# ENGINEERING GRAPHICS
# FUNDAMENTALS

## SECOND CUSTOM EDITION

Taken from:

*Engineering Graphics with AutoCAD® 2009\**
by James D. Bethune

*Technical Drawing,* Thirteenth Edition
by Frederick E. Giesecke

*Geo-Metrics III: The Application of Geometric Dimensioning and
Tolerancing Techniques (Using the Customary Inch Systems)*
by Lowell W. Foster

*Blueprint Reading for the Machine Trades,* Sixth Edition
by Russ Schultz and Larry Smith

\*Material from this book has been customized to reflect AutoCAD® 2010 updates

**Learning Solutions**

New York  Boston  San Francisco
London  Toronto  Sydney  Tokyo  Singapore  Madrid
Mexico City  Munich  Paris  Cape Town  Hong Kong  Montreal

Cover Art: Cover images courtesy of Photodisc/Getty

Taken from:

*Engineering Graphics with AutoCAD® 2009*
by James D. Bethune
Copyright © 2009 by Pearson Education, Inc.
Published by Prentice Hall
Upper Saddle River, New Jersey 07458

*Technical Drawing*, Thirteenth Edition
by Frederick E. Giesecke
Copyright © 2008 by Pearson Education, Inc.
Published by Prentice Hall

*Geo-Metrics III: The Application of Geometric Dimensioning and
Tolerancing Techniques (Using the Customary Inch Systems)*
by Lowell W. Foster
Copyright © 1994 by Prentice Hall
A Pearson Education Company

*Blueprint Reading for the Machine Trades,* Sixth Edition
by Russ Schultz and Larry Smith
Copyright © 2008, 2004, 2001, 1996, 1988, 1981 by Pearson Education, Inc.
Published by Prentice Hall

This special edition published in cooperation with Pearson Learning Solutions.

All trademarks, service marks, registered trademarks, and registered service marks are the property of their respective
owners and are used herein for identification purposes only.

Pearson Learning Solutions, 501 Boylston Street, Suite 900, Boston, MA 02116
A Pearson Education Company
www.pearsoned.com

Printed in the United States of America

1 2 3 4 5 6 7 8 9 10  V312  14 13 12 11 10

2009760012

KW

**PEARSON**

ISBN 10: 0-558-56414-3
ISBN 13: 978-0-558-56414-8

# Contents

Taken from: *Engineering Graphics with AutoCAD® 2009* by James D.
Bethune and *Geo-Metrics III: The Application of Geometric Dimensioning
and Tolerancing Techniques (Using the Customary Inch Systems)* by
Lowell W. Foster

## Chapter 2—Sectional Views and GD&T

Taken from: *Engineering Graphics with AutoCAD® 2009* by James D.
Bethune and *Geo-Metrics III: The Application of Geometric Dimensioning
and Tolerancing Techniques (Using the Customary Inch Systems)* by
Lowell W. Foster

## Chapter 3—Auxiliary Views and GD&T

Taken from: *Engineering Graphics with AutoCAD® 2009* by James D.
Bethune and *Geo-Metrics III: The Application of Geometric Dimensioning
and Tolerancing Techniques (Using the Customary Inch Systems)* by
Lowell W. Foster

## Chapter 4—Dimensioning and Tolerancing in AutoCAD and GD&T

Taken from: *Engineering Graphics with AutoCAD® 2009* by James D. Bethune and *Geo-Metrics III: The Application of Geometric Dimensioning and Tolerancing Techniques (Using the Customary Inch Systems)* by Lowell W. Foster

## Chapter 5—Geometric Tolerancing in AutoCAD and GD&T

Taken from: *Engineering Graphics with AutoCAD® 2009* by James D. Bethune and *Geo-Metrics III: The Application of Geometric Dimensioning and Tolerancing Techniques (Using the Customary Inch Systems)* by Lowell W. Foster

# Chapter 6—Threads, Fasteners, Springs, and GD&T

Taken from: *Engineering Graphics with AutoCAD® 2009* by James D. Bethune, *Geo-Metrics III: The Application of Geometric Dimensioning and Tolerancing Techniques (Using the Customary Inch Systems)* by Lowell W. Foster, and *Technical Drawing*, Thirteenth Edition by Frederick E. Giesecke

# Chapter 7—Working Drawings

Taken from: *Engineering Graphics with AutoCAD® 2009* by James D. Bethune and *Technical Drawing*, Thirteenth Edition by Frederick E. Giesecke

# Chapter 8—Gears, Bearings, and Cams

Taken from: *Engineering Graphics with AutoCAD® 2009* by James D. Bethune and *Technical Drawing*, Thirteenth Edition by Frederick E. Giesecke

# Chapter 9—Axonometric and Oblique Projections

Taken from: *Technical Drawing*, Thirteenth Edition by Frederick E. Giesecke

## Chapter 10—Welding Representation

Taken from: *Technical Drawing*, Thirteenth Edition by Frederick E. Giesecke

## Index 767

## Appendix

Taken from: *Blueprint Reading for the Machine Trades*, Sixth Edition by Russ Schultz and Larry Smith

# C H A P T E R   1

# Orthographic Projection and GD&T

## 1-1  INTRODUCTION

This chapter introduces orthographic views. *Ortho-graphic views* are two-dimensional views of three-dimensional objects. Orthographic views are created by projecting a view of an object onto a plane that is usually positioned so it is parallel to one of the planes of the object. See Figure 1-1.

Orthographic projection planes may be positioned at any angle relative to an object, but in general the six views parallel to the six sides of a rectangular object are used. See Figure 1-2. Technical drawings usually include only the front, top, and right-side orthographic views because together they are considered sufficient to completely define an object's shape.

## 1-2  THREE VIEWS OF AN OBJECT

Orthographic views are positioned on a technical drawing so that the top view is located directly over the front view, and the side view is directly to the right side of the front view. See Figure 1-2.

The location of the front, top, and right-side views relative to each other is critical. Correct relative positioning of views allows information to be projected between the views. Vertical lines are used to project information between the front and top views, and horizontal lines are used to project information between the front and right-side views.

Orthographic views are two-dimensional, so they cannot show depth. In Figure 1-2 plane A-B-C-D appears as a straight line in both the front and top orthographic views. Edge A-B appears as a vertical line in the top view, as a horizontal line in the side view, and as point AB in the front view. The end view of a plane is a straight line, and the end view of a line is a point. See Figure 1-3.

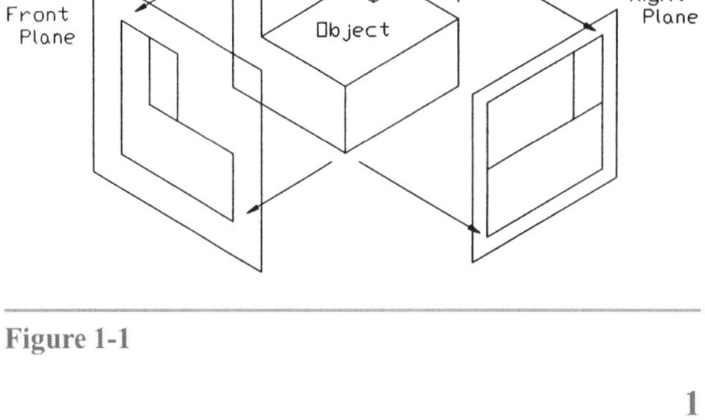

Figure 1-1

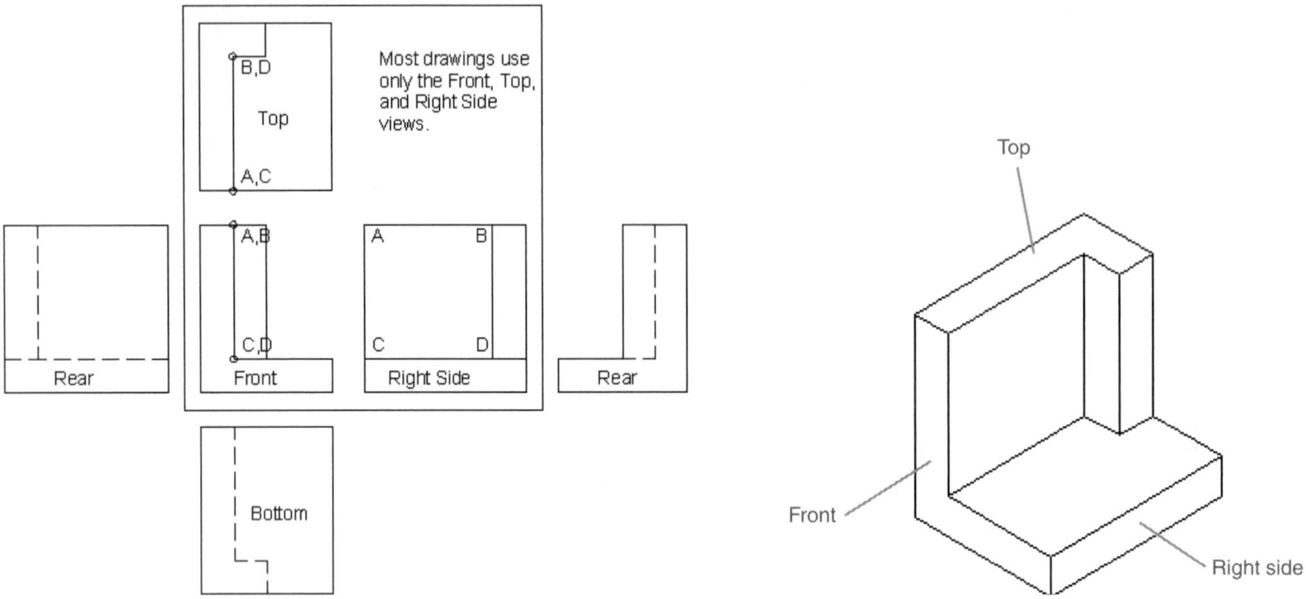

**Figure 1-2**

Figure 1-4 shows an object and the front, top, and right-side orthographic views of the object. Plane A-B-C-D appears as a rectangle in the front view and as a line in both the top and right-side views.

To help you better understand the relationship between orthographic views and in particular the front, top, and side views of an object, place a book on a table as shown in Figure 1-5. Note how the front, top, and side views were projected and then arranged to create a technical drawing of the book.

## 1-3  VISUALIZATION

*Visualization* is the ability to look at a three-dimensional object and mentally see the appropriate three orthographic views.

It is an important skill for a designer and an engineer to develop. Visualization also includes the ability to look at two-dimensional orthographic views and mentally picture the three-dimensional object the views represent.

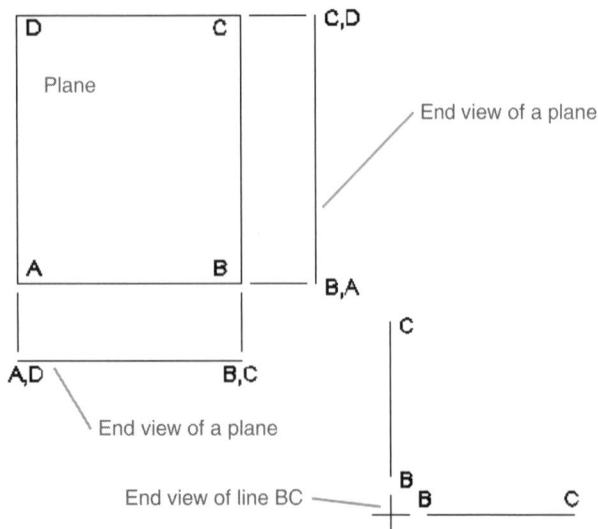

**Figure 1-3**

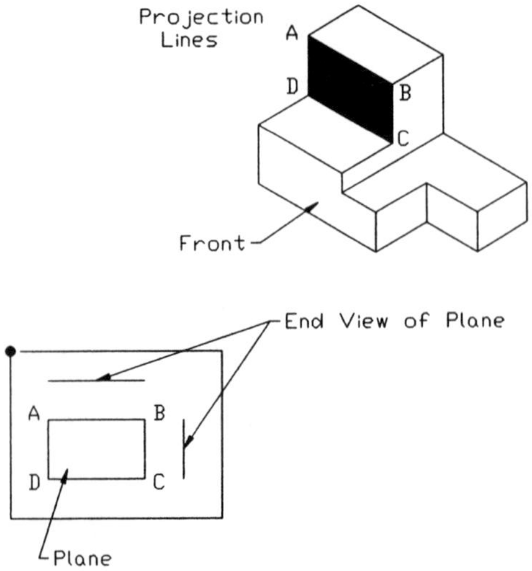

**Figure 1-4**

Figure 1-5

Figure 1-6

Figure 1-6 shows a 5 × 3.5 × 2-inch box and three orthographic views of the box. This section presents a short exercise to help you better understand how orthographic views are related to three-dimensional objects.

**To draw a three-dimensional box**

1. Set the screen for **inches,** then select the **3D View** command from the **View** menu accessed by clicking the **Menu Browser** in the upper left corner of the screen, then **SE Isometric.**

   The origin axis reference icon will change to the 3D alignment.

2. Type the word **box** to a command prompt.

   *Specify corner of box or [Center] <0,0,0>:*

3. Press **Enter.**

   *Specify corner or [Cube/Length]:*

4. Type **L**; press **Enter.**

   *Specify length:*

5. Type **5**; press **Enter.**

   *Specify width:*

6. Type **3.5**; press **Enter.**

   *Specify height:*

7. Type **2**; press **Enter.**

   A box will appear on the screen.

8. Select the **3D Orbit** command from the **View** menu. Move the cursor around the drawing screen holding the left mouse button down. Position the box so as to create each of the three orthographic views shown in Figure 1-6.

   Figures 1-7 and 1-8 show two additional objects and their orthographic views.

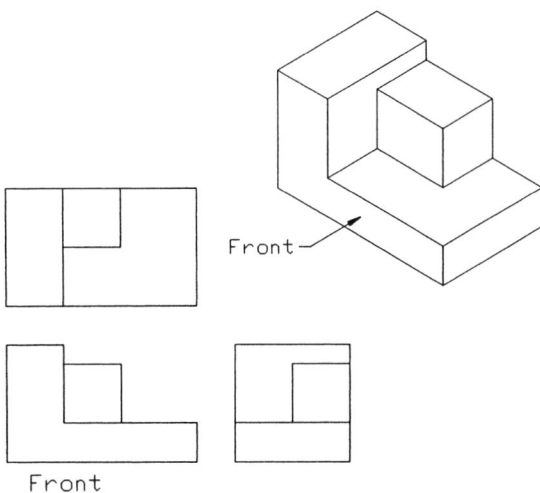

Figure 1-7

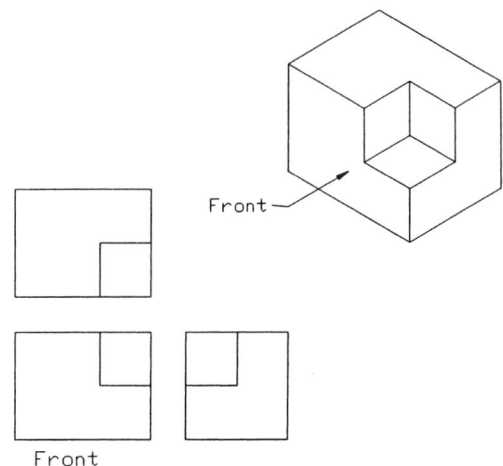

Figure 1-8

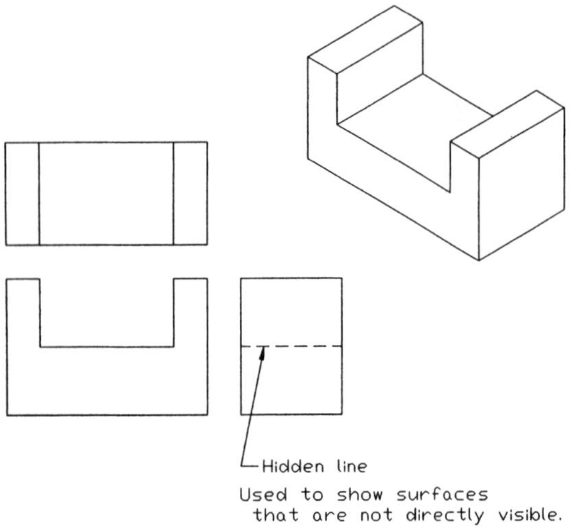

**Figure 1-9**

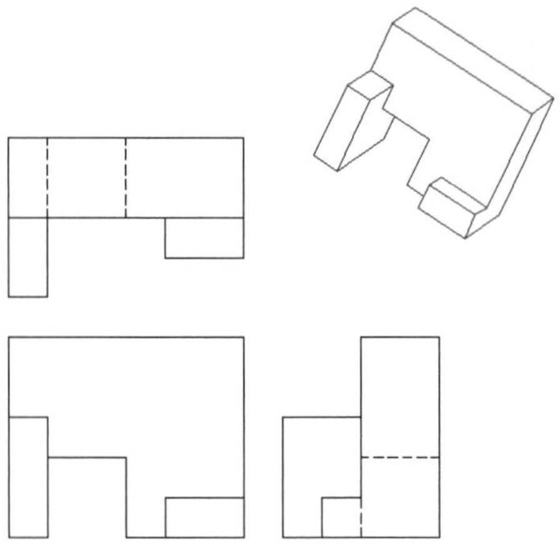

## 1-4  HIDDEN LINES

Hidden lines are used to represent surfaces that are not directly visible in an orthographic view. Figure 1-9 shows an object that contains a surface A-B-C-D that is not directly visible in the side view. A hidden line is used to represent the end view of plane A-B-C-D in the side view.

Figure 1-10 shows an object and three views of the object. The top and side views of the object contain hidden lines.

Figure 1-11 shows an object that contains an edge line A-B. In the top view of the object, line A-B is partially hidden and partially visible. When the line is directly visible through the square hole it is drawn as a continuous line; when it is not directly visible it is drawn as a hidden line.

**Figure 1-11**

**Figure 1-10**

## 1-5  HIDDEN LINE CONVENTIONS

Figure 1-12 shows several conventions associated with drawing hidden lines. Whenever possible, show intersections of hidden lines as touching lines, not as open gaps. Corners should also be shown as touching lines.

The **Linetype Scale** option on the **Properties** dialog box can be used to change the spacing of hidden lines. Figure 1-13 shows two hidden lines and a continuous line that are aligned. A small gap should be included between the two types of lines to prevent confusion as to where one type of line ends and the other begins. A visual distinction can also be created by selecting a different color for hidden lines. If continuous and hidden lines are drawn using different colors, then gaps are not necessary.

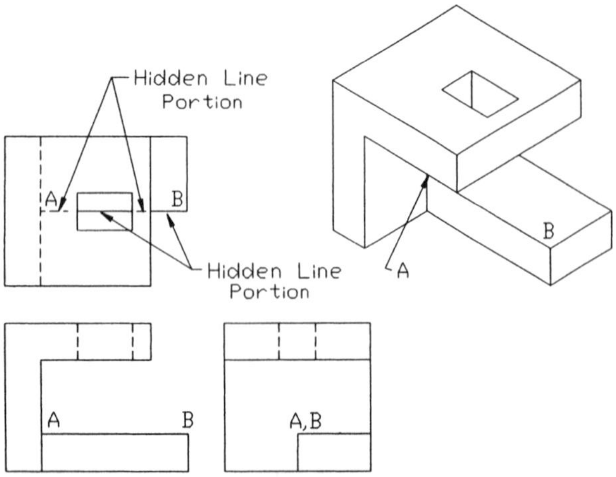

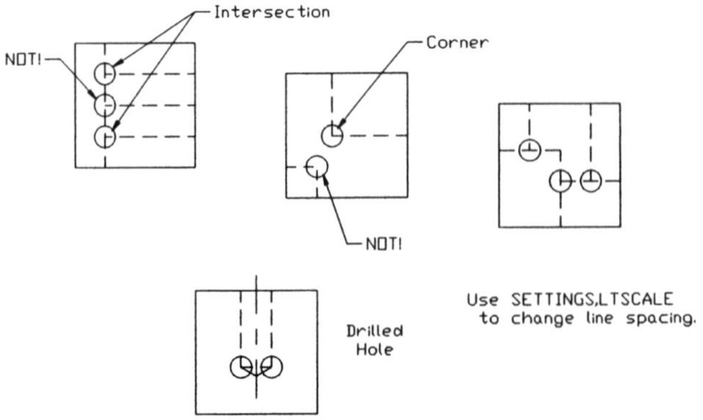

**Figure 1-12**

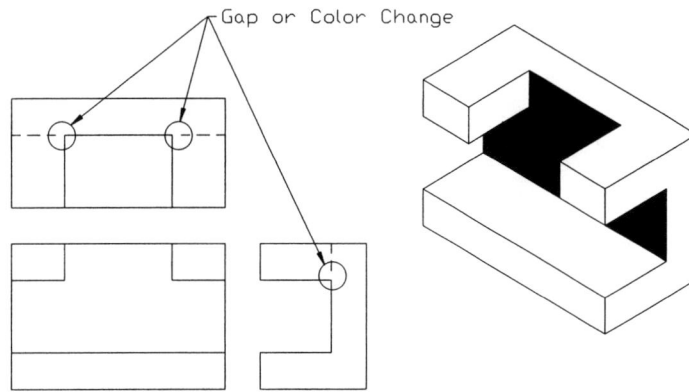

**Figure 1-13**

## 1-6 DRAWING HIDDEN LINES

Hidden lines may be created in several different ways but two of the easiest are to modify an existing continuous line or to transfer an existing continuous line to a special layer created specifically for hidden lines. In the example presented both linetype and line color will be changed.

**To change a continuous line to a hidden line**

1. Select the line.

The grip points will appear with a dialog box showing some of the line's properties. See Figure 1-14.

2. Click the **Linetype** box on the dialog box.
3. Click the arrow on the right end on the dialog box.

A listing of available linetypes will cascade down.

4. Click the **Hidden linetype**.

See Figure 1-15. The selected line will be changed to a hidden line. If the linetype is not listed, type the word linetype on the **Command:** line and press Enter. The **Linetype Manager** dialog box will appear. Select the **Load** option, and the **Load or Reload Linetypes** dialog box will appear.

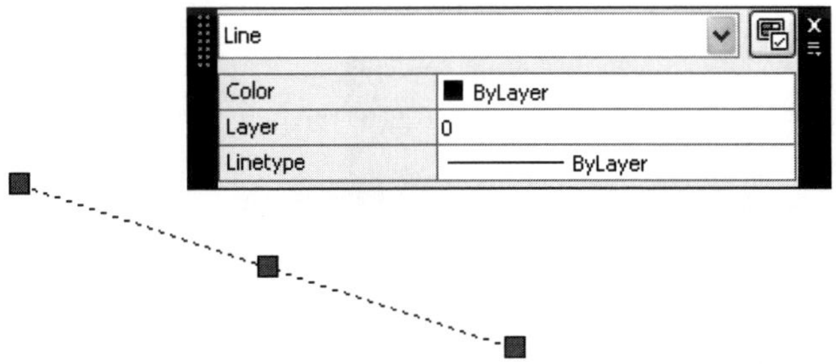

**Figure 1-14**

**Figure 1-15**

Figure 1-16

See Figure 1-16. Scroll down and select the **Hidden** line option. Select **OK** and return to the drawing screen. Access the **Modify Line** dialog box as explained above. This time the **Hidden** line pattern will be listed as an option.

5.  Click the Color box on the dialog box
6.  Click the arrow at the right end of the Color box.

A listing of available colors will cascade down.

7.  Select the color blue.

8.  Press the <Esc> key.

If the line changes color but the broken-line pattern does not appear, or if the line segments are too short or too long, use the **Linetype Scale** option found in the **Properties** palette to change the length of the hidden line segments. See Figure 1-14. It may be necessary to use a trial-and-error process to obtain the most appropriate hidden line pattern.

### To use Layer to create a hidden line

Linetype can be changed by creating a layer specifically set up for hidden lines. Both the line pattern and color can be preset on the layer. This procedure may seem a bit lengthy when compared with the procedure outlined above, but it is very efficient if there are many lines to change.

1.  Select the Layer Properties tool from the Layer panel under the Home tab.

The **Layer Properties Manager** dialog box will appear. See Figure 1-18. Create a new layer called Hidden.

2.  Select the New option.
3.  Create a layer named **Hidden** using the **hidden** line pattern and the color **blue,** then return to the drawing screen.
4.  Click an existing continuous line.

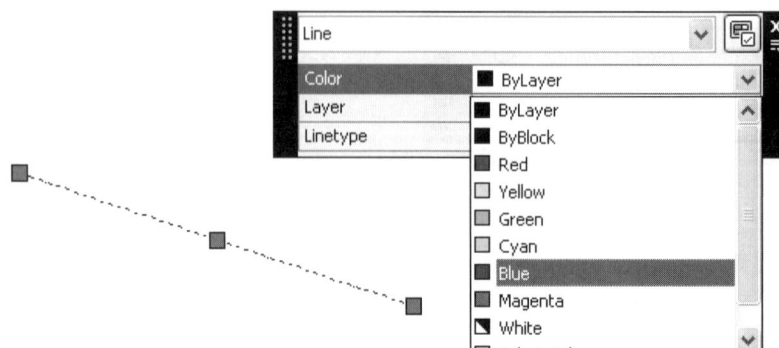

Figure 1-17

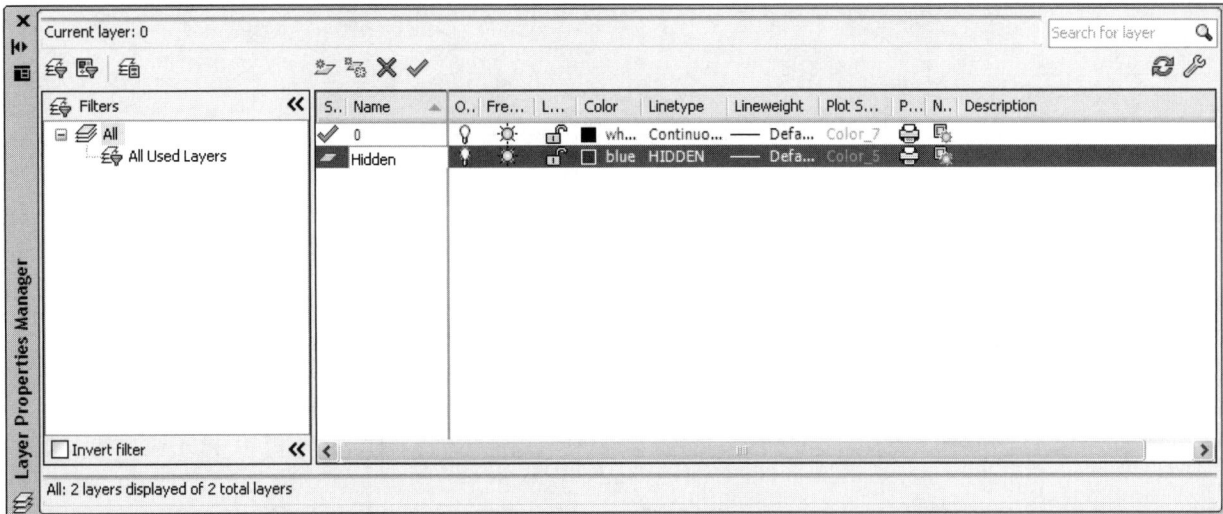

Figure 1-18

The Grip points will appear with a dialog box listing some of the line's properties. See Figure 1-19.

5. Click the Layer box.
6. Click the arrow at the right of the Layer box.

A listing of available layers will cascade down.

7. Select the Hidden layer.
8. Press the <Esc> key.

The line will now have a hidden linetype and be blue in color.

Figure 1-19

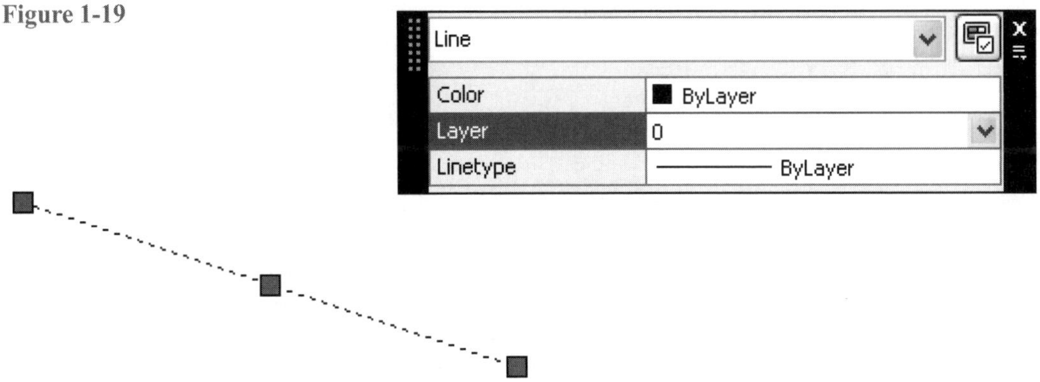

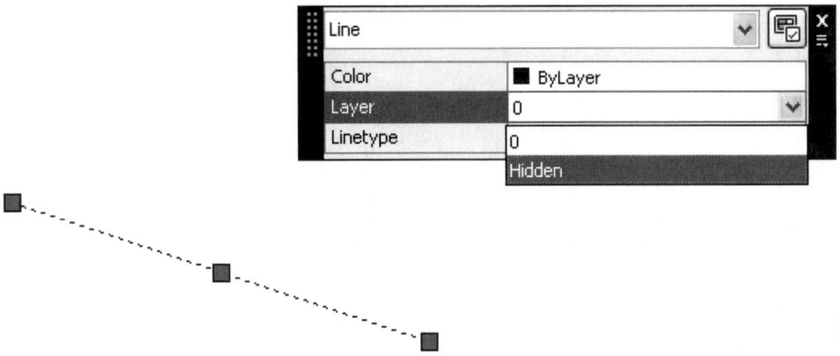

Figure 1-20

**Figure 1-21**

**Figure 1-22**

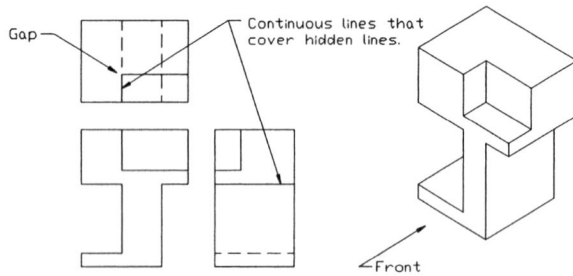

**Figure 1-23**

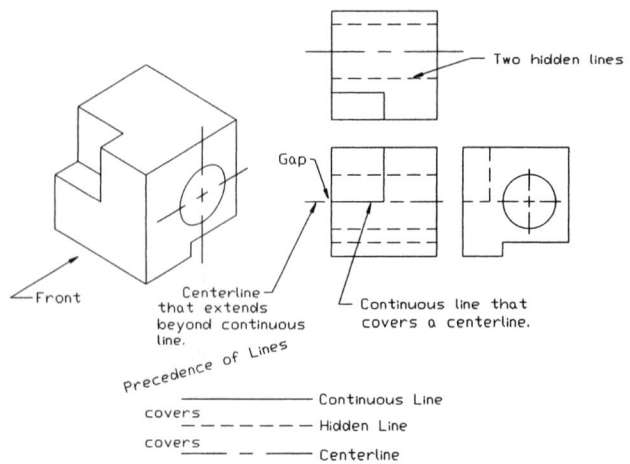

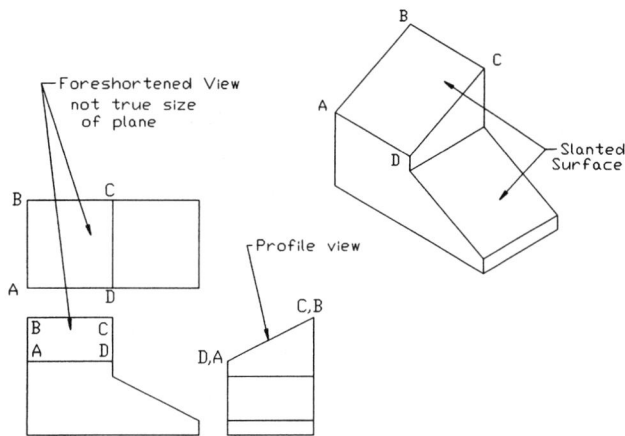

Figure 1-24

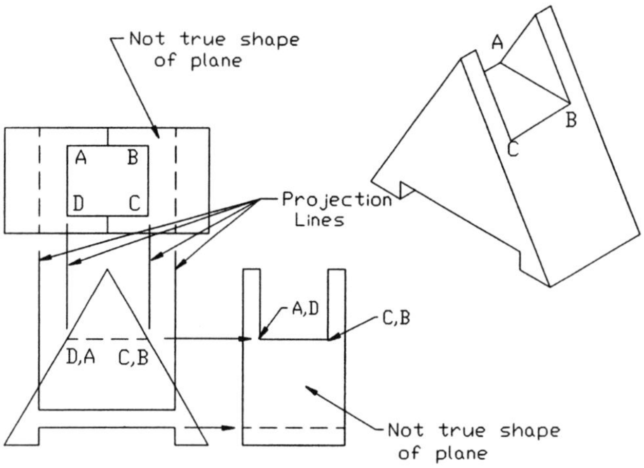

Figure 1-25

Figures 1-21 and 1-22 show an object and three views of that object. There are hidden lines in several of the orthographic views.

## 1-7 PRECEDENCE OF LINES

When orthographic views are prepared, it is not unusual for one type of line to be drawn over another type: a continuous line over a centerline, for example. See Figure 1-23. Drawing convention has established a precedence of lines: A continuous line takes precedence over a hidden line, and a hidden line takes precedence over a centerline.

If different linetypes are of different lengths, include a gap where the line changes from continuous to either hidden or center to create a visual distinction between the lines. Color changes may also be used to create visual distinctions between lines.

## 1-8 SLANTED SURFACES

*Slanted surfaces* are surfaces that are not parallel to either the horizontal or vertical axis. Figure 1-24 shows a slanted surface A-B-C-D. Slanted surface A-B-C-D ap-

pears as a straight line in the side view and as a plane in both the front and top views. It is important to note that neither the front nor top view of surface A-B-C-D is a true representation of the surface. Both orthographic views are actually smaller than the actual surface. Lines B-C and A-D in the top view are true length, but lines A-B and C-D are shorter than the actual edge lengths. The same is true for the front view of the surface. The side view shows the true length of lines A-B and D-C. True lengths of lines and true shapes of planes are discussed again in Chapter 3, Auxiliary Views and GDT.

Figure 1-25 shows another object that contains slanted surfaces. Note how projection lines are used between the views. As objects become more complex the shape of a surface or the location of an edge line will not always be obvious, so projecting information between views, and therefore exact, correct relative view location, becomes critical. Information can be projected only between views that are correctly positioned and accurately drawn.

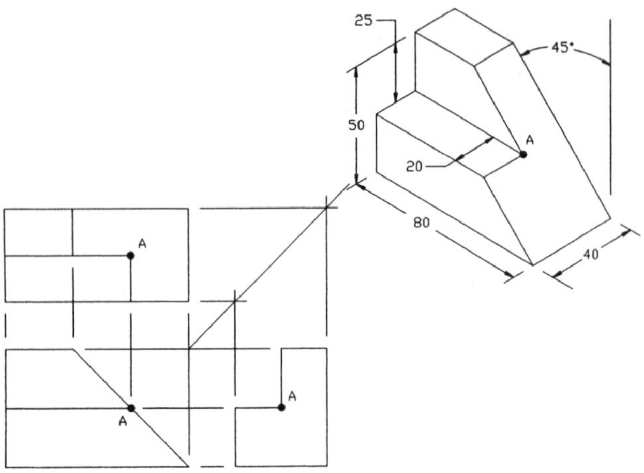

**Figure 1-26**

## 1-9 PROJECTION BETWEEN VIEWS

Information is projected between the front and side views using horizontal lines and between the front and top views using vertical lines. Information can be projected between the top and side views using a combination of horizontal and vertical lines that intersect a 45° miter line. See Figure 1-26.

A 45° miter line is constructed between the top and side views. This line allows the project lines to change direction, turn a corner, and change from horizontal to vertical lines. To go from the top view to the side view, construct horizontal projection lines from the top view so that they intersect the miter line. Construct vertical lines from the intersection points on the miter line into the side view. The reverse process is used to go from the side view to the top view.

The following sample problem shows how projection lines can be used to create orthographic views using AutoCAD.

## 1-10 SAMPLE PROBLEM SP1-1

Draw the front, top, and right-side orthographic views of the object shown in Figure 1-26. Set up the drawing as follows.

1. Create a separate layer for hidden lines.

Limits = **297 × 210** (Metric setup)
Grid  = **10**
Snap = **5**
Ortho = **ON** (F8)
Layer = **HIDDEN** (LType = **HIDDEN**, Color = **Green**)

2. Use the overall dimensions (length = **80,** height = **50,** and depth = **40**) to define the overall size requirements of the three orthographic views. See Figure 1-27. The **20** spacing between the views is arbitrary. The distance between the front and top views does not have to equal the distance between the front and side views. In this example the two distances are equal.

3. Draw a **45°** miter line starting from the upper right corner of the front view. This step is possible only if the distances between the views are equal. If the distances are not equal, draw the front and top views, add the miter line, and project the overall size of the side view from the front and top views.

3a. (Optional) Trim away the excess lines to clearly define the areas of the front, top, and side views.

3b. (Optional) The **Layers** tool may be used to create a layer for construction lines and another layer for final drawing lines. The lines for the final drawing may then be transferred to the final drawing layer and the construction layer turned off. This method eliminates the need for extensive erasing or trimming.

4. Draw the **45°** slanted surface in the front view as shown in Figure 1-27. Project the intersection of the slanted surface and the top edge of the front view into the top view. Add a horizontal line across the front and right-side views **25** from the bottom edge line. The 25 value comes from the given dimension. Label the intersection of the slanted line and the horizontal line as point **A.**

5. Draw a horizontal line in the top view **20** from the top edge as shown in Figure 1-27. Continue the line so that it intersects the miter line. Project the line into the side view. Use **Osnap, Intersection** with **Ortho** on to ensure an accurate projection. Label point **A** in the side view.

6. Project point **A** in the front view into the top view (vertical line). Label the intersection of the projection line and the horizontal line in the top view as point **A.**

7. Use **Erase** and **Trim** to remove the excess lines.

8. Save the drawing if desired.

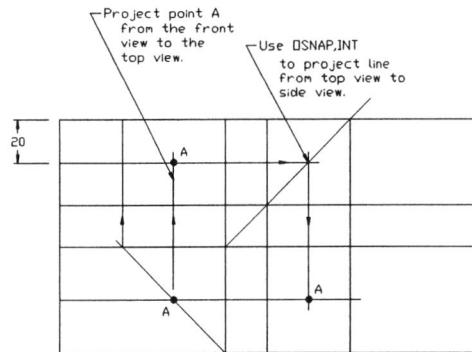

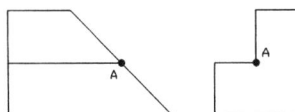

Figure 1-28

## 1-11 COMPOUND LINES

A *compound line* is formed when two slanted surfaces intersect. See Figure 1-28. The true length of a compound line is not shown in the front, top, or side views.

Figure 1-29 shows another object that contains compound lines. The three orthographic views of a compound line are sometimes difficult to visualize, making it more important to be able to project information accurately between views. Usually, part of an object can be visualized and part of the orthographic views created. The remainder of the drawing can be added by projecting from the known information.

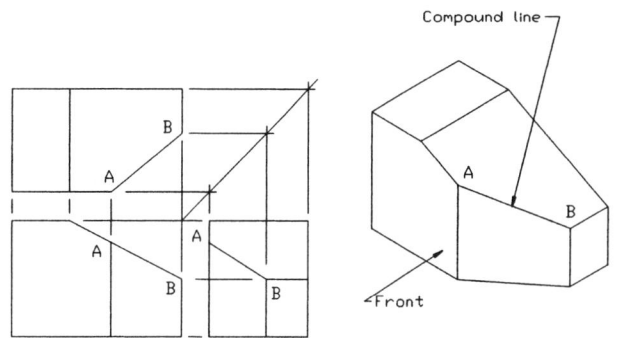

Figure 1-27

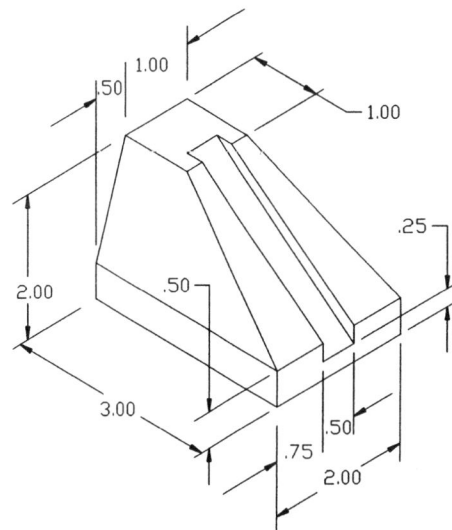

Figure 1-29

## 1-12  SAMPLE PROBLEM SP1-2

Figure 1-30 shows how the front, top, and side views of the object shown in Figure 1-29 were created by projecting information. The procedure is as follows.

1. Set the drawing up as follows.

   Limits = **Default**
   Grid   = **.5**
   Snap   = **.25**

2. Use the overall dimensions to define the space and location requirements for the views. Draw the **45°** miter line.

3. Use the given dimensions to define the starting and end points on the compound lines in the top and front views. One of the lines has been labeled **A-B.**

4. Draw the compound line in the top view and then project it into the side view using information from both the top and front views.

5. Add the dovetailed shape using the given dimensions. Draw the appropriate hidden lines for the dovetail.

6. Erase and trim the excess lines.

7. Save the drawing if desired.

Figure 1-31 shows the three orthographic views and pictorial view of another example of an object that includes compound lines. Some of the projection lines have also been included.

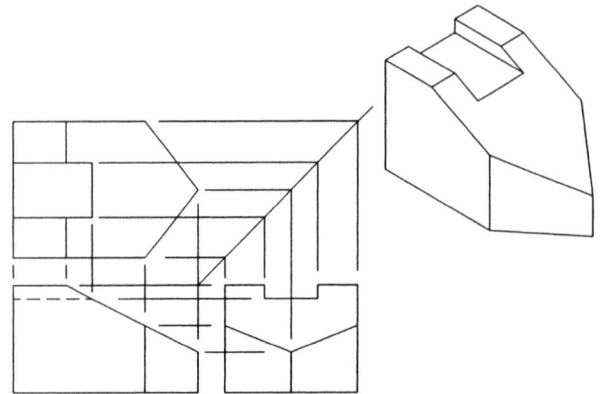

**Figure 1-31**

**Figure 1-30**

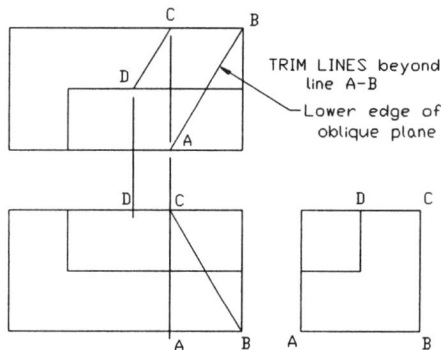

**Figure 1-32**

## 1-13  OBLIQUE SURFACES

*Oblique surfaces* are surfaces that do not appear correctly shaped in the front, top, or side views. Figure 1-32 shows an object that contains oblique surface A-B-C-D. Figure 1-32 also shows three views of just surface A-B-C-D. Oblique surfaces are projected by first projecting their corner points and then joining the points with straight lines. The orthographic views of oblique surfaces are sometimes visually abstract. This makes the development of their orthographic views more dependent on projection than other types of surfaces.

Figure 1-33 shows another object that contains an oblique surface. Figure 1-34 shows how the three orthographic views were developed. The oblique surface is defined using only two angles, but this is sufficient to create the three views because the surface is flat. The edge lines are parallel. Lines A-B, C-D, and E-F are parallel to each other, and lines C-B, D-E, and F-A are also parallel to each other. Figure 1-34 also shows three views of just the oblique surface along with the appropriate projection lines.

**Figure 1-33**

**Figure 1-34**

## 1-14 SAMPLE PROBLEM SP1-3

The three views of the object shown in Figure 1-35A may be drawn as follows. (The given dimensions are in millimeters.)

1. Use the **Line** and the **Offset** commands to set up the sizes and locations of the three views. Use **Osnap, Intersection** to draw the projection lines. See Figure 1-35B.
2. Draw an oblique surface based on the given dimensions. The edge lines of the oblique surface are parallel.
3. Draw the slanted surface in the side view and project the surface into the other views. Use the intersection between the slanted surface and flat sections on the top and side of the object to determine the shape of the oblique surface.
4. Use **Erase** and **Trim** to remove any excess lines. Save the drawing if desired.

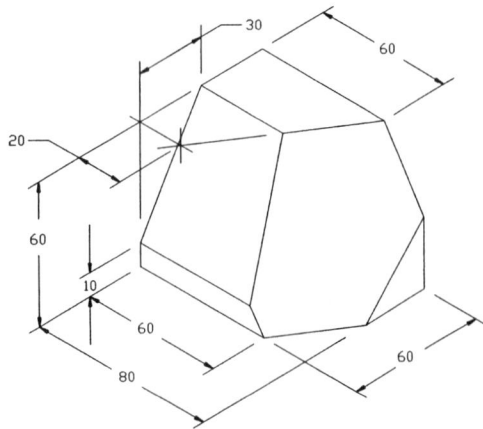

Figure 1-35A

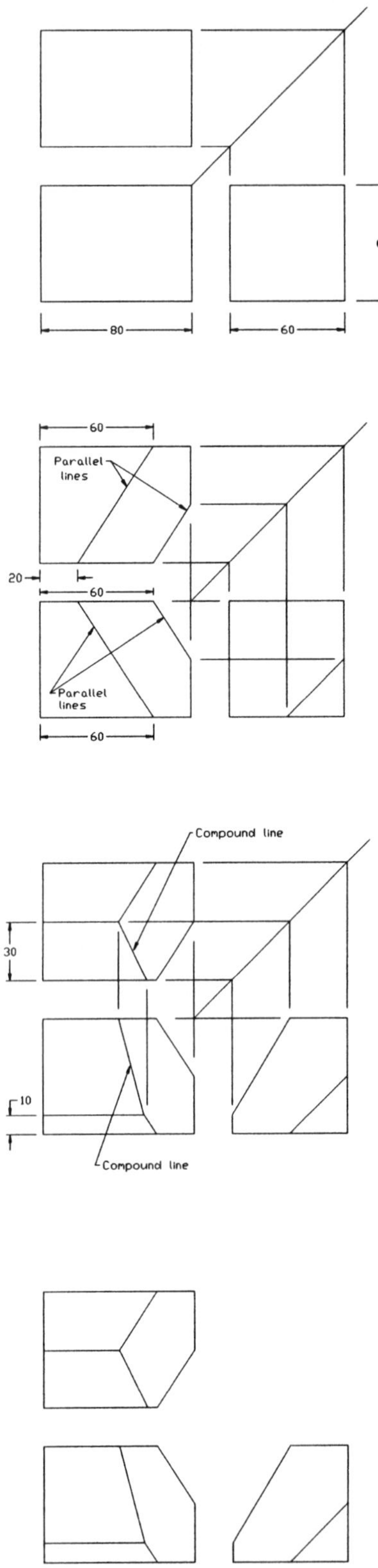

Figure 1-35B

## 1-15  ROUNDED SURFACES

*Rounded surfaces* are surfaces that have constant radii, such as arcs or circles. Surfaces that do not have constant radii are classified as *irregular surfaces.* See Section 1-24.

Figure 1-36 shows an object with rounded surfaces. Surface A-B-C-D is tangent to both the top and side surfaces of the object, so no edge line is drawn. This means that the top and side orthographic views of the object are ambiguous. If only the top and side views are considered, then other interpretations of the views are possible. Figure 1-36 shows another object that generates the same top and side orthographic views but does not include rounded surface A-B-C-D. The front view is needed to define the rounded surfaces and limit the views to only one interpretation.

Surfaces perpendicular to an orthographic view always produce lines in that orthographic view. The vertical surface shown in the front view of Figure 1-37A requires an edge line in the top view. The vertical line in the front view is perpendicular to the top views.

The object shown in Figure 1-37B has no lines perpendicular to the top view, so no lines are drawn in the top view. Figure 1-37C shows two rounded surfaces that intersect at points tangent to their centerlines. The intersection point is considered sufficient to require an edge line in the top view. No line would be visible on the actual object, but drawing convention prescribes that a line be drawn.

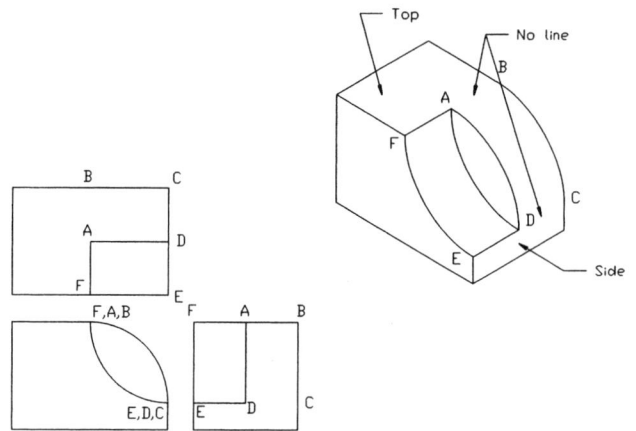

Figure 1-36

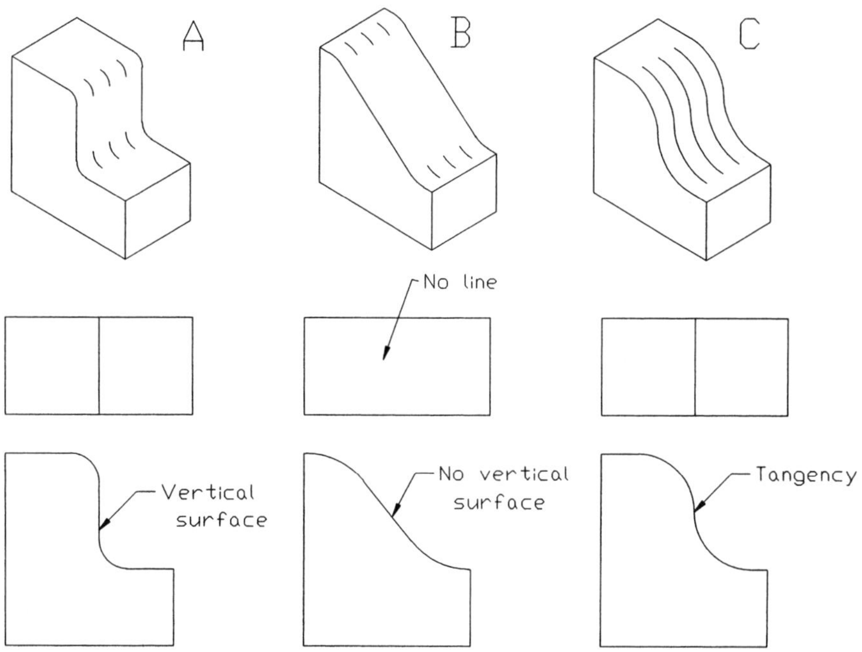

Figure 1-37

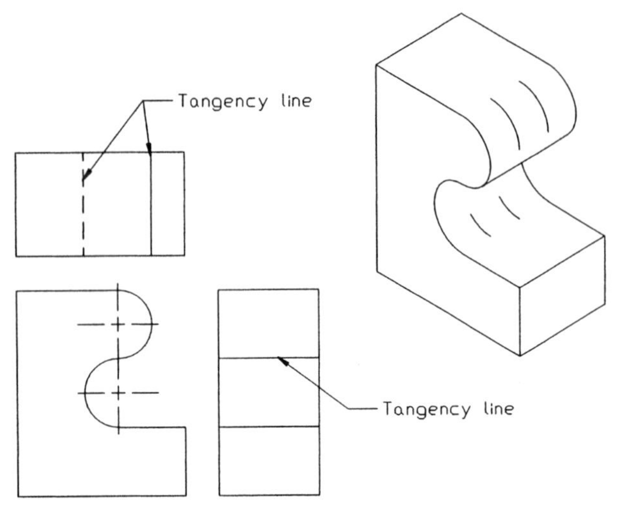

**Figure 1-38**

**Figure 1-39**

Figure 1-38 shows an object that contains two semi-circular surfaces. Note how lines representing the widest and deepest point of the surfaces generate lines in the top view. No line would actually appear on the object. Figure 1-39 shows further examples of objects with curved surfaces.

## 1-16  SAMPLE PROBLEM SP1-4

Figure 1-40 shows an object that includes rounded surfaces, and Figure 1-41 shows how the three views of the object were developed.

The procedure is as follows.

1. Use the given overall dimensions to lay out the size and location of the three views.
2. Draw in the external details, the cutout, and the circular extension in all three views.
3. Add the appropriate hidden lines. Use either the modify line properties method, or create a layer for hidden lines.
4. **Erase** and **Trim** any excess lines and save the drawing if desired.

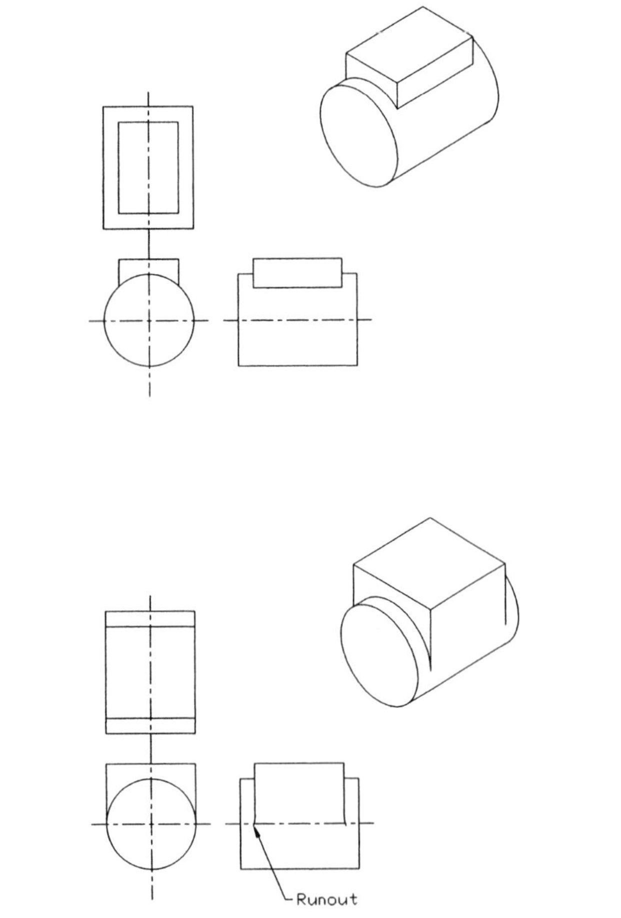

## 1-17  HOLES

Holes are represented in orthographic views by circles and parallel hidden lines. All views of a hole always include centerlines. See Figure 1-42. It is suggested that a layer titled **Center** be added to any drawing being created. Specify the linetype **Center** and the color **Red.**

Holes that go completely through an object are dimensioned using only a diameter. The notation **THRU** may be added to the diameter specification for clarity.

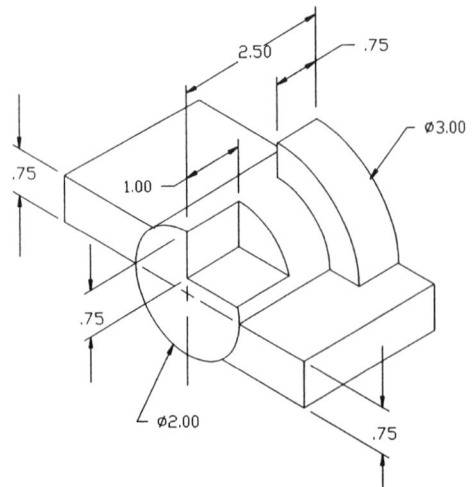

**Figure 1-40**

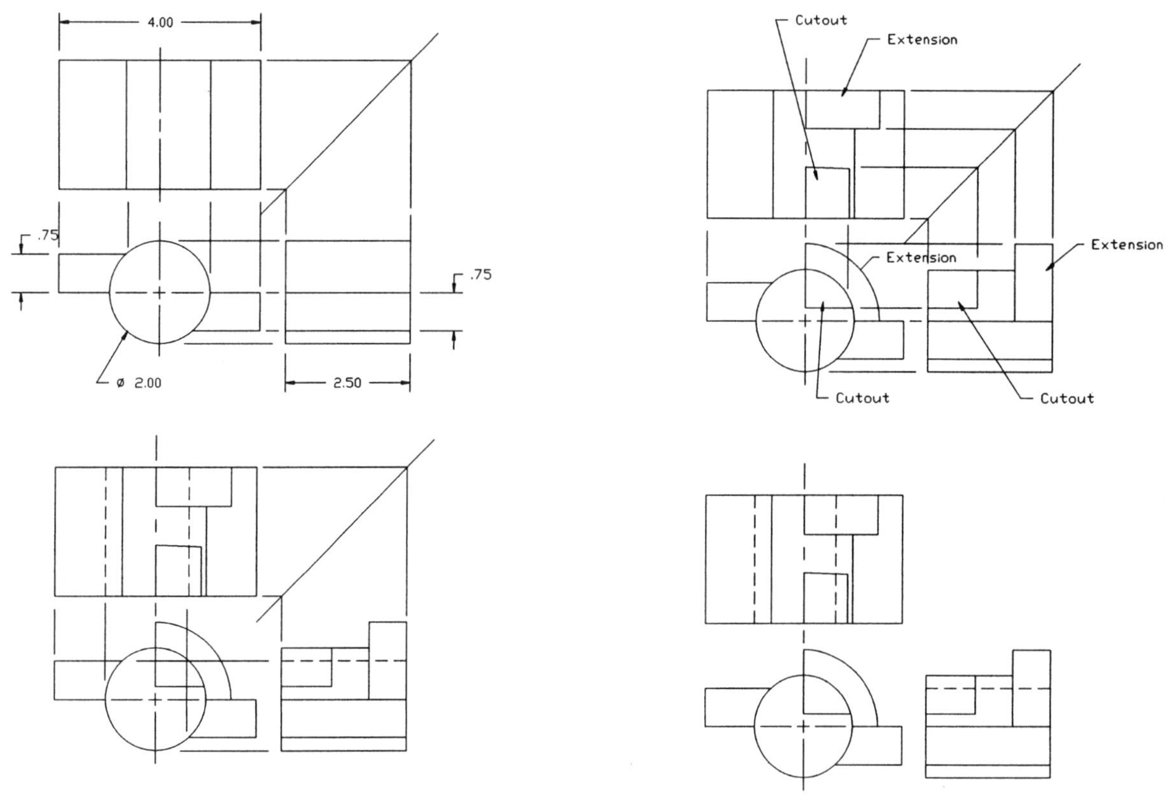

**Figure 1-41**

Holes that do not go completely through an object include a depth specification in their dimension. The depth specification is interpreted as shown in Figure 1-42. Holes that do not go completely through an object must include a conical point. The conical point is not included in the depth specification. Conical points must be included because most holes are produced using a twist drill with conically shaped cutting edges.

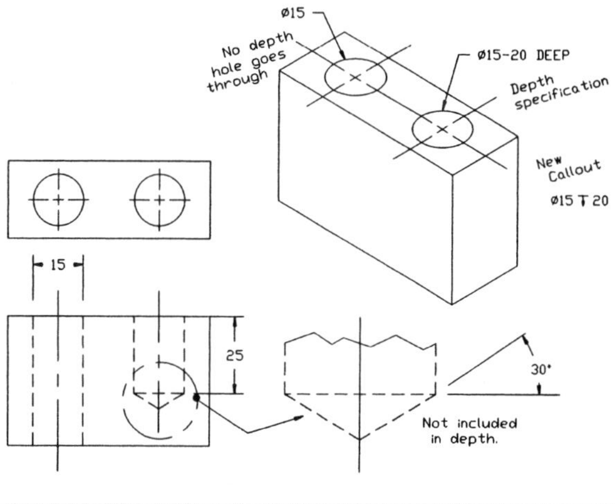

**Figure 1-42**

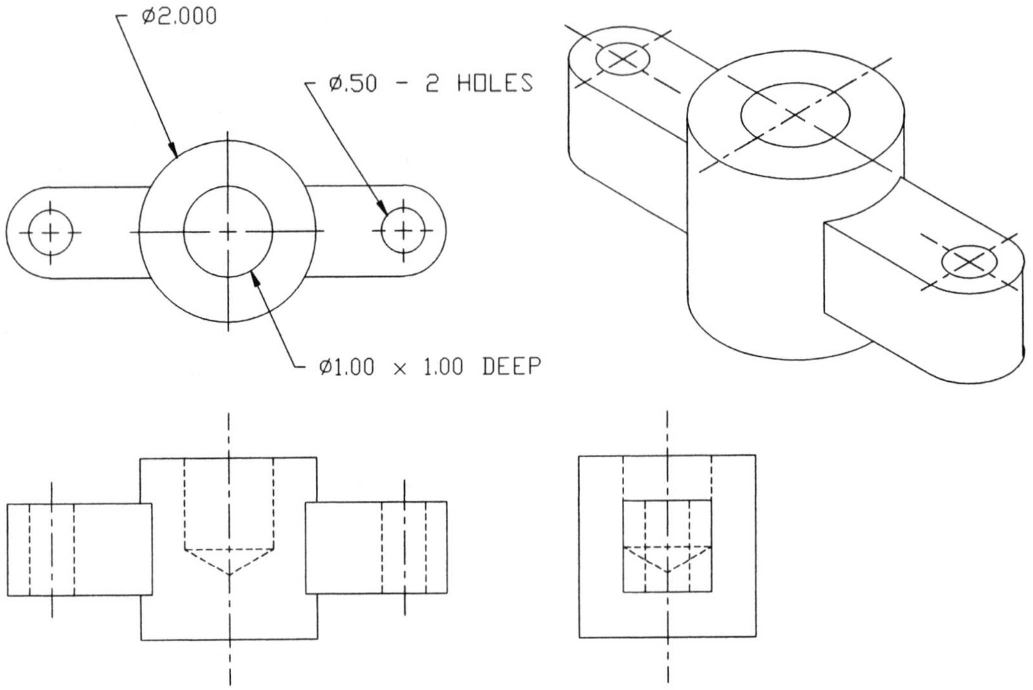

Ø2.000

Ø.50 - 2 HOLES

Ø1.00 × 1.00 DEEP

**Figure 1-43**

The cutting angles of twist drills vary, but they are always represented by a 30° angle as shown.

Figure 1-43 shows an object that contains two through holes and one hole with a depth specification. Note how these holes are represented in the orthographic views and how centerlines are used in the different views.

Figure 1-44 is another example of an object that includes a hole. The hole is centered about the edge line between the two normal surfaces. Figure 1-45 shows the same object with the hole's center point offset from the edge line between the two normal surfaces. Compare the differences between Figures 1-44 and 1-45.

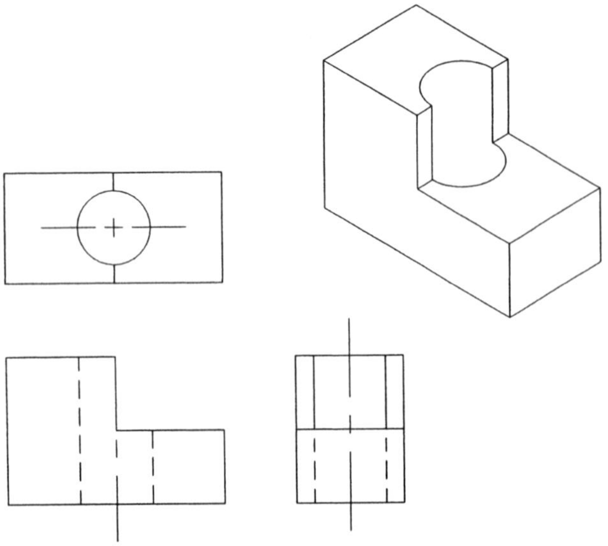

**Figure 1-44**

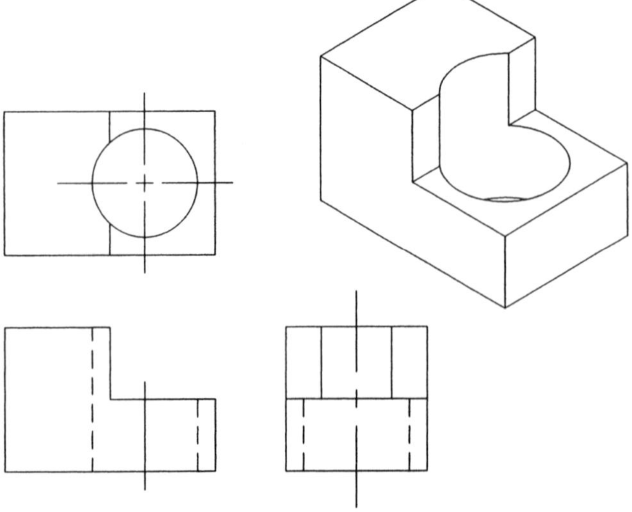

**Figure 1-45**

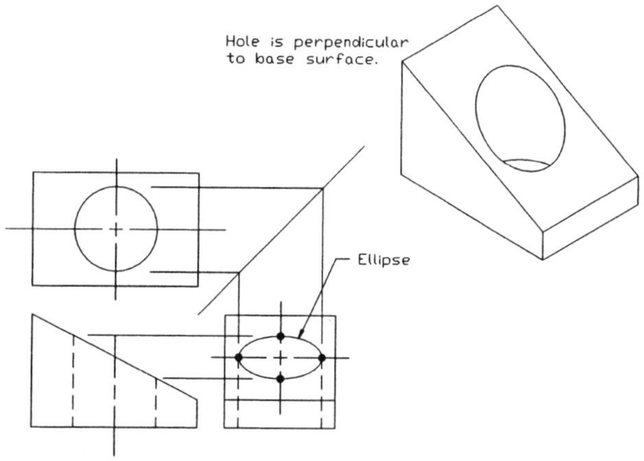

Figure 1-46

## 1-18 HOLES IN SLANTED SURFACES

Figure 1-46 shows a hole that penetrates a slanted surface. The hole is perpendicular to the bottom surface of the object. The top view of the hole appears as a circle because we are looking straight down into it. The side view shows a distorted view of the circle and is represented using an ellipse.

The shape of the ellipse in the side view is defined by projecting information from the front and top views. Points 1 and 2 are known to be on the vertical centerline, so their location can be projected from the front view to the side view using horizontal lines. Points 3 and 4 are located on the horizontal centerline, so their location can be projected from the top view into the side view using the 45° miter line.

The elliptical shape of the hole in the side view is drawn connecting the projected points using the **Ellipse** command.

**To draw an ellipse representing a projected hole**

See Figure 1-47.

1. Define the **first point** as one of the points where the major axis intersects one of the centerlines.
2. Define the **second point** as the other point on the major axis that intersects the same centerline.
3. Define the **third point** as one of the points on the minor axis that intersects the centerline perpendicular to the centerline used in steps 1 and 2.

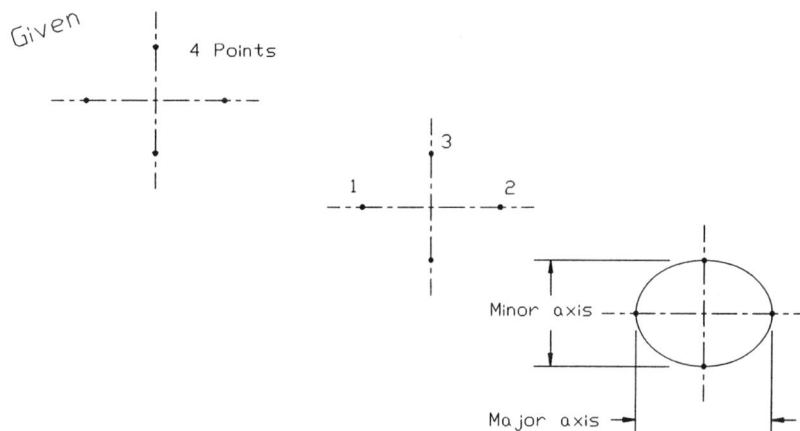

Figure 1-47

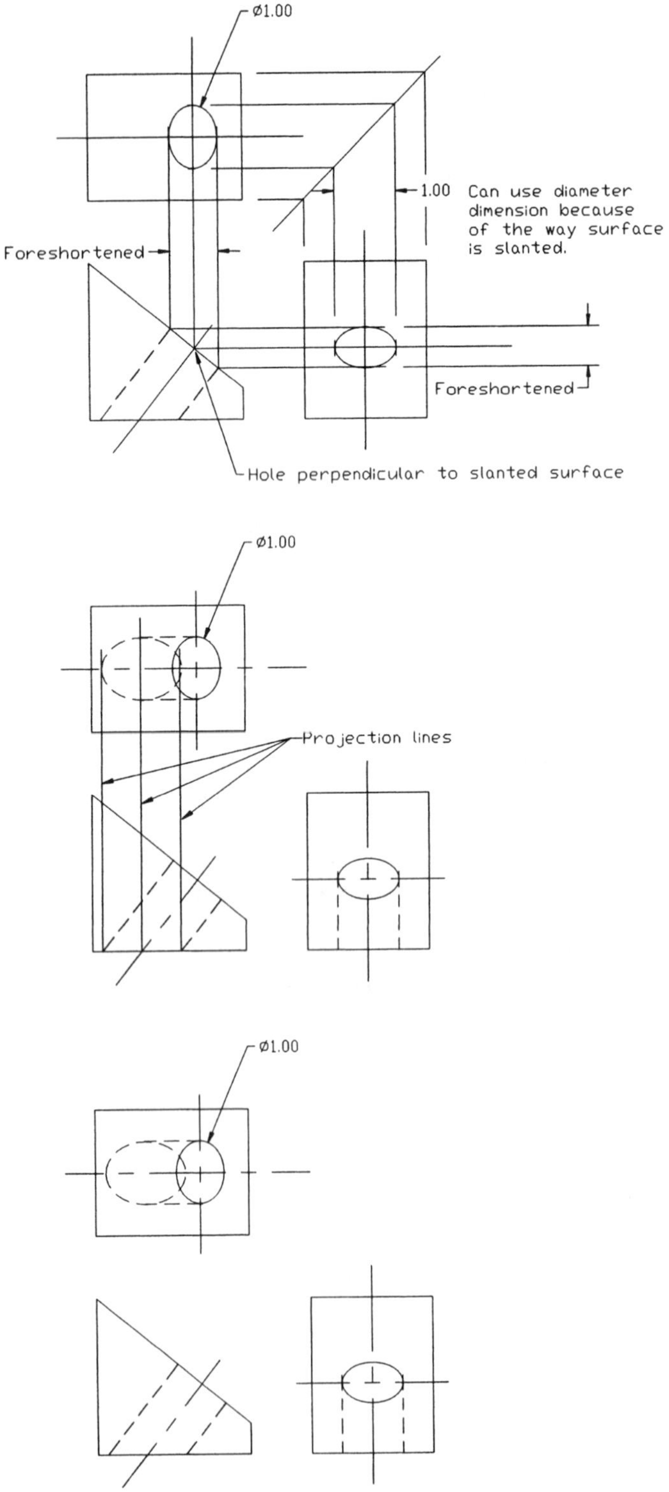

**Figure 1-48**

## To draw three views of a hole in a slanted surface

Figure 1-48 shows an object that includes a hole drilled perpendicular to the slanted surface. The hole is 1.00 in diameter. The hidden lines in the front view that represent the edges of the hole are drawn parallel to the centerline at a distance of .50. The horizontal and vertical centerlines of the hole are located in the top and right views. The intersection of the hole edges with the slanted surface shown in the front view is projected into the top and side views as shown.

The slanted surface shown in Figure 1-48 appears as a straight line in the front view. This means that the hole representations in the top and side views are rotated about only one axis. This in turn means that the hole representation is foreshortened in only one direction. The other direction, the other axis length, remains at the original diameter distance on 1.00. The four points needed to define the projected elliptical shape can be defined by the projection lines and the measured 1.00 distance.

The hole also penetrates the bottom surface, forming a different-sized ellipse. The hole's penetration can be shown in the side view by hidden lines parallel to the given centerline. The elliptical shape in the top view is defined by drawing horizontal parallel lines from the intersection of the hole's projected shape with the vertical centerline, and by projection lines from the front view that define the hole's length.

## To draw three views of a hole through an oblique surface

Figure 1-49A shows a hole drilled horizontally through the exact center of an oblique surface of an object. The oblique surface means that both axes in the top and

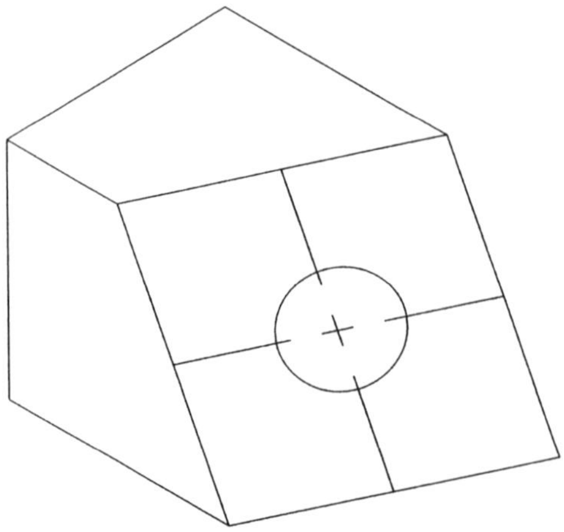

**Figure 1-49A**

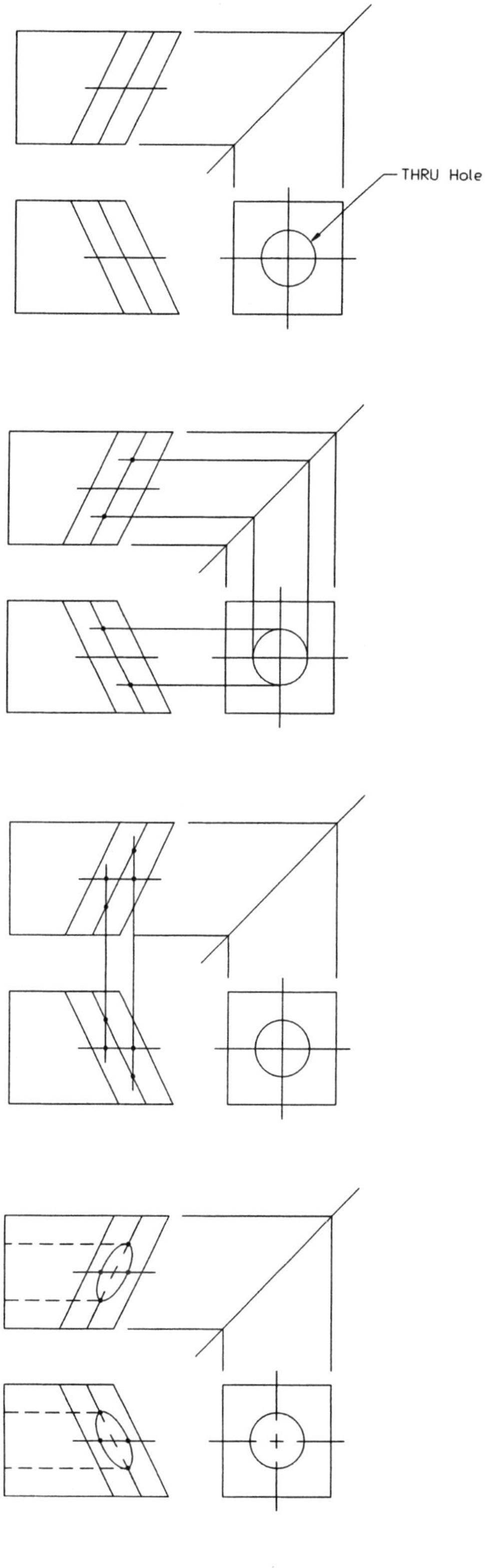

front views will be foreshortened. The procedure used to create the hole's projection in the front and top views is as follows. (See Figure 1-49B.)

1. Draw the hole in the side view. The hole will appear as a circle because it is drilled horizontally.
2. Draw the hole's centerlines in the front and top views. The hole is located exactly in the center of the oblique surface, so lines parallel to the surface's edge lines located midway across the surface may be drawn.
3. Project the intersection of the hole and the horizontal centerline into the top view, and project the intersection of the hole with the vertical centerline into the front view.
4. Project the intersections created in step 3 with the horizontal centerline, represented by the slanted centerline, onto the horizontal centerline in the front view. Likewise, project the intersection points created in step 3 on the horizontal centerline in the front view to the slanted centerline in the top view.
5. Draw the ellipses as required, add the appropriate hidden lines, and erase and trim any excess lines. Save the drawing if desired.

## 1-19 CYLINDERS

Figure 1-50 shows three views of a cylinder that is cut along the centerline by surface A-B-C-D. Note that each of the three views includes centerlines. The front and

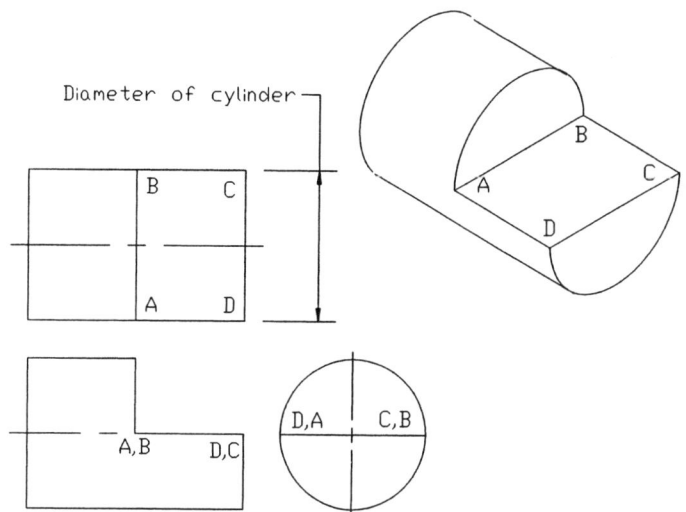

**Figure 1-49B**

**Figure 1-50**

**Figure 1-51**

**Figure 1-52**

top views of a cylinder are called the *rectangular views,* and the side view (end view) is called the *circular view.* The width of the top view is equal to the diameter of the cylinder.

Figure 1-51 shows a cylinder that has a second surface located above the centerline in the front view. The top view of the cylinder shows the flat surface and the rounded portion of the cylinder. Both the flat surface and the two rounded surfaces appear as rectangles.

Figure 1-52 shows a cylinder with a surface J-K-L-M that is located below the centerline in the front view. The top

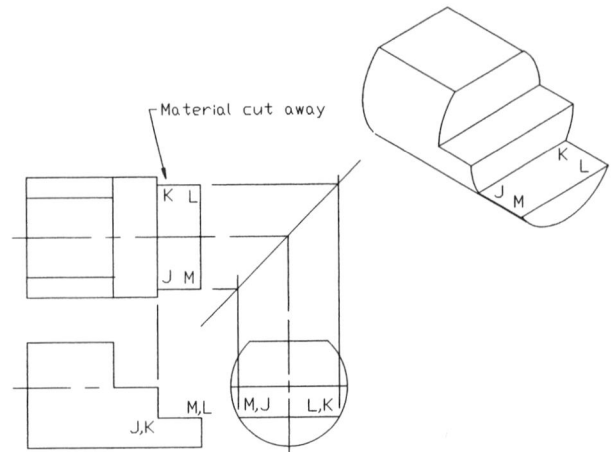

view of this surface does not include any rounded surfaces because they have been cut away.

## 1-20 SAMPLE PROBLEM SP1-5

Figure 1-53 shows a cylindrically shaped object that has three surface cuts: one above the centerline, one below the centerline, and one directly on the centerline. Figure 1-54 shows how the three views of the object were developed. The procedure is as follows.

**Figure 1-53**

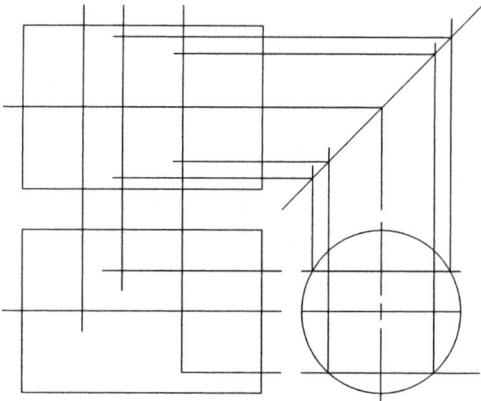

1. Use the given overall length of the cylinder and its diameter to create the two rectangular top and front views and the circular end views.
2. Draw horizontal lines in the side view that define the end views of the three surfaces. Project the size and location of the surfaces into the front and top views. Use **Osnap, Intersection,** with **Ortho** on, for accurate projection.
3. **Erase** and **Trim** the excess lines.
4. Save the drawing if desired.

## 1-21  CYLINDERS WITH SLANTED AND ROUNDED SURFACES

Figure 1-55 shows a front and side view of a cylindrical object that includes a slanted surface. Slanted surfaces are projected by defining points along their edges in known orthographic views and then projecting the edge points. At least two known views are needed for accurate projection. Sample Problem SP1-6 shows how to define and project a slanted cylindrical surface into the top view given the front and side views.

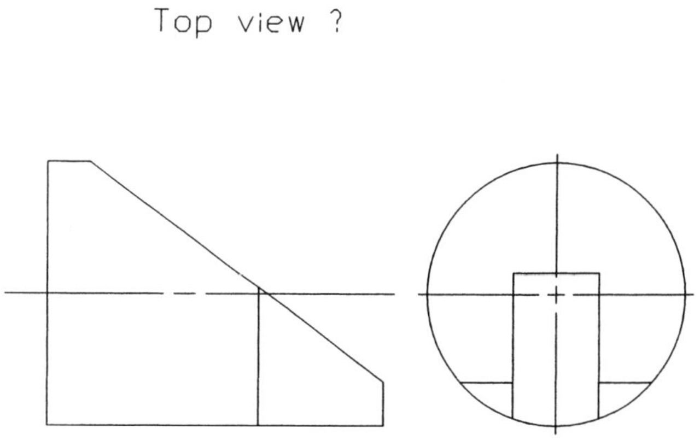

Top view ?

**Figure 1-54**

**Figure 1-55**

## 1-22 SAMPLE PROBLEM SP1-6

(See Figure 1-56.)

1. Draw the rectangular shape of the top view by projecting the length from the front view and the width from the side view.
2. Project the centerlines, and label points 1, 2, and 3 in the front and side views. The locations of these points are known from the given information.
3. Draw three horizontal lines across the front and side views. The location of the lines is arbitrary. Label the intersections of the horizontal lines with the edge of the slanted surface as shown.
4. The three horizontal lines serve to define point locations along the surface's edge.

A line contains an infinite number of points, so any six can be used.

5. Draw vertical projection lines (**Osnap, Intersection**) from the six point locations in the front view into the area of the top view.
6. Project the six point locations from the side view into the area of the top view. Use the 45° miter line to make the turn between the two views. The intersection of the vertical projection lines of step 2 and the projection lines from the side view defines the location of the points in the top view. Label the six points.
7. Use **Draw, Polyline, Edit Polyline, Fit** to create a smooth line between the six projected points.
8. Remove any excess lines and save the drawing if desired.

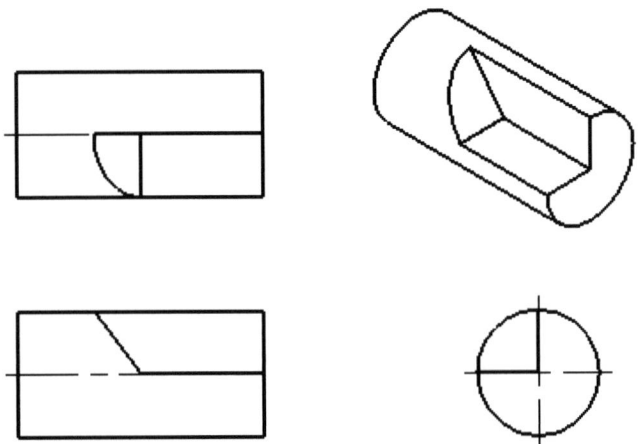

**Figure 1-56**

**Figure 1-57**

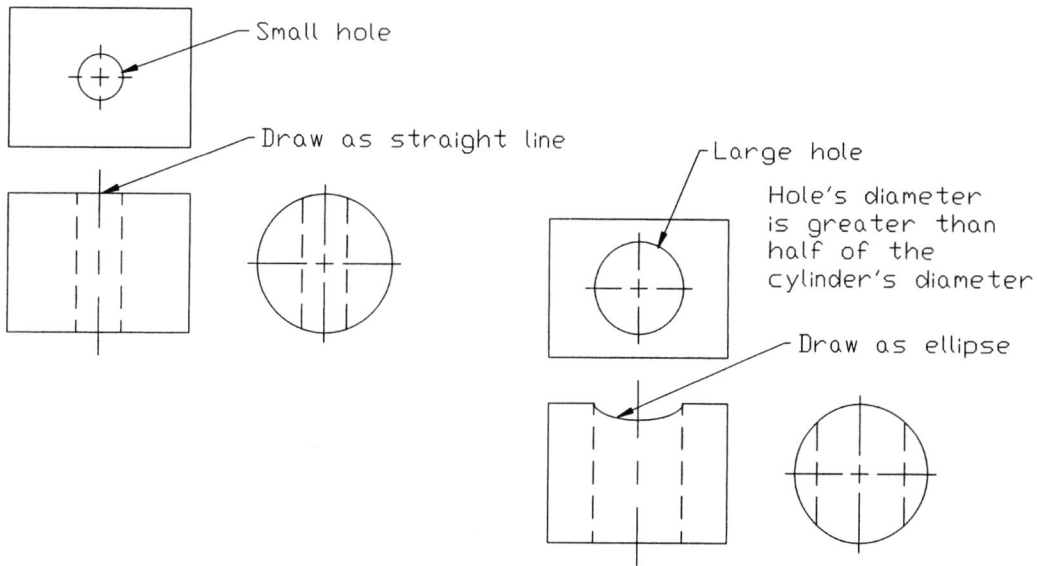

**Figure 1-58**

Figure 1-57 shows another cylindrical object that contains a slanted surface along with some of the projection lines.

## 1-23 DRAWING CONVENTIONS AND CYLINDERS

A hole drilled into a cylinder will produce an elliptically shaped edge line in a profile view of the hole. See Figure 1-58. Drawing convention allows a straight line to be drawn in place of the elliptical shape for small holes. The elliptical shape should be drawn for large holes in cylinders, that is, one whose diameter is greater than half the diameter of the cylinder.

Drawing convention also permits straight lines to be drawn for the profile view of a keyway cut into a cylinder. See Figure 1-59. As with holes, if a keyway is large relative to a cylinder, the correct offset shape should be drawn. A keyway is considered large if its width is greater than half the diameter of the cylinder.

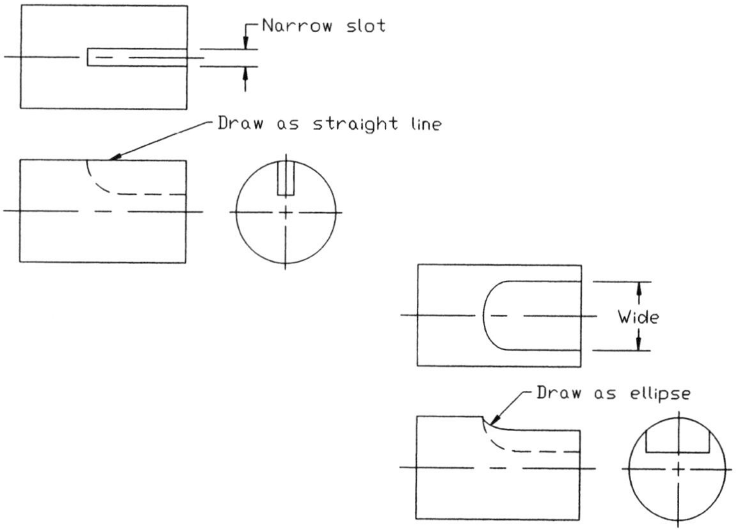

**Figure 1-59**

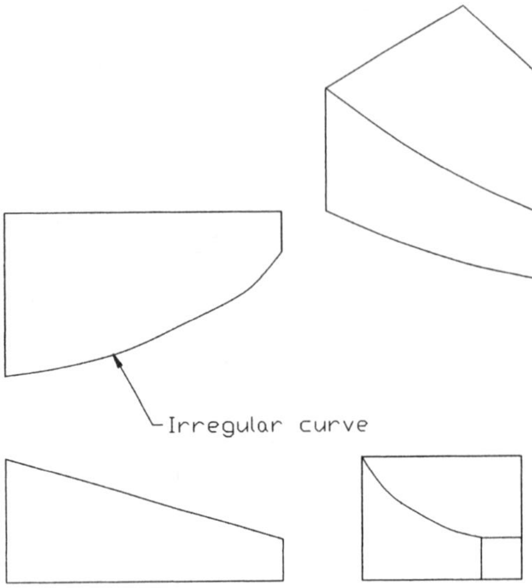

Figure 1-60

## 1-24 IRREGULAR SURFACES

*Irregular surfaces* are curved surfaces that do not have constant radii. See Figures 1-60 and 1-61. Irregular surfaces are defined using the X,Y coordinates of points located on the edge of the surface. The points may be dimensioned directly on the view but are often presented in chart form, as shown in Figure 1-61. Figure 1-61 also shows a wing surface defined relative to a given X,Y coordinate system.

Irregular surfaces are projected by defining points along their edges and then projecting the points. At least two known views are needed for accurate projection. The more points used to define the curve, the more accurate the final curve shape will be.

## 1-25 SAMPLE PROBLEM SP1-7

Figure 1-62 shows an object that includes an irregular surface. The irregular surface is defined by points referenced to an X,Y coordinate system. The point values are listed in a chart. Draw three views of the object shown in Figure 1-62. Your views should look similar to those in Figure 1-63.

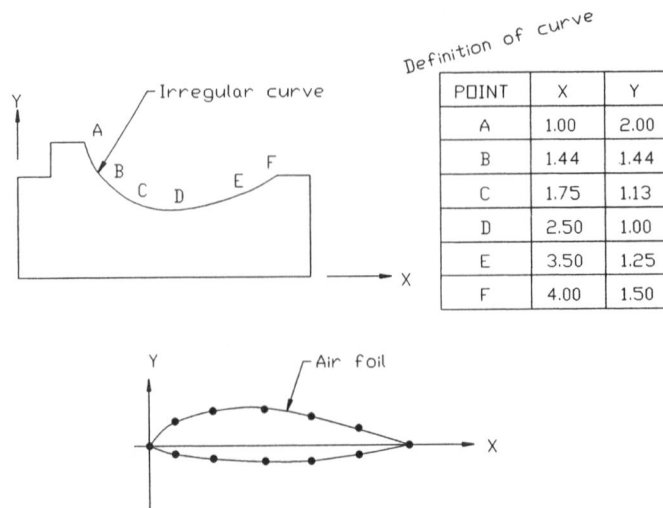

Definition of curve

| POINT | X | Y |
|-------|------|------|
| A | 1.00 | 2.00 |
| B | 1.44 | 1.44 |
| C | 1.75 | 1.13 |
| D | 2.50 | 1.00 |
| E | 3.50 | 1.25 |
| F | 4.00 | 1.50 |

Figure 1-61

1. Draw the front and top views using the given dimensions and chart values. Draw the outline of the side view using the given dimensions.
2. Project the points that define the irregular curve in the top view into the front view. Label the points.
3. Project the points for the curve from both the top and front views into the side view. Label the intersection points.
4. Use **Polyline, Edit Polyline,** and **Fit** to create a curve that represents the side view of the irregular surface.
5. **Erase** and **Trim** any excess lines. Save the drawing if desired.

Remember, it is possible to use **Layer** to create a layer for construction lines and for the final drawing. After the initial drawing is laid out, the final drawing lines may be copied onto another layer and the construction layer turned off. This method eliminates the need for extensive erasing and trimming.

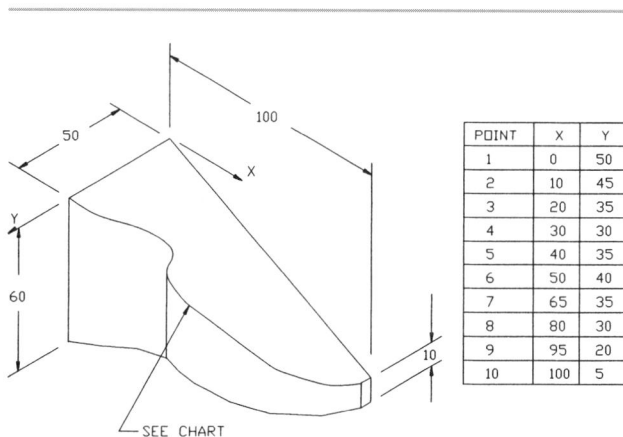

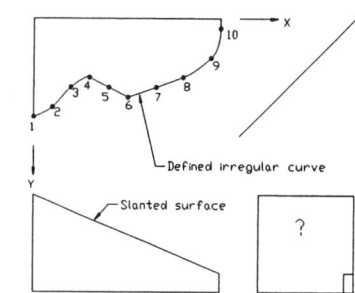

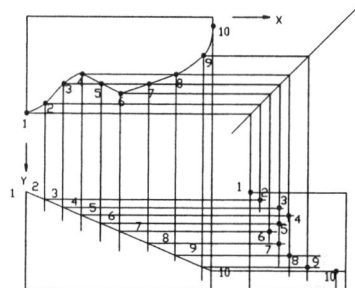

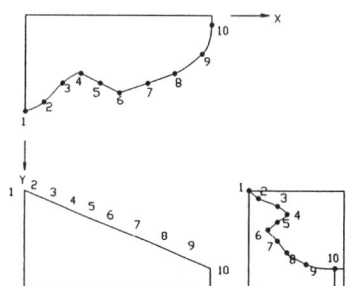

| POINT | X | Y |
|---|---|---|
| 1 | 0 | 50 |
| 2 | 10 | 45 |
| 3 | 20 | 35 |
| 4 | 30 | 30 |
| 5 | 40 | 35 |
| 6 | 50 | 40 |
| 7 | 65 | 35 |
| 8 | 80 | 30 |
| 9 | 95 | 20 |
| 10 | 100 | 5 |

**Figure 1-62**

**Figure 1-63**

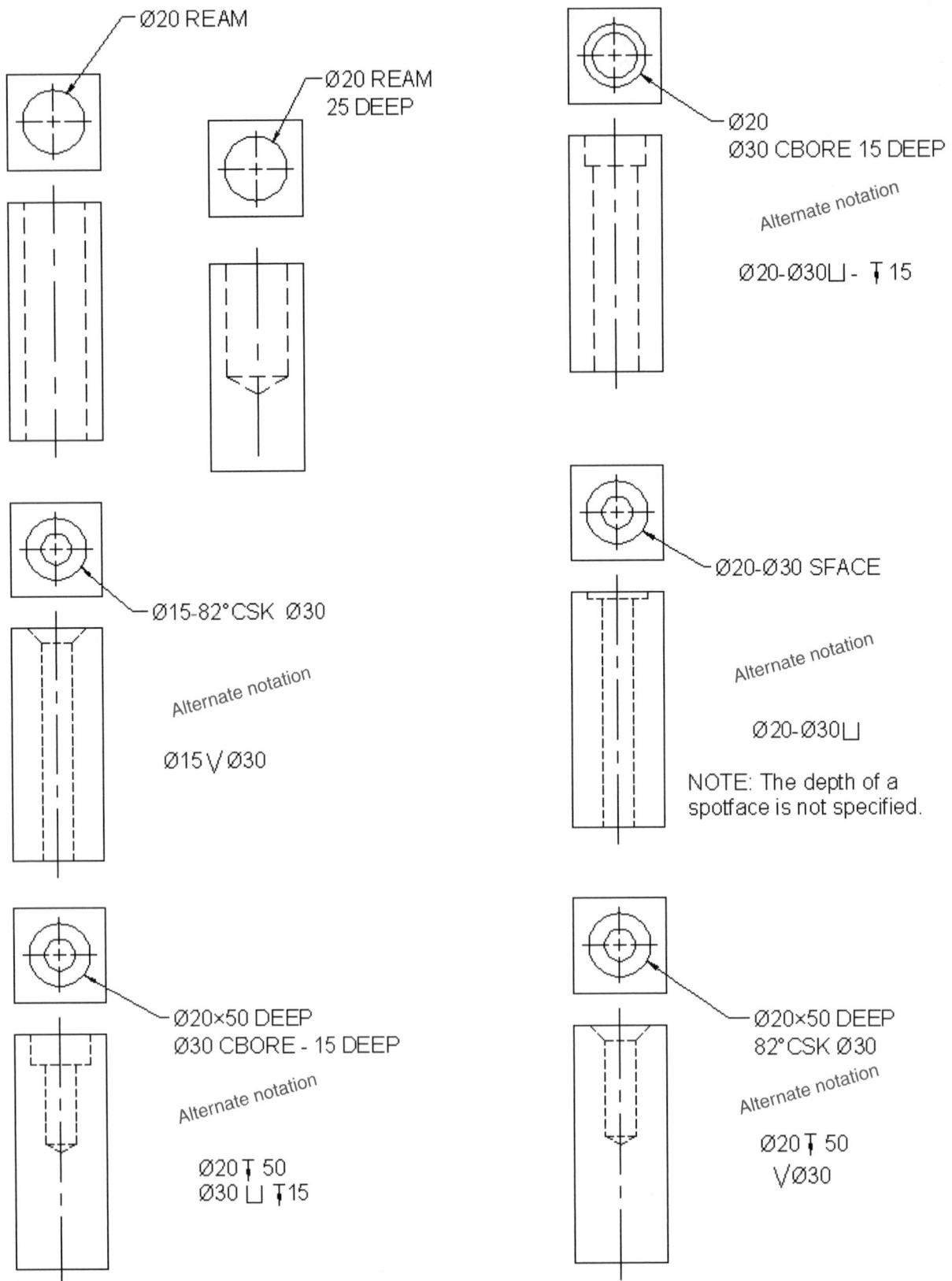

Ø20 REAM

Ø20 REAM
25 DEEP

Ø20
Ø30 CBORE 15 DEEP

Alternate notation

Ø20-Ø30⊔ - ⊤ 15

Ø15-82°CSK  Ø30

Alternate notation

Ø15 ∨ Ø30

Ø20-Ø30 SFACE

Alternate notation

Ø20-Ø30⊔

NOTE: The depth of a
spotface is not specified.

Ø20×50 DEEP
Ø30 CBORE - 15 DEEP

Alternate notation

Ø20 ⊤ 50
Ø30 ⊔ ⊤15

Ø20×50 DEEP
82°CSK Ø30

Alternate notation

Ø20 ⊤ 50
∨ Ø30

**Figure 1-64**

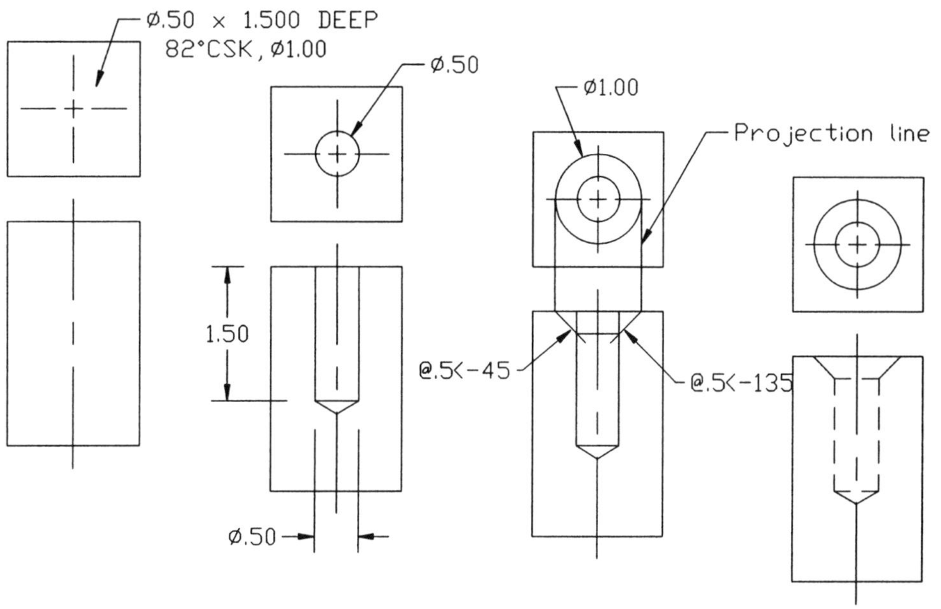

Ø.50 × 1.500 DEEP
82°CSK, Ø1.00

Ø.50

Ø1.00

Projection line

1.50

@.5<-45

@.5<-135

Ø.50

**Figure 1-65**

## 1-26 HOLE CALLOUTS

There are four hole-shape manufacturing processes that are used so often that they are defined using a standardized drawing callout: ream, counterbore, countersink, and spotface. See Figure 1-64.

*A ream* is a process that smooths out the inside of a drilled hole. Holes created using twist drills have spiral-shaped machine marks on their surfaces. Reaming is used to remove the spiral machine marks and to increase the roundness of the hole. Ream callouts generally include a much tighter tolerance than do drill diameter callouts.

A *counterbore* consists of two holes drilled along the same centerline. Counterbores are used to allow fasteners or other objects to be recessed, thus keeping the top surface uniform in height.

A counterbore drawing callout specifies the diameter of the small hole, the diameter of the large hole, and the depth of the large hole.

The information is given in this sequence because it is also the sequence the manufacturer uses. Note that the hidden lines used in the front view of the counterbored hole clearly show the intersection between the two holes.

Figure 1-64 shows an alternative notation for drawing callouts as defined by ISO standards (see Chapter 4) that is intended to remove language from drawings. As parts are often designed in one country and manufactured in another, it is important that drawing callouts be universally understood.

A *countersink* is a conical-shaped hole used primarily for flat-head fasteners. The drawing callout specifies the diameter of the hole, the included angle of the countersink, and the diameter of the countersink as measured on the surface of the object. Almost all countersinks are 82°, but some are drawn at 45°.

**To draw a countersunk hole (See Figure 1-65.)**

1. Draw a **0.50**-diameter hole in the top view. Project the hole's diameter into the front view.
2. Draw a **1.00**-diameter hole in the top view and project its diameter into the front view. In the front view draw **45°** lines from the intersection of the 1.00-diameter hole and the top surface of the front view so that the lines intersect the 0.50-diameter hole's projection lines. Use **Osnap, Intersection** to ensure accuracy.
3. Draw a horizontal line between the intersection created in step 2. Again use **Osnap, Intersection.**
4. **Erase** and **Trim** the excess lines.

A *spotface* is a very shallow counterbored hole generally used on cast surfaces. Cast surfaces are more porous than machine surfaces. Rather than machine an entire surface flat, it is cheaper to manufacture a spotface because only a small portion of the surface is machined. Spotfaces are usually used for bearing surfaces of fasteners.

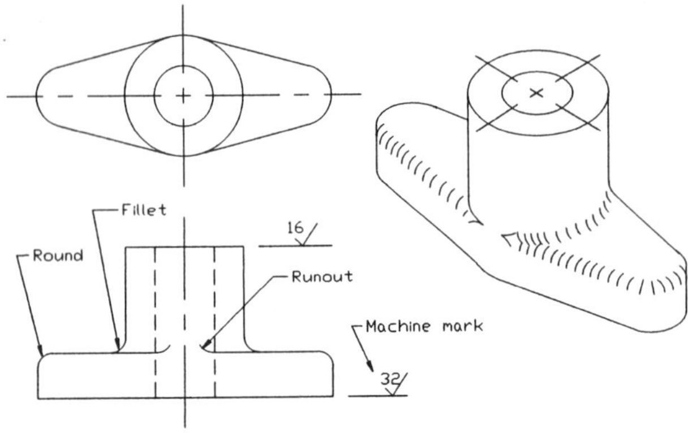

**Figure 1-66**

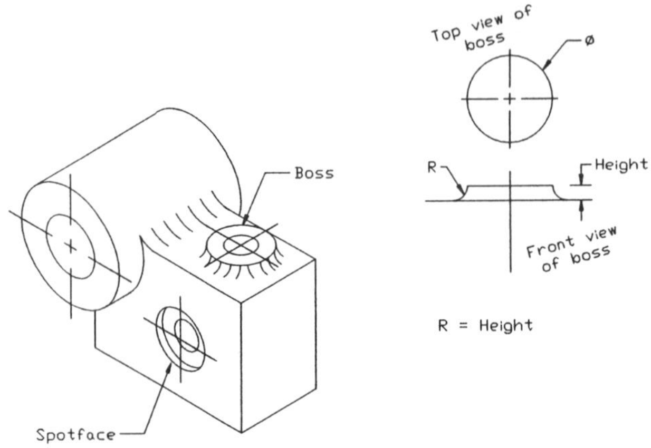

**Figure 1-67**

A spotface callout defines the diameter of the hole and the diameter of the spotface. A depth need not be given. When machining was done mostly by hand, the machinist would make the spotface just deep enough to produce a shiny surface (most cast surfaces are more gray in color), so no depth was needed. Automated machines require a spotface depth specification. Usually the depth is very shallow.

## 1-27 CASTINGS

Casting is one of the oldest manufacturing processes. Metal is heated to liquid form, then poured into molds and allowed to cool. The resulting shapes usually include many rounded edges and surface tangencies because it is very difficult to cast square edges. Concave edges are called *rounds,* and convex edges are called *fillets.* See Figure 1-66. A runout is used to indicate that two rounded surfaces have become tangent to one another. A *runout* is a short arc of arbitrary radius, that is, arbitrary as long as the runout is visually clear.

Cast objects are often partially machined to produce flatter surfaces than can be produced by the casting process. Machined surfaces are defined using machine marks (check marks), as shown in Figure 1-66. The numbers within the machine marks specify the flatness requirements in terms of microinches or micrometers. Machine marks are explained in greater detail in Chapter 4.

A *boss* is a turretlike shape that is often included on castings to localize and minimize machining. A boss is defined by its diameter and its height. The sides of a boss are

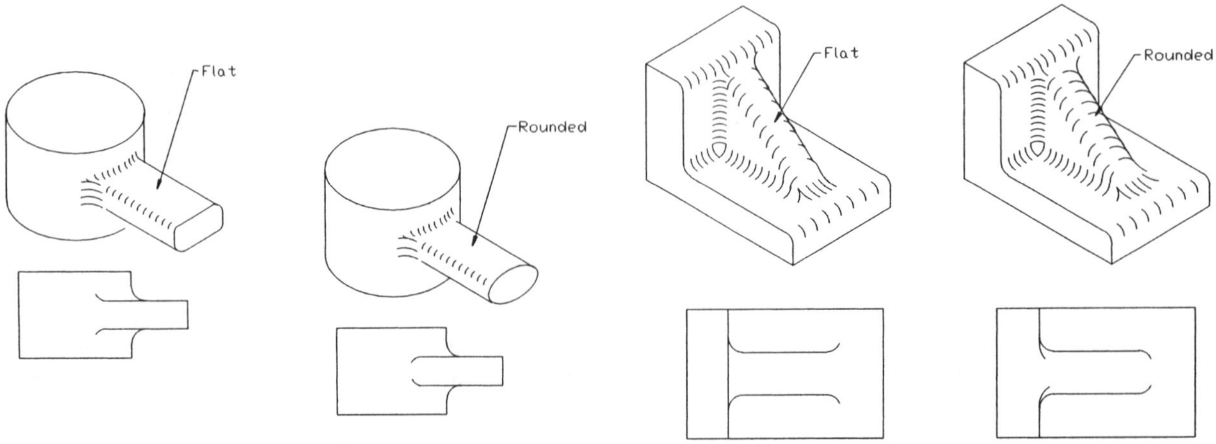

**Figure 1-68**

Figure 1-69

rounded and defined by a radius usually equal to the height of the boss. See Figure 1-67.

Spotfacing is a machine process often associated with castings. As with bosses, spotfaces help localize and minimize machining requirements. Spotfaces were defined in Section 1-26.

The direction of runouts is determined by the shape of the two surfaces involved. Flat surfaces intersecting rounded surfaces generate runouts that turn out. Rounded surfaces that intersect rounded surfaces generate runouts that turn in. See Figure 1-68.

The same convention is followed for flat and rounded surfaces that intersect flat surfaces.

The location of a runout is determined by the location of the tangent it represents. See Figure 1-69. The location of a tangency point can be determined by first drawing the top view of the tangency line using **Osnap, Tangent.** A line can then be drawn from the center point of the circular top view to the end of the tangency line using **Osnap, Endpoint.** The point of tangency can then be projected from the top view to the front view using **Osnap, Intersection** with **Ortho** on.

Use **Arc** (type the word **arc** in response to a command prompt) to draw runouts. Any convenient radius may be used, providing the runout is clearly visible and visually distinct from the straight line. Figure 1-70 shows some further examples of cast surfaces that include runouts.

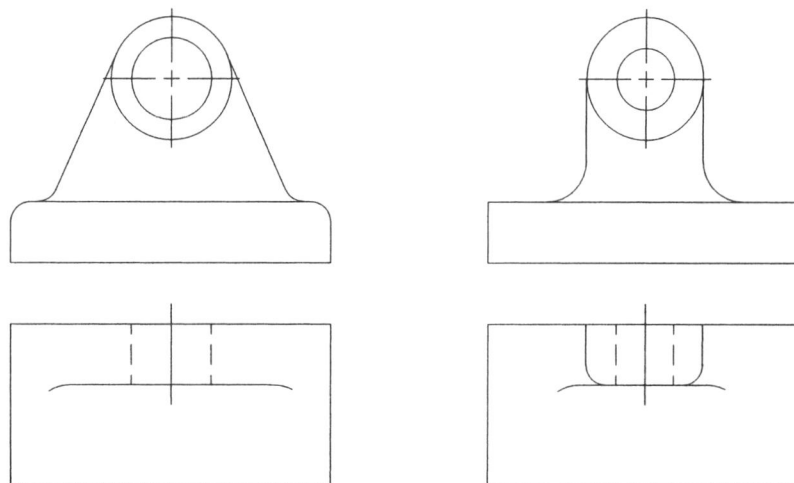

Figure 1-70

## 1-28  SAMPLE PROBLEM SP1-8

Draw three views of the object shown in Figure 1-71. The solution was drawn as follows:

1. Use the given overall dimensions and draw the outline of the front, top, and side views.
2. Modify the appropriate lines to hidden lines. Draw the fillets and rounds. The **Fillet** command may erase needed lines. This cannot always be avoided, so simply redraw the needed lines after the fillet or round is complete.
3. Draw the spotface and boss in each of the views. The hole in the boss is **.500** deep. Clearly show the bottom of the hole, including the conical point. Draw the hole in the front view and use **Copy** to copy the hole into the side view.
4. Use **Draw, Circle** to draw the **1.00**-diameter hole in the front view. Draw the appropriate hidden and centerlines in the other views.
5. Save the drawing if desired.

See Figure 1-72.

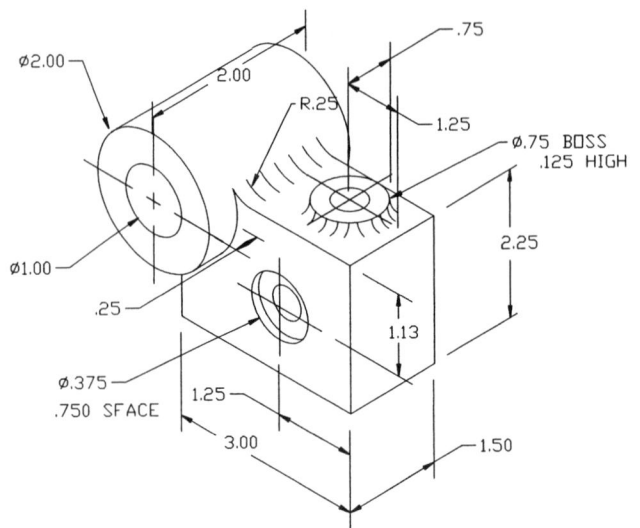

**Figure 1-71**

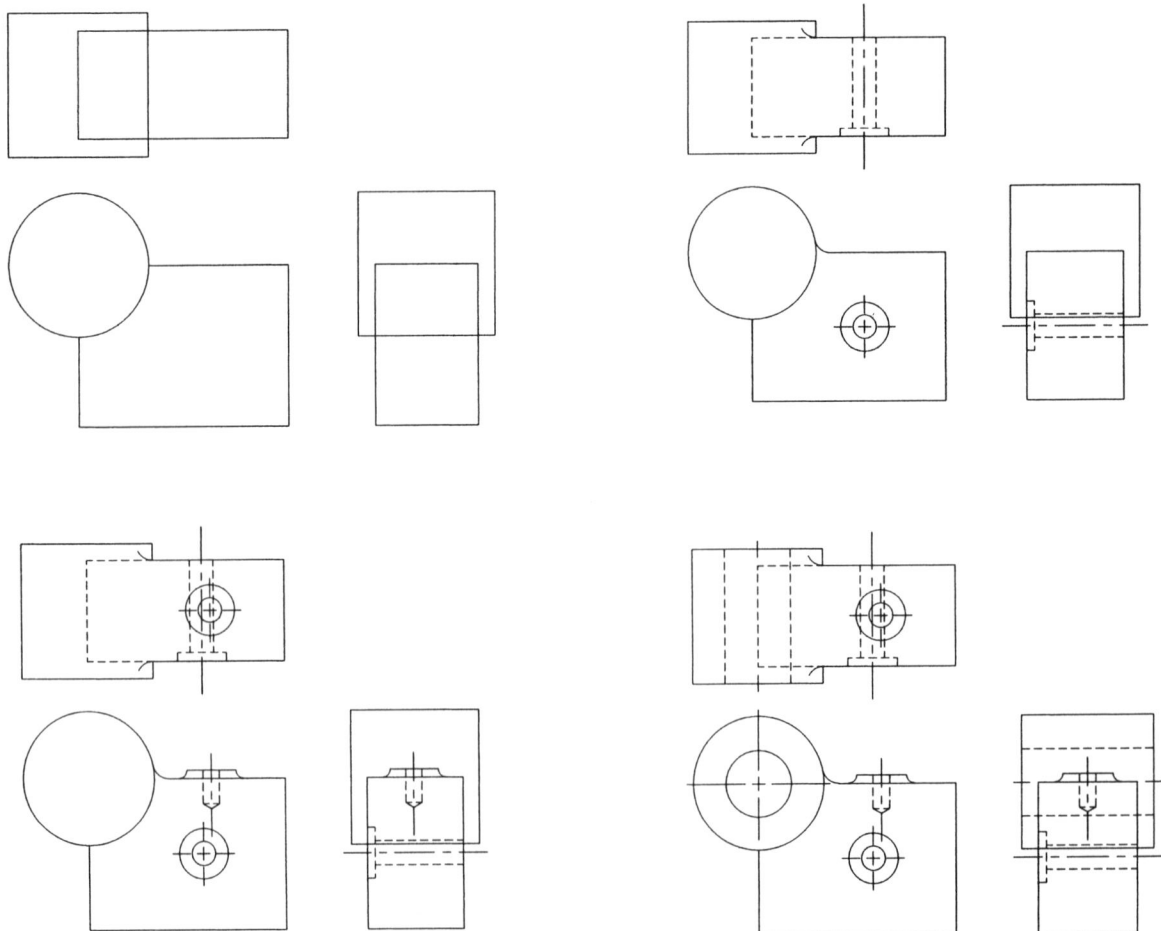

**Figure 1-72**

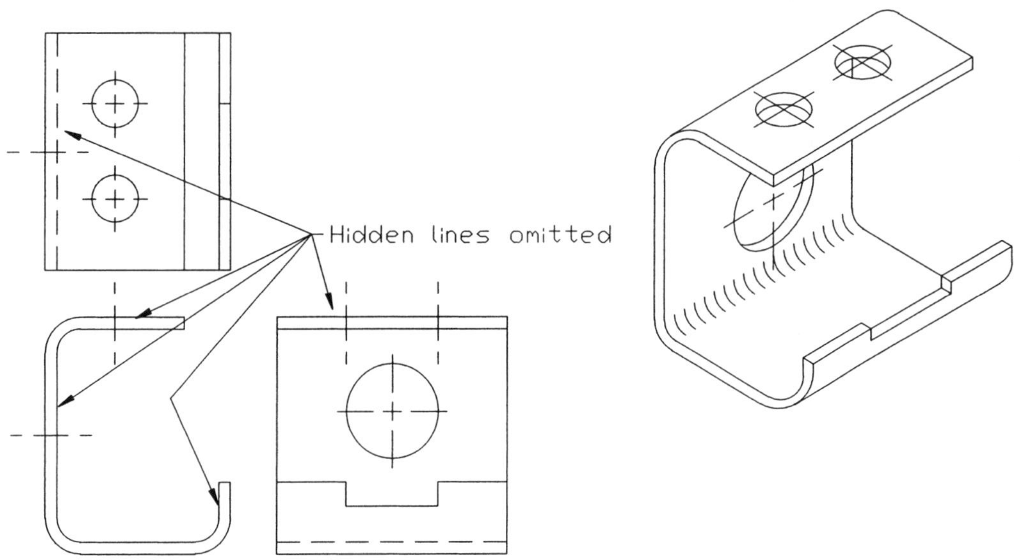

**Figure 1-73**

## 1-29 THIN-WALLED OBJECTS

Thin-walled objects such as parts made from sheet metal or tubing present some unique drawing problems. For example, the distance between surfaces is usually so small that hidden lines can't be used to define holes, and there isn't enough distance to draw a broken-line pattern. See Figure 1-73.

Hidden lines may be applied to thin-walled objects in one of three ways: draw the lines as continuous lines but use a different color from that used for the actual continuous lines; use an enlarged detail of the area including the hidden lines; or omit the hidden lines and include only a centerline.

Each of these techniques is shown in Figure 1-73. All holes must include a centerline.

Companies often have a drawing manual that states their policy on drawing hidden lines in thin-walled objects. The most important goal is to be consistent in the representation.

Many sheet-metal parts are manufactured by bending. Bending produces an inside bend radius and an outside bend radius. The inside bend radius plus the material thickness should equal the outside bend radius. See Figure 1-74. The same radius should not be used for both the inside and outside bends.

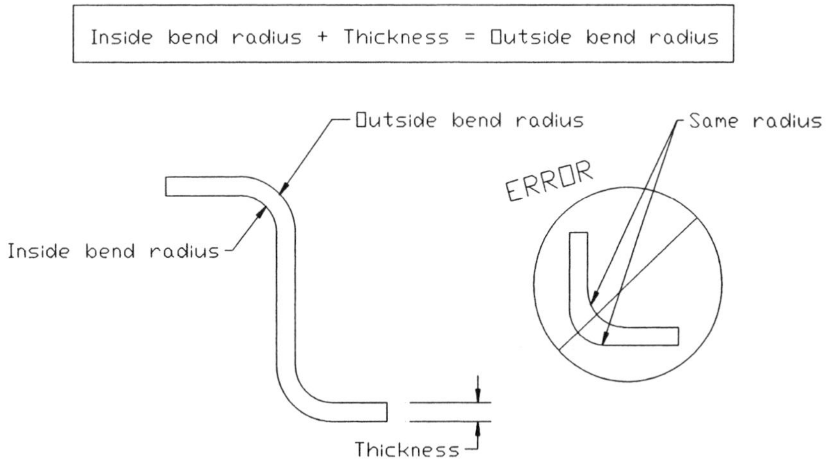

Inside bend radius + Thickness = Outside bend radius

**Figure 1-74**

# 1-30 SAMPLE PROBLEM SP1-9

Draw three views of the object shown in Figure 1-75. Figure 1-76 shows how the three views were developed.

1.  Use the given overall dimensions to draw the outline of the three views. Draw the object as if all bends are 90°.
2.  Use **Offset,** set to a distance of **3 mm,** to draw the thickness of the object. Use **Extend** and **Trim** to add or remove lines as needed. The thickness can also be drawn by setting the **Snap** spacing equal to the material thickness.

    Change the appropriate lines to hidden lines.

3.  Draw fillets using a radius of **5 mm** for the inside bend radii and **8 mm** for the outside bend radii.
4.  Use **Draw, Circle** to draw the holes in the appropriate view and add centerlines as shown.
5.  **Save** the drawing if desired.

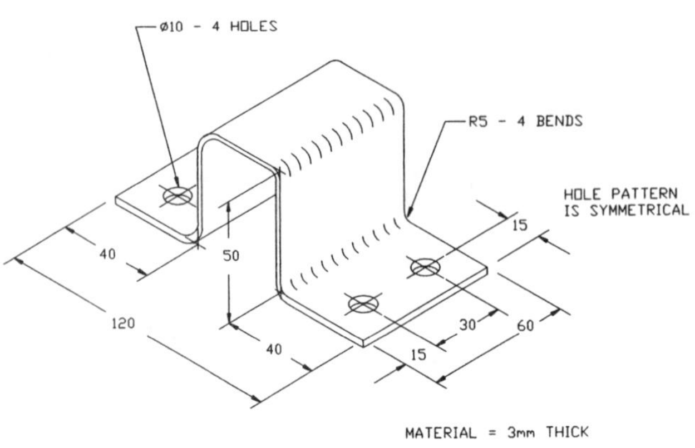

**Figure 1-75**

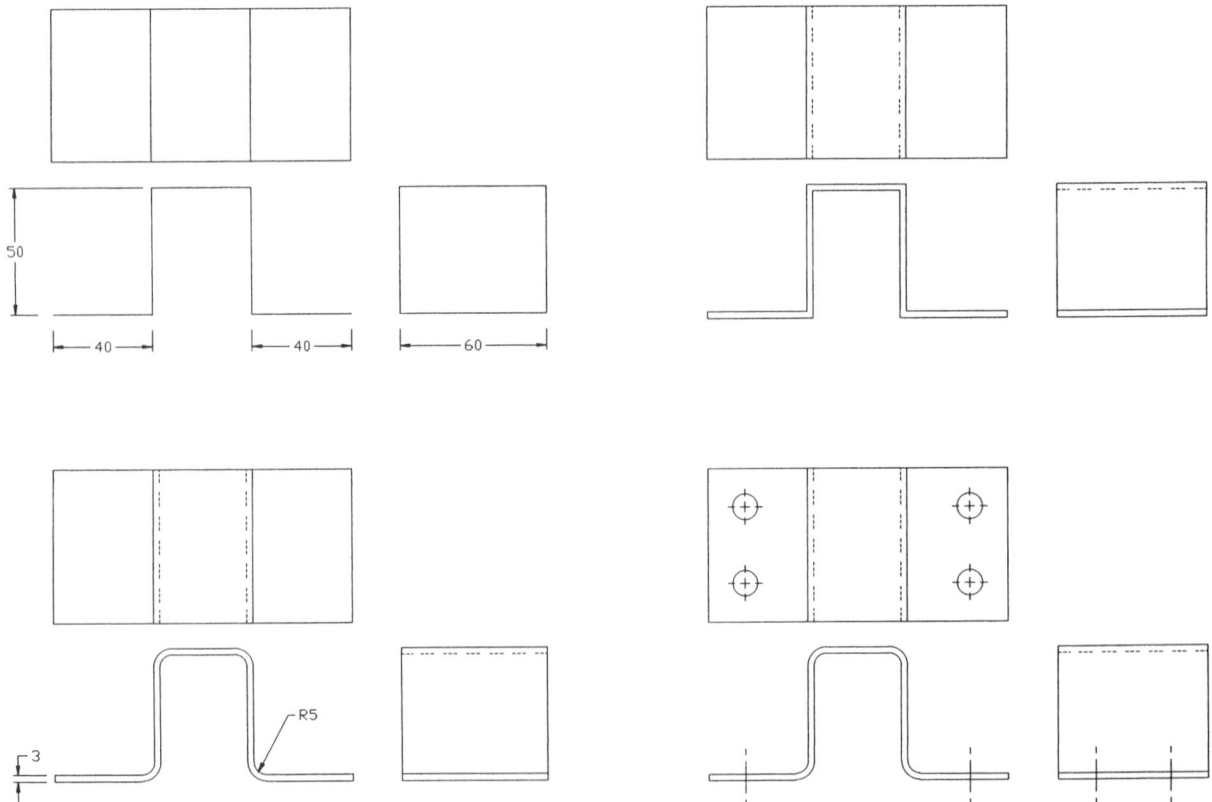

**Figure 1-76**

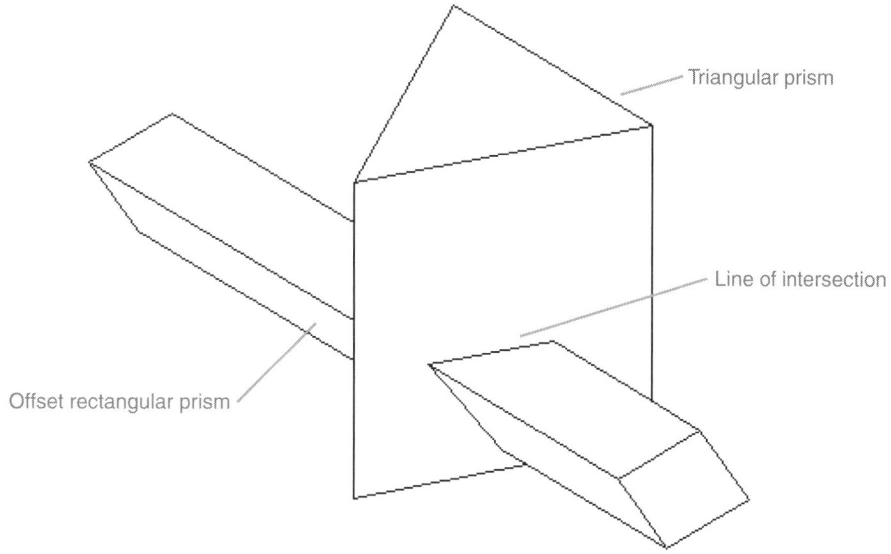

Triangular prism

Line of intersection

Offset rectangular prism

Figure 1-77

## 1-31 INTERSECTIONS

*Intersection drawings* are drawings that show the intersection of two objects. Figure 1-77 shows the intersection between two offset squares. A discussion of intersections has been included at this point in the book because they rely heavily on projection of information between views. They require not only a knowledge of the principles of projection but also an understanding about what the various lines represent. Intersections will be discussed again in Chapter 16, Solid Modeling.

Three intersection problems are presented in the following three sample problems.

## 1-32 SAMPLE PROBLEM SP1-10

Given the side and top views of a smaller circle intersecting a larger circle as shown in Figure 1-78, draw the front view. All dimensions are in inches. Because the object is symmetrical, the intersection for one side will be developed and then mirrored.

1. Draw the outline of the front view by projecting lines from the given top and front views. See Figure 1-79.
2. Define points on the edge of the smaller circle. In this example, 17 points were defined.

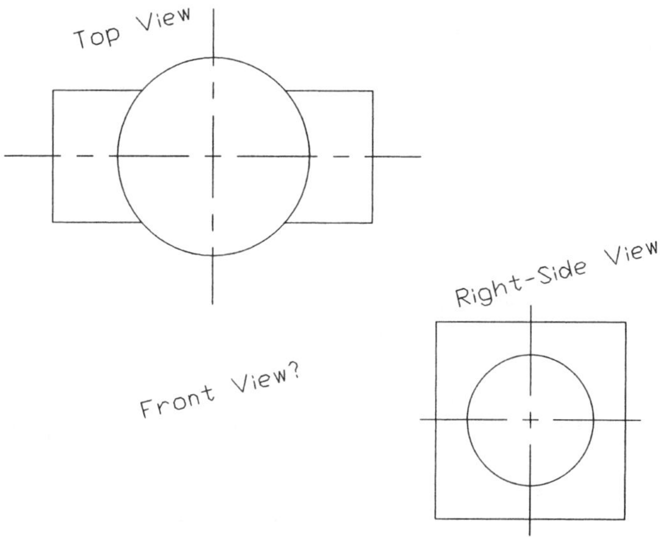

Top View

Front View?

Right-Side View

Figure 1-78

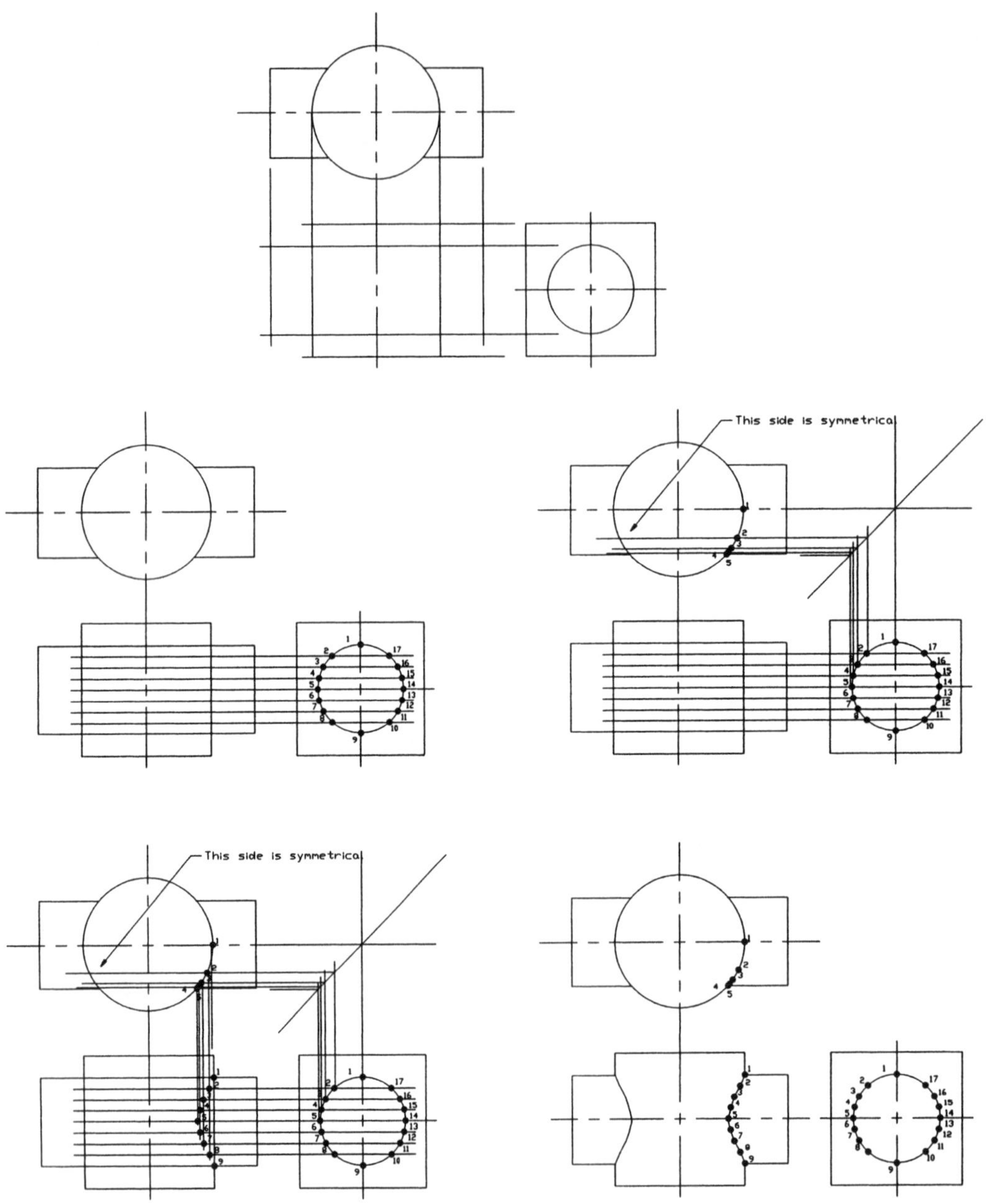

**Figure 1-79**

3. Extend the horizontal centerline of the smaller circle in the side view into the area of the front view. Use **Draw, Offset** and add six lines parallel to the horizontal centerline in the side view. Label the intersection of the horizontal lines with the edge of the smaller circle as shown.

4. Project points **5, 6, 7, 8, 9,** and **10** into the top view. Use **Draw, Line** along with **Osnap, Intersection** to draw vertical lines from the side view so that they intersect the 45° miter line. Then, project the intersections on the miter line into the

top view using horizontal lines. Label the points as shown.

5. Project points **5, 6, 7, 8, 9,** and **10** from the top view into the area of the front view so they intersect the horizontal lines from the side view. Label the points as shown. Use **Draw, Polyline, Edit Polyline,** and **Fit** to draw the curve that represents the intersection.

6. Add the lines needed to complete the front view, remove all excess lines, and save the drawing if desired.

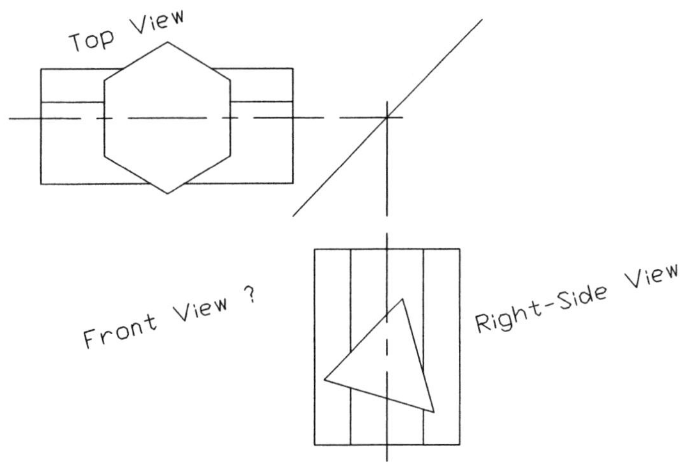

**Figure 1-80**

## 1-33  SAMPLE PROBLEM SP1-11

Figure 1-80 shows the top and right-side views of a tri-angular-shaped piece intersecting a hexagonal-shaped piece. What is the shape of the front view?

Figure 1-81 shows the solution obtained by projecting lines from the two given views into the front view, and shows an enlargement of the intersecting surfaces. Use a straight-edge and verify the location of each labeled point by project-ing horizontal lines from the front view and vertical lines from the top view into the front view.

## 1-34  SAMPLE PROBLEM SP1-12

Given the side and top views of a circle intersecting a cone as shown in Figure 1-82, draw the front view.

As in the previous sample problems, the problem is solved by defining intersection points in the given two views and then projecting them into the front view. The cone, how-ever, presents a unique problem because it is both round and tapered. There are few edge lines to work with.

The solution requires that there be a more precise definition of the cone's surface than is presented by the

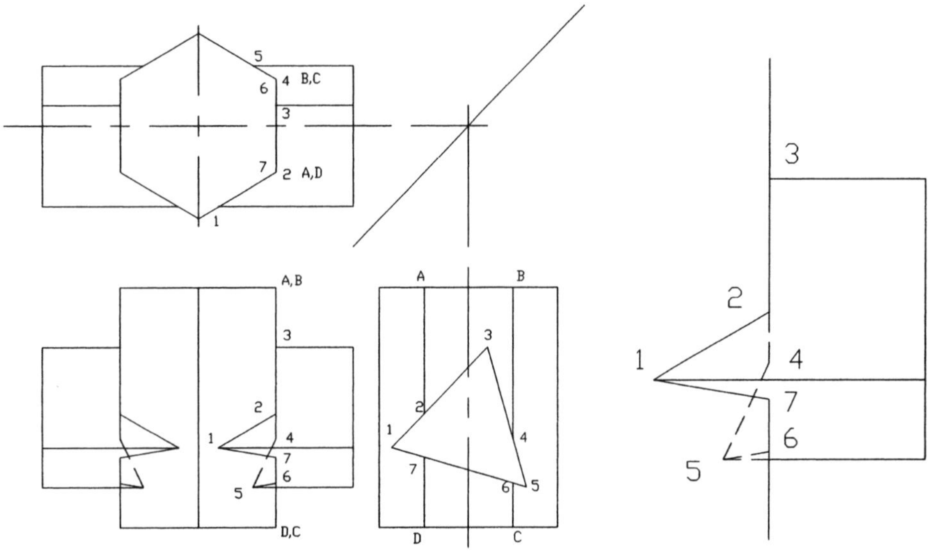

**Figure 1-81**

circle and centerlines in the top view and the profile front view.

1. Figure 1-83, step 1, shows how points **1** and **3,** located on the vertical centerline of the cylinder, are projected from the side view to the front view and then to the top view. The vertical centerline is aligned with the right-side profile line of the cone so that it can be used for projection.

This is not true for points 2 and 4, located on the horizontal centerline in the side view. Currently the location of points 2 and 4 is unknown in both the front and top views, so projection lines cannot be drawn. The location of points 2 and 4 can be determined in the front and top views.

2. Extend the horizontal centerline in the side view so that it intersects the edge lines of the cone. Use **Osnap, Intersection,** with **Ortho** on and project the intersection of the extended horizontal centerline and the cone's edge line so that it intersects the vertical centerline in the top view. Draw a circle in the top view using the distance between the intersection with the vertical centerline and the projection line and the center point of the cone as the radius.

The circle in the top view represents a slice of the cone located at exactly the same height as points 2 and 4. Points 2 and 4 must be located somewhere in the slice.

3. Project points **2** and **4** into the top view from the side view so that the projection line intersects the circular slice drawn in step 2. Label the intersections **2** and **4.**

4. Project the locations of points **2** and **4** in the top view into the front view using a vertical line. Project the locations of the points from the side view into the front view using a horizontal line. The intersection of the vertical and horizontal projection lines defines the locations of points 2 and 4 in the front view.

5. Expand the procedure explained in steps 2 and 3 by drawing a series of horizontal projection lines between the front and side views as shown. Use **Draw** for the first line and **Offset** for the other lines. The location of these additional lines is random.

Use **Osnap, Intersection** with **Ortho** on to project the intersections of the horizontal lines with the cone's edge lines in the top view.

6. Draw circles using the distance between the cone's center point and the projection lines' intersections with the vertical centerline as radii. Use **Osnap, Intersection** to locate the circles' center point and radii distances.

7. Use **Draw, Polyline, Edit Polyline,** and **Fit** to draw the required curves in the top and front views. Use **Osnap, Intersection** to ensure accurate curve point locations. The **Move** option located on the **Edit Polyline** command options line can be used to move vertex points on the curves to make them appear smoother and more continuous.

8. Trim and erase all excess lines and save the drawing if desired.

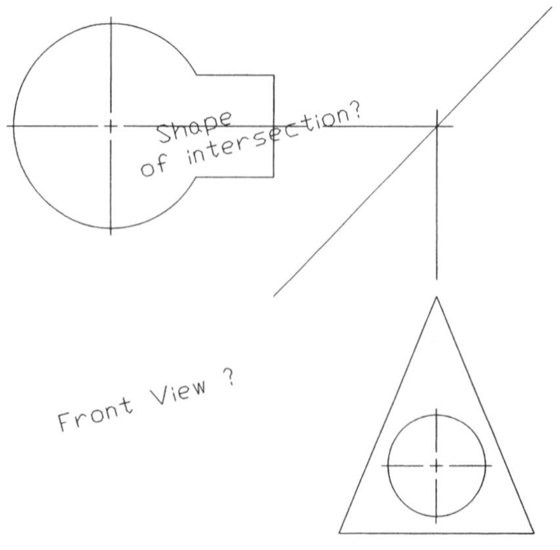

Figure 1-82

Step 1

Step 2

Slice of cone
that contains points
2 and 4.

Project intersection
point into the top
view.

Radius for
circular
cone slice.

EXTEND centerline
to the edge of
the cone.

Step 3

Step 4

Project the location
of points 2 and 4
into the top view.

Project point
from the top to
front view.

Project point location
from the side to the
front view.

Horizontal projection lines

Step 5

Step 6

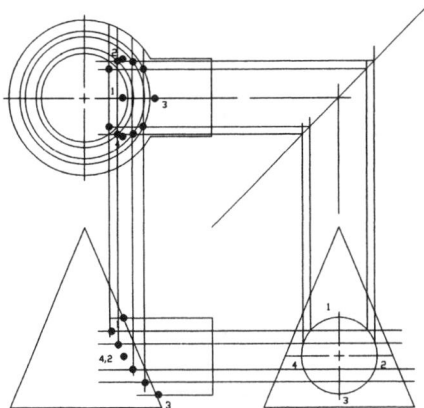

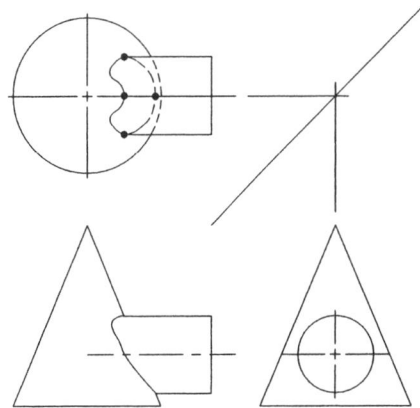

**Figure 1-83**

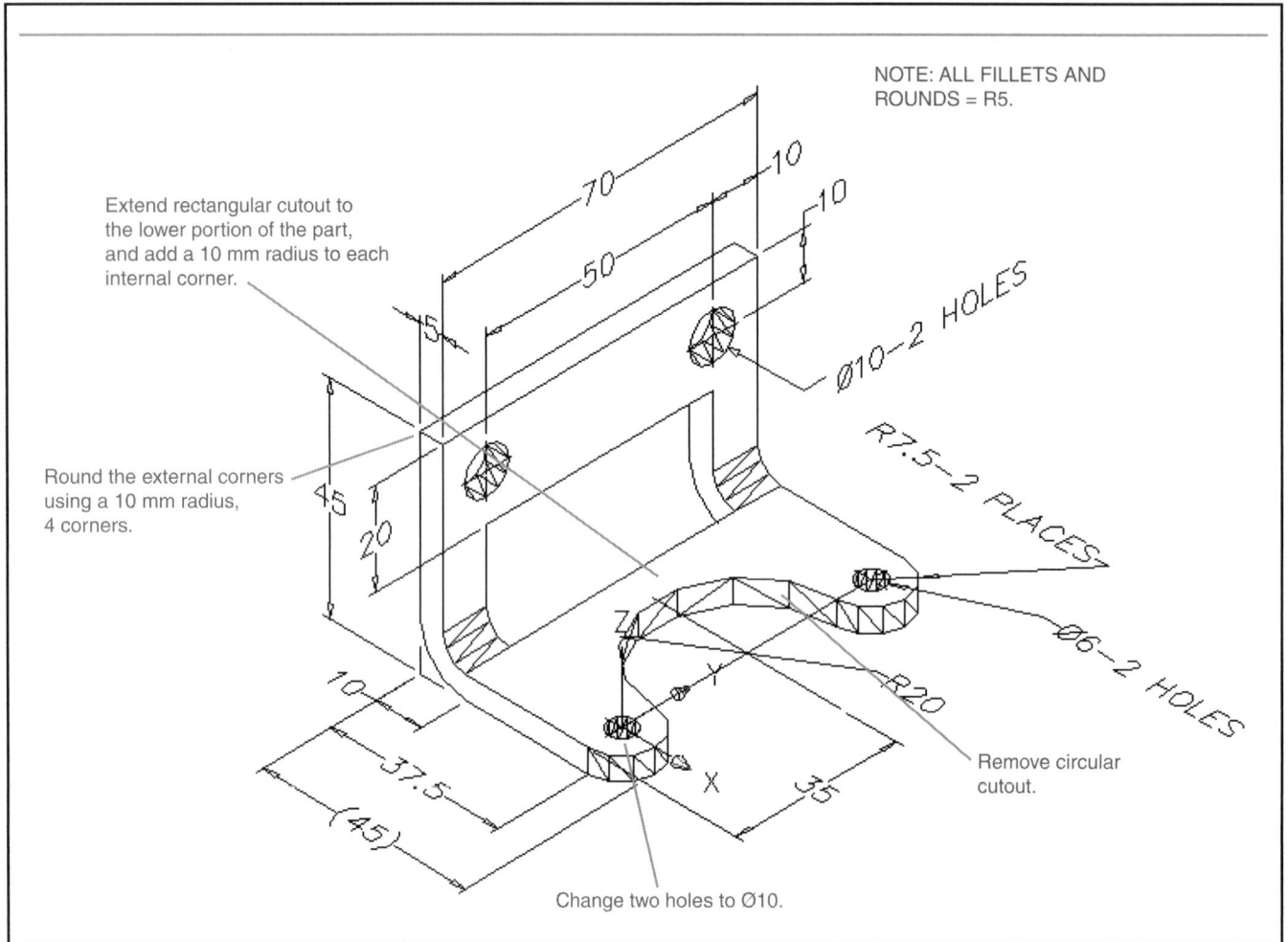

NOTE: ALL FILLETS AND
ROUNDS = R5.

Extend rectangular cutout to
the lower portion of the part,
and add a 10 mm radius to each
internal corner.

Round the external corners
using a 10 mm radius,
4 corners.

Ø10–2 HOLES

R7.5–2 PLACES

Ø6–2 HOLES

70

50

10

10

45

20

45

10

37.5

(45)

35

R20

Remove circular
cutout.

Change two holes to Ø10.

**Figure 1-84**

## 1-35  DESIGNING BY MODIFYING AN EXISTING PART

Many beginning design assignments require that an existing part be modified to meet a new set of requirements. Designers must therefore create something different from the original drawing. This ability to look beyond the drawing is an important design skill.

Figure 1-84 shows a dimensioned part, and Figure 1-85 shows the three orthographic views of the part. The part is to be redesigned as follows.

1. Replace the two **Ø7.5** holes with **Ø10** holes.
2. Remove the **R20** cutout.
3. Modify the horizontal portion of the part so its overall length is **45** millimeters; make the entire part symmetrical about its **90°** axis.

Original views

**Figure 1-85**

4.  Round the four external and four internal corners using a **10**-millimeter radius.

Figure 1-86 shows the resulting three orthographic views.

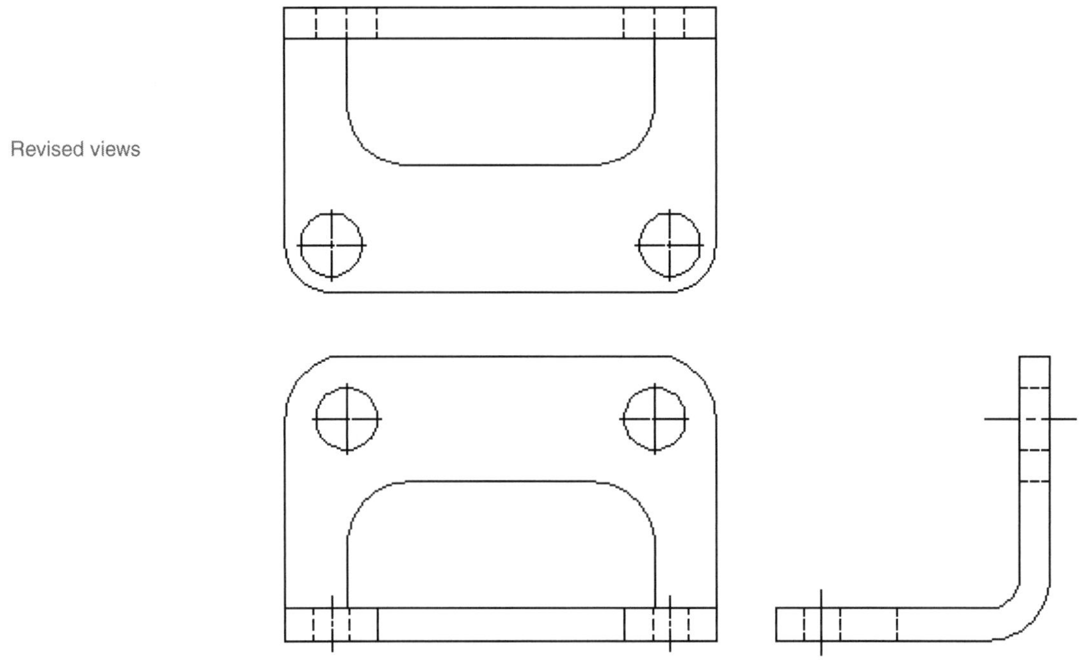

Revised views

**Figure 1-86**

# 1-36 EXERCISE PROBLEMS

Draw a front, top, and right-side orthographic view of each of the objects in Exercise Problems EX1-1 to EX1-94. Do not include dimensions.

### EX1-1 INCHES

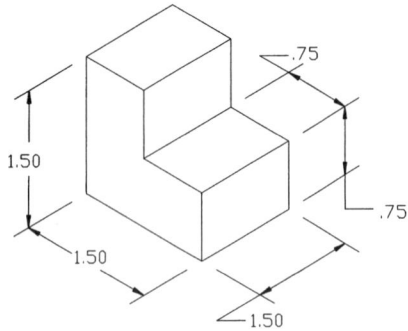

### EX1-2 MILLIMETERS

### EX1-3 MILLIMETERS

### EX1-4 INCHES

### EX1-5 MILLIMETERS

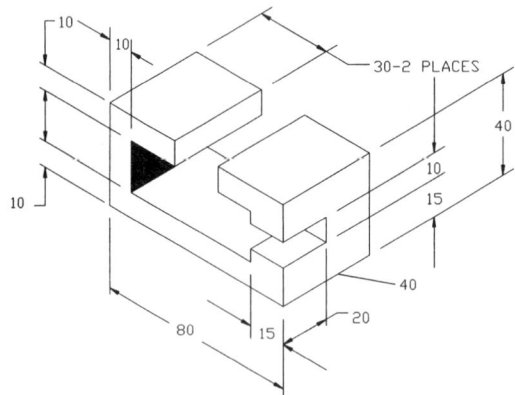

### EX1-6 MILLIMETERS

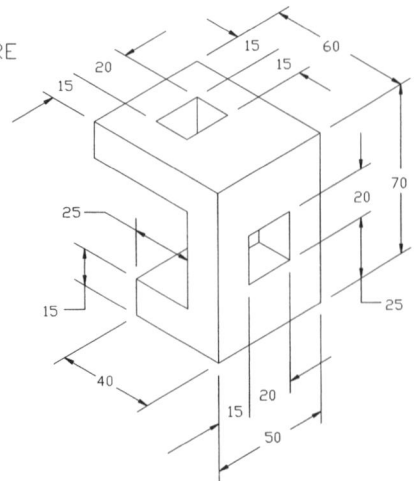

## EX1-7 MILLIMETERS

SETTER BRACKET

50

10 – ALL AROUND

30

40 –2 PLACES

25

15

100

30 –2 PLACES

## EX1-10 MILLIMETERS

10 BOTH SIDES

15

15

10

25

20

STANDOFF – ONE SIDE ONLY

10

60

20

30

80

## EX1-8 MILLIMETERS

10 – BOTH SIDES

40

15

15

S-CLIP

15

40

10 – BOTH SIDES

50

40

MATL = 10mm SAE 1020 STEEL

## EX1-11 MILLIMETERS

10

40

30°

15

20

5 All Around

30

8

50

2 Places

## EX1-9 INCHES

KEY CLIP

.75 BOTH SIDES

.50 BOTH SIDES

.50 ALL AROUND

.50

.25

1.50

1.00

1.50

2.50

2.00

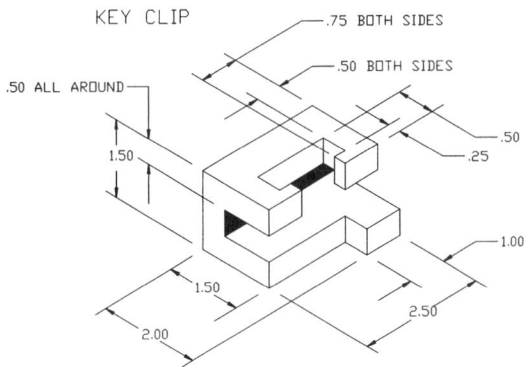

## EX1-12 INCHES

.50

.50

.25

30°

.75

.25

.75

.25

2.38

1.25

1.50

## EX1-13 MILLIMETERS

## EX1-16 MILLIMETERS

## EX1-17 MILLIMETERS

## EX1-14 MILLIMETERS

## EX1-18 MILLIMETERS

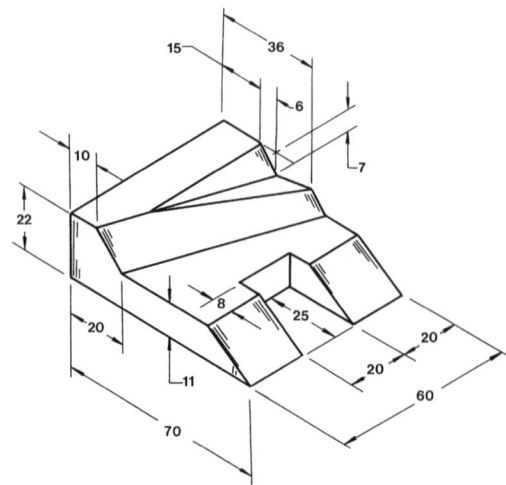

## EX1-15 MILLIMETERS

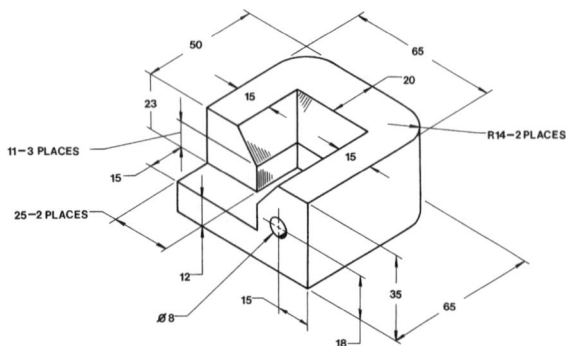

## EX1-19 MILLIMETERS

## EX1-20 MILLIMETERS

## EX1-21 INCHES

## EX1-22 MILLIMETERS

## EX1-23 MILLIMETERS

## EX1-24 INCHES

## EX1-25 MILLIMETERS

CYLINDRICAL
KEY

20
10 DEEP
Ø50
10
20
70
20
20
Ø80

## EX1-28 MILLIMETERS

15
25
15
12
Ø 14　2 HOLES
22° - BOTH
SLANTED SURFACES
60
27
38
100
30
7

## EX1-26 MILLIMETERS

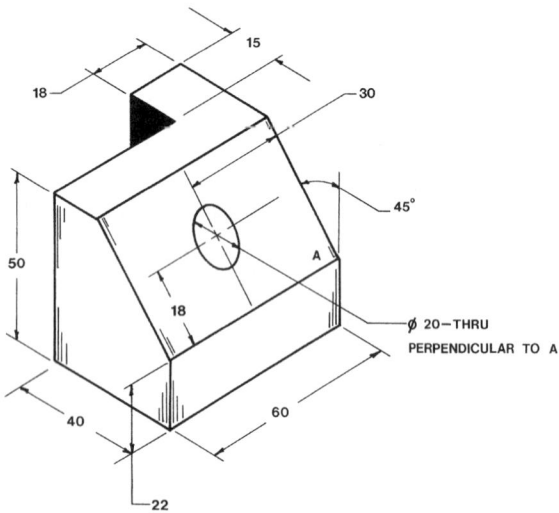

15
18
30
45°
50
A
18
Ø 20—THRU
PERPENDICULAR TO A
40
60
22

## EX1-29 MILLIMETERS

40
15
15
50
R13 – 4 Corners
40
10
R13 – Both Sides
25
R8

## EX1-27 MILLIMETERS

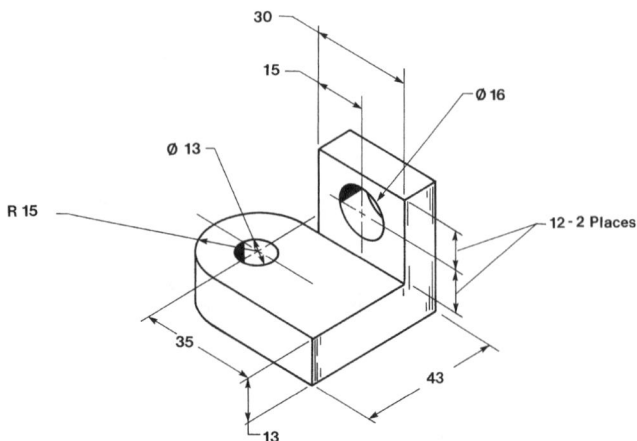

30
15
Ø 16
Ø 13
R 15
12 - 2 Places
35
43
13

## EX1-30 MILLIMETERS

18
15 BOTH SIDES
Ø14
25
60
30
35
15
20
20
10
12
8
55
30
25

## EX1-31 INCHES

2.00R
1.00
.69
.50 DIA
2.50
1.25
.63
.63
.63
1.13
.88
.75
.75
2.38
3.88


## EX1-34 MILLIMETERS

Ø 13
15
30
70
15
80
75°
35
50
15
R
12
10
15
50
8


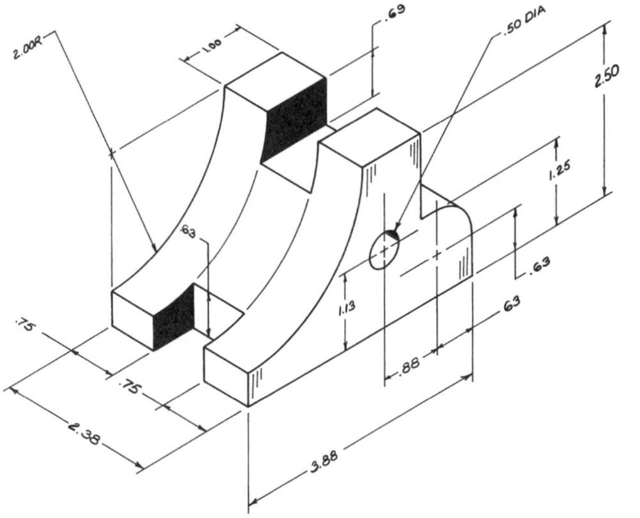

## EX1-32 MILLIMETERS

45°
17
CENTERLINE OF CIRCLE IS A VERTICAL LINE
52
Ø 24
80
HEXAGON 60 ACROSS THE FLATS


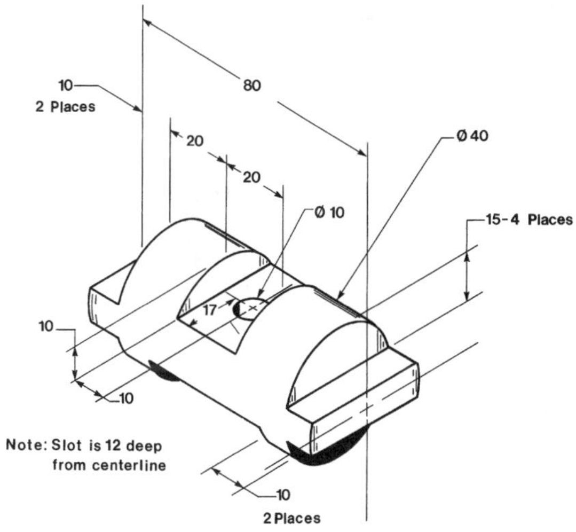

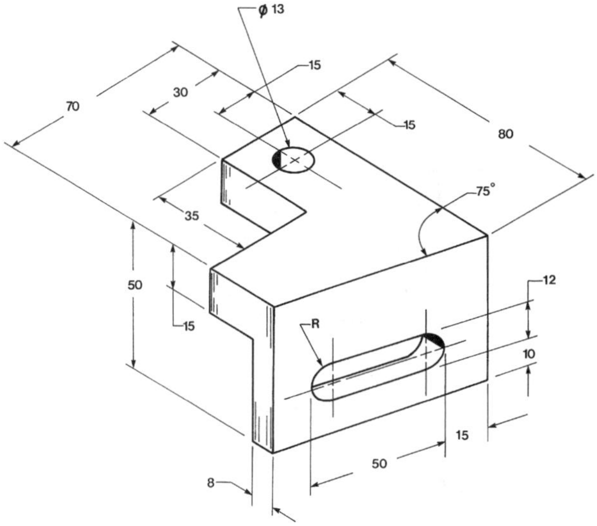

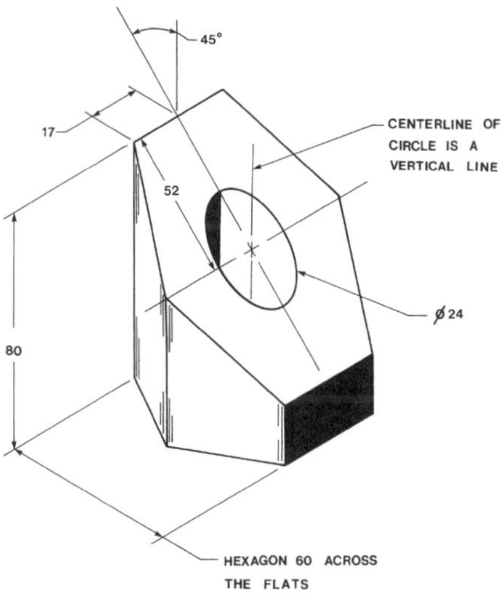

## EX1-35 MILLIMETERS

10
2 Places
80
20
20
Ø 40
Ø 10
15-4 Places
17
10
10
Note: Slot is 12 deep from centerline
10
2 Places


## EX1-33 INCHES

1.13
.19
.75
.25
.50
3.00
1.75 DIA


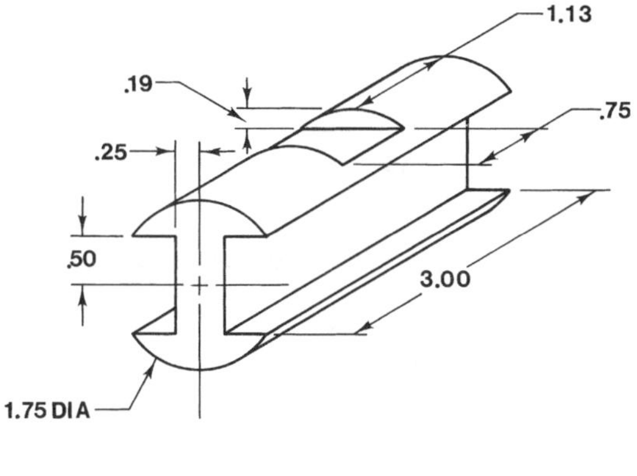

## EX1-36 INCHES

2.75
1.13
1.38 DIA
.25
.25
.31
1.13

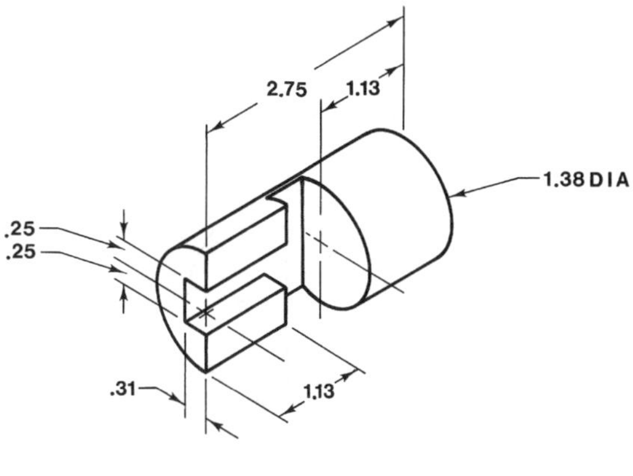

## EX1-37 MILLIMETERS

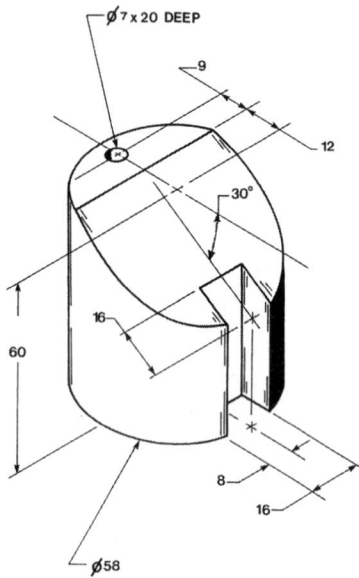

## EX1-40 MILLIMETERS

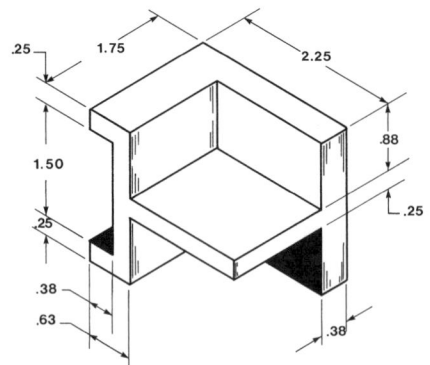

## EX1-38 MILLIMETERS

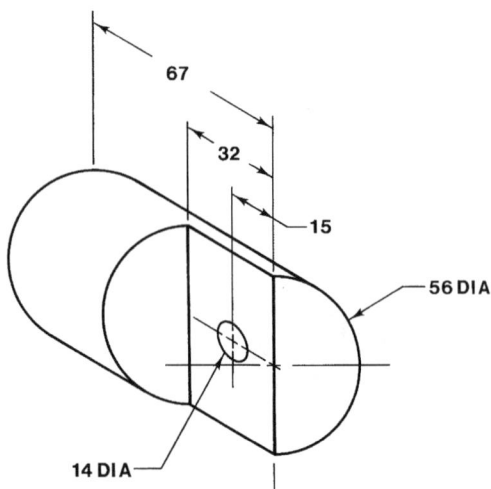

## EX1-41 INCHES

## EX1-39 MILLIMETERS

## EX1-42 INCHES

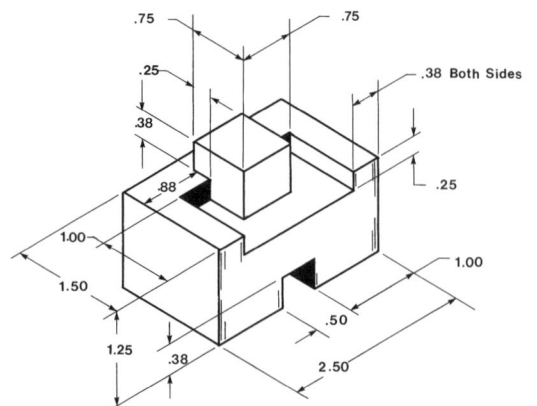

**EX1-43 MILLIMETERS**

**EX1-46 MILLIMETERS**

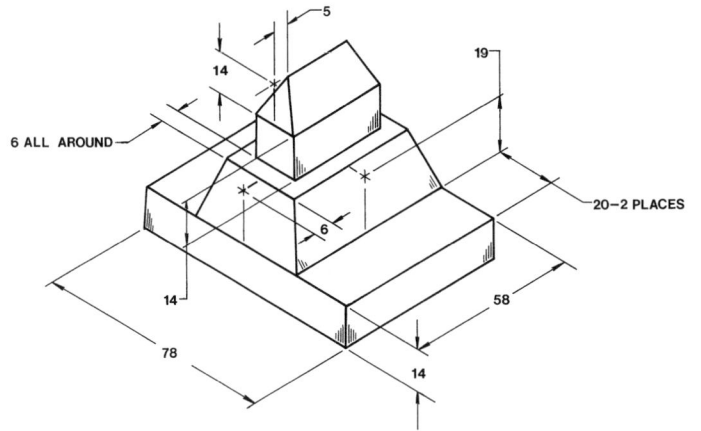

**EX1-47 INCHES**

**EX1-44 MILLIMETERS**

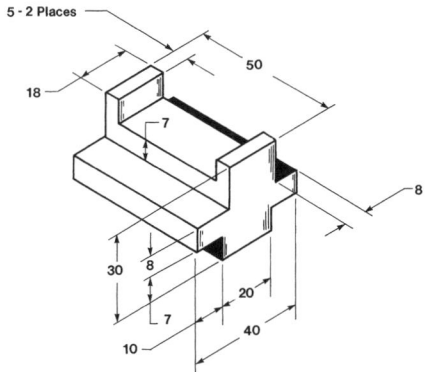

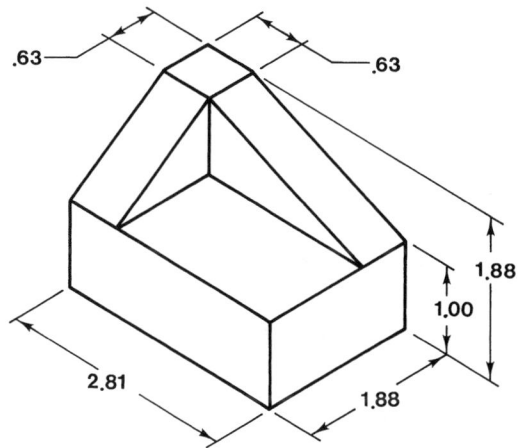

**EX1-48 INCHES**

**EX1-45 MILLIMETERS**

NOTE: THE SLOT
IS 15 LONG

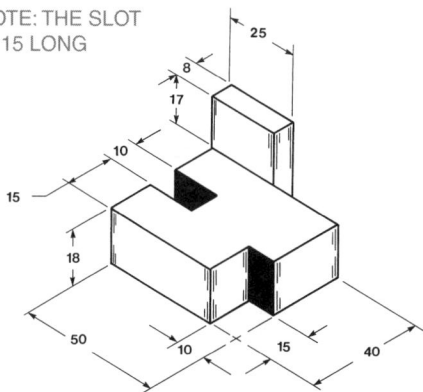

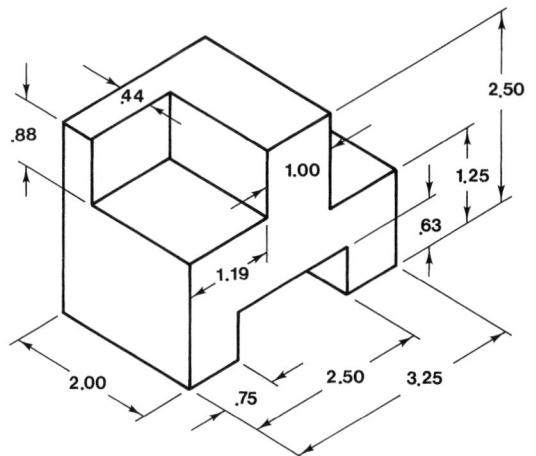

## EX1-49 MILLIMETERS

## EX1-52 MILLIMETERS

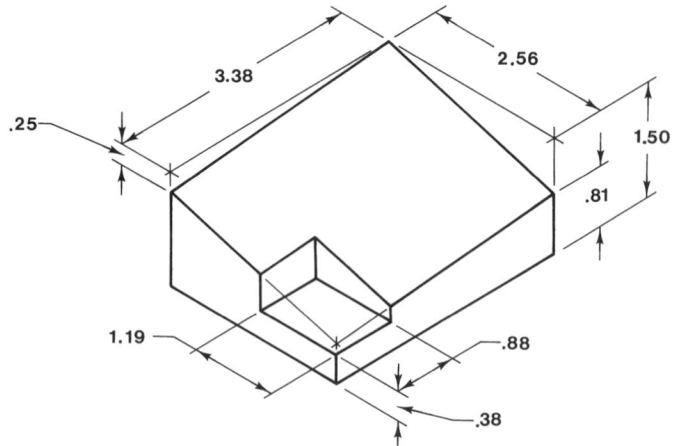

## EX1-53 INCHES

## EX1-50 INCHES

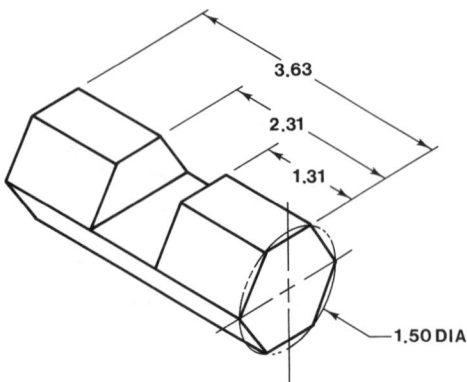

## EX1-54 MILLIMETERS

## EX1-51 INCHES

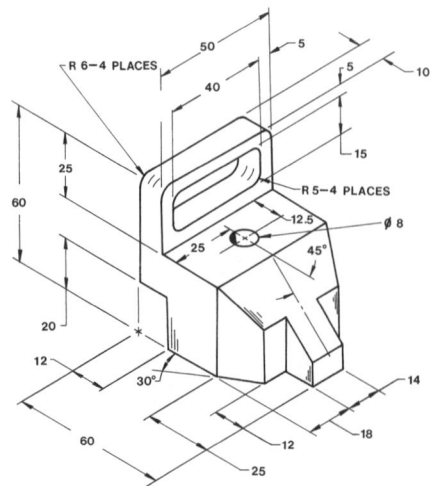

## EX1-55 MILLIMETERS

## EX1-58 MILLIMETERS

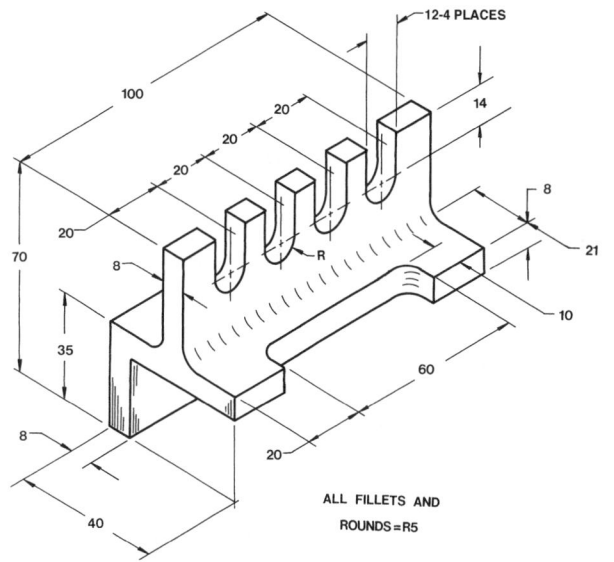

## EX1-56 MILLIMETERS

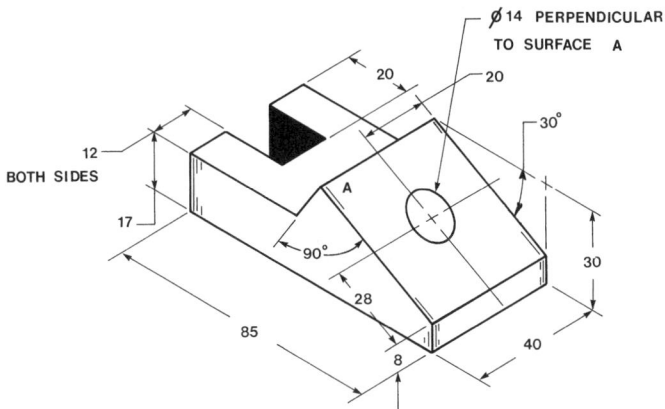

## EX1-59 MILLIMETERS

## EX1-57 MILLIMETERS

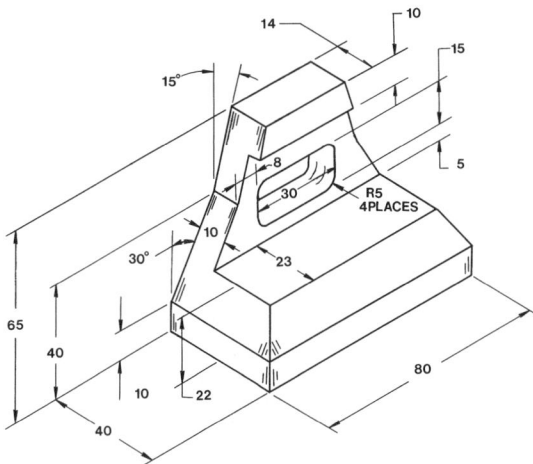

## EX1-60 INCHES

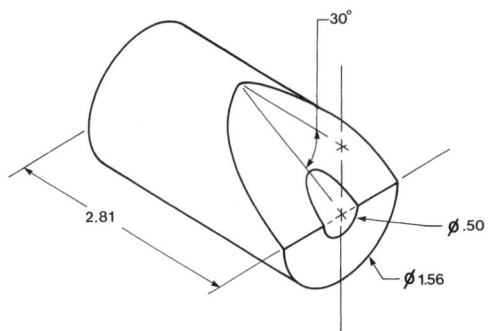

## EX1-61 MILLIMETERS

ALL FILLETS AND ROUNDS = R3

## EX1-64 MILLIMETERS

## EX1-62 MILLIMETERS

## EX1-65 MILLIMETERS

## EX1-63 MILLIMETERS

## EX1-66 MILLIMETERS

## EX1-67 MILLIMETERS

## EX1-68 MILLIMETERS

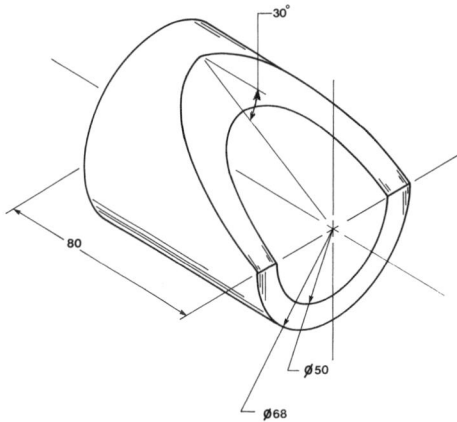

## EX1-69 MILLIMETERS

## EX1-70 MILLIMETERS

## EX1-71 MILLIMETERS

## EX1-72 MILLIMETERS

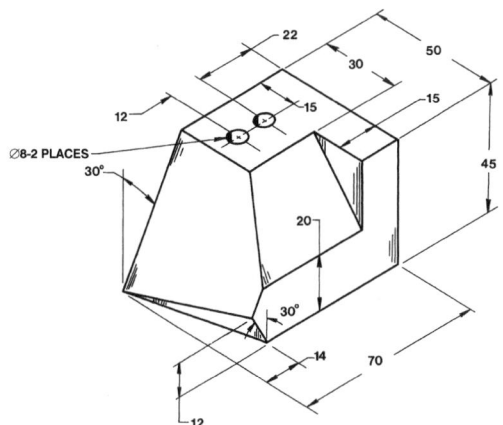

## EX1-73 MILLIMETERS

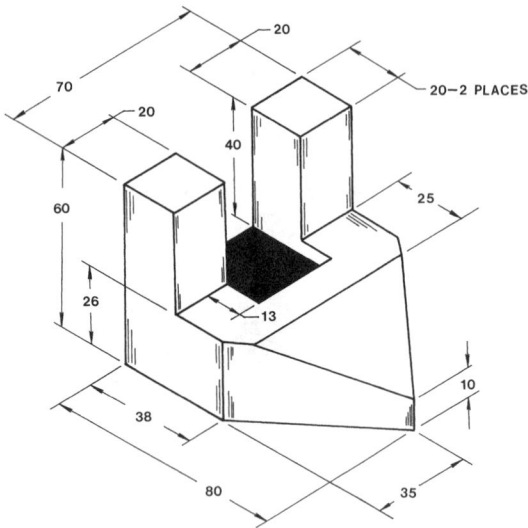

## EX1-74 INCHES

## EX1-75 INCHES

## EX1-76 MILLIMETERS

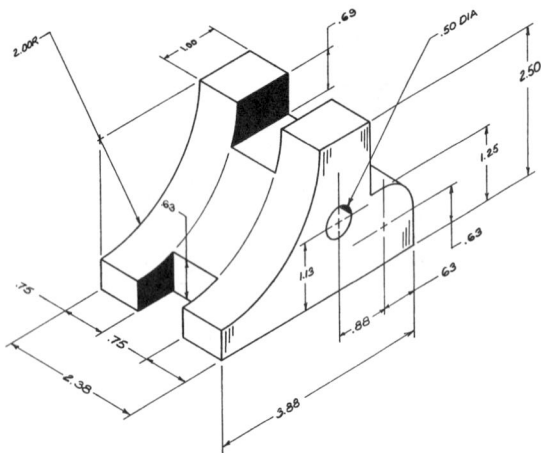

## EX1-77 MILLIMETERS

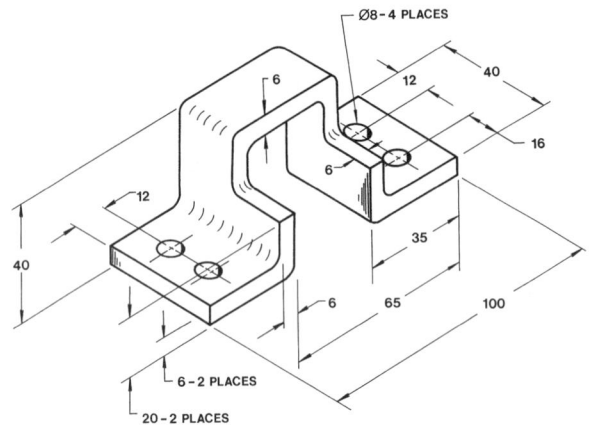

NOTE: ALL FILLET AND ROUNDS=R3

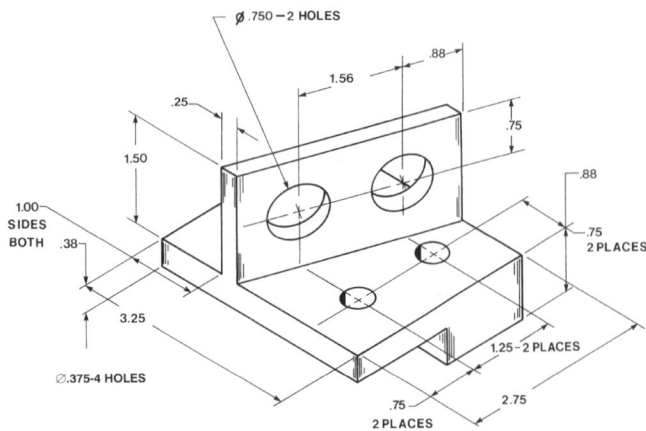

## EX1-78 MILLIMETERS

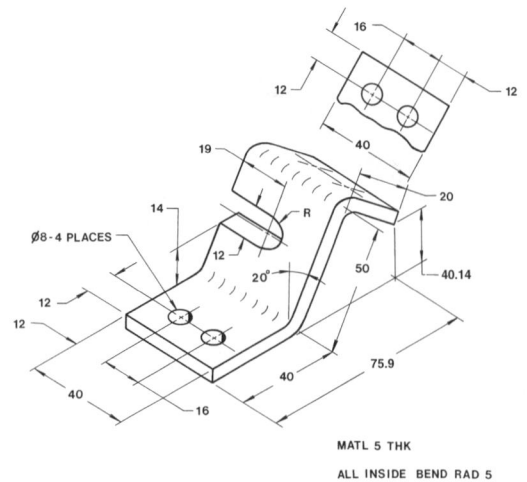

MATL 5 THK

ALL INSIDE BEND RAD 5

## EX1-79 MILLIMETERS

70
50
10
45
50
Ø10 - 2 PLACES
32.5
10 - 2 PLACES
20
5
Ø6 - 2 PLACES
27.5
7.5 - 2 PLACES
27.5
R 20
70
R 7.5 - 2 PLACES

ALL FILLETS AND ROUNDS = R5
MATL 5 THK

## EX1-80 MILLIMETERS

8 – 2 PLACES
60
28
15
25
16 – 2 PLACES
8 – 2 PLACES
Ø8 – 2 PLACES
R 5 – 2 PLACES
32
48
60
60 REF
Ø6 – 2 PLACES
8 – 2 PLACES
60
Ø12 – 4 PLACES
5 – 2 PLACES
20
16 – 2 PLACES
10
28

MATL 5 THK

## EX1-81 MILLIMETERS

35
25
50
15
35
12
18
10
45
80
20 – 4 PLACES
40 – 2 PLACES
12
10 – 4 PLACES
Ø8 – 4 PLACES
8 – 2 PLACES
12 – 2 PLACES

ALL FILLETS AND ROUNDS = R3
MATL 5 THK

## EX1-82 MILLIMETERS

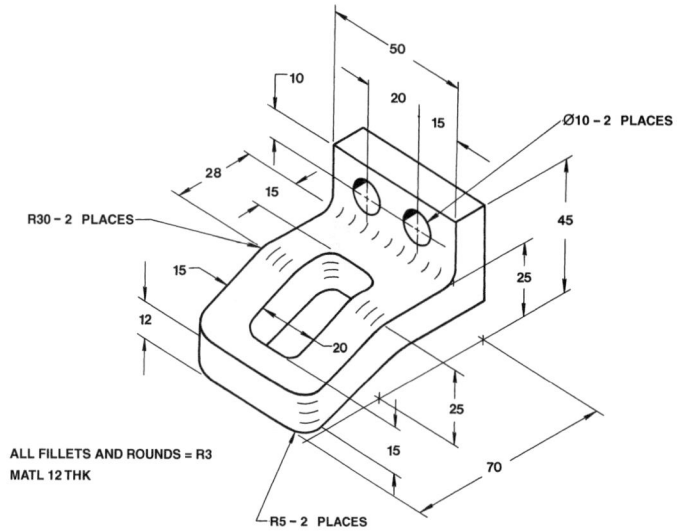

50
10
20
15
Ø10 – 2 PLACES
28
15
R30 – 2 PLACES
45
15
25
12
20
25
15
70
R5 – 2 PLACES

ALL FILLETS AND ROUNDS = R3
MATL 12 THK

## EX1-83 MILLIMETERS

40
12
20
16
10
53
30°
10
40
13
50
20
15
20
30°
10
Ø10 - 2 HOLES

## EX1-84 INCHES

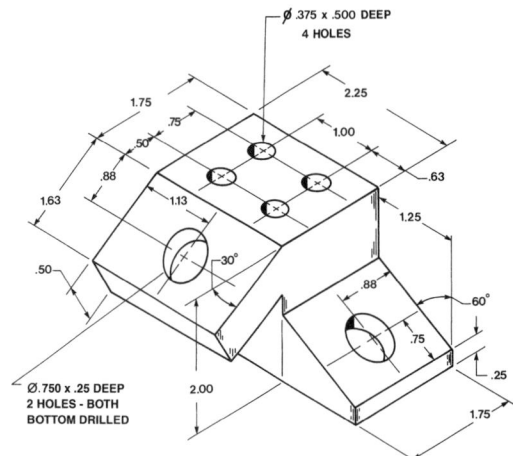

Ø .375 x .500 DEEP
4 HOLES
1.75
2.25
.75
.50
1.00
.88
.63
1.63
1.13
1.25
.50
30°
.88
.60°
.75
2.00
.25
1.75
Ø .750 x .25 DEEP
2 HOLES - BOTH
BOTTOM DRILLED

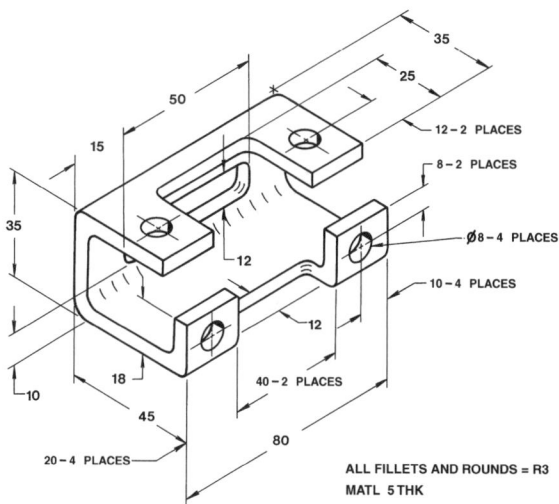

## EX1-85 MILLIMETERS

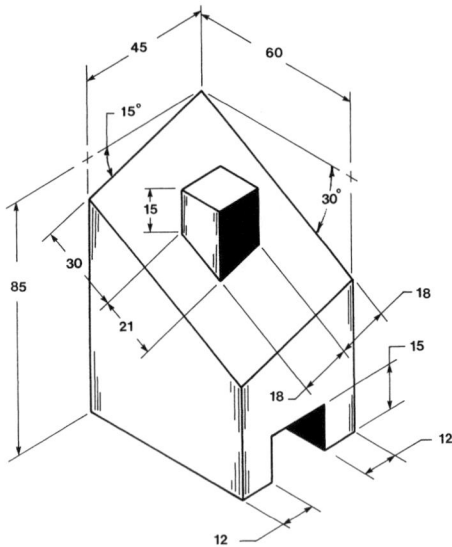

## EX1-87 INCHES

## EX1-86 MILLIMETERS

## EX1-88 MILLIMETERS

## EX1-89 MILLIMETERS

35
80
17.5
14
15
Ø12
5−4 PLACES
10−2 PLACES    20
14−4 PLACES
15−2 PLACES
40
10−2 PLACES
8
25−2 PLACES
13
Ø22
28 − 2 PLACES
R10
60°
12−2 PLACES
ALL FILLETS AND ROUNDS = R3

## EX1-90 MILLIMETERS

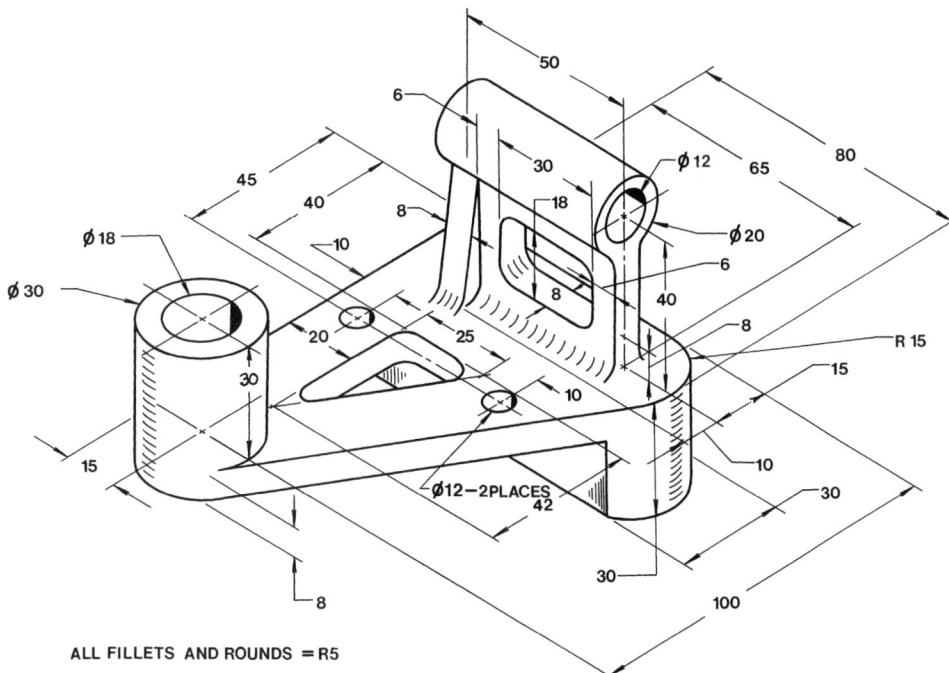

50
6
45
40
30
Ø 12    65
80
Ø 18
8
10
18
Ø 20
Ø 30
6
20
8
25
40
8
30
R 15
10    15
15
Ø12−2PLACES
10
30
42
8
30
100
ALL FILLETS AND ROUNDS = R5

**EX1-91 MILLIMETERS**

**EX1-92 MILLIMETERS**

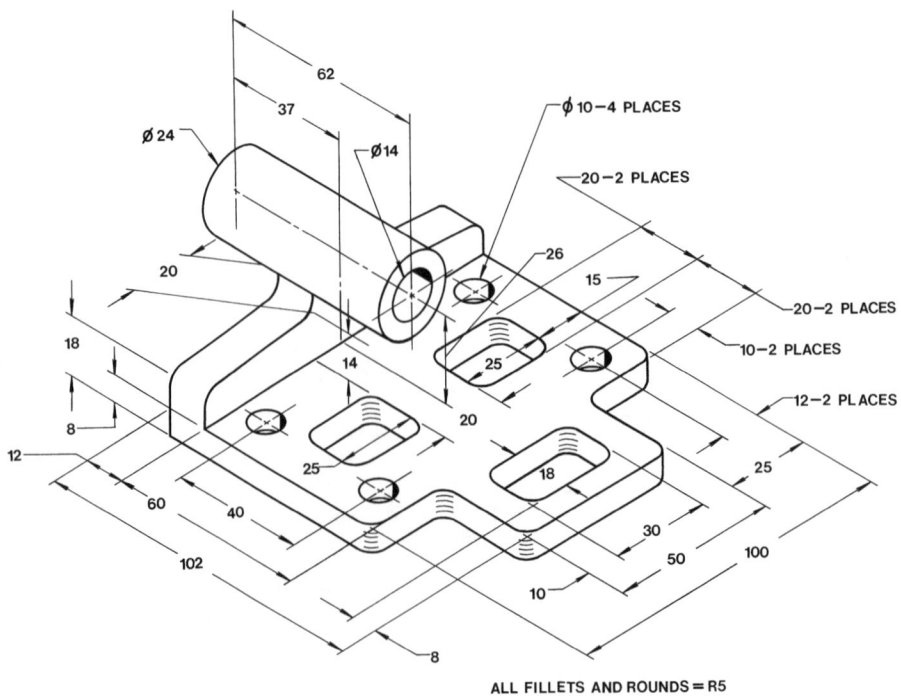

ALL FILLETS AND ROUNDS = R5

**EX1-93 MILLIMETERS**

M12-2 PLACES

56

30

28

60

30

30

15

16

19

22

40

22

40

60

20

M7
2 PLACES

20

4

12

25

8

3

3

5

7

7

10

10

5

12

25

8

8

3

20

8

4.4

100

**EX1-94 MILLIMETERS**

10    50

40    30

Ø10-4 HOLES

59.21    R77.5

5

Ø16-8 HOLES

R16-8 PLACES

30

15

55    35

30

10    R8
4 PLACES    24

15    160

20    22

10    24

50    30

80    15

For Exercises EX1-95 to EX1-100:

A. Sketch the given orthographic views and add the top view so that the final sketch includes a front, top, and right-side view.
B. Prepare a three-dimensional sketch of the object.

**EX1-95**

**EX1-98**

**EX1-96**

**EX1-99**

**EX1-97**

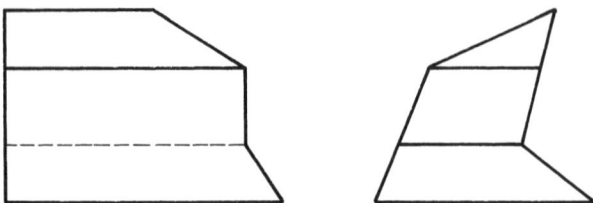

**EX1-100**

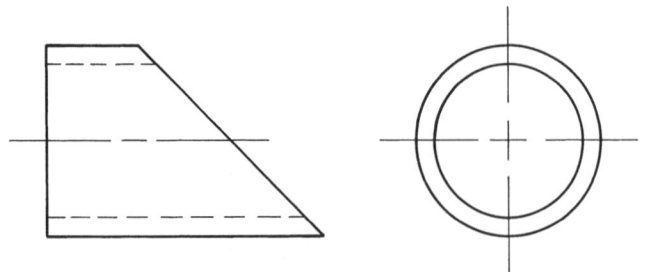

For Exercise Problems EX1-101 to EX1-128:

A. Redraw the given views and draw the third view.
B. Prepare a three-dimensional sketch of the object.

**EX1-101 INCHES**

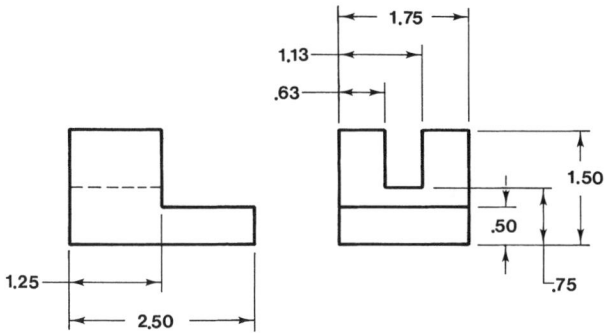

**EX1-104 INCHES**

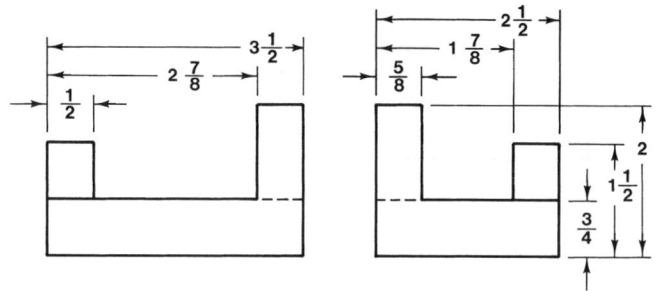

**EX1-102 INCHES**

**EX1-105 INCHES**

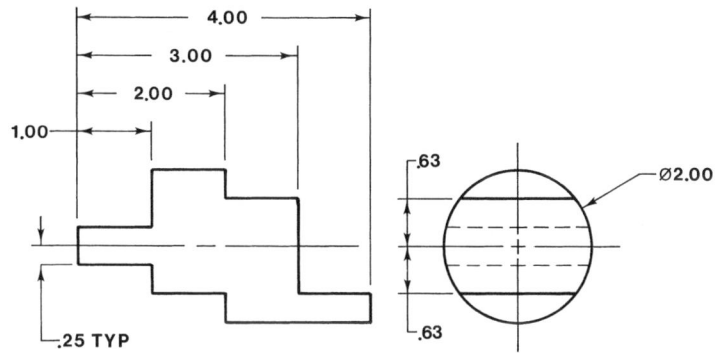

**EX1-103 INCHES**

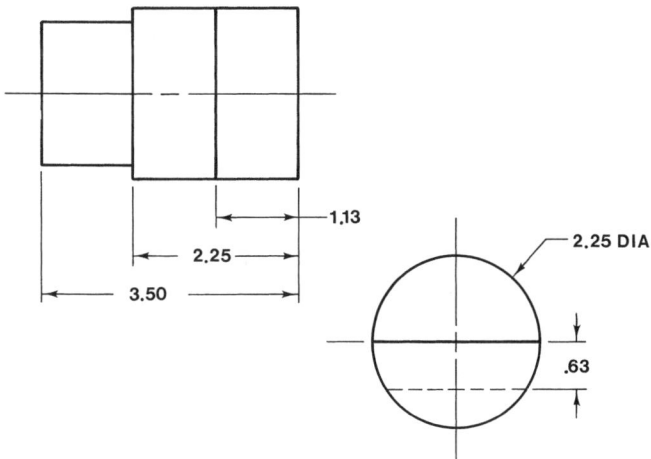

**EX1-106 INCHES**

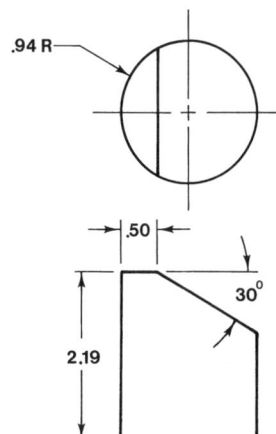

## EX1-107 INCHES

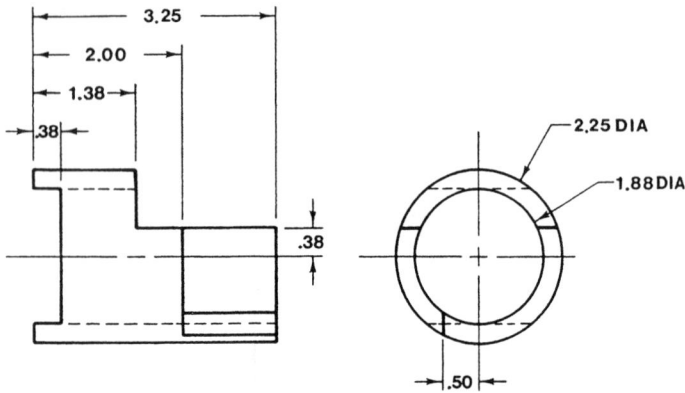

## EX1-110 INCHES

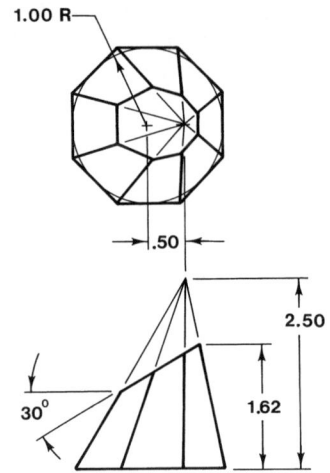

## EX1-108 INCHES

## EX1-111 INCHES

## EX1-109 INCHES

All fillets and rounds = $\frac{1}{8}$ R

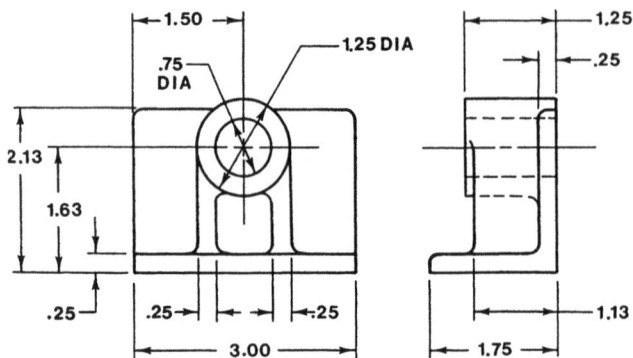

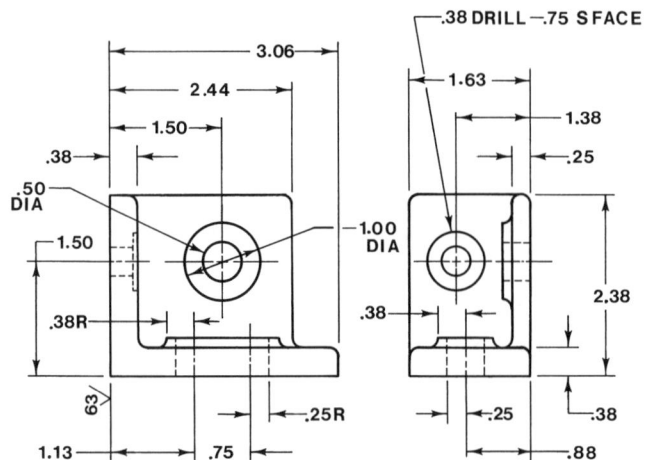

## EX1-112 INCHES

All fillets and rounds = $\frac{1}{8}$ R

Each exercise on this page is presented on a $10 \times 10$–mm grid.

**EX1-113**

**EX1-115**

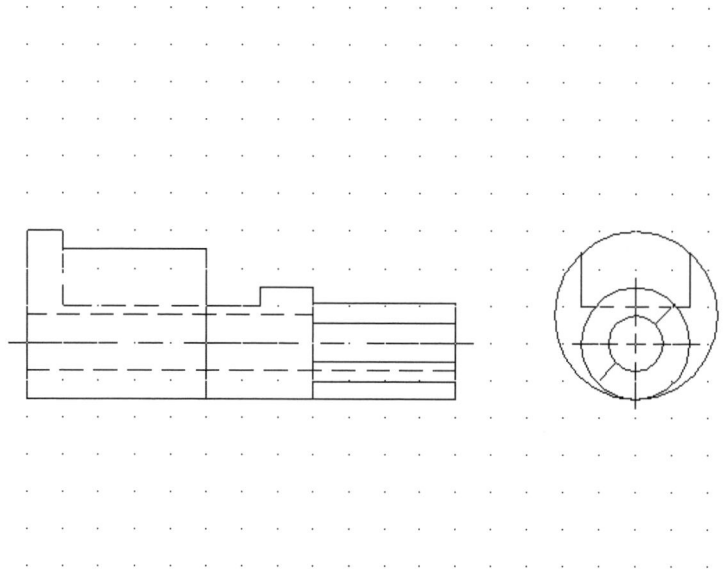

**EX1-114**

**EX1-116**

Each exercise on this page is presented on a .50″ × .50″ grid.

**EX1-117**

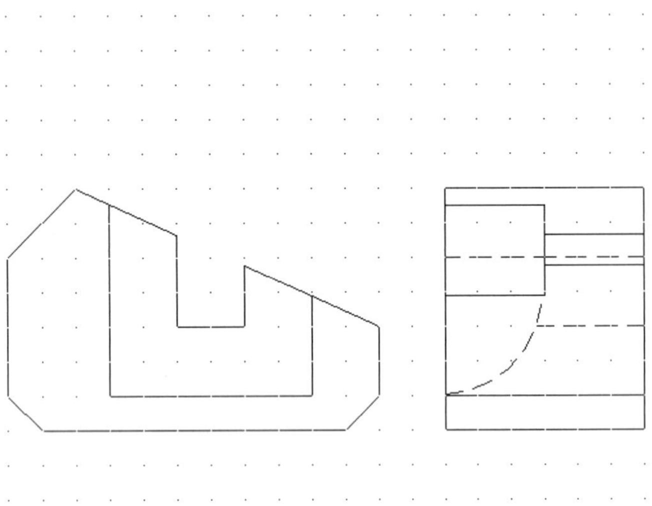

**EX1-119**

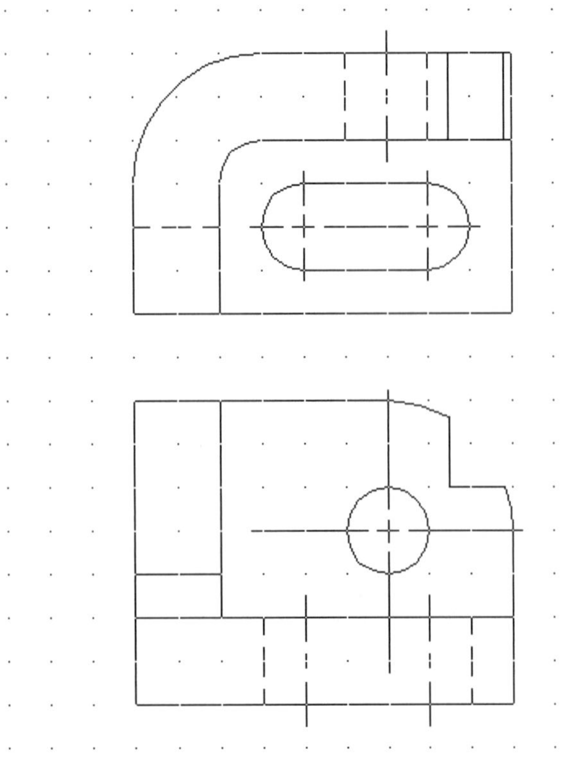

**EX1-118**

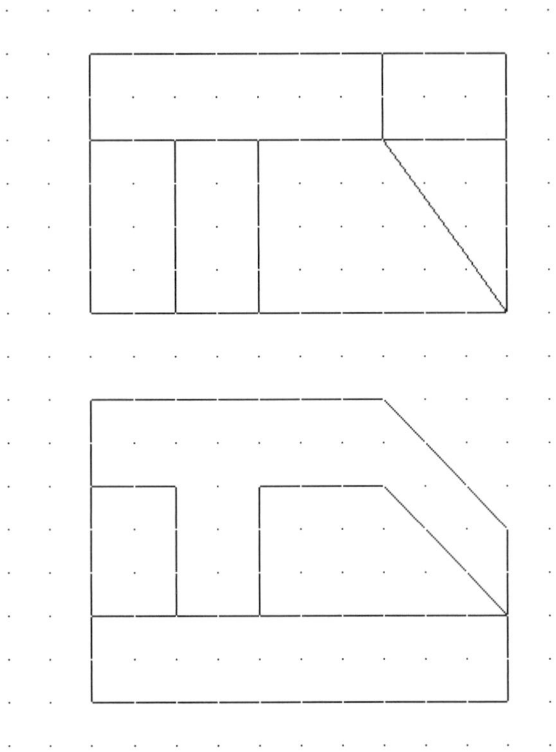

**EX1-120**

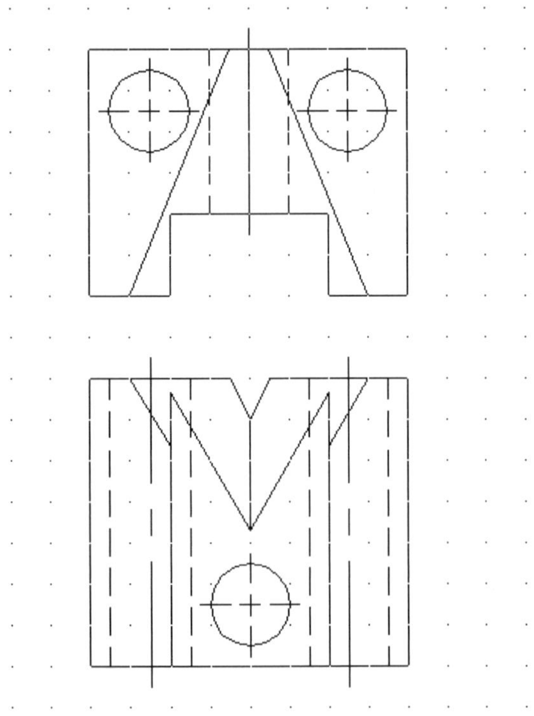

Each exercise on this page is presented on a .50″ × .50″ grid.

**EX1-121**

**EX1-123**

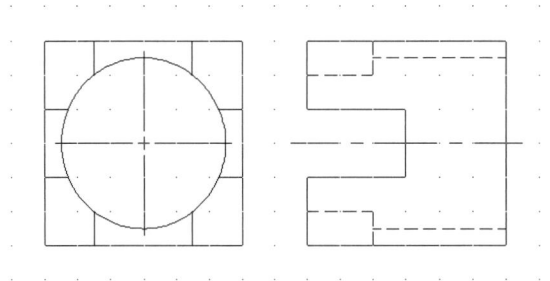

**EX1-122**

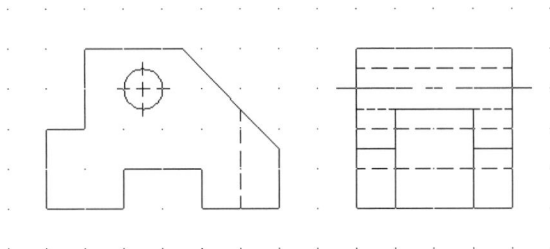

**EX1-124**

Each exercise on this page is presented on a 10 × 10–mm grid.

**EX1-125**

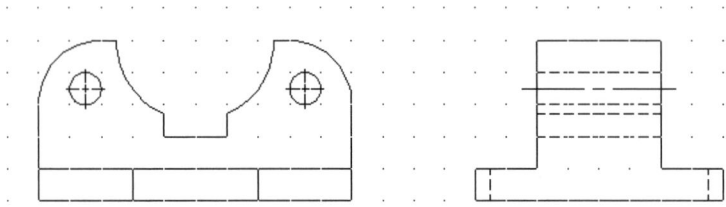

**EX1-127**

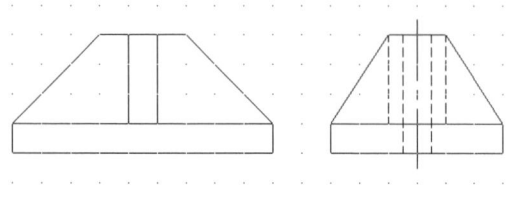

**EX1-126**

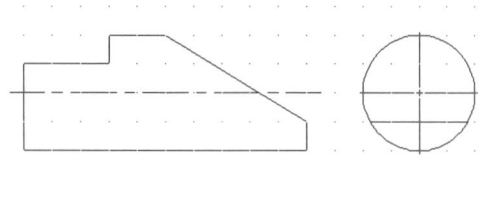

**EX1-128**

Draw the complete front, top, and side views of the two intersecting objects given in Exercise Problems EX1-129 to EX1-134 based on the given complete and partially complete orthographic views.

### EX1-129 MILLIMETERS

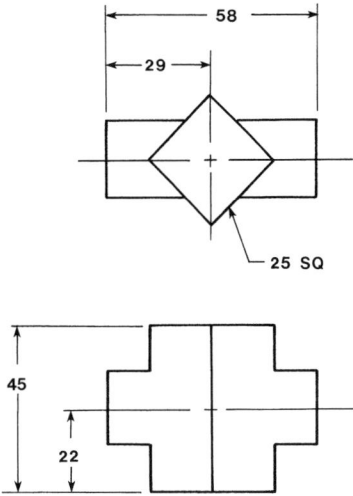

### EX1-131 INCHES

### EX1-130 MILLIMETERS

### EX1-132 INCHES

Draw the front, top, and side orthographic views of the objects given in Exercise Problems EX1-135 to EX1-138 based on the partially complete isometric drawings.

## EX1-133 MILLIMETERS

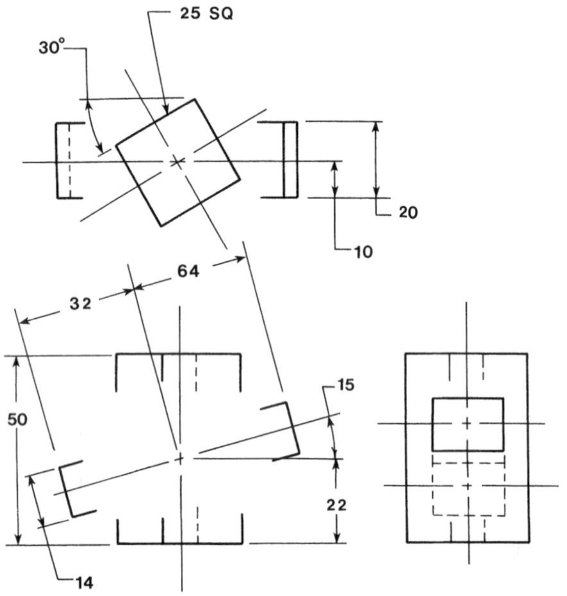

25 SQ
30°
64
32
50
14
20
10
15
22

## EX1-135 MILLIMETERS

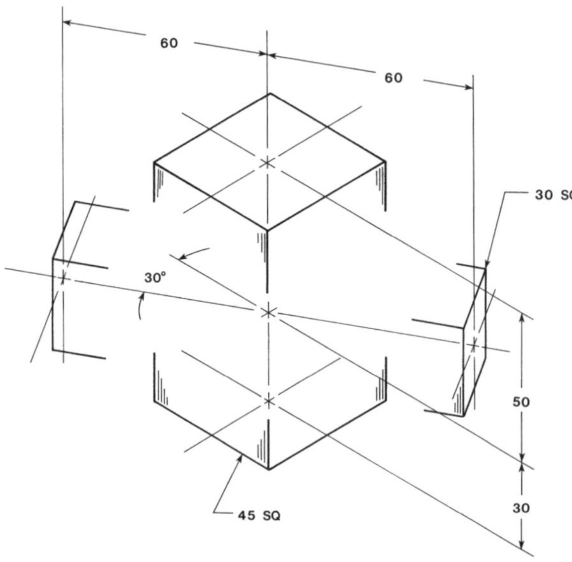

60
60
30 SQ
30°
50
30
45 SQ

## EX1-134 MILLIMETERS

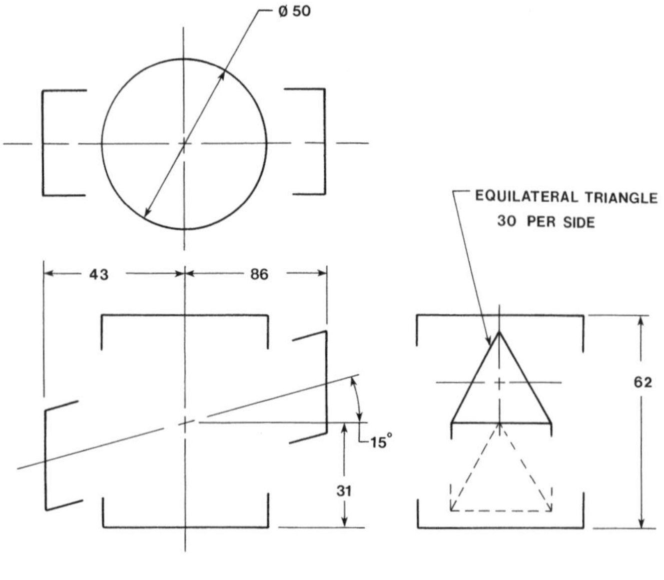

Ø 50
43
86
15°
31
EQUILATERAL TRIANGLE
30 PER SIDE
62

## EX1-136 MILLIMETERS

36
36
20 SQ
41
28
Ø 30

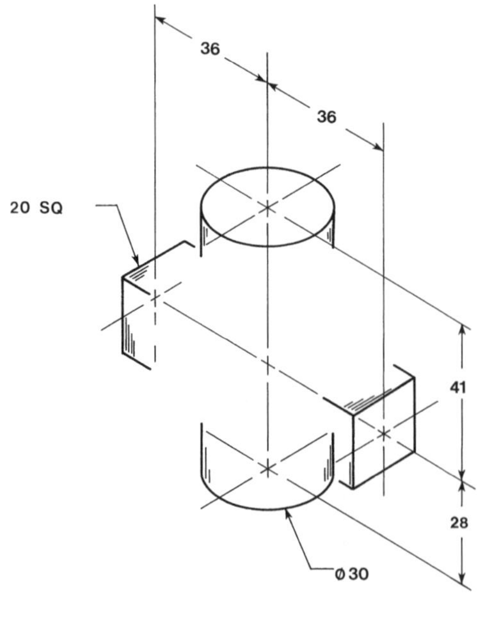

**EX1-137 INCHES**

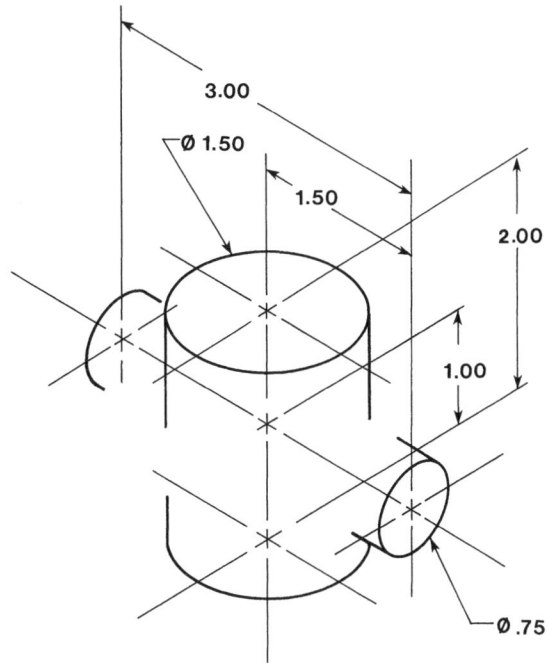

**EX1-138 INCHES (SCALE: 5=1)**

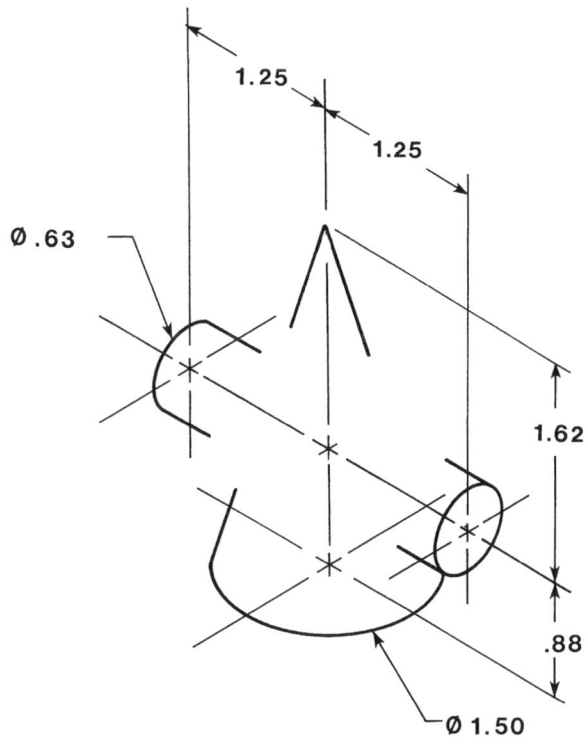

Redesign the existing objects as indicated in Exercise Problems EX1-139 to EX1-142, then prepare a front, top, and right-side view of the object.

## EX1-139 MILLIMETERS

1. Replace the four existing 12-wide slots with seven slots 16 wide.
2. Increase the distance between the slots from 20 to 24.
3. Modify the overall length as needed.
4. Increase the size of the 60 slot so that it is 20 from both ends.

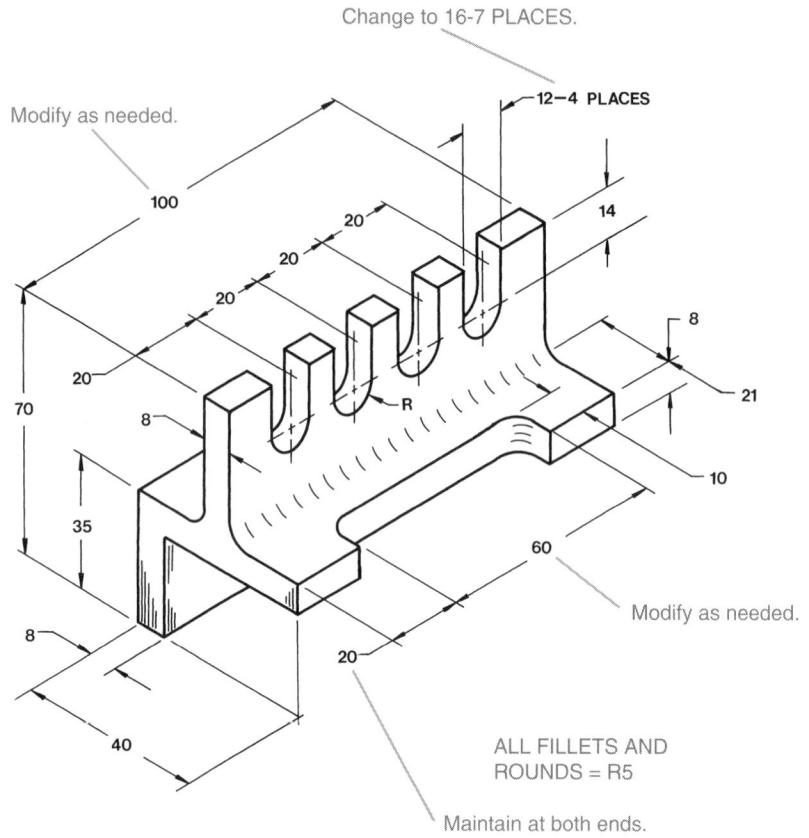

Change to 16-7 PLACES.

12—4 PLACES

Modify as needed.

100

20

20

20

20

14

8

21

70

8

R

35

8

10

60

Modify as needed.

8

20

40

ALL FILLETS AND ROUNDS = R5

Maintain at both ends.

## EX1-140 MILLIMETERS

Ø16

23 R

Modify to Ø10 Ø16CBORE - 8 DEEP.

35

Modify to Ø16-23 DEEP Ø30-82°CSK.

Ø18

15

35

30

27.5

50

Modify to 80.

110

20
25

20   15   55

Modify to 130.

**EX1-141 MILLIMETERS**

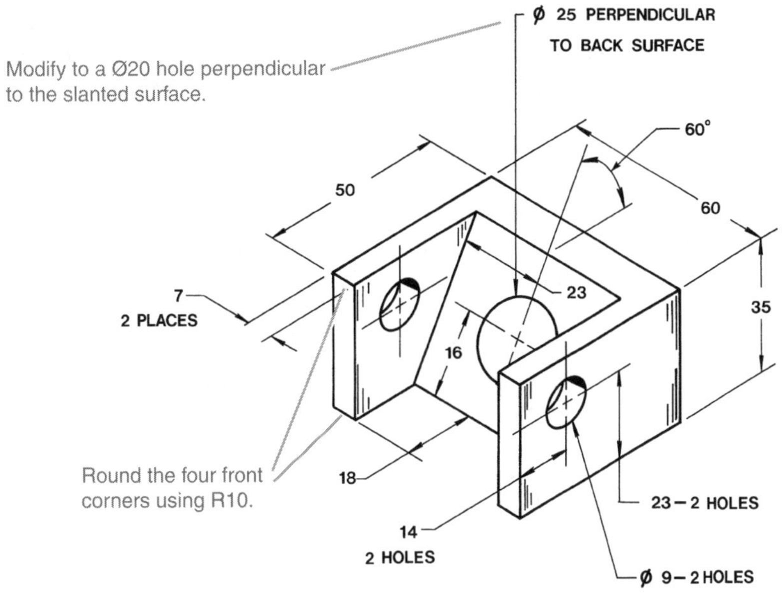

Modify to a Ø20 hole perpendicular to the slanted surface.

Ø 25 PERPENDICULAR TO BACK SURFACE

60°

50

60

7
2 PLACES

23

35

16

Round the four front corners using R10.

18

14
2 HOLES

23 – 2 HOLES

Ø 9 – 2 HOLES

**EX1-142 MILLIMETERS**

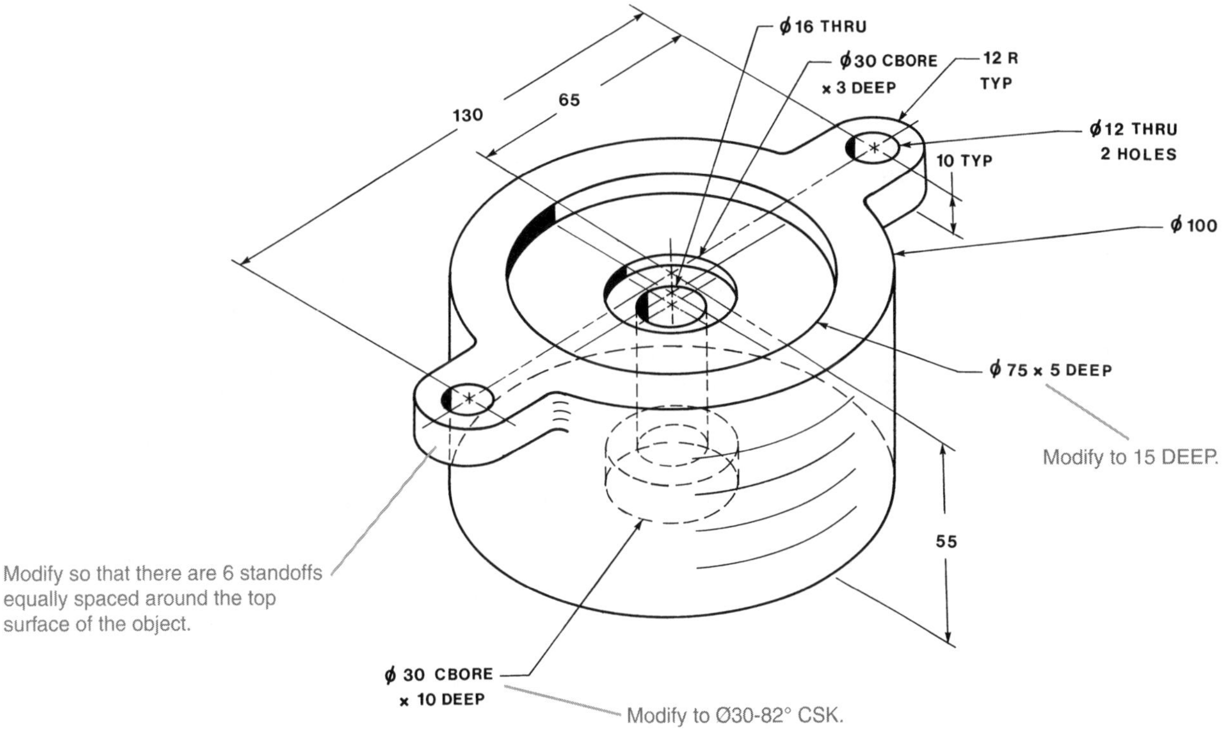

Ø 16 THRU

Ø 30 CBORE
x 3 DEEP

12 R
TYP

Ø 12 THRU
2 HOLES

130

65

10 TYP

Ø 100

Ø 75 x 5 DEEP

Modify to 15 DEEP.

55

Modify so that there are 6 standoffs equally spaced around the top surface of the object.

Ø 30 CBORE
x 10 DEEP

Modify to Ø30-82° CSK.

# 1-37  BRIEF OVERVIEW OF BASIS AND STATUS OF GD&T

Geometric dimensioning and tolerancing (GD&T) is a means of specifying engineering design and drawing requirements with respect to actual "function" and "relationship" of part features. Furthermore, it is a technique which, properly applied, ensures the most economical and effective production of these features. Thus geometric dimensioning and tolerancing can be considered both an engineering design drawing language and a functional production and inspection technique. Uniform understanding and interpretation among design, production, and inspection groups are the major objectives of the system. This text discusses the subject step by step, focusing on practical application. Before presenting these details, however, we wish to provide the reader with a brief overview of the basis and status of geometric dimensioning and tolerancing.

The authoritative document governing the use of geometric dimensioning and tolerancing in the United States is ANSI/ASME Y14.5M-1994, "Dimensioning and Tolerancing." This standard evolved out of a consolidation of earlier standards, ANSI Y14.5M–1982, ANSI Y14.5–1973, USASI Y14.5–1966, ASA Y14.5–1957, SAE Automotive Aerospace Drawing Standards (Sections A6, 7, and 8–September 1963) and MIL–STD-8C, October 1963. This consolidation was accomplished over many years by committee actions representing military, industrial, and educational interests in developing national and international standards. The work of these committees has had and continues to have three prime objectives:

1. to provide a single standard for practices in the United States and harmonize as much as possible USA practices with International Standards Organization (ISO) standards.

2. to update existing practices in keeping with technological advances and extend the principles into new areas of application.

3. to establish a single basis and "voice" for the United States in the interest of international trade, in keeping with the United States' desire to be more active, gain greater influence, and pursue a more extensive exchange of ideas with other nations in the area of international standards development.

The historical evolution of geometric dimensioning and tolerancing in the United States is an interesting story which, however, is not discussed in this text. It suffices to say that the early introduction of functional gaging, giving rise to the possibility of new techniques, along with the growing need for more specifically and economically stated engineering design requirements, has caused its growth. Advancing product sophistication and complexity, computerization, rapid industrial expansion, diversification, etc., have created an environment in which more exacting engineering drawing communication is not only desirable but mandatory for competitive and effective operation.

Updated and expanded practices have been initiated in the latest Y14.5 standard. Further expansion will no doubt occur as growth in the area continues. In the process of extending into new areas, this expansion is confronted by the challenge of ensuring progress without upsetting stability. Rapid advances in this subject, although desirable, must be tempered by the ability to make the transition with no loss of continuity or understanding.

United States coordination and compatibility with international dimensioning and tolerancing practices have been extended significantly in the latest Y14.5M standard.

The symbology and influence from ISO (International Standards Organization) and earlier ABCA (America, Britain, Canada, Australia) documents, activities, and committees have found their way into the latest Y14.5 standard. This influence continues in the ongoing development and use of geometric tolerancing in the United States and as the USA plays a major role in developing, coordinating, and adopting international standards.

Many of the United States industrial concerns and the military have overseas affiliations and markets. Thus, the increasing need for understanding and for more uniform practices throughout the world is evident as Total Quality Management (TQM), World Class Engineering, ISO 9000 quality standards and USA competitiveness require urgent consideration.

This text presents the subject of dimensioning and tolerancing in order of complexity of the details, and attempts to clarify and promote the use of Y14.5. It also emphasizes the importance of the ongoing effort to expand the principles and to more closely incorporate international practices.

### The New GD&T—A Pervasive Technology

Much has been written about GD&T, Geometric Dimensioning and Tolerancing. The annals of literature on the subject would fill many volumes and still brim over. The author alone has written papers on the subject more times than can be remembered. Many other authors and standards writers of various persuasions and objectives have also contributed over the years; no doubt, all with noble intentions in a desire to be noted, published, and to play a role in the parade of industrial progress. All such efforts are to be applauded for their contributions and many for enriching the sphere of knowledge in significant measure.

It is not without some hesitation that again an effort is made in this introduction to promote the subject. An apology is in order, if the reader deems it appropriate, if this appears as another "sales pitch" GD&T missile. Such an effort sometimes tends to present the material in the sunshine of creation as if introducing GD&T as a new technology and new idea for the first time. The subject has been, of course, propounded long before. Yet by wearing another hat, it must also be remembered that what might be "warmed-over words" to one person may be a dawn of awakening to another. Admittedly, the string must go on with the basic raw materials of GD&T presented sufficiently and clearly in the standards of authority (presently ANSI/ASME Y14.5M) and in supporting papers, texts, and the many necessary other training resources also required.

As in almost any subject, an aura of stagnation can appear to predominate if the same timeworn old GD&T story, as valid as it may be, is continually put forth in editorials, published texts, and papers. We do *not* need to rediscover GD&T again. What is needed is to find ways and means to better implement it and make it the core of our engineering design and manufacturing culture and its integrated communication network. First, there is a need for a new accounting of the subject in a more dynamic light. This is both as the dependable well-adapted and long-established engineering communication tool that it is, but also as the catalyst and mechanism for a real breakthrough of new technology. It could be identified as "The New GD&T—A Pervasive Technology."

Over many years of observing, working with, and promoting GD&T, tremendous advances have been made. The subject has progressed from one of early nominal use by those primarily in military work, using the MIL–STD–8 series of standards, to a modern engineering tool absolutely essential to the technical and economic well-being of the company or organization. ANSI/ASME Y14.5M, ISO/1101, ISO standards, and standards of other countries, have made GD&T the technical drawing language of the world. It has bridged local and international barriers of communication and provided that tool to "say what we mean and mean what we say" better than ever before. Its advantages can best be realized from the experiences in one's own facility, dealing with sub-contractors, multi-divisional coordination, military contracts, and reaching out the arm of communication anywhere in the world. There yet remain some minor differences between the latest ANSI/ASME Y14.5M (USA) standard and the latest ISO (International) standards. However, it can be roughly estimated that there is about 90 to 95% compatibility between these standards now for all practical purposes, and work continually goes on to improve that compatibility. The graphic language of symbols and numbers can "speak for itself" and surmount any language barrier around the world. Later in this introduction the harmonizing of USA and ISO standards is discussed.

The term "pervasive technology" is not new to industrial parlance. It has been used as a term of reference in a number of ways. Used here it implies that geometric dimensioning and tolerancing, GD&T, can now be considered a new and pervasive technology in two distinct, albeit related, ways. One, the standards and the contained universal GD&T techniques and language have "arrived"; it is no longer considered an optional approach. Its presence, influence and use is mandated to fill the void of communication which has

persisted in many areas for years. It is a new day and age for GD&T and in many ways it is coming to the rescue; rescue from the doldrums of poor quality, complacency, archaism and resistance to the changing world; a life-ring to the future.

## *A Renaissance of GD&T Use*

In every sense there is occurring a renaissance of use and recognition of GD&T. All should take advantage of the new frontiers of use of the subject. One cannot afford to be left behind. Those that do not have at least some speaking acquaintance with these standards may be left incommunicado.

This renaissance of new life and recognition is not a circumstantial quirk of fate. It has been a continual evolutionary trend and development, based upon hard-earned experience, economics, state of the art, and a desperate need for improved technical communication. Continually, more attention has been drawn to these standards as an essential authoritative base of credibility as we face the advantages and realities of the world in which we now live. The second reason, and probably the predominant one for the "new" GD&T age, is the advent and growth of the computer with all it offers and demands. The computer has revolutionized our world. The use and development of this technology has advanced to the point where the limits of communication and innovation have virtually been obliterated. The standards and the computer have provided the ideal marriage; one begets the other.

## *Standard ANSI Y14.5M (GD&T). The Catalyst and Base for the New Technology*

Active and numerous projects are underway which are exploring and developing new frontiers relative to computer application and research in mechanical engineering. This is in addition to the hundreds of CAD, CAM, CIM, CADD, CNC, CAI, etc., program applications already in place and functioning as a part of everyday operations. What has been discovered and is being increasingly recognized, is that for nearly all such programs, the fundamental "standard" or language must be used if these efforts are to have a broad base of credibility, authority and be usable and coherent to others. ANSI Y14.5M (GD&T) *is the standard and language* which provides that base.

Further developments are underway in the USA for example ASME B89.1.12 committee which deal with coordinate measuring machines (CMM's) to better implement computerized inspection operations as based upon the concepts of the Y14.5 standard. ASME, industry, government, and academia are also cooperating on developmental and standards projects which deal with "Mathematical Definition of Dimensioning and Tolerancing Principles" (the ASME Y14.5.1 committee), "Certification of GD&T Professionals" (the ASME Y14.5.2 committee) and national and internationally coordinated projects dealing with such subjects as "Mechanical Tolerancing." The basic standards are required as the essential key ingredients to give such research and development the common bond for consolidation and departure to new areas of analysis and technological enlightenment. There are also many additional computer explorations underway that would seem to need that same "home-base" anchor or ground zero origin.

### The Technology Gap

While the renaissance of GD&T provides the necessary communication link in the computer revolution and the future appears revitalized, there is a "catch-22" to threaten the successes envisioned. Experience tends to convince that wherever there is "good news" there probably will also be a down side or price to pay somewhere. In this case, no search for the negatives is needed. Personal experience in daily contact with a cross section of academia, government, industry and the international arena will adequately convey the impression that there is a serious growing technology gap on this subject. Unfortunately, many are not even aware of the standard, do not understand it, misuse it, don't want to use it, fear it, etc.

There is a dilemma here; a direction is already charted and in place to address this technology gap, but all too many are not yet on course. Standard ANSI/ASME Y14.5M is the course and direction provided; on a "silver platter" just waiting to be used. All we have to do is use it. Many have spent a good deal of an industrial life sharing in the experience and development of the standard and its forerunners. Now is the best time in history to "bite the bullet", face facts, and take advantage of the technology we ourselves have created. History also reveals that where supposed disadvantages seem to prevail, there must also be advantages therein which can evolve out of the turmoil.

### Implementing the New GD&T—A Pervasive Technology

The foregoing commentary can really be summed up by saying it another way. Our own technology is out-pacing us. Many cannot or will not make that effort to "learn something new" or reach for that higher plane of communication. It can also be added that many possibly have not had a chance or may not even be aware of the "bridge" needed to survive in the world of engineering communication today. As stated before, continual progress is being made but tends to be like walking in sand. Two steps forward with one step backward. The technology gap continually widens.

The down side of this situation then is that there is an engineering standard technology well designed and adapted to the present and future; but, its understanding and use is left wanting by a monumental share of the industrial populace. As the technology expands and the industrial population increases, the technology gap increases commensurately. The chasm between those knowledgeable and conversant with the standard and the GD&T state-of-the-art are being further separated from the pack. At the present pace there will never be enough highly-qualified persons to fill the need. There must be devised some way to package the technology where at least some of the needed intelligence can be supplied in simplified form. If somehow the implementation of the product design-to-manufacture cycle could be facilitated through "GD&T smart" aids, this could be accomplished. These simplified "tools" could bring the technology to a more nominal-user level. This approach could also be implemented in manufacturing and ensure correct and efficient inspection follow-through. The technology gap is then dramatically narrowed. The need-to-know depth becomes more identifiable and manageable.

The foregoing obviously describes computer programs and the present new technology already in place at many levels. This again introduces the pervasive technology theme. The Y14.5 standard (GD&T) provides the mechanism to fill the void and provides the authoritative language and technique for specific communication. The catch is: How well is the standard understood and used in design and in the computer programs developed? Those accepting the design-model data in integrated computer application place great "trust" in its correctness and build on or integrate the data directly. If the data is not clearly representative of the requirements, done well, and in the language of the standard, communication again is headed for failure. Worse yet, its shortcomings are transmitted to other operations.

The above scenario is put on-track and becomes a positive pervasive technology when the product designer and programmers responsible for creating the computer output are qualified in the language and dynamics of the standard. Then truly the bridge is constructed across the technology gap and each side is brought closer with limits and responsibilities better defined. There are high hopes and achievable goals then set down. Progress, of course, is already well down this road in many of the activities before mentioned.

The key to successful application of GD&T lies in education, training, and proper use of the standard and its contained techniques. It is the bridge needed to span the fate of further slipping into the chaos of our own doing. As before mentioned, certification requirements are being considered as a means of ensuring that those who play a key role in initially applying GD&T are qualified and where such input provides a dependable base for all involved. Emphasizing the bright side, the renaissance of GD&T use and prominence, provides the tools, advantage, and the language for both manual and automated communication in defining and building products. GD&T is playing a timely role as a partner to the computer. It is almost as if it was planned that way from the beginning; and who is to say that it was not ordained to happen.

## Automated Intelligence—GD&T Spoken Here

Relative to the state of things as cited in the preceding commentary, what is being done to build the bridge over the technology gap? Many things; some have already been mentioned. Any observer or participant in the changing industrial scene, will detect many notable efforts now in place or underway to couple the computer to daily tasks. When GD&T is married to the computer great things can happen. With the type of quantum leaps forward, represented by the creative and innovative breakthroughs now being realized, the technology gap and crisis is addressed in great measure. The new GD&T, as the required engineering language, and the pervasive avenue provided by the computer programs, opens the heretofore locked doors of communication. Now everyone can, and everyone must, participate in the exciting new GD&T and the age of pervasive technology. This text modestly tries to play a role in this effort by providing the daily "ready reference" tool and the teaching mechanism for classroom use or self-study.

## Harmonizing ANSI Y14.5 and ISO standards

The ultimate objective of facilitating world-wide technical design standards, using the language of geometric dimensioning and tolerancing (GD&T), has been, and is, a lofty goal to which many have aspired for decades. As the world "gets smaller," the need for such standards becomes more proven as ways and means are sought to better communicate product requirements on a global scale. Thus, the dedicated efforts by many over the years have come around full circle and now appear to be able to help fulfill far-reaching communication needs.

Geometric dimensioning and tolerancing, as provided in ANSI/ASME Y14.5M and comparable ISO standards, has now evolved to be the major tool to achieve such global communication. Its role in the industrial world has matured to a point where these techniques are called upon as a mandatory avenue for defining a product for most universal meaning, effective production, and economic advantage.

## Emphasis on quality

There is a major emphasis today also upon all aspects of "quality" while referencing TQM, World Class Engineering, SPC, ISO 9000, etc. These are the "buzzwords" that are commonly heard and used as indicators of desirable goals and representative of progress for any given organization. GD&T supplies a fundamental basis and ground zero for most of these allied disciplines; quality seems to start with a well-defined product.

It is not a new idea, but a more commonly conceived one in recent years, that a company or organization decides to also "go to ISO standards." Apparently, it is the trend; sometimes in necessity as a multinational mandate or as a move toward the "big picture" emphasis in the organization's future. It seems that the "quality" and international emphases often become integrated into a common objective. On the surface, this move to ISO appears as a significant, yet readily achievable one, by simply shifting to a higher level standard usage. This is not so easy, but, it is possible to an extent, as described below. This goal has also been the ultimate objective as the long range plan for the national (USA) GD&T standards developers for at least forty-five years (possibly longer). However, to "what extent" does, and can, the aspiration to adopt and use ISO standards go?

## What are ISO standards?

Since ISO standards are developed via an international forum, compromises and differences prevail. The finalization and approval of ISO standards, as well as their original initiation and progressive formulation, are based upon the very principles of cooperation and compromise by the countries involved. Therefore, ISO standards achieve the highest level of stature in the spirit of international cooperation and the common good. It follows then that each country attempts, as the obvious, to influence and input into the standards development process, their own preferences, philosophy and practices as much as possible. In this way their own national standards and habits are preserved as a higher order standard to which compliance is nicely accommodated. Volumes could be written on the details of the USA and its representatives' involvement and contribution to ISO/GD&T standards development over the years. Suffice it to say, it has been a very significant and successful endeavor reaching back about thirty years.

## ISO; is it "them" or "us"?

A major misunderstanding seems to prevail when there is consideration of adopting ISO standards on GD&T. That is, that such an action seems to require adoption of "their" standards which could involve ideas totally alien to us. This is not so; it is "our" standard just as much, if not more, in some cases, than any other country could claim. Of course, the spirit of ISO standards is really a joint "ownership" by all signatories of the final standard. The point being made is, that the USA has been active as member "Experts" and/or "Conveners" (Chair) on all GD&T-related standards that have been developed in the ISO arena since the beginning of the evolution of ISO standards on this subject.

## When the USA adopts ISO standards, it's "us" & "them"!

When ISO standards are adopted on any GD&T subject, it can be said that the USA has had previous input, contribution, voting influence, and privilege on that standard. The author alone has been convener (chairman) of six of the committees that developed the later listed ISO standards and, in addition, served on *all* of the committees which developed these standards. USA delegates have convened or served on all Working Groups (WG's). The WG's are ISO committees which do the preliminary spade work to develop the ISO technical standards from inception. The ISO/GD&T standards have USA fingerprints all over them. That does not mean, of course, that we (the USA) were totally successful in winning every point. We have learned from, and gained great respect for, our ISO colleagues from other countries and their contribution, as well, to the ISO. Together, the ISO standards efforts have borne fruit. The USA influence has been significant. When there is an adoption of an ISO/GD&T standard, it constitutes an adoption of something of our own doing; perhaps not totally satisfactory, but a product of cooperative effort in which USA influence was included. Our "batting average" in promoting our proposals to ultimately become ISO standards is a "ball park" .750 in the opinion of the author.

### ISO standards versus ANSI/ASME Y14.5M

In adopting GD&T ISO standards it will be discovered that about fifteen to twenty ISO standards are directly or indirectly involved to approximate the coverage of USA ANSI/ASME Y14.5M. In the ISO standards development agenda it is, "one subject, one standard." This philosophy and history alone could deserve volumes of coverage. Yet, it can be said in brief, that this method evolves as the only reasonable way when being exposed to the lengthy travel, involved time consumed, cost outlays, national pride, language differences, due process required, etc. One subject at a time surfaces as the only practical and achievable approach. Obviously, related subjects must be discussed simultaneously (such as position tolerance and datums), but the detailed coverage on each ends up in separate documents of minimum size and very basic coverage.

### ISO standards necessary for GD&T coverage

The following documents must be considered when adopting ISO/GD&T standards:

1. ISO/1101–    Technical Drawings Geometical tolerancing
2. ISO/5458–    Technical Drawings Positional tolerancing
3. ISO/5459–    Technical Drawings Datums and Datum Systems
4. ISO/2692–    Technical Drawings Maximum material principle
5. ISO/3040–    Technical Drawings Cones
6. ISO/1660–    Technical Drawings Profiles
7. ISO/129–    Technical Drawings General principles
8. ISO/406–    Technical Drawings Linear and angular dimensions
9. ISO/10578    Technical Drawings Projected tolerance zones
10. ISO/2692: 1988/DAM 1    Technical Drawings Least material principle
11. ISO/8015    Technical Drawings Fundamental tolerance principle
12. ISO/7083    Technical Drawings Symbols proportions
13. ISO/10579    Technical Drawings Non-rigid parts

Additional ISO standards involved:

1. ISO/1000 – SI Units
2. ISO/286 – Limits & Fits
3. ISO/TR5460 Technical Drawings–Verification principles
4. ISO/2768–2 General geometrical tolerances
5. ISO/1302 – Surface Texture
6. ISO/2768–1 Tolerances for linear and angular dimensions
7. Other peripheral standards on screw threads, gears, drills, welding, etc., may also be required for coverage beyond Y14.5 for product design.

### USA contribution to ISO standard concepts

Some examples of USA-originated concepts, principles or standard methodology which are now contained within ISO standards:

The datum reference frame (three plane concept)
Datum precedence
Datum targets, point, lines, areas
Cylindricity
Projected Tolerance zone
Total runout
Composite positional tolerancing
Least material condition
Non-rigid part specification
Multiple/compound datums
Conicity using profile tolerance
Spherical diameter symbol
Formulas for positional tolerancing

### USA delegates attendance at ISO meetings

Since 1967, USA delegates have attended ISO/TC10/SC5 plenary or working group meetings in:

| | | | |
|---|---|---|---|
| Moscow | Stockholm | Vilnius | Carmel, CA |
| London | Ottawa | Zurich | Beijing |
| Berlin | New York | Prague | Paris |

Cologne    The Hague    Gothenburg
Oslo       Orlando      Copenhagen

Cities such as New York, London, Paris, Berlin, Zurich, Ottawa have been common repeat locations.

## ANSI/ASME Y14.5M and ISO compatibility

As of this date, in the opinion of the author, the ANSI Y14.5 standard and ISO principles are 90–95% in agreement. This is based upon major concepts and methodology, and not a myriad of minor details of preference where some discrepancies exist and probably will continue to exist. Of the differences that did exist, the major concerns were the datum feature symbol and methodology, the RFS symbol (US used only), and projected tolerance zone placement within the feature control frame. Other differences of some note are: Concentricity applied on MMC basis (USA allows only RFS), ISO permits position tolerance application to non-size feature (USA uses profile), and ISO uses flatness tolerance on a coplanarity application (USA uses profile). There are some other differences of minor note which may continue to exist.

Current considerations, in the new revision to USA ANSI/ASME Y14.5 subcommittee activities, have addressed all of these foregoing matters. Some have been completely resolved. The ISO datum feature symbol and methodology is now the USA standard. The RFS symbol has been removed and implies the RFS principle (by default) with no indicator (RFS symbol may be used by those who desire to retain temporarily the redundant modifier), the projected tolerance zone symbol is placed within the feature control frame.

Since ISO 1101, ISO 5458 and ISO 2692 are work items for possible future revision within ISO/TC10/SC5, some of the remaining USA/ISO differences have been already discussed in meetings and via correspondence. Resolution of these matters favorable to USA preferences, are possible in the future. This could eliminate many more variations between the USA Y14.5 standard and the ISO family of GD&T standards.

By completion of the new revision to Y14.5M, there is a very close harmonization between Y14.5 and ISO standards up to the 95% range (in the opinion of the author) in those matters of major consequence.

## Unpredictables and new concepts

Another phase of the effort to harmonize concepts between ISO and new USA GD&T technology, brings new challenges.

The USA standards occasionally reach out beyond the ISO work agenda with newer concepts and thus some more sophisticated differences may result for a period of time. The impact of those differences can be minimized, however, by the now established, improved communication atmosphere between the USA and ISO participants. That is, the newer concepts are now being discussed somewhat earlier due to improved modern communication networks (FAX, computer, phone). Such concepts as the "axis of the actual mating envelope" using position tolerance RFS; the "center-plane of the actual mating envelope" using position tolerance RFS; coaxial use of position tolerance RFS; coaxial use of position tolerance MMC; modified definitions to accommodate the "mathematizing" of GD&T; "tangent plane" application and symbol; added symbols for countersink, counterbore, depth, arc length, composite position tolerancing, composite profile tolerancing, Resultant Condition, Inner and Outer Boundary, Coplanarity; and other refinements may cause the standards to struggle some to stay abreast (i.e., Y14.5 & ISO). However, the USA will be working these matters jointly through US/Technical Advisory Groups (TAG's) and committees. The USA plays a major role in both USA and ISO interests; they are both on our work agenda. The USA has, since 1992, held the Secretariat of the ISO/TC10/SC5 international committee (Secretariat = Chairmanship). As a new development, the ISO/TC10/SC5 committee has been disbanded and is now integrated into a new technical committee ISO/TC213 with the Secretariat and Chairmanship shifted to Denmark.

## Third angle projection versus First angle projection

This matter is not directly a GD&T concern but has, nevertheless, a major impact on international engineering drawings. No permanent resolution of this matter seems visible; we have "agreed to disagree." So long as the symbolic logo to designate one or the other is found on the drawing, the problem is minimized and the two systems can coexist comfortably; at least for the foreseeable future.

## Future meetings of ISO/TC10/SC5

Future meetings, to further advance standardization of GD&T, both nationally and internationally, and its compatibility with electronic (computer) application, will be scheduled as necessary.

New activities to further coordination and cooperation between related ISO committees is underway. ISO committees ISO/TC10/SC5, ISO/TC3 and ISO/TC57, under the banner of the Joint Harmonization Group (JHG), have been pursuing efforts to better harmonize the work of the subject matter found in these standards and their committees' work agenda. Subjects such as limits and fits, technical drawings, product definition, dimensioning and tolerancing, properties of surfaces, and metrology are involved. A new technical committee, ISO/TC213, is now established to merge these previous separate committees. The USA will actively participate in this new ISO mission.

*Conclusion*

The future seems very bright for the use of GD&T as a major factor in the years ahead. Its use as a technical language for global communication, enabling quality programs to thrive and improve international trade, economics, and understanding seems ensured. Its use will be a necessity and a rewarding culmination of the years of dedication to its purpose. Its continuance, maintenance, and continued success in the future lies in our collective hands.

Following in this text is our best effort to assist this important endeavor. That is, to present GD&T in a digestible manner and with the flexibility necessary to accommodate varying needs, interests, and background to pursue the desired educational goals on this important subject.

## 1-38  WHY USE GEOMETRIC DIMENSIONING AND TOLERANCING?

Why is it that we should be so interested in this subject?

FIRST AND FOREMOST ITS USE SAVES MONEY!
It saves money directly by providing for maximum producibility of the part through maximum production tolerances. It provides "bonus" or extra tolerances in many cases.
It ensures that design dimensional and tolerance requirements, as they relate to actual function, are specifically stated and thus carried out.
It adapts to, and assists, computerization techniques in design and manufacture.
It ensures interchangeability of mating parts at assembly.
It provides uniformity and convenience in drawing delineation and interpretation, thereby reducing controversy and guesswork.

Aside from these primary reasons there are others of a more general nature:

The intricacies of today's sophisticated engineering design demand new and better ways of accurately and reliably communicating requirements. Old methods simply no longer suffice.
Diversity of product line and manufacture makes considerably more stringent demands of the completeness, uniformity, and clarity of drawings.
It is increasingly becoming the "spoken word" throughout industry, the military, and, internationally, on engineering drawing documentation. Every engineer or technician involved in

originating or reading a drawing should have a working knowledge of this new state of the art.

## 1-39  WHAT IS GEOMETRIC DIMENSIONING AND TOLERANCING?

In particular, it is a means of dimensioning and tolerancing a drawing with respect to the actual function or relationship of part features which can be most economically produced. *Function* and *relationship* are the key words.

In general, it is a system of building blocks for good drawing practice which provides the means of stating necessary dimensional or tolerance requirements on the drawing not otherwise covered by implication or standard interpretation.

## 1-40  WHEN SHOULD GEOMETRIC DIMENSIONING AND TOLERANCING BE USED?

When part features are critical to function or interchangeability;
when functional gaging techniques are desirable;
when datum references are desirable to ensure consistency between design, manufacturing and verification operations;
when computerization techniques in design and manufacture are desirable;
when standard interpretation or tolerance is not already implied.

## 1-41  GEOMETRIC CHARACTERISTICS AND SYMBOLS

The geometric characteristics and symbols that are used as the building blocks for geometric dimensioning and tolerancing are:

|  |  |
|---|---|
| ⟱ | Flatness |
| — | Straightness |
| ○ | Circularity (Roundness) |
| ⌭ | Cylindricity |
| ⊥ | Perpendicularity (Squareness) |
| ∠ | Angularity |
| // | Parallelism |
| ⌒ | Profile of a Surface |

⌒    Profile of a Line

↗    Circular Runout

↗↗   Total Runout

⊕    Position

◎    Concentricity

═    Symmetry

# 1-42  USING SYMBOLS

The general use of symbols instead of notes on a drawing provides a number of advantages. The illustrations below incorporate the geometric characteristic symbols with datum and feature control symbols. Some of the advantages of symbols over notes are

1. The symbol has uniform meaning. A note can be stated inconsistently, with a possibility of misunderstanding.

2. Symbols are compact, quickly drawn, and can be placed on the drawing where the control applies; symbols adapt readily to computer applications.
   Notes require much more time and space, tend to be scattered on the drawing, often appear as footnotes which separate the note from the feature to which it applies.

3. Symbols are the international language and surmount individual language barriers.
   Notes may require translation if the drawing is used in another country.

4. Symbols can be applied with drafting templates or computer techniques and retain better legibility in various forms of copy reproduction.

5. Geometric tolerancing symbols follow the established precedent of other well-known symbol systems, e.g., electrical and electronic, welding, surface texture, etc.

## Using Symbols

## Using Notes

# 1-43  MAXIMUM MATERIAL CONDITION PRINCIPLE

**Symbol Ⓜ Abbreviation (MMC)**

One of the fundamental and most important principles of geometric dimensioning and tolerancing is MAXIMUM MATERIAL CONDITION. A thorough understanding of its meaning is therefore essential.

Note in the figure below that the "maximum material condition" size of the .250 ± .005 diameter hole is .245 or its *low* limit size. The hole at its low limit obviously retains more material than if it were at its *high* limit or larger size; thus the term "maximum material condition" defines the *low* limit when it applies to a hole or similar feature.

Note similarly that the .235 diameter pin is at its "maximum material condition" size when it is at its *high* limit of size of .240. In this instance it is more readily seen that more material exists in the pin when it is at its maximum permissible size. However, the same principle exists in both hole and pin MMC situations. Relating mating part features in this manner ensures their functional relationships, and as will be seen later in the text, establishes the criteria for determining necessary form, orientation, and position tolerances.

The symbol for "maximum material condition," the M enclosed in a circle, and the occasionally used abbreviation MMC are shown above. The symbolic method is to be used with feature control frames only. The abbreviation MMC may be used with note callouts but not with symbolic representations. We shall discuss later the application of the "maximum material condition" principle and illustrate it with practical examples.

Generally, the use of the "maximum material condition" principle permits greater possible tolerance as part feature sizes vary from their calculated "maximum material condition" limits. It also ensures interchangeability and permits functional gaging techniques. It is one of the fundamental principles upon which the system of geometric dimensioning and tolerancing is

2X ⌀.250 ±.005

⌀.245 MAXIMUM MATERIAL
CONDITION OF HOLE
(LOW LIMIT OF HOLE TOL)

2X ⌀.235 ±.005

⌀.240 MAXIMUM MATERIAL
CONDITION OF PIN
(HIGH LIMIT OF PIN TOL)

based. Below is the definition of maximum material condition and the usual prerequisites for application. We shall later expand the use of the principle by means of examples.

**Definition.** The condition in which a feature of size contains the maximum amount of material within the stated limits of size: for example, minimum hole diameter and maximum shaft diameter.

The "maximum material condition" principle is normally valid only when both of the following conditions exist:

1. Two or more features are interrelated with respect to location or orientation (e.g., a hole and an edge or surface, two holes, etc.). At least *one* of the related features is to be a feature of size.
2. The feature (or features) to which the MMC principle is to apply must be a feature of size (e.g., a hole, slot, pin, etc.) with an axis or center plane.

"Maximum material condition" has two connotations. One as seen in the preceding figure; the ⌀.245 MAXIMUM MATERIAL CONDITION HOLE further means that there is a Perfect Form at MMC boundary at ⌀.245 and the pin likewise at ⌀.240. (See Rule 1). The second connotation is where the material condition symbol Ⓜ is used in a feature control frame to apply the principle of MMC to the geometric tolerance specified to locate the pins and holes. See numerous figures in this text.

"Maximum material condition" might also be considered as a "new" term for an "old" situation, such as the familiar terms "worst condition," "critical size," etc., used in the past for relating mating part features.

Where the maximum material condition principle is not appropriate, the "regardless of feature size" principle or "least material principle" may be applied. (See below.)

# 1-44  REGARDLESS OF FEATURE SIZE

### Abbreviation (RFS)

**Definition.** The term used to indicate that a geometric tolerance or datum reference applies at any increment of size of the feature within its size tolerance.

"Regardless of feature size" is another principle of geometric dimensioning and tolerancing which must be well understood. Unlike maximum material condition, the "regardless of feature size" principle permits *no* additional positional, form or orientation tolerance, no matter to which size the related features are produced. It is really the independent form of dimensioning and tolerancing which has always been used prior to the introduction of the MMC principle.

The abbreviation for "regardless of feature size" is RFS. We shall later clarify the principle by means of examples.

The RFS principle is valid only when applied to features of size (for example, a hole, slot, pin, etc., with an axis or center plane). The size connotation cannot be applied to a feature which does not have "size."

## 1-45 LEAST MATERIAL CONDITION PRINCIPLE

### Symbol Ⓛ Abbreviation (LMC)

**Definition.** The condition in which a feature of size contains the least amount of material within the stated limits of size: for example, maximum hole diameter and minimum shaft diameter.

The least material condition principle may be desirable as an alternative to MMC or RFS in certain design considerations.

Note that the actual local sizes of the hole at Ø.255 and the pin at Ø.230 of the figures on page 14 are also their least material condition sizes.

## 1-46 BASIC AND DATUM

The terms BASIC and DATUM are most important. Proper application of the principles implied by these terms greatly contributes to effective geometric dimensioning and tolerancing.

### Basic

**Definition.** A numerical value used to describe the theoretically exact size, profile, orientation, or location of a feature or datum target. It is the basis from which permissible variations are established by tolerances in feature control frames or on other dimensions or notes.

Use of a BASIC dimension, which is a theoretical and exact value, requires also a *tolerance* stating the permissible variation from this exact value (most often relative to a position, angularity or profile requirement). A BASIC dimension states only *half* the requirement. To complete it, a tolerance must be associated with the features involved in the BASIC dimension.

In the past, a BASIC dimension was identified on the drawing by the word BASIC (or the accepted abbreviation BSC) adjacent to, or below, the dimension, or by a general note on the drawing. The symbolic method that follows on page 82 is the recommended method in keeping with latest standards and international practice.

### Example

Some companies use a "naked" or untoleranced dimension instead of BASIC. The same meaning as BASIC is invoked by adding a drawing footnote, title block notation, or company standards.

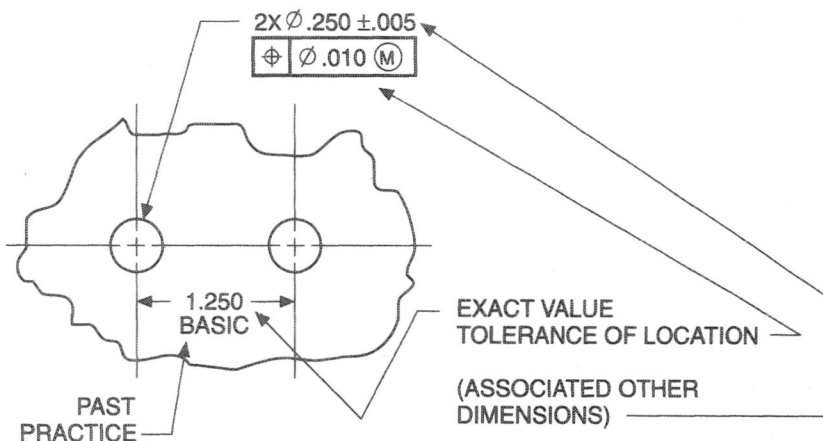

The term TP (for True Position) derived from British standards has also been used in the past. It has the same meaning as BASIC.

BASIC dimensions are also used in other applications such as tapers for which tolerances must be derived from associated size dimensions.

The use of BASIC dimensions on datum targets assumes standard tooling or gagemaker's tolerances (see DATUM section for more detail). BASIC dimensions used to indicate a limited area or portion of a surface where a tolerance (e.g., Runout) applies, assumes standard inspection set-up precision.

# 1-47 SYMBOLIC METHOD OF STATING A BASIC OR THEORETICAL EXACT VALUE—RECOMMENDED

The preferred method of stating an "exact" value replacing BASIC, BSC, TP, etc., is the international (ISO) method recommended by ANSI Y14.5. According to this method, the exact value is enclosed in a frame or box (see example below).

### Example

The symbolic method for "exact" values may be used with symbolic *or* notated geometric tolerancing. Because of the need to standardize U.S. practices and encourage compatibility with international practices, using the symbolic or boxed "exact" value is strongly recommended.

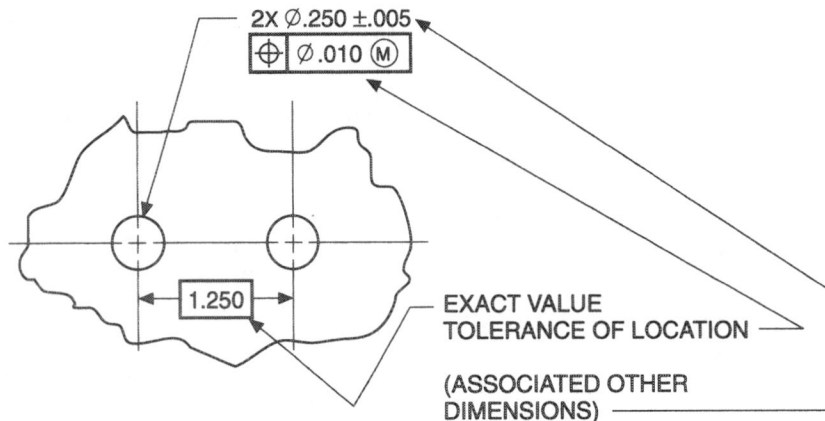

# 1-48 DATUMS AND DATUM FEATURE SYMBOL

**Definitions.** A theoretically exact point, axis, or plane derived from the true geometric counterpart of a specified datum feature. A datum is the origin from which the location or geometric characteristics of features of a part are established.

Datum surfaces and datum features are actual part surfaces or features used to establish datums. They include all the surface or feature inaccuracies.

### Datum Feature Symbol

To identify a feature as a datum, the following datum feature symbol is used:

(Datum feature triangle may be filled or open. Leader may be appropriately directed to a feature.)

The datum feature symbol consists of a capital letter enclosed in a square frame, a leader line extending from the frame to the concerned feature and terminating with a triangle.

Each datum requiring identification is assigned a different reference letter. Do not use letters I, O, Q. If the single letter alphabet is exhausted, double letters may be used, i.e., AA, AB, etc.

Where datum feature symbol is repeated to identify the same feature in other locations of a drawing, it need not be identified as reference.

### Placement of the Datum Feature Symbol Application to Plane Surfaces

The datum feature symbol is applied to the concerned feature surface outline, extension line, dimension line or feature control frame as follows:

1. Placed on the outline of a feature surface or an extension line of the feature outline (but clearly separated from the dimension line when the datum feature is represented by the extension line or feature surface itself.)

## Placement of the Datum Feature Symbol Application to size Features

2. Placed on an extension of the dimension line of a size feature when the datum is the axis or median center plane. If there is insufficient space for the two arrows, one of them may be replaced by the datum feature triangle.

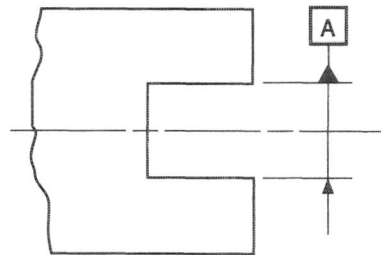

3. Placed on the outline of a cylindrical feature surface or an extension line of the feature outline, separated from the size dimension, when the datum is the axis. For CAD systems, the triangle may be tangent to the feature.

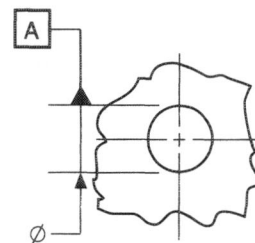

4. Placed below or above and attached to the feature control frame when the feature, or group of features, controlled is the datum axis or datum centerplane.

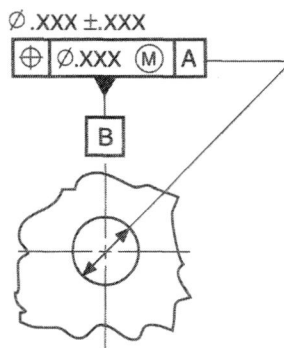

5. Placed on the planes established by datum targets on complex or irregular datum features, equalizing datums, etc., as an option for clarification, or to re-identify previously established datum axes or planes, as reference, on repeated or multi-sheet drawing requirements.

6. Placed on a dimension leader line to the feature size dimension where no geometrical tolerance and feature control frame is used.

## 1-49 FEATURE AND FEATURE CONTROL FRAME

### Feature

The general term applied to a physical portion of a part, such as a surface, hole or slot.

Features are specific component portions of a part and may include one or more surfaces such as holes, faces, screw threads, profiles, or slots. Features may be "individual" or "related."

### Feature Control Frame

The feature control frame consists of a box containing the geometric characteristic symbol, datum references, tolerance, and the material condition symbol (e.g., for MMC) if applicable.

The example below shows this feature control frame as used on a part drawing.

### Example

### Placement of the Feature Control Frame

The feature control frame is associated with the feature(s) being toleranced by one of the following methods:

1. Attaching a side, end, or corner of the symbol box to an EXTENSION LINE or leader from the feature (used on most form tolerances). See Fig. 1-87.

2. Attaching a side or end of the symbol box to the DIMENSION LINE or EXTENSION LINE pertaining to the feature when it is cylindrical. See Fig. 1-88.

3. Placing the symbol box below or closely adjacent to the dimension or note pertaining to that feature. See Fig. 1-89.

4. Running a leader line from the symbol to the feature. See Fig. 1-90.

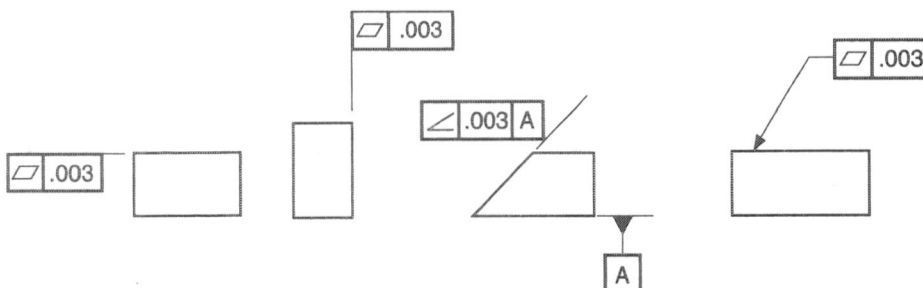

**Figure 1-87**

**Figure 1-88**

Ø X.XXX ± .XXX

**Figure 1-89**

**Figure 1-90**

# 1-50  COMBINED FEATURE CONTROL FRAME AND DATUM FEATURE SYMBOL

When a feature serves as a datum and is also controlled by a geometric tolerance, the feature control frame and the datum feature symbol may be combined as shown.

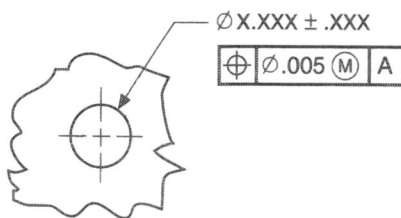

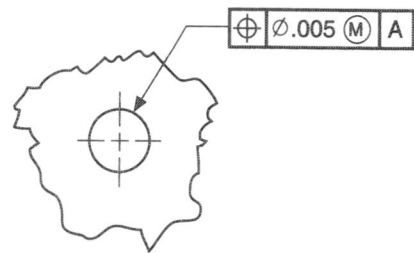

# 1-51  REFERENCE TO DATUM

When an orientation, profile, runout, or location tolerance must be related to a datum, this relationship is stated by placing the datum reference letter following the geometric characteristic symbol and the tolerance.

The illustrations show additional examples of the feature control frames with reference to datums.

Figure 1-91 is a typical feature control frame using a single datum reference. The symbol reads "This feature shall be within a .002 tolerance zone perpendicular to datum A."

Figure 1-92 shows a feature control frame with *two* datums. The symbol reads "This feature shall be located at true position within .005 diameter at maximum material condition with respect to both datums A and B."

Note that vertical lines are used to separate the characteristic symbol, the feature tolerance, and the datum references. These vertical lines are used on all feature control frames to ensure clarity. One reason for this is illustrated in Fig. 1-93, in which the maximum material condition symbol is used. The vertical lines clearly show that MMC condition symbols apply only to those datums or tolerances with which they appear in the subdivision of the symbol box.

Figure 1-94 illustrates primary, secondary, and tertiary datums showing the order of precedence. When the order of precedence of datums is significant to function, datum references should be classified as primary, secondary, and tertiary. The datum precedence is shown by placing each datum reference letter in the proper order. The first datum letter (left to right) is considered the primary datum, the second letter secondary, and the third letter tertiary. Thus the datum reference letters will not necessarily be in alphabetical order. See section on DATUMS for further explanation.

Figure 1-95 illustrates a feature control frame in which multiple datum features are used simultaneously to establish a single datum reference (equal precedence of datum features) e.g., to establish a common datum axis. See section on DATUMS and RUNOUT for further details.

Figure 1-96 illustrates a possibly questionable use of a datum reference. Note that datum A applies at MMC, whereas the feature controlled applies at RFS. This means that the datum reference is subject to variation and cannot serve as a fixed reference for any RFS relationship. Al-though there may be exceptions under special circumstances, generally, wherever MMC is used on any datum, the feature controlled should also be controlled at MMC.

## Diameter Symbol Ø

The symbol used to designate a diameter (DIA) or cylindrical feature or tolerance zone is shown above. The symbol (Ø) precedes the specified tolerance. See Figs. 1-92, 1-93, 1-94, and 1-96. The symbol is used elsewhere on the drawing to indicate that the feature is circular or cylindrical. The symbol (Ø) precedes the stated nominal size value of the concerned feature.

**Figure 1-91**

**Figure 1-92**

**Figure 1-93**

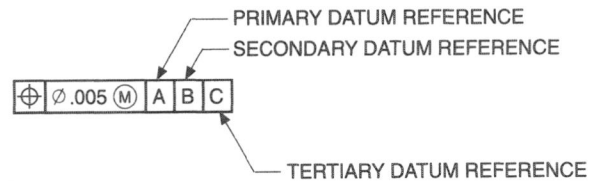

**Figure 1-94**

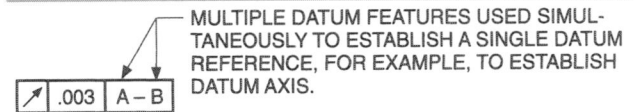

**Figure 1-95**

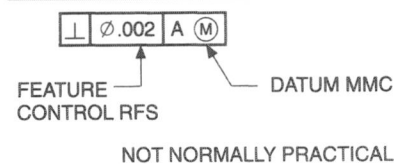

**Figure 1-96**

## 1-52 GEOMETRIC CHARACTERISTICS—FORM, ORIENTATION, PROFILE, RUNOUT, AND LOCATIONAL TOLERANCE— OTHER SYMBOLS AND TERMS

Illustrated at right are the geometric characteristics and symbols which are the basis for the language of geometric dimensioning and tolerancing.

### Five Types of Geometric Characteristics

Expanding on preceding text explanation, it is seen that the geometric characteristics are of five types (see "5-TYPES" column at right):

1. FORM tolerance—A form tolerance states how far an actual surface or feature is permitted to vary from the desired form implied by the drawing.

2. ORIENTATION tolerance—An orientation tolerance states how far an actual surface or feature is permitted to vary relative to a datum or datums.

3. PROFILE tolerance—A profile tolerance states how far an actual surface or feature is permitted to vary from the desired form on the drawing and/or vary relative to a datum or datums.

4. RUNOUT tolerance—A runout tolerance states how far an actual surface or feature is permitted to vary from the desired form implied by the drawing during full (360°) rotation of the part on a datum axis.

5. LOCATION tolerance—A location tolerance states how far an actual size feature is permitted to vary from the perfect location implied by the drawing as related to a datum, or datums, or other features.

## Kinds of Features to which a Geometric Characteristics is Applicable

The geometric characteristics are also divisible into three "kinds" of features to which a particular characteristic is applicable. (See "KIND OF FEATURE" column at below.)

1. *INDIVIDUAL* feature—A single surface, element, or size feature which relates to a perfect geometric counterpart of itself as the desired form; no datum is proper nor used. (characteristics ▱,—,○,⌭ ).

2. *RELATED* feature—A single surface or element feature which relates to a datum, or datums, in form and orientation. (characteristics ⊥,∠,∥ ). A size feature (e.g., hole, slot, pin, shaft) which relates to a datum, or datums, in form, attitude (orientation), runout, or location. (characteristics ⊥,∠,∥,↗,↗,⊕,◎,≡ ).

3. *INDIVIDUAL* or *RELATED* feature—A single surface or element feature whose perfect geometric profile is described which may, or may not, relate to a datum, or datums, (characteristics ⌒,⌓ ).

### Other Symbols and Terms

For review, other symbols and characteristics are shown at right. The Projected Tolerance Zone, Least Material Condition, and Datum Target symbols are explained in later text.

## 1-53 GEOMETRIC CHARACTERISTICS, SYMBOLS, AND TERMS

| SYMBOL | CHARACTERISTIC | 5 TYPES | KIND OF FEATURE |
|---|---|---|---|
| ▱ | Flatness | | |
| — | Straightness | | |
| ○ | Circularity (Roundness) | FORM | INDIVIDUAL |
| ⌭ | Cylindricity | | |
| ⌒ | Profile of a Line | | |
| ⌓ | Profile of a Surface | PROFILE | INDIVIDUAL OR RELATED |
| ⊥ | Perpendicularity (Squareness) | | |
| ∠ | Angularity | ORIENTATION | |
| ∥ | Parallelism | | |
| ↗ | Circular Runout | | |
| ↗ | Total Runout | RUNOUT | RELATED |
| ⊕ | Position | | |
| ◎ | Concentricity | LOCATION | |
| ≡ | Symmetry | | |
| Ⓜ | Maximum Material Condition  MMC | | |
| | Regardless of Feature Size  RFS* | | |
| Ⓛ | Least Material Condition  LMC | | |
| Ⓟ | Projected Tolerance Zone | | |
| Ⓣ | Tangent Plane | | |
| ⌀ | Diametrical (Cylindrical) Tol Zone or Feature | | |
| Ⓕ | Free State | | |
| .605 | Basic, or Exact, Dimension | | |
| A◀ | Datum Feature Symbol | | |
| ⊕ ⌀.005 Ⓜ A | Feature Control Frame | | |
| A1 | Datum Target | | |

---

\* (RFS implied unless otherwise specified under Material Condition Rule , See Rule 2 and Alternate Rule 2A.)

## 1-54 GENERAL RULES

Like any discipline, geometric dimensioning and tolerancing is based on certain fundamental rules. Some of these follow from standard interpretation of the various characteristics, some govern specification, and some are general rules applying across the entire system.

The various rules appropriate to given geometric characteristics and related nomenclature will be discussed later. The general rules are described below and on succeeding pages.

### Applicability of General Rules

ANSI Y14.5M contains four general rules. ANSI Y14.5M must be referenced whenever these rules are to be applied.

### Individual Features of Size
### Limits of Size

Unless otherwise specified the limits of size of a feature prescribe the extent within which variations of geometric form, as well as size, are allowed. This control applies solely to individual features of size.

---

## RULE 1—LIMITS OF SIZE RULE

### Individual Feature of Size

Where only a tolerance of size is specified, the limits of size of an individual feature prescribe the extent to which variations in its geometric form as well as size are allowed.

### Variations of Size

The actual local size of an individual feature at any cross-section shall be within the specified tolerance of size.

### Variations of Form (Envelope Principle)

The form of an individual feature is controlled by its limits of size to the extent prescribed in the following paragraph and illustration:

a. The surface, or surfaces, of a feature shall not extend beyond a boundary (envelope) of perfect form at MMC. This boundary is the true geometric form represented by the drawing. No variation in form is permitted if the feature is produced at its MMC limit of size.

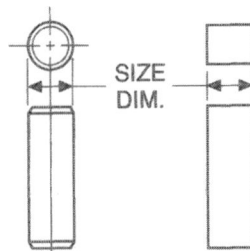

SIZE DIM.

INDIVIDUAL SIZE FEATURES

# RULE 1—LIMITS OF SIZE RULE (CONTINUED)

EXTERNAL FEATURE                    INTERNAL FEATURE

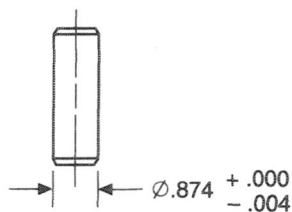

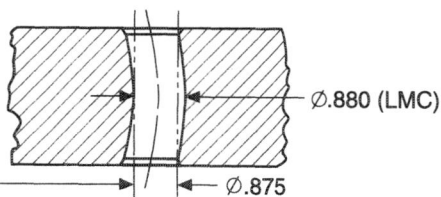

Ø.874 $^{+\ .000}_{-\ .004}$        Ø.875 $^{+\ .005}_{-\ .000}$

Ø.874 (MMC)        Ø.880 (LMC)

Ø.875

Ø.874 — BOUNDARY OF PERFECT FORM AT MMC

Ø.870 (LMC)        Ø.875 (MMC)

Ø.870 (LMC)        Ø.880 (LMC)

Ø.874 — BOUNDARY OF PERFECT FORM AT MMC

Ø.875

b. Where the actual local size of a feature has departed from MMC toward LMC, a variation in form is allowed equal to the amount of such departure.

c. There is no requirement for a boundary of perfect form at LMC. Thus, a feature produced at its LMC limit of size is permitted to vary from true form to the maximum variation allowed by the boundary of perfect form at MMC.

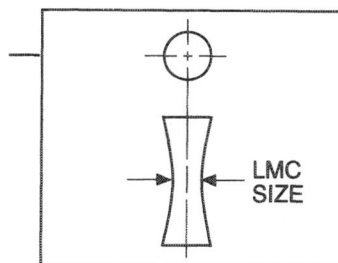

LMC SIZE

## When Perfect Form at MMC (Rule 1) Does *not* Apply:

The control of geometric form prescribed by limits of size does not apply to the following:

a.  Stock such as bars, sheets, tubing, structural shapes, and other items produced to established industry or government standards that prescribe limits for straightness, flatness, and other geometric characteristics. Unless geometric tolerances are specified on the drawing of a part made from these items, standards for the items govern the surfaces that remain in the "as furnished" condition on the finished part.

b.  Parts subject to free state variation in the unrestrained condition.

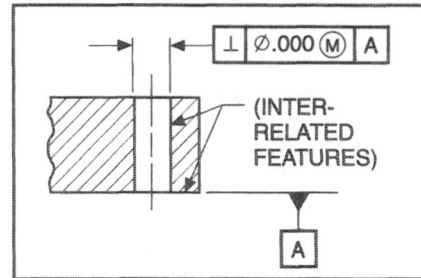

## Perfect Form at MMC *not* Required (Rule 1 Removed)

Where it is desirable to permit a surface, or surfaces, of a feature to exceed the boundary of perfect form at MMC, a note such as PERFECT FORM AT MMC NOT REQUIRED is specified exempting the pertinent size dimension from the provision of RULE 1.

## Relationship  Between Individual Features

The limits of *size do not* control the runout, orientation, or location relationship *between* individual features.

Features shown perpendicular, coaxial, or symmetrical or otherwise geometrically related to each other must be controlled for runout, orientation, or location to avoid incomplete drawing requirements. These controls may be specified by the use of appropriate geometric tolerances.

If it is necessary to establish a boundary of perfect form at MMC to control the relationship between features, the following methods are used:

a.  Specify a zero tolerance of orientation at MMC, including a datum reference (at MMC if applicable), to control perpendicularity, parallelism, or angularity of the feature.

b.  Specify a zero positional tolerance at MMC, including a datum reference to control locational, coaxial, or symmetrical features.

c.  Relate the dimensions to a datum reference frame, or by a local or general note indicating datum precedence.

d.  Indicate this control for the features involved by a note such as PERFECT ORIENTATION (OR COAXIALITY OR SYMMETRY) AT MMC REQUIRED FOR RELATED FEATURES.

## Applicability of RFS, MMC, and LMC

Applicability of RFS, MMC, or LMC is *limited to features subject to variations in size*. They may be datum features or other features whose axes or center planes are controlled by geometric tolerances. In such cases the following practices apply:

---

# RULE 2 —MATERIAL CONDITION RULE

For *all* applicable geometric tolerances, RFS applies with respect to the individual tolerance, datum reference, or both, where no modifying symbol is specified. MMC, Ⓜ , or LMC, Ⓛ , must be specified on the drawing where it is required. Such as:

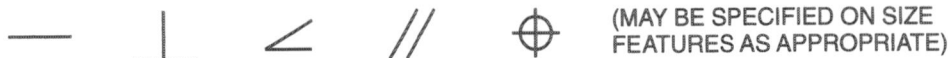

—    ⊥    ∠    //    ⊕    (MAY BE SPECIFIED ON SIZE FEATURES AS APPROPRIATE)

**NOTE:**

The below characteristics and controls are *always* applicable at RFS and due to the nature of the control cannot be applied at MMC or LMC.

# PITCH DIAMETER RULE

Each tolerance of orientation or position and datum reference specified for a screw thread applies to the axis of the thread derived from the pitch cylinder. Where an exception to this practice is necessary, the specific feature of the screw thread (such as MAJOR Ø or MINOR Ø) shall be stated beneath the feature control frame or beneath the datum feature symbol, as applicable.

$$\boxed{\oplus \ \vert \ \emptyset.005 \ \textcircled{M} \ \vert \ A}$$

MAJOR Ø

$$\boxed{A} \blacktriangleleft$$

MAJOR Ø

Each tolerance of orientation or location and datum reference specified for gears, splines, etc., must designate the specific feature of the gear, spline, etc., to which it applies (such as PITCH Ø, PD, MAJOR Ø, or MINOR Ø). This information is stated beneath the feature control frame or beneath the datum feature symbol.

$$\boxed{\oplus \ \vert \ \emptyset.005 \ \textcircled{M} \ \vert \ A}$$

PD

$$\boxed{A} \blacktriangleleft$$

PD

# DATUM/VIRTUAL CONDITION RULE

A virtual condition exists for a datum feature of size where its axis or center plane is controlled by a geometric tolerance. In such cases, the datum feature applies at its virtual condition even though it is referenced in a feature control frame at MMC or LMC.

**NOTE:**

Alternate practice (Rule 2A). For a tolerance of position RFS may be specified on the drawing with respect to the individual tolerance, datum reference, or both, as applicable. (Past Practices.) (Not compatible with ISO practices.)

# APPLICABILITY OF MMC, RFS, OR LMC

| | Characteristic | Applicability to feature | Applicability to datum reference |
|---|---|---|---|
| ▱ | Flatness | Not applicable | |
| — | Straightness | MMC Ⓜ or RFS applicable if tolerance applies to axis or center plane of a feature with size; not applicable if considered feature is a single plane surface | No datum reference |
| ○ | Circularity | Not applicable | |
| ⌀ | Cylindricity | | |
| ⌒ ⌒ | Profile of a surface Profile of a line | Not applicable | RFS on datum reference applicable if datum feature has size and has an axis or center plane; not applicable if datum feature is a single plane surface. MMC Ⓜ not applicable† |
| ⊥ // ∠ | Perpendicularity Parallelism Angularity | MMC Ⓜ, RFS, or LMC Ⓛ applicable if tolerance applies to axis or center plane of a feature with size; not applicable if considered feature is a single plane surface | MMC Ⓜ, RFS, or LMC Ⓛ on datum reference applicable if datum feature has size and has an axis or center plane; not applicable if datum feature is a single plane surface |
| ⊕ | Position | MMC Ⓜ, RFS, or LMC applicable if tolerance applies to axis or center plane of a feature with size; not applicable if considered feature is a single plane surface | MMC Ⓜ, RFS, or LMC Ⓛ on datum reference applicable if datum feature has size and has an axis or center plane; not applicable if datum feature is a single plane surface |
| ↗ ↗↗ ◎ ⹀ | Circular Runout Total Runout Concentricity Symmetry | Only RFS applicable | Only RES applicable |

† May have exceptions under special conditions. (See PROFILE section)

## Shape of Tolerance Zone for Form, Orientation, Profile, Runout, and Locational Tolerances

Where the specified tolerance value represents the diameter of a cylindrical zone, the diameter symbol Ø shall be included in the feature control frame.

Where the tolerance zone is other than a diameter, the tolerance value represents the distance between two parallel straight lines or planes or the distance between two uniform boundaries.

## Examples

TOLERANCE ZONE SHAPE IS: Where the diameter (cylindrical) symbol Ø is specified, the tolerance is a diameter (or cylindrical) shape. Where no Ø is specified, the tolerance zone is between two parallel lines or planes in the direction of the dimension arrows. The tolerance indicated is the TOTAL tolerance permitted.

| POSITION | CONCENTRICITY | SYMMETRY | FLATNESS |
|---|---|---|---|
| ⊕ Ø.XXX Ⓜ A<br>— DIAMETER (CYL) TOL ZONE<br><br>⊕ .XXX Ⓜ A<br>— TOTAL WIDTH TOL ZONE | ◎ Ø.XXX A<br>— DIAMETER (CYL) TOL ZONE | ⬄ .XXX A<br>— TOTAL WIDTH TOL ZONE | ▱ .XXX<br>— TOTAL WIDTH TOL ZONE |
| STRAIGHTNESS | CIRCULARITY | CYLINDRICITY | PERPENDICULARITY |
| — .XXX<br>— TOTAL WIDTH TOL ZONE<br><br>— Ø.XXX Ⓜ<br>— DIAMETER (CYL) TOL ZONE | ○ .XXX<br>— TOTAL WIDTH TOL ZONE BETWEEN TWO CONCENTRIC CIRCLES | ⌀ .XXX<br>— TOTAL WIDTH OR THICKNESS TOL ZONE BETWEEN TWO CONCENTRIC CYLINDERS | ⊥ .XXX A<br>— TOTAL WIDTH TOL ZONE<br><br>⊥ Ø.XXX A<br>— DIAMETER TOL ZONE |
| PARALLELISM | ANGULARITY | RUNOUT | PROFILE OF SURFACE PROFILE OF LINE |
| // .XXX A<br>— TOTAL WIDTH TOL ZONE<br><br>// Ø.XXX A<br>— DIAMETER (CYL) TOL ZONE | ∠ .XXX A<br>— TOTAL WIDTH TOL ZONE<br><br>∠ Ø.XXX A<br>— DIAMETER (CYL) TOL ZONE | ↗ .XXX A–B<br>— CIRCULAR RUNOUT (FIM) TOL ZONE<br><br>↗↗ .XXX A–B<br>— TOTAL RUNOUT (FIM) TOL ZONE | ⌓ .XXX A    ⌒ .XXX<br>— TOTAL WIDTH OR THICKNESS TOL ZONE AS APPLICABLE |

# 1-55 VIRTUAL CONDITION (MMC)

It is necessary to understand Virtual Condition as it applies to features. The definition below and the examples below and throughout the text will clarify its meaning.

**Definition.** Virtual Condition. A constant boundary generated by the collective effects of a size feature's specified MMC or LMC material condition and the geometric tolerance for that material condition.

## Where MMC is Specified:

The virtual condition of a feature is the extreme boundary of that feature which represents the "worst case"

### Example

for, typically, such concerns as a clearance or fit possibility relative to a mating part or situation.

- FOR PIN: In the case of an external feature such as a pin or shaft, the virtual condition is determined by: MMC + TOL = VC; e.g., Pin MMC + stated orientation or position tolerance = Pin virtual condition.

Virtual condition of a pin or shaft is always a "constant value" and can also be referred to as the "outer boundary (locus)" in worst case analysis calculations.

### Meaning

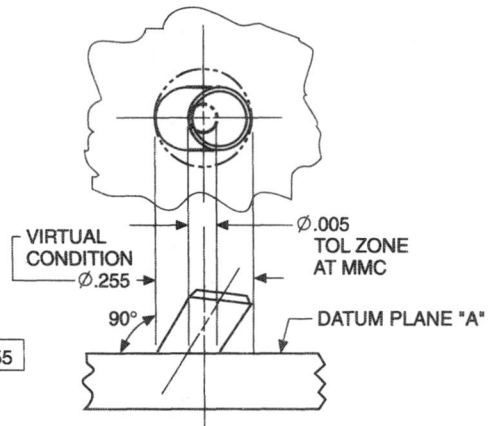

- FOR HOLE: In the case of an internal feature such as a hole, the virtual condition is determined by: MMC − TOL = VC; e.g., Hole MMC − stated orientation or position tolerance = Hole virtual condition.

Virtual condition of a hole is always a "constant value" and can also be referred to as the "inner boundary (locus)" in worst case analysis calculations.

### Example

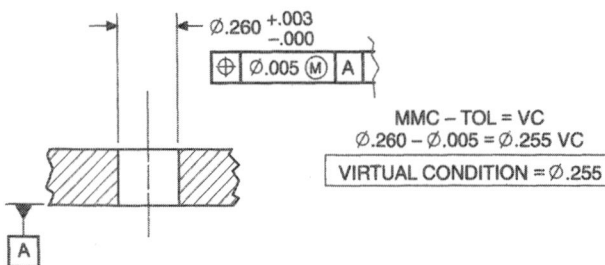

### Meaning

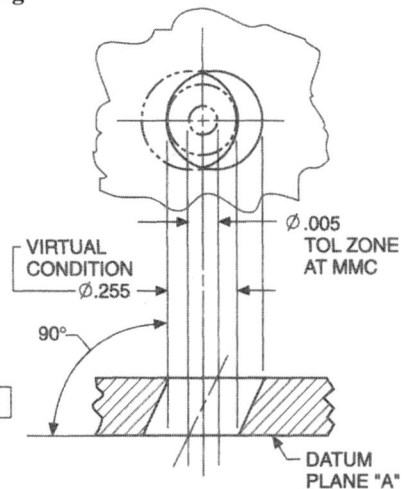

## 1-56 VIRTUAL CONDITION (LMC)

**Definition.** Virtual Condition. A constant boundary generated by the collective effects of a size feature's specified MMC or LMC material condition and the geometric tolerance for that material condition.

### Where LMC is Specified:

The virtual condition of a feature is the extreme boundary of that feature which represents the "worst case"

for such concerns as cross-sectional mass, strength, alignment, wall thickness, compensating effects, interference, etc., relative to a mating part or situation.

- FOR PIN: In the case of an external feature such as a pin or shaft, the virtual condition is determined by: LMC − TOL = VC; e.g., Pin LMC − stated orientation position tolerance = Pin virtual condition.

Virtual condition of a pin or shaft is always a "constant value" and can also be referred to as the "inner boundary (locus)" in worst case analysis calculations.

**Example**

**Meaning**

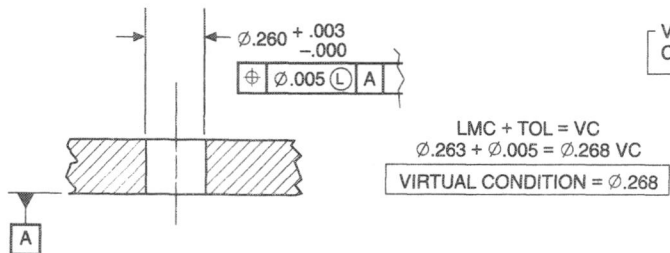

$$LMC - TOL = VC$$
$$\varnothing.247 - \varnothing.005 = \varnothing.242\ VC$$

VIRTUAL CONDITION = ∅.242

- FOR HOLE: In the case of an internal feature, such as a hole, the virtual condition is determined by: LMC + TOL = VC; e.g., Hole LMC + stated orientation or position tolerance = Hole virtual condition.

Virtual condition of a hole or shaft is always a "constant value" and can also be referred to as the "outer boundary (locus)" in worst case analysis calculations.

**Example**

**Meaning**

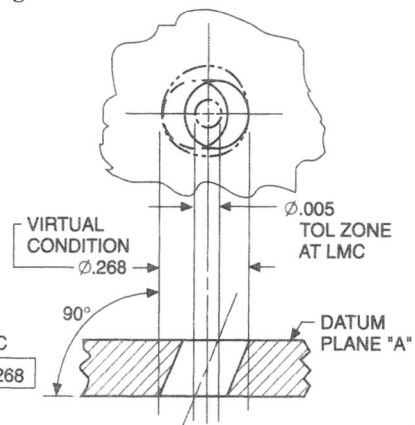

$$LMC + TOL = VC$$
$$\varnothing.263 + \varnothing.005 = \varnothing.268\ VC$$

VIRTUAL CONDITION = ∅.268

## Resultant Condition — (MMC)

It is necessary to understand Resultant Condition as it applies to features. The definition below and the examples shown will clarify its meaning.

**Definition.** Resultant Condition.

The variable boundary generated by the collective effects of a size feature's specified MMC or LMC material condition, the geometric tolerance for that material condition, the size tolerance, and the additional geometric tolerance derived from the feature's departure from its specified MMC or LMC material condition.

### Where MMC is Specified:

The resultant condition of a feature is the extreme boundary which represents the "worst case" of that feature (in the opposite direction from its virtual condition).

**Example**

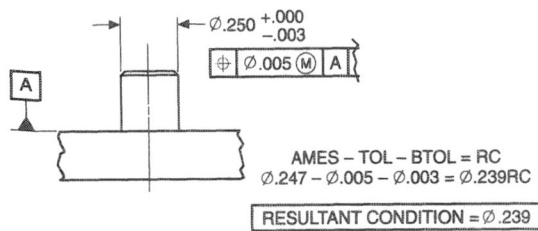

AMES − TOL − BTOL = RC
Ø.247 − Ø.005 − Ø.003 = Ø.239RC

RESULTANT CONDITION = Ø.239

(NOTE: Above example uses  .247 (LMC) for explanation.)

- FOR PIN: In the case of an external feature such as a pin or shaft, the resultant condition is determined by: AMES − TOL − BTOL = RC; e.g., Pin Actual Mating Envelope Size (AMES) − stated position or orientation Tolerance (TOL) − Bonus Tolerance (BTOL) = Resultant Condition (RC).

Resultant condition of a pin is a variable "worst case value" and can also be referred to as the "inner boundary (locus)" in worst case analysis calculations.

| AMES | − TOL | − BTOL | = RC |
|---|---|---|---|
| Ø.250 MMC | Ø.005 | 0 | Ø.245 |
| Ø.249 | Ø.005 | Ø.001 | Ø.243 |
| Ø.248 | Ø.005 | Ø.002 | Ø.241 |
| Ø.247 LMC | Ø.005 | Ø.003 | Ø.239 |

**Meaning**

Ø.239 RESULTANT CONDITION

Ø.008 AT LMC

Ø.005 TOL ZONE AT MMC

Ø.247 LMC PIN

(ALL PERPENDICULAR TO DATUM PLANE "A")

- FOR HOLE: In the case of an internal feature such as a hole, the resultant condition is determined by: AMES + TOL + BTOL = RC; e.g., Hole Actual Mating Envelope Size (AMES) + stated position or orientation Tolerance (TOL) + Bonus Tolerance (BTOL) = Resultant Condition (RC).

Resultant Condition of a hole is a variable "worst case value" and can also be referred to as the "outer boundary (locus)" in worst case analysis calculations.

| AMES | + TOL + | BTOL | = RC |
|---|---|---|---|
| Ø.260 MMC | Ø.005 | Ø 0 | Ø.265 |
| Ø.261 | Ø.005 | Ø.001 | Ø.267 |
| Ø.262 | Ø.005 | Ø.002 | Ø.269 |
| Ø.263 LMC | Ø.005 | Ø.003 | Ø.271 |

**Example**

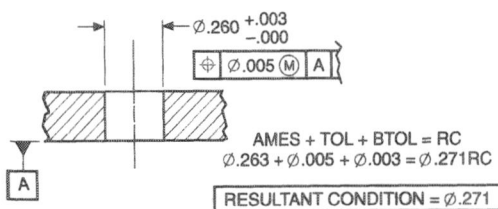

AMES + TOL + BTOL = RC
Ø.263 + Ø.005 + Ø.003 = Ø.271RC

RESULTANT CONDITION = Ø.271

(NOTE: Above example uses Ø.263 (LMC) for explanation.)

**Meaning**

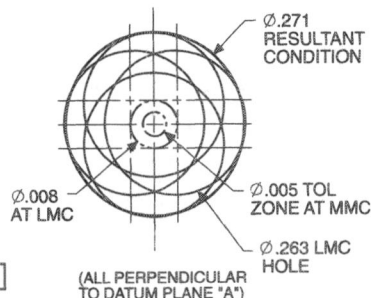

Ø.271 RESULTANT CONDITION

Ø.008 AT LMC

Ø.005 TOL ZONE AT MMC

Ø.263 LMC HOLE

(ALL PERPENDICULAR TO DATUM PLANE "A")

## Resultant Condition — (LMC)

It is necessary to understand Resultant Condition as it applies to features. The definition below and the examples shown will clarify its meaning.

**Definition.** Resultant Condition.
The variable boundary generated by the collective effects of a size features specified MMC or LMC material condition, the geometric tolerance for that material condition, the size tolerance, and the additional geometric tolerance derived from the feature's departure from its specified material condition.

## Where LMC is Specified:

The resultant condition of a feature is the extreme boundary which represents the "worst case" of that feature (in the opposite direction from its virtual condition).

**Example**

Ø.250 $^{+.000}_{-.003}$

⊕ | Ø.005 Ⓛ | A

A

AMES + TOL + BTOL = RC
Ø.250 + Ø.005 + Ø.003 = Ø.258 RC

RESULTANT CONDITION = Ø.258

- FOR HOLE: In the case of an internal feature such as a hole, the resultant condition is determined by: AMES − TOL − BTOL = RC; e.g., Hole Actual Mating Envelope Size (AMES) − stated position or orientation Tolerance (TOL) − Bonus Tolerance (BTOL) = Resultant Condition (RC).

- FOR PIN: In the case of an external feature such as a pin or shaft, the resultant condition is determined by: AMES + TOL + BTOL = RC; e.g., Pin Actual Mating Envelope Size (AMES) + stated position or orientation Tolerance (TOL) + Bonus Tolerance (BTOL) = Resultant Condition (RC).

Resultant condition of a pin is a variable "worst case value" and can also be referred to as the "outer boundary (locus)" in worst case analysis calculations.

| AMES | + TOL | + BTOL | = RC |
|---|---|---|---|
| Ø.247 LMC | Ø.005 | 0 | Ø.252 |
| Ø.248 | Ø.005 | Ø.001 | Ø.254 |
| Ø.249 | Ø.005 | Ø.002 | Ø.256 |
| Ø.250 MMC | Ø.005 | Ø.003 | Ø.258 |

**Meaning**

Ø.258 RESULTANT CONDITION

Ø.008 AT MMC

Ø.005 TOL ZONE AT LMC

Ø.250 MMC PIN

(ALL PERPENDICULAR TO DATUM PLANE "A")

Resultant Condition of a hole is a variable "worst case value" and can also be referred to as the "outer boundary (locus)" in worst case analysis calculations.

| AMES | − TOL | − BTOL | = RC |
|---|---|---|---|
| Ø.263 LMC | Ø.005 | Ø 0 | Ø.258 |
| Ø.262 | Ø.005 | Ø.001 | Ø.256 |
| Ø.261 | Ø.005 | Ø.002 | Ø.254 |
| Ø.260 MMC | Ø.005 | Ø.003 | Ø.252 |

**Example**

Ø.260 $^{+.003}_{-.000}$

⊕ | Ø.005 Ⓛ | A

A

AMES − TOL − BTOL = RC
Ø.260 − Ø.005 − Ø.003 = Ø.252 RC

RESULTANT CONDITION = Ø.252

**Meaning**

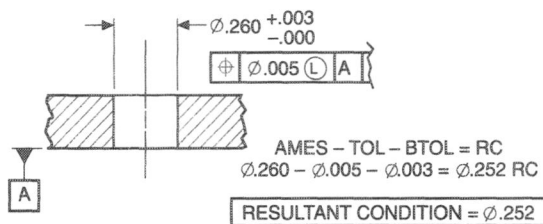

Ø.252 RESULTANT CONDITION

Ø.008 AT MMC

Ø.005 TOL ZONE AT LMC

Ø.260 MMC HOLE

(ALL PERPENDICULAR TO DATUM PLANE "A")

(NOTE: Above example uses Ø.260 (MMC) for explanation.)

# Sectional Views and GD&T

## 2-1 INTRODUCTION

*Sectional views* are used in technical drawing to expose internal surfaces. They serve to present additional orthographic views of surfaces that appear as hidden lines in the standard front, top, and side orthographic views.

Figure 2-1A shows an object intersected by a cutting plane. Figure 2-1B shows the same object with its right side and the cutting plane removed. Hatch lines are drawn on the surfaces that represent where the cutting plane passed through solid material. Also shown are the front and right-side orthographic views and a sectional view.

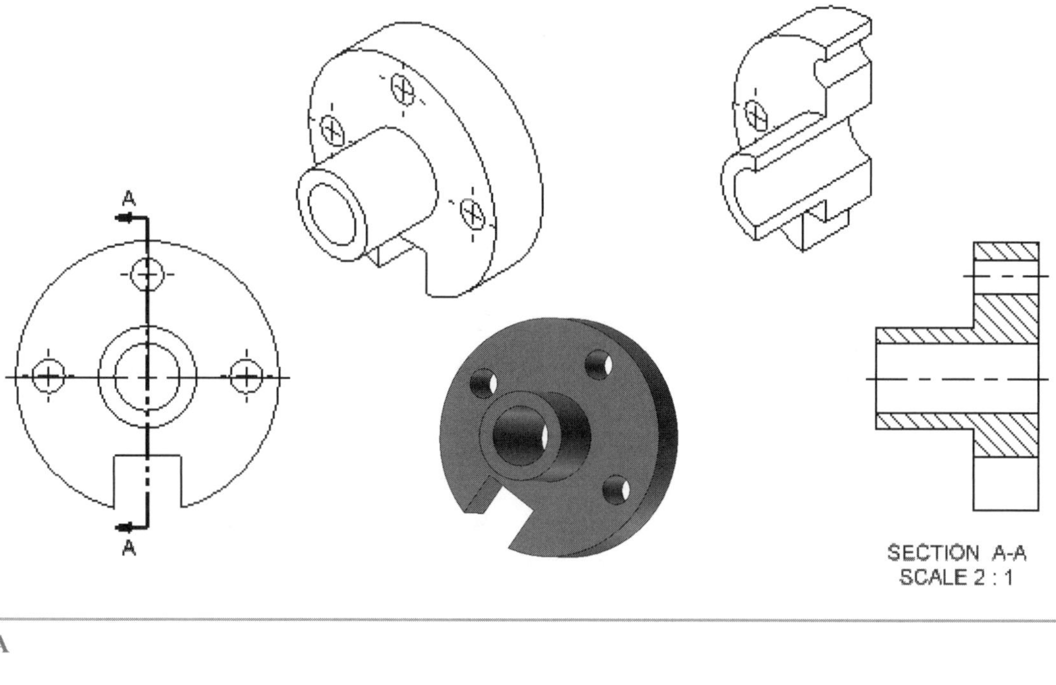

SECTION A-A
SCALE 2 : 1

**Figure 2-1A**

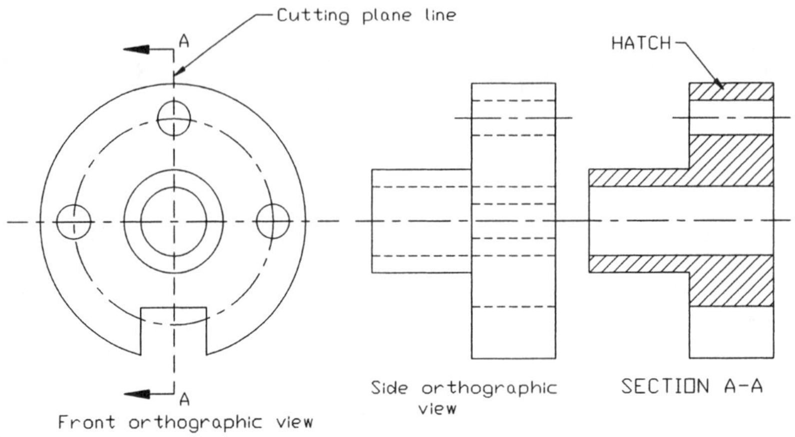

Figure 2-1B

Note the similarity between the sectional view and the cut pictorial view.

Sectional views do not contain hidden lines, so they are used to clarify orthographic views that are difficult to understand because of excessive hidden lines. Figure 2-2 shows an object with a complex internal shape. The standard right view contains many hidden lines and is difficult to follow. Note how much easier it is to understand the object's internal shape when it is presented as a sectional view.

Sectional views *do* include all lines that are directly visible. Figure 2-3 shows an object, a cutting plane line, and a sectional view taken along the cutting plane line. Surfaces that are directly visible are shown in the sectional view. For example, part of the large hole in the back left surface is, from the given sectional view's orientation, blocked by the shorter rectangular surface. The part of the hole that appears above the blocking surface is shown; the part behind the surface is not.

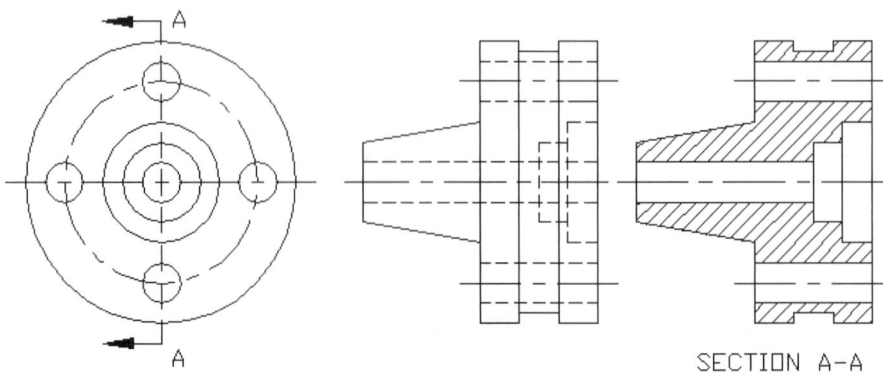

Figure 2-2

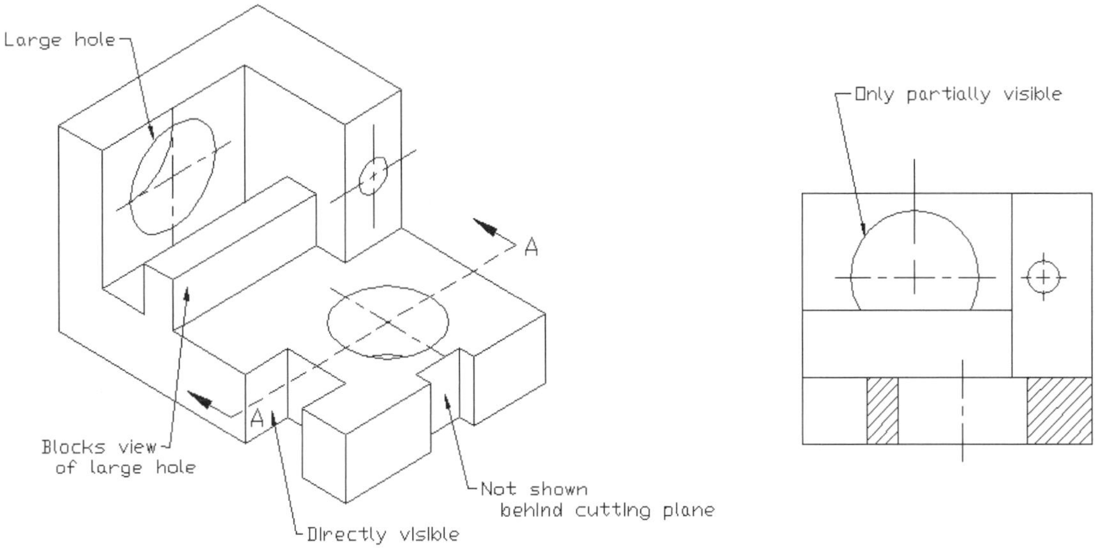

Figure 2-3

Sectional views are always viewed in the direction defined by the cutting plane arrows. Any surface that is behind the cutting plane is not included in the sectional view. Sectional views are aligned and oriented relative to the cutting plane lines as a side orthographic view is to the front view. The orientation of a sectional view may be better understood by placing your right hand on the cutting plane line so that your thumb is pointing up. Move your hand to the right and place it palm down. Your thumb should now be pointing to the left. Your thumb indicates the top of the view.

Drafters and designers often refer to sectional views as "sectional cuts" or simply "cuts." The terminology is helpful in understanding how sectional views are defined and created.

## 2-2  CUTTING PLANE LINES

Cutting plane lines are used to define the location for the sectional view's cutting plane. An object is "cut" along a cutting plane line.

Figure 2-4 shows two linetype patterns for cutting plane lines. Either pattern is acceptable, although some companies prefer to use only one linetype for all drawings to ensure a uniform appearance in all their drawings. The dashed line pattern will be used throughout this book.

The two patterns shown in Figure 2-4 are included in AutoCAD's linetype library as **Dashed** and **Phantom** styles. The arrow portion of the cutting plane line is created using

one of the tools listed within the **Dimension** panel and **Exploding** the dimension, then positioning the arrowhead onto a line or by creating a wblock that includes an arrowhead and extension line.

### To draw a cutting plane line—Method I

Change a given continuous line, A-A, to a cutting plane line. See Figure 2-5.

1. Select **Format** (click the **Menu Browser** in the upper left corner of the screen), then **Linetype**.

   The **Linetype Manager** dialog box will appear.

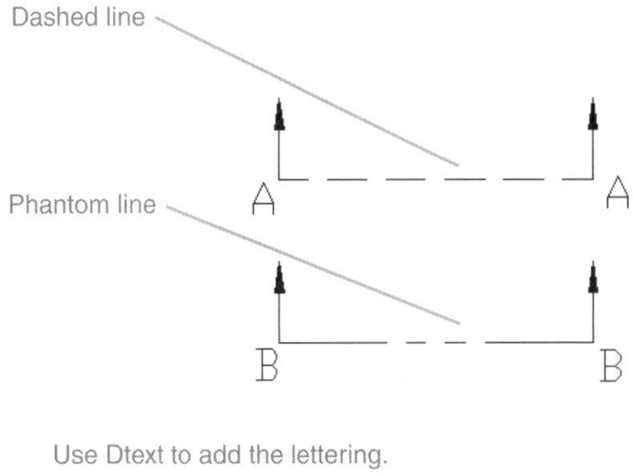

Use Dtext to add the lettering.

Figure 2-4

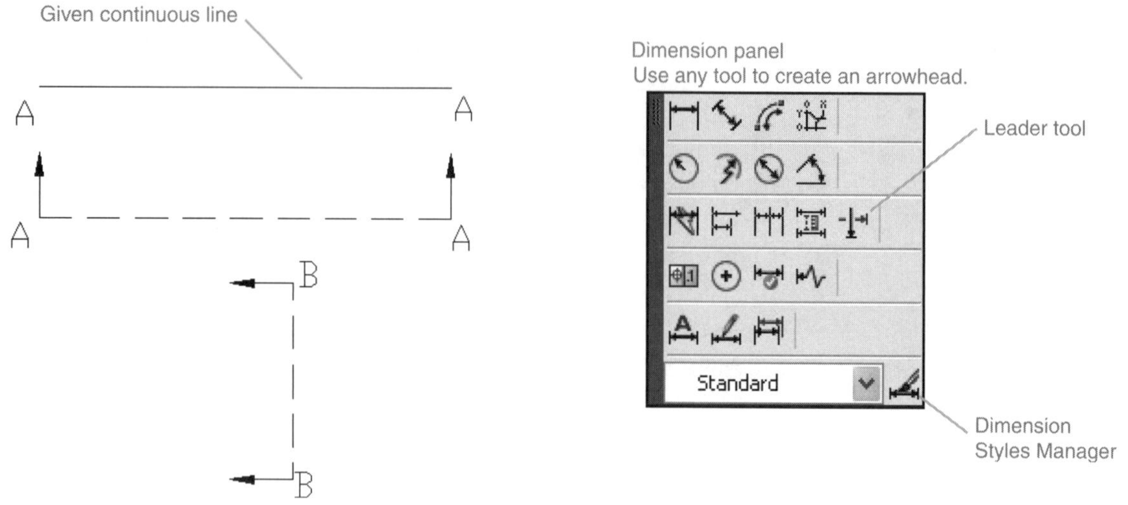

Given continuous line

Dimension panel
Use any tool to create an arrowhead.

Leader tool

Standard

Dimension
Styles Manager

Figure 2-5

2. Select **Load**.

The **Load or Reload Linetypes** dialog box will appear. See Figure 2-6.

3. Select **Dashed** or **Phantom**, then return to the drawing screen.
4. Select **Modify** (pull-down), **Properties**.

*Select object(s):*

5. Select line **A-A**.

The **Change Properties** dialog box will appear.

6. Select **Linetype**.

The **Linetype Manager** dialog box will appear. See Figure 2-7.

7. Select **Dashed** and return to the drawing screen.

Select the Dashed linetype.

Click here to access the Load or Reload Linetypes dialog box.

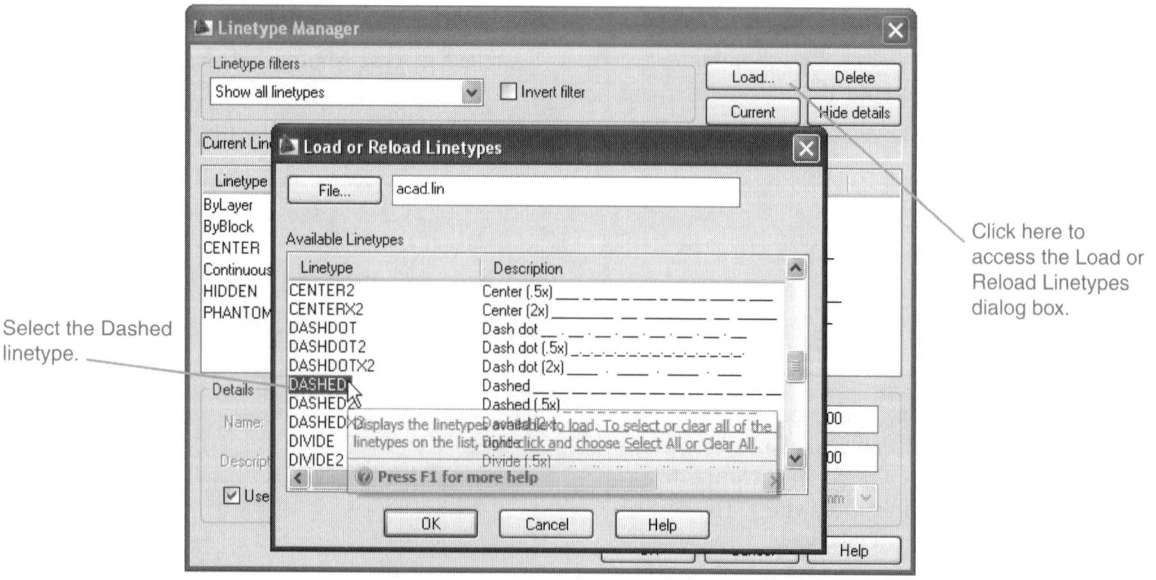

Figure 2-6

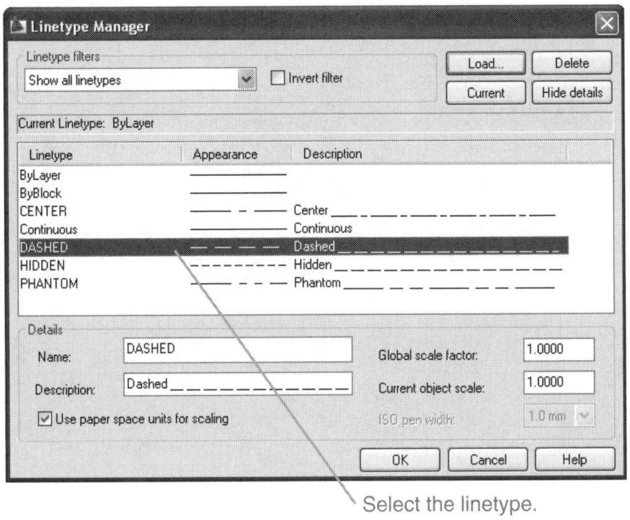

Select the linetype.

Figure 2-7

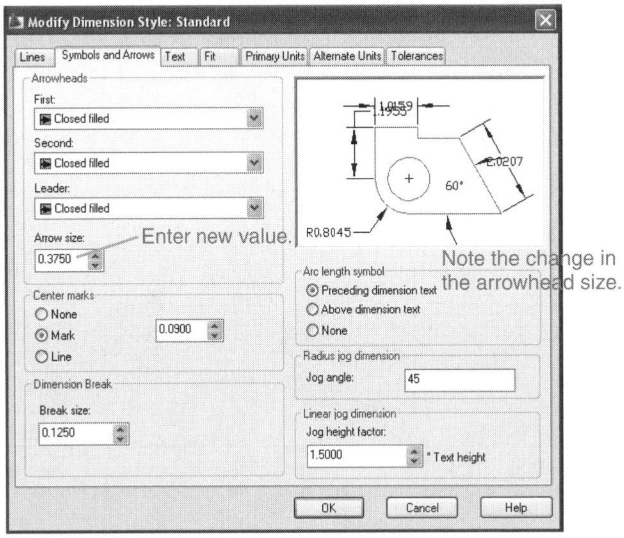

Enter new value.

Note the change in the arrowhead size.

Figure 2-8

## To draw an arrowhead

1. Use the **Rectangle** command on the **Draw** toolbar and draw a rectangle.
2. Select the **Linear** tool on the **Dimension panel** and dimension the vertical end of the rectangle.
3. Use the **Explode** tool and explode the dimension.

   The dimension will now be broken into parts.

4. Select the arrowhead and a short vertical line segment and use the **Move** tool to locate the arrowhead and line to the end of the dashed section line.

   If necessary, use the **Rotate** command to align the arrowhead and line segment with the sectional line.

## To change the size of an arrowhead

Arrowheads used for cutting plane lines are usually drawn larger than arrowheads used for dimension lines. This serves to make them more distinctive and easier to find. The normal arrowhead scale factor is 0.19. In this example, the scale factor is 0.375.

1. Select the **Dimension Style Manager** from the Dimension panel.
2. Select **Modify**.

The **Modify Dimension Style: Standard** dialog box will appear. See Figure 2-8.

3. Select the **Symbols and Arrows** tab and use the **Arrow size** box within the **Arrowheads** portion of the dialog box to change the arrowhead's scale factor (change to **0.3750**).
4. Return to the drawing screen and use **Leader** to create the needed arrowheads.

The arrowhead size can also be changed by typing **dimasz** in response to a command prompt and typing in the new scale factor.

## To draw a cutting plane line—Method II

A cutting plane can also be created by first defining a separate layer setup for cutting plane lines. The linetype will be dashed or phantom, and a color can be assigned if desired.

## To draw cutting plane lines

Cutting plane lines should extend beyond the edges of the object. See Figure 2-9. A cutting plane line should extend far enough beyond the edges of the object so that there is a clear gap between the arrowhead and the edge of the object.

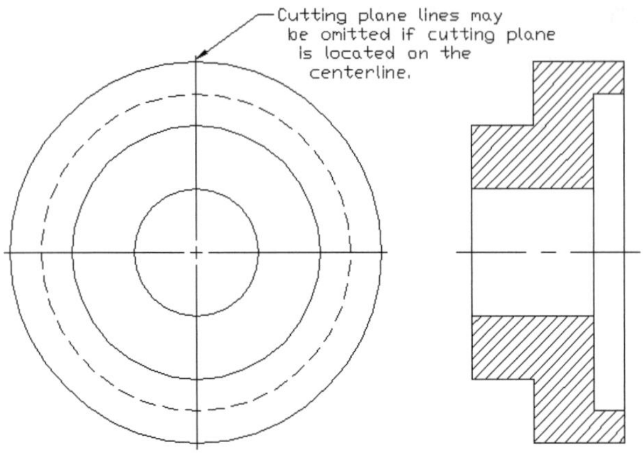

**Figure 2-9**

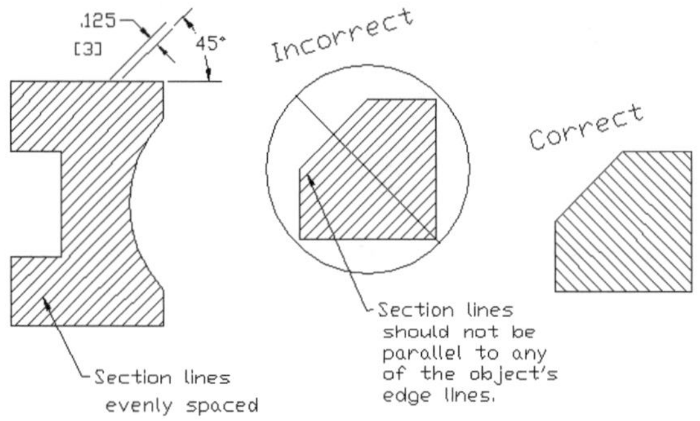

**Figure 2-10**

If an object is symmetrical about the centerline and only one sectional view is to be taken exactly aligned with the centerline, the cutting plane line may be omitted.

## 2-3 SECTION LINES

*Section lines* are used to define areas that represent where solid material has been cut in a sectional view. Section lines are evenly spaced at any inclined angle that is not parallel to any existing edge line and should be visually distinct from the continuous lines that define the boundary of the sectional view.

Figure 2-10 shows an area that includes uniform section lines evenly spaced at 45°. The other area shown in

Figure 2-10 includes a 45° edge line; therefore, the section lines cannot be drawn at 45°. Lines at 135° (0° is horizontal to the right) were drawn instead.

Figure 2-11 shows an object that contains edge lines at both 45° and 135°. The section lines within this area were drawn at 60°.

If two or more parts are included within the same sectional view, each part must have visually different section lines. Figure 2-12 shows a sectional view that contains two parts. Part one's section lines were spaced 3 apart at 45°, and part two's were spaced 5 apart at 135° (−45).

The recommended spacing for sectional lines is 0.125 inch or 3 millimeters, but smaller areas may use section lines spaced closer together than larger areas. See Figure 2-13. Section lines should never be spaced so close together as to

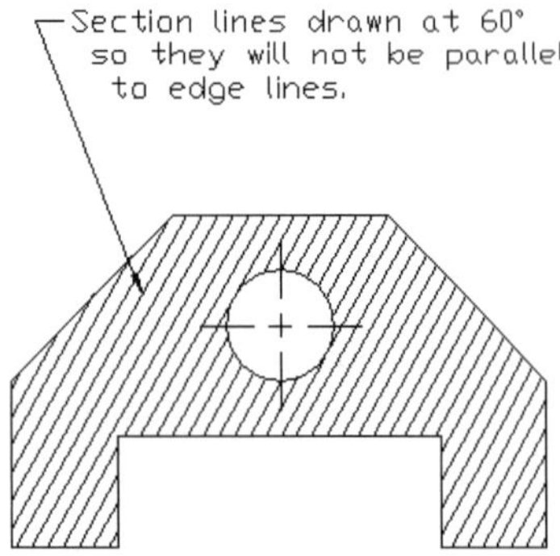

**Figure 2-11**

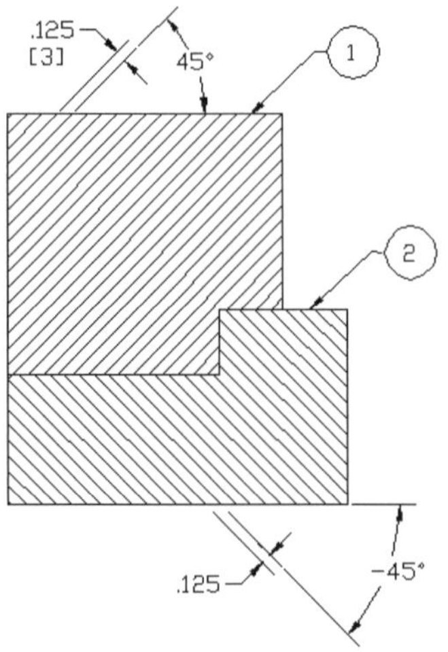

**Figure 2-12**

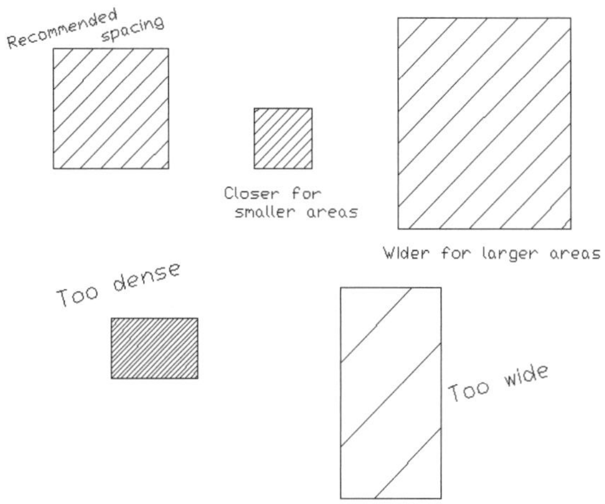

**Figure 2-13**

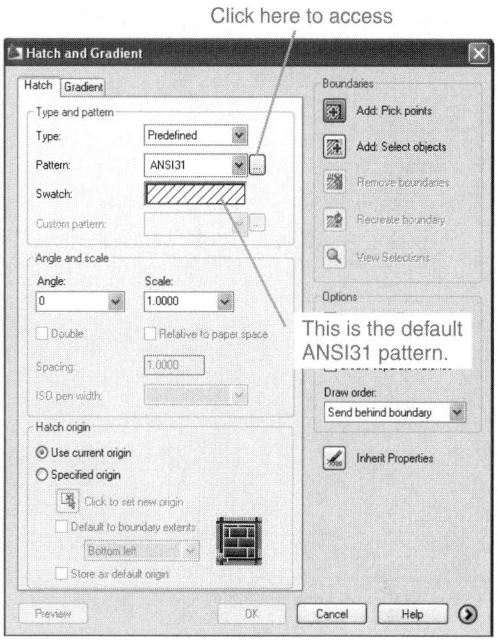

**Figure 2-14**

look blurry or be so far apart that they are not clearly recognizable as section lines.

Evenly spaced sectional line patterns drawn at 45° are called *general,* or *uniform,* patterns; other patterns are available. Different section line patterns are used to help distinguish different materials. The different patterns allow the drawing reader to see what materials are used in a design without having to refer to the drawing's parts list. It should be noted that not all companies use different patterns to define material differences. When you are in doubt, the general pattern is usually acceptable.

How to draw different section line patterns is explained in Section 2-4.

## 2-4  HATCH

Section lines are drawn in AutoCAD using the **Hatch** tool, which is located on the **Draw** toolbar. The **Hatch** tool offers many different hatch patterns and spacings. The general pattern of evenly spaced lines at 45° is defined as pattern ANSI31 and is the default setting for the **Hatch** tool.

**To hatch a given area**

Given an area, use **Hatch** to draw section lines.

1. Select the **Hatch** tool from the **Draw** toolbar.

The **Hatch and Gradient** dialog box will appear. See Figure 2-14. Note that **ANSI31** is the default pattern.

2. Select the **Pick Points** option.

   *Command: _bhatch*
   *Select internal point:*

3. Select a point within the area to be hatched, then right-click the mouse, then select **Enter**.

   The **Hatch and Gradient** dialog box will reappear.

4. Select the **OK** option.

   The area will be hatched. See Figure 2-15.

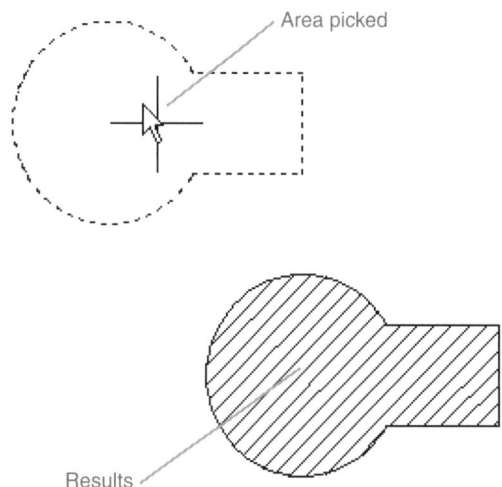

**Figure 2-15**

A list of available hatch patterns

**Figure 2-16**

**To change hatch patterns**

1. Select the **Hatch** tool from the **Draw** toolbar.

   The **Hatch and Gradient** dialog box will appear.

2. Click the arrow to the right of the **Pattern ANSI31** designation.

   A list of available ANSI patterns will appear. See Figure 2-16.

3. Select the desired pattern and apply it as described above.

**Figure 2-17**

New values

**Figure 2-18**

A swatch of the selected pattern will appear on the **Hatch and Gradient** dialog box. Figure 2-17 shows three areas with three different hatch patterns applied.

**To change the spacing and angle of a hatch pattern**

1. Select the **Hatch** tool from the **Draw** toolbar.

   The **Hatch and Gradient** dialog box will appear. See Figure 2-18. The **Angle** and **Scale** option boxes are located just below the **Custom pattern** box.

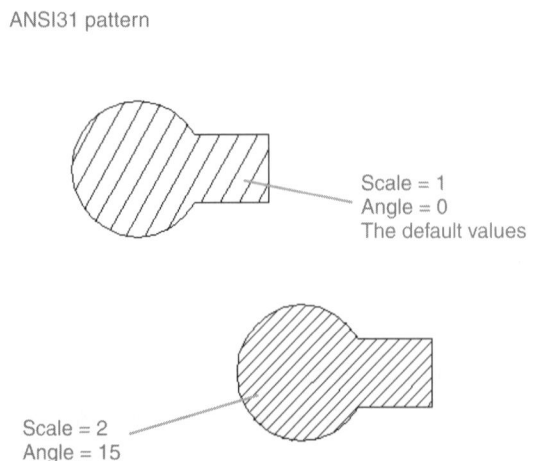

**Figure 2-19**

2. Change the **Angle** to **15** and the **Scale** to **2**.

Figure 2-19 shows a comparison between an ANSI31 pattern created using the default values of 1 and 0°, and the adjusted pattern of 2 and 15°.

## 2-5  SAMPLE PROBLEM SP2-1

Figure 2-20 shows an object with a cutting plane line. Figure 2-21 shows how a front view and a sectional view of the object are created.

1. Draw the front orthographic view of the object and lay out the height and width of the object based on the given dimensions. Sectional views must be directly aligned with the angle of the cutting plane line. Sectional views are in fact orthographic views that are based on the angle of the cutting plane line. Information is projected to sectional views as it was projected to top and side views. In this example the cutting plane line is a vertical line, so horizontal projection lines are used to draw the sectional view.
2. Draw the object features. No hidden lines are drawn.
3. Use **Hatch** to draw sectional lines within the appropriate areas.
4. Erase or trim any excess lines.
5. Modify the color of the sectional lines if desired.
6. Save the drawing if desired.

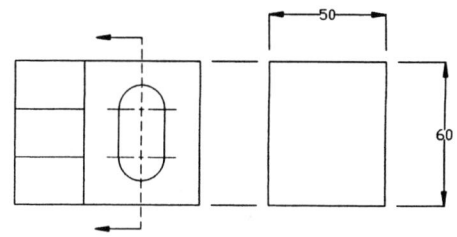

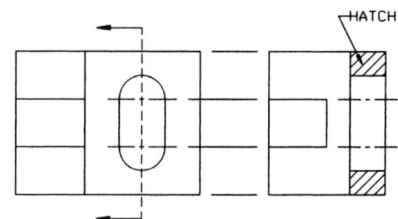

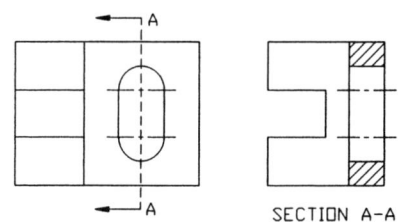

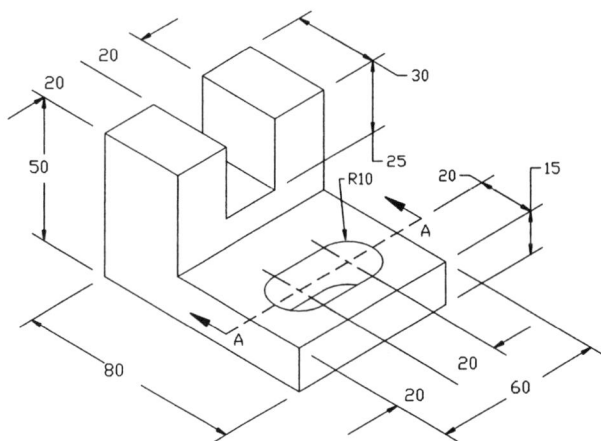

**Figure 2-20**

**Figure 2-21**

## 2-6 STYLES OF SECTION LINES

AutoCAD has more than 50 different hatch patterns. The different patterns can be previewed in the **Swatch** option in the **Hatch and Gradient** dialog box.

Most of the available options are for architectural use. The patterns can be used to create elevation drawings and to add texture patterns to drawings of houses, buildings, and other structures. Technical drawings usually refer to objects made from steel, aluminum, or a composite material. There are hundreds of variations of each of these materials.

In general, if you decide to assign a particular pattern to a material, clearly state which pattern has been assigned to which material on the drawing. This is best done by including a note on the drawing that includes a picture of the pattern and the material that it is to represent. Figure 2-22 shows a drawing note for a hatch pattern used to represent SAE 1040 steel. The representation is unique for the drawing shown.

## 2-7 SECTIONAL VIEW LOCATION

Sectional views should be located on a drawing behind the arrows. The arrows represent the viewing direction for the sectional view. See Figure 2-23. If it is impossible to locate sectional views behind the arrows, they may be located above or below, but still behind, the arrowed portion of the cutting plane line. Sectional views should never be located in front of the arrows.

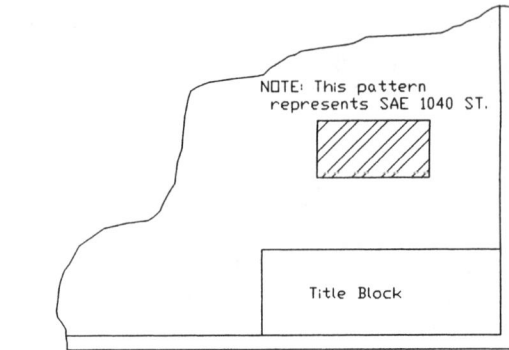

**Figure 2-22**

Sectional views located on a different drawing sheet than the cutting plane line must be cross-referenced to the appropriate cutting plane line. See Figure 2-24. The boxed notation C/3 SHT2 next to the cutting plane line means that the sectional view may be found in zone C/3 on sheet 2 of the drawing.

The location of the sectional view must be referenced to the cutting plane line if the view's location is not near the cutting plane. The notation C/3 SHT2 defines the location of the cutting plane line for the sectional view.

The reference numbers and letters are based on a drawing area charting system similar to that used for locating

**Figure 2-23**

Figure 2-24

Figure 2-26

features on maps. A sectional view located at C/3 SHT2 can be found by going to sheet 2 of the drawing, then drawing a vertical line from the box marked 3 at the top or bottom of the drawing and a horizontal line from the box marked C on the left or right edge of the drawing. The sectional view should be located somewhere near the intersection of the two lines.

## 2-8  HOLES IN SECTIONS

Figure 2-25 shows a sectional view of an object that contains three holes. As with orthographic views, a conical

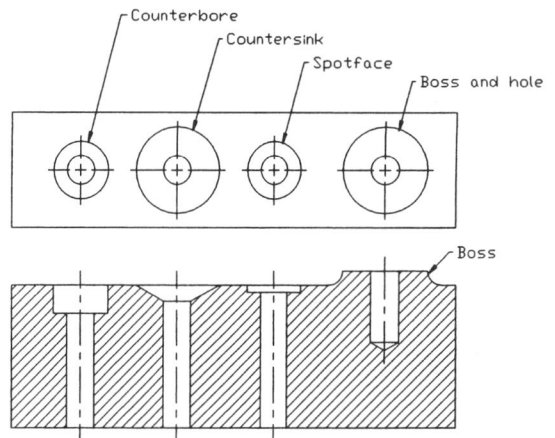

point must be included on holes that do not completely penetrate the object.

A common mistake is to omit the back edge of a hole when drawing a sectional view. Figure 2-25 shows a hole drilled through an object. Note that the sectional view includes a straight line across the top and bottom edges of the view that represents the back edges of the hole.

Figure 2-26 shows a countersunk hole, a counterbored hole, a spotface, and a hole through a boss. Again, any hole that does not completely penetrate the object must include a conical point.

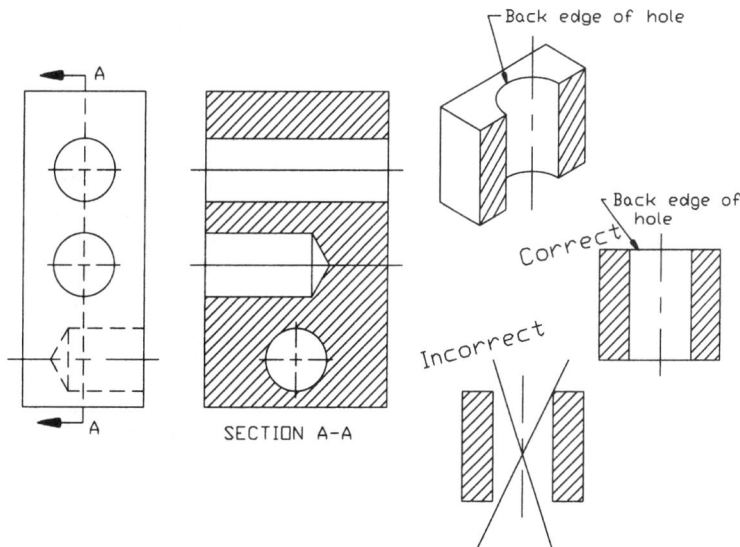

Figure 2-25

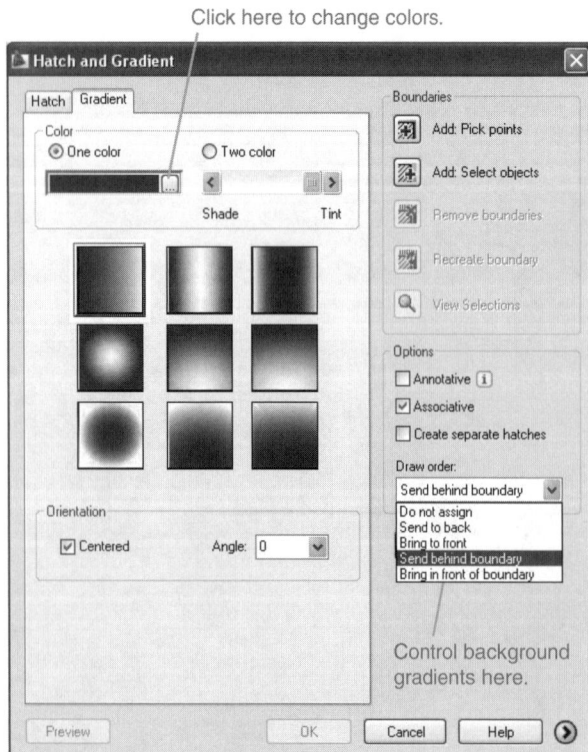

Original shape | Shape with gradients added

**Figure 2-27**

## 2-9 GRADIENTS

In AutoCAD a *gradient* is shading that varies in intensity. Figure 2-27 shows a rectangular shape with two internal circles. The figure also shows the same shape with various gradients added.

### To create a gradient

1. Select the **Gradient** tool from the **Draw** toolbar.

The **Hatch and Gradient** dialog box will appear. See Figure 2-28.

2. Select the **Add: Select objects** button.

The dialog box will disappear.

3. Select one of the circles by clicking its edge line.
4. Right-click the mouse and select the **Enter** option.

The **Hatch and Gradient** dialog box will appear.

5. Select a gradient pattern, then click **OK**.

A gradient will appear within the circle. Use the **Send behind boundary** options to control drawings that include more than one gradient.

Click on the box to the right of the **One color** bar to change gradient colors. Select a different color by clicking on the desired color, as shown within the square field of color on the **Select Color** dialog box.

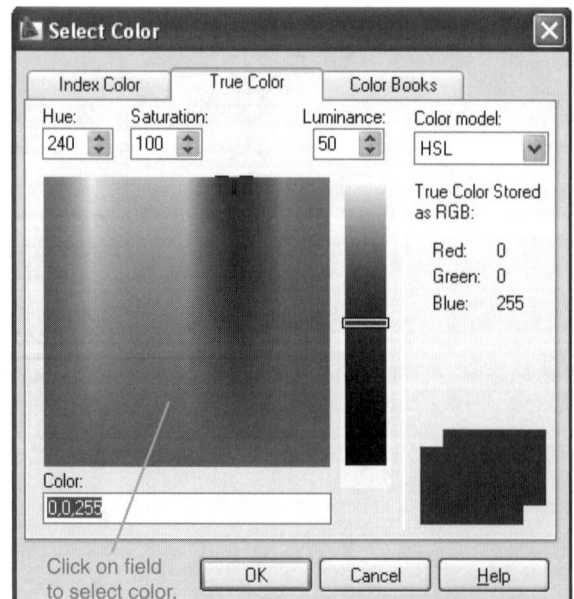

**Figure 2-28**

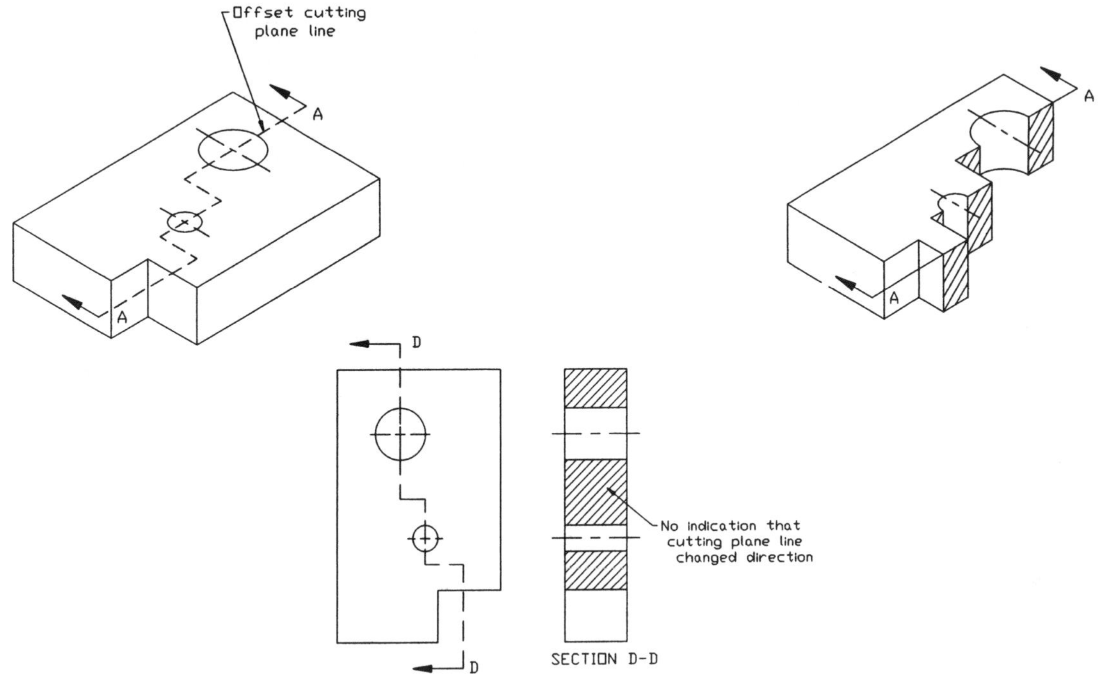

**Figure 2-29**

## 2-10  OFFSET SECTIONS

Cutting plane lines need not be drawn as straight lines across the surface of an object. They may be stepped so more features can be included in the sectional view. See Figure 2-29. In the object shown in Figure 2-29, the cutting plane line crosses two holes and a directly visible open surface that contains a hole. There is no indication in the sectional view that the cutting plane line has been offset.

Cutting plane lines should be placed to include as many features as possible without causing confusion. The intent of using sectional views is to simplify views and to clarify the drawing. Several features can be included on the same cutting plane line so that fewer views are used, giving the drawing a less cluttered look and making it easier for the reader to understand the shape of the object's features.

Figure 2-30 shows another example of an offset cutting plane line.

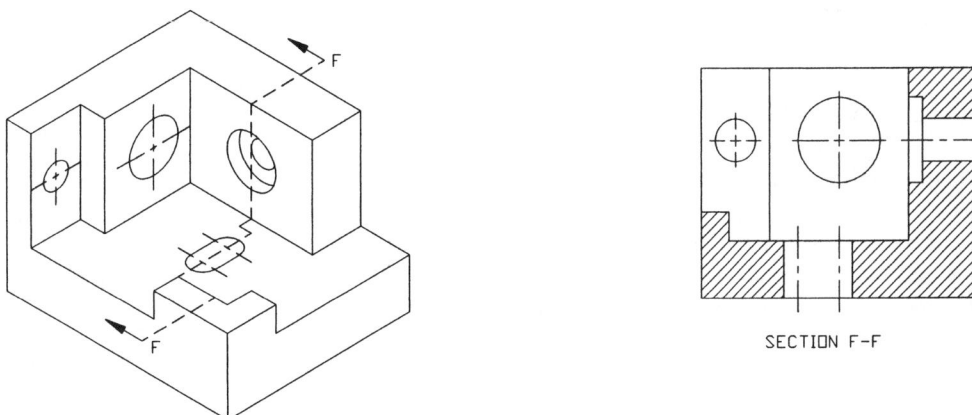

**Figure 2-30**

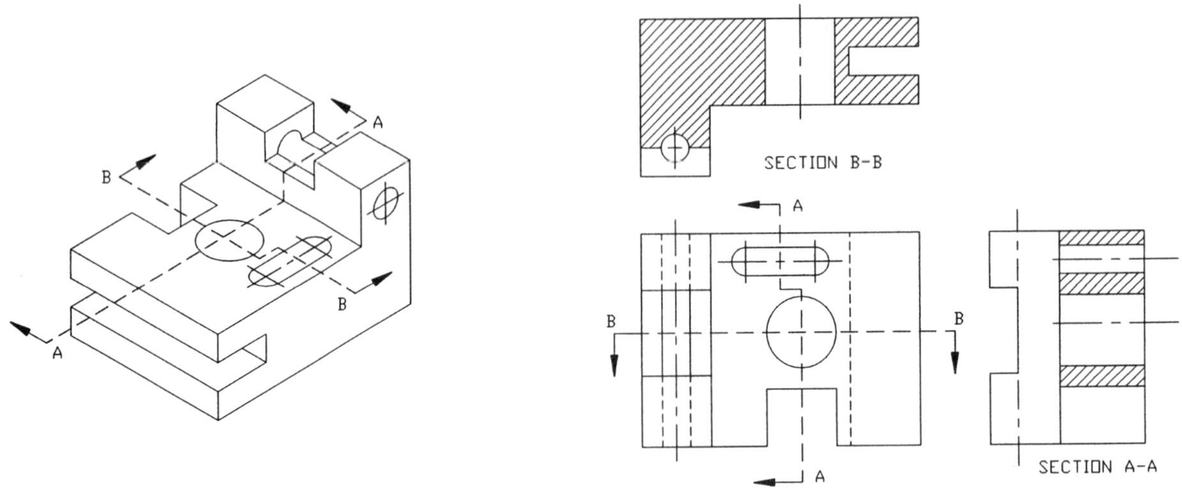

**Figure 2-31**

## 2-11 MULTIPLE SECTIONS

More than one sectional view may be taken off the same orthographic view. Figure 2-31 shows a drawing that includes three sectional views, all taken off a front view. Each cutting plane line is labeled with a letter. The letters are in turn used to identify the appropriate sectional view. Identifying letters are also placed at the ends of the cutting plane lines, behind the arrowheads. Identifying letters are also placed below the sectional view and written using the format SECTION A-A, SECTION B-B, and so on. The abbreviation SECT may also be used: SECT A-A, SECT B-B.

The letters I, O, and X are generally not used to identify sectional views because they can easily be misread. If more than 23 sectional views are used, the lettering starts again with double letters: AA-AA, BB-BB, and so on.

## 2-12 ALIGNED SECTIONS

Cutting plane lines taken at angles on circular shapes may be aligned as shown in Figure 2-32. Aligning the sectional views prevents the foreshortening that would result if the view were projected from the original cutting plane line location. A foreshortened view would not present an accurate picture of the object's surfaces.

## 2-13 DRAWING CONVENTIONS IN SECTIONS

Slots and small holes that penetrate cylindrical surfaces may be drawn as straight lines, as shown in Figure 2-33. Larger holes, that is, holes whose diameters are greater than the radii of their cylinders, should be drawn showing an elliptical curvature.

Intersecting holes are represented by crossed lines, as shown in Figure 2-34. The crossed lines are drawn from the intersecting corners of the holes.

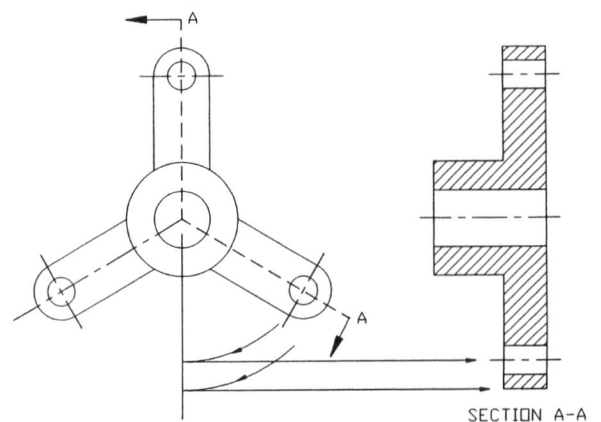

**Figure 2-32**

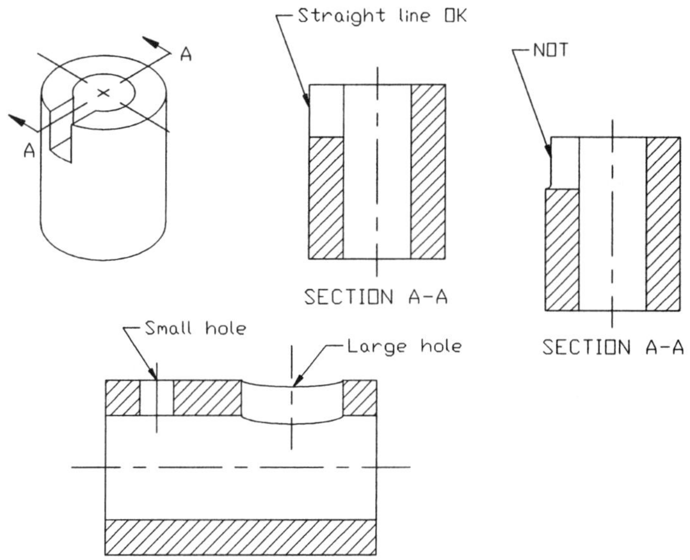

**Figure 2-33**

# 2-14 HALF, PARTIAL, AND BROKEN-OUT SECTIONAL VIEWS

Half and partial sectional views allow a designer to show an object using an orthographic view and a sectional view within one view. A half-sectional view is shown in Figure 2-35. Half the view is a sectional view; the other half is a normal orthographic view including hidden lines. The cutting plane line is drawn as shown and includes an arrowhead at only one end of the line.

Figure 2-36 shows a partial sectional view. It is similar to a half-sectional view, but the sectional view is taken at a location other than directly on a centerline or one defined by a cutting plane line. A broken line is used to separate the sectional view from the orthographic view. A broken line is a freehand line drawn using the **Sketch** command.

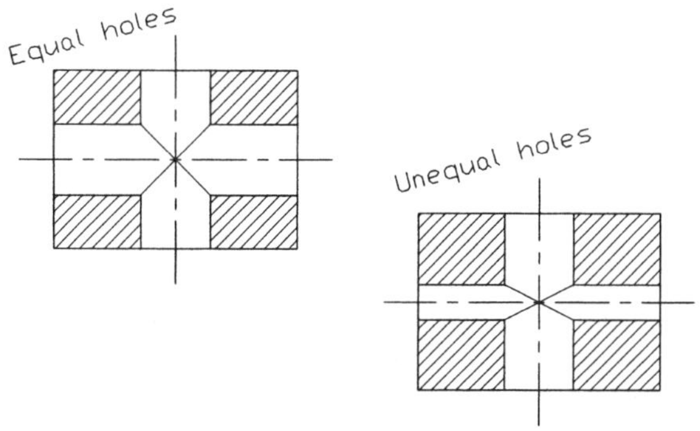

**Figure 2-34**

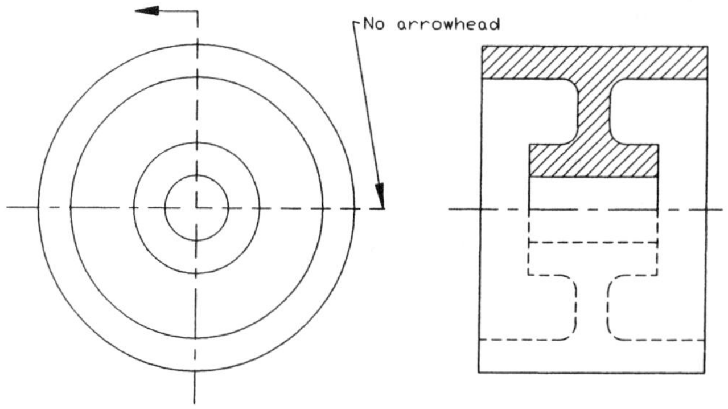

**Figure 2-35**

**Figure 2-36**

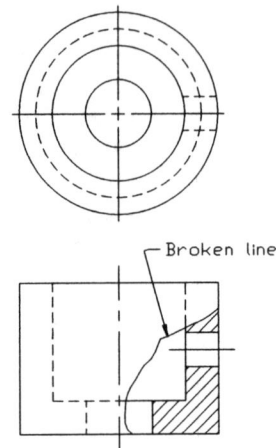

**Figure 2-37**

Broken-out sectional views are like small partial views. They are used to show only small internal portions of an object. Figure 2-37 shows a broken-out sectional view. Broken lines are used to separate the broken-out section from the rest of the orthographic views.

**To draw a broken line**

1. Type **Sketch** in response to a command prompt.

   *Command:_sketch*
   *Record Increment <0.1000>:*

2. Press **Enter**.

   *Sketch. Pen/eXit/Quit/Record/Erase/Connect:*

3. Press the left mouse button.

   *<Pen down>*

4. Move the cursor across the screen. A green free-hand line will emerge from the crosshairs as it is moved.

5. Press the left mouse button again.

   *<Pen up>*

   The crosshairs can now be moved without producing a sketched line.

6. Select **RECORD** from the menu.

   *23 lines recorded*

   The number of lines generated and recorded will depend on the length of the line and the size of the line increment.

7. Select **eXit** from the menu.

   *Command:*

Do *not* use the right mouse button to exit the **Sketch** command. If the right mouse button is pressed, a line will be drawn from the end of the sketched line to the present location of the crosshairs. Use **eXit** to end the **Sketch** command.

## 2-15 REMOVED SECTIONAL VIEWS

Removed sectional views are used to show how an object's shape changes over its length. Removed sectional views are most often used with long objects whose shape changes continuously over its length. See Figure 2-38. The sectional views are not positioned behind the arrowheads but are positioned across the drawing as shown; however, the view orientation is the same as it would be if the view were projected from the arrowheads; it is simply located in a different position in the drawing.

It is good practice to identify the cutting plane lines and the sectional views in alphabetical order. This will make it easer for the drawing's readers to find the sectional views.

## 2-16 BREAKS

It is often convenient to break long continuous shapes so that they take up less drawing space. There are two drawing conventions used to show breaks: *freehand lines* used for rectangular shapes, and *S-breaks* used for cylindrical shapes. See Figure 2-39. Freehand break lines are drawn using the **Sketch** command as explained in Section 2-14. Instructions for drawing S-breaks follow.

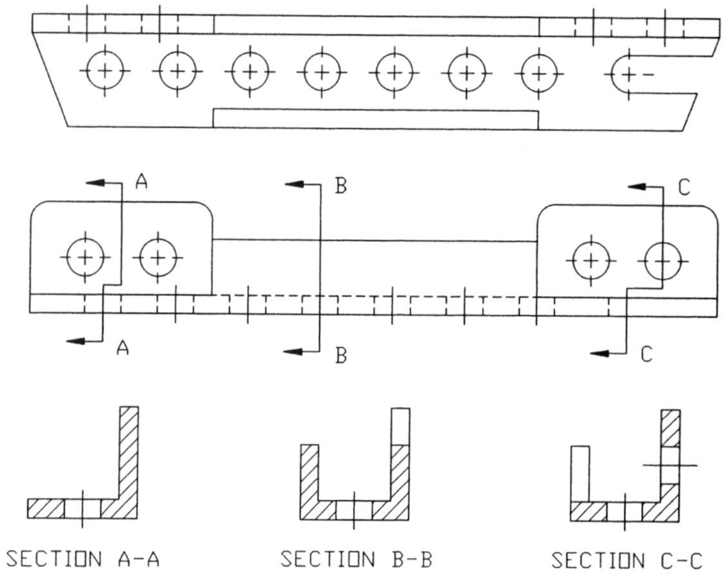

A
B
C
A
B
C
SECTION A-A
SECTION B-B
SECTION C-C

Figure 2-38

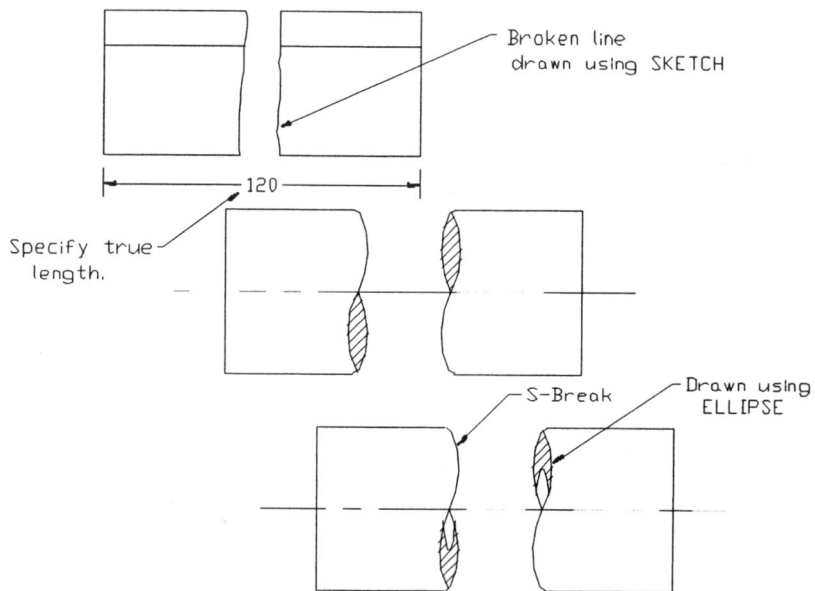

Broken line
drawn using SKETCH
120
Specify true
length.
S-Break
Drawn using
ELLIPSE

Figure 2-39

**To draw an S-break (See Figure 2-40.)**

1. Draw a rectangular view of the cylindrical object and draw a construction line where the break is to be located. The rectangular view should include a centerline.
2. Draw two **30°** lines: one from the intersection of the construction line and the outside edge line of the view, and the other from the intersection of the construction line and the centerline as shown.
3. Draw an arc using the intersection created in step 2 as the center point. Mirror the arc about the centerline and then about the construction line.
4. Use **Fillet,** set to a small radius, to smooth the corners between the arc and the edge lines. Any small radius may be used for the fillet, provided it produces a smooth visual transition between the arc and the edge lines.
5. Erase and trim any excess lines.
6. Hatch the area created by the two arcs and fillet as shown.

The **Copy, Mirror,** and **Move** commands may be used to create the opposing S-break as shown in Figure 2-39. The internal shapes of the tubular S-breaks are created using **Ellipse.** Position the end of the ellipse according to the thickness specifications, and determine the elliptical shape by eye. Trim the sectional lines from the inside of the ellipse.

## 2-17 SECTIONAL VIEWS OF CASTINGS

Cast objects are usually designed to include a feature called a **rib.** See Figure 2-41. Ribs add strength and rigidity to an object. Sectional views of ribs do not include complete section lines because this is considered misleading to the reader. Ribs are usually narrow, and a large sectioned area gives the impression of a denser and stronger area than is actually on the casting.

There are two conventions used to present sectional views of cast ribs: one that does not draw any section lines on the ribs, and one that puts every other section line on the rib. Sectional views of castings that do not include section lines on ribs are created using **Hatch** for those areas that are to be hatched.

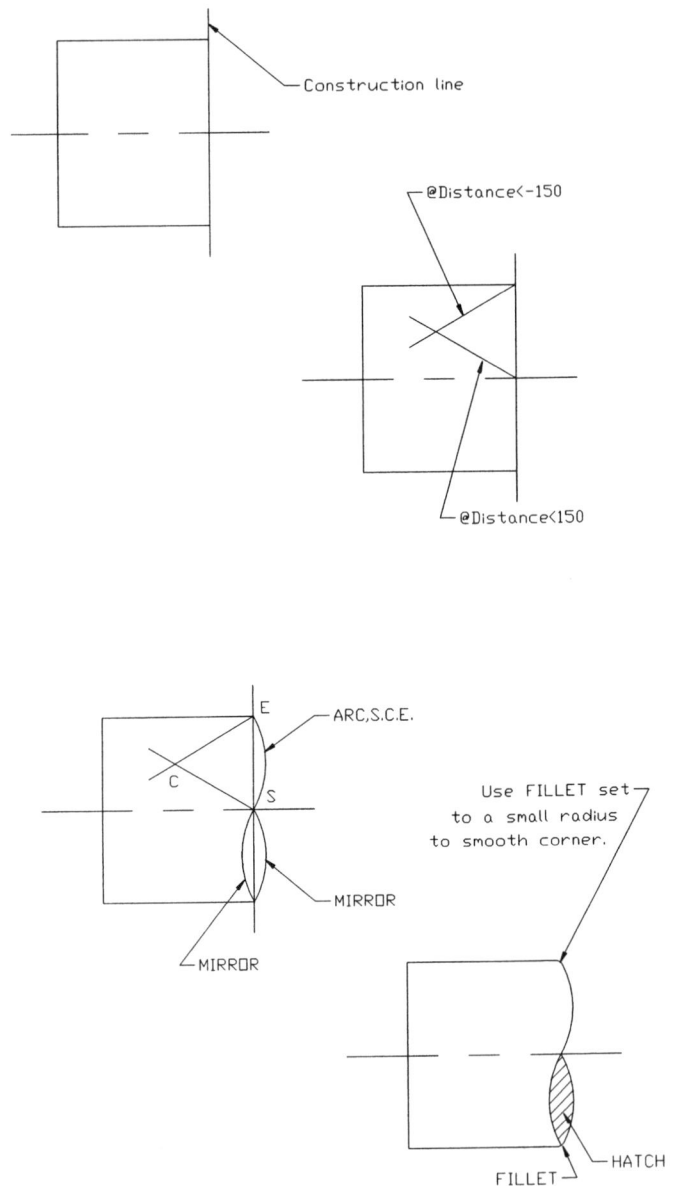

Figure 2-40

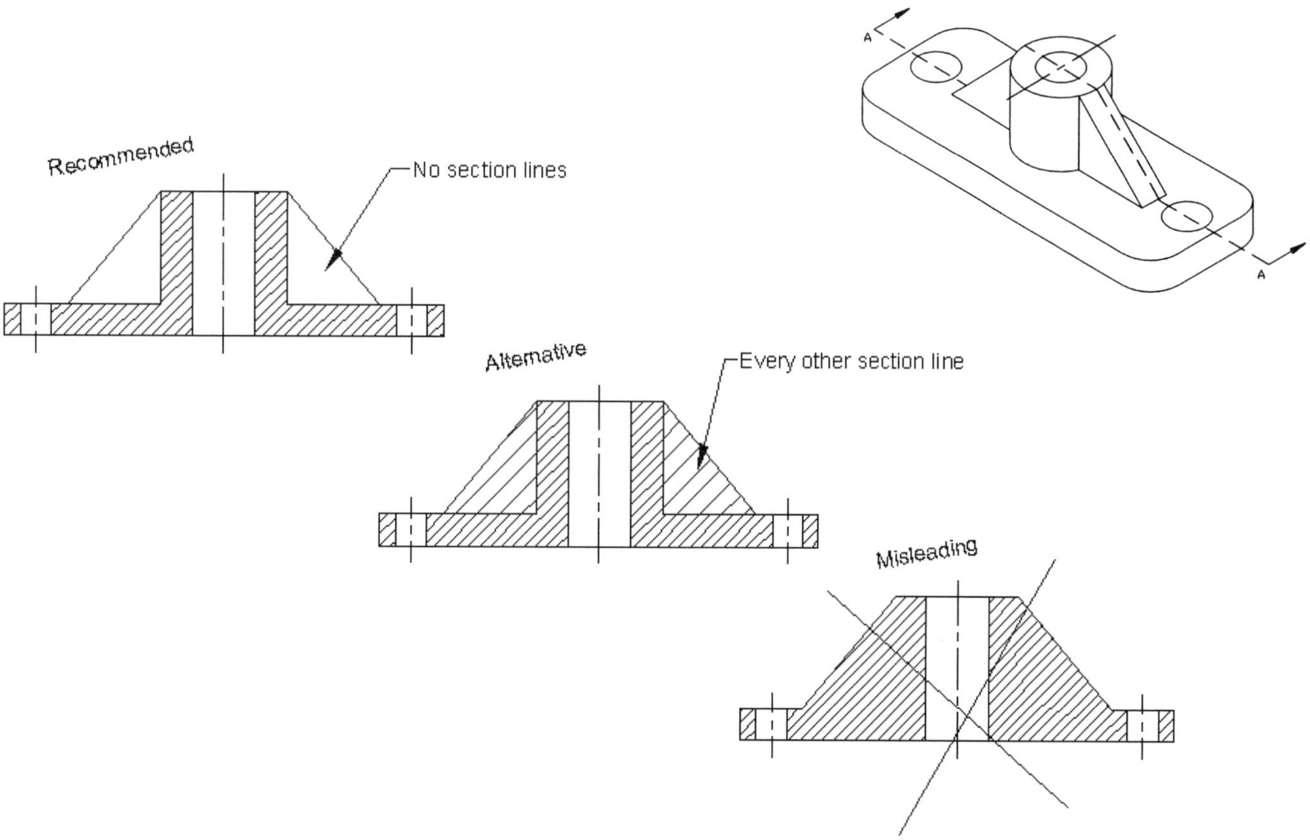

Recommended

No section lines

Alternative

Every other section line

Misleading

**Figure 2-41**

## 2-18 EXERCISE PROBLEMS

Draw a complete front view and a sectional view of the objects in Exercise Problems EX2-1 through EX2-4. The cutting plane line is located on the vertical centerline of the object.

### EX2-1 MILLIMETERS

### EX2-3 INCHES

### EX2-2 MILLIMETERS

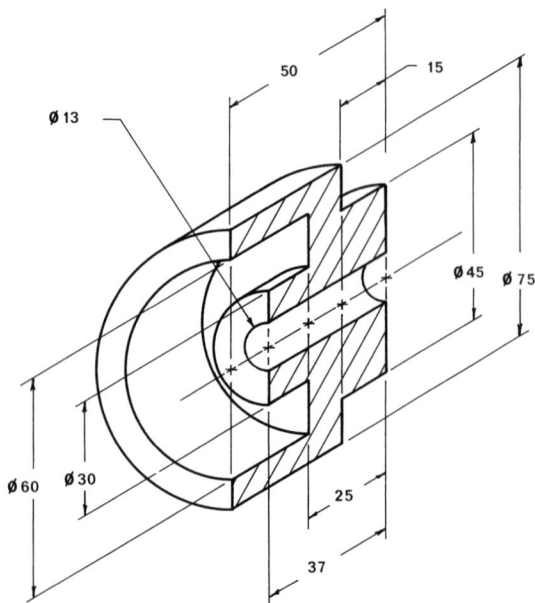

### EX2-4 MILLIMETERS

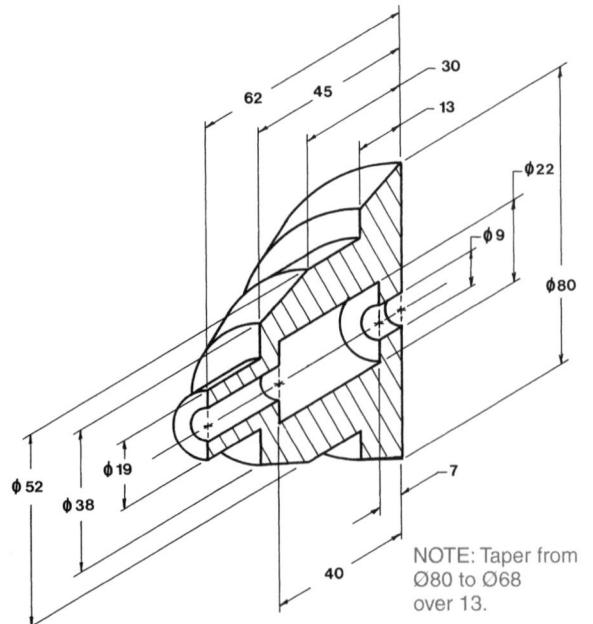

NOTE: Taper from Ø80 to Ø68 over 13.

Draw the following sectional views using the given
dimensions.

EX2-5

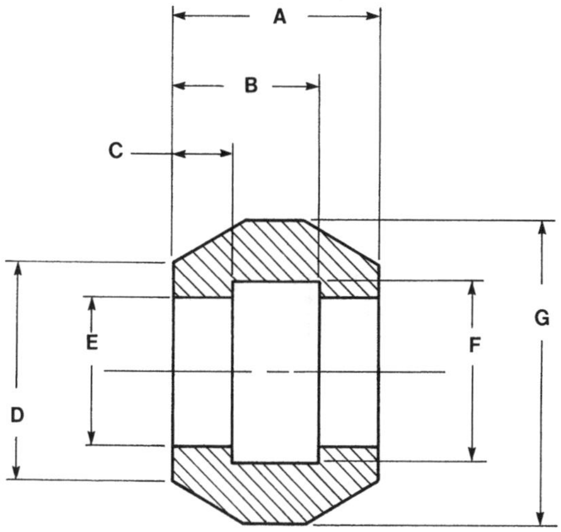

| DIMENSIONS | INCHES | mm |
|------------|--------|-----|
| A | .50 | 12 |
| B | 1.25 | 32 |
| C | 1.75 | 44 |
| D | Ø1.75 | 44 |
| E | Ø 1.25 | 3 2 |
| F | Ø1.50 | 38 |
| G | Ø2.50 | 64 |

EX2-6

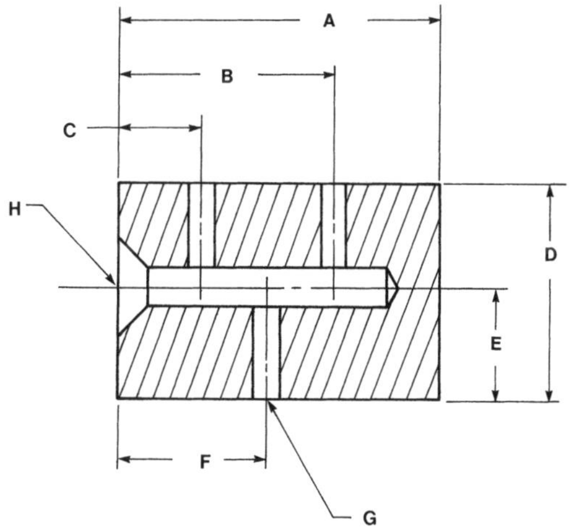

| DIMENSIONS | INCHES | mm |
|------------|--------|-----|
| A | 3.00 | 72 |
| B | 2.00 | 48 |
| C | .75 | 18 |
| D | 2.00 | 48 |
| E | 1.00 | 24 |
| F | 1.38 | 33 |
| G | Ø.25 | 6 |
| H | Ø .375 X 2.50  DEEP  Ø.875  X 82° CSINK | Ø 10 X 60  DEEP  Ø 24 X 82° CSINK |

Resketch the given top view and Section A-A, then sketch Sections B-B, C-C, and D-D.

**EX2-7**

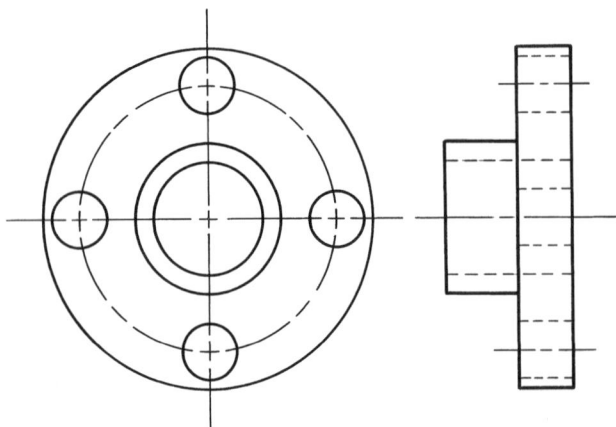

Section   A-A

Resketch the given front views in Exercise Problems EX2-8 through EX2-11 and replace the given side orthographic view with the appropriate sectional view.

**EX2-9**

**EX2-8**

EX2-10 INCHES

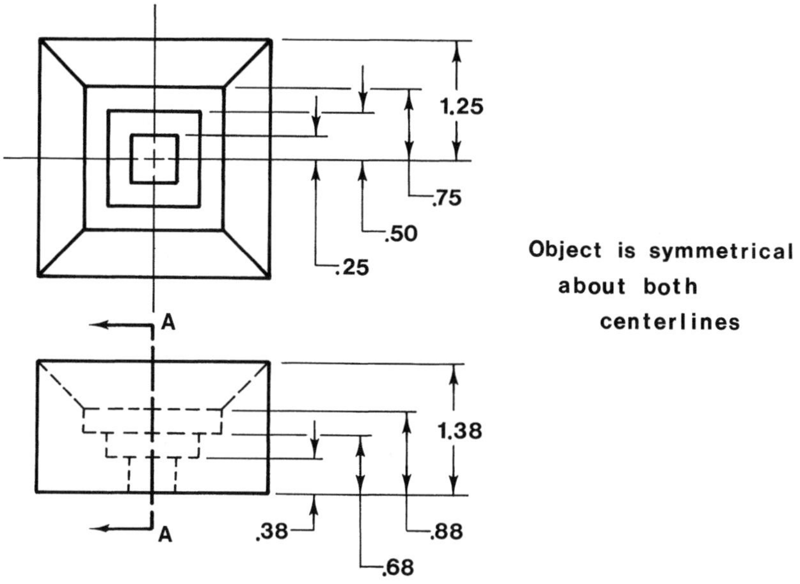

1.25
.75
.50
.25
A
A
1.38
.38
.88
.68
Object is symmetrical
about both
centerlines

EX2-11

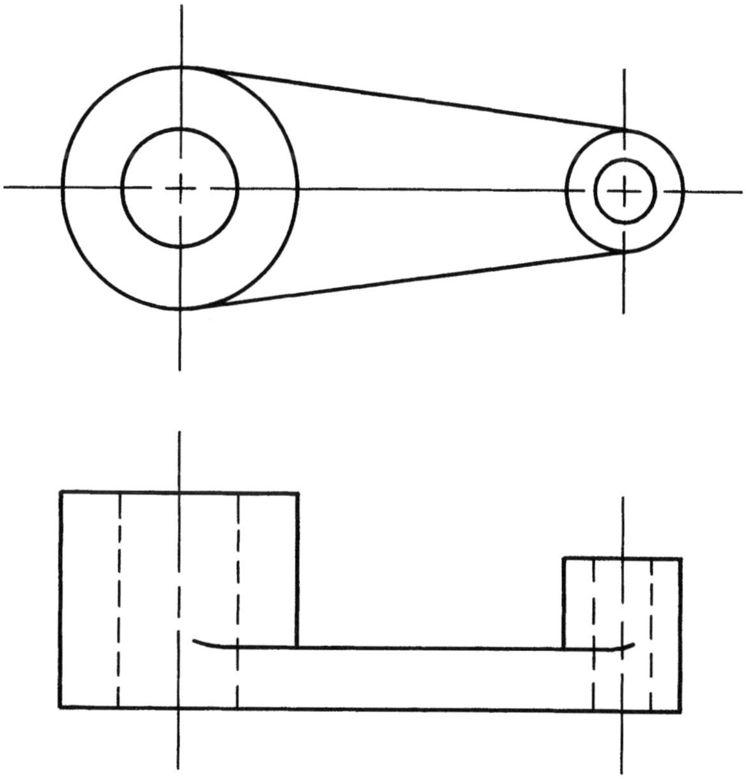

Redraw the given front views in Exercise Problems EX2-12 through EX2-15 and replace the given side orthographic view with the appropriate sectional view. The cutting plane is located on the vertical centerlines of the objects.

**EX2-12 INCHES**

**EX2-14 INCHES**

**EX2-13 INCHES**

**EX2-15 MILLIMETERS**

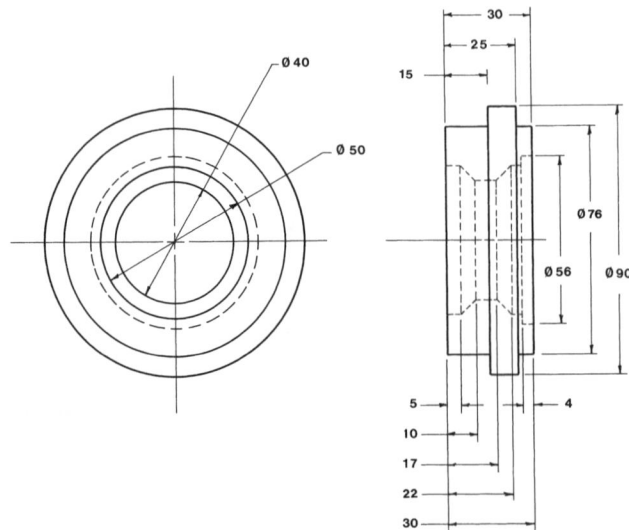

Draw the top orthographic view and the indicated sectional view.

## EX2-16 MILLIMETERS

45

45

Ø25
2 Places

Ø30

Ø38
2 Places

20 – 2 Places

6
Both Keyseats

A

27

Both Keyseats

25

A

50

10

Ø 5 – 2 Places
Second hole at other
end of object

## EX2-18 MILLIMETERS

120

50

Ø 12
3 Places

A   R 12
2 Places

6

30

A

14

Ø25

30

6

Centerline
of object

## EX2-17 INCHES

Ø 1.125

2.25

1.38

2.250

A   1.25

.63

R.31

R .15

Ø .31

A

1.00

.75

## EX2-19 MILLIMETERS

7

Ø 32

6

3

Ø 22   A

6

60

Ø 6
3 Holes

25

40

8

70

*10

30

20

30

20

10

10

A   ø12

3

ø12

14

40

30

10

60

15

Redraw the given views and add the specified sectional views in Exercise Problems EX2-20 through EX2-34.

**EX2-20 INCHES**

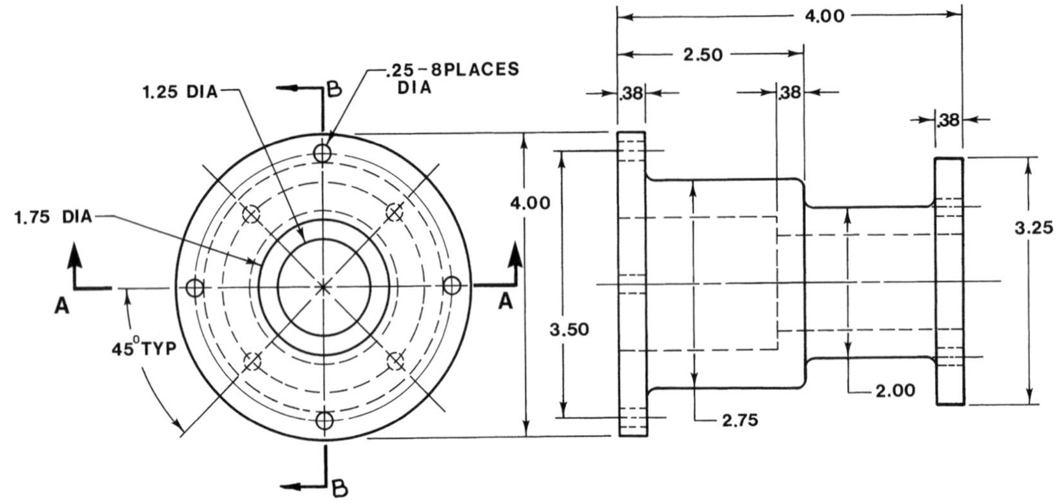

ALL FILLETS AND ROUNDS = R.125

**EX2-21 INCHES**

| HOLE | X | Y | DIA |
|------|------|------|-----|
| A | 1.63 | 2.00 | .44 |
| B | 1.13 | 1.00 | .56 |
| C | 2.50 | 2.00<br>1.00 | .63 |
| D | 3.88 | 2.00<br>1.00 | .50 |

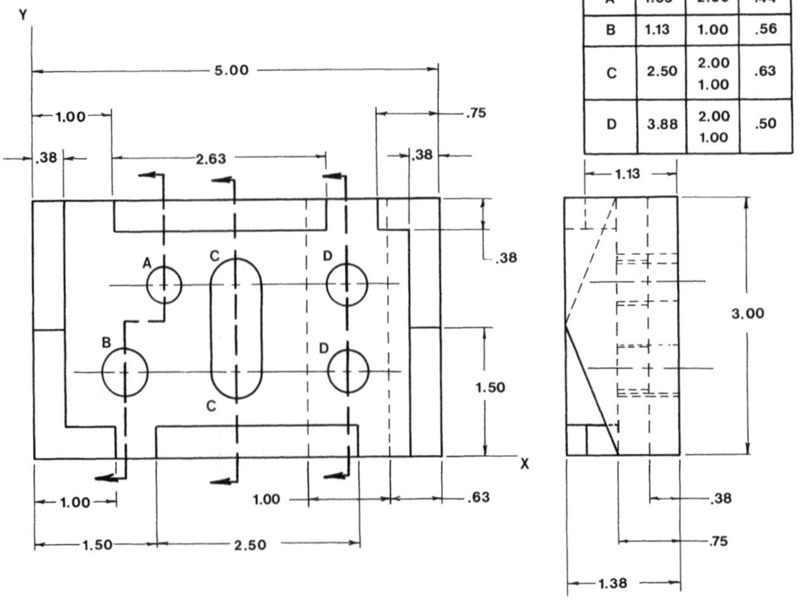

**EX2-22 INCHES**

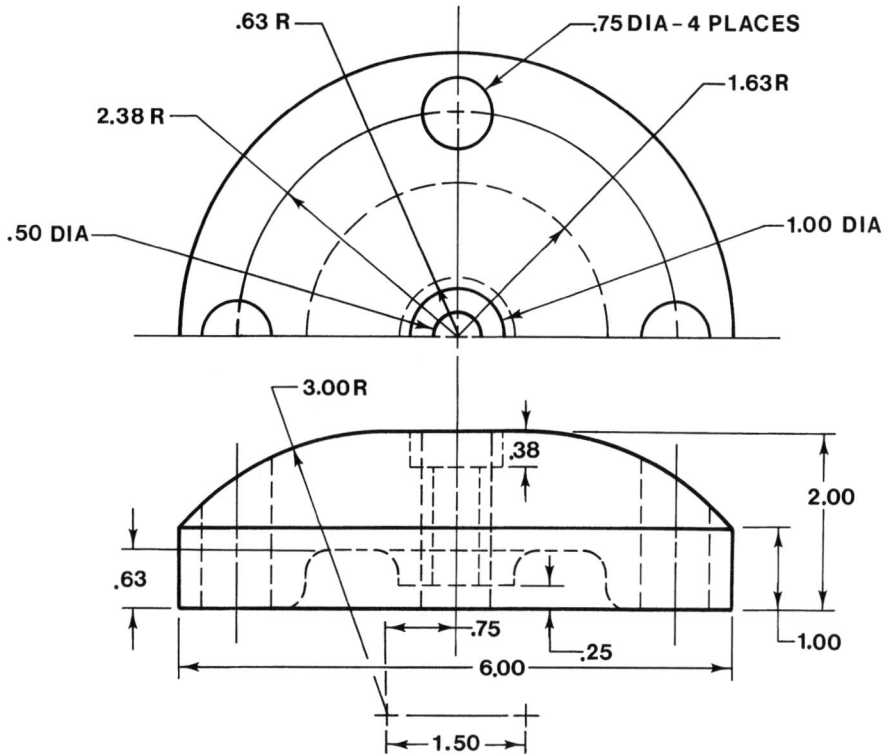

**EX2-23 INCHES**

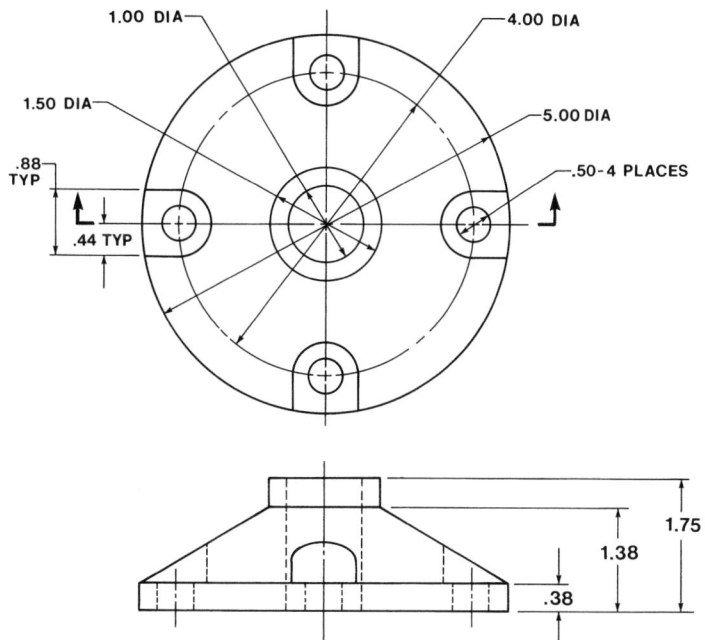

## EX2-24 INCHES

Redraw the given front and top views and add Sections
A-A, B-B, and C-C as indicated.

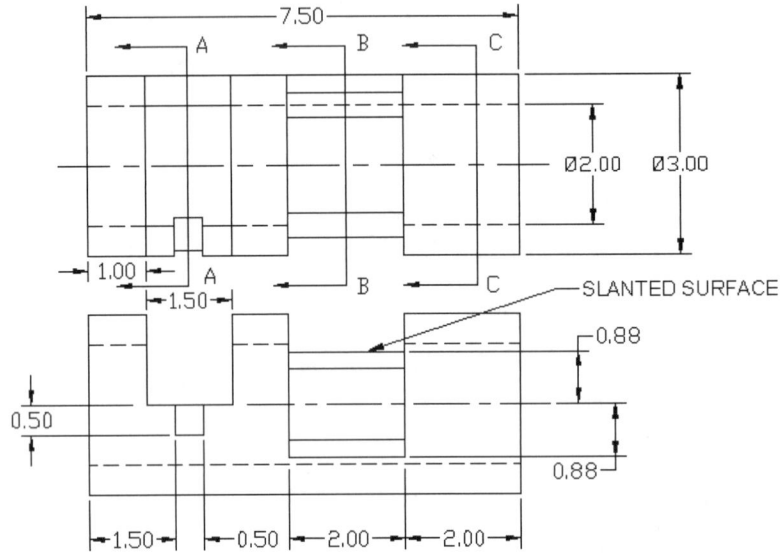

## EX2-25 MILLIMETERS

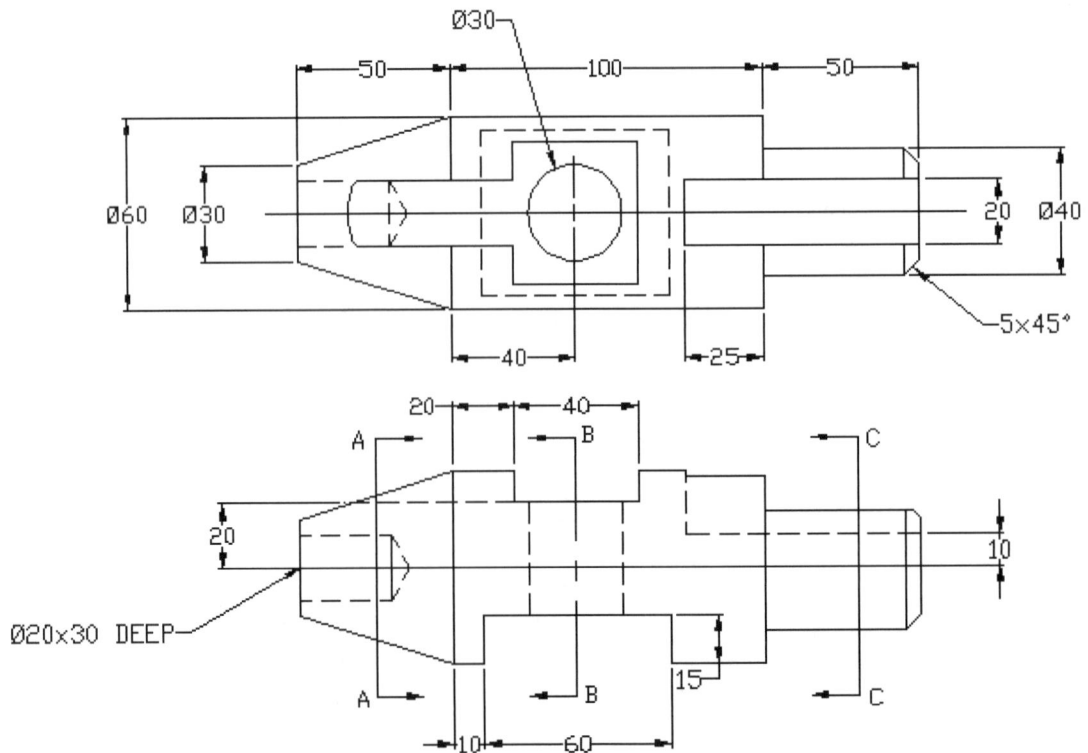

**EX2-26 INCHES**

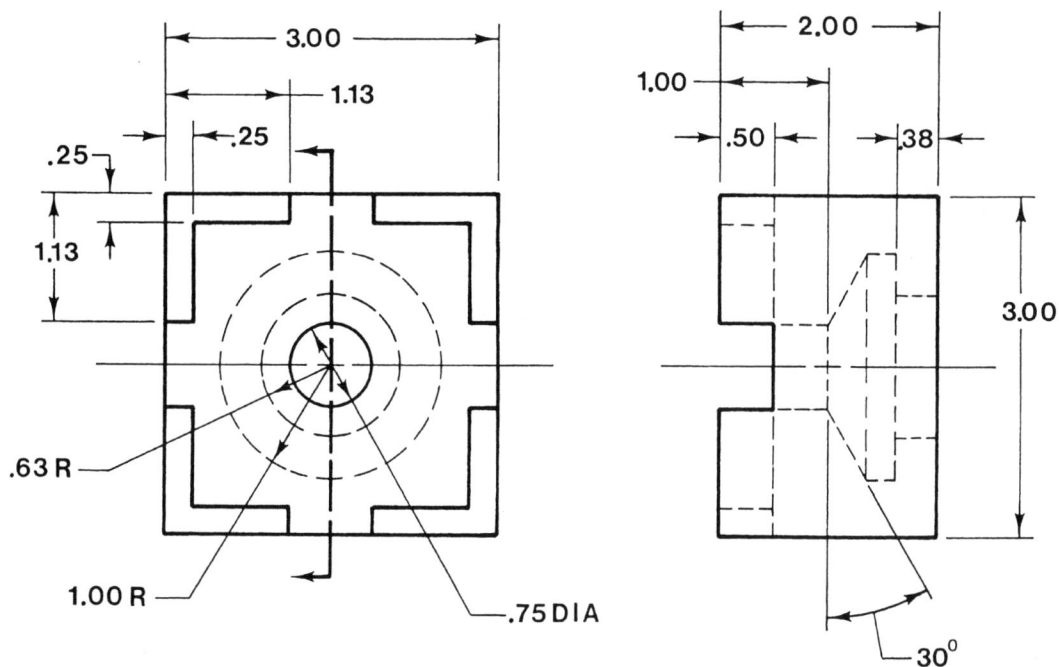

**EX2-27 INCHES**

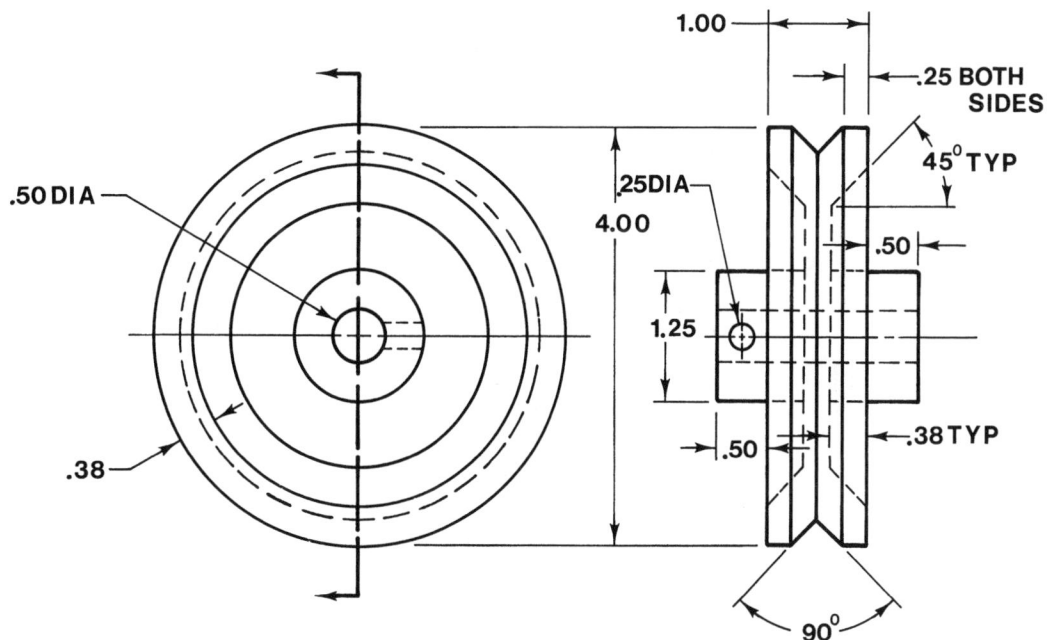

**EX2-28 INCHES**

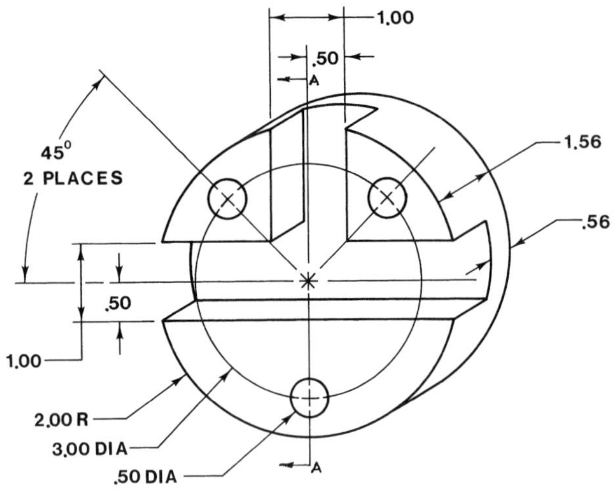

- 1.00
- .50
- A
- 45° 2 PLACES
- 1.56
- .56
- .50
- 1.00
- 2.00 R
- 3.00 DIA
- .50 DIA
- A

**EX2-30 MILLIMETERS**

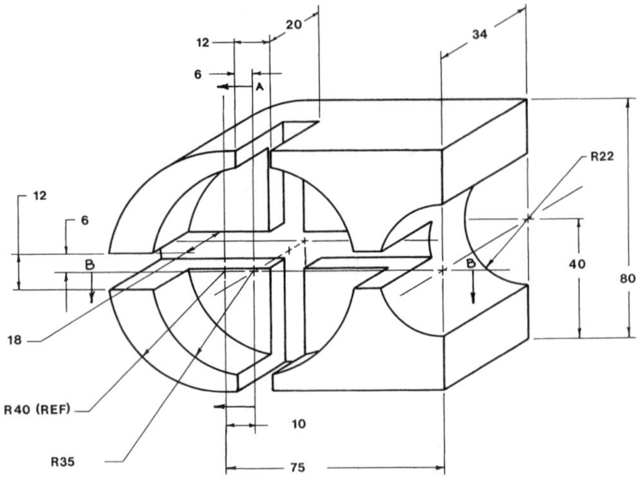

- 12
- 20
- 6
- A
- 34
- 12
- R22
- 6
- B
- 40
- 80
- 18
- B
- R40 (REF)
- 10
- R35
- 75

**EX2-29 MILLIMETERS**

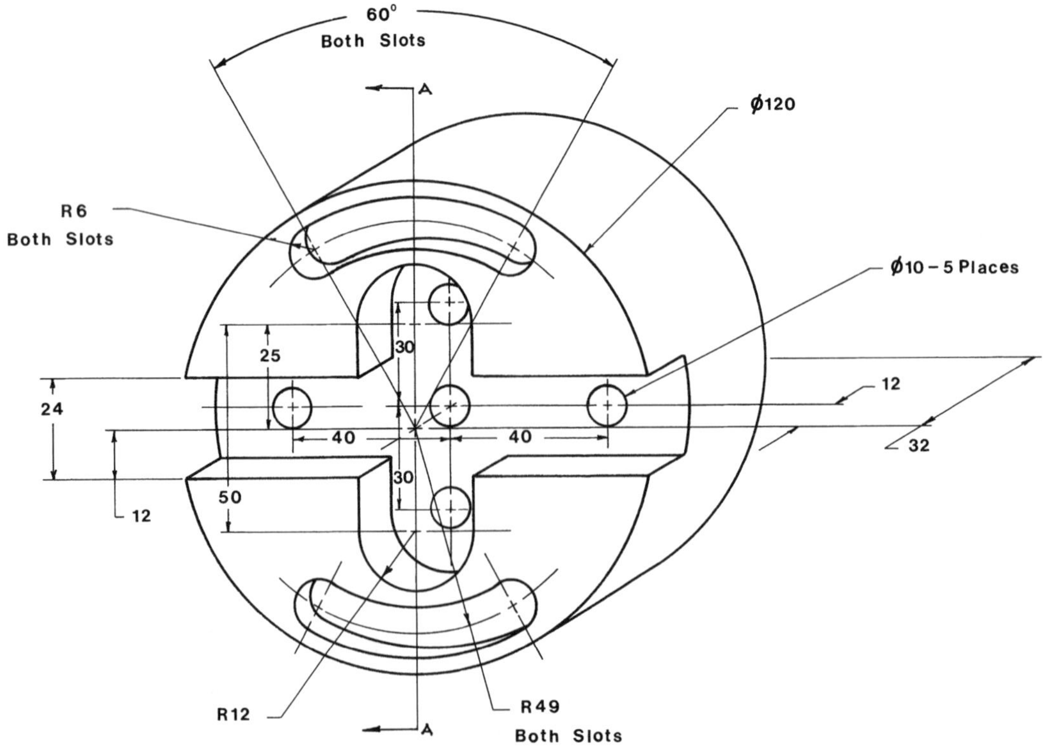

- 60° Both Slots
- A
- Ø120
- R 6 Both Slots
- Ø10 – 5 Places
- 25
- 30
- 24
- 12
- 40
- 40
- 50
- 30
- 12
- R12
- R49 Both Slots
- A

**EX2-31 MILLIMETERS**

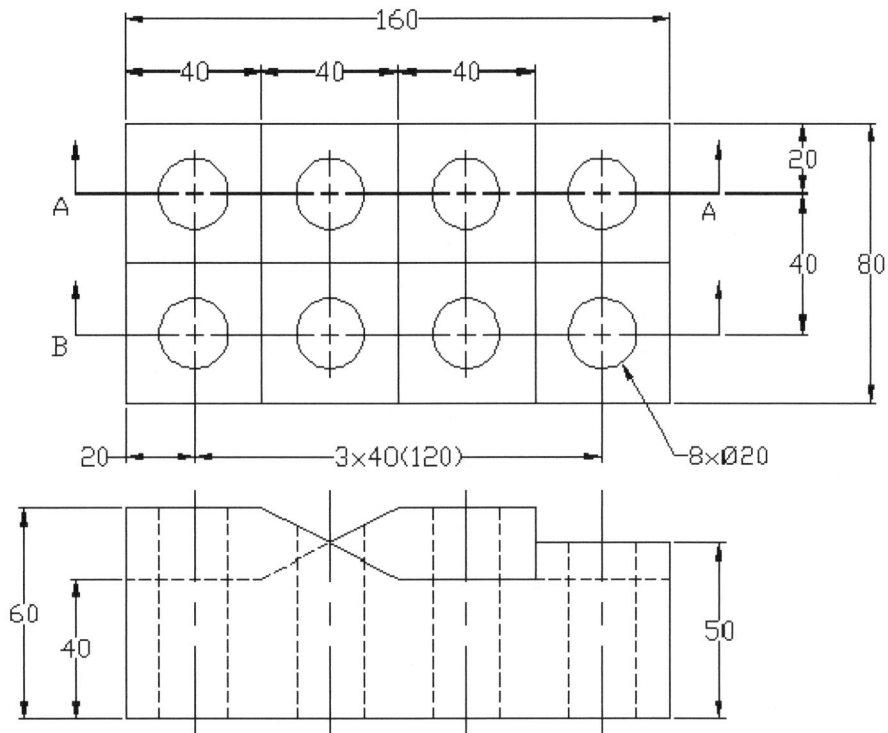

**EX2-32 MILLIMETERS**

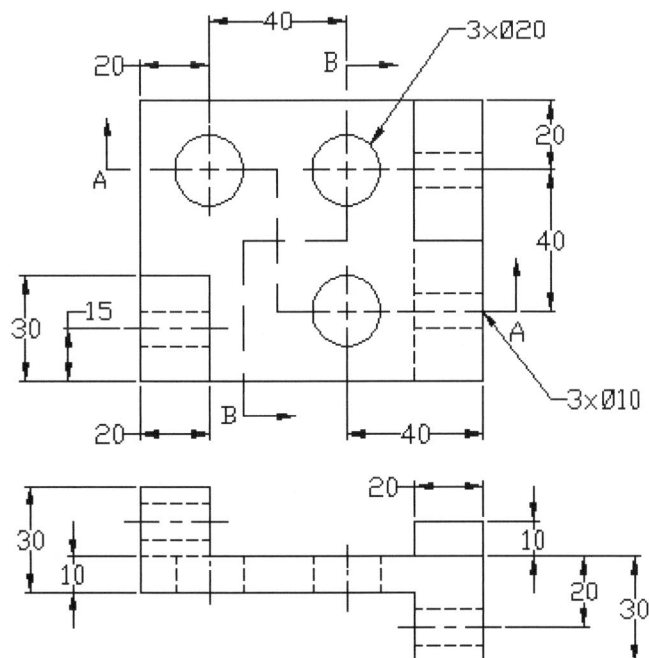

**EX2-33 MILLIMETERS**

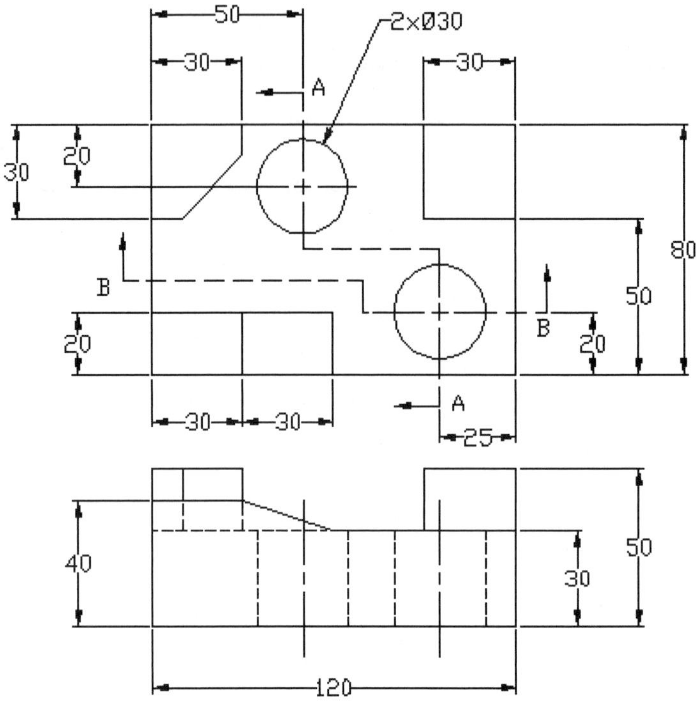

**EX2-34 INCHES**

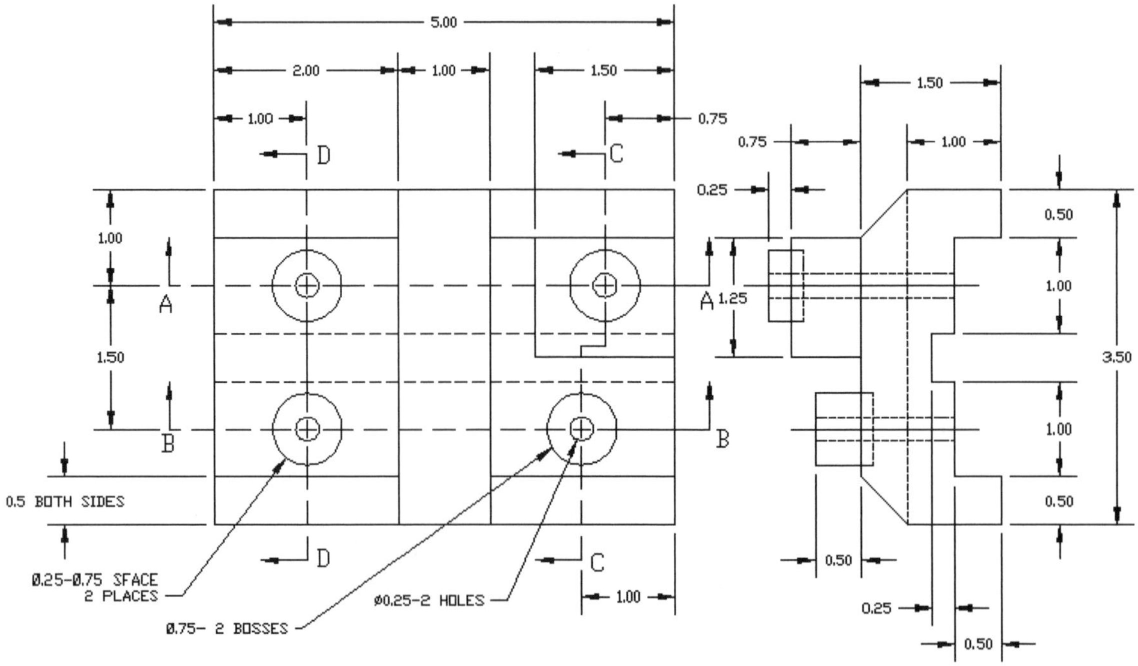

**EX2-35**

New customer requirements require that the following part be redesigned as follows. Draw a front and a sectional view of the redesigned part.

1.  The diameters of the Ø9 internal access holes are to be increased to Ø12.
2.  The diameter of the internal cavity is to be increased to Ø30.
3.  The length of the internal cavity is to be increased from 33 to 50.
4.  All other sizes and distances are to be increased to maintain the same wall thickness.

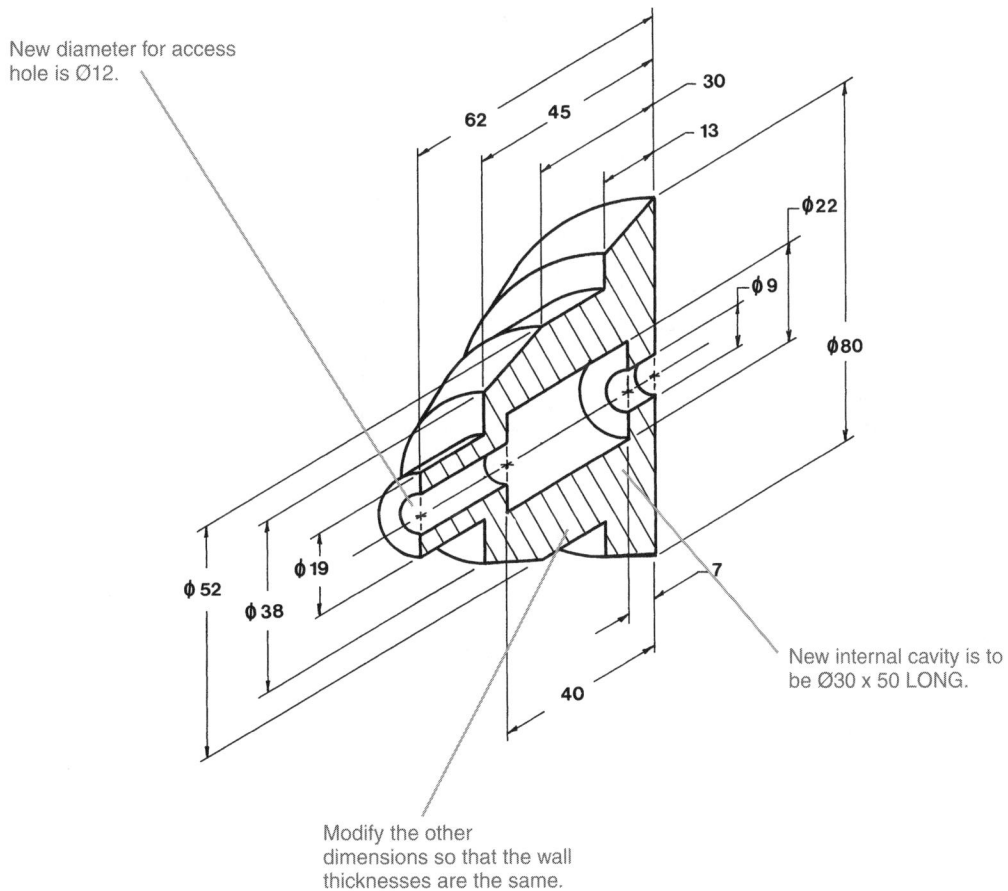

New diameter for access hole is Ø12.

New internal cavity is to be Ø30 x 50 LONG.

Modify the other dimensions so that the wall thicknesses are the same.

NOTE: Taper from Ø80 to Ø68 over 13.

**EX2-36**

The following object is to be redesigned as follows.

1. Extend the 1.00-inch vertical slot all the way through the object, creating four corner sections.
2. Locate countersunk holes in each of the four corner sections. Note that two of the corner sections presently have Ø.50 holes. The countersink specifications are Ø.50 - 82°, Ø.75.
3. Locate a Ø.625 hole in the center of the object.

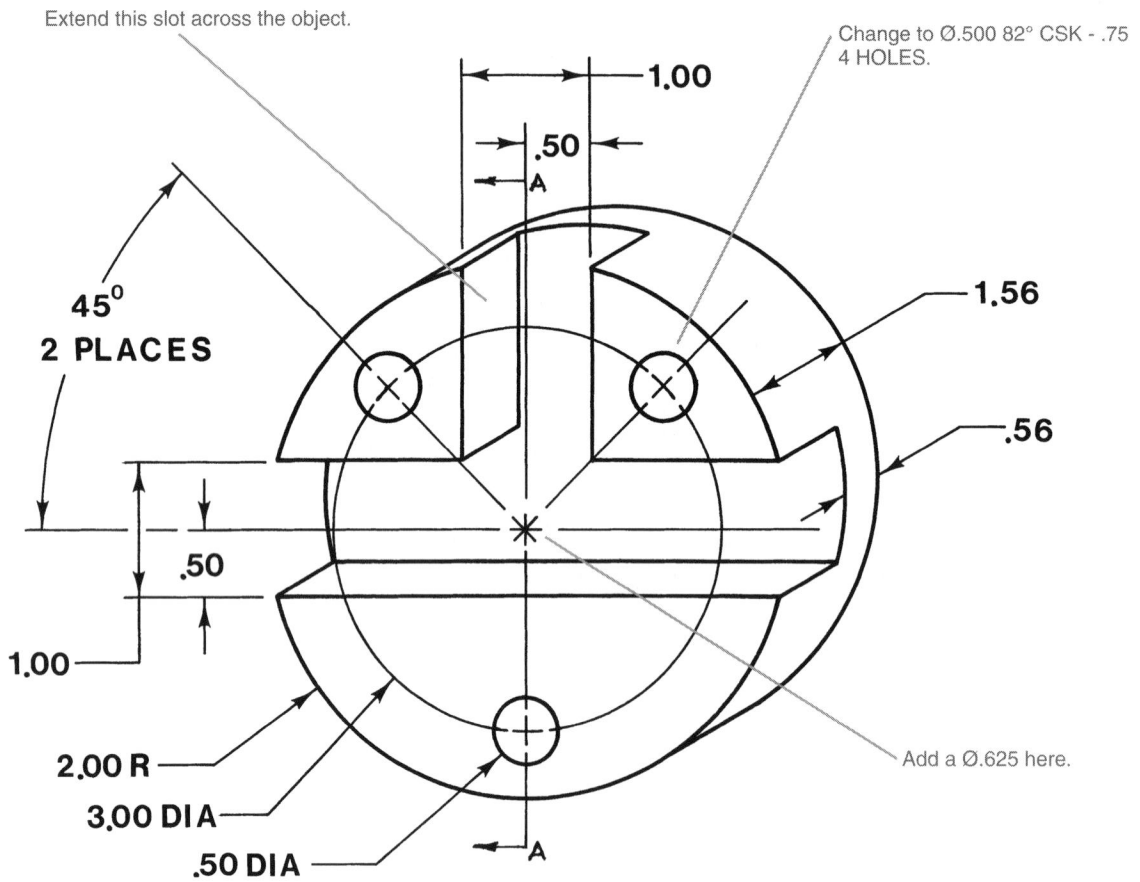

Extend this slot across the object.

Change to Ø.500 82° CSK - .75 4 HOLES.

1.00

.50

A

45°
2 PLACES

1.56

.56

.50

1.00

2.00 R

3.00 DIA

.50 DIA

A

Add a Ø.625 here.

# 2-19 TOLERANCES OF FORM, ORIENTATION, PROFILE, AND RUNOUT

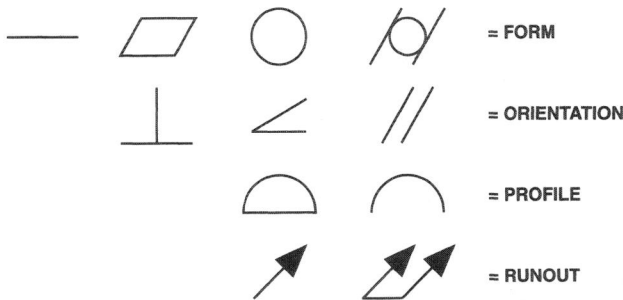

Tolerances of form, orientation, profile and runout state how far actual surfaces or features are permitted to vary from those implied by the drawing. Expressions of form and orientation tolerances refer to flatness, straightness, circularity, cylindricity, parallelism, perpendicularity and angularity. Profile of a surface, profile of a line, circular runout and total runout tolerances are unique variations and combinations of form, orientation and sometimes location and are considered as separate types of characteristics.

Form, orientation, profile, or runout tolerances should be specified for all features critical to the design requirements:

a. where established workshop practices cannot be relied upon to provide the required accuracy;
b. where documents establishing suitable standards of workmanship cannot be prescribed;
c. where tolerances of size and location do not provide the necessary control.

The various tolerances of form, orientation, profile and runout often have an effect on one another; that is, parallelism could include flatness or straightness, and runout could include circularity, straightness, or cylindricity.

The following series of form, orientation, profile, and runout tolerance examples address each of these characteristics and their individual purpose and meaning, in order to explain the basic principles.

# 2-20 TOLERANCES OF FORM, ORIENTATION, AND PROFILE

Tolerances of form, orientation, and profile provide methods by which to control part geometry where size or location does not adequately do so. Such tolerances state how

far an actual surface or feature is permitted to vary from the desired geometric shape implied by the drawing.

### Kinds of Features to which a Form, Orientation, or Profile Tolerance is Applicable

To correctly apply form, orientation, or profile tolerances, an understanding of the kind of features, INDIVIDUAL OR RELATED, upon which each characteristic can be used is required. Such tolerances can be applied as follows:

*INDIVIDUAL* feature—A *single surface, element*, or *size* feature which relates to a perfect geometric counterpart of itself as the desired form; no datum is proper or used.

*Form Characteristics* which can be applied:

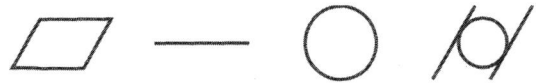

INDIVIDUAL FEATURES

*RELATED* surface feature—A *single surface* or *element* feature which relates to a datum, or datums, in orientation.*

*Orientation Characteristics* which can be applied:

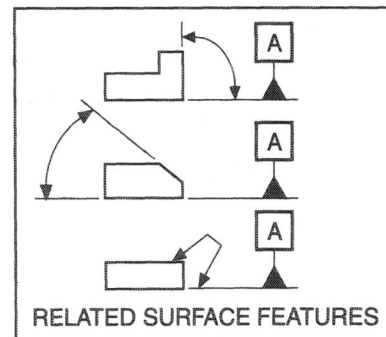

RELATED SURFACE FEATURES

---

*Those tolerances involving related features and datums are sometimes also referred to as attitude tolerances.

*RELATED* size feature—A single *size* feature which relates to a datum, or datums, in form and orientation.*

*Orientation Characteristics* which can be applied:

RELATED SIZE FEATURES

*INDIVIDUAL* or *RELATED* surface or size feature— A *single surface, element* or *size* feature whose perfect geometric profile is described, which may or may not relate to a datum, or datums, in form, orientation, and profile.*

*Profile Characteristics* which can be applied:

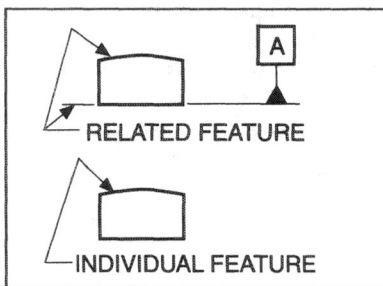

RELATED FEATURE

INDIVIDUAL FEATURE

## 2-21 TOLERANCES OF FORM— INDIVIDUAL FEATURES—NO DATUM

Tolerances of form used on individual features where no datum is proper nor used involve the characteristics below:

(FLATNESS)   (STRAIGHTNESS)   (CIRCULARITY)   (CYLINDRICITY)

*Those tolerances involving related features and datums are sometimes also referred to as attitude tolerances.

These characteristics are used to describe form tolerances of *single surface, element,* or *size* features and relate to a perfect geometric counterpart of itself.

See following pages for details of application.

## 2-22 FLATNESS

**Definition.** Flatness is the condition of a surface having all elements in one plane.

### FLATNESS TOLERANCE

Flatness tolerance specifies a tolerance zone confined by two parallel planes within which the surface must lie.

## 2-23 FLATNESS TOLERANCE APPLICATION

The example below shows how a flatness symbol is applied.

The symbol is interpreted to read "This surface shall be flat within .002 total tolerance zone over entire surface." Note that the .002 tolerance zone is *total* variation. To be acceptable, the entire actual surface must fall within the parallel plane extremities of the .002 tolerance zone. The .002 flatness tolerance is based upon the design requirement which is derived from the precision determined necessary for mating part interface, seal-off, appearance, etc.

A flatness tolerance is a form control of all elements of a surface as it compares to a simulated perfect geometric counterpart of itself. The perfect geometric counterpart of a flat surface is a plane. The tolerance zone is established as a width or thickness zone relative to this plane as established from the actual part surface.

**Example**

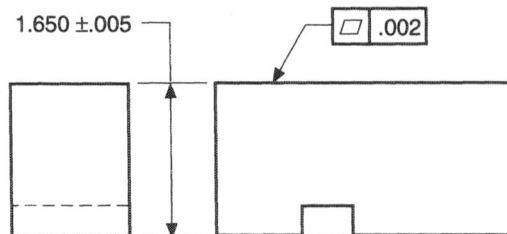

1.650 ±.005

**Meaning**

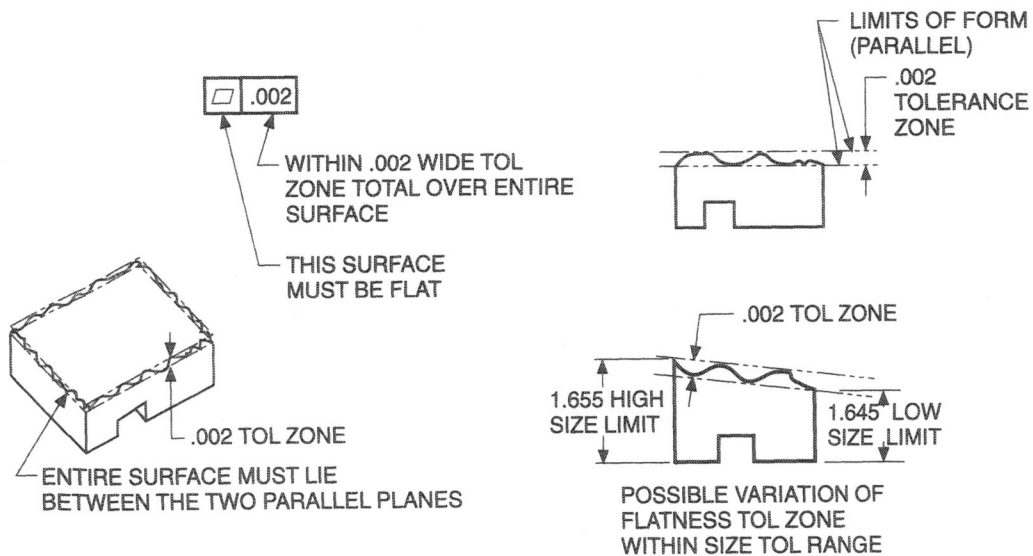

WITHIN .002 WIDE TOL ZONE TOTAL OVER ENTIRE SURFACE

THIS SURFACE MUST BE FLAT

.002 TOL ZONE

ENTIRE SURFACE MUST LIE BETWEEN THE TWO PARALLEL PLANES

LIMITS OF FORM (PARALLEL)

.002 TOLERANCE ZONE

.002 TOL ZONE

1.655 HIGH SIZE LIMIT

1.645 LOW SIZE LIMIT

POSSIBLE VARIATION OF FLATNESS TOL ZONE WITHIN SIZE TOL RANGE

Note that the extremities or high points of the surface determine one limit or plane of the tolerance zone, with the other limit or plane being established .002 (the specified tolerance) parallel to it.

Since flatness tolerancing control is essentially a relationship of a feature to itself, no datum references are required or proper.

Also, note that since flatness is a form tolerance controlling surface elements only, it is not applicable to RFS or MMC considerations.

In the absence of a flatness tolerance specification, the size tolerance and method of manufacture of a part will exercise some control over its flatness. However, when a flatness tolerance is specified, as applicable to a single surface, the flatness tolerance zone must be contained *within* the size tolerance limits.* It cannot be additive to the size tolerance. Therefore, a flatness tolerance should normally be *less than* the part size tolerance.

Where necessary the terms "MUST NOT BE CONCAVE" or "MUST NOT BE CONVEX" may be added beneath the feature control frame.

Author advisory—as a "rule-of-thumb," as based upon the norms of production and probabilities, it may be well to consider that a design calculated flatness tolerance requirement on a stable rigid part, should be equal to, or less than, one-half of the overall size tolerance (i.e. .010 on illustrated part) for justification as a specified flatness tolerance. The .002 tolerance illustrated being less than .005 (1/2 of .010 size tolerance) is an example of this logic.

## 2-24 STRAIGHTNESS —

**Definition.** Straightness is a condition where an element of a surface or an axis is a straight line.

### Straightness Tolerance

A straightness tolerance specifies a tolerance zone within which the considered element or derived median line must lie.

### Straightness Tolerance Application

A straightness tolerance is applied in the view where the elements to be controlled are represented by a straight line.

### Straightness Tolerance—Surface Element Control

Straightness tolerance is typically used as a form control of individual surface elements such as those on cylindrical or conical surfaces. Since surfaces of this kind are made up of an infinite number of longitudinal elements, a straightness requirement applies to the entire surface as controlled in single line elements in the direction specified.

The example following illustrates straightness control of individual longitudinal surface elements on a cylindrical part. Note that the symbol is directed to the feature surface (or extension line) and not to the dimension lines. The straightness tolerance must be less than the size tolerance.

Author advisory—as a "rule-of-thumb," as based upon the norms of production and probabilities, it may be well to consider that a design calculated surface element straightness tolerance requirement on a stable rigid part, should be equal to, or less than, one half of the overall diameter size

---

*Under special circumstances, a note specifically exempting the pertinent size dimension, i.e., PERFECT FORM AT MMC NOT REQUIRED, may be specified. In such cases the flatness may be greater than the size tolerance.

tolerance (i.e. .010 on illustrated part) for justification as a specified straightness tolerance.

All actual local size (circular elements) of the surface must be within the specified size tolerance and the boundary of perfect form at MMC.* Also, each longitudinal element of the surface must lie in a tolerance zone defined by two parallel lines spaced apart by the amount of the prescribed tolerance where the two lines and the nominal axis of the part share a common plane.

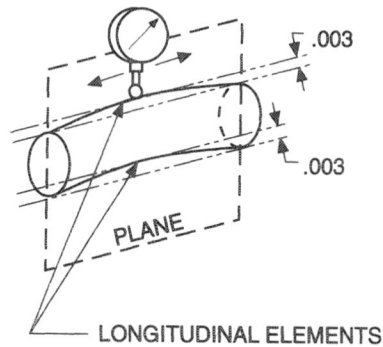

> **NOTE:**
>
> Since surface element control is specified, the tolerance zone applies uniformly whether the part is of a bowed, waisted, or barreled shape.

Note the absence of any datum reference when straightness tolerancing is used. Straightness is a form control of a single element, or an axis, as it relates to a perfect geometric counterpart of itself. Therefore, datum references are neither required nor proper.

## 2-25 STRAIGHTNESS TOLERANCE APPLIED TO FLAT SURFACE

Straightness tolerancing may be applied to a flat surface to provide element surface control in a specific direction as a refinement of size tolerance or other form tolerance such as flatness. If so used, the straightness tolerance must be less than the size tolerance or the refined tolerance.

The example below illustrates straightness tolerance of surface elements as a refinement of size tolerance. Note that the tolerance is applied in the view in which the elements to be controlled appear as a straight line.

The individual straightness elements must be within both the size* tolerance and the straightness tolerance zone of .003, whereas element to element in the other view, variation within the size tolerance may occur.

**Example**

$\varnothing.500 \pm.005$

—|.003

**SYMBOL MEANING**

—|.003 ← WITHIN .003 WIDE TOL ZONE

EACH LONGITUDINAL ELEMENT MUST BE STRAIGHT

THE FEATURE MUST BE WITHIN THE SPECIFIED TOL OF SIZE AND THE BOUNDARY OF PERFECT FORM AT MMC (.505). EACH LONGITUDINAL ELEMENT OF THE SURFACE MUST LIE BETWEEN TWO PARALLEL LINES (.003 APART) WHERE THE TWO LINES AND THE NOMINAL AXIS OF THE PART SHARE A COMMON PLANE.

**Meaning**

$\varnothing.505^{*}$

.003 WIDE TOL ZONE

$\varnothing.505^{*}$

.003 WIDE TOL ZONE

$\varnothing.505^{*}$

.003 WIDE TOL ZONE

*BOUNDARY OF PERFECT FORM AT MMC.

.003

.003

PLANE

LONGITUDINAL ELEMENTS

---

*Under special circumstances a note specifically exempting the pertinent size dimensioning, i.e., PERFECT FORM AT MMC NOT REQUIRED, may be specified. In such cases, the straightness tolerance may be greater than the size tolerance.

**Example**

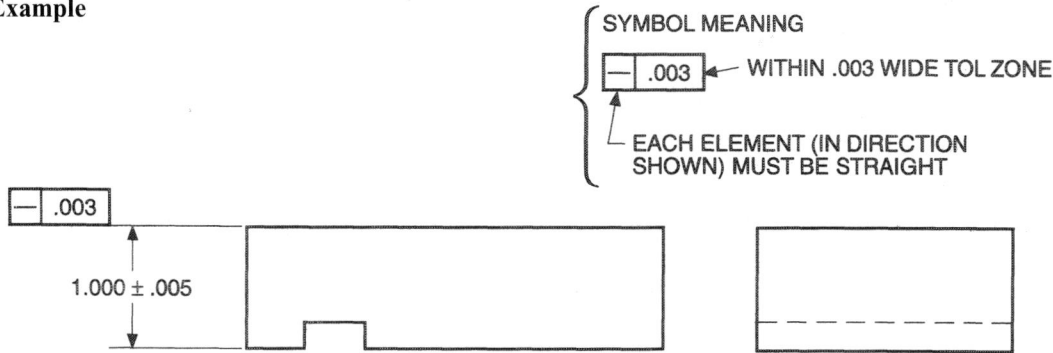

SYMBOL MEANING

— | .003 ← WITHIN .003 WIDE TOL ZONE

EACH ELEMENT (IN DIRECTION SHOWN) MUST BE STRAIGHT

— | .003

1.000 ± .005

**Meaning**

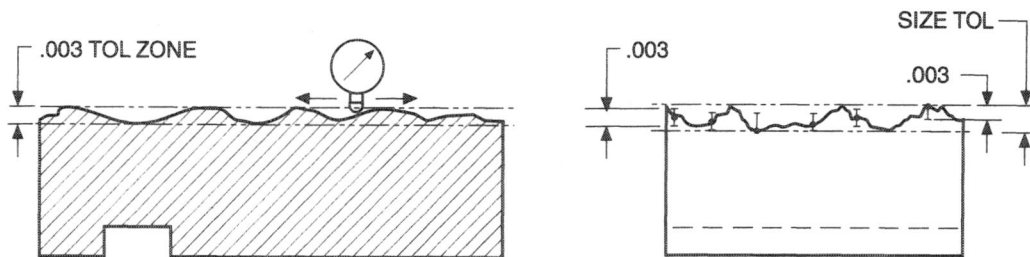

.003 TOL ZONE

.003

SIZE TOL

.003

## Straightness Tolerance RFS and MMC to Axis

Where function of a size feature permits a collective result of *size* and *form* variation and the perfect form at MMC boundary may be exceeded, the RFS or MMC principles may be used.

In this instance, where the appropriate symbology and specifications are stated, the part is not confined to the perfect form at MMC boundary. All actual local sizes (cross-sectional elements) of the surface are to remain within the specified size tolerance, but the total part surface may exceed the perfect form at MMC boundary to the extent of the straightness tolerance.

This principle may be applied to individual size features such as pins, shafts, bars, etc., where the longitudinal elements are to be specified with a straightness tolerance

independent of, or in addition to, the size tolerance as a design reality or manufacturing necessity. In such a case, a new outer boundary or virtual condition is developed which represents the collective effect of the size and form error that must be considered in determining the fit, mating feature relationship clearance or between parts.

### Straightness—RFS

Where a cylindrical feature is to be controlled on an RFS basis, as below, the feature control symbol must be located with the size dimension or attached to the dimension line, and the diameter symbol must precede the straightness tolerance.

**Example**

Ø.500 ± .002

SYMBOL MEANING

| Ø.015 |  ←— OF .015, RFS

└── WITHIN Ø (CYL) TOL ZONE

└── THE DERIVED MEDIAN LINE
MUST BE STRAIGHT

| Ø.015 |

**Meaning**

DERIVED MEDIAN LINE

Ø.498 – .502   ACTUAL LOCAL SIZES

EACH ACTUAL LOCAL SIZE OF THE FEATURE MUST BE
WITHIN THE SPECIFIED TOLERANCE OF SIZE. THE DERIVED
MEDIAN LINE OF THE FEATURE ACTUAL LOCAL SIZES
MUST LIE WITHIN A CYLINDRICAL TOLERANCE
ZONE OF .015 REGARDLESS OF FEATURE SIZE.

.015 DIA (CYL) ZONE, RFS

.517 DIA  OUTER BOUNDARY

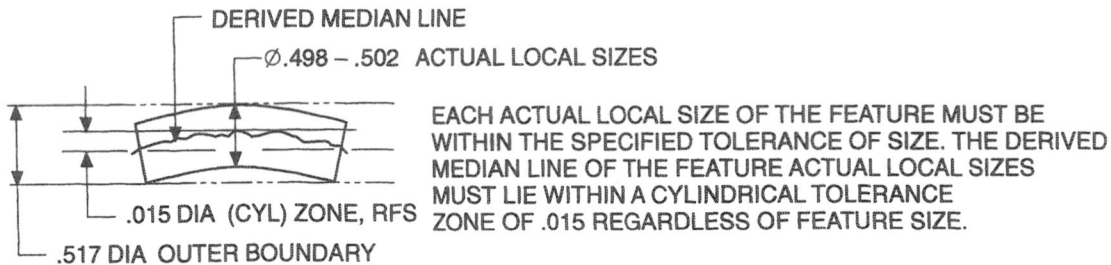

Author advisory—In such cases, the straightness tolerance need *not* be greater than the size tolerance; however, it is a common situation. Where the part length may approach *ten times* the part diameter or more, maintaining the perfect form at MMC boundary may be impractical. Therefore, this method could be a necessary alternative in the design considerations.

### Straightness—MMC—Virtual Condition

Where a cylindrical feature has a functional relationship with another feature, such as a pin or shaft and a hole, the control of straightness on an MMC basis may be desirable. If the pin or shaft, for example, is to fit into a hole of a given diameter, the collective effect of the pin size and its straightness error must be considered in relationship to the hole size minimum, i.e., their virtual conditions must be considered relative to one another.

On the part below, the size must be maintained at all actual local sizes (cross-sectional elements) within the stated limits. Likewise its straightness, using the axis as the criteria, must be within tolerance, but only when the part is at MMC. Therefore, the part develops (or is based upon) a virtual condition of .517, which represents the extreme condition the part can have and yet perform its function, or fit the mating part.

By stating the requirements on an MMC basis, the allowable straightness tolerance may increase an amount equal to the actual local size departure from MMC. The feature control symbol must be located with the size dimension, or attached to the dimension line; the diameter symbol must precede the straightness tolerance; and the MMC symbol must be inserted following the tolerance. In this manner maximum tolerance is achieved, part fit is guaranteed, and functional gaging techniques may be used.

**Example**

Ø.500 ± .002

SYMBOL MEANING

| Ø.015 Ⓜ |  ←— OF .015 AT MMC

└── WITHIN Ø (CYL) TOL ZONE

└── THE DERIVED MEDIAN LINE
MUST BE STRAIGHT

| Ø.015 Ⓜ |

**Meaning**

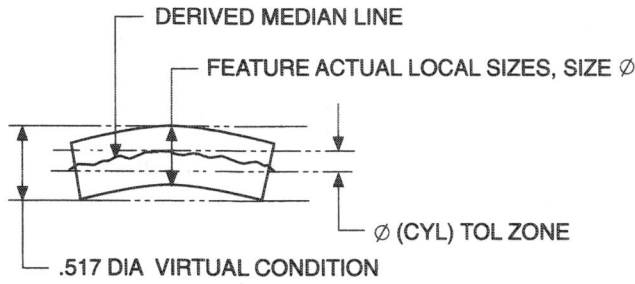

DERIVED MEDIAN LINE

FEATURE ACTUAL LOCAL SIZES, SIZE ∅

∅ (CYL) TOL ZONE

.517 DIA VIRTUAL CONDITION

| ACTUAL LOCAL SIZE ∅ | ∅ TOL ZONE IS |
|---|---|
| .502 MMC | .015 |
| .501 | .016 |
| .500 | .017 |
| .499 | .018 |
| .498 LMC | .019 |

EACH ACTUAL LOCAL SIZE OF THE FEATURE MUST BE
WITHIN THE SPECIFIED TOLERANCE OF SIZE. THE DERIVED
MEDIAN LINE OF THE FEATURE MUST LIE WITHIN A CYLINDRICAL
TOLERANCE ZONE OF .015 AT MMC. AS THE ACTUAL LOCAL
SIZES OF THE FEATURE DEPART FROM MMC, AN INCREASE
IN THE STRAIGHTNESS TOLERANCE IS ALLOWED WHICH
IS EQUAL TO THE AMOUNT OF SUCH DEPARTURE.

Where straightness tolerance on an MMC basis is specified, functional gaging techniques may be used. The below gage and conditions demonstrate how these principles can be applied to the preceding part.

From the above it is seen that straightness applied on an MMC basis provides control of mating part conditions to facilitate design and provides maximum production tolerance. Functional gaging techniques are permissible. Note that the above gage simulates both the extreme permissible

**Gage**

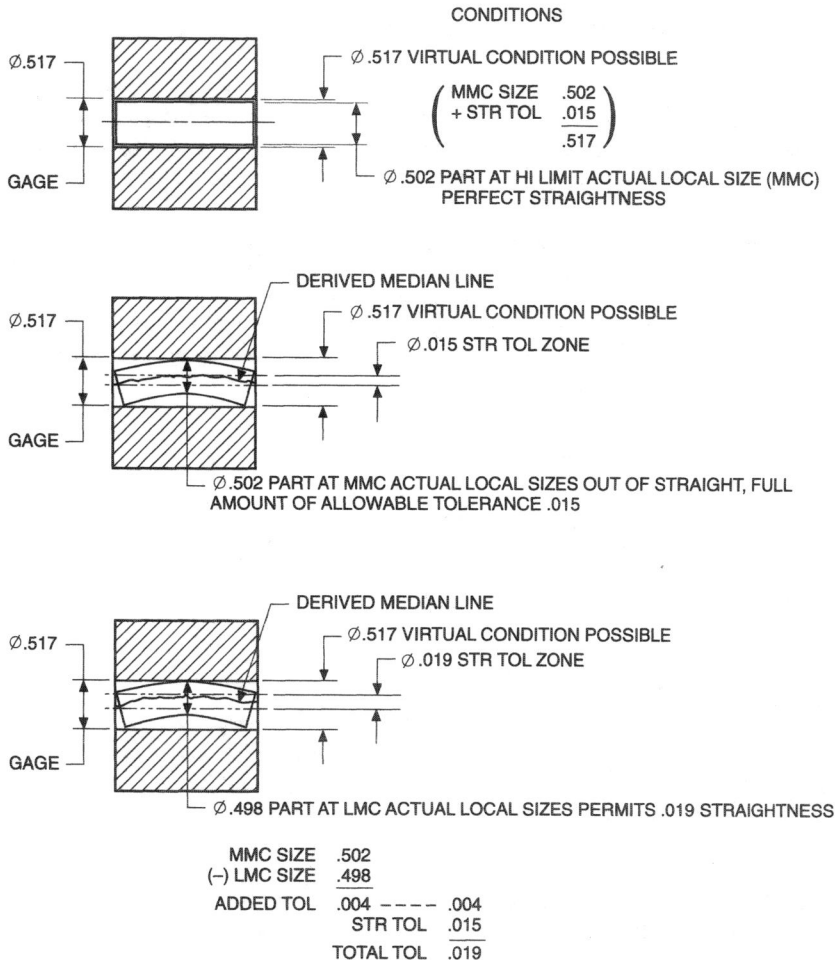

condition of the part as well as the condition of the mating part.

Author Advisory—As a practical matter it is not realistic to try to determine *different* resulting straightness tolerance values at each actual local size. It is, therefore, necessary to base the results of the varying sizes on a collective basis. As the individual actual local sizes of the feature depart from maximum material condition (MMC), they develop an actual mating envelope. This actual mating envelope may be of any varying size to the maximum value which would then reach "virtual condition" size. Thus, the straightness tolerance of such a part is gradually increased as the centers of the actual local sizes derive the median line resulting from the feature size and form deviations (i.e., the shaft gets smaller and more bow is permitted). This simultaneously develops a gradually increasing actual mating envelope size as the collective result of the various involved surface elements. Thus, the specific value of the amount of the increased tolerance would be difficult to state since that amount would be equal the collective result of these conditions on any one part. The illustrated functional gage shows this principle and demonstrates the validity of the functional gage method with the "virtual condition" gage hole size representing the "worst case" condition.

## Straightness on Unit Length Basis

Where required, straightness may be applied on a unit length basis. This method is occasionally used to facilitate special design requirements and to prevent abrupt surface variations within a relatively short length of the feature. To prevent extreme variations of bow over the total length of the part, the amount of permissible straightness variation allowable on the total length of the part should be specified. If the unit variations are permitted to continue along the length of the part with no maximum tolerance indicated, an unsatisfactory part could result.

The example below illustrates a part with unit straightness specified on an RFS basis (MMC principles could also be used if desired.)

**Example**

**Meaning**

EACH ACTUAL LOCAL SIZE OF THE FEATURE MUST BE WITHIN THE SPECIFIED TOLERANCE OF SIZE. THE DERIVED MEDIAN LINE OF THE FEATURE MUST LIE WITHIN A CYLINDRICAL TOL ZONE OF .020 FOR THE TOTAL 4.00 LENGTH, AND WITHIN A .005 CYLINDRICAL TOL ZONE FOR EACH 1.000 LENGTH, RFS

# 2-26 CIRCULARITY ◯

**Definition.** Circularity is the condition of a surface where:

1. for a feature other than a sphere, all points of the surface intersected by any plane perpendicular to an axis are equidistant from that axis;
2. for a sphere, all points of the surface intersected by any plane passing through a common center are equidistant from that center.

## Circularity Tolerance

A circularity tolerance specifies a tolerance zone bounded by two concentric circles within which each circular element of the surface must lie and applies independently at any plane as described above.

## Circularity Tolerance Application

Limits of size exercise control of circularity within the size tolerance. Often this provides adequate control. However, where necessary to further refine form control, circularity tolerancing can be used on any figure of revolution or circular cross section.

The example illustrates a part with a circularity tolerance of .002 specified on a cylindrical part.

The interpretation shows how one establishes the .002 tolerance zone. Note that the tolerance zone is the width of the annular zone between the two concentric circles.

A circularity tolerance zone is established relative to the actual local size of the part when measured at the surface periphery at any cross section perpendicular to the part axis. It should be noted that the circularity tolerance applies only at the cross-sectional point of measurement, and is relative to the *size* at that point. Therefore, a cylindrical part with circularity tolerance control could taper or otherwise vary in its surface contour within its size tolerance range, yet still meet circularity requirements if it is within the circularity tolerance at that point.

The part size in this example has been assumed to measure .503 at its largest point at the cross section selected for measurement. The .002 circularity tolerance zone is then established by two theoretically perfect concentric circles, one at the .503 diameter and the other .004 *smaller* at the .499 diameter. This establishes the tolerance zone of .002 *width* between the concentric circles. To be acceptable, the part surface at that cross section must fall within the .002 wide tolerance zone.

As is seen, the tolerance zone is established relative to the part actual local size wherever it may fall in its size tolerance range. That is, the part *size* is first determined and its circularity is then defined as a refinement of the part *form* relative to that *size*. Unless otherwise specified, any established size at any point along the surface can be used to determine the circularity tolerance zone. It is therefore seen that the circularity tolerance may be based on *different* sizes on the same part. The circularity tolerance zone, however, remains constant.

Author Advisory—as a "rule-of-thumb," in determining a circularity tolerance, the calculated value should be equal to, or less than, one-half of the feature's total size tolerance; i.e., in the illustrated figure, the total size tolerance is .010, the .002 determined is less than .005. This rule-of-thumb would obviously be applicable only to rigid parts not subject to free state variation.

Note again that the circularity tolerance must always be contained *within* the part *size* tolerance range. The circularity tolerance cannot exceed the size tolerance limits. Furthermore, a circularity tolerance cannot be modified to an MMC application since it controls surface elements only.

A circularity tolerance is a form control of a single part element as it compares to a perfect counterpart of itself. Therefore, no datum references are required nor proper. A circularity tolerance does have a reference center, but this is considered only as a part of the perfect frame of reference (concentric circles and their common center) for measurement, just as a straightness tolerance refers to a perfect line of reference for its measurement. You may compare a circularity tolerance to a straightness tolerance zone curled around a circle.

## Circularity of a Cylinder

**Example**

Ø.500 ± .005

.002 — WITHIN .002 WIDE TOL ZONE

THIS FEATURE MUST BE CIRCULAR

.002

A

90°

A

.499          .503

.002 TOL ZONE
SECTION A - A
4X

SURFACE PERIPHERY AT ANY CROSS SECTION PERPENDICULAR TO THE FEATURE AXIS MUST BE WITHIN THE SPECIFIED TOLERANCE OF SIZE AND MUST LIE BETWEEN TWO CONCENTRIC CIRCLES, ONE HAVING A RADIUS .002 LARGER THAN THE OTHER.

(ABOVE SIZES ARBITRARILY SELECTED FOR ILLUSTRATION)

## Circularity of a Cone

The example below illustrates a cone-shaped part for which a circularity tolerance of .003 is specified. As previously discussed, the periphery at any cross section perpendicular to the axis must be within the specified tolerance of size and must lie between the two concentric circles (one having a radius .003 larger than the other).

**Example**

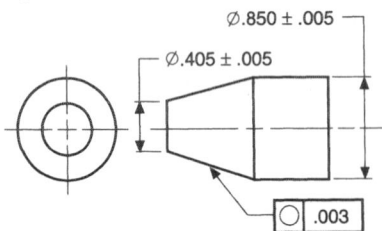

Ø.850 ± .005

Ø.405 ± .005

.003

**Meaning**

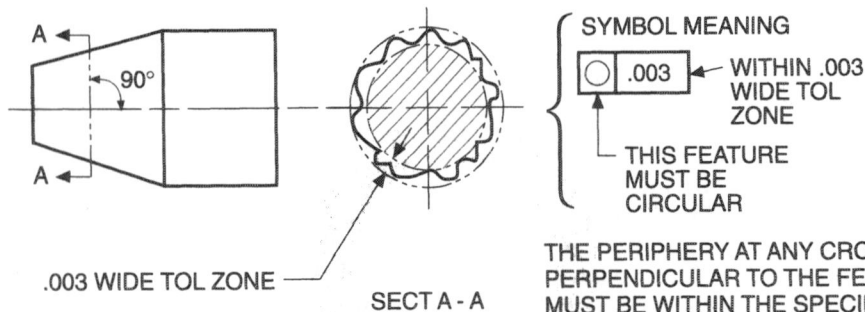

A

90°

A

SYMBOL MEANING

.003 — WITHIN .003 WIDE TOL ZONE

THIS FEATURE MUST BE CIRCULAR

.003 WIDE TOL ZONE

SECT A - A

THE PERIPHERY AT ANY CROSS SECTION PERPENDICULAR TO THE FEATURE AXIS MUST BE WITHIN THE SPECIFIED TOL OF SIZE AND MUST LIE BETWEEN TWO CONCENTRIC CIRCLES (ONE HAVING A RADIUS .003 LARGER THAN THE OTHER).

## Circularity of a Sphere

Circularity of a spherical part is based upon the same principles as those for a cylinder or cone preceding, except that the tolerance control reference is to *any cross section* *passing through a common center* rather than to *any cross section perpendicular to the axis,* as in the conventional application of circularity tolerancing.

**Example**

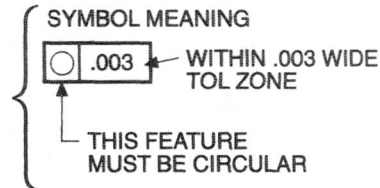

S∅.500 ± .005

| ◯ | .003 |

SYMBOL MEANING

| ◯ | .003 | ← WITHIN .003 WIDE TOL ZONE

└ THIS FEATURE MUST BE CIRCULAR

THE PERIPHERY AT ANY CROSS SECTION PASSING THROUGH A COMMON CENTER MUST BE WITHIN THE SPECIFIED TOLERANCE OF SIZE AND MUST BE BETWEEN TWO CONCENTRIC CIRCLES (ONE HAVING A RADIUS .003 LARGER THAN THE OTHER).

**Meaning**

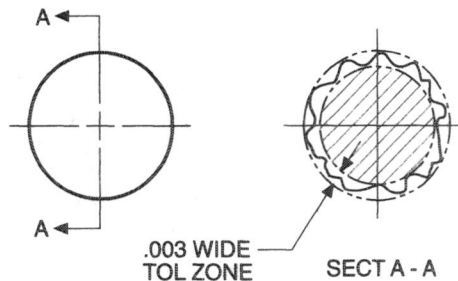

A

A

.003 WIDE TOL ZONE

SECT A - A

# 2-27 CYLINDRICITY

**Definition.** Cylindricity is a condition of a surface of revolution in which all points of the surface are equidistant from a common axis.

## Cylindricity Tolerance

A cylindricity tolerance specifies a tolerance zone bounded by two concentric cylinders within which the surface must lie.

## Cylindricity Tolerance Application

Limits of size exercise control of cylindricity within the size tolerance. This control is often adequate. However, where more refined form control is required, cylindricity tolerancing can be used. Note that in cylindricity, unlike circularity, the tolerance applies simultaneously to both circular and longitudinal elements of the entire surface.

The example illustrates a part with a cylindricity tolerance of .003. A cylindricity tolerance is interpreted as "on the radius" or, as in this case, within the .003 wide tolerance zone defined by two concentric cylinders .003 apart. A cylindricity tolerance can be considered as circularity tolerancing extended to control the *entire* surface of a cylinder.

**Example**

SYMBOL MEANING

∅| .003 ← WITHIN .003 WIDE TOL ZONE

↑ THIS FEATURE MUST BE CYLINDRICAL

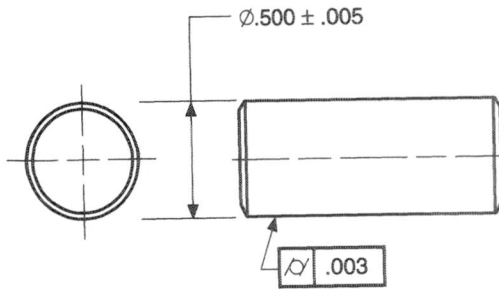

THE FEATURE MUST BE WITHIN THE SPECIFIED TOLERANCE OF SIZE AND MUST LIE BETWEEN TWO CONCENTRIC CYLINDERS (ONE HAVING A RADIUS .003 LARGER THAN THE OTHER).

**Meaning**

Cylindricity tolerancing can be applied *only* to cylindrical forms. The leader from the feature control symbol may be directed to either view.

It should be noted that a cylindricity tolerance simultaneously controls circularity, straightness, and parallelism of the elements of the cylindrical surface.

A cylindricity tolerance applies to the *entire* cylindrical surface as opposed to the cross-sectional or diametral measurement considered in circularity. Also, in measuring cylindricity, the concentric cylinders defining the tolerance zone are always based on the actual mating size of the part produced. The part size in the example has been assumed to measure .503 at its largest diameter. This .503 is the actual mating size diameter of the largest of the concentric cylinders defining the tolerance zone. The smaller of the concentric cylinders is .503 minus the amount of the cylindricity tolerance (.003 on R = .006 on dia), i.e., .497. The cylindricity tolerance zone is therefore .003 *wide* between the con-

centric cylinders. The entire part surface must then fall within this tolerance zone to be acceptable.

It should be noted that the cylindricity tolerance must always be contained *within* the part *size* tolerance range. The cylindricity tolerance cannot exceed the size tolerance limits.

As is seen, the cylindricity tolerance zone is established relative to the part actual mating size wherever it may fall in its *size* tolerance range. That is, the part size is first determined and its *cylindricity* is then defined as a refinement of the part *form* relative to that *size*.

If the largest measurement of the produced part had been near the *low* limit, for example at .497, the cylindricity tolerance could not have been more than .001 on R (= .002 on dia). The cylindricity tolerance cannot exceed *size* tolerance limits. Therefore, the form (cylindricity) variation could not be less than the low limit size of .495.

A cylindricity tolerance (similar to flatness) is a form control of a surface element as it compares to a perfect counterpart of itself. Therefore, no datum references are required nor proper. One can compare a cylindricity tolerance zone to a flatness tolerance zone by visualizing that flatness zone curled around a cylinder. A cylindricity tolerance does have a reference axis, but this is considered only a part of the perfect frame of reference (concentric cylinders and their common axis) for measurement, just as a flatness tolerance refers to a perfect plane of reference for its measurement.

Since cylindricity is a form tolerance controlling surface elements only, it cannot be modified to an MMC application.

Author advisory—as a rule-of-thumb in determining a cylindricity tolerance, the calculated value should be equal to or less than, one-half of the feature's size tolerance; i.e., in the illustrated figure the total size tolerance is .010, the .003 determined is less than .005. This rule-of-thumb would obviously be applicable only to rigid parts not subject to free state variation.

# 2-28 EVALUATION OF CIRCULARITY AND CYLINDRICITY

Cylindricity is checked or measured by the same basic techniques that are used to check circularity except that cylindricity involves a cylindrical tolerance zone of uniform thickness over the entire surface and is based on a single size reference. This must be considered in any measuring procedure concerned with cylindricity. For example, in the following discussions of vee block, between centers, and polargraph methods, the measurements are made at cross sections only; whereas in cylindricity, one must consider the entire surface as controlled by the one tolerance zone.

Vee block or between centers methods are often used in open set-up inspection of circularity and cylindricity.

The vee block method provides an approximation or a rough check which may be adequate for some applications. Variables, such as the varying number and arrangement of lobes on the part surface and the angle of the vee block, may affect the resultant measurement sufficiently for the method to become rather inaccurate. In addition, the procedure itself contributes to inaccuracies, because one portion of the part surface (in the vee) is used as the basis for checking another portion of the same surface, thus compounding the chances for error. With certain lobe and vee block combinations, a visibly out-of-circularity part can be rotated with no evidence of error on the dial indicator.

When the vee block method is used, the total error read is roughly the error on the *diameter*; it must be halved, as an approximation, to compare the result with the specified circularity tolerance which is implied as the radial separation or total width between annular circles.

The between centers method more nearly establishes the reference axis for the geometric relationship. Any error in the centers, however, will affect the accuracy of the resultant measurement. Since the part is rotated about its nominal center or axis, the reading will be on the radius and representative of the total width zone between concentric circles. This method, however, is more correctly termed a runout relationship of a surface and centers and is technically not a circularity analysis. This may be used as a method to support the part recognizing that other considerations are necessary to isolate the circularity error.

$\frac{FIM}{2}$ = CIRCULARITY ERROR

ROTATE 360°

VEE BLOCK

INDICATOR STATIONARY, MEASURED AT ONE LINE ON SURFACE WHILE PART ROTATED.

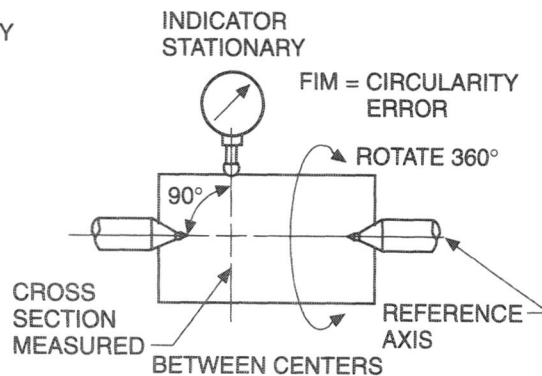

INDICATOR STATIONARY

FIM = CIRCULARITY ERROR

ROTATE 360°

90°

CROSS SECTION MEASURED

BETWEEN CENTERS

REFERENCE AXIS

Figures 2-42 and 2-43 demonstrate that certain vee block and lobed-part conditions do not register the correct part circularity or cylindricity error.

Obviously these parts have noticeable form errors. However, when the parts are rotated in the 60° vee block, the true error will not register. In fact, in these hypothetical examples of a five-lobed part an oval shape, *no* error registers in any position. When the number of lobes of the part is known, certain vee block angles can be used to obtain accurate diametral readings. However, the above indicated variables and the difficulty of accurately predicting conditions on any given part make this method impractical.

Figures 2-44 and 2-45 illustrate how "miking" a lobed part (particularly a part with an uneven number of lobes) will not pick up error. The parts shown, although obviously not true in form, register the same measurements at any diametrical location. Suppose the parts in Fig. 2-42 (five lobes) and Fig. 2-44 (three lobes) were "miked" at 1.999 since they are intended to go through the Ø2.000 hole shown in. Fig. 2-46. The parts will not pass through the hole although they were "miked" at the lower size measurement of 1.999.

Figures 2-47 and 2-48 illustrate the parts shown in. Figs. 2-42 and 2-43 evaluated on the basis of radial values. Using appropriate measuring techniques, the error can be determined directly.

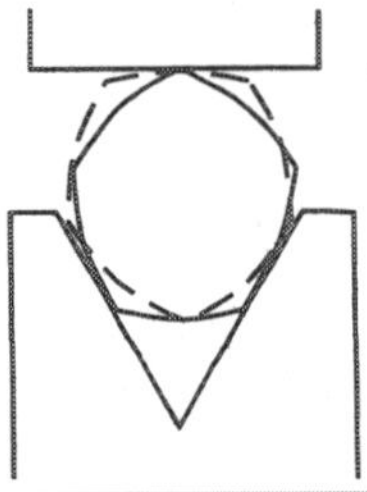

**Figure 2-42** 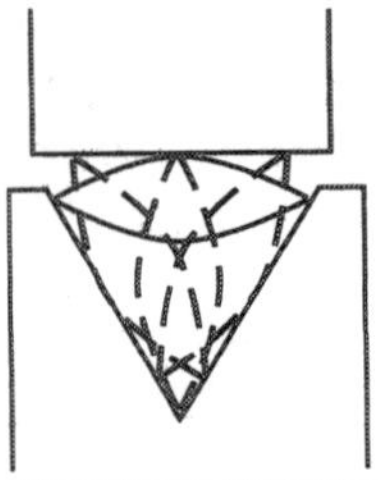          **Figure 2-43** 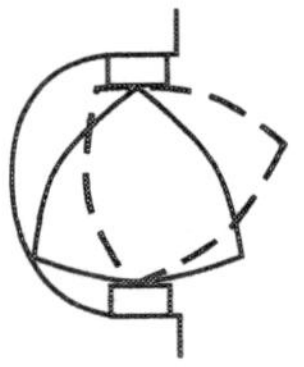          **Figure 2-44**

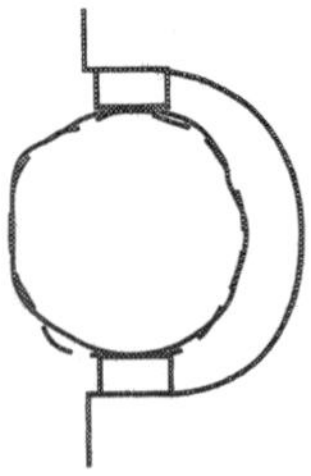

Ø 2.000 HOLE

**Figure 2-45**       **Figure 2-46** 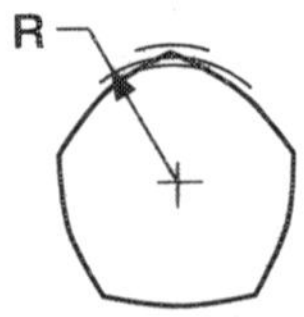          **Figure 2-47** 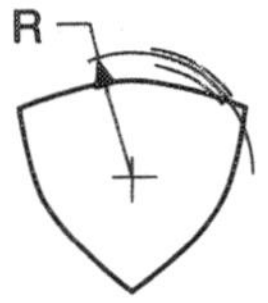          **Figure 2-48**

### Precision Method of Evaluating Circularity and Cylindricity

To evaluate circularity and cylindricity precisely, one must relate the part surface periphery to the geometry of a perfectly round or cylindrical form as constructed from a reference axis. Several kinds of special gaging equipment utilizing optical, mechanical, electronic, and pneumatic principles are available. One method uses an electronic probe which travels around the periphery of the part while the part is chuck-mounted on an extremely accurate spindle and transcribes an enlarged profile of the part periphery on a polargraph. This profile is then, compared with a transparent overlay gage which contains circles at various increments. Note that the final basis for comparison is the part profile only and the reference axis is merely a means of constructing the geometry for measurement. Although more costly and time consuming than other methods, this method utilizes geometric relationships which more directly evaluate circularity and cylindricity where necessary.

POLARGRAPH

TRANSCRIBED PART PROFILE

### Other Methods of Evaluating Circularity and Cylindricity

Due to the increasing sophistication of design requirements, manufacturing processes, and measuring processes, such control as circularity and cylindricity may be accomplished by yet further methods.

Such methods as "Least Mean Square," which determines the center of a circular form by mathematical formula, or a computer program based on ordinate and radius measurements may be used. Other methods, such as "minimum circumscribed circle," "maximum inscribed circle," and "minimum radial separation," which utilize precision spindle techniques can be used in appropriate circumstances.

GAGE

SELECTED LINES REPRESENT WIDTH OF TOLERANCE ZONE

More specific callout of the drawing requirement indicating one of the above methods (or stylus tip radius, cycles per revolution, etc.) may occasionally be necessary.

## 2-29 DATUMS

### Basis for Relating Features to One Another
### Meaning of Datum

Flatness, straightness, circularity, cylindricity (and occasionally profile) tolerancing are characteristics or controls which are applied to single features. It might be said that in order to define controls, the feature is compared to a perfect counterpart of itself with a stated tolerance to indicate the variation permissible from perfect for that feature. It might also be said that the feature serves as its own "datum"; a datum being a feature from which relationships originate. The following definition will further clarify the meaning and intention for a datum.

**Definition.** A datum is a theoretically exact point, axis, or plane derived from the true geometric counterpart of a specified datum feature. A datum is the origin from which the location or geometric characteristics of features of a part are established.

### Datum vs. Datum Surface and Datum Feature

A distinction must be made between a "datum" (theoretically exact—as on a drawing), which represents design requirement and function, and the actual "datum surface" or "datum feature" on the produced part.

**Definition.** A datum feature is an actual feature of a part that is used to establish a datum.

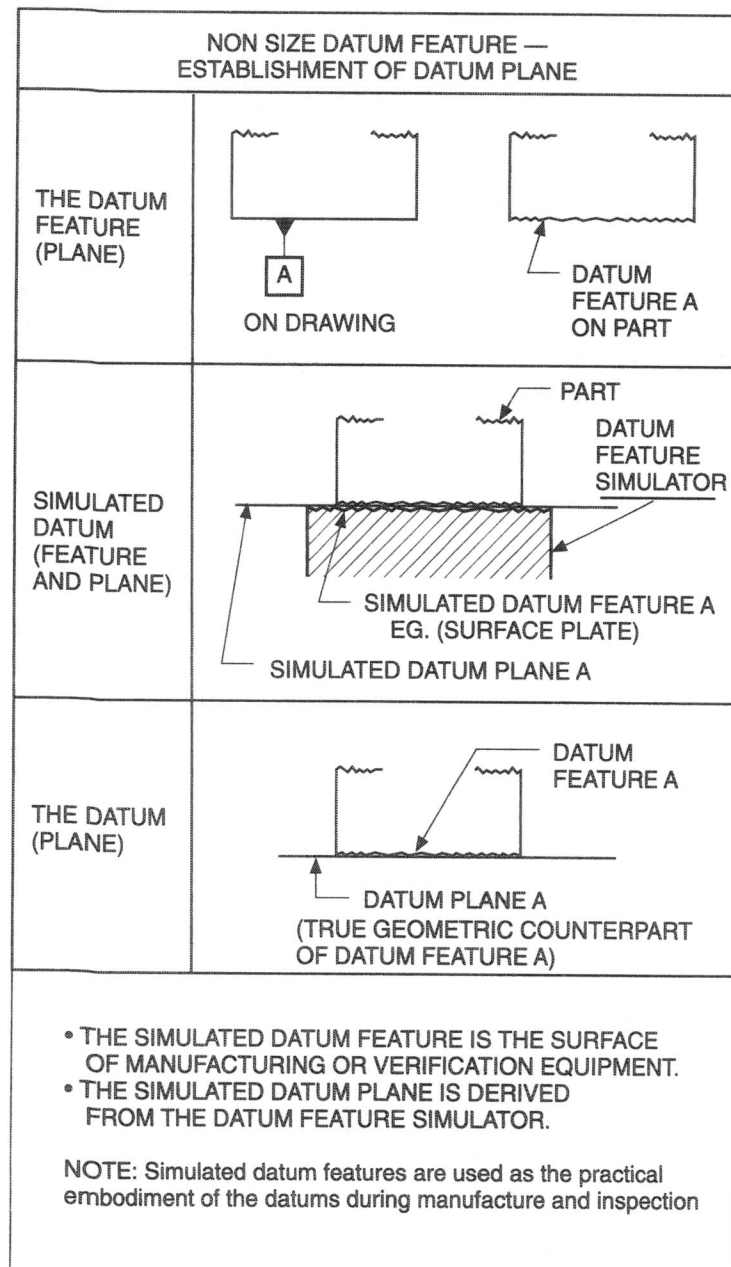

| | NON SIZE DATUM FEATURE —<br>ESTABLISHMENT OF DATUM PLANE | |
|---|---|---|
| THE DATUM FEATURE (PLANE) | ON DRAWING | DATUM FEATURE A ON PART |
| SIMULATED DATUM (FEATURE AND PLANE) | PART — DATUM FEATURE SIMULATOR — SIMULATED DATUM FEATURE A EG. (SURFACE PLATE) — SIMULATED DATUM PLANE A | |
| THE DATUM (PLANE) | DATUM FEATURE A — DATUM PLANE A (TRUE GEOMETRIC COUNTERPART OF DATUM FEATURE A) | |

• THE SIMULATED DATUM FEATURE IS THE SURFACE OF MANUFACTURING OR VERIFICATION EQUIPMENT.
• THE SIMULATED DATUM PLANE IS DERIVED FROM THE DATUM FEATURE SIMULATOR.

NOTE: Simulated datum features are used as the practical embodiment of the datums during manufacture and inspection

SIZE DATUM FEATURE

| | |
|---|---|
| THE DATUM FEATURE (CYLINDER) | ON DRAWING — DATUM FEATURE A (CYL) ON PART |
| SIMULATED DATUM (FEATURE AND AXIS) | SIMULATED DATUM CYLINDER A — DATUM FEATURE SIMULATOR — DATUM CYLINDER A. TRUE GEOMETRIC COUNTERPART OF DATUM FEATURE A (SMALLEST TRUE CYLINDER) — SIMULATED DATUM AXIS A — DATUM FEATURE A — SIMULATED DATUM FEATURE A — DATUM AXIS A |
| THE DATUM (AXIS) | DATUM CYLINDER A. TRUE GEOMETRIC COUNTERPART OF DATUM FEATURE A — DATUM FEATURE A — DATUM AXIS A |

- THE SIMULATED DATUM FEATURE IS THE SURFACE OF MANUFACTURING OR VERIFICATION EQUIPMENT.
- THE SIMULATED DATUM AXIS IS DERIVED FROM THE DATUM FEATURE SIMULATOR.

NOTE: Simulated datum features are used as the practical embodiment of the datums during manufacture and inspection.

Author advisory: Any physical differences between the "true geometric counterpart of the datum feature," the "simulated datum feature" and the "simulated datum plane" represent a necessary compromise between unpredictable part, tool and gage surface precision variables versus practical application within human capabilities to approach such finite exactness. To attempt to bridge this gap between the theory and the realities involved, the "simulated datum feature," the "simulated datum plane" and the "true geometric counterpart of the datum feature" are, for all practical purposes, normally considered consolidated into one composite result when applied under typical every day conditions. This means that when the "simulated datum feature (via the datum feature simulator)" and the part "datum feature" are brought into contact, "the datum (point, axis or plane)" is considered to be, thus, established about as well as it can be done under normal operating conditions and constraints. Where sophisticated computerized electronic verification or evaluation methods are undertaken, it may be possible and/or necessary to detect and respect such minute differences. Under these conditions, results are based upon recognition, detection and distinction of these values. Because of the above referenced electronic options and more sophisticated requirements arising in industry, the standards must recognize this more detailed nomenclature to adequately describe the datum concepts where necessary.

## Datum Required to Relate Features

When relating form, orientation, profile, runout, or location tolerances to features in relation to one another, a datum (or datums), must be used. Datum precedence also must be considered when necessary. (See below and Datum section.)

Further examples of datum application using form, orientation, profile, runout, and location characteristics are given in the DATUM section and in following examples throughout this text.

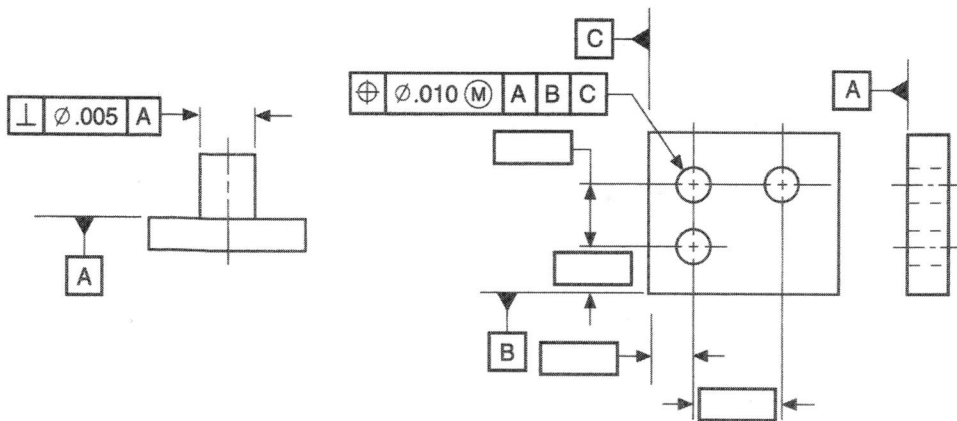

# CHAPTER 3

# Auxiliary Views and GD&T

## 3-1 INTRODUCTION

*Auxiliary views* are orthographic views used to present true-shaped views of slanted and oblique surfaces. Slanted and oblique surfaces appear foreshortened or as edge views in normal orthographic views. Holes in the surfaces are elliptical, and other features are also distorted.

Figure 3-1 shows an object with a slanted surface that includes a hole drilled perpendicular to that surface. Note how the slanted surface is foreshortened in both the normal

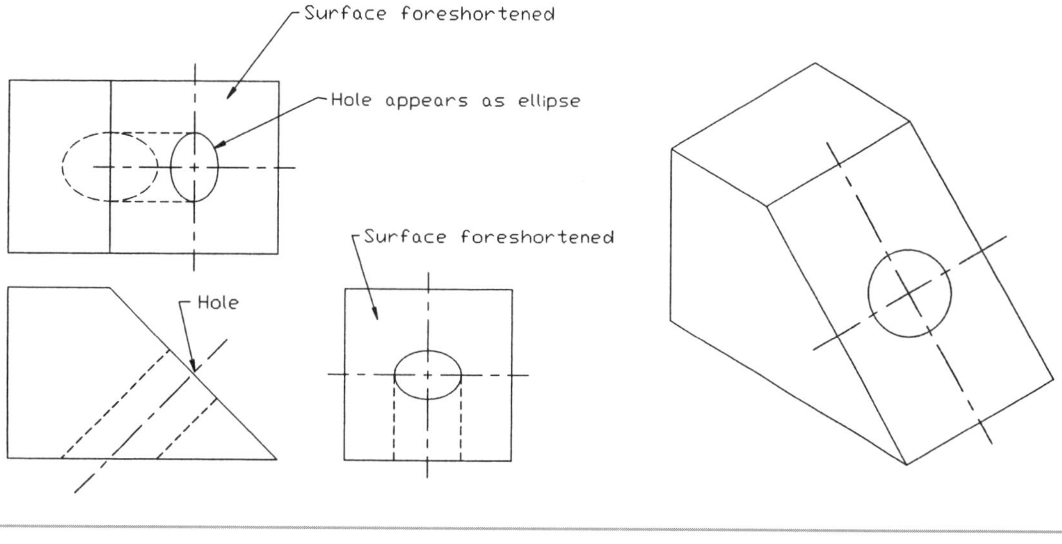

Figure 3-1

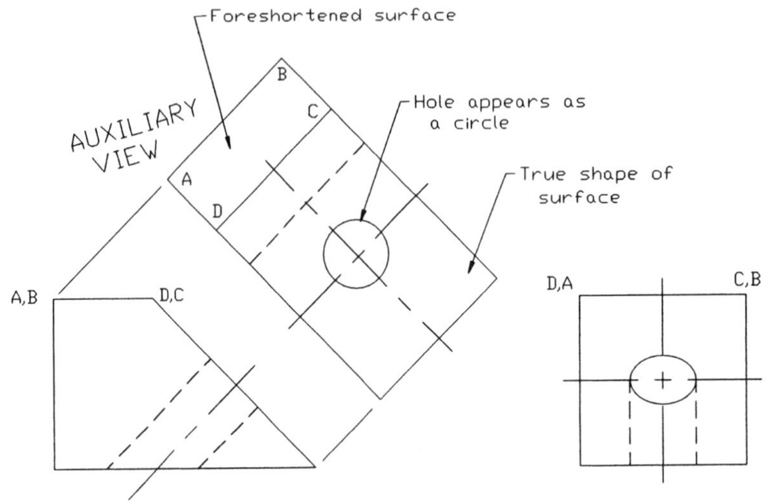

**Figure 3-2**

top and side views and that the hole appears as an ellipse in both of these views. The hole appears as an edge view with hidden lines in the front view, so none of the normal views show the hole as a circle.

Figure 3-2 shows the same object shown in Figure 3-1. The top and side views are the same, but the top view is replaced with an auxiliary view. The auxiliary view shows the true shape of the slanted surface, and the hole appears as a circle.

The auxiliary view in Figure 3-2 shows the true shape of the slanted surface but a foreshortened view of surface A-B-C-D. Positioning the auxiliary view to generate the true shape of the slanted surface foreshortened the other surfaces.

## 3-2 PROJECTION BETWEEN NORMAL AND AUXILIARY VIEWS

Information about an object's features can be projected between the normal views and any additional auxiliary views by establishing reference planes. A reference plane RPT is located between the top and front views of Figure 3-3. The plane appears as a horizontal line, and its location is arbitrary. In the example shown, the plane was located 10 millimeters from the top view.

A second reference plane, RPA, was established parallel to the slanted surface at a location that prevents the

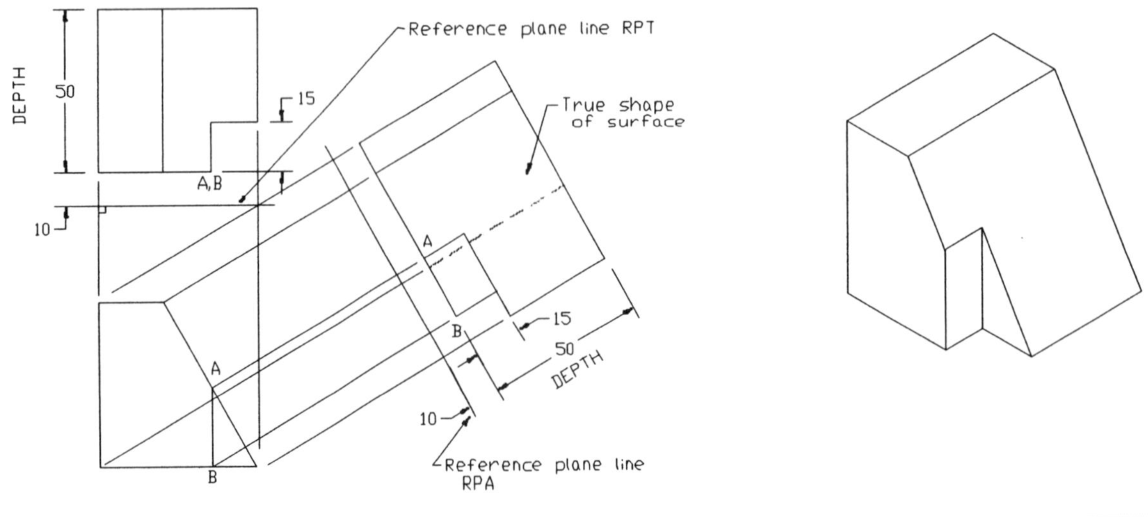

**Figure 3-3**

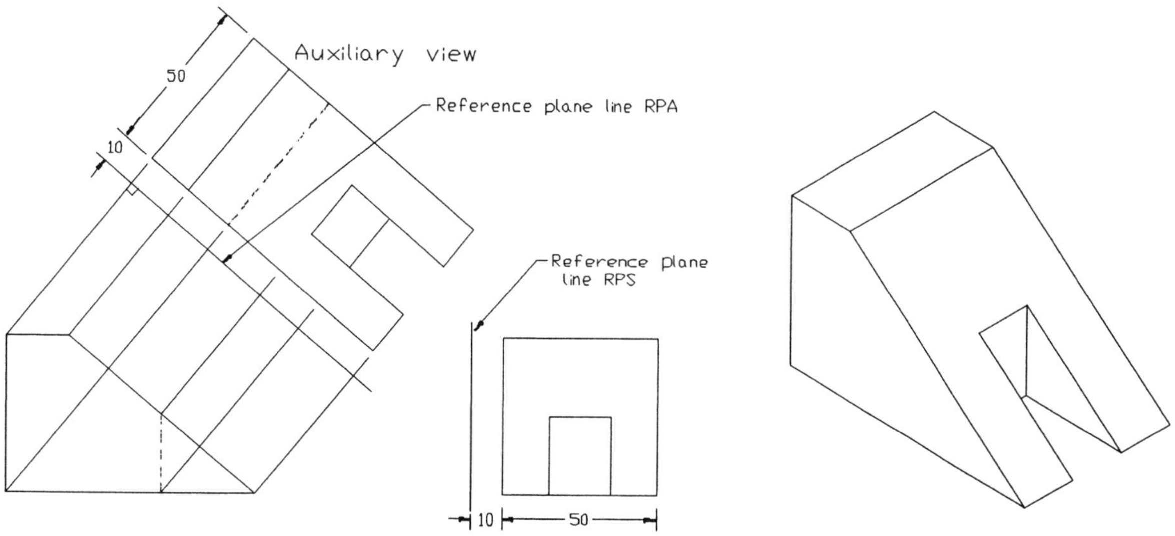

**Figure 3-4**

auxiliary view from interfering with the top view. The **Move** command can be used to move the top view farther away from the front view if necessary.

Information was projected from the front view into the auxiliary view by using projection lines perpendicular to reference plane RPA.

The depth of the object was transferred from the top view to the auxiliary view. Figure 3-3 shows the 50-millimeter depth and 15-millimeter slot depth dimensioned in both the top and auxiliary views.

Figure 3-4 shows information projected from given front and side views into an auxiliary view. Reference planes RPS and RPA were located 10 millimeters from and parallel to the side and auxiliary views, and information was projected from the front view into the auxiliary view using lines perpendicular to plane RPA. The depth information was transferred from the side view to the auxiliary view.

Figure 3-5 shows an object that has a slanted surface in the top view. The reference plane line, RPA, for the auxiliary view was located parallel to the slanted surface and

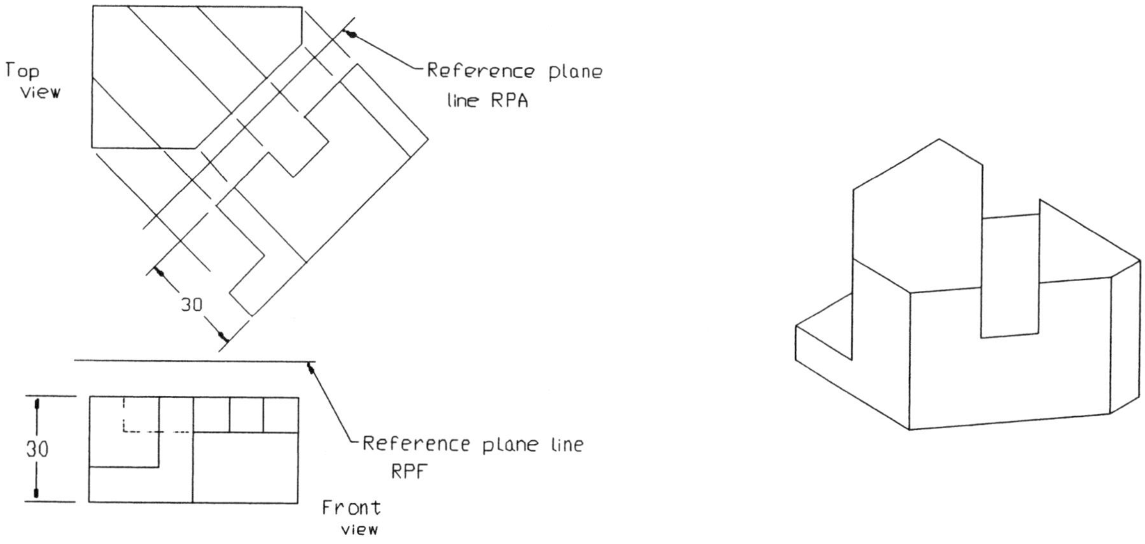

**Figure 3-5**

another reference plane, RPF (horizontal line), was established between the front and top views. Information was projected from the top view into the auxiliary view using lines perpendicular to RPA. The 30-millimeter height measurement was transferred from the front view into the auxiliary view.

Reference plane lines and projection lines drawn perpendicular to them are best drawn by rotating the drawing's axis system, indicated by the crosshairs, so that it is parallel to the slanted surface. A separate layer may be created for projection lines.

**To rotate the drawing's axis system (See Figure 3-6.)**

1. Type **Snap** in response to a command prompt.

   The command prompts are as follows.

   *Command: _snap*
   *Specify snap spacing or [ON/OFF/Aspect/*
   *Rotate/Style/Type]<0.5000>:*

2. Type **r**; press **Enter.**

   *Specify base point <0.0000,0.0000>:*

3. Accept the default value by pressing **Enter.**

   *Specify rotation angle <0>:*

4. Type **45**; press **Enter.**

   The cursor rotates 45° counterclockwise. Any angle value, including negative values, may be used.

   The grid also is rotated, and the **Snap** command will limit the crosshairs to spacing aligned with the rotated axis. **Ortho** will limit lines to horizontal and vertical relative to the rotated axis.

   The angle value entered for **Snap, Rotate** is always interpreted as an absolute value. For example, if the above

procedure is repeated and a value of 60 entered, the crosshairs will advance only 15° from its present location, or 60° from the system's absolute 0° axis system.

## 3-3  SAMPLE PROBLEM SP3-1

Figure 3-7 shows an object that includes a slanted surface. The front, top, and auxiliary views projected from the slanted surface were developed as follows. See Figure 3-8.

1. Draw the front and top orthographic views as explained in Chapter 1.

   Use the **Move** command to position the top view so that it will not interfere with the auxiliary view.

2. Establish two reference plane lines: RPT parallel to the top view (horizontal line) and RPA parallel to the slanted surface (45° line).

   In this example the reference plane line for the top view is located along the lower edge of the view.
   Use **Snap, Rotate** to establish an axis system parallel to the slanted surface. The slanted surface is 45° to the horizontal. Turn **Ortho** on and draw a line parallel to the slanted surface.
   Option: Create a layer for projection lines.

3. Project lines perpendicular to RPA from the drawing's feature presented in the front view into the area of the auxiliary view. Use **Osnap, Intersection** to ensure accuracy.

4. Type **Dist** in response to a command prompt to determine the distance from the reference plane line RPT to the object's features.

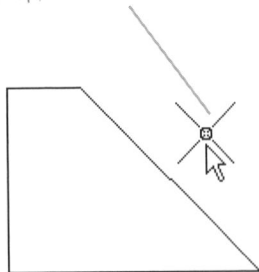

Figure 3-6

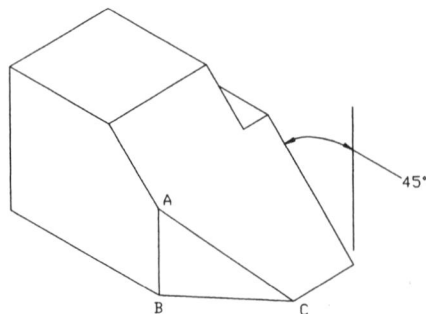

Figure 3-7

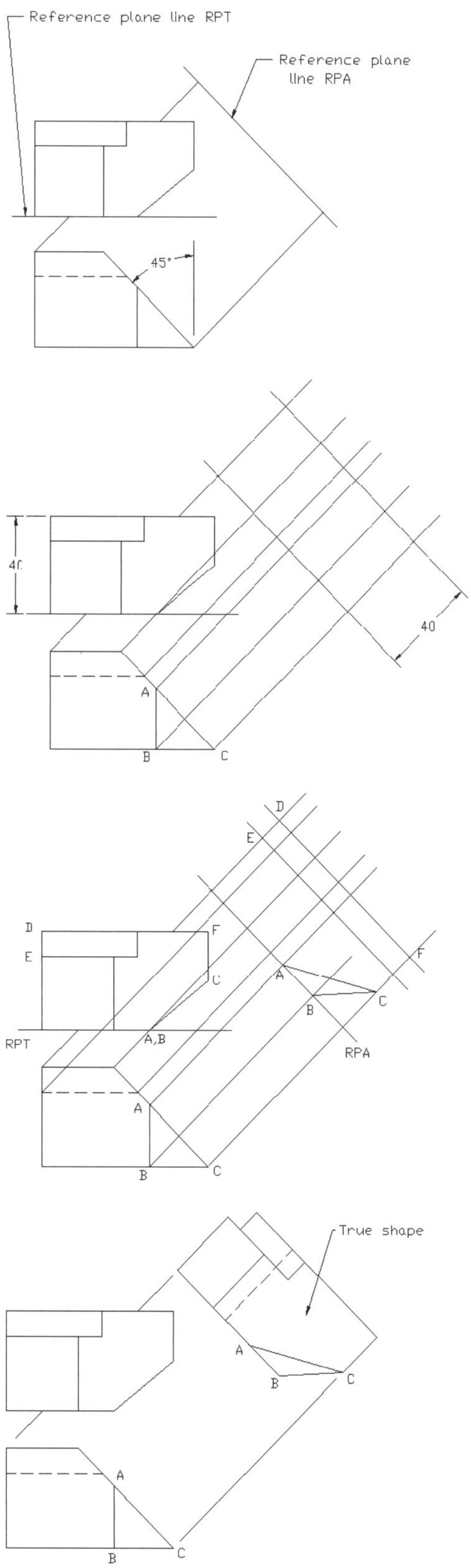

**Figure 3-8**

5. Use **Offset** to draw a line parallel to RPA at a distance equal to the distances of the object's features.

6. Erase and trim any excess lines and save the drawing, if desired.

## 3-4 TRANSFERRING LINES BETWEEN VIEWS

Objects are often dimensioned so that only some of their edge lines are dimensioned. This means that **Offset** may not be used to transfer distances to auxiliary views until the line length is known. There are three possible methods that can be used to transfer an edge line of unknown length: measure the line length using **Dist** and then use the determined distance with the **Offset** command; use **Grips,** and **Rotate** and **Move** the line; or use **Copy, Modify, Rotate** and relocate the line. The procedures are as follows.

**To measure the length of a line**

1. Type **Dist** in response to a command prompt.

   The command prompts are as follows:

   *Command: Dist*
   *Specify first point:*

2. Select one end of the line. The **Osnap, Endpoint** option will help identify the endpoint of the line. The intersection of one of the projection lines with a reference plane is usually used.

   *Specify second point:*

3. Select the other end of the line.

   The following style display will appear in the screen's prompt area.

   *Distance=40, Angle in X-Y Plane=45, Angle from the X-Y Plane=0*
   *Delta X=28.2885, Delta Y=28.2885, Delta Z=0.0000*

   Figure 3-9 shows the meaning of the displayed distance information.

   Record the distance and use it with the **Offset** command to locate the line's endpoints relative to a reference plane. See Figure 3-9.

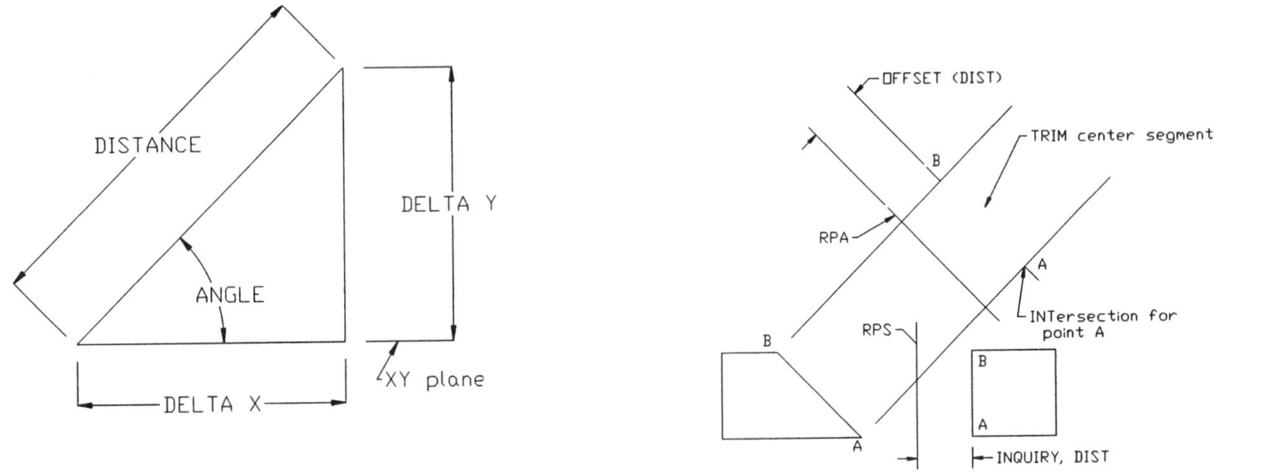

Figure 3-9

**To Grip and Move a line (See Figure 3-10.)**

1. Copy the line and move the copy to an open area of the drawing.
2. Click the line to highlight its grip points.
3. Select the line's lower endpoint; press the right mouse button.
4. Select the **Rotate** option from the menu.

    *\*\*ROTATE\*\**
    *Specify rotation angle or [Base point/Copy/ Undo/Reference/eXit]:*

5. Select the endpoint as the base point.
6. Type **45**; press **Enter.**

    The **Dist** command can also be used to determine the angle of a slanted surface if it is not known.

7. Select the line's lower endpoint again.
8. Select **Move** from the menu.
9. Select the lower line point as the base point.
10. Move the line to a new location or use dynamic input to define a new location.

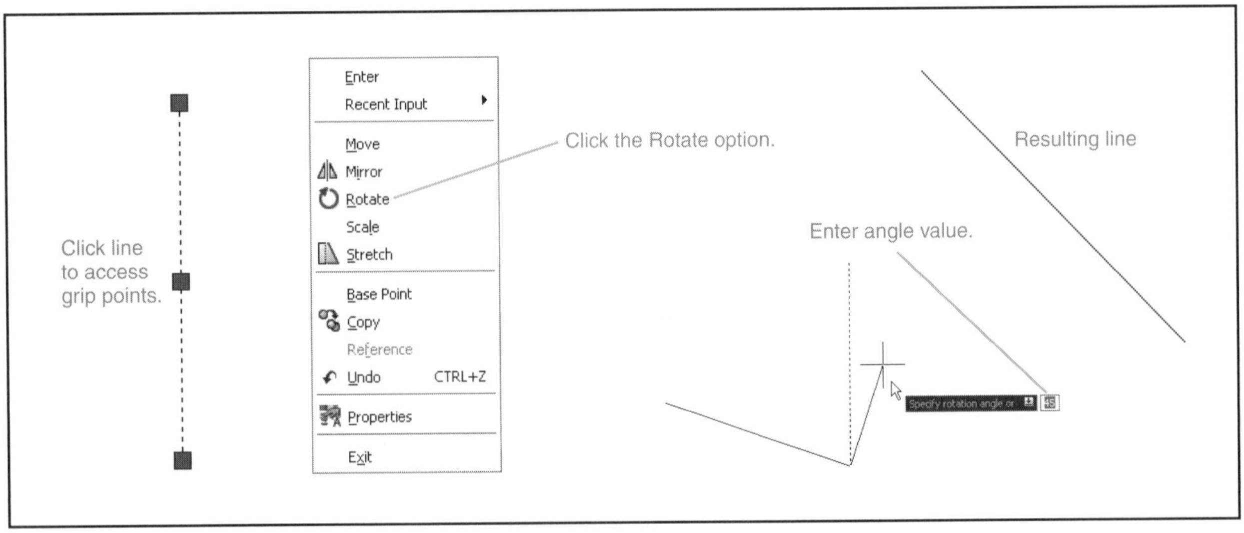

Figure 3-10

## To rotate and move a line

1. Copy the line and move the line to an open area of the drawing.
2. Select **Modify** (pull-down menu), **Rotate** and rotate the line **45°**.
3. Select **Modify** (pull-down menu), **Move** and relocate the line to the auxiliary view. Use **Osnap** if necessary to ensure accuracy.

## 3-5  SAMPLE PROBLEM SP3-2

Figure 3-11 shows an object that contains two slanted surfaces and two cutouts. Draw a front and right-side orthographic view, and an auxiliary view. See Figure 3-12.

1. Draw the front and side orthographic views as described in Chapter 1.
2. Define a reference plane, **RPS,** between the front and side views.
3. Use **Snap, Rotate (37.5)** to align the crosshairs with the slanted surface. The 37.5° value was determined using the **Dist** command.
4. Draw a reference plane line, **RPA,** parallel to the slanted surface and project the features of the object into the auxiliary view area.
5. Transfer the distance measurements from RPS and the side view into the auxiliary view.

   Label the various points as needed.

6. Erase and trim any excess lines and save the drawing, if desired.

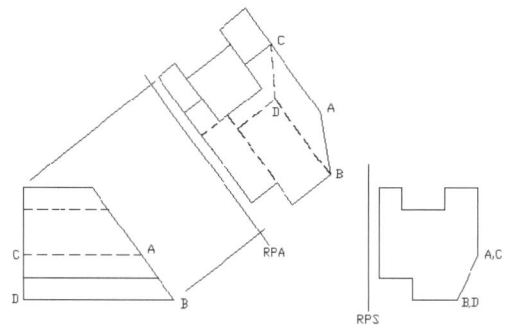

Figure 3-11

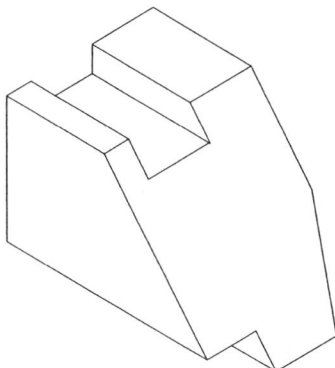

Figure 3-12

## 3-6  PROJECTING ROUNDED SURFACES

Rounded surfaces are projected into auxiliary views as they were projected into the normal orthographic views, as described in Section 1-15. Additional lines are added to one of the normal orthographic views and then projected into the other normal view. The additional lines define a series of points that define the outside shape of the surface. The two views with the added lines are used to project and transfer the points into the auxiliary view.

## 3-7  SAMPLE PROBLEM SP3-3

Figure 3-13 shows a cylindrical object that includes a slanted surface. Draw the front, side, and auxiliary views.
The procedure is as follows. See Figure 3-14.

1. Draw the front and side orthographic views.
2. Draw a reference plane line, **RPS.**

In this example the reference plane line was located on the side view's vertical centerline.

3. Draw a reference plane line, **RPA,** parallel to the slanted surface. The reference plane line will also be used as the centerline for the auxiliary view.

In this example the endpoints of both the major and minor axes of the projected elliptical auxiliary view are

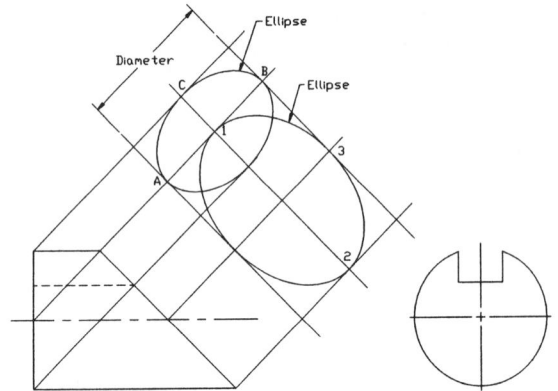

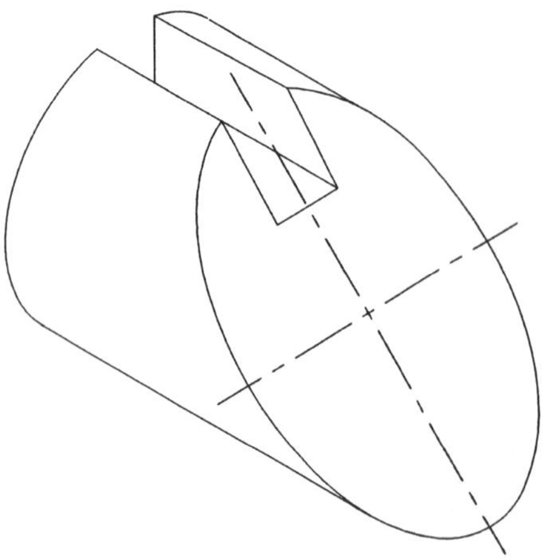

Figure 3-13

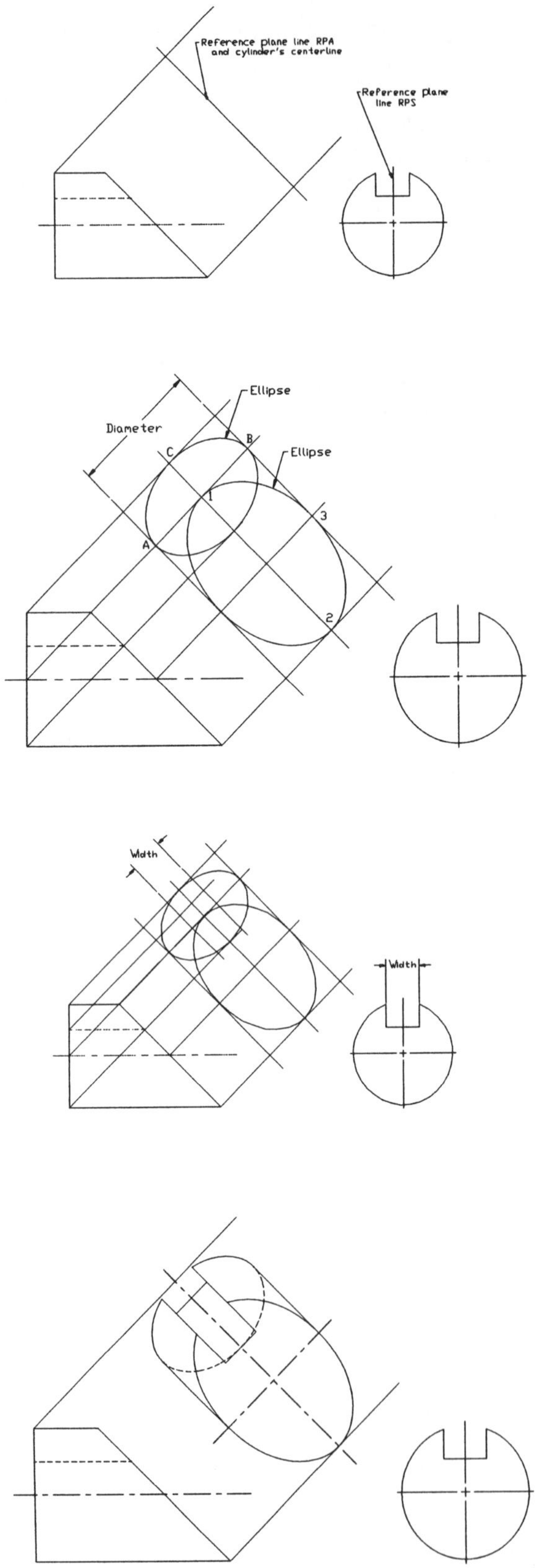

Figure 3-14

known, as well as the angle of the auxiliary view. It is therefore not necessary to define points in the circular side view, as was done in Section 1-21. Enough information is present to draw the elliptical shape directly in the auxiliary view.

4. Draw the elliptical surfaces in the auxiliary view as shown.
5. Transfer the width and location of the slot from the side view to the auxiliary view.
6. Erase and trim any excess lines and save the drawing, if desired.

## 3-8 PROJECTING IRREGULAR SURFACES

Auxiliary views of irregular surfaces are created by projecting information from given normal orthographic views into the auxiliary views in a manner similar to the way information was projected between orthographic views. See Section 1-28. The irregular surface is defined by a series of points along its edge line. The location of the points is random, although more points should be used when the curve's shape is changing sharply than when the curve tends to be smoother.

## 3-9 SAMPLE PROBLEM SP3-4

Figure 3-15 shows an object that includes an irregular surface. Draw front and side orthographic and auxiliary views of the object. The procedure is described below. See Figure 3-16.

1. Draw the front and side views as defined in Chapter 1.

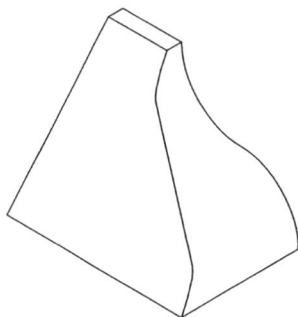

Figure 3-15

2. Draw two reference lines: RPF between the front and side views, and RPA parallel to the slanted surface in the side view.

Use **Snap, Rotate** to align the crosshairs with the slanted surface.

3. Define points along the irregular surface edge line in the front view and project the points into the side view using horizontal lines.
4. Project the points into the auxiliary view using lines perpendicular to line RPA. Label the points and their projection lines.
5. Transfer the depth measurements from RPS and the object's features and the points defining the irregular curve to the auxiliary view.

Type **Dist** in response to a command prompt to determine the distance from RPF to the points. Record the distances.

A = 3.00
B = 2.91
C = 2.62
D = 1.50
E = 0.30
F = 0.09
G = 0

Use **Offset** to draw lines parallel to RPA.

6. Use **Polyline, Edit Polyline, Fit** as explained in Section 1-26 to draw the irregular curve required in the auxiliary view. Use **Edit Vertex** to smooth the curve, if necessary.
7. Project the points defined in the front view to the back surface (far right vertical line) in the side view, and project the points into the auxiliary view.
8. Use **Polyline, Edit Polyline, Fit** to draw the required irregular curve.
9. Erase and trim any excess lines and save the drawing, if desired.

## 3-10 SAMPLE PROBLEM SP3-5

Figure 3-17 shows a top view and an auxiliary view of an object. Redraw the given views and add the front and right-side views. The procedure is described below.

1. Draw a reference plane line, **RPA,** parallel to the auxiliary view and a reference plane line, **RPT,** a horizontal line.

The object contains a dihedral angle, and the auxiliary view was aligned with the vertex line of the angle. Reference

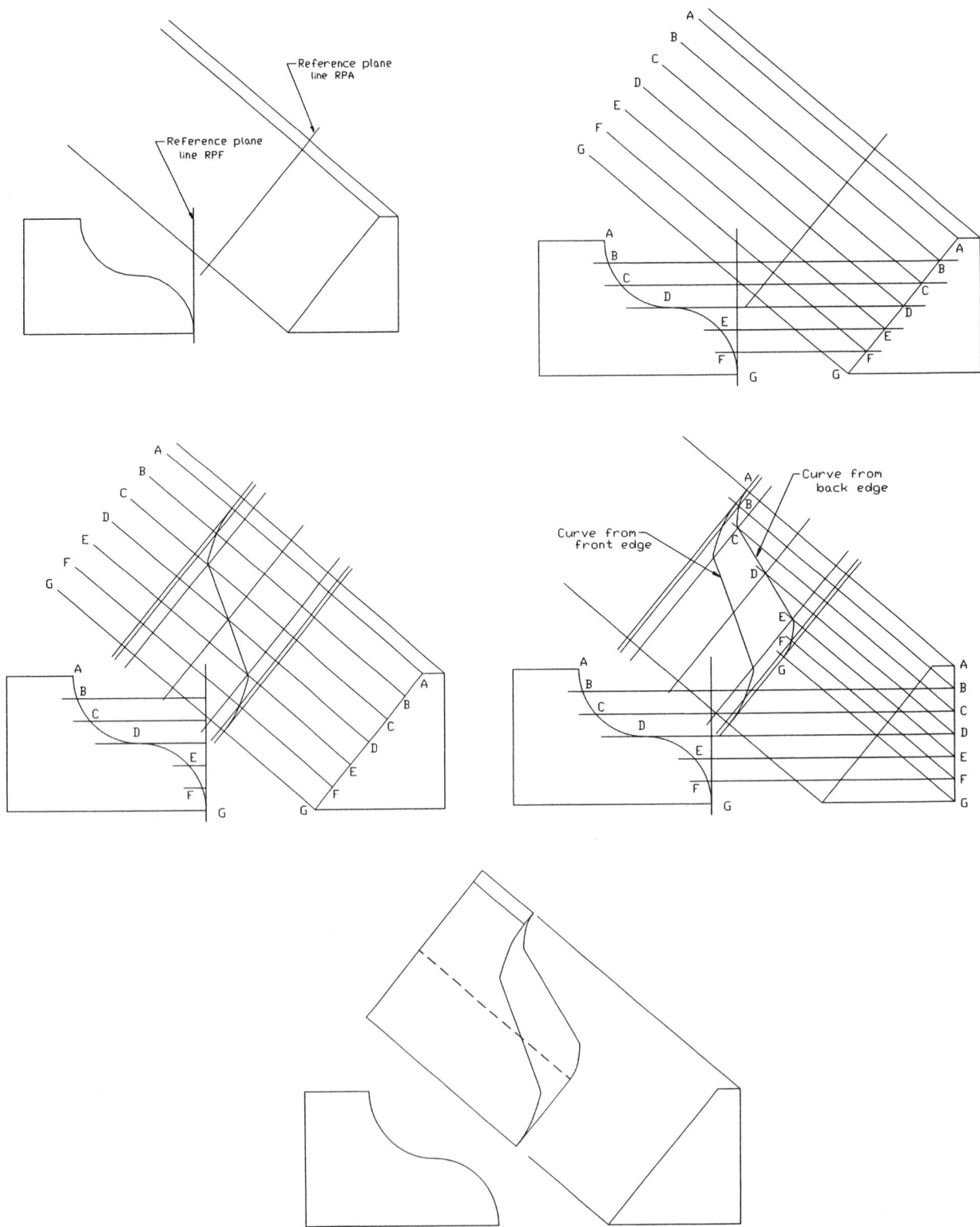

Figure 3-16

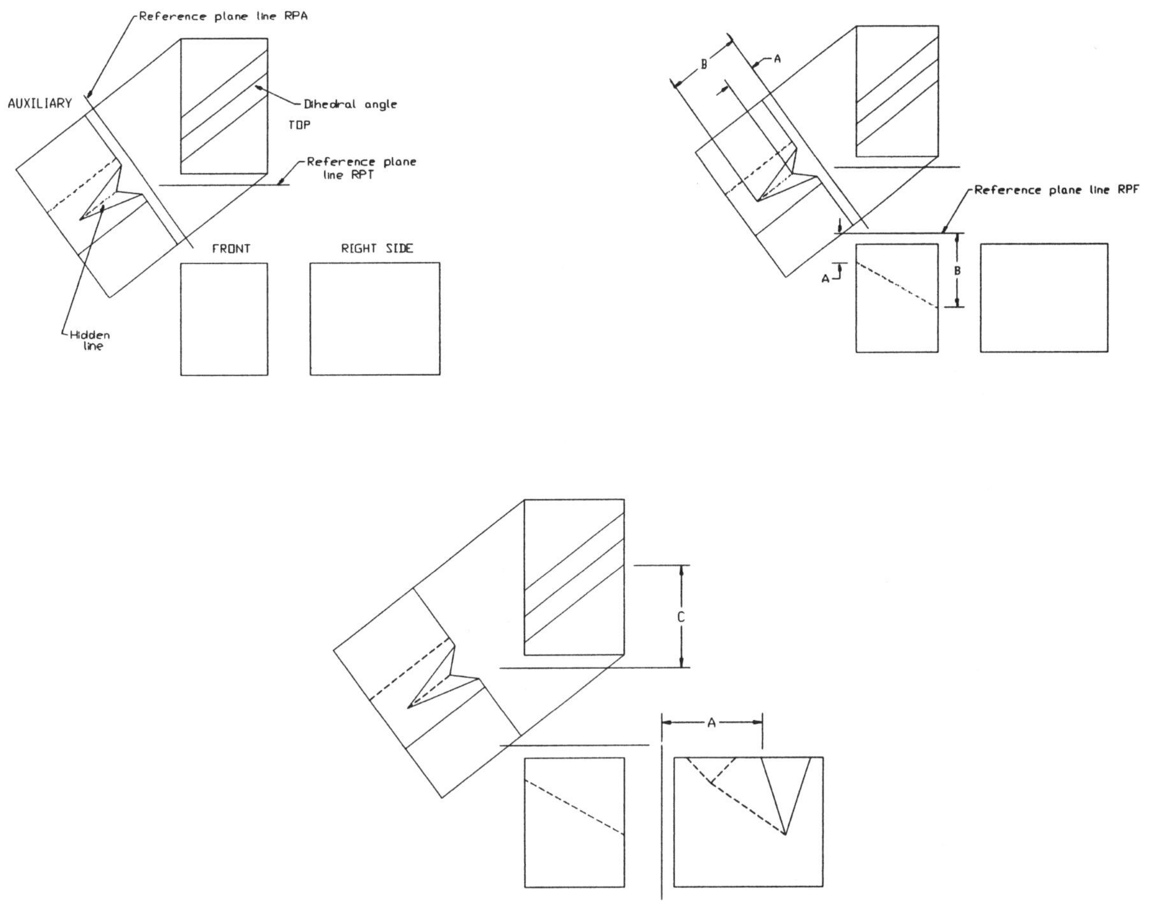

**Figure 3-17**

plane RPA is perpendicular to the vertex of the dihedral angle's vertex.

2. Project information from the top view and transfer information from the auxiliary view into the front view.
3. Project information from the front view and transfer information from the top view into the side view.
4. Erase and trim any excess lines and save the drawing, if desired.

## 3-11  PARTIAL AUXILIARY VIEWS

Auxiliary views help present true-shaped views of slanted and oblique surfaces, but in doing so generate foreshortened views of other surfaces. It is often clearer to create an auxiliary view of just the slanted surface and omit the surfaces that would be foreshortened. Auxiliary views that show only one surface of an object are called ***partial auxiliary views.***

Figure 3-18 shows a front view and three partial views of the object: a partial top view and two partial auxiliary views. A broken line (see Section 2-14) may be used to show that the partial auxiliary view is part of a larger view that has been omitted. Likewise, hidden lines may or may not be included. Figure 3-19 shows a front, side, and two partial auxiliary views of an object. Both the hidden lines and broken lines were omitted. If you are unsure about the interpretation of a view with hidden lines omitted, add a note to the drawing next to the partial views: ALL HIDDEN LINES OMITTED FOR CLARITY.

## 3-12  SECTIONAL AUXILIARY VIEWS

Sectional views may also be drawn as auxiliary views. Figure 3-20 shows an auxiliary sectional view. The cutting plane line is positioned across the object. A reference plane line is drawn parallel to the cutting plane line, and information is projected into the auxiliary sectional view, using lines perpendicular to the cutting plane line.

Broken line

Partial
Auxiliary
Views

**Figure 3-18**

PARTIAL
AUXILIARY

PARTIAL
AUXILIARY

FRONT        RIGHT SIDE

**Figure 3-19**

Figure 3-21 shows front, partial auxiliary, and auxiliary sectional views of an object. In this example the partial sectional view was used to help clarify the shape of the object's feature that would appear vague in the normal top or side views.

## 3-13 AUXILIARY VIEWS OF OBLIQUE SURFACES

The true shape of an oblique surface cannot be determined by a single auxiliary view taken directly from the oblique surface. An auxiliary view shows the true shape of a

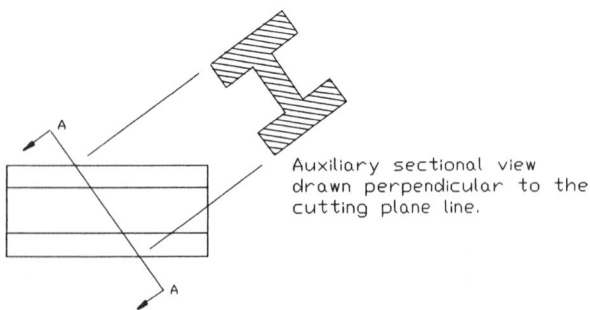

A

Auxiliary sectional view
drawn perpendicular to the
cutting plane line.

A

**Figure 3-20**

Auxiliary sectional
view

Front
orthographic
view

Partial auxiliary
view

**Figure 3-21**

**Figure 3-22**

**Figure 3-23**

surface only when it is taken at exactly 90° to the surface. The auxiliary views taken for Sample Problems 1 through 5 were taken off slanted surfaces.

*Slanted surfaces* are surfaces that are rotated about only one axis. See Figure 3-22. One of the orthographic views must be an edge view of the surface (the surface appears as a line) for an auxiliary view created by projecting lines perpendicular to that view to show the true shape of the surface.

*Oblique surfaces* are surfaces rotated about two axes. See Figures 3-22 and 3-23. This means that none of the normal orthographic views will show the oblique surface as an edge view; there is no given end view of the surface. This in turn means that an auxiliary view taken directly from one of the given views will not be perpendicular to the surface and therefore will not show the true shape of the surface. An auxiliary view that shows an edge view of the surface must first be created, then a second auxiliary taken perpendicular to one of the other normal orthographic views is needed to show the true shape of the surface.

## 3-14 SECONDARY AUXILIARY VIEWS

Consider the object shown in Figure 3-24. What is the true shape of surface A-B-C? An auxiliary view taken perpendicular to the surface will show its true shape, but what is the angle for a plane perpendicular to the surface?

Figure 3-25 shows the normal orthographic views of the object.

Start by choosing an edge line in the plane that is perpendicular to either the X or Y axis. Edge line A-B in the front view is a horizontal line, so it is parallel to the X axis. Take an auxiliary view aligned with the top view of the edge line (you're looking straight down the line). This auxiliary view will be perpendicular to the edge line and will generate an end view of the plane. A second auxiliary view can then be taken perpendicular to the end view that will show the true shape of the surface.

Figure 3-26 shows how a secondary auxiliary view is created for the object shown in Figure 3-24. The specific procedure is as follows.

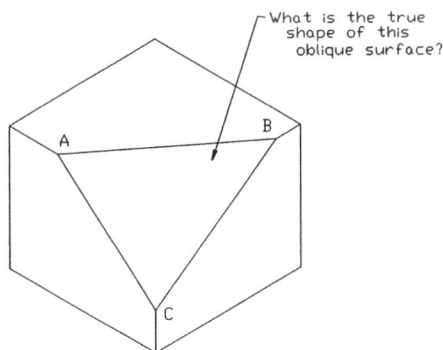

**Figure 3-24**

**Figure 3-25**

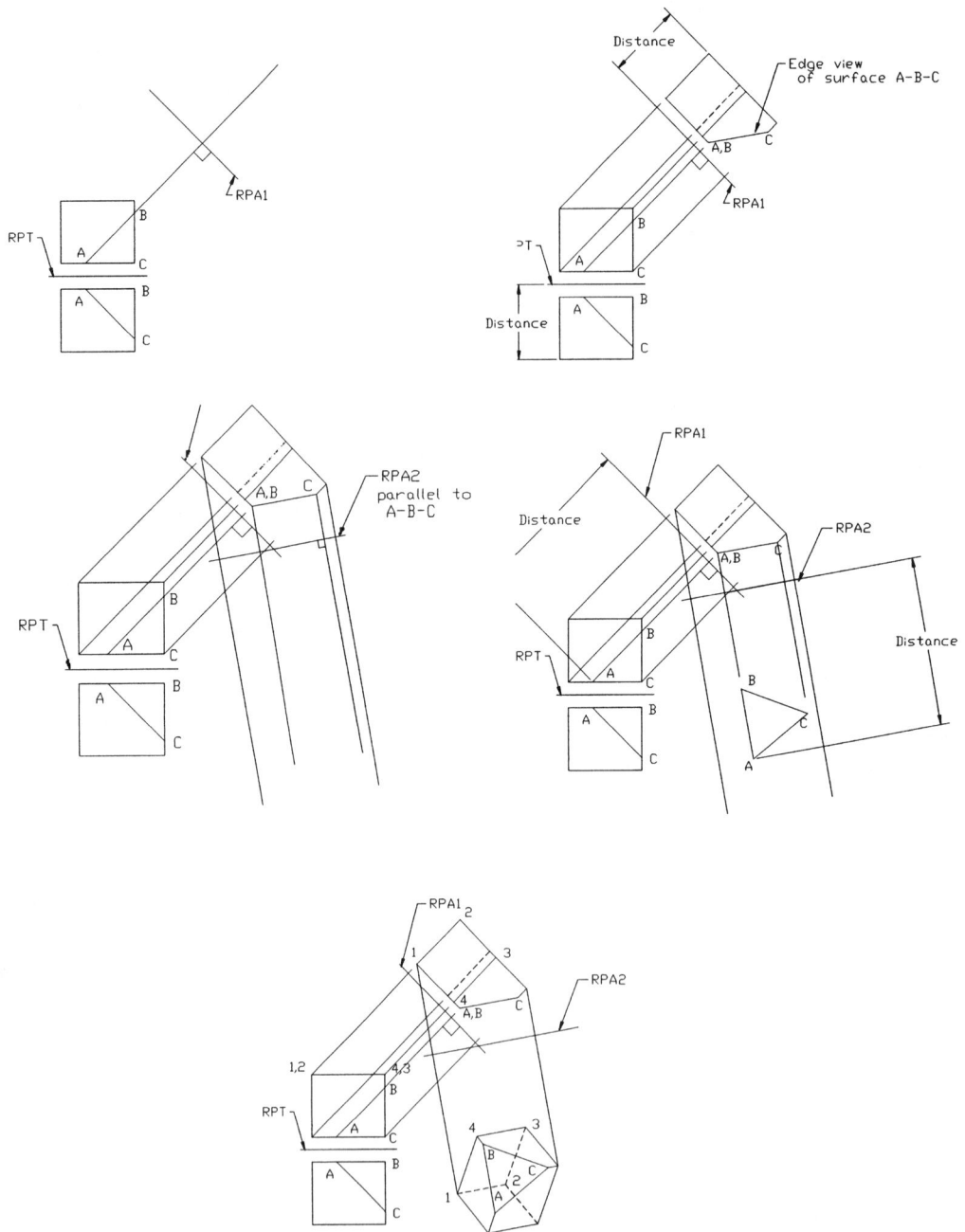

**Figure 3-26**

## To draw the first auxiliary view

1. Draw the normal orthographic views of the object.
2. Draw a reference plane, **RPT,** between the front and top views.
3. Extend a line from the top view of line A-B in the area for the first auxiliary view.

The **Extend** command can be used by drawing a construction line slightly beyond the expected area of the first auxiliary view, then using the construction line as an **Extend** boundary line, and extending line A-B to the boundary. The construction line can then be erased.

4. Draw a reference plane line, **RPA1,** perpendicular to the line A-B extension.

Use **Snap, Rotate** to align the crosshairs with the extension line. The **Dist** command can be used to determine the line's angle if it is not known.

5. Project the feature of the object into the auxiliary view using lines perpendicular to RPA1.
6. Transfer the distance measurements from RPT and the front view as shown.
7. Erase and trim any excess lines and create the first auxiliary view.

The surface A-B-C should appear as a straight line. This is an edge view of surface A-B-C.

## To draw the secondary auxiliary view

1. Use **Snap, Rotate** and align the crosshairs with the end view of surface A-B-C. **Dist** can be used to determine the angle of the edge view line.
2. Draw a reference plane line, **RPA2,** parallel to the edge view of the surface.
3. Project the features of the object into the second auxiliary view using lines perpendicular to RPA2.
4. Transfer the distance measurements from RPA1 and the top view as shown.
5. Erase and trim any excess lines and save the drawing, if desired.

The secondary auxiliary view shows the true shape of surface A-B-C.

Figure 3-27 shows another example of a secondary auxiliary view used to show the true shape of an oblique surface. In this example the first auxiliary view was taken from line F-A in the front view because it is a vertical line in the side view. The distance measurements came from the top view. The procedure used is the same as described above.

The auxiliary view shown in Figure 3-27 includes only the oblique surface. This is a partial auxiliary view. The purpose of the auxiliary view is to determine the true shape of the oblique surface. All the other surfaces would be foreshortened, not their true shape, if included in the auxiliary views.

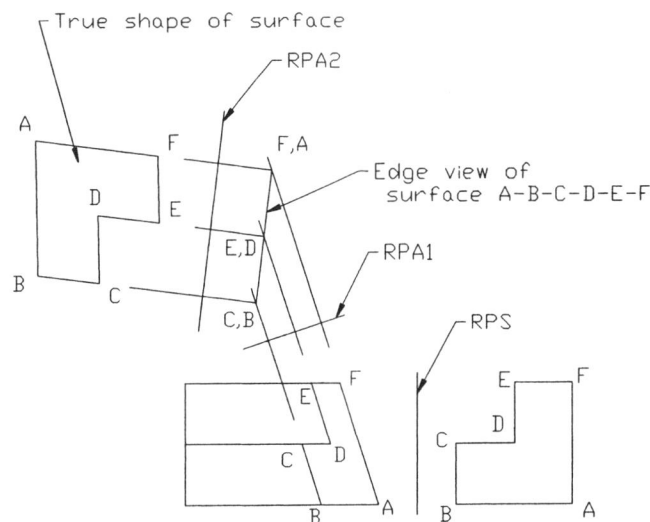

Figure 3-27

## 3-15  SAMPLE PROBLEM SP3-6

What is the true shape of the plane shown in Figure 3-28?

None of the edge lines are horizontal or vertical lines, so a line must be defined within the plane that is either horizontal or vertical. The added line can then be used to generate the two auxiliary views needed to define the true shape of the plane.

The procedure is as follows. See Figure 3-29.

1.  Draw a horizontal line from point A in the front view.

Use **Osnap, Intersection** with **Ortho** on and draw the line across the view. Use **Trim** to remove the excess portion of the line.

2.  Define the intersection of the new line with one of the plane's edge lines (D-C) as x.
3.  Project line A-x into the top view.

Option: Create a layer for projection lines.

The location of point A is already known in the top view. The location of point x in the top view can be found by drawing a vertical line from point x in the front view so that the line intersects line D-C in the top view. Use **Osnap, Intersection, Ortho** on to ensure accuracy.

4.  Extend line A-x into the area for the first auxiliary view.

Draw a construction line to use as a boundary line for the **Extend** command.

5.  Draw two reference plane lines: **RPT** between the front and top views, and **RPA1** perpendicular to the extension of line A-x.

Use **Snap, Rotate** to align the crosshairs with the extension of line A-x.

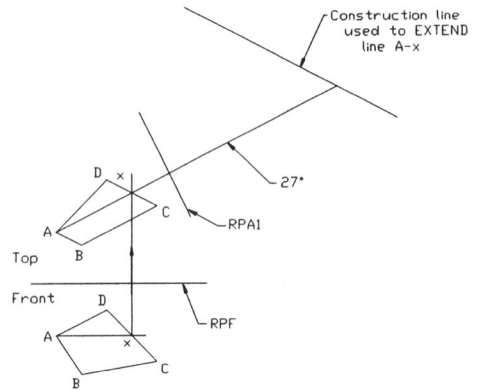

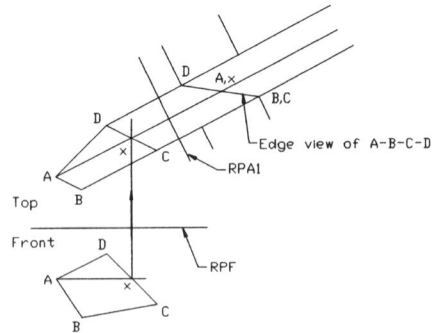

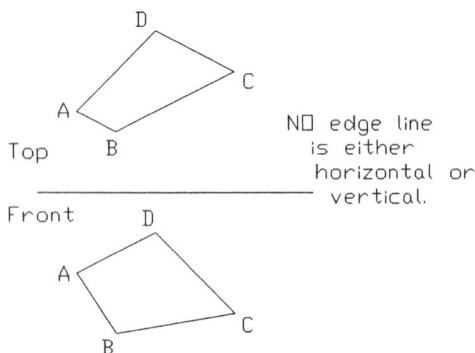

**Figure 3-28**

**Figure 3-29**

6. Project the plane's corner points into the area of the first auxiliary view from the top view. Transfer the distance measurements from RPT and the front view.

Use **Dist** to determine the distance from RPF and the corner points of the surface. Use **Offset** to transfer the point distance from RPF to RPA1. Use **Trim** to remove the internal portion of the offset lines. This will help clarify the drawing by removing excess lines from the auxiliary view but will retain intersection points needed to draw lines using **Osnap, Intersection.**

The plane should appear as a straight line, the end view of the plane.

7. Draw a third reference plane line, **RPA2,** parallel to the end view of the plane and project the plane's corner points into the secondary auxiliary view area.
8. Transfer the depth distances from RPA1 and the top view to RPA2 and the secondary auxiliary view.
9. Erase and trim any excess lines and save the drawing, if desired.

The secondary auxiliary view shows the true shape of surface A-B-C-D.

## 3-16 SECONDARY AUXILIARY VIEW OF AN ELLIPSE

Figure 3-30 shows the front and side views of an oblique surface and includes a foreshortened view of a hole, that is, an ellipse. Also included are two auxiliary views, the second of which shows the hole as a circle.

The ellipse is projected by first determining the correct projection angle that will produce an edge view of surface A-B-C-D. In this example, line B-C appears as a vertical line in the side view, so its front view can be used to project the required edge view.

It is known that the secondary view of the surface will show the hole as a circle, so only a radius value need be carried between the views. In this example, point 1 in the front view was projected into the first auxiliary view and then into the second auxiliary view, thereby defining the radius of the circle.

The hole is located at the center of the surface, which means that its centerlines are located on the midpoints of the edge lines in the secondary auxiliary views. If the hole's center point was not in the center of the surface, the intersections of the hole's centerlines with the surface's edge lines would also have to be projected into the secondary auxiliary view to accurately locate the hole.

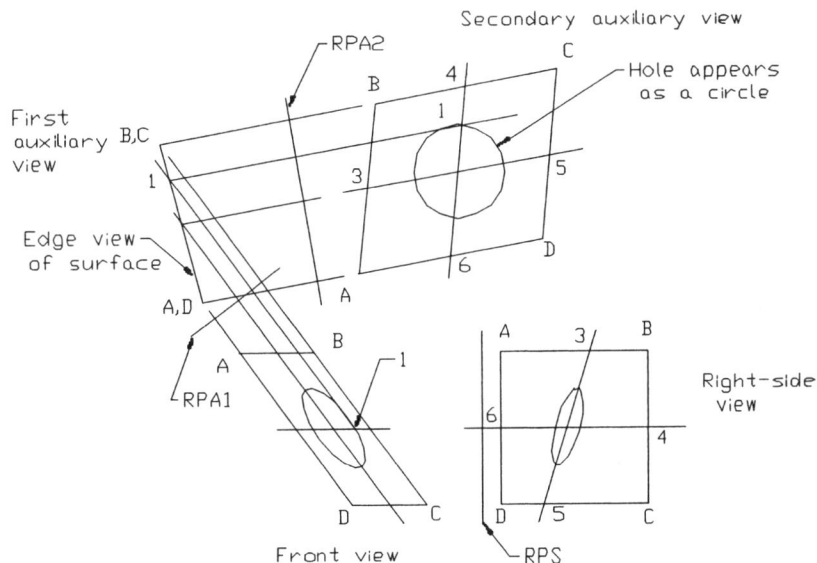

Figure 3-30

## 3-17 EXERCISE PROBLEMS

Given the outlines for front, side, and auxiliary views as shown, substitute one of the side views shown in Exercise Problems EX3-1A through EX3-1D and complete the front and auxiliary views. All dimensions are in millimeters.

**EX3-1A**

**EX3-1B**

NOTE: This surface is 10 from and parallel to the top surface.

**EX3-1C**

**EX3-1D**

Given the outlines for front, side, and auxiliary views as shown, substitute one of the side views shown in Exercise Problems EX3-2A through EX3-2D and complete the front and auxiliary views. All dimensions are in millimeters.

Draw two orthographic views and an auxiliary view for each of the objects shown.

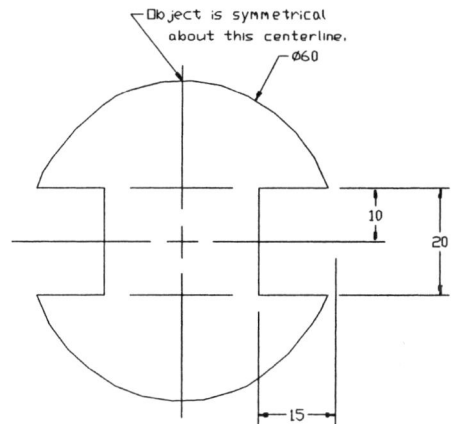

**EX3-2A**

**EX3-2B**

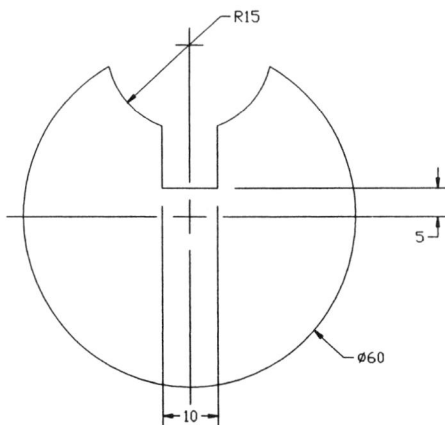

**EX3-2C**

**EX3-2D**

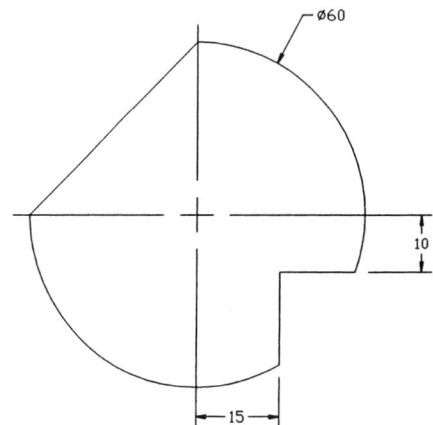

Draw two orthographic views and an auxiliary view for each of the objects in Exercise Problems EX3-3 through EX3-6.

**EX3-3 MILLIMETERS**

**EX3-5 INCHES**

**EX3-4 MILLIMETERS**

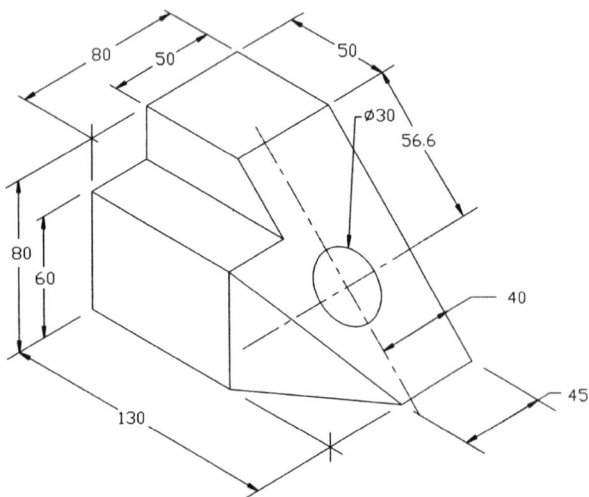

**EX3-6 MILLIMETERS**

Redraw the given orthographic views in Exercise Problems EX3-7 through EX3-10 and add the appropriate auxiliary view.

## EX3-7 MILLIMETERS

## EX3-9 INCHES

## EX3-8 MILLIMETERS

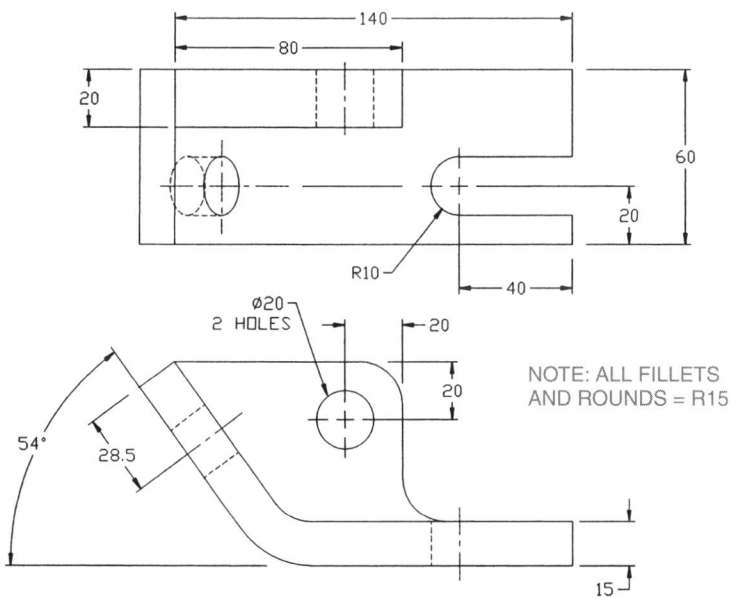

NOTE: ALL FILLETS
AND ROUNDS = R15

## EX3-10 MILLIMETERS

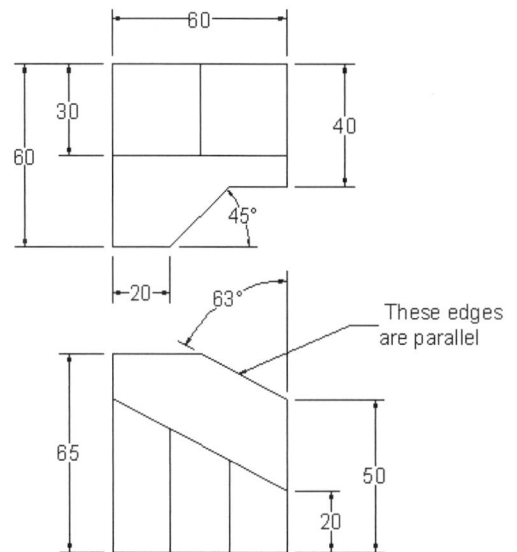

Draw two orthographic views and an auxiliary view for each of the objects in Exercise Problems EX3-11 through EX3-40.

## EX3-11 MILLIMETERS

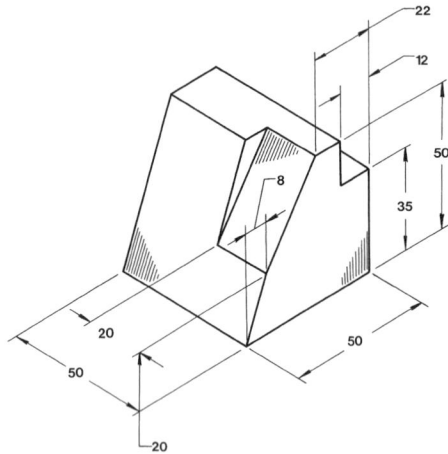

## EX3-14 MILLIMETERS

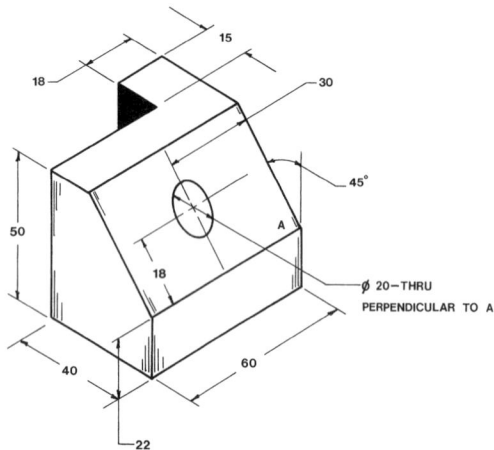

## EX3-12 MILLIMETERS

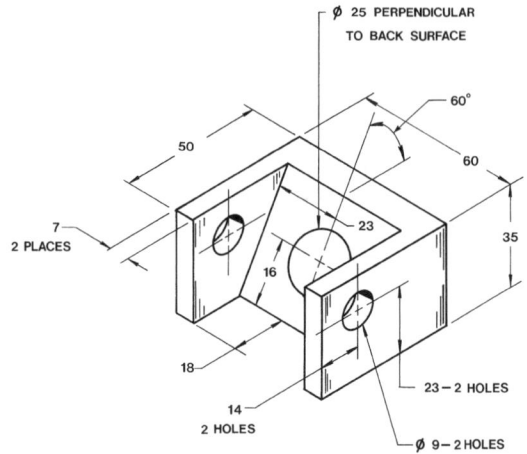

## EX3-15 INCHES

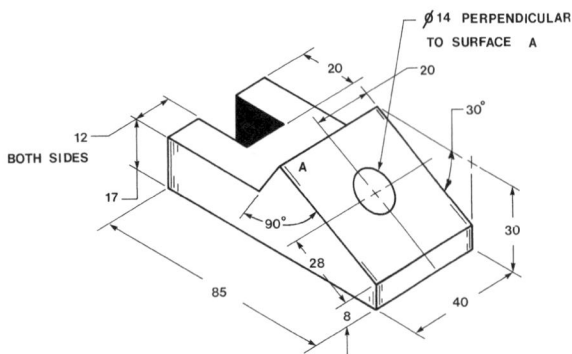

## EX3-13 MILLIMETERS

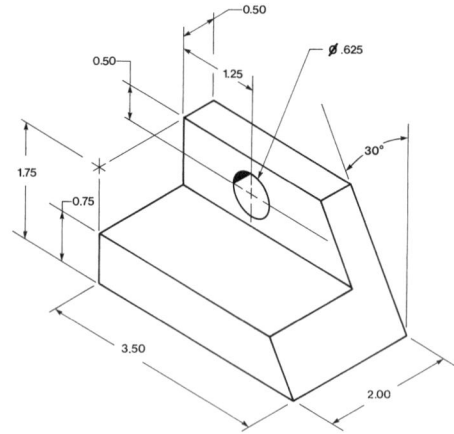

## EX3-16 INCHES

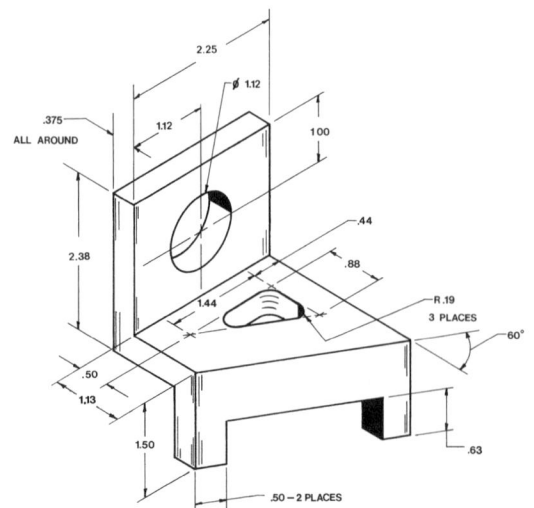

## EX3-17 MILLIMETERS

45°

17

52

80

CENTERLINE OF
CIRCLE IS A
VERTICAL LINE

Ø24

HEXAGON 60 ACROSS
THE FLATS

## EX3-20 INCHES

2.69

1.00

1.00

2 LARGE HOLES—.75 DIA
1 SMALL HOLE—.50 DIA

.75

1.94

.88

.38

.75

1.25    2.75

1.50

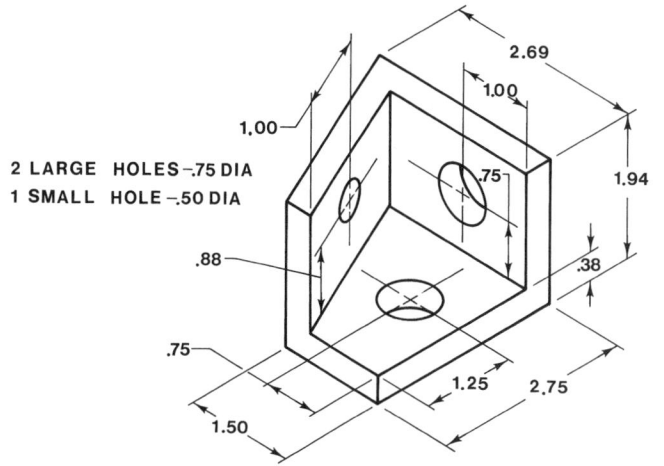

## EX3-18 MILLIMETERS

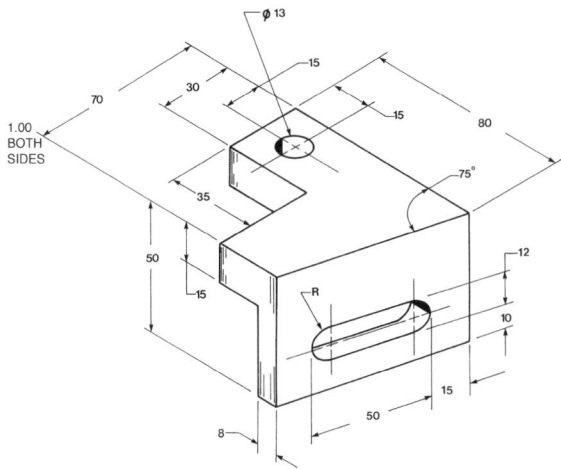

Ø13

15

70

30

15

1.00
BOTH
SIDES

35

80

75°

50

12

15

R

10

8

50

15

## EX3-21 MILLIMETERS

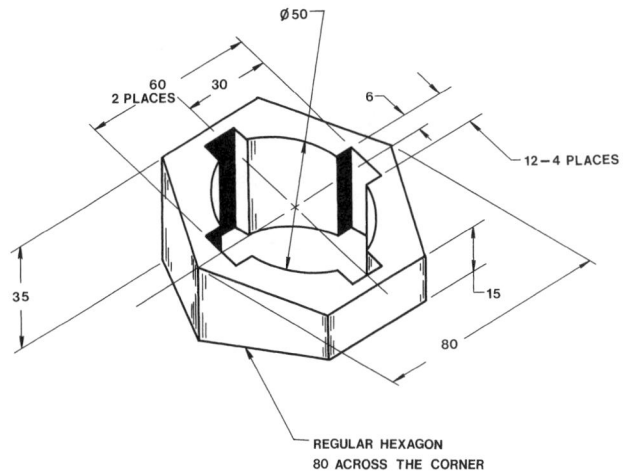

Ø50

60
2 PLACES

30

6

12—4 PLACES

35

15

80

REGULAR HEXAGON
80 ACROSS THE CORNER

## EX3-19 MILLIMETERS

8 – 2 PLACES

60

28

25

15

16—2 PLACES

60

8—2 PLACES

Ø12—4 PLACES

Ø8—2 PLACES

R5—2 PLACES

32

20

48

16 – 2 PLACES

10

60

28

60 REF

Ø6—2 PLACES

MATL 5 THK

## EX3-22 MILLIMETERS

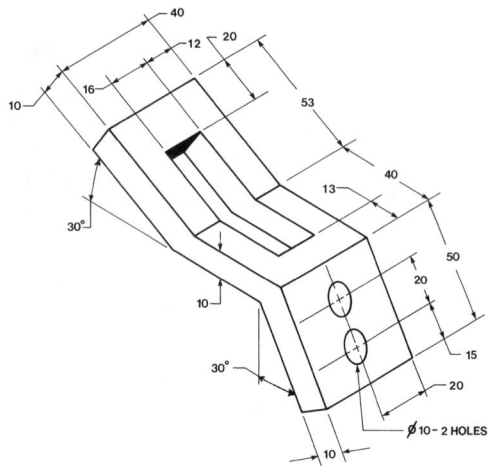

## EX3-25 INCHES

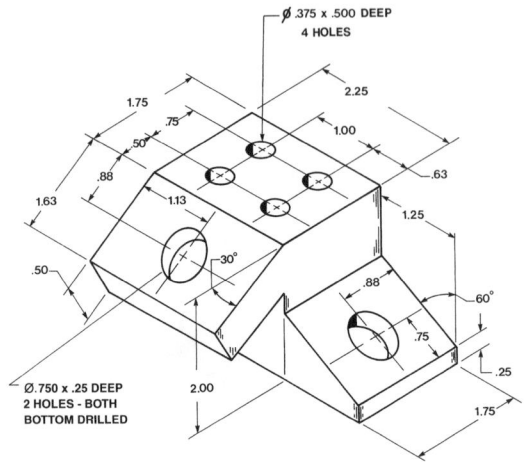

## EX3-23 INCHES

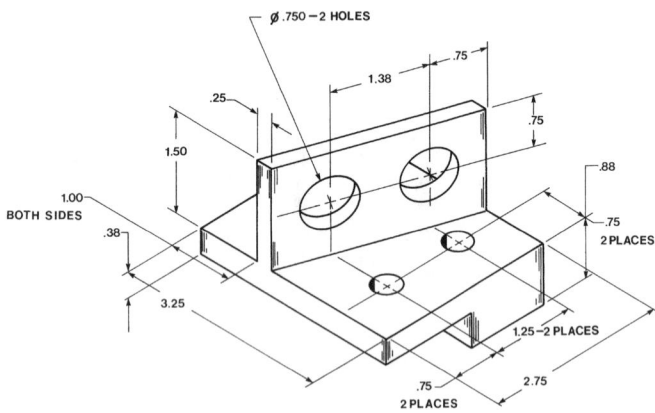

## EX3-26 MILLIMETERS

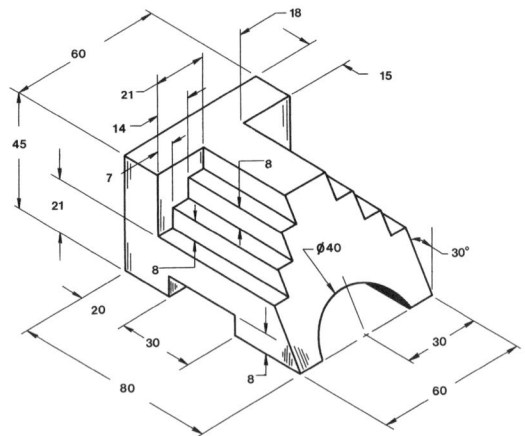

## EX3-24 MILLIMETERS

ALL FILLETS AND ROUNDS = R3

## EX3-27 MILLIMETERS

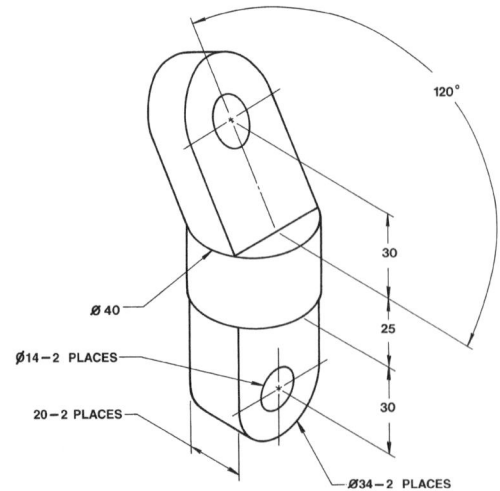

**EX3-28 MILLIMETERS**

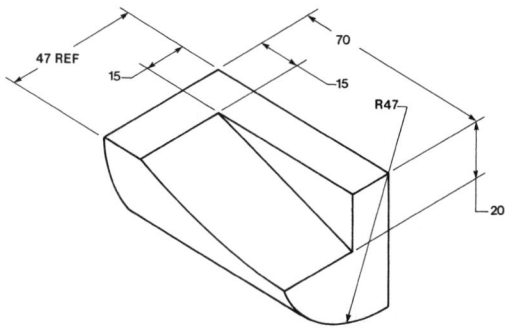

47 REF
15
70
15
R47
20

**EX3-31 INCHES**

-30°
2.81
∅.50
∅1.56

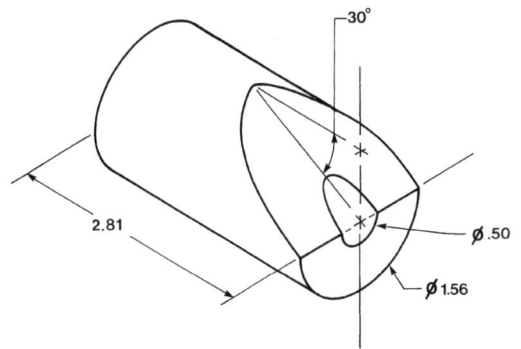

**EX3-29 MILLIMETERS**

-30°
80
∅50
∅68

**EX3-32 MILLIMETERS**

∅7 x 20 DEEP
9
12
-30°
60
16
∅58
8
16

**EX3-30 MILLIMETERS**

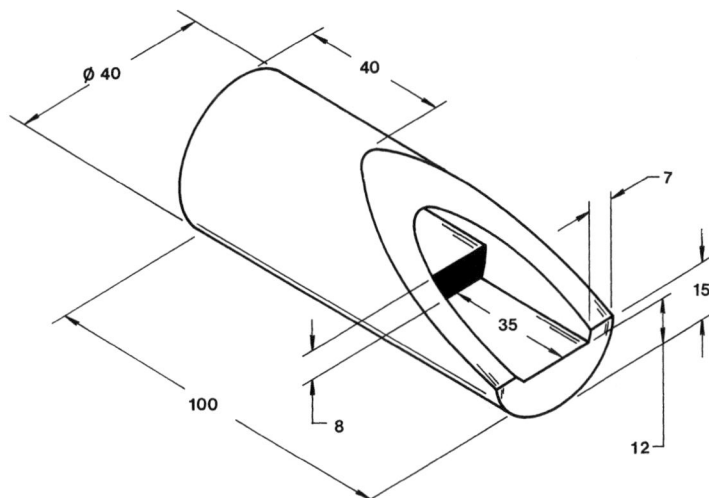

∅ 40
40
7
35
15
100
8
12

## EX3-33 MILLIMETERS

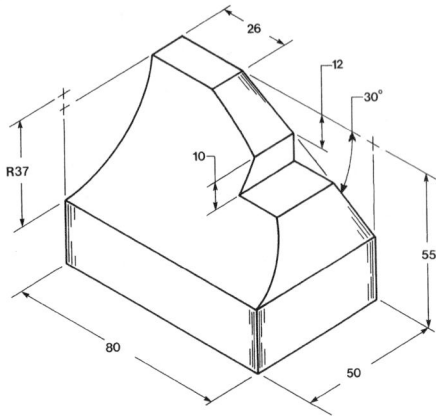

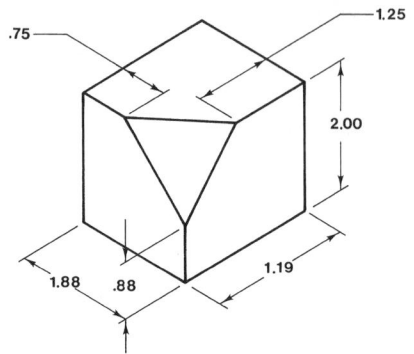

## EX3-34 MILLIMETERS

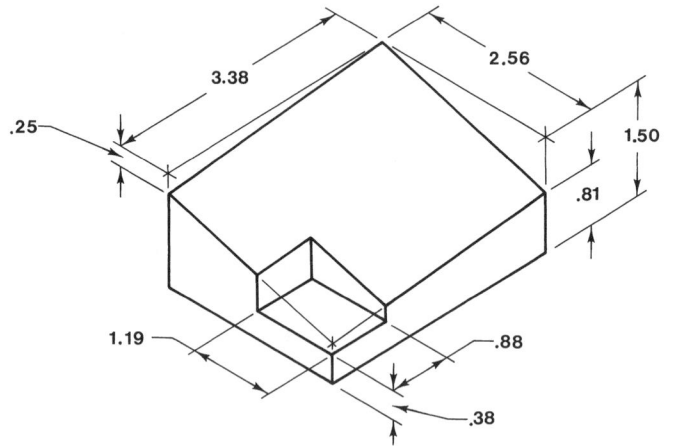

## EX3-35 MILLIMETERS

## EX3-36 INCHES

## EX3-37 INCHES

## EX3-38 INCHES

## EX3-39 MILLIMETERS

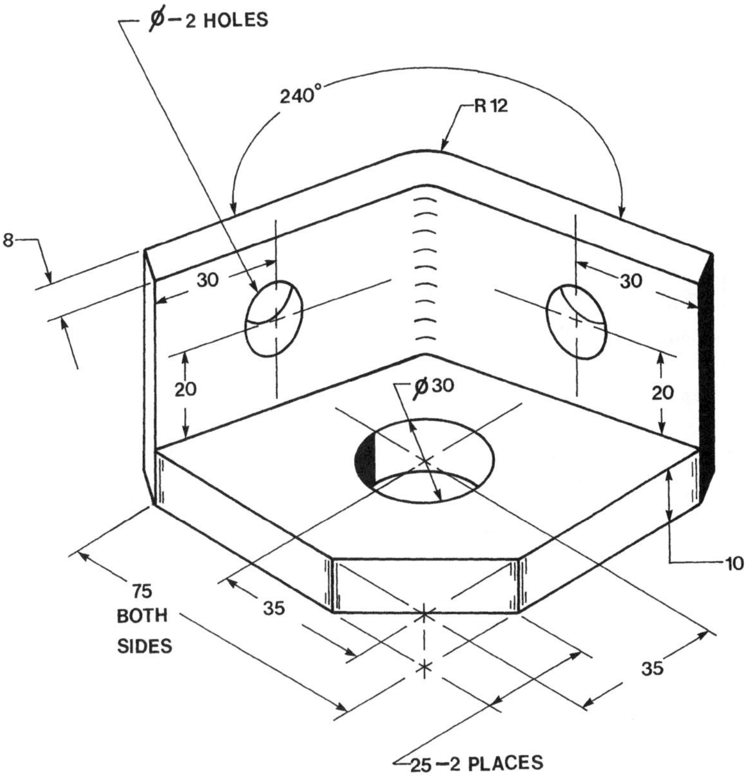

∅−2 HOLES

240°

R 12

8

30

30

20

∅ 30

20

75
BOTH
SIDES

35

35

10

25−2 PLACES

## EX3-40 MILLIMETERS

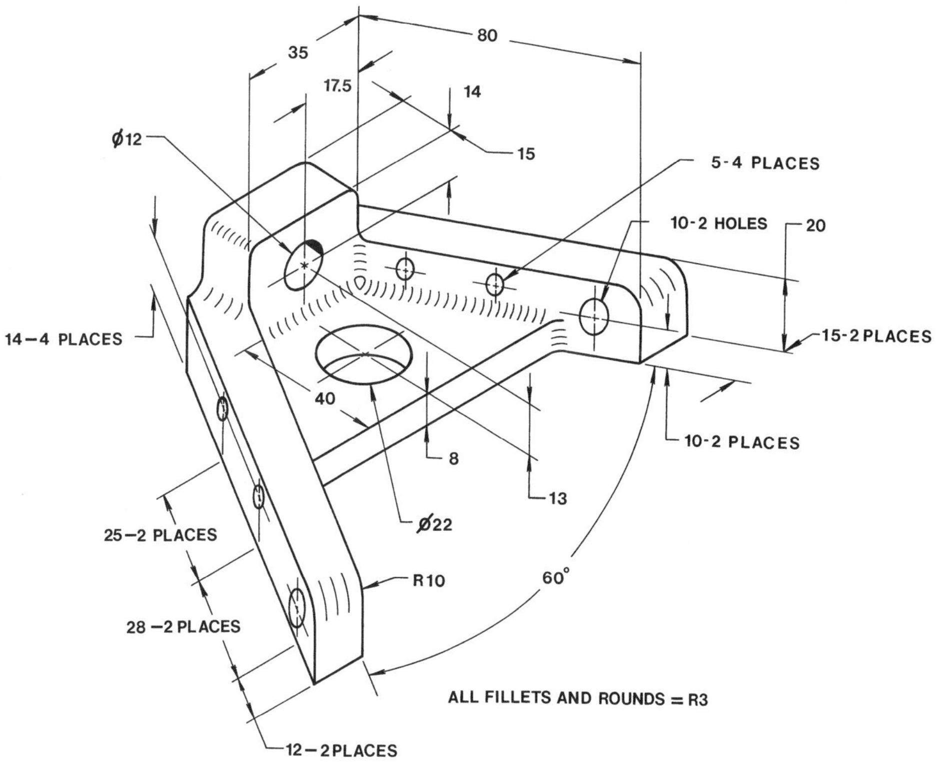

35

80

17.5

14

15

∅12

5-4 PLACES

10-2 HOLES    20

14−4 PLACES

15-2 PLACES

40

10-2 PLACES

8

25−2 PLACES

∅22

13

28−2 PLACES

R 10

60°

12−2 PLACES

ALL FILLETS AND ROUNDS = R3

Draw at least two orthographic views and one auxiliary view for each of the following objects.

**EX3-41 INCHES**

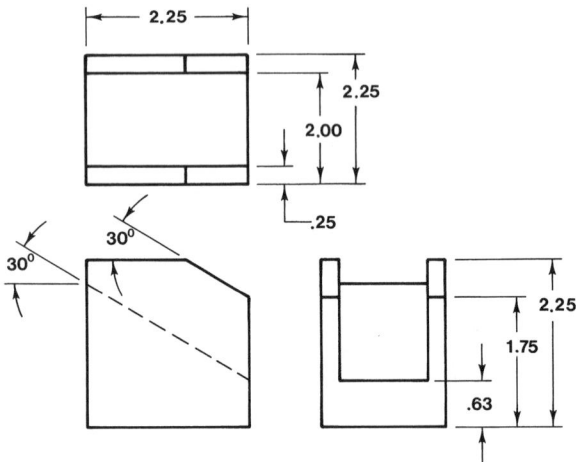

**EX3-43 INCHES**

**EX3-42 INCHES**

**EX3-44 INCHES**

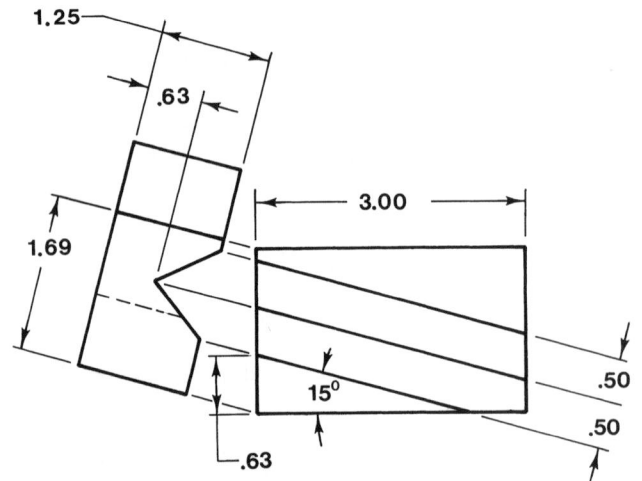

Use a secondary auxiliary view to find the true shape of the planes in Exercise Problems EX3-45 through EX3-50.

**EX3-45 MILLIMETERS**

C = −15,20,20

A = 70,30,50

B = 50,20,20

**EX3-46 MILLIMETERS**

A = 0,−5,60

C = −20,−10,10

B = 50,−15,−20

**EX3-47 MILLIMETERS**

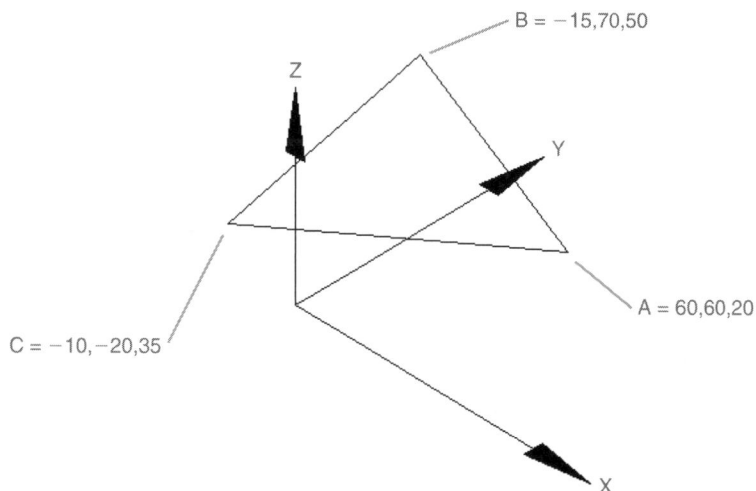

B = −15,70,50

A = 60,60,20

C = −10,−20,35

## EX3-48 INCHES (SCALE: 4:1)

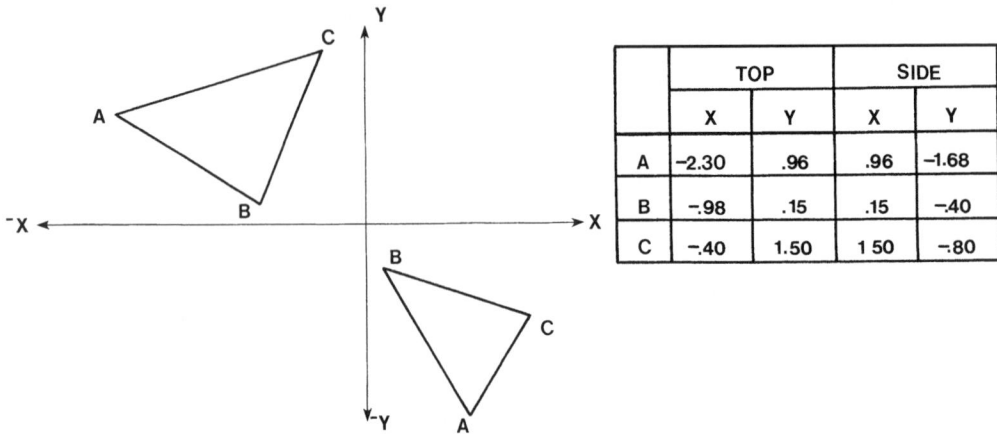

|   | TOP | | SIDE | |
|---|------|------|------|------|
|   | X | Y | X | Y |
| A | −2.30 | .96 | .96 | −1.68 |
| B | −.98 | .15 | .15 | −.40 |
| C | −.40 | 1.50 | 1 50 | −.80 |

## EX3-49 MILLIMETERS (SCALE: 2:1)

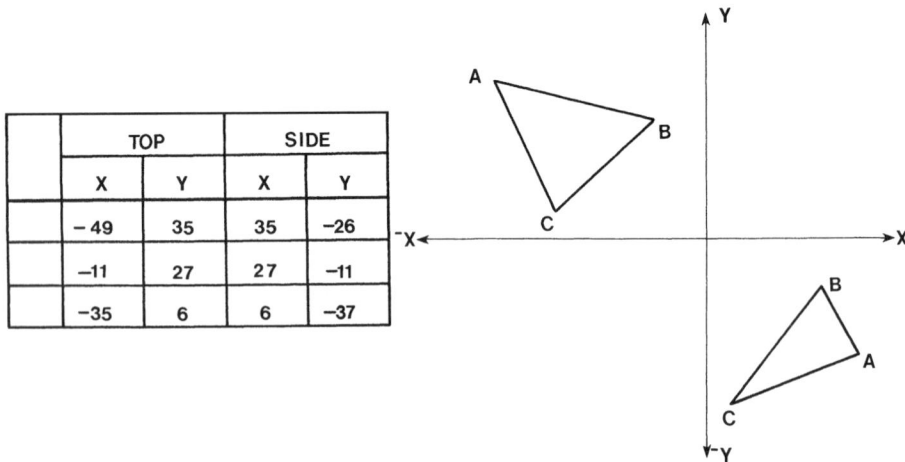

|   | TOP | | SIDE | |
|---|------|------|------|------|
|   | X | Y | X | Y |
|   | − 49 | 35 | 35 | −26 |
|   | −11 | 27 | 27 | −11 |
|   | −35 | 6 | 6 | −37 |

## EX3-50 INCHES (SCALE: 3:1)

|   | FRONT | | TOP | |
|---|------|------|------|------|
|   | X | Y | X | Y |
| A | −2.80 | −.60 | −2.80 | 1.05 |
| B | −1.17 | −.20 | −1.17 | .35 |
| C | −.61 | −1.62 | −.61 | 1.62 |

## EX3-51 MILLIMETERS

Redesign the following object to include two Ø10 holes in the slanted surface. The holes should be centered along the longitudinal axis, and spaced so that the distance between the holes' centers equals the distance from the holes' centers to the upper and lower edges of the slanted surface. The holes should be perpendicular to the slanted surface.

Draw the front, top, and side views of the object plus an auxiliary view of the slanted surface.

47 REF

15

70

15

R47

20

Add two evenly spaced and perpendicular Ø10 holes to the slanted surface.

## EX3-52 MILLIMETERS

Redesign the following object so that the outside shape is a regular heptagon (seven-sided polygon) 80 across the flats, and the inside Ø50 hole is intersected by six evenly spaced slots each 12 wide. Draw front and top orthographic views and an auxiliary view of the object.

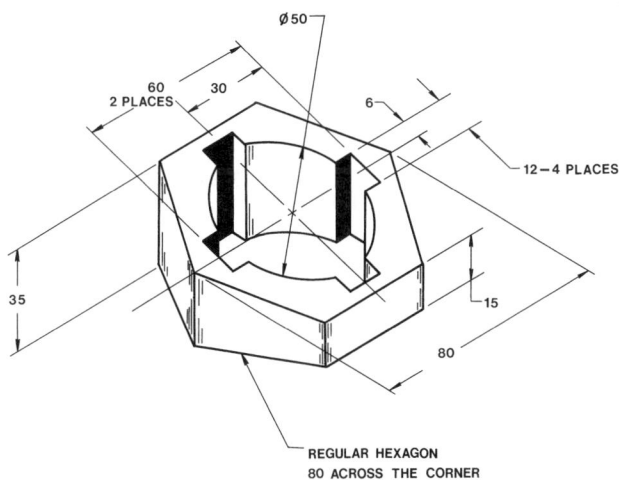

Ø50

60
2 PLACES

30

6

12–4 PLACES

35

15

80

REGULAR HEXAGON
80 ACROSS THE CORNER

## EX3-53 MILLIMETERS

Calculate the area of the following plane. Redesign the plane so that the new plane is congruent to the original but has an area equal to 1.5 times the original area. Draw and label the true shape of the new plane relative to the X, Y, Z axes.

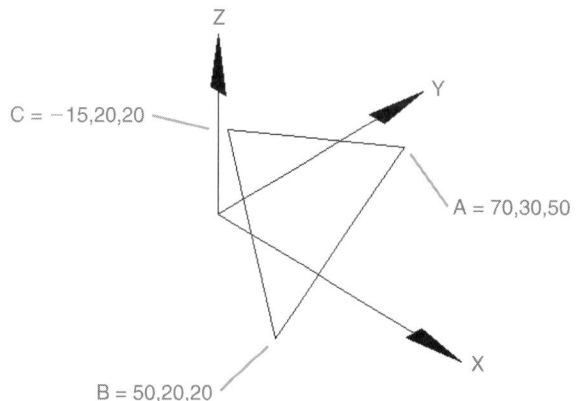

Z

Y

C = −15,20,20

A = 70,30,50

X

B = 50,20,20

# 3-18 ORIENTATION TOLERANCES—RELATED FEATURES USING DATUMS

Orientation tolerances used on related features and which require a datum involve the characteristics below:

(PERPENDICULARITY)    (ANGULARITY)    (PARALLELISM)

These characteristics are used to describe orientation tolerances of *single surface, element,* or *size* features and are always related to a *datum.*

Orientation tolerance when applied to a plane surface, is a refinement added to the design which would not be controlled by the size dimensions. Orientation tolerance when applied to a plane surface controls the "flatness" as well as its orientation.

Orientation tolerance when applied to a size feature (hole or pin) is a refinement of orientation (perpendicularity, etc.) within the stated locational tolerance; therefore, it must be *less than* the governing locational tolerance of the features involved. Orientation tolerance and the values determined are based upon design requirements for part relationships, fit, function, strength, appearance, etc., as deemed necessary by the designer.

See following pages for details of application.

# 3-19 PERPENDICULARITY ⊥

## (Squareness, Normality)

**Definition.** Perpendicularity is the condition of a surface, center plane, or axis at a right angle (90°) to a datum plane or axis.

## Perpendicularity Tolerance

A perpendicularity tolerance specifies one of the following:

1. a tolerance zone defined by two parallel planes perpendicular to a datum plane or axis within which
   a. the surface of a feature must lie (see Fig. 3-31);
   b. the center plane of a feature must lie (see Fig. 3-32);
2. a tolerance zone defined by two parallel planes perpendicular to a datum axis within which the axis of the considered feature must lie (see Fig. 3-33);
3. a cylindrical tolerance zone perpendicular to a datum plane within which the axis of the considered feature must lie (see Fig. 3-34);
4. a tolerance zone defined by two parallel lines perpendicular to a datum plane or axis within which a line element of the surface must lie (see Fig. 3-35—radial perpendicularity).

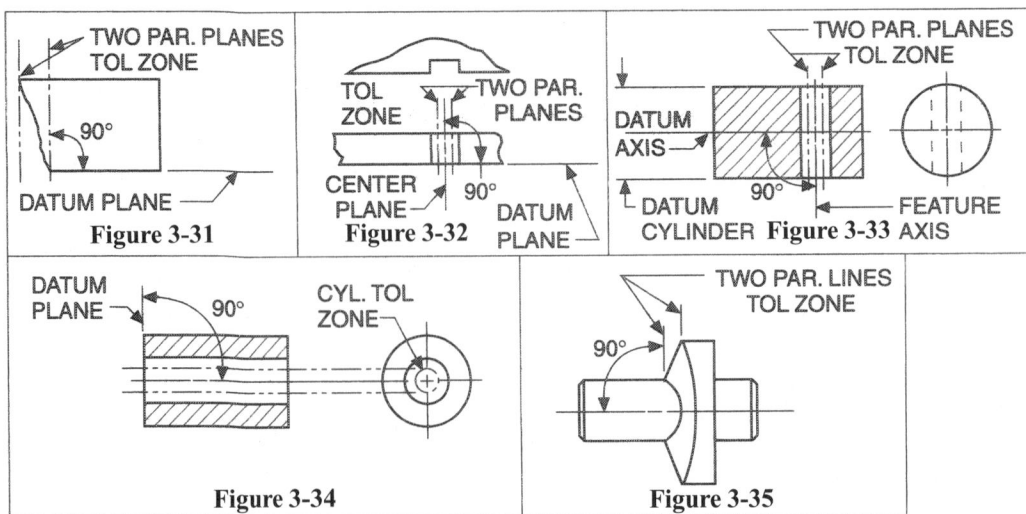

**Figure 3-31**

**Figure 3-32**

**Figure 3-33**

**Figure 3-34**

**Figure 3-35**

## Perpendicularity Application

The example below illustrates perpendicularity tolerance as applied to a surface.

**Example**

| ⊥ | .005 | A |

SYMBOL MEANING

| ⊥ | .005 | A | ← TO DATUM PLANE A

└─ WITHIN .005 WIDE TOL ZONE

└─ THIS SURFACE MUST BE PERPENDICULAR

| A |

**Meaning**

⊥ TOL ZONE

.005 WIDE TOL ZONE

DATUM PLANE

90°

DATUM PLANE

THE SURFACE MUST BE WITHIN THE SPECIFIED TOLERANCE OF SIZE AND MUST LIE BETWEEN TWO PARALLEL PLANES (.005 APART) WHICH ARE PERPENDICULAR TO THE DATUM PLANE.

Note that the perpendicularity tolerance applied to a plane surface controls flatness if a flatness tolerance is not specified (that is, the flatness will be at least as good as the perpendicularity).

The examples below and on the following pages show perpendicularity tolerancing under various conditions. Note also how MMC applications permit greater tolerance.

## Noncylindrical Feature at RFS, Datum a Plane

### Example

NONCYLINDRICAL FEATURE AT
RFS. DATUM A PLANE

SYMBOL MEANING

⊥  .005  A ◄— TO DATUM
                PLANE A

WITHIN .005 WIDE
TOL ZONE, RFS

THIS FEATURE MUST
BE PERPENDICULAR

### Meaning

POSSIBLE DIRECTION FOR THE
FEATURE CENTER PLANE

DATUM PLANE

.005 WIDE
TOL ZONE

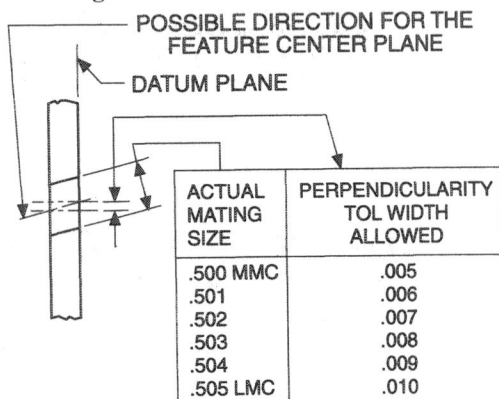

THE FEATURE CENTER PLANE MUST BE WITHIN THE
SPECIFIED TOLERANCE OF LOCATION. REGARDLESS
OF THE ACTUAL MATING ENVELOPE SIZE OF THE
FEATURE,  ITS CENTER PLANE MUST LIE BETWEEN
TWO PARALLEL PLANES (.005 APART) WHICH ARE
PERPENDICULAR TO THE DATUM PLANE.

## Noncylindrical Feature at MMC, Datum a Plane

### Example

⊥  .005 Ⓜ  A

NONCYLINDRICAL FEATURE AT
MMC. DATUM A PLANE

SYMBOL MEANING

⊥  .005 Ⓜ  A ◄— TO DATUM
                  PLANE A

WITHIN .005 WIDE
TOL ZONE AT MMC

THIS FEATURE MUST
BE PERPENDICULAR

.500 +.005
     −.000

### Meaning

POSSIBLE DIRECTION FOR THE
FEATURE CENTER PLANE

DATUM PLANE

| ACTUAL MATING SIZE | PERPENDICULARITY TOL WIDTH ALLOWED |
|---|---|
| .500 MMC | .005 |
| .501 | .006 |
| .502 | .007 |
| .503 | .008 |
| .504 | .009 |
| .505 LMC | .010 |

THE FEATURE CENTER PLANE MUST BE
WITHIN THE SPECIFIED TOLERANCE OF
LOCATION. WHEN THE FEATURE IS AT
MAXIMUM MATERIAL CONDITION (.500),
THE MAXIMUM PERPENDICULARITY
TOLERANCE IS .005 WIDE. WHERE THE
ACTUAL MATING ENVELOPE SIZE OF THE
FEATURE IS LARGER THAN ITS MAXIMUM
MATERIAL CONDITION SIZE, AN INCREASE
IN THE PERPENDICULARITY TOLERANCE
IS ALLOWED EQUAL TO THAT AMOUNT.
VIRTUAL CONDITION IS .495.

## Cylindrical Feature at RFS, Datum a Cylinder RFS

**Example**

⊥ | .005 | A

A

SYMBOL MEANING

⊥ | .005 | A — TO DATUM
AXIS A, RFS

WITHIN .005 WIDE
TOL ZONE, RFS

THIS FEATURE MUST
BE PERPENDICULAR

**Meaning**

.005 WIDE TOL ZONE

DATUM AXIS

POSSIBLE DIRECTION
FOR THE FEATURE AXIS

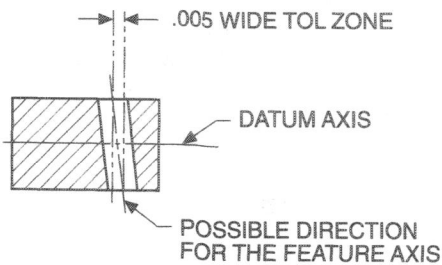

THE FEATURE AXIS MUST BE WITHIN THE SPECIFIED
TOLERANCE OF LOCATION. REGARDLESS OF THE
ACTUAL MATING ENVELOPE SIZE, ITS AXIS MUST
LIE BETWEEN TWO PLANES (.005 APART) WHICH ARE
PERPENDICULAR TO THE DATUM AXIS.

NOTE: This tolerance applies only to the view on which
it is specified.

## Cylindrical Feature at MMC, Datum a Plane

Note that the Ø symbol is required to indicate a diameter (cylindrical) tolerance zone.

**Example**

A

$\emptyset.250 \begin{smallmatrix} +.000 \\ -.001 \end{smallmatrix}$

⊥ | Ø.003 Ⓜ | A

SYMBOL MEANING

⊥ | Ø.003 Ⓜ | A — TO DATUM
PLANE A

WITHIN Ø.003 TOL
ZONE AT MMC

THIS FEATURE MUST
BE PERPENDICULAR

**Meaning**

| ACTUAL MATING SIZE Ø | PERP TOL Ø ALLOWED |
|---|---|
| .250 MMC | .003 |
| .2495 | .0035 |
| .249 LMC | .004 |

DATUM PLANE

Ø.253

FUNCTIONAL GAGE

FLUSH

PART

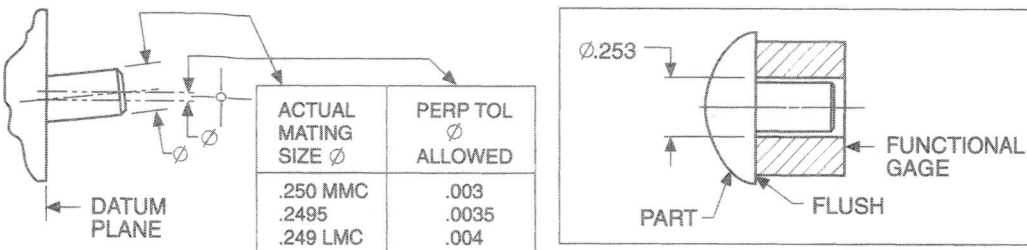

THE FEATURE MUST BE WITHIN THE SPECIFIED TOLERANCE OF
LOCATION. WHEN THE FEATURE IS AT MMC .250, THE PERPENDICULARITY
TOLERANCE IS .003 DIAMETER.  AS THE FEATURE ACTUAL MATING
ENVELOPE SIZE DEPARTS FROM MMC (GETS SMALLER), AN INCREASE
IN TOLERANCE IS PERMITTED EQUAL TO THAT AMOUNT  OF THAT
DEPARTURE. VIRTUAL CONDITION IS Ø.253.

## Cylindrical Feature at MMC, Datum a Plane

Note that the Ø symbol is required to indicate a diameter (cylindrical) tolerance zone.

### Example

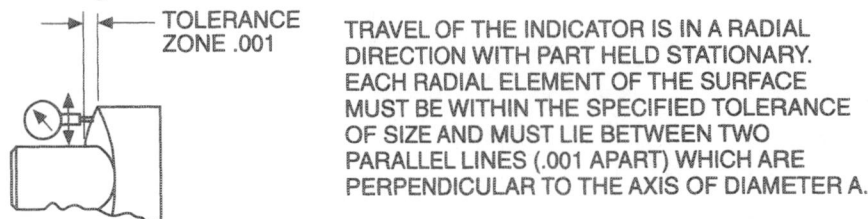

SYMBOL MEANING

⊥ | Ø.010 Ⓜ | A ← TO DATUM PLANE A

WITHIN Ø.010 TOL ZONE AT MMC

THIS FEATURE MUST BE PERPENDICULAR

Ø 2.000 +.005 −.000

⊥ | Ø.010 Ⓜ | A

### Meaning

POSSIBLE DIRECTION FOR THE FEATURE AXIS

DATUM PLANE

| ACTUAL MATING SIZE Ø | PERPENDICULARITY TOLERANCE DIA Ø ALLOWED |
|---|---|
| 2.000 MMC | .010 |
| 2.001 | .011 |
| 2.002 | .012 |
| 2.003 | .013 |
| 2.004 | .014 |
| 2.005 LMC | .015 |

WHEN THE FEATURE IS AT MMC (2.000), THE MAXIMUM PERPENDICULARITY TOLERANCE FOR ITS AXIS IS .010 DIAMETER. AS THE FEATURE ACTUAL MATING ENVELOPE SIZE DEPARTS FROM MMC (GETS LARGER), AN INCREASE IN TOLERANCE IS PERMITTED EQUAL TO THE AMOUNT OF THAT DEPARTURE. VIRTUAL CONDITION IS Ø1.990.

## Radial Perpendicularity

### Example

⊥ | .001 | A

EACH RADIAL ELEMENT

A

SYMBOL MEANING

⊥ | .001 | A ← TO DATUM AXIS A, RFS

WITHIN .001 WIDE TOL ZONE

EACH RADIAL ELEMENT MUST BE PERPENDICULAR

### Meaning

TOLERANCE ZONE .001

TRAVEL OF THE INDICATOR IS IN A RADIAL DIRECTION WITH PART HELD STATIONARY. EACH RADIAL ELEMENT OF THE SURFACE MUST BE WITHIN THE SPECIFIED TOLERANCE OF SIZE AND MUST LIE BETWEEN TWO PARALLEL LINES (.001 APART) WHICH ARE PERPENDICULAR TO THE AXIS OF DIAMETER A.

Where "element" control is required and the perfect form at MMC requirement of Rule 1 is to be removed, the method below may be used.

**Example**

⊥ | .005 | A

EACH ELEMENT

1.000 ± .010

△1

— | .040

△1

48.0 ± .1

A

△1 PERFECT FORM AT MMC NOT REQUIRED

**Meaning**

.005 WIDE (⊥)
TOL ZONE
(EACH ⊥
ELEMENT)

90°

DATUM PLANE A

.040 (−) TOL ZONE

Each perpendicular element of the surface at any location along the entire length of the surface must lie between two parallel lines (.005 apart) which are perpendicular to datum plane A. Each longitudinal element of the surface must lie between two parallel straight lines (.040 apart). The part perfect form of the MMC envelope may be exceeded but the stated tolerances are the maximum form error permissible on the surface.

When perpendicularity tolerancing is critical, it may be necessary to limit the *tolerance* deviation to an amount equal to the feature *size* deviation from MMC. This assumes that the part form must be perfect at MMC size and that the virtual condition (size) can be no greater than that at MMC. The only permissible form tolerance must be acquired from the variation in part size (see Example below) in the increase of the feature hole size.

Example 2 limits the tolerance acquired from the feature size increase to a MAX amount.

## Zero Tolerance at Maximum Material Condition

Note that the Ø symbol is required to indicate a diameter (cylindrical) tolerance zone.

**Example 1**

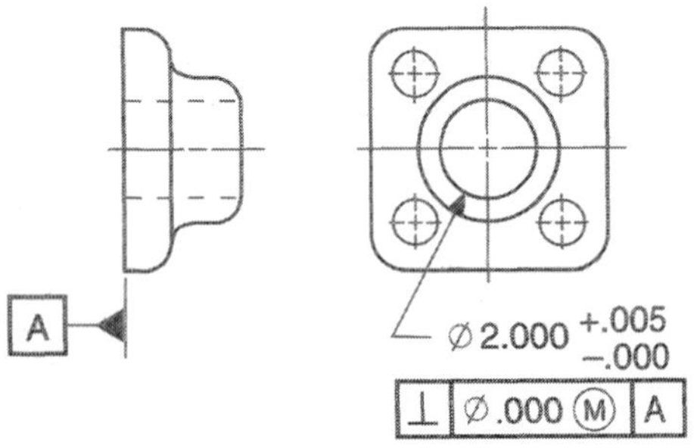

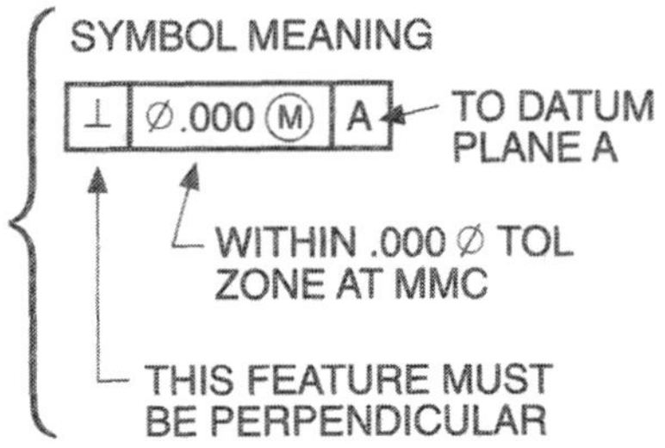

**Meaning**

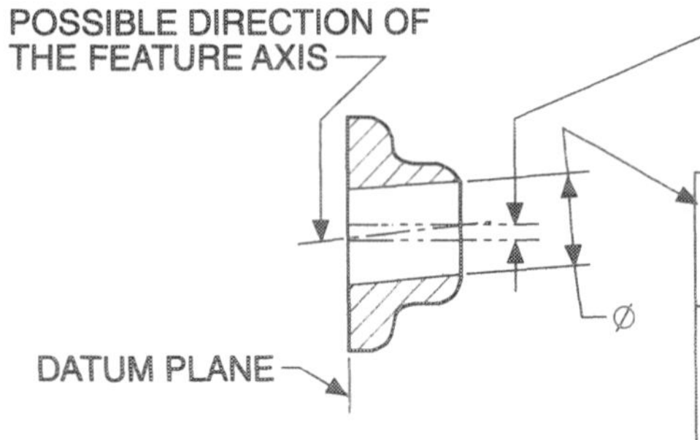

| ACTUAL MATING SIZE Ø | PERPENDICULARITY TOLERANCE DIA Ø ALLOWED |
|---|---|
| 2.000 MMC | .000 |
| 2.001 | .001 |
| 2.002 | .002 |
| 2.003 | .003 |
| 2.004 | .004 |
| 2.005 LMC | .005 |

WHEN THE FEATURE IS AT MMC 2.000, ITS AXIS MUST BE PERPENDICULAR TO THE DATUM PLANE. AS THE FEATURE ACTUAL MATING ENVELOPE SIZE DEPARTS FROM MMC (GETS LARGER), AN INCREASE IN TOL IS PERMITTED EQUAL TO THE AMOUNT OF THAT DEPARTURE. VIRTUAL CONDITION AND MMC ARE BOTH 2.000 IN THIS CASE.

Zero Tolerance at MMC, Max Deviation

**Example 2**

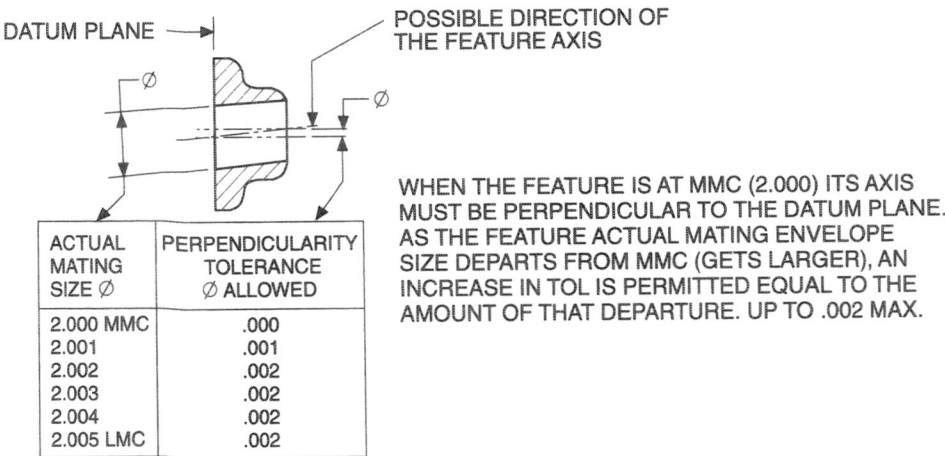

THIS FEATURE MUST BE PERPENDICULAR

WITHIN .000 TOL ZONE ⌀ AT MMC

| ⌖ | ⌀.000 Ⓜ | ⌀.002 MAX | A |

TO DATUM PLANE A

.002 ⌀ MAX TOL AS FEATURE DEPARTS FROM MMC

SYMBOL MEANING

⌀ 2.000 $^{+.005}_{-.000}$

| ⌖ | ⌀.000 Ⓜ | ⌀.002 MAX | A |

**Meaning**

DATUM PLANE

POSSIBLE DIRECTION OF THE FEATURE AXIS

| ACTUAL MATING SIZE ⌀ | PERPENDICULARITY TOLERANCE ⌀ ALLOWED |
|---|---|
| 2.000 MMC | .000 |
| 2.001 | .001 |
| 2.002 | .002 |
| 2.003 | .002 |
| 2.004 | .002 |
| 2.005 LMC | .002 |

WHEN THE FEATURE IS AT MMC (2.000) ITS AXIS MUST BE PERPENDICULAR TO THE DATUM PLANE. AS THE FEATURE ACTUAL MATING ENVELOPE SIZE DEPARTS FROM MMC (GETS LARGER), AN INCREASE IN TOL IS PERMITTED EQUAL TO THE AMOUNT OF THAT DEPARTURE. UP TO .002 MAX.

# 3-20  ANGULARITY ∠

**Definition.** Angularity is the condition of a surface, center plane, or axis which is at a specified angle (other than 90°) from a datum plane or axis.

## Angularity Tolerance

Angularity is the condition of a surface, center plane, or axis at a specified angle (other than 90°) from a datum plane or axis. An angularity tolerance specifies one of the following:

1. a tolerance zone defined by two parallel planes at the specified basic angle from one or more datum planes or a datum axis, within which the surface or center plane of the considered feature must lie.

2. a tolerance zone defined by two parallel planes at the specified basic angle from one or more datum planes or a datum axis, within which the axis of the considered feature must lie.

3. a cylindrical tolerance zone whose axis is at the specified basic angle from one or more datum planes, or a datum axis, within which the axis of the considered feature must lie.

4. a tolerance zone defined by two parallel lines at the specified basic angle from a datum plane, or axis, within which a line element of the surface must lie.

## Angularity Application

The example shows a part with a surface angular requirement. Note that the symbol is interpreted as "This surface must be at 45° angle in relation to datum plane A within .005 wide tolerance zone in relation to datum plane A."

The interpretation shows how the tolerance zone is established. Note that the angular tolerance zone is at 45° BASIC (exact) from the datum plane A. To be acceptable, the entire angular surface must fall within this tolerance zone. The angular surface must be contained within the limits of part size.

Note in the lower portion of the interpretation that the tolerance zone is also affected and established by the surface itself. That is, the surface extremities actually determine one plane of the tolerance zone as it bottoms out while the plane is inclined at an exact 45° angle with reference to the datum plane.

## Plane Surfaces

**Example**

SYMBOL MEANING

IN RELATION TO DATUM PLANE A

WITHIN .005 WIDE TOL ZONE

THIS SURFACE MUST BE AT 45°

**Meaning**

.005 TOL ZONE
PARALLEL PLANES

SEE NOTE*

45° BASIC

DATUM PLANE A

SEE NOTE*

LENGTH TOL AS ESTABLISHED BY LIMITS OF SIZE

.005 TOL ZONE

PARALLEL PLANES

45° BASIC

DATUM PLANE A

### NOTE*:

Part must be within *size* tolerance limits. The angular tolerance zone, composed of two parallel planes .005 apart, is 45° BASIC to the datum plane A. This tolerance zone is established by contact of the outermost of the two planes with the extremities of the angular surface and with the other plane parallel and inward at .005 distance. The entire surface must fall within this tolerance zone to be acceptable. Actually, this also controls the *flatness* of the surface to .005. Note that the angular surface extremities must be within both size and angular tolerance. See the examples below.

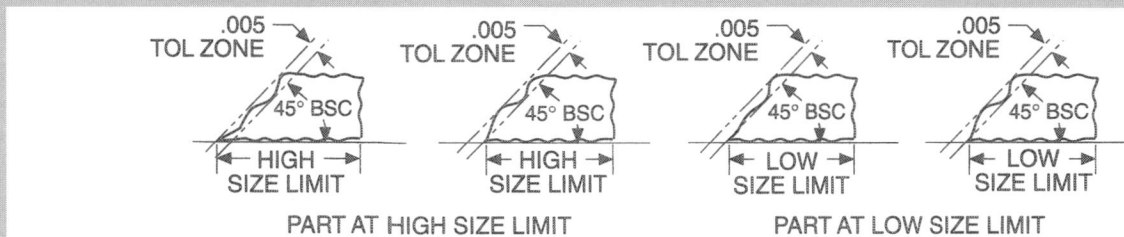

.005 TOL ZONE — 45° BSC — HIGH SIZE LIMIT

.005 TOL ZONE — 45° BSC — HIGH SIZE LIMIT

PART AT HIGH SIZE LIMIT

.005 TOL ZONE — 45° BSC — LOW SIZE LIMIT

.005 TOL ZONE — 45° BSC — LOW SIZE LIMIT

PART AT LOW SIZE LIMIT

### Cylindrical Feature at MMC, Datum a Plane

Note that the Ø symbol is required to indicate a diameter (cylindrical) tolerance zone. In this example an angularity tolerance has been used to control the axis of a hole at the specified angle. A cylindrical tolerance zone is used. Such an application would normally be used as a refinement of a locational tolerance to ensure that the angular relationship of the hole is maintained to the datum plane surface (in the view shown). The use of the cylindrical tolerance zone requires that the hole axis also remains within the angularity tolerance in a 360° relationship as well as within positional tolerance specified. MMC has been used as based upon design requirements and could gain additional tolerance as the feature is produced (see tabulation).

If it was desired to control the hole in its angularity relationship to the datum feature but only within a total wide zone, the diameter symbol (Ø) would be omitted in the feature control frame. The positional tolerance on such a part would then provide the feature control in the other directions.

**Example**

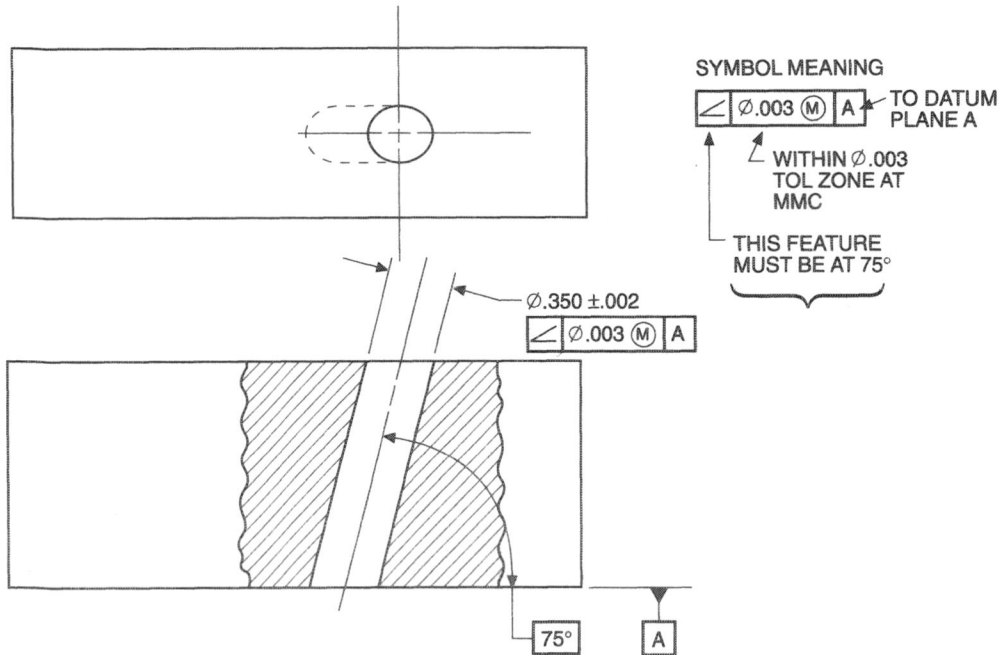

**Meaning**

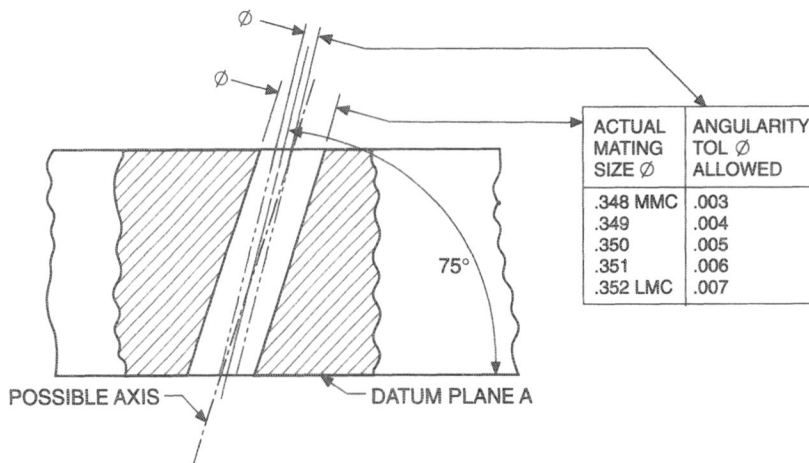

| ACTUAL MATING SIZE Ø | ANGULARITY TOL Ø ALLOWED |
|---|---|
| .348 MMC | .003 |
| .349 | .004 |
| .350 | .005 |
| .351 | .006 |
| .352 LMC | .007 |

The feature must be within the specified tolerance of location. When the feature is at MMC .348, the angularity tolerance is .003 diameter. As the feature actual mating envelope size departs from MMC (gets larger), an increase in tolerance is permitted equal to the amount of that departure. Virtual condition is Ø.345. Note: This tolerance applies only to the view on which it is specified.

# 3-21 PARALLELISM //

**Definition.** Parallelism is the condition of a surface or center plane, equidistant at all points from a datum plane; or an axis, equidistant along its length from one or more datum planes or a datum axis.

## Parallelism Tolerance

A parallelism tolerance specifies:

1. a tolerance zone defined by two parallel planes parallel to a datum plane or axis within which the surface or center plane of the considered feature must lie (see Fig. 3-36);
2. a tolerance zone defined by two parallel planes parallel to a datum plane or axis, within which the axis of the considered feature must lie (see Fig. 3-37);
3. a cylindrical tolerance zone parallel to one or more datum planes or a datum axis, within which the axis of the feature must lie (see Fig. 3-38);
4. a tolerance zone defined by two parallel lines parallel to a datum plane or axis, within which the line element of the surface must lie.

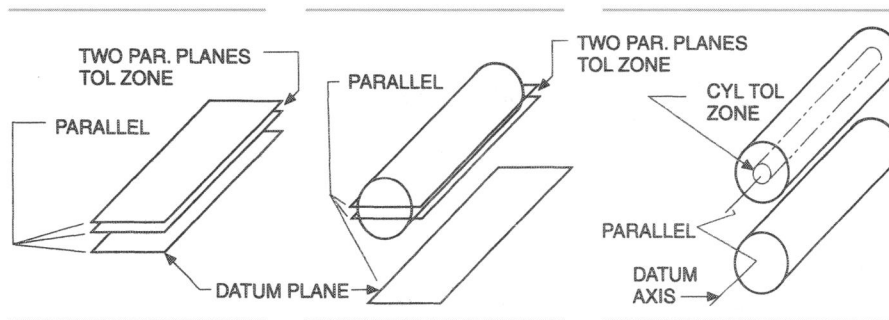

**Figure 3-36**          **Figure 3-37**          **Figure 3-38**

## Parallelism Application

Note in the following example that the bottom surface has been selected as the datum and the top surface is to be parallel to datum plane A within .002.

The Meaning beneath the example clarifies the symbol: It reads, "This feature must be parallel within .002 to datum plane A."

The lower example illustrates the tolerance zone and the manner in which the surface must fall within the tolerance zone to be acceptable. Note that the tolerance zone is established parallel to the datum plane A. Note also that the parallelism tolerance, when applied to a plane surface, controls flatness if a flatness tolerance is not specified (that is, the implied flatness will be *at least* as good as the parallelism).

## Surface to Datum Plane

**Example**

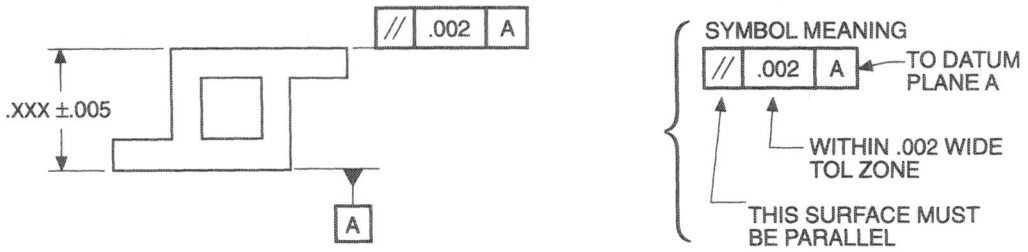

**Meaning**

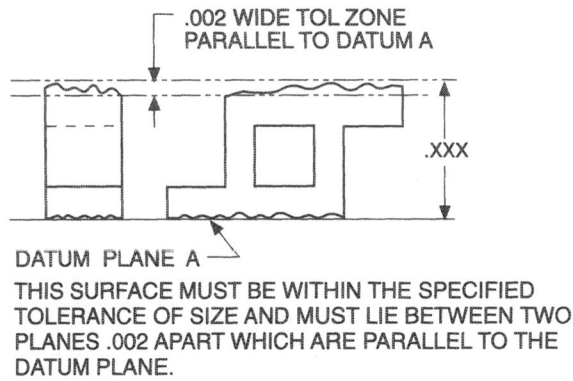

DATUM PLANE A

THIS SURFACE MUST BE WITHIN THE SPECIFIED
TOLERANCE OF SIZE AND MUST LIE BETWEEN TWO
PLANES .002 APART WHICH ARE PARALLEL TO THE
DATUM PLANE.

**Example**                                    **Meaning**

.005                     .005                     .005
TOL ZONE                 TOL ZONE                 TOL ZONE

// | .005 | A

.300 ±.010

A

.310                     .300                     .290
(PART AT HIGH            (PART AT                 (PART AT LOW
SIZE LIMIT)              NOMINAL SIZE)            SIZE LIMIT)
DATUM PLANE A
(POSSIBLE VARIATION OF PARALLELISM TOLERANCE ZONE
WITHIN ACTUAL LOCAL SIZE TOLERANCE RANGE)

## Parallelism—Features of Size

This page and the one following cover the application of
parallelism tolerancing to features of size. Note how the
MMC applications give the possibility of greater tolerance.

**Feature at RFS, Datum a Plane**

**Example**

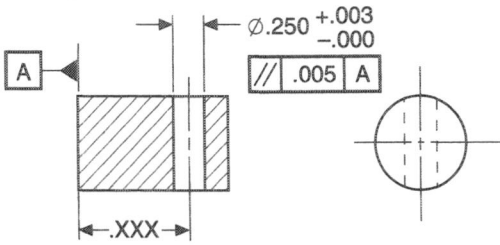

SYMBOL MEANING

// .005 A ← TO DATUM PLANE A

WITHIN .005 WIDE TOL ZONE, RFS

THIS FEATURE MUST BE PARALLEL

**Meaning**

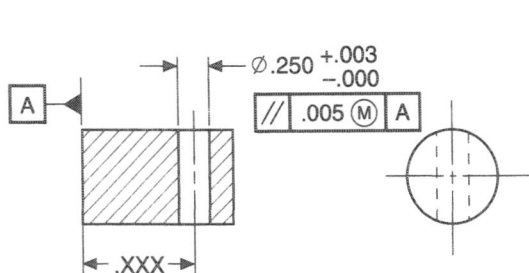

.005 TOL ZONE

THE FEATURE AXIS MUST BE WITHIN THE SPECIFIED TOLERANCE OF LOCATION. REGARDLESS OF THE ACTUAL MATING ENVELOPE SIZE OF THE FEATURE, IT'S AXIS MUST LIE BETWEEN TWO PARALLEL PLANES (.005 APART) WHICH ARE PARALLEL TO THE DATUM PLANE A.

PARALLEL
DATUM PLANE A

**Feature at MMC, Datum a Plane**

**Example**

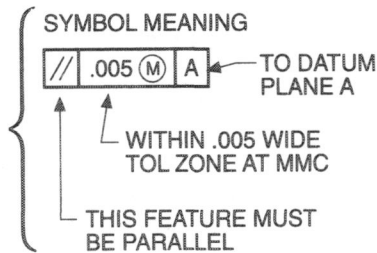

SYMBOL MEANING

// .005 Ⓜ A ← TO DATUM PLANE A

WITHIN .005 WIDE TOL ZONE AT MMC

THIS FEATURE MUST BE PARALLEL

**Meaning**

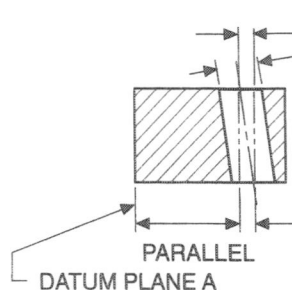

WIDE TOL ZONE

PARALLEL
DATUM PLANE A

| ACTUAL MATING SIZE ⌀ | PARALLELISM TOL WIDTH ALLOWED |
|---|---|
| .250 MMC | .005 |
| .2505 | .0055 |
| .251 | .006 |
| .2515 | .0065 |
| .252 | .007 |
| .2525 | .0075 |
| .253 LMC | .008 |

FEATURE AXIS MUST BE WITHIN SPECIFIED TOL OF LOCATION. HOLE (AT MMC) AXIS MUST LIE BETWEEN TWO PARALLEL PLANES (.005 APART) WHICH ARE PARALLEL TO THE DATUM PLANE A.

WHEN FEATURE ACTUAL MATING ENVELOPE SIZE DEPARTS FROM MMC (GETS LARGER), AN INCREASE IN THE PARALLELISM TOL IS ALLOWED EQUAL TO THAT DEPARTURE.

Feature at RFS, Datum Feature at RFS

**Example**

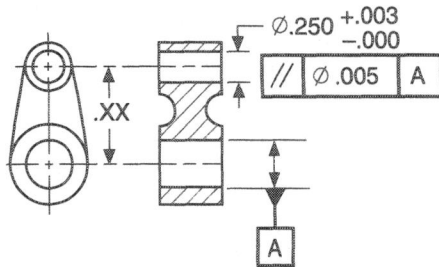

SYMBOL MEANING

// | Ø.005 | A → TO DATUM AXIS A, RFS

WITHIN Ø.005 TOL ZONE, RFS

THIS FEATURE MUST BE PARALLEL

**Meaning**

POSSIBLE DIRECTION FOR THIS FEATURE AXIS

.005 DIAMETER PARALLELISM TOL ZONE

Ø

PARALLEL

DATUM AXIS

THE FEATURE AXIS MUST BE WITHIN THE SPECIFIED TOL OF LOCATION. REGARDLESS OF THE ACTUAL MATING ENVELOPE SIZE OF THE FEATURE, ITS AXIS MUST LIE WITHIN A .005 DIAMETER (CYLINDRICAL) TOLERANCE ZONE WHICH IS PARALLEL TO THE DATUM AXIS.

Feature at MMC, Datum Feature at RFS

(Note that the Ø symbol is required to indicate a diameter (cylindrical) tolerance zone.)

**Example**

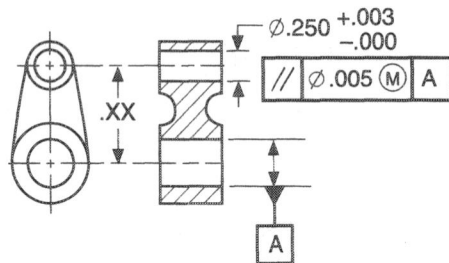

SYMBOL MEANING

// | Ø.005 Ⓜ | A → TO DATUM AXIS A, RFS

WITHIN .005 Ø TOL ZONE AT MMC

THIS FEATURE MUST BE PARALLEL

**Meaning**

POSSIBLE DIRECTION FOR THIS FEATURE AXIS

Ø

PARALLEL

DATUM AXIS

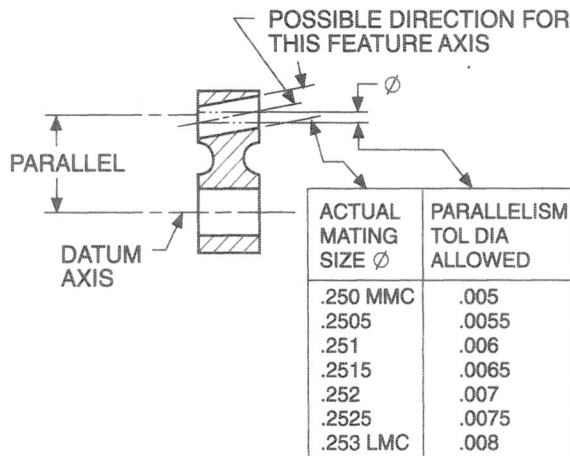

| ACTUAL MATING SIZE Ø | PARALLELISM TOL DIA ALLOWED |
|---|---|
| .250 MMC | .005 |
| .2505 | .0055 |
| .251 | .006 |
| .2515 | .0065 |
| .252 | .007 |
| .2525 | .0075 |
| .253 LMC | .008 |

THE FEATURE AXIS MUST BE WITHIN THE SPECIFIED TOL OF LOCATION. WHEN THE FEATURE IS AT MAXIMUM MATERIAL CONDITION, .250, THE MAXIMUM PARALLELISM TOLERANCE IS .005 DIAMETER. WHEN THE FEATURE ACTUAL MATING ENVELOPE SIZE DEPARTS FROM MMC (GETS LARGER), AN INCREASE IN THE PARALLELISM TOL IS ALLOWED EQUAL TO THE AMOUNT OF THAT DEPARTURE.

# Dimensioning and Tolerancing in AutoCAD and GD&T

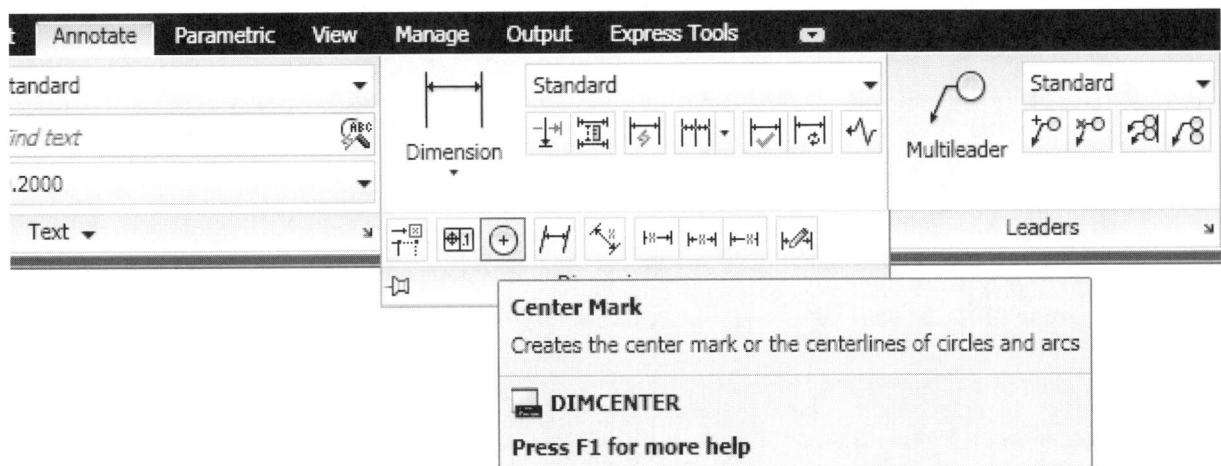

Figure 4-1

# 4-1  INTRODUCTION

This chapter explains the **Dimension** panel. See Figure 4-1. The chapter first explains dimensioning terminology and conventions then presents an explanation of each tool within the **Dimension** panel. The chapter also demonstrates how dimensions are applied to drawings and gives examples of standard drawing conventions and practices.

# 4-2  TERMINOLOGY AND CONVENTIONS

**Some common terms (See Figure 4-2.)**

**Dimension lines:** Mechanical drawings contain lines between extension lines that end with arrowheads and include a numerical dimensional value located within the line; architectural drawings contain lines between extension lines that end with tick marks and include a numerical dimensional value above the line.

**Extension lines:** Lines that extend away from an object and allow dimensions to be located off the surface of an object.

**Leader lines:** Lines drawn at an angle, not horizontal or vertical, that are used to dimension specific shapes such as holes. The start point of a leader line includes an arrowhead. Numerical values are drawn at the end opposite the arrowhead.

**Linear dimensions:** Dimensions that define the straight-line distance between two points.

**Angular dimensions:** Dimensions that define the angular value, measured in degrees, between two straight lines.

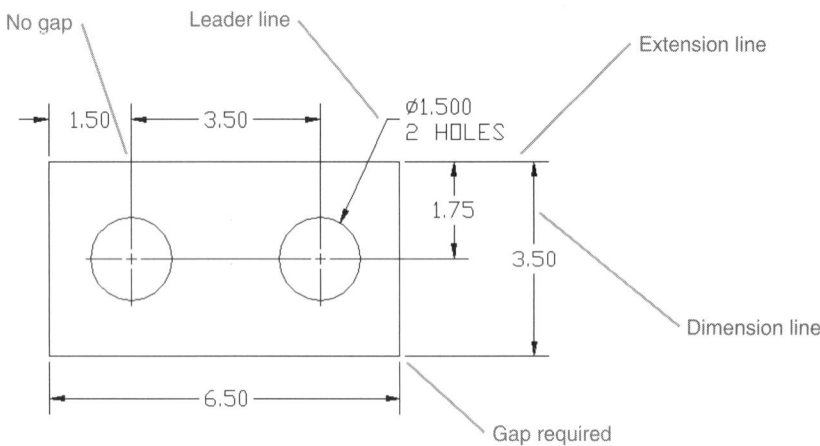

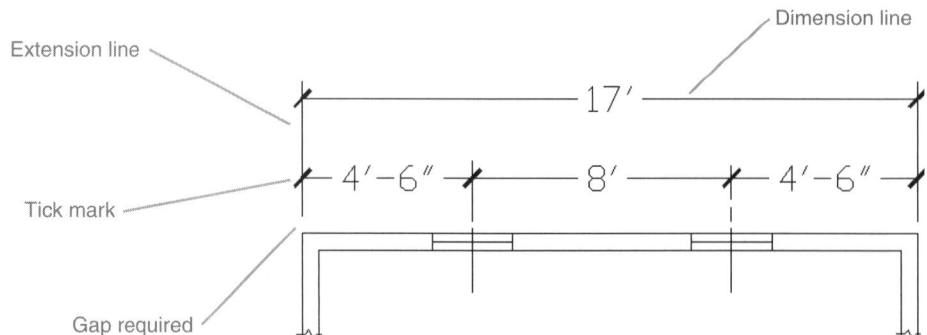

Figure 4-2

## Some dimensioning conventions (See Figure 4-3.)

1. Dimension lines are drawn evenly spaced; that is, the distance between dimension lines is uniform. A general rule of thumb is to locate dimension lines about 1/2 inch, or 15 millimeters, apart.
2. There should be a noticeable gap between the edge of a part and the beginning of an extension line. This serves as a visual break between the object and the extension line. The visual difference between the linetypes can be emphasized by using different colors for the two types of lines.
3. Leader lines are used to define the size of holes and should be positioned so that the arrowhead points at the center of the hole.
4. Centerlines may be used as extension lines. No gap is used when a centerline is extended beyond the edge lines of an object.
5. Dimension lines should be aligned whenever possible to give the drawing a neat, organized appearance.

## Some common errors to avoid (See Figure 4-4.)

1. Avoid crossing extension lines. Place longer dimensions farther away from the object than shorter dimensions.
2. Do not locate dimensions within cutouts; always use extension lines.
3. Do not locate any dimension too close to the object. Dimension lines should be at least 1/2 inch, or 15 millimeters, from the edge of the object.
4. Avoid long extension lines. Locate dimensions in the same general area as the feature being defined.

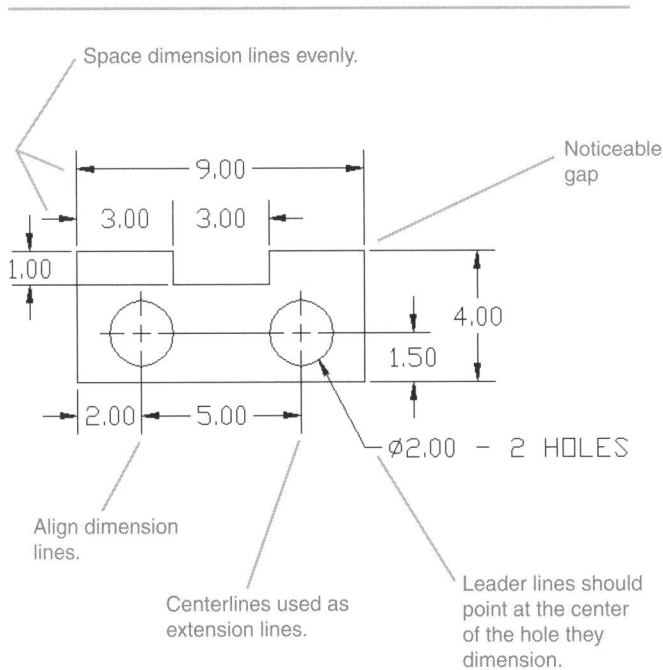

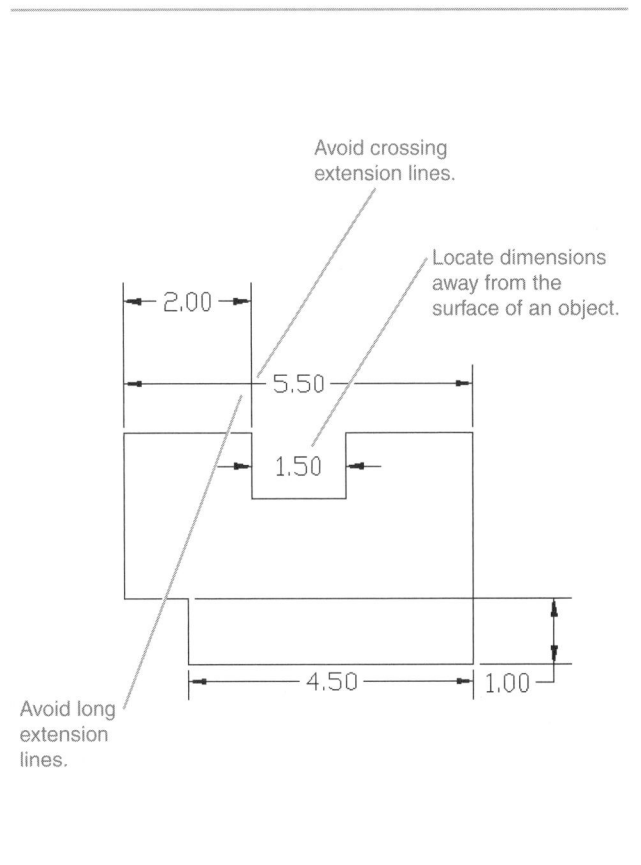

Figure 4-3

Figure 4-4

First extension line origin

Dimension line location

Linear Dimension tool

7.00

2.50

Second extension line origin

Figure 4-5

## 4-3 LINEAR DIMENSION

The **Linear Dimension** command is used to create horizontal and vertical dimensions.

**To create a horizontal dimension by selecting extension line locations or origins (See Figure 4-5.)**

1. Select the **Linear** tool from the **Dimension** panel.

   *Command: _dimlinear*
   *Specify first extension line origin or <select object>:*

2. Select the starting point for the first extension line.

   *Specify second extension line origin:*

3. Select the starting point for the second extension line.

   *[MText/Text/Angle/Horizontal/Vertical/Rotated]:*

4. Locate the dimension line by moving the crosshairs to the desired location.
5. Press the left mouse button to place the dimension.

The dimensional value locations shown in Figure 4-5 are the default settings locations. The location and style may be changed using the **Dimension Style** command discussed in Section 4-4.

**To create a vertical dimension**

The vertical dimension shown in Figure 4-5 was created using the same procedure demonstrated for the hori-

zontal dimension, except different extension line origin points were selected. AutoCAD will automatically switch from horizontal to vertical dimension lines as you move the cursor around the object.

If there is confusion between horizontal and vertical lines when adding dimensions, that is, you don't seem to be able to generate a vertical line, type **V** and press **Enter** in response to the following prompt:

*[MText/Text/Angle/Horizontal/Vertical/Rotated]:*

The system will now draw vertical dimension lines.

**To create a horizontal dimension by selecting the object to be dimensioned (See Figure 4-6.)**

1. Select the **Linear** tool from the **Dimension** panel.

   *Command: _dimlinear*
   *Specify first extension line origin or <select object>:*

7.00

2.50

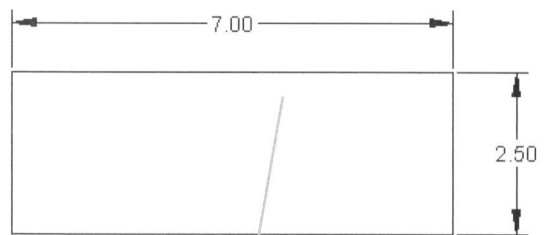

Select this line for dimensioning; extension lines will be added automatically.

Figure 4-6

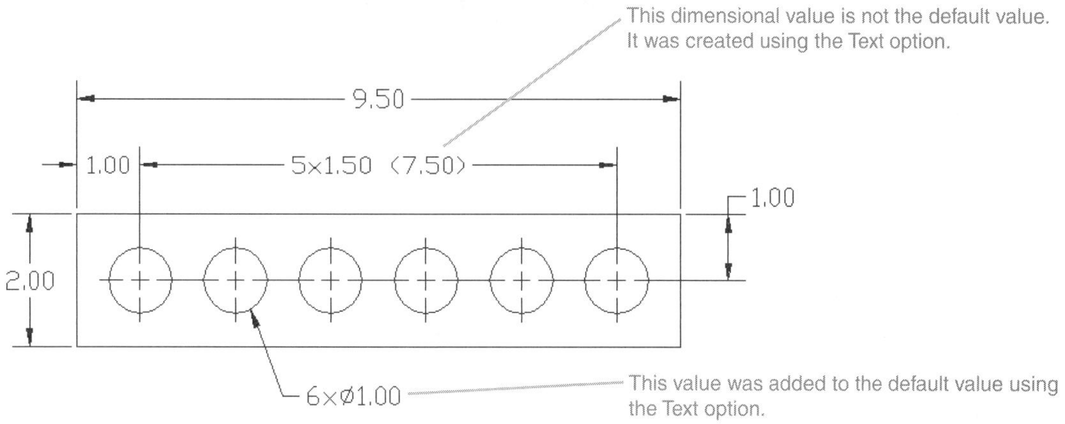

This dimensional value is not the default value. It was created using the Text option.

This value was added to the default value using the Text option.

Figure 4-7

2. Press the right mouse button.

   *Select object to Dimension:*

3. Select the line to be dimensioned.

   This option allows you to select the distance to be dimensioned directly. The option applies only to horizontal and vertical lengths. Aligned dimensions, although linear, are created using the **Aligned** tool.

**To change the default dimension text—Text option**

   AutoCAD will automatically create a text value for a given linear distance. A different value or additional information may be added as follows. See Figure 4-7.

1. Select the **Linear** tool from the **Dimension** panel.

   *Command: _dimlinear*
   *Specify first extension line origin or <select object>:*

2. Select the starting point for the first extension line.

   *Specify second extension line origin:*

3. Select the starting point for the second extension line.

   *[MText/Text/Angle/Horizontal/Vertical/Rotated]:*

4. Type **t**; press **Enter.**

   *Enter dimension text <7.50>:*

   The value given will be the linear value of the distance selected. In this example more information is required, so the default distance value must be modified.

5. Type **5 × 1.50 (7.50)**; press **Enter.**

   The typed dimension will appear on the screen and can be located by moving the cursor.

**To change the default dimension text—Mtext option**

1. Select the **Linear** tool from the **Dimension** panel.

   *Command: _dimlinear*
   *Specify first extension line origin or <select object>:*

2. Select the starting point for the first extension line.

   *Specify second extension line origin:*

3. Select the starting point for the second extension line.

   *[MText/Text/Angle/Horizontal/Vertical/Rotated]:*

4. Type **m**; press **Enter.**

   The **Text Formatting** dialog box will appear. See Figure 4-8. A box will appear around the dimension text and the box will include a colored background. Press the Del key to delete the existing text. Type in the new text. In the example shown, the default text value of 7.0000 was replaced with a value of 7.00.

5. Click the screen.

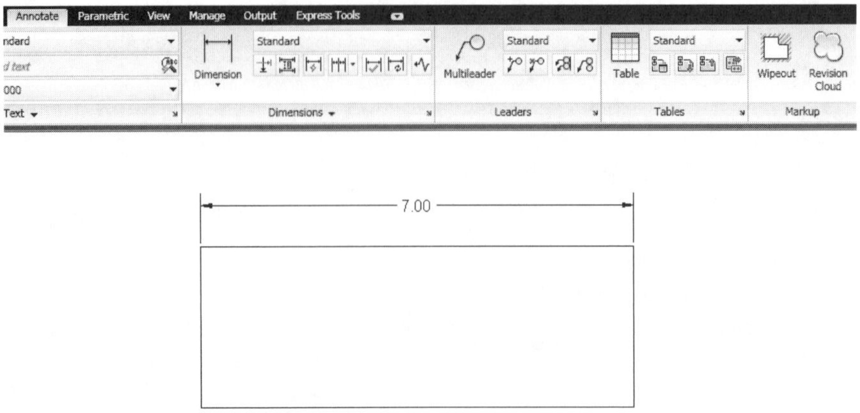

**Figure 4-8**

Information can be added before or after the default text by placing the cursor in the appropriate place and typing the additional information. For example, placing the cursor to the right of the text value box and typing **-2 PLACES** would produce the text 7.00-2 PLACES on the drawing.

**To edit an existing dimension**

Figure 4-9 shows a figure with an existing 7.00 dimension.

1. Click the existing dimension.

    A dialog box will appear. See Figure 4-10

2. Move the cursor to the bottom edge of the dialog box to expand the box's listings.

3. Click the Text override box.

    The dimension text value of 7.00 will be highlighted. See Figure 4-11

4. Type in the new text value.

    The ± sign was created using the character map – alt 0177.

5. Delete the dialog box and press the <Esc> key.

    Figure 4-12 shows the edited dimension.

    Dimension text can also be edited using the Properties dialog box. See Figure 4-13. To access the Properties dialog box, double click the dimension.

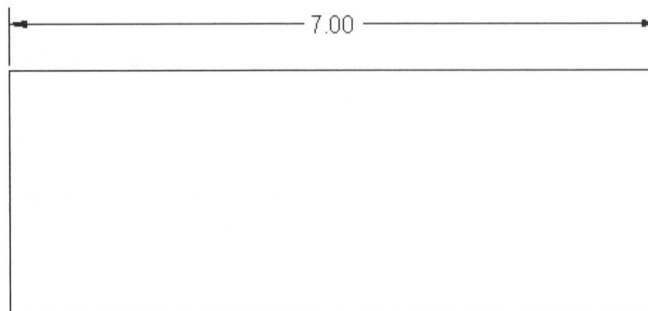

**Figure 4-9**

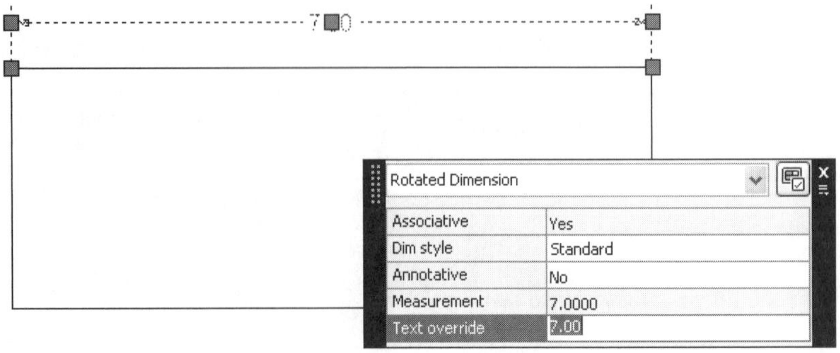

**Figure 4-10**

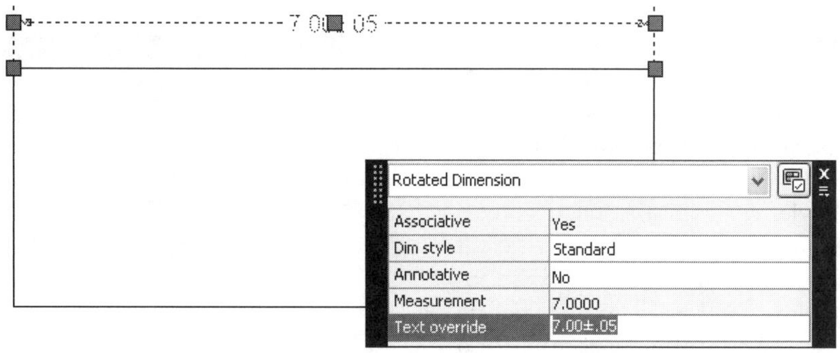

7.00±.05
Rotated Dimension
Associative
Yes
Dim style
Standard
Annotative
No
Measurement
7.0000
Text override
7.00±.05

**Figure 4-11**

7.00±.05

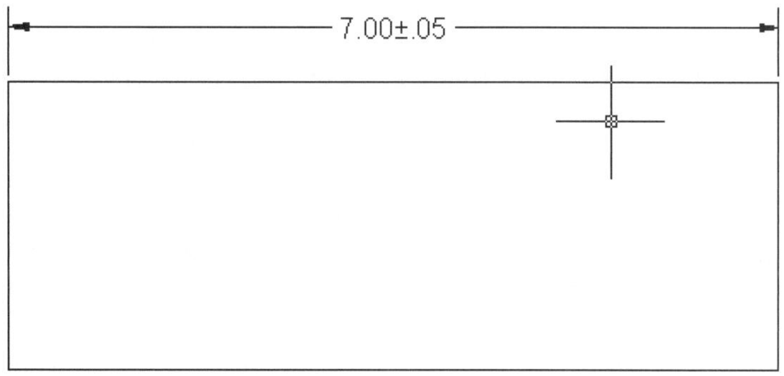

**Figure 4-12**

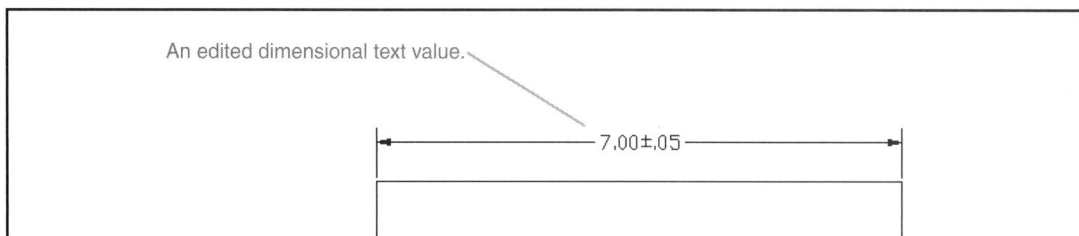

An edited dimensional text value.
7.00±.05

**Figure 4-13**

# 4-4 DIMENSION STYLES

The Dimension Style Manager is used to control the appearance and format of dimensions. The Dimension Style Manager is accessed by clicking the arrow in the lower right corner of the Dimensions Panel under the Annotation tab. See Figure 4-14.

A great variety of styles are used to create technical drawings. The style difference may be the result of different drawing conventions. For example, architects locate dimensions above the dimension lines, and mechanical engineers locate the dimensions within the dimension lines. AutoCAD works in decimal units for either millimeters or inches, so parameters set for inches would not be usable for millimeter drawings. The **Dimension Style** Manager allows you to conveniently choose and set dimension parameters that suit your particular drawing requirements.

Figure 4-15 shows the **Dimension Style Manager** dialog box. This section will explain how to use the **Modify** option to change the **Standard** style settings to suit a specific drawing requirement. The **Set Current, New, Override,** and **Compare** options are used to create a new custom dimension style designed to meet specific applications. Figure 4-16 shows the **Primary Units** option of the **Modify Dimension Style: Standard** dialog box.

### To change the scale of a drawing

Drawings are often drawn to scale because the actual part is either too big to fit on a sheet of drawing paper or too small to be seen. For example, a microchip circuit must be drawn at several thousand times its actual size to be seen.

**Figure 4-14**

Drawing scales are written using the following formats:

SCALE: 1 = 1
SCALE: FULL
SCALE: 1000 = 1
SCALE: .25 = 1

In each example the value on the left indicates the scale factor. A value greater than 1 indicates that the drawing is larger than actual size. A value less than 1 indicates that the drawing is smaller than actual size.

Regardless of the drawing scale selected, the dimension values must be true size. Figure 4-17 shows the same rectangle drawn at two different scales. The top rectangle is drawn at a scale of 1 = 1, or its true size. The bottom rectangle is drawn at a scale of 2 = 1, or twice its true size. In both examples the 3.00 dimension remains the same.

The **Measurement scale** box on the **Primary Units** option on the **Modify Dimension Style: Standard** dialog

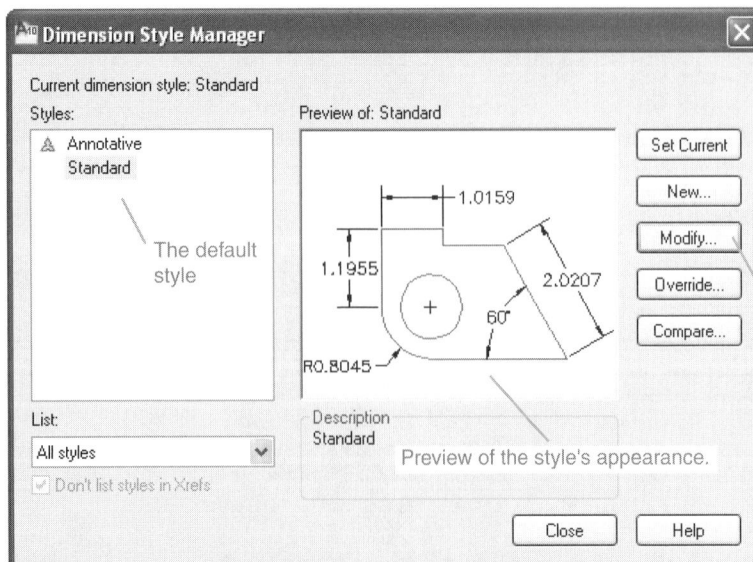

The Dimension Style Manager dialog box can be used to create and modify custom dimension styles designed to meet specific applications.

The Modify box is used to make changes to the existing standard values.

Preview of the style's appearance.

**Figure 4-15**

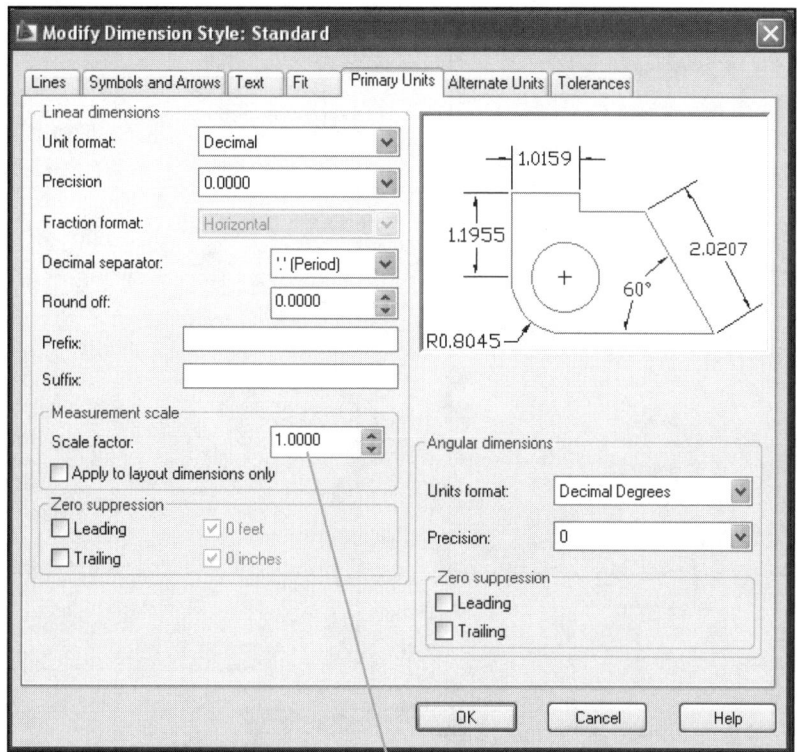

Change this value to change the scale of the drawing.

**Figure 4-16**

box is used to change the dimension values to match different drawing scales. Figure 4-18 shows the measurement scale set to a factor of 0.5000. If the drawing scale is 2 = 1, as shown in Figure 4-17, then the scale factor for the **Measurement scale** must be 0.5000. Compare the preview

in Figure 4-16 with the preview in Figure 4-18, where the scale factor has been changed.

### To use the Text option

Figure 4-19 shows the **Text** option on the **Modify Dimension Style: Standard** dialog box. This option can be

**Figure 4-17**

**Figure 4-18a**

**Figure 4-18b**

**Figure 4-19**

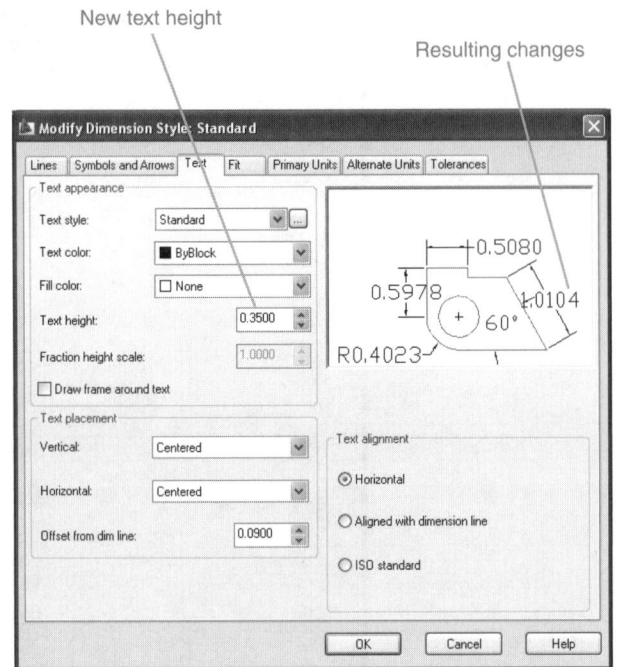

New text height

Resulting changes

**Figure 4-20**

used to change the height of dimension text or the text placement. In Figure 4-20 the text height was changed from the default value of 0.1800 to a value of 0.3500. The preview box shows the resulting changes in both the text and how it will be positioned on the drawing.

Figure 4-21 shows text located above the dimension lines and positioned nearer the first extension line. These changes were created using the **Vertical** and **Horizontal** options within the **Text placement** box.

Figure 4-22 shows text aligned with the direction of the dimension lines in accordance with ISO standards. This change was created using the ISO standard radio button within the **Text alignment** box.

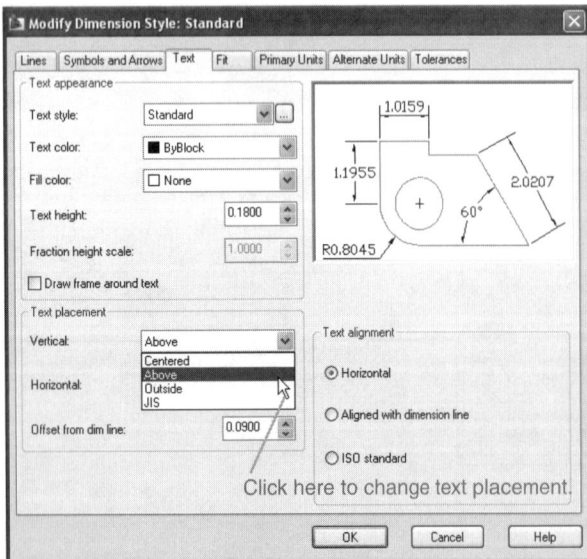

**Figure 4-21**

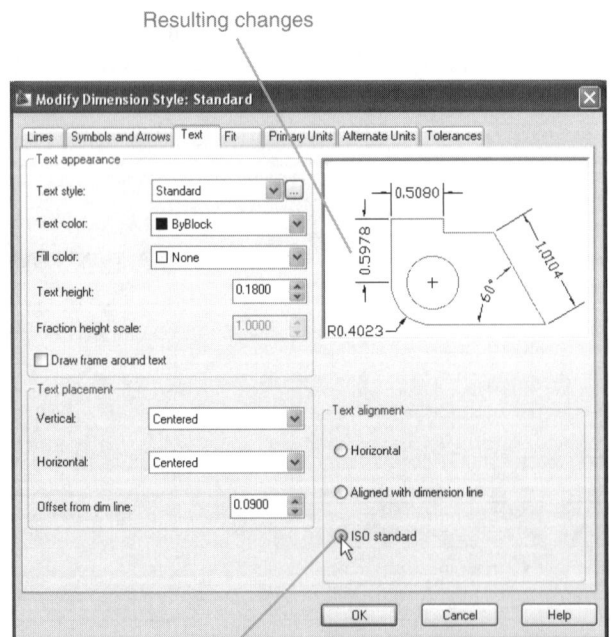

Resulting changes

Change in text alignment—use ISO Standard.

**Figure 4-22**

STANDARD TOLERANCES

X = ±1
X.X = ±0.1
X.XX = ±0.01
X.XXX = ±0.001
X.XXXX = ±0.0005

X° = ±0.1°

THESE TOLERANCES APPLY
UNLESS OTHERWISE STATED.

Figure 4-23

| MILLIMETERS | | Zero required |
|---|---|---|
| 0.25 | 0.5 | 0.033 |
| 32 | 1.45 | 3 |
| INCHES | | No zero required |
| 25 | 5 | .003 |
| 32.00 | 145.0 | 3.000 |
| ARCHITECTURAL UNITS (feet and inches) | | |
| 0"-0 1/2" | 8" | 2'-8" |

Figure 4-25

## 4-5 UNITS

It is important to understand that dimensional values are not the same as mathematical units. Dimensional values are manufacturing instructions and always include a tolerance, even if the tolerance value is not stated. Manufacturers use a predefined set of standard dimensions that are applied to any dimensional value that does not include a written tolerance. Standard tolerance values differ from organization to organization. Figure 4-23 shows a chart of standard tolerances.

In Figure 4-24 a distance is dimensioned twice: once as 5.50 and a second time as 5.5000. Mathematically these two values are equal, but they are not the same manufacturing instruction. The 5.50 value could, for example, have a standard tolerance of ±0.01, whereas the 5.5000 value could have a standard tolerance of ±0.0005. A tolerance of ±0.0005 is more difficult and therefore more expensive to manufacture than a tolerance of ±0.01.

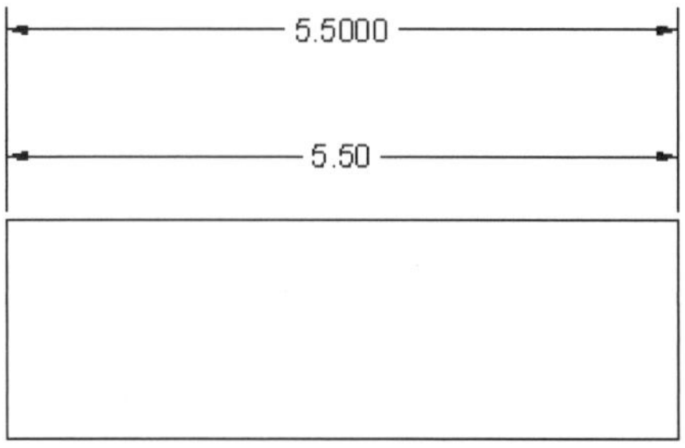

Figure 4-25 shows examples of units expressed in millimeters, decimal inches, and architectural units. A zero is not required to the left of the decimal point for decimal inch values less than one. Millimeter values do not require zeros to the right of the decimal point. Architectural units should always include the feet (') and inch (") symbols. Millimeter and decimal inch values never include symbols; the units will be defined in the title block of the drawing.

**To prevent a 0 from appearing to the left
of the decimal point**

1. Access the Dimension Style Manager.

   The **Dimension Style** dialog box will appear.

2. Select **Modify**.

   The **Modify Dimension Style: Standard** dialog box will appear.

3. Select the **Primary Units** option.

   The **Primary Units** dialog box will appear. See Figure 4-26.

4. Click the box to the left of the word **Leading** within the **Zero suppression** box.

   A check mark will appear in the box indicating that the function is on.

5. Select **OK** to return to the drawing.

   Save the change, if desired. You can now dimension using any of the dimension commands; no zeros will appear to the left of the decimal point. Figure 4-27 shows the results.

Figure 4-24

Turn the Leading option on to prevent zeros to the left of the decimal point.

Turn the Trailing option on to prevent zeros to the right of the decimal point.

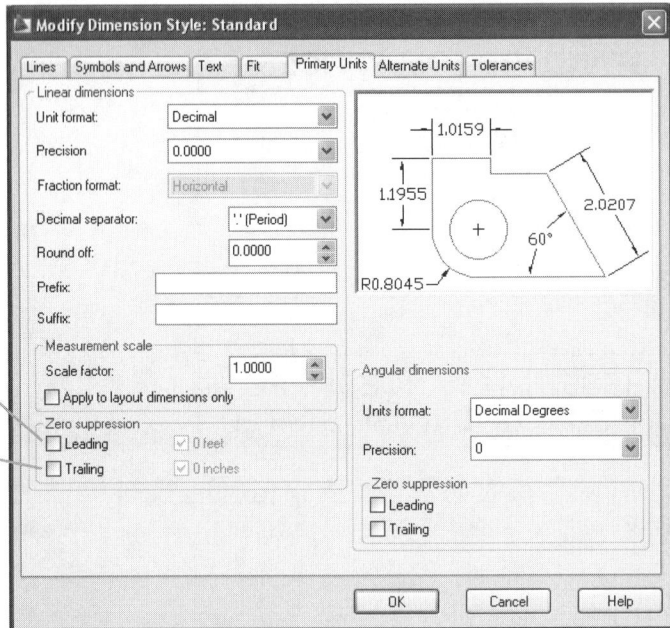

**Figure 4-26**

**To change the number of decimal places in a dimension value**

1. Access the Dimension Style Manager.

   The **Dimension Style Manager** dialog box will appear.

2. Select **Modify**.

   The **Modify Dimension Style: Standard** dialog box will appear.

3. Select the **Primary Units** option.

4. Select the arrow to the right of the **Precision** box.

   A list of precision options will cascade down. See Figure 4-28.

5. Select the desired value.

   Save the changes, if desired. You can now dimension using any of the dimension commands, and the resulting values will be expressed using the selected precision.

Leading Zero Suppression on

0.50

.50

Trailing Zero Suppression on

2.00

2

**Figure 4-27**

Select the unit precision here.

**Figure 4-28**

# 4-6  ALIGNED DIMENSIONS (See Figures 4-29 and 4-30.)

## To create an aligned dimension

1. Select the **Aligned** tool from the **Dimension** panel.

   *Command: _dimaligned*
   *Specify first extension line origin or <select object>:*

2. Select the first extension line origin point.

   *Specify second extension line origin:*

3. Select the second extension line origin point.

   *[Mtext/Text/Angle]:*

4. Select the location for the dimension line.

## The Select Object option

1. Select the **Aligned** tool from the **Dimension** panel.

   *Command: _dimaligned*
   *Specify first extension line origin or <select object>:*

2. Press the **Enter** key.

   *Select object to dimension:*

3. Select the line.

   *[Mtext/Text/Angle]:*

4. Select the dimension line location.

A response of **M** to the last prompt line will activate the **Multiline Text** option. The **Text Formatting** dialog box will appear. The **Text** option can be used to replace or supplement the default text generated by AutoCAD.

A dimension aligned using the Aligned with Dimension line option on the Modify Dimension Style: Standard dialog box.

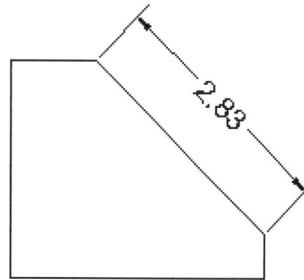

This format can also be achieved using the Angle option within the Aligned Dimension command.

**Figure 4-29**

A response of **A** to the prompt will activate the **Angle** option. The **Angle** option allows you to change the angle of the text within the dimension line. See Figure 4-30. The default angle value is **0°**, or horizontal. The example shown in Figure 4-29 used an angle of −45°. The prompt responses are as follows.

*Specify angle of dimension text:*
*[MTExt/Text/Angle]:*

1. Type **-45;** press **Enter.**
2. Select the location for the dimension.

# 4-7  RADIUS AND DIAMETER DIMENSIONS

Figure 4-31 shows an object that includes both arcs and circles. The general rule is to dimension arcs using a radius dimension, and circles using diameter dimensions. This convention is consistent with the tooling required to produce

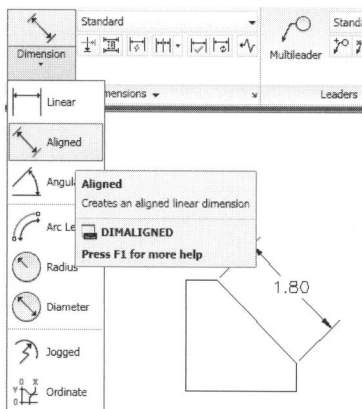

**Figure 4-30**

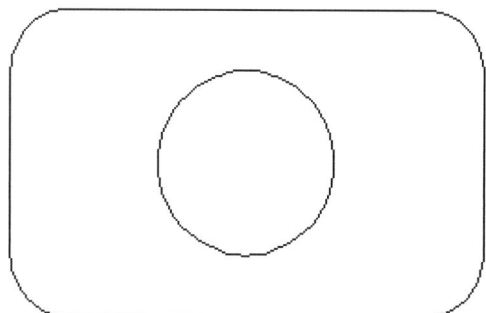

**Figure 4-31**

Figure 4-32

Figure 4-34

the feature shape. Any arc greater than 180° is considered a circle and is dimensioned using a diameter.

**To create a radius dimension**

1. Select the **Radius** tool from the **Dimension** panel.

   *Command: _dimradius*
   *Select arc or circle:*

2. Select the arc to be dimensioned.

   *Specify dimension line location or [Mtext/Text/ Angle]:*

3. Position the radius dimension so that its leader line is not horizontal or vertical.

   Figure 4-32 shows the resulting dimension. The dimension text can be altered by clicking the text. A dialog box will appear. See Figure 4-33. Access the Text override box, delete the existing text and type in the new text. Delete the dialog box and press the <Esc> key.

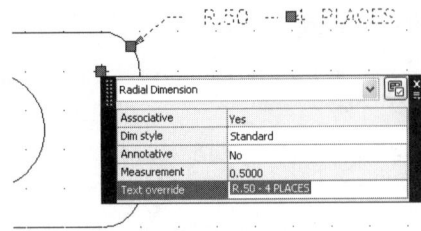

**To alter the default dimensions**

1. Select the **Radius** tool from the **Dimension** panel.

   *Command: _dimradius*
   *Select arc or circle:*

2. Select the arc to be dimensioned.

   *Specify dimension line location or [Mtext/Text/ Angle]:*

3. Type **M;** press **Enter.**

   The **Text Formatting** dialog box will appear.

4. Place the flashing cursor just to the right of the dimension value box and type **-4 PLACES.**

5. Select **OK.**

   Figure 4-34 shows the resulting dimension. The **Radius** dimension command will automatically include a center point with the dimension. The center point can be excluded from the dimension as follows.

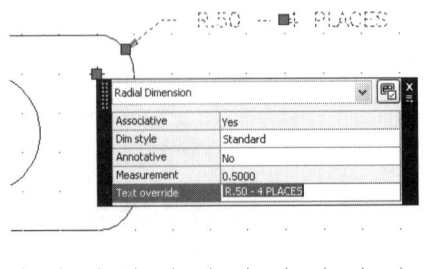

Figure 4-33

Figure 4-35

**To remove the center mark from a radius dimension**

1. Acces the **Dimension Style Manager**.

   The **Dimension Style Manager** dialog box will appear.

2. Select **Modify.**

   The **Modify Dimension Style: Standard** dialog box will appear.

3. Select the **Symbols and Arrows** option.
4. Select the **None** radio button in the **Center marks** area.

   A solid circle will appear in the preview box, indicating that it is on. See Figure 4-35.

5. Select **OK** to return to the drawing.

   You will have to redimension the arc, including the text alteration. Figure 4-36 shows the results.

**To create a diameter dimension**

   Circles require three dimensions: a diameter value plus two linear dimensions used to locate the circle's center point. AutoCAD can be configured to automatically add horizontal and vertical centerlines as follows.

1. Access the **Dimension Style Manager**.

   The **Dimension Style Manager** dialog box will appear.

2. Select **Modify.**

   The **Modify Dimension Style: Standard** dialog box will appear.

3. Select the **Symbols and Arrows** option.
4. Select the **Line** option from the **Center marks** box.
5. Select **OK** to return to the drawing.

   Centerlines will be added to the existing radius dimension.

6. Explode the radius dimension, then erase the radius centerlines.
7. Select the **Diameter** tool from the **Dimension** panel.

   *Command: _dimdiameter*
   *Select arc or circle:*

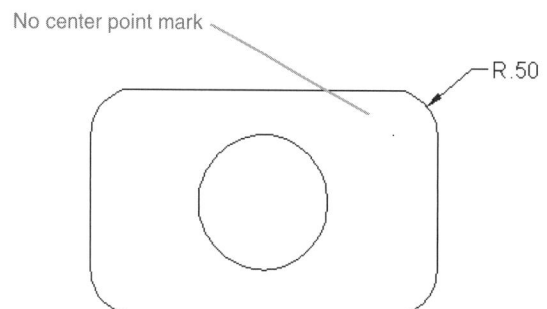

Figure 4-36

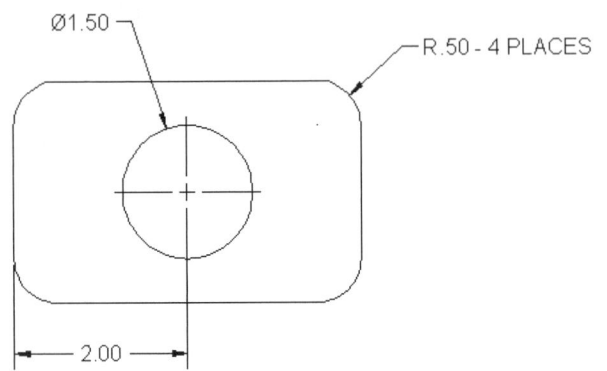

Figure 4-37

Figure 4-38

8.  Select the circle.

    *Specify dimension line location or [MText/ Text/Angle]:*

9.  Locate the dimension away from the object so that the leader line is neither horizontal nor vertical.

    Figure 4-37 shows the results.

**To add linear dimensions to given centerlines**

1.  Select the **Linear** tool on the **Dimension** panel.

    *Command: _dimlinear*
    *Specify first extension line origin or <select object>:*

2.  Select the lower endpoint of the circle's vertical centerline.

    *Specify second extension line origin:*

3.  Select the endpoint of the vertical edge line (the endpoint that joins with the corner arc).

    Figure 4-38 shows the results.

4.  Repeat the above procedure to add the vertical dimension needed to locate the circle's center point.

5.  Add the overall dimensions using the **Linear** dimension tool.

    Figure 4-39 shows the results. Radius and diameter dimensions are usually added to a drawing after the linear dimensions because they are less restricted in their locations. Linear dimensions are located close to the distance they are defining, whereas radius and diameter dimensions can be located farther away and use leader lines to identify the appropriate arc or circle.

    Avoid crossing extension and dimension lines with leader lines. See Figure 4-40.

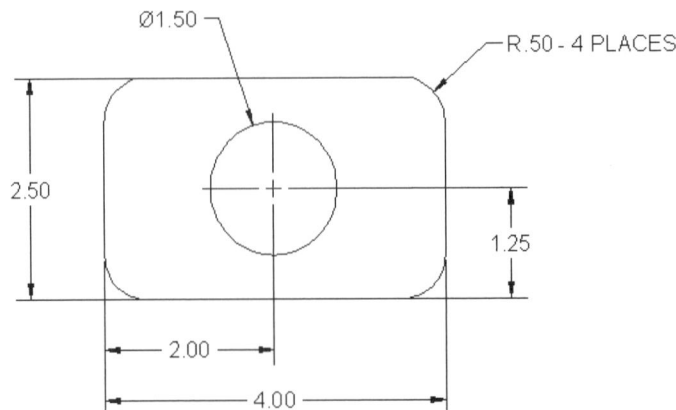

Figure 4-39

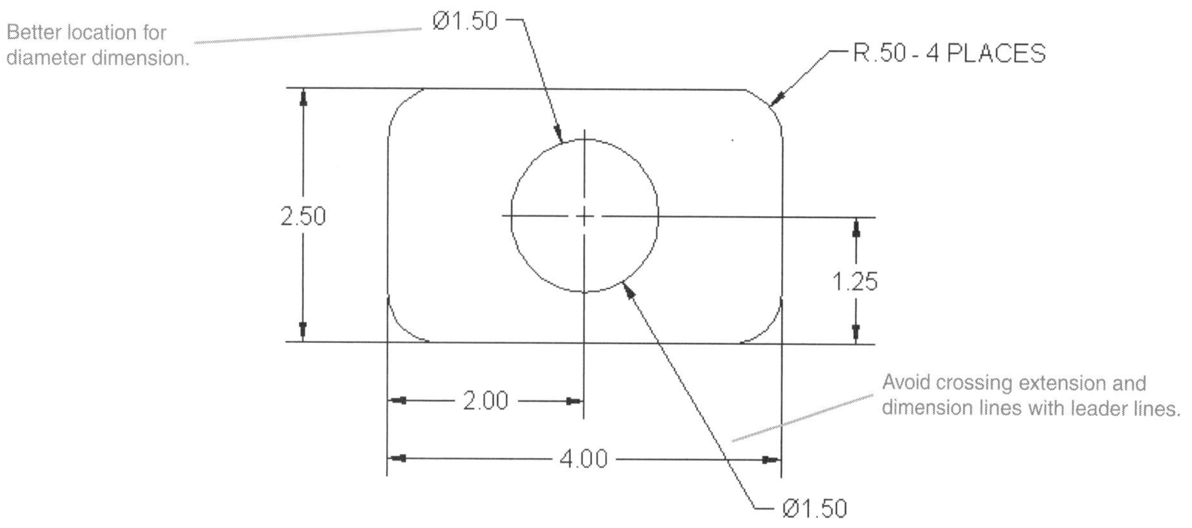

Better location for
diameter dimension.

Ø1.50

R.50 - 4 PLACES

2.50

1.25

Avoid crossing extension and
dimension lines with leader lines.

2.00

4.00

Ø1.50

**Figure 4-40**

51.34°

128.66°

128.66°

51.34°

Angular dimensions

Aligned

Angular

**Angular**
Creates an angular dimension

Arc L

DIMANGULAR
Press F1 for more help

Radi

Diameter

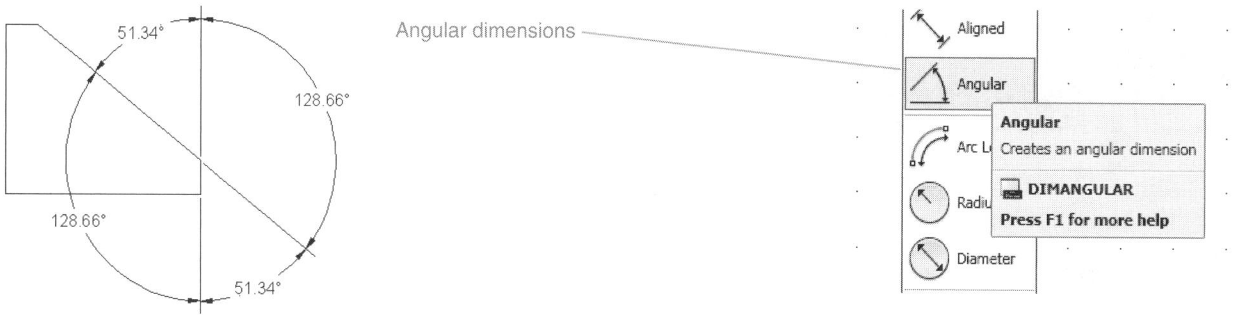

**Figure 4-41**

> **NOTE:**
>
> The diameter symbol can be added when us-
> ing the **Text Formatting** dialog box by typing
> **%%c** or by using the **Symbols** option. See
> Figure 4-40. The characters **%%c** will appear
> on the **Text Formatting** screen but will be con-
> verted to the diameter symbol Ø when the text
> is applied to the drawing.

## 4-8  ANGULAR DIMENSIONS

Figure 4-41 shows four possible angular dimensions
that can be created using the **Angular** tool on the **Dimension**
toolbar. The extension lines and degree symbol will be added
automatically.

Dimension this slanted surface using
an angular dimension.

**To create an angular dimension (See Figure 4-42.)**

1. Select the **Angular** tool on the **Dimension** panel.

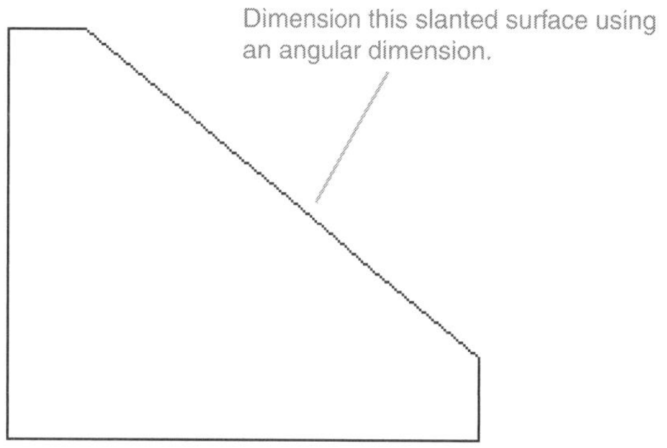

**Figure 4-42**

*Command: _dimangular*
*Select arc, circle, line, or <specify vertex>:*

2. Select the short vertical line on the lower right side of the object.

   *Select second line:*

3. Select the slanted line.

   *Specify dimension arc line location [MText/Text/ Angle]:*

4. Locate the text away from the object.

Figure 4-43 shows the results. It is considered better to use two extension lines for angular dimensions and to not have an arrowhead touch the surface of the part.

> **NOTE:**
>
> The degree symbol can be added when using the **Text Formatting** dialog box by using the **Symbols** option or by typing **%%d.** The characters **%%d** will appear on the **Text Formatting** screen but will be converted to ° when the text is applied to the drawing.

### Avoid overdimensioning

Figure 4-44 shows a shape dimensioned using an angular dimension. The shape is completely defined. Any additional dimension would be an error. It is tempting, in an effort to make sure a shape is completely defined, to add more dimensions, such as a horizontal dimension for the short horizontal edge at the top of the shape. This dimension is not needed and is considered double dimensioning.

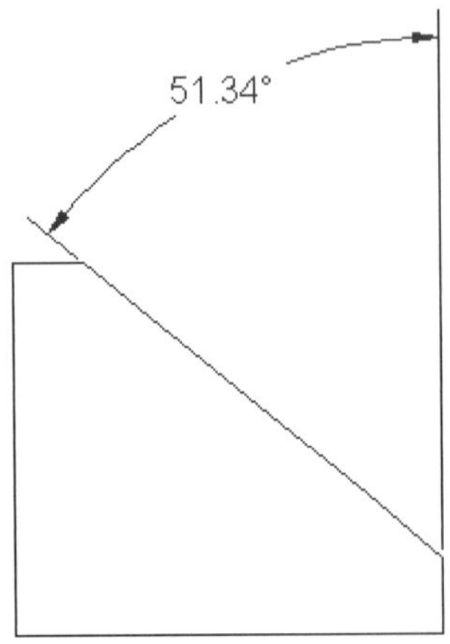

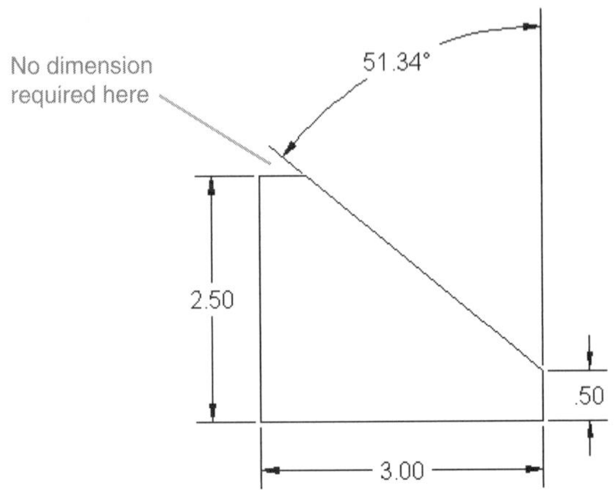

**Figure 4-44**

## 4-9 ORDINATE DIMENSIONS

Ordinate dimensions are dimensions based on an X,Y coordinate system. Ordinate dimensions do not include extension, or dimension lines, or arrowheads but simply horizontal and vertical leader lines drawn directly from the features of the object. Ordinate dimensions are particularly useful when dimensioning an object that includes many small holes.

Figure 4-45 shows an object that is to be dimensioned using ordinate dimensions. Ordinate dimensions are automatically calculated from the X, Y origin, or in this example, the lower left corner of the screen. If the object had been drawn with its lower left corner on the origin, you could proceed directly to the **Ordinate** tool on the **Dimension** toolbar; however, the lower left corner of the object is currently located at X = 3, Y = 4. First move the origin to the corner of the object, then use the **Ordinate** dimension tool.

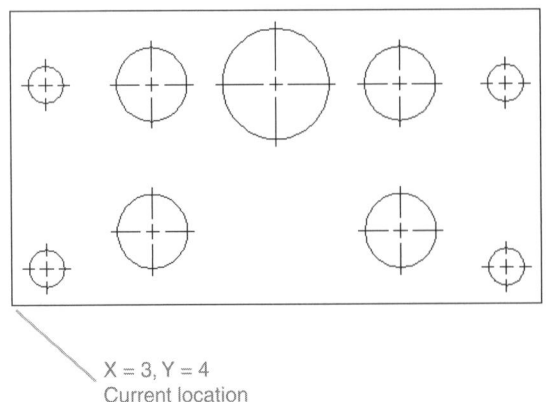

**Figure 4-43**

**Figure 4-45**

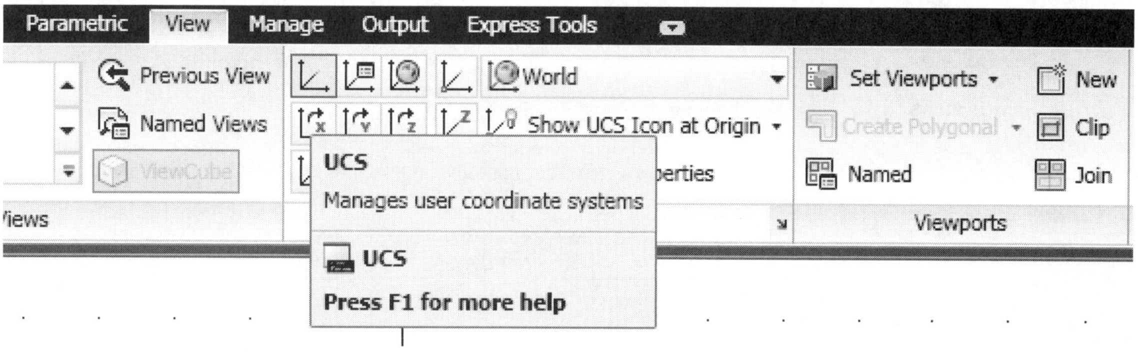

Figure 4-46

## To move the origin

1.  Access the UCS tool located on the Coordinates panel under the View tab.

    Specify origin of UCS or [Face/Named/Object/Previous/ View/World/X/Y/ZAxis]<World>:

2.  Click the lower left corner of the object.
3.  Right click the mouse.
4.  Select the lower left corner of the object.

The origin (0,0) is now located at the lower left corner of the object. This can be verified by looking at the coordinate display at the lower left corner of the screen.

The origin icon may move to the lower left corner as shown in Figure 4-47, depending on your computer's settings. The tool can be moved back to the original screen location as follows.

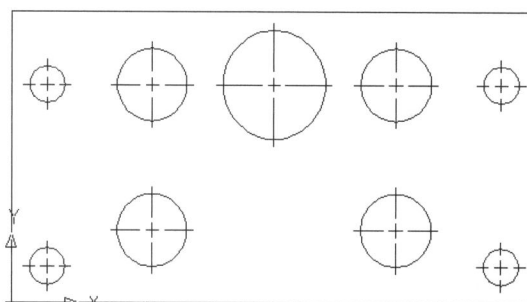

Origin has been relocated to the lower-left corner of the object.

Figure 4-47

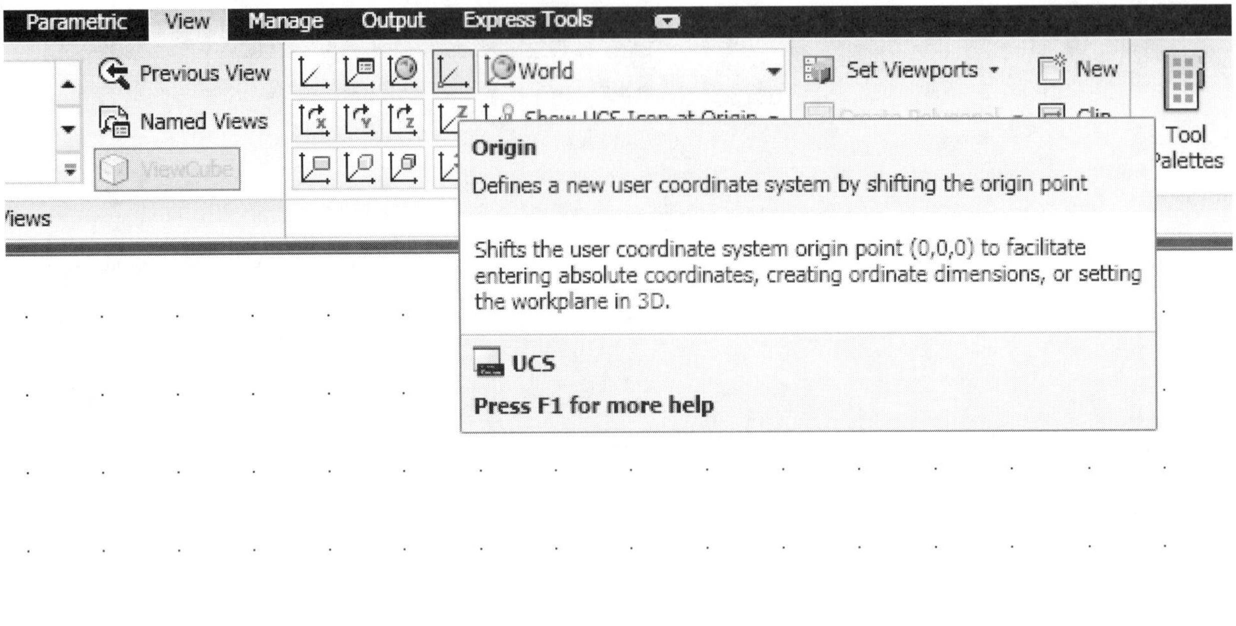

Figure 4-48

## To move the Origin icon

The Origin icon can also be moved using the Origin tool located on the Coordinates panel under the View tab.

1. Click the Origin tool

   Specify new origin point <0,0,0>:

2. Click the lower left corner of the object.
3. Right click the mouse.

## To add ordinate dimensions to an object

The following procedure assumes that you have already used the **Dimension Style Manager** (Section 4-4) to set the desired dimension style and that you have moved the origin to the lower left corner of the object as shown.

1. Turn the **Ortho Mode** command on (click the **Ortho Mode** box at the bottom of the screen). See Figure 4-49.
2. Select the **Ordinate** tool from the **Dimension** toolbar.

   *Command: _dimordinate*
   *Select feature location:*

3. Select the lower endpoint of the first circle's vertical centerline.

   *Specify leader endpoint or [Xdatum/Ydatum/ MText/Text/Angle]:*

4. Select a point along the X axis directly below the vertical centerline of the circle.

   The ordinate value of the point will be added to the drawing. This point should have a **0.50** value. The text value may be modified using either the **MText** or **Text** option, or by using the **Dimension Style Manager** dialog box to define the precision of the text.

5. Press the right mouse button to restart the command and dimension the object's other features.
6. Extend the centerlines across the object and add the diameter dimensions for the holes.

   Figure 4-49 shows the completed drawing. The **Text** option of the prompt shown in step 3 can be used to modify or remove the default text value.

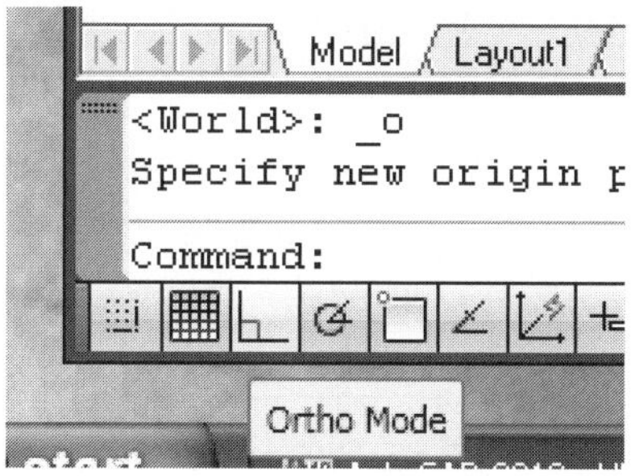

Ordinate dimensions

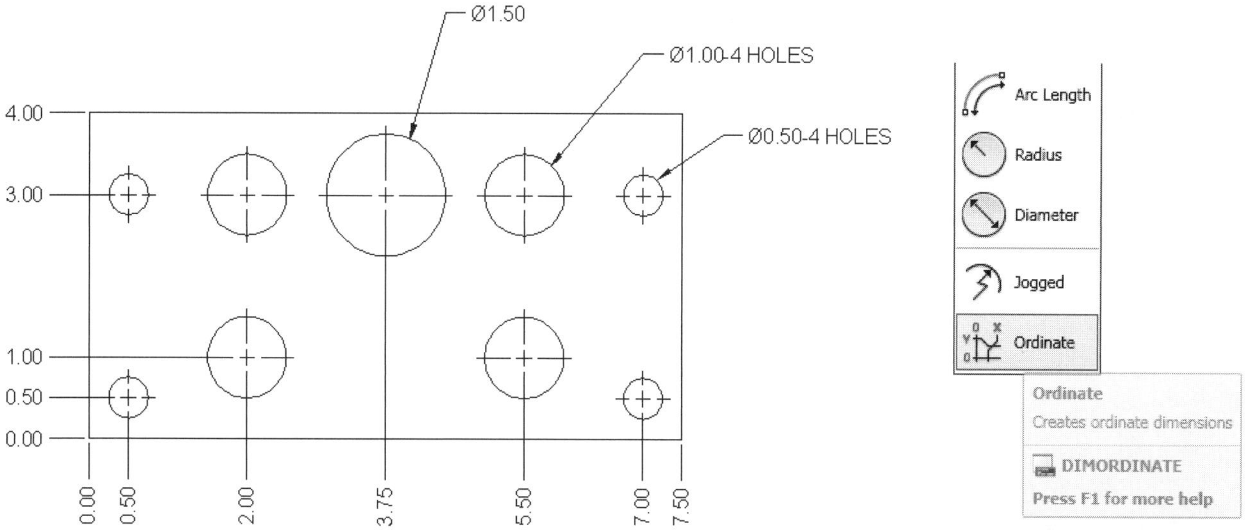

Figure 4-49

## 4-10  BASELINE DIMENSIONS

***Baseline dimensions*** are a series of dimensions that originate from a common baseline or datum line. Baseline dimensions are very useful because they help eliminate tolerance buildup associated with chain-type dimensions.

The **Baseline** tool can be used only after an initial dimension has been drawn. AutoCAD will define the first extension line origin of the initial dimension selected as the baseline for all baseline dimensions.

**To use the Baseline dimension tool (See Figure 4-50.)**

1. Select the **Linear** tool on the **Dimension** panel.

   *Command: _dimlinear*
   *Specify first extension line origin or <select object>:*

2. Select the upper left corner of the object.

   This selection determines the baseline.

   *Specify second extension line origin:*

3. Select the endpoint of the first circle's vertical centerline.

*Specify dimension line location or [Text/Angle/ Horizontal/Vertical/Rotated]:*

4. Select a location for the dimension line.

   *Command:*

5. Select the **Baseline** tool on the **Dimension** panel.

   *Specify a second extension line origin or [Undo/ Select] <Select>:*

6. Select the endpoint of the next circle's vertical centerline.

   *Specify a second extension line origin or [Undo/Select] <Select>:*

7. Continue to select the circle centerlines until all circles are located.

   *Specify a second extension line origin or [Undo/Select] <Select>:*

8. Select the upper right corner of the object.
9. Press the right mouse button, then press **Enter**.

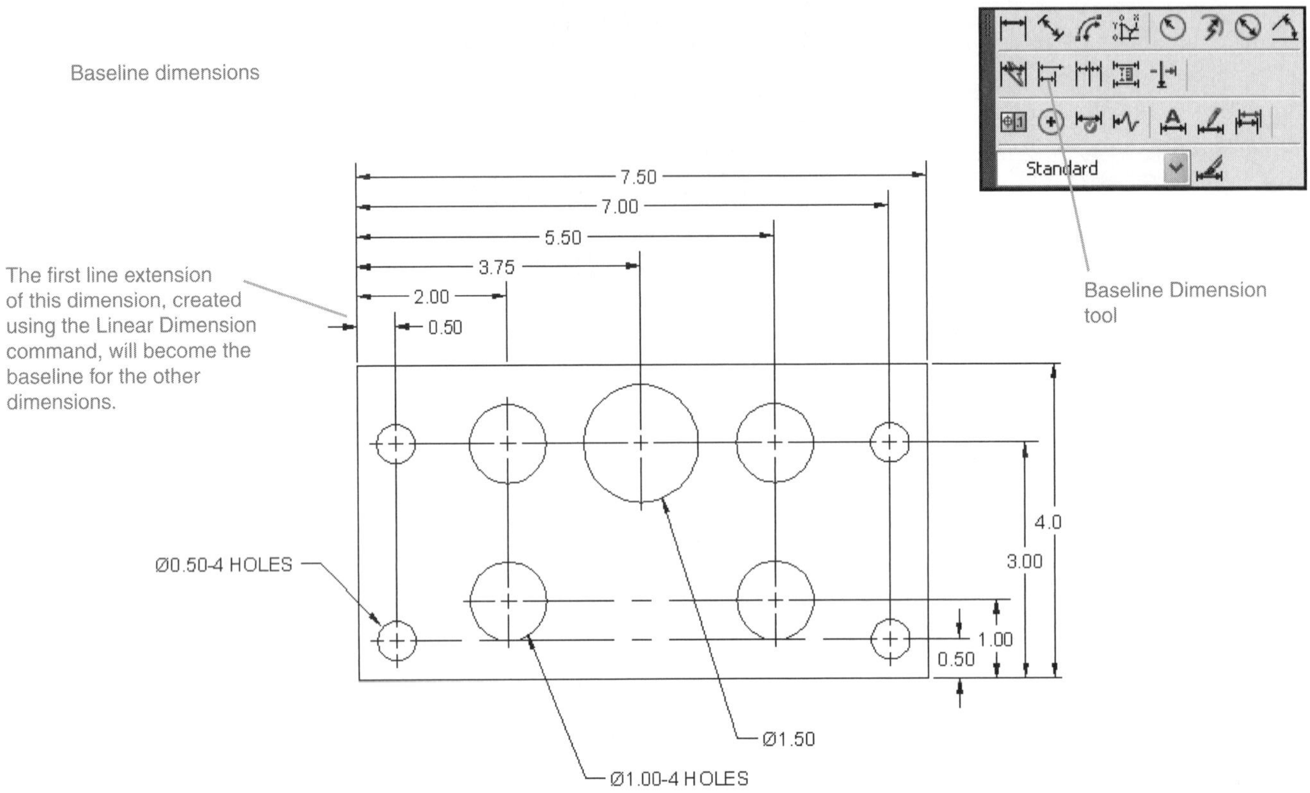

Baseline dimensions

The first line extension of this dimension, created using the Linear Dimension command, will become the baseline for the other dimensions.

Baseline Dimension tool

Ø0.50-4 HOLES

Ø1.50

Ø1.00-4 HOLES

**Figure 4-50**

This will end the **Baseline** dimension command.

10. Repeat the preceding procedure for the vertical baseline dimensions.
11. Add the circles' diameter values.

The **Baseline** dimension option can also be used with the **Angular** dimension option.

## 4-11 CONTINUE DIMENSION

The **Continue** tool on the **Dimension** panel is used to create chain dimensions based on an initial linear, angular, or ordinate dimension. The second extension line's origin becomes the first extension line origin for the continued dimension.

**To use the Continue dimension command (See Figure 4-51.)**

1. Select the **Linear** tool on the **Dimension** panel.

   *Command: dimcontinue*
   *Specify first extension line origin or <select object>:*

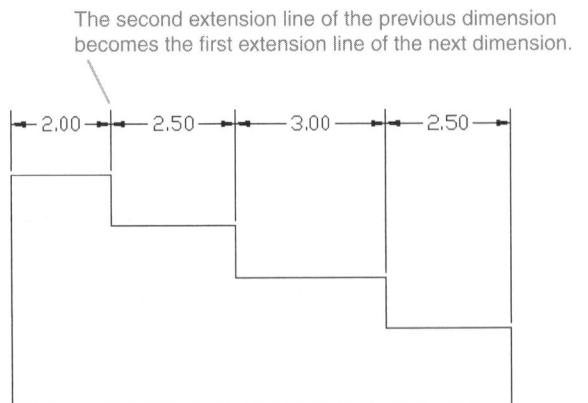

2. Select the upper left corner of the object.

   *Specify second extension line origin:*

3. Select the right endpoint of the uppermost horizontal line.

   *Dimension line location (Text/Angle/Horizontal/Vertical/Rotated):*

The second extension line of the previous dimension becomes the first extension line of the next dimension.

**Figure 4-51**

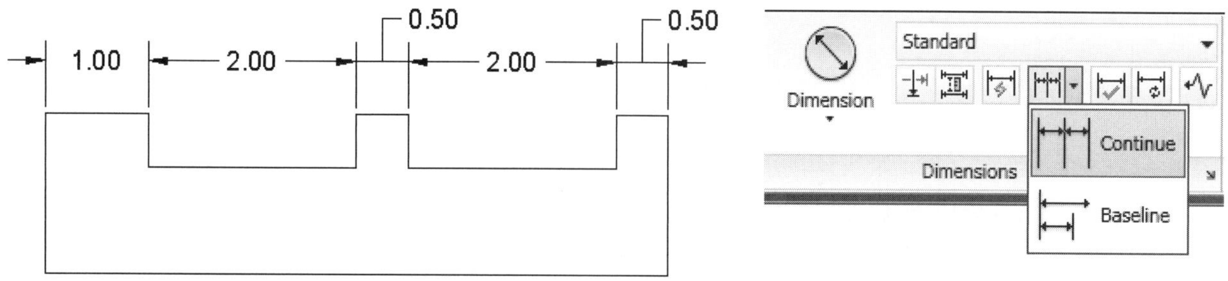

Figure 4-52

4. Select a dimension line location.

   *Command:*

5. Select the **Continue** tool on the **Dimension** panel.

   *Command: _dimcontinue*
   *Specify a second extension line origin or [Undo/Select] <Select>:*

6. Select the next linear distance to be dimensioned.

   *Specify a second extension line origin or [Undo/Select] <Select>:*

7. Continue until the object's horizontal edges are completely dimensioned.

   AutoCAD will automatically align the dimensions. Figure 4-52 shows how the **Continue** tool dimensions

distances that are too small for both the arrowhead and dimension value to fit within the extension lines.

## 4-12  QUICK DIMENSION

The **Quick Dimension** tool is used to add a series of dimensions. See Figure 4-53.

**To use the Quick Dimension command**

1. Select the **Quick Dimension** tool from the **Dimension** panel.

   *Command: _qdim*
   *Select geometry to dimension:*

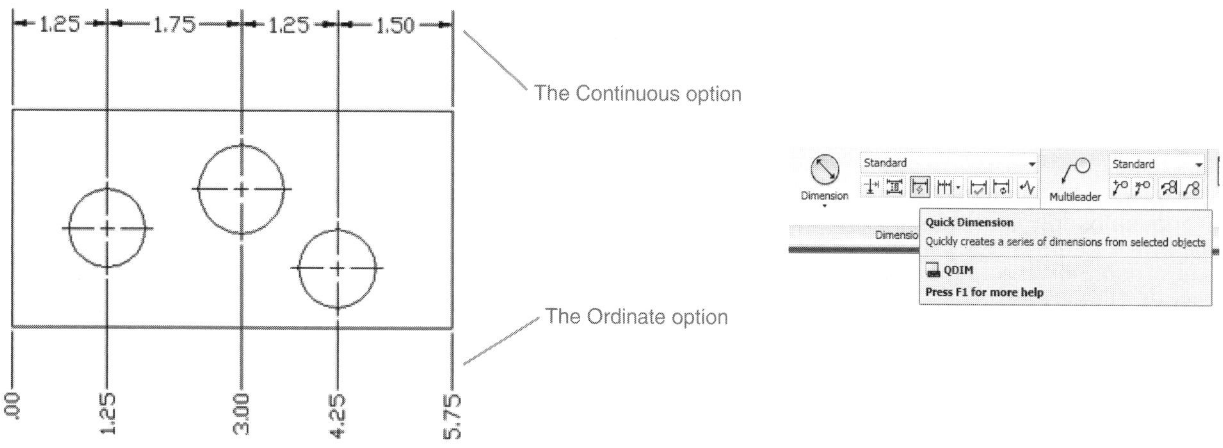

Figure 4-53

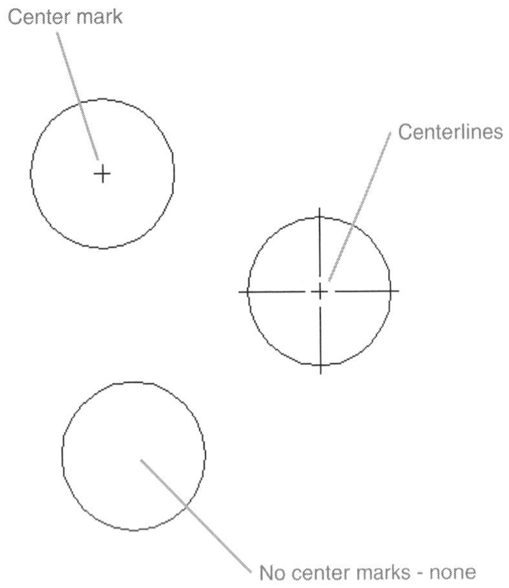

Center mark

Centerlines

No center marks - none

**Figure 4-54**

2. Select the left vertical edge line of the object to be dimensioned; press **Enter.**

   *Specify dimension line position, or [Continuous/ Staggered/Baseline/Ordinate/Radius/Diameter/ datumPoint/Edit]<Baseline>:*

3. Select the vertical centerline of the first hole.

   *Select geometry to dimension:*

4. Select the vertical centerline of the second hole.

   *Select geometry to dimension:*

5. Select the vertical centerline of the third hole.
6. Press the right mouse button.

   *Specify dimension line position, or [Continuous/ Staggered/Baseline/Ordinate/Radius/Diameter/ datumPoint/Edit]<Baseline>:*

7. Type **c;** press **Enter.**
8. Position the dimension lines, press the right mouse button, and enter the position.

   The ordinate dimensions along the bottom of the object in Figure 4-53 were created by typing **o** rather than **c** in step 7.

## 4-13  CENTER MARK

When AutoCAD first draws a circle or arc, a center mark appears on the drawing; however, these marks will disappear when the **Redraw View** or **Redraw All** command is applied. See Figure 4-54.

**To add centerlines to a given circle**

1. Select the **Dimension Style** panel.

   The **Dimension Style** dialog box appears.

2. Select **Modify,** then the **Symbols and Arrows** option.

   The **Modify Dimension Style: Standard** dialog box will appear.

3. Select the **Line** option.

   The preview display will show a horizontal and a vertical centerline.

4. Select **OK** to return to the drawing.
5. Select the **Center Mark** tool.

   *Select arc or circle:*

6. Select the circle.

   Horizontal and vertical centerlines will appear. The size of the center mark can be controlled using the **Size** box in the **Center marks** box. If the centerline's size appears unacceptable, try different sizes until you get an acceptable size.

## 4-14  MLEADER AND QLEADER

*Leader lines* are slanted lines that extend from notes or dimensions to a specific feature or location on the surface of a drawing. They usually end with an arrowhead or dot. The **Radius** and **Diameter** tools on the **Dimension** panel automatically create a leader line. The **MLEADER** and **QLEADER** commands can be used to add leader lines not associated with radius and diameter dimensions.

**To create a leader line with text**

1. Type **qleader** in response to a command prompt.

   *Command: _qleader
   Specify first leader point, or [Settings]<Settings>:*

2. Select the starting point for the leader line.

   This is the point where the arrowhead will appear. In the example shown in Figure 4-55 the upper right corner of the object was selected.

   *Specify next point:*

3. Select the location of the endpoint of the slanted line segment.

   *Specify next point:*

Figure 4-55

4. Draw a short horizontal line segment; press **Enter.**

   *Specify text width <0.0000>:*

5. Press **Enter.**

   *Enter first line of annotation text <MText>:*

6. Press **Enter.**

   The **Text Formatting** dialog box will appear.

7. Use the cursor and extend the text box as needed.
8. Type the desired text. See Figure 4-56.
9. Click the **OK** button.

   The text will appear next to the horizontal line segment of the leader line.

**To draw a curved leader line**

   The **Leader** command can be used to draw curved leader lines and leader lines that end with dots. See Figure 4-57.

1. Type **qleader** in response to a command prompt.

   *Command: _qleader*
   *Specify first leader point, or [Settings] <Settings>:*

2. Type **s;** press **Enter.**
3. Select the **Leader Line & Arrow** option.

   The **Leader Settings** dialog box will appear. See Figure 4-58.

4. Select the **Spline** option in the **Leader Line** box, then **OK.**

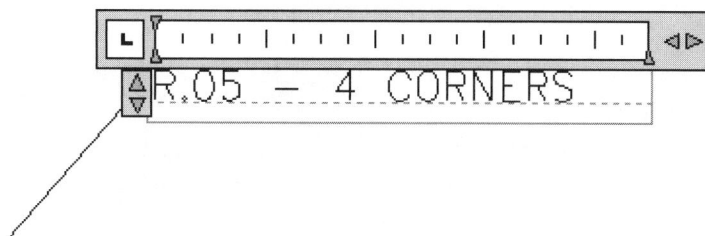

Figure 4-56

Curved leader line

SERVICE SPACE

5 WASHERS

Figure 4-57

*Specify first leader point, or [Settings] <Settings>:*

5.  Select the starting point for the leader line.

    This is the point where the arrowhead will appear.

    *Specify next point:*

6.  Select the next point.

    *Select next point:*

AutoCAD will shift to the **Drag** mode, which allows you to move the cursor around and watch the changes in shapes of the leader line. More than one point may be selected to define the shape.

7.  Complete the leader line as explained previously.

**To draw a leader line with a dot at its end**

To draw a leader with a dot at its end

1.  Click the arrow in the lower left corner of the Leaders panel located under the Annotation panel.

    The Multileader style Manager dialog box will appear. See Figure 4-58.

2.  Click the Modify option.

    The Modify Multileader Style: Standard dialog box will appear. See Figure 4-59.

3.  Click the arrow on the right end of the Symbol box under the Arrowhead heading.

4.  Select the Dot option

5.  Click OK.

## 4-15  DIMENSION EDIT

The **Dimension Edit** tool is used to edit existing dimensioning text. Existing text can be moved or rotated, or the text value can be changed. See Figure 4-60.

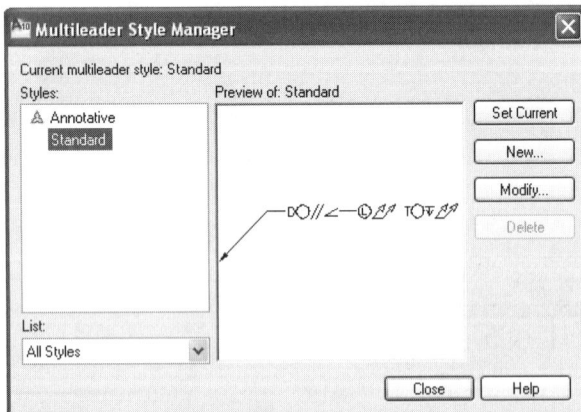

Figure 4-58

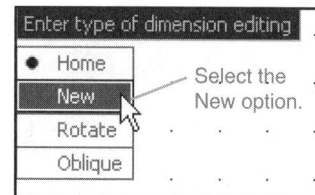

Figure 4-59

**To use the Dimension Edit option to change a text value**

1. Type dimtedit in response to a command prompt.

Select dimension:

2. Select the dimension.

Select new location for dimension or [Left/Right/Center/Home/Angle]:

3. Use the cursor and move the dimension text to the left.
4. Click the mouse.

See Figure 4-60.

**To change the angle of existing dimension text**

1. Type dimtedit in response to a command prompt.

Select dimension:

2. Select the dimension.

Select new location for dimension or [Left/Right/Center/Home/Angle]:

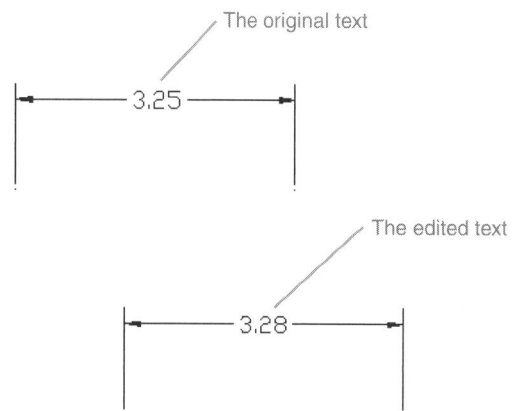

Figure 4-60

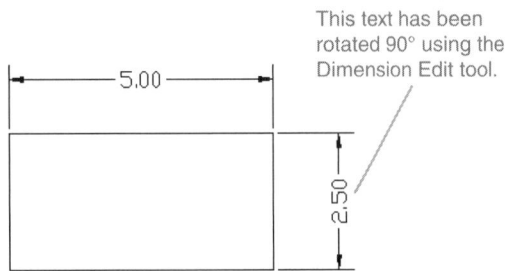

**Figure 4-61**

3. Type A

   Specify angle for dimension text:

4. Type 90; press Enter.

   See Figure 4-61.

## 4-16 TOLERANCES

*Tolerances* are numerical values assigned with the dimensions that define the limits of manufacturing acceptability for a distance. AutoCAD can create four types of tolerances: symmetrical, deviation, limits, and basic. See Figure 4-62. Many companies also use a group of standard

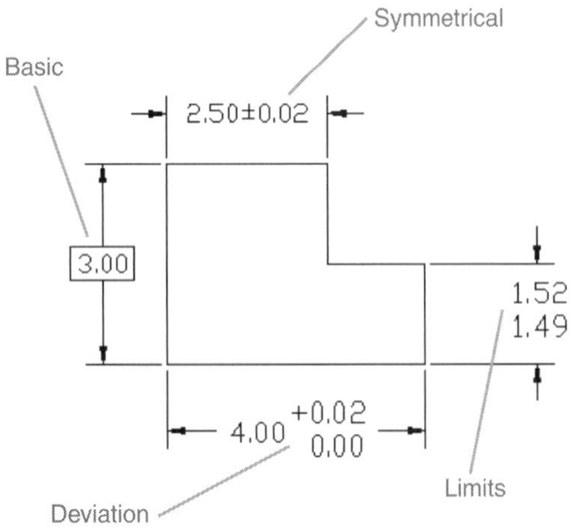

**Figure 4-62**

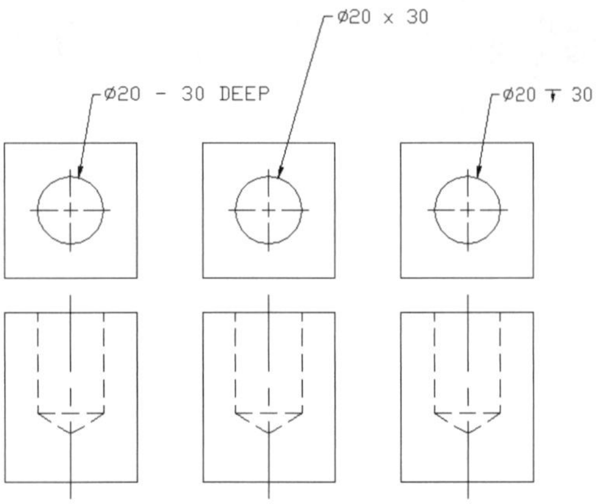

**Figure 4-63**

tolerances that are applied to any dimensional value that is not assigned a specific tolerance. See Figure 4-23. Tolerances for numerical values expressed in millimeters are applied using a different convention than that used for inches. Tolerances are discussed in Chapters 4 and 5. The discussion of the appropriate dimensioning tools will be covered in those chapters.

## 4-17 DIMENSIONING HOLES

Holes are dimensioned by stating their diameter and depth, if any. The symbol Ø is used to represent diameter. It is considered good practice to dimension a hole using a diameter value because the tooling used to produce the hole is also defined in terms of diameter values. A notation like 12 DRILL is considered less desirable because it specifies a machining process. Manufacturing processes should be left, whenever possible, to the discretion of the shop.

**To dimension individual holes**

Figure 4-63 shows three different methods that can be used to dimension a hole that does not go completely through an object. If the hole goes completely through, only the diameter need be specified. The **Radius** and **Diameter** dimension tools were covered in Section 4-7. Depth values may be added using the **Text Formatting** dialog box or by using the **Text** option of any of the dimension commands.

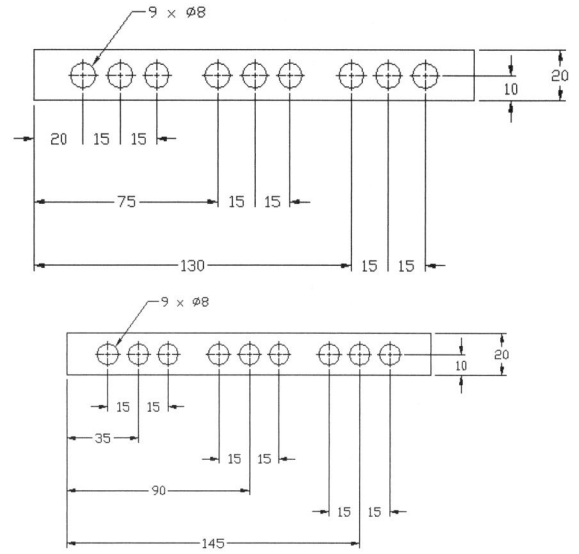

**Figure 4-66**

**Figure 4-64**

Figure 4-64 shows two methods of dimensioning holes in sectional views. The single-line note version is the preferred method.

### To dimension hole patterns

Figure 4-65 shows two different hole patterns dimensioned. The circular pattern includes the note **Ø10-4 HOLES.** This note serves to define all four holes within the object.

Figure 4-65 also shows a rectangular object that contains five holes of equal diameter, equally spaced from one another. The notation **5 × Ø10** specifies five holes of 10 diameter. The notation **4 × 20 (=80)** means 4 equal spaces of 20. The notation **(=80)** is a reference dimension and is included for convenience. Reference dimensions are explained in Chapter 9.

Figure 4-66 shows two additional methods for dimensioning repeating hole patterns. Figure 4-67 shows a circular hole pattern that includes two different hole diameters. The hole diameters are not noticeably different and could be confused. One group is defined by indicating letter (A); the other is dimensioned in a normal manner.

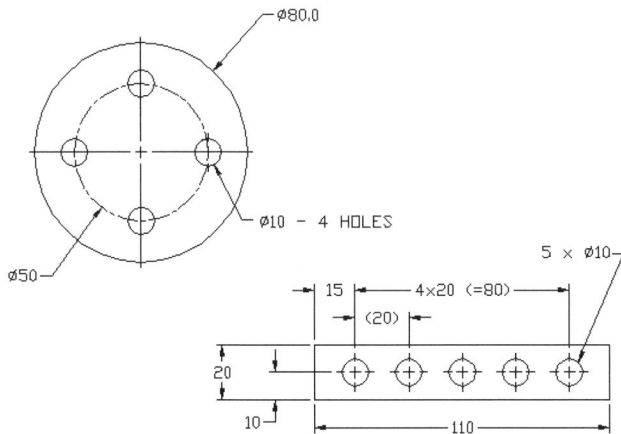

**Figure 4-65**

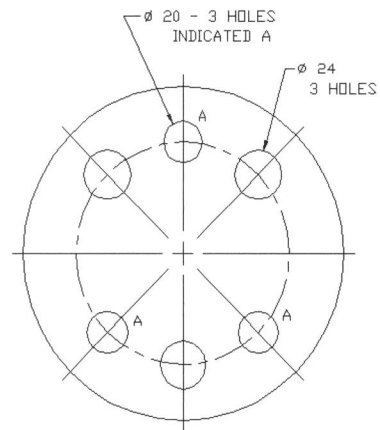

**Figure 4-67**

Place shorter dimensions closer to the object than longer ones.

Place dimensions near the features they are defining.

DO NOT PLACE DIMENSIONS ON THE SURFACE OF THE OBJECT.

Use the Explode, Erase, and Move commands to reconstruct and relocate inappropriate dimensions.

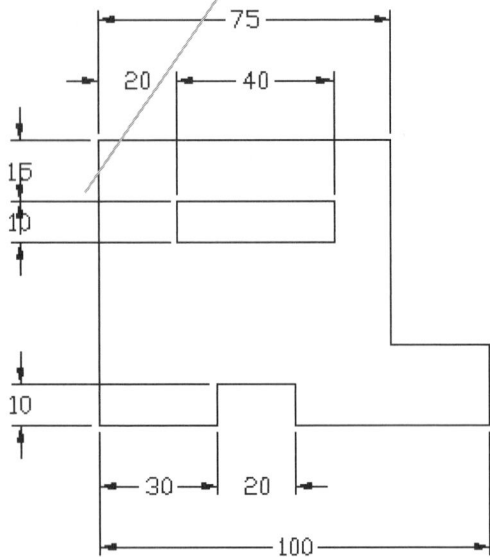

Align groups of dimensions.

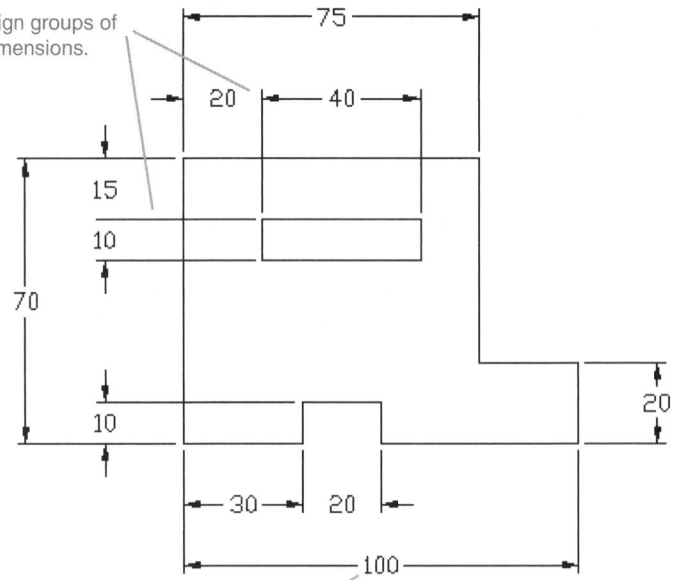

Place overall dimensions the farthest away from the object.

Figure 4-68

## 4-18  PLACING DIMENSIONS

There are several general rules concerning the placement of dimensions. See Figure 4-68.

1. Place dimensions near the features they are defining.
2. Do not place dimensions on the surface of the object.
3. Align and group dimensions so that they are neat and easy to understand.
4. Avoid crossing extension lines.

Sometimes it is impossible not to cross extension lines because of the complex shape of the object, but whenever possible, avoid crossing extension lines.

5. Place shorter dimensions closer to the object than longer ones.
6. Always place overall dimensions the farthest away from the object.
7. Do not dimension the same distance twice. This is called *double dimensioning* and will be discussed in Chapter 4.

## 4-19  FILLETS AND ROUNDS

Fillets and rounds may be dimensioned individually or by a note. In many design situations all the fillets and rounds are the same size, so a note as shown in Figure 4-69 is used. Any fillets or rounds that have a different radius from that specified by the note are dimensioned individually.

## 4-20  ROUNDED SHAPES (INTERNAL)

Internal rounded shapes are called *slots.* Figure 4-70 shows three different methods for dimensioning slots. The end radii are indicated by the note **R - 2 PLACES,** but no numerical value is given. The width of the slot is dimensioned, and it is assumed that the radius of the rounded ends is exactly half of the stated width.

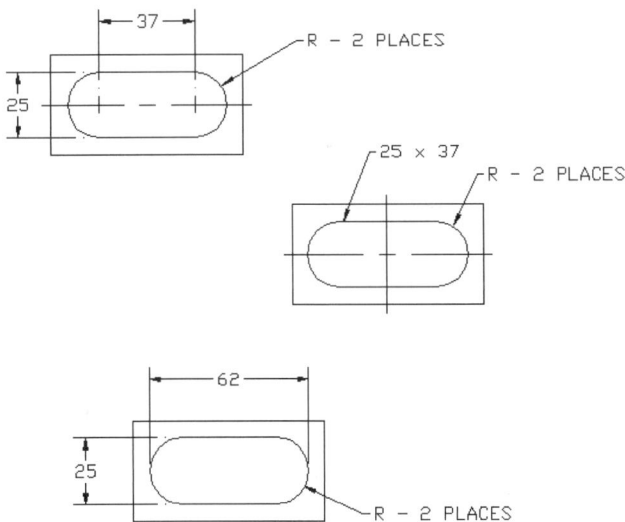

Figure 4-69

Figure 4-70

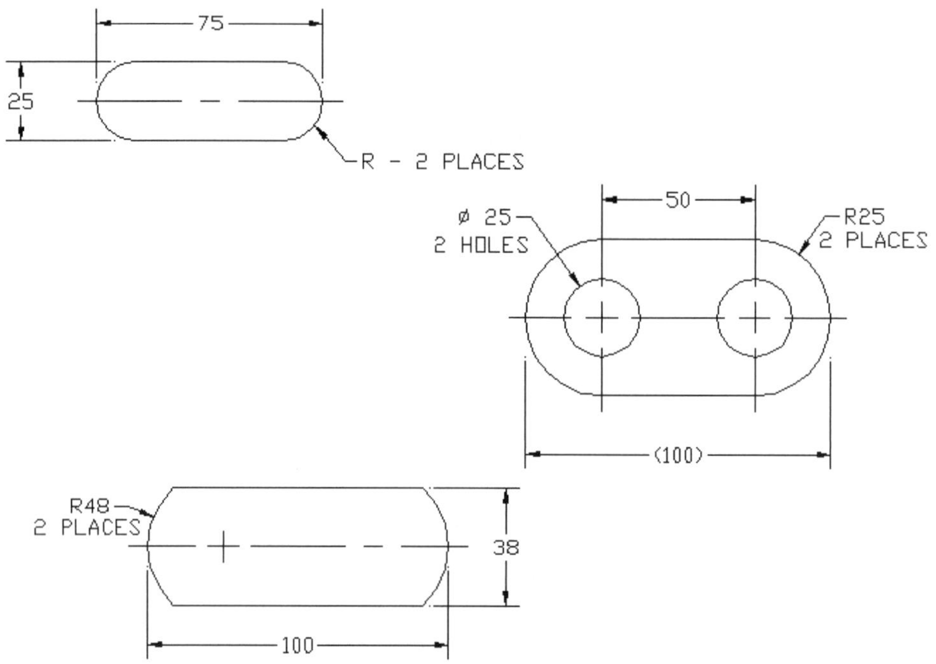

Figure 4-71

## 4-21 ROUNDED SHAPES (EXTERNAL)

Figure 4-71 shows two example shapes with external rounded ends. As with internal rounded shapes, the end radii are indicated but no value is given. The width of the object is given, and the radius of the rounded end is assumed to be exactly half of the stated width.

The second example shown in Figure 4-72 shows an object dimensioned using the object's centerline. This type of dimensioning is used when the distance between the hole is more important than the overall length of the object; that is, the tolerance for the distance between the holes is more exact than the tolerance for the overall length of the object.

The overall length of the object is given as a reference dimension (100). This means the object will be manufactured based on the other dimensions, and the 100 value will be used only for reference.

Objects with partially rounded edges should be dimensioned as shown in Figure 4-72. The radii of the end features are dimensioned. The center point of the radii is implied to be on the object centerline. The overall dimension is given; it is not referenced unless specific radii values are included.

## 4-22 IRREGULAR SURFACES

There are three different methods for dimensioning irregular surfaces: tabular, baseline, and baseline with oblique extension lines. Figure 4-72 shows an irregular surface dimensioned using the tabular method. The X,Y axes are defined using the edges of the object. Points are then defined relative to the X,Y axes. The points are assigned reference numbers, and the reference numbers and X,Y coordinate values are listed in chart form as shown.

Figure 4-73 shows an irregular curve dimensioned using baseline dimensions. The baseline method references all dimensions back to specified baselines. Usually there are two baselines, one horizontal and one vertical.

It is considered poor practice to use a centerline as a baseline. Centerlines are imaginary lines that do not exist on the object and would make it more difficult to manufacture and inspect the finished objects.

Baseline dimensioning is very common because it helps eliminate tolerance buildup (see Section 4-12) and is easily adaptable to many manufacturing processes. Auto-CAD has a special **Baseline** dimension tool for creating baseline dimensions.

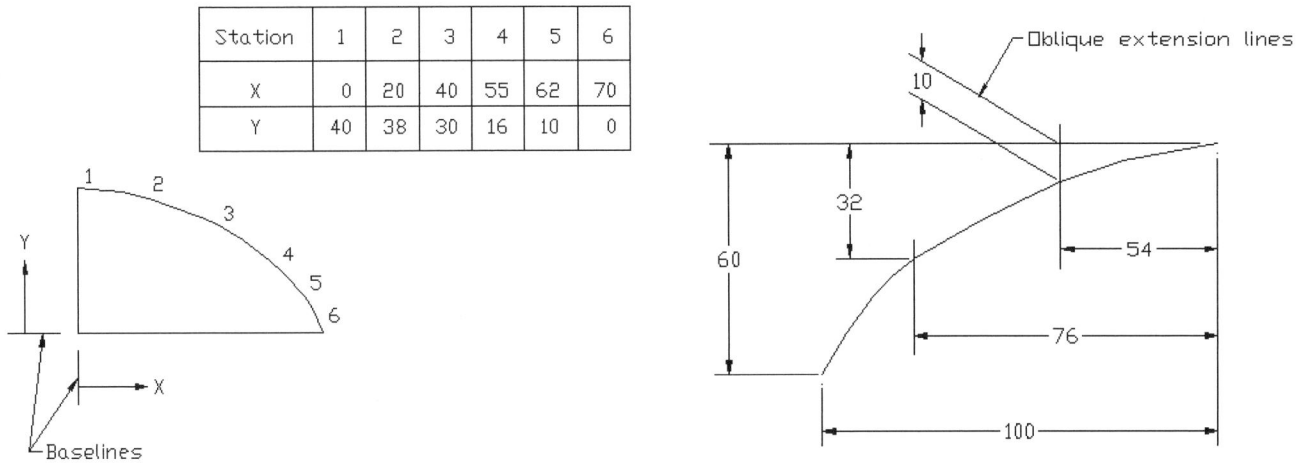

| Station | 1 | 2 | 3 | 4 | 5 | 6 |
|---------|---|---|---|---|---|---|
| X | 0 | 20 | 40 | 55 | 62 | 70 |
| Y | 40 | 38 | 30 | 16 | 10 | 0 |

**Figure 4-72**

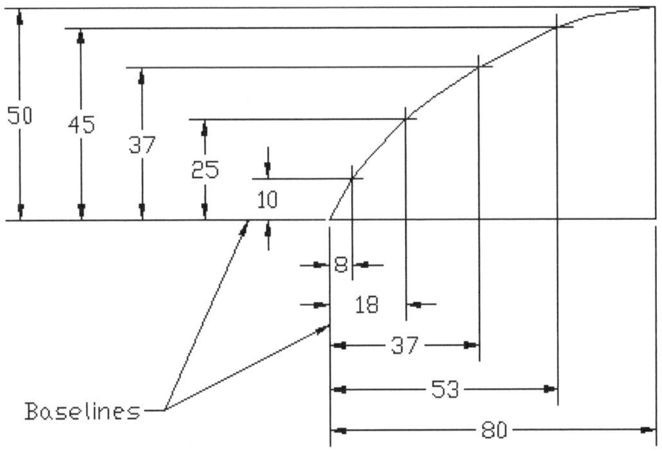

**Figure 4-73**

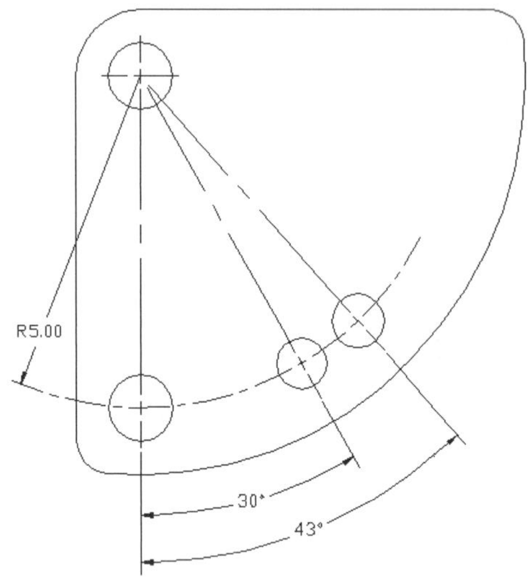

Figure 4-74

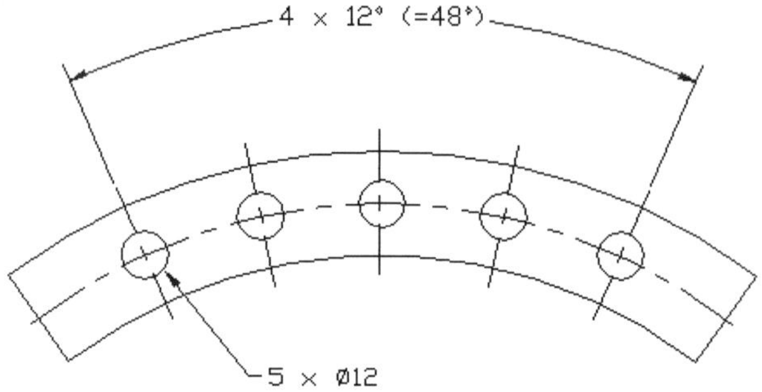

Figure 4-75

## 4-23  POLAR DIMENSIONS

Polar dimensions are similar to polar coordinates. A location is defined by a radius (distance) and an angle. Figure 4-74 shows an object that includes polar dimensions. The holes are located on a circular centerline, and their positions from the vertical centerline are specified using angles.

Figure 4-75 shows an example of a hole pattern dimensioned using polar dimensions.

## 4-24  CHAMFERS

**Chamfers** are angular cuts made on the edges of objects. They are usually used to make it easier to fit two parts together. They are most often made at 45° angles but may be made at any angle. Figure 4-76 shows two objects with chamfers between surfaces 90° apart and two examples between surfaces that are not 90° apart. Either of the two types of dimensions shown for the 45° dimension may be used. If an angle other than 45° is used, the angle and setback distance must be specified.

Figure 4-77 shows two examples of internal chamfers. Both define the chamfers using an angle and diameter. Internal chamfers are very similar to countersunk holes. See Section 1-26.

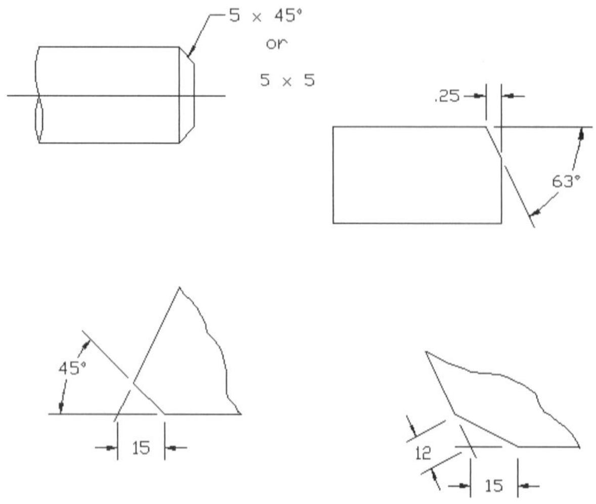

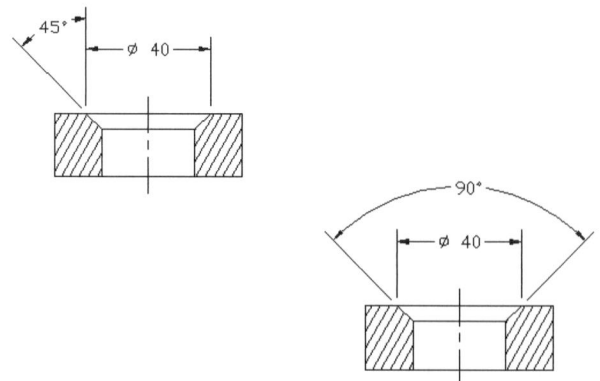

Figure 4-76

Figure 4-77

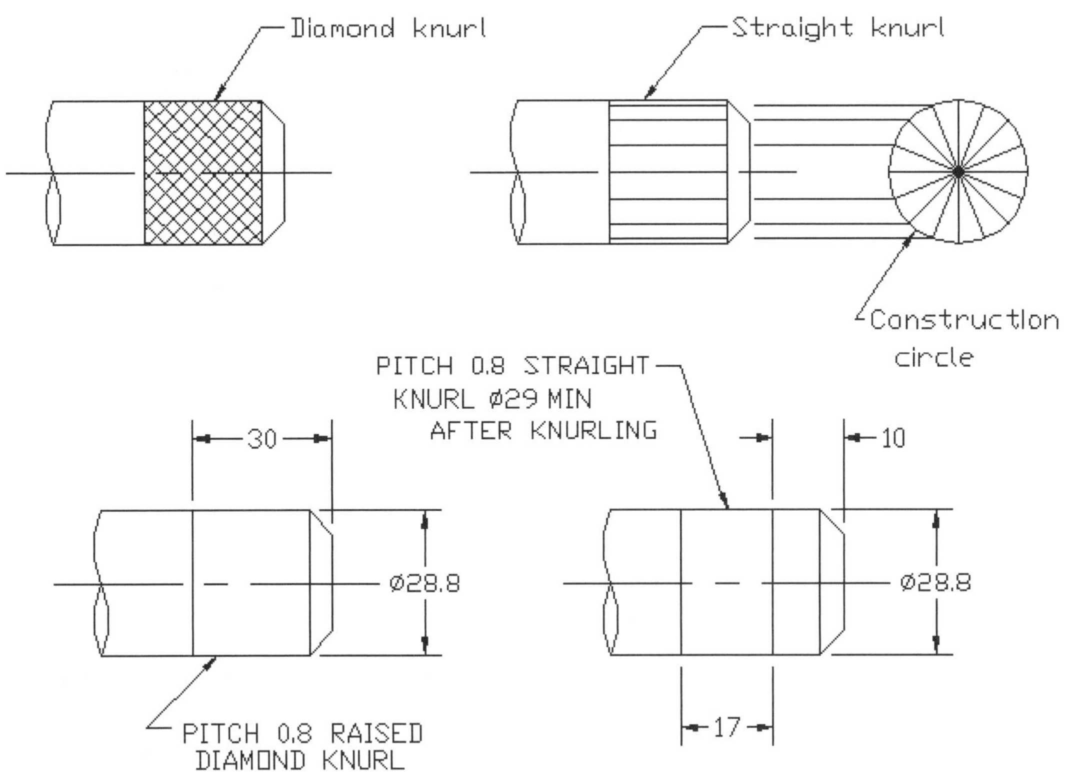

Figure 4-78

## 4-25 KNURLING

*Knurls* are used to make it easier to grip a shaft, or to roughen a surface before it is used in a press fit. There are two types of knurls: diamond and straight.

Knurls are defined by their pitch and diameter. See Figure 4-78. The *pitch* of a knurl is the ratio of the number of grooves on the circumference to the diameter. Standard knurling tools sized to a variety of pitch sizes are used to manufacture knurls for both English and metric units.

Diamond knurls may be represented by a double hatched pattern or by an open area with notes. The **Hatch** command is used to draw the double-hatched lines. See Section 2-4.

Straight knurls may be represented by straight lines in the pattern shown or by an open area with notes. The straight-line pattern is created by projecting lines from a construction circle. The construction points are evenly spaced on the circle. Once drawn, the straight-line knurl pattern can be saved as a wblock for use on other drawings.

## 4-26 KEYS AND KEYSEATS

*Keys* are small pieces of material used to transmit power. For example, Figure 4-79 shows how a key can be fitted between a shaft and a gear so that the rotary motion of the shaft can be transmitted to the gear.

There are many different styles of keys. The key shown in Figure 4-79 has a rectangular cross section and is called a square key. Keys fit into grooves called *keyseats,* or *keyways.*

Keyways are dimensioned from the bottom of the shaft or hole as shown.

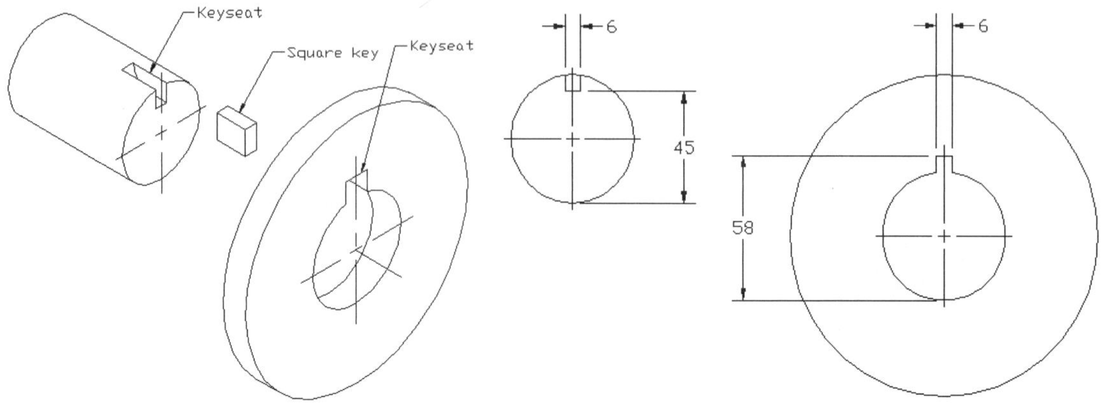

Figure 4-79

## 4-27 SYMBOLS AND ABBREVIATIONS

Symbols are used in dimensioning to help accurately display the meaning of the dimension. Symbols also help eliminate language barriers when reading drawings. Figure 4-80 shows a list of dimensioning symbols and their meanings. The height of a symbol should be the same as the text height.

Abbreviations should be used very carefully on drawings. Whenever possible, write out the full word including correct punctuation. Figure 4-81 shows several standard abbreviations used on technical drawings.

## 4-28 SYMMETRY AND CENTERLINE

An object is symmetrical about an axis when one side is an exact mirror image of the other. Figure 4-82 shows a symmetrical object. The two short parallel lines symbol or the note **OBJECT IS SYMMETRICAL ABOUT THIS AXIS** (centerline) may be used to designate symmetry.

If an object is symmetrical, only half the object need be dimensioned. The other dimensions are implied by the symmetry note or symbol.

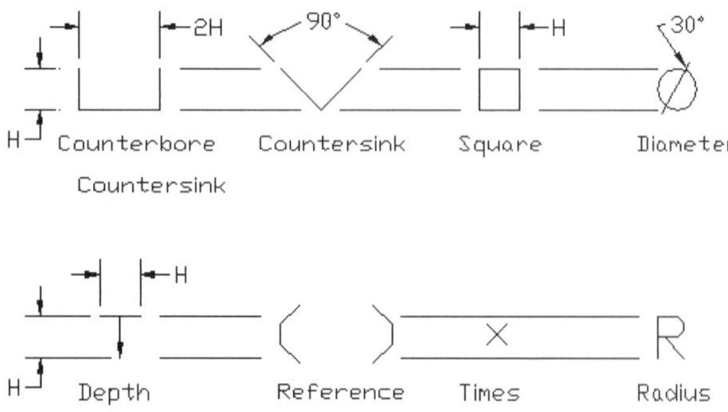

Figure 4-80

| | | |
|---|---|---|
| AL | = | Aluminum |
| C'BORE | = | Counterbore |
| CRS | = | Cold Rolled Steel |
| CSK | = | Countersink |
| DIA | = | Diameter |
| EQ | = | Equal |
| HEX | = | Hexagon |
| MAT'L | = | Material |
| R | = | Radius |
| SAE | = | Society of Automotive Engineers |
| SFACE | = | Spotface |
| ST | = | Steel |
| SQ | = | Square |
| REQD | = | Required |

Figure 4-81

**Figure 4-82**

**Figure 4-83**

Centerlines are slightly different from the axis of symmetry. An object may or may not be symmetrical about its centerline. See Figure 4-82. Centerlines are used to define the center of both individual features and entire objects. Use the centerline symbol when a line is a centerline, but do not use it in place of the symmetry symbol.

## 4-29  DIMENSIONING TO POINTS

Curved surfaces can be dimensioned using theoretical points. See Figure 4-83. There should be a small gap between the surface of the object and the lines used to define the theoretical point. The point should be defined by the intersection of at least two lines.

There should also be a small gap between the extension lines and the theoretical point used to locate the point.

## 4-30  COORDINATE DIMENSIONS

Coordinate dimensions are used for objects that contain many holes. Baseline dimensions can also be used, but when there are many holes, baseline dimensions can create a confusing appearance and will require a large area on the drawing. Coordinate dimensions use charts that simplify the appearance, use far less space on the drawing, and are easy to understand.

Figure 4-84 shows an object that has been dimensioned using coordinate dimensions without dimension lines. Holes are identified on the drawing using letters. Holes of equal diameter use the same letter. The hole diameters are presented in chart form.

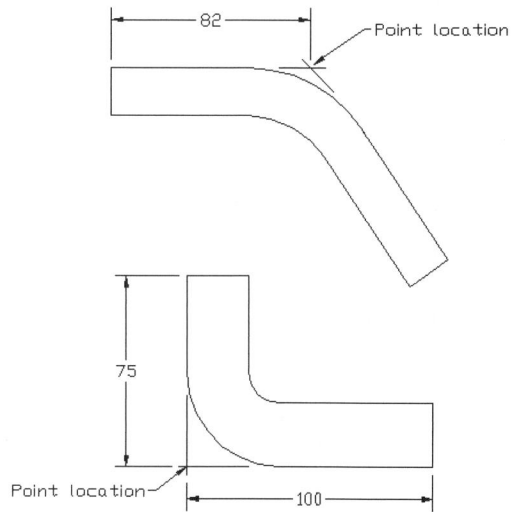

Hole locations are defined using a series of centerlines referenced to baselines. The distance from the baseline to the centerline is written below the centerline as shown.

Figure 4-85 shows an object that has been dimensioned using coordinate dimensions in tabular form. Each hole is assigned both a letter and a number. Holes of equal diameter are assigned the same letter. A chart is used to define the diameter values for each hole letter.

Hole locations are defined relative to X,Y axes. A Z axis is used for depth dimensions. A chart lists each hole by its letter–number designation and specifies its distance from the X, Y, or Z axis. The overall dimensions are given using extension and dimension lines.

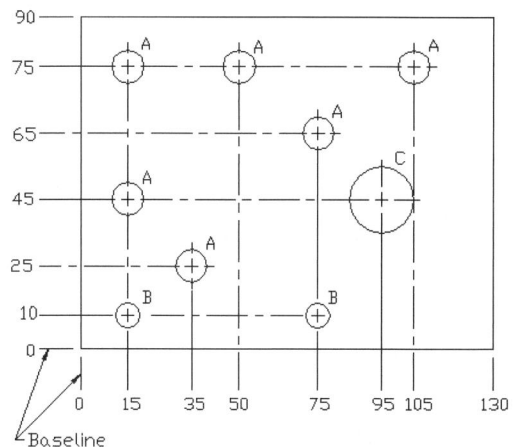

| SIZE SYMBOL | A | B | C |
|---|---|---|---|
| HOLE DIA | 10 | 7.5 | 20 |

**Figure 4-84**

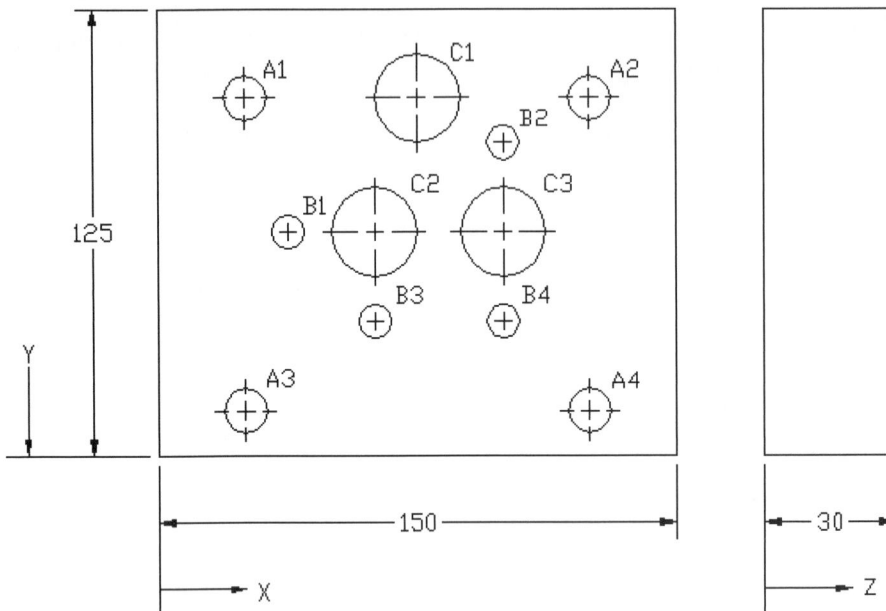

| HOLE | FROM | X | Y | Z |
|------|------|-----|-----|------|
| A1 | XY | 15 | 65 | THRU |
| A2 | XY | 80 | 65 | THRU |
| A3 | XY | 15 | 10 | THRU |
| A4 | XY | 80 | 10 | THRU |
| B1 | XY | 25 | 40 | 12 |
| B2 | XY | 65 | 56 | 12 |
| B3 | XY | 40 | 25 | 12 |
| B4 | XY | 65 | 25 | 12 |
| C1 | XY | 48 | 65 | THRU |
| C2 | XY | 40 | 40 | THRU |
| C3 | XY | 65 | 40 | THRU |

| HOLE | DESCRIPTION | QTY |
|------|-------------|-----|
| A | ⌀8 | 4 |
| B | ⌀5 | 4 |
| C | ⌀16 | 3 |

**Figure 4-85**

The side view does not show any hidden lines because if all the lines were shown, it would be too confusing to understand. A note **THIS VIEW LEFT BLANK FOR CLARITY** may be added to the drawing.

## 4-31 SECTIONAL VIEWS

Sectional views are dimensioned, as are orthographic views. See Figure 4-86. The sectional lines should be drawn at an angle that allows the viewer to clearly distinguish between the sectional lines and the extension lines.

## 4-32 ORTHOGRAPHIC VIEWS

Dimensions should be added to orthographic views where the features appear in contour. Holes should be dimensioned in their circular views. Figure 4-87 shows three views of an object that has been dimensioned.

The hole dimensions are added to the top view where the hole appears circular. The slot is also dimensioned in the top view because it appears in contour. The slanted surface is dimensioned in the front view.

The height of surface A is given in the side view rather than run along extension lines across the front view. The length of surface A is given in the front view. This is a contour view of the surface.

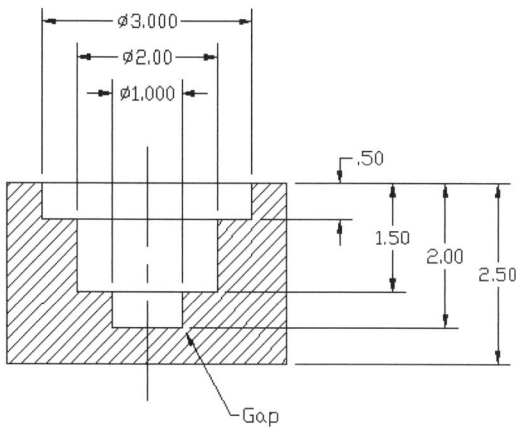

**Figure 4-86**

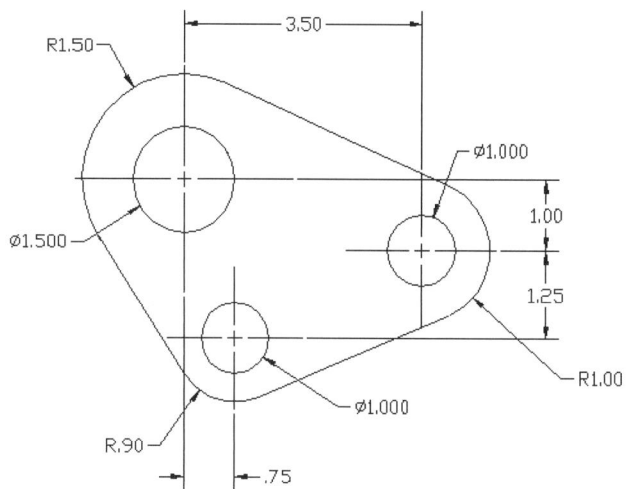

**Figure 4-88**

It is considered good practice to keep dimensions in groups. This makes it easier for the viewer to find dimensions.

Be careful not to double-dimension a distance. A distance should be dimensioned only once per view. If a 30 dimension were added above the 25 dimension on the right-side view, it would be an error. The distance would be double-dimensioned: once with the 25 + 30 dimension and again with the 55 overall dimension. The 25 + 30 dimensions are mathematically equal to the 55 overall dimension, but there is a distinct difference in how they affect the manufacturing tolerances. Double dimensions are explained more fully in Chapter 4.

Figure 4-88 shows an object dimensioned from its centerline. This type of dimensioning is used when the distance between the holes relative to each other are critical.

## 4-33 VERY LARGE RADII

Some radii are so large that it is not practical to draw the leader for the radius dimension at full size. Figure 4-89

shows an example of an object that uses foreshortened leader lines.

**To create a radius for large radii**

1. Click the **Jogged** tool on the **Dimension** panel.

   *Select arc or circle:*

2. Click the large arc.

   *Specify center location override:*

3. Select a center point for the dimension.

   The point selected is not the arc's true center point, as that is probably off the screen. The **Jogged** tool will override the true center point location and substitute the one selected. See Figure 4-89.

4. Right-click the mouse and click **Enter.**
5. Select a text location by moving the cursor and then clicking the left mouse button.

   Add a center mark to the end of the leader line.

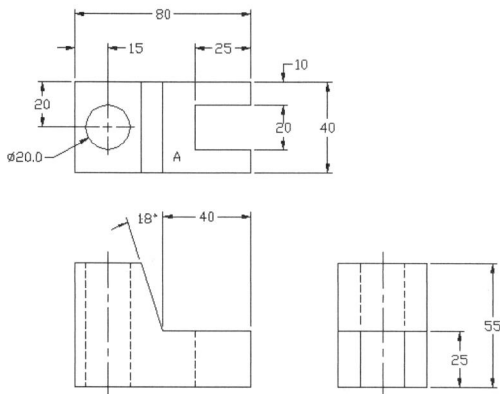

**Figure 4-87**

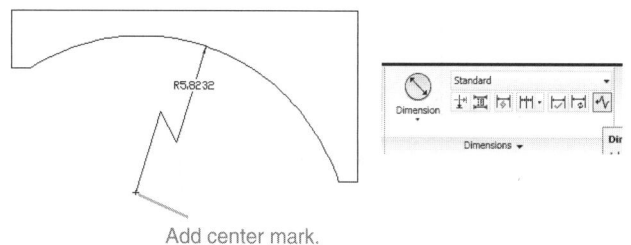

Add center mark.

**Figure 4-89**

# 4-34 EXERCISE PROBLEMS

Redraw the shapes shown in Exercise Problems EX4-1 through EX4-6. Locate the dimensions and tolerances as shown.

## EX4-1 INCHES

1. 3.00
2. 1.56

3. 46°

4. .750
5. 2.75
6. 3.625

7. 45°

8. 2.250

## EX4-2 MILLIMETERS

1. 38
2. 10
3. 5

4. 45°

5. 40
6. 22
7. 12
8. 25
9. 51
10. 76

## EX4-3 MILLIMETERS

1. 34.0
2. 17.0
3. 25.0
4. 15.00
5. 50.0
6. 80.0
7. R5 - 8 PLACES
8. 45
9. 60
10. Ø14 - 3 HOLES
11. 15.00
12. 30.00

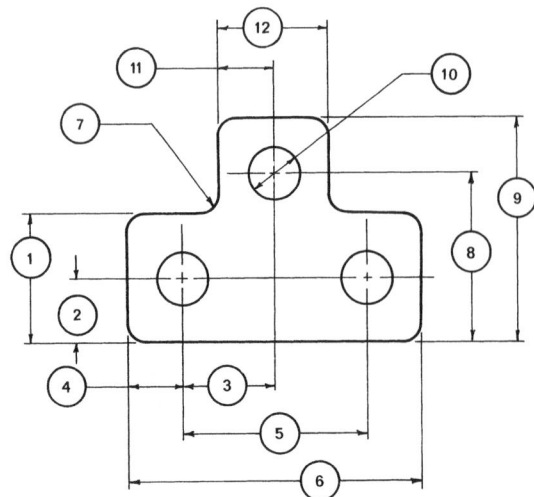

**EX4-4 INCHES**

1. 1.50

2. 1.50

3. .625

4. .750

5. .625

6. 2.250

7. ∅.500

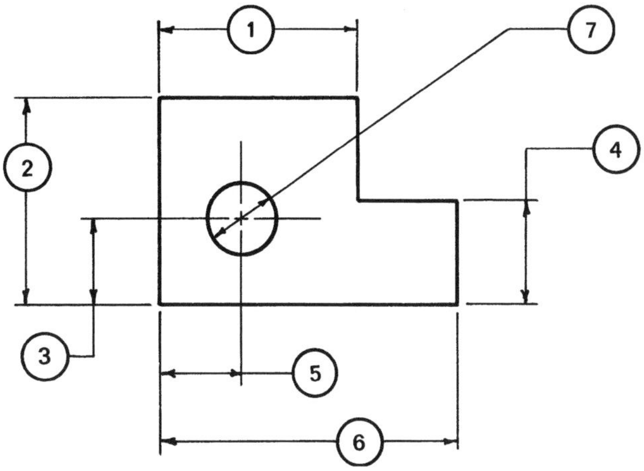

**EX4-5 MILLIMETERS**

1. ∅30.0

2. ∅15.00

3. 10.0

4. 20.0

5. 66.2

6. 15.1

7. 35.02

8. 70.00

   NOTE: ALL FILLETS AND
   ROUNDS = R5.0 UNLESS
   OTHERWISE STATED.

Note : ⑨

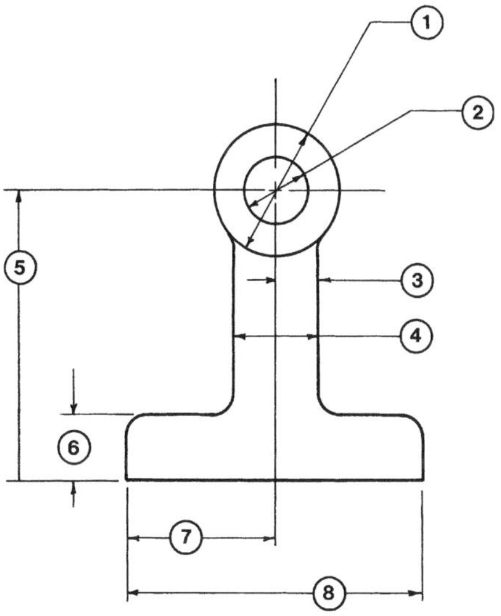

## EX4-6 MILLIMETERS

| | | | |
|---|---|---|---|
| 1. 184.5 | 7. 28.0 | 13. 83.2 | 19. 120.0 |
| 2. 91.5 | 8. 16.00 | 14. 63.00 | 20. ⟨184.0⟩ |
| 3. 44.2 | 9. 16.00 | 15. 50.00 | 21. 12 × 31 |
| 4. 22.00 | 10. 28.0 | 16. 28.5 | R - 3 SLOTS |
| 5. 13.00 | 11. 12.5 | 17. 32.0 | 22. 6.00 |
| 6. 6.51 | 12. Ø6.00 | 18. 76.0 | 23. 6.00 |

Measure and redraw the shapes in Exercise Problems EX4-7 through EX4-49. Add the appropriate dimensions. Specify the units and scale of the drawing. The dotted grid background has either .50-inch or 10-millimeter spacing.

**EX4-7**

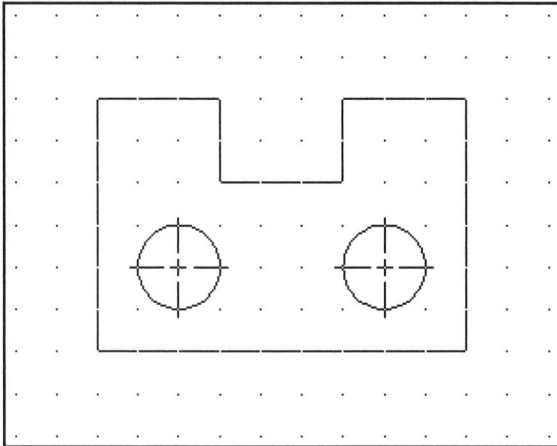

**EX4-9**

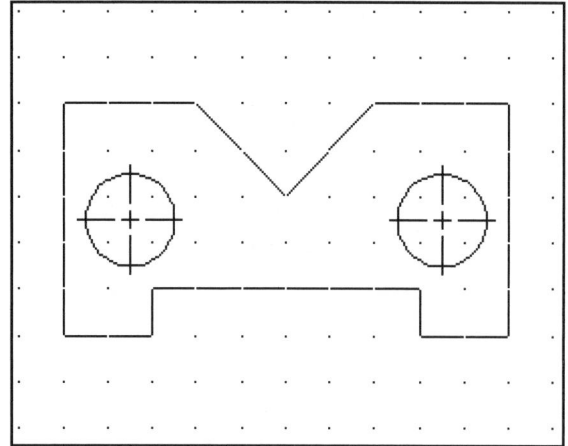

**EX4-8**

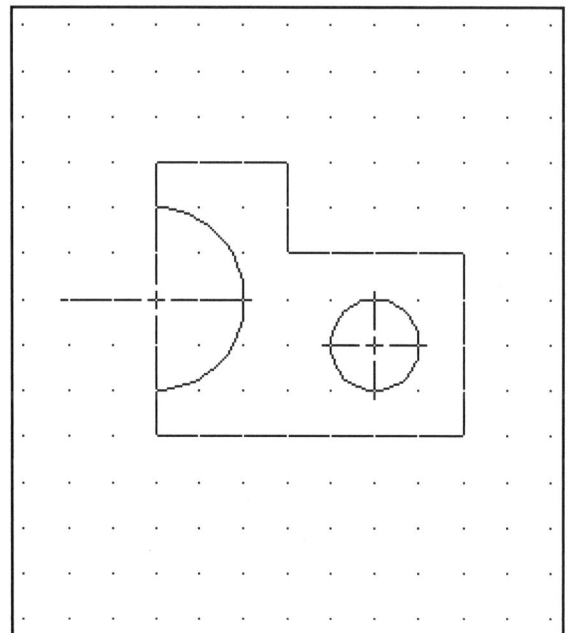

**EX4-10**

**EX4-11**

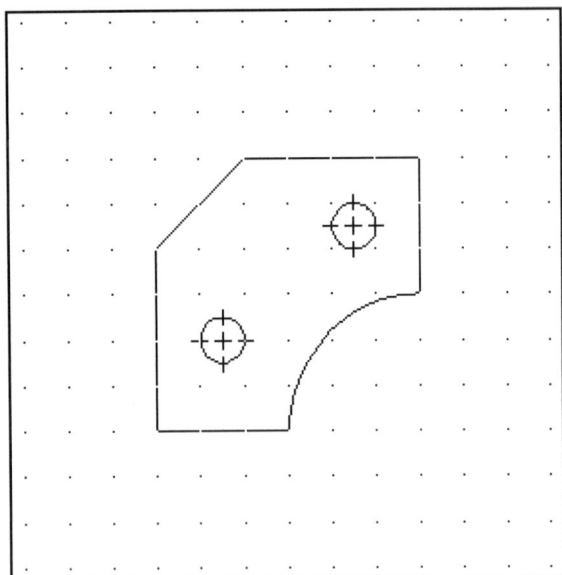

**EX4-13**

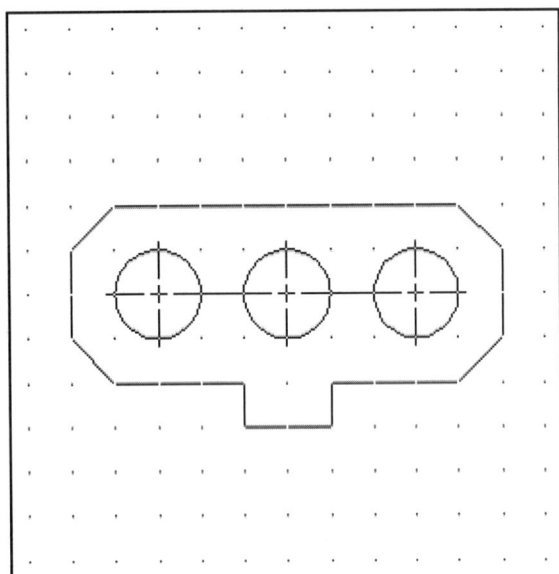

**EX4-12**

**EX4-14**

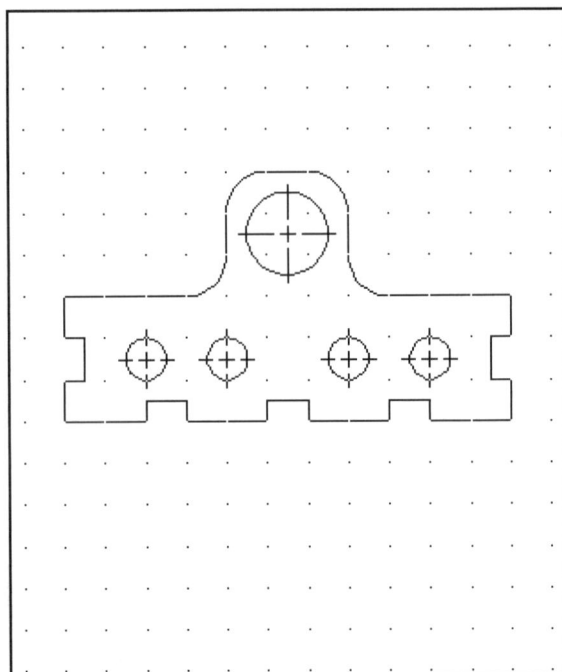

EX4-15

EX4-17

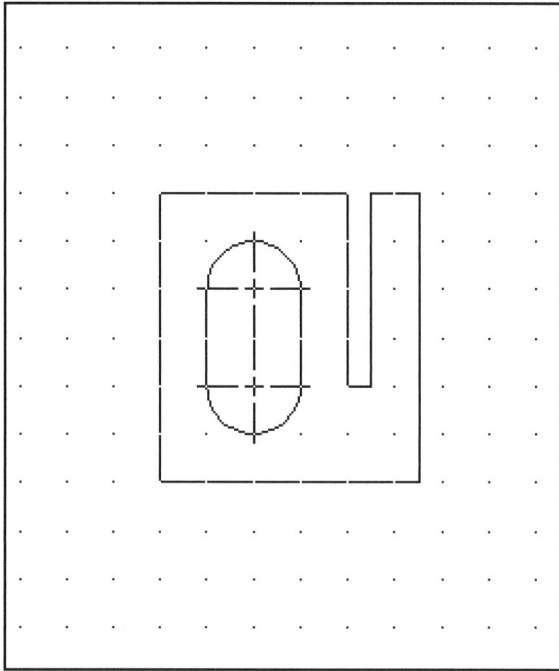

EX4-16

EX4-18

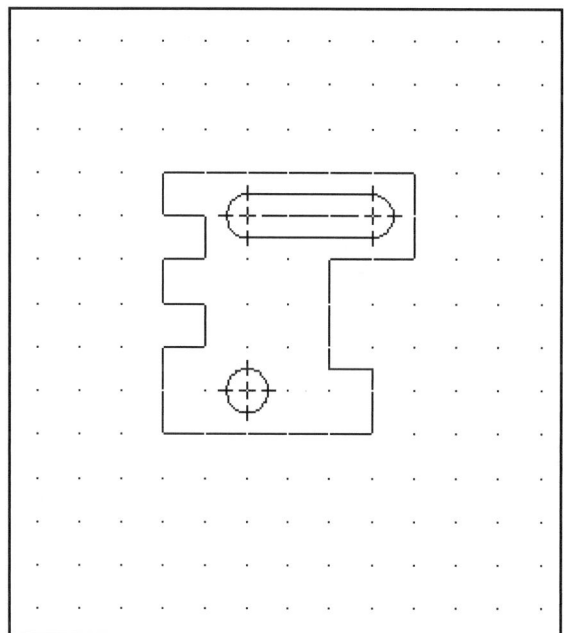

**EX4-19**

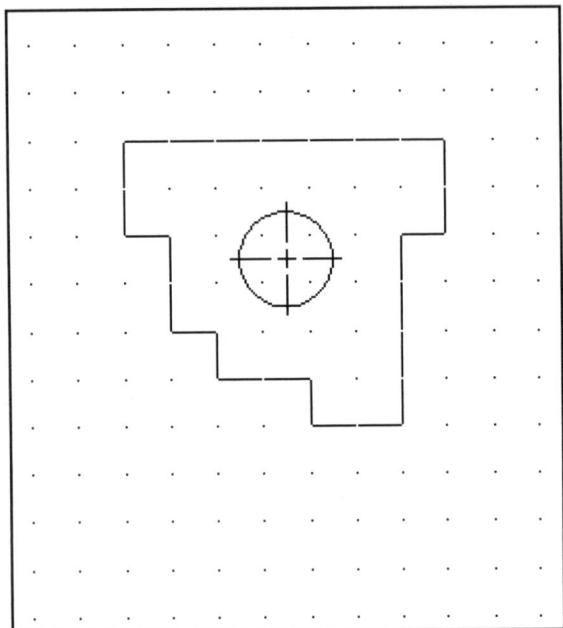

**EX4-21**

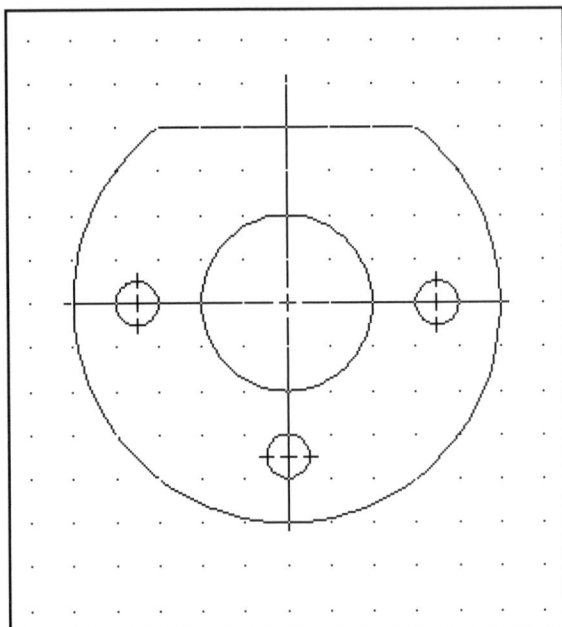

**EX4-20**

**EX4-22**

EX4-23

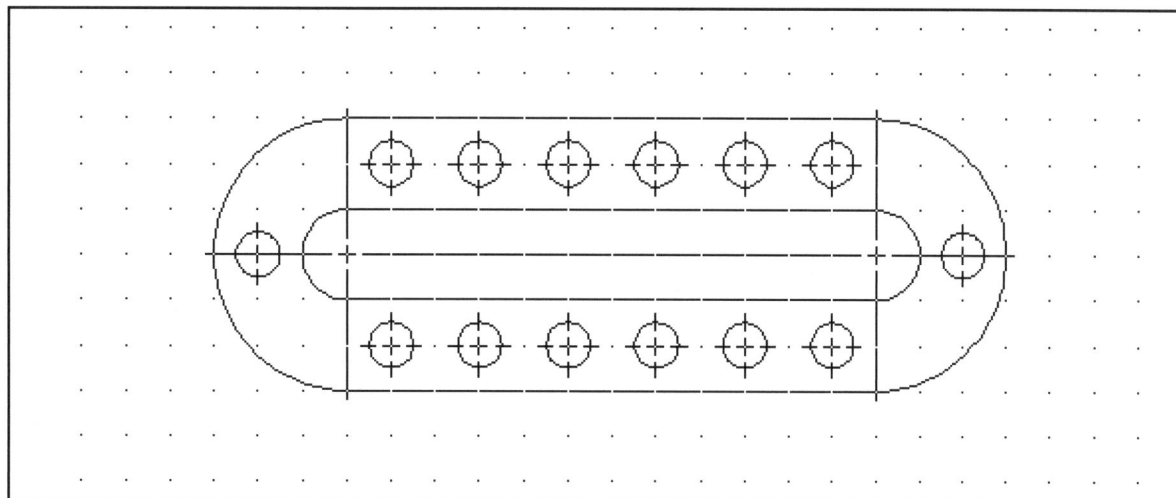

EX4-24

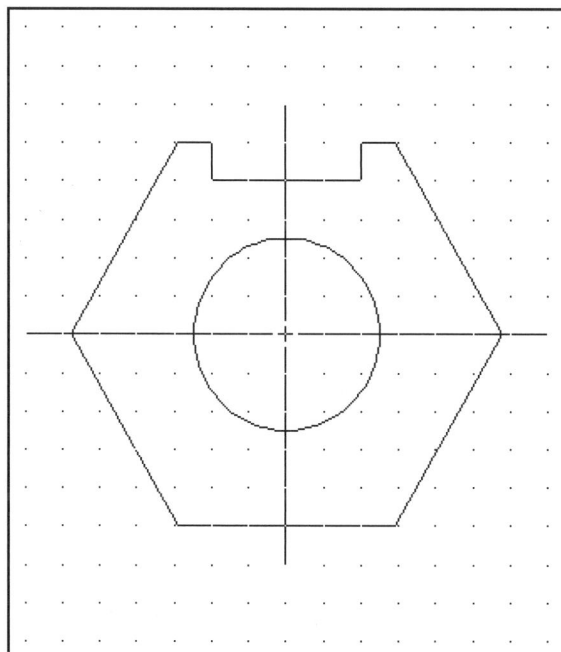

EX4-25

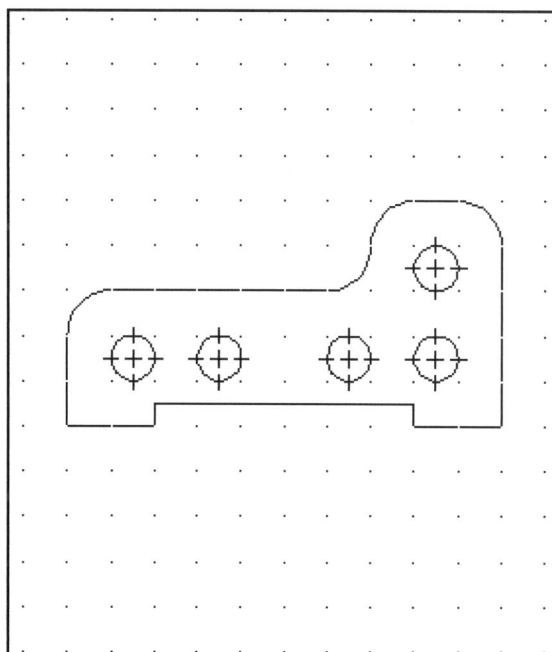

**EX4-26**

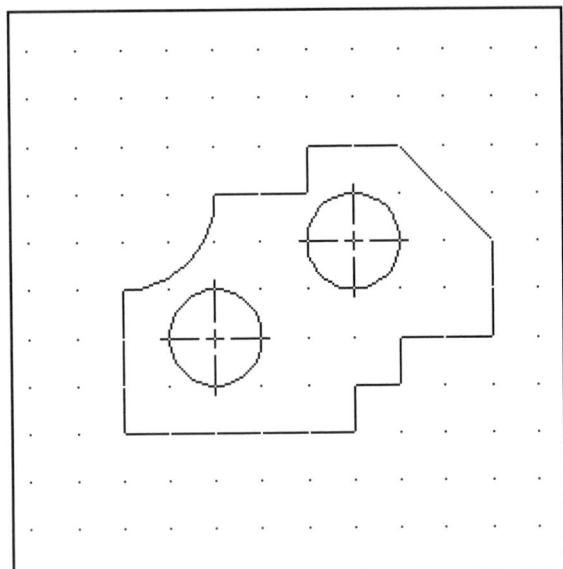

**EX4-28**

**EX4-27**

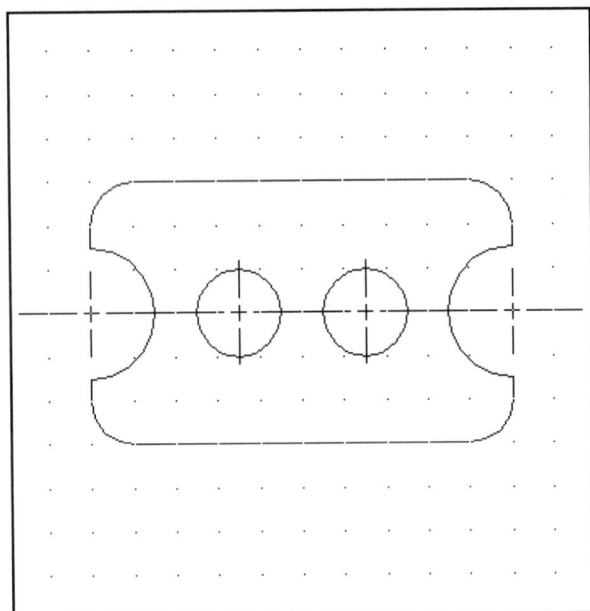

**EX4-29**

EX4-30

EX4-32

EX4-31

EX4-33

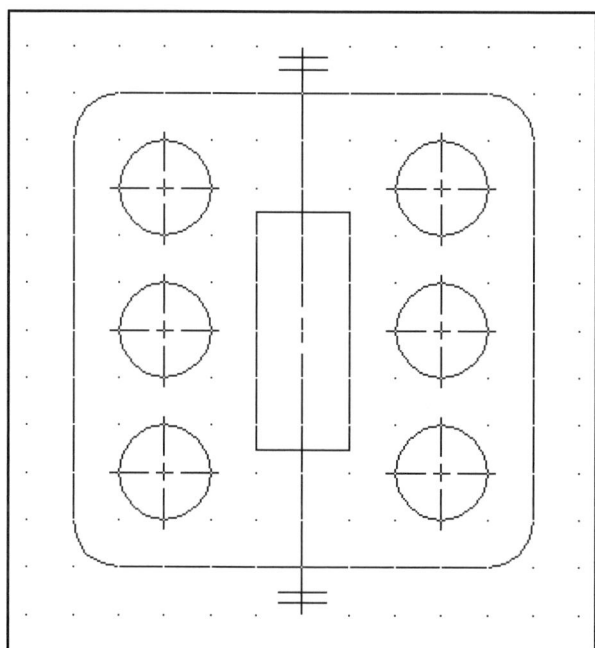

EX4-34

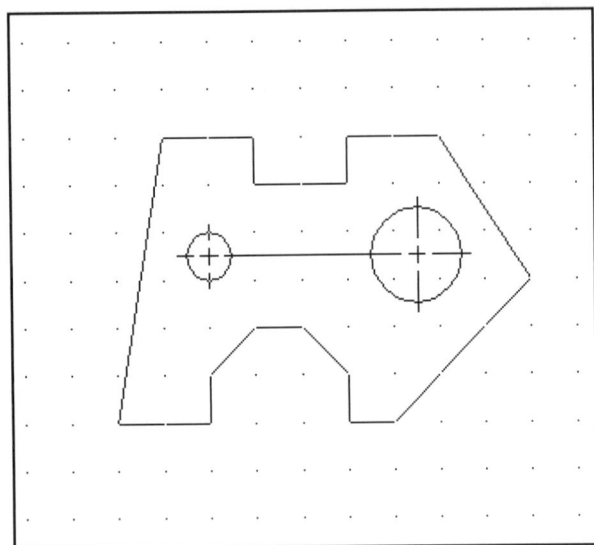

EX4-36

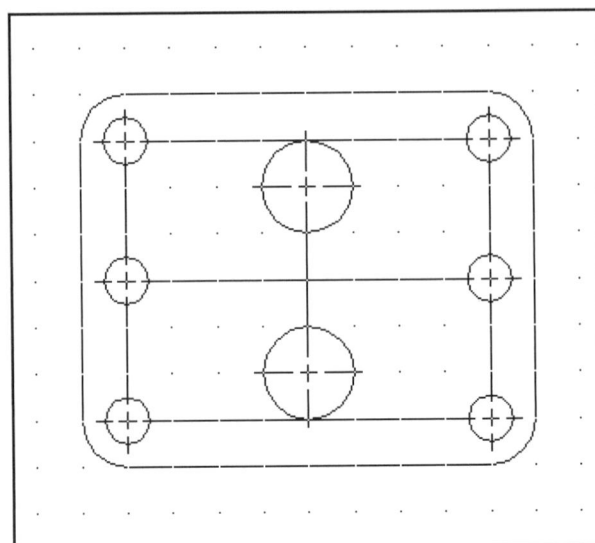

EX4-35

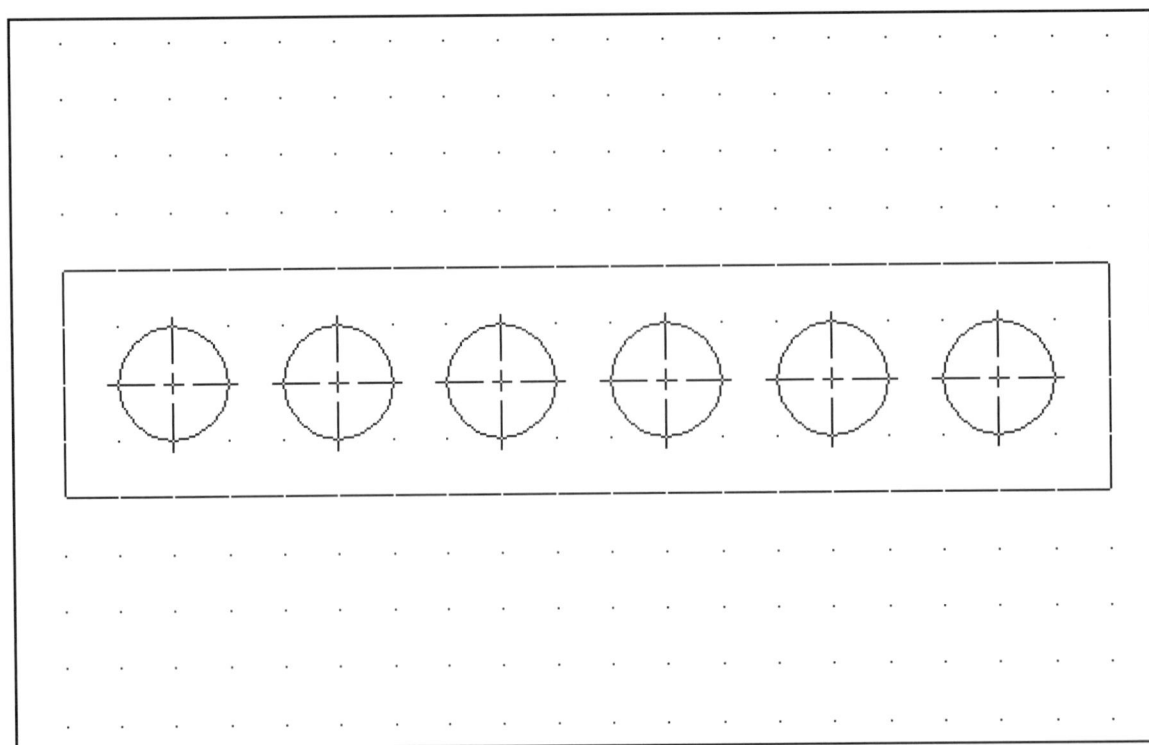

EX4-37

EX4-38

EX4-39

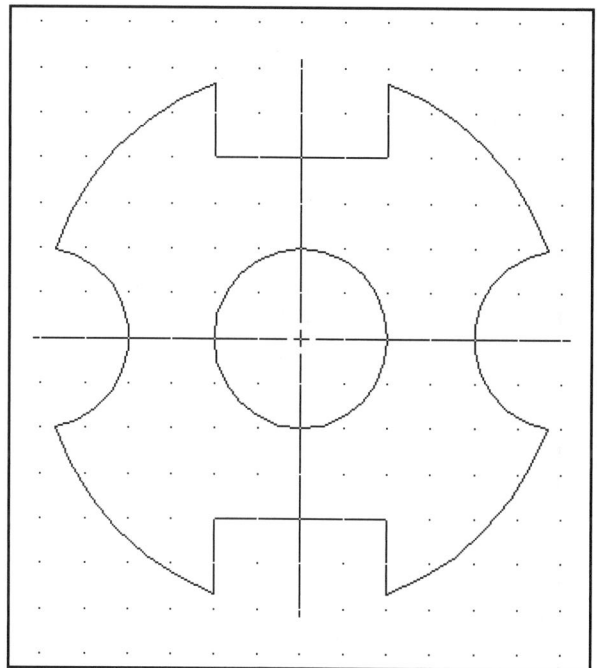

**EX4-40**

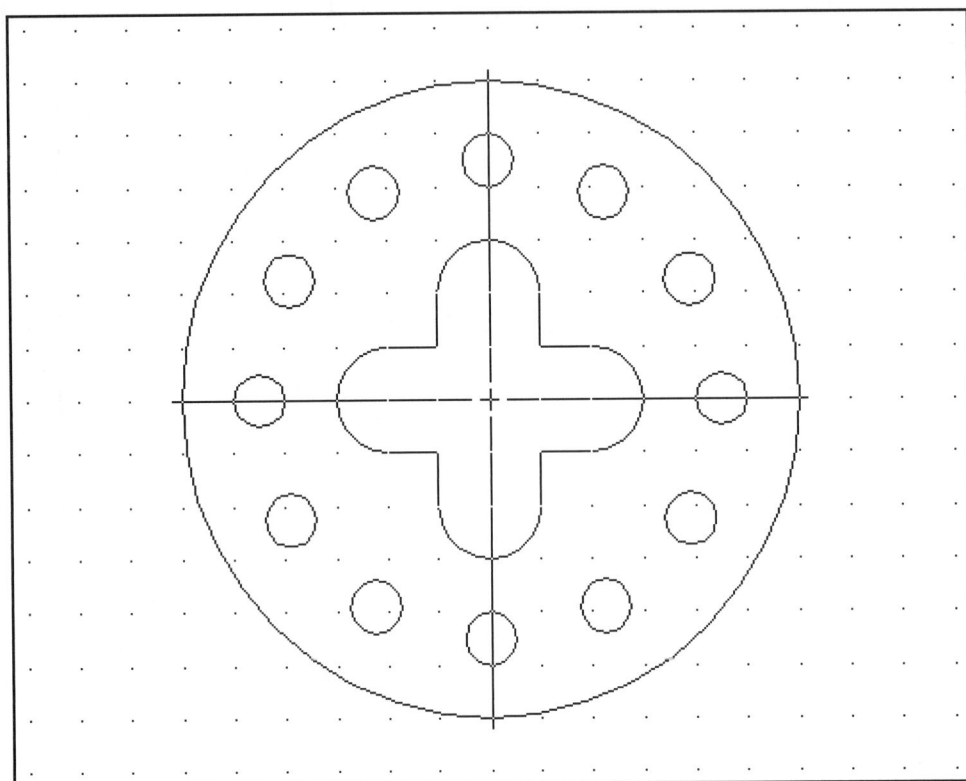

**EX4-41**

**EX4-42A**

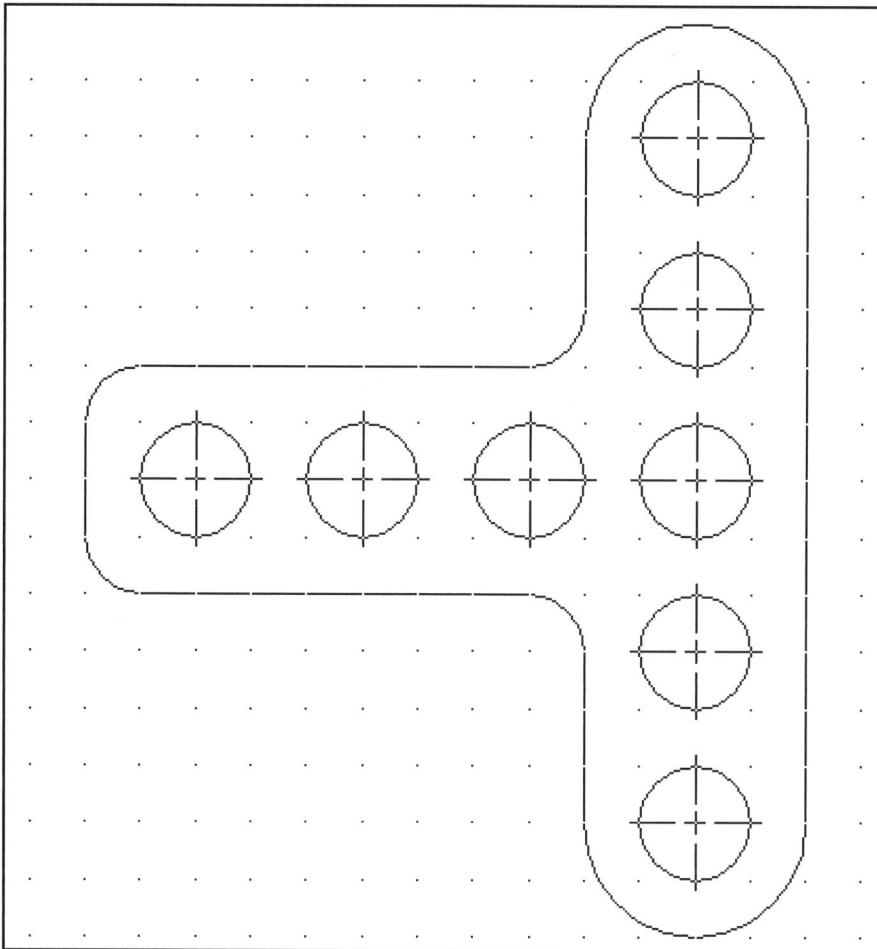

**EX4-43A**

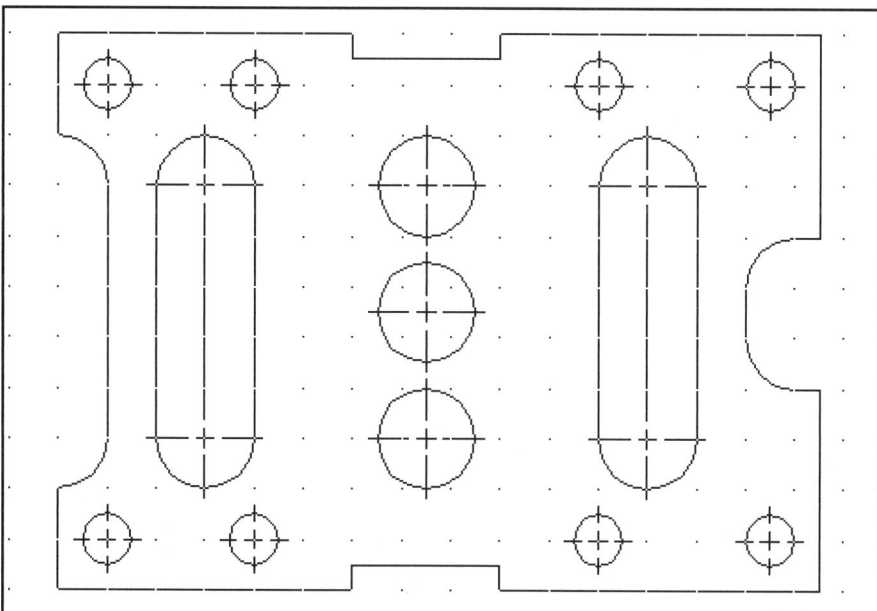

**EX4-42B**

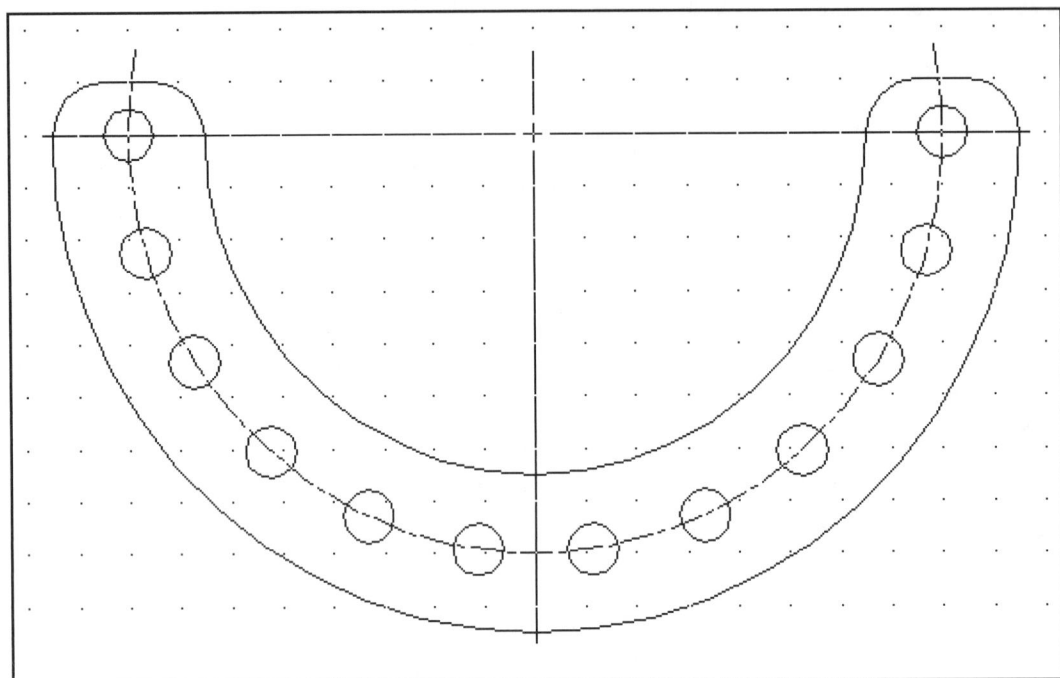

**EX4-43B**

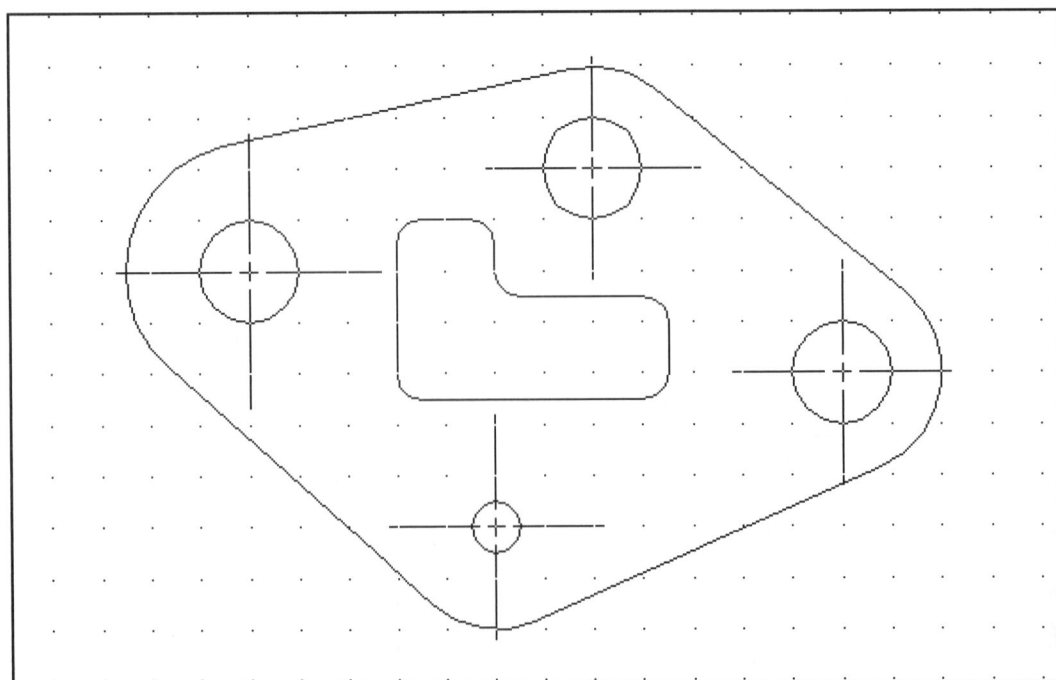

**EX4-42C**

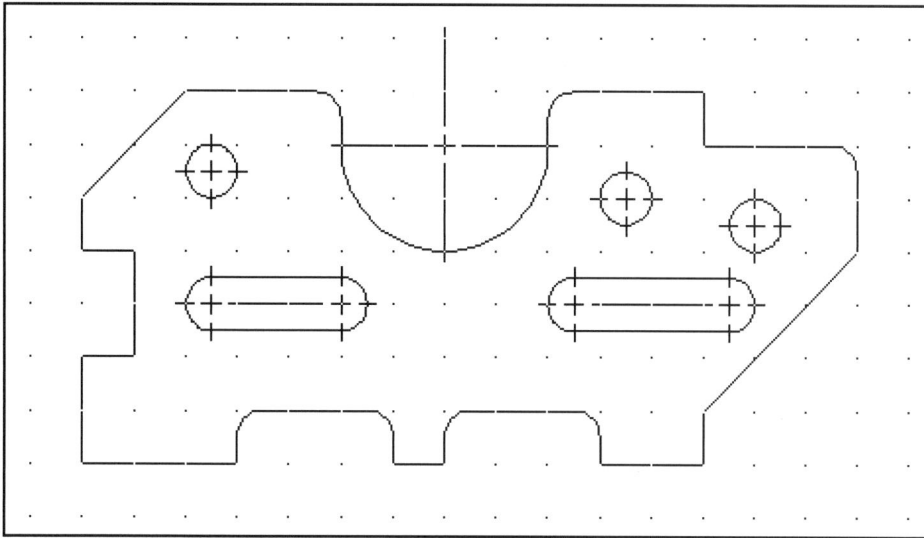

**EX4-43C**

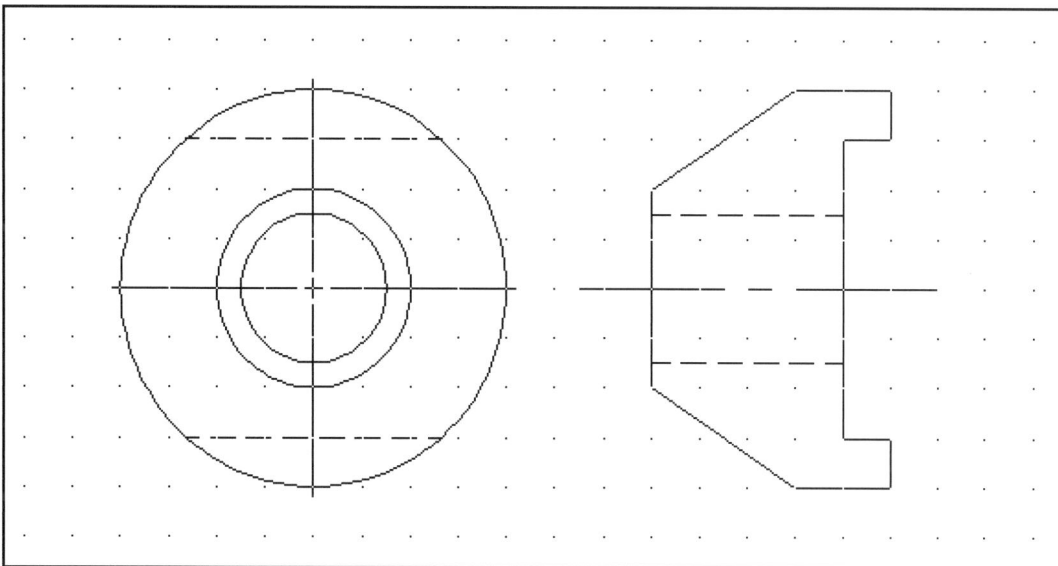

Front View                    Right Side View

**EX4-44**

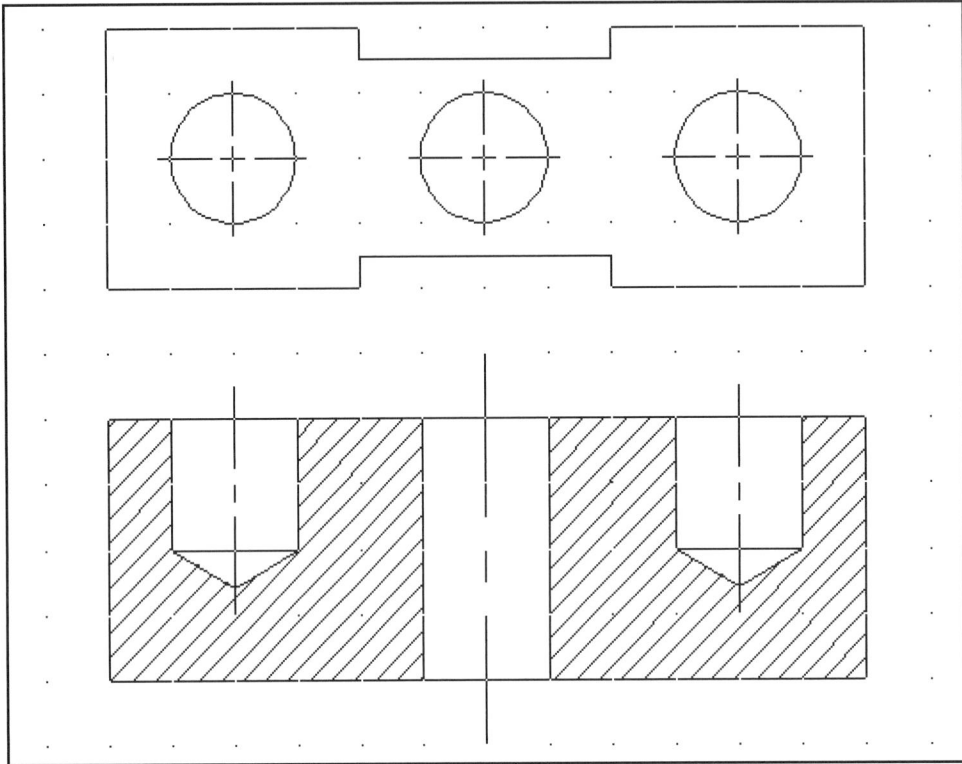

**EX4-45**

EX4-46

EX4-48

EX4-47

EX4-49

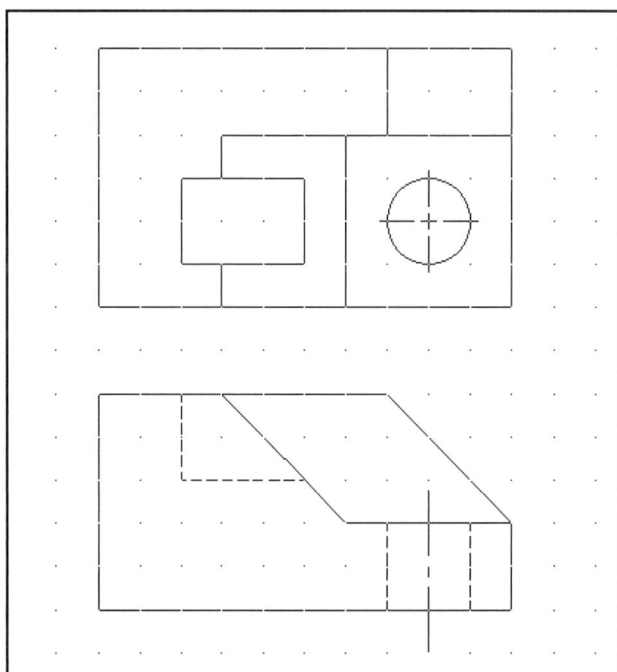

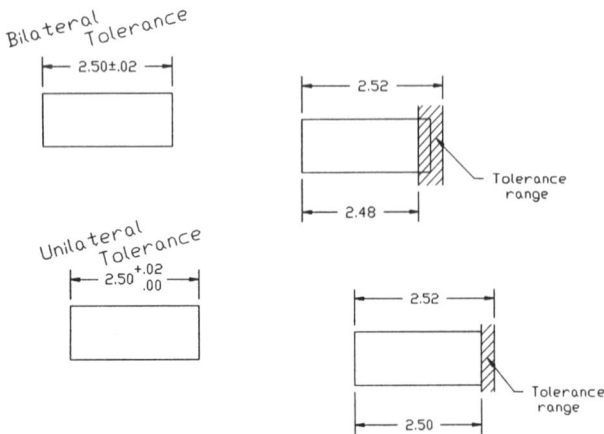

**Figure 4-90**

## 4-35 INTRODUCTION

*Tolerances* define the manufacturing limits for dimensions. All dimensions have tolerances either written directly on the drawing as part of the dimension or implied by a predefined set of standard tolerances that apply to any dimension that does not have a stated tolerance.

This chapter explains general tolerance conventions and how they are applied using AutoCAD. It includes a sample tolerance study and an explanation of standard fits and surface finishes.

Chapter 5 explains geometric tolerances.

## 4-36 DIRECT TOLERANCE METHODS

There are two methods used to include tolerances as part of a dimension: plus and minus, and limits. Plus and minus tolerances can be expressed in either bilateral or unilateral forms.

A bilateral tolerance has both a plus and a minus value. A unilateral tolerance has either the plus or minus

value equal to 0. Figure 4-90 shows a horizontal dimension of 60 millimeters that includes a bilateral tolerance of plus or minus 0.1 and another dimension of 60 millimeters that includes a bilateral tolerance of plus 0.20 or minus 0.10. Figure 4-90 also shows a dimension of 65 millimeters that includes a unilateral tolerance of plus 1 or minus 0.

Plus or minus tolerances define a range for manufacturing. If inspection shows that all dimensioned distances on an object fall within their specified tolerance range, the object is considered acceptable; that is, it has been manufactured correctly.

The dimension and tolerance $60 \pm 0.1$ means that the distance must be manufactured within a range no greater than 60.1 nor less than 59.9. The dimension and tolerance $65 + 1, -0$ defines the tolerance range as 65 to 66.

Figure 4-91 shows some bilateral and unilateral tolerances applied using decimal inch values. Inch dimensions and tolerances are written using a slightly different format from that used for millimeter dimensions and tolerances, but they also define manufacturing ranges for dimension values. The horizontal bilateral dimension and tolerance $2.50 \pm .02$ defines the longest acceptable distance as 2.52 inches and the shortest as 2.48. The unilateral dimension $2.50 + .02, -.00$ defines the longest acceptable distance as 2.52 and the shortest as 2.50.

## 4-37 TOLERANCE EXPRESSIONS

Dimension and tolerance values are written differently for inch and millimeter values. See Figure 4-92. Unilateral dimensions for millimeter values specify a zero limit with a single 0. A zero limit for inch values must include the same number of decimal places given for the dimension value. In the example shown in Figure 4-92, the dimension value .500 has a unilateral tolerance with minus zero tolerance. The zero limit is written as .000, three decimal places for both the dimension and the tolerance.

Both values in a bilateral tolerance must contain the same number of decimal places, although for millimeter values the tolerance values need not include the same number

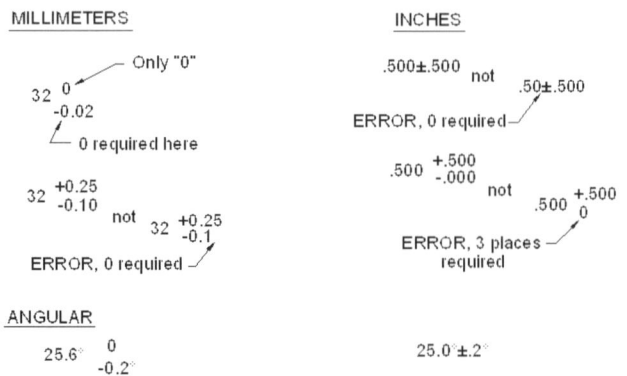

**Figure 4-91**　　　　　　　　　　　　　　　　　　　　**Figure 4-92**

**Figure 4-93**

**Figure 4-94**

of decimal places as the dimension value. In Figure 4-92 the dimension value 32 is accompanied by tolerances of +0.25 and −0.10. This form is not acceptable for inch dimensions and tolerances. An equivalent inch dimension and tolerance would be written 32.00 + .25/−.10.

Degree values must include the same number of decimal places in both the dimension value and the tolerance values for bilateral tolerances. A single 0 may be used for a unilateral tolerance.

## 4-38 UNDERSTANDING PLUS AND MINUS TOLERANCES

A millimeter dimension and tolerance of 12.0 +0.2/−0.1 means the longest acceptable distance is 12.2000...0, and the shortest, 11.9000...0. The total range is .3000...0.

After an object is manufactured, it is inspected to ensure that the object has been manufactured correctly. Each dimensioned distance is measured and, if it is within the specified tolerance, is accepted. If the measured distance is not within the specified tolerance, the part is rejected. Some rejected objects may be reworked to bring them into the specified tolerance range, whereas others are simply scrapped.

Figure 4-93 shows a dimension with a tolerance. Assume that five objects were manufactured using the same 12 +0.2/−0.1 dimension and tolerance. The objects were then inspected and the results were as listed. Inspected measurements are usually at least one more decimal place than that specified in the tolerance. Which objects are acceptable and which are not? Object 3 is too long, and object 5 is too short because their measured distances are not within the specified tolerances.

Figure 4-94 shows a dimension and tolerance of 3.50±.02 inches. Object 3 is not acceptable because it is too short, and object 4 is too long.

## 4-39 CREATING PLUS AND MINUS TOLERANCES USING AUTOCAD

Plus and minus tolerances may be created using AutoCAD in four ways: using the **Text** option or the **Mtext** option, typing the tolerances directly using **Dtext**, or by setting the plus and minus values using the **Dimension Style Manager**.

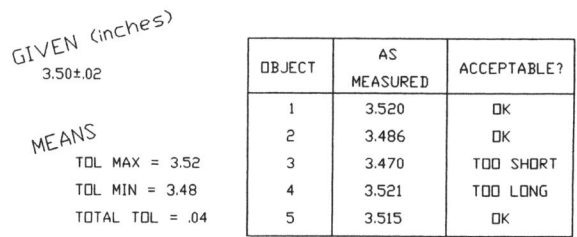

## To create plus and minus tolerances using the Text option

The example given is for a horizontal dimension, but the procedure is the same for any linear or radial dimension.

1. Select the **Linear** tool from the **Dimension** panel.
2. Select the extension line origins as explained in Section 4-3 and locate the dimension line.

   *Specify dimension line location or [Mtext/Text/Angle/Horizontal/Vertical/Rotated]:*

3. Type **t**; press **Enter**.

   *Dimension text <x.xxxx>:*

4. Type the appropriate text value; press **Enter**.

Type the dimension value, then **%%p**, then the tolerance value.

*Dimension text <x.xxxx>:5.00%%p.02.*

In this example, the resulting dimension will be 5.00±.02.

## To create plus and minus tolerances using the Mtext option

To add plus and minus tolerances to an existing dimension.

Given an existing dimension text.

1. Click the dimension text.

   A dialog box will appear. See Figure 4-95.

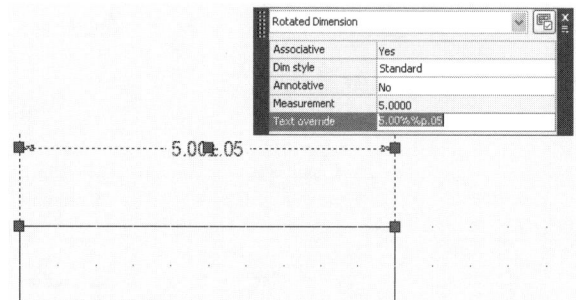

**Figure 4-6**

2. Click the Text override box.

3. Type %%p.05 next to the referenced text in the Text override box.

4. Delete the dialog box.

5. Press the <Esc> key.

The modified dimension will appear on the drawing screen and can be located by moving the mouse.

The **Dimension Style Manager** on the **Dimension** panel can be used to change the precision of the inital Auto-CAD select dimension values. (The default value is four places; 1.0000.) If, for example, two decimal places are desired, use the **Precision** box on the **Primary Units** tab on the **Modify Dimension Style: Standard** dialog box to reset the system for two decimal places. See Section 4-5.

## To use Dtext to create a plus and minus tolerance

1. Type **Dtext** in response to a command prompt.
2. Place the starting point for the text, then define the appropriate height and angle.
3. Type the desired dimension.

   *Text: 5.00%%p.02*

4. Press **Enter.**

In the example shown, the resulting dimension will be **5.00±.02.** The symbol will initially appear on the screen as %%p but will change to ± when the last text line is entered. The symbol ± can also be created using the Windows character map by holding down the <**Alt**> key and typing **0177.**

## To use the Dimension Style Manager

1. Access the **Dimension Style Manager**. The **Dimension Style Manager** box will appear.
2. Select the **Modify** option, then the **Tolerances** tab.
3. Select the arrow to the right of the **Method** box.

A list of available tolerancing methods will cascade. See Figure 4-96. AutoCAD offers two options for plus and minus tolerancing: **Symmetrical** and **Deviation.**

## The symmetrical method

4a. Select the **Symmetrical** method.

The **Tolerance format** box within the **Modify Dimension Style: Standard** dialog box is used to create a symmetrical tolerance by entering a value in the **Upper value** box. See Figure 4-97.

5a. Type a value.

In this example, a value of **0.0300** was entered. Only an upper value needed to be entered, as the tolerance value is symmetrical.

6a. Return to the drawing screen.

Dimensions created using the **Dimension** panel will now automatically include a ±0.03000 tolerance.

## The deviation method

4b. Select the **Deviation** method.

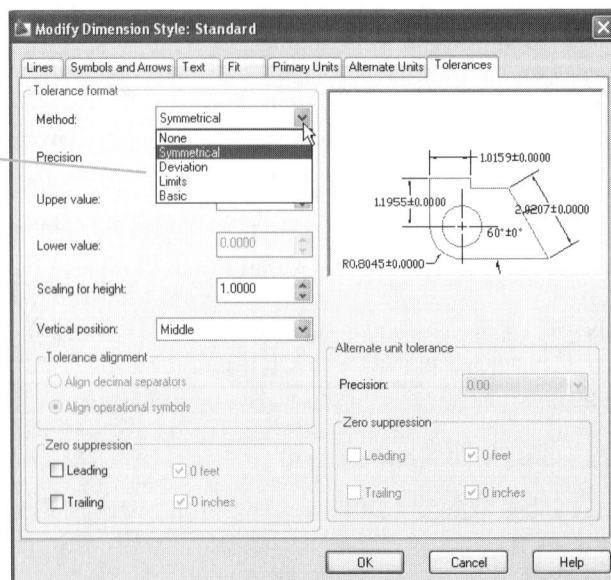

Select the tolerance method here.

Figure 4-96

Resulting dimension

— 2.50±0.03 —

Define the tolerance limits here.

NOTE: The precisions of the dimension value is changed using the Primary Units option.

**Modify Dimension Style: Standard**

Lines | Symbols and Arrows | Text | Fit | Primary Units | Alternate Units | Tolerances

Tolerance format

Method: Symmetrical

Precision 0.00

Upper value: 0.0300

Lower value: 0.0300

Scaling for height: 1.0000

Vertical position: Middle

1.02±0.03
1.20±0.03
2.02±0.03
60°±0°
R0.80±0.03

**Figure 4-97**

The **Tolerance format** dialog box is used to create a deviation tolerance by entering values in the **Upper value** and **Lower value** boxes. See Figure 4-98.

5b. Type in values.

In this example values of **0.0100** and **0.0300** were entered.

6b. Return to the drawing screen.

Dimensions created using the **Dimension** toolbar will now automatically include a +0.0100, −0.0300 tolerance.

The **Dimension Style Manager** can also be used to change the precision of the initial AutoCAD selected dimension values. (The default value is four places: 1.0000.) If, for example, two decimal places are desired, use the **Precision**

box on the **Primary Units** tab on the **Dimension Style Manager** dialog box to reset the system for two decimal places. See Section 4-5.

## 4-40 LIMIT TOLERANCES

Figure 4-99 shows examples of limit tolerances. Limit tolerances replace dimension values. Two values are given: the upper and lower limits for the dimension value. The limit tolerance of 62.1 and 59.9 is mathematically equal to $62 \pm 0.1$, but the stated limit tolerance is considered easier to read and understand.

Limit tolerances define a range for manufacture. Final distances on an object must fall within the specified range to be acceptable.

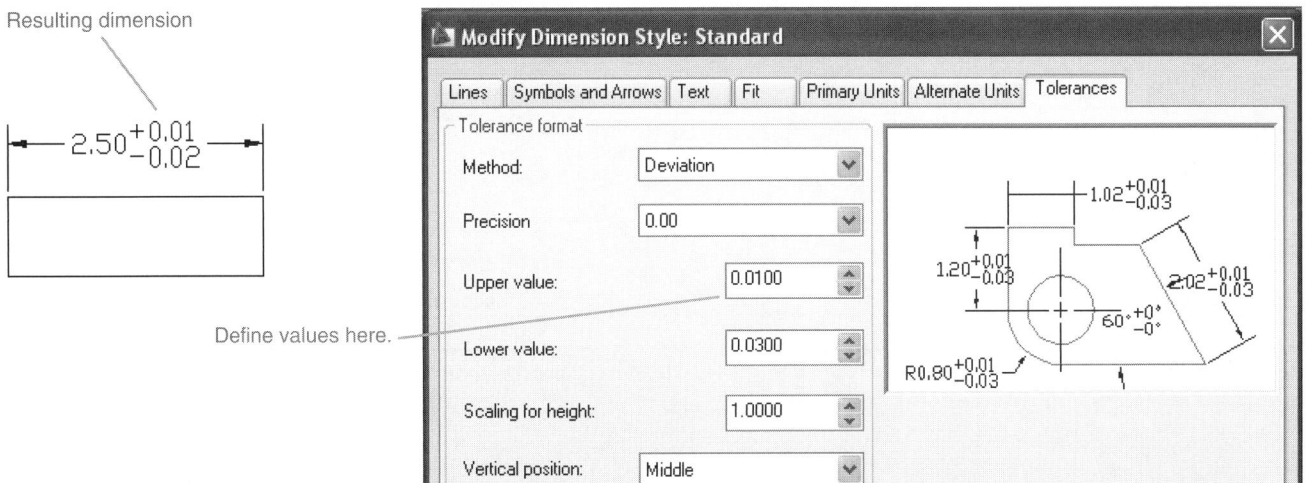

Resulting dimension

— 2.50$^{+0.01}_{-0.02}$ —

Define values here.

**Modify Dimension Style: Standard**

Lines | Symbols and Arrows | Text | Fit | Primary Units | Alternate Units | Tolerances

Tolerance format

Method: Deviation

Precision 0.00

Upper value: 0.0100

Lower value: 0.0300

Scaling for height: 1.0000

Vertical position: Middle

1.02$^{+0.01}_{-0.03}$
1.20$^{+0.01}_{-0.03}$
2.02$^{+0.01}_{-0.03}$
60°$^{+0°}_{-0°}$
R0.80$^{+0.01}_{-0.03}$

**Figure 4-98**

**Figure 4-99**

# 4-41 CREATING LIMIT TOLERANCES USING AUTOCAD

Limit tolerances may be created using AutoCAD in two ways: by using the **Dimension Style Manager** or the edit tools.

**To create a limit tolerance using the Dimension Style Manager**

1. Select the **Dimension Style Manager** from the **Dimension** panel.

The **Dimension Style Manager** dialog box will appear.

2. Select the **Modify** option, then the **Tolerances** tab.

The **Modify Dimension Style: Standard** dialog box will appear.

3. Select the arrow to the right of the **Method** option box.

A list of available tolerance methods will cascade. See Figure 4-96.

4. Select the **Limits** option.

The **Tolerance format** dialog box will reappear with the headings **Upper value** and **Lower value** now in black letters, meaning that they can be accessed.

5. Type in the upper and lower values.

In this example values of **0.01000** and **0.03000** were chosen. See Figure 4-100.

6. Return to the drawing screen.

Every dimension created using the **Linear** dimension tool will now automatically create a limit tolerance based on the selected distance and the selected upper and lower values. Figure 4-100 shows the results of the **Linear** dimension tool applied to a distance of 3.2500. The upper limit is 3.2500 + 0.0100 and the lower limit is 3.2500 - 0.0300.

If only a few limit-type tolerances are to be used on a drawing, it is sometimes easier simply to modify an

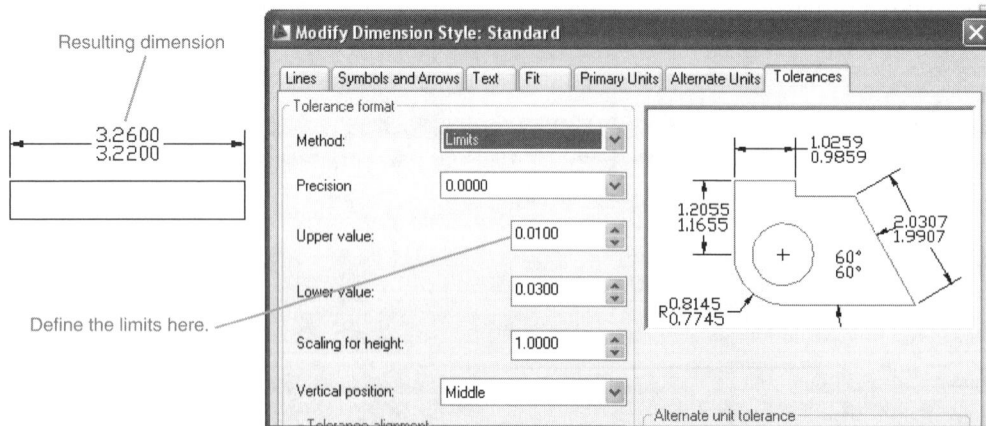

**Figure 4-100**

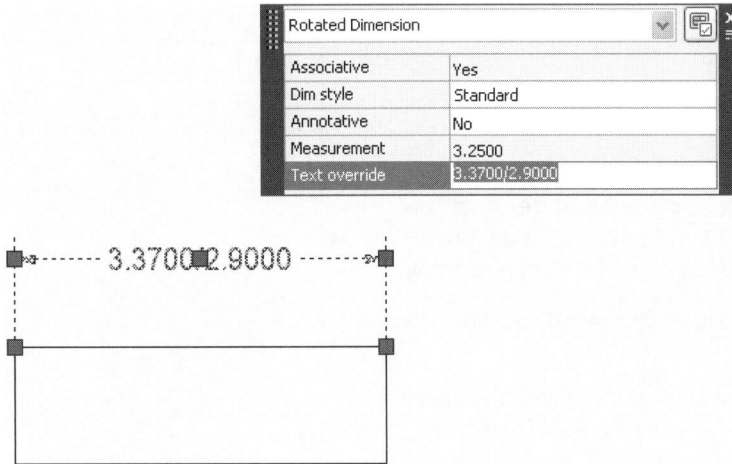

Figure 4-101

existing dimension rather than to change the **Tolerance** settings.

**To modify an existing dimension into a limit tolerance**

1. Click the existing dimension

   A dialog box will appear. See Figure 4-101.

2. Click the Text override box.
3. Type 3.3700/2.9000 in the Text override box.

   AutoCAD does allow two lines of text in the override box so write the limit text on a single line as shown.

4. Delete the dialog box.
5. Press the <Esc> key.

## 4-42  ANGULAR TOLERANCES

Figure 4-102 shows an example of an angular dimension with a tolerance. The procedure explained for plus and minus tolerances applies to angular as well as to linear dimensions and tolerances.

An angular dimension with a tolerance.

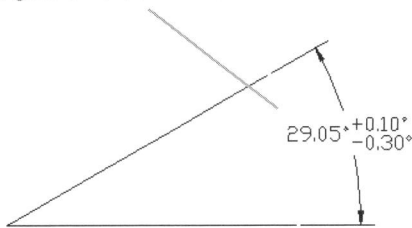

Select the dimension precision here.

Select the angular tolerance precision here.

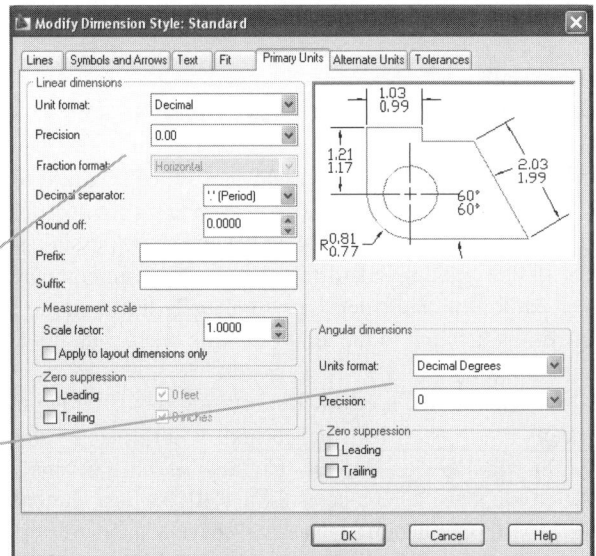

Figure 4-102

The precision of angular dimensions is set using the **Primary Units** dialog box. In the following example, a deviation tolerance of +0.10, −0.30 was also specified.

**To set the precision for angular dimensions and tolerances**

1. Access the **Dimension Style Manager**, then the Modify option, then the **Primary Units** option.
2. Select the arrow to the right of the **Precision** box.

A list of available precision factors will cascade. See Figure 4-102.

3. Select two significant figures for both **Dimension** and **Tolerance** and return to the drawing screen.

The precision for the angular value can be changed using the **Precision** option under **Angular Dimensions.**

**To create an angular dimension and tolerance**

1. Select the **Angular** tool from the **Dimension** panel.

   *Select arc, circle, line, or <specify vertex>:*

2. Select one of the lines that defines the angle.

   *Select second line:*

3. Select the second line.

   *Specify dimension arc line location or [Mtext/ Text/Angle]:*

4. Select a location for the dimension; press **Enter.**

The **Mtext** and **Text** options are used as explained for linear dimensions and tolerance in Section 4-39. Symmetrical, deviation, and limit tolerances can be applied to angular tolerances in the same way they were to linear tolerances.

## 4-43 STANDARD TOLERANCES

Most manufacturers establish a set of standard tolerances that are applied to any dimension that does not include a specific tolerance. Figure 4-103 shows some possible standard tolerances. Standard tolerances vary from company to company. Standard tolerances are usually listed on the first page of a drawing to the left of the title block, but this location may vary.

The X value used when specifying standard tolerances means any X stated in that format. A dimension value of 52.00 will have an implied tolerance of ±.01 because the stated standard tolerance is .XX ±.01, so any dimension value with two decimal places will have a standard implied tolerance of ±.01. A dimension value of 52.000 will have an implied tolerance of ±.005.

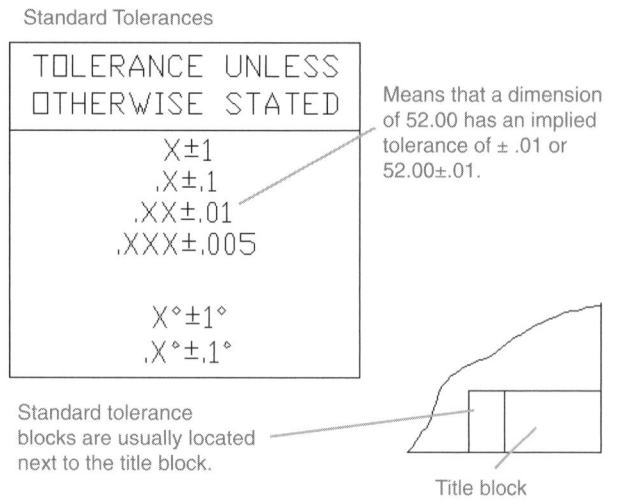

**Figure 4-103**

## 4-44 DOUBLE DIMENSIONING

It is an error, called *double dimensioning,* to dimension the same distance twice. Double dimensioning is an error because it does not allow for tolerance buildup across a distance.

Figure 4-104 shows an object that has been dimensioned twice across its horizontal length; once using three 30-millimeter dimensions and a second time using the 90-millimeter overall dimension. The two dimensions are mathematically equal but are not equal when tolerances are considered. Assume that each dimension has a standard tolerance of ±1 millimeter. The

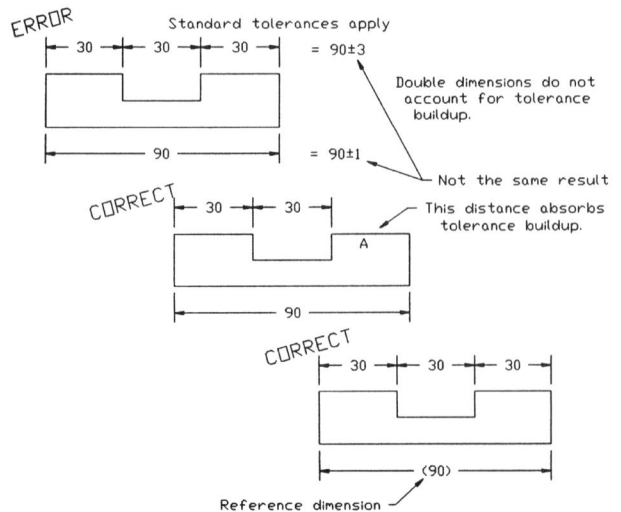

**Figure 4-104**

three 30-millimeter dimensions could create an acceptable distance of 90±3 millimeters, or a maximum distance of 93 and a minimum distance of 87. The overall dimension of 90 millimeters allows a maximum distance of 91 and a minimum distance of 89. The two dimensions yield different results when tolerances are considered.

The size and location of a tolerance depends on the design objectives of the object, how it will be manufactured, and how it will be inspected. Even objects that have similar shapes may be dimensioned and toleranced very differently.

One possible solution to the double dimensioning shown in Figure 4-104 is to remove one of the 30-millimeter dimensions and to allow that distance to "float," that is, absorb the cumulated tolerances. The choice of which 30-millimeter dimension to eliminate depends on the design objectives of the part. For this example the far-right dimension was eliminated to remove the double-dimensioning error.

Another possible solution to the double-dimensioning error is to retain the three 30-millimeter dimensions and to

change the 90-millimeter overall dimension to a reference dimension. A reference dimension is used only for mathematical convenience. It is not used during the manufacturing or inspection process. A reference dimension is designated on a drawing with parentheses (90).

If the 90-millimeter dimension were referenced, then only the three 30-millimeter dimensions would be used to manufacture and inspect the object. This would eliminate the double-dimensioning error.

## 4-45 CHAIN DIMENSIONS AND BASELINE DIMENSIONS

There are two systems used to apply dimensions and tolerances to a drawing: *chain* and *baseline*. Figure 4-105 shows examples of both systems. Chain dimensions relate each feature to the feature next to it; baseline dimensions relate all features to a single baseline or datum.

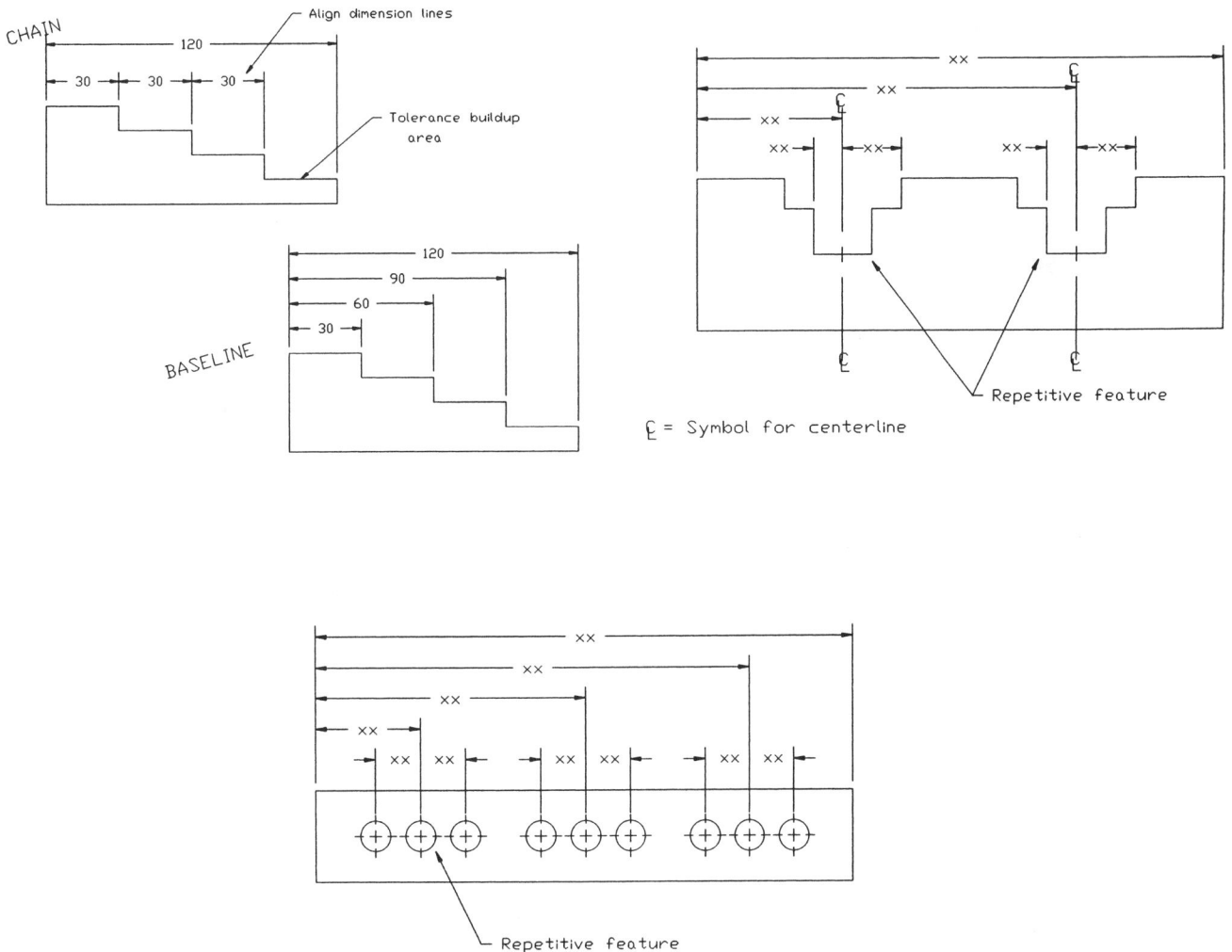

**Figure 4-105**

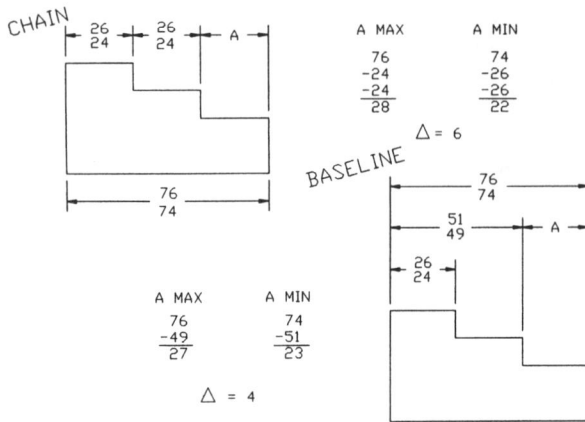

Figure 4-106

Chain and baseline dimensions may be used together. Figure 4-105 also shows two objects with repetitive features: one object includes two slots, and the other three sets of three holes. In each example, the center of the repetitive feature is dimensioned to the left side of the object, which serves as a baseline. The sizes of the individual features are dimensioned using chain dimensions referenced to centerlines.

Baseline dimensions eliminate tolerance buildup and can be related directly to the reference axis of many machines. They tend to take up much more area on a drawing than do chain dimensions.

Chain dimensions are useful in relating one feature to another, such as the repetitive hole pattern shown in Figure 4-105. In this example the distance between the holes is more important than the distance of the individual hole from the baseline.

Figure 4-106 shows the same object dimensioned twice, once using chain dimensions and once using baseline dimensions. All distances are assigned a tolerance range of 2 millimeters stated using limit tolerances. The maximum distance for surface A is 28 millimeters using the chain system and 27 millimeters using the baseline system. The 1-millimeter difference comes from the elimination of the first 26–24 limit dimension found on the chain example but not on the baseline.

The total tolerance difference is 6 millimeters for the chain and 4 millimeters for the baseline. The baseline reduces the tolerance variations for the object simply because it applies the tolerances and dimensions differently. So why not always use baseline dimensions? For most applications, the baseline system is probably better, but if the distance between the individual features is more critical than the distance from the feature to the baseline, use the chain system.

### To create baseline dimensions using AutoCAD

The **Baseline** tool can be applied only after a linear dimension has been created. The object shown in Figure 4-107 was dimensioned by first creating the 1.00 dimension using the **Linear** dimension tool, then using the **Baseline** tool as follows.

1. Create a linear dimension using the **Linear** tool from the **Dimension** panel.

The point selected as the origin for the first extension line will become the origin of the baseline.

2. Select the **Baseline** tool from the **Dimension** panel.

*Specify a second extension line origin or [Undo/<Select>]:*

3. Select the origin for the second extension line of the baseline dimension.

*Specify a second extension line origin or [Undo/Select] <Select>:*

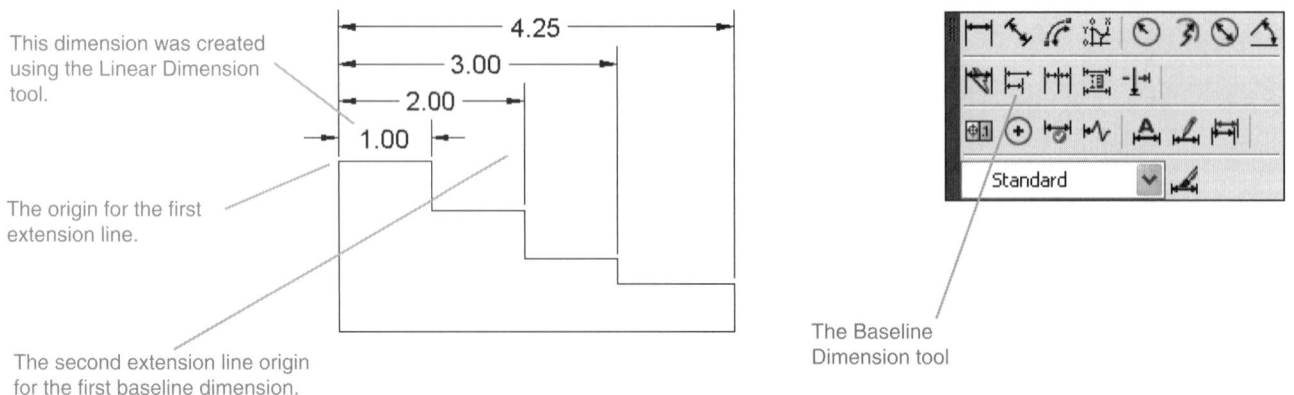

Figure 4-107

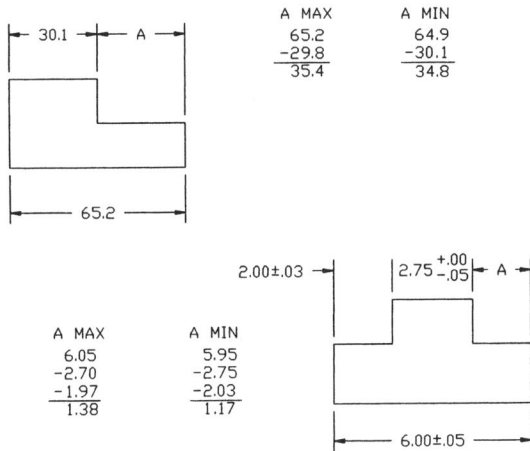

**Figure 4-108**

**Figure 4-109**

4. Repeat the process until the baseline dimensioning is complete.
5. Press the right mouse button and **Enter** the dimensions.

## 4-46 TOLERANCE STUDIES

The term *tolerance study* is used when analyzing the effects of a group of tolerances on each other and on an object. Figure 4-108 shows an object with two horizontal dimensions. The horizontal distance A is not dimensioned. Its length depends on the tolerances of the two horizontal dimensions.

### To calculate A's maximum length

Distance A will be longest when the overall distance is at its longest and the other distance is at its shortest:

$$\begin{array}{r} 65.2 \\ -29.8 \\ \hline 35.4 \end{array}$$

### To calculate A's minimum length

Distance A will be shortest when the overall length is at its shortest and the other length is at its longest:

$$\begin{array}{r} 64.9 \\ -30.1 \\ \hline 34.8 \end{array}$$

Figure 4-108 also shows an object that includes three horizontal dimensions. Surface A is at its maximum length

when the overall dimension is at its longest and the other dimensions are at their shortest. Surface A is at its minimum when the overall length is at its shortest and the other dimensions are at their longest.

## 4-47 RECTANGULAR DIMENSIONS

Figure 4-109 shows an example of rectangular dimensions referenced to baselines. Figure 4-110 shows a circular object for which dimensions are referenced to a circle's centerlines. Dimensioning to a circle's centerline is critical to accurate hole location.

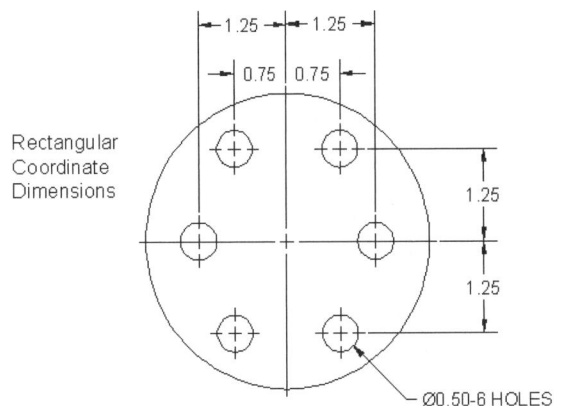

**Figure 4-110**

**Figure 4-111**

## 4-48  HOLE LOCATIONS

When rectangular dimensions are used, the location of a hole's center point is defined by two linear dimensions. The result is a rectangular tolerance zone whose size is based on the linear dimension's tolerances. The shape of the center point's tolerance zone may be changed to circular using positional tolerancing, as described in Section 5-16.

Figure 4-111 shows the location and size dimensions for a hole. Also shown is the resulting tolerance zone and the overall possible hole shape. The center point's tolerance is 0.2 by 0.3, based on the given linear locating tolerances.

The hole diameter has a tolerance of ±0.05. This value must be added to the center point location tolerances to define the maximum overall possible shape of the hole. The maximum possible hole shape is determined by drawing the maximum radius from the four corner points of the tolerance zone.

This means that the left edge of the hole could be as close to the vertical baseline as 12.75 or as far as 13.25. The 12.75 value was derived by subtracting the maximum hole diameter value, 12.05, from the minimum linear distance, 24.80 (24.80 − 12.05 = 12.75). The 13.25 value was derived by subtracting the minimum hole diameter, 11.95, from the maximum linear distance, 25.20 (25.20 − 11.95 = 13.25).

Figure 4-112 shows a hole's tolerance zone based on polar dimensions. The zone has a sector shape, and the possible hole shape is determined by locating the maximum radius at the four corner points of the tolerance zone.

## 4-49  CHOOSING A SHAFT FOR A TOLERANCED HOLE

Given the hole location and size shown in Figure 4-112, what is the largest diameter shaft that will always fit into the hole?

Figure 4-113 shows the hole's center point tolerance zone based on the given linear locating tolerances. Four circles have been drawn centered at the four corners of the linear tolerance zone that represent the smallest possible hole diameter. The circles define an area that represents the maximum shaft size that will always fit into the hole, regardless of how the given dimensions are applied.

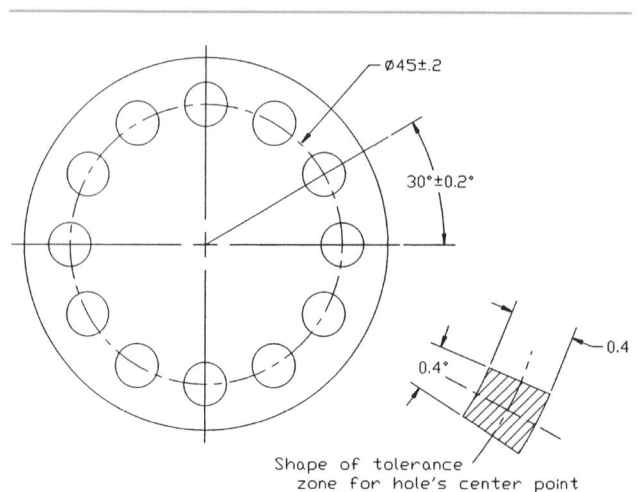

**Figure 4-112**

**Figure 4-113**

**Figure 4-114**

The diameter of this circular area can be calculated by subtracting the maximum diagonal distance across the linear tolerance zone (corner to corner) from the minimum hole diameter.

The results can be expressed as a formula.

**For linear dimensions and tolerances**

$$S_{max} = H_{min} - DTZ$$

where

$S_{max}$ = Maximum shaft diameter

$H_{min}$ = Minimum hole diameter

DTZ = Diagonal distance across the tolerance zone

In the example shown the diagonal distance is determined using the Pythagorean theorem:

$$DTZ = \sqrt{(.4)^2 + (.6)^2}$$
$$= \sqrt{.16 + .36}$$
$$DTZ = .72$$

This means that the maximum shaft diameter that will always fit into the given hole is 11.23.

$$S_{max} = H_{min} - DTZ$$
$$= 11.95 - .72$$
$$S_{max} = 11.23$$

This procedure represents a restricted application of the general formula presented in Chapter 5 for positional tolerances. For a more complete discussion see Section 5-16.

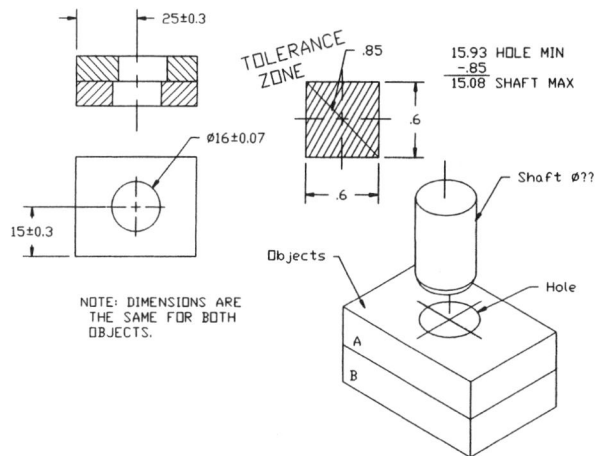

Once the maximum shaft size has been established, a tolerance can be applied to the shaft. If the shaft had a total tolerance of .25, the minimum shaft diameter would be 11.23 − .25, or 10.98. Figure 4-113 shows a shaft dimensioned and toleranced using these values.

The formula presented is based on the assumption that the shaft is perfectly placed on the hole's center point. This assumption is reasonable if two objects are joined by a fastener and both objects are free to move. When both objects are free to move about a common fastener, they are called *floating objects*.

## 4-50 SAMPLE PROBLEM SP4-1

Parts A and B in Figure 4-114 are to be joined by a common shaft. The total tolerance for the shaft is to be 0.05. What are the maximum and minimum shaft diameters?

Both objects have the same dimensions and tolerances and are floating relative to each other.

$$S_{max} = H_{min} - DTZ$$
$$= 15.93 - .85$$
$$S_{max} = 15.08$$

The shaft's minimum diameter is found by subtracting the total tolerance requirement from the calculated maximum diameter.

$$15.08 - .05 = 15.03$$

Therefore,

$$\text{Shaft max} = 15.08$$
$$\text{Shaft min} = 15.03$$

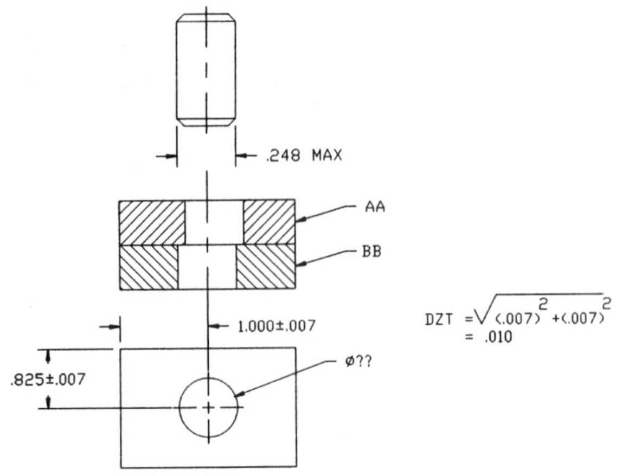

Figure 4-115

## 4-51  SAMPLE PROBLEM SP4-2

The procedure presented in Sample Problem SP4-1 can be worked in reverse to determine the maximum and minimum hole size based on a given shaft size.

Objects AA and BB as shown in Figure 4-115 are to be joined using a bolt whose maximum diameter is .248. What is the minimum hole size for the objects that will always accept the bolt? What is the maximum hole size if the total hole tolerance is .007?

$$S_{max} = H_{min} - DZT$$

In this example the $H_{min}$ is the unknown factor, so the equation is rewritten:

$$H_{min} = S_{max} + DZT$$
$$= .248 + .010$$
$$H_{min} = .258$$

This is the minimum hole diameter, so the total tolerance requirement is added to this value:

$$.258 + .007 = .265$$

Therefore,

Hole max = .265
Hole min = .258

## 4-52  STANDARD FITS (METRIC VALUES)

Calculating tolerances between holes and shafts that fit together is so common in engineering design that a group of standard values and notations has been established. These values are listed in tables in the appendix.

There are three possible types of fits between a shaft and a hole: clearance, transition, and interference. See Figure 4-116. There are several subclassifications within each of these categories.

A *clearance fit* always defines the maximum shaft diameter as smaller than the minimum hole diameter. The difference between the two diameters is the amount of clearance. It is possible for a clearance fit to be defined with zero clearance; that is, the maximum shaft diameter is equal to the minimum hole diameter.

An *interference fit* always defines the minimum shaft diameter as larger than the maximum hole diameter, or more simply said, the shaft is always bigger than the hole. This definition means that an interference fit is the converse of a clearance fit. The difference between the diameter of the shaft and the hole is the amount of interference.

An interference fit is primarily used to assemble objects together. Interference fits eliminate the need for threads, welds, or other joining methods. Using an interference for joining two objects is generally limited to light-load applications.

It is sometimes difficult to visualize how a shaft can be assembled into a hole with a diameter smaller than that of the shaft. It is sometimes done using a hydraulic press that slowly forces the two parts together. The joining process can be augmented by the use of lubricants or heat. The hole is heated, causing it to expand, the shaft is inserted, and the hole is allowed to cool and shrink around the shaft.

A *transition fit* may be either a clearance or an interference fit. It may have a clearance between the shaft and the hole or an interference.

Figure 4-116 shows two graphic representations of 20 different standard hole/shaft tolerance ranges. The figure shows ranges for hole tolerances, shaft tolerances, and the amount of clearance or interference for each classification. The notations are based on Standard International Tolerance values. A specific description for each category of fit is as follows.

**Clearance Fits**

H11/c11 or C11/h11 = Loose running fit
H9/d9 or D9/h9 = Free running fit
H8/f7 or F8/h7 = Close running fit
H7/g6 or G7/h6 = Sliding fit
H7/h6 = Locational clearance fit

**Transition Fits**

H7/k6 or K7/h6 = Locational transition fit
H7/n6 or N7/h6 = Locational transition fit

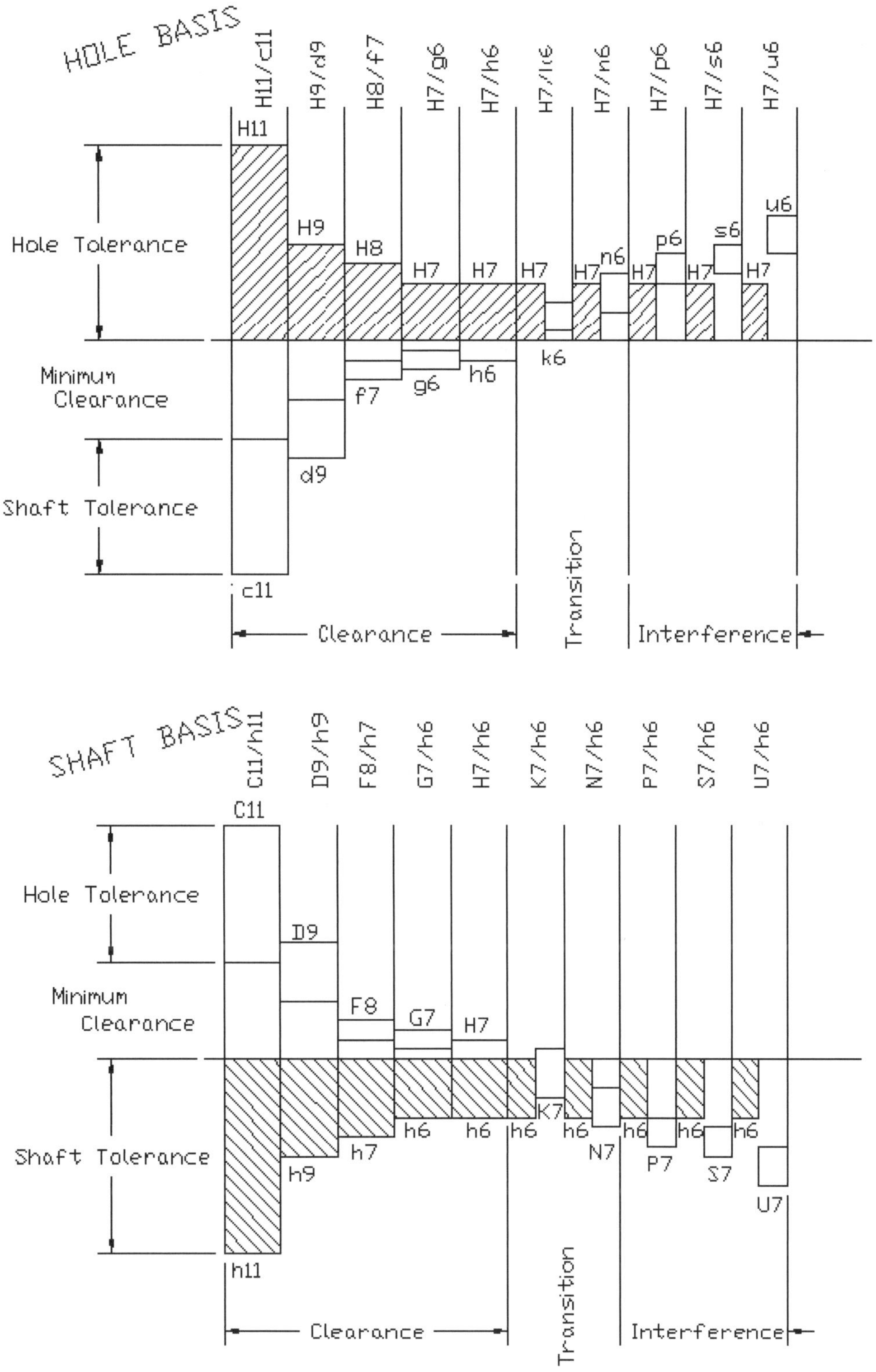

**Figure 4-116**

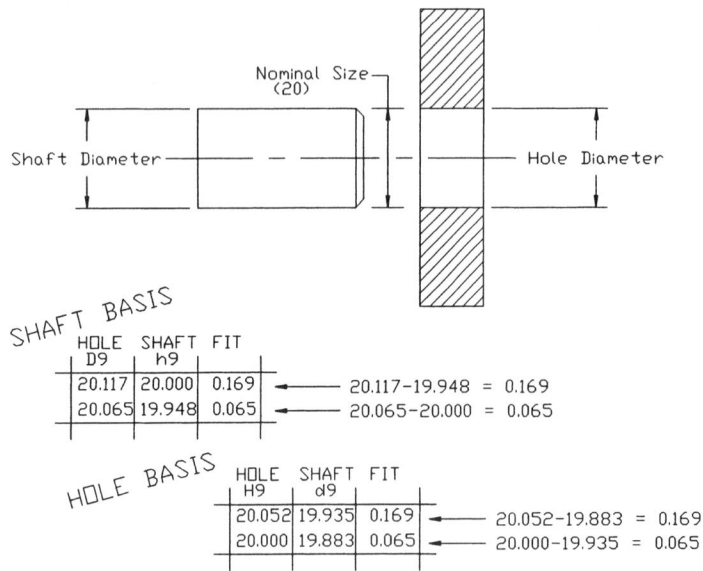

**Figure 4-117**

### Interference Fits

H7/p6 or P7/h6 = Locational transition fit
H7/s6 or S7/h6 = Medium drive fit
H7/u6 or U7/h6 = Force fit

Not all possible sizes are listed in the tables in the appendix. Only preferred sizes are listed. Tolerances for sizes between the stated sizes are derived by going to the next nearest given size. The values are not interpolated. A basic size of 27 would use the tolerance values listed for 25. Sizes that are exactly halfway between two stated sizes may use either set of values depending on the design requirements.

## 4-53 NOMINAL SIZES

The term *nominal* refers to the approximate size of an object that matches a common fraction or whole number. A shaft with a dimension of 1.500±.003 is said to have a nominal size of "one and a half inches." A dimension of 1.500 +.000/-0.005 is still said to have a nominal size of one and a half inches. In both examples, 1.5 is the closest common fraction.

## 4-54 HOLE AND SHAFT BASIS

One of the charts shown in Figure 4-116 applies tolerances starting with the nominal hole sizes, called *hole basis tolerances*; the other applies tolerances starting with the shaft nominal sizes, called *shaft basis tolerances.* The choice of which set of values to use depends on the design application. In general, hole basis numbers are used more often because it is more difficult to vary hole diameters manufactured using specific drill sizes than shaft sizes manufactured using a lathe. Shaft sizes may be used when a specific fastener diameter is used to assemble several objects.

Figure 4-117 shows a hole, a shaft, and a set of sample values taken from the tables found in the appendix. One set of values is for hole basis tolerance and the other for shaft basis tolerance. The fit values are the same for both sets of values. The hole basis values were derived starting with a nominal hole size of 20.000, whereas the shaft basis values were derived starting with a shaft nominal size of 20.000. The letters used to identify holes are always written using capital letters, and the letters for shaft values use lowercase.

Additional fit tolerances may be found on the Web. Search for fits and tolerances.

## 4-55 SAMPLE PROBLEM SP4-3

Dimension a hole and a shaft that are to fit together using a preferred clearance fit. Use hole basis values based on a nominal size of 12 mm.

Figure 4-118 shows values taken from the appropriate table in the appendix. The values may be applied directly to the shaft and hole as shown.

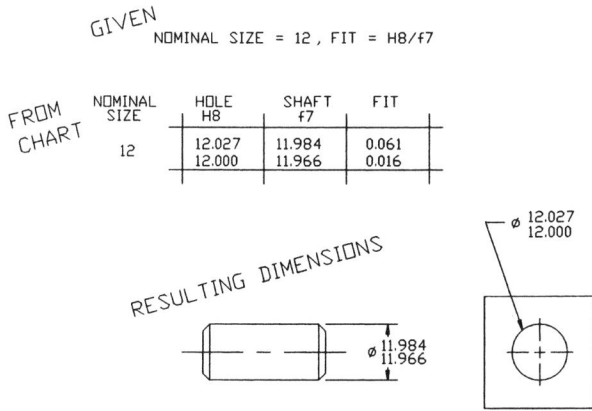

Figure 4-118

## 4-56 STANDARD FITS (INCH VALUES)

The appendix also includes tables of standard fit tolerances for inch values. The tables for inches are presented for a range of nominal values and are not for specific values, as are the metric value tables. The values may be hole or shaft basis.

**Fits defined using inch values are classified as follows:**

RC = Running and sliding fits
LC = Clearance locational fits
LT = Transitional locational fits
LN = Interference fits
FN = Force fits

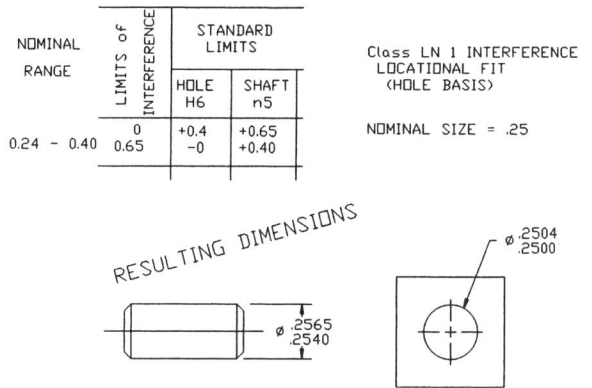

Figure 4-119

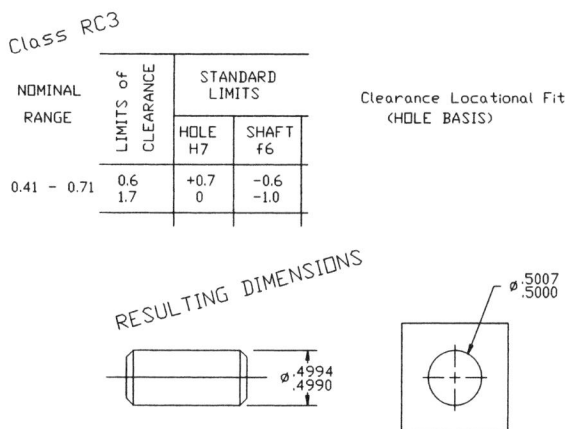

Figure 4-120

Each of these general categories has several subclassifications within it defined by a number, for example, Class RC1, Class RC2, through Class RC9. The letter designations are based on International Tolerance Standards, as are metric designations.

The values are listed in thousandths of an inch. A table value of 1.1 means .0011 inch. A table value of .5 means .0005 inch.

Figure 4-119 shows a set of values for a Class RC3 clearance fit hole basis taken from the appendix. If the values are applied to a nominal size of .5 inch, the resulting hole and shaft sizes will be as shown. Plus table values are added to the nominal value; minus values are subtracted from the nominal value.

Nominal values that are common to two nominal ranges (0.71) may use values from either range.

## 4-57 SAMPLE PROBLEM SP4-4

Dimension a hole and shaft for a Class LN1 Interference fit based on a nominal diameter of .25 in. Use hole basis values.

Figure 4-120 shows the values for the 0.24–0.40 nominal range as listed in the appendix. The values are in thousandths of an inch. Plus values are added to the nominal size. The resulting shaft and hole dimensions are as shown. The diameter of the shaft is larger than that of the hole because this example calls for an interference fit.

| PREFERRED SIZES (mm) | | | |
|---|---|---|---|
| First Choice | Second Choice | First Choice | Second Choice |
| 1 | 1.1 | 12 | 14 |
| 1.2 | 1.4 | 16 | 18 |
| 1.6 | 1.8 | 20 | 22 |
| 2 | 2.2 | 25 | 28 |
| 2.5 | 2.8 | 30 | 35 |
| 3 | 3.5 | 40 | 45 |
| 4 | 4.5 | 50 | 55 |
| 5 | 5.5 | 60 | 70 |
| 6 | 7 | 80 | 90 |
| 8 | 9 | 100 | 110 |
| 10 | 11 | 120 | 140 |

Figure 4-121

| Fraction | Decimal Equivalent | Fraction | Decimal Equivalent | Fraction | Decimal Equivalent |
|---|---|---|---|---|---|
| 7/64 | .1094 | 21/64 | .3281 | 11/16 | .6875 |
| 1/8 | .1250 | 11/32 | .3438 | 3/4 | .7500 |
| 9/64 | .1406 | 23/64 | .3594 | 13/16 | .8125 |
| 5/32 | .1562 | 3/8 | .3750 | 7/8 | .8750 |
| 11/64 | .1719 | 25/64 | .3906 | 15/16 | .9375 |
| 3/16 | .1875 | 13/32 | .4062 | 1 | 1.0000 |
| 13/64 | .2031 | 27/64 | .4219 | | |
| 7/32 | .2188 | 7/16 | .4375 | PARTIAL LIST | |
| 1/4 | .2500 | 29/64 | .4531 | of standard | |
| 17/64 | .2656 | 15/32 | .4688 | Twist Drill Sizes | |
| 9/32 | .2812 | 1/2 | .5000 | (Fractional | |
| 19/64 | .2969 | 9/16 | .5625 | Sizes) | |
| 5/16 | .3125 | 5/8 | .6250 | | |

Figure 4-122

## 4-58 PREFERRED AND STANDARD SIZES

It is important that designers always consider preferred and standard sizes when selecting sizes for designs. Most tooling is set up to match these sizes, so manufacturing is greatly simplified when preferred and standard sizes are specified. Figure 4-121 shows a list of preferred sizes for metric values.

Consider the case of design calculations that call for a 42-mm-diameter hole. A 42-mm-diameter hole is not a preferred size. A diameter of 40 mm is the closest preferred size, and a 45-mm diameter is a second choice. A 42-mm hole could be manufactured but would require an unusual drill size that may not be available. It would be wise to reconsider the design to see if a 40-mm diameter hole could be used, and if not, possibly a 45-mm diameter hole.

A very large quantity production run could possibly justify the cost of special tooling, but for smaller runs it is probably better to use preferred sizes. Machinists will have the required drills, and maintenance people will have the appropriate tools for these sizes.

Figure 4-122 shows a list of standard fractional drill sizes. Most companies now specify metric units or decimal inches; however, many standard items are still available in fractional sizes, and many older objects may still require fractional-sized tools and replacement parts. A more complete listing is available in the appendix.

## 4-59 SURFACE FINISHES

The term *surface finish* refers to the accuracy (flatness) of a surface. Metric values are measured using micrometers (μm), and inch values are measured in microinches (μin.).

The accuracy of a surface depends on the manufacturing process used to produce the surface. Figure 4-123 shows a list of several manufacturing processes and the quality of the surface finish they can be expected to produce.

Surface finishes have several design applications. *Datum surfaces,* or surfaces used for baseline dimensioning, should have fairly accurate surface finishes to help assure accurate measurements; bearing surfaces should have good-quality surface finishes for better load distribution, and parts that operate at high speeds should have smooth finishes to help reduce friction.

Figure 4-124 shows a screw head sitting on a very wavy surface. Note that the head of the screw is in contact with only two wave peaks, meaning all the bearing load is concentrated on the two peaks. This situation could cause

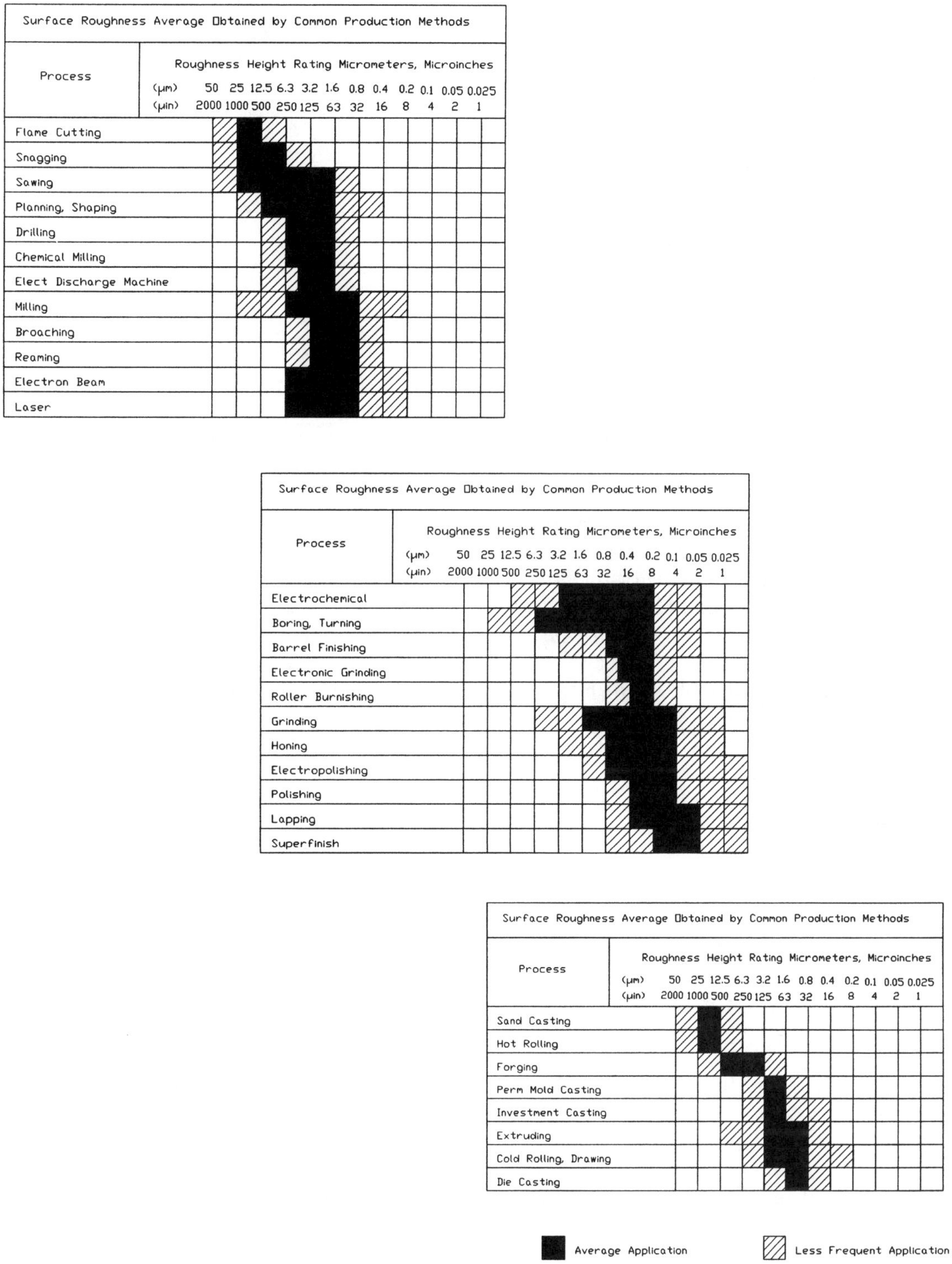

**Surface Roughness Average Obtained by Common Production Methods**

| Process | Roughness Height Rating Micrometers, Microinches |
|---------|---------------------------------------------------|
| | (μm) 50 25 12.5 6.3 3.2 1.6 0.8 0.4 0.2 0.1 0.05 0.025 |
| | (μin) 2000 1000 500 250 125 63 32 16 8 4 2 1 |
| Flame Cutting | |
| Snagging | |
| Sawing | |
| Planning, Shaping | |
| Drilling | |
| Chemical Milling | |
| Elect Discharge Machine | |
| Milling | |
| Broaching | |
| Reaming | |
| Electron Beam | |
| Laser | |

**Surface Roughness Average Obtained by Common Production Methods**

| Process | Roughness Height Rating Micrometers, Microinches |
|---------|---------------------------------------------------|
| | (μm) 50 25 12.5 6.3 3.2 1.6 0.8 0.4 0.2 0.1 0.05 0.025 |
| | (μin) 2000 1000 500 250 125 63 32 16 8 4 2 1 |
| Electrochemical | |
| Boring, Turning | |
| Barrel Finishing | |
| Electronic Grinding | |
| Roller Burnishing | |
| Grinding | |
| Honing | |
| Electropolishing | |
| Polishing | |
| Lapping | |
| Superfinish | |

**Surface Roughness Average Obtained by Common Production Methods**

| Process | Roughness Height Rating Micrometers, Microinches |
|---------|---------------------------------------------------|
| | (μm) 50 25 12.5 6.3 3.2 1.6 0.8 0.4 0.2 0.1 0.05 0.025 |
| | (μin) 2000 1000 500 250 125 63 32 16 8 4 2 1 |
| Sand Casting | |
| Hot Rolling | |
| Forging | |
| Perm Mold Casting | |
| Investment Casting | |
| Extruding | |
| Cold Rolling, Drawing | |
| Die Casting | |

■ Average Application        ▨ Less Frequent Application

Figure 4-123

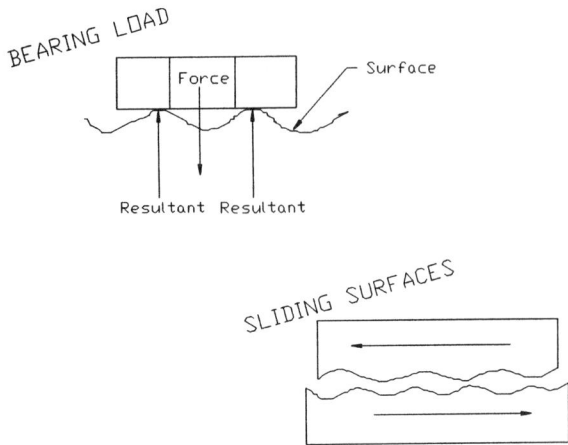

**Figure 4-124**

**Figure 4-125**

stress cracks and greatly weaken the surface. A better-quality surface finish would increase the bearing contact area.

Figure 4-124 also shows two very rough surfaces moving in contact with each other. The result will be excess wear to both surfaces because the surfaces touch only on the peaks, and these peaks will tend to wear faster than flatter areas. Excess vibration can also result when interfacing surfaces are too rough.

Surface finishes are classified into three categories: surface texture, roughness, and lay. *Surface texture* is a general term that refers to the overall quality and accuracy of a surface.

*Roughness* is a measure of the average deviation of a surface's peaks and valleys. See Figure 4-125.

*Lay* refers to the direction of machine marks on a surface. See Figure 4-126. The lay of a surface is particularly

**Figure 4-126**

**Figure 4-127**

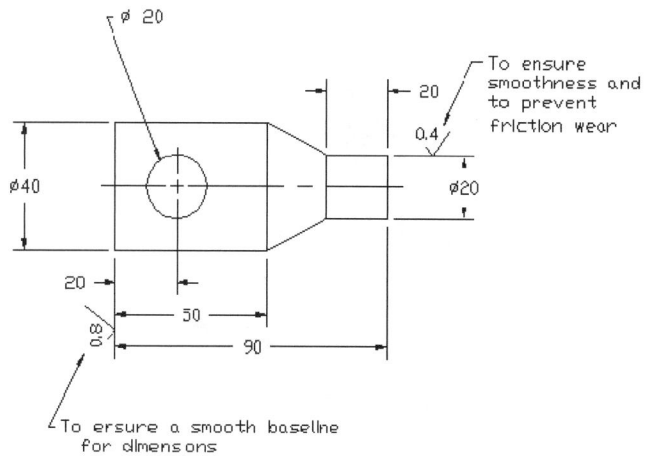

**Figure 4-128**

important when two moving objects are in contact with each other, especially at high speeds.

## 4-60 SURFACE CONTROL SYMBOLS

Surface finishes are indicated on a drawing with surface control symbols. See Figure 4-127. The general surface control symbol looks like a check mark. Roughness values may be included with the symbol to specify the required accuracy. Surface control symbols can also be used to specify the manufacturing process that may or may not be used to produce a surface.

Figure 4-128 shows two applications of surface control symbols. In the first example, a 0.8-μm (32-μin.) surface finish is specified on the surface that serves as a datum for several horizontal dimensions. A 0.8-μm surface finish is generally considered the minimum acceptable finish for datums.

A second finish mark with a value of 0.4 μm is located on an extension line that refers to a surface that will be in contact with a moving object. The extra flatness will help prevent wear between the two surfaces.

It is suggested that a general finish mark be drawn and saved as a wblock so it can be inserted as needed on future drawings. Add the machine mark wblock to any prototype drawings created.

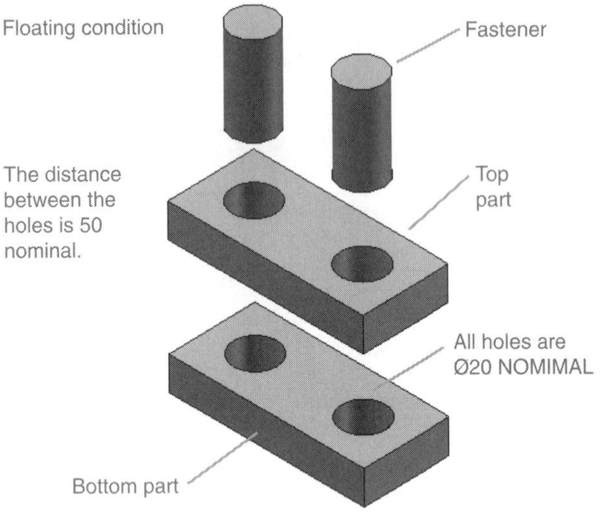

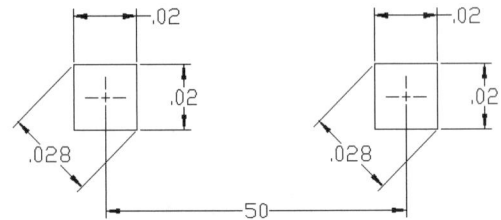

Figure 4-129

Figure 4-130

## 4-61 DESIGN PROBLEMS

Figure 4-129 shows two objects that are to be fitted together using a fastener such as a screw-and-nut combination. For this example a cylinder will be used to represent a fastener. Only two nominal dimensions are given. The dimensions and tolerances were derived as follows.

The distance between the centers of the holes is given as 50 NOMINAL. The term *nominal* means that the stated value is only a starting point. The final dimensions will be close to the given value, but do not have to equal it.

Assigning tolerances is an iteration process; that is, a tolerance is selected and other tolerance values are calculated from the selected initial values. If the results are not satisfactory, go back and modify the initial value and calculate the other values again. As your experience grows, you will become better at selecting realistic initial values.

In the example shown in Figure 4-129, start by assigning a tolerance of ±0.01 to both the TOP and BOTTOM parts for both the horizontal and vertical dimensions used to locate the holes. This means that there is a possible center point variation of 0.02 for both parts. The parts must always fit together, so tolerances must be assigned based on the worst-case condition, or when the parts are made at the extreme ends of the assigned tolerances.

Figure 4-130 shows a greatly enlarged picture of the worst-case condition created by a tolerance of ±0.01. The center points of the two holes could be as much as 0.028 apart if they were located at opposite corners of the tolerance zones. This means that the minimum hole diameter must always be at least 0.028 larger than the maximum stud diameter. See Section 4-49. In addition, there should be a clearance

tolerance assigned so that the hole and stud are never exactly the same size. Figure 4-131 shows the resulting tolerances.

### Floating condition

The TOP and BOTTOM parts shown in Figure 4-129 are to be joined by two independent fasteners; that is, the location of one fastener does not depend on the location of the other. This is called a *floating condition.* This means that the tolerance zones for both the TOP and BOTTOM parts can be

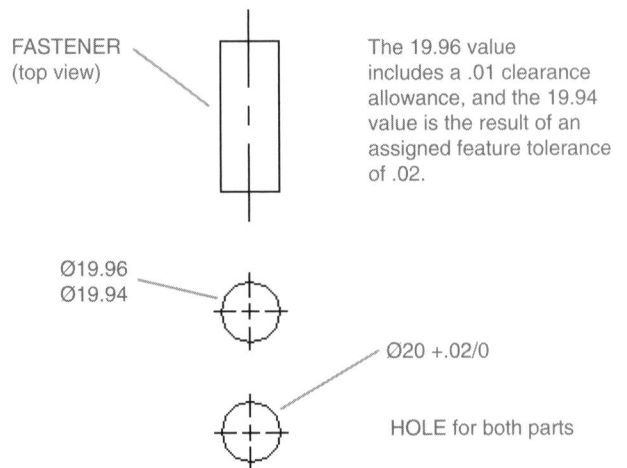

Figure 4-131

assigned the same values, and that a fastener diameter selected to fit one part will also fit the other part.

The final tolerances were developed by first defining a minimum hole size of 20.00. An arbitrary tolerance of .02 was assigned to the hole and was expressed as 20.00 +.02/−0, so that the hole can never be any smaller than 20.00.

The 20.00 minimum hole diameter dictates that the maximum fastener diameter can be no greater than 19.97, or .03 (the rounded-off diagonal distance across the tolerance zone −.028) less than the minimum hole diameter. A .01 clearance was assigned. The clearance ensures that the hole and fastener are never exactly the same diameter. The resulting maximum allowable diameter for the fastener is 19.96. Again, an arbitrary tolerance of .02 was assigned to the fastener. The final fastener dimensions are therefore 19.96 to 19.94.

The assigned tolerances ensure that there will always be at least .01 clearance between the fastener and the hole. The other extreme condition occurs when the hole is at its largest possible size (10.02) and the fastener is at its smallest (19.94). This means that there could be as much as .08 clearance between the parts. If this much clearance is not acceptable, then the assigned tolerances will have to be re evaluated.

Figure 4-132 shows the TOP and BOTTOM parts dimensioned and toleranced. Any dimensions that do not have assigned tolerances are assumed to have standard tolerances. See Figure 4-103.

Note, in Figure 4-132, that the top edge of each part was assigned a surface finish. This was done to help ensure the accuracy of the 20±.01 dimension. If this edge surface were rough, it could affect the tolerance measurements.

This example will be repeated in Chapter 5 using geometric tolerances. Geometric tolerance zones are circular rather than rectangular.

## Fixed condition

Figure 4-133 shows the same nominal conditions presented in Figure 4-129, but the fasteners are now fixed to the TOP part. This is called the *fixed condition.* In analyzing the tolerance zones for the fixed condition, one must consider two positional tolerances: the positional tolerances for the holes in the BOTTOM part and the positional tolerances for the fixed fasteners in the TOP part. This may be expressed in an equation as follows:

$$S_{max} + DTSZ = H_{min} - DTZ$$

where

$S_{max}$ = Maximum shaft (fastener) diameter

$H_{min}$ = Minimum hole diameter

$DTSZ$ = Diagonal distance across the shaft's center point tolerance zone

$DTZ$ = Diagonal distance across the hole's center point tolerance zone

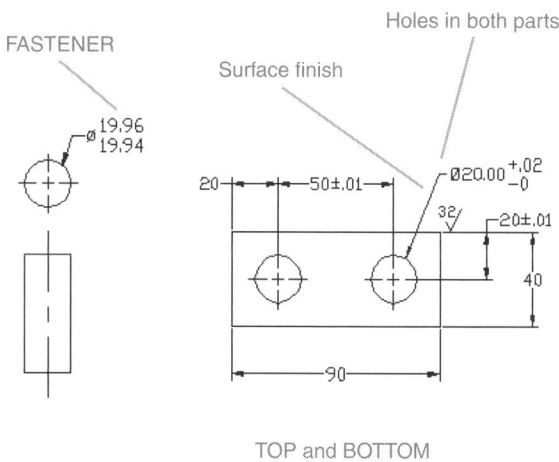

FASTENER

Surface finish

Holes in both parts

Ø 19.96 / 19.94

20   50±.01

Ø20.00 +.02/−0

32/   20±.01

40

90

TOP and BOTTOM

All dimensions not assigned a tolerance will be assumed to have a standard tolerance. See Figure 4-14.

**Figure 4-132**

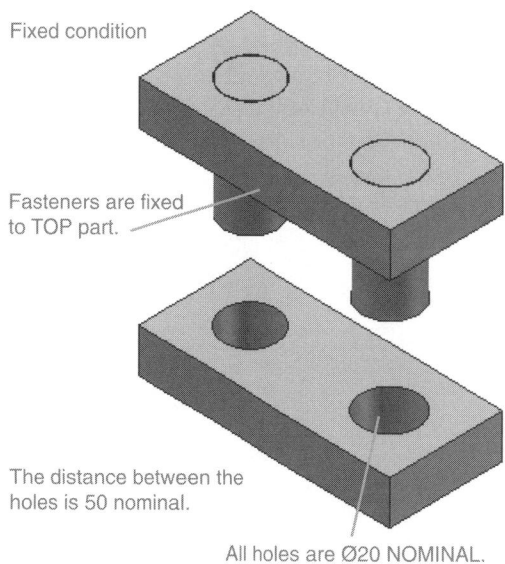

Fixed condition

Fasteners are fixed to TOP part.

The distance between the holes is 50 nominal.

All holes are Ø20 NOMINAL.

**Figure 4-133**

If a dimension and tolerance of 50±.01 and 20±.01 are assigned to both the center distance between the holes and the center distance between the fixed fasteners, the values for DTSZ and DTZ will be equal. The formula can then be simplified as follows:

$$S_{max} = H_{min} - 2(DTZ)$$

where DTZ equals the diagonal distance across the tolerance zone. If a hole tolerance of 20.00 +.02/−0 is also defined, the resulting maximum shaft size can be determined, assuming that the calculated distance of .028 is rounded off to .03. See Figure 4-134.

$$\begin{aligned} S_{max} &= 20.00 - 2(.03) \\ &= 19.94 \end{aligned}$$

This means that the largest possible shaft diameter that will just fit equals 19.94. If a clearance tolerance of .01 is assumed to ensure that the shaft and hole are never exactly the same size, the maximum shaft diameter becomes 19.93. See Figure 4-135.

A feature tolerance of .02 on the shaft will result in a minimum shaft diameter of 19.91. Note that the .01 clearance tolerance and the .02 feature tolerance were arbitrarily chosen. Other possible values could have been used.

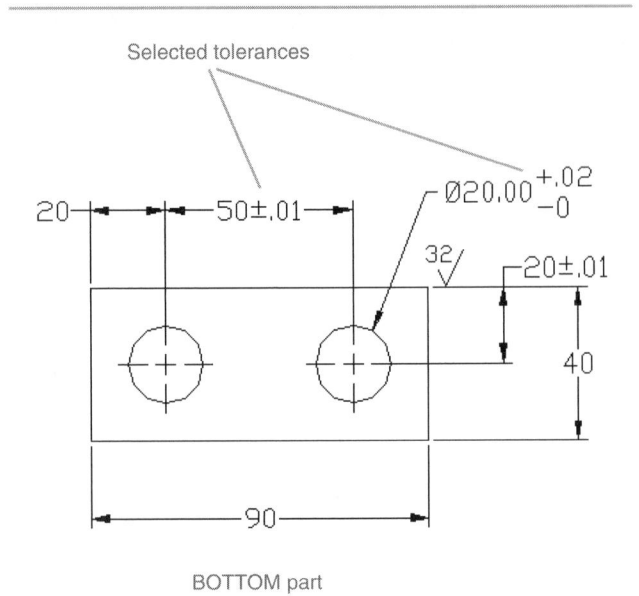

Figure 4-134

The shaft values were derived as follows.

| 20.00 | The selected value for the minimum hole diameter |
|---|---|
| −.03 | The rounded-off value for the hole positional tolerance |
| −.03 | The rounded-off value for the shaft positional tolerance |
| −.01 | The selected clearance value |
| 19.93 | The maximum shaft value |
| −.02 | The selected tolerance value |
| 19.91 | The minimum shaft diameter |

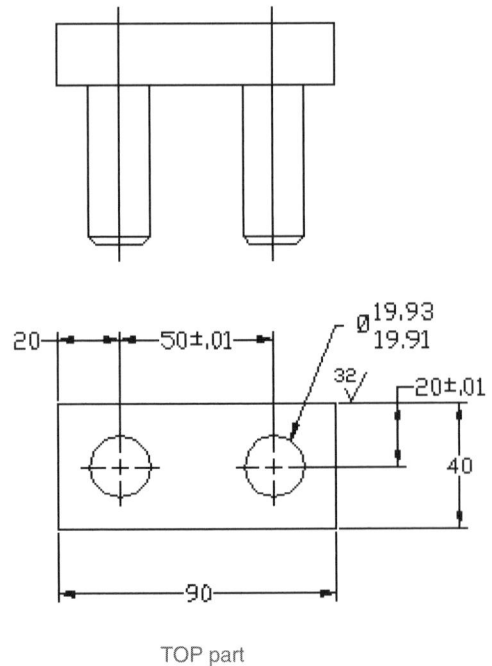

Figure 4-135

## To design a hole given a fastener size

The previous two examples started by selecting a minimum hole diameter and then calculating the resulting fastener size. Figure 4-136 shows a situation in which the fastener size is defined, and the problem is to determine the appropriate hole sizes. Figure 4-137 shows the dimensions and tolerances for both top and bottom parts.

Requirements:
Clearance, minimum = .003
Hole tolerances = .005
Positional tolerance = .002

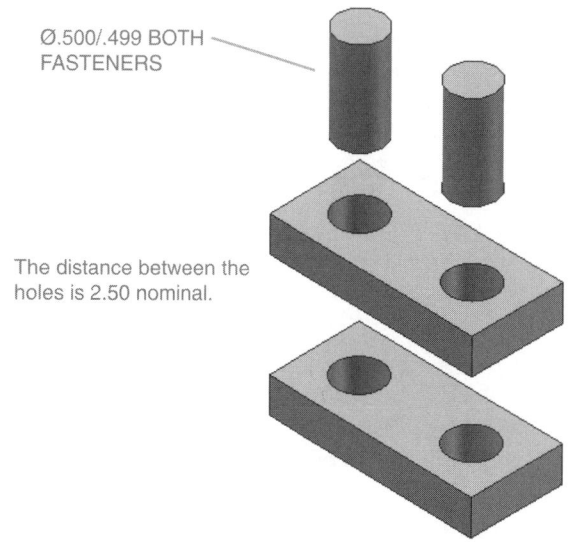

Ø.500/.499 BOTH FASTENERS

The distance between the holes is 2.50 nominal.

Figure 4-136

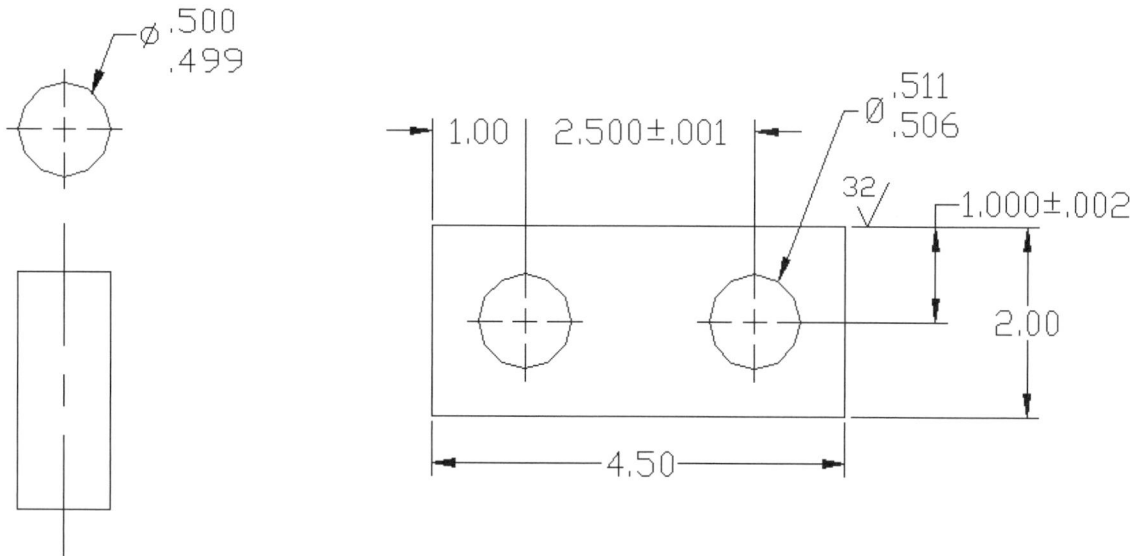

Figure 4-137

## 4-62 EXERCISE PROBLEMS

Redraw the objects shown in Exercise Problems through EX4-50 through EX4-53 using the given dimensions. Include the listed tolerances.

### EX4-50 MILLIMETERS

1. 38±0.05
2. 10±0.1
3. 5±0.05
4. 45.50°
   44.50°
5. 40±0.1
6. 22±0.1
7. $25 \begin{smallmatrix} +0 \\ -0.1 \end{smallmatrix}$
8. $25 \begin{smallmatrix} +0.05 \\ -0 \end{smallmatrix}$
9. 51.50
   50.75
10. 76±0.1

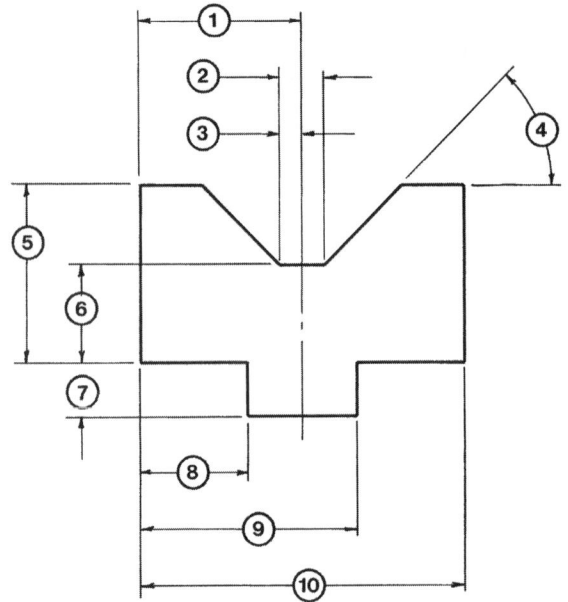

### EX4-51 MILLIMETERS

1. 34±0.25
2. 17±0.25
3. 25±0.05
4. 15.00
   14.80
5. 50±0.05
6. 80±0.1
7. R5±0.1-8 PLACES
8. 45±0.25
9. 60±0.1
10. Ø14-3 HOLES
11. 15.00
    14.80
12. 30.00
    29.80

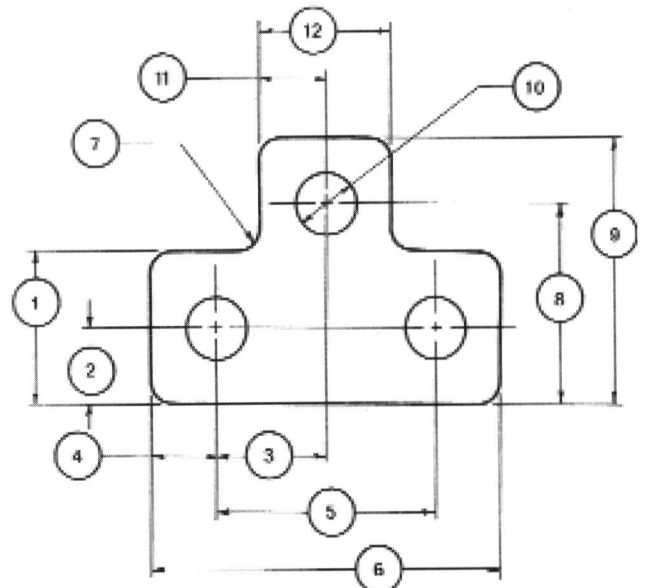

## EX4-52 INCHES

1. 3.00±0.1
2. 1.56±.01
3. $\dfrac{46.50}{45.50}$
4. .750±.005
5. $\dfrac{2.75}{2.70}$
6. 3.625±.010
7. 45°±.5°
8. 2.250±.005

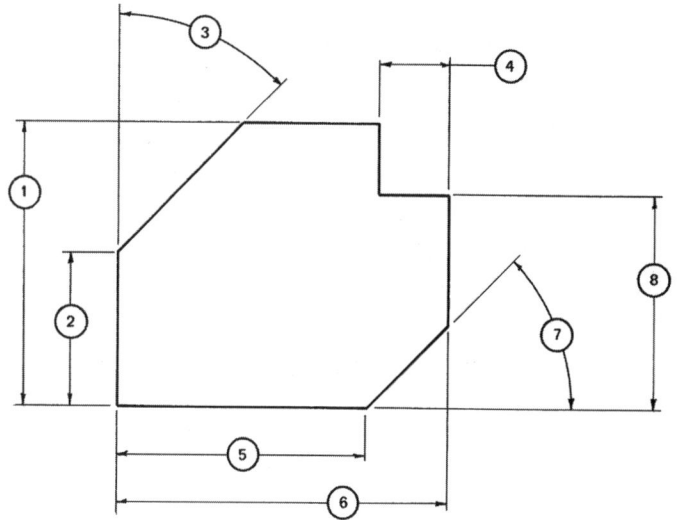

## EX4-53 INCHES

1. $50 \, {}^{+.2}_{\ \ 0}$
2. R45±.1-2 PLACES
3. $63.5 \, {}^{\ \ 0}_{0.2}$
4. 76±.1
5. 38±.1
6. Ø12.00 $\,{}^{+.05}_{\ \ \ 0}$ -3 HOLES
7. 30±.03
8. 30±.03
9. $100 \, {}^{+.4}_{\ \ 0}$

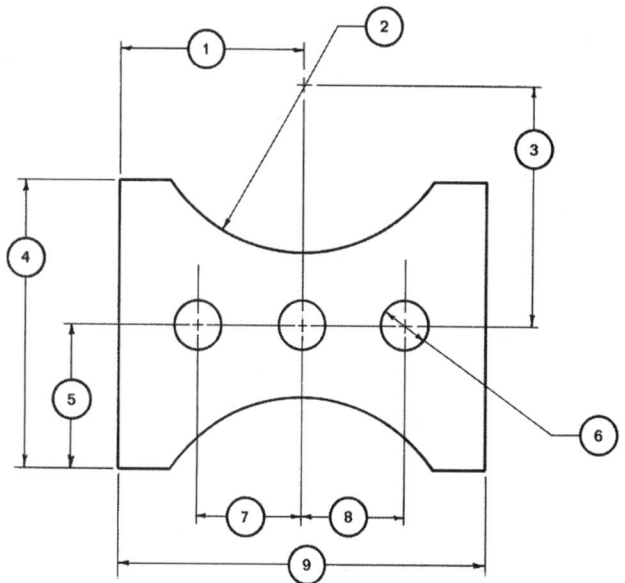

## EX4-54 MILLIMETERS

Redraw the following object, including the given dimensions and tolerances. Calculate and list the maximum and minimum distances for surface A.

## EX4-56 MILLIMETERS

Redraw the following object, including the dimensions and tolerances. Calculate and list the maximum and minimum distances for surfaces D and E.

## EX4-55 INCHES

A. Redraw the following object, including the dimensions and tolerances. Calculate and list the maximum and minimum distances for surface A.
B. Redraw the given object and dimension it using baseline dimensions. Calculate and list the maximum and minimum distances for surface A.

## EX4-57 MILLIMETERS

Dimension the following object twice: once using chain dimensions and once using baseline dimensions. Calculate and list the maximum and minimum distances for surface D for both chain and baseline dimensions. Compare the results.

## EX4-58 INCHES

Redraw the following shapes, including the dimensions and tolerances. Also list the required minimum and maximum values for the specified distances.

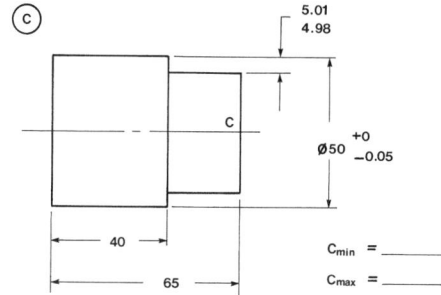

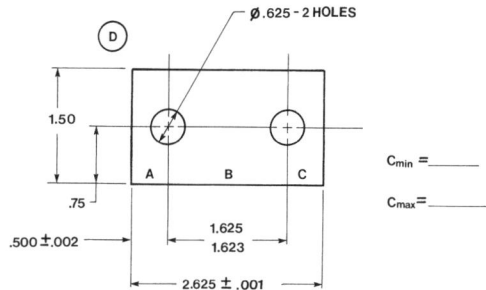

**A**

$2.00 {}^{+.05}_{-.02}$

1.02
.97

A

B

1.25

.75

$B_{min} =$ _____
$B_{max} =$ _____

**B**

$60 {}^{+0.1}_{-0}$

$20 \pm 0.2$

21.0
19.5

B

A        C

30

20

$C_{min} =$ _____
$C_{max} =$ _____

**C**

5.01
4.98

C

$\varnothing 50 {}^{+0}_{-0.05}$

40

65

$C_{min} =$ _____
$C_{max} =$ _____

**D**

$\varnothing .625 - 2$ HOLES

1.50

A        B        C

.75

$.500 \pm .002$

1.625
1.623

$2.625 \pm .001$

$C_{min} =$ _____
$C_{max} =$ _____

## EX4-59

Redraw and complete the following inspection report. Under the Results column classify each "AS MEASURED" value as OK if the value is within the stated tolerances, REWORK if the value indicates that the measured value is beyond the stated tolerance but can be reworked to bring it into the acceptable range, or SCRAP if the value is not within the tolerance range and cannot be reworked to make it acceptable.

INSPECTION REPORT

PART NAME AND NO:   *1075500 2*

1.00  3 PLACES

INSPECTOR:

DATE:

| BASE DIMENSION | TOLERANCES | | AS MEASURED | RESULTS |
|---|---|---|---|---|
| | MAX | MIN | | |
| ① $100 \pm 0.5$ | | | 99.8 | |
| ② $\varnothing {}^{57}_{56}$ | | | 57.01 | |
| ③ $22 \pm 0.3$ | | | 21.72 | |
| ④ $40.05$ $39.95$ | | | 39.98 | |
| ⑤ $22 \pm 0.3$ | | | 21.68 | |
| ⑥ $R52 {}^{+0}_{-0.2}$ | | | 51.99 | |
| ⑦ $35 {}^{+0.2}_{-0.3}$ | | | 35.20 | |
| ⑧ $30 {}^{+0.4}_{0}$ | | | 30.27 | |
| ⑨ $6.0 {}^{+.1}_{-.2}$ | | | 5.85 | |
| ⑩ $12.0 \pm 0.2$ | | | 11.90 | |

.50 — 10 PLACES

**EX4-60 MILLIMETERS**

Redraw the following charts and complete them based on the following information. All values are in millimeters.

A. Nominal=16, Fit=H9/d9
B. Nominal=30, Fit=H11/c11
C. Nominal=22, Fit=H7/g6
D. Nominal=10, Fit=C11/h11
E. Nominal=25, Fit=F8/h7
F. Nominal=12, Fit=H7/k6
G. Nominal=3, Fit=H7/p6
H. Nominal=19, Fit=H7/s6
I. Nominal=27, Fit=H7/u6
J. Nominal=30, Fit=N7/h6

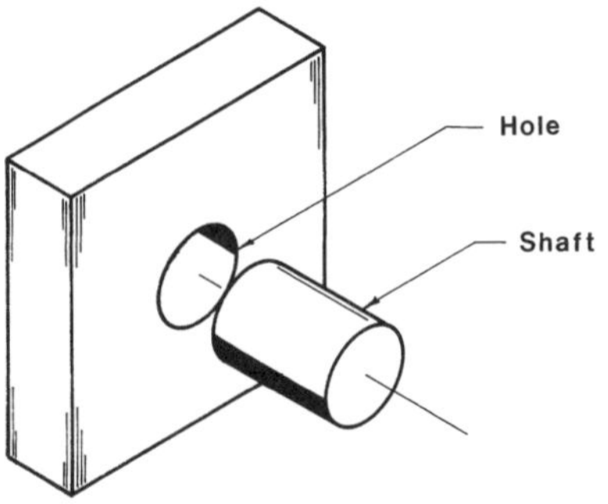

Hole

Shaft

half space

| NOMINAL | HOLE | | SHAFT | | CLEARANCE | |
|---|---|---|---|---|---|---|
| | MAX | MIN | MAX | MIN | MAX | MIN |
| A | | | | | | |
| B | | | | | | |
| C | | | | | | |
| D | | | | | | |
| E | | | | | | |

3.75
6 equal spaces

1.5

6.0 – 6 equal spaces

| NOMINAL | HOLE | | SHAFT | | INTERFERENCE | |
|---|---|---|---|---|---|---|
| | MAX | MIN | MAX | MIN | MAX | MIN |
| F | | | | | | |
| G | | | | | | |
| H | | | | | | |
| I | | | | | | |
| J | | | | | | |

Use the same dimensions given above

## EX4-61

Redraw the following charts and complete them based on the following information. All values are in inches.

A. Nominal=0.25, Fit=Class LC5
B. Nominal=1.00, Fit=Class LC7
C. Nominal=1.50, Fit=Class LC10
D. Nominal=0.75, Fit=Class RC3
E. Nominal=2.50, Fit=Class RC6
F. Nominal=.500, Fit=Class LT2
G. Nominal=1.25, Fit=Class LT5
H. Nominal=3.00, Fit=Class LN3
I. Nominal=1.625, Fit=Class FN1
J. Nominal=2.00, Fit=Class FN4

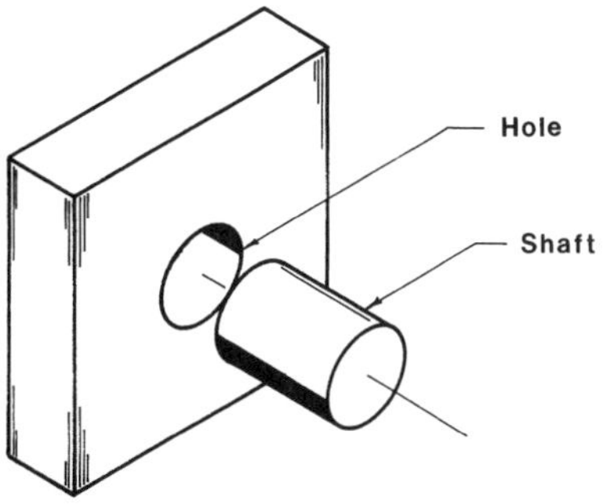

Hole

Shaft

half space

3.75
6 equal
spaces

| NOMINAL | HOLE | | SHAFT | | CLEARANCE | |
|---------|------|-----|-------|-----|-----------|-----|
|  | MAX | MIN | MAX | MIN | MAX | MIN |
| A |  |  |  |  |  |  |
| B |  |  |  |  |  |  |
| C |  |  |  |  |  |  |
| D |  |  |  |  |  |  |
| E |  |  |  |  |  |  |

1.5        6.0 – 6 equal spaces

| NOMINAL | HOLE | | SHAFT | | INTERFERENCE | |
|---------|------|-----|-------|-----|--------------|-----|
|  | MAX | MIN | MAX | MIN | MAX | MIN |
| F |  |  |  |  |  |  |
| G |  |  |  |  |  |  |
| H |  |  |  |  |  |  |
| I |  |  |  |  |  |  |
| J |  |  |  |  |  |  |

Use the same dimensions given above

Draw the chart shown and add the appropriate values based on the dimensions and tolerances given in Exercise Problems EX4-62 through EX4-65.

### EX4-62 MILLIMETERS

```
PART NO: 9-M53A
    A.  20±0.1
    B.  30±0.2
    C.  Ø20±0.05
    D.  40
    E.  60
```

MAXIMUM SHAFT DIAMETER
THAT WILL ALWAYS FIT
= _____ ?

| PART NO: | | | | | | | | | |
|---|---|---|---|---|---|---|---|---|---|
| A | | B | | C | | D | | E | |
| MAX | MIN | MAX | MIN | MAX | MIN | MAX | MIN | MAX | MIN |
| | | | | | | | | | |

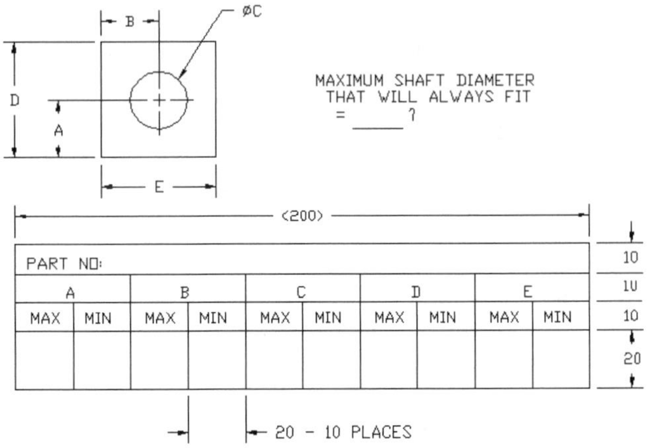

### EX4-64 MILLIMETERS

```
PART NO: 9-M53B
    A.  32.02
        31.97
    B.  47.52
        47.50
    C.  Ø18 +0.05
             0
    D.  64±0.05
    E.  100±0.05
```

### EX4-63 MILLIMETERS

```
PART NO: 9-M53B
    A.  32.02
        31.97
    B.  47.52
        47.50
    C.  Ø18 +0.05
             0
    D.  64±0.05
    E.  100±0.05
```

### EX4-65 MILLIMETERS

```
PART NO: 9-E47B
    A.  18  +0
        -0.02
    B.  26  +0
        -0.04
    C.  Ø  24.03
           23.99
    D.  52±0.04
    E.  36±0.02
```

## EX4-66 MILLIMETERS

Prepare front and top views of Parts 4A and 4B based on the given dimensions. Add tolerances to produce the stated clearances.

PART 4A

ø 33

Maximum allowable mismatch

10

21

Maximum allowable clearance

0.5

0.3

21

ø 13

PART 4B

## EX4-67 INCHES

Redraw parts A and B and add dimensions and tolerances to meet the "UPON ASSEMBLED" requirements.

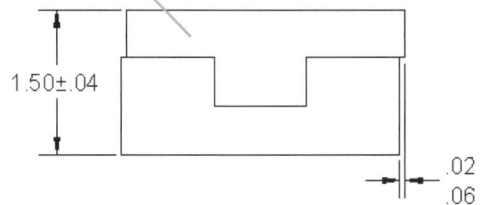

Material is 1.00 thick.

Nominal sizes

0.50

0.50

1.00    1.00    1.00

0.50

1.00

The UPON ASSEMBLED requirements.

1.50±.04

.02
.06

## EX4-68 MILLIMETERS

Draw a front and a top view of both given objects. Add dimensions and tolerances to meet the "FINAL CONDITION" requirements.

7

Ø 38

Ø 12

12

12

Ø 38

FINAL CONDITION

MAX =0.03
MIN =0.01

MIN =0.00
MAX =0.04

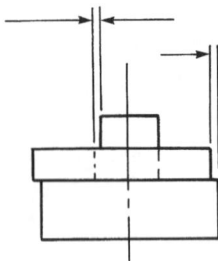

## EX4-69 INCHES

Given the following nominal sizes, dimension tolerance parts AM311 and AM312 so that they always fit together regardless of orientation. Further, dimension the overall lengths of each part so that, in the assembled condition, they will always pass through a clearance gauge with an opening of 90.00±.02.

In the assembled condition, both parts must always pass through the clearance gauge.

20

10

90.05±.02
CLEARANCE

AM312

25
BOTH PARTS

20-3 PLACES
BOTH PARTS

AM311

15-2 SLOTS

10

25

CLEARANCE GAGE
SURFACES

All given dimensions, except for the clearance gauge, are nominal.

## EX4-70 MILLIMETERS

Design a bracket that will support the three Ø100 wheels shown. The wheels will utilize three Ø5.00±0.01 shafts attached to the bracket. The bottom of the bracket must have a minimum of 10 millimeters from the ground. The wall thickness of the bracket must always be at least 5 millimeters, and the minimum bracket opening must be at least 15 millimeters.

1. Prepare a front and a side view of the bracket.
2. Draw the wheels in their relative positions using phantom lines.
3. Add all appropriate dimensions and tolerances.

ALL SIZES ARE NOMINAL, UNLESS OTHERWISE STATED.

SHAFT Ø = 5.00±.01
3 REQUIRED

ROLLER BLADE ASSEMBLY
PART NUMBER  BU110-44

Given a TOP and BOTTOM part in the floating condition as shown in Figure EX4-71 add dimensions and tolerances to satisfy the following conditions. Size the TOP and BOTTOM parts and FASTENER length as needed.

### EX4-71 INCHES–CLEARANCE FIT

A. The distance between the holes is 2.00 nominal.
B. The diameter of the fasteners is Ø.375 nominal.
C. The fasteners have a total tolerance of .001.
D. The holes have a tolerance of .002.
E. The minimum allowable clearance between the fastener and the holes is .003.
F. The material is .375 inch thick.

### EX4-72 MILLIMETERS–CLEARANCE FIT

A. The distance between the holes is 80 nominal.
B. The nominal diameter of the fasteners is Ø12.
C. The fasteners have a total tolerance of 0.05.
D. The holes have a tolerance of 0.03.
E. The minimum allowable clearance between the fastener and the holes is 0.02.
F. The material is 12 millimeters thick.

### EX4-73 INCHES–CLEARANCE FIT

A. The distance between the holes is 3.50 nominal.
B. The diameter of the fasteners is Ø.625.
C. The fasteners have a total tolerance of .005.
D. The holes have a tolerance of .003.
E. The minimum allowable clearance between the fastener and the holes is .002.
F. The material is .500 inch thick.

### EX4-74 MILLIMETERS–CLEARANCE FIT

A. The distance between the holes is 120 nominal.
B. The diameter of the fasteners is Ø24 nominal.
C. The fasteners have a total tolerance of 0.01.
D. The holes have a tolerance of 0.02.
E. The minimum allowable clearance between the fastener and the holes is 0.04.
F. The material is 20 millimeters thick.

### EX4-75 INCHES–INTERFERENCE FIT

A. The distance between the holes is 2.00 nominal.
B. The diameter of the fasteners is Ø.250 nominal.
C. The fasteners have a total tolerance of .001.
D. The holes have a tolerance of .002.
E. The maximum allowable interference between the fastener and the holes is .0065.
F. The material is .438 inch thick.

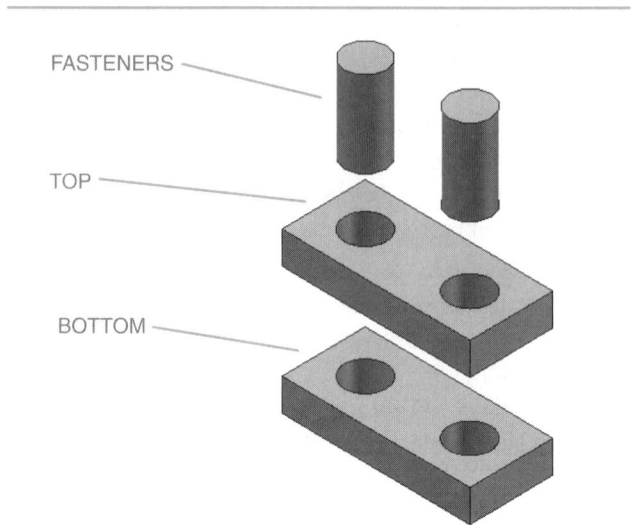

Figure EX4-71

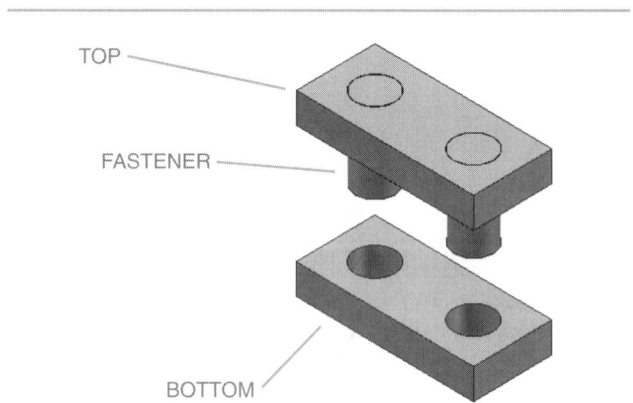

Figure EX4-72

### EX4-76 MILLIMETERS–INTERFERENCE FIT

A. The distance between the holes is 80 nominal.
B. The diameter of the fasteners is Ø10 nominal.
C. The fasteners have a total tolerance of 0.01.
D. The holes have a tolerance of 0.02.
E. The maximum allowable interference between the fastener and the holes is 0.032.
F. The material is 14 millimeters thick.

### EX4-77 INCHES–LOCATIONAL FIT

A. The distance between the holes is 2.25 nominal.
B. The diameter of the fasteners is Ø.50 nominal.

C. The fasteners have a total tolerance of .001.
D. The holes have a tolerance of .002.
E. The minimum allowable clearance between the fastener and the holes is .0010.
F. The material is .370 inch thick.

## EX4-78 MILLIMETERS–TRANSITIONAL FIT

A. The distance between the holes is 100 nominal.
B. The diameter of the fasteners is Ø16 nominal.
C. The fasteners have a total tolerance of 0.01.
D. The holes are to have a tolerance of 0.02.
E. The minimum allowable clearance between the fastener and the holes is 0.01.
F. The material is 20 millimeters thick.

Given a TOP and BOTTOM part in the fixed condition as shown in Figure EX4-72 add dimensions and tolerances to satisfy the following conditions. Size the TOP and BOTTOM parts and FASTENER length as needed.

## EX4-79 INCHES–CLEARANCE FIT

A. The distance between the holes is 2.00 nominal.
B. The diameter of the fasteners is Ø.375 nominal.
C. The fasteners have a total tolerance of .001.
D. The holes have a tolerance of .002.
E. The minimum allowable clearance between the fastener and the holes is .003.
F. The material is .375 inch thick.

## EX4-80 MILLIMETERS–CLEARANCE FIT

A. The distance between the holes is 80 nominal.
B. The nominal diameter of the fasteners is Ø12.
C. The fasteners have a total tolerance of 0.05.
D. The holes have a tolerance of 0.03.
E. The minimum allowable clearance between the fastener and the holes is 0.02.
F. The material is 12 millimeters thick.

## EX4-81 INCHES–CLEARANCE FIT

A. The distance between the holes is 3.50 nominal.
B. The diameter of the fasteners is Ø.625.
C. The fasteners have a total tolerance of .005.
D. The holes have a tolerance of .003.
E. The minimum allowable clearance between the fastener and the holes is .002.
F. The material is .500 inch thick.

## EX4-82 MILLIMETERS–CLEARANCE FIT

A. The distance between the holes is 120 nominal.
B. The diameter of the fasteners is Ø24 nominal.
C. The fasteners have a total tolerance of 0.01.
D. The holes have a tolerance of 0.02.
E. The minimum allowable clearance between the fastener and the holes is 0.04.
F. The material is 20 millimeters thick.

## EX4-83 INCHES–INTERFERENCE FIT

A. The distance between the holes is 2.00 nominal.
B. The diameter of the fasteners is Ø.250 nominal.
C. The fasteners have a total tolerance of .001.
D. The holes have a tolerance of .002.
E. The maximum allowable interference between the fastener and the holes is .0065.
F. The material is .438 inch thick.

## EX4-84 MILLIMETERS–INTERFERENCE FIT

A. The distance between the holes is 80 nominal.
B. The diameter of the fasteners is Ø10 nominal.
C. The fasteners have a total tolerance of 0.01.
D. The holes have a tolerance of 0.02.
E. The maximum allowable interference between the fastener and the holes is 0.032.
F. The material is 14 millimeters thick.

## EX4-85 INCHES–LOCATIONAL FIT

A. The distance between the holes is 2.25 nominal.
B. The diameter of the fasteners is Ø.50 nominal.
C. The fasteners have a total tolerance of .001.
D. The holes have a tolerance of .002.
E. The minimum allowable clearance between the fastener and the holes is .0010.
F. The material is .370 inch thick.

## EX4-86 MILLIMETERS–TRANSITIONAL FIT

A. The distance between the holes is 100 nominal.
B. The diameter of the fasteners is Ø16 nominal.
C. The fasteners have a total tolerance of 0.01.
D. The holes are to have a tolerance of 0.02.
E. The minimum allowable clearance between the fastener and the holes is 0.01.
F. The material is 20 millimeters thick.

## EX4-87 MILLIMETERS

Given the following two assemblies, size the individual parts so that they always fit together. Create a drawing for each part including dimensions and tolerances.

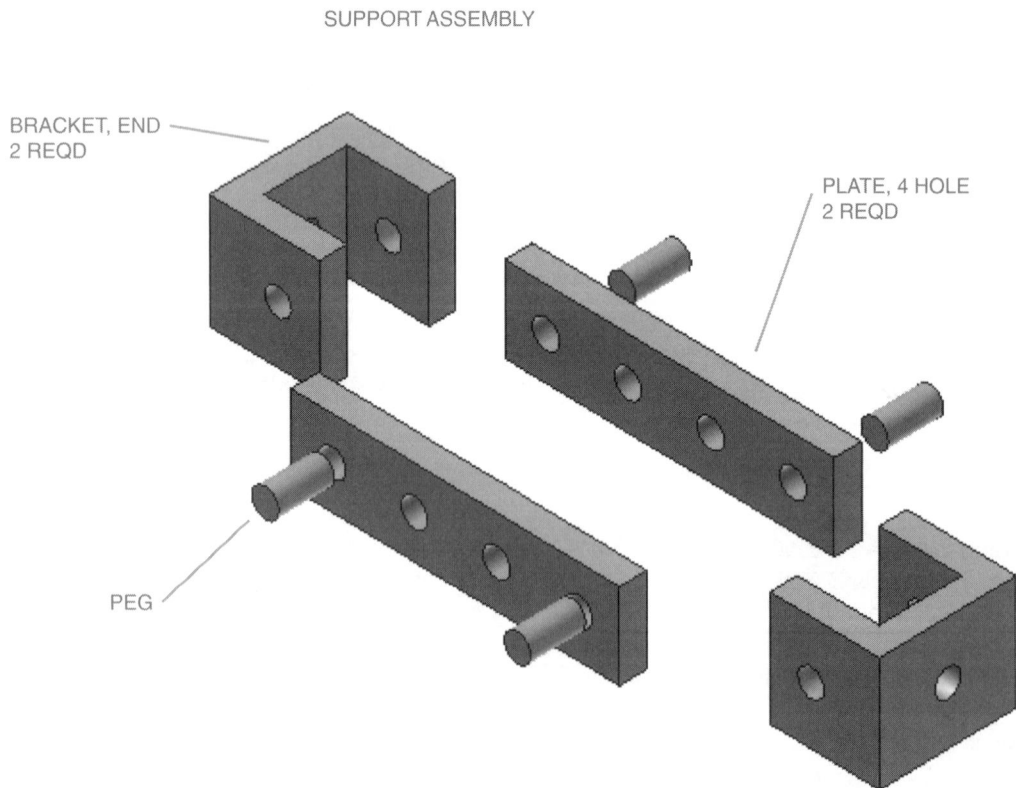

BLOCK ASSEMBLY

BLOCK, BASE

BLOCK, TOP

PEG

SUPPORT ASSEMBLY

BRACKET, END
2 REQD

PLATE, 4 HOLE
2 REQD

PEG

# 4-63 TOLERANCES OF FORM— PROFILE TOLERANCING

## Applied to Individual Features (No Datum) or Related Features (Using Datums)

Profile tolerancing is of two varieties and involves the characteristics below. According to the design requirement, these characteristics may be applied to an individual feature, such as a *single surface* or *element*, or to related features, such as a *single surface* or *element* relative to a datum or datums.

(PROFILE OF A LINE)   (PROFILE OF A SURFACE)

See following pages for details of application.

# 4-64 PROFILE

## Method of Specifying

**Definition.** Profile tolerancing specifies a uniform boundary along the true profile within which the elements of the surface must lie.

## Profile Tolerance

A profile tolerance (either bilateral or unilateral) specifies a tolerance zone, always* intended and measured normal to the basic profile (applicable to the view in which drawn) at all points of the profile, within which the true part surface profile or line profile must lie.

## Profile Tolerance Application

Profile tolerancing is an effective method for controlling lines, arcs, irregular surfaces, or other unusual part profiles. Profile tolerances are usually applied to surface features but may also be applied to a line (element on a feature surface). Profile tolerances can also be used as a locational control to size features to specify their shape, form, orientation and relationship to datum features. In any case, these requirements must be specified in association with the desired profile in a plane of projection (view) on the drawing as follows:

a. An APPROPRIATE VIEW or SECTION is drawn which shows the desired basic profile in true shape.
b. The profile is defined by basic dimensions. This dimensioning may be in the form of located radii

and angles, coordinate dimensioning, formulas, or undimensioned drawing.
c. Depending on design requirements, the tolerance may be divided bilaterally to both sides of the true profile or applied unilaterally to either side of the true profile.
d. To control location of a size feature relative to datum planes.

*Unless otherwise specified.

Where an equally disposed bilateral tolerance is intended, it is only necessary to show the feature control frame with a leader directed to the surface. For an unequally disposed or unilateral tolerance, phantom lines are drawn parallel to the true profile to clearly indicate the tolerance zone inside or outside the true profile. One end of a dimension line is extended to the feature control frame.

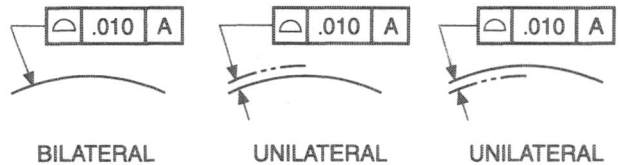

BILATERAL   UNILATERAL   UNILATERAL

e. Other appropriate dimensions, as well as the applicable feature control frame, are added. The FRAME should be applied in a view in which the surface or lines to be controlled is shown as a profile and which pictorially represents the desired feature orientation and relationship.
f. If some limits on a drawing are expressed by a profile tolerance and others by regular tolerance dimensions, the extent or limitation of the profile tolerance must be clearly indicated by reference letters applied at the extremities of the profile controlled portion and a symbolic notation added beneath the profile tolerance feature control frame. Where a profile tolerance applies *all around* the profile of the part, the symbolic representation ⌒ is specified or the notation ALL AROUND is placed beneath the feature control frame.

Common surface profile tolerancing includes a combination of FORM, ORIENTATION, and SIZE controls, with the profile basic dimensions amounting to overall dimensions as well. In this situation, the standard limits of size interpretation, which require that "form" control must be contained within "size" control, do not apply.

Surface profile tolerancing may also be associated with conventional size dimensions and tolerances. The profile itself, however, must be controlled by basic dimensions and the profile tolerance zone. Under this condition the standard limits of size interpretation do apply and the profile tolerance zone must be contained *within* the *size* tolerance zone. If, for example, the bilateral profile method is used, some portions of the profile tolerance zone may be sacrificed if the controlled feature is at its extreme size limit at that point of the profile.

Since profile tolerancing is a control of *surface form* and *orientation,* no modifier (MMC) can be used on the feature controlled. Where a datum reference is used, it is usually intended to apply RFS. However, there may be unique applications permitting the use an MMC datum to expedite gaging provided there is no detrimental effect on the design requirement (e.g., a functional shaft size as related to a cam profile).

## Gaging

The example meaning that follows the basic gaging principle of measurement *normal to the basic profile.* Gage traverse or part rotation techniques could be used. Where overall size permits, optical comparator (or similar) techniques are usually the most economical and effective method of inspecting this type of part. These techniques make it possible to compare a blown-up (10X, 20X, etc.) projected shadow or profile of the part with an optical gage chart containing accurately scribed profile tolerance zone limits.

Note, however, that shadow image of the part profile which is seen on the comparator screen will be the extreme profile and may not represent the entire surface. Focus adjustment and surface illumination methods (available on some comparators) may be able to account for the entire surface. If necessary, further inspection may be required to account for surface irregularities which are below the extreme profile or not made visible by the method used.

A larger part exceeding the normal limitations of the optical chart size might be checked in segments or with the use of tracer and reticle.

Computerization techniques, either independently applied or in conjunction with optical methods, also provide possible verification options.

## Two Types of Profile Tolerance

In practice, a profile tolerance may be applied to either an entire surface or to individual line element profiles taken at various cross sections through the part. The two types or methods of controlling profile are:

a. Profile of a surface ⌓

The tolerance zone established by profile of a surface tolerance is a three-dimensional zone or total control across the entire length and width or circumference of the feature; it may be applied to parts having a constant cross section or to parts having a surface of revolution. Usually profile of a surface requires datum references.

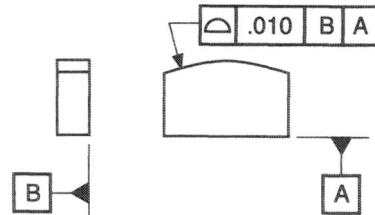

b. Profile of a line ⌒

The tolerance zone established by profile of a line tolerance is a two-dimensional zone extending along the length of the considered feature; it may apply to the profiles of parts having a varying cross section such as a propeller, aircraft wing, nose cone, or to random cross sections of parts where is not desired to control the entire surface as a single entity. Profile of a line may, or may not, require datum references.

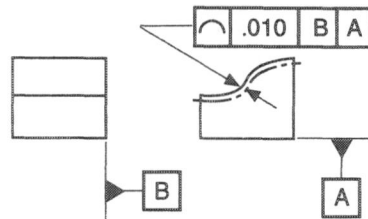

⌒ Profile of a Surface

**Example**

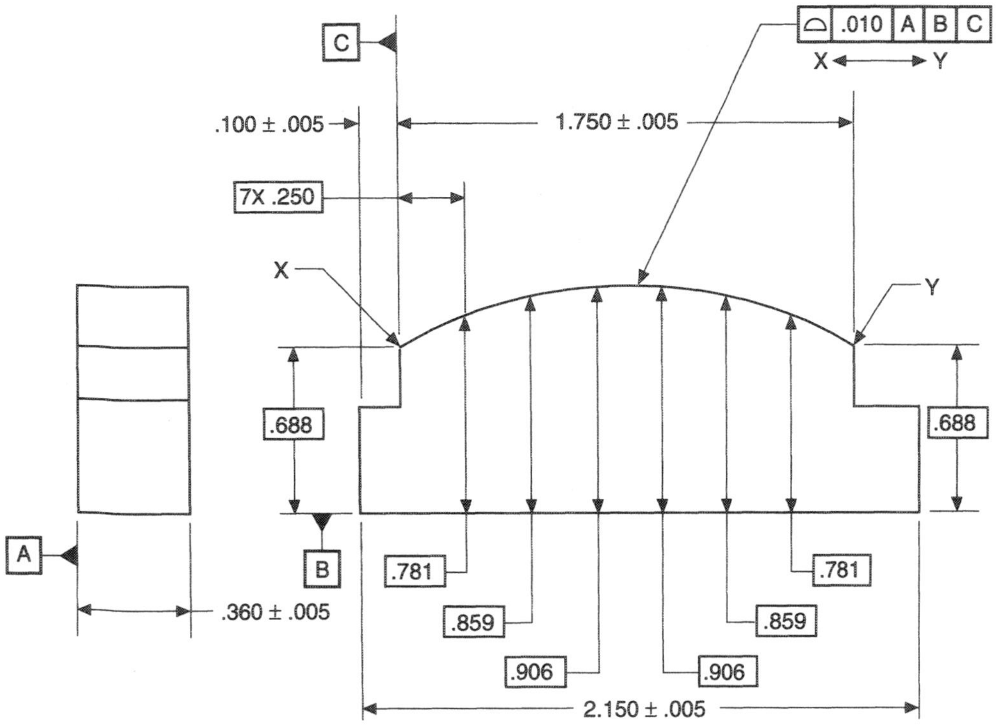

| ⌒ | .010 | A | B | C |

X ◄──────► Y

.100 ± .005

1.750 ± .005

7X .250

X

Y

.688

.688

A

B

.360 ± .005

.781

.781

.859

.859

.906

.906

2.150 ± .005

**Meaning**

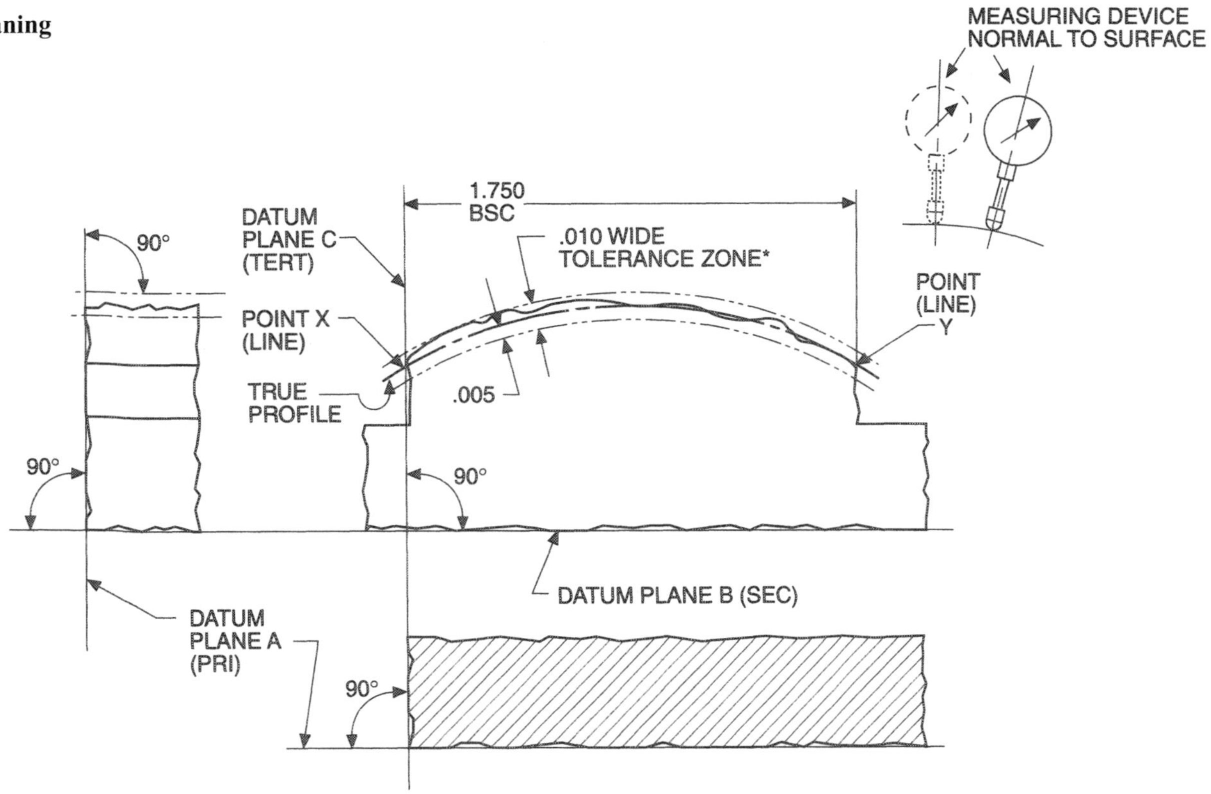

MEASURING DEVICE
NORMAL TO SURFACE

DATUM
PLANE C
(TERT)

90°

1.750
BSC

.010 WIDE
TOLERANCE ZONE*

POINT X
(LINE)

POINT
(LINE)
Y

TRUE
PROFILE

.005

90°

90°

90°

DATUM PLANE B (SEC)

DATUM
PLANE A
(PRI)

90°

*.010 total wide zone (.005 each side of true profile). The surface is between the two profile boundaries .010 apart, equally disposed about the true profile, which are perpendicular to datum plane A and positioned with respect to datum planes B and C.

## 4-65  PROFILE

### Profile of a Surface ⌒

Where the profile tolerance is to be used in controlling the entire surface profile, the symbolic representation may be specified or the words "ALL AROUND" should be added below the profile tolerance.

### Zone Tolerance Around Entire Profile

**Example**

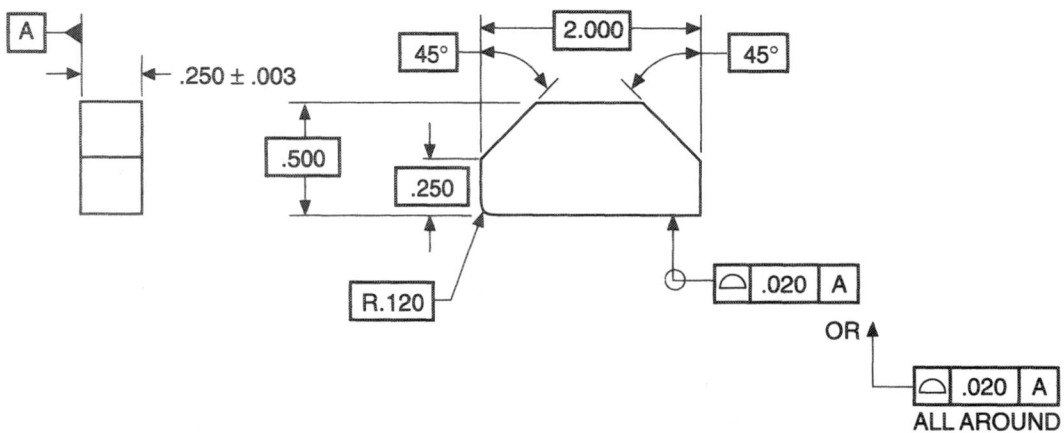

**Meaning**

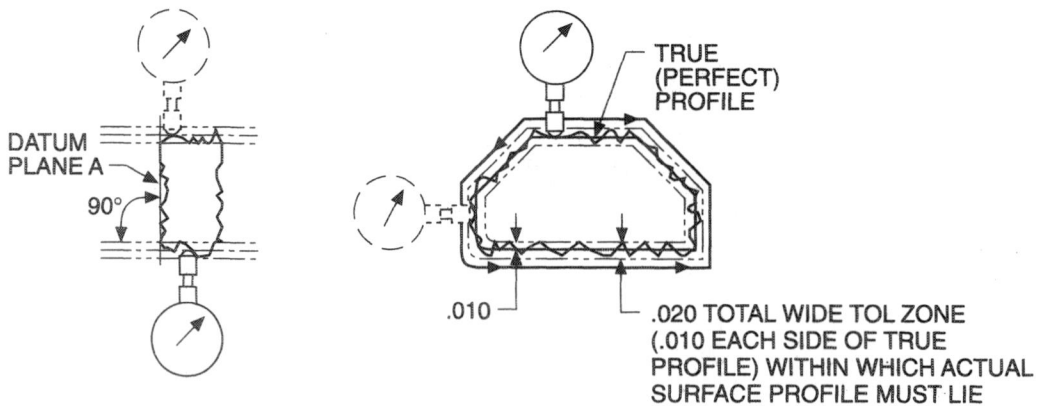

DATUM PLANE A
90°

TRUE (PERFECT) PROFILE

.010

.020 TOTAL WIDE TOL ZONE (.010 EACH SIDE OF TRUE PROFILE) WITHIN WHICH ACTUAL SURFACE PROFILE MUST LIE

### Gaging

The example illustrates the basic gaging principle of measurement *normal to the true profile* as oriented from datum A. Where overall size permits, optical "comparator" (or similar) techniques are usually the most economical and effective method of inspecting this type of part (see also GAGING of preceding example).

Where the profile tolerance is to be used in controlling all, or large portions, of the surface profile, specific notations (such as angular degrees on a cam) should be added with the profile tolerance symbol or note.

This type of tolerance control is particularly useful on cams or similar parts.

## Different Zone Tolerance Around Entire Profile

**Example**

| CAM DEGREES BASIC | CAM RADIUS BASIC |
|---|---|
| 0° | .390 |
| 52° | .390 |
| 59° | .376 |
| 67° | .362 |
| 75° | .359 |
| 92.5° | .355 |
| 145° | .338 |
| 180° | .323 |
| 213° | .313 |
| 269° | .297 |
| 299° | .287 |
| 336° | .194 |

**Meaning**

TOTAL WIDE TOL ZONE
.002 FROM 67° TO 299°
.004 FROM 0° TO 67°
.004 FROM 299° TO 360°
(.001 & .002 RESPECTIVELY
EACH SIDE OF TRUE PROFILE)
WITHIN WHICH ACTUAL
SURFACE PROFILE MUST LIE

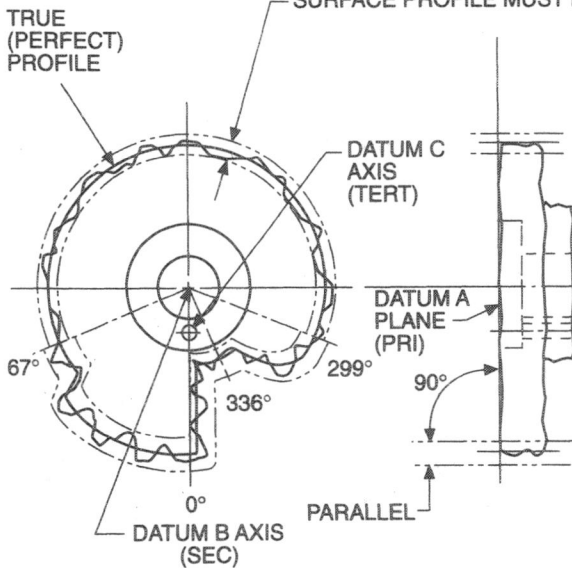

**Gaging**

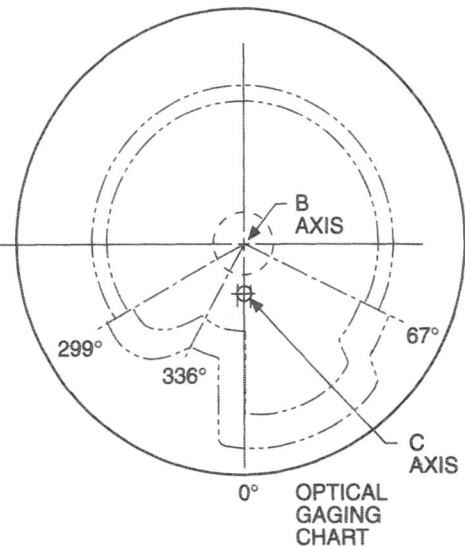

Optical comparator (or similar) techniques are ideal for this part. A transparent chart made to 20X, 50X, etc., size with accurately scribed lines defining the tolerance zones permits a direct surface profile check of the magnified part with the chart. If the surface profile of the part lies within the scribed tolerance zones when oriented to the datum B and C axes and while bottoming on datum A, it is acceptable.

## 4-66 PROFILE

### Profile of a Line

The profile of a surface controls the entire surface, or all the elements of the surface, within a uniform tolerance zone as established from the true profile. However, where line elements of a surface are to be specifically controlled or controlled as a refinement of size or surface profile control, the profile of a line characteristic may be used. Line profile control is applied in a manner similar to the application of surface profile.

The example below illustrates line profile control as a refinement of size. As with surface profile, line profile must be shown in the drawing view in which the control applies. The tolerance zone is established in the same way as surface profile. However, its tolerance zone is disposed about each element of the surface. Therefore the tolerance zone applies for the full length of each element of the surface (in the view in which it is shown), but only for the width or height at a cutting plane bisecting the element. This cutting plane must be considered to be perpendicular or parallel to the part ori-entation within its size tolerances. Where datum references are specified, the orientation is with respect to the datums to give a more specific relationship. Each actual line element of the controlled profile may vary within its prescribed tolerance zone, but the element-to-element control may vary within the entire size tolerance.

Line profile control is illustrated in the example below. Accuracy and consistency of the individual elements at the part surface are critical, but variation from element to element is less critical. More lenient control within the size tolerances is adequate in the direction perpendicular to the line profile. The requirement to hold an accurate line profile on an irregularly shaped part within a more lenient size control is similar to holding straightness within a more lenient size control on a square, rectangular, or cylindrical part.

Surface profile and line profile may be applied to the same feature, where applicable—for example, in situations in which the line elements are to be controlled more closely than the surface as a whole. Combined surface and line profile control may be used and specified as shown in the figure below.

**Example**

SYMBOL MEANING

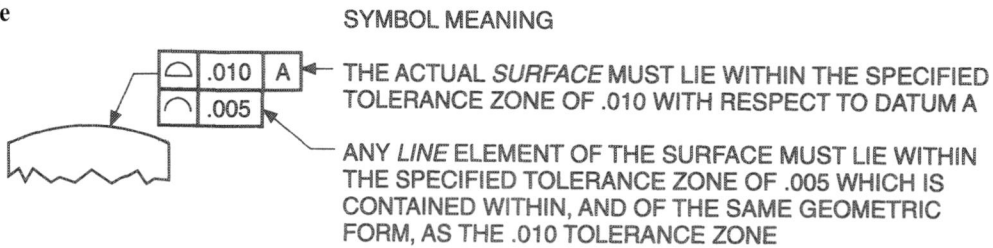

THE ACTUAL *SURFACE* MUST LIE WITHIN THE SPECIFIED TOLERANCE ZONE OF .010 WITH RESPECT TO DATUM A

ANY *LINE* ELEMENT OF THE SURFACE MUST LIE WITHIN THE SPECIFIED TOLERANCE ZONE OF .005 WHICH IS CONTAINED WITHIN, AND OF THE SAME GEOMETRIC FORM, AS THE .010 TOLERANCE ZONE

### Profile of a Line

**Example**

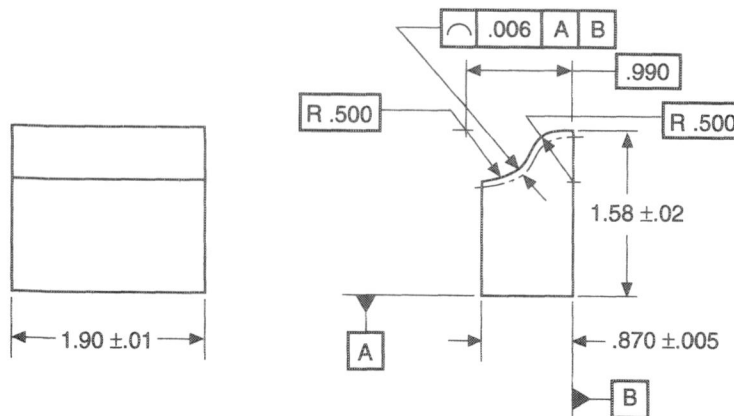

**Meaning**

**Gaging**

The illustration above shows the basic gaging principle of measurement *normal to the basic profile* and in an imaginary cutting plane which bisects the line element while oriented basically perpendicular and parallel within the size tolerances.

Dial indicator, master template, and optical comparator (or similar) techniques can be used (see also GAGING in preceding examples).

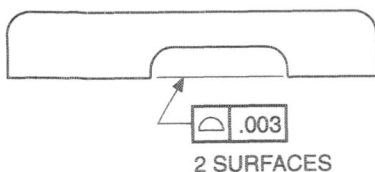

**Coplanar Surfaces**

Profile tolerancing may be used to specify coplanarity of two or more surfaces where it is desired to treat these surfaces as a single interrupted or noncontinuous surface. In such an application, a control is provided similar to that achieved by a flatness tolerance applied to a single plane surface. Coplanarity is the condition of two or more surfaces having all elements in one plane.

Any profile line element of the surface (elements in the view in which the symbol is shown and elements represented by the true profile) must be within the specified .006 tolerance zone which is unilaterally disposed from the true profile. The *profile* tolerance zone must be contained within the *size* tolerance zone.

As shown in Example 1, the profile of a surface tolerance establishes a tolerance zone defined by two parallel planes within which the considered surfaces must lie. No datum reference is stated (as in the case of flatness) since the orientation of the tolerance zone is established from the contact of the part against a reference standard; the datum is established by the considered surfaces themselves.

Where more than two surfaces are involved, it may be desirable to identify which specific surface (or surfaces) is to be used as a datum and to establish the tolerance zone. Where necessary, datum target methods could also be used. Where a datum (or datums) is used, it is understood that the tolerance zone established applies to all coplanar surfaces, including the datum surfaces, unless otherwise specified as shown in Example 2.

**Offset Surfaces**

As a variation of coplanar surfaces, the principles described above may be extended to offset surfaces as shown in the example. In this instance, the variation from coplanar is a desired amount (.105) and is stated by a basic dimension.

**Example 1**

**Coplanar Surfaces**        **Meaning**

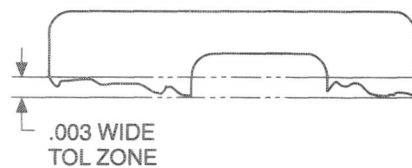

THE SURFACES MUST BE WITHIN
THE SPECIFIED TOL OF SIZE AND
MUST BOTH LIE BETWEEN TWO
PARALLEL PLANES .003 APART.

**Example 2**

**Meaning**

.003
WIDE
TOL
ZONE

DATUM PLANE A

THE SURFACE MUST BE WITHIN
THE SPECIFIED TOL OF SIZE AND
MUST LIE BETWEEN TWO
PARALLEL PLANES .003 APART
AS ESTABLISHED RELATIVE TO
DATUM PLANE A

**Offset Surfaces**

**Meaning**

.0015

DATUM PLANE A
.105
(PARALLEL
TO A)
.003
WIDE
TOL
ZONE

THE SURFACE MUST BE WITHIN
THE SPECIFIED TOL OF SIZE AND
MUST LIE BETWEEN TWO
PARALLEL PLANES .003 APART
AS ESTABLISHED RELATIVE TO
DATUM A

## Conicity

Profile of a surface tolerance can be applied to conical parts to control conicity. It may be applied as an independent control of form (Example 1) or as a combined control of form and orientation with respect to a datum feature and datum axis (Example 2). It could also be used, when necessary, as profile ALL AROUND control, which can control size and form simultaneously using basic dimensions to define the conical shape.

In Example 1, profile of a surface is used as a *form* control (of the 20° cone) as a refinement within the size control. This is comparable to a cylindricity tolerance control-

ling *form* of a cylinder as a refinement within its size control. No datum is required nor proper in this case.

In Example 2, profile of a surface is used as a *form* and *orientation* control (of the 15° cone) as a refinement within the size control and relative to datums A and B. The added notation ALL AROUND (or symbolic version) could be used if desired for clarification but is not necessary here. In this case, profile tolerance is applied to a surface of revolution as a design preference (hypothetically) over other methods (i.e., runout).

See DATUM section for more information on datums and datum precedence (primary, secondary, tertiary datums).

**Conical Surfaces**

**Example 1**

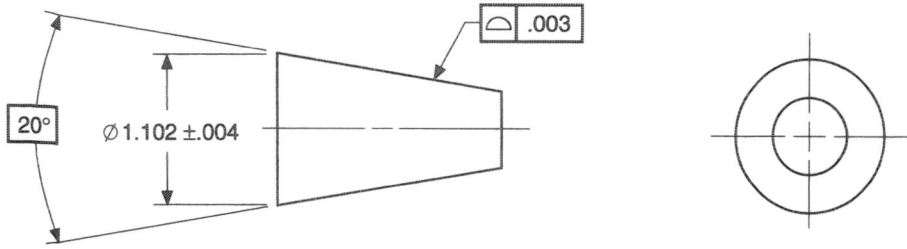

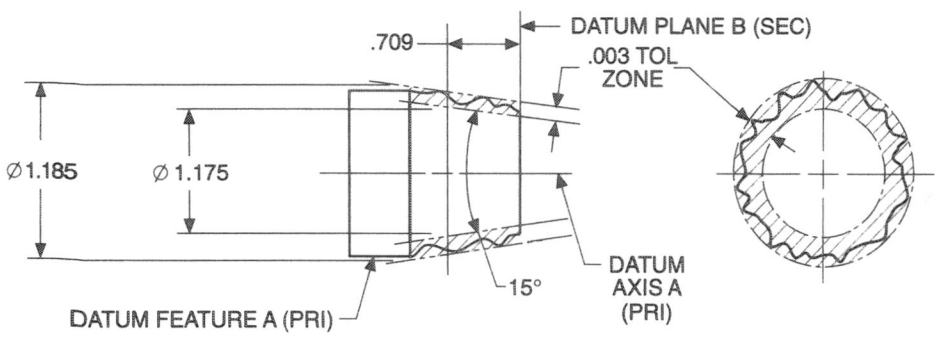

# CHAPTER 5

# Geometric Tolerances in AutoCAD and GD&T

## 5-1 INTRODUCTION

*Geometric tolerancing* is a dimensioning and tolerancing system based on the geometric shape of an object. Surfaces may be defined in terms of their flatness or roundness, or in terms of how perpendicular or parallel they are to other surfaces.

Geometric tolerances allow a more exact definition of the shape of an object than do conventional coordinate-type tolerances. Objects can be toleranced in a manner more closely related to their design function, or so that their features and surfaces are more directly related to each other.

Figure 5-1 shows a square shape dimensioned and toleranced using plus and minus tolerances. The resulting tolerance zone has an outside length of 51 and an inside length of 49 square. The defined tolerance zone allows any shape that falls within in it to be deemed acceptable, or correctly manufactured. Figure 5-1 shows an exaggerated shape that fits within the defined tolerance zone and is not square yet would be acceptable under the specified dimensions and tolerances. Geometric tolerancing can be used to more precisely define the tolerance zone when a more nearly square shape is required.

It should be pointed out that geometric tolerancing is not a panacea for all dimensioning and tolerancing problems. In many cases coordinate tolerancing, as presented in Chapter 4, is sufficient to accurately define an object. Unnecessary or excessive use of geometric tolerances can increase production costs. Most objects are toleranced using a combination of coordinate and geometric tolerances, depending on the design function of the object.

The key to using tolerances and types of tolerances may be simply stated as "decimal points cost money." Every tolerance should be made as loose as possible while still

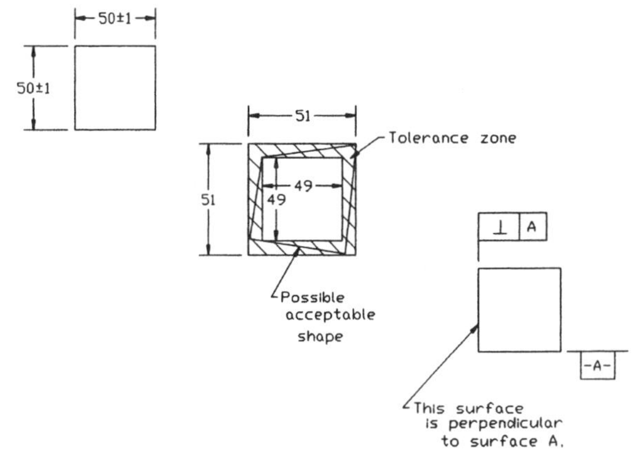

Figure 5-1

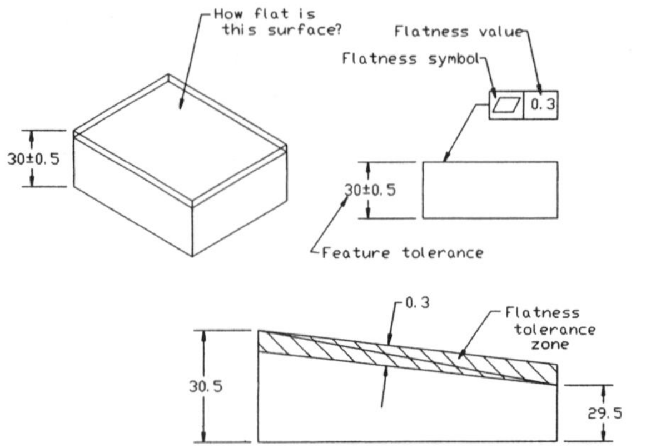

Figure 5-2

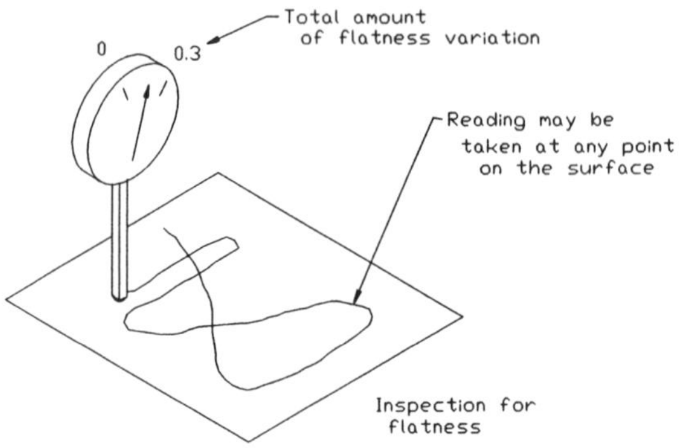

Figure 5-3

maintaining the design integrity of the object. If a surface flatness is critical to the correct functioning of the object, then, of course, it will require a very close tolerance. But every tolerance should be considered individually and loosened wherever possible to make the object's manufacture easier and therefore less expensive.

## 5-2 TOLERANCES OF FORM

Tolerances of form are used to define the shape of a surface relative to itself. There are four classifications: flatness, straightness, roundness, and cylindricity. Tolerances of form are not related to other surfaces but apply only to an individual surface.

## 5-3 FLATNESS

Flatness tolerances are used to define the amount of variation permitted in an individual surface. The surface is thought of as a plane not related to the rest of the object.

Figure 5-2 shows a rectangular object. How flat is the top surface? The given plus or minus tolerances allow a variation of ±0.5 across the surface. Without additional tolerances the surface could look like a series of waves that vary between 30.5 and 29.5.

If the example in Figure 5-2 is assigned a flatness tolerance of 0.3, the height of the object, the feature tolerance could continue to vary based on the 30±0.5 tolerance, but the surface itself could not vary by more than 0.3. In the most extreme condition, one end of the surface could be 30.5 above the bottom surface and the other end 29.5, but the surface would still be limited to within two parallel planes 0.3 apart as shown.

To better understand the meaning of flatness, consider how the surface would be inspected. The surface would be acceptable if a gauge could be moved all around the surface and never varied by more than 0.3. See Figure 5-3. Every point in the plane must be within the specified tolerance.

## 5-4 STRAIGHTNESS

Straightness tolerances are used to measure the variation of an individual feature along a straight line in a specified direction. Figure 5-4 shows an object with a straightness tolerance applied to its top surface. Straightness differs from flatness because straightness measurements are checked by moving a gauge directly across the surface in a single direction. The gauge is not moved randomly about the surface, as is required by flatness.

Figure 5-4

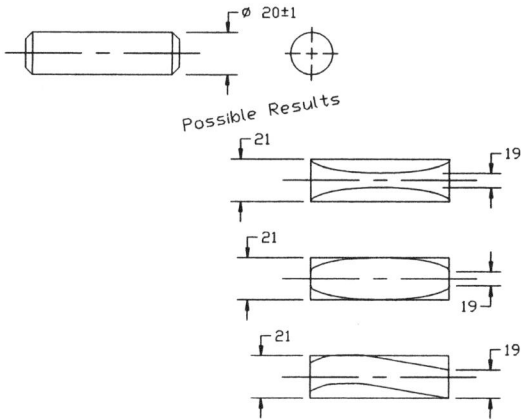

**Figure 5-5**

**Figure 5-6**

Straightness tolerances are most often applied to circular or matching objects to help ensure that the parts are not barreled or warped within the given feature tolerance range and therefore not fit together well. Figure 5-5 shows a cylindrical object dimensioned and toleranced using a standard feature tolerance. The surface of the cylinder may vary within the specified tolerance range as shown.

Figure 5-6 shows the same object shown in Figure 5-5 dimensioned and toleranced using the same feature tolerance, but also including a 0.05 straightness tolerance. The straightness tolerance limits the surface variation to 0.05 as shown.

## 5-5 STRAIGHTNESS (RFS AND MMC)

Figure 5-7 again shows the same cylinder shown in Figures 5-5 and 5-6. This time the straightness tolerance is applied about the cylinder's centerline. This type of tolerance permits the feature tolerance and geometric tolerance to be used together to define a virtual condition. A **virtual condition** is used to determine the maximum possible size variation of the cylinder or the smallest diameter hole that will always accept the cylinder. See Section 5-17.

The geometric tolerance specified in Figure 5-7 is applied to any circular segment along the cylinder, regardless of the cylinder's diameter. This means that the 0.05 tolerance is applied equally when the cylinder's diameter measures 19 or when it measures 21. This application is called *RFS, regardless of feature size.* RFS conditions are specified in a tolerance either by an S with a circle around it or implied tacitly when no other symbol is used. In

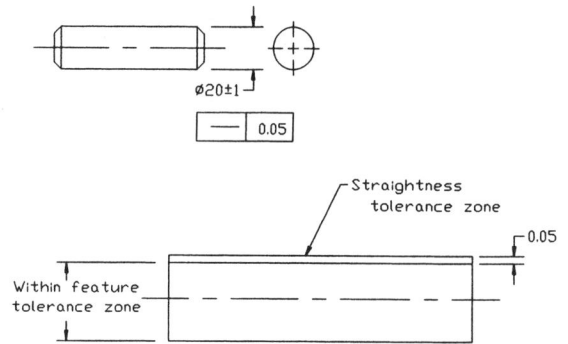

Figure 5-7 no symbol is listed after the 0.05 value, so it is assumed to be applied RFS.

Figure 5-8 shows the cylinder dimensioned with an *MMC* condition applied to the straightness tolerance. MMC stands for *maximum material condition* and means that the specified straightness tolerance (0.05) is applied only at the MMC condition or when the cylinder is at its maximum diameter size (21).

A shaft is an external feature, so its largest possible size or MMC occurs when it is at its maximum diameter. A hole is an internal feature. A hole's MMC condition occurs when it is at its smallest diameter. The MMC condition for holes will be discussed later in the chapter along with positional tolerances.

**Figure 5-7**

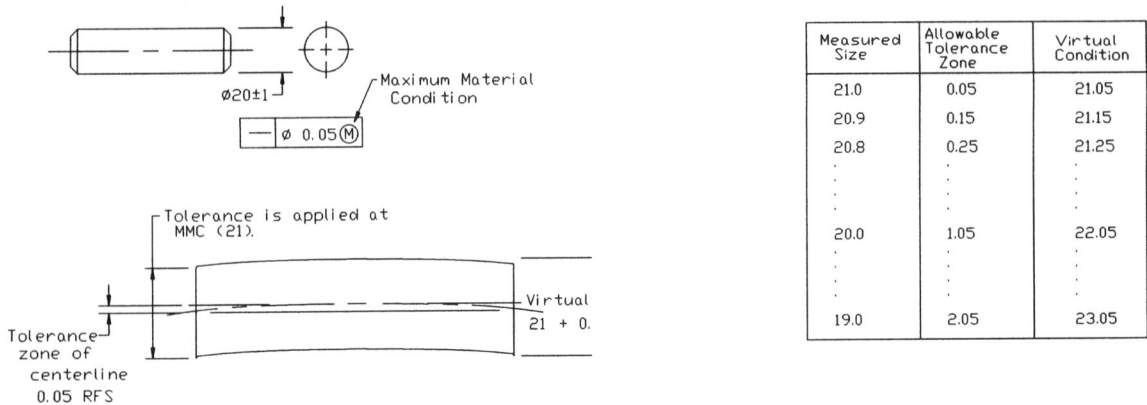

| Measured Size | Allowable Tolerance Zone | Virtual Condition |
|---|---|---|
| 21.0 | 0.05 | 21.05 |
| 20.9 | 0.15 | 21.15 |
| 20.8 | 0.25 | 21.25 |
| . | . | . |
| . | . | . |
| . | . | . |
| 20.0 | 1.05 | 22.05 |
| . | . | . |
| . | . | . |
| . | . | . |
| 19.0 | 2.05 | 23.05 |

**Figure 5-8**

Applying a straightness tolerance at MMC allows for a variation in the resulting tolerance zone. Because the 0.05 flatness tolerance is applied at MMC, the virtual condition is still 21.05, the same as with the RFS condition; however, the tolerance is applied only at MMC. As the cylinder's diameter varies within the specified feature tolerance range the acceptable tolerance zone may vary to maintain the same virtual condition.

Figure 5-8 lists how the tolerance zone varies as the cylinder's diameter varies. When the cylinder is at its largest size, or MMC, the tolerance zone equals 0.05, or the specified flatness variation. When the cylinder is at its smallest diameter the tolerance zone equals 2.05, or the total feature size plus the total flatness size. In all variations the virtual size remains the same, so at any given cylinder diameter value, the size of the tolerance zone can be determined by subtracting the cylinder's diameter value from the virtual condition.

Figure 5-9 shows a comparison between different methods used to dimension and tolerance a .750 shaft. The first example uses only a feature tolerance. This tolerance sets an upper limit of .755 and a lower limit of .745. Any variations within that range are acceptable.

The second example in Figure 5-9 sets a straightness tolerance of .003 about the cylinder's centerline. No conditions are defined, so the tolerance is applied RFS. This limits the variations in straightness to .003 at all feature sizes. For example, when the shaft is at its smallest possible feature size, .745, the .003 still applies. This means that a shaft measuring .745 that had a straightness variation greater than .003 would be rejected. If the tolerance

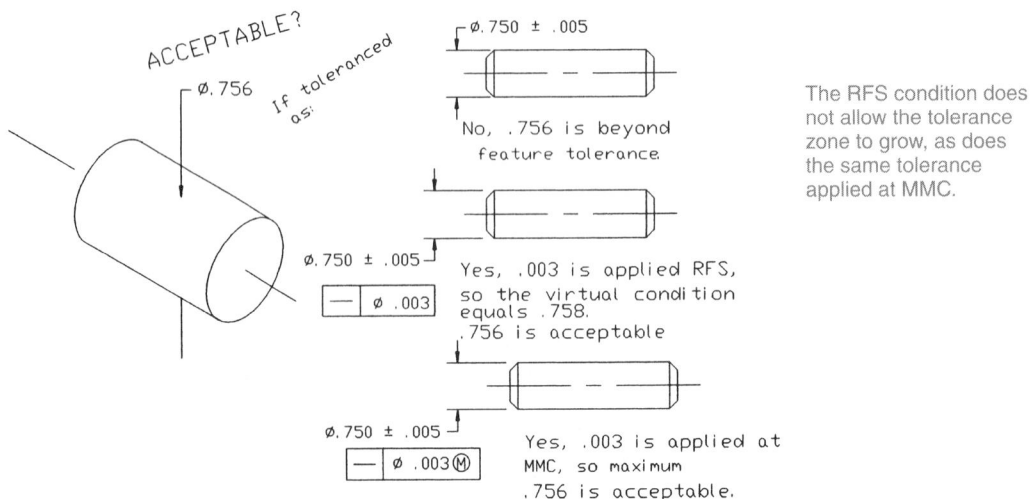

The RFS condition does not allow the tolerance zone to grow, as does the same tolerance applied at MMC.

**Figure 5-9**

**Figure 5-10**

**Figure 5-11**

has been applied at MMC, the part would be accepted. This does not mean that straightness tolerances should always be applied at MMC. If straightness is critical to the design integrity or function of the part, then straightness should be applied in the RFS condition.

The third example in Figure 5-9 applies the straightness tolerance about the centerline at MMC. This tolerance creates a virtual condition of .758. The MMC condition allows the tolerance to vary as the feature tolerance varies, so when the shaft is at its smallest feature size, .745, a straightness tolerance of .013 is acceptable (.010 feature tolerance +.003 straightness tolerance).

If the tolerance specification for the cylinder shown in Figure 5-9 were to have a 0.000 tolerance applied at MMC, it would mean that the shaft would have to be perfectly straight at MMC, or when the shaft was at its maximum value (.755); however, the straightness tolerance could vary as the feature size varied, as discussed for the other tolerance conditions. A 0.000 tolerance means that the MMC and the virtual conditions are equal.

Figure 5-10 shows a very long .750-diameter shaft. Its straightness tolerance includes a length qualifier that serves to limit the straightness variations over each inch of the shaft length and prevents excess waviness over the full length. The tolerance Ø.002/1.000 means that the total straightness may vary over the entire length of the shaft by .003 but that the variation is limited to .002 per 1.000 of shaft length.

## 5-6  CIRCULARITY

Circularity tolerances are used to limit the amount of variation in the roundness of a surface of revolution and are measured at individual cross sections along the length of the

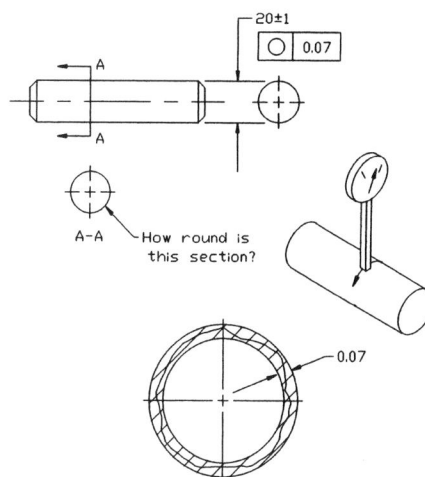

object. The measurements are limited to the individual cross sections and are not related to other cross sections. This means that in extreme conditions the shaft shown in Figure 5-11 could actually taper from a diameter of 21 to a diameter of 19 and never violate the circularity requirement. It also means that qualifications such as MMC cannot be applied.

Figure 5-11 shows a shaft that includes a feature tolerance and a circularity tolerance of 0.07. To understand circularity tolerances, consider an individual cross section, or slice, of the cylinder. The shape of the outside edge of the slice varies around the slice. The difference between the maximum diameter and the minimum diameter of the slice can never exceed the stated circularity tolerance.

Circularity tolerances can be applied to tapered sections and spheres, as shown in Figure 5-12. In both applications,

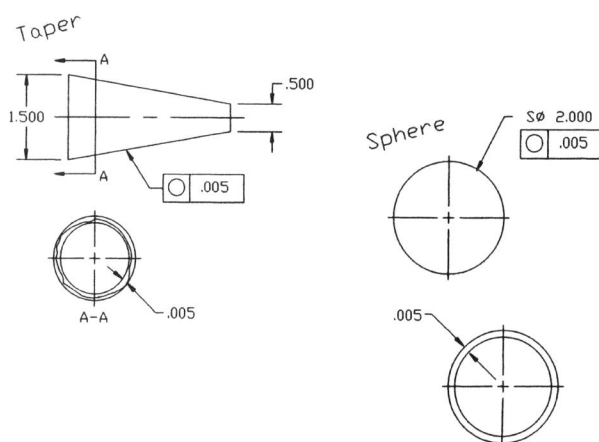

**Figure 5-12**

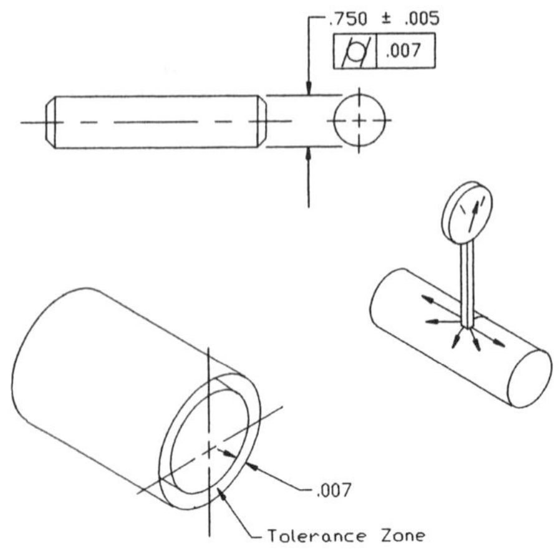

**Figure 5-13**

**Figure 5-14**

circularity is measured around individual cross sections, as it was for the shaft shown in Figure 5-11.

## 5-7 CYLINDRICITY

Cylindricity tolerances are used to define a tolerance zone both around individual circular cross sections of an object and also along its length. The resulting tolerance zone looks like two concentric cylinders.

Figure 5-13 shows a shaft that includes a cylindricity tolerance that establishes a tolerance zone of .007. This means that if the maximum measured diameter is determined to be .755, the minimum diameter cannot be less than .748 anywhere on the cylindrical surface. Figure 5-14 shows how to draw the feature control frames that surround the tolerance symbol and size specifications.

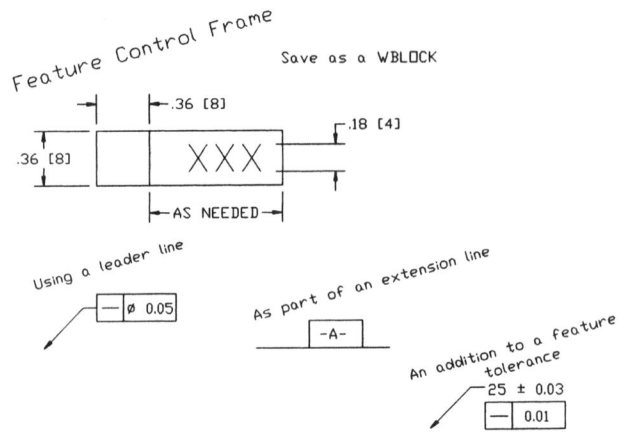

Cylindricity and circularity are somewhat analogous to flatness and straightness. Flatness and cylindricity are concerned with variations across an entire surface or plane. In the case of cylindricity, the plane is shaped like a cylinder. Straightness and circularity are concerned with variations of a single element of a surface, that is, a straight line across the plane in a specified direction for straightness, and a path around a single cross section for circularity.

## 5-8 GEOMETRIC TOLERANCES USING AUTOCAD

Geometric tolerances are tolerances that limit dimensional variations based on the geometric properties of an object. Figure 5-15 shows a list of geometric tolerance symbols. Figure 5-16 shows an object dimensioned using

| | TYPE OF TOLERANCE | CHARACTERISTIC | SYMBOL |
|---|---|---|---|
| FOR INDIVIDUAL FEATURES | FORM | STRAIGHTNESS | — |
| | | FLATNESS | ▱ |
| | | CIRCULARITY | ○ |
| | | CYLINDRICITY | ⌀ |
| INDIVIDUAL OR RELATED FEATURES | PROFILE | PROFILE OF A LINE | ⌒ |
| | | PROFILE OF A SURFACE | ⌓ |
| RELATED FEATURES | ORIENTATION | ANGULARITY | ∠ |
| | | PERPENDICULARITY | ⊥ |
| | | PARALLELISM | // |
| | LOCATION | POSITION | ⊕ |
| | | CONCENTRICITY | ◎ |
| | RUNOUT | CIRCULAR RUNOUT | ⟋ |
| | | TOTAL RUNOUT | ⫽ |

| TERM | SYMBOL |
|---|---|
| AT MAXIMUM MATERIAL CONDITION | Ⓜ |
| REGARDLESS OF FEATURE SIZE | Ⓢ |
| AT LEAST MATERIAL CONDITION | Ⓛ |
| PROJECTED TOLERANCE ZONE | Ⓟ |
| DIAMETER | ⌀ |
| SPHERICAL DIAMETER | S⌀ |
| RADIUS | R |
| SPHERICAL RADIUS | SR |
| REFERENCE | ( ) |
| ARC LENGTH | ⌒ |

**Figure 5-15**

Figure 5-16

geometric tolerances. The geometric tolerances were created as follows.

### To define a datum

1. Select the **Tolerance** tool from the **Dimension** panel.

   The **Geometric Tolerance** dialog box will appear. See Figure 5-17.

2. Click the **Datum Identifier** box and type **A**, then click **OK.**

   *Command: _tolerance*
   *Enter tolerance location:*

3. Position the datum identifier and press the left mouse button. See Figure 5-16.

### To define a straightness value

1. Select the **Tolerance** tool from the **Dimension** panel.

   The **Geometric Tolerance** dialog box will appear.

2. Select the top open box under the heading **Sym.**

   The **Symbol** dialog box will reappear. See Figure 5-18.

3. Select the straightness symbol, then **OK.**

   The **Geometric Tolerance** dialog box will reappear with the straightness symbol in the first box under the heading **Sym.**

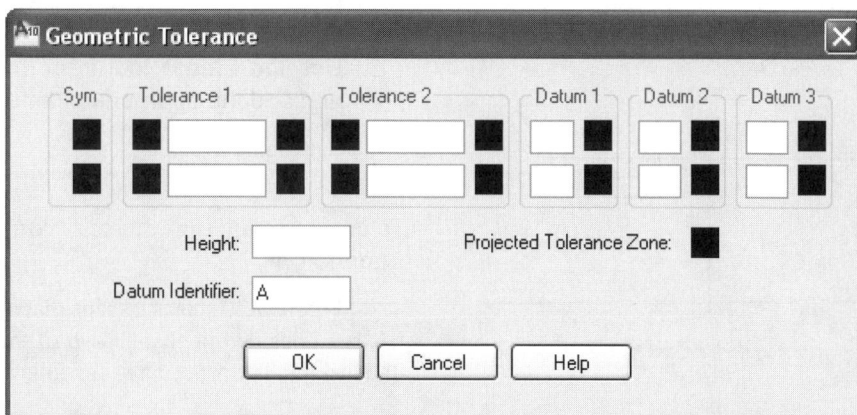

Figure 5-17

Figure 5-18

4. Click the open box in the **Tolerance 1** box and type .**003**; click **OK.**

*Command: _tolerance*
*Enter tolerance location:*

5. Position the straightness tolerance and press the left mouse button

See Figure 5-19. Use the **Move** and **Osnap** tools if necessary to reposition the tolerance box. See Figure 5-16.

### To create a positional tolerance

A *positional tolerance* is used to locate and tolerance a hole in an object. Positional tolerances require base locating dimensions for the hole's center point. Positional toler-ances also require a feature tolerance to define the diameter tolerances of the hole, and a geometric tolerance to define the position tolerance for the hole's center point.

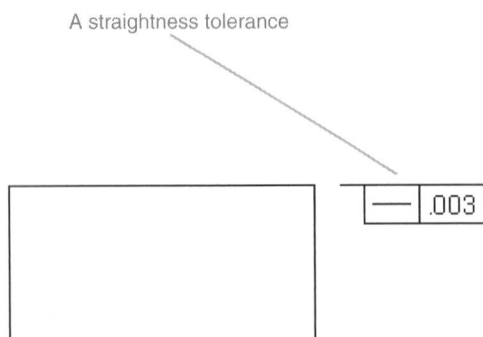

### To create a basic dimension

See the two 1.50 dimensions in Figure 5-16 used to lo-cate the center position of the hole.

1. Access the **Dimension Style Manager** or type **DDIM** in response to a command prompt.

The **Dimension Style Manager** dialog box will ap-pear.

2. Select **Modify.**

The **Modify Dimension Style: Standard** dialog box will appear. See Figure 4-9.

3. Click the Tolerance tab and select the **Basic** op-tion next to the heading **Method** in the **Tolerance** box.
4. Return to the drawing screen.
5. Use the **Linear** tool from the **Dimension** panel and add the appropriate dimensions.

See Figure 5-16.

### To create basic dimensions from existing dimensions

Figure 5-20 shows a shape that includes dimensions. It has been decided to change two of the dimensions to basic dimensions. The procedure is as follows.

Figure 5-19

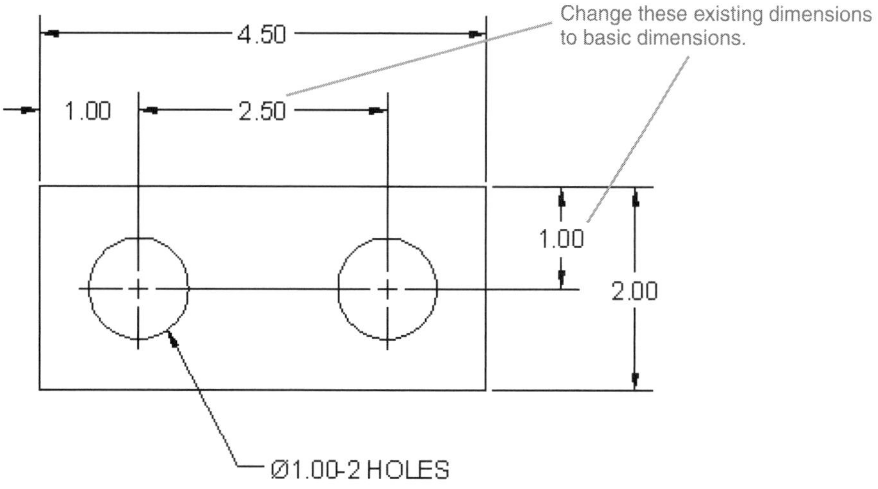

**Figure 5-20**

1. Click on the **2.50** dimension.

   Blue squares will appear on the dimension.

2. Right-click the mouse.

   A dialog box will appear.

3. Select the **Properties** option.

   The **Properties** dialog box will appear. See Figure 5-21.

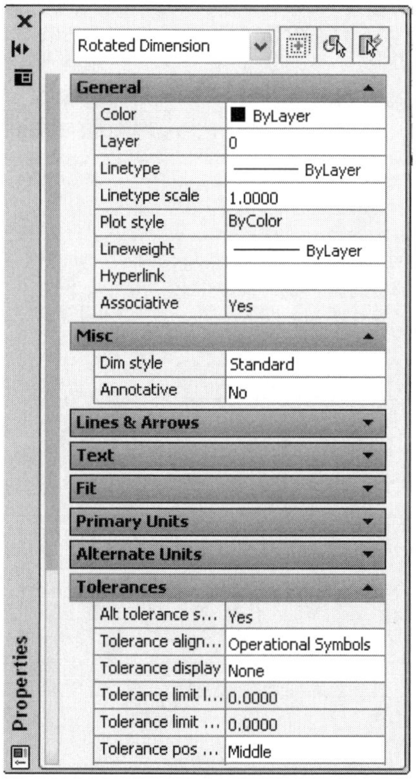

**Figure 5-21**

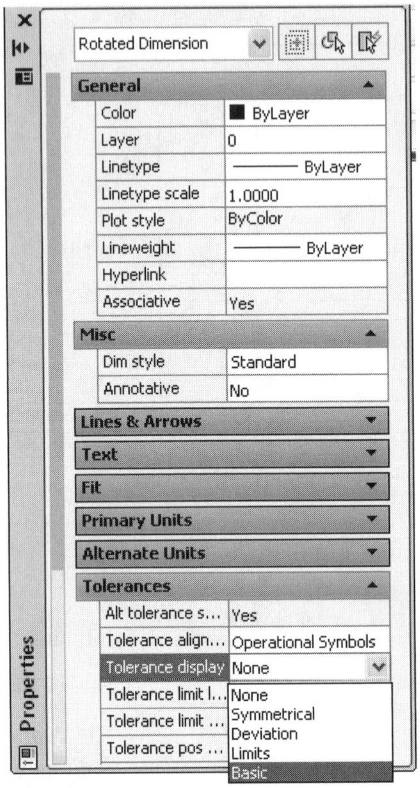

**Figure 5-22**

4. Use the scroll arrow next to the **P** in **Properties** and locate the **Tolerances** option.
5. Click the open space next to the word **None** and scroll down to the **Basic** option.

   See Figure 5-22.

6. Select the **Basic** option, then click the return **X,** then press the **<Esc>** key.
7. Repeat the procedure for the vertical 1.00 dimension.

   Figure 5-23 shows the final results.

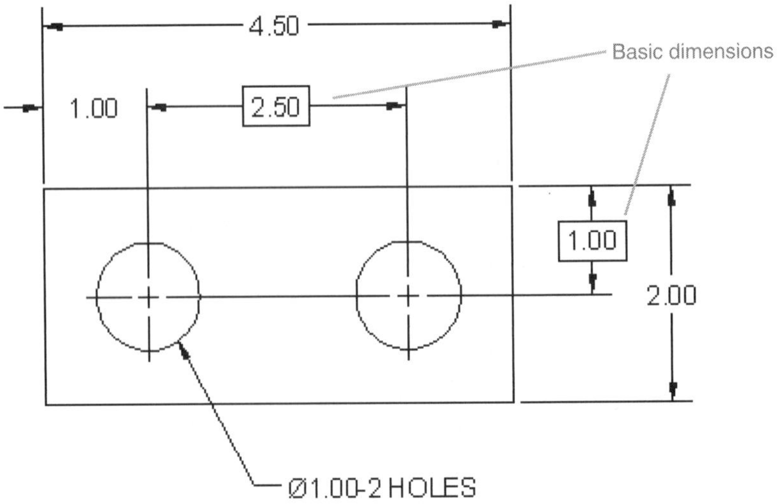

**Figure 5-23**

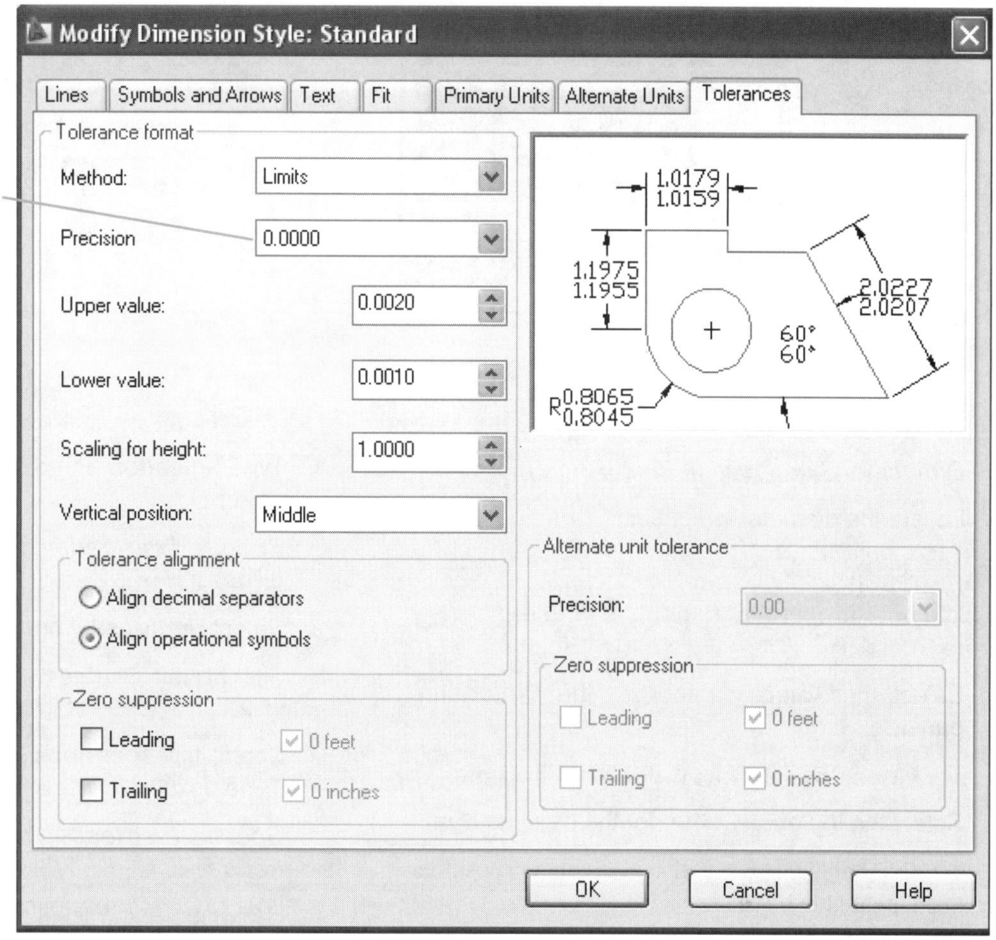

The tolerance precision must contain the same number of decimal places as the tolerances.

Figure 5-24

**To add a limit feature tolerance to a hole**

1. Access the **Dimension Style Manager** or type **DDIM** in response to a command prompt.

   The **Dimension Style Manager** dialog box will appear.

2. Select **Modify,** then the **Tolerance** tab.

   The **Modify Dimension Style: Standard** dialog box will appear.

3. Select the **Limits** option in the **Method** box located in the **Tolerance format** box.

   See Figure 5-24.

4. Change the upper value to **0.0002** by placing the cursor within the **Upper value** box, backspacing out the existing value, and typing in the new value.

5. Change the lower value to **0.0001** by placing the cursor within the **Lower value** box, backspacing

out the existing value, and typing in the new value.

6. Select the arrow to the right of the **Precision** box.

   The **Precision** options will cascade down.

7. Change the precision for both **Dimension** and **Tolerance** boxes to three decimal places **(0.000)**.

   See Figure 5-24. AutoCAD will truncate any input according to the number of decimal places allowed by the precision settings. If the precision settings had been two decimal places (0.00), the resulting limit dimensions would have both been 1.50. The values defined in the third decimal place would have been ignored.

8. Return to the drawing screen.
9. Select the **Diameter** tool on the **Dimension** panel.

   *Select arc or circle:*

**Figure 5-25**

**Figure 5-26**

10. Select the hole.

   *Dimension line location (Text/Angle):*

11. Locate the diameter dimension.

**To add a positional tolerance to the hole's feature tolerance**

1. Select the **Tolerance** tool on the **Dimension** panel.

   The **Geometric Tolerance** dialog box will appear.

2. Select the top open box under the heading **Sym**.

   The **Symbol** dialog box will appear with the positioning symbol highlighted. See Figure 5-25.

3. Select the top left open box under the heading **Tolerance 1.**

   A diameter symbol will appear.

4. Select the **Value** box and type **0.0005.**

   The numbers will appear in the box.

5. Select the top far right open box under the heading **Tolerance 1.**

   The **Material Condition** dialog box will appear. See Figure 5-25.

6. Select the maximum material condition (MMC) symbol (the circle with an M in it).

7. Select **OK.**

   The MMC symbol will appear in the material condition box in the **Geometric Tolerance** dialog box. Figure 5-26 shows the resulting **Geometric Tolerance** dialog box.

8. Select **OK.**

   *Enter tolerance location:*

9. Locate the tolerance box.

   Use the **Move** and **Osnap** commands to position the box, if necessary.

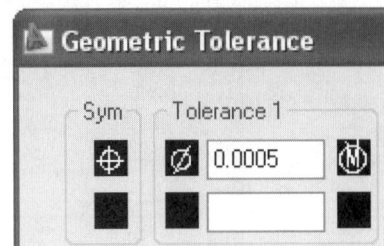

**To add a geometric tolerance with a leader line**

1. Type **qleader** in response to a command prompt.

   *Specify first leader point, or [Settings] <Settings>:*

2. Select **Settings** by pressing the **Enter** key.

   The **Leader Settings** dialog box will appear. See Figure 5-27.

3. Select the **Tolerance** option in the **Annotation Type** box, then **OK.**

   *Specify first leader point, or [Settings] <Settings>:*

4. Select a starting point for the leader line.

   *Specify next point:*

5. Draw a short horizontal segment.

   The **Geometric Tolerance** dialog box will appear. See Figure 5-27.

6. Select the top **Sym** box.

   The **Symbol** dialog box will appear.

7. Select the parallel symbol.
8. Select the open box under the heading **Tolerance 1** and type **0.0010.**
9. Select the open box under the heading **Datum 1,** type **A,** then **OK.**

   See Figure 5-27.

## 5-9 TOLERANCES OF ORIENTATION

Tolerances of orientation are used to relate a feature or surface to another feature or surface. Tolerances of orientation include perpendicularity, parallelism, and angularity.

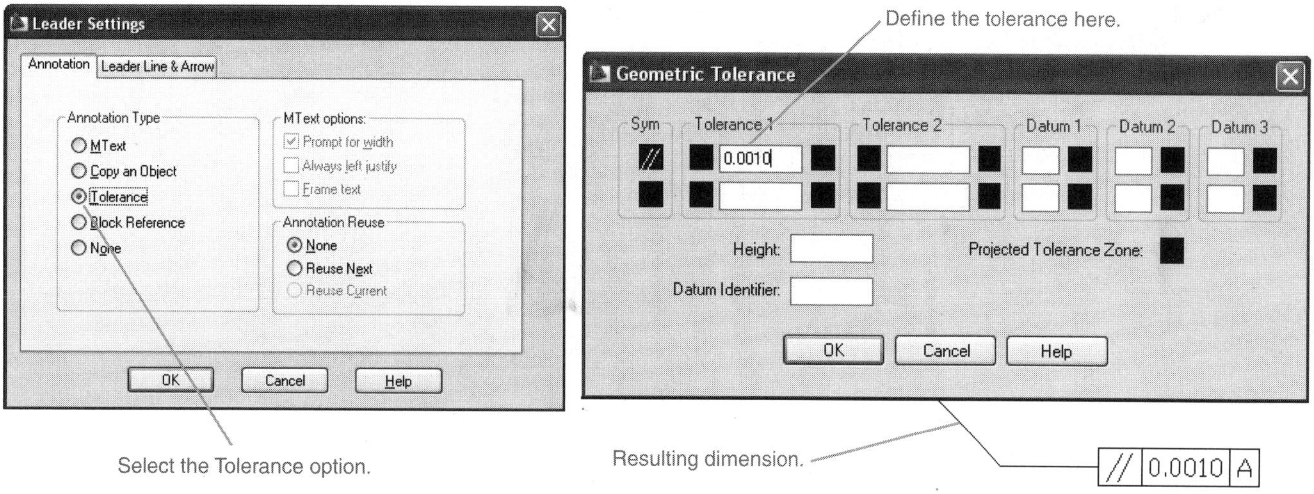

Define the tolerance here.

**Leader Settings**

Annotation | Leader Line & Arrow

Annotation Type
- ○ MText
- ○ Copy an Object
- ⦿ Tolerance
- ○ Block Reference
- ○ None

MText options:
- ☑ Prompt for width
- ☐ Always left justify
- ☐ Frame text

Annotation Reuse
- ⦿ None
- ○ Reuse Next
- ○ Reuse Current

[ OK ]  [ Cancel ]  [ Help ]

**Geometric Tolerance**

Sym | Tolerance 1 | Tolerance 2 | Datum 1 | Datum 2 | Datum 3

0.0010

Height:

Datum Identifier:

Projected Tolerance Zone:

[ OK ]  [ Cancel ]  [ Help ]

Select the Tolerance option.

Resulting dimension.

// | 0.0010 | A

**Figure 5-27**

They may be applied using RFS or MMC conditions, but they cannot be applied to individual features by themselves. To define a surface as parallel to another surface is very much like assigning a flatness value to the surface. The difference is that flatness applies only within the surface; every point on the surface is related to a defined set of limiting parallel planes. Parallelism defines every point in the surface relative to another surface. The two surfaces are therefore directly related to each other, and the condition of one affects the other.

Orientation tolerances are used with locational tolerances. A feature is first located, then it is oriented within the locational tolerances. This means that the orientation toler-

ance must always be less than the locational tolerances. The next four sections will further explain this requirement.

## 5-10 DATUMS

*A datum* is a point, axis, or surface used as a starting reference point for dimensions and tolerances. Figure 5-28 shows a rectangular object with three datum planes labeled –A–, –B–, and –C–. The three datum planes are called the primary, secondary, and tertiary datums, respectively. The three datum planes are, by definition, oriented exactly 90° to one another.

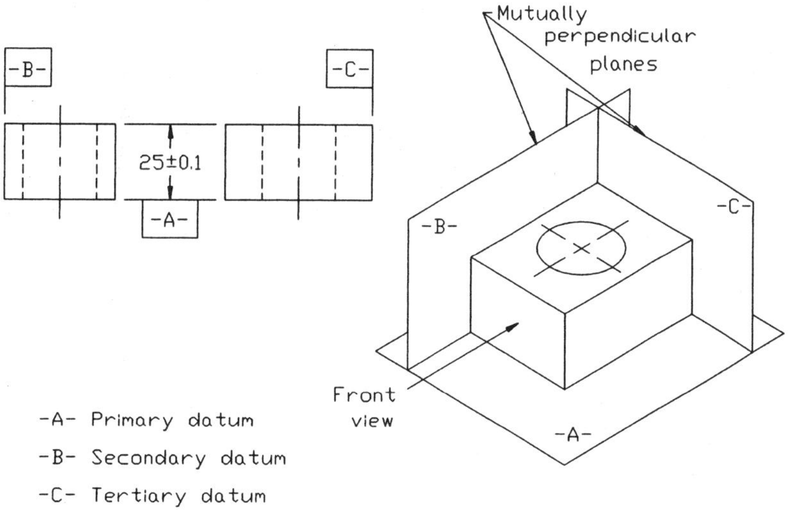

-B-    -C-

25±0.1

-A-

Mutually perpendicular planes

-B-    -C-

Front view

-A-

-A- Primary datum

-B- Secondary datum

-C- Tertiary datum

**Figure 5-28**

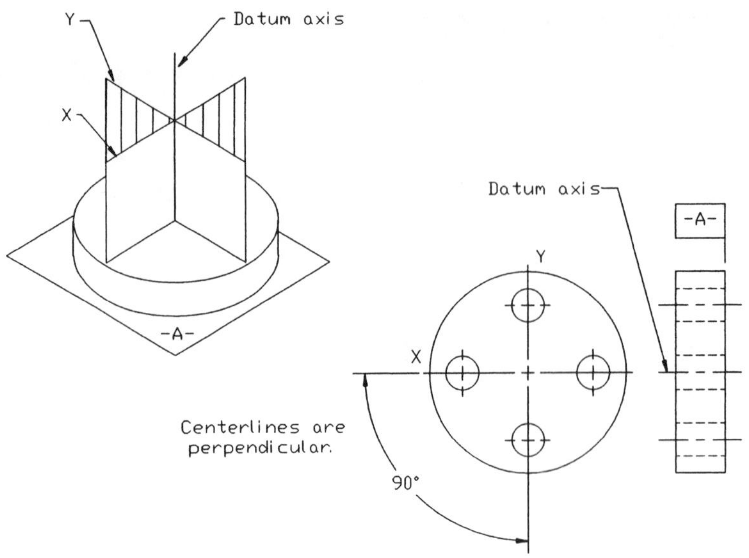

**Figure 5-29**

Figure 5-29 shows a cylindrical datum frame that includes three datum planes. The X and Y planes are perpendicular to each other, and the base A plane is perpendicular to the datum axis between the X and Y planes.

Datums are defined on a drawing by using letters enclosed in rectangular boxes, as shown. The defining letters are written between dash lines: –A– , –B– , and –C–.

Datum planes are assumed to be perfectly flat. When assigning a datum status to a surface, be sure that the surface is reasonably flat. This means that datum surfaces should be toleranced using surface finishes, or created using machine techniques that produce flat surfaces.

## 5-11  PERPENDICULARITY

Perpendicularity tolerances are used to limit the amount of variation for a surface or feature within two planes perpendicular to a specified datum. Figure 5-30 shows a rectangular object. The bottom surface is assigned as datum A, and the right vertical edge is toleranced so that it must be perpendicular within a limit of 0.05 to datum A. The perpendicularity tolerance defines a tolerance zone 0.05 wide between two parallel planes that are perpendicular to datum A.

The object also includes a horizontal dimension and tolerance of 40±1. This tolerance is called a ***locational tol-***

***erance*** because it serves to locate the right edge of the object. As with rectangular coordinate tolerances discussed in Chapter 4, the 40±1 controls the location of the edge, how far away or how close it can be to the left edge, but does not directly control the shape of the edge. Any shape that falls within the specified tolerance range is acceptable. This may, in fact, be sufficient for a given design, but if a more controlled shape is required, a perpendicularity tolerance must be added. The perpendicularity tolerance works within the

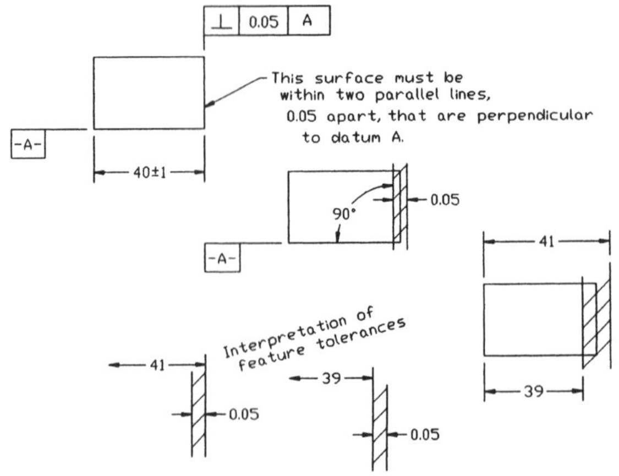

**Figure 5-30**

**Figure 5-31**

**Figure 5-32**

locational tolerance to ensure that the edge is not only within the locational tolerance but is also perpendicular to datum A.

Figure 5-30 shows the two extreme conditions for the $40 \pm 1$ locational tolerance. The perpendicularity tolerance is applied by first measuring the surface and determining its maximum and minimum lengths. The difference between these two measurements must be less than 0.05. Thus, if the measured maximum distance is 41, then no other part of the surface may be less than $41 - 0.05 = 40.95$.

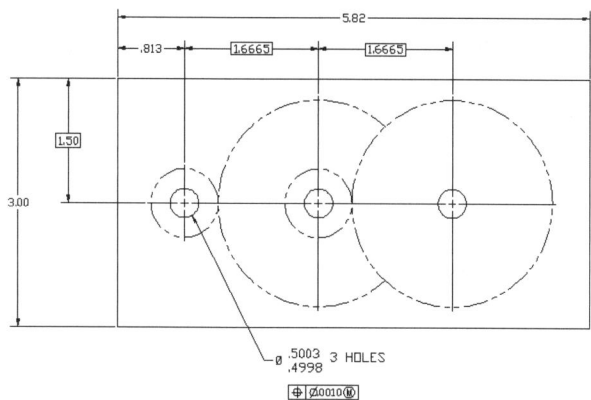

**Figure 5-33**

Tolerances of perpendicularity serve to complement locational tolerances, to make the shape more exact, so tolerances of perpendicularity must always be smaller than tolerances of location. It would be of little use, for example, to assign a perpendicularity tolerance of 1.5 for the object shown in Figure 5-30. The locational tolerance would prevent the variation from ever reaching the limits specified by such a large perpendicularity tolerance.

Figure 5-31 shows a perpendicularity tolerance applied to cylindrical features: a shaft and a hole. The figure includes examples of both RFS and MMC applications. As with straightness tolerances applied at MMC, perpendicularity tolerances applied about a hole or shaft's centerline allow the tolerance zone to vary as the feature size varies.

The inclusion of the Ø symbol in a geometric tolerance is critical to its interpretation. See Figure 5-32. If the Ø symbol is not included, the tolerance applies only to the view in which it is written. This means that the tolerance zone is shaped like a rectangular slice, not a cylinder, as would be the case if the Ø symbol were included. In general, it is better always to include the Ø symbol for cylindrical features because it generates a tolerance zone more like that used in positional tolerancing.

Figure 5-33 shows a perpendicularity tolerance applied to a slot, a noncylindrical feature. Again, the MMC specification is always for variations in the tolerance zone.

Figure 5-34

Figure 5-35

## 5-12  PARALLELISM

Parallelism is used to ensure that all points within a plane are within two parallel planes that are parallel to a referenced datum plane. Figure 5-34 shows a rectangular object that is toleranced so that its top surface is parallel to the bottom surface within 0.02. This means that every point on the top surface must be within a set of parallel planes 0.02 apart. These parallel tolerancing planes are located by determining the maximum and minimum distances from the datum surface. The difference between the maximum and minimum values may not exceed the stated 0.02 tolerance.

In the extreme condition of maximum feature size, the top surface is located 40.5 above the datum plane. The parallelism tolerance is then applied, meaning that no point on the surface may be closer than 40.3 to the datum. This is an RFS condition. The MMC condition may also be applied, thereby allowing the tolerance zone to vary as the feature size varies.

## 5-13  ANGULARISM

Angularism tolerances are used to limit the variance of surfaces and axes that are at an angle relative to a datum. Angularism tolerances are applied like perpendicularity and parallelism tolerances as a way to better control the shape of locational tolerances.

Figure 5-35 shows an angularism tolerance and several ways in which it is interpreted at extreme conditions.

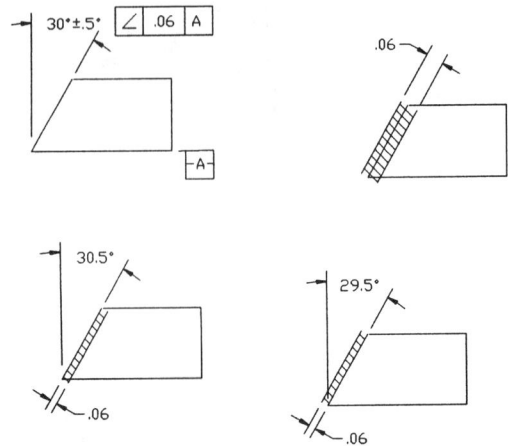

## 5-14  PROFILES

Profile tolerances are used to limit the variations of irregular surfaces. They may be assigned as either bilateral or unilateral tolerances. There are two types of profile tolerances: surface and line. **Surface profile tolerances** limit the variation of an entire surface, whereas a **line profile tolerance** limits the variations along a single line across a surface.

Figure 5-36 shows an object that includes a surface profile tolerance referenced to an irregular surface. The tolerance is considered a bilateral tolerance because no other specification is given. This means that all points on the surface must be located between two parallel planes 0.08 apart that are centered about the irregular surface. The measurements are taken perpendicular to the surface.

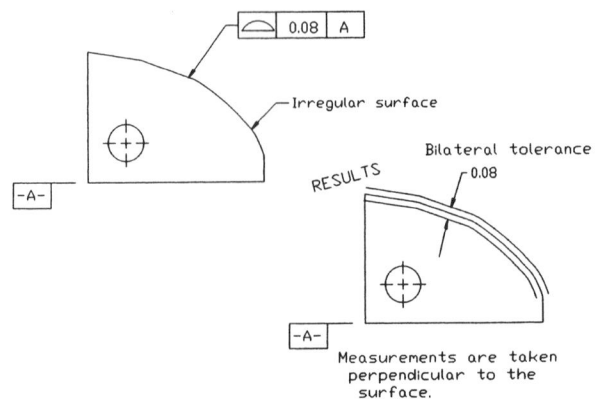

Figure 5-36

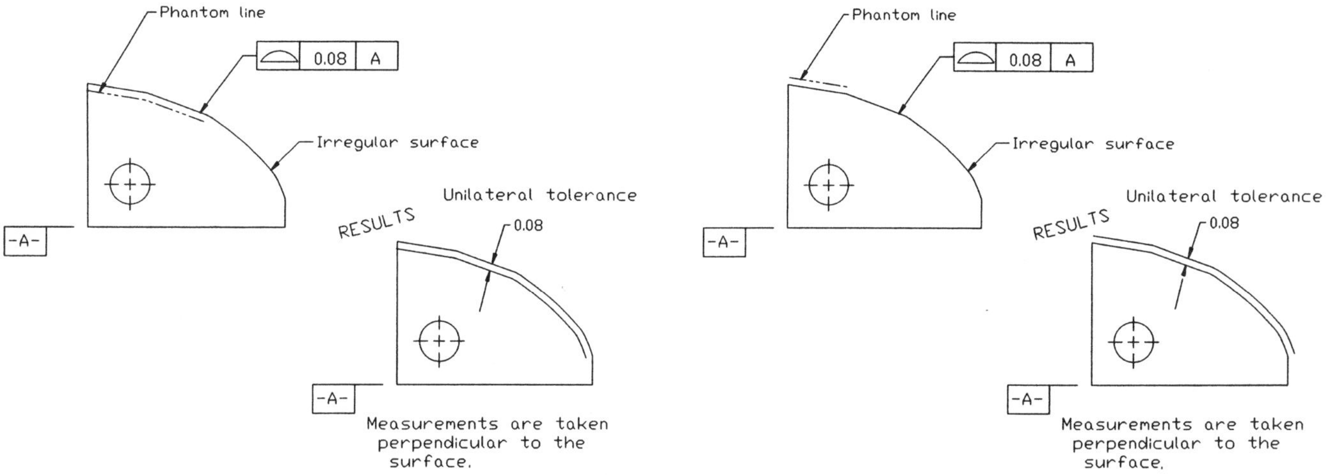

Figure 5-37

variation of an entire surface, whereas a *line profile tolerance* limits the variations along a single line across a surface.

Figure 5-36 shows an object that includes a surface profile tolerance referenced to an irregular surface. The tolerance is considered a bilateral tolerance because no other specification is given. This means that all points on the surface must be located between two parallel planes 0.08 apart that are centered about the irregular surface. The measurements are taken perpendicular to the surface.

Unilateral applications of surface profile tolerances must be indicated on the drawing using phantom lines. The phantom line indicates on which side of the true profile line of the irregular surface the tolerance is to be applied. A phantom line above the irregular surface indicates that the tolerance is to be applied using the true profile line as 0 and then adding a specified tolerance range above that line. See Figures 5-37 and Figure 5-38.

Profiles of line tolerances are applied to irregular surfaces, as shown in Figure 5-37. Profiles of line tolerances are

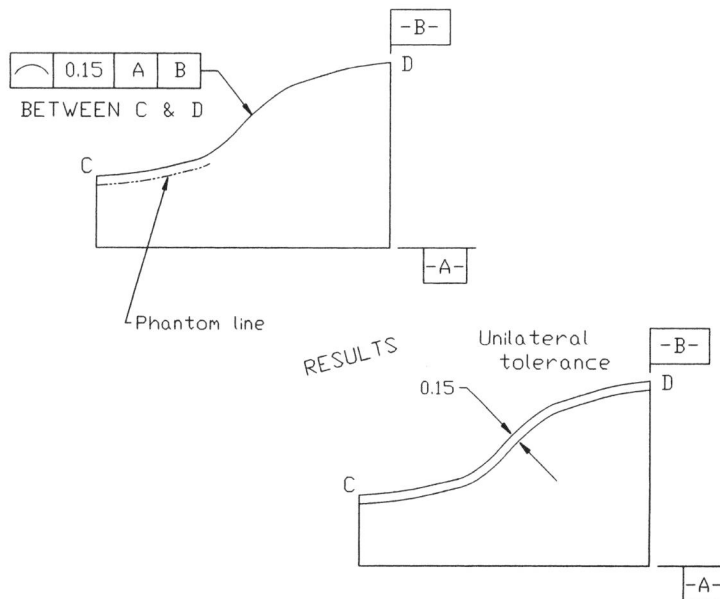

Figure 5-38

**Figure 5-39**

## 5-15  RUNOUTS

A *runout tolerance* is used to limit the variations between features of an object and a datum. More specifically, runout tolerances are applied to surfaces around a datum axis such as a cylinder or to a surface constructed perpendicular to a datum axis. There are two types of runout tolerances: circular and total.

Figure 5-39 shows a cylinder that includes a circular runout tolerance. The runout requirements are checked by rotating the object about its longitudinal axis or datum axis while holding an indicator gauge in a fixed position on the surface of the object.

Runout tolerances may be either bilateral or unilateral. A runout tolerance is assumed to be bilateral unless otherwise indicated. If a runout tolerance is to be unilateral, a phantom line is used to indicate to which side of the object's true surface the tolerance is to be applied. See Figure 5-40.

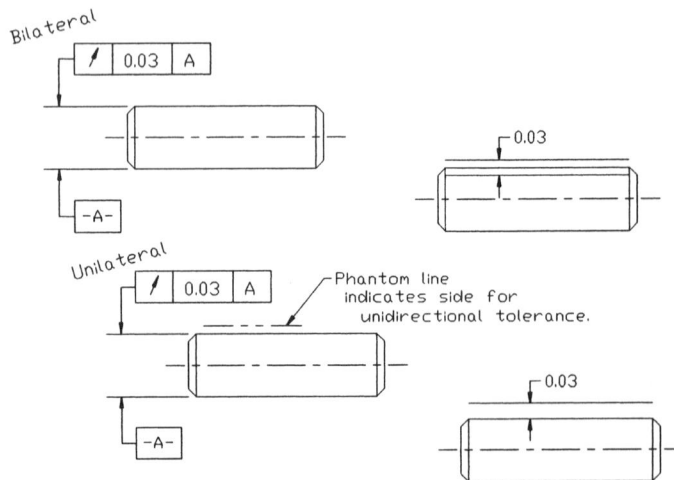

**Figure 5-40**

**Figure 5-41**

**Figure 5-42**

Runout tolerances may be applied to tapered areas of cylindrical objects, as shown in Figure 5-41. The tolerance is checked by rotating the object about a datum axis while holding an indicator gauge in place.

A total runout tolerance limits the variation across an entire surface. See Figure 5-42. An indicator gauge is not held in place while the object is rotated, as it is for circular runout tolerances, but is moved about the rotating surface.

Figure 5-43 shows a circular runout tolerance that references two datums. The two datums serve as one datum. The object can then be rotated about both datums simultaneously as the runout tolerances are checked.

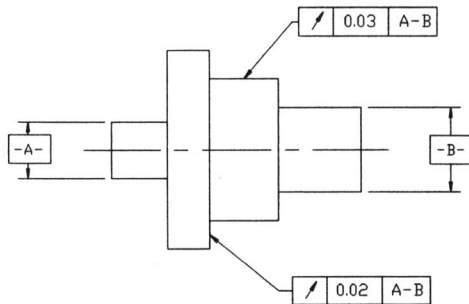

## 5-16 POSITIONAL TOLERANCES

Positional tolerances are used to locate and tolerance holes. Positional tolerances create a circular tolerance zone for hole center point locations that differs from the rectangular-shaped tolerance zone created by linear coordinate dimensions. See Figure 5-44. The circular tolerance zone allows for an increase in acceptable tolerance variation without compromising the design integrity of the object. Note that some of the possible hole center points fall in an area outside the rectangular tolerance zone but are still within the circular tolerance zone. If the hole had been located using linear coordinate dimensions, center points located beyond the

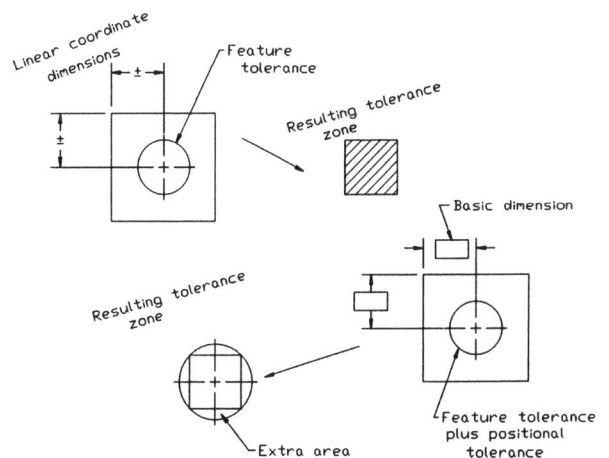

**Figure 5-43**

**Figure 5-44**

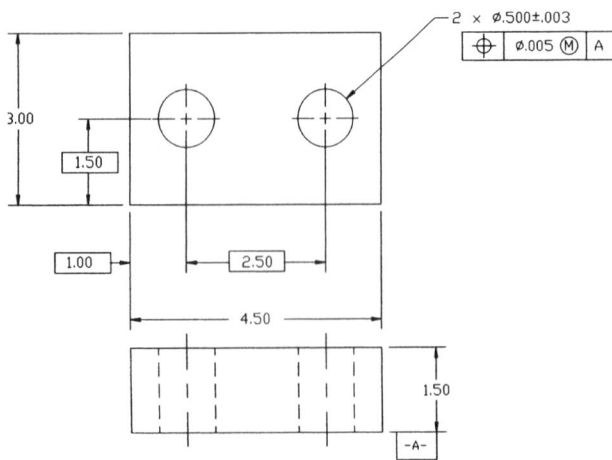

**Figure 5-45**

rectangular tolerance zone would have been rejected as beyond tolerance, and yet holes produced using these locations would function correctly from a design standpoint. The center point locations would be acceptable if positional tolerances had been specified. The finished hole is round, so a round tolerance zone is appropriate. The rectangular tolerance zone rejects some holes unnecessarily.

Holes are dimensioned and toleranced using geometric tolerances by a combination of locating dimensions, feature dimensions and tolerances, and positional tolerances. See Figure 5-45. The locating dimensions are enclosed in

rectangular boxes and are called **basic dimensions.** Basic dimensions are assumed to be exact.

The feature tolerances for the hole are as presented in Chapter 4. They can be presented using plus or minus or limit-type tolerances. In the example shown in Figure 5-45 the diameter of the hole is toleranced using a plus and minus 0.05 tolerance.

The basic locating dimensions of 45 and 50 are assumed to be exact. The tolerances that would normally accompany linear locational dimensions are replaced by the positional tolerance. The positional tolerance also specifies that the tolerance be applied at the centerline at maximum material condition. The resulting tolerance zones are as shown in Figure 5-45.

Figure 5-46 shows an object containing two holes that are dimensioned and toleranced using positional tolerances. There are two consecutive horizontal basic dimensions. Because basic dimensions are exact, they do not have tolerances that accumulate; that is, there is no tolerance buildup.

## 5-17  VIRTUAL CONDITION

*Virtual condition* is a combination of a feature's MMC and its geometric tolerance. For external features (shafts) it is the MMC plus the geometric tolerance; for internal features (holes) it is the MMC minus the geometric tolerance.

The following calculations are based on the dimensions shown in Figure 5-47.

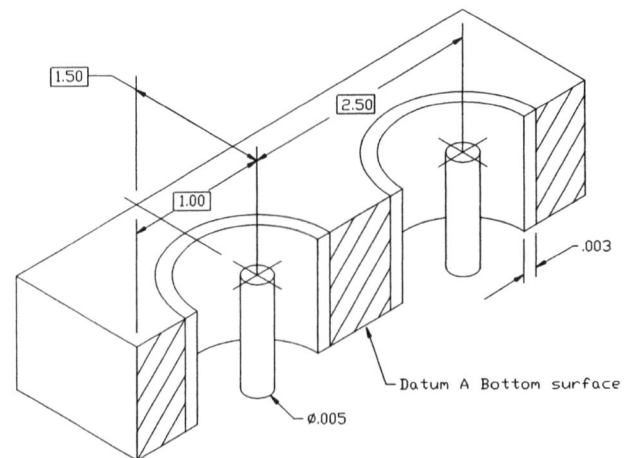

**Figure 5-46**

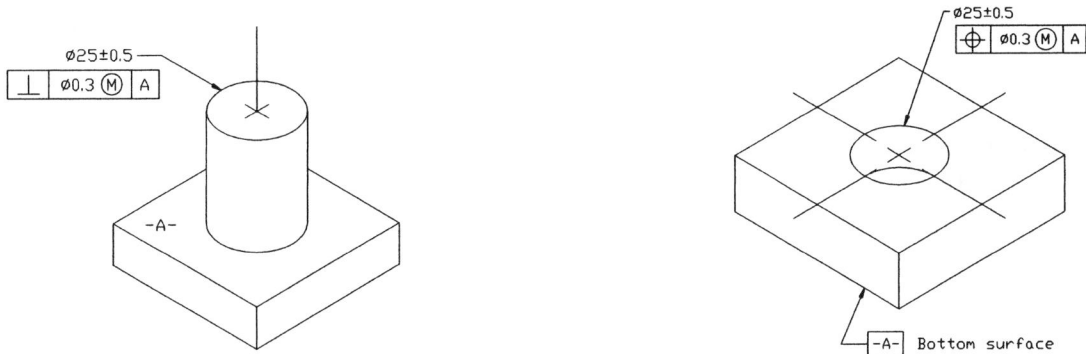

**Figure 5-47**

**To calculate the virtual condition for a shaft**

    25.5  MMC for shaft—maximum diameter
    +0.3  Geometric tolerance
    25.8  Virtual condition

**To calculate the virtual condition for a hole**

    24.5  MMC for hole—minimum diameter
    −0.3  Geometric tolerance
    24.2  Virtual condition

## 5-18 FLOATING FASTENERS

Positional tolerances are particularly helpful when dimensioning matching parts. Because basic locating dimensions are considered exact, the sizing of mating parts is dependent only on the MMC of the hole and shaft and the geometric tolerance between them.

The relationship for floating fasteners and holes in objects may be expressed as a formula:

$$H - T = F$$

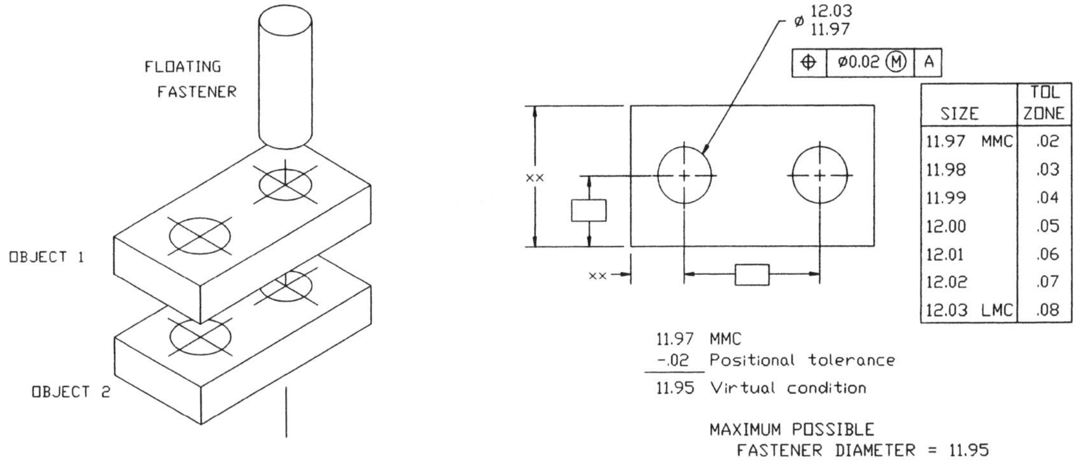

**Figure 5-48**

where

$H$ = Hole at MMC
$T$ = Geometric tolerance
$F$ = Shaft at MMC

A *floating fastener* is one that is free to move in either object. It is not attached to either object and it does not screw into either object. Figure 5-48 shows two objects that are to be joined by a common floating shaft, such as a bolt or screw. The feature size and tolerance and the positional geometric tolerance are both given. The minimum size hole that will always just fit is determined using the preceding formula:

$$H - T = F$$
$$11.97 - .02 = 11.95$$

Therefore, the shaft's diameter at MMC, the shaft's maximum diameter, equals 11.95. Any required tolerance would have to be subtracted from this shaft size.

The 0.02 geometric tolerance is applied at the hole's MMC. Thus, as the hole's size expands within its feature tolerance, the tolerance zone for the acceptable matching parts also expands. See the table in Figure 5-48.

## 5-19  SAMPLE PROBLEM SP5-1

The situation presented in Figure 5-48 can be worked in reverse; that is, hole sizes can be derived from given shaft sizes.

The two objects shown in Figure 5-49 are to be joined by a .250-inch bolt. The parts are floating; that is, they are both free to move, and the fastener is not joined to either ob-

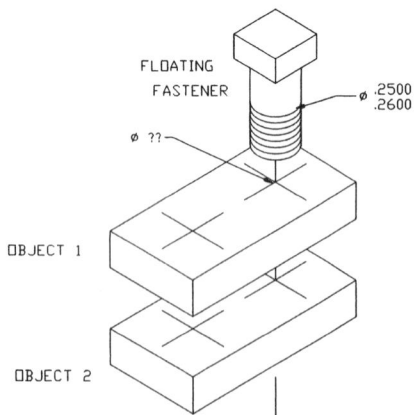

**Figure 5-49**

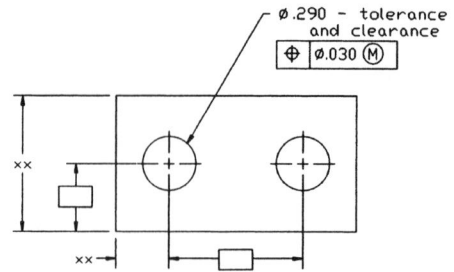

**Figure 5-50**

ject. What is the MMC of the holes if the positional tolerance is to be .030?

A manufacturer's catalog specifies that the tolerance for a .250 bolt is .2500 to .2600.

Rewriting the formula

$$H - T = F$$

to isolate the H, we have

$$H = F + T$$
$$= .260 + .030$$
$$= .290$$

The .290 value represents the minimum hole diameter, MMC, for all four holes that will always accept the .250 bolt. Figure 5-50 shows the resulting drawing callout.

Any clearance requirements or tolerances for the hole would have to be added to the .290 value.

## 5-20  SAMPLE PROBLEM SP5-2

Repeat the problem presented in SP5-1 but be sure that there is always a minimum clearance of .002 between the hole and the shaft and assign a hole tolerance of .0010.

Sample Problem SP5-1 determined that the maximum hole diameter that would always accept the .250 bolt was .290 based on the .030 positional tolerance. If the minimum clearance is to be .002, the maximum hole diameter is found as follows:

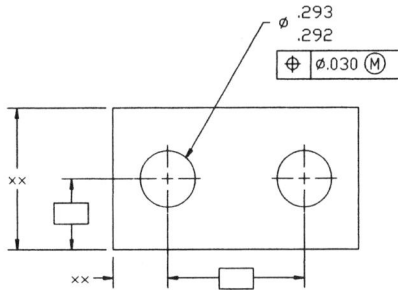

Figure 5-51

Figure 5-53

.290  Minimum hole diameter that will always
      accept the bolt (0 clearance at MMC)
+.002  Minimum clearance
.292  Minimum hole diameter including clearance

Now assign the tolerance to the hole.

.292  Minimum hole diameter
+.001  Tolerance
.293  Minimum hole diameter

See Figure 5-51 for the appropriate drawing callout.
The choice of clearance size and hole tolerance varies with
the design requirements for the objects.

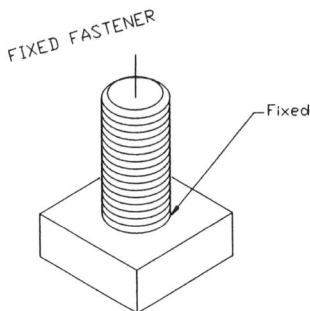

## 5-21  FIXED FASTENERS

A *fixed fastener* is one that is attached to one of the
mating objects. See Figure 5-52. Because the fastener is
fixed to one of the objects, the geometric tolerance zone
must be smaller than that used for floating fasteners. The
fixed fastener cannot move without moving the object it is
attached to. The relationship between fixed fasteners and
holes in mating objects is defined by the formula

$$H - 2T = F$$

The tolerance zone is cut in half, as can be demonstrated by the objects shown in Figure 5-53. The same feature sizes that were used in Figure 5-48 are assigned, but in
this example the fasteners are fixed. Solving for the geometric tolerance, we obtain the following value:

$$H - F = 2T$$
$$11.97 - 11.95 = 2T$$
$$.02 = 2T$$
$$.01 = T$$

The resulting positional tolerance is half of that obtained for floating fasteners.

Figure 5-52

Figure 5-54

## 5-22  SAMPLE PROBLEM SP5-3

This problem is similar to Sample Problem SP5-1, but the given conditions are applied to fixed fasteners rather than to floating fasteners. Compare the resulting shaft diameters for the two problems. See Figure 5-54.

A. What is the minimum-diameter hole that will always accept the fixed fasteners?
B. If the minimum clearance is .005 and the hole is to have a tolerance of .002, what are the maximum and minimum diameters of the hole?

$$H - 2T = F$$
$$H = F + 2T$$
$$= .260 + 2(.030)$$
$$= .260 + .060$$
$$= .320 \text{ Minimum diameter that will}$$
always accept the fixed fastener

If the minimum clearance = .005 and the hole tolerance is .002,

.320  Virtual condition
+.005  Clearance
.325  Minimum hole diameter

.325  Minimum hole diameter
+.002  Tolerance
.327  Maximum hole diameter

The maximum and minimum values for the hole's diameter can then be added to the drawing of the object that fits over the fixed fasteners. See Figure 5-55.

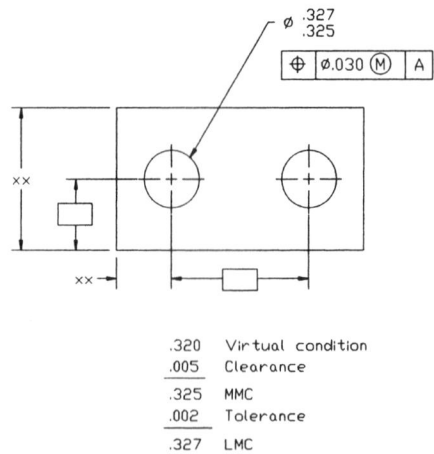

Figure 5-55

## 5-23  DESIGN PROBLEMS

This problem was originally done in Section 4-61 using rectangular tolerances. It is done in this section using positional geometric tolerances so that the two systems can be compared. It is suggested that you review Section 4-61 before reading this section.

Figure 5-56 shows top and bottom parts that are to be joined in the floating condition. A nominal distance of 50 between hole centers and 20 for the holes has been assigned. In Section 4-61 a rectangular tolerance of $\pm.01$ was selected and there was a minimum hole diameter of 20.00. Figure 5-57 shows the resulting tolerance zones.

The diagonal distance across the rectangular tolerance zone is .028 and was rounded off to .03 to yield a maximum possible fastener diameter of 19.97. If the same .03 value is used to calculate the fastener diameter using positional tolerance, the results will be as follows:

$$H - T = F$$
$$20.00 - .03 = 19.97$$

The results seem to be the same, but because of the circular shape of the positional tolerance zone, the manufactured results are not the same. The minimum distance between the inside edges of the rectangular zones is 49.98, or .01 from the center point of each hole. The minimum distance from the innermost points of the circular tolerance zones is 49.97, or .015 (half of the rounded-off .03 value)

Fastener

TOP

All holes are
Ø20 NOMINALs

The distance between
the holes is 50 nominal.

Floating condition

BOTTOM

**Figure 5-56**

from the center point of each hole. The same value difference also occurs for the maximum distance between center points, where 50.02 is the maximum distance for the rectangular tolerances, and 50.03 is the maximum distance for the circular tolerances. The size of the circular tolerance zone increased more because the hole tolerances are assigned at

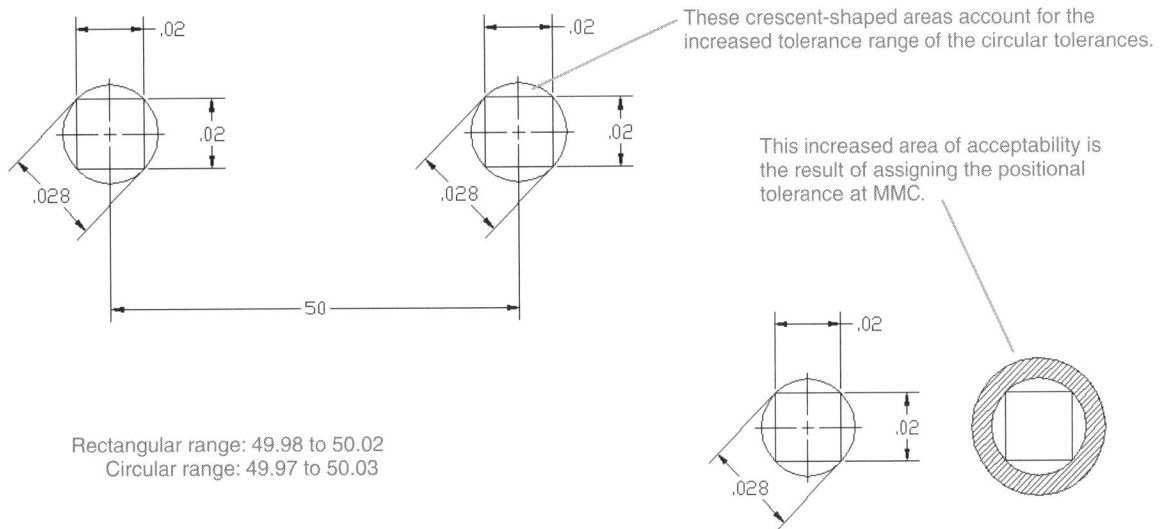

.02

.02

.028

50

These crescent-shaped areas account for the increased tolerance range of the circular tolerances.

This increased area of acceptability is the result of assigning the positional tolerance at MMC.

.02

.02

.028

Rectangular range: 49.98 to 50.02
Circular range: 49.97 to 50.03

**Figure 5-57**

**Figure 5-58**

**Figure 5-59**

MMC. Figure 5-57 shows a comparison between the tolerance zones, and Figure 5-58 shows how the positional tolerances would be presented on a drawing of either the top or bottom part.

Figure 5-59 shows the same top and bottom parts joined together in the fixed condition. The initial nominal values are the same. If the same .03 diagonal value is assigned as a positional tolerance, the results are as follows:

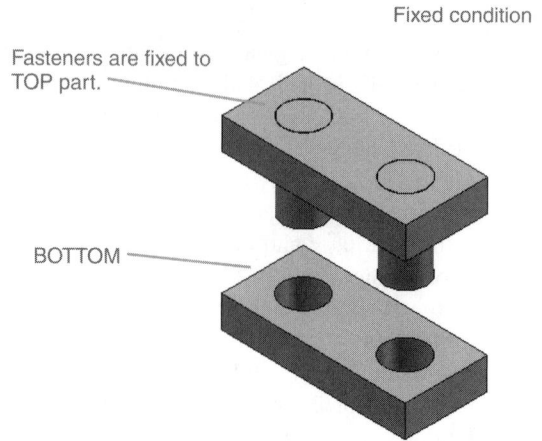

$$H - 2T = F$$
$$20.00 - .06 = 19.94$$

These results appear to be the same as those generated by the rectangular tolerance zone, but the circular tolerance zone allows a greater variance in acceptable manufactured parts. Figure 5-60 shows how the positional tolerance would be presented on a drawing.

**Figure 5-60**

# 5-24  EXERCISE PROBLEMS

## EX5-1

Redraw the object shown. Include all dimensions and tolerances.

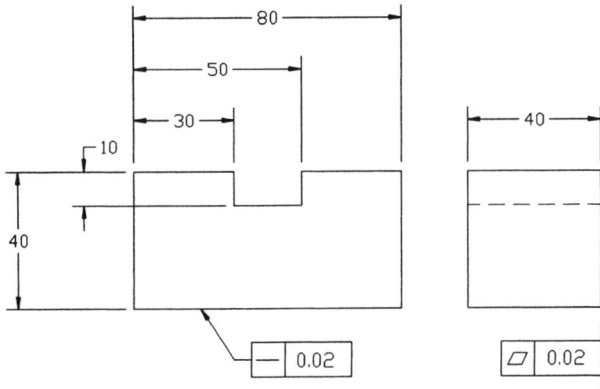

## EX5-2

Redraw the following shaft and add a feature dimension and tolerance of 36±0.1 and a straightness tolerance of 0.07 about the centerline at MMC.

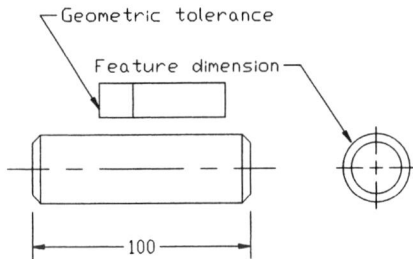

## EX5-3

A.  Given the shaft shown, what is the minimum hole diameter that will always accept the shaft?
B.  If the minimum clearance between the shaft and a hole is equal to 0.02, and the tolerance on the hole is to be 0.6, what are the maximum and minimum diameters for the hole?

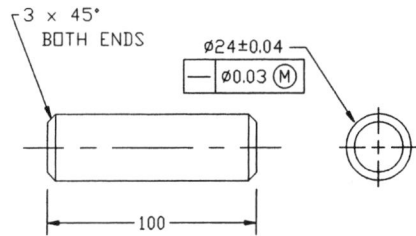

## EX5-4

A.  Given the shaft shown, what is the minimum hole diameter that will always accept the shaft?
B.  If the minimum clearance between the shaft and a hole is equal to .005, and the tolerance on the hole is to be .007, what are the maximum and minimum diameters for the hole?

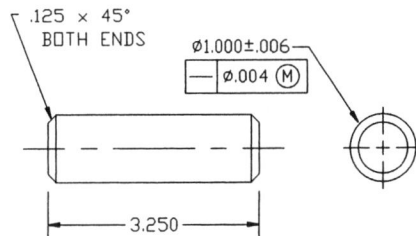

**EX5-5**

Draw a front and a right-side view of the object shown in Figure EX5-5 and add the appropriate dimensions and tolerances based on the following information. Numbers located next to an edge line indicate the length of the edge.

A. Define surfaces A, B, and C as primary, secondary, and tertiary datums, respectively.
B. Assign a tolerance of ±0.5 to all linear dimensions.
C. Assign a feature tolerance of 12.07 – 12.00 to the protruding shaft.
D. Assign a flatness tolerance of 0.01 to surface A.
E. Assign a straightness tolerance of 0.03 to the protruding shaft.
F. Assign a perpendicularity tolerance to the centerline of the protruding shaft of 0.02 at MMC relative to datum A.

**Figure EX5-5**

**EX5-6**

Draw a front and right-side view of the object shown in Figure EX5-6 and add the following dimensions and tolerances.

A. Define the bottom surface as datum A.
B. Assign a perpendicularity tolerance of 0.4 to both sides of the slot relative to datum A.
C. Assign a perpendicularity tolerance of 0.2 to the centerline of the 30-diameter hole centerline at MMC relative to datum A.
D. Assign a feature tolerance of ±0.8 to all three holes.
E. Assign a parallelism tolerance of 0.2 to the common centerline between the two 20-diameter holes relative to datum A.
F. Assign a tolerance of ±0.5 to all linear dimensions.

**Figure EX5-6**

## EX5-7

Draw a circular front and the appropriate right-side view of the object shown in Figure EX5-7 and add the following dimensions and tolerances.

A. Assign datum A as indicated.
B. Assign the object's longitudinal axis as datum B.
C. Assign the object's centerline through the slot as datum C.
D. Assign a tolerance of ±0.5 to all linear tolerances.
E. Assign a tolerance of ±0.5 to all circular-shaped features.
F. Assign a parallelism tolerance of 0.01 to both edges of the slot.
G. Assign a perpendicularity tolerance of 0.01 to the outside edge of the protruding shaft.

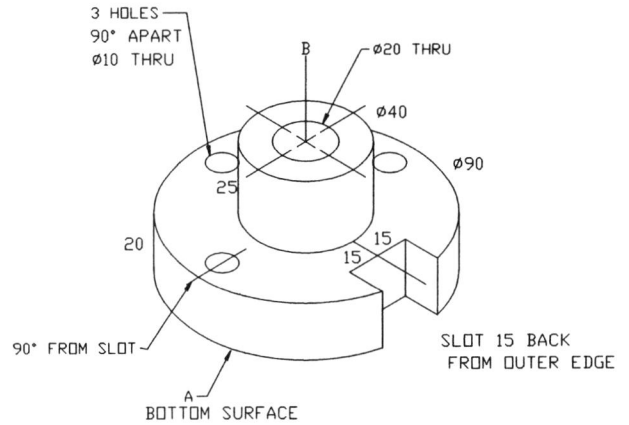

**Figure EX5-7**

## EX5-8

Given the two objects shown in Figure EX5-8, draw a front and a side view of each. Assign a tolerance of ±0.5 to all linear dimensions. Assign a feature tolerance of ±0.4 to the shaft, and also assign a straightness tolerance of 0.2 to the shaft's centerline at MMC.

Tolerance the hole so that it will always accept the shaft with a minimum clearance of 0.1 and a feature tolerance of 0.2. Assign a perpendicularity tolerance of 0.05 to the centerline of the hole at MMC.

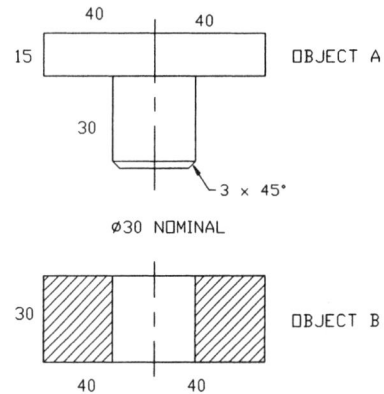

**Figure EX5-8**

## EX5-9

Given the two objects shown in Figure EX5-9, draw a front and a side view of each. Assign a tolerance of ±.005 to all linear dimensions. Assign a feature tolerance of ±.004 to the shaft, and also assign a straightness tolerance of .002 to the shaft's centerline at MMC.

Tolerance the hole so that it will always accept the shaft with a minimum clearance of .001 and a feature tolerance of .002.

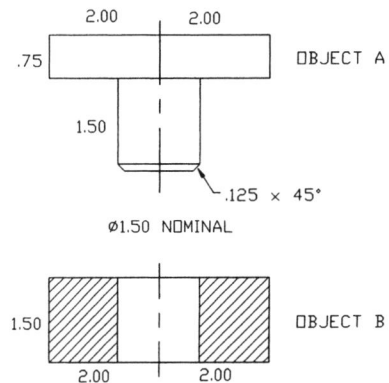

**Figure EX5-9**

**EX5-10**

Refer to parts A through G for this exercise problem. Use the format shown in Figure EX5-10 and redraw the given geometric tolerance symbols and frame as shown in the sample. Express in words (**Dtext**) the meaning of each tolerance callout.

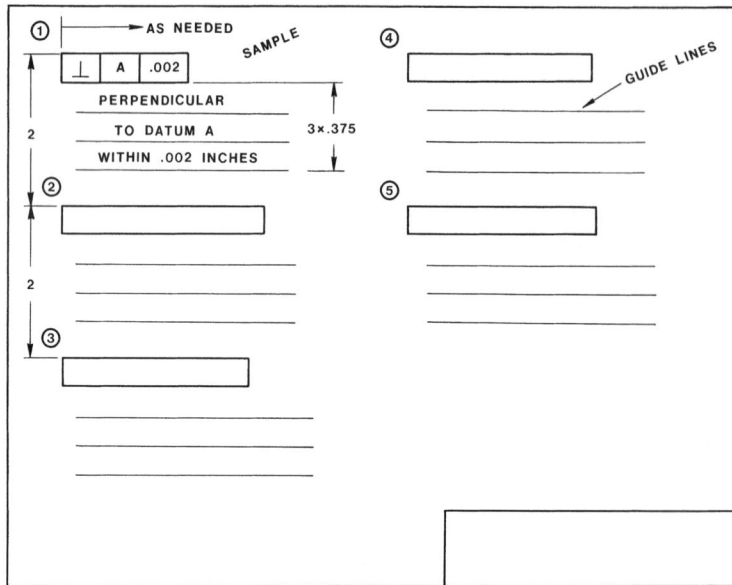

**Figure EX5-10**

A.

B.

**EX5-10 continued**

**C.**

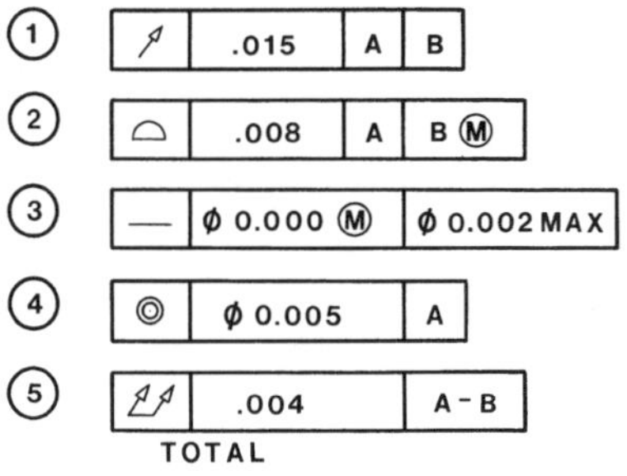

| ① | ⌰ | .015 | A | B | |
| ② | ⌓ | .008 | A | B Ⓜ | |
| ③ | — | ⌀ 0.000 Ⓜ | ⌀ 0.002 MAX | |
| ④ | ◎ | ⌀ 0.005 | A | |
| ⑤ | ⌰⌰ | .004 | A – B | |

TOTAL

**F.**

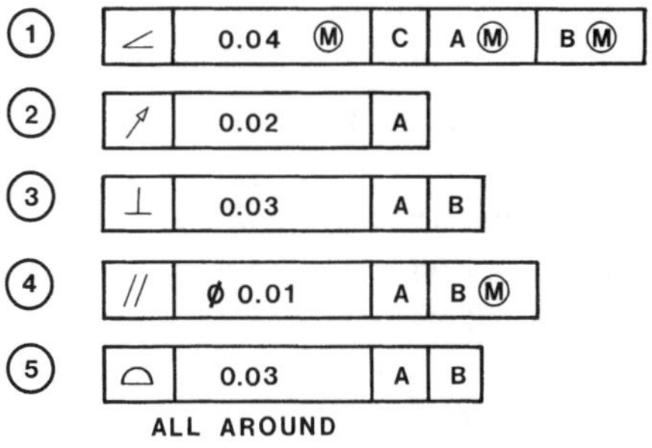

| ① | ∠ | 0.04 Ⓜ | C | A Ⓜ | B Ⓜ |
| ② | ⌰ | 0.02 | A | |
| ③ | ⊥ | 0.03 | A | B |
| ④ | // | ⌀ 0.01 | A | B Ⓜ |
| ⑤ | ⌓ | 0.03 | A | B |

ALL AROUND

**D.**

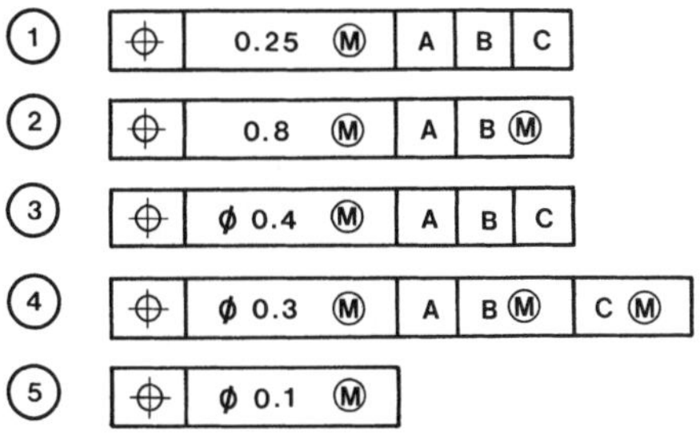

| ① | ⊕ | 0.25 Ⓜ | A | B | C |
| ② | ⊕ | 0.8 Ⓜ | A | B Ⓜ | |
| ③ | ⊕ | ⌀ 0.4 Ⓜ | A | B | C |
| ④ | ⊕ | ⌀ 0.3 Ⓜ | A | B Ⓜ | C Ⓜ |
| ⑤ | ⊕ | ⌀ 0.1 Ⓜ | | |

**G.**

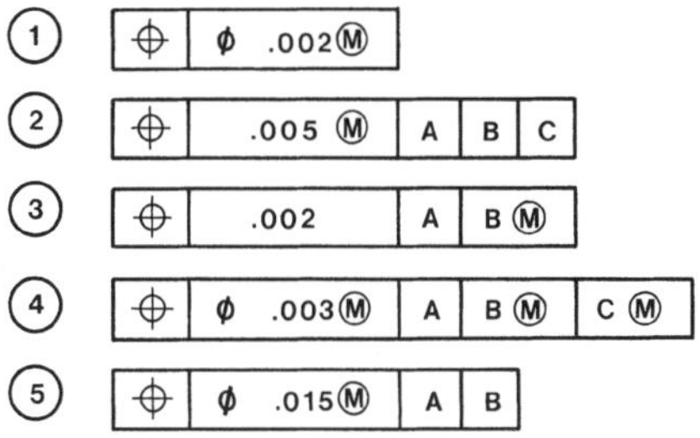

| ① | ⊕ | ⌀ .002 Ⓜ | | |
| ② | ⊕ | .005 Ⓜ | A | B | C |
| ③ | ⊕ | .002 | A | B Ⓜ | |
| ④ | ⊕ | ⌀ .003 Ⓜ | A | B Ⓜ | C Ⓜ |
| ⑤ | ⊕ | ⌀ .015 Ⓜ | A | B |

**E.**

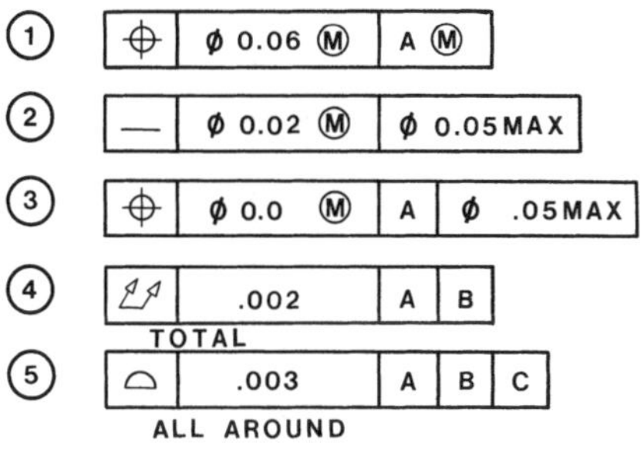

| ① | ⊕ | ⌀ 0.06 Ⓜ | A Ⓜ | |
| ② | — | ⌀ 0.02 Ⓜ | ⌀ 0.05 MAX | |
| ③ | ⊕ | ⌀ 0.0 Ⓜ | A | ⌀ .05 MAX |
| ④ | ⌰⌰ | .002 | A | B |

TOTAL

| ⑤ | ⌓ | .003 | A | B | C |

ALL AROUND

## EX5-11 MILLIMETERS

Draw front, top, and right-side views of the object in Figure EX5-11, including dimensions. Add the following tolerances and specifications to the drawing.

A. Surface 1 is datum A.
B. Surface 2 is datum B and is perpendicular to datum A within 0.1 millimeter.
C. Surface 3 is datum C and is parallel to datum A within 0.3 millimeter.
D. Locate a 16-millimeter-diameter hole in the center of the front surface that goes completely through the object. Use positional tolerances to locate the hole. Assign a positional tolerance of 0.02 at MMC perpendicular to datum A.

Figure EX5-11

## EX5-12

Draw front, top, and right-side views of the object in Figure EX5-12, including dimensions. Add the following tolerances and specifications to the drawing.

A. Surface 1 is datum A.
B. Surface 2 is datum B and is perpendicular to datum A within .003 inch.
C. Surface 3 is parallel to datum A within .005 inch.
D. The cylinder's longitudinal centerline is to be straight within .001 inch at MMC.
E. Surface 2 is to have circular accuracy within .002 inch.

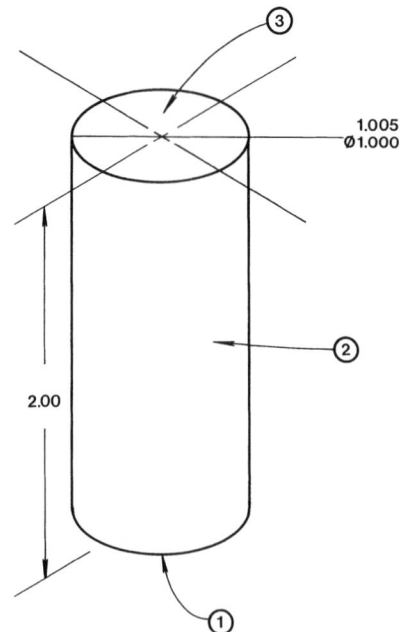

Figure EX5-12

## EX5-13

Draw front, top, and right-side views of the object in Figure EX5-13, including dimensions. Add the following tolerances and specifications to the drawing.

A.  Surface 1 is datum A.
B.  Surface 4 is datum B and is perpendicular to datum A within 0.08 millimeter.
C.  Surface 3 is flat within 0.03 millimeter.
D.  Surface 5 is parallel to datum A within 0.01 millimeter.
E.  Surface 2 has a runout tolerance of 0.2 millimeter relative to surface 4.
F.  Surface 1 is flat within 0.02 millimeter.
G.  The longitudinal centerline is to be straight within 0.02 millimeter at MMC and perpendicular to datum A.

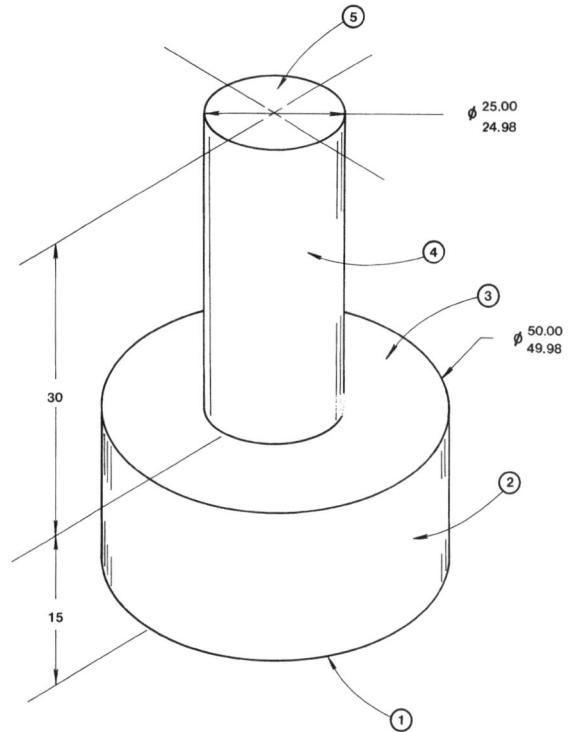

Figure EX5-13

## EX5-14

Draw front, top, and right-side views of the object in Figure EX5-14, including dimensions. Add the following tolerances and specifications to the drawing.

A.  Surface 2 is datum A.
B.  Surface 6 is perpendicular to datum A with 0.000 allowable variance at MMC but with a .002 inch MAX variance limit beyond MMC.
C.  Surface 1 is parallel to datum A within .005 inch.
D.  Surface 4 is perpendicular to datum A within .004 inch.

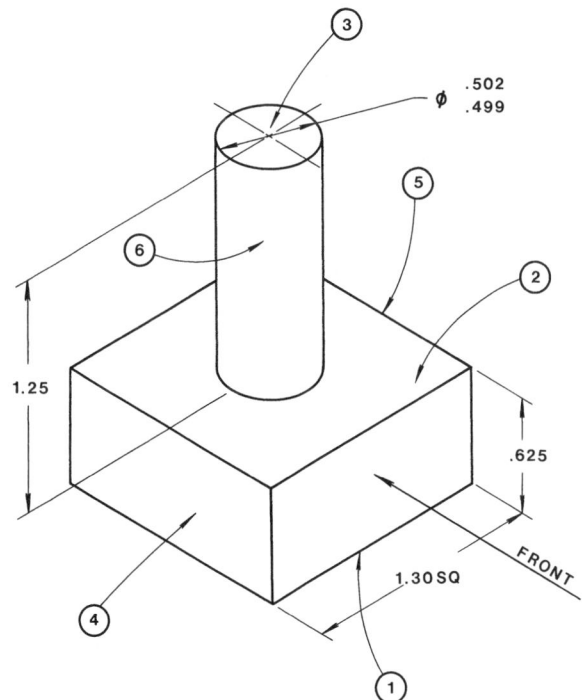

Figure EX5-14

## EX5-15

Draw front, top, and right-side views of the object in Figure EX5-15, including dimensions. Add the following tolerances and specifications to the drawing.

A.  Surface 1 is datum A.
B.  Surface 2 is datum B.
C.  The hole is located using a true position tolerance value of 0.13 millimeter at MMC. The true position tolerance is referenced to datums A and B.
D.  Surface 1 is to be straight within 0.02 millimeter.
E.  The bottom surface is to be parallel to datum A within 0.03 millimeter.

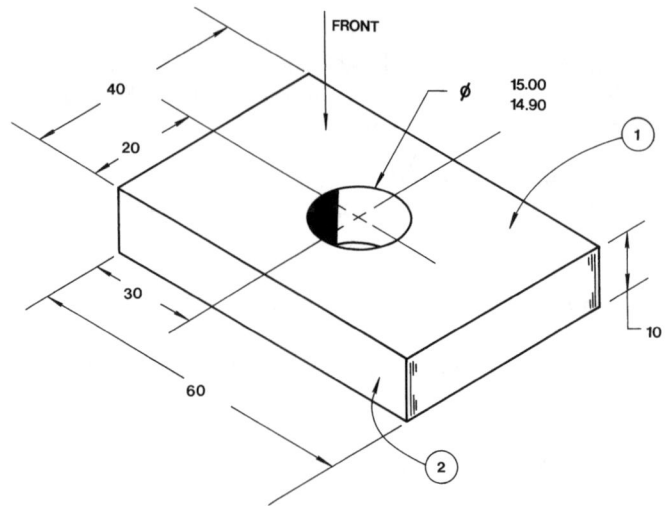

Figure EX5-15

## EX5-16

Draw front, top, and right-side views of the object in Figure EX5-16, including dimensions. Add the following tolerances and specifications to the drawing.

A.  Surface 1 is datum A.
B.  Surface 2 is datum B.
C.  Surface 3 is perpendicular to surface 2 within 0.02 millimeter.
D.  The four holes are to be located using a positional tolerance of 0.07 millimeter at MMC referenced to datums A and B.
E.  The centerlines of the holes are to be straight within 0.01 millimeter at MMC.

Figure EX5-16

## EX5-17

Draw front, top, and right-side views of the object in Figure EX5-17, including dimensions. Add the following tolerances and specifications to the drawing.

A. Surface 1 has a dimension of .378−.375 inch and is datum A. The surface has a dual primary runout with datum B to within .005 inch. The runout is total.

B. Surface 2 has a dimension of 1.505−1.495 inch. Its runout relative to the dual primary datums A and B is .008 inch. The runout is total.

C. Surface 3 has a dimension of 1.000±.005 and has no geometric tolerance.

D. Surface 4 has no circular dimension but has a total runout tolerance of .006 inch relative to the dual datums A and B.

E. Surface 5 has a dimension of .500−.495 inch and is datum B. It has a dual primary runout with datum A within .005 inch. The runout is total.

Figure EX5-17

## EX5-18

Draw front, top, and right-side views of the object in Figure EX5-18, including dimensions. Add the following tolerances and specifications to the drawing.

A. Hole 1 is datum A.

B. Hole 2 is to have its circular centerline parallel to datum A within 0.2 millimeter at MMC when datum A is at MMC.

C. Assign a positional tolerance of 0.01 to each hole's centerline at MMC.

Figure EX5-18

## EX5-19

Draw front, top, and right-side views of the object in Figure EX5-19, including dimensions. Add the following tolerances and specifications to the drawing.

A.  Surface 1 is datum A.
B.  Surface 2 is datum B.
C.  The six holes have a diameter range of .502−.499 inch and are to be located using positional tolerances so that their centerlines are within .005 inch at MMC relative to datums A and B.
D.  The back surface is to be parallel to datum A within .002 inch.

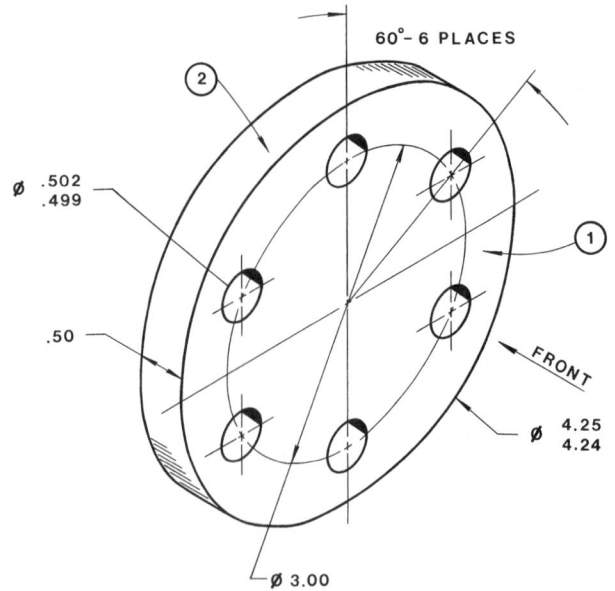

Figure EX5-19

## EX5-20

Draw front, top, and right-side views of the object in Figure EX5-20, including dimensions. Add the following tolerances and specifications to the drawing.

A.  Surface 1 is datum A.
B.  Hole 2 is datum B.
C.  The eight holes labeled 3 have diameters of 8.4–8.3 millimeters with a positional tolerance of 0.15 millimeter at MMC relative to datums A and B. Also, the eight holes are to be counterbored to a diameter of 14.6–14.4 millimeters and to a depth of 5.0 millimeters.
D.  The large center hole is to have a straightness tolerance of 0.2 at MMC about its centerline.

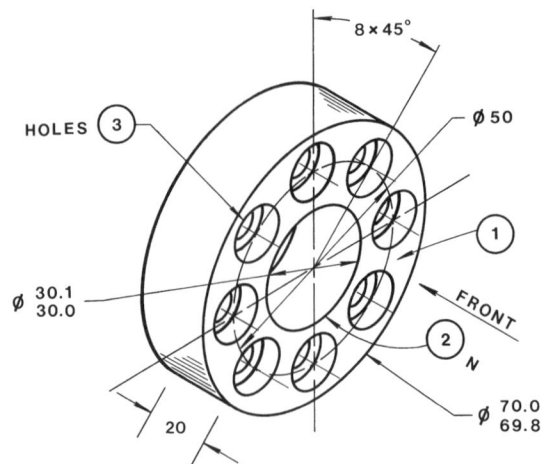

Figure EX5-20

## EX5-21

Draw front, top, and right-side views of the object in Figure EX5-21, including dimensions. Add the following tolerances and specifications to the drawing.

A. Surface 1 is datum A.
B. Surface 2 is datum B.
C. Surface 3 is datum C.
D. The four holes labeled 4 have a dimension and tolerance of 8 +0.3, −0 millimeters. The holes are to be located using a positional tolerance of 0.05 millimeter at MMC relative to datums A, B, and C.
E. The six holes labeled 5 have a dimension and tolerance of 6 +0.2, −0 millimeters. The holes are to be located using a positional tolerance of 0.01 millimeter at MMC relative to datums A, B, and C.

## EX5-22

The objects on page 434 labeled A and B, are to be toleranced using four different tolerances as shown. Redraw the charts shown in Figure EX5-22 and list the appropriate allowable tolerance for "as measured" increments of 0.1 millimeter or .001 inch. Also include the appropriate geometric tolerance drawing called out above each chart.

Figure EX5-21

Figure EX5-22

**EX5-22, CONTINUED**

A.

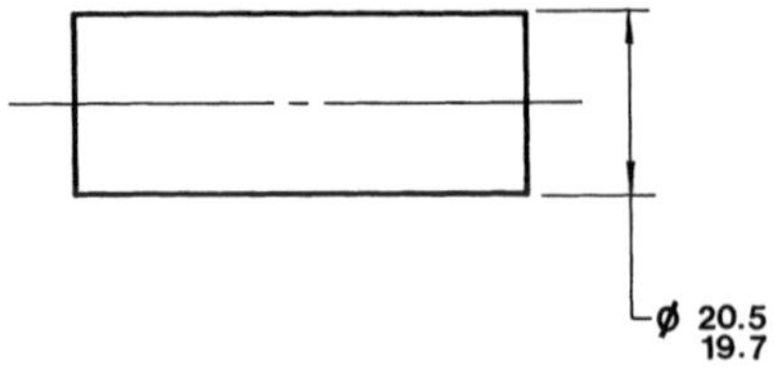

Ø 20.5
19.7

① | — | Ø 0.3 |

② | — | Ø 0.3 Ⓜ |

③ | — | Ø 0.0 Ⓜ |

④ | — | Ø 0.0 Ⓜ | Ø0.04 MAX |

B.

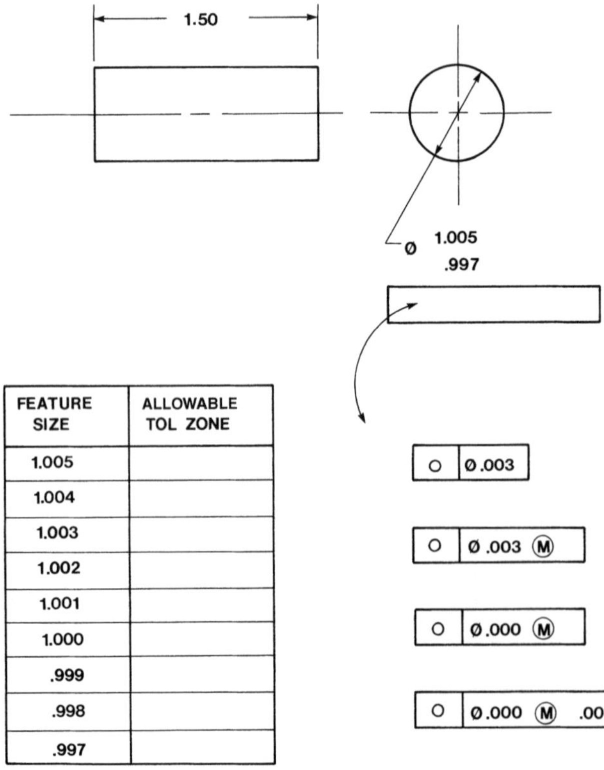

1.50

Ø 1.005
.997

| FEATURE SIZE | ALLOWABLE TOL ZONE |
|---|---|
| 1.005 | |
| 1.004 | |
| 1.003 | |
| 1.002 | |
| 1.001 | |
| 1.000 | |
| .999 | |
| .998 | |
| .997 | |

| ○ | Ø .003 |

| ○ | Ø .003 Ⓜ |

| ○ | Ø .000 Ⓜ |

| ○ | Ø .000 Ⓜ | .004 MAX |

## EX5-23

Dimension and tolerance parts 1 and 2 of Figure EX5-23 so that part 1 always fits into part 2 with a minimum clearance of .005 inch. The tolerance for part 1's outer matching surface is .006 inch.

**Figure EX5-23**

## EX5-24

Dimension and tolerance parts 1 and 2 of Figure EX5-24 so that part 1 always fits into part 2 with a minimum clearance of 0.03 millimeter. The tolerance for part 1's diameter is 0.05 millimeter. Take into account the fact that the interface is long relative to the diameters.

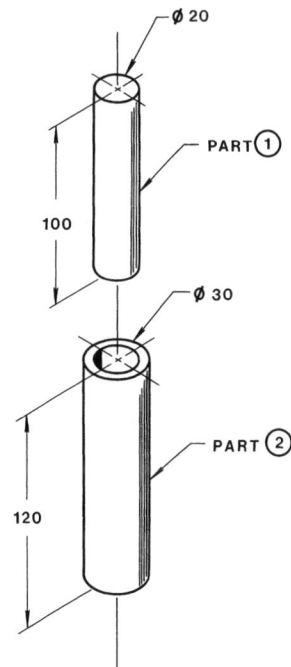

**Figure EX5-24**

## EX5-25

Prepare front and top views of parts 4A and 4B of Figure EX5-25 based on the given dimensions. Add geometric tolerances to produce the stated maximum clearance and mismatch.

## EX5-26

Redraw parts A and B of Figure EX5-26 and add dimensions and tolerances to meet the "UPON ASSEMBLY" requirements.

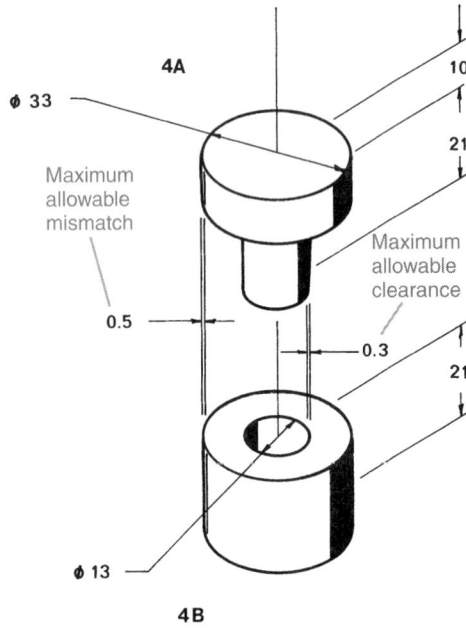

Figure EX5-25

The UPON ASSEMBLY condition.

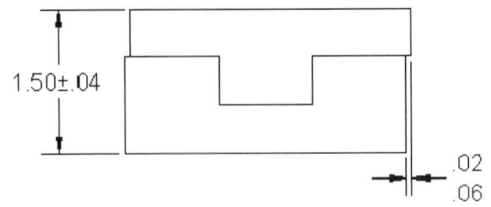

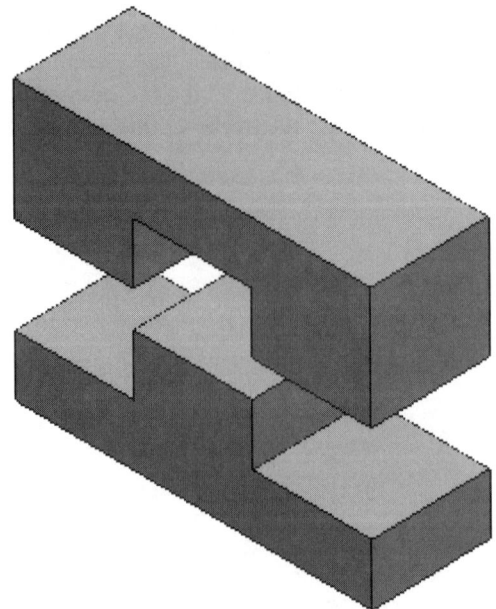

Figure EX5-26

EX5-27

Draw front and top views of both objects in Figure EX5-27. Add dimensions and geometric tolerances to meet the "FINAL CONDITION" requirements.

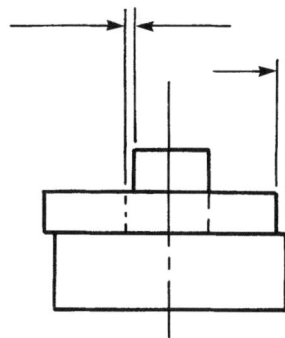

**FINAL CONDITION**

MAX = 0.03
MIN = 0.01

MIN = 0.00
MAX = 0.04

Figure EX5-27

Given a TOP and BOTTOM part in the floating condition as shown in Figure EX5-28 add dimensions and tolerances to satisfy the following conditions. Size the TOP and BOTTOM parts and FASTENER length as needed. Use geometric and positional tolerances.

### EX5-28 INCHES–CLEARANCE FIT

A. The distance between the holes is 2.00 nominal.
B. The diameter of the fasteners is Ø.375 nominal.
C. The fasteners have a total tolerance of .001.
D. The holes have a tolerance of .002.
E. The minimum allowable clearance between the fastener and the holes is .003.
F. The material is .375 inch thick.

### EX5-29 MILLIMETERS–CLEARANCE FIT

A. The distance between the holes is 80 nominal.
B. The nominal diameter of the fasteners is Ø12.
C. The fasteners have a total tolerance of 0.05.
D. The holes have a tolerance of 0.03.
E. The minimum allowable clearance between the fastener and the holes is 0.02.
F. The material is 12 millimeters thick.

### EX5-30 INCHES–CLEARANCE FIT

A. The distance between the holes is 3.50 nominal.
B. The diameter of the fasteners is Ø.625.
C. The fasteners have a total tolerance of .005.
D. The holes have a tolerance of .003.
E. The minimum allowable clearance between the fastener and the holes is .002.
F. The material is .500 inch thick.

### EX5-31 MILLIMETERS–CLEARANCE FIT

A. The distance between the holes is 120 nominal.
B. The diameter of the fasteners is Ø24 nominal.
C. The fasteners have a total tolerance of 0.01.
D. The holes have a tolerance of 0.02.
E. The minimum allowable clearance between the fastener and the holes is 0.04.
F. The material is 20 millimeters thick.

### EX5-32 INCHES–INTERFERENCE FIT

A. The distance between the holes is 2.00 nominal.
B. The diameter of the fasteners is Ø.250 nominal.
C. The fasteners have a total tolerance of .001.
D. The holes have a tolerance of .002.

**Figure EX5-28**

E. The maximum allowable interference between the fastener and the holes is .0065.
F. The material is .438 inch thick.

### EX5-33 MILLIMETERS–INTERFERENCE FIT

A. The distance between the holes is 80 nominal.
B. The diameter of the fasteners is Ø10 nominal.
C. The fasteners have a total tolerance of 0.01.
D. The holes have a tolerance of 0.02.
E. The maximum allowable interference between the fastener and the holes is 0.032.
F. The material is 14 millimeters thick.

### EX5-34 INCHES–LOCATIONAL FIT

A. The distance between the holes is 2.25 nominal.
B. The diameter of the fasteners is Ø.50 nominal.
C. The fasteners have a total tolerance of .001.
D. The holes have a tolerance of .002.
E. The minimum allowable clearance between the fastener and the holes is .0010.
F. The material is .370 inch thick.

### EX5-35 MILLIMETERS–TRANSITIONAL FIT

A. The distance between the holes is 100 nominal.
B. The diameter of the fasteners is Ø16 nominal.
C. The fasteners have a total tolerance of 0.01.
D. The holes are to have a tolerance of 0.02.
E. The minimum allowable clearance between the fastener and the holes is 0.01.
F. The material is 20 millimeters thick.

Given a TOP and BOTTOM part in the fixed condition as shown in Figure EX5-36 add dimensions and tolerances to satisfy the following conditions. Size the TOP and BOTTOM parts and FASTENER length as needed. Use geometric and positional tolerances.

**Figure EX5-36**

## EX5-36 INCHES–CLEARANCE FIT

A. The distance between the holes is 2.00 nominal.
B. The diameter of the fasteners is Ø.375 nominal.
C. The fasteners have a total tolerance of .001.
D. The holes have a tolerance of .002.
E. The minimum allowable clearance between the fastener and the holes is .003.
F. The material is .375 inch thick.

## EX5-37 MILLIMETERS–CLEARANCE FIT

A. The distance between the holes is 80 nominal.
B. The nominal diameter of the fasteners is Ø12.
C. The fasteners have a total tolerance of 0.05.
D. The holes have a tolerance of 0.03.
E. The minimum allowable clearance between the fastener and the holes is 0.02.
F. The material is 12 millimeters thick.

## EX5-38 INCHES–CLEARANCE FIT

A. The distance between the holes is 3.50 nominal.
B. The diameter of the fasteners is Ø.625.
C. The fasteners have a total tolerance of .005.
D. The holes have a tolerance of .003.
E. The minimum allowable clearance between the fastener and the holes is .002.
F. The material is .500 inch thick.

## EX5-39 MILLIMETERS–CLEARANCE FIT

A. The distance between the holes is 120 nominal.
B. The diameter of the fasteners is Ø24 nominal.
C. The fasteners have a total tolerance of 0.01.
D. The holes have a tolerance of 0.02.
E. The minimum allowable clearance between the fastener and the holes is 0.04.
F. The material is 20 millimeters thick.

## EX5-40 INCHES–INTERFERENCE FIT

A. The distance between the holes is 2.00 nominal.
B. The diameter of the fasteners is Ø.250 nominal.
C. The fasteners have a total tolerance of .001.
D. The holes have a tolerance of .002.

E. The maximum allowable interference between the fastener and the holes is .0065.
F. The material is .438 inch thick.

## EX5-41 MILLIMETERS–INTERFERENCE FIT

A. The distance between the holes is 80 nominal.
B. The diameter of the fasteners is Ø10 nominal.
C. The fasteners have a total tolerance of 0.01.
D. The holes have a tolerance of 0.02.
E. The maximum allowable interference between the fastener and the holes is 0.032.
F. The material is 14 millimeters thick.

## EX5-42 INCHES–LOCATIONAL FIT

A. The distance between the holes is 2.25 nominal.
B. The diameter of the fasteners is Ø.50 nominal.
C. The fasteners have a total tolerance of .001.
D. The holes have a tolerance of .002.
E. The minimum allowable clearance between the fastener and the holes is .0010.
F. The material is .370 inch thick.

## EX5-43 MILLIMETERS–TRANSITIONAL FIT

A. The distance between the holes is 100 nominal.
B. The diameter of the fasteners is Ø16 nominal.
C. The fasteners have a total tolerance of 0.01.
D. The holes are to have a tolerance of 0.02.
E. The minimum allowable clearance between the fastener and the holes is 0.001.
F. The material is 20 millimeters thick.

**EX5-44**

Given the following two assemblies, size the parts so that they always fit together. Create individual drawings of each part including dimensions and tolerances. Use geometric and positional tolerances.

SUPPORT ASSEMBLY

BRACKET, END
2 REQD

PLATE, 4 HOLE
2 REQD

PEG
4 REQD

## EX5-45

Assume that there are two copies of the part in Figure EX5-45 and that these parts are to be joined together using four fasteners in the floating condition. Draw front and top views of the object, including dimensions and tolerances. Add the following tolerances and specifications to the drawing, then draw front and top views of a shaft that can be used to join the two objects. The shaft should be able to fit into any of the four holes.

A. Surface 1 is datum A.
B. Surface 2 is datum B.
C. Surface 3 is perpendicular to surface 2 within 0.02 millimeter.
D. Specify the positional tolerance for the four holes applied at MMC.
E. The centerlines of the holes are to be straight within 0.01 millimeter at MMC.
F. The clearance between the shafts and the holes is to be 0.05 minimum and 0.10 maximum.

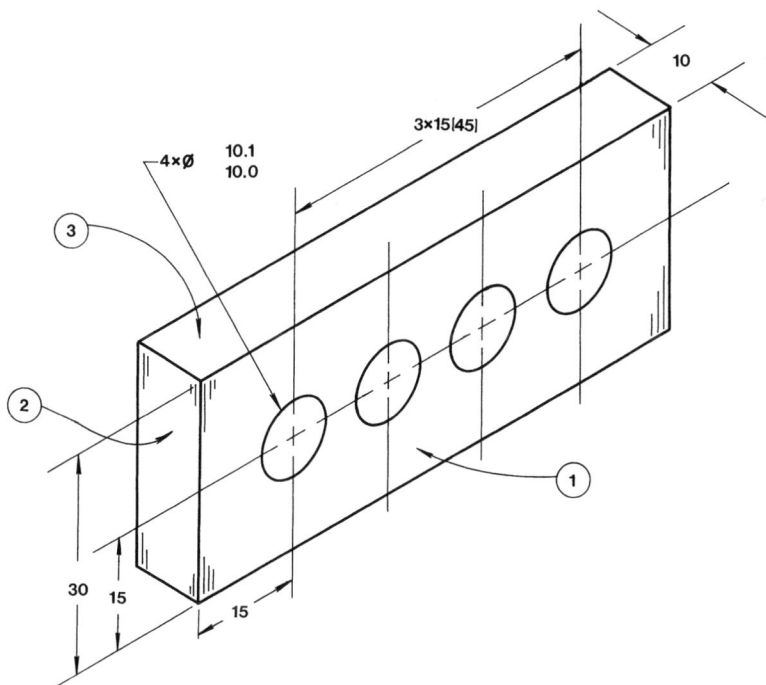

**Figure EX5-45**

## EX5-46

Assume that there are two copies of the part in Figure EX5-46 and that these parts are to be joined together using six fasteners in the floating condition. Draw front and top views of the object, including dimensions and tolerances. Add the following tolerances and specifications to the drawing, then draw front and top views of a shaft that can be used to join the two objects. The shaft should be able to fit into any of the six holes.

A. Surface 1 is datum A.
B. Surface 2 is round within .003.
C. Specify the positional tolerance for the six holes applied at MMC.
D. The clearance between the shafts and the holes is to be .001 minimum and .003 maximum.

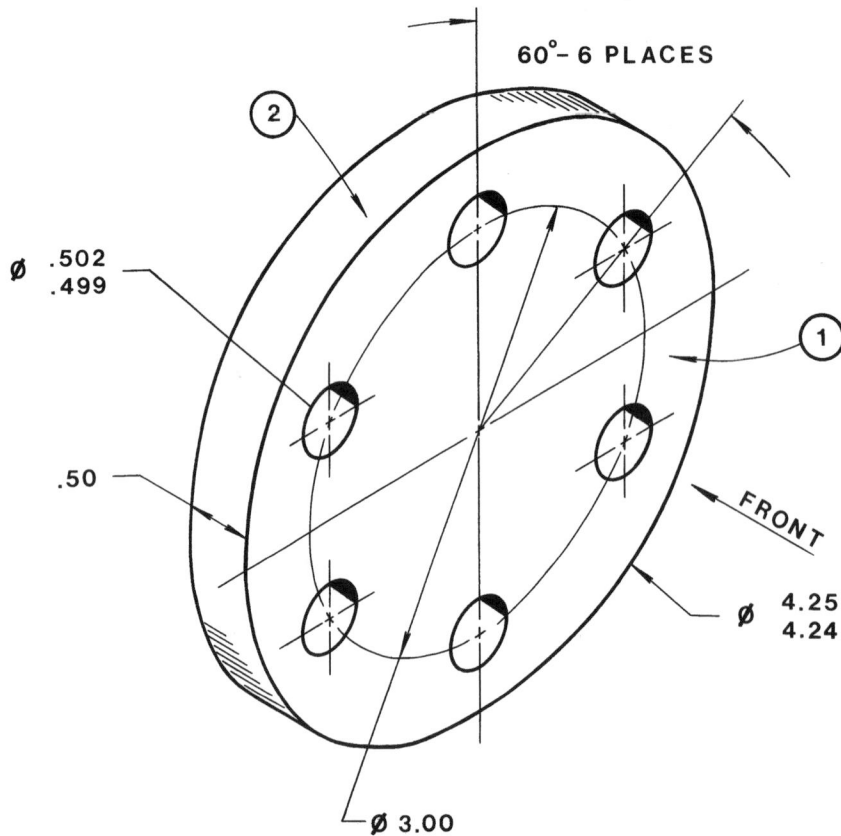

Figure EX5-46

# 5-25 RUNOUT TOLERANCES—RELATED FEATURES USING DATUMS

A runout tolerance is a relationship between surfaces or features, therefore, a datum (or datums) is required.

Runout tolerance may be applied in two different ways using the characteristics shown below.

(CIRCULAR RUNOUT)          (TOTAL RUNOUT)

See following pages for details of application.

# 5-26 RUNOUT (CIRCULAR AND TOTAL)

**Definition.** Runout is the composite deviation from the desired form and orientation of a part surface of revolution during full rotation (360°) of the part on a datum axis.

## Runout Tolerance

Runout tolerance states how far an actual surface or feature is permitted to deviate from the desired form and orientation implied by the drawing during full rotation (360°) of the part on a datum axis.

## Runout Application

Runout tolerancing is a method used to control the composite surface effect of one or more features of a part relative to a datum axis. Runout tolerance is applicable to rotating parts in which this composite surface control is based on the part function and design requirement. A runout tolerance always applies on an RFS basis; namely, size variation has no effect upon the runout tolerance compliance.

Each considered feature must be within its individual runout tolerance when rotated 360° about the datum axis. The tolerance specified for a controlled surface is the total tolerance or full indicator movement (FIM) in terms of common inspection criteria. Former terms, full indicator reading (FIR) and total indicator reading (TIR), have the same meaning as FIM.

As is seen on the following page, the basis of runout tolerance control is the datum axis of the part. Surfaces controlled may be those constructed *around* a datum axis, or those constructed at *right angles* to a datum axis. As is also seen, the datum axis is established from a datum feature.

> **NOTE:**
>
> On runout tolerance, where it is desired to consider a worst case "outer boundary" or "inner boundary," it is determined by the maximum material condition of the feature plus (if outside diameter) or minus (if inside diameter) the runout tolerance.

## Basis of Control of Runout Tolerance

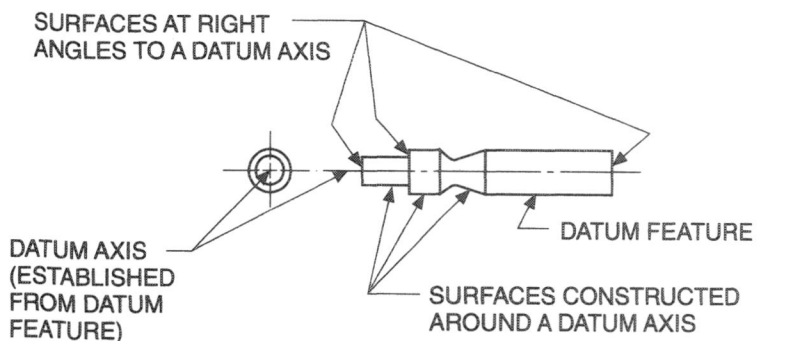

TYPES OF FEATURES APPLICABLE TO A RUNOUT TOLERANCE

⟋ **Circular Runout**

**Example**

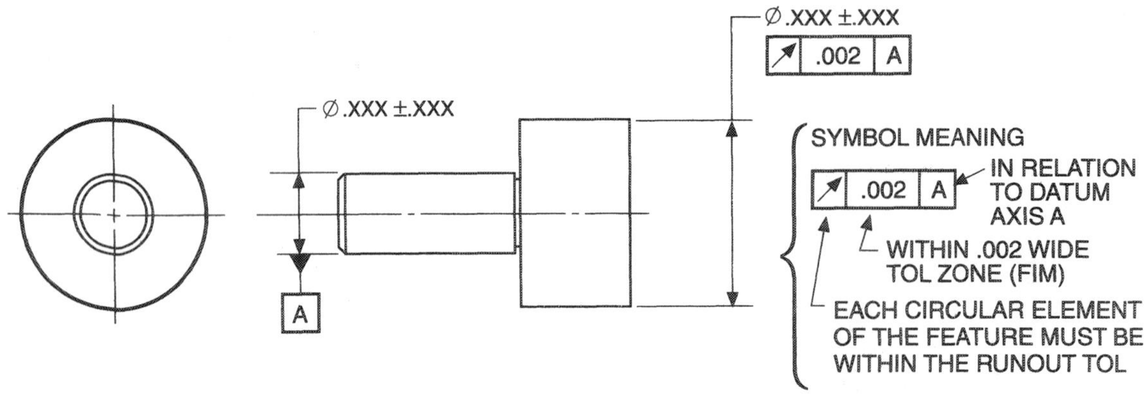

⌀ .XXX ±.XXX

| ⟋ | .002 | A |
|---|------|---|

⌀ .XXX ±.XXX

A

SYMBOL MEANING

| ⟋ | .002 | A |
|---|------|---|

IN RELATION
TO DATUM
AXIS A

WITHIN .002 WIDE
TOL ZONE (FIM)

EACH CIRCULAR ELEMENT
OF THE FEATURE MUST BE
WITHIN THE RUNOUT TOL

**Meaning**

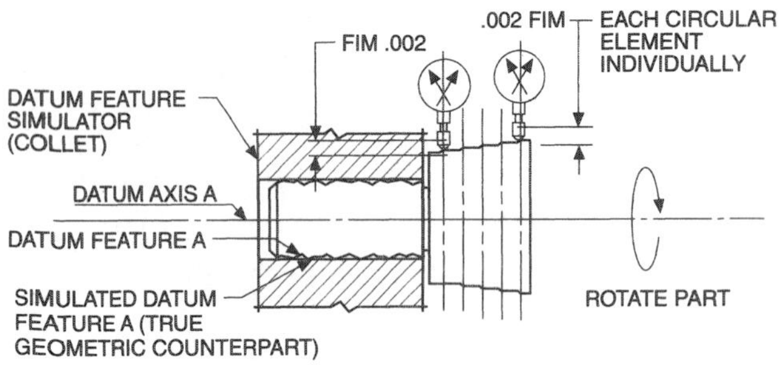

FIM .002

.002 FIM — EACH CIRCULAR
ELEMENT
INDIVIDUALLY

DATUM FEATURE
SIMULATOR
(COLLET)

DATUM AXIS A

DATUM FEATURE A

SIMULATED DATUM
FEATURE A (TRUE
GEOMETRIC COUNTERPART)

ROTATE PART

⟋⟋ **Total Runout**

**Example**

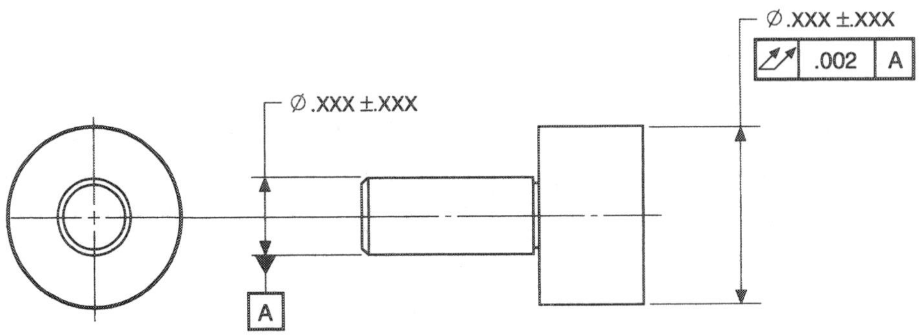

⌀ .XXX ±.XXX

| ⟋⟋ | .002 | A |
|----|------|---|

⌀ .XXX ±.XXX

A

**Meaning**

.002 FIM — ALL ELEMENTS TOGETHER

SYMBOL MEANING

IN RELATION TO DATUM AXIS A

WITHIN .002 WIDE TOL ZONE (FIM)

ALL SURFACE ELEMENTS, TOTAL, ACROSS ENTIRE SURFACE MUST BE WITHIN THE RUNOUT TOL

DATUM FEATURE SIMULATOR (COLLET)

DATUM AXIS A

DATUM FEATURE A

SIMULATED DATUM FEATURE A (TRUE GEOMETRIC COUNTERPART)

ROTATE PART

The datum axis may be established by a single diameter (cylinder) of sufficient length, two diameters with sufficient axial separation, or a diameter and a face surface which is at right angles to it. Features selected as datums should, as much as possible, be functional to the part requirement (e.g., bearing mounting diameters, etc.).

**Types of Runout Control**

The two types of runout control are *circular* runout and *total* runout. Selection of the proper type is based upon the design requirement and manufacturing considerations. The fundamental difference between the two types is illustrated on pages 349-350 and above with detailed examples following in this section. Note that with either method, the collective or composite control of various form and orientation variations of the part provides a more direct representation of part functions, integrates manufacturing operations, and minimizes inspection setup requirements. If necessary to predict a worst case possibility on such parts, an "outer boundary" or "inner boundary" is determined; see definitions.

## 5-27 COAXIAL FEATURES— SELECTION OF PROPER CONTROL

There are four characteristics for controlling interrelated coaxial features:

1. RUNOUT TOLERANCE          (RFS)
   (circular or total)
2. POSITION TOLERANCE        (MMC or RFS)

3. CONCENTRICITY             (RFS)
   TOLERANCE
4. PROFILE OF A SURFACE      (RFS DATUM)

Any of the above methods provides effective control. However, it is important to select the *most appropriate* one to both meet the design requirements and provide the most economical manufacturing conditions. (See also details of preceding and following sections.)

Below are recommendations to assist in selecting the proper control:

If the need is to control only CIRCULAR cross-sectional elements in a composite relationship to the datum axis, RFS, e.g., multidiameters on a shaft, use:

**Circular Runout**  ↗  **Example**   | ↗ | .005 | A – B |

(This method controls any composite error effect of circularity, concentricity, and circular cross-sectional profile variations.)

If the need is to control the TOTAL cylindrical or profile surface in composite relative to the datum axis, RFS, e.g., multi-diameters on a shaft, bearing mounting diameters, etc., use:

**Total Runout**  ⤢  **Example**   | ⤢ | .005 | A – B |

(This method controls any composite error effect of circularity, cylindricity, straightness, coaxiality, angularity, and parallelism.)

If the need is to control the total cylindrical or profile surface and its actual mating envelope axis relative to the datum axis on an MMC or RFS basis, e.g., on mating parts to assure inter-changeability or assemblability, use:

> **NOTE:**
>
> Runout is always implied as an RFS application. It cannot be applied on an MMC basis, since an MMC situation involves functional interchangeability or assemblability (probably of mating parts), in which case POSITION tolerance would be used. See below.

**Total Runout**   ⊕   **(IF MMC)**   **Example**   | ⊕ | ⌀.005 Ⓜ | A Ⓜ |

**(IF RFS)**   **Example**   | ⊕ | ⌀ .005 | A |

**OR RFS Datum**   | ⊕ | ⌀ .005 Ⓜ | A |

If the need is to control the *axis* of one or more features in composite relative to a *datum axis*, RFS, e.g., to control such as balance of a rotating part, use:

**Concentricity**   ◎   **Example**   | ◎ | ⌀.005 | A – B |

> **NOTE:**
>
> Concentricity is always implied as an RFS application. Variations in size (departure from MMC size, out-of-circularity, out-of-cylindricity, etc.) do not in themselves conclude *axis* error.

If the need is to control the total cylindrical or profile surface simultaneously with the size dimension(s) (using basic dimensions for both), relative to a datum axis, e.g., precise fit, multi-diameters, etc., use:

**Profile of a Surface**   ⌓   **Example**   | ⌓ | .005 | A |

## Part Mounted on Functional Diameter (Datum)

Circular runout provides a composite control of circular elements of a surface. Circular runout is normally a less complex requirement than total runout. The tolerance is applied independently at any circular cross section or measuring position on the part as it rotates through 360°. Circular runout should be considered when the part function and manufacturing requirements are satisfied by this type of control. Where more complete control of all elements in composite is necessary, total runout should be considered.

Circular runout controls composite variations of circularity and cross-sectional size and form variations of the surface at each circular element where applied to surfaces constructed around a datum axis. It controls circular elements of the surface (wobble) where applied to surfaces constructed at right angles (perpendicular) to a datum axis.

The example below utilizes circular runout control. It applies circular runout to the angular surface controlling the individual circular elements of the part as it rotates.

When circular runout is to be applied at specified locations, it must be so stated on the drawing. The example on the next page illustrates this application.

Please see further pages and examples in this section for further applications of and considerations on circular runout and for circular runout and total runout use on the same part.

**Example**

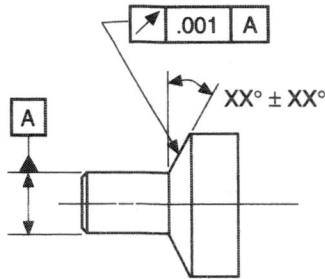

SYMBOL MEANING

┌─────┬──────┬───┐
│ ↗  │ .001 │ A │ ◄── IN RELATION TO
└─────┴──────┴───┘         DATUM AXIS A

WITHIN .001 WIDE
TOL ZONE (FIM)

EACH CIRCULAR ELEMENT
MUST BE WITHIN THE
RUNOUT TOLERANCE

**Meaning**

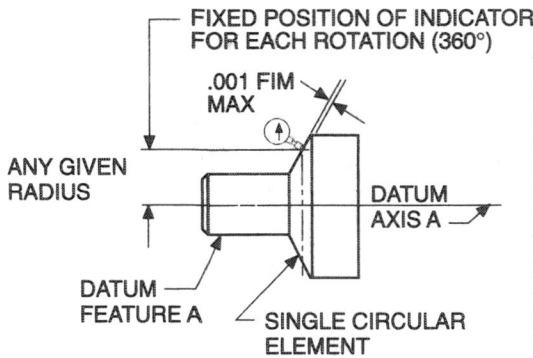

THE FEATURE MUST BE WITHIN THE SPECIFIED TOLERANCE OF SIZE AT ANY MEASURING POSITION. EACH CIRCULAR ELEMENT OF THE SURFACE MUST BE WITHIN .001 FULL INDICATOR MOVEMENT WHEN THE PART IS ROTATED ONE FULL ROTATION ABOUT THE SPECIFIED DATUM AXIS WITH THE INDICATOR FIXED IN A POSITION NORMAL TO THE SURFACE. (THIS DOES NOT CONTROL FORM OF THE TOTAL SPECIFIED SURFACE AREA, BUT ONLY CONTROLS THE RUNOUT OF EACH CIRCULAR ELEMENT.)

**Example**

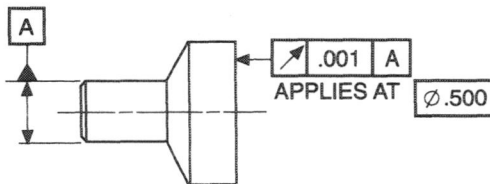

SYMBOL MEANING

┌─────┬──────┬───┐
│ ↗  │ .001 │ A │ ◄── IN RELATION TO
└─────┴──────┴───┘         DATUM AXIS A

WITHIN .001 WIDE
TOL ZONE (FIM)

THE CIRCULAR ELEMENT
AT ⌀.500 MUST BE WITHIN
THE RUNOUT TOLERANCE

**Meaning**

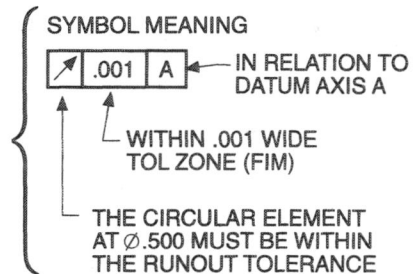

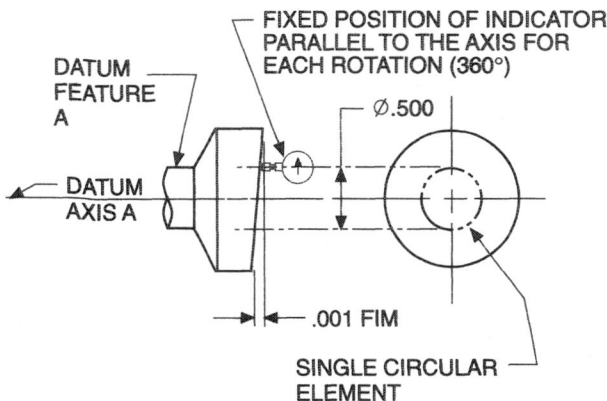

THE CIRCULAR ELEMENT OF THE SURFACE AT ⌀.500 MUST BE WITHIN .001 FULL INDICATOR MOVEMENT WHEN THE PART IS ROTATED ONE FULL ROTATION ABOUT THE SPECIFIED DATUM AXIS WITH THE INDICATOR FIXED IN A POSITION PARALLEL TO THE AXIS. (THIS DOES NOT CONTROL PERPENDICULARITY, BUT CONTROLS ONLY THE LATERAL RUNOUT (WOBBLE) OF EACH CIRCULAR ELEMENT AT THE SPECIFIED SURFACE LOCATION.)

## Part Mounted on Two Functional Diameters (Datums)

Where multiple diameters of a rotating part are to be controlled relative to a datum axis, the datum axis can be established by the features which will provide the functional mounting of the part at assembly. These are called multiple datums.

Two diameters (cylinders) are used on the part to establish the datum axis of relationship. These diameters could represent the bearing mounting features and thus they establish the datum axis of rotation. Note that two datum features are selected (C and D). By stating the two datums simultaneously with a dash line between them, the relationship of all of the other features so designated relate to their common datum axis C-D. The precedence of datums C and D are equal in this case.

Any two surfaces given runout tolerances about the datum axis are to be individually within their stated runout tolerance; collectively they are related to each other within the sum of their indicator readings.

A runout tolerance specified for the datum feature has no effect on the considered features related to it.

### Part Mounted on Two Functional Diameters (Datums)

**Example**

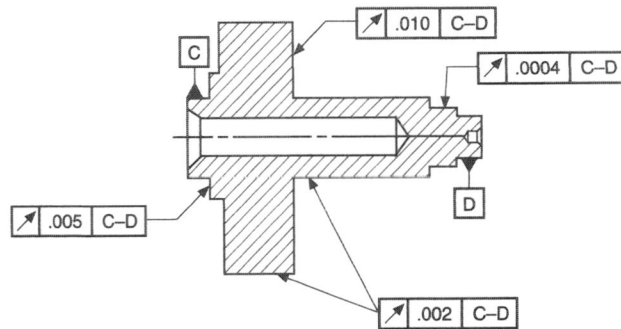

(SIZE DIMENSIONS & TOLERANCES NOT SHOWN FOR SIMPLICITY)

**Meaning**

WHEN MOUNTED ON DATUMS C AND D.
DESIGNATED SURFACES MUST BE WITHIN CIRCULAR
RUNOUT ( ↗ ) TOLERANCE SPECIFIED, RFS

# 5-28  CIRCULAR RUNOUT ↗ AND TOTAL RUNOUT ↗↗

## Part Mounted on Two Functional Diameters (Datums) Including Runout Tolerance on Datums

The part shown extends the principles of the previous example. It is a shaft of multi-diameters about a common datum axis C–D with each feature, *including the datum features*, stating an individual runout tolerance. It utilizes both circular runout and total runout control.

Total runout controls composite surface variations of circularity, cylindricity, parallelism, straightness, angularity, taper, and profile of a surface where applied to features constructed around a datum axis. Total runout applied to surfaces constructed perpendicular to a datum axis controls composite variations of perpendicularity and flatness. The tolerance is applied *simultaneously* at all circular and profile measuring positions as the part is rotated through 360°.

Due to the design requirements involved in this example, certain diameters (the datums) must be given total runout control, whereas the remaining diameters and face surfaces may be controlled with circular runout.

Since in the end assembly of this part, (mount to bearings), the relationship to both datums simultaneously is the result desired, all the features including each datum (C and D) must meet their individual runout tolerances. Note that the runout tolerance of each datum (C and D) individually is relative to their common axis C–D.

Multiple leaders may be used in controlling two or more features with a common runout tolerance as is shown. Runout tolerance may be specified individually or in groups, as is convenient, without affecting the runout tolerance.

## Part Mounted on Two Functional Diameters

**Example**

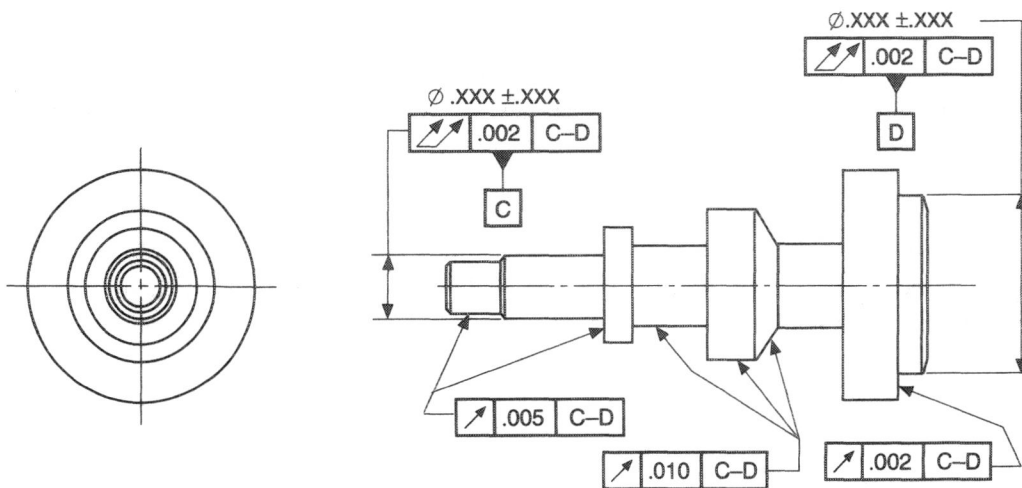

**Meaning**

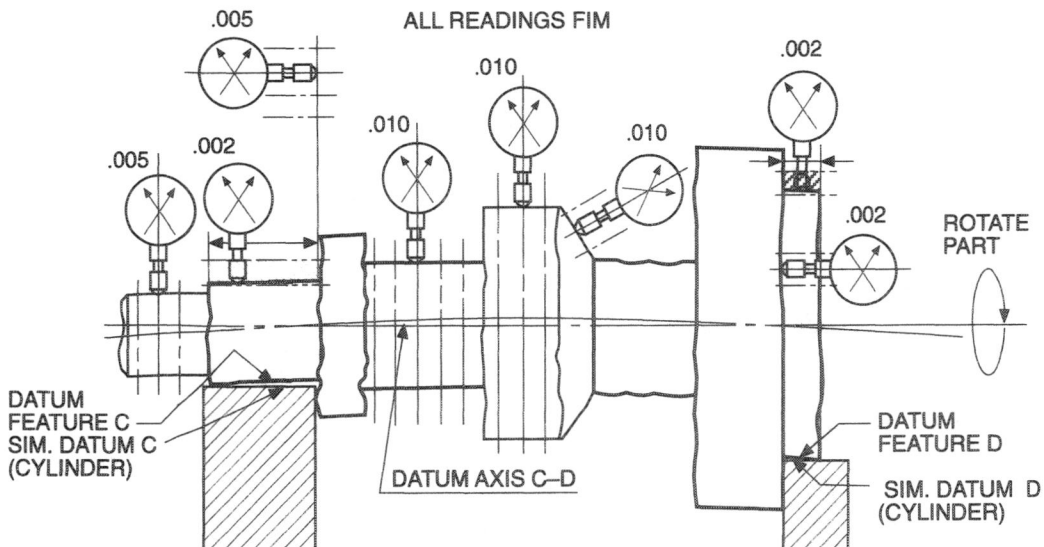

—EACH CIRCULAR ELEMENT INDIVIDUALLY—
—DATUM DIAMETERS C AND D ALL ELEMENTS TOGETHER (TOTAL)—

## Part Mounted on Two Functional Diameters (Datums) including Runout and Cylindricity Tolerance on Datum

The part shown further extends the principles of the previous examples. It is a shaft of multi-diameters about a common datum axis C–D, with each feature including the datum features stating an individual runout tolerance. In addition to having both circular and total runout control, each of the datum features has a cylindricity tolerance.

It should be noted that the datum features of this part have controls of three orders of magnitude: size, total runout,

and cylindricity (form). In this example, an accurate fit to bearings as controlled by size and form was required in addition to the runout relationship of the datum features to the datum axis (C–D) of rotation.

Features with a specific relationship to another feature rather than to the common datum axis may be so indicated by addition of appropriate datum references (such as datum E on page at right). Note that the inside diameter on the right end of the part relates the runout requirement to datum E.

## Part Mounted on Two Functional Diameters (Datums)

**Example**

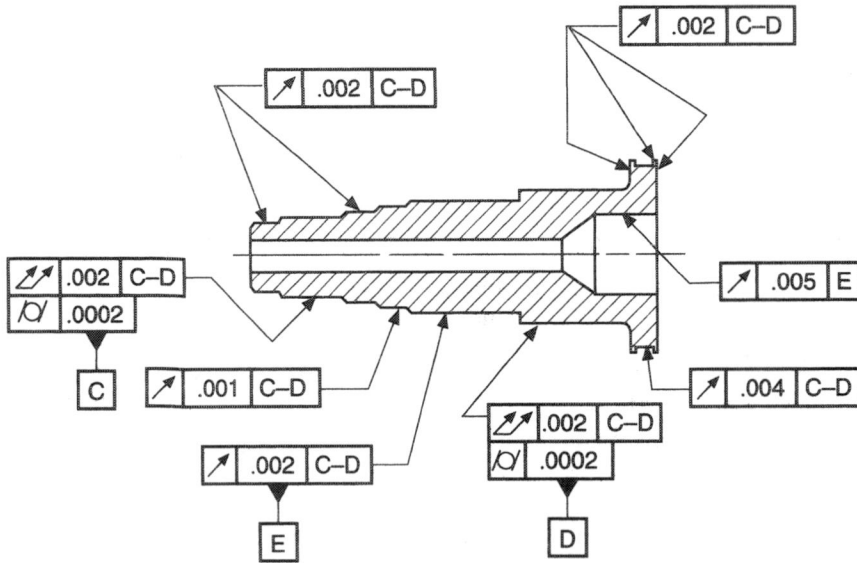

(SIZE DIMENSIONS & TOLERANCES NOT SHOWN FOR SIMPLICITY)

**Meaning**

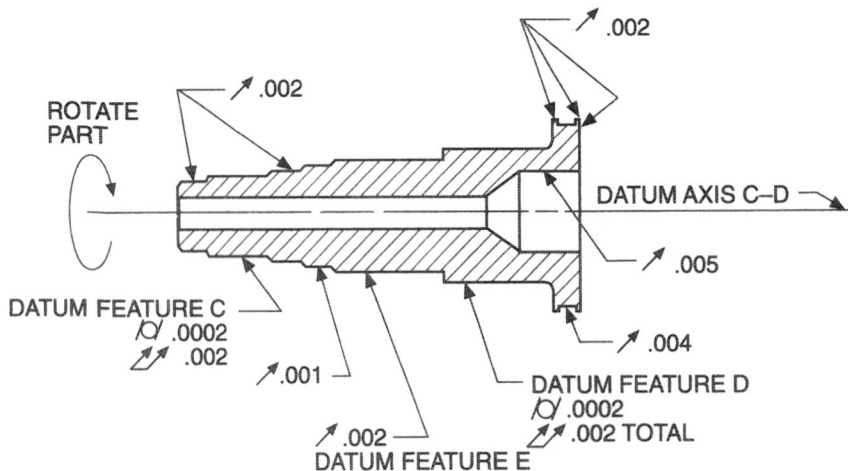

ALL VALUES FIM (TIR, FIR)

WHEN MOUNTED ON DATUMS C AND D, DESIGNATED SURFACES MUST BE WITHIN CIRCULAR RUNOUT (↗) TOLERANCE SPECIFIED. DATUMS C AND D MUST ALSO BE WITHIN TOTAL RUNOUT (↗↗) TOLERANCE SPECIFIED AND CYLINDRICAL WITHIN .0002.

WHEN MOUNTED ON DATUM E, DESIGNATED SURFACE MUST BE WITHIN CIRCULAR RUNOUT (↗) TOLERANCE SPECIFIED.

## Part Mounted on Functional Face Surface (Datum) and Diameter (Datum)

Runout tolerancing may be applied to features of rotation where the feature datum references are an axis and a face surface perpendicular to the axis. In such an application, datum precedence is usually considered necessary. The influence of the appropriate feature as a primary datum is determined relative to the design requirement.

In the example, the face surface perpendicular to the axis of rotation is considered the functionally important surface (bottoms on the inner race of bearing) and is thus selected as the primary datum A. Secondary datum diameter (cylinder) B is in contact with the extremities of datum feature B. With the part in this orientation and located functionally on its datums, all related features must meet their individual runout tolerances when the part is rotated about datum axis B.

Note that the primary datum A and secondary datum B features are indicated with that precedence by separate enclosures in the feature control frame (reading left to right).

## Part Mounted on Functional Face Surface (Datum) and Diameter (Datum)

**Example**

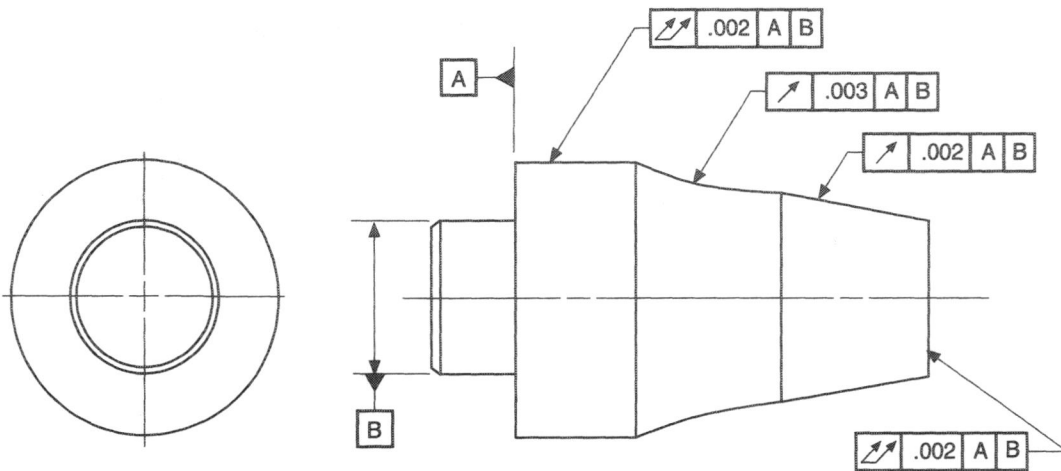

**Meaning**

# 5-29  CIRCULAR RUNOUT ↗ AND FLATNESS ▱

### Part Mounted on Functional Face Surface (Datum) and Diameter (Datum)

The part shown extends the principles of runout tolerancing. It illustrates a situation in which the large flat mounting surface requires flatness control to assure desired accuracy of the primary datum surface as it mounts and orients the remainder of the part to proper functional relationships. In this instance it can be assumed that this method was most representative of part end function.

The remainder of the part relates to the datum reference frame established by datum features C and D, with C being the primary datum for attitude of the part, and D the secondary datum for establishing axis of rotation. Note again that the other features *including datum D* relate to the datum system C and D.

### Part Mounted on Large Flat Surface (Datum) and Diameter (Datum)

**Example**

**Meaning**

WHEN MOUNTED ON DATUMS C AND D, DESIGNATED SURFACES MUST BE WITHIN CIRCULAR RUNOUT ( ↗ ) TOLERANCE SPECIFIED.

## 5-30  TOTAL RUNOUT ⟋⟋

### Runout Application and Analysis

The following example illustrates a functional design requirement where the part mounts upon live centers in assembly. Suppose components such as a gear, pulley, clutch, and brake mount upon the four diameters shown. The datum axis A–B is established for the runout relationships. The size and runout tolerance are determined by the calculated values permissible as based upon the design requirement. Another possibility might be where the datums A and B are used as "temporary datums" so that the Ø.449 and Ø.351 features could be established as shown with these features then specified as datums (e.g. C and D) for the Ø.940 and Ø.7450 relationship to the new datums axis "C–D."

Typical total runout applications, inspection analysis, and meaning are discussed below and in the following text. An understanding of the fundamentals involved in these examples will provide the reader with the basis for total runout characteristic and use. The methods and principles shown may be readily adapted to circular runout as well, except for the substitution, where appropriate in the text and illustrations, of the circular elements (only) principles. In checking total runout, the datum feature, or features, is normally centered or positioned in an appropriate inspection device (e.g., between centers, collet, chuck, mandrel, vee block, or other centering device) to establish the datum axis from which the total runout relationships are to occur.

### Part Mounted on Machining Centers

**Example**

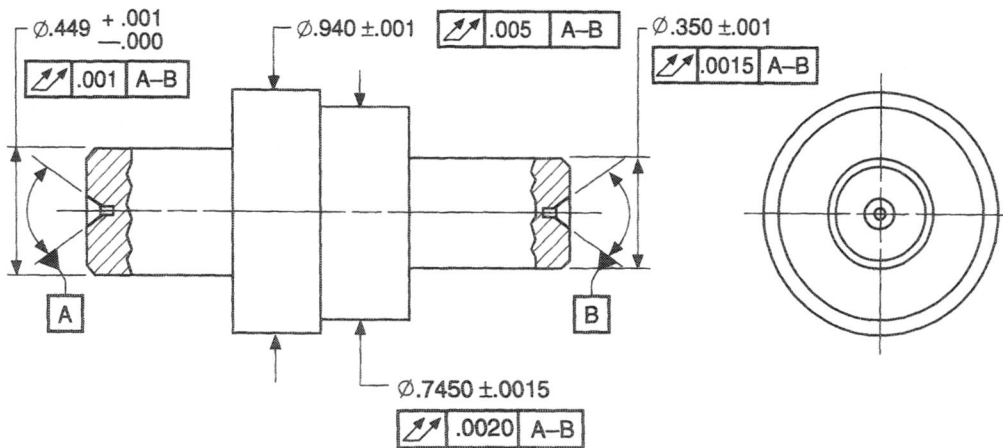

**Meaning**

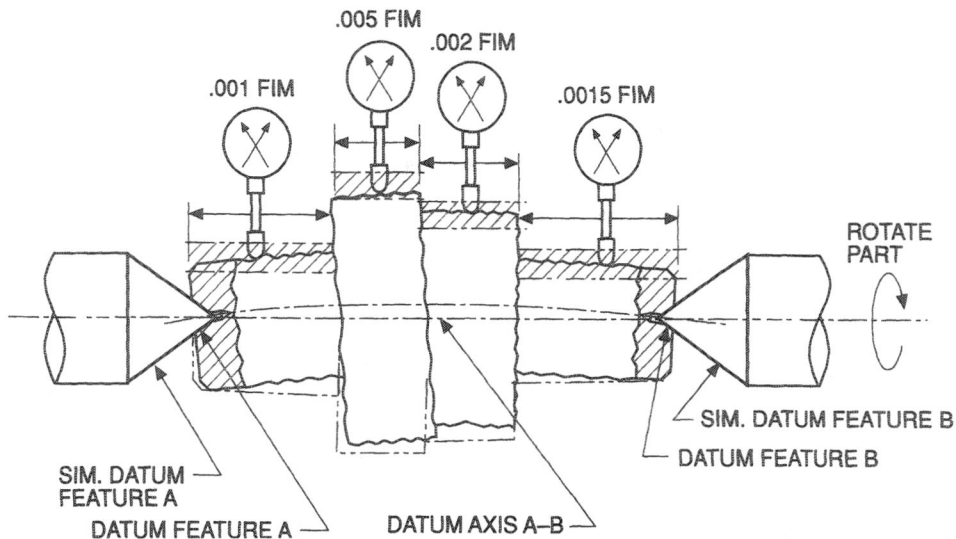

## Part Mounted on Two Functional Diameters

**Example**

Ø.605 ±.003

[⟋⟋ | .002 | A–B]

[⟋⟋ | .002 | A–B]

[⟋⟋ | .001 | A–B]

[B]

Ø.750 ±.001

[A]

Ø.375 ±.002

[⟋⟋ | .001 | A–B]

[R]

Ø.8800 ±.0005

[⟋⟋ | .001 | A–B]

Ø.500 ±.001

[⟋⟋ | .0005 | R]

Ø1.200 ±.003

Ø.810 ±.001

**Meaning 1**

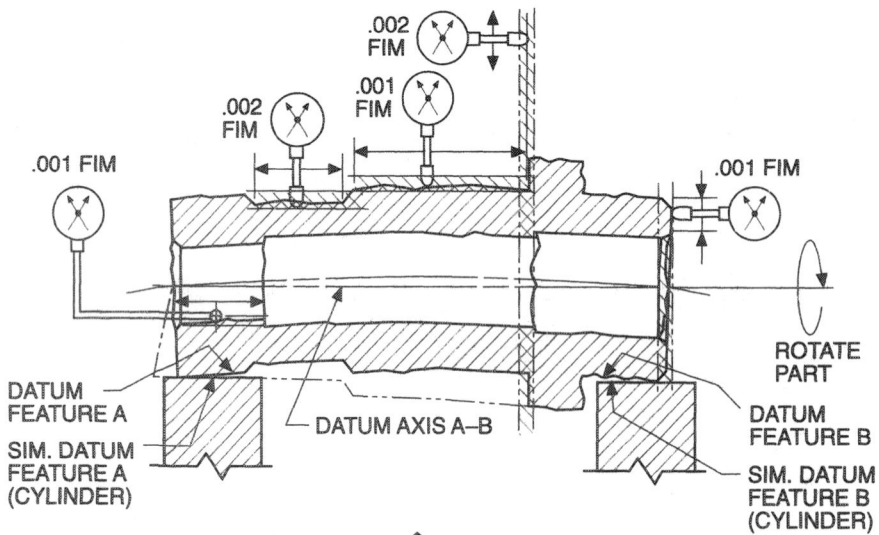

.002 FIM

.002 FIM

.001 FIM

.001 FIM

.001 FIM

ROTATE PART

DATUM FEATURE A

DATUM AXIS A–B

DATUM FEATURE B

SIM. DATUM FEATURE A (CYLINDER)

SIM. DATUM FEATURE B (CYLINDER)

**Meaning 2**

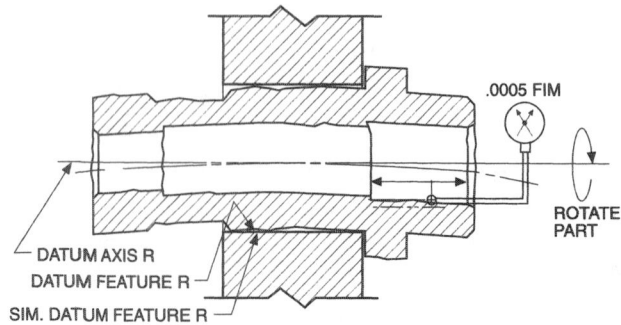

.0005 FIM

ROTATE PART

DATUM AXIS R

DATUM FEATURE R

SIM. DATUM FEATURE R

## Total Surface Runout (Wobble)

Note that when flat or face surface runout (wobble) is part of the specification as in the case of the example on the preceding page, the runout analysis must consider the need for restricting end movement of the part while checking it. If collets or other holding devices are applied to establish the datum axis, end movement of the part is controlled. Also, specified secondary or tertiary datums (specified faces, shoulders, etc.) may indicate the end location or "stop" surface. However, in the absence of the above conditions and when the flat or face surface runout is to be checked, one should, if possible, "stop" on the surface being checked (see Meaning 1 on the preceding page and continued below), in order to avoid adding the error of *another* surface to the reading. When this is done, however, dependent on the location or placement of the stop, the dial indicator runout reading may need to be *halved* (divided by 2) before comparison with the stated part drawing runout tolerance to determine whether the requirement has been met. Referring to the illustration below, note that to use the fixed factor 2 the stop should be as close as possible to the O.D. of the surface being checked. As the placement of the stop approaches the axis of the mounting datum diameters (see also Meaning 2 and 3), the factor approaches a 1:1 ratio.

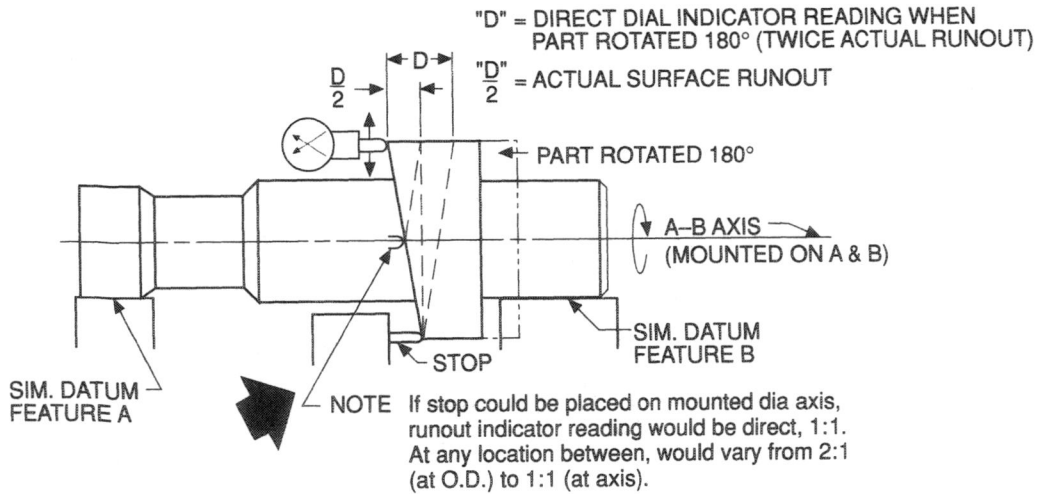

"D" = DIRECT DIAL INDICATOR READING WHEN PART ROTATED 180° (TWICE ACTUAL RUNOUT)

"$\frac{D}{2}$" = ACTUAL SURFACE RUNOUT

PART ROTATED 180°

A–B AXIS (MOUNTED ON A & B)

SIM. DATUM FEATURE B

STOP

SIM. DATUM FEATURE A

NOTE   If stop could be placed on mounted dia axis, runout indicator reading would be direct, 1:1. At any location between, would vary from 2:1 (at O.D.) to 1:1 (at axis).

## Meaning 2

As an alternative method a full end stop may be used. The end stop must, however, be extremely accurate at 90° (exact within gage tolerance) to the axis of the mounting diameters. See below.

In this case the dial indicator reading taken is direct and can be compared immediately with the stated part drawing runout (wobble) tolerance.

Note also that no error of the part surface or face which is used for "stopping" is introduced into the indicator reading, since the high point or extremity of this surface remains in consistent contact with the full stop gage as the part is rotated.

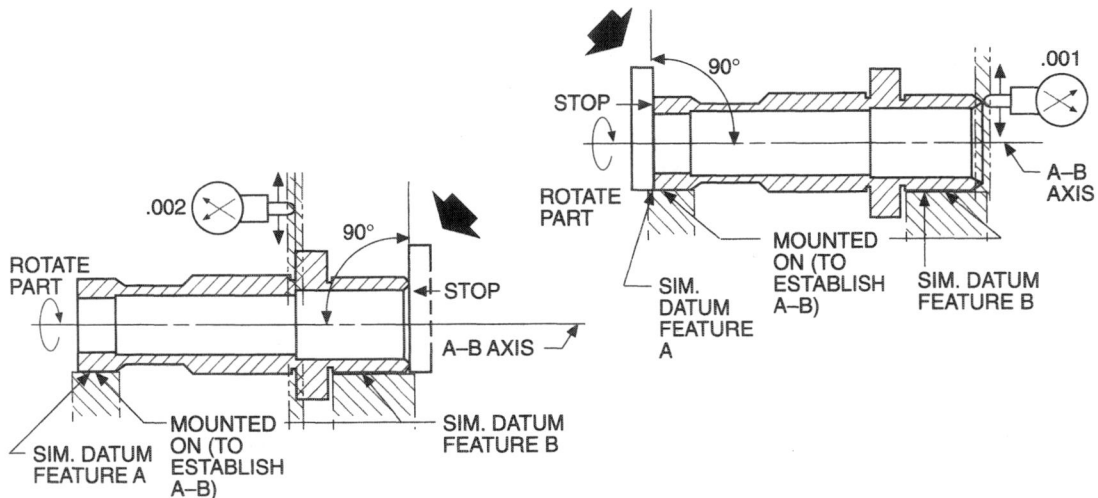

.002

90°

ROTATE PART

STOP

A–B AXIS

SIM. DATUM FEATURE A

MOUNTED ON (TO ESTABLISH A–B)

SIM. DATUM FEATURE B

90°

.001

STOP

A–B AXIS

ROTATE PART

SIM. DATUM FEATURE A

MOUNTED ON (TO ESTABLISH A–B)

SIM. DATUM FEATURE B

## Meaning 3

If the part has a *closed end* and a stop can be placed at the *axis* of the mounting datum diameters (see illustration below), the runout requirement may be checked without regard for extreme accuracy of the stop. The dial indicator reading taken is direct and can be compared immediately with the stated part drawing runout (wobble) tolerance.

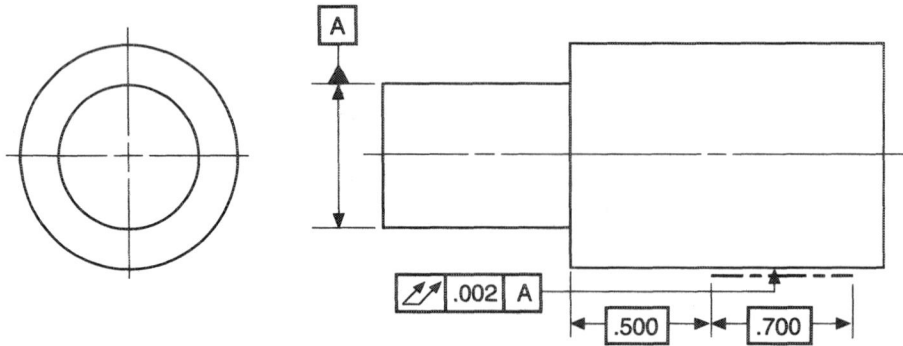

# 5-31 RUNOUT ↗↗
# PARTIAL SURFACE*

Where it is desirable to indicate that a runout tolerance is applicable only at a specific place or portion of a surface feature, it may be shown as indicated in the example below. Basic (exact) dimensions are shown indicating the portion of the surface to which the runout tolerance applies.

A thick chain line is shown slightly off the part with the leader from the feature control frame extended to the concerned surface. The thick chain line means that the stated tolerance applies only at the designated area. The basic dimensions invoke standard inspection accuracy in verifying the requirements. Otherwise, standard interpretation is given to the total runout requirement shown (see previous pages).

**Example**

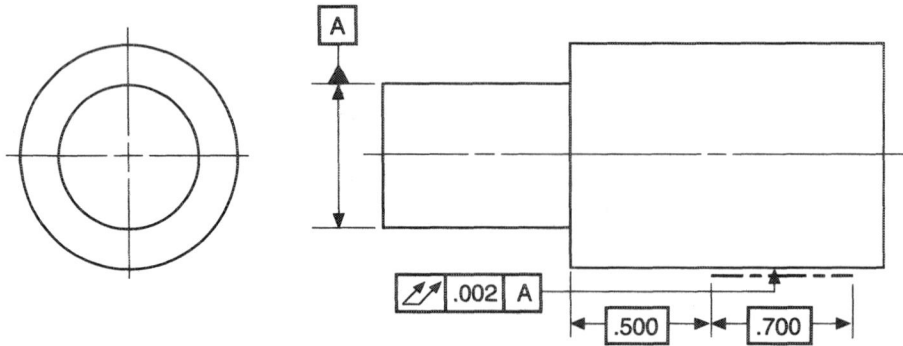

**Meaning**

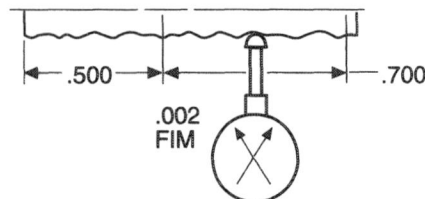

---

*This principle may be applied to other geometric controls as well.

# CHAPTER 6

# Threads, Fasteners, Springs, and GD&T

## 6-1 INTRODUCTION

This chapter explains how to draw threads, washers, keys, and springs. It explains how to use fasteners to join parts together and design uses for washers, keys, and springs.

Throughout the chapter it will be suggested that blocks and wblocks be created of the various thread and fastener shapes. Thread representations, fastener head shapes, setscrews, and both internal and external thread representations for orthographic views and sectional views are so common in technical drawings that it is good practice to create a set of wblocks that can be used on future drawings to prevent having to redraw a thread shape every time it is needed.

## 6-2 THREAD TERMINOLOGY

Figure 6-1 shows a thread. The peak of a thread is called the *crest*, and the valley portion is called the *root*. The *major diameter* of a thread is the distance across the thread from crest to crest. The *minor diameter* is the distance across the thread from root to root.

The *pitch* of a thread is the linear distance along the thread from crest to crest. Thread pitch is usually referred to in terms of a unit of length such as 20 threads per inch or 1.5 threads per millimeter.

Detailed representation

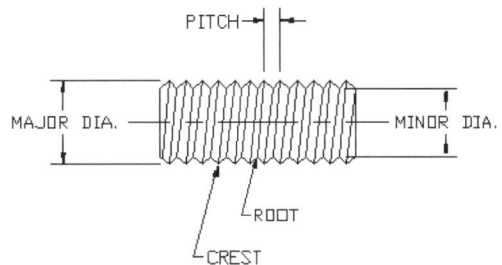

Figure 6-1

## 6-3 THREAD CALLOUTS (METRIC UNITS)

Threads are specified on a drawing using drawing callouts. See Figure 6-2. The M preceding a drawing callout specifies that the callout is for a metric thread. Holes that are not threaded use the Ø symbol.

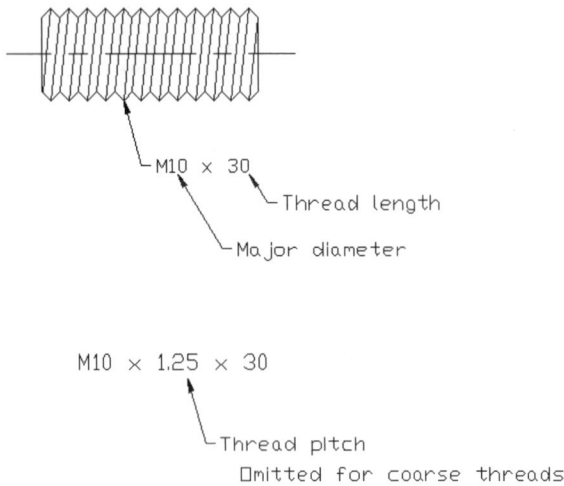

M10 × 30

Thread length

Major diameter

M10 × 1.25 × 30

Thread pitch
Omitted for coarse threads

**Figure 6-2**

| Major Dia | Coarse | | Fine | |
|---|---|---|---|---|
| | Pitch | Tap Drill Dia | Pitch | Tap Drill Dia |
| 1.6 | 0.35 | 1.25 | | |
| 2 | 0.4 | 1.6 | | |
| 2.5 | 0.45 | 2.05 | | |
| 3 | 0.5 | 2.5 | | |
| 4 | .7 | 3.3 | | |
| 5 | 0.8 | 4.2 | | |
| 6 | 1 | 5.0 | | |
| 8 | 1.25 | 6.7 | 1 | 7.0 |
| 10 | 1.5 | 8.5 | 1.25 | 8.7 |
| 12 | 1.75 | 10.2 | 1.25 | 10.8 |
| 16 | 2 | 14 | 1.5 | 14.5 |
| 20 | 2.5 | 17.5 | 1.5 | 18.5 |
| 24 | 3 | 21 | 2 | 22 |
| 30 | 3.5 | 26.5 | 2 | 28 |
| 36 | 4 | 32 | 3 | 33 |
| 42 | 4.5 | 37.5 | 3 | 39 |
| 48 | 5 | 43 | 3 | 45 |

**Figure 6-3**

The number following the M is the major diameter of the thread; for example, an M10 thread has a major diameter of 10 millimeters. The pitch of a metric thread is assumed to be a coarse thread unless otherwise stated. The callout M10 × 30 assumes a coarse thread, or 1.5 threads per millimeter. The number 30 is the thread length in millimeters. The "×" is read as "by," so the thread is called a "ten by thirty."

The callout M10 × 1.25 × 30 specifies a pitch of 1.25 threads per millimeter. This is not a standard coarse thread size, so the pitch must be specified.

Figure 6-3 shows a list of preferred thread sizes. These sizes are similar to the standard sizes shown in Figure 4-121. A list of other metric thread sizes is included in the appendix.

Whenever possible use preferred thread sizes for designing. Preferred thread sizes are readily available and are usually cheaper than nonstandard sizes. In addition, tooling such as wrenches is also readily available for preferred sizes.

## 6-4  THREAD CALLOUTS (ENGLISH UNITS)

English unit threads always include a thread form specification. Thread form specifications are designated by capital letters, as shown in Figure 6-4, and are defined as follows:

UNC—Unified National Coarse
UNF—Unified National Fine
UNEF—Unified National Extra Fine
UN—Unified National, or constant pitch threads

An English unit thread callout starts by defining the major diameter of the thread followed by the pitch specification. The callout .500 –13 UNC means a thread whose major diameter is .500 inch with 13 threads per inch and is manufactured to the Unified National Coarse standards.

There are three possible classes of fit for a thread: 1, 2, and 3. The different class specifications specify a set of manufacturing tolerances. A class 1 thread is the loosest and a class 3, the most exact. A class 2 fit is the most common.

The letter A designates an external thread, B an internal thread. The symbol × means "by" as in 2 × 4, "two by four." The thread length (3.00) may be followed by the word LONG to prevent confusion about which value represents the length.

Drawing callouts for English unit threads are sometimes shortened, as in Figure 6-4. The callout .500–13UNC–2A × 3.00 LONG is shortened to .500–13 × 3.00. Only a coarse thread has 13 threads per inch, and it should be obvious whether a thread is internal or external, so these specifications may be dropped. Most threads are class 2, so it is tacitly accepted that all threads are class 2 unless otherwise specified. The shortened callout form is not universally accepted. When in doubt, use a complete thread callout.

A partial list of standard English unit threads is shown in Figure 6-5. A more complete list is included in the appendix. Some of the drill sizes listed use numbers and letters. The decimal equivalents to the numbers and letters are listed in the appendix.

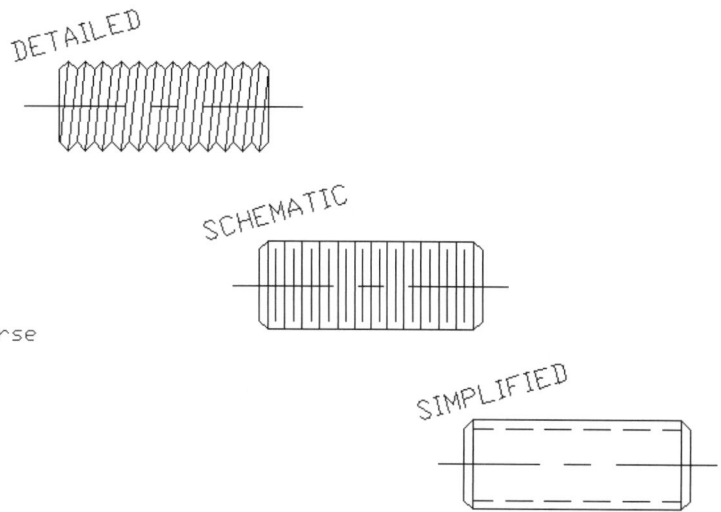

Class of fit
Optional
.500-13UNC-2A×3.00 LONG
Thread length
External thread
Thread form
Unified National Coarse
Threads per inch
Major diameter

SHORTENED VERSION
.500-13×3.00

**Figure 6-4**

DETAILED

SCHEMATIC

SIMPLIFIED

**Figure 6-6**

## 6-5 THREAD REPRESENTATIONS

There are three ways to graphically represent threads on a drawing: detailed, schematic, and simplified. Figure 6-6 shows the three representations.

Detailed representations look the most like actual threads but are time-consuming to draw. Creating a wblock of a detailed shape will help eliminate this time constraint.

Schematic and simplified thread representations are created using a series of straight lines. The simplified repre-

sentation uses only two hidden lines and can be mistaken for an internal hole if it is not accompanied by a thread specification callout. The choice of which representation to use depends on individual preferences. The resulting drawing should be clear and easy to understand. All three representations may be used on the same drawing, but in general, only very large threads (those over 1.00 in., or 25 mm) are drawn using the detailed representation.

Ideally, thread representations should be drawn with each thread equal to the actual pitch size. This is not practical

| Major Dia | Decimal | UNC Thread/in | UNC Tap drill Dia. | UNF Thread/in | UNF Tap drill Dia. | UNEF Thread/in | UNEF Tap drill Dia. |
|---|---|---|---|---|---|---|---|
| #6 | .138 | 40 | #38 | 44 | #37 | | |
| #8 | .164 | 32 | #29 | 36 | #29 | | |
| #10 | .190 | 24 | #25 | 32 | #21 | | |
| 1/4 | .250 | 20 | 7 | 28 | 3 | 32 | .219 |
| 5/16 | .312 | 18 | F | 24 | 1 | 32 | .281 |
| 3/8 | .375 | 16 | .312 | 24 | Q | 32 | .344 |
| 7/16 | .438 | 14 | U | 20 | .391 | 28 | Y |
| 1/2 | .500 | 13 | .422 | 20 | .453 | 28 | .469 |
| 9/16 | .562 | 12 | .484 | 18 | .516 | 24 | .516 |
| 5/8 | .625 | 11 | .531 | 18 | .578 | 24 | .578 |
| 3/4 | .750 | 10 | .656 | 16 | .688 | 20 | .703 |
| 7/8 | .875 | 9 | .766 | 14 | .812 | 20 | .828 |
| 1 | 1.000 | 8 | .875 | 12 | .922 | 20 | .953 |
| 1 1/4 | 1.250 | 7 | 1.109 | 12 | 1.172 | 18 | 1.188 |
| 1 1/2 | 1.500 | 6 | 1.344 | 12 | 1.422 | 18 | 1.438 |

UNC = Unified National Coarse

UNF = Unified National Fine

UNEF = Unified National Extra Fine

**Figure 6-5**

for smaller threads and not necessary for larger ones. Thread representations are not meant to be exact duplications of the threads but representations, so convenient drawing distances are acceptable.

**To draw a detailed thread representation**

Draw a detailed thread representation for a 1.00-inch-diameter thread that is 3.00 inches long. See Figure 6-7.

1. Set **GRID= .5** and **Snap= .125.**
2. Draw a **4.00-inch** centerline near the center of the screen.
3. Zoom the area around the centerline.
4. Draw a zigzag pattern **.375** above the centerline using the .125 snap points. Start the zigzag line **.50** from the left end of the centerline.
5. Access the **Array** command.

    The **Array** dialog box will appear.

6. Select the **Rectangular** option.
7. Click the **Select objects** button, then select the zigzag pattern.
8. Set the **Row value** for **1** and the **Column value** for **12.**
9. Set the **Row offset** for **0.0000.**
10. Set the **Column offset** for **0.25,** then select the **OK** button.
11. Mirror the arrayed zigzag line about the centerline.
12. Draw vertical lines at both ends of the thread and two slanted lines between the thread's roots and crests as shown.
13. Array both slanted lines using the same array parameters used for the zigzag line: **12** columns **.25** apart.
14. Save the thread representation as a block and wblock named **DETLIN.** Define the insertion point as shown.

Another technique for drawing a detailed thread representation is to draw a single thread completely and then use the **Copy** or **Array** command to generate as many additional threads as is necessary.

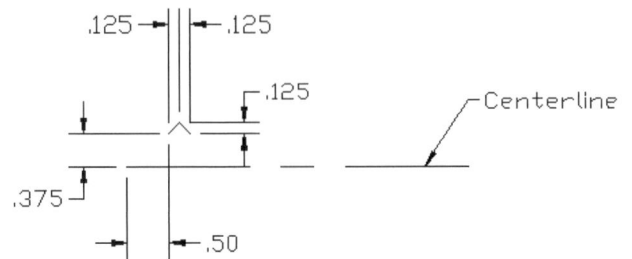

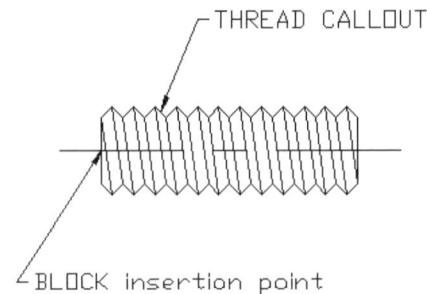

Figure 6-7

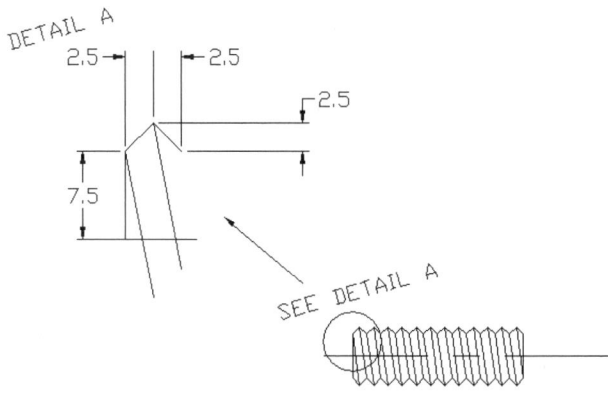

Figure 6-8

It is recommended that you save all thread wblocks on a separate disk. This disk will become a reference disk that you can use when creating other drawings that require threads.

Figure 6-8 shows a metric unit detailed thread representation. It was created using the procedure outlined. **Grid** was set at **10, Snap** was set at **2.5,** and the distance from the centerline to the zigzag pattern was **10.** Draw and save the metric detailed thread representation shown in Figure 6-8 as a wblock named **DETLMM.**

### To create an internal detailed thread representation in a sectional view

Figure 6-9 shows how to create a 1.00-inch internal detailed thread representation from the wblock DETLIN of the external detailed thread created above.

1. Use **Block, Insert** and locate the detailed wblock on the drawing screen at the indicated insert point. In this example the thread is to be drawn in a vertical orientation, so the block is rotated **90°** when it is inserted. The same scale size is used for the wblock as was drawn for a 1.00-inch diameter thread.
2. The thread created from wblock DETLIN is longer than needed, so explode the block and then erase the excess lines.
3. Define the hatch pattern as **ANSI31** and apply hatching to the areas outside the thread as shown.

Figure 6-9

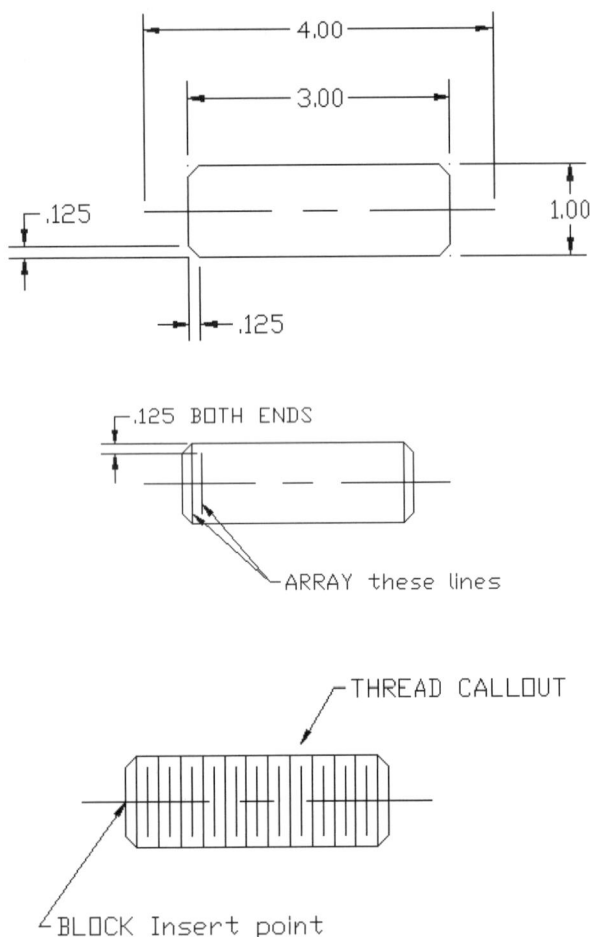

Figure 6-10

## To create a schematic thread representation

Draw a schematic representation of a 1.00-inch-diameter thread. See Figure 6-10.

1. Set Grid to **.50** and **Snap** to **.125.**
2. Draw a **4.00-inch** centerline near the middle of the drawing screen.
3. Draw the outline of the thread, including a chamfer, using the dimensions shown.

The chamfer was drawn in this example using the .125 snap points, but the **Chamfer** command could also have been used.

The 1.00 diameter was chosen because it will make it easier to determine scale factors for the wblock when inserting the representation into other drawings.

4. Draw three vertical lines as shown.
5. Use the **Array, Rectangular** command to draw **11** vertical lines across the thread. The distance between the 11 lines (COLUMNS) is **.25.**

A distance of –.25 would create lines to the left of the original line.

6. Save the representation as a wblock named **SCHMINCH.** Define the insertion point as shown.

Figure 6-11 shows a metric unit version of a schematic thread representation in a sectional view. The procedure used to create the representation is the same as explained previously but with different drawing limits and different values. The major diameter is **20,** and the spacing between lines is **2.5** and **5** as shown. Draw and save the representation as a wblock named **SCHMMM.**

## To create an internal schematic thread representation

Figure 6-12 shows a 36-millimeter-diameter internal schematic thread representation. It was developed from the wblock SCHMMM created previously.

1. Set **Grid** = **10**
   **Snap** = **5**
   **Limits** = **297,210**
   **Zoom** = **ALL**

2. Insert the wblock SCHMMM at the indicated insert point.

The required thread diameter is **36** millimeters. The wblock SCHMMM was drawn using a diameter of **20** mil-

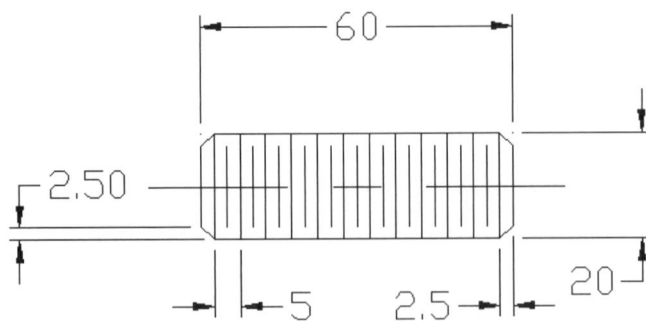

Figure 6-11

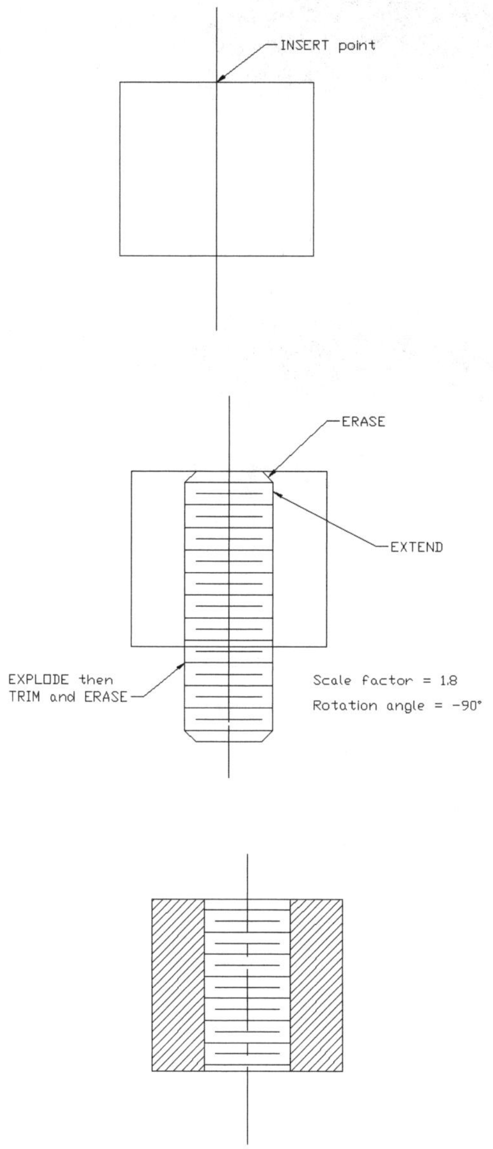

**Figure 6-12**

limeters. This means that the block must be enlarged by using a scale factor. The scale factor is determined by dividing the desired diameter by the wblock's diameter.

36/20 = 1.8

The wblock must also be rotated **90°** to give it the correct orientation.

3. The inserted thread shape is longer than desired, so first explode the wblock and then use the **Erase, Trim,** and **Extend** commands as needed.
4. Draw the sectional lines using **Hatch ANSI31.**
5. Save the drawing as a wblock if desired.

## To create a simplified thread representation

The simplified representation looks very similar to the orthographic view of an internal hole, so it is important to always include a thread callout with the representation. In the example shown, a leader line was included with the representation. See Figure 6-13. The leader serves as a reminder to add the appropriate drawing callout. If the leader line is in an inconvenient location when the wblock is inserted into a drawing, the leader line can be moved or simply erased.

1. Set **Grid** = **.5**
   **Snap** = **.125**
2. Draw a **4.00** centerline near the center of the screen.
3. Draw the thread outline using the given dimensions.
4. Draw the hidden lines using the given dimensions.
5. Save the thread representation as a wblock named **SIMPIN**. Define the insertion point as shown.

Figure 6-14 shows an internal simplified thread representation in a sectional view. Note how hidden lines that cross over the sectional lines are used. The hatch must be drawn first and the hidden lines added over the pattern. If the hidden lines are drawn first, the **Hatch** command may not add section lines to the portion between the hidden line and the solid line that represents the edge of the threaded hole.

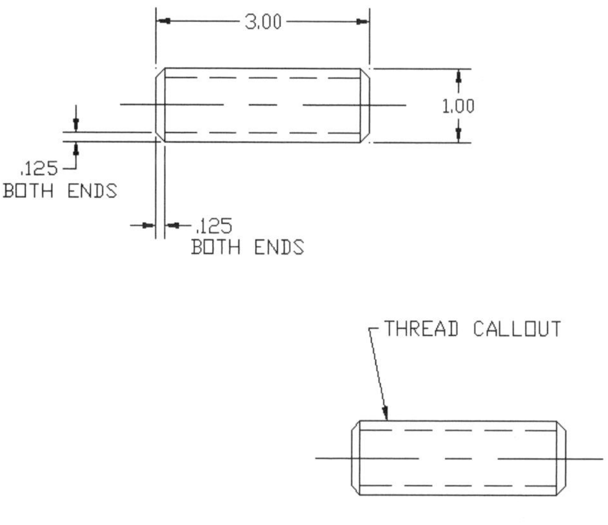

**Figure 6-13**

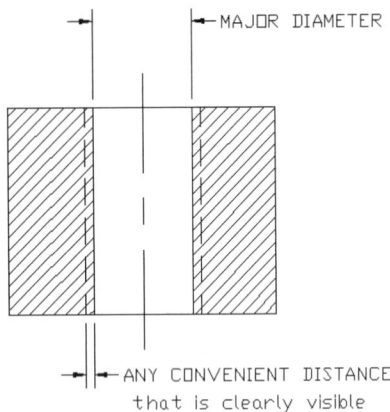

Figure 6-14

Figure 6-15

## 6-6 ORTHOGRAPHIC VIEWS OF INTERNAL THREADS

Figure 6-15 shows top and front orthographic views of internal threads. One thread goes completely through the object, the other only partially.

Internal threads are represented in orthographic views using parallel hidden lines. The distance between the lines should be large enough so that there is a clear distinction between the lines; that is, the lines should not appear to blend together or become a single, very thick line.

Circular orthographic views of threaded holes are represented by two circles: one drawn using a continuous line and the other drawn using a hidden line. The distance between the circles should be large enough to be visually distinctive. The two circles shown should both be clearly visible.

Threaded holes are created by first drilling a tap hole and then tapping (cutting) the threads using a tapping bit. Tapping bits have cutting surfaces on their side surfaces, not on the bottom. This means that if the tapping bit were forced all the way to the bottom of the tap hole, the bit could be damaged or broken. It is good design practice to make the tap hole deep enough so that a distance equivalent to at least two thread lengths ($2P$) extends beyond the tapped portion of the hole.

Threaded holes that do not go completely through an object must always show the unused portion of the tap hole. The unused portion should also include the conical point. See Chapter 1.

If an internal .500–13 UNC thread does not go completely through an object, the length of the unused portion of the tap hole is determined as follows:

Inches/Thread
$$= \text{Pitch length} = 1.00/13$$
$$= .077 \text{ inch}$$

so

$$2P = 2(.077) = .15 \text{ inch}$$

The distance .15 represents a minimum. It would be acceptable to specify a pilot hole depth greater than .15 depending on the specific design requirements.

If an internal M12 × 1.75 thread does not go completely through an object, the length of the unused portion of the tap hole is determined as follows:

$$2(\text{Pitch length}) = 2(1.75) = 3.50 \text{ mm}$$

## 6-7 SECTIONAL VIEWS OF INTERNAL THREAD REPRESENTATIONS

Figure 6-16 shows sectional views of internal threads that do not go completely through an object. Each example was created from wblocks of the representations. The hatch pattern used was ANSI32. Each example includes both a threaded portion and an untapped pilot hole that extends approximately $2P$ beyond the end of the tapped portion of the hole.

When drawing a simplified representation in a sectional view, draw the hidden lines that represent the outside edges of the threads after the section lines have been added. If the lines are drawn before the section lines are added, the section lines will stop at the outside line. Section lines should be drawn up to the solid line, as shown.

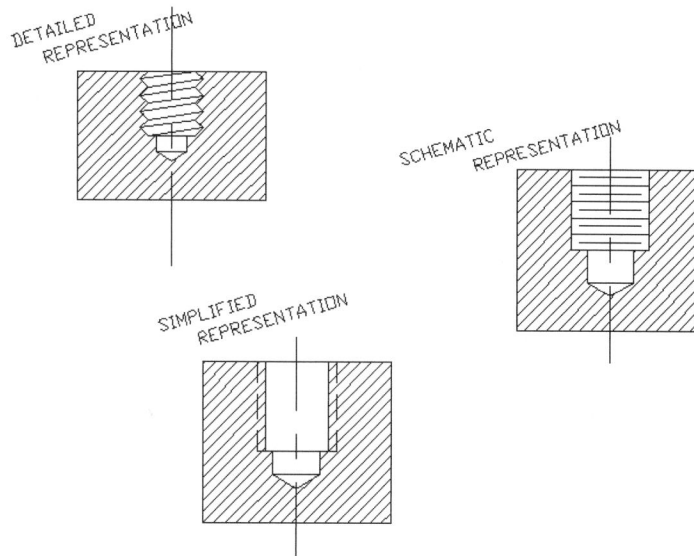

Figure 6-16

## 6-8 TYPES OF THREADS

Figure 6-17 shows the profiles of four different types of thread: American National, square, acme, and knuckle. There are many other types of threads. In general, square and acme threads are used when heavy loading is involved. A knuckle thread can be manufactured from sheet metal and is most commonly found on a lightbulb.

The American National thread is the thread shape most often used in mechanical design work. All threads in this chapter are assumed to be American National threads unless otherwise stated.

Figure 6-17

# 6-9 HOW TO DRAW AN EXTERNAL SQUARE THREAD

Figure 6-18 shows how to draw a 4.00-inch-long external square thread that has a major diameter of 5.25 and 2 threads per inch. The procedure is as follows.

1. Set **Grid** = **.50**
   **Snap** = **.125**
2. Draw two **5.00**-long horizontal lines **2.00** apart and a **4.50** centerline between them.
3. Draw a rhomboid centered about the centerline using the given dimensions.

   In this example $P = .5$, so $.5P = .25$.

4. Use the **Array, Rectangular** command to create eight columns (2 threads per inch) **.50** apart.
5. Draw slanted lines 1-2 and 3-4.
6. Zoom the upper right portion of the thread as needed.
7. Draw a horizontal line **.25** (.5P) from the outside edge of the thread as shown.
8. Draw line 5-6 from the intersection of the horizontal line drawn in step 7 with line 1-2, labeled point 5, to the intersection of the toothline 1-6 and the thread's centerline.
9. Trim the horizontal line to create line 5-7, and trim line 1-2 below point 5.
10. Array lines 1-5, 5-6, and 5-7 using **Array, Rectangular** with eight columns −**.25** apart.

    The minus sign will generate a right-to-left array.

11. Repeat steps 7 through 10 for the lower portion of the thread. Array eight columns, +**.25** apart. The plus sign generates a left-to-right array.
12. Erase any excess lines and add any necessary shaft information to the drawing.
13. Save the drawing as a wblock named **SQIN** using the indicated insertion point.

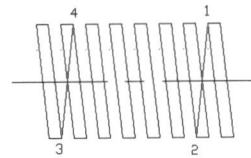

**Figure 6-18**

# 6-10 HOW TO DRAW AN INTERNAL SQUARE THREAD

Figure 6-19 shows an external 2.25 × 2 square thread. The drawing was developed from wblock SQIN created in Section 6-9. The wblock was rotated to the correct orientation. **Erase** and **Trim** were used to fit the thread within the required depth. No scale factor was needed.

# 6-11 HOW TO DRAW AN EXTERNAL ACME THREAD

Draw a 2.25 × 2 × 4.00–long external acme thread. The procedure is as follows. See Figure 6-20.

1.  Set **Grid** = **.5**
    **Snap** = **.125**
2.  Draw a **5.00**-long horizontal centerline.
3.  Draw two **.5**-long horizontal lines **1.125** above and below.

These lines establish the major diameter of the thread.

4.  Draw a single acme thread using the given dimensions. Use **Zoom** to help create an enlarged working area. The first thread should start at the right end of the short horizontal line above the centerline.

The width dimensions are taken along the horizontal centerline of the individual thread. Each side of the thread is slanted at **14.5°** [.5(29)]. In this example $P = .5$, so $.5P = .25$, and $.25P = .125$.

5.  Use the **Array** command and draw eight columns (two threads per inch) **.5** apart to develop the top portion of the thread.
6.  Copy and move the top portion of the thread to create the lower portion.

Do not use **Mirror.** The lower portion of the thread is not a mirror image of the upper portion.

7.  Erase and extend lines as necessary along the lower left portion of the thread to blend the thread into the shaft.

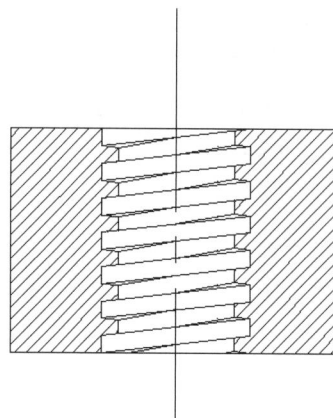

**Figure 6-19**

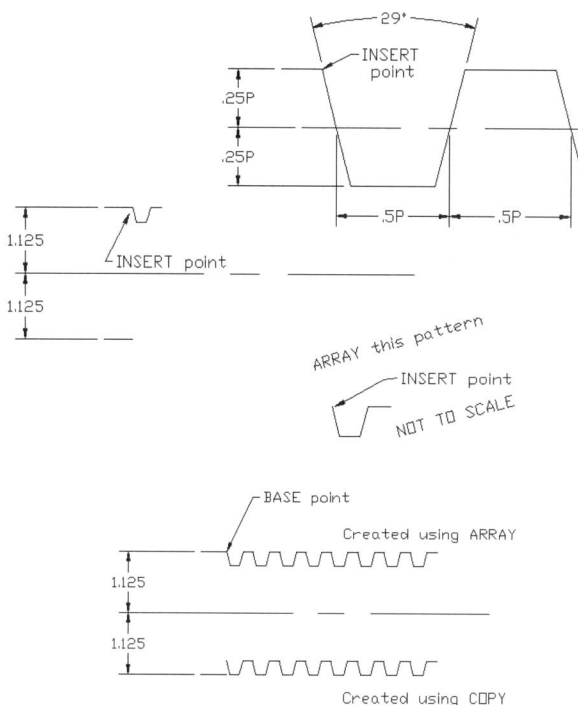

**Figure 6-20, Part 1**

8. Draw two slanted lines between the thread's root lines at the left end of the thread.
9. Array the lines drawn in step 8 so that there are eight columns, **.5** apart.
10. Draw two slanted lines across the thread's crest lines at the left end of the thread.
11. Array the lines drawn in step 10 so that there are eight columns, **.5** apart.

12. Draw a vertical line at each end of the thread, establishing the thread's 4.00-inch length.
13. Erase and trim the excess lines from the left end of the thread.
14. Copy and trim a crest-to-crest line to complete the left end of the thread.
15. Save the drawing as a wblock named **ACMEIN** using the indicated insertion point.

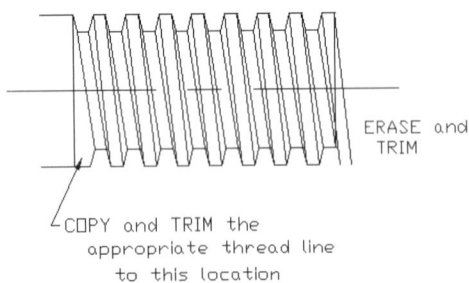

**Figure 6-20, Part 2**

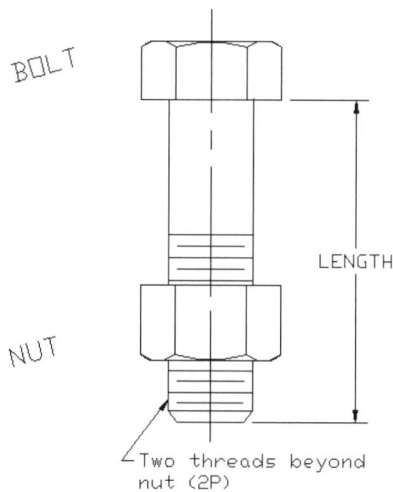

**Figure 6-21**

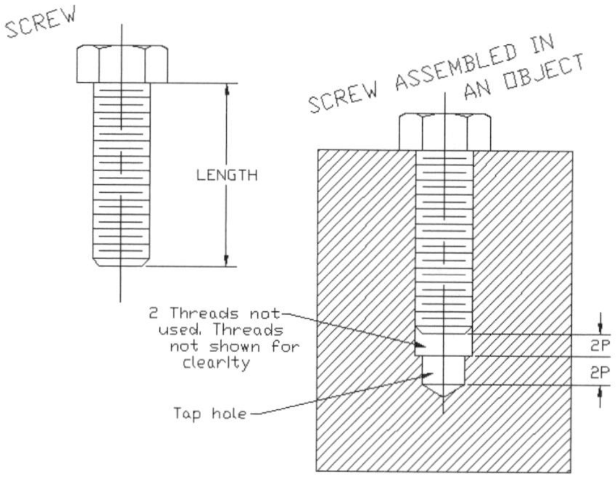

**Figure 6-22**

## 6-12 BOLTS AND NUTS

A *bolt* is a fastener that passes through a clearance hole in an object and is joined to a nut. There are no threads in the object. See Figure 6-21. Note that there are no hidden lines within the nut to indicate that the bolt is passing through. Drawing convention allows for nuts to be drawn without hidden lines.

Threads on a bolt are usually made just long enough to correctly accept a nut. This is done to minimize the amount of contact between the edges of threads and the inside surfaces of the clearance holes. The sharp, knifelike thread edges could cut into the object, particularly if the application involves vibrations.

It is considered good design practice to specify a bolt length long enough to allow at least two threads to extend beyond the end of the nut. This ensures that the nut is fully attached to the bolt.

Bolt drawings can be created from wblocks of threads. Remember that a block must be exploded before it can be edited.

## 6-13 SCREWS

A *screw* is a fastener that assembles into an object. It does not use a nut. The joining threads are cut into the assembling object. See Figure 6-22.

Screws may or may not be threaded over their entire length. If a screw passes through a clearance hole in an object before it assembles into another object, it is good design practice to minimize the number of threads that contact the sides of the clearance hole.

It is also considered good design practice to allow a few unused threads in the threaded hole beyond the end of an assembled screw. If a screw was forced to the bottom of a tapped hole, it might not assemble correctly or could possibly be damaged.

It is good design practice to allow at least two unused threads beyond the end of the screw. Figure 6-22 shows a schematic thread representation of a screw mounted in a threaded hole.

There is a distance of $2P$ (two threads) between the end of the screw and the end of the threaded portion of the hole. There should also be a $2P$ distance between the end of the threaded hole and the end of the pilot hole, plus the conical point of the tap hole, as described in Section 6-6.

Threads are usually not drawn in a threaded hole beyond the end of an assembling screw. This makes it easier to visually distinguish the end of the screw.

Figure 6-23 shows a screw assembled into a hole drawn using the detailed, schematic, and simplified representations in a sectional view. An orthographic view of a screw in a threaded hole is also shown.

The top view shown in Figure 6-23 applies to all three representations and the orthographic view.

## 6-14 STUDS

A *stud* is a threaded fastener that both screws into an object and accepts a nut. See Figure 6-24. The thread callouts and representations for studs are the same as they are for bolts and screws.

**Figure 6-23**

## 6-15 HEAD SHAPES

Bolts and screws are manufactured with a variety of different head shapes, but hexagon (hex) and square head shapes are the most common. There are many different head sizes available for different applications. Extra-thick heads are used for heavy-load applications, and very thin heads are used for applications where space is limited. The exact head size specifications are available from fastener manufacturers.

This section shows how to draw hex and square heads based on accepted average sizes that are functions of the major diameters of both the bolt and the screw. It is suggested that hexagon and square head drawings be saved as wblocks for both inch and millimeter values so that they can be combined with the thread wblocks to form fasteners.

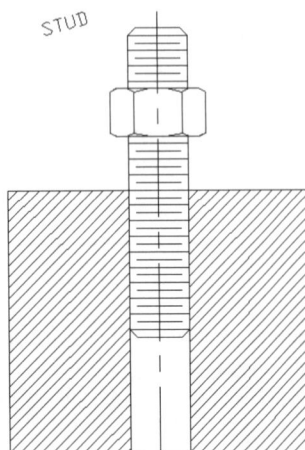

**Figure 6-24**

**To draw a hexagon (hex) shaped head**

Draw front and top orthographic views of a hex head based on a thread with an M24 major diameter. See Figure 6-25.

1. Set **Limits = 297,210**
   **Grid = 10**
   **Snap = 5**
2. Draw a vertical line and two horizontal lines using the given dimensions.

The two horizontal lines are used to locate the center of the head in the top view and the bottom of the head in the front view.

3. Use **Polygon** and draw a hexagon distance across the flats equal to **1.5D,** where **D** is the major diameter of the thread.

In this example a radius of **18** was used to draw the hexagon. $1.5D = 1.5(24) = 36$ is the distance across the flats of the hexagon. Use the **Polygon** command to circumscribe a six-sided polygon around a circle of radius 18.

4. Offset a line **.67D** from the lower horizontal.

This line defines the thickness of the head. The head thickness is .67D, where $D$ is the major diameter of the thread. In this example $D = 24$, so $.67D = .67(24) = 16.08$, which can be rounded off to 16.

5. Draw projection lines from the corners of the hexagon in the top view into the front view.
6. **Zoom** the front view portion of the drawing if necessary. Draw two **60°** lines from the corners of the front view as shown so that they intersect on the vertical centerline.

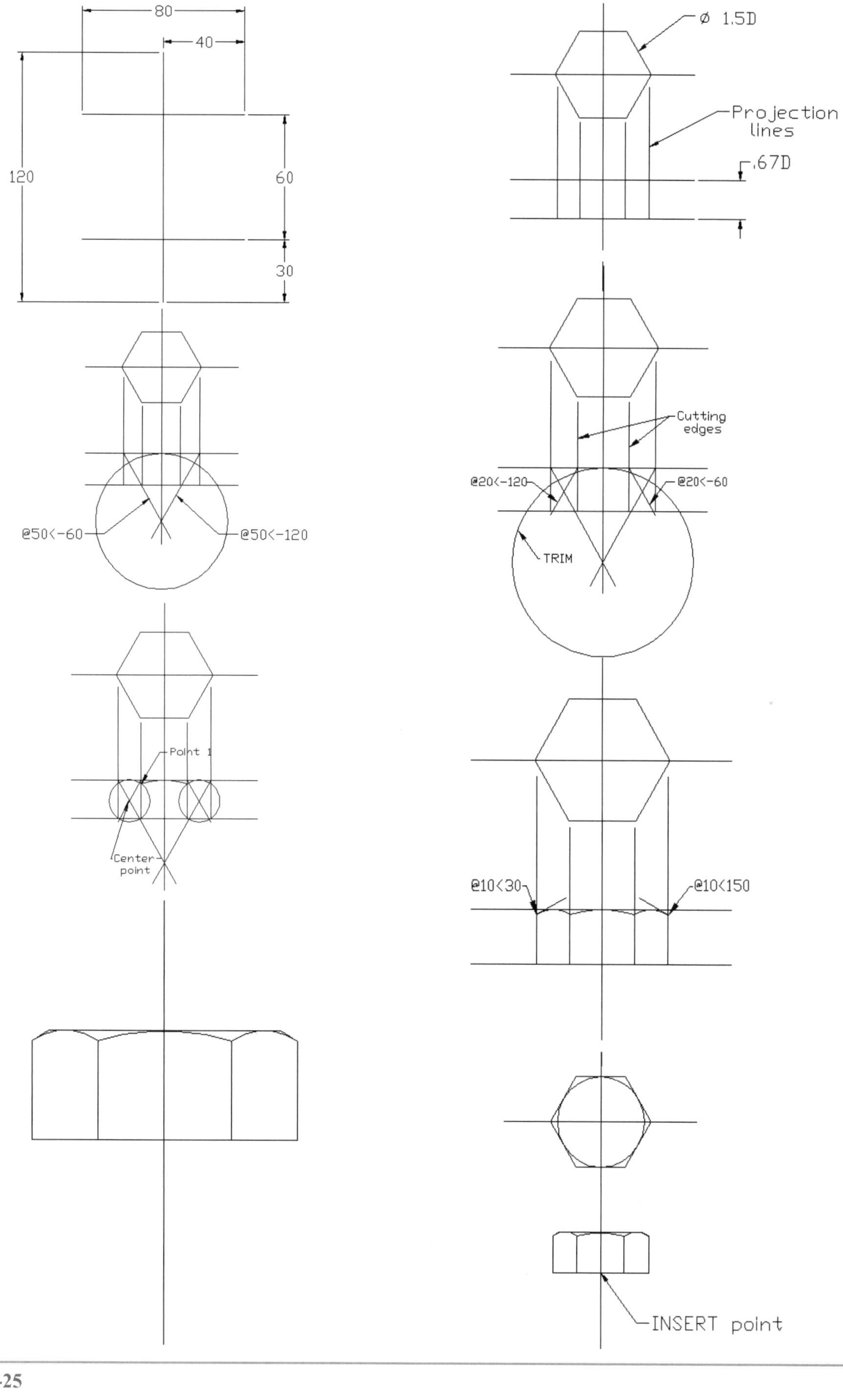

Figure 6-25

Use **Osnap, Intersection** to accurately locate the corner points. The line from the left corner uses an input of @50<−60; the line from the right corner uses the input @50<−120.

7. Draw a circle whose center point is the intersection of the two 60° lines drawn in step 6 and the vertical centerline, and whose radius equals the distance from the center point to the line at the top of the front view.
8. Trim the circle so that only the arc between the inside projection lines remains.
9. Draw two **60°** lines as shown using the inputs @**20**<−**120** and @**20**<−**60**.
10. Draw two circles centered about the intersection of the slanted lines drawn in steps 6 and 8. The radius of each circle equals the distance from the center point to the intersection of the circle drawn in step 7 and the inside projection lines labeled point 1.
11. Trim and erase as needed.
12. Draw two lines from the intersections of the smaller arcs with the outside edge of the head. The input for the lines is @**10**<**30** and @**10**<**150**.

These lines could have been generated using the **Chamfer** command.

13. Trim the chamfer lines to the top of the head.
14. **Zoom All** and draw a circle that is circumscribed within the hexagon as shown.
15. Save the drawing as a wblock named **HEXHEAD**, using the indicated insertion point.

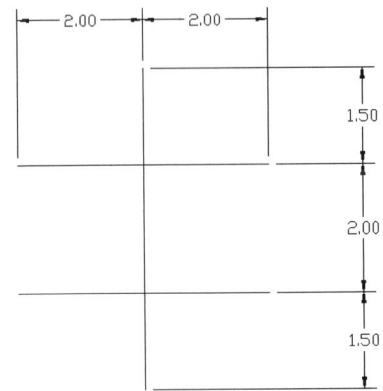

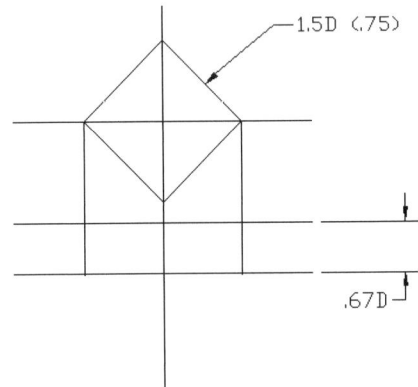

### To draw a square-shaped head

Draw front and top orthographic views of a hex head based on a thread with a 1.00-inch major diameter. See Figure 6-26.

1. Set **Grid = .50**
   **Snap = .25**
2. Draw two horizontal lines and a vertical line using the given dimensions.

The intersection of the top horizontal line and the vertical line will be the center point of the top view, and the lower horizontal line will be the bottom edge of the head.

3. Use **Draw, Polygon** and **Modify, Rotate (45°)** to create a square oriented as shown.

The distance across the square equals 1.50*D* or 1.50 inches. This means that the radius for the polygon is **.75.**

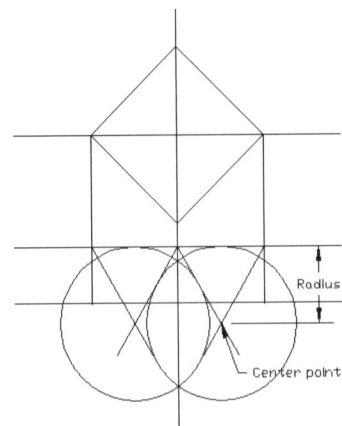

**Figure 6-26, Part 1**

between the arc and the vertical side line of the head.

10. Trim any excess lines.
11. Zoom the drawing back to its original size and draw a circle in the top view tangent to the inside edges of the square.
12. Save the drawing as a wblock named **SQHEAD** using the insertion point indicated.

## 6-16 NUTS

This section explains how to draw hexagon- and square-shaped nuts. Both construction methods are based on the head shape wblocks created in Section 6-15.

There are many different styles of nuts. A *finished nut* has a flat surface on one side that acts as a bearing surface when the nut is tightened against an object. A finished nut has a thickness equal to .88*D*, where *D* is the major diameter of the nut's thread size.

A *locknut* is symmetrical, with the top and bottom surfaces identical. A locknut has a thickness equal to .5*D*, where *D* is the major diameter of the nut's thread size.

**To draw a hexagon-shaped finished nut**

Draw a hexagon-shaped finished nut for an M36 thread. See Figure 6-27.

1. Set **Limits = 297,210**
   **Grid = 10**
   **Snap = 2.5**
2. Construct a vertical line that is intersected by two horizontal lines **32** millimeters apart.

The thickness of a finished nut is .88*D*. In this example, .88(36) = 31.68, or 32.

3. Use **Block, Insert** and insert the HEXHEAD wblock created in Section 6-15.

The wblock HEXHEAD was created for an M24 thread, so the X and Y scale factors must be increased to accommodate the larger thread size. The scale factor is determined by dividing the desired size by the wblock size. In this example, 36/24 = 1.5. The scale factor is **1.5.** Respond to the command prompts as follows:

*Command: INSERT*

The **Insert** dialog box will appear.

4. Type or select **HEXHEAD**; press **Enter**.
5. Select the indicated insertion point.

**Figure 6-26, Part 2**

4. Offset a line **.67** (.67*D*) from the lower horizontal line.

The offset distance is equal to the thickness of the head.

5. Project the corners of the square in the top view into the front view.
6. Zoom the front view and draw four **60°** lines from the head's upper corners and intersection of the centerline and the top surface of the head as shown. The inputs for the lines are **@1.5<−120** and **@1.5<−60**. Use **Osnap, Intersection** to ensure accuracy.
7. Use **Draw, Circle, Center, Radius** and draw two circles about the center points created in step 6. Trim the excess portions of the circle.
8. Trim and erase any excess lines.
9. Add the **30°** chamfer lines as shown. The line inputs are **@.5<150** and **@.5<30**. Use **Osnap, Intersection** to accurately locate the intersection

6. Specify an X and Y scale of **1.5** and a **0** rotation angle; select **OK**.

*Specify insertion point or [Scale/X/Y/Z/Rotate/ PScale/PX/PY/PZ/PRotate]:*

7. Press **Enter**.

If you have not created a wblock, refer to Section 6-15 and draw a hexagon-shaped head.

8. Explode the wblock.
9. Move the front view of the nut so that the top sur-

face aligns with the top parallel horizontal line.
10. Erase the bottom line of the front view of the nut and extend the vertical line to the lower horizontal line. Erase the top horizontal line in the front view.
11. Use **Offset** to draw a horizontal line **2** millimeters below the bottom of the nut.

This line defines the shoulder surface of the nut. Any offset distance may be used as long as the line is clearly

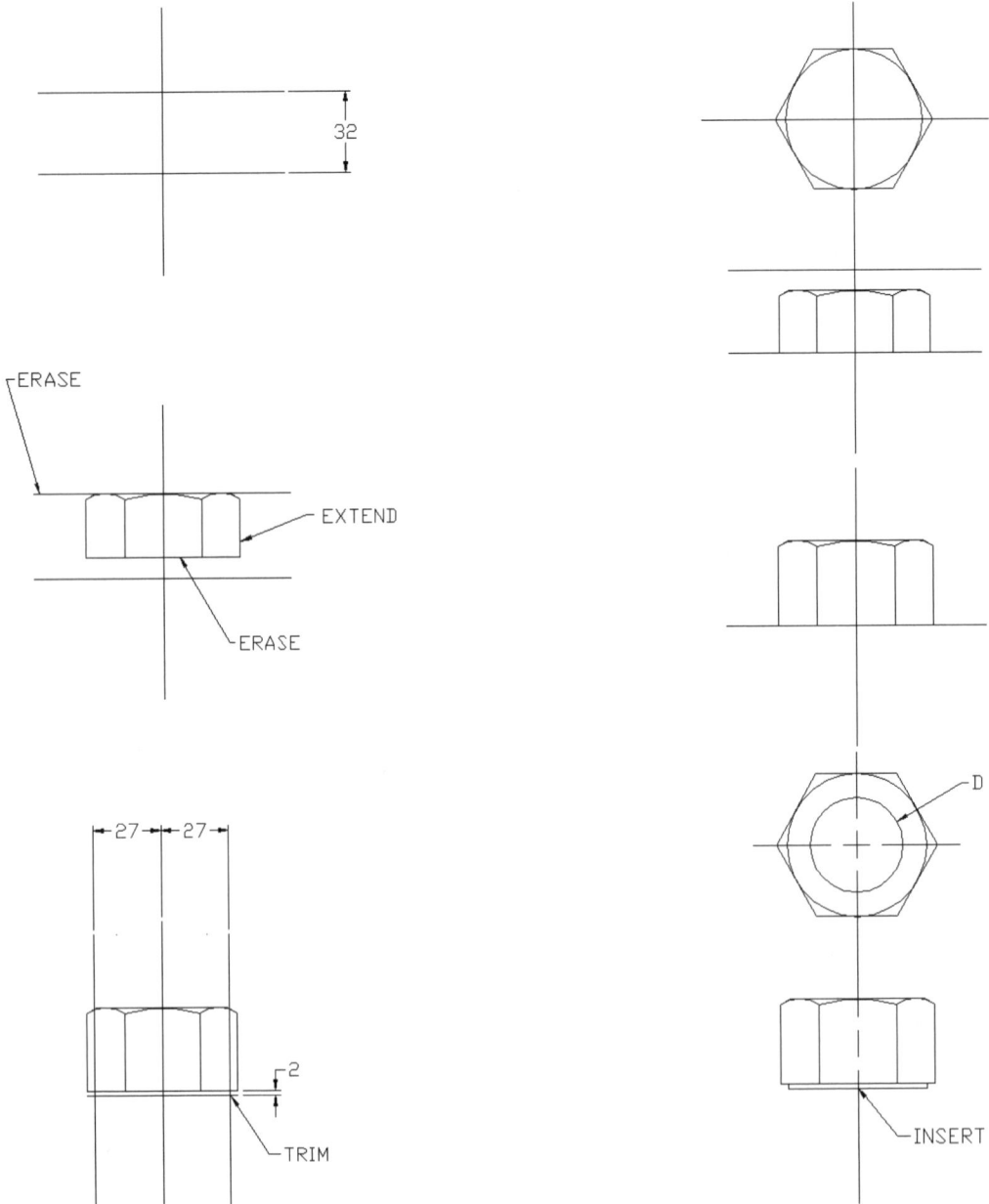

Figure 6-27

visible. The actual shoulder surface is less than 1 millimeter deep and would not appear clearly on the drawing.

The shoulder surface has a diameter equal to 1.5D [1.5(36) = 54] or, in this example, 54.

12. Offset the centerline **27** (half of 54) to each side and trim the excess lines.
13. Save the drawing as a wblock named **FNUTHEX.** Define the insertion point as shown.

## To draw a locking nut

Draw a locking nut for an M24 thread. See Figure 6-28. The HEXHEAD wblock was originally drawn for an M24 thread, so no scale factor is needed.

1. Set **Limits** = **297,210**
   **Grid** = **10**
   **Snap** = **5**
2. Draw a vertical line and two horizontal lines using the given dimensions.

The 6 distance is half the .5D [.5(24) = 12] thickness distance recommended for locknuts. Locknuts are symmetrical, so half the nut will be drawn and then mirrored.

3. Insert the HEXHEAD wblock at the indicated insert point.
4. Explode the wblock.
5. Move the front view of the hex head so that it aligns with the horizontal line as shown.
6. Trim and erase the lines that extend beyond the lower horizontal line.
7. Mirror the remaining portion of the front view about the lower horizontal line, then erase the horizontal line.
8. Draw a circle of diameter **D** in the top view as shown.
9. Save the drawing as a wblock named **LNUTHEX** using the indicated insertion point.

The procedure explained previously for hexagon-shaped finish nuts and locknuts is the same for square-shaped nuts. Use the wblock SQHEAD in place of the HEXHEAD wblock.

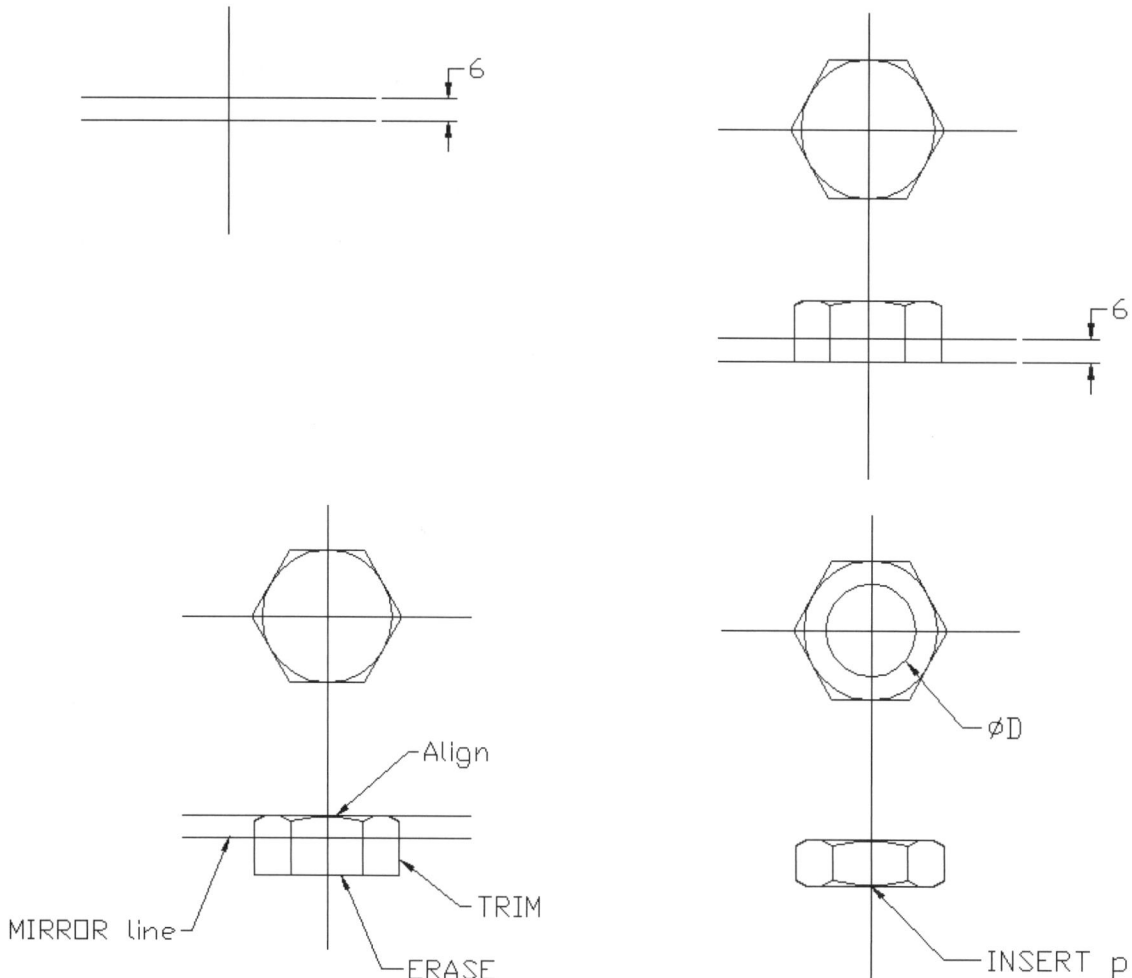

**Figure 6-28**

## 6-17 SAMPLE PROBLEM SP6-1

Draw and specify the mimimum threaded hole depth and pilot hole depth for an M12 × 1.75 × 50 hex head screw. Assemble the screw into the object shown in Figure 6-29. Use the schematic thread representation and a sectional view.

The SCHMMM wblock was originally drawn for an M20 thread, so a scale factor is needed. The scale factor to reduce an M20 diameter to an M12 is 12/20 = .6.

1. Insert the SCHMMM wblock and insert it at the indicated point. Use a **.6** X and Y scale factor.
2. Explode the block and trim any excess threads.
3. Insert the HEXHEAD wblock as indicated. The HEXHEAD wblock was also created for an M20 thread, so the scale factor is again **.6.**

The thread pitch equals 1.75, so $2P = 3.5$. This means that the threaded hole should be at least $50 + 3.5$, or 53.5, deep to allow for two unused threads beyond the end of the screw.

The unused portion of the pilot hole should also extend $2P$ beyond the end of the threaded portion of the hole, so the minimum pilot hole depth equals $53.5 + 3.5 = 57$.

The diameter of the tap hole for the thread is given in a table in the appendix as 10.3. For this example a diameter of 10 was used for drawing purposes. If required, the 10.3 value would be given in the hole's drawing callout.

4. Draw the unused portion of the thread hole and pilot hole as shown. Omit the thread representa-

tion in the portion of the threaded hole beyond the end of the screw for clarity.

## 6-18 SAMPLE PROBLEM SP6-2

Determine the length of a .750–10UNC square head bolt needed to pass through objects 1 and 2 shown in Figure 6-30. Include a hexagon-shaped locknut on the end of the bolt. Allow at least two threads beyond the end of the nut. Use the closest standard bolt length and draw a schematic representation as a sectional view.

This construction is based on the wblocks SCHMINCH, HEXHEAD, and LNUTHEX created in Sections 6-5, 6-15, and 6-16. If no wblocks are available, refer to these sections for an explanation of how to create the required shapes.

The wblocks SCHMINCH, HEXHEAD, and LNUTHEX were created for a major diameter of 1.00 inch, so the scale factor needed to create a major diameter of .750 is .75.

1. Insert the SCHMINCH wblock at the indicated point. Use an X and Y scale factor of **.75.** Rotate the block −**90°** to the correct orientation.

The wblock SCHMINCH is not long enough to go completely through the objects, so a second wblock is inserted below the first.

2. Insert the SQHEAD wblock at the indicated point. Use an X and Y scale factor of **.75.**

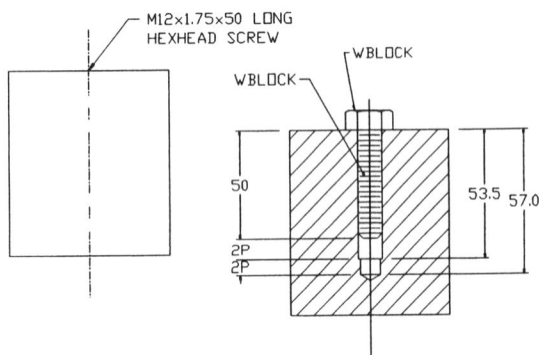

**Figure 6-29**

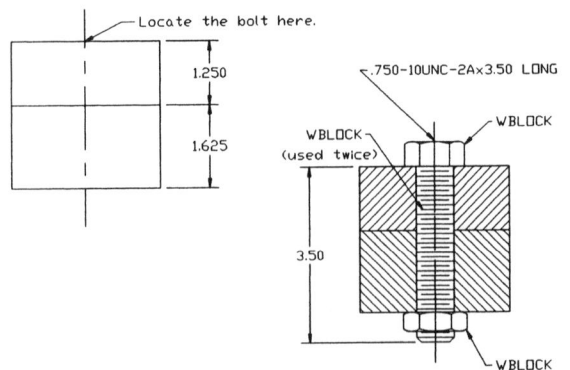

**Figure 6-30**

3. Insert the wblock LNUTHEX at the indicated point. Use an X and Y scale factor of **.75.**
4. Explode the wblocks.
5. Trim the excess lines within the nut.
6. Calculate the minimum length for the bolt.

The total depth of objects 1 and 2 equals 1.250 + 1.625 = 2.875.

The thickness of the nut equals .88$D$ or .88(.75) = .375.

The pitch length of the thread equals 1.00/10 = .1, so 2$P$ = .2.

The minimum bolt length equals 2.875 + .375 + .200 = 3.450.

From the table of standard bolt lengths in the appendix, the standard bolt length that is greater than 3.45 is 3.50, so the bolt callout can now be completed:

$$.750 - 10UNC - 2A \times 3.50 \text{ LONG}$$

The completed drawing with the appropriate bolt callout is shown in Figure 6-30.

# 6-19  STANDARD SCREWS

Figure 6-31 shows a group of standard screw shapes. The proportions given in Figure 6-30 are acceptable for general drawing purposes and represent average values. The exact dimensions for specific screws are available from manufacturers' catalogs. A partial listing of standard screw sizes is included in the appendix.

The given head shape dimensions are all in terms of $D$, the major diameter of the screw's thread. Information about the available standard major diameters and lengths is included in the appendix.

The choice of head shape is determined by the specific design requirements. For example, a flat head mounted flush with the top surface is a good choice when space is critical, when two parts butt against each other, or when aerodynamic considerations are involved. A round head can be assembled using a common blade screwdriver, but it is more susceptible to damage than a hex head. The hex head, however, requires a specific wrench for assembly.

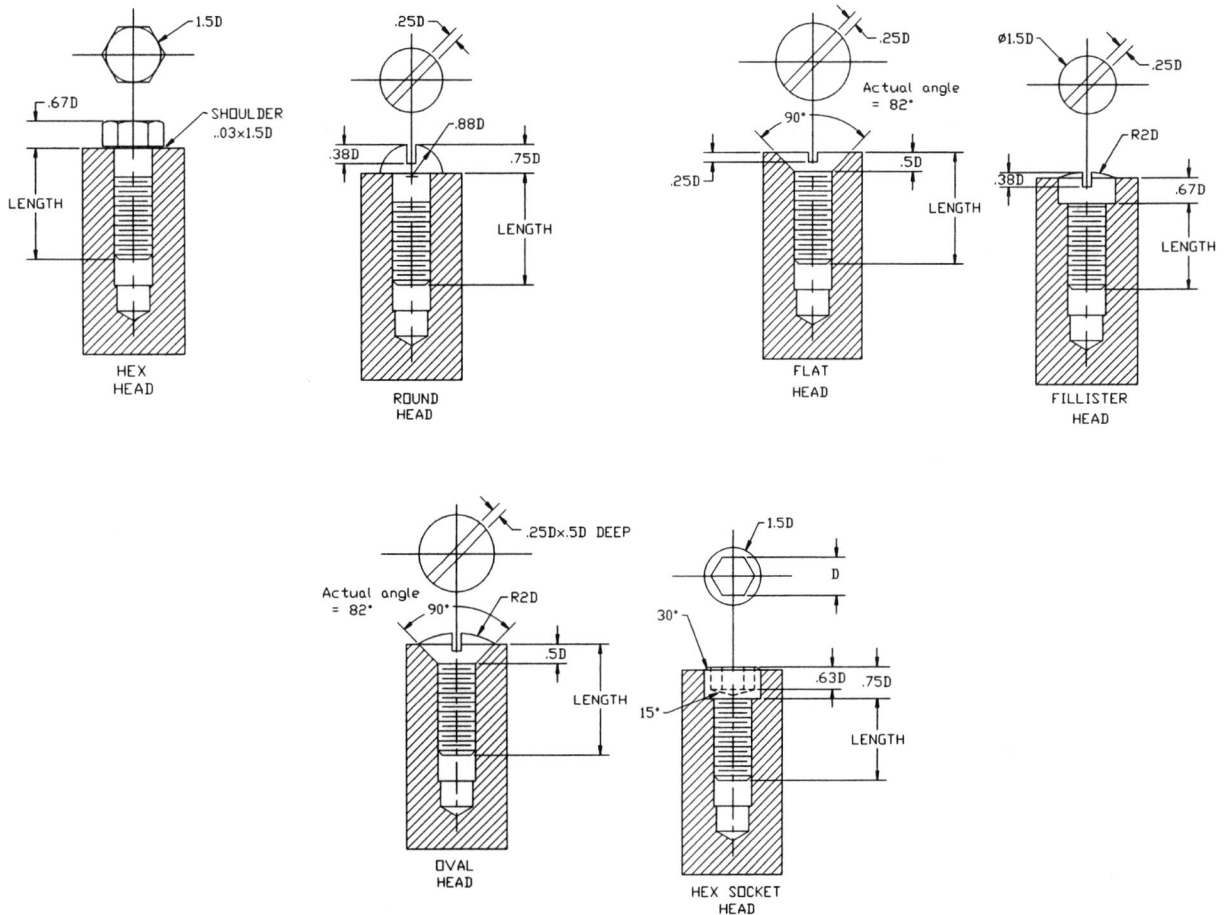

**Figure 6-31**

## 6-20  SETSCREWS

*Setscrews* are fasteners used to hold parts like gears and pulleys to rotating shafts or other objects to prevent slippage between the two objects. See Figure 6-32.

Most setscrews have recessed heads to help prevent interference with other parts. Many different head styles and point styles are available. See Figure 6-33.

Setscrews are referenced on a drawing using the following format.

THREAD SPECIFICATION
HEAD  SPECIFICATION  POINT  SPECIFICA-
    TION
SET SCREW

.250 – 20UNC – 2A 1.00 LONG
SLOT HEAD FLAT POINT
SET SCREW

The words LONG, HEAD, and POINT are optional.

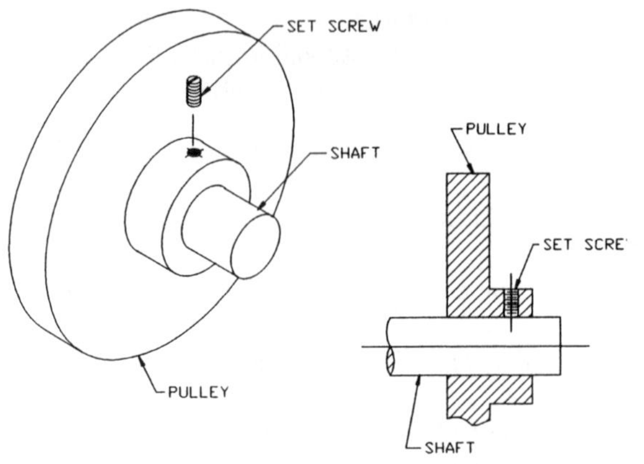

Figure 6-32

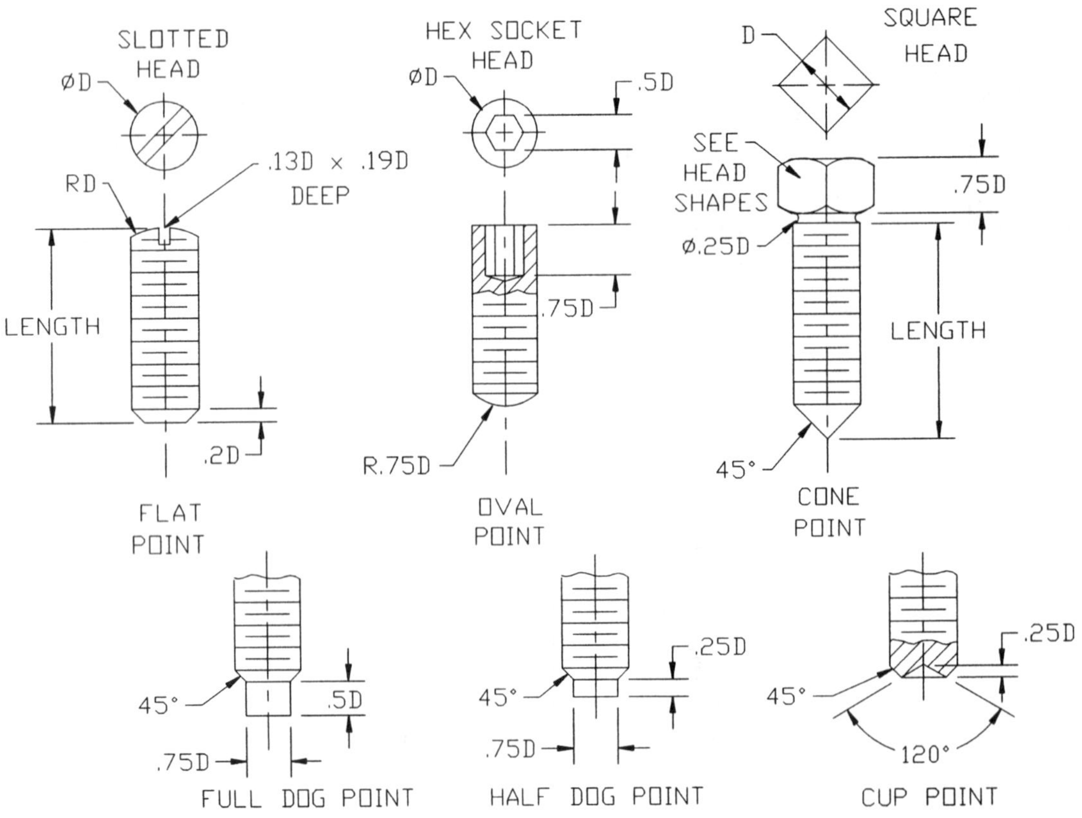

Figure 6-33

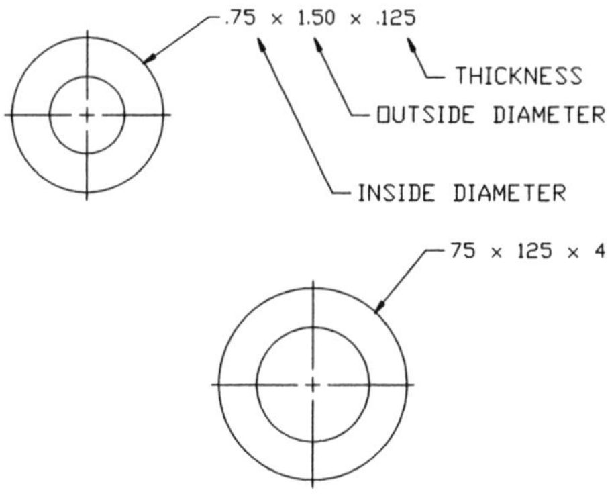

.75 × 1.50 × .125

THICKNESS

OUTSIDE DIAMETER

INSIDE DIAMETER

75 × 125 × 4

**Figure 6-34**

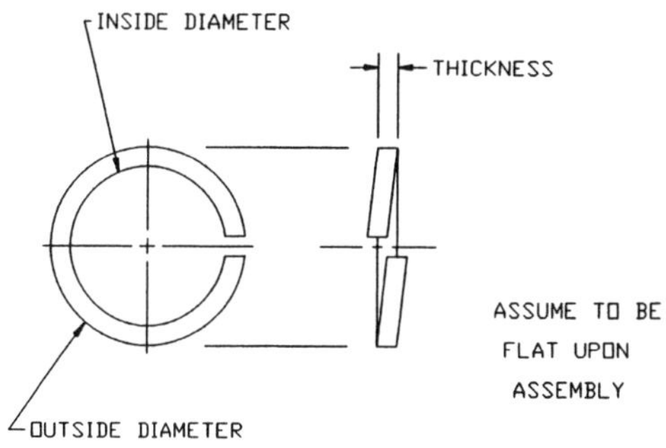

INSIDE DIAMETER

THICKNESS

ASSUME TO BE
FLAT UPON
ASSEMBLY

OUTSIDE DIAMETER

**Figure 6-36**

## 6-21  WASHERS

There are many different styles of washers available for different design applications. The three most common types of washers are *plain, lock,* and *star.* Plain washers can be used to help distribute the bearing load of a fastener or used as a spacer to help align and assemble objects. Lock and star washers help absorb vibrations and prevent fasteners from loosening prematurely. All washers are identified as follows:

Inside diameter × Outside diameter × Thickness

Examples of plain washers and their callouts are shown in Figure 6-34. A listing of standard washer sizes is included in the appendix.

**To draw a plain washer (See Figure 6-35.)**

1. Draw concentric circles for the inside and outside diameters.
2. Project lines from the circular view and draw a line that defines the width of the washer.
3. Use **Offset** to define the thickness.

Figure 6-36 shows two views of a lock washer. As the lock washer is compressed during assembly it tends to flatten, so the slanted end portions are not usually included on the drawing. The drawing callout should include the words LOCK WASHER.

Figure 6-37 shows an internal and an external tooth lock-type washer. They may also be called *star washers.*

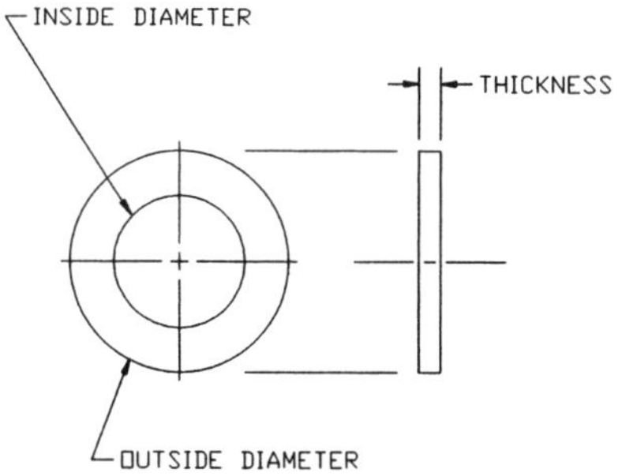

INSIDE DIAMETER

THICKNESS

OUTSIDE DIAMETER

**Figure 6-35**

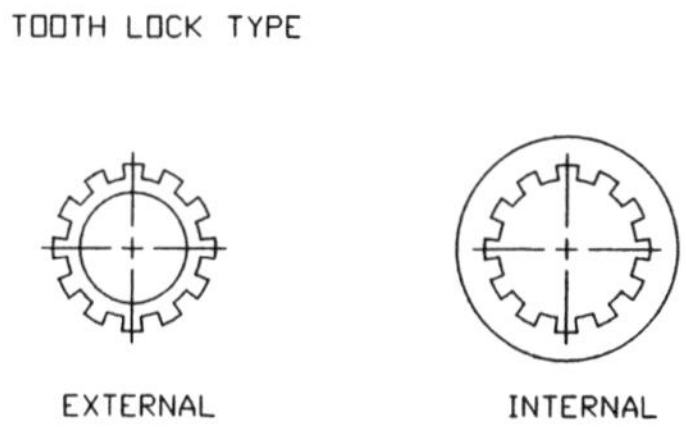

TOOTH LOCK TYPE

EXTERNAL          INTERNAL

**Figure 6-37**

They are best drawn by first drawing an individual tooth, then using **Array** to draw a total of 12 teeth. Because there are 12 teeth, each tooth and the space between the tooth and the next tooth require a total of $30° - 15°$ for the tooth and $15°$ for the space between.

## 6-22 KEYS

*Keys* are used to help prevent slippage in power transmission between parts, for example, a gear and a drive shaft. Grooves called *keyways* are cut into both the gear and the drive shaft and a key is inserted between, as shown in Figure 6-38.

There are four common types of keys: *square, Pratt & Whitney, Woodruff,* and *gib head*. Each has design advantages and disadvantages. See Figure 6-39. A list of standard key sizes is included in the appendix.

Square keys are called out on a drawing by specifying the length of one side of the square cross section and the overall length; Pratt & Whitney and Woodruff keys are specified by numbers; and gib head keys are defined by a group of dimensions. See the appendix for the appropriate tables and charts of standard key sizes.

Keyways are dimensioned as shown in Figure 4-46. Note that the depth of a keyway in a shaft is dimensioned from the bottom of the shaft. Because material has been cut away, the intersection between the shaft's centerline and top

outside edge does not exist. It is better to dimension from a real surface than from a theoretical one. The same dimensioning technique also applies to the gear.

## 6-23 RIVETS

*Rivets* are fasteners that hold adjoining or overlapping objects together. A rivet starts with a head at one end and a straight shaft at the other. The rivet is then inserted into the object, and the headless end is "bucked" or otherwise forced into place. A force is applied to the headless end that changes its shape so that another head is formed holding the objects together.

There are many different shapes and styles of rivets. Figure 6-40 shows five common head shapes for rivets. Aircraft use hollow rivets because they are very lightweight. A design advantage for using rivets is that they can be drilled out and removed or replaced without damaging the objects they hold together.

Rivet types are represented on technical drawings using a coding system. See Figure 6-41. The lines used to code a rivet must be clearly visible on the drawing. Because rivets are sometimes so small and the material they hold together so thin that it is difficult to clearly draw the rivets, companies draw only the rivet's centerline in the side view and identify the rivet using a drawing callout.

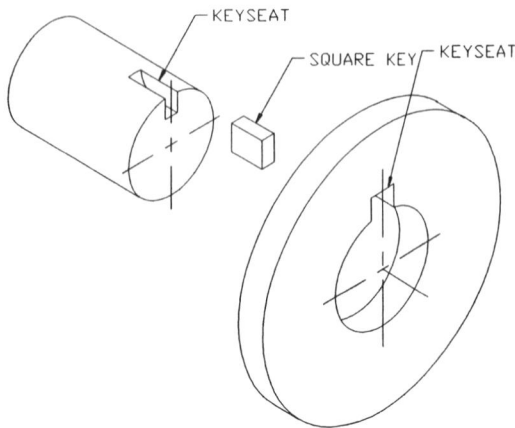

**Figure 6-38**

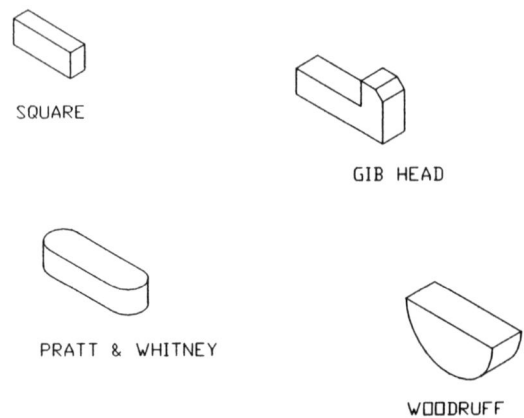

**Figure 6-39**

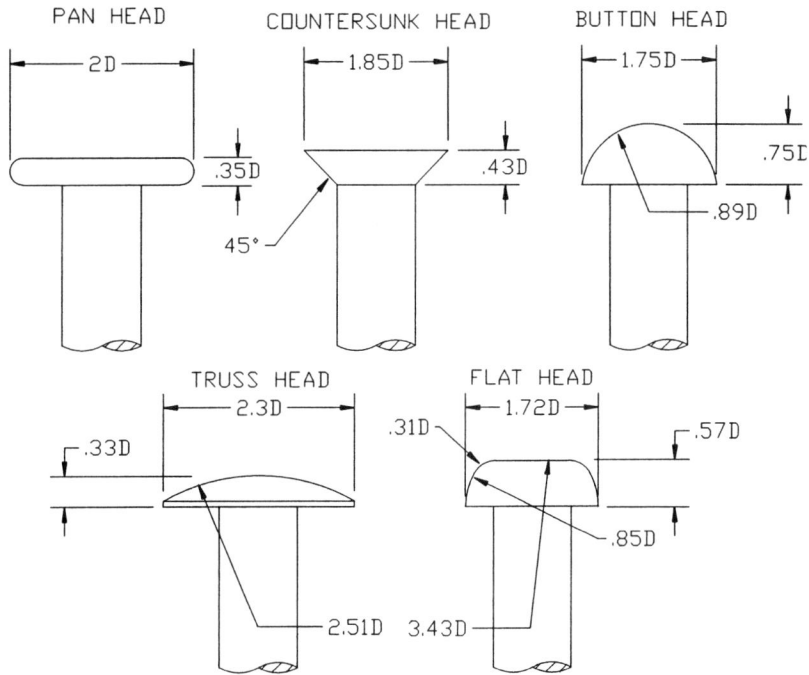

**Figure 6-40**

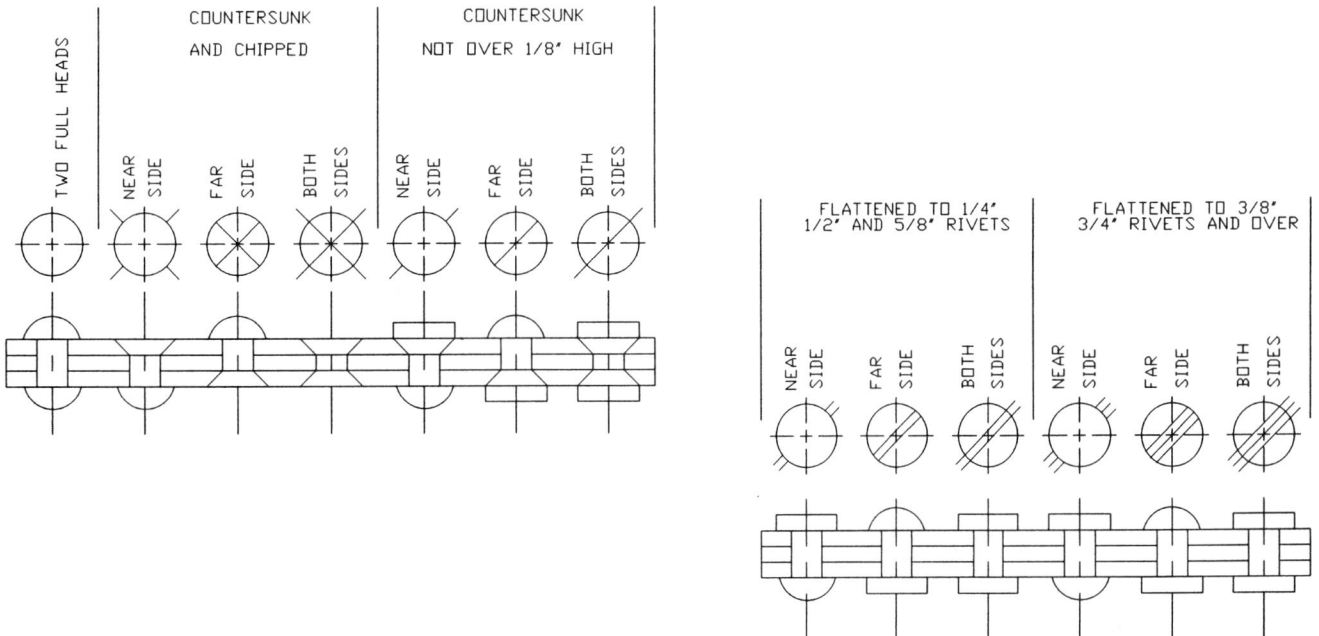

**Figure 6-41**

# 6-24 SPRINGS

The most common types of springs used on technical drawings are compression, extension, and torsional. This section explains how to draw detailed and schematic representations of springs. Springs are drawn in their relaxed position (not expanded or compressed). Phantom lines may be used to show several different positions for springs as they are expanded or compressed.

Springs are defined by the diameter of their wire, the direction of their coils, their outside diameter, their total number of coils, and their total relaxed length. Information about their loading properties may also be included. The pitch of a spring equals the distance from the center of one coil to the center of the next coil.

As with threads, the detailed representations are difficult and time consuming to draw, but the **Array** command shortens the drawing time considerably. Saving the representation as a wblock also helps avoid having to redraw the representation for each new drawing.

Ideally, the distance between coils should be exactly equal to the pitch of the spring, but this is not always practical for smaller springs. Any convenient distance between coils may be used that presents a clear, easy-to-understand representation of the spring.

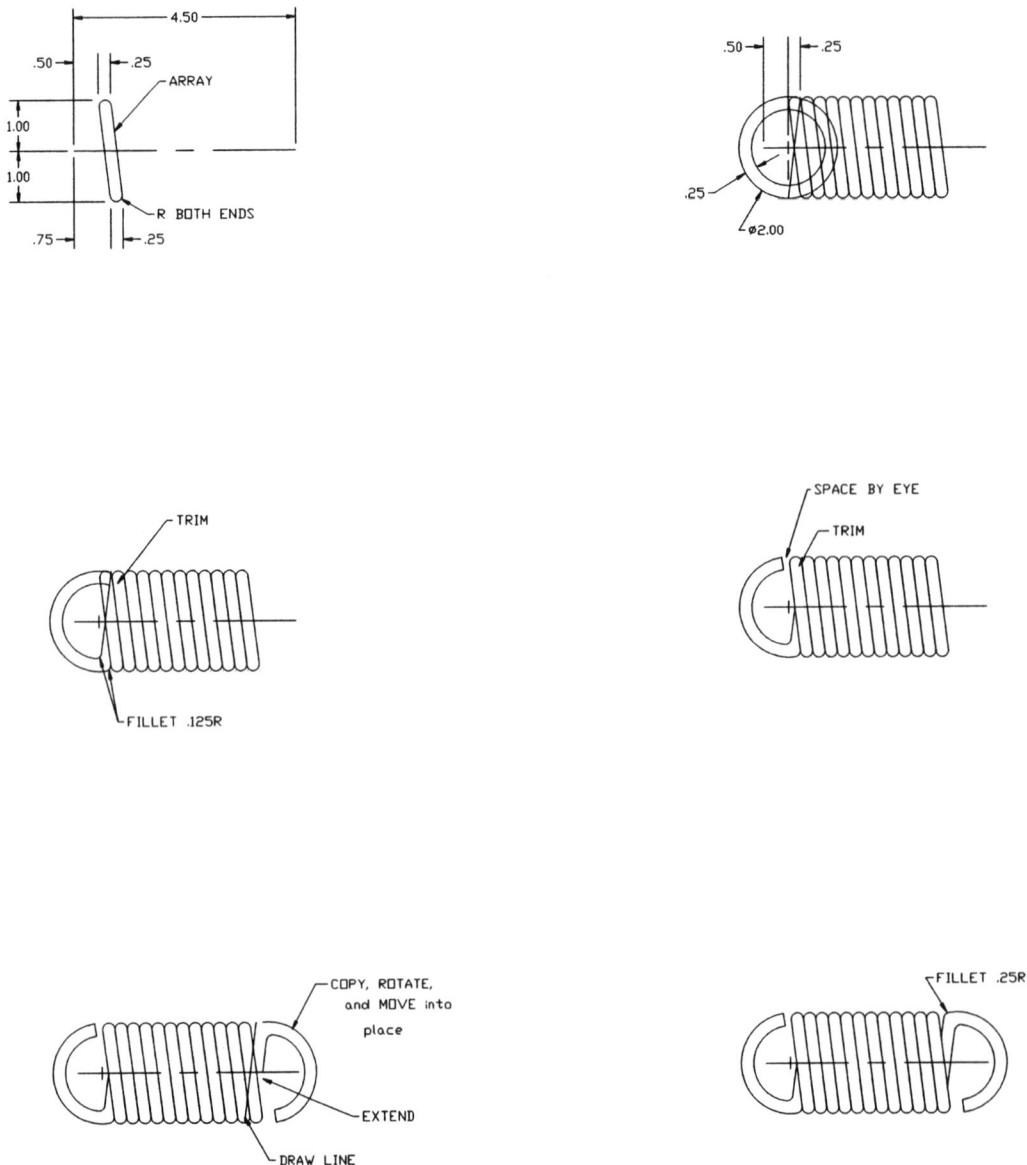

**Figure 6-42**

## To draw a detailed representation of an extension spring

Draw an extension spring 2.00 inches in diameter with right-hand coils made from .250-inch wire. The spring's pitch equals .25, and the total length of the coils is 3.00 inches. See Figure 6-42.

Because the pitch of the spring equals the wire diameter, the coils will touch each other.

1. Set **Grid** = **.5**
   **Snap** = **.25**
2. Draw the rounded shape of one of the coils.

The distance between the lines equals the diameter of the spring wire. The diameter of the rounded ends also equals the diameter of the wire. The coil's offset equals the spring's pitch.

3. Array the coil shape.

A rectangular array was used with one row, and 12 columns **.25** apart.

4. Draw two concentric circles at the left end of the spring so that the larger circle's diameter equals the spring's diameter. The vertical centerline of the circles is tangent to the left edge of the first coil.
5. Draw a line across the thread offset a distance equal to one pitch. The endpoints of the line are on snap points.
6. Trim the excess portion of the circles and fillet the circles as shown.
7. Trim the excess portion of the slanted line and cut back the end of the circles so that there is a noticeable gap between the first coil and the end of the circles. The gap distance is determined by eye.
8. **Copy, Rotate,** and **Move** the circular end portion to the right end of the spring.
9. Draw a slanted line across the farthest right coil as shown.
10. Trim and extend the lines as necessary and add the **FILLET .25R** as shown.
11. Save the drawing as a wblock named **EXTSPRNG.**

## To draw a detailed representation of a compression spring

Draw a compression spring with six right-hand coils made from .25-inch-diameter wire. The pitch of the spring equals .50 inch. The diameter of the spring is 2.00 inches. See Figure 6-43.

1. Set **Grid** = **.50**
   **Snap** = **.25**

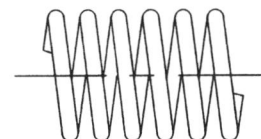

2. Draw the rounded shape of the first left coil as shown. The diameter of the rounded ends of the coil equals the diameter of the spring's wire.
3. Array the coil shape.

In this example a rectangular array was used to draw one row, and six columns **.50** inch apart.

**Figure 6-43**

4. Draw two slanted lines as shown.

These lines represent the back portion of the coil. They are most easily drawn from a snap point next to the rounded portion of the coil shape (this point would be the corner point of the coil if the coil were not rounded) to a point tangent to the opposite rounded end of the coil.

5. Trim and array the slanted lines as shown.

Use a rectangular array with one row, and six columns .50 inch apart.

6. Draw the right end of the spring so that it appears to end just short of the spring's centerline. Any convenient distance may be used.

7. Draw the left end of the spring using **Copy,** and copy one of the existing slanted lines. Use **Zoom,** if necessary, to align the copied line with the coil. Draw the left end of the spring so that it appears to end just short of the centerline.

8. Save the drawing as a wblock named **COMPSPNG.**

Figure 6-44 shows schematic representations of a compression and an extension spring. The distance between the peaks of the slanted lines should equal the pitch of the spring.

## 6-25 DESIGNCENTER

AutoCAD includes a **DesignCenter** that includes, among other things, a set of fastener blocks. The **DesignCenter** is accessed using the **DesignCenter** tool located on the Content panel under the insert tab.

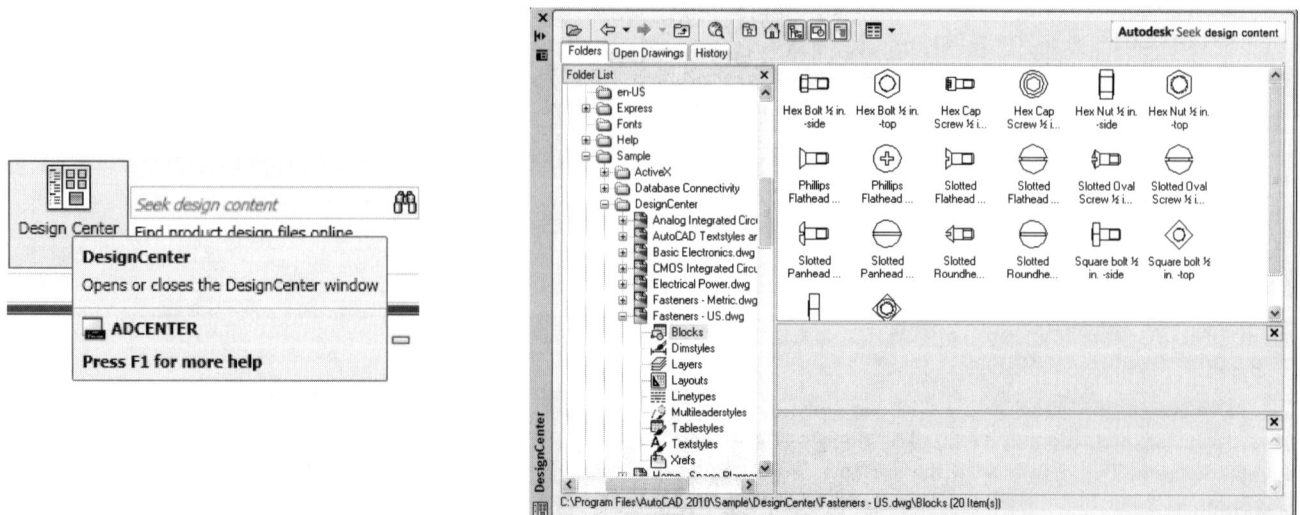

**Figure 6-44**

**To access U.S. fastener blocks**

1. Click the **DesignCenter** tool on the Insert panel.

The **DesignCenter** dialog box will appear. See Figure 6-45.

2. Select the **Blocks** tool (double-click) under the **Fasteners - US.dwg** heading.

A grouping of available blocks will appear in the dialog box.

3. Double-click the **Hex Bolt 1/2 in. -side** option in the display.

**Figure 6-45**

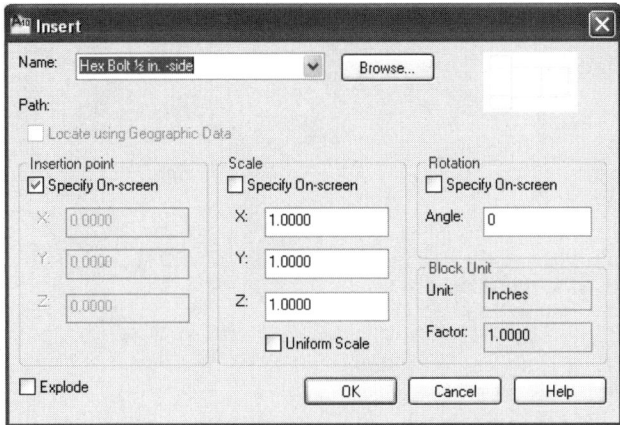

Figure 6-46

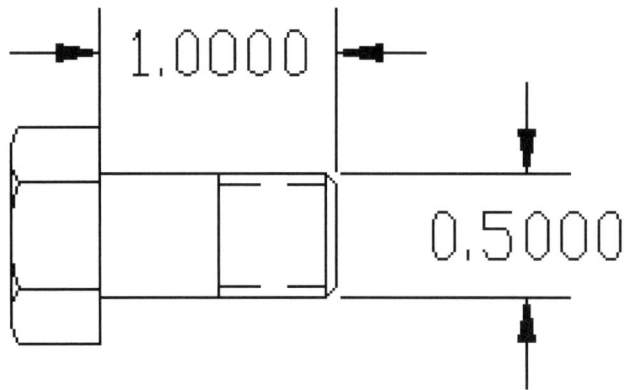

Given block dimensions

Figure 6-48

The **Insert** dialog box will appear. See Figure 6-46.

4.  Accept the default scale values and click **OK.**

A side view of a hex bolt will appear on the screen. See Figure 6-47.

### To change the size of a fastener block

The available blocks are common sizes and can easily be modified to different sizes using the **Scale** options on the **Insert** dialog box. Figure 6-48 shows the dimension values for the **Hex Bolt** block. The dimensions were created using the **Linear** dimension tool. The dimensions for any block can be determined by first inserting a drawing of the block and then using the **Dimension** tool.

### To create a hex bolt that has a .75-inch diameter and a length of 2.00 inches

1.  Select the **Hex Bolt 1/2 in. -side** block from the **DesignCenter.**

The **Insert** dialog box will appear. See Figure 6-49.

Hex Bolt 1/2 in.-side

Figure 6-47

2.  Set the **X** scale factor for **2.0000** and the **Y** scale factor for **1.5000,** then click **OK.**

Figure 6-50 shows the resulting fastener drawing. The dimensions were added for demonstration purposes. Dimensions will not appear on the initial drawing.

### To use the Scale tool

See Figure 6-49.

1.  Click the **Scale** tool located on the **Modify** panel.

    *Command: _scale*
    *Select objects:*

2.  Window the fastener; press the right mouse button.

    Specify base point.

3.  Select a base point.

    The base point need not be on the fastener.

    *Specify scale factor or [Copy/Reference]*
    *<1.0000>:*

4.  Type **2.50;** press **Enter.**

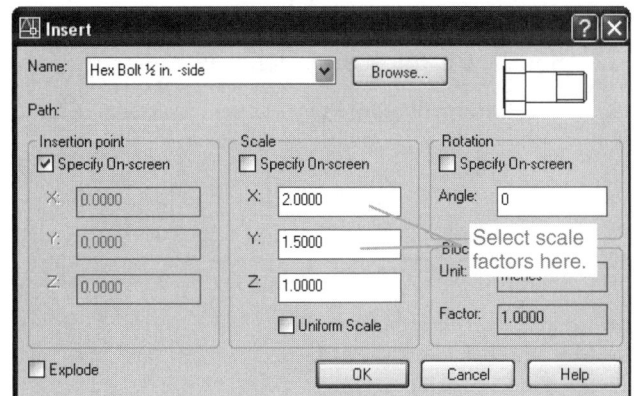

Figure 6-49

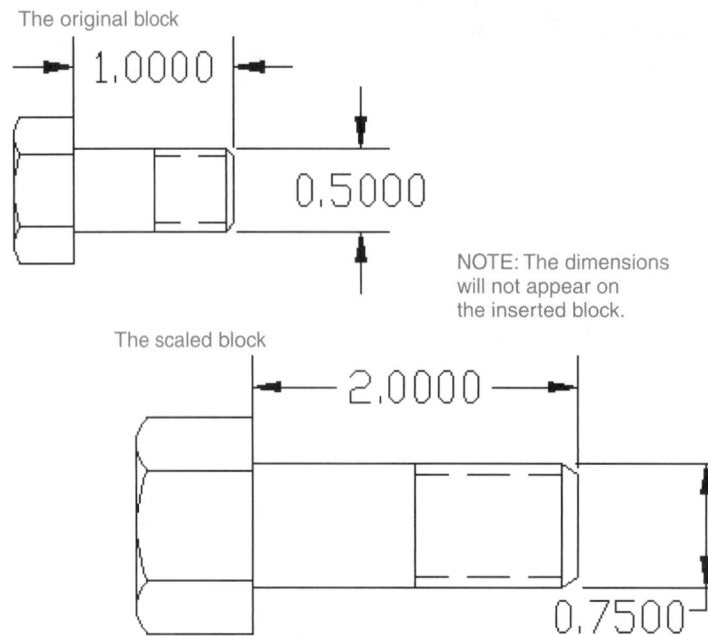

The original block

1.0000

0.5000

NOTE: The dimensions will not appear on the inserted block.

The scaled block

2.0000

0.7500

**Figure 6-50**

The scale tool can be used to increase the size of an object, but the object's proportions remain the same.

**To change a fastener's proportions**

Figure 6-51 shows a Hex Bolt ½in. -side view taken from the **DesignCenter.** The original proportions are shown in Figure 6-48. Draw a bolt that has a Ø0.375 inch and a length of 2.00 inches.

1. Click the **Scale** tool on the **Modify** panel.

The **Scale** tool is used because the proportions of a fastener's head are generally related to its major diameter. The given major diameter is 0.5000 inch. Scaling this dimension down to 0.375 will also reduce the proportions of the fastener's head.

2. Specify a **0.750** scale factor.

This factor will reduce the fastener's major diameter to 0.375.

3. Use the **Explode** tool from the **Modify** panel and explode the fastener.

The fastener as presented from the **DesignCenter** is a block. It must be exploded so that individual lines may be modified.

4. Use the **Move** tool and move the right portion of the fastener 1.00 inch to the right as shown in Figure 6-52.
5. Use the **Extend** tool and extend lines from the fastener's head to the extended protion.
6. Use the **Move** and **Extend** tools to define any required thread length.
7. Save the fastener as a new block. See Figure 6-52.

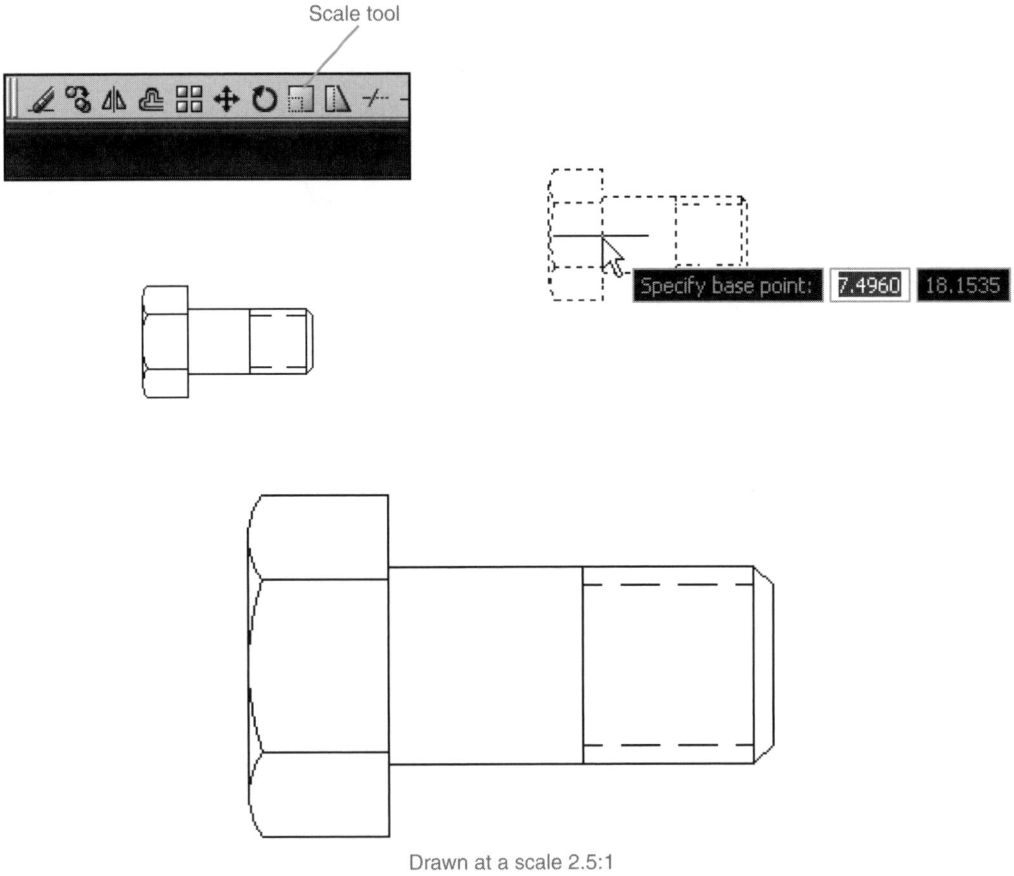

Scale tool

Specify base point:  7.4960   18.1535

Drawn at a scale 2.5:1

**Figure 6-51**

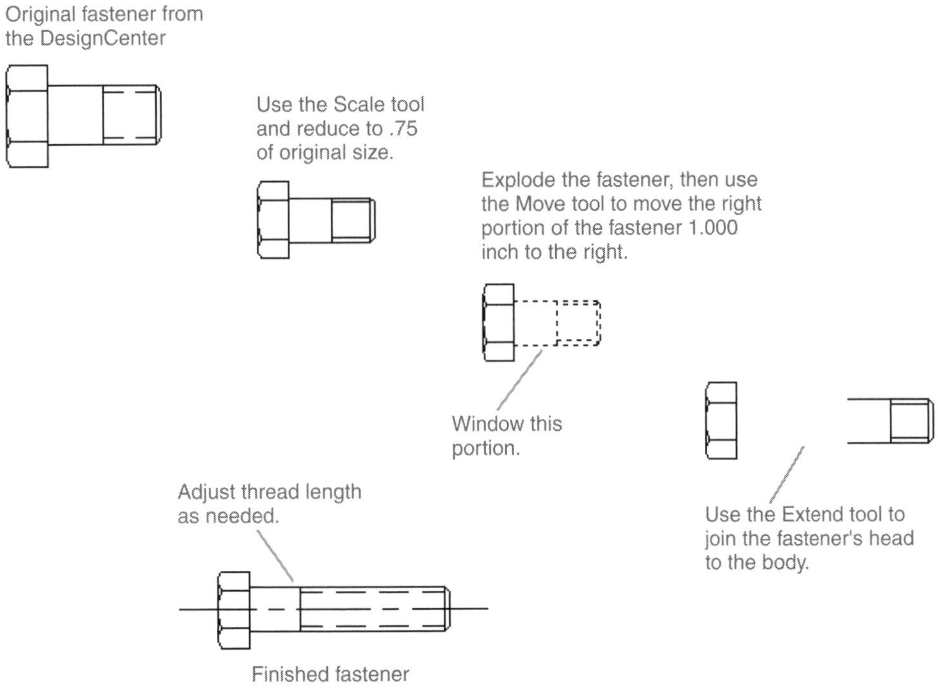

Original fastener from
the DesignCenter

Use the Scale tool
and reduce to .75
of original size.

Explode the fastener, then use
the Move tool to move the right
portion of the fastener 1.000
inch to the right.

Window this
portion.

Use the Extend tool to
join the fastener's head
to the body.

Adjust thread length
as needed.

Finished fastener

**Figure 6-52**

# 6-26 EXERCISE PROBLEMS

### EX6-1

Create wblocks for the following thread major diameters.

A. 1.00-in. detailed representation
B. 1.00-in. schematic thread representation
C. 1.00-in. simplified thread representation
D. 100-mm detailed representation
E. 100-mm schematic thread representation
F. 100-mm simplified thread representation

### EX6-2

Draw a .750–10UNC–2_ × _____ LONG thread. Include the thread callout.

A. External –3.00 LONG.
B. Internal to fit into the object shown in Figure EX6-2.
C. Use a detailed representation.
D. Use a schematic representation.
E. Use a simplified representation.

### EX6-3

Draw a .250–28UNF–2_ × _____ LONG thread. Include the thread callout.

A. External –1.50 LONG.
B. Internal to fit into the object shown in Figure EX6-2.
C. Use a detailed representation.
D. Use a schematic representation.
E. Use a simplified representation.

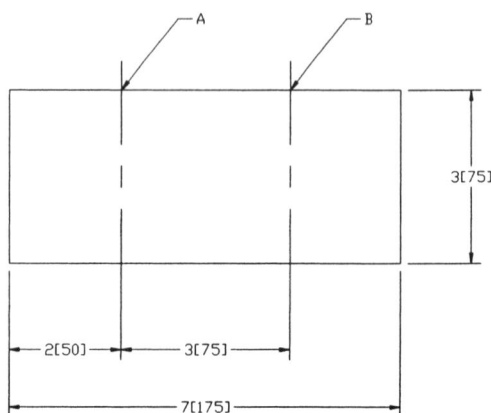

**Figure EX6-2**

### EX6-4

Draw an M36 × 4 × _____ LONG thread. Include the thread callout.

A. External –100 LONG.
B. Internal to fit into the object shown in Figure EX6-2.
C. Use a detailed representation.
D. Use a schematic representation.
E. Use a simplified representation.

### EX6-5

Draw an M12 × 1.75 × _____ LONG thread. Include the thread callout.

A. External –40 LONG.
B. Internal to fit into the object shown in Figure EX6-2.
C. Use a detailed representation.
D. Use a schematic representation.
E. Use a simplified representation.

### EX6-6

Draw a 2.75 × 2 × 5.00 inch–long external square thread.

### EX6-7

Draw a 50 × 2 × 100 millimeter–long external square thread.

### EX6-8

Draw a 3 × 1.5 × 6 inch–long external acme thread.

### EX6-9

Draw an 80 × 1.5 × 200 millimeter–long external acme thread.

Draw the bolts, nuts, and screws in Exercise Problems EX6-10 through EX6-13 assembled into the object shown in Figure EX6-2. If not given, determine the length of the fasteners using the tables given in this chapter or in the appendix. Include the appropriate drawing callout.

### EX6-10

A. .500–13UNC × _____ LONG HEX HEAD BOLT. Include a finished nut on the end of the bolt.
B. 5/8(.625)–18UNF × 1.5 LONG HEX HEAD SCREW. Specify the diameter and length of the tap hole.
C. Draw a detailed representation.

D. Draw a schematic representation.
E. Draw a simplified representation.
F. Draw an orthographic view.
G. Draw a sectional view.

## EX6-11

A. M24– _____ LONG HEX HEAD BOLT. Include a finished nut on the end of the bolt.
B. M16–30 LONG HEX HEAD SCREW. Specify the diameter and length of the tap hole.
C. Draw a detailed representation.
D. Draw a schematic representation.
E. Draw a simplified representation.
F. Draw an orthographic view.
G. Draw a sectional view.

## EX6-12

A. M30 × _____ LONG SQUARE HEAD BOLT. Include two locknuts on the end of the bolt.
B. M12 × 1.4 × 24 LONG SQUARE HEAD SCREW. Specify the diameter and length of the tap hole.
C. Draw a detailed representation.
D. Draw a schematic representation.
E. Draw a simplified representation.
F. Draw an orthographic view.
G. Draw a sectional view.

## EX6-13

A. 1.25–7UNC × _____ LONG SQUARE HEAD BOLT. Include two locknuts on the end of the bolt.
B. .375–32UNEF × 1.5 LONG SQUARE HEAD SCREW. Specify the diameter and length of the tap hole.
C. Draw a detailed representation.
D. Draw a schematic representation.
E. Draw a simplified representation.
F. Draw an orthographic view.
G. Draw a sectional view.

## EX6-14

Redraw the sectional view shown in Figure EX6-14. Include a drawing callout that defines the threads as M12. Specify the depth of the threaded hole and the diameter and depth of the tap hole.

## EX6-15

Redraw the drawing shown in Figure EX6-15. Add a drawing callout for the bolt and nut based on the following thread major diameters. Use only standard bolt lengths as defined in the appendix.

A. .375–24UNF–2A × _____ LONG
B. M16 × 2 × _____ LONG

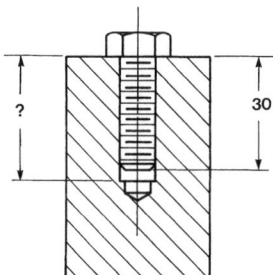

Figure EX6-14

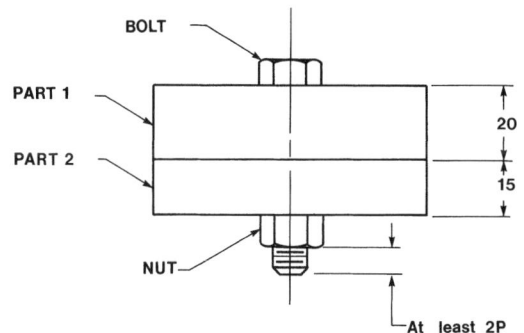

Figure EX6-15

**EX6-16**

Draw a front sectional view and a top orthographic view based on the following drawing and table information. Include the following setscrews in the indicated holes. Include the appropriate drawing callout for each setscrew.

FOR INCH VALUES:
1. .250–20UNC–2A × 1.00
   SLOT HEAD, FLAT POINT
   SET SCREW
2. .375–24UNF–2A × .750
   HEX SOCKET, FULL DOG
   SET SCREW
3. #10–28UNF–2A × .625
   SQUARE, OVAL
   SET SCREW
4. #6–32UNC–2A × .50
   SLOT, CONE POINT
   SET SCREW

FOR MILLIMETER VALUES:
1. M12 × 20
   SLOT HEAD, FLAT POINT
   SET SCREW
2. M16 × 2 × 30
   SQUARE HEAD, HALF DOG
   SET SCREW
3. M6 × 20
   HEX SOCKET, CONE
   SET SCREW
4. M10 × 1.5 × 20
   CUP, SLOT
   SET SCREW

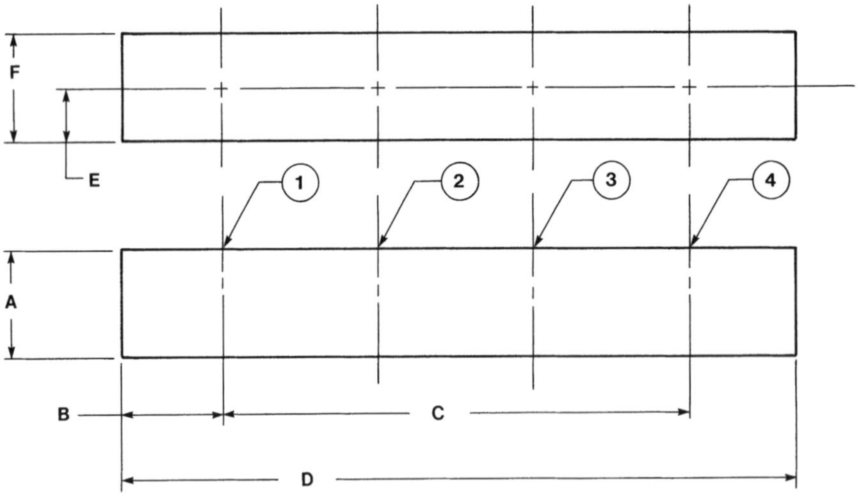

| DIMENSION | INCHES | mm |
|-----------|--------|--------|
| A | 2.00 | 50 |
| B | 1.00 | 25 |
| C | 3 X 1.00 | 3 X 40 |
| D | 6.5 | 170 |
| E | .75 | 20 |
| F | 1.50 | 40 |

**EX6-17**

Draw a front sectional view and a top orthographic view based on the following drawing and table information.

A.  Use inch values.
B.  Use millimeter values.

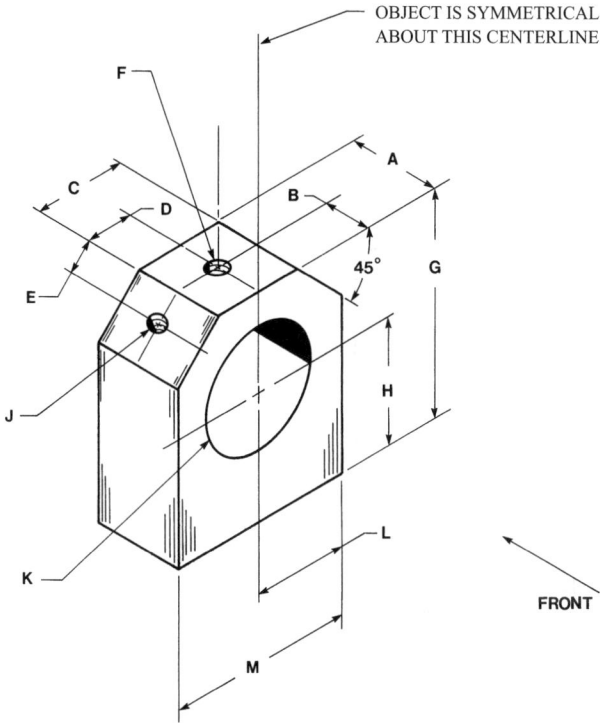

OBJECT IS SYMMETRICAL
ABOUT THIS CENTERLINE

F

C

D

B

A

E

45°

G

H

J

L

K

FRONT

M

| DIMENSION | INCHES | mm |
|-----------|--------|-----|
| A | 1.00 | 26 |
| B | .50 | 13 |
| C | 1.00 | 26 |
| D | .50 | 13 |
| E | .38 | 10 |
| F | .190-32 UNF | M8X1 |
| G | 2.38 | 60 |
| H | 1.38 | 34 |
| J | .164-36 UNF | M6 |
| K | Ø1.25 | Ø30 |
| L | 1.00 | 26 |
| M | 2.00 | 52 |

**EX6-18**

Redraw the following drawing as a sectional view. Select and include the drawing callouts for the nut and washer based on the following bolts. Specify the bolt lengths based on standard lengths listed in the appendix.

A. .625–11UNC–2A × _____ LONG HEX HEAD BOLT
B. M16 × 2 × _____ LONG HEX HEAD

C. .500-20 UNF × _____ LONG SQUARE HEAD
D. M20 × _____ SQUARE HEAD
E. .438-28 UNEF × _____ LONG HEX HEAD
F. M12 × 1.25 × _____ LONG HEX HEAD
G. .750-10 UNC × _____ LONG HEX HEAD
H. M24 × 3 × _____ LONG HEX HEAD

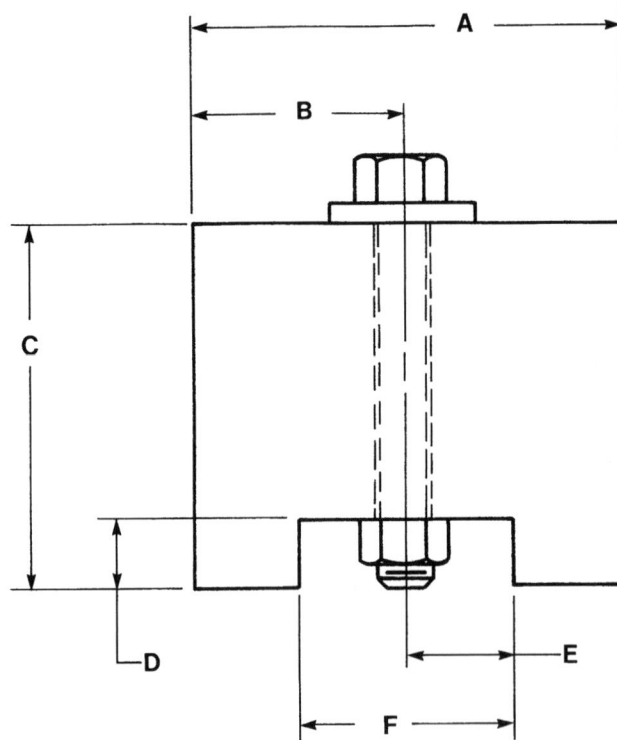

| DIMENSION | INCHES | mm |
|:---:|:---:|:---:|
| A | 3.00 | 76 |
| B | 1.50 | 38 |
| C | 2.50 | 64 |
| D | .50 | 13 |
| E | .75 | 19 |
| F | 1.50 | 38 |

**EX6-19**

Draw a front sectional view and a top orthographic view based on the following drawing and table information. Add fasteners to the labeled holes based on the following information. Include the appropriate drawing callouts. Use only standard sizes as listed in the appendix.

FOR INCH VALUES:
1. Nominal diameter = .250, UNC
   Square head bolt and nut
   A washer between the bolt head and part 23
   A washer between the nut and part 24
2. .375–24UNF–2A × 1.00
   SLOT, OVAL
   SETSCREW
3. Nominal diameter = .375, UNF
   Flat head screw, 1.25 LONG

FOR MILLIMETER VALUES:
1. Nominal diameter = 12, coarse
   Square head bolt and nut
   A washer between the bolt head and part 23
   A washer between the nut and part 24
2. M10 × .5 × 20
   SQUARE HEAD, FULL DOG
   SET SCREW
3. Nominal diameter = 12, fine
   Flat head screw, 20 LONG

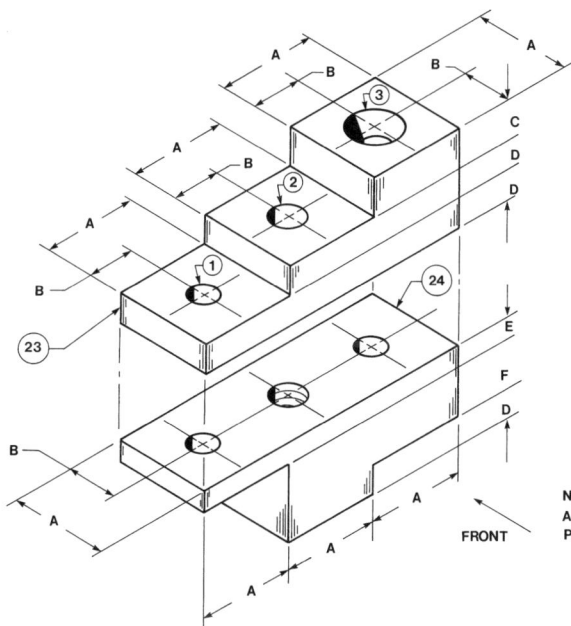

NOTE: HOLES IN PART 24 ALIGN WITH THOSE IN PART 23.

| DIMENSION | INCHES | mm |
|-----------|--------|-----|
| A | 1.25 | 32 |
| B | .63 | 16 |
| C | .50 | 13 |
| D | .38 | 10 |
| E | .25 | 7 |
| F | .63 | 16 |

## EX6-20

Redraw the following sectional view and include the appropriate bolts and nuts at locations W, X, Y, and Z so that the selected bolt heads sit flat on their bearing surfaces. Select the bolts from the standard sizes listed in the appendix.

(W)(X)(Y)(Z) INDICATES FASTENER LOCATIONS

| DIMENSION | INCHES | mm |
|-----------|--------|------|
| A | 1.00 | 50 |
| B | 1.63 | 41 |
| C | .38 | 10 |
| D | 2.00 | 50 |
| E | .75 | 19 |
| F | 1.50 | 38 |
| G | 1.00 | 25 |
| H | .25 | 6 |
| J | .38 | 10 |
| K | 2.00 | 50 |
| L | 3.63 | 92 |
| M | 4.00 | 100 |
| N | R.13 | 3 |

**EX6-21**

Redraw the following drawing based on the following information. Include all bolt, nut, washer, and spring callouts. Use only standard sizes as listed in the appendix.

FOR INCH VALUES:
A.  Major diameter of bolt = .375 coarse thread.
B.  Add the appropriate nut.
C.  Washer is .125 thick and allows a minimum clearance from the bolt of at least .125.
D.  Compression springs are made from .125-diameter wire and are 1.00 long. They have an inside diameter that always clears the bolt by at least .125.
E.  Parts 1 and 2 are 1.00 high and 4.00 wide.
F.  Locate the bolts at least 1.00 from each end.

FOR MILLIMETER VALUES:
A.  Major diameter of bolt = 12 coarse thread.
B.  Add the appropriate nut.
C.  Washer is 3 thick and allows a minimum clearance from the bolt of at least 2.
D.  Compression springs are made from 4-diameter wire and are 24 long. They have an inside diameter that always clears the bolt by at least 3.
E.  Parts 1 and 2 are 25 high and 100 wide.
F.  Locate the bolts at least 25 from each end.

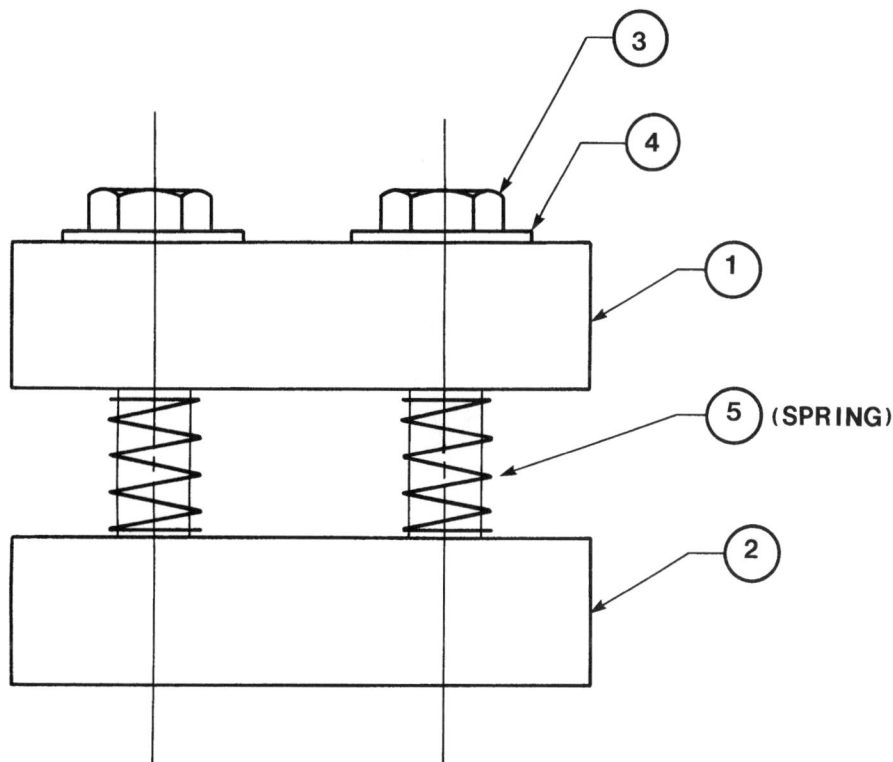

**EX6-22**

Figure EX6-22 shows two identical blocks that are to be held together using a bolt and a nut. Two washers are also used. Specify the appropriate fastener, nut, and washer for the following block sizes.

Block and nominal hole sizes:
A. 1.50 × 1.50 × 0.75 inches: Ø.375 inch
B. 20 × 20 × 12 mm; Ø8 mm
C. 2.00 × 2.00 × 0.625 inches; Ø.500 inch
D. 30 × 30 × 15 mm; Ø10 mm

Figure EX6-22

## EX6-23

Figure EX6-23 shows three parts that are to be held to-gether using six screws. Given the part sizes, specify the six fasteners needed to assemble the parts.

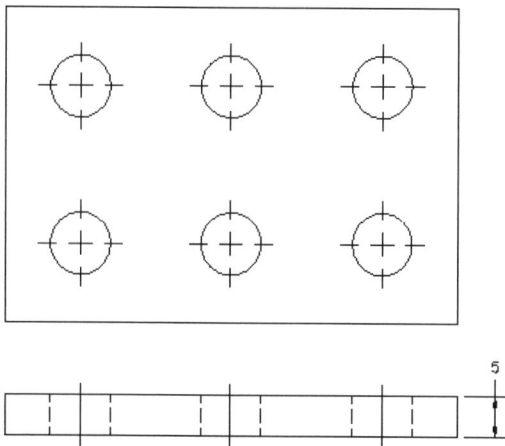

**Figure EX6-23**

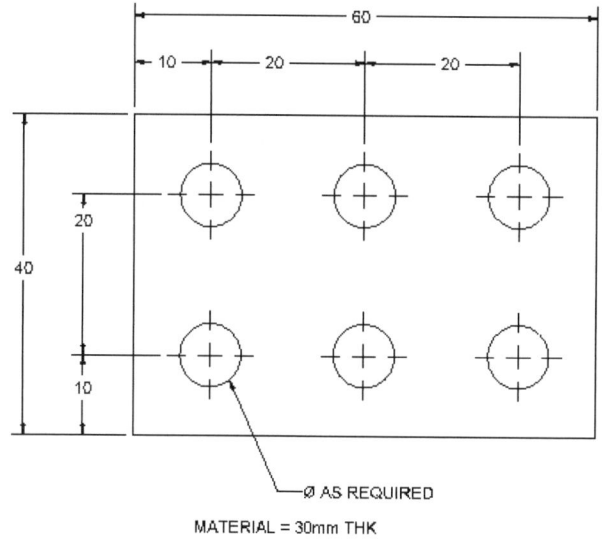

Ø AS REQUIRED

MATERIAL = 30mm THK

**A.**

**B.**

**C.**

Fasteners. *Courtesy of TwinNut Corp., Germany.*

## 6-27 THREADED FASTENERS

Threaded fasteners are the principal devices used for assembling components. To speed production time and reduce costs, many new types of fasteners are created every year. Existing fasteners are also modified to improve their insertion in mass production. Many companies provide CAD drawings of their fasteners on the Web. When you are using standard fasteners in your designs, save time by downloading drawings or models. Thread is usually dimensioned by giving a thread note in the drawing. This allows you to combine more information in a compact space.

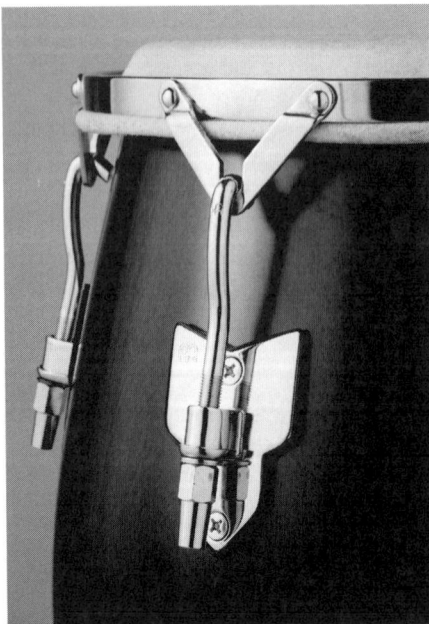

**Figure 6-54** Thread Used for Adjustment

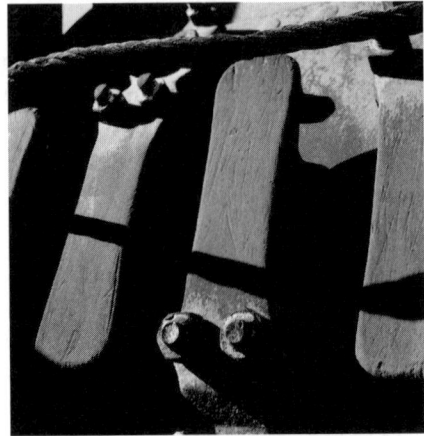

**Figure 6-53** Thread Used for Attachment. *Courtesy of Arthur S. Aubry/Getty Images, Inc.-Photodisc.*

The information in this chapter will prepare you to specify various types of thread and fasteners and use the standard methods of representing them in your drawings.

Web sites related to this chapter

- http://www.mcmaster.com/ A great site for standard parts including fasteners
- http://www.pemnet.com/fastening_products PEM fasteners
- https://sdp-si.com/eStore/ More standard parts with downloadable CAD files

## 6-28 UNDERSTANDING THREADS AND FASTENERS

Screw threads are vital to industry. They are designed for hundreds of different purposes. The three basic applications are as follows:

1. To hold parts together (Figure 6-53).
2. To provide for adjustment between parts (Figure 6-54).
3. To transmit power (Figure 6-55).

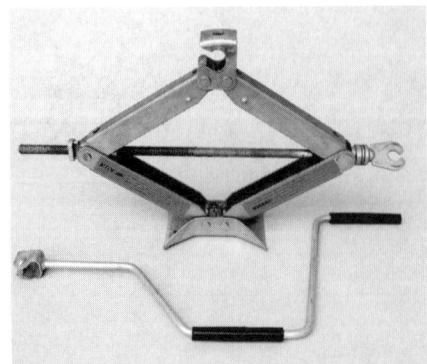

**Figure 6-55** Thread Used to Transmit Power

# THE STANDARDIZATION OF SCREW THREADS

At one time, there was no such thing as standardization. Nuts made by one manufacturer would not fit the bolts of another. In 1841 Sir Joseph Whitworth started crusading for a standard screw thread, and soon the Whitworth thread was accepted throughout England.

In 1864 the United States adopted a thread proposed by William Sellers of Philadelphia, but the Sellers nuts would not screw onto a Whitworth bolt or vice versa. In 1935 the American standard thread, with the same 60° V form of the old Sellers thread, was adopted in the United States.

Still there was no standardization among countries. In peacetime it was a nuisance; in World War I it was a serious inconvenience; and in World War II the obstacle was so great that the Allies decided to do something about it. Talks began among the Americans, British, and Canadians, and in 1948 an agreement was reached on the unification of American and British screw threads. The new thread was called the Unified screw thread, and it represented a compromise between the American standard and Whitworth systems, allowing complete interchangeability of threads in three countries.

In 1946 a committee called the International Organization for Standardization (ISO) was formed to establish a single international system of metric screw threads.

Sir Joseph Whitworth. *Courtesy of National Park Service.*

Consequently, through the cooperative efforts of the Industrial Fasteners Institute (IFI), several committees of the American National Standards Institute, and the ISO representatives, a metric fastener standard was prepared.*

The shape of the helical (spiral shaped) thread is called the **thread form.** The metric thread form is the international standard, although the unified thread form is common in the United States. Other thread forms are used in specific applications.

CAD drawing programs are often used to automatically depict threads. The thread specification is a special leader note that defines the type of thread or fastener. This is an instruction for the shop technician so the correct type of thread is created during the manufacturing process.

(a) External thread

(b) Internal thread

**Figure 6-56** Screw Thread Nomenclature

## Screw Thread Terms

The following definitions apply to screw threads in general and are shown on the illustration Figure 6-56. For additional information regarding specific Unified and metric screw thread terms and definitions, refer to the appropriate standards.

**Screw Thread:** A ridge of uniform cross section in the form of a helix on the external or internal surface of a cylinder.

**External Thread:** A thread on the outside of a member, as on a shaft.

**Internal Thread:** A thread on the inside of a member, as in a hole.

**Major Diameter:** The largest diameter of a screw thread (for both internal and external threads).

**Minor Diameter:** The smallest diameter of a screw thread (for both internal and external threads).

**Pitch:** The distance from a point on a screw thread to a corresponding point on the next thread measured parallel to the axis. In the U.S., the pitch is equal to 1 divided by the number of threads per inch.

**Pitch Diameter:** The diameter of an imaginary cylinder passing through the threads where the widths of the threads and the widths of the spaces would be equal.

**Lead:** The distance a screw thread advances axially in one turn.

**Angle of Thread:** The angle between the sides of the thread measured in a plane through the axis of the screw.

**Crest:** The top surface joining the two sides of a thread.

**Root:** The bottom surface joining the sides of two adjacent threads.

**Side:** The surface of the thread that connects the crest with the root.

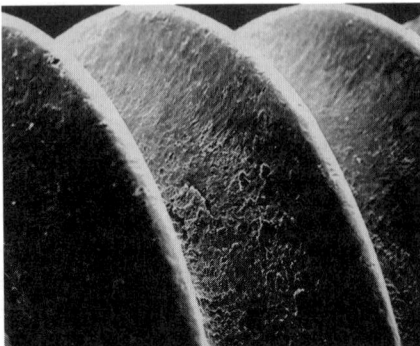

Electron Microscope View of a Thread Surface. *Courtesy of David Gnizak/Phototake NYC.*

**Axis of Screw:** The longitudinal centerline through the screw.

**Depth of Thread:** The distance between the crest and the root of the thread measured normal to the axis.

**Form of Thread:** The cross section of thread cut by a plane containing the axis.

**Series of Thread:** The standard number of threads per inch for various diameters.

## Screw Thread Forms

The thread form is the cross sectional shape of the thread. Various forms of threads are used for different purposes. The main uses for threads are to hold parts together, to adjust parts with reference to each other, and to transmit power. Figures 6-57–6-58 show some of the typical thread forms.

**Sharp-V thread** (60 degrees) useful for certain adjustments because of the increased friction resulting from the full thread face. It is also used on brass pipe work (Figure 6-57a).

**American national thread,** with flattened roots and crests, is a stronger thread. This form replaced the sharp- V thread for general use. (Figure 6-57b)

**Unified thread** is the standard thread agreed upon by the United States, Canada, and Great Britain in 1948. It has replaced the American national form. The crest of the external thread may be flat or rounded, and the root is rounded; otherwise, the thread form is essentially the same as the American national. Some earlier American national threads are still included in the new standard, which lists 11 different numbers of threads per inch for the various standard diameters, together with selected combinations of special diameters and pitches. The 11 series includes: the coarse thread series (UNC or NC), recommended for general use; the fine thread series (UNF or NF), for general use in automotive and aircraft work and in applications where a finer thread is required; the extra fine series (UNF or NF), which is the same as the SAE extra fine series, used particularly in aircraft and aeronautical equipment and generally for threads in thin walls; and the eight series of 4, 6, 8, 12, 16, 20, 28, and 32 threads with constant pitch. The 8UN or 8N, 12UN or 12N, and 16UN or 16N series are recommended for the uses corresponding to the old 8-, 12-, and 16-pitch American national threads. In addition, there are three special thread series—UNS, NS, and UN—that involve special combinations of diameter, pitch, and length of engagement (Figure 6-57c).

*Unified extra fine thread series* (UNEF) has many more threads per inch for given diameters than any series of the American national or unified. The form of thread is the same as the American national. These small threads are used in thin metal where the length of thread engagement is small, in cases where close adjustment is required, and where vibration is great.

*Metric thread* is the standard screw thread agreed upon for international screw thread fasteners. The crest and root are flat, but the external thread is often rounded if formed by a rolling process. The form is similar to the American national and unified threads but with less depth of thread. The preferred metric thread for commercial purposes conforms to the ISO basic profile M for metric threads. This M profile design is comparable to the unified inch profile, but the two are not interchangeable. For commercial purposes, two series of metric threads are preferred—coarse (general purpose) and fine—much fewer than previously used (Figure 6-58a).

*Square thread* is theoretically the ideal thread for power transmission, since its face is nearly at right angles to the axis, but due to the difficulty of cutting it with dies and because of other inherent disadvantages (such as the fact that split nuts will not readily disengage), square thread has been displaced to a large extent by the acme thread. Square thread is not standardized (Figure 6-58b).

*Acme thread* is a modification of the square thread and has largely replaced it. It is stronger than the square thread, is easier to cut, and has the advantage of easy disengagement from a split nut, as on the lead screw of a lathe (Figure 6-58c).

*Standard worm thread* (not shown) is similar to the acme thread but is deeper. It is used on shafts to carry power to worm wheels.

*Whitworth thread* was the British standard and has been replaced by the unified thread. The uses of Whitworth thread correspond to those of the American national thread (Figure 6-59a).

*Knuckle thread* is usually rolled from sheet metal but is sometimes cast. In modified forms knuckle thread is used in electric bulbs and sockets, bottle tops, etc. (Figure 6-59b).

*Buttress thread* is designed to transmit power in one direction only. It is commonly used in large guns, in jacks, and in other mechanisms that have high strength requirements (Figure 6-59c).

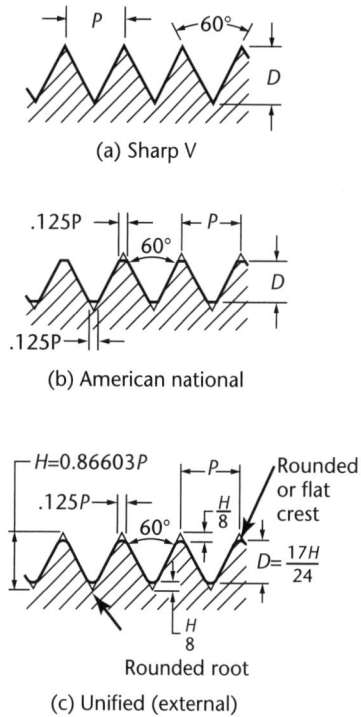

**Figure 6-57** Sharp, American National and Unified Screw Thread Forms

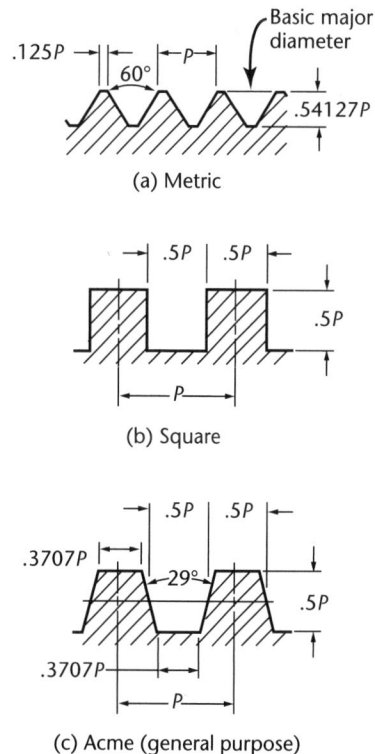

**Figure 6-58** Metric, Square, and Acme Screw Thread Forms

Knuckle Thread is Often Used on Electric Bulbs. *Stephen Oliver © DORLING KINDERSLEY.*

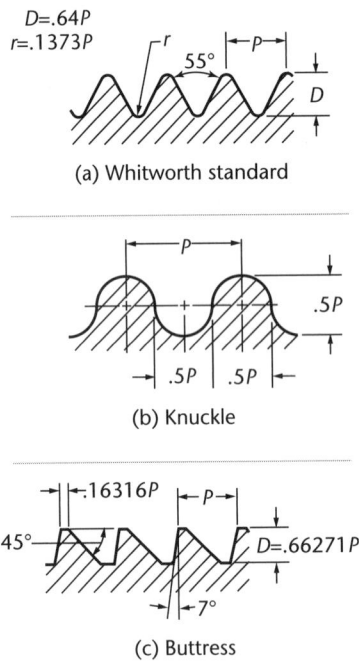

**Figure 6-59**  Whitworth Standard, Knuckle, and Buttress Screw Thread Forms

A number of different thread forms are defined in various ASME standards which specify requirements for the design and selection of screw threads. For example, the old N thread series has been superseded by the UN series.

**Thread Pitch**

The pitch of any thread form is the distance parallel to the axis between corresponding points on adjacent threads, as shown in Figure 6-60.

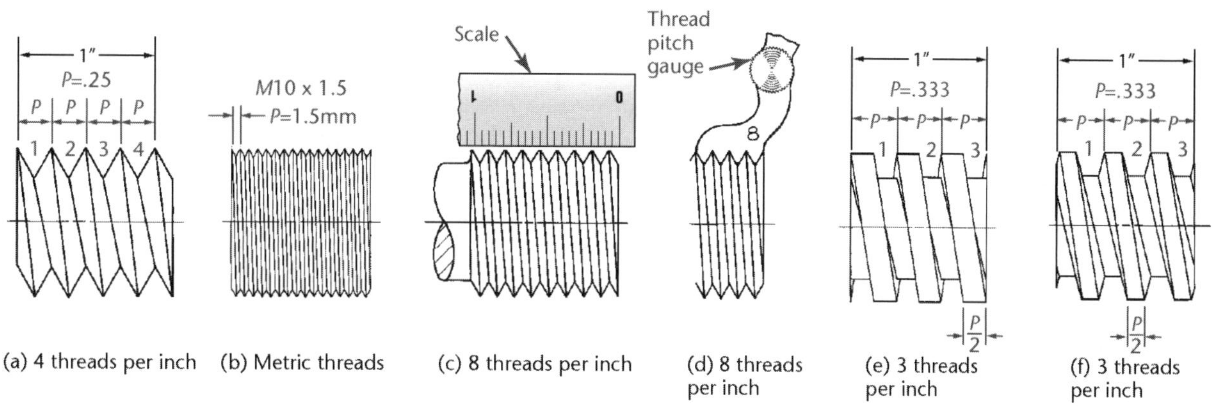

(a) 4 threads per inch   (b) Metric threads   (c) 8 threads per inch   (d) 8 threads per inch   (e) 3 threads per inch   (f) 3 threads per inch

**Figure 6-60**   Pitch

For metric threads, this distance is specified in millimeters. The pitch for a metric thread that is included with the major diameter in the thread designation determines the size of the thread—for example, as shown in Figure 6-60b.

For threads dimensioned in inches, the pitch is equal to 1 divided by the number of threads per inch. For example, a unified coarse thread of 1" diameter has eight threads per inch, and the pitch P equals 1/8" (.125").

If a thread has only four threads per inch, the pitch and the threads themselves are quite large, as shown in Figure 6-60a. If there are 16 threads per inch, the pitch is only 1/16" (.063"), and the threads are relatively small, similar to those in Figure 6-60b.

The pitch or the number of threads per inch can be measured with a scale or with a **thread pitch gage.**

## A HISTORICAL THREAD

The concept of the screw thread seems to have occurred first to Archimedes, the third-century-B.C. mathematician who wrote briefly on spirals and designed

An Archimedean Screw. © 2007 Jupiterimages Corporation.

several simple devices applying the screw principle. By the first century B.C., the screw was a familiar element but was crudely cut from wood or filed by hand on a metal shaft. Not much was heard of the screw thread until the 15th century.

Leonardo da Vinci understood the screw principle, and created sketches showing how to cut screw threads by machine. In the 16th century, screws appeared in German watches and were used to fasten suits of armor. In 1669, the Frenchman, Besson, invented the screw-cutting lathe, but this method of production did not take hold for another century and a half; nuts and bolts continued to be made largely by hand. Screw manufacturing began in 18th century England, during the Industrial Revolution.

## Thread Series

ASME/ANSI Y14.6-2001, Screw Thread Representation, is a standard for drawing, specifying, and dimensioning threads on drawings.

The thread series is the detail of the shape and number of threads per inch composing different groups of fasteners. Table 6-1 shows the thread series for UN thread.

Five series of threads were used in the old ANSI standards:

*Coarse thread* A general-purpose thread used for holding. It is designated NC (national coarse).

*Fine thread* A greater number of threads per inch it is used extensively in automotive and aircraft construction. It is designated NF (national fine).

*8-pitch thread* All diameters have eight threads per inch. It is used on bolts for high-pressure pipe flanges, cylinder-head studs, and similar fasteners. It is designated 8N (national form, 8 threads per inch).

*12-pitch thread* All diameters have 12 threads per inch. It is used in boiler work and for thin nuts on shafts and sleeves in machine construction. It is designated 12N (national form, 12 threads per inch).

*16-pitch thread* All diameters have 16 threads per inch. It is used where necessary to have a fine thread regardless of diameter, as on adjusting collars and bearing retaining nuts. It is designated 16N (national form, 16 threads per inch).

**Table 6-1** Thread Series of UN Thread.

| Basic Thread Series | Constant Pitch | Coarse | Fine | Extra Fine | Special Diameter |
|---|---|---|---|---|---|
| UN | UN | UNC | UNF | UNEF | UNS |
| UNJ | UNJ | UNJC | UNJF | UNJEF | UNJS |
| N | N | NC | NF | NEF | NS |
| UNR | UNR | UNRC | UNRF | UNREF | UNRS |

*This series is superseded by the UN series.*

### Right-Hand and Left-Hand Threads

A right-hand thread is one that advances into a nut when turned clockwise, and a left-hand thread is one that advances into a nut when turned counterclockwise, as shown in Figure 6-61. A thread is always considered to be right-handed (RH) unless otherwise specified. A left-hand thread is always labeled LH on a drawing.

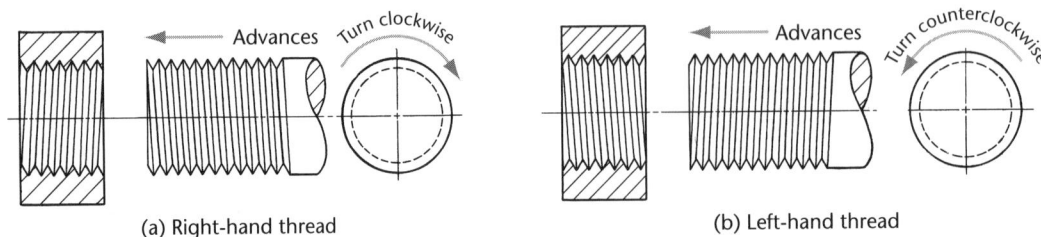

(a) Right-hand thread

(b) Left-hand thread

**Figure 6-61** Right-Hand and Left-Hand Threads

### Single and Multiple Threads

A **single thread**, as the name implies, is composed of one ridge, and the lead is therefore equal to the pitch. Multiple threads are composed of two or more ridges running side by side. As shown in Figures 6-62a–c, the slope line is the hypotenuse of a right triangle whose short side equals .5*P* for single threads, *P* for double threads, 1.5*P* for triple threads, and so on. This applies to all forms of threads. In double threads, the lead is twice the pitch; in triple threads, the lead is three times the pitch, and so on. On a drawing of a single or triple thread, a root is opposite a crest; in the case of a dou-

ble or quadruple thread, a root is drawn opposite a root. Therefore, in one turn, a double thread advances twice as far as a single thread, and a triple thread advances three times as far. RH double square and RH triple acme threads are shown in Figures 6-62d and 6-62e, respectively.

**Multiple threads** are used wherever quick motion, but not great power, is desired, as on ballpoint pens, toothpaste caps, valve stems, and so on. The threads on a valve stem are frequently multiple threads to impart quick action in opening and closing the valve. Multiple threads on a shaft can be recognized and counted by observing the number of thread starts on the end of the screw.

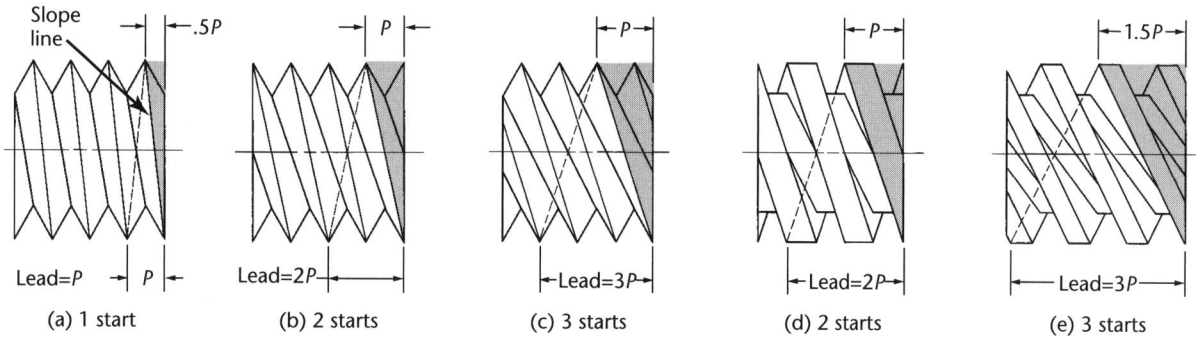

**Figure 6-62**    Multiple Threads

## American National Thread Fits

For general use, three classes of screw thread fits between mating threads (as between bolt and nut) have been established by ANSI.

These fits are produced by the application of tolerances listed in the standard and are as follows:

*Class 1 fit*  Recommended only for screw thread work where clearance between mating parts is essential for rapid assembly and where shake or play is not objectionable.

*Class 2 fit*  Represents a high quality of commercial thread product and is recommended for the great bulk of interchangeable screw thread work.

*Class 3 fit*  Represents an exceptionally high quality of commercially threaded product and is recommended only in cases where the high cost of precision tools and continual checking are warranted.

The standard for unified screw threads specifies tolerances and allowances defining the several classes of fit (degree of looseness or tightness) between mating threads. In the symbols for fit, the letter A refers to the external threads and B to internal threads. There are three classes of fit each for external threads (1A, 2A, 3A) and internal threads (1B, 2B, 3B). Classes 1A and 1B have generous tolerances, facilitating rapid assembly and disassembly. Classes 2A and 2B are used in the normal production of screws, bolts, and nuts, as well as in a variety of general applications. Classes 3A and 3B provide for applications needing highly accurate and close-fitting threads.

## Metric and Unified Thread Fits

Some specialized metric thread applications are specified by tolerance grade, tolerance position, class, and length of engagement. There are two general classes of metric thread fits. The first is for general-purpose applications and has a tolerance class of 6H for internal threads and a class of 6g for external threads. The second is used where closer fits are necessary and has a tolerance class of 6H for internal threads and a class of 5g6g for external threads. Metric thread tolerance classes of 6H/6g are generally assumed if not otherwise designated and are used in applications comparable to the 2A/2B inch classes of fits.

The single-tolerance designation of 6H refers to both the tolerance grade and position for the pitch diameter and the minor diameter for an internal thread. The single-tolerance designation of 6g refers to both the tolerance grade and position for the pitch diameter and the major diameter of the external thread. A double designation of 5g6g indicates separate tolerance grades for the pitch diameter and for the major diameter of the external thread.

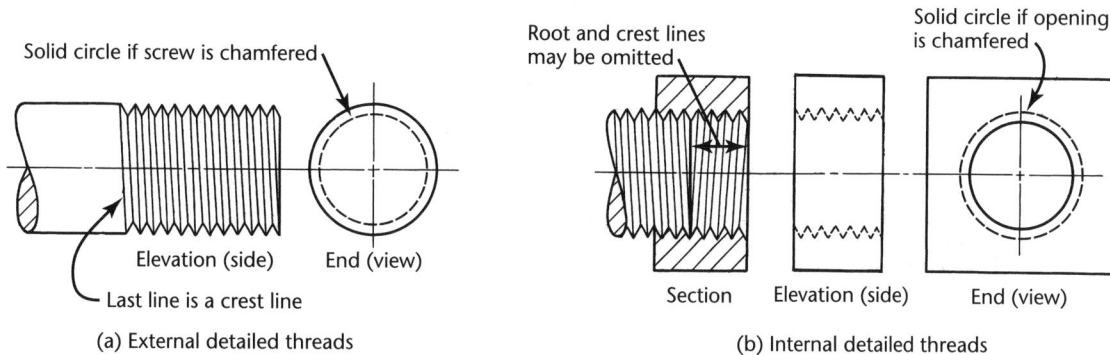

**Figure 6-63**    Detailed Metric, American National, and Unified Threads

## Three Methods for Drawing Thread

There are three methods of representing screw threads on drawings—the schematic, simplified, and detailed methods. Schematic, simplified, and detailed thread symbols may be combined on a single drawing.

Schematic and the more common simplified representations are used to show threads. The symbols are the same for all forms of threads, such as metric, unified, square, and acme, but the thread specification identifies which is to be used.

Detailed representation is a closer approximation of the exact appearance of a screw thread, where the true profiles of the thread's form are drawn; but the helical curves are replaced by straight lines. The true projection of the heli-cal curves of a screw thread takes too much time to draw, so it is rarely used in practice.

Do not use detailed representation unless the diameter of the thread on the drawing is more than 1" or 25 mm and then only to call attention to the thread when necessary. Schematic representation is much simpler to draw and still presents the appearance of thread. Detailed representation is shown in Figure 6-63. Whether the crests or roots are flat or rounded, they are represented by single lines and not double lines. American national and unified threads are drawn the same way. Figure 6-64 shows schematic thread symbols, and Figure 6-65 shows simplified thread symbols.

Figure 6-66 shows detailed directions for drawing schematic and simplified thread.

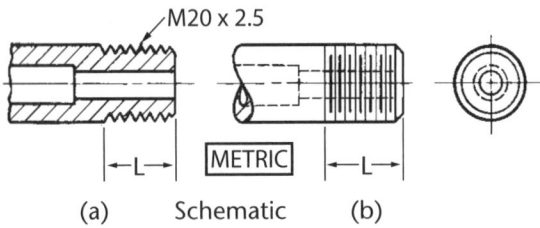

**Figure 6-64**  Schematic Thread Symbols

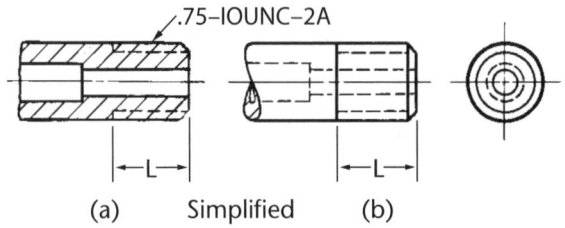

**Figure 6-65**  Simplified Thread Symbols

# STEP BY STEP

## Showing Detailed Thread

1. Make centerline and lay out length and major diameter as shown at right.

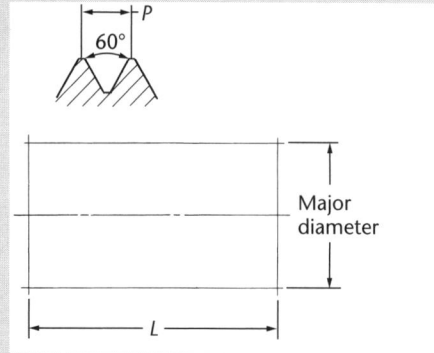

2. Find the number of threads per inch for American National and Unified threads. This number depends on the major diameter of the thread and whether the thread is internal or external.

   Find $P$ (pitch) by dividing 1 by the number of threads per inch. The pitch for metric threads is given directly in the thread designation. For example, the thread has a pitch of 2 mm.

   Establish the slope of the thread by offsetting the slope line .5$P$ for single threads, $P$ for double threads, 1.5$P$ for triple threads, and so on. For right-hand external threads, the slope line slopes upward to the left; for left-hand external threads, the slope line slopes upward to the right.

   By eye, mark off even spacing for the pitch. If using CAD, make a single thread and array the lines using the pitch as the spacing.

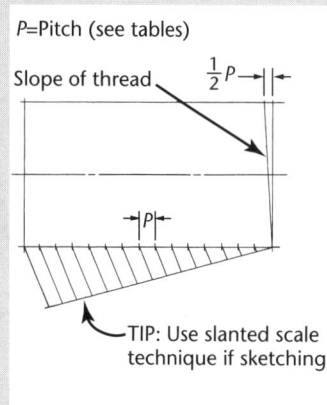

3. From the pitch points, make crest lines parallel to the slope line. These should be dark, thin lines. Make two V's to establish the depth of thread, and sketch light guidelines for the root of thread, as shown.

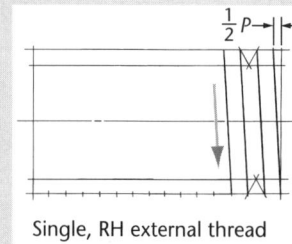

4. Finish the final 60° V's. The V's should be vertical; they should not lean with the thread.

   Make root lines. Root lines will not be parallel to crest lines, but should appear parallel to each other.

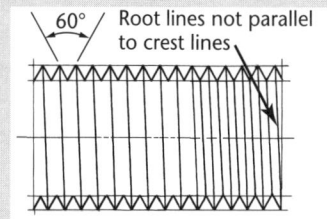

5. When the end is chamfered (usually 45° with end of shaft, sometimes 30°), the chamfer extends to the thread depth. The chamfer creates a new crest line, which you make between the two new crest points. It is not parallel to the other crest lines. When finished, all thread root and crest lines should be shown thin, but dark.

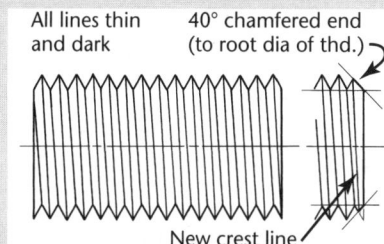

**TIP**

The thread depth table within Figure 6-66 gives approximate thread depths for a variety of common diameters.

| MAJOR DIAMETER | #5 (.125) TO #12 (.216) | .25 | .3125 | .375 | .4375 | .5 | .5625 | .625 | .6875 | .75 | .8125 | .875 | .9375 | I. |
|---|---|---|---|---|---|---|---|---|---|---|---|---|---|---|
| DEPTH, D | .03125 | .03125 | .03125 | .0468 | .0468 | .0625 | .0625 | .0625 | .0625 | .0781 | .0937 | .0937 | .0937 | .0937 |
| PITCH, P | .0468 | .0625 | .0625 | .0625 | .0625 | .0937 | .0937 | .0937 | .0937 | .125 | .125 | .125 | .125 | .125 |

Approximate thread depth table        *(For metric values: 1″ = 25.4mm or see inside front cover)*

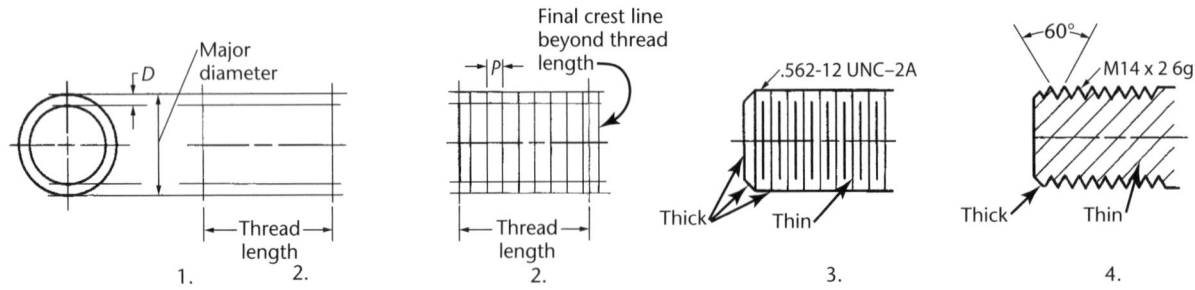

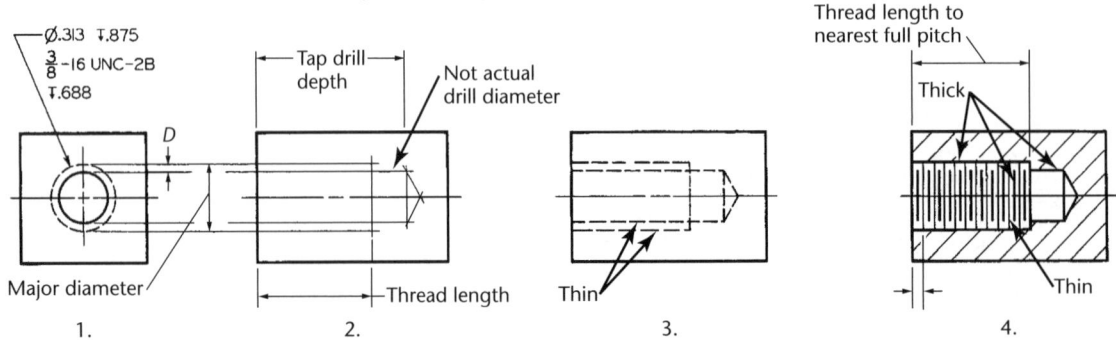

**Figure 6-66**   Steps to Draw Thread Symbols—Simplified and Schematic

# 6-29 THREAD NOTES

ASME/ANSI Y14.6-2001, Screw Thread Representation, is a standard for representing, specifying, and dimensioning screw threads on drawings. Thread notes for metric, unified, and American national screw threads are shown in Figure 6-67 and 6-68. These same notes or symbols are used in correspondence, on shop and storeroom records, and in specifications for parts, taps, dies, tools, and gages.

Metric screw threads are designated basically by the letter M for metric thread symbol followed by the thread form and nominal size (basic major diameter) in millimeters and separated by the symbol × followed by the pitch, also in millimeters. For example, the basic thread note M10 × 1.5 is adequate for most commercial purposes, as shown in Figure 6-67. If needed, the class of fit and LH for left-hand designation is added to the note. (The absence of LH indicates a RH thread.)

If necessary, the length of the thread engagement is added to the thread note. The letter S stands for short, N means normal, and L means long. For example, the single note M10 × 1.5-6H/6g-N-LH combines the specifications for internal and external mating of left-hand metric threads of 10 mm diameter and 1.5 mm pitch with general-purpose tolerances and normal length of engagement.

If the thread is a multiple thread, the word STARTS, with the number of thread starts all contained in parentheses, should precede the thread form; otherwise, the thread is understood to be single. For example:

(2 STARTS) UNC

would indicate that this is double thread.

A thread note for a blind tapped hole is shown in Figure 6-67b. A tap drill is sized to form a hole that will leave enough material for thread to be cut using a tap in order to form a threaded hole. In practice the tap drill size and depth are omitted and left up to the shop. At times it is desirable to state a tolerance range for the size of the hole prior to threading. This can be stated as follows:

Ø.656–.658 BEFORE THD .75–20–NEF–2B

For tap drill sizes.

Thread notes for holes are preferably attached to the circular views of the holes. Thread notes for external threads are preferably given where the threaded shaft appears rectangular, as shown in Figures 6-67c–g. A sample special thread designation is 1.50-7N-LH.

General-purpose acme threads are indicated by the letter G, and centralizing acme threads by the letter C. Typical thread notes are 1-4 ACME-2G or 1-6 ACME-4C.

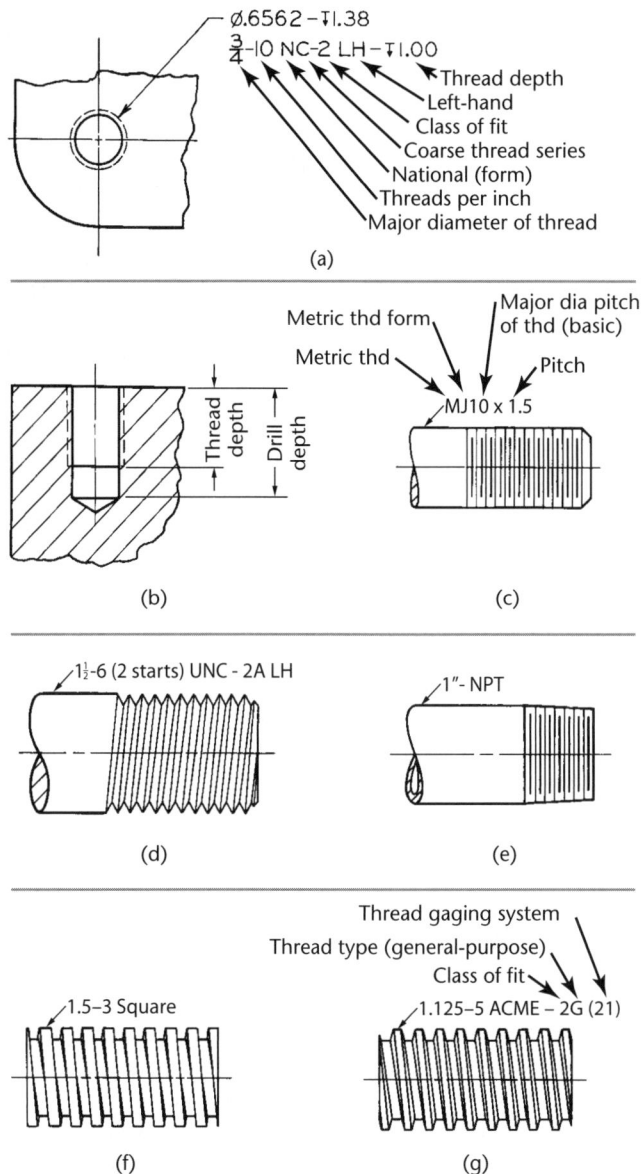

**Figure 6-67**  Thread Notes

Thread notes for unified threads are shown in Figures 6-67d and e. The letters A and B designate external or internal, respectively, after the numeral designating the class of fit. If the letters LH are omitted, the thread is understood to be right hand. Some typical thread notes are:

-20 (3 STARTS) UNC-2A
-18 UNF-2B
1-16 UN-2A

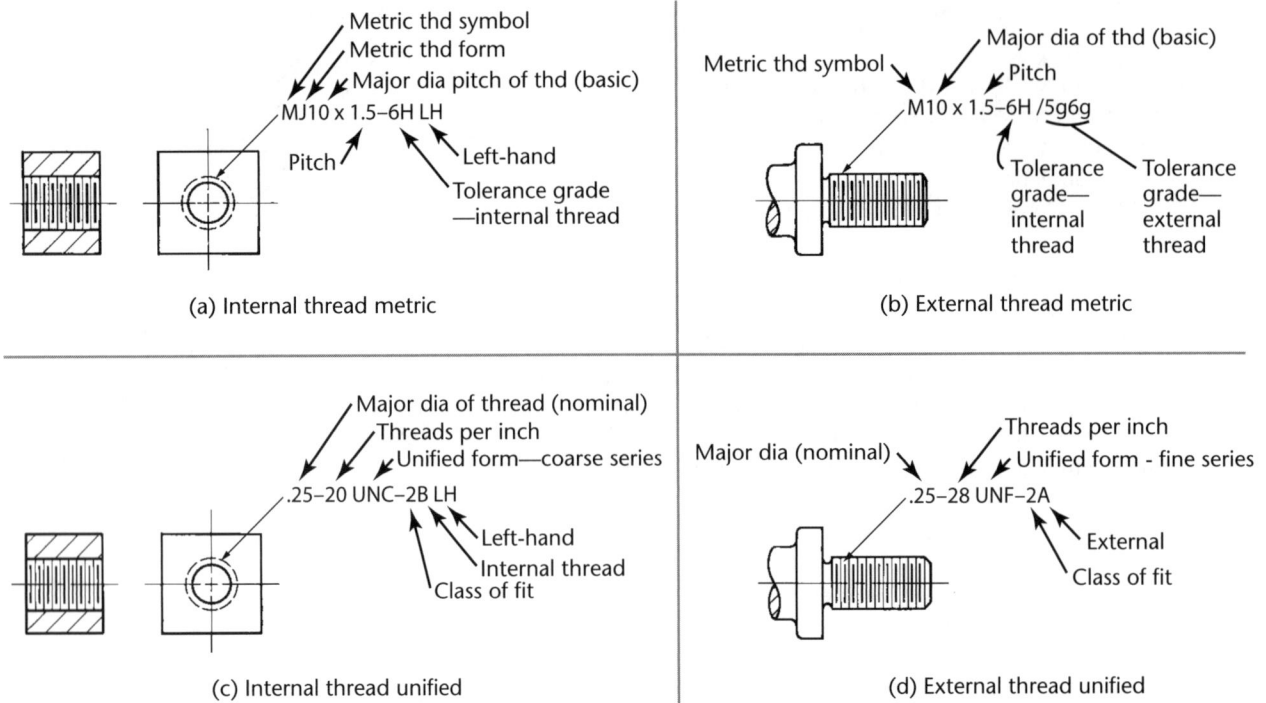

(a) Internal thread metric

Metric thd symbol
Metric thd form
Major dia pitch of thd (basic)
MJ10 x 1.5–6H LH
Pitch
Left-hand
Tolerance grade
—internal thread

(b) External thread metric

Metric thd symbol
Major dia of thd (basic)
Pitch
M10 x 1.5–6H /5g6g
Tolerance grade— internal thread
Tolerance grade— external thread

(c) Internal thread unified

Major dia of thread (nominal)
Threads per inch
Unified form—coarse series
.25–20 UNC–2B LH
Left-hand
Internal thread
Class of fit

(d) External thread unified

Major dia (nominal)
Threads per inch
Unified form - fine series
.25–28 UNF–2A
External
Class of fit

**Figure 6-68**  Thread Notes

## 6-30  EXTERNAL THREAD SYMBOLS

Simplified representation for external threads are shown in Figures 6-69a and b. The threaded portions are indicated by hidden lines parallel to the axis at the approximate depth of the thread, whether the cylinder appears rectangular or circular. The depth shown is not always the actual thread depth, just a representation of it. Use the table in Figure 6-66 for the general appearance of these lines.

When the schematic form is shown in section, as in Figure 6-70a, show the V's of the thread to make the thread obvious. It is not necessary to show the V's to scale or to the actual slope of the crest lines. To draw the V's, use the schematic thread depth, as shown in Figure 6-67, and determine the pitch by drawing 60° V's.

Schematic threads are indicated by alternate long and short lines, as shown in Figure 6-70b. The short lines representing the root lines are thicker than the long crest lines. Theoretically, the crest lines should be spaced according to actual pitch, but this would make them crowded and tedious to draw, defeating the purpose, which is to save time in sketching them. Space the crest lines carefully by eye, then add the heavy root lines halfway between the crest lines. Generally, lines closer together than about 1/16" are hard to distinguish. The spacing should be proportionate for all diameters. You do not need to use these actual measurements in sketching schematic threads, just use them to get a feel for how far apart to make the lines.

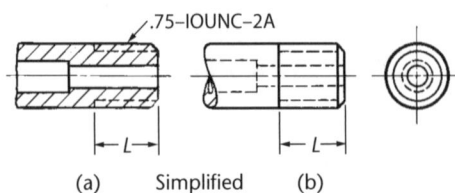

.75–IOUNC–2A

(a)    Simplified    (b)

**Figure 6-69**  External Thread Symbols for Simplified Thread

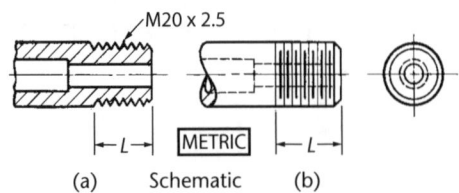

M20 x 2.5

METRIC

(a)    Schematic    (b)

**Figure 6-70**  External Thread Symbols for Schematic Thread

# 6-31 INTERNAL THREAD SYMBOLS

Internal thread symbols are shown in Figure 6-71. Note that the only differences between the schematic and simplified internal thread symbols occur in the sectional views. The representation of the schematic thread in section in Figures 6-71k, 6-71m, and 6-71n is exactly the same as the external representation shown in Figure 6-70b. Hidden threads, by either method, are represented by pairs of hidden lines. The hidden dashes should be staggered, as shown.

In the case of blind tapped holes, the drill depth normally is drawn at least three schematic pitches beyond the thread length, as shown in Figures 6-71d, 6-71e, 6-71l, and 6-71m. The symbols in Figures 6-71f and 6-71n represent the use of a bottoming tap, when the length of thread is the same as the depth of drill. The thread length you sketch may be slightly longer than the actual given thread length. If the tap drill depth is known or given, draw the drill to that depth. If the thread note omits this information, as is often done in practice, sketch the hole three schematic thread pitches beyond the thread length. The tap drill diameter is represented approximately, not to actual size.

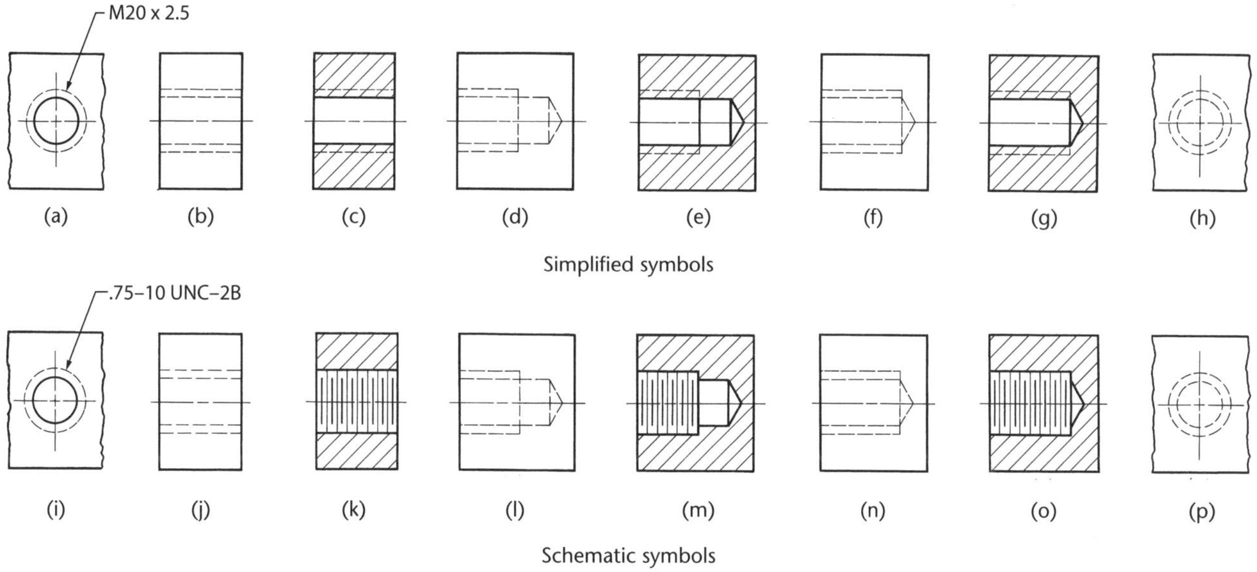

**Figure 6-71**    Internal Thread Symbols

# 6-32 DETAILED REPRESENTATION: METRIC, UNIFIED, AND AMERICAN NATIONAL THREADS

The detailed representation for metric, unified, and American national threads is the same, since the flats are disregarded.

Internal detailed threads in section are drawn as shown in Figure 6-72. Notice that for left hand threads the lines slope upward to the left (Figure 6-72a to 6-72c), while for right hand threads the lines slope upward to the right (Figures 6-72d to 6-72f).

## Detailed External Square Thread

Figure 6-73 is an assembly drawing showing an external square thread partly screwed into a nut. When the exter-

nal and internal threads are assembled, the thread in the nut overlaps and covers up half of the V, as shown at B.

Sometimes in assemblies the root and crest lines may be omitted from the nut only portion of the drawing so that it is easier to identify the inserted screw.

## Detailed Internal Square Thread

The internal thread construction is the same as in Figure 6-74. Note that the thread lines representing the back half of the internal threads (since the thread is in section) slope in the opposite direction from those on the front side of the screw.

Steps in drawing a single internal square thread in section are shown in Figure 6-74. Note in Figure 6-74b that a crest is drawn opposite a root. This is the case for both single and triple threads. For double or quadruple threads, a crest is opposite a crest. Thus, the construction in Figures 6-74a and b is the same for any multiple of thread. The differences ap-

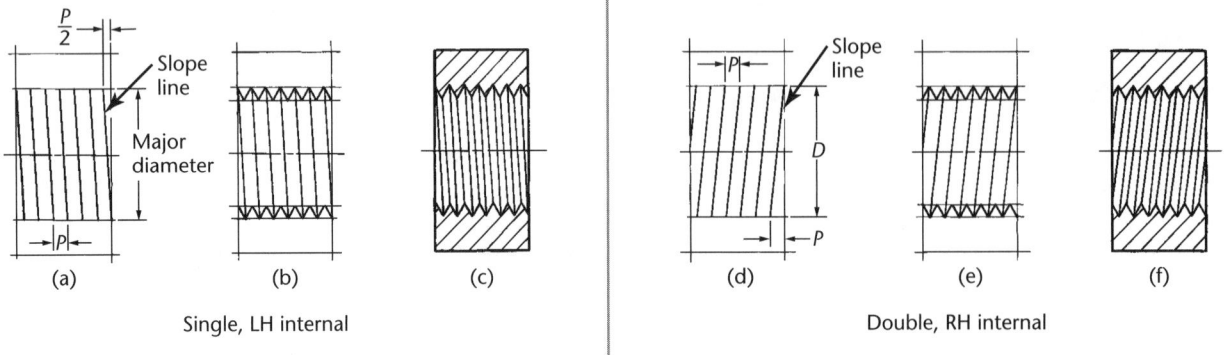

**Figure 6-72**    Detailed Representation—Internal Metric, Unified, and American

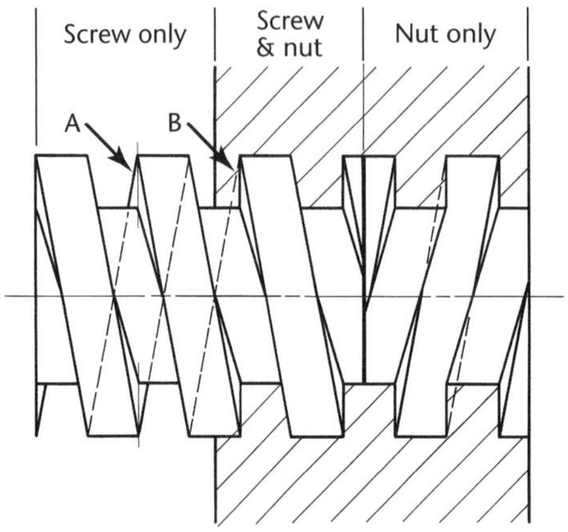

**Figure 6-73**    Square Threads in Assembly

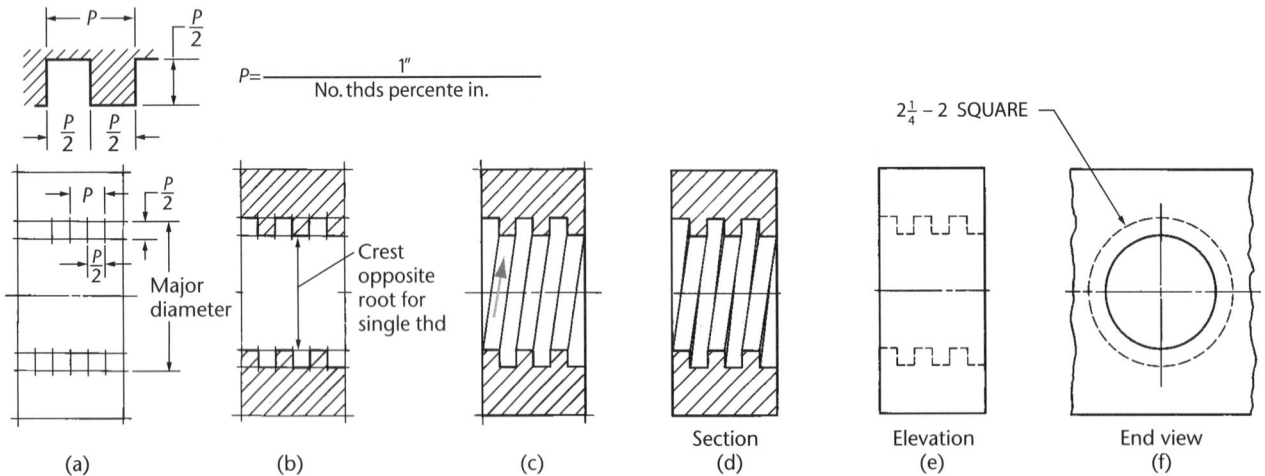

**Figure 6-74**    Detailed Representation—Internal Square Threads

pear in Figure 6-74c, where the threads and spaces are distinguished and outlined.

The same internal thread is shown in Figure 6-74e from an external view. The profiles of the threads are drawn in their normal position, but with hidden lines, and the slop-

ing lines are omitted for simplicity. The end view of the same internal thread is shown in Figure 6-74f. Note that the hidden and solid circles are opposite those for the end view of the shaft.

# STEP BY STEP

## Detailed Representation of Square Threads

Detailed representation of external square threads is only used when the major diameter is over about 1" or 25 mm, and it is important to show the detail of the thread on the finished sketch or plotted drawing. The steps to create detailed square thread are as follows.

1. Make a centerline and lay out the length and major diameter of the thread. For U.S. drawings, determine the pitch (P) by dividing 1 by the number of threads per inch. For a single right-hand thread, the lines slope upward to the left, and the slope line is offset as for all single threads of any form. On the upper line, use spacing equal to P/2, as shown.

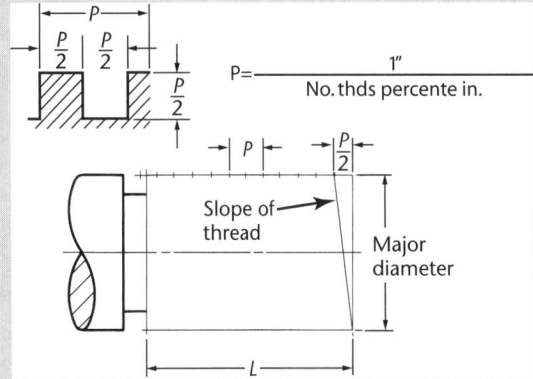

$$P = \frac{1''}{\text{No. thds percent in.}}$$

2. From the points on the upper line, draw guidelines for root of thread, making the depth as shown.

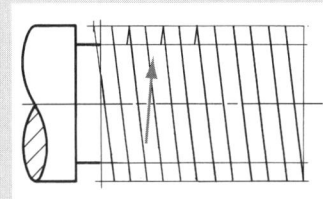

Single, R H external thread

3. Make parallel visible back edges of threads.

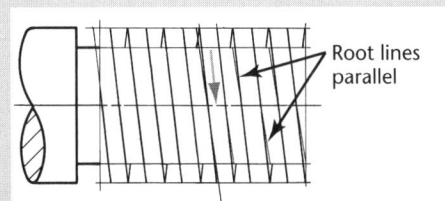

4. Make parallel visible root lines.

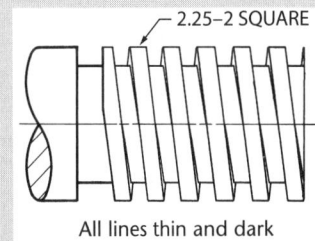

Root lines parallel

5. All lines should be thin and dark.

2.25–2 SQUARE

All lines thin and dark

## TIP

### End view of a shaft

The end view of the shaft illustrated in this Step by Step feature is shown below. Note that the root circle is hidden. When sketching, no attempt is made to show the true projection of any but the major diameter.

If the end of a shaft is chamfered, a solid circle would be drawn instead of the hidden circle.

End view

# STEP BY STEP

## Detailed Representation of Acme Thread

Detailed representation of acme threads is used only to call attention when details of the thread are important and the major diameter is larger than 1" or 25 mm on the drawing. The steps are as follows.

1. Make a centerline and lay out the length and major diameter of the thread, as shown. For U.S. drawings, determine the pitch by dividing 1 by the number of threads per inch. Make construction lines for the root diameter, making the thread depth P/2. Make construction lines halfway between crest and root guidelines.

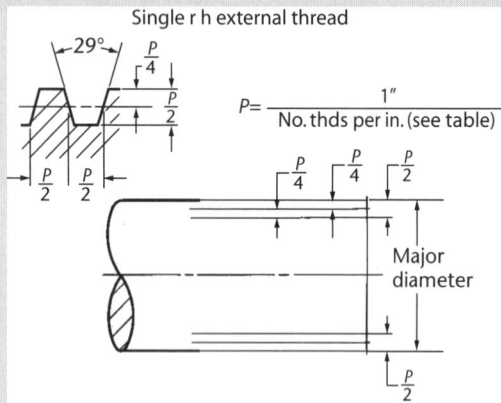

2. Mark off spaces on the intermediate construction lines.

3. Through alternate points, make construction lines for the sides of the threads at 15° (instead of 141/2°).

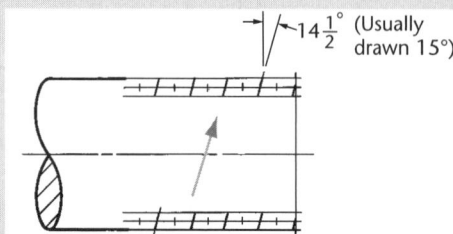

4. Make construction lines for the other sides of the threads, as shown. For single and triple threads, a crest is opposite a root, while for double and quadruple threads, a crest is opposite a crest. Finish tops and bottoms of threads.

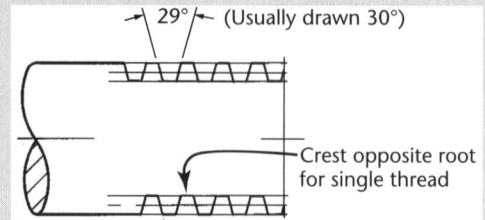

5. Make parallel crest lines.

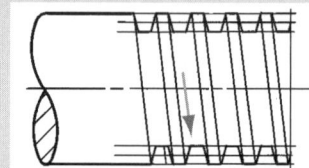

6. Make parallel root lines, and finish the thread profiles. All lines should be thin and dark. The internal threads in the back of the nut will slope in the opposite direction to the external threads on the front side of the screw.

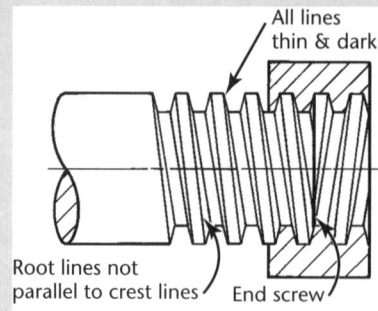

*End views of acme threaded shafts and holes are drawn exactly like those for the square thread.*

## 6-33  USE OF PHANTOM LINES

Use phantom lines to save time when representing identical features, as shown in Figure 6-75. Threaded shafts and springs may be shortened without using conventional breaks, but must be correctly dimensioned.

**Figure 6-75**  Use of Phantom Lines

## 6-34  THREADS IN ASSEMBLY

Threads in an assembly drawing are shown in Figure 6-76. It is customary not to section a stud or a nut or any solid part unless necessary to show some internal shapes.

Show these items "in the round," as they would look if they were set in the hole after the assembly was cut to form the section. When external and internal threads are sectioned in assembly, the V's are required to show the threaded connection.

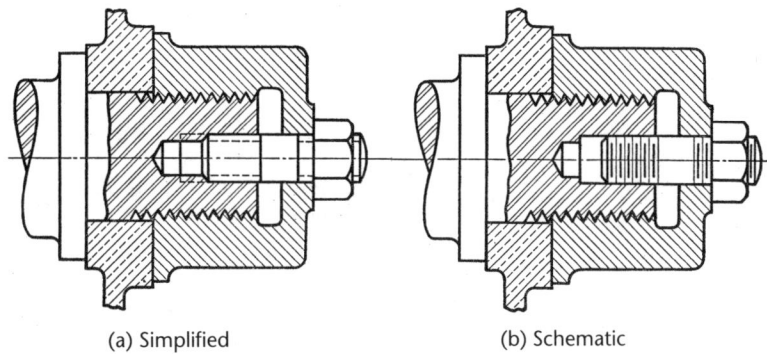

(a) Simplified    (b) Schematic

**Figure 6-76**  Threads in Assembly

## 6-35  AMERICAN NATIONAL STANDARD PIPE THREADS

The American National Standard for pipe threads, originally known as the Briggs standard, was formulated by Robert Briggs in 1882. Two general types of pipe threads have been approved as American National Standard: taper pipe threads and straight pipe threads.

The profile of the tapered pipe thread is illustrated in Figure 6-77. The taper of the standard tapered pipe thread is 1 in 16, or .75" per foot measured on the diameter and along the axis. The angle between the sides of the thread is 60°. The depth of the sharp V is $.8660p$, and the basic maximum depth of the thread is $.800p$, where $f$ = pitch. The basic pitch diameters, $E_0$ and $E_1$, and the basic length of the effective external taper thread, $L_2$ are determined by the formulas

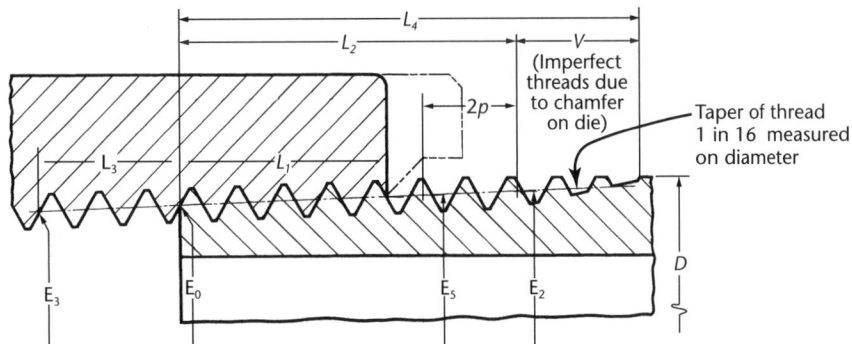

**Figure 6-77**  American National Standard Taper Pipe Thread, ANSI/ASME B1.20.1-1983 (R1992). *Courtesy of The American Society of Mechanical Engineers. All rights reserved.*

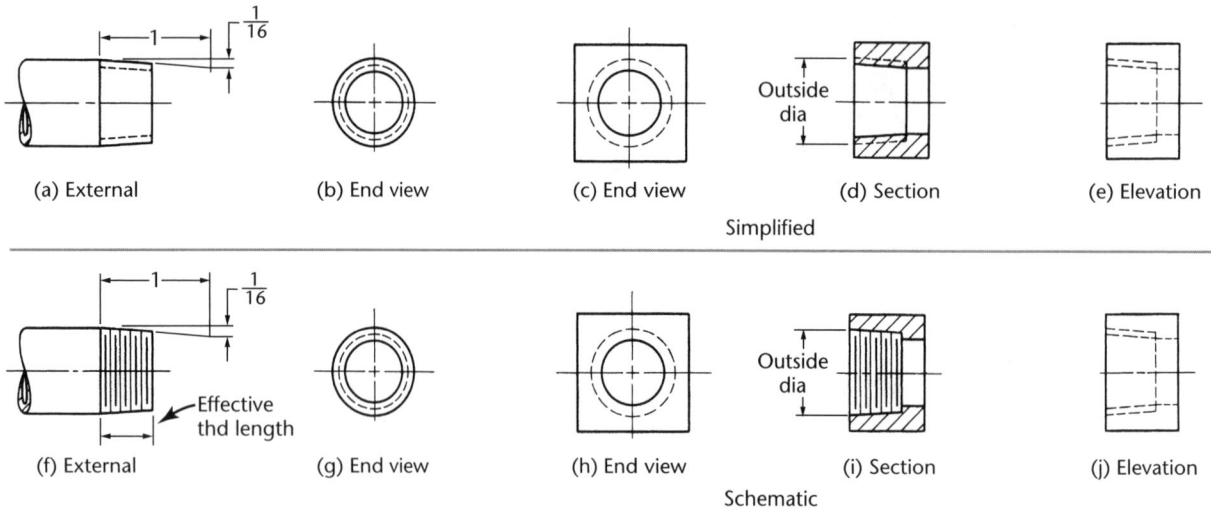

(a) External    (b) End view    (c) End view    (d) Section    (e) Elevation

Simplified

(f) External    (g) End view    (h) End view    (i) Section    (j) Elevation

Schematic

**Figure 6-78**   Conventional Pipe Thread Representation

$$E_0 = D - (.050D + 1.1)\frac{1}{n}$$
$$E_1 = E + .0625L_1$$
$$L_2 = (.80D + 6.8)\frac{1}{n}$$

where $D$ = outer diameter of pipe, $E_0$ = pitch diameter of thread at end of pipe, $E_1$ = pitch diameters of thread at large end of internal thread, $L_1$ = normal engagement by hand, and $n$ = number of threads per inch.

ANSI also recommends two modified tapered pipe threads for (1) dry seal pressure-tight joints (.880 per foot taper) and (2) rail fitting joints. The former is used for metal-to-metal joints, eliminating the need for a sealer, and is used in refrigeration, marine, automotive, aircraft, and ordnance work. The latter is used to provide a rigid mechanical thread joint as is required in rail fitting joints.

While tapered pipe threads are recommended for general use, there are certain types of joints in which straight pipe threads are used to advantage. The number of threads per inch, the angle, and the depth of thread are the same as on the tapered pipe thread, but the threads are cut parallel to the axis. Straight pipe threads are used for pressure-tight joints for pipe

couplings, fuel and oil line fittings, drain plugs, free-fitting mechanical joints for fixtures, loose-fitting mechanical joints for locknuts, and loose-fitting mechanical joints for hose couplings.

Pipe threads are represented by detailed or symbolic methods in a manner similar to the representation of unified and American national threads. The symbolic representation (schematic or simplified) is recommended for general use regardless of the diameter, as shown in Figure 6-78. The detailed method is recommended only when the threads are large and when it is desired to show the profile of the thread, as for example, in a sectional view of an assembly.

As shown in Figure 6-78, it is not necessary to draw the taper on the threads unless there is some reason to emphasize it, since the thread note indicates whether the thread is straight or tapered. If it is desired to show the taper, it should be exaggerated, as shown in Figure 6-79, where the taper is drawn 1/16" per 1" on radius (or 6.75" per 1" on diameter) instead of the actual taper of 1/16" on diameter. American National Standard tapered pipe threads are indicated by a

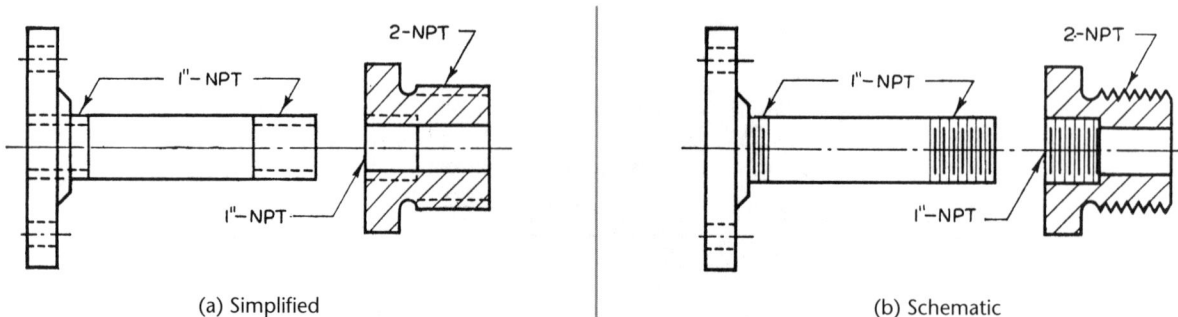

(a) Simplified    (b) Schematic

**Figure 6-79**   Conventional Representation of American National Standard Tapered Pipe Threads

note giving the nominal diameter followed by the letters NPT (national pipe taper), as shown in Figure 6-79. When straight pipe threads are specified, the letters NPS (national pipe straight) are used. In practice, the tap drill size is normally not given in the thread note.

## 6-36  BOLTS, STUDS, AND SCREWS

The term bolt is generally used to denote a "through bolt" that has a head on one end, is passed through clearance holes in two or more aligned parts, and is threaded on the other end to receive a nut to tighten and hold the parts together, as shown in Figure 6-80a. A hexagon head cap screw, shown in Figure 6-80b, is similar to a bolt except it often has greater threaded length. It is often used when one of the parts being held together is threaded to act as a nut. The cap screw is screwed on with a wrench. Cap screws are not screwed into thin materials if strength is desired.

A stud, shown in Figure 6-80c, is a steel rod threaded on one or both ends. If threaded on both ends, it is screwed into place with a pipe wrench or with a stud driver. If threaded on one end, it is force fitted into place. As a rule, a stud is passed through a clearance hole in one member, is screwed into another member, and uses a nut on the free end, as shown in Figure 6-80c.

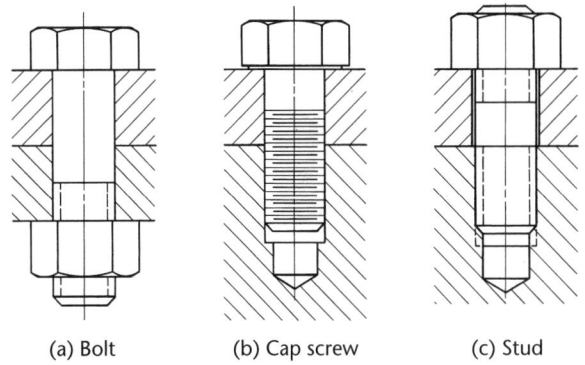

(a) Bolt          (b) Cap screw          (c) Stud

**Figure 6-80**   Bolt, Cap Screw, and Stud

A machine screw is similar to a slotted head cap screw but usually smaller. It may be used with or without a nut. Figure 6-81 shows different screw head types.

A set screw is a screw, with or without a head, that is screwed through one member and whose special point is forced against another member to prevent motion between the two parts.

Do not section bolts, nuts, screws, and similar parts when drawn in assembly because they do not have interior detail that needs to be shown.

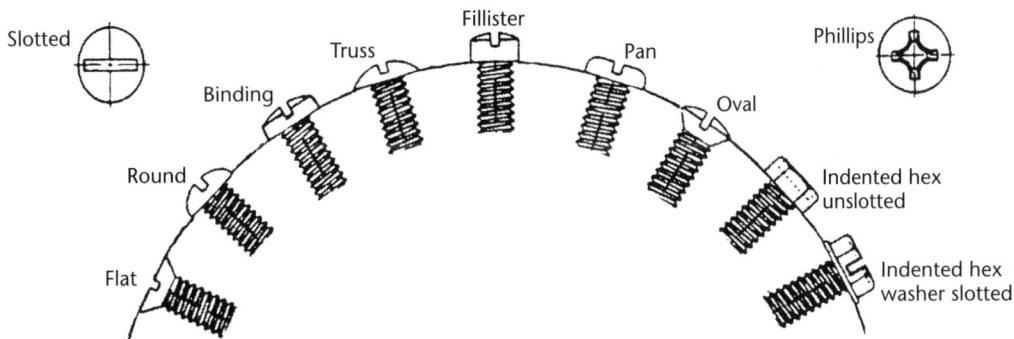

**Figure 6-81**   Types of Screw Heads

## 6-37  TAPPED HOLES

The bottom of a drilled hole, formed by the point of a twist drill, is cone-shaped, as shown in Figures 6-82a and 6-82b. When an ordinary drill is used to make holes that will be tapped, it is referred to as a tap drill. When drawing the drill point, use an angle of 30° to approximate the actual 31° slope of the drill bit.

The thread length is the length of full or perfect threads. The tap drill depth does not include the cone point of the drill. In Figure 6-82c and 6-82d, the drill depth shown beyond the threads (labeled A) includes several imperfect threads produced by the chamfered end of the tap. This distance varies according to drill size and whether a plug tap or a bottoming tap is used to finish the hole.

A drawing of a tapped hole finished with a bottoming tap is shown in Figure 6-82e. Blind bottom-tapped holes are hard to form and should be avoided when possible. Instead, a relief with its diameter slightly greater than the major diameter of the thread is used, as shown in Figure 6-82f. Tap drill sizes and lengths may be given in the thread note, but are generally left to the manufacturer to determine. Since the tapped thread length contains only full threads, it is necessary to make this length only one or two pitches beyond the end of the engaging screw. In simplified or schematic representation, don't show threads in the bottoms of tapped holes. This way the ends of the screw show clearly.

The thread length in a tapped hole depends on the major diameter and the material being tapped. The minimum engagement length ($X$), when both parts are steel, is equal to the

**Table 6-2** Thread Engagement Lengths for Different Materials.

| Screw Material | Material of Parts | Thread Engagement* |
|---|---|---|
| Steel | Steel | D |
| Steel | Cast iron | 1 1/2D |
| Steel | Aluminum | 2D |

*Requirements for thread engagement vary based on the specific materials. Use these rules of thumb only as guidelines.*

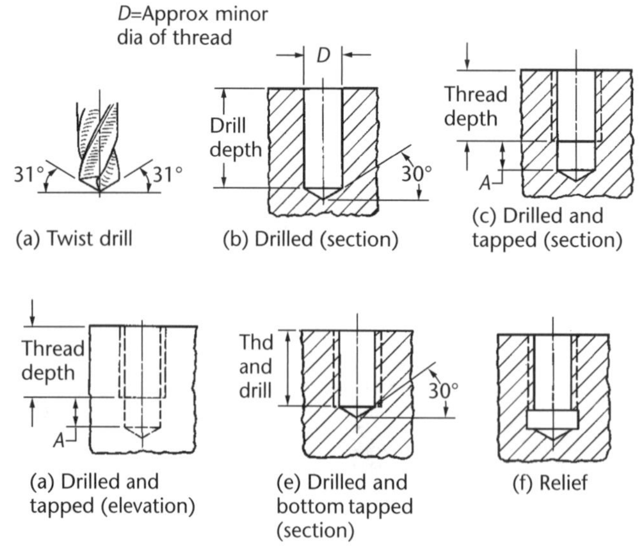

diameter ($D$) of the thread. Table 6-2 shows different engagement lengths for different materials.

**Figure 6-82**    Drilled and Tapped Holes

**TIP**

*Prevent tap breakage:* A chief cause of tap breakage is insufficient tap drill depth. When the depth is too short, the tap is forced against a bed of chips in the bottom of the hole. Don't specify a blind hole when a through hole of not much greater length can be used. When a blind hole is necessary, the tap drill depth should be generous.

*Clearance holes:* When a bolt or a screw passes through a **clearance hole**, the hole is often drilled 0.8 mm larger than the screw for screws of 3/8" (10 mm) diameter and 1.5 mm larger for larger di-

ameters. For more precise work, the clearance hole may be only 1/64" (0.4 mm) larger than the screw for diameters up to 10 mm and 0.8 mm larger for larger diameters.

Closer fits may be specified for special conditions. The clearance spaces on each side of a screw or bolt need not be shown on a drawing unless it is necessary to show clearly that there is no thread engagement. When it is necessary to show that there is no thread engagement, the clearance spaces should be drawn about 3/64" (1.2 mm) wide.

# 6-38  STANDARD BOLTS AND NUTS

American National Standard hexagon bolts and nuts are made in both metric and inch sizes. Square bolts and nuts, shown in Figure 6-83, are only produced in inch sizes. Metric bolts, cap screws, and nuts also come in hexagon form. Square heads and nuts are chamfered at 30°, and hexagon heads and nuts are chamfered at 15–30°. Both are drawn at 30° for simplicity.

**Bolt Types**  Bolts are grouped into bolt types according to use: regular bolts for general use and heavy bolts for heavier use or easier wrenching. Square bolts come only in the regular type; hexagon bolts, screws, nuts, and square nuts are available in both regular and heavy.

Metric hexagon bolts are grouped according to use: regular and heavy bolts and nuts for general service and high-strength bolts and nuts for structural bolting.

**Finish**  Square bolts and nuts, hexagon bolts, and hexagon flat nuts are unfinished. Unfinished bolts and nuts are not machined on any surface except for the threads. Hexagon cap screws, heavy hexagon screws, and all hexagon nuts, except hexagon flat nuts, are considered finished to some degree and have a "washer face" machined or otherwise formed on the bearing surface. The washer face is 1/64" thick (drawn 1/32" so that it will be visible on the plotted drawing), and its diameter is 1.5 times the body diameter for the inch series.

For nuts, the bearing surface may also be a circular surface produced by chamfering. Hexagon screws and hexagon nuts have closer tolerances and a more finished appearance but are not completely machined. There is no difference in the drawing for the degree of finish on finished screws and nuts.

**Proportions**  Proportions for both inch and metric are based on the diameter (D) of the bolt body. These are shown in Figure 6-84.

For regular hexagon and square bolts and nuts, proportions are:

$$W = 1\tfrac{1}{2}D \quad H = \tfrac{2}{3}D \quad T = \tfrac{7}{8}D$$

where $W$ = width across flats, $H$ = head height, and $T$ = nut height.

© *Dorling Kindersley.*

For heavy hexagon bolts and nuts and square nuts, the proportions are:

$$W = 1\tfrac{1}{2}D + \tfrac{1}{8}\text{" (or + 3 mm)}$$
$$W = \tfrac{2}{3}D \quad T = D$$

The washer face is always included in the head or nut height for finished hexagon screw heads and nuts.

**Threads**  Square and hex bolts, hex cap screws, and finished nuts in the inch series are usually Class 2 and may have coarse, fine, or 8-pitch threads. Unfinished nuts have coarse threads and are Class 2B.

**Thread** lengths For bolts or screws up to 6" (150 mm) long.

Thread length $= 2D + \tfrac{1}{4}$" (or + 6 mm)

For bolts or screws over in length,

Thread length $= 2D + \tfrac{1}{2}$" (or + 12 mm)

Fasteners too short for these formulas are threaded as close to the head as practical. For drawing purposes, use approximately three pitches. The threaded end may be rounded or chamfered, but it is usually drawn with a 45° chamfer from the thread depth, as shown in Figure 6-84.

**Bolt Lengths**  Have not been standardized because of the endless variety required by industry. Short bolts are typically available in standard length increments of 1/4" (6 mm), while long bolts come in increments of 1/2" to 1 inch (12 to 25 mm).

**Figure 6-83** Standard Bolt and Nut.

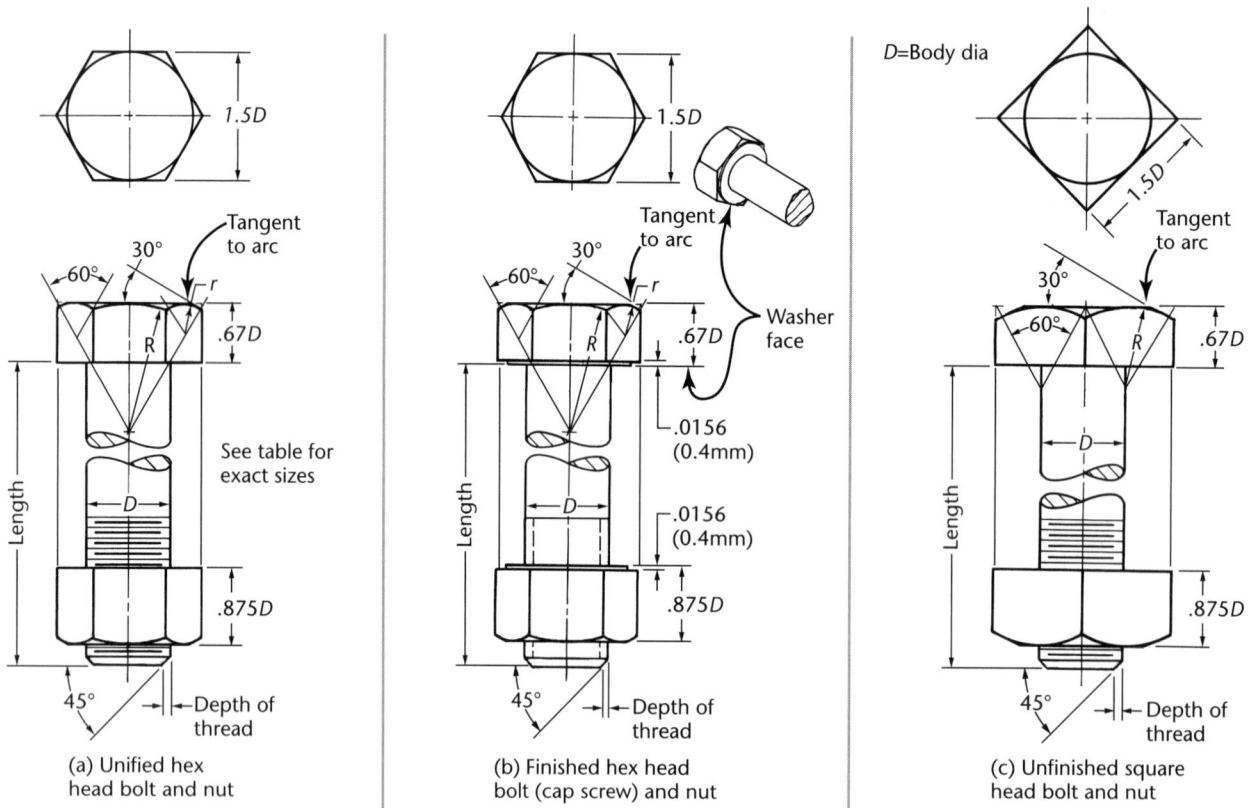

**Figure 6-84** Bolt Proportions (Regular)

## 6-39 DRAWING STANDARD BOLTS

Detail drawings show all of the necessary information defining the shape, size, material, and finish of a part. Standard bolts and nuts do not usually require detail drawings unless they are to be altered (for example, by having a slot added through the end of a bolt), because they are usually stock parts that can easily be purchased. But you often need to show them on assembly drawings.

Templates are available to help you add bolts quickly to sketches. In most cases, a quick representation, where proportions are based on the body diameter, is sufficient. Three typical bolts illustrating the use of these proportions are shown in Figure 6-84.

Many CAD systems have fastener libraries that you can use to add a wide variety of nuts and bolts to your drawings. Often these symbols are based on a diameter of 1 unit so that you can quickly figure a scale at which to insert them. Other systems prompt for the diameter and lengths and create a symbol to your specifications. In 3D models, when nuts and bolts are represented, the thread is rarely shown because it adds to the complexity and size of the drawing and is difficult to model. The thread specification is annotated in the drawing.

Generally, bolt heads and nuts should be drawn "across corners" in all views, regardless of projection. This conventional violation of projection is used to prevent confusion between the square and hexagon heads and nuts and to show actual clearances. Only when there is a special reason should bolt heads and nuts be drawn across flats, as shown in Figure 6-85.

## 6-40 SPECIFICATIONS FOR BOLTS AND NUTS

In specifying bolts in parts lists, in correspondence, or elsewhere, the following information must be covered in order:

1. Nominal size of bolt body
2. Thread specification or thread note
3. Length of bolt
4. Finish of bolt
5. Style of head
6. Name

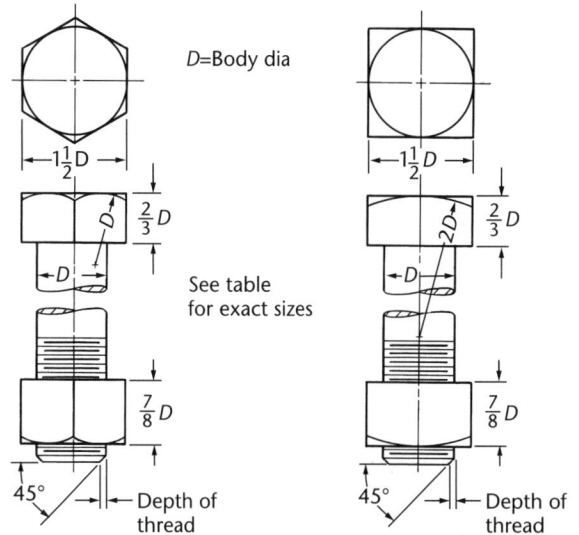

**Figure 6-85**  Bolts "Across Flats"

Example (complete decimal inch)

.75-10 UNC- HEXAGON CAP SCREW

Example (abbreviated decimal inch)

HEXCAP SCR

Example (metric)

HEXCAP SCR

Nuts may be specified as follows:
Example (complete)

$\frac{5}{8}$–11 UNC–2B SQUARE NUT

Example (abbreviated)

$\frac{5}{8}$ SQ NUT

Example (metric)

HEX NUT

For either bolts or nuts, REGULAR or GENERAL PURPOSE are assumed if omitted from the specification. If the heavy series is intended, the word HEAVY should appear as the first word in the name of the fastener. Likewise, HIGH STRENGTH STRUCTURAL should be indicated for such metric fasteners. However, the number of the specific ISO standard is often included in the metric specifications—for example, HEXAGON NUT ISO. Finish need not be mentioned if the fastener or nut is correctly named.

# STEP BY STEP

## Sketching Hexagonal Bolts, Cap Screws, and Nuts

1. Determine the diameter of the bolt, the length (from the underside of the bearing surface to the tip), the style of head (square or hexagon), the type (regular or heavy), and the finish before starting to draw.

(Use $1\frac{1}{2}D + \frac{1}{8}$" or 3mm for heavy bolt and nut)

$1\frac{1}{2}D$

$R$

Usually drawn as $\frac{1}{32}$" (1mm)

$\frac{1}{64}$"

$\frac{2}{3}D$

$D$

Length

$\frac{7}{8}D$

2. Lightly sketch the top view as shown, where $D$ is the diameter of the bolt. Project the corners of the hexagon or square to the front view. Sketch the head and nut heights. Add the 1/64" (0.4 mm) washer face if needed. Its diameter is equal to the distance across flats of the bolt head or nut. Only the metric and finished hexagon screws or nuts have a washer face. The washer face is 1/64" thick, but is shown at about 1/32" (1 mm) for clarity. The head or nut height includes the washer face.

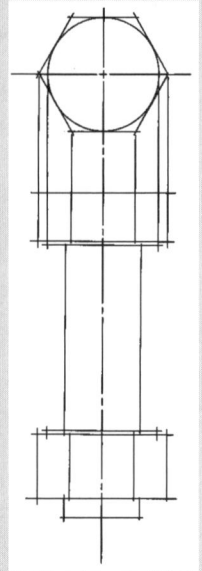

3. Represent the curves produced by the chamfer on the bolt heads and nuts as circular arcs, although they are actually hyperbolas. On drawings of small bolts or nuts under approximately 1/2" (12 mm) in diameter, where the chamfer is hardly noticeable, omit the chamfer in the rectangular view.

4. Chamfer the threaded end of the screw at 45° from the schematic thread depth.

5. Show threads in simplified or schematic form for diameters of 1" (25 mm) or less on the drawing. Detailed representation is rarely used because it clutters the drawing and takes too much time.

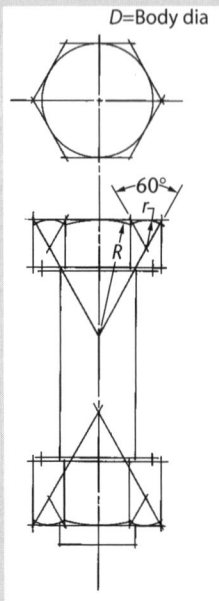

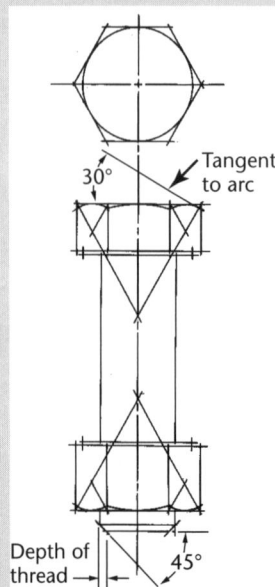

$D$=Body dia

60°

$r$

$R$

30°

Tangent to arc

Depth of thread

45°

Enlarged view of chamfer

Thread length

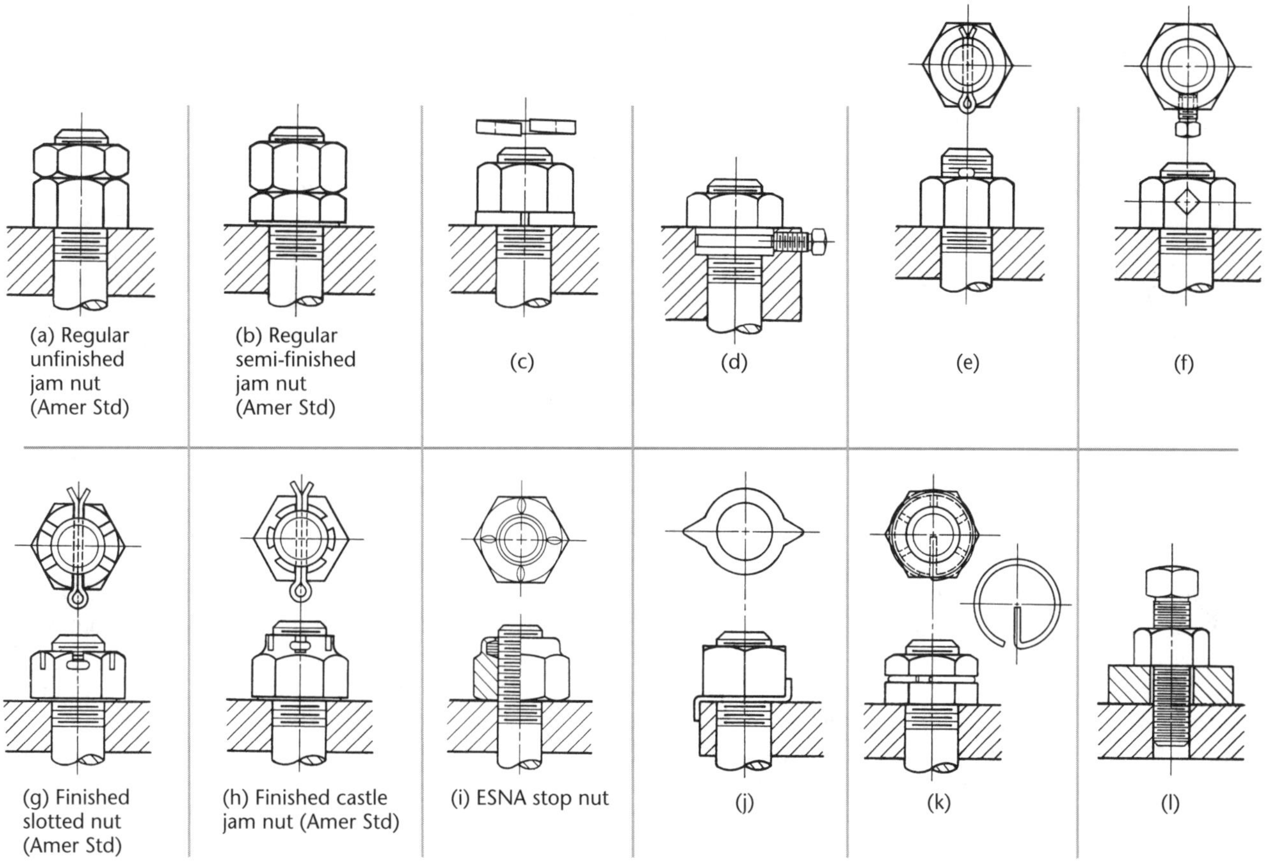

**Figure 6-86**    Locknuts and Locking Devices

# 6-41  LOCKNUTS AND LOCKING DEVICES

Many types of special nuts and devices to prevent nuts from unscrewing are available, and some of the most common are shown in Figure 6-86. The American National Standard jam nuts, as shown in Figures 6-86a and 6-86b, are the same as the hexagon or hexagon flat nuts, except that they are thinner. The application shown in Figure 6-86b, where the larger nut is on top and is screwed on more tightly, is recommended. They are the same distance across flats as the corresponding hexagon nuts ($1\ 1/2D$ or $1''$). They are slightly over $1/2D$ in thickness but are drawn $1/2D$ for simplicity. They are available with or without the washer face in the regular and heavy types. The tops of all are flat and chamfered at 30°, and the finished forms have either a washer face or a chamfered bearing surface.

The lock washer, shown in Figure 6-86c, and the cotter pin, shown in Figure 6-86e, are very common. The set screw, shown in Figure 6-86f, is often made to press against a plug of softer material, such as brass, which in turn presses against the threads without deforming them. For use with cotter pins, it is recommended to use a hex slotted nut (Figure 6-86g), a hex castle nut (Figure 6-86h), or a hex thick slotted nut or a heavy hex thick slotted nut.

Similar metric locknuts and locking devices are available. See fastener catalogs for details.

---

Reid Tool is one company that has a free download of its catalog available as CAD files at

• http://www.reidtool.com/download.htm.

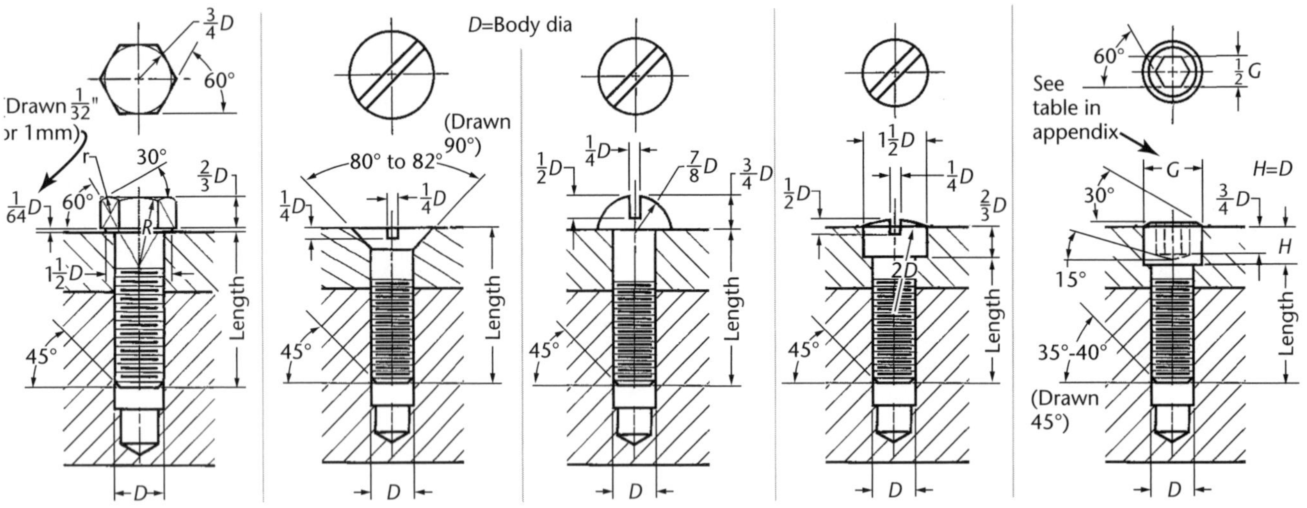

(a) Hexagon head  (b) Flat head  (c) Round head  (d) Fillister head  (e) Hex socket

*Hexagon Head Screws* Coarse, fine, or 8-thread series, 2A. Thread length = $2D + \frac{1}{4}$" up to 6" long and $2D + \frac{1}{2}$" if over 6" long. For screws too short for formula, threads extend to within $2\frac{1}{2}$ threads of the head for diameters up to 1". Screw lengths not standardized.

*Slotted Head Screws* Coarse, fine, or 8-thread series, 2A. Thread length = $2D + \frac{1}{4}$". Screw lengths not standardized. For screws too short for formula, threads extend to within $2\frac{1}{2}$ threads of the head.

*Hexagon Socket Screws* Coarse or fine threads, 3A. Coarse thread length = $2D + \frac{1}{2}$" where this would be over $\frac{1}{2}L$; otherwise thread length = $\frac{1}{2}L$. Fine thread length = $1\frac{1}{2}D + \frac{1}{2}$" where this would be over $\frac{3}{4}L$; otherwise thread length $+ \frac{3}{8}L$. Increments in screw lengths = $\frac{1}{8}$" for screws $\frac{1}{4}$" to 1" long, $\frac{1}{4}$" for screws 1" to 3" long, and $\frac{1}{2}$" for screws $3\frac{1}{2}$" to 6" long.

**Figure 6-87**   Standard Cap Screws

## 6-42 STANDARD CAP SCREWS

Five types of American National Standard cap screws are shown in Figure 6-87. The first four of these have standard heads, while the socket head cap screws, as shown in Figure 6-87e, have several different shapes of round heads and sockets. Cap screws are normally finished and are used on machine tools and other machines when accuracy and appearance are important. The hexagon head cap screw and hex socket head cap screw are also available in metric.

Cap screws ordinarily pass through a clearance hole in one member and screw into another. The clearance hole need not be shown on the drawing when the presence of the unthreaded clearance hole is obvious.

Cap screws are inferior to studs when frequent removal is necessary. They are used on machines requiring few adjustments. The slotted or socket-type heads are used for crowded conditions.

Actual dimensions may be used in drawing cap screws when exact sizes are necessary. Figure 6-87 shows the propor-

tions in terms of body diameter (*D*) that are usually used. Hexagonal head cap screws are drawn similar to hex head bolts. The points are chamfered at 45° from the schematic thread depth.

Note that screwdriver slots are drawn at 45° in the circular views of the heads, without regard to true projection, and that threads in the bottom of the tapped holes are omitted so that the ends of the screws may be clearly seen. A typical cap screw note is:

Example (complete)

.375-16 UNC-2A × 2.5 HEXAGON HEAD CAP SCREW

Example (abbreviated)

.375 × 2.5 HEXHD CAP SCR

Example (metric)

M20 × 2.5 × 80 HEXHD CAP SCR

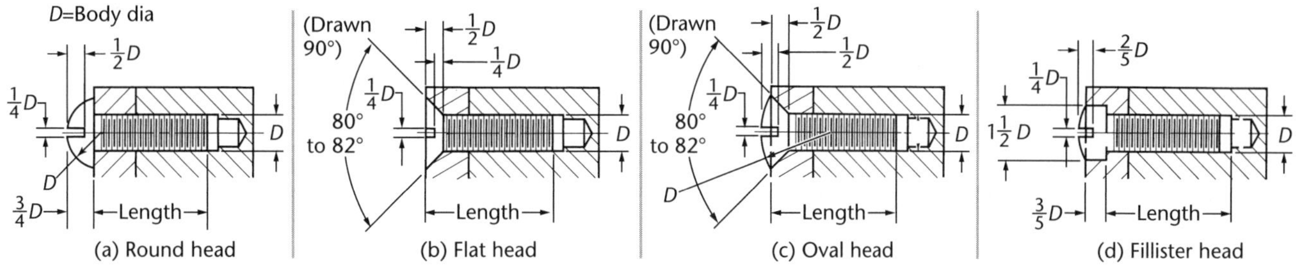

**Figure 6-88** Standard Machine Screws

## 6-44 STANDARD MACHINE SCREWS

Machine screws are similar to cap screws but are usually smaller (.060" to .750" diameter) and the threads generally go all the way to the head. There are eight ANSI-approved forms of heads. The hexagonal head may be slotted if desired. All others are available in either slotted or recessed-head forms. Standard machine screws are produced with a naturally bright finish, not heat treated, and have plain-sheared ends, not chamfered.

Machine screws are used for screwing into thin materials, and the smaller-numbered screws are threaded nearly to the head. They are used extensively in firearms, jigs, fixtures, and dies. Machine screw nuts are used mainly on the round head, pan head, and flat head types and are usually hexagonal.

The four most common types of machine screws are shown in Figure 6-88, with proportions based on the diameter ($D$). Clearance holes and counterbores should be made slightly larger than the screws.

Typical machine screw notes are:

Example (complete)
NO. 10 (.1900) $-32$ NF$-3 \times \frac{5}{8}$ FILLISTER HEAD MACHINE SCREW

Example (abbreviated)
NO. 10 (.1900) $\times \frac{5}{8}$ FILH MSCR

Example (metric)
M8 $\times$ 1.25 $\times$ 30 SLOTTED PAN HEAD MACHINE SCREW

## 6-45 STANDARD SET SCREWS

Set screws, shown in Figure 6-89, are used to prevent motion, usually rotary, between two parts, such as the movement of the hub of a pulley on a shaft. A set screw is screwed into one part so that its point bears firmly against another part. If the point of the set screw is cupped, or if a flat is milled on the shaft, the screw will hold much more firmly. Obviously, set screws are not efficient when the load is heavy or when it is suddenly applied. Usually they are manufactured of steel and case hardened.

Headless set screws have come into greater use because the projecting head of headed set screws has caused many industrial casualties; this has resulted in legislation prohibiting their use in many states.

Metric hexagon socket headless set screws with the full range of points are available. Nominal diameters of metric hex socket set screws are 1.6, 2, 2.5, 3, 4, 5, 6, 8, 10, 12, 16, 20, and 24 mm.

Square head set screws have coarse, fine, or 8-pitch threads and are Class 2A, but are usually furnished with coarse threads since the square head set screw is generally used on the rougher grades of work. Slotted headless and socket set screws have coarse or fine threads and are Class 3A.

Nominal diameters of set screws range from number 0 up through 2" set-screw lengths are standardized in increments of 1/32" to 1" depending on the overall length of the set screw.

Metric set screw length increments range from 0.5 to 4 mm, again depending on overall screw length.

Set screws are specified as follows:

Example (complete)
.375- 16UNC-2A $\times$ .75 SQUARE HEAD FLAT POINT SET SCREW

Example (abbreviated)
.375- $\times$ 1.25 SQH FP SSCR
.438 $\times$ .750 HEXSOC CUP PT SSCR
$\frac{1}{4}$ $-20$ UNC 2A $\times \frac{1}{2}$ SLTD HDLS CONE PT SSCR

Example (metric)
M10 $\times$ 1.5 12 HEX SOCKET HEAD SET SCREW

**Figure 6-89** Set Screws. *Courtesy of Penninsula Components Inc.*

# 6-46 AMERICAN NATIONAL STANDARD WOOD SCREWS

Wood screws with three types of heads—flat, round, and oval—have been standardized. The approximate dimensions sufficient for drawing purposes are shown in Figure 6-90.

The Phillips style recessed head is also available on several types of fasteners, as well as on wood screws. Three styles of cross recesses have been standardized by ANSI. A special screwdriver is used, as shown in Figure 6-43q, and this results in rapid assembly without damage to the head.

**Figure 6-90**  American National Standard Wood Screws

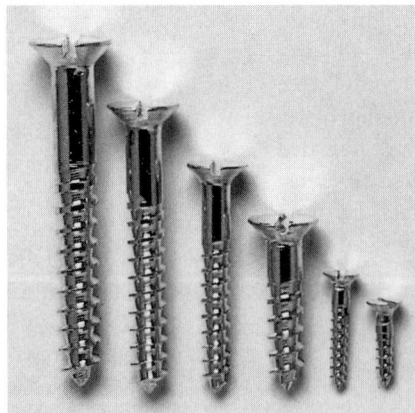

*Courtesy of Michael Newman/PhotoEdit Inc.*

# 6-47 MISCELLANEOUS FASTENERS

Many other types of fasteners have been devised for specialized uses. Some of the more common types are shown in Figure 6-91. A number of these are American National Standard round head bolts, including carriage, button head, step, and countersunk bolts.

Helical-coil-threaded inserts, as shown in Figure 6-91p, are shaped like a spring except that the cross section of the wire conforms to threads on the screw and in the hole. These are made of phosphor bronze or stainless steel, and they provide a hard, smooth protective lining for tapped threads in soft metals and plastics.

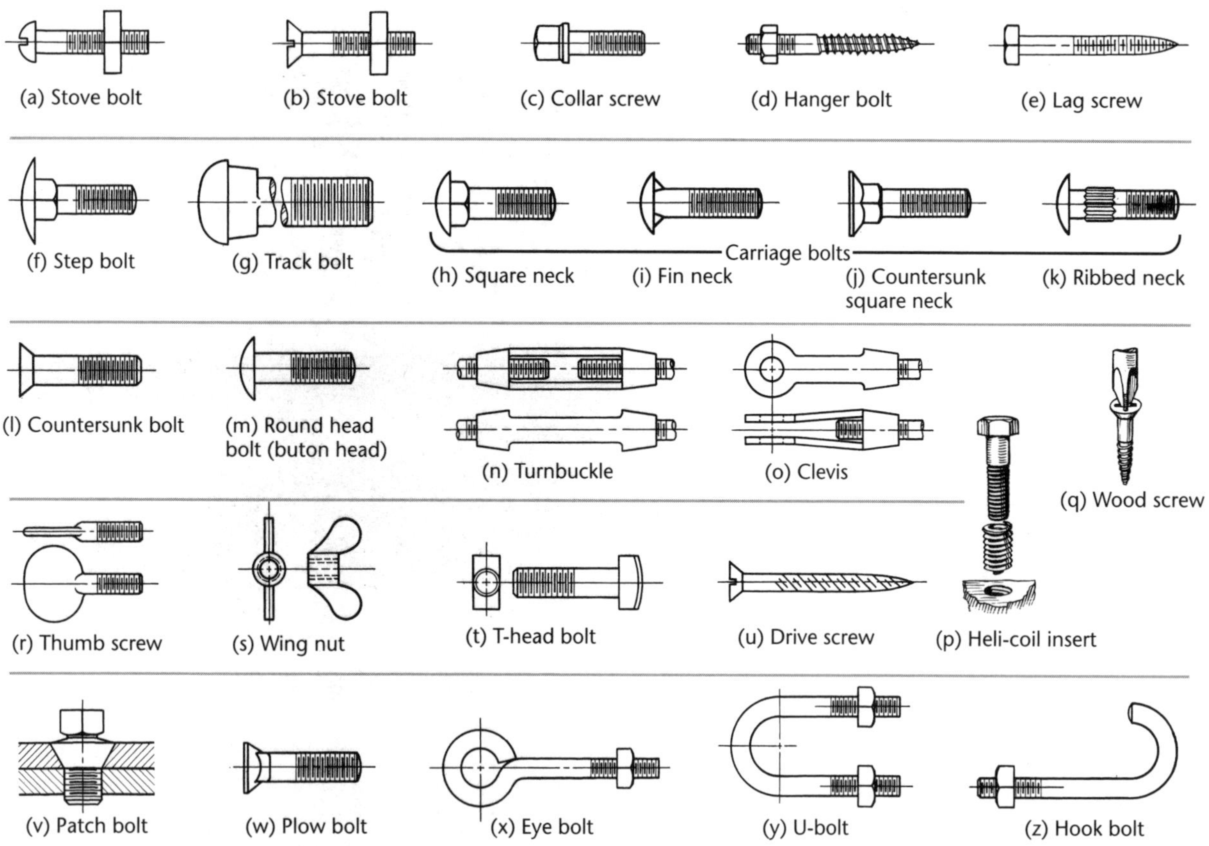

**Figure 6-91**  Miscellaneous Bolts and Screws

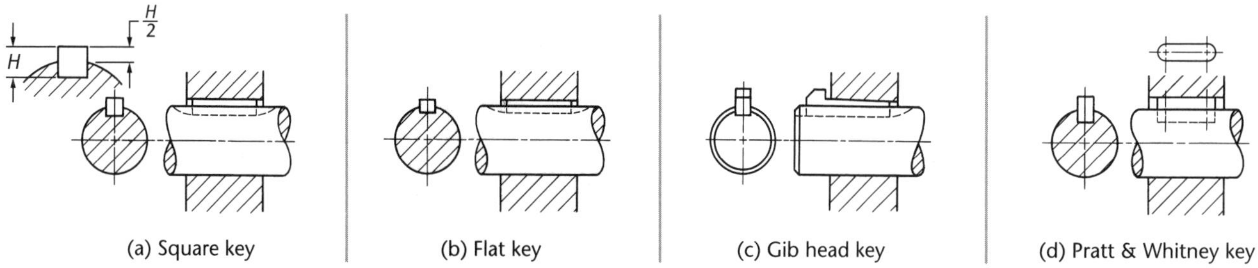

**Figure 6-92**  Square and Flat Keys

## 6-48  KEYS

Keys are used to prevent movement between shafts and wheels, couplings, cranks, and similar machine parts attached to or supported by shafts, as shown in Figure 6-92. A keyseat is in a shaft; a keyway is in the hub or surrounding part.

For heavy-duty functions, rectangular keys (flat or square) are used, and sometimes two rectangular keys are necessary for one connection. For even stronger connections, interlocking splines may be machined on the shaft and in the hole.

A *square key* is shown in Figure 6-92a.

A *flat key* is shown in Figure 6-92b. The widths of keys are generally about one fourth the shaft diameter. In either case, one half the key is sunk into the shaft. The depth of the keyway or the keyseat is measured on the side—not the center—as shown in Figure 6-92a. Square and flat keys may have the top surface tapered 1/8" per foot, in which case they become square taper or flat taper keys.

Woodruff key-slot cutter →

(a)          (b)          (c)

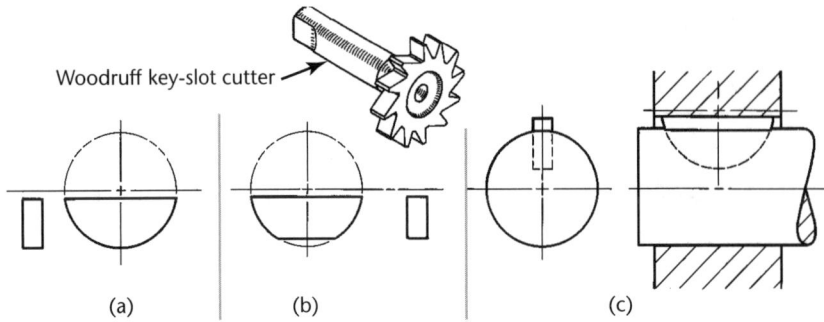

**Figure 6-93**   Woodruff Keys and Key-Slot Cutter

A *feather key* is rectangular to prevent rotary motion, but permits relative longitudinal motion. Usually feather keys have gib heads, or are fastened so they cannot slip out of the keyway.

A *gib head key* (Figure 6-92c) is the same as a square taper or flat taper key except that a gib head allows its easy removal. Square and flat keys are made from cold-finished stock and are not machined.

A *Pratt & Whitney key* (P&W key) is shown in Figure 6-92d. It is rectangular, with semicylindrical ends. Two-thirds of its height is sunk into the shaft keyseat.

*Woodruff keys* are semicircular, as shown in Figure 6-93. This key fits into a semicircular key slot cut with a Woodruff cutter, as shown, and the top of the key fits into a plain rectangular keyway. Sizes of keys for given shaft diameters are not standardized. For average conditions, select a key whose diameter is approximately equal to the shaft diameter. See manufacturers' catalogs for specifications for metric counterparts.

Typical specifications for keys are:

.25 × 1.50 SQ KEY
No. 204 WOODRUFF KEY

1/4 × 1/6 × 1-FLAT KEY
No. 10 P&W KEY

## 6-49  MACHINE PINS

Machine pins include taper pins, straight pins, dowel pins, clevis pins, and cotter pins. For light work, taper pins can be used to fasten hubs or collars to shafts. Figure 6-94 shows the use of a taper pin where the hole through the collar and shaft is drilled and reamed when the parts are assembled. For slightly heavier duty, a taper pin may be used parallel to the shaft as for square keys.

Dowel pins are cylindrical or conical and usually used to keep two parts in a fixed position or to preserve alignment. They are usually used where accurate alignment is essential. Dowel pins are generally made of steel and are hardened and ground in a centerless grinder.

Clevis pins are used in a clevis and held in place by cotter pins.

$D$

$L$ (max)

Taper .25 per ft

**Figure 6-94**   Taper Pin

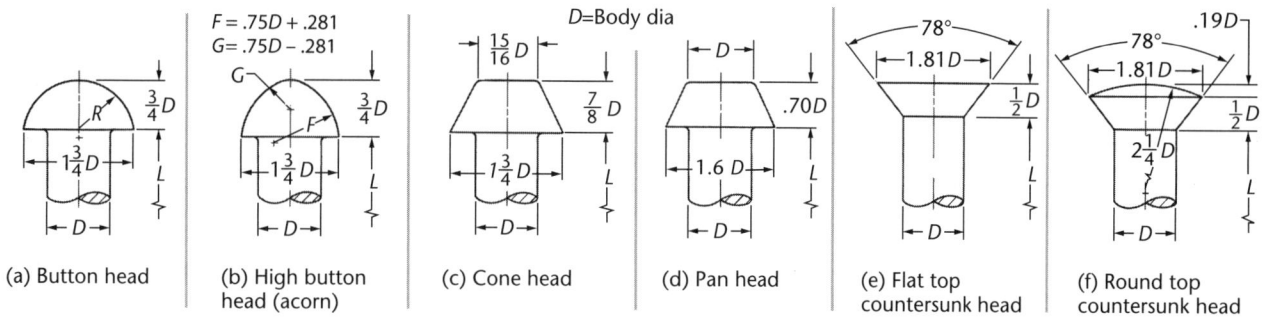

**Figure 6-95**  Standard Large Rivets

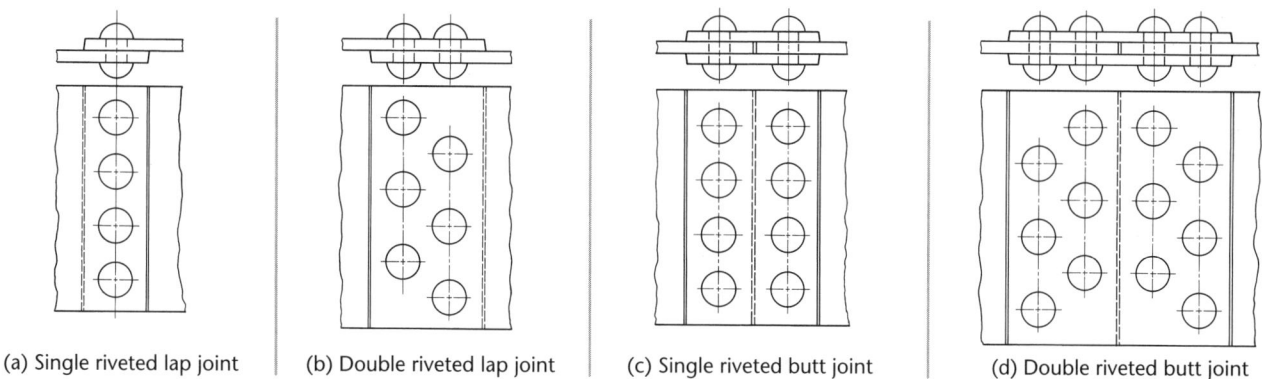

**Figure 6-96**  Common Riveted Joints

# 6-50 RIVETS

Rivets are regarded as permanent fastenings, unlike removable fastenings, such as bolts and screws. Rivets are generally used to hold sheet metal or rolled steel together and are made of wrought iron, carbon steel, copper, or occasionally other metals.

To fasten two pieces of metal together, holes are punched, drilled, or punched and then reamed, all slightly larger in diameter than the shank of the rivet. Rivet diameters are made from $d = 1.2 \sqrt{t}$ to $d = 1.4\sqrt{t}$, where $d$ is the rivet diameter and $t$ is the metal thickness. The larger rivet diameter size is used for steel and single-riveted joints, and the smaller may be used for multiple-riveted joints. In structural work it is common to make the hole 1.6 mm (1/16") larger than the rivet.

When the red-hot rivet is inserted, a "dolly bar" with a depression the shape of the driven head is held against the head. A riveting machine is used to drive the rivet and forms the head on the plain end. This causes the rivet to swell and fill the hole tightly.

Large rivets or heavy hex structural bolts are often used in structural work of bridges and buildings and in ship and boiler construction. They are shown in their exact formula proportions in Figure 6-95. Button heads (Figure 6-95a), and countersunk heads (Figure 6-95e), are the rivets most commonly used in structural work. The button head and cone head are commonly used in tank and boiler construction.

### Riveted Joints

Typical riveted joints are shown in Figure 6-96. Note that the rectangular view of each rivet shows the shank of the rivet with both heads made with circular arcs, and the circular view of each rivet is represented by only the outside circle of the head.

### Rivet Symbols

Since many engineering structures are too large to be built in the shop, they are built in the largest units possible and then are transported to the desired location. Trusses are common examples of this. The rivets driven in the shop are called shop rivets, and those driven on the job are called field rivets. However, heavy steel bolts are commonly used on the job for structural work. Solid black circles are used to represent field rivets, and other standard symbols are used to show other features, as shown in Figure 6-97.

**Figure 6-97**   Conventional Rivet Symbols

## Small Rivets

Small rivets are used for light work. American National Standard small solid rivets are illustrated with dimensions that show their standard proportions in Figure 6-98, ANSI/ASME B18.1.1–1972 (R1995). Included in the same standard are tinners', coppers', and belt rivets. Metric rivets are also available. Dimensions for large rivets are in ANSI/ASME B18.1.2-1972 (R1995). See manufacturers' catalogs for additional details.

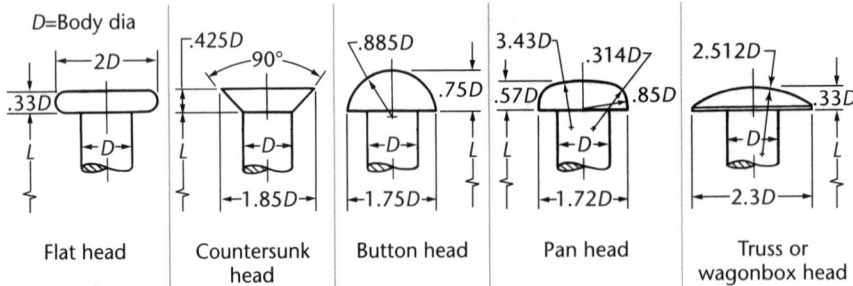

**Figure 6-98**   American National Standard Small Solid Rivet Proportions

## Blind Rivets

Blind rivets, commonly known as pop rivets (Figure 6-99), are often used for fastening together thin sheet-metal assemblies. Blind rivets are hollow and are installed with manual or power-operated rivet guns which grip a center pin or mandrel, pulling the head into the body and expanding the rivet against the sheet metal. They are available in aluminum, steel, stainless steel, and plastic. As with any fastener, the designer should be careful to choose an appropriate material to avoid corrosive action between dissimilar metals.

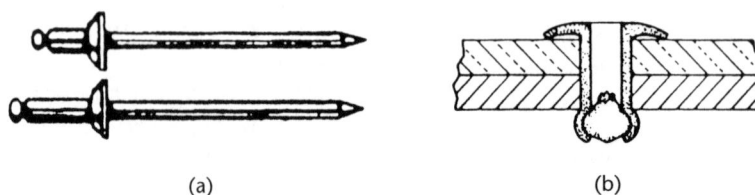

**Figure 6-99**   Blind Rivets (a) Before Installation, and (b) Installed

## 6-51 SPRINGS

A spring is a mechanical device designed to store energy when deflected and to return the equivalent amount of energy when released, ANSI Y14.13M-1981 (R1992). Springs are commonly made of spring steel, which may be music wire, hard-drawn wire, or oil-tempered wire. Other materials used for compression springs include stainless steel, beryllium copper, and phosphor bronze. Urethane plastic is used in applications where conventional springs would be affected by corrosion, vibration, or acoustic or magnetic forces.

Springs are classified as *helical springs* (Figure 6-100), or *flat springs*.

### Helical Springs

Helical springs are usually cylindrical but may also be conical. There are three types of helical springs.

- **Compression springs** offer resistance to a compressive force.
- **Extension springs** offer resistance to a pulling force.
- **Torsion springs** offer resistance to a torque or twisting force.

On working drawings, true projections of helical springs are not drawn because of the labor involved. Like screw threads, they are drawn in detailed and schematic methods, using straight lines to replace helical curves, as shown in Figure 6-100.

A square wire spring is similar to the square thread with the core of the shaft removed, as in Figure 6-100b. Use standard cross-hatching if the areas in section are large, as in

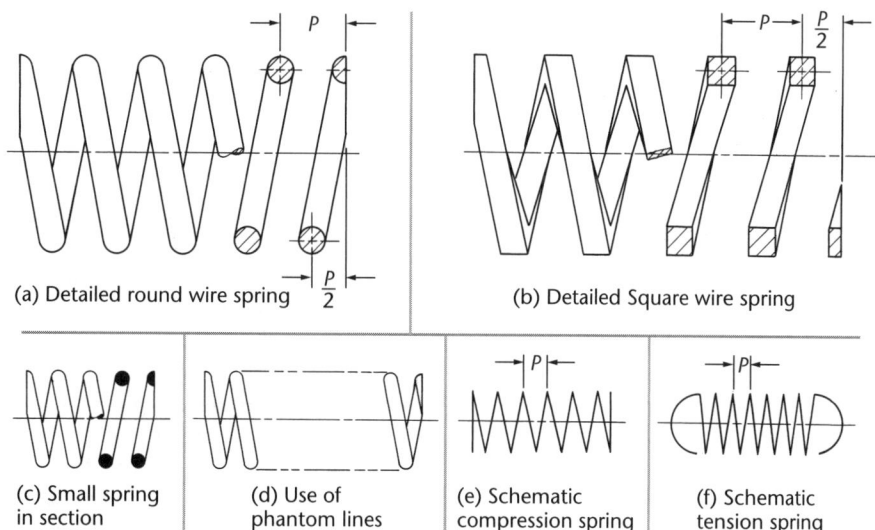

Springs. *NORTON, ROBERT L. MACHINE DESIGN: AN INTEGRATED APPROACH, 3rd Edition, © 2006. Reprinted by permission of Pearson Education, Inc., Upper Saddle River, NJ.*

Figure 6-100a and b. Small sectioned areas may be made solid black, as in Figure 6-100c.

In cases where a complete picture of the spring is not necessary, use phantom lines to save time in drawing the coils, as in Figure 6-100d. If the drawing of the spring is too small to be represented by the outlines of the wire, use schematic representation, shown in Figure 6-100e and f.

Compression springs have plain ends, as in Figure 6-101a, or squared (closed) ends, as in Figure 6-101b. The ends may be ground, as shown in Figure 6-101c, or both squared and ground, as in Figure 6-101d. Required dimensions are indicated in the figure. When required, RH or LH is specified for right-hand or left-hand coil direction.

(a) Detailed round wire spring

(b) Detailed Square wire spring

(c) Small spring in section

(d) Use of phantom lines

(e) Schematic compression spring

(f) Schematic tension spring

**Figure 6-100**  Helical Springs

FL = Free length  D = Controlling dia inside or outside
t = Dia of wire  L1 = Comp length (min)
L2 = Comp length (max)

(a) No. of coils Plain ends

(b) Square ends

(c) Plain end ground

(d) Squared and ground

**Figure 6-101** Compression Springs

Extension springs may have many types of ends, so it is necessary to draw the spring or at least the ends and a few adjacent coils, as shown in Figure 6-102.

A typical torsion spring drawing is shown in Figure 6-103. A typical flat spring drawing is shown in Figure 6-104. Other types of flat springs are power springs (or flat coil springs),

Belleville springs (like spring washers), and leaf springs (commonly used in automobiles).

Many companies use a printed specification form to provide the necessary spring information, including data such as load at a specified deflected length, load rate, finish, and type of service.

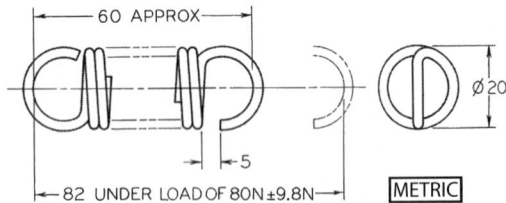

60 APPROX

Ø20

5

82 UNDER LOAD OF 80N ±9.8N

METRIC

MATERIAL: 2.00 OIL TEMPERED SPRING STEEL WIRE
14.5 COILS RIGHT HAND
MACHINE LOOP AND HOOK IN LINE
SPRING MUST EXTEND TO 110 WITHOUT SET
FINISH: BLACK JAPAN

**Figure 6-102** Extension Spring Drawing

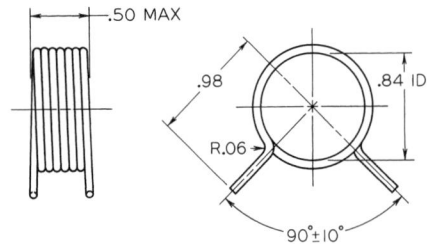

.50 MAX

.98   .84 ID

R.06

90° ±10°

MATERIAL : .059 MUSIC WIRE
6.75 COILS RIGHT HAND NO INITIAL TENSION
TORQUE : 2.50 INCH LB AT 155° DEFLECTION SPRING MUST
DEFLECT 180° WITHOUT PERMANENT SET AND
MUST OPERATE FREELY ON .75 DIAMETER SHAFT
FINISH : CADMIUM OR ZINC PLATE

**Figure 6-103** Torsion Spring Drawing

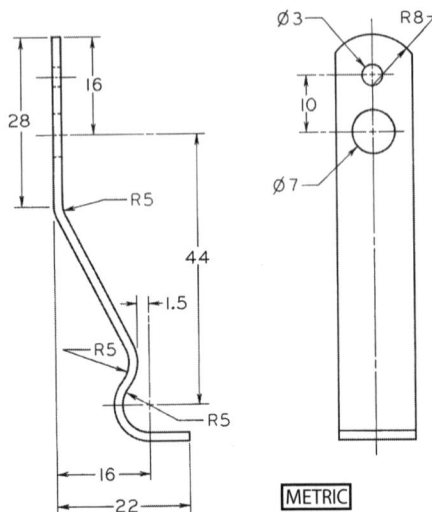

## 6-52 DRAWING HELICAL SPRINGS

The construction for a schematic elevation view of a compression spring with six total coils is shown in Figure 6-105a. Since the ends are closed, or squared, two of the six coils are "dead" coils, leaving only four full pitches to be set off along the top of the spring.

If there are six total coils, as shown in Figure 6-105, the spacings will be on opposite sides of the spring. The construction of an extension spring with six active coils and loop ends is shown in Figure 6-105c.

Figure 6-106 shows the steps in drawing a detailed section and elevation view of a compression spring. The spring is shown pictorially in Figure 6-106a. Figure 6-106b shows the cutting plane through the centerline of the spring. Figure 6-106c shows the section with the cutting plane removed. Steps to construct the sectional view are shown in

Ø3   R8

16

28

10

Ø7

R5

44

1.5

R5

R5

16

22

METRIC

MATERIAL : 1.20 X 14.0 SPRING STEEL
HEAT TREAT : 44-48 C ROCKWELL
FINISH : BLACK OXIDE AND OIL

**Figure 6-104** Flat Spring

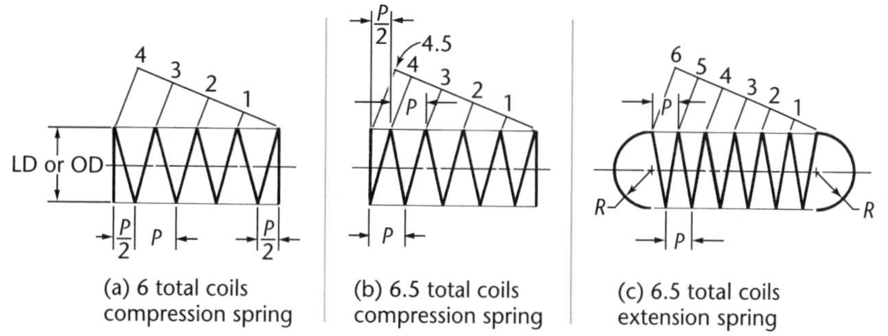
LD or OD
(a) 6 total coils
compression spring
(b) 6.5 total coils
compression spring
(c) 6.5 total coils
extension spring

**Figure 6-105** Schematic Spring Representation.

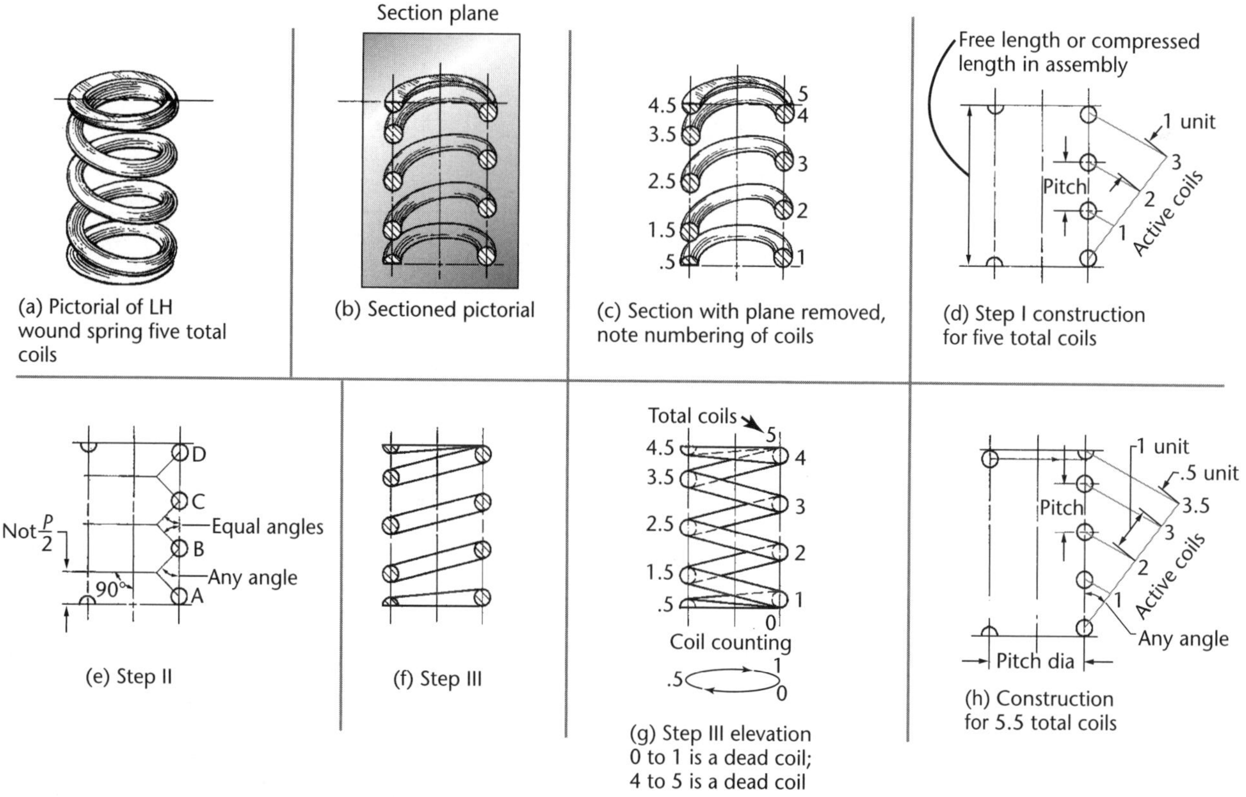
Section plane
(a) Pictorial of LH
wound spring five total
coils
(b) Sectioned pictorial
(c) Section with plane removed,
note numbering of coils
Free length or compressed
length in assembly
1 unit
Pitch
Active coils
(d) Step I construction
for five total coils
Equal angles
Any angle
(e) Step II
(f) Step III
Total coils
Coil counting
(g) Step III elevation
0 to 1 is a dead coil;
4 to 5 is a dead coil
.5 unit
Pitch dia
Any angle
(h) Construction
for 5.5 total coils

**Figure 6-106** Steps in Detailed Representation of Spring

Figures 6-106d–f. Figure 6-106g shows the corresponding elevation view.

If there is a fractional number of coils, such as the five coils in Figure 6-106h, the half-rounds of sectional wire are placed on opposite sides of the spring.

## 6-53 COMPUTER GRAPHICS

Standard representations of threaded fasteners and springs, in both detailed and schematic forms, are available in CAD symbol libraries. Use of computer graphics frees the drafter from the need to draw time-consuming repetitive features by hand and also makes it easy to modify drawings if required.

In 3D modeling, thread is not usually represented because it can be difficult to create and computer intensive to view and edit. Instead, the nominal diameter of a threaded shaft or hole is usually created along with notation calling out the thread. Sometimes the depth of the thread is shown in the 3D drawing to call attention to the thread and to help in determining fits and clearances.

# CAD AT WORK

## Downloading Standard Fasteners

Many stock fasteners in standard CAD formats are available to download, ready to use in CAD drawings and models. One example of such a site is http://www.pemnet.com/fastening_products, from the Penn Engineering company. Their PEM brand fasteners are often used to fasten sheet metal parts, such as the part shown in Figure B.

The CAD files that you download can be inserted into your drawings to save drawing time. The stock fastener type is specified in a drawing note as shown in Figure B.

Data sheets in PDF format are also available from the Pemnet site. Figure C shows the cover of a 12-page PDF booklet that lists material, thread size, performance data, and other key data for unified and metric self-clinching nuts that the company carries.

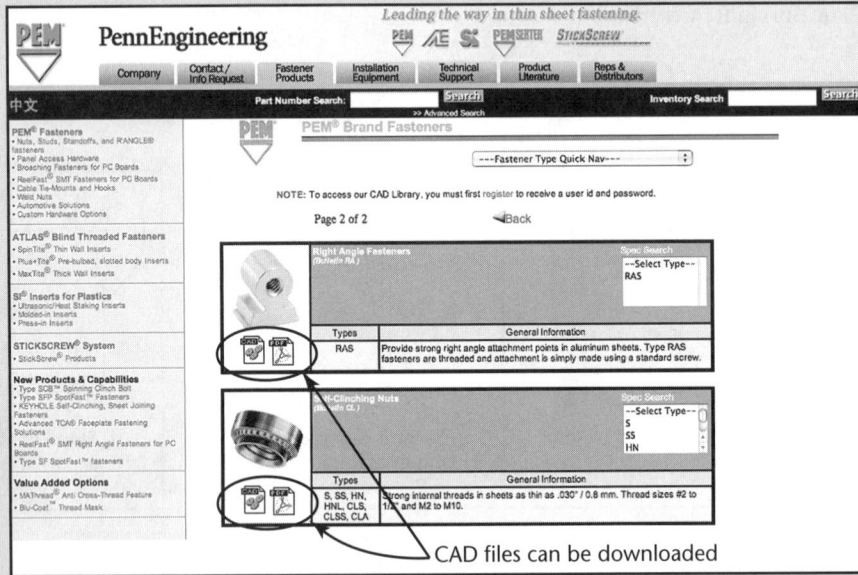

CAD files can be downloaded

(a) PEM brand fasteners, available from Penn Engineering are an example of the many stock fasteners that you can download in CAD file formats for easy insertion into drawings. *Courtesy of PennEngineering.*

(b) PEM stock fasteners shown on a drawing. *Courtesy of Dynojet Research, Inc.*

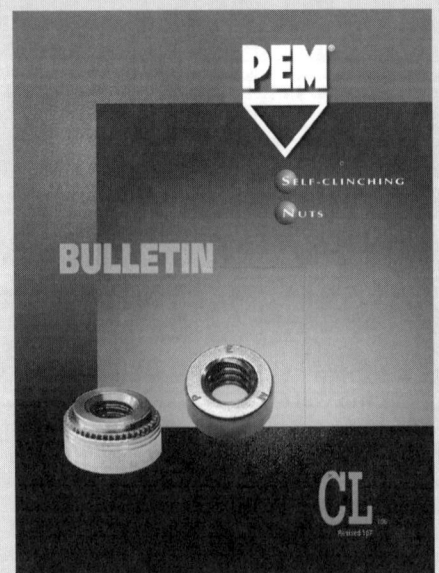

(c) Data sheets are often available in PDF format, as in this example from PEM. *Courtesy of PennEngineering.*

# Portfolio

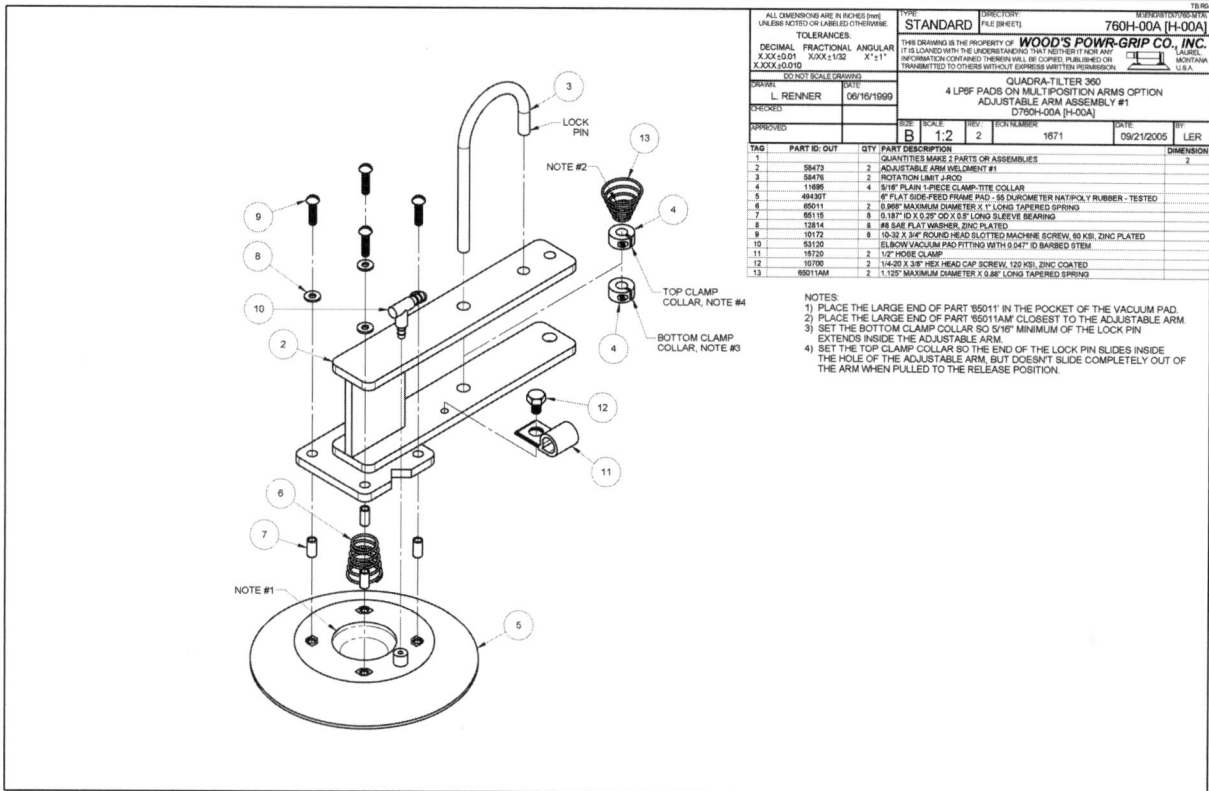

Assembly Drawing Showing Fasteners and Springs. *Courtesy of Wood's Power-Grip Co. Inc.*

Part Drawing for a Special Purpose Threaded Part (Scale 4:1). *Courtesy of Dynojet Research, Inc.*

## Portfolio

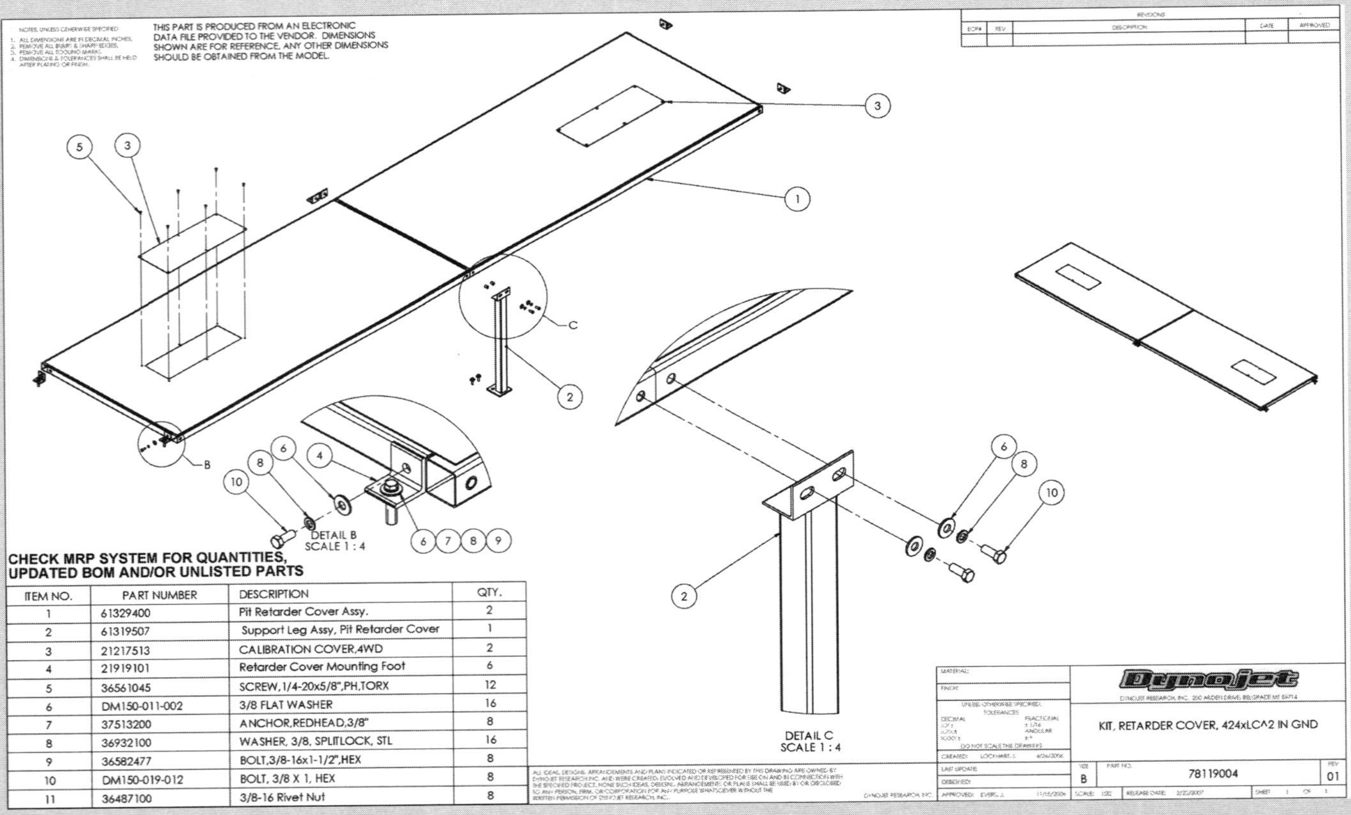

THIS PART IS PRODUCED FROM AN ELECTRONIC DATA FILE PROVIDED TO THE VENDOR. DIMENSIONS SHOWN ARE FOR REFERENCE. ANY OTHER DIMENSIONS SHOULD BE OBTAINED FROM THE MODEL.
DETAIL B
SCALE 1 : 4
DETAIL C
SCALE 1 : 4
CHECK MRP SYSTEM FOR QUANTITIES, UPDATED BOM AND/OR UNLISTED PARTS
ITEM NO. | PART NUMBER | DESCRIPTION | QTY.
1 | 61329400 | Pit Retarder Cover Assy. | 2
2 | 61319507 | Support Leg Assy, Pit Retarder Cover | 1
3 | 21217513 | CALIBRATION COVER,4WD | 2
4 | 21919101 | Retarder Cover Mounting Foot | 6
5 | 36561045 | SCREW,1/4-20x5/8",PH,TORX | 12
6 | DM150-011-002 | 3/8 FLAT WASHER | 16
7 | 37513200 | ANCHOR,REDHEAD,3/8" | 8
8 | 36932100 | WASHER, 3/8, SPLITLOCK, STL | 16
9 | 36582477 | BOLT,3/8-16x1-1/2",HEX | 8
10 | DM150-019-012 | BOLT, 3/8 X 1, HEX | 8
11 | 36487100 | 3/8-16 Rivet Nut | 8
Dynojet
KIT, RETARDER COVER, 424xLC^2 IN GND
B
78119004
01

Enlarged Details Show the Fasteners in this Assembly Drawing. *Courtesy of Dynojet Research, Inc.*

# 6-54 EXERCISES

## Thread and Fastener Projects

Use the information in this chapter and in various manufacturers' catalogs in connection with the working drawings at the end of the next chapter, where many different kinds of threads and fasteners are required. Several projects are included here (Ex 6-24 to 6-28).

## Design Project

Design a system that uses thread to transmit power, for use in helping transfer a handicapped person from a bed to a wheelchair. Use either schematic or detailed representation to show the thread in your design sketches.

## EX 6-24

Draw specified detailed threads arranged as shown. Omit all dimensions and notes given in inclined letters. Letter only the thread notes and the title strip. (Some dimensions are given to help you match the sheet layout.)

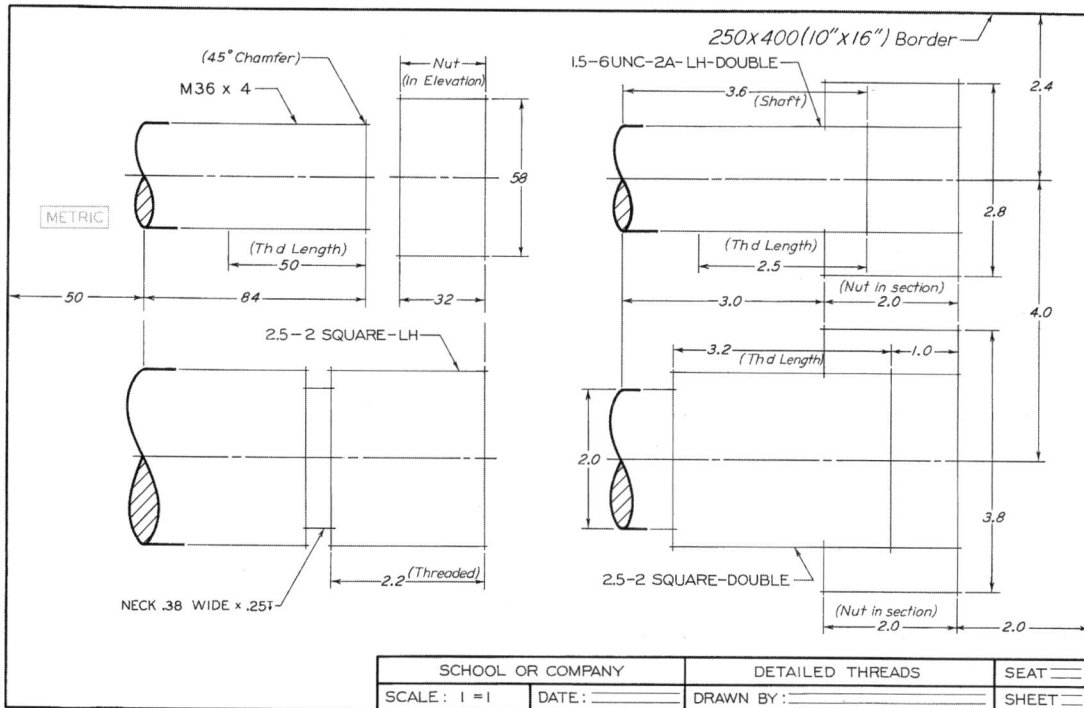

**EX 6-25**

Draw specified detailed notes given in inclined letters. Letter only the thread notes and the title strip. (Some dimensions are given to help you match the sheet layout.)

10"x16" Border

1.50   3.00   1.50   .90   1.50   .90   2.75

(Nut in sect)   (Nut in el)   (End view of nut)

2.50

1.75–5UNC–3A

(45° Chamfer)

2.76

(Thread Length)
3.00   .76

2.25   2.00   3.00   2.50   1.00   3.50   4.40

(Nut in section)   (End view of nut)

2–2.5 ACME
LH–DOUBLE

3.50

4.25 (Threaded)

| SCHOOL OR COMPANY | | DETAILED THREADS | SEAT |
|---|---|---|---|
| SCALE : 1 =1 | DATE : | DRAWN BY : | SHEET |

## EX 6-26

Draw fasteners, arranged as shown. At (a) draw 7/8-9 UNC-2A × 4 Hex Cap Screw. At (b) draw 7 UNC-2A × 41/4 Sq Hd Bolt. At (c) draw 3/8-16 UNC-2A × 11/2 Flat Hd Cap Screw. At (d) draw 7/16-14 UNC-2A × 1 Fill Hd Cap Screw. At (e) draw 1/2 × 1 Headless Slotted Set Screw. At (f) draw front view of No. 1010 Woodruff Key. Draw simplified or schematic thread symbols as assigned. Letter titles under each figure as shown. (Some dimensions are given to help you match the sheet layout.)

## EX 6-27

Draw specified thread symbols, arranged as shown. Draw simplified or schematic symbols, as assigned by instructor, using Layout B-5 or A3-5. Omit all dimensions and notes given in inclined letters. Letter only the drill and thread notes, the titles of the views, and the title strip. (Some dimensions are given to help you match the sheet layout.)

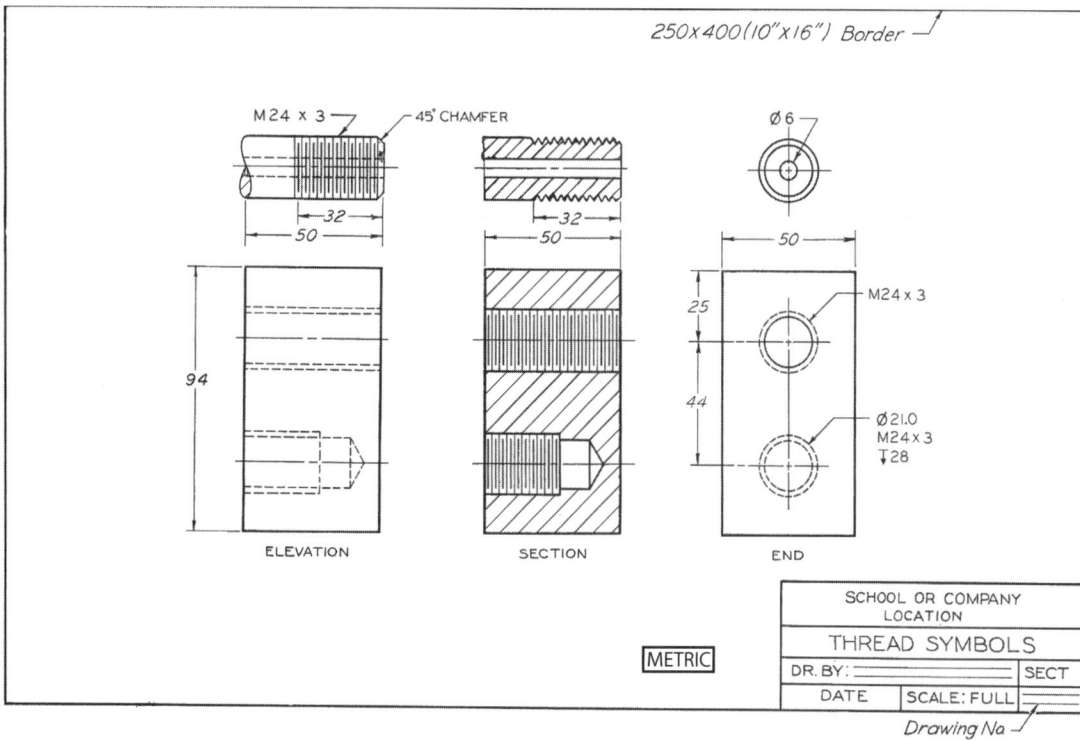

## EX 6-28

Specify fasteners for attaching the sheet metal and standard electrical components shown. Use the web to research power and electrical connectors.

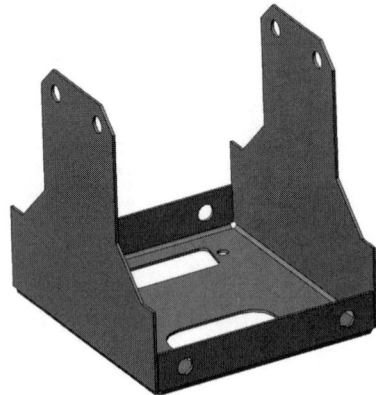

## EX 6-29

Captive hardware.

Captive hardware is a term for fasteners that, once installed, cannot be easily or accidentally removed. This is typically achieved by removing a portion of the threading on a screw shaft and then threading the captive hardware into a special sleeve that, once installed, prevents the hardware from backing out. Captive hardware is useful in situations where many fasteners are necessary—for example in sheet metal covers and large panels and when the fastener must be repeatedly unfastened and refastened.

For this exercise, modify a standard 6-32 × .75" socket head cap screw (SHCS) so it can be installed into a sleeve that you will design. The dimensions for the lock and flat washer are provided on the facing page, and the dimensions from the screw to be modified are provided from a CAD file downloaded from the McMaster-Carr Web site. Provide dimensions for turning down the thread on the 6-32 SHCS and the missing callouts on the captive screw sleeve. Ensure the dimensions provided for the threaded portion of the screw and the clearance for the captive screw sleeve allow the screw to be fully inserted, and allow it to be completely removed from the threaded blank material without causing interference.

LOCATION OF LOCKING
CUT NOT CRITICAL

Ø.250
Ø.148

.010

.031

**#6 LOCK WASHER**

Ø.141
Ø.375

.024

**#6 FLAT WASHER**

**6-32 X .75"
SOCKET HEAD CAP SCREW
REF McMASTER-CARR PART#
92196A151 FOR DIMENSIONS**

SPECIFY WHAT ID TO
TURN THREAD OFF OF
THE SCREW

SPECIFY

SPECIFY UNDERCUT

SPECIFY INTERNAL
6-32 THREAD

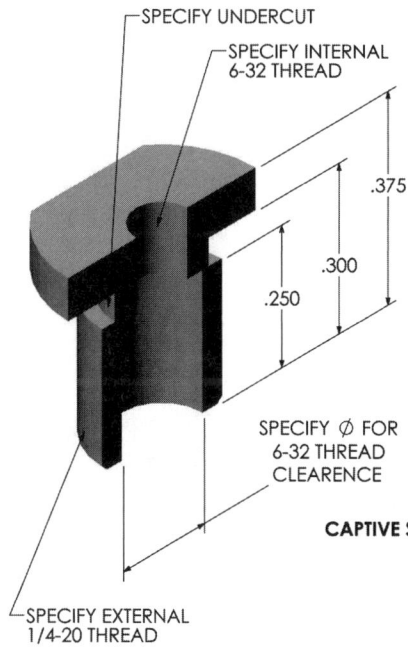

.375

.300

.250

SPECIFY Ø FOR
6-32 THREAD
CLEARENCE

SPECIFY EXTERNAL
1/4-20 THREAD

.313

2X Ø.391

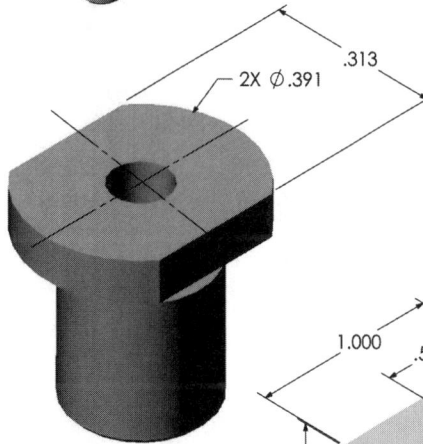

**CAPTIVE SCREW SLEEVE**

1.000    .500    .500    1.000

.375

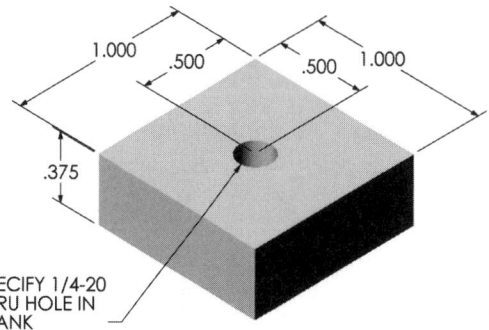

SPECIFY 1/4-20
THRU HOLE IN
BLANK

VIEWS SHOWN WITH
CAPTIVE 6-32 SCREW
ENGAGED⌐

VIEWS SHOWN WITH
CAPTIVE 6-32 SCREW
RETRACTED⌐

# 6-55 TOLERANCES OF LOCATION

$$\oplus \quad = \quad \bigodot$$

Tolerances of location state the permissible variation in the specified location of a feature in relation to some other feature or datum. Tolerances of location refer to the geometric characteristics: position, symmetry, and concentricity.

In the course of the discussion on location tolerancing, more detail on maximum material condition, datums, basic dimensioning, and the interrelationship of location and orientation tolerancing will be introduced.

Location tolerances involve features of size and relationships to center planes and axes. At least two features, one of which is a size feature, are required before location tolerancing is valid. Where function or interchangeability of mating part features is involved, the MMC principle may be introduced to great advantage. Perhaps the most widely used and best example of the application of this principle is position tolerancing.

The use of the position concept in conjunction with the maximum material condition concept provides some of the major advantages of the geometric tolerancing system.

# 6-56 POSITION $\oplus$

**Definition.** Position is a term used to describe the perfect (exact) location of a point, line, or plane of a feature in relationship with a datum reference or other feature.

## Position Tolerance

A position tolerance is the total permissible variation in the location of a feature about its exact true position. For cylindrical features (holes and bosses) the position tolerance is the *diameter (cylinder)* of the tolerance zone within which the axis of the feature must lie, the center of the tolerance zone being at the exact true position. For other features (slots, tabs, etc.) the position tolerance is the total width of the tolerance zone within which the center plane of the feature must lie, the center plane of the zone being at the exact true position.

## Position Theory

The illustration at right examines the position theory as typically applied to a part for purposes of function or interchangeability. As a means of describing this theory it is helpful to first compare the position system with the bilateral or coordinate system.

Imagine a part with four holes in a pattern which must line up with a mating part to accept screws, pins, rivets, etc., to accomplish assembly, or four holes in a pattern to accept the pins, dowels, or studs of a mating part to accomplish assembly.

The top figure on the next page shows the part with a hole pattern dimensioned and toleranced using a coordinate system. The bottom figure shows the same part dimensioned using the position system. Comparing the two approaches, the following differences are noted:

1. The derived tolerance zones for the hole centers are square in the coordinate system and round in the position system.
2. The hole center location tolerance in the top figure is part of the coordinates (the 2.000 and 1.750 dimensions). In the bottom figure, however, the location tolerance is associated with the hole size dimension and is shown in the feature control frame on the next page. The 2.000 and 1.750 coordinates are retained in the position application, but are stated as BASIC or exact values.

For this comparison, the .005 square coordinate tolerance zone has been converted to an equivalent .007 position tolerance zone. The two tolerance zones are superimposed on each other in the enlarged detail.

The black dots represent possible inspected centers of this hole on eight separate piece parts. Note that if the coordinate zone is applied, only three of the eight parts are acceptable. However, with the position zone applied, six of the eight parts appear immediately acceptable.

## Coordinate System

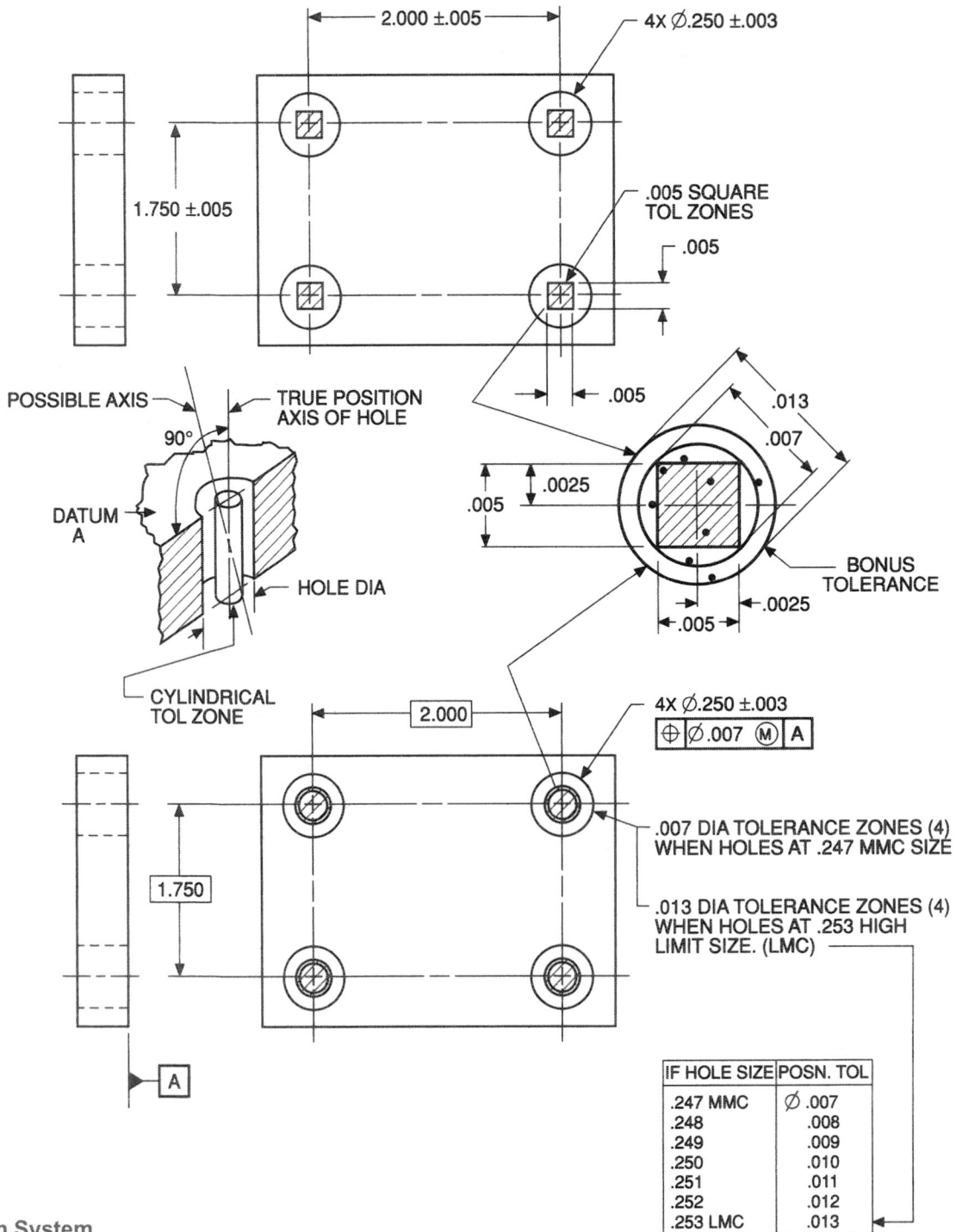

2.000 ±.005

4X ⌀.250 ±.003

1.750 ±.005

.005 SQUARE
TOL ZONES

.005

POSSIBLE AXIS

TRUE POSITION
AXIS OF HOLE

90°

DATUM
A

HOLE DIA

CYLINDRICAL
TOL ZONE

.013

.007

.0025

.005

.005

BONUS
TOLERANCE

.0025

.005

2.000

4X ⌀.250 ±.003

⊕ ⌀.007 Ⓜ A

.007 DIA TOLERANCE ZONES (4)
WHEN HOLES AT .247 MMC SIZE

1.750

.013 DIA TOLERANCE ZONES (4)
WHEN HOLES AT .253 HIGH
LIMIT SIZE. (LMC)

A

## Position System

| IF HOLE SIZE | POSN. TOL |
|---|---|
| .247 MMC | ⌀ .007 |
| .248 | .008 |
| .249 | .009 |
| .250 | .010 |
| .251 | .011 |
| .252 | .012 |
| .253 LMC | .013 |

The position diameter shaped zone can be justified by recognizing that the .007 diagonal is unlimited in orientation. Also, a cylindrical hole should normally have a cylindrical tolerance zone.

A closer analysis of the representative black dots and their position with respect to the desired exact location clearly illustrates the fallacies of the coordinate system when applied to a part such as that illustrated.

The dot in the upper left diagonal corner of the square zone and the dot on the left outside the square zone are in reality at nearly the same distance from the desired exact center. However, in terms of the square coordinate zone, the hole on the left is unacceptable by a wide margin, whereas the upper left hole is acceptable.

Note, then, that a hole produced off center under the coordinate system has *greater* tolerance if the shift is on the diagonal and not in the horizontal or vertical direction.

Realizing that the normal function of a hole relates to its mating feature in *any* direction (i.e., a hole vs. a round pin), we see that the square zone restriction seems unreasonable and incorrect. Thus the position tolerance zone, which recognizes and accounts for unlimited orientation of round or cylindrical features as they relate to one another, is more realistic and practical.

In normal applications of position principles, the tolerance is derived, of course, from the design requirement, *not* from converted coordinates. The maximum material sizes of the features (hole and mating component) are used to determine this tolerance.

Thus the .007 position tolerance of the example on the preceding page (see also Fig. 6-107) would normally be based on the MMC size of the hole (.247). As the hole size deviates from the MMC size, the position of the hole is permitted to shift off its "true position" beyond the original tolerance zone to the extent of that departure. The "bonus tolerance" of .013 illustrates the possible position tolerance should the hole be produced, for example, to its high limit size of .253. The tabulation on the preceding page shows the enlargement of the position tolerance zone as the hole size departs from MMC.

Although we have considered only one hole to this point in the explanation, the same reasoning applies to all the holes in the pattern. Note that position tolerancing is also a noncumulative type of control in which each hole relates to its own true or exact position and no error is accumulated from the other holes in the pattern.

Position tolerancing is usually applied on mating parts in cases where fit, function, and interchangeability are the considerations. It provides greater production tolerances, ensures design requirements, and provides the advantages of functional inspection practices as desired.

Functional gaging techniques, familiar to a large segment of industry through many years of application, are fundamentally based on the MMC position concept. It should be clearly understood, however, that functional gages are not mandatory in fulfilling MMC position requirements.

Functional gages are used and discussed in this text for the dual purpose of explaining the principles involved in position tolerancing and of introducing the functional gage technique as a valuable tool. A functional gage can be considered as a simulated master mating part at its worst condition.

Position, although a locational tolerance, also includes form and orientation tolerance elements in composite. For example, as shown in the illustration on page 90, perpendicularity is invoked as part of the control to the extent of the diameter zone, actually as a "cylindrical" zone, for the depth of the hole. Further, the holes in the pattern are parallel to one another within the position tolerance. Various other elements of form are included as a part of the composite functional control provided by position tolerancing. Use of datum references and the relation of the hole pattern to specific surfaces or other features further ensures that the holes (and their tolerance zones) are related to part function and will be uniformly interpreted. For purposes of simplicity in explaining the theory, only *one* datum feature was specified in this example. A complete specification of this requirement would include *two* additional datums (edges of the part).

## Position System

The example below further clarifies the position theory; two of the holes on the part shown in the example are enlarged to illustrate the actual effect of feature *size* variation on the previous positional location of the features.

## Position Theory

Figure 6-107 shows the two .250 ± .003 holes at MMC size (or the low limit of their size tolerance) of .247 and with their centers perfectly located in the .007 diameter position tolerance zone. The drawing illustrates the mating part situation represented by a functional or fixed pin. The gage pins are shown undersize an amount equal to the positional tolerance of .007, i.e., at .240 diameter. This represents the maximum permissible offset of the holes within their stated positional tolerance when the hole is at MMC size of .247.

Figure 6-108 shows the two .250 MMC holes offset in opposite directions to the maximum permissible limits of the .007 position tolerance zone. Note that we illustrate the worst condition: the edges of the holes are tangent to the diameters of the simulated mating part or gage pins. The holes are within tolerance and, as can be seen, would satisfactorily

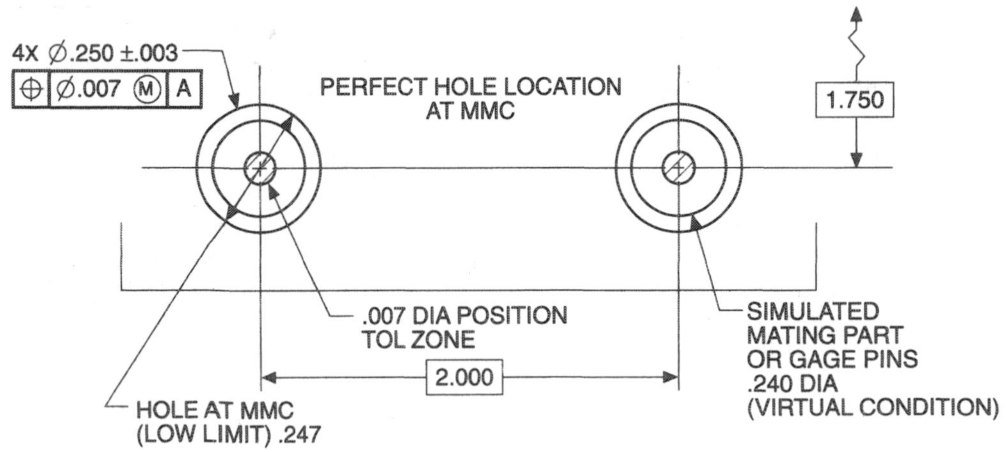

**Figure 6-107**

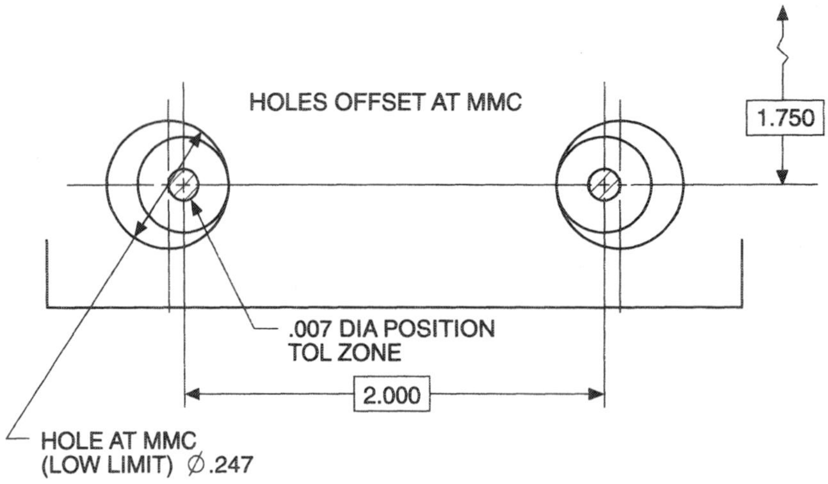

**Figure 6-108**

pass the simulated mating part condition as represented by the gage pins.

In Fig. 6-109 the .250 ± .003 holes have been produced to the opposite, or *high* limit (least material condition) size of .253. It can now be seen that when we retain the same offset and tangency of the holes and mating part of the gage pins as shown in Fig. 6-108, the produced centers of the holes are allowed to shift *beyond* the original .007 tolerance zone to a resulting .013 diameter tolerance zone still providing an acceptable situation.

The foregoing illustrates the interrelationship of size and location tolerances which is utilized in position dimensioning and tolerancing.

Although in this example we have used only two of the holes, the same reasoning applies to *all* the holes in the pattern; similarly, each individual hole could be offset within its tolerance zone in any direction around 360° and provide an acceptable situation.

It should be noted that a functional or fixed pin gage such as the one used here to explain the position theory can

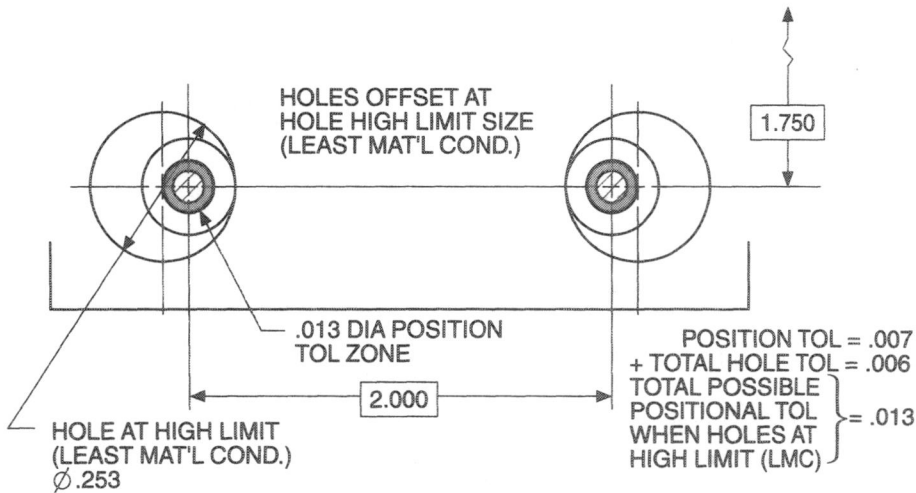

HOLES OFFSET AT
HOLE HIGH LIMIT SIZE
(LEAST MAT'L COND.)

1.750

.013 DIA POSITION
TOL ZONE

2.000

HOLE AT HIGH LIMIT
(LEAST MAT'L COND.)
$\varnothing$.253

POSITION TOL = .007
+ TOTAL HOLE TOL = .006
TOTAL POSSIBLE
POSITIONAL TOL
WHEN HOLES AT
HIGH LIMIT (LMC) } = .013

**Figure 6-109**

be used *only* to check the *positional* location of the holes. *Positional* tolerance can be added as the holes increase in size or depart from MMC size within their size tolerance range. Hole size tolerance, however, must be held within the tolerances specified on the drawing and must be checked individually and separately from the positional check.

### Mating Parts—Floating Fastener

Position tolerancing techniques are most effective and appropriate in mating part situations. The illustrations on page 457, in addition to demonstrating the calculations required, also emphasize the importance of decisions at the design stage to recognize and initiate the position principles.

The mating parts shown in the illustration are to be interchangeable. Thus, the calculation of their position tolerances should be based on the two parts and their interface with the fastener in terms of MMC sizes.

The two parts are to be assembled with four .190 screws. The holes in the two parts are to line up sufficiently to pass the four screws at assembly. Since the four screws ("fasteners") are separate components, they are considered to have some "float" with respect to one another. The colloquial term, "floating fastener" application, has been popularly used to describe this situation.

The calculations are shown in the upper right corner of the illustration. Also, note that, in this case, the same basic dimensions and position tolerances are used on both parts. They are, of course, separate parts and are on separate drawings.

The position tolerance calculations are based on the MMC sizes of the holes and the screws. The maximum material basis then sets the stage for maximum producibility, interchangeability, functional gaging (if desired), etc., at production. As seen from the illustration, part acceptance tolerances will increase as the hole actual mating sizes in the parts are actually produced and vary in size as a departure from MMC. From the .016 diameter tolerance calculated, the tolerance may increase to as much as .022 dependent upon the actually produced hole size. It should be noted that clearance between the mating features (in this case hole and screw) is the criterion for establishing the position tolerances.

Simultaneously with these production advantages, the design is protected since it has been based upon the realities of the hole and screw sizes as they interrelate at assembly and in their function. Thus, as parts are produced, assembly is ensured, and the design function is carried out specifically as planned.

A possible functional gage is also shown in the illustration on page 457. The .190 gage pin diameters are determined by the MMC size of the hole, .206, minus the stated position tolerance of .016. In our example, the same functional gage can be used on both parts. Functional gages are, of course, not required with position application, but they do, however, provide an effective method of evaluation where desired. Functional gaging principles may be carried out by the use of coordinate measuring machines (CMM's), and the variables data derived, through the use of calculators, computer programs or tabulated mathematical equivalents. This technique is often referred to as "soft gaging."

### Mating Parts—Floating Fastener

Referring to the position tolerance calculations, if more than two parts are assembled in a floating fastener application, we must determine the position tolerance to ensure that any two parts and the fastener will mate properly. Calculate each part to mate with the fastener using the illustrated formula and MMC sizes.

Only the primary datums (the interface surfaces) are shown on the illustrations for simplicity of explanation of this method. The hole pattern alignment is emphasized between the two parts. A complete specification of these requirements on both parts would include two additional datums to locate the holes and the pattern from the part outside surfaces (datums).

The clearance holes on these parts are all specified as same size. Where they are specified as *different sizes,* each part can be calculated separately for the allowable positional displacement based on the difference between the MMC of the hole and fastener. If one part in our example had .203 ± .003–4 HOLES specified, the method below would be used:

| MMC hole | $\varnothing$.200 |
| MMC fastener | $(-)\varnothing$.190 |
| | $\varnothing$.010 |

$$\boxed{\oplus \;\; \varnothing.010 \;\; \text{Ⓜ} \;\; | \; A \;}$$

| Other part | |
| MMC hole | $\varnothing$.206 |
| MMC fastener | $(-)\varnothing$.190 |
| | $\varnothing$.016 |

$$\boxed{\oplus \;\; \varnothing.016 \;\; \text{Ⓜ} \;\; | \; A \;}$$

The position tolerance calculation method illustrated assumes the possibility of a zero interference-zero clearance

**NOTE:**

Where mating parts are to be assembled with rivets, dowel pins, press fits, etc., other considerations or methods may be necessary; such as, drilling or fabricating of the holes, pins or other features at assembly.

condition of the mating part features at extreme tolerance limits. Additional compensation of the calculated tolerance values should be considered as necessary relative to the particular application.

Formulas used as a basis for the position floating fastener calculations are:

To calculate position tolerance with fastener and hole size known:

$$T = H - F$$

where T = tolerance, H = MMC hole, and F = MMC fastener.

Where the hole size or fastener size is to be derived from an established position tolerance, the formula is altered to:

$$H = T + F$$

$$F = H - T$$

**Mating Parts—Floating Fastener**

## Example

4X ⌀.206 $^{+.006}_{-.000}$

| ⊕ | ⌀.016 Ⓜ | A |

1.500

1.250

A

4X ⌀.206 $^{+.006}_{-.000}$

| ⊕ | ⌀.016 Ⓜ | A |

1.500

1.250

A

**CALCULATIONS**

MMC OF HOLE        =⌀.206

MMC OF FASTENER
SCREW MAX        =(−)⌀.190

POSITIONAL TOL        =⌀.016
FOR ALL HOLES ON
BOTH PARTS

PART ACCEPTANCE TOL
PART HOLES AT LOW
LIMIT .206 (MMC)
POSITIONAL TOL = .016 DIA

PART HOLES AT HIGH
LIMIT .212 (LMC)
POSITIONAL TOL = .022 DIA

| AT SIZE | TOL |
|---------|-----|
| .206 MMC | .016 |
| .207 | .017 |
| .208 | .018 |
| .209 | .019 |
| .210 | .020 |
| .211 | .021 |
| .212 LMC | .022 |

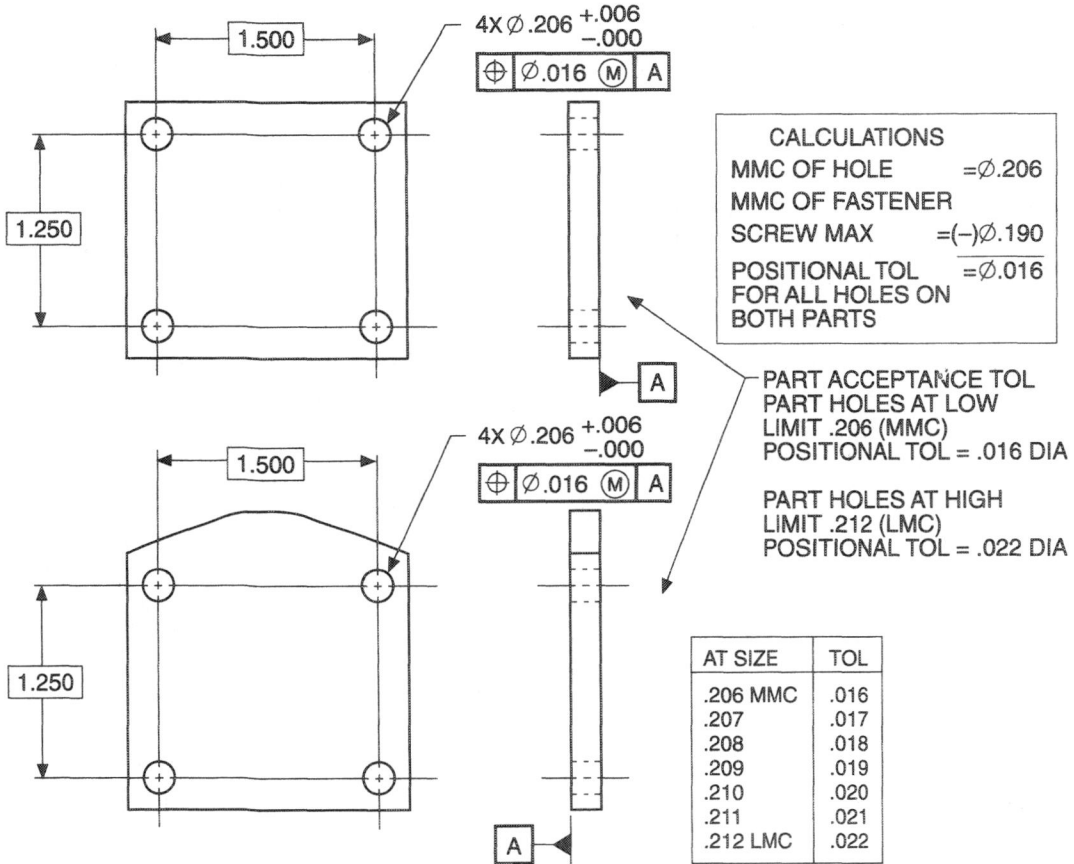

## Gage for above Parts

1.500

1.250

4X .190 { .206 MMC / (−).016 TOL / .190 PIN }

## Mating Parts—Fixed Fastener

When one of two mating parts has "fixed" features, such as the threaded studs in this example, the "fixed fastener" method is used in calculating position tolerances.

The term "fixed fastener" is a colloquialism popularly used to describe this application. Both the term and the technique are applied to numerous other manufacturing situations such as locating dowels and holes, tapped holes, etc.

The advantages of the MMC principle as described in the foregoing "floating fastener" application also apply here. However, with a "fixed fastener" application, the difference between the MMC sizes of mating features must be divided between the two features, since the total position tolerance must be shared by the two mating features. In this example, the two mating features (actually four of each in each pattern) are the studs and the clearance holes. The studs must fit through the holes at assembly.

Again, we see that the clearance of the mating features as they relate to each other at assembly determines the position tolerances. When one feature is to be assembled *within* another on the basis of the MMC sizes and "worst" condition of assembly, the clearance, or total tolerance, must be divided for assignment to *each* of the mating part features. In this case, the derived .016 was divided equally, with .008 diameter position tolerance assigned to each mating part feature (stud and hole). The total tolerance of .016 can be distributed to the two parts as desired, so long as the total is .016 (e.g., .010 + .006, .012 + .004, etc.). This decision is made at the design stage, however, and must be fixed on the drawing before release to production.

Only the primary datums (the interface surfaces) are shown on the illustrations for simplicity of explanation of the fixed fastener method.

Application of the MMC principle to situations of this type guarantees functional interchangeability, design integrity, maximum production tolerance, functional gaging (if desired), and uniform understanding of the requirements.

As the part features of both parts are produced, any departure in actual mating size from MMC will increase the calculated position by an amount equal to that departure. Thus, for example, the position tolerance of the upper part could possibly increase up to .014 and that of the lower part up to .013 dependent upon the amount of departure from their MMC sizes. However, parts must actually be produced and sizes established before the *amount* of increase in tolerance can be determined.

Functional gages (shown below each part in the illustration) can be used for checking, and, although their use is not a must, they provide a very effective method of evaluation if desired. Note that the functional gages resemble the mating parts; as a matter of fact, functional gages simulate mating parts at their worst condition.

The functional gage pins of the upper part are determined by the MMC hole size minus the stated position tolerance. Gage tolerances are not shown, although they may be imagined to be on the order of $.1981 \, {}^{+\,.0002}_{-\,.0000}$ for pin size, and ± .0002 on between pin locations. Local gage practices would prevail using a minimum amount of part tolerance (up to 10%) for gage making tolerances as assigned.

The functional gage on the lower part of the illustration contains holes instead of pins. The gage hole sizes are determined by the MMC (O.D.) size of the .190 pins plus the stated position tolerance. The tolerances are similar to those of the above pin gage. Tolerances on the order of $.1979 \, {}^{+\,.0000}_{-\,.0002}$ for hole size, and ± .0002 between holes could be applied, depending on local gage practices using a minimum amount of part tolerance (up to 10%) for gage making tolerances as assigned.

**Example**

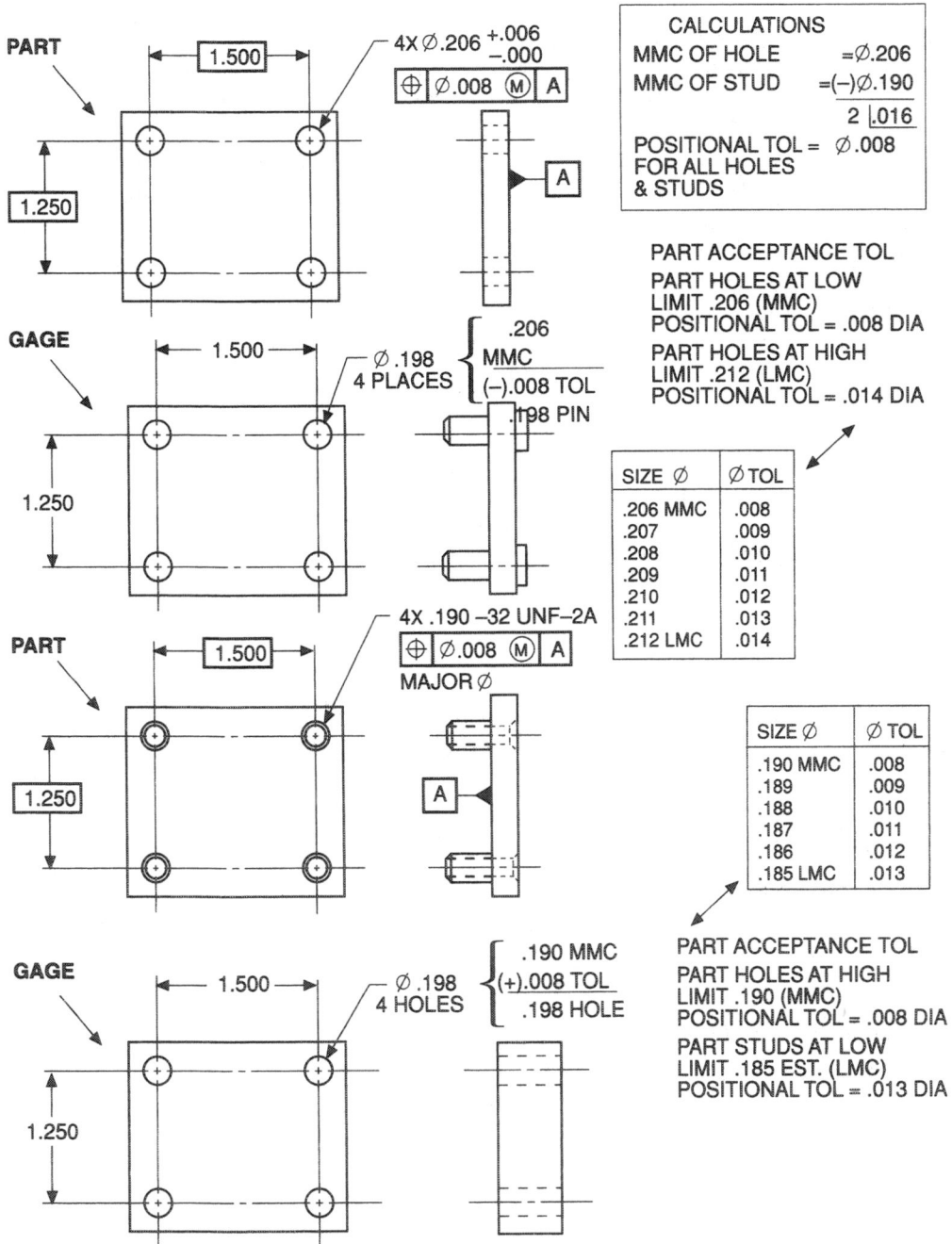

**PART**

1.500

1.250

4X Ø.206 $^{+.006}_{-.000}$

⊕ | Ø.008 Ⓜ | A

A

| CALCULATIONS | |
|---|---|
| MMC OF HOLE | =Ø.206 |
| MMC OF STUD | =(−)Ø.190 |
| | 2 ⌊.016 |
| POSITIONAL TOL = Ø.008 FOR ALL HOLES & STUDS | |

PART ACCEPTANCE TOL

PART HOLES AT LOW
LIMIT .206 (MMC)
POSITIONAL TOL = .008 DIA

PART HOLES AT HIGH
LIMIT .212 (LMC)
POSITIONAL TOL = .014 DIA

**GAGE**

1.500

1.250

Ø .198
4 PLACES

.206
MMC
(−).008 TOL
.198 PIN

| SIZE Ø | Ø TOL |
|---|---|
| .206 MMC | .008 |
| .207 | .009 |
| .208 | .010 |
| .209 | .011 |
| .210 | .012 |
| .211 | .013 |
| .212 LMC | .014 |

**PART**

1.500

1.250

4X .190 −32 UNF−2A

⊕ | Ø.008 Ⓜ | A

MAJOR Ø

A

| SIZE Ø | Ø TOL |
|---|---|
| .190 MMC | .008 |
| .189 | .009 |
| .188 | .010 |
| .187 | .011 |
| .186 | .012 |
| .185 LMC | .013 |

PART ACCEPTANCE TOL

PART HOLES AT HIGH
LIMIT .190 (MMC)
POSITIONAL TOL = .008 DIA

PART STUDS AT LOW
LIMIT .185 EST. (LMC)
POSITIONAL TOL = .013 DIA

**GAGE**

1.500

1.250

Ø .198
4 HOLES

.190 MMC
(+).008 TOL
.198 HOLE

It should be noted that the term MAJOR Ø is used beneath the position callout on the lower part on page 459. In the absence of this special notation of exception, the ANSI Y14.5, Pitch Diameter Rule would have invoked the tolerance on the basis of the pitch diameter of the threads. The major diameter (or O.D.) of the thread was the desired criterion in this example. See POSITION—EXTENDED PRINCIPLES section for additional examples of fixed fastener applications.

The calculations on these parts illustrate a balanced tolerance application in which the total permissible position tolerance of the two parts is equally divided, for example, .008 on each part. The total position tolerance can, however, be distributed as desired, as discussed earlier.

If more than two parts are assembled in a fixed fastener application, each part containing clearance holes must be calculated to mate with the part with the fixed features.

The position tolerance calculation method illustrated assumes the possibility of a zero interference-zero clearance condition of the mating part features at extreme tolerance limits. Additional compensation of the calculated tolerance values should be considered as necessary relative to the particular application.

Formulas used as a basis for the position fixed fastener (or locator) calculations are:

$$T = \frac{H - F}{2}$$

| MMC hole | = H |
| MMC fastener (or pin, dowel, etc.) | = F |
| Tolerance | = T |

Where the hole size or fastener (or pin, dowel, etc.) size is to be derived from an established tolerance, the formula is altered to:

$$H = F + 2T$$
$$F = H - 2T$$

The illustrations on page 461 show position tolerancing applied to two mating parts with a circular hole pattern. The same reasoning applies here as in the preceding examples except that the basic dimensions are angular (45° angles, 8 places) and a diameter (the 1.500 diameter).

These two parts again are of the fixed fastening type, the studs of the lower part being the fixed elements. To determine the positional tolerances for each part, the MMC of the hole and the MMC of the stud are used to determine the total positional tolerance. This is divided by two to give the positional tolerance value for each part. The total value may be divided as desired, as previously described.

Note again how the positional tolerance *increases* as the actual mating sizes of the holes in the upper part and the studs in the lower part depart from their MMC sizes, that is, when the holes get larger and the pins get smaller during the production process.

Functional gages are shown in the illustration for each of the parts. Note that the pins in the upper gage are calculated to the MMC or low limit of the holes in the part (which is .187 in this case) *minus* the positional tolerance (.0025), resulting in the .1845 gage pin size.

## Example

8X Ø.190 ±.003

| ⊕ | Ø.0025 Ⓜ | A |

8X 45°

PART

Ø1.500

A

**CALCULATIONS**

MMC OF HOLE = Ø.187
MMC OF STUD = (−) Ø.182

$$2\overline{|.005}$$

POSITIONAL    = Ø.0025
TOL FOR ALL
HOLES AND STUDS

PART ACCEPTANCE TOL

PART HOLES AT LOW
  LIMIT .187 (MMC)
  POSITIONAL TOL = Ø.0025

PART HOLES AT HIGH
  LIMIT .193
  POSITIONAL TOL = Ø.0085

| Ø SIZE | Ø TOL |
|---|---|
| .187 MMC | .0025 |
| .1875 | .003 |
| .188 | .0035 |
| .189 | .0045 |
| .190 | .0055 |
| .191 | .0065 |
| .192 | .0075 |
| .193 LMC | .0085 |

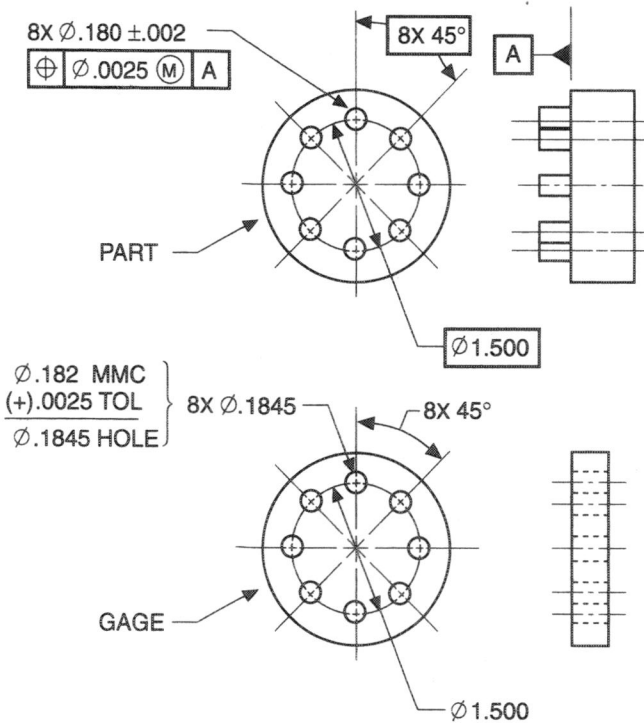

Ø.187 MMC
(−).0025 TOL
Ø.1845 PIN
8X Ø.1845

8X 45°

GAGE

Ø1.500

8X Ø.180 ±.002

| ⊕ | Ø.0025 Ⓜ | A |

8X 45°

A

PART

Ø1.500

PART ACCEPTANCE TOL

PART STUDS AT HIGH
  LIMIT .182 (MMC)
  POSITIONAL TOL = Ø.0025

PART STUDS AT LOW
  LIMIT .178
  POSITIONAL TOL = Ø.0065

Ø.182 MMC
(+).0025 TOL
Ø.1845 HOLE
8X Ø.1845

8X 45°

GAGE

Ø1.500

| Ø SIZE | Ø TOL |
|---|---|
| .182 MMC | .0025 |
| .1815 | .003 |
| .181 | .0035 |
| .180 | .0045 |
| .179 | .0055 |
| .178 LMC | .0065 |

The lower gage is calculated in reverse, using the MMC on high limit of the stud, .182, plus the positional tolerance (.0025), resulting in the .1845 gage hole size.

These calculations illustrate a balanced tolerance application in which the total permissible position tolerance of the two parts is equally divided, for example, .0025 on each part. The total position tolerance can, however, be distributed as desired, for example, .002 on one part, .003 on the other, etc., so long as it totals the tolerance calculated (in this case .005).

The position tolerance calculation method illustrated here and in preceding examples assumes the possibility of a zero interference-zero clearance condition of the mating part features at extreme tolerance limits. Additional compensation of the calculated tolerance value should be considered as necessary relative to the particular application.

A primary datum (A) only is shown in the examples for simplicity of explanation. A secondary datum (i.e., the outside diameter) would be required to complete the specifications on both parts.

## Relation to Specified Datum Surfaces

Datum planes or surfaces as the basis for position relationships must be *specified* on the drawing. This illustration introduces the datum reference frame and datum precedence as it relates to hole patterns. The primary datum A placed first in the feature control frame is probably established because it is the most important stabilizing feature for the part interface, the corresponding feature, with a mating part. The secondary datum B is placed next in the feature control frame indicating the next most important orientation of the part to a mating surface and the third, or tertiary, datum C is then added in the third segment of the feature control frame to complete the hole pattern relationship to the datum planes, the three mutually perpendicular datum planes. Very often the selection of which features is to be selected as the secondary datum is arbitrary. That is, the part function and interface with a mating assembly may suggest that either of the second or third necessary datums could be equally selected as secondary datum. However, to give uniform meaning to design intent and manufacturing and verification follow-through, a decision is made to remove the either/or ambiguity.

The extremities or high points of the datum surfaces from which the $\boxed{.750}$ dimensions are taken establish the datum planes from which the position hole pattern is oriented. These datum planes are functional to the part requirement and may also be used for tooling or fixturing reference and in establishing measuring planes to inspect the part.

The position tolerance zone is .010 when the holes are produced at MMC or the low limit of .245. The tolerance zone increases up to .020, if the hole actual mating envelope size is produced larger to the high (LMC) limit of .255. Note that the position tolerance applies while the part is in contact with the datum surface; thus the position tolerance stated also controls the pattern location from the edges and provides the tolerance for the $\boxed{.750}$ pattern locating dimensions.

Note that in this example, the precision of the datum feature surfaces will be the result of manufacturing processes used to produce the rather lenient part overall size dimensions. The question is then raised, are these surfaces of sufficient precision to ensure the part relationship to the mating part situation is adequate. It is the decision of the design to determine this. Note also that the squareness (perpendicularity) of the two outside surfaces selected as datum features (B and C) have no ensured specific perpendicularity requirement. The result, unless further refined by form and/or orientation controls will be "whatever is produced." This depends on what can be considered the "four factors", i.e., good workmanship, discretion of the production operations, probabilities and any unless otherwise specified controls (such as title block, industry specs, milspecs, etc.).

**Example**

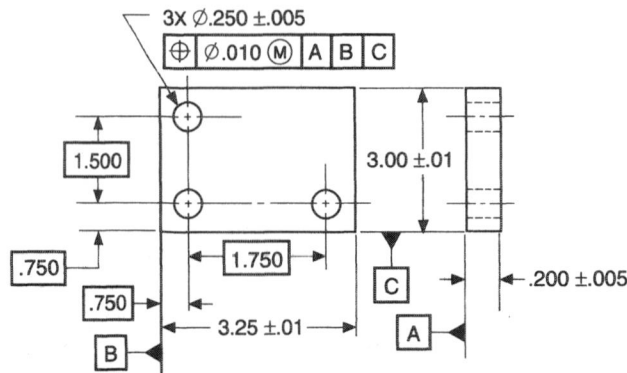

**Meaning**

| SIZE ⌀ | ⌀ TOL |
|---|---|
| .245 MMC | .010 |
| .246 | .011 |
| .247 | .012 |
| .250 | .015 |
| .255 LMC | .020 |

DATUM PLANE B (SEC)

.245/.255 3 HOLES

HOLE (3) LOCATED AT TRUE POSITION WITHIN .010 DIA. IF HOLE AT .245 MMC, .020 DIA IF HOLE AT .255 LMC

90°

90°

90°

90°

90°

1.500

DATUM SURFACE B

.750

DATUM PLANE C (TERT)

.750

1.750

.195 .205

DATUM PLANE A (PRI)

DATUM SURFACE A

DATUM SURFACE C

90°

(DATUM PLANES DETERMINED BY EXTREMITIES OF DATUM SURFACES)

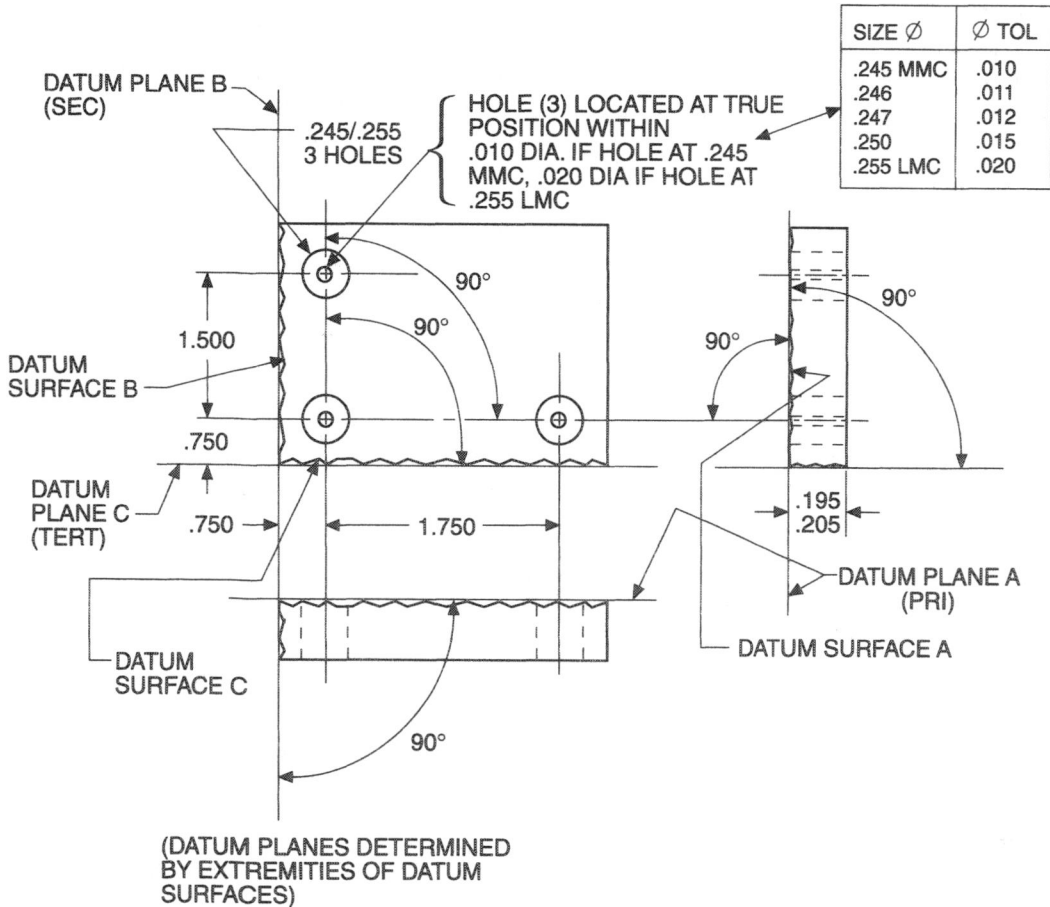

## Relation to Specified Datum Surfaces—with Form and Orientation Tolerances

The illustration on the right shows the same part as that shown on the preceding page, but with added form and orientation tolerances specified to control the precision of the datum features.

Where part function, and thus the stated drawing requirements, are more critical, specified datums and greater geometric control are essential to the design.

In this example, it was considered necessary to control the accuracy of the datum surfaces in their specific relationship to each other. To accomplish this, identification of the specific surfaces as datum references was required. Further, since the hole position pattern was critical in this orientation to the surfaces, datum identification was required. With specification of the datums, precedence of the datum surfaces is established, and the part and the hole pattern are stabilized relative to the datum reference frame. Datum precedence is also established.

Datum surface A (top surface of the part) is to be held to a flatness of .001 total. Datum surface B is to be perpendicular to datum plane A within .001 total. Datum surface C is to be perpendicular to datum plane A within .001 total and also perpendicular to datum plane B within .002 total.

The extremities or high points of the datum surfaces from which the .750 dimensions are taken establish the secondary and tertiary datum planes (B and C). The primary datum plane (A) is established by the extremities of datum surface A. The part orientation with respect to the position pattern is thus fixed. These datum planes are functional to the part requirement and may also be used for tooling or fixturing reference.

The position tolerance zone is .010 when the holes are produced at MMC or the low limit of .245. The tolerance zone increases up to .020 if the holes are produced larger to the high limit of .255. Note that the position tolerance applies while the part is in contact with the datum surfaces according to their stated precedence or sequence. The position tolerance stated also controls the pattern location from the edges and provides the tolerance for the .750 pattern locating dimensions.

In this example all geometric controls are specifically stated, removing all doubt as to design intent and follow-through manufacture and inspection requirements. The tools provided by specified datums and greater geometric control can be very effectively applied and will protect design integrity.

It should be noted that the form and orientation controls shown are not necessarily required when specifying the datum planes (A,B,C). The designer should consider the precision of the surfaces being used as datum features (as controlled by size), and then only as necessary, per the design requirements, add any form or orientation controls. See the preceding illustration and explanation where no form and orientation tolerances were required.

## Position Tolerance—Specified Datums—Form and Orientation Tolerances

**Example**

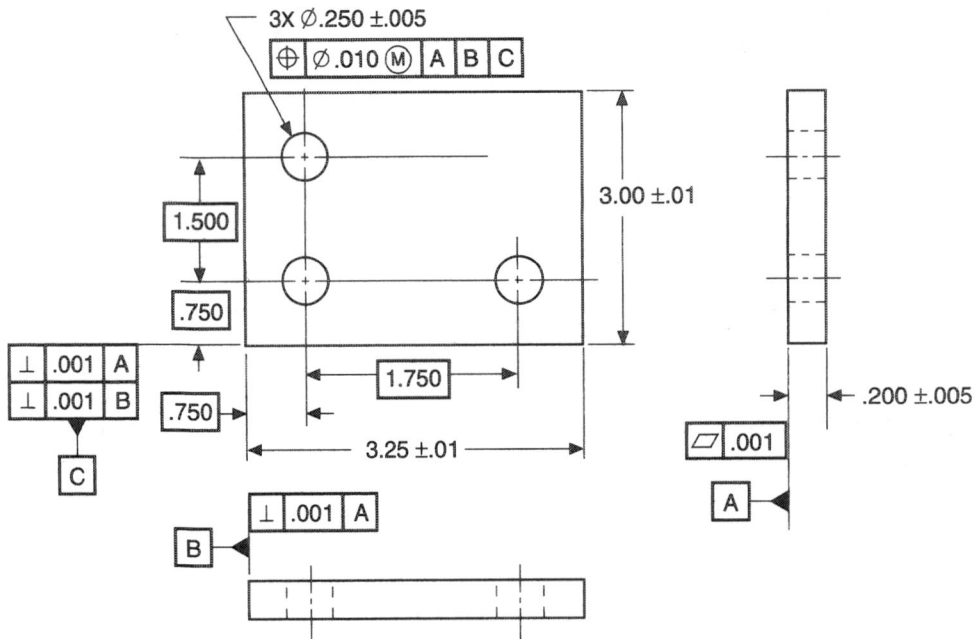

**Meaning**

| SIZE ⌀ | ⌀ TOL |
|---|---|
| .245 MMC | .010 |
| .246 | .011 |
| .247 | .012 |
| .250 | .015 |
| .255 LMC | .020 |

DATUM PLANE B (SEC)

.245/.255 3 HOLES

HOLE (3) LOCATED AT POSITION WITHIN .010 DIA. IF HOLE AT .245 MMC, .020 DIA IF HOLE AT .255 LMC.

DATUM PLANE A (PRI)

90°

90°

1.500

90°

90°

.750

90°

.750

1.750

DATUM PLANE C (TERT)

.195 .205

DATUM SURFACE "B" PERPENDICULAR TO "A" WITHIN .001

DATUM SURFACE "C" PERPENDICULAR TO "A" WITHIN .001 & PERPENDICULAR TO "B" WITHIN .002

DATUM SURFACE "A" FLAT WITHIN .001 TOTAL

90°

(DATUM PLANES DETERMINED BY EXTREMITIES OF DATUM SURFACES)

## Relation to Datum Surfaces—Composite Position Tolerancing

When the location of a pattern of features from datum surfaces is less important than the accuracy required *within* the pattern of features, composite position tolerancing may be used.

Composite position tolerancing also extends the use of specified datum relationships and geometric tolerance control.

Composite position tolerancing incorporates a dual feature control frame with two positional controls. One, the upper segment in the symbol, specifies the applicable datums and the pattern locating position tolerance. The lower segment specifies the applicable datum and the feature relating position tolerance. A single position tolerance symbol is used.

The composite position tolerance method utilizes the full advantages of MMC (and RFS or LMC if desired) and extends the principles to control of patterns of features as well as of the individual feature interrelationship.

In the upper right pattern of eight mounting holes in the example, the theoretical center of the pattern is located by the 2.008 and 3.978 basic dimensions and is related to datums A, B, and C. This center is the axis for the 1.990 basic diameter. The individual theoretical centers of the eight .188 holes are established by the intersection of the 45° basic angles at the 1.990 basic diameter. The pattern as a unit, yet actually determined by the holes themselves, is to be located within a position tolerance of .030 when the holes are at MMC. As the individual holes in the pattern depart from MMC toward LMC, additional tolerance to that hole (and thus to the pattern) is acquired equal to that departure.

Within the hole pattern itself, the feature relating position tolerance is established at .005 diameter. Then, as previously described, each hole in the pattern may increase its position tolerance an amount equal to the departure from MMC as it is produced to a maximum of .009 at LMC. Note that the hole-to-hole interrelationship in the pattern, as well as the relationship to datum A, is maintained. The attitude (perpendicularity) of each individual hole simultaneously must be within the .005 diameter tolerance zone as well as within position of .005 at MMC.

As can be seen in the typical hole cross-sectional view, the axes of both the large (pattern) and small (feature) tolerance zones are parallel. The axes of the holes must lie within both the larger and smaller tolerance zones. Portions of the smaller zones may fall outside the peripheries of the larger tolerance. However, this portion of the smaller zone is not usable since the axis of each hole must fall within both zones.

The lower left four hole pattern follows the same reasoning as described above and as seen in the illustration.

**Example**

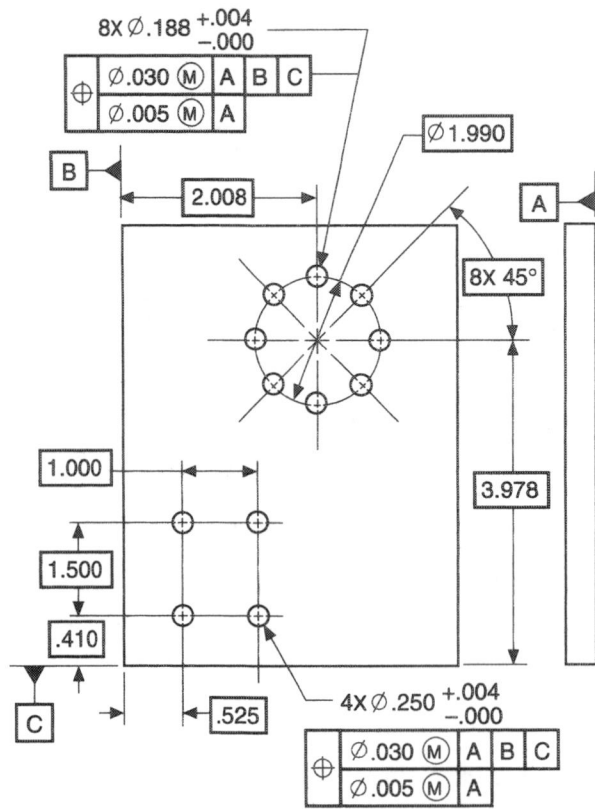

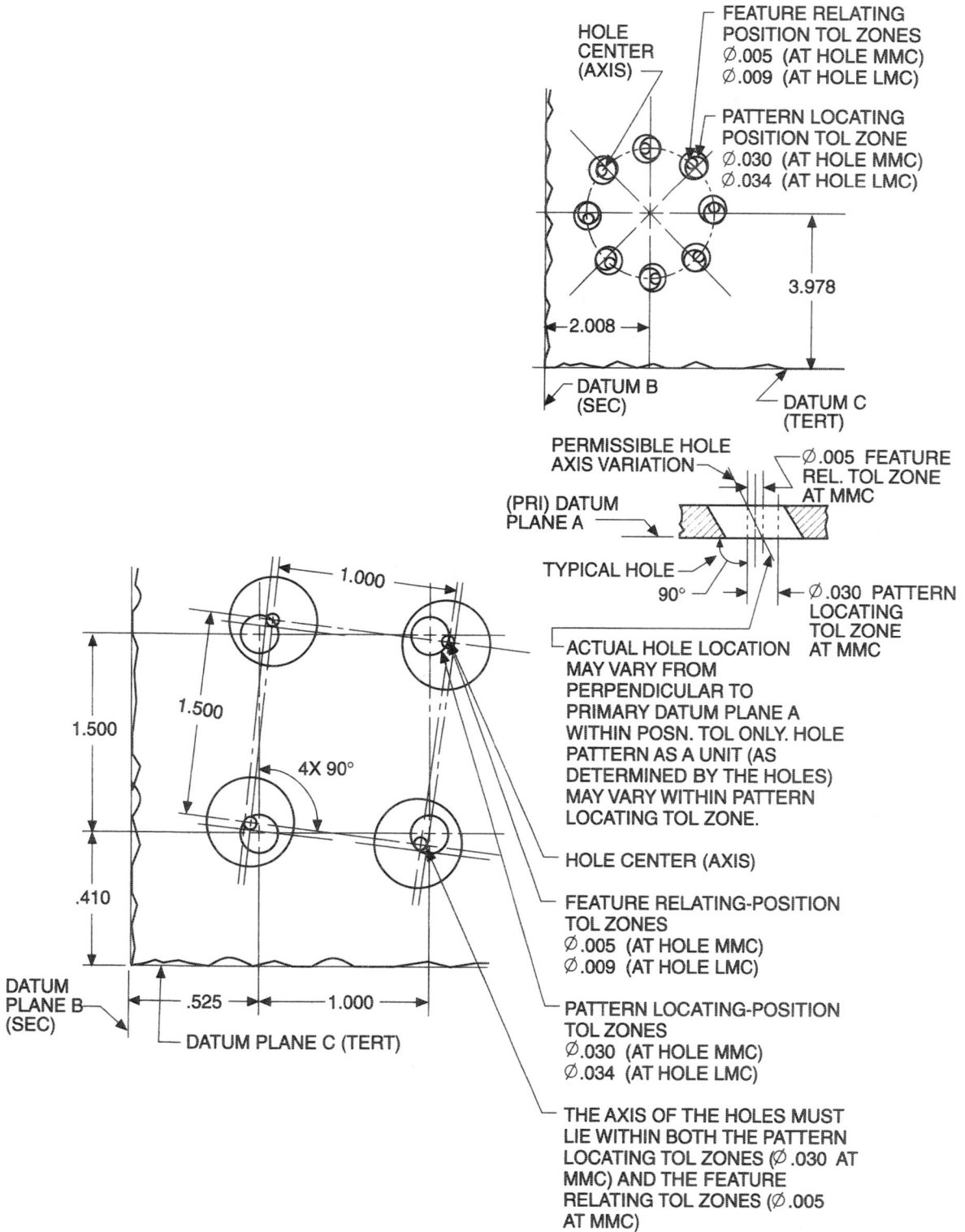

**Meaning**

HOLE CENTER (AXIS)

FEATURE RELATING POSITION TOL ZONES
$\emptyset$.005 (AT HOLE MMC)
$\emptyset$.009 (AT HOLE LMC)

PATTERN LOCATING POSITION TOL ZONE
$\emptyset$.030 (AT HOLE MMC)
$\emptyset$.034 (AT HOLE LMC)

3.978

2.008

DATUM B (SEC)

DATUM C (TERT)

PERMISSIBLE HOLE AXIS VARIATION

$\emptyset$.005 FEATURE REL. TOL ZONE AT MMC

(PRI) DATUM PLANE A

TYPICAL HOLE

90°

$\emptyset$.030 PATTERN LOCATING TOL ZONE AT MMC

ACTUAL HOLE LOCATION MAY VARY FROM PERPENDICULAR TO PRIMARY DATUM PLANE A WITHIN POSN. TOL ONLY. HOLE PATTERN AS A UNIT (AS DETERMINED BY THE HOLES) MAY VARY WITHIN PATTERN LOCATING TOL ZONE.

1.000

1.500

1.500

4X 90°

.410

HOLE CENTER (AXIS)

FEATURE RELATING-POSITION TOL ZONES
$\emptyset$.005 (AT HOLE MMC)
$\emptyset$.009 (AT HOLE LMC)

PATTERN LOCATING-POSITION TOL ZONES
$\emptyset$.030 (AT HOLE MMC)
$\emptyset$.034 (AT HOLE LMC)

THE AXIS OF THE HOLES MUST LIE WITHIN BOTH THE PATTERN LOCATING TOL ZONES ($\emptyset$.030 AT MMC) AND THE FEATURE RELATING TOL ZONES ($\emptyset$.005 AT MMC)

DATUM PLANE B (SEC)

DATUM PLANE C (TERT)

.525

1.000

Relation to Datum Surfaces Composite Position
Tolerancing—Functional Gages

**Example**                                                  **Gages**

8X Ø.188 +.004 −.000

| ⊕ | Ø.030 Ⓜ A B C |
| | Ø.005 Ⓜ A |

B       2.008       Ø1.990       A

8X 45°

1.000

1.500

3.978

C   .410      .525      4X Ø.250 +.004 −.000

| ⊕ | Ø.030 Ⓜ A B C |
| | Ø.005 Ⓜ A |

SIMULATED DATUM A

Ø1.990

45° (8)

90°

8x Ø.183 PINS

Ø.188 +.004 −.000 HOLE

Ø.188 MMC
(−)Ø.005 TOL AT MMC
Ø.183 GAGE PIN SIZE

FEATURE RELATING

SIMULATED DATUM A

1.500

90°

1.000          4x Ø.245 PINS

Ø.250 +.004 −.000 HOLE

Ø.250 MMC
(−)Ø.005 TOL AT MMC
Ø.245 GAGE PIN SIZE

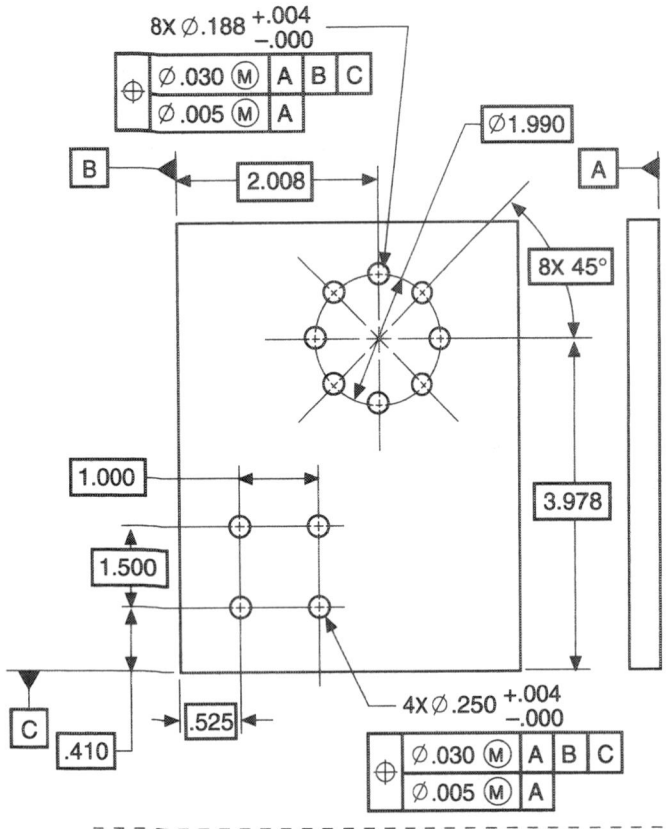

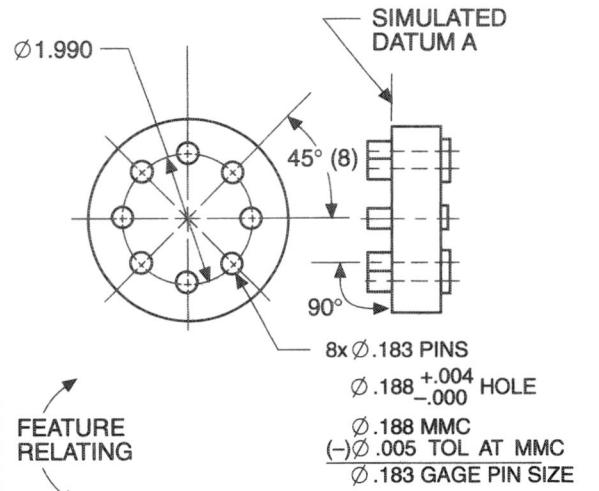

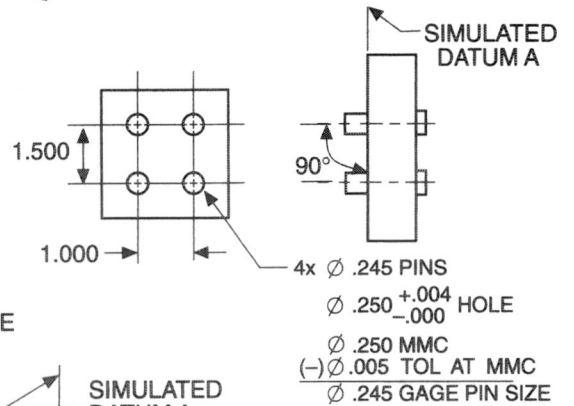

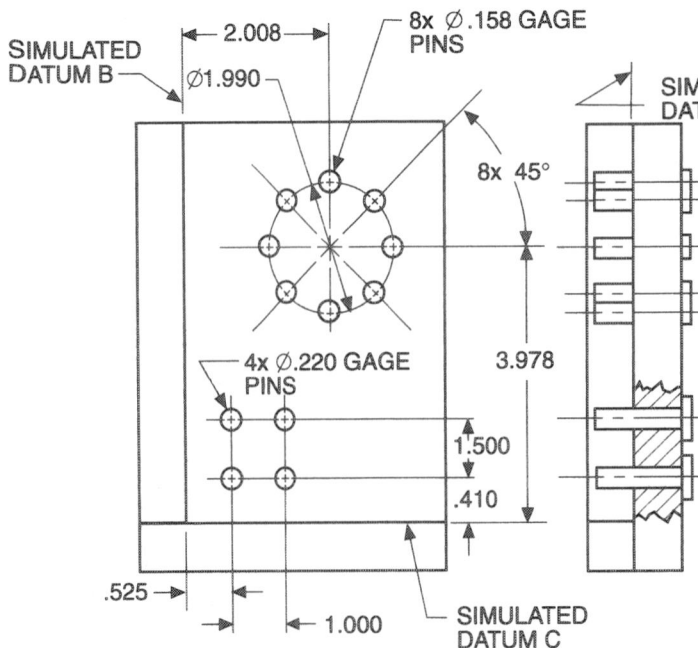

PATTERN LOCATING

SIMULATED DATUM B       Ø1.990       2.008       8x Ø.158 GAGE PINS

8x 45°

4x Ø.220 GAGE PINS

3.978

1.500

.410

.525       1.000       SIMULATED DATUM C

SIMULATED DATUM A

GAGE PIN CALCULATIONS

Ø.188 +.004 −.000 HOLE

Ø.188 MMC
(−)Ø.030 TOL AT MMC
Ø.158 GAGE PIN SIZE

Ø.250 +.004 −.000 HOLE

Ø.250 MMC
(−)Ø.030 TOL AT MMC
Ø.220 GAGE PIN SIZE

## Composite Position Tolerancing—Functional and Paper Gaging, Graphic Analysis, Computation

The previous page illustrates functional gage principles for the four hole pattern of the part on page 466–67.

Paper gaging methods are shown to demonstrate actual usable techniques which can also be used to quantify position tolerance principles pictorially. The mathematical method may be used in lieu of paper gaging or functional gaging directly as backup verification of borderline cases, or for further analytical work. Paper gaging, also referred to as graphic analysis, provides a useful tutorial tool to explore principles and better understand the mechanics of positional tolerancing. Once understood, mathematical or computer manipulation of the data in such a situation is more readily achieved.

## Paper Gaging, Graphic Analysis

Paper gaging is accomplished through plotting an enlarged scale of coordinately measured feature positions onto a piece of standard graph paper and then plotting the resulting differentials (actual position versus true position) to a selected scale (e.g., one square = .001) with a dot on the graph. An overlay chart (gage) of tracing paper or other transparent material containing a series of graph-scale circles of desired increments is placed over the graph to depict the position tolerance zones. Note that the paper gaging method simulates part function and functional gaging. However, the individual tolerance zones are each assumed to be represented by the one exact (true) position on the graph. The exact (basic) dimensions of the pattern are assumed as 0 in the X and Y directions.

**Step 1** The four .250 holes are located both as a pattern (Ø.030) and hole-to-hole within the pattern (Ø.005). In open set-up inspection methods, two steps are required to determine whether both requirements have been met. Step 1, using coordinate measuring and paper gaging, will determine whether the hole pattern as a unit (as based upon individual hole location) has met the Ø.030 tolerance requirement. The part is set up according to the datum surfaces and is measured to holes 1, 2, and 3 (see illustration on page 470–71); only three of the four holes are necessary for this evaluation.

The resulting X and Y measurements are compared to the specified coordinate dimensions on the drawing resulting in a differential (off position) value. This differential is plotted on the paper gage graph according to the above stated scale (i.e., one square = .001).

When the actual location of the three holes are plotted, an overlay gage, with the circles, the scale of the graph plot, and a representation of the tolerances, is placed over the graph plot. The center of the overlay must be placed on the center of the plot. If the plotted centers fall within the position zone (in this case the Ø.030 zone), the pattern, via the

holes, meets the requirement. Since hole #1 in our hypothetical example exceeds the Ø.030 tolerance zone, MMC principles may be invoked. The size of hole #1 is found to be .254 which is .004 departure from MMC; thus that hole has Ø.034 tolerance, and as seen in the illustration on the following page, is now acceptable. The hole pattern is also thus found acceptable within the Ø.030 tolerance.

**Step 2** To evaluate the accuracy of the four holes in the pattern (part shown on page 466–67) relative to the individual .005 position tolerances, it is necessary to consider the hole-to-hole relationship in the pattern, exclusively. The four-hole grid pattern from which the individual hole positions can be compared is established by selecting two of the holes as a basis. In our example, hole #1 is selected as the origin for the X- and Y- coordinate measurements.

Since one additional hole must be selected to give orientation or square-up to the pattern, hole #3 is selected for the X orientation.

From the part orientation now established, each hole is measured in X and Y from this set-up. From the illustration on page 470–71, resultant differentials are derived from our hypothetical measurements. Note, of course, that #1 is zero in X and Y (as the origin for measurement), and hole #3 is zero in X as the square-up for the basic pattern.

The differentials are plotted on the graph paper to the desired scale (in this case every five squares = .001) using an origin on the graph as the hole #1 position. Each hole is plotted in the appropriate value and quadrant in X and Y from the hole #1 position.

With the holes plotted on the graph, the overlay gage with the circles representing the tolerances scaled to the graph plot is placed over the graph. The overlay is moved at random to try to encompass all the plotted hole centers simultaneously within the stated tolerance .005 circles. After trial and error, let us assume that the illustrated location of the overlay gage can successfully encompass holes #1, #2, and #3 but hole #4 is outside the .005 circle. MMC principles may now be invoked. By checking the actual mating size of hole #4, it is determined to be .2515, which is a departure of .0015 from MMC; thus hole #4 has .0065 position tolerance, and as is seen in the illustration, is now acceptable.

All holes in the pattern have met their position tolerance requirements. It is extremely important to note that hole #1, even though used as a zero location origin, is actually assumed as an equal partner in the pattern and must likewise be treated as imperfect in its relationship to its desired position.

It should be noted that other methods of deriving coordinate data could be used. For example, via computer analysis determine the centroid of the hole pattern and then derive X and Y values. Any method chosen is attempting to duplicate a situation which most closely represents the best fit manipulation of the mating part pins (or representative hard gage). The method shown would represent only one position, or rotation, of the pattern of the many possible options.

Paper gaging simulates hard gaging and part function and thus is an effective technique. This method is further advantageous in that it visually detects error trends through periodic inspections, gives a permanent record, and requires no gaging tolerance.

## Computation

Mathematical verification is possible using a calculator or computer program, proving the holes within the ∅.030 and ∅.005 tolerance zones.

**Step 1.** Mathematically proven using a calculator or computer program.

- Hole #1 calculates to ∅.0328024–.004 (bonus tol) = ∅.0288024. Hole #1 is *good* using the ∅.004 bonus tolerance (less than ∅.030).
- Hole #2 calculates to = ∅.0268328. Hole #2 is *good* (less than ∅.030).
- Hole #3 calculates to = .0233238. Hole #3 is *good* (less than ∅.030).

**Step 2.** Mathematically proven using a suitable computer program.

- Hole—hole location *is* within tolerance as calculated.
- Hole—hole position tolerance calculated for *all four holes*, considering bonus tolerance of hole #4 (∅.0015), is ∅.0047472. That is, the smallest circle (tolerance zone) which will encompass all the four hole locations (including bonus tol ∅ .0015 of hole #4) is ∅.0047472. This is less than ∅.005, the stated tolerance. Therefore, all four holes are acceptable.

**STEP 1**

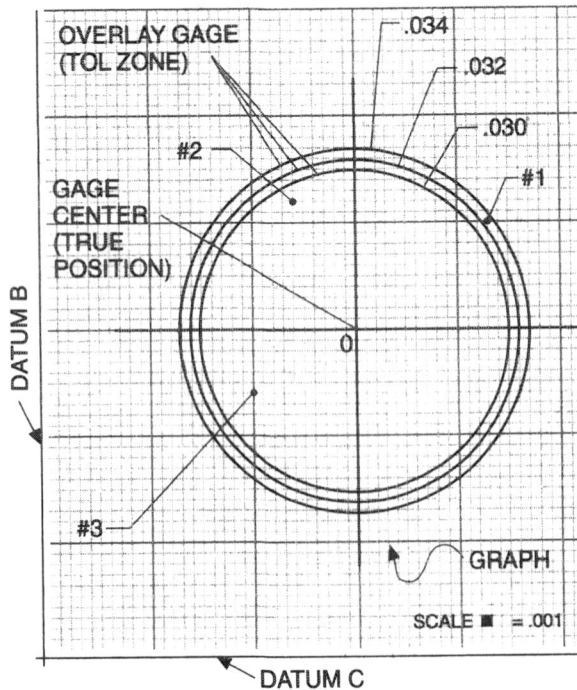

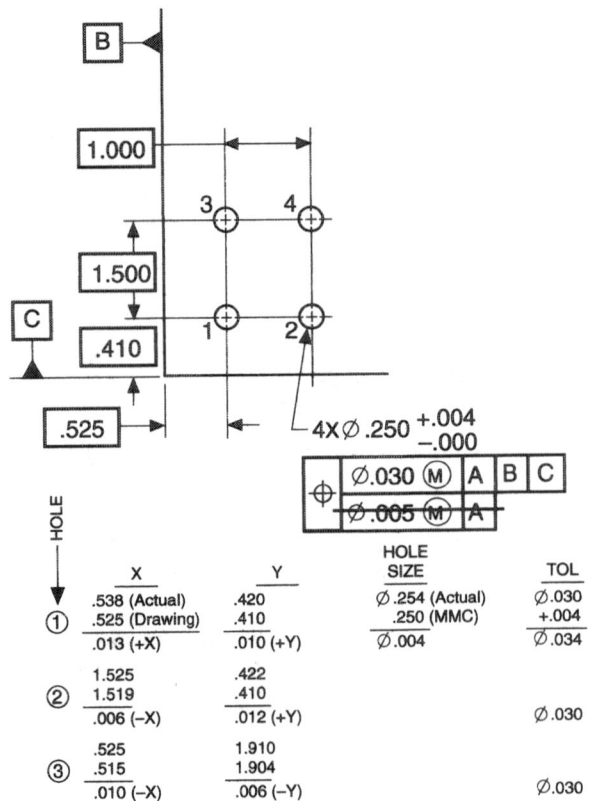

**STEP 2**

OVERLAY GAGE (TOL ZONE)

.005
.0055
.006
.0065

#1

SCALE ▨ = .001
■ = .0002

GRAPH

0

#2
#3
#4

(OVERLAY GAGE)

.005
.0055
.006
.0065

B

1.000

3    4

1.500

C

.410    1    2

.525    4X Ø.250 +.004 / −.000

⌖ | Ø.030 Ⓜ A B C |
  | Ø.005 Ⓜ A |

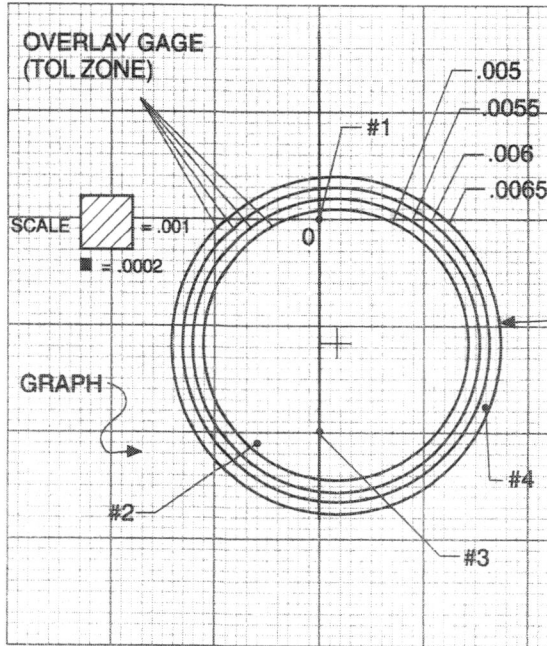

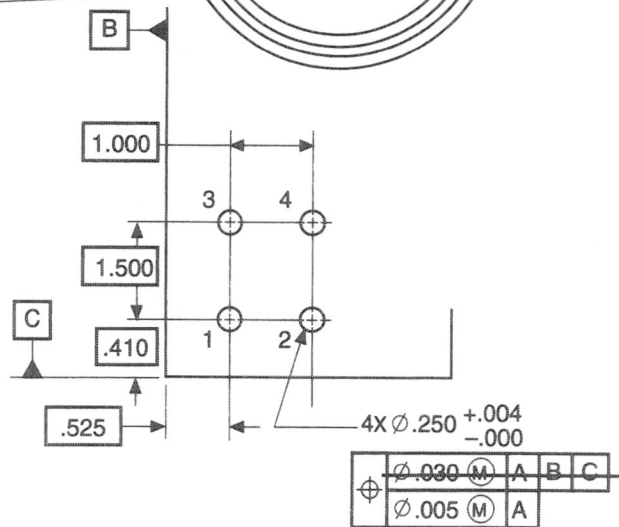

|   | X | Y | HOLE SIZE | TOL |
|---|---|---|---|---|
| ① | X = O (ORIGIN) | Y = O (ORIGIN) | GO/NO GO | Ø.005 |
| ② | 1.0000 / .9988 / .0012 (−X) | (−).0042 / .0000 / .0042 (−Y) | GO/NO GO | Ø.005 |
| ③ | X = O (SQUARE-UP) | 1.5000 / 1.4960 / .0040 (−Y) | GO/NO GO | Ø.005 |
| ④ | 1.0032 / 1.0000 / .0032 (+X) | 1.5000 / 1.4965 / .0035 (−Y) | Ø.2515 (Actual) / Ø.2500 (MMC) / Ø.0015 | Ø.005 / +.0015 / Ø.0065 |

## Composite Position Tolerancing

The composite position tolerancing principle can be applied in numerous ways to meet similar, yet different, design requirements from that shown on previous pages. For example, where the two patterns of features are in fact *one* pattern, or are to be treated as *one* pattern for both the pattern locating and feature relating (upper *and* lower segments), the method shown in the upper example could be used. One callout addressed to both patterns and the number designation 12X (twelve holes) will convey the one pattern meaning.

Functional gaging, coordinate measuring, and evaluation of the upper segment, pattern locating tolerance, would be exactly as shown in previous pages. The lower segment, feature relating tolerance, would also be handled as in previous pages except that it would be put into a *single gage* (accommodating all twelve holes at once) or into a *single measuring process* where all twelve holes are measured and evaluated as *one* pattern.

The lower example illustrates the possibility where one pattern (the four Ø.250 holes) is to serve as the datum reference. Although feasible, this method includes some complications in verification. Where the design requirement is that the diametrical pattern (the eight Ø.188 holes) be located with respect to the four holes (designated datum D) location, this method can be used. Production is so guided. However, verification from the four holes required either some form or functional gage which can pick up the four holes simultaneously (and thus their functional centroid), or open set-up measuring by coordinate methods selecting two of the four holes or mathematically deriving the centroid of the four holes via computer. Methods shown previously can then be applied with some trial and error attempts possibly necessary. Where a pattern of features is used as a single datum reference, the centroid of the pattern establishes the datum axis from which the feature relationship originates.

If a functional gage method is to be used in this application, due consideration must be given to the Datum Virtual Rule as it would apply to the datum D pattern gage pins (i.e., Ø.245). Should the design require more stringent control or relationship of the eight hole pattern to the datum D pattern, zero (.000) tolerancing of the feature relating hole pattern could be considered (see below). In such a case, the Datum Virtual Condition Rule would not apply since the gage "pins" would be at MMC size.

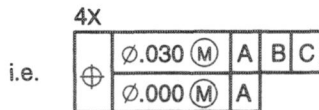

i.e.

4X

| ⊕ | Ø.030 Ⓜ | A | B | C |
|---|---------|---|---|---|
|   | Ø.000 Ⓜ | A |   |   |

Consideration could be given to whether selection of two of the four holes might better serve as secondary and tertiary datum features (arbitrarily selected by design). This method would remove most of the complexities of using a group of features as a single datum. If RFS reference to the datum as a group is considered, the foregoing method (select two holes) should seriously be considered. Otherwise, the complexities become deeper and can only be resolved by sophisticated analytical or computer methods.

As discussed in later sections of this text, it can be argued whether using a pattern of features (such as holes as shown) as a single datum, and where the datum is referenced on an MMC basis, is any different than if the datum feature pattern of holes and the features related to it were all stated as a single requirement as one pattern. When the datum virtual condition rule is invoked, it can be proven that they are exactly the same. However, where extenuating circumstances may enter in, or surely where RFS datum reference is used, there is a difference. Where CMM processing is used, the RFS method is automatic in measurement. Whether the MMC latitude is included in the analysis can bring differing results.

## Composite Position Tolerancing—One Pattern

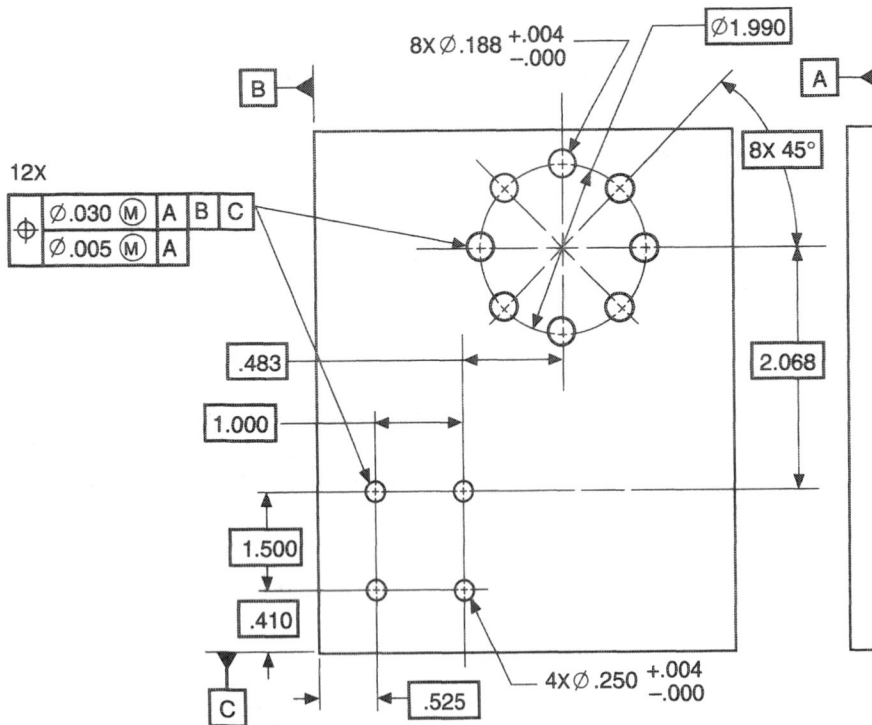

## Composite Position Tolerancing—Hole Pattern as Datum

See later sections in this text for further discussion on two hole datum applications, coordinate measuring, relationship to gaging, and mathematical (computer) simulation of hole patterns.

### Relation to Datum Surfaces Composite Position Tolerancing

When the location of a pattern of features from datum surfaces is less important that the accuracy required *within* the pattern of feature composite position tolerancing may be used. Composite position tolerancing also extends the use of specified datum relationships and geometric tolerance control.

Composite position tolerancing incorporates a dual feature control frame with two positional controls. One, the upper segment in the symbol, specifies the applicable datums and the pattern locating position tolerance. The lower segment specifies the applicable datum and the feature relating position tolerance. A single position tolerance symbol is used.

The composite position tolerance method utilizes the full advantages of MMC (and RFS or LMC if desired) and extends the principles to control of patterns of features as well as of the individual feature interrelationship.

Extending the composite tolerancing principles, this example incorporates a *secondary* datum in the features relating tolerance frame. Such a specification permits the patterns to float within the restrictions of the pattern locating tolerance but maintain a parallel orientation of the feature relating

tolerance framework to datum plane B. Details of the four hole pattern meaning are shown; the bolt circle pattern would float independently in the same manner. Note that since the purpose is to *orient* each pattern framework only (*location* of the pattern from the datum reference frame is specified in the pattern location frame), the basic dimensions from the datums B and C do not apply to the lower segment feature relating tolerance. The basic dimensions, hole-to-hole, in the pattern yet control the hole locations to one another.

Functional gaging principles would be as described before in previous examples relative to the pattern locating tolerance. Possible functional gages for the feature-relating requirements are shown at right. Note how the gages stabilize against datums A and B with movement permitted of the gage pin pattern parallel to datum plane B. Graphic analysis and CMM processing with computer soft-gaging techniques are again possible.

Within the hole pattern itself, the feature relating position tolerance is established at .005 diameter. Then, as previously described, each hole in the patterns may increase its position tolerance an amount equal to the departure from MMC as it is produced to a maximum of .009 at LMC. Note that the hole-to-hole interrelationship in the pattern, as well as the relationship to datum A, is maintained. The attitude (perpendicularity) of each individual hole simultaneously must be within the .005 diameter tolerance zone as well as within position of .005 at MMC.

As can be seen in the typical hole cross-sectional view, the axes of both the large (pattern) and small (feature) tolerance zones are parallel. The axes of the holes must lie within *both* the larger and smaller tolerance zones. Portions of the smaller zones may fall outside the peripheries of the larger tolerance. However, this portion of the smaller zone is not usable since the axis of each hole must fall within both zones.

See page 468 for gage pin size calculations.

## Composite Position Tolerancing

**Example**

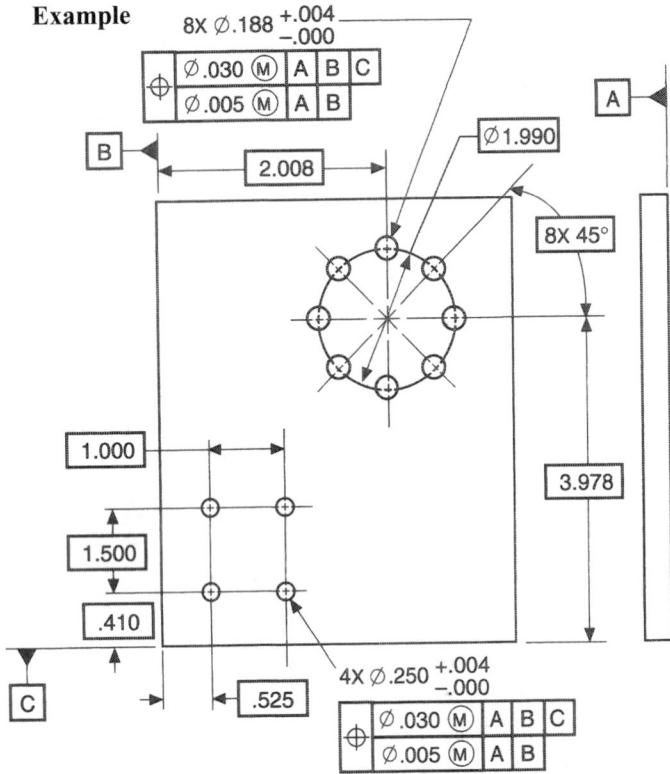

**Functional Gages**

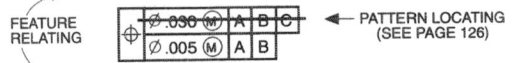

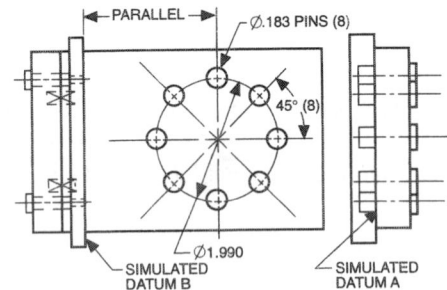

**Meaning**

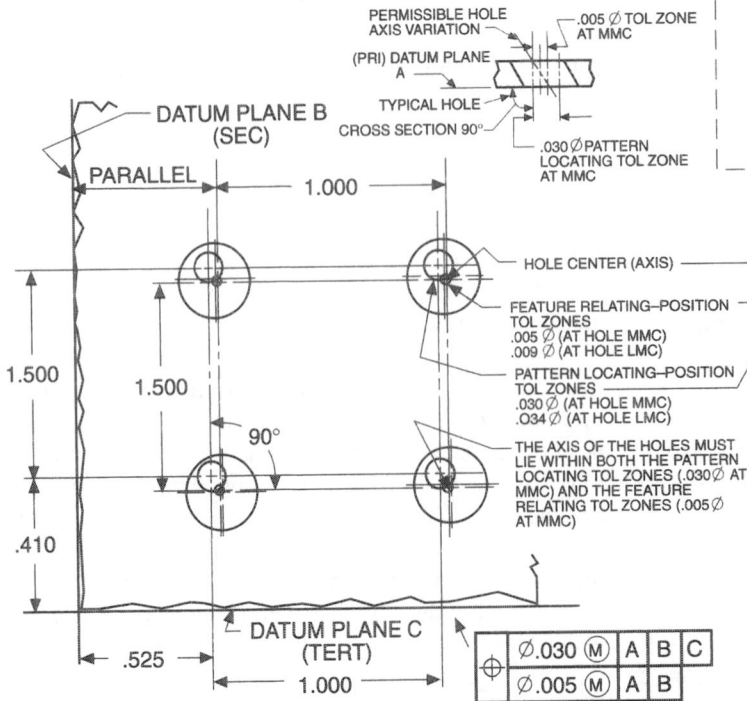

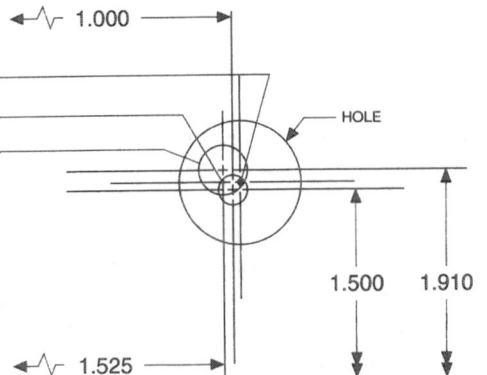

(NOTE: 8X ⌀.188 HOLE PATTERN, USE SAME PRINCIPLES)

## MMC Related to MMC Datum Feature

When a pattern of holes is dimensioned relative to the location of another hole, this hole is identified as a datum and the hole pattern is located dimensionally with respect to it.

In the example, the four holes are related to the center datum hole. As the position of the center datum hole shifts, the position of the 4 hole pattern itself must follow as dictated by the function of the part. Imagine that this part has a mating part with a shaft and four pins which must assemble with the five holes in the illustrated part.

Note that the .500 datum center hole is also located from datums (surfaces) A, B, and C. It is given a position tolerance of Ø.010 at MMC and a refined orientation, or perpendicularity, tolerance of Ø.003 at MMC. This properly controls the center hole relationship to the edge surfaces with a rather lenient position tolerance while the orientation to the primary datum A is maintained to closer tolerance. The center hole is identified as datum D so that the four .380 holes can be located with respect to it. Under MMC principles, the position and perpendicularity tolerances increase an amount equal to the hole size departure from MMC as shown in the illustration.

Wherever the datum hole D position varies in the design considerations, or in actual production, the four .380 hole pattern must follow. Note that the positional pattern dimensions originate at datum D to carry out this intent.

The four .380 holes are located by a position tolerance of Ø.005 at MMC with respect to datum A (for orientation), datum D at MMC (virtual condition) (for location), and

datum B (for rotation). Reference to the lower portion of the illustration under Meaning will assist in the understanding of the effect of the three datums and the importance of datum precedence. Although not illustrated, imagine the four .380 hole position tolerances have been calculated relative to a mating part using the "fixed fastener" method previously explained.

The four holes individually with respect to their own true positions can vary in location up to Ø.005 at MMC. While at any actual size in a departure from MMC a hole can vary an additional amount (enlarges positional tolerance) equal to that departure. For example (as shown), if the hole size is produced at .381, the position tolerance becomes Ø.006, at .382 it becomes Ø.007, etc., up to Ø.008 if the hole is produced at .383 LMC.

Since the datum feature is a size feature, its variation in size (.500 to .502) has an effect on the four hole pattern relationship which locates from it. That is, as the actual size of the datum feature increases, its relationship to the mating part corresponding feature (e.g., a shaft) changes; if the imagined mating part shaft and corresponding pins (at MMC) will insert into holes larger than those we used to calculate the position tolerances, greater latitude (off position) of the pattern as a unit can be realized. This latitude is, however, to the hole pattern as a unit and not relative to the four holes individually or hole-to-hole in the four hole pattern.

Please see the following gaging illustrations and text describing evaluation of this part with representative techniques.

## Examples

**Meaning**

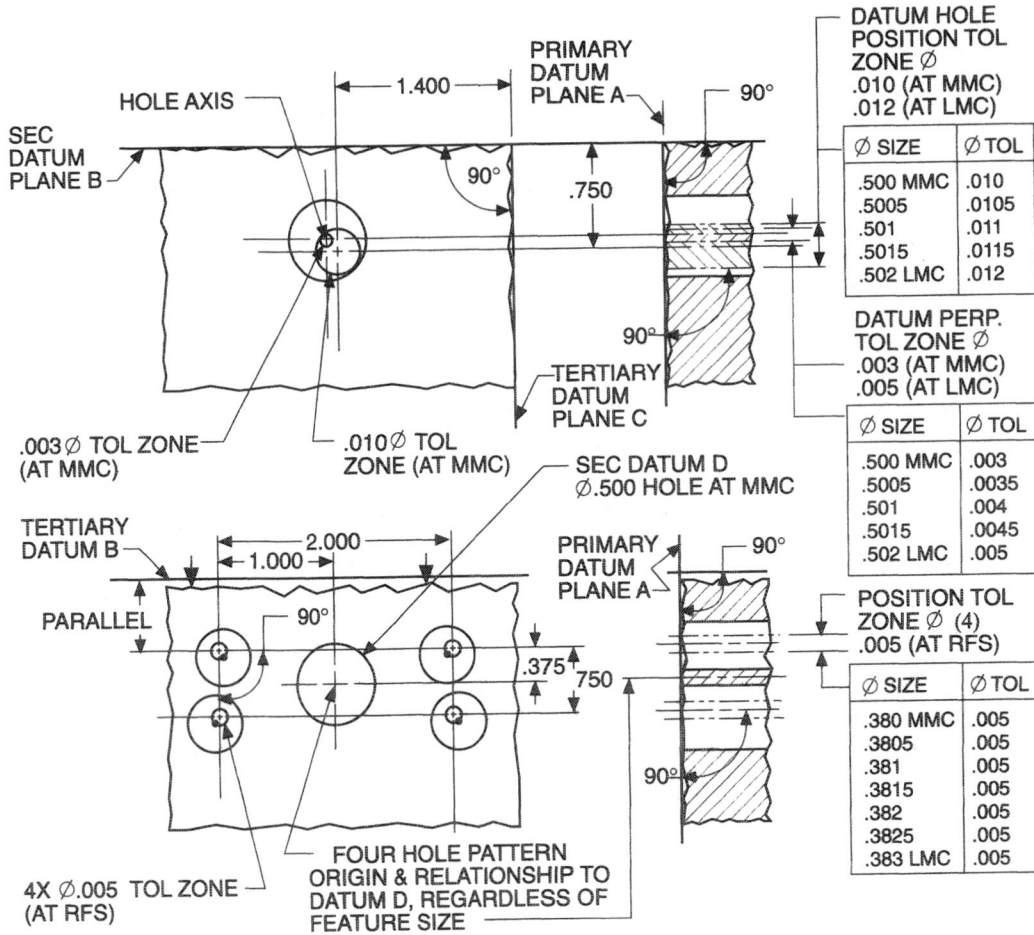

## MMC Related to MMC Datum Feature

### Functional Gaging Principles

Functional gaging principles which can be applied to the preceding example (duplicated at the upper left of the opposite page) are shown in the illustration.

It must first be clarified that functional gaging is not required when position tolerancing is used. Other techniques (i.e., CMM, open set-up coordinate methods, optical methods, etc.) can also be used. For example, the paper gaging and computation methods, illustrated earlier in this section could effectively be applied using the datum D feature as zero and orientation and rotation from datums A and B.

The purpose of this illustration is to depict representative functional gages and show actual methods as well as to demonstrate principle.

The illustration is intended to be self-descriptive and to explain the necessary details. Note that three gages (or operations) are required to fulfill the requirements.

Note specifically the clarity afforded by the clearly specified design requirements via the datums. Also, note the manner in which the gages pattern after the part function and simulate mating part relationships.

Also worthy of note is the manner in which the datum D pick-up pin size virtual condition is established based upon the Datum Virtual Condition Rule. The virtual condition applicable (.497) is derived from the smallest orientation or position tolerance controlling the hole (i.e., Ø.003) as subtracted from MMC size .500.

**Example**

4X $\varnothing$.380 $^{+.003}_{-.000}$

$\oplus$ | $\varnothing$.005 Ⓜ | A | D Ⓜ | B

2.75 ±.01

2.000

1.000

.750

.375

.750

1.75±.01

1.400

.560 ±.005

$\varnothing$.500 $^{+.002}_{-.000}$

$\oplus$ | $\varnothing$.010 Ⓜ | A | B | C
$\perp$ | $\varnothing$.003 Ⓜ | A

B

C

A

D

PERPENDICULARITY GAGE
PIN CALCULATIONS ( $\perp$ )

DATUM
A
SIM

90°

$\varnothing$.497 GAGE
PIN

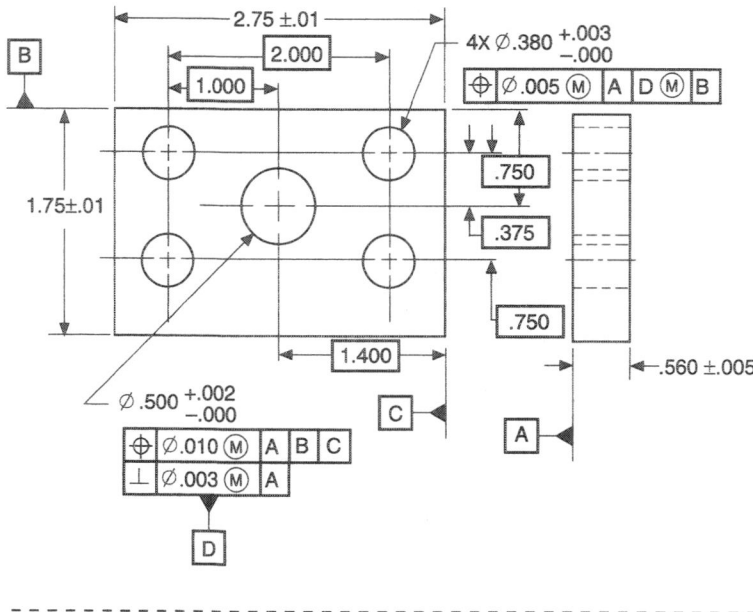

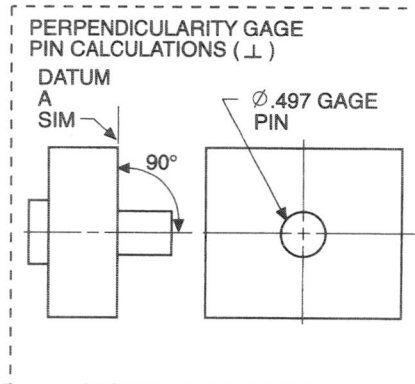

GAGE PIN CALCULATION

$\varnothing$.500 $^{+.002}_{-.000}$ HOLE

$\varnothing$.500 MMC
$(-)$ $\varnothing$.003 $\perp$ TOL AT MMC
$\varnothing$.497 GAGE PIN SIZE

$\varnothing$.500 HOLE LOCATING ($\oplus$)
GAGE PIN CALCULATIONS

$\varnothing$.500 $^{+.002}_{-.000}$ HOLE

$\varnothing$ .500 MMC
$(-)$ $\varnothing$.010 $\oplus$ TOL AT MMC
$\varnothing$.490 GAGE PIN SIZE

**Functional Gages**

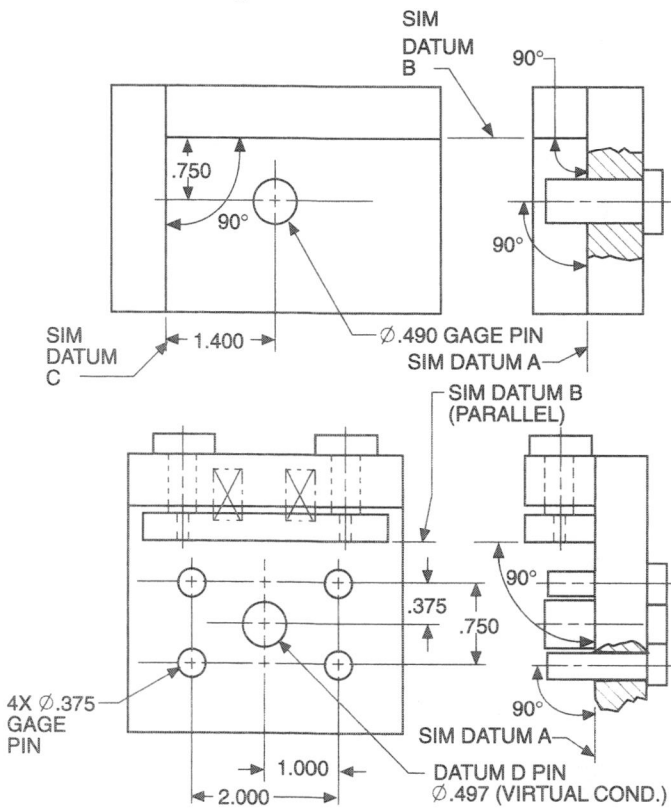

SIM
DATUM
B

90°

.750

90°

90°

SIM
DATUM
C

1.400

$\varnothing$.490 GAGE PIN
SIM DATUM A

SIM DATUM B
(PARALLEL)

90°

.375

.750

90°

4X $\varnothing$.375
GAGE
PIN

1.000

2.000

SIM DATUM A

DATUM D PIN
$\varnothing$.497 (VIRTUAL COND.)

PATTERN LOCATING ($\oplus$)
GAGE PIN CALCULATIONS

$\varnothing$.500 $^{+.002}_{-.000}$ HOLE (DATUM D)

$\varnothing$.500 MMC
$(-)$ $\varnothing$.003 $\perp$ TOL AT MMC
$\varnothing$.497 DATUM PIN SIZE
(VIRTUAL CONDITION)

$\varnothing$.380 $^{+.003}_{-.000}$ HOLE

$\varnothing$ .380 MMC
$(-)$ $\varnothing$.005 $\oplus$ TOL AT MMC
$\varnothing$ .375 GAGE PIN SIZE

### MMC Related to RFS Datum Feature

When a pattern of holes is dimensioned relative to the location of another hole, this hole is identified as a datum and the hole pattern is located dimensionally with respect to it.

In the illustrated example, the four holes are related to the center datum hole. As the position of the center datum hole shifts, the position of the four hole pattern itself must follow as dictated by the function of the part. Imagine that this part has a mating part with a shaft and four pins which must assemble with the five holes in the illustrated part.

Differing from the preceding illustration, assume there is a closer precision fit required between the shaft of the imagined part as it fits into the .500 hole. The four pins of the imagined part, however, are to relate to the four .380 holes on an MMC basis as in the preceding illustration. Therefore, since in this case the relationship between the four holes and their datum is critical or more precise, the datum D is referenced regardless of feature size (RFS).

Note that the .500 datum center hole is located from surface datums A, B and C. It is given a position tolerance relative to these edges of Ø.010 at MMC since the relationship to the edges on this basis can have a rather lenient position tolerance. The orientation of the datum hole relative to the primary datum A is, however, to be maintained to a closer tolerance. Since datum D position and orientation is controlled on an MMC basis, the tolerances (position and perpendicularity) increase an amount equal to the produced size departure from MMC shown in the illustration. It should be noted here, however, that the reference to datum D in the relationship of the four .380 holes is on an RFS basis. This means the four hole pattern takes reference from the exact center (axis) of the datum hole at whatever size it is produced (RFS) within .500 to .502.

Wherever the datum hole D position varies in the design considerations, or in actual production, the four .380 hole pattern must follow. Note that the positional pattern dimensions originate at datum D to carry out this intent.

The four .380 holes are located by a position tolerance of Ø.005 at MMC with respect to datum A (for orientation), datum D at RFS (for location), and datum B (for rotation). Reference to the lower portion of the illustration under Meaning will assist in understanding the effect of the three datums and the importance of datum precedence. Although not illustrated, imagine that the four .380 hole position tolerances have been calculated relative to a mating part using the "fixed fastener" method previously explained.

The four holes individually, with respect to their own true or exact positions, can vary in location up to Ø.005 at MMC. While at any actual size in a departure from MMC, a hole can vary an additional amount (enlarge positional tolerance) equal to that departure. For example (as shown), if the hole size is produced at .381, the position tolerance becomes Ø.006, at .382 it becomes Ø.007, etc., up to Ø.008, if the hole is produced at .383 (LMC).

Although the datum feature is a size feature as in the preceding illustration, its variation in size (.500 to .502) will have *no* effect on the four hole pattern relationship. This is because the pattern relationship is to the center (axis) of the datum D, .500 hole, no matter to which size it is produced in its size tolerance range (i.e., RFS).

In this instance, it is seen that a more critical or precise relationship is maintained between the four holes and their datum. The design requirements and mating part functional interface determined this approach in our example.

Please see the following gaging illustrations and text describing evaluation of this part with representative techniques.

**Example**

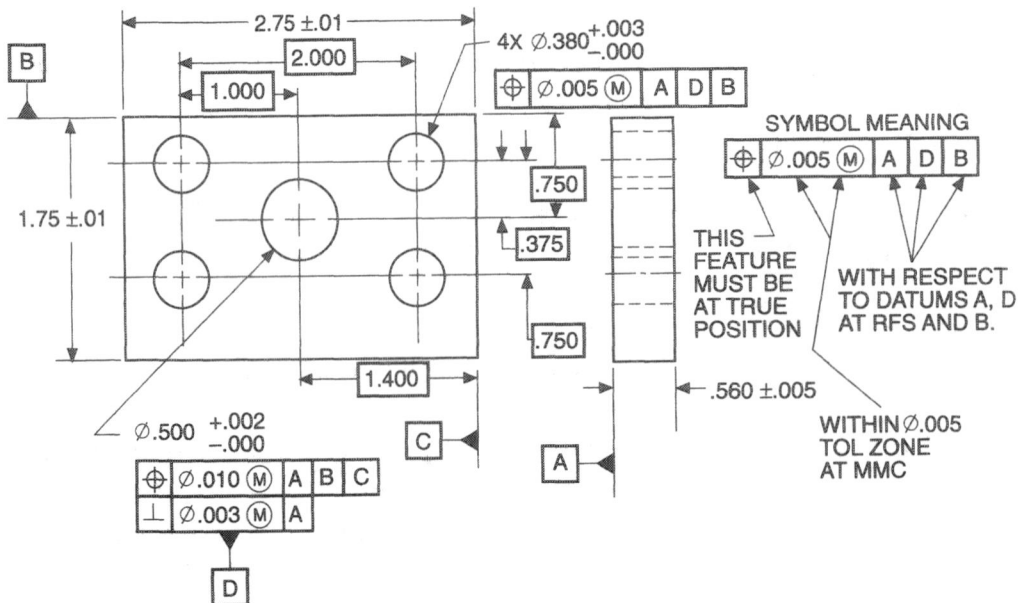

**Meaning**

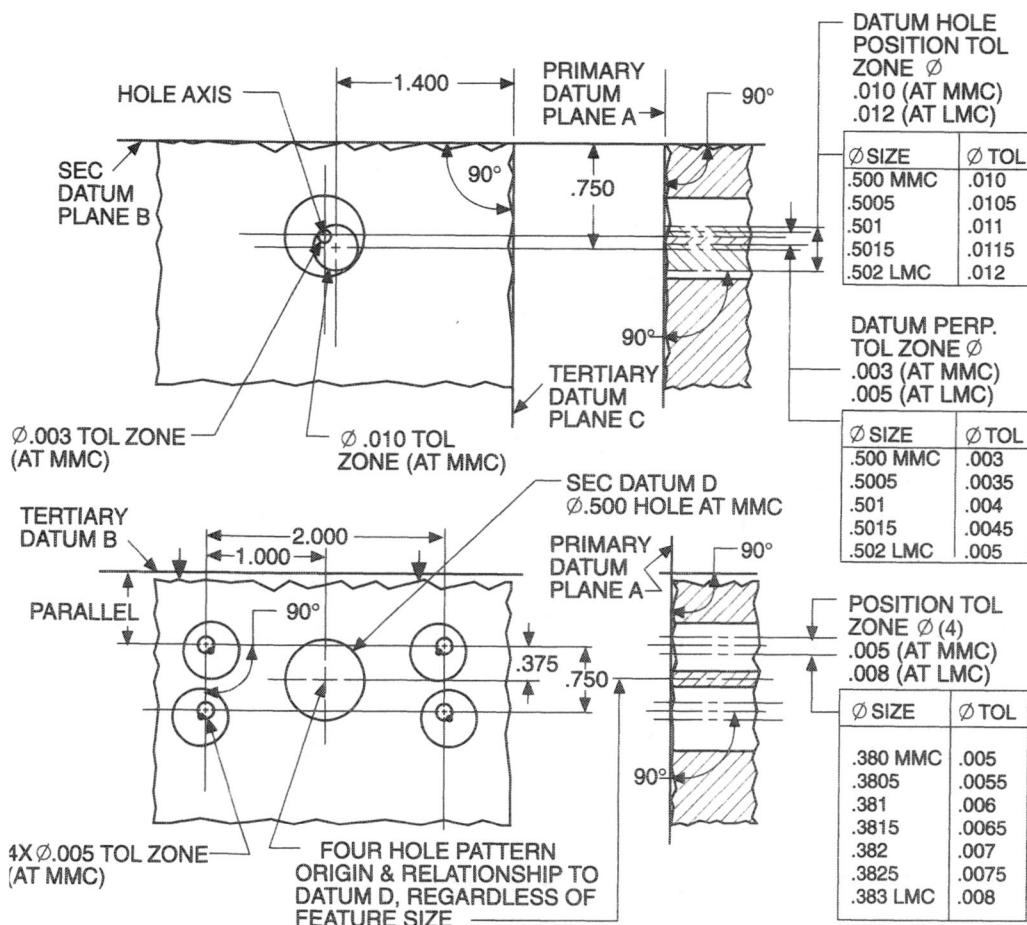

## MMC Related to RFS Datum Feature Functional Gaging Principles

Functional gaging principles which can be applied to the preceding example (duplicated at the upper left of the opposite page) are shown in the illustration.

It must first be clarified that functional gaging is not required when position tolerancing is used. Other techniques (i.e., CMM, open set-up coordinate methods, optical methods, etc.) can also be used. For example, the paper gaging and computation methods illustrated earlier in this section could effectively be applied using the datum D feature as zero and orientation and rotation from datums A and B.

The purpose in this illustration is to depict representative functional gages to show actual methods as well as to illustrate principle. Note that functional principles, identical to the preceding part, can be used except for the RFS pickup of the datum feature.

The illustration is intended to be self-descriptive and to explain the necessary details. Note that three gages (or operations) are required to fulfill the requirements.

Note specifically the clarity afforded by the clearly specified design requirements via the datums. Also, note the manner in which the gages pattern after the part function and simulate mating part relationships.

In this example, the use of the RFS datum has tightened requirements. Costs in manufacturing will probably be higher than with total use of MMC as in the preceding example; yet, functional principles can be applied as shown.

**Example**

**4X** $\varnothing.380 \, ^{+.003}_{-.000}$

| $\oplus$ | $\varnothing.005$ Ⓜ | A | D | B |

2.75 ±.01

2.000

1.000

1.75 ±.01

.750

.375

.750

1.400

.560 ±.005

$\varnothing.500 \, ^{+.002}_{-.000}$

| $\oplus$ | $\varnothing.010$ Ⓜ | A | B | C |
| $\perp$ | $\varnothing.003$ Ⓜ | A | | |

B

C

A

D

PERPENDICULARITY GAGE
PIN CALCULATIONS ( $\perp$ )

SIM
DATUM
A

90°

$\varnothing.497$ GAGE
PIN

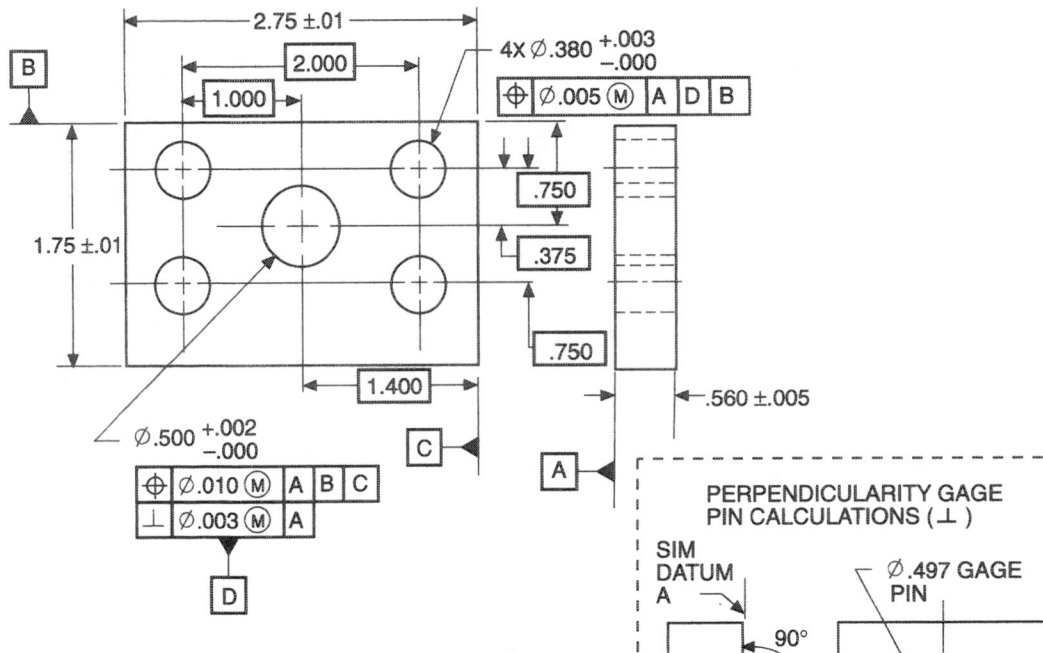

**Functional Gages**

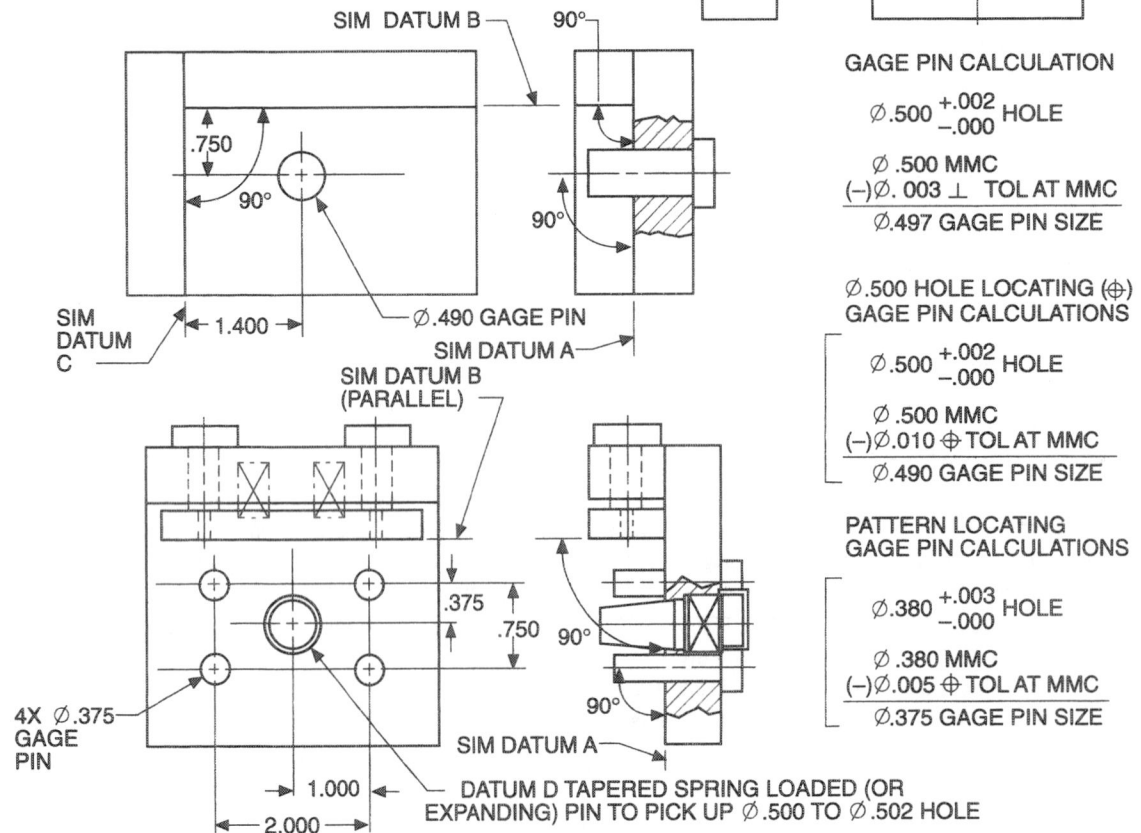

SIM DATUM B

90°

.750

90°

90°

SIM
DATUM
C

1.400

$\varnothing.490$ GAGE PIN

SIM DATUM A

SIM DATUM B
(PARALLEL)

.375

.750

90°

90°

4X $\varnothing.375$
GAGE
PIN

1.000

2.000

SIM DATUM A

DATUM D TAPERED SPRING LOADED (OR
EXPANDING) PIN TO PICK UP $\varnothing.500$ TO $\varnothing.502$ HOLE

GAGE PIN CALCULATION

$\varnothing.500 \, ^{+.002}_{-.000}$ HOLE

$\varnothing.500$ MMC
$(-)\varnothing.003 \perp$ TOL AT MMC
_____
$\varnothing.497$ GAGE PIN SIZE

$\varnothing.500$ HOLE LOCATING ($\oplus$)
GAGE PIN CALCULATIONS

$\varnothing.500 \, ^{+.002}_{-.000}$ HOLE

$\varnothing.500$ MMC
$(-)\varnothing.010 \oplus$ TOL AT MMC
_____
$\varnothing.490$ GAGE PIN SIZE

PATTERN LOCATING
GAGE PIN CALCULATIONS

$\varnothing.380 \, ^{+.003}_{-.000}$ HOLE

$\varnothing.380$ MMC
$(-)\varnothing.005 \oplus$ TOL AT MMC
_____
$\varnothing.375$ GAGE PIN SIZE

## RFS Related to RFS Datum Feature

When a pattern of holes is dimensioned relative to the location of another hole, this hole is identified as a datum and the hole pattern is located dimensionally with respect to it.

In the illustrated example, the four holes are related to the center datum hole. As the position of the center datum hole shifts, the position of the four hole pattern itself must follow as dictated by the function of the part.

In this instance, note that the four holes of the pattern are assigned an RFS position tolerance and that they are related to the datum hole D also on an RFS basis. Imagine that this part has precision requirements between the holes either to provide accurate relationship with a mating part or to maintain accuracy for a mating situation, such as a semicritical gear plate mounting.

Note that the .500 datum center hole is located from surface datums A, B, and C. It is given a position tolerance relative to these edges of Ø.010 at MMC since the relationship to the edges can be on this basis with a rather lenient position tolerance. The orientation of the datum hole relative to the primary datum A is, however, to be maintained to a closer tolerance. Since datum D position and orientation are controlled on an MMC basis, the tolerances (position and perpendicularity) increase an amount equal to the produced size departure from MMC as shown in the illustration. Note, however, that the reference to datum D in the relationship of the four .380 holes is on an RFS basis. The four hole pattern, therefore, takes its positional reference from the exact center (axis) of the datum hole at whatever size it is produced (RFS) within .500 to .502.

Wherever the datum hole D position varies in the design considerations or in actual production, the four .380 pattern must follow. Note that the positional pattern dimensions originate at datum D to carry out this intent.

The four .380 holes are located by a position tolerance of Ø.005 at RFS with respect to datum A (for attitude), datum D at RFS (for location), and datum B (for rotation). Reference to the explanation portion of the illustration (under Meaning) will assist in understanding the effect of the three datums and the importance of datum precedence. Although not illustrated, imagine that, as previously stated, the four .380 hole position tolerances have been determined to relate to a mating part or to maintain accuracy in a mating situation where other features or components must relate with precision regardless of the produced sizes (RFS) of the .380 holes (.380 to .383).

The four holes individually, with respect to their own true or exact positions, can vary in location up to Ø.005. Under the RFS method, however, this tolerance applies to each hole regardless of the size to which it is produced. That is to say (as shown in the lower right corner of the illustration), the Ø.005 position tolerance is the maximum allowable to each hole no matter to which size it is produced (.380 to .383). If, as was shown in previous illustrations, the MMC method had been applied, the position tolerance would increase to the extent of departure of the .380 holes from MMC. Not so, however, in this example since the RFS method has been invoked. The choice of proper approach, be it MMC or RFS, is of course decided by the design requirements. MMC methods are obviously recommended wherever possible due to the added functional (interchangeability) and tolerance advantages.

In this instance, it is seen that a more critical or precise relationship is maintained between the four holes and their datum. The design requirements based on part function determined this approach in our example.

## Example

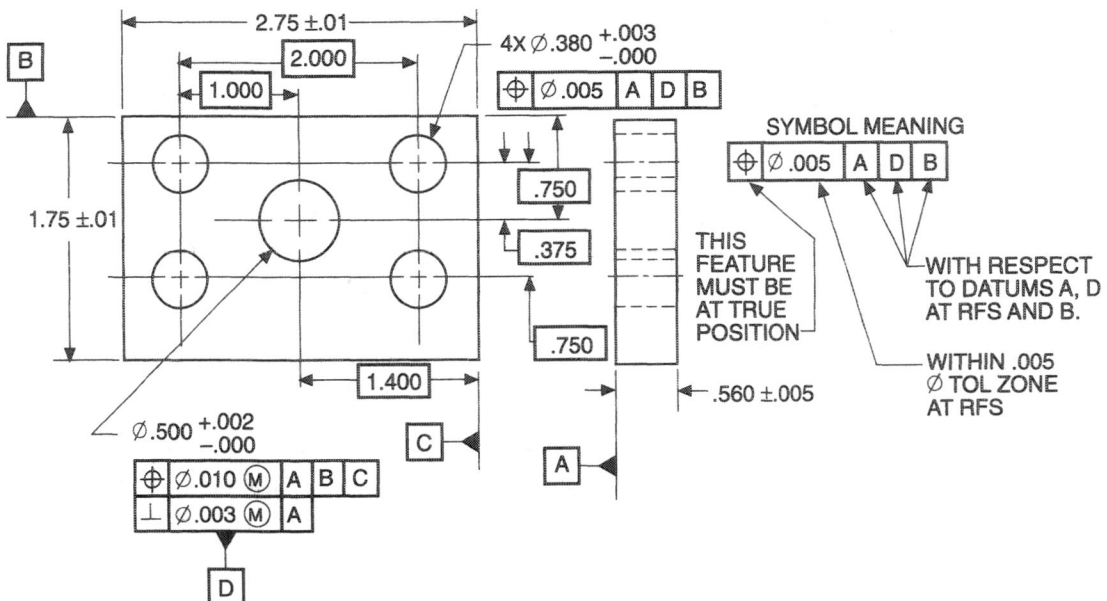

**Meaning**

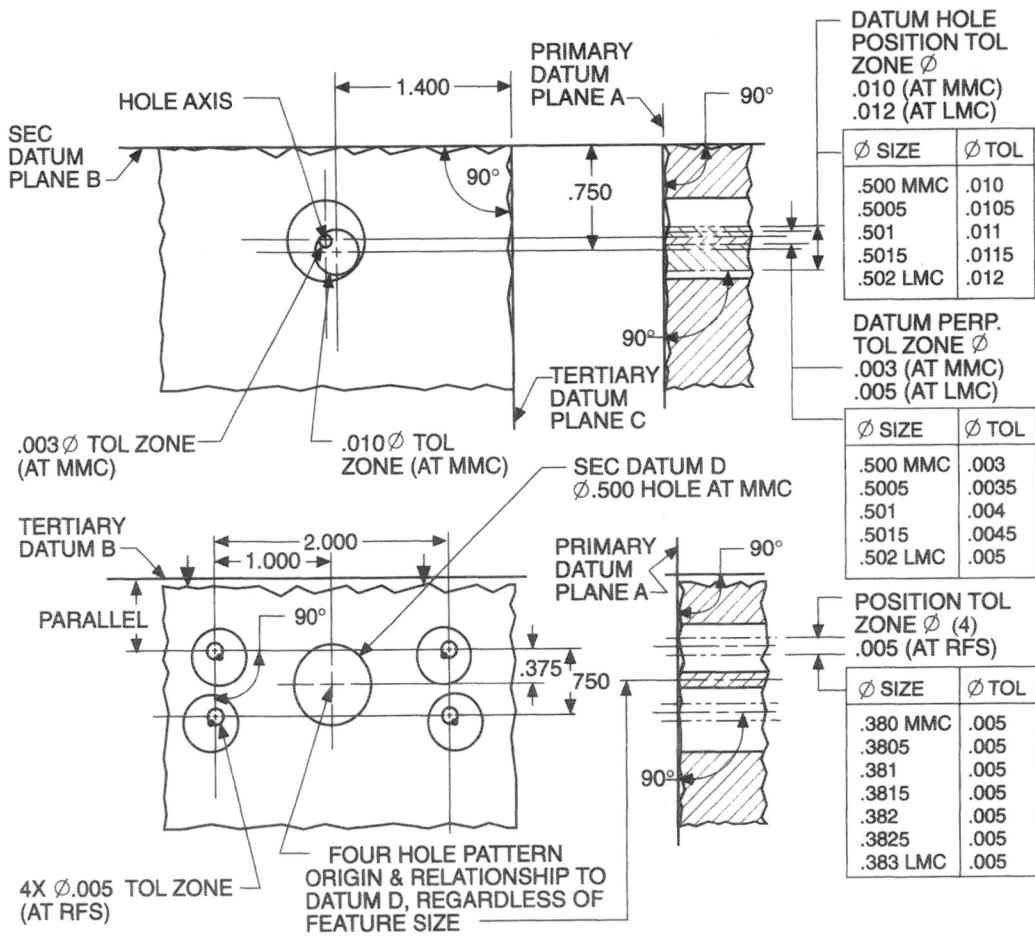

As may be seen in the previous examples in this section, functional gaging principles can be used to evaluate the datum hole location and relationships. However, the four hole (.380) pattern relative to datum D in this example cannot utilize such methods because it is an RFS application; open set-up or CMM methods would be necessary. The techniques of "paper gaging and computation," described earlier in the text could, of course, be used as desired.

## MMC Related to MMC Datum Feature

### Projected Tolerance Zone

This illustration shows a part with four tapped mounting holes. A cover plate mechanism is assumed as a mating part (not shown). Position tolerancing is used to assure assembly of the mating parts. The projected tolerance zone requirement is also added.

We wish to establish a relationship between the mounting surface, the mounting hole pattern, and the 1.506 counterbored seat diameter (identified as datum B). The mating part has a seat mechanism which must fit within this counterbore and attach and locate with screws to the four mounting holes on the flange surface.

The flange surface itself is established as the primary datum and is identified as datum A to ensure clarity of the hole pattern positional relationship with the top surface.

We calculate the position tolerances using the "fixed fastener" method as based on the .250–28 screw and the clearance hole in the mating part cover plate. Assuming that the cover plate has .256 clearance holes and .250 maximum thickness, we calculate the position tolerance and assign .003 to each part (.006/2 = .003).

The four .250–28 holes are thus designated as shown. The feature control symbol specifies "at true position within Ø.003 at MMC size of the holes with respect to datums A and B (at MMC) and to datum C."

Conventional position tolerancing discussed previously reveals that a position (MMC) application is affected by the actual mating size departure of the involved features from MMC. However, in this instance, the peculiarity of thread assembly requires further consideration and may cause an exception to the usually inferred interpretation.

The centering effect of a screw as it tightens in the tapped hole tends to negate or diminish any *added* position tolerance based on greater clearance between the pitch diameters of the screw and tapped hole. The screw seeks center as, in tightening, the flank of the mating thread forms (screw and hole) come into contact or bottom out. Thus additional position tolerance relative to the tapped hole increase in pitch diameter size (departure from MMC) may not be fully realized. There may be some additional tolerance derived in actual assembly but it cannot be predicted.

The same reasoning as in the foregoing paragraph applies to attempts to assign position tolerance to countersinks. Position tolerance would be appropriate to the through hole, but, as a rule, is not practical for the countersink itself.

Added tolerance may be acquired, however, for the tapped hole pattern as a unit relative to datum B (counterbore) as it departs from MMC size (gets larger). Datum B is a straight sided feature, and size deviation from MMC will have the effect previously discussed in other examples.

## Example

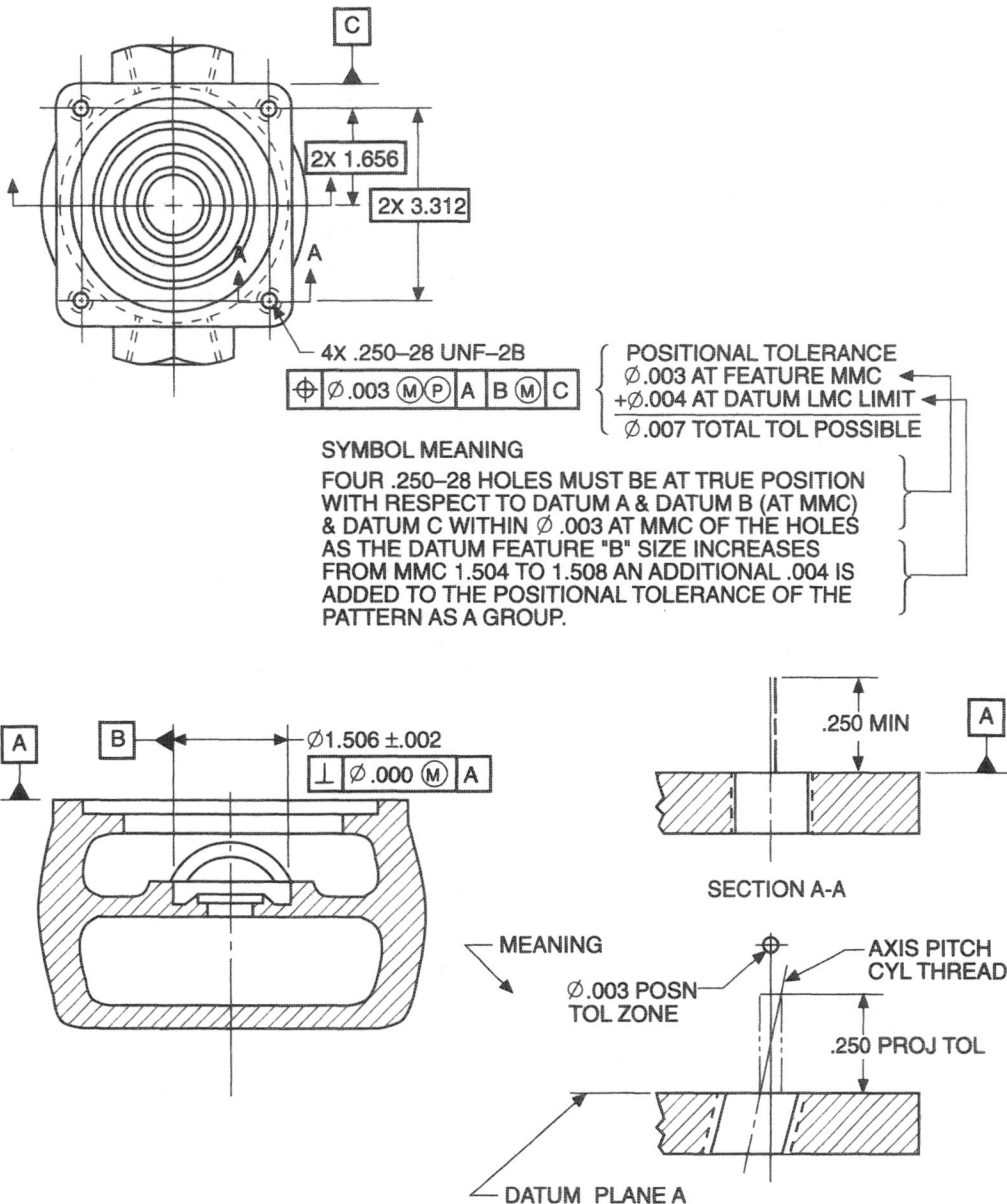

## Functional Gage (Principles)

## MMC Related to MMC Datum Feature

### Projected Tolerance Zone

Since the threaded holes are to attach a mating part, we see that in assembly the critical position location of these holes will actually concern the inserted screws or the actual projection from these holes which must accommodate the mating part. Therefore, the location of the threaded holes is controlled by a *projected* position tolerance zone to represent the projected screw locations. As is seen in the callout and sectioned projected tolerance zone view, the tolerance zone is cylindrical and extends above, and perpendicular to, the datum plane A. The drawing specification and view using the thick chain line clearly shows the direction and extent of the projected tolerance zone.

Since in this case the threaded hole is a through hole, a drawing view is required to clearly show the direction from datum plane A that the tolerance zone projects. The length of extent of the tolerance zone is conveniently shown in this view as well. If projected tolerance zone is applied on a blind hole, the extent of the projected tolerance zone may be placed in the feature control frame as ⊕ Ø.003 Ⓜ Ⓟ .250 A BⓂ C. A drawing sectional view may then be unnecessary.

The height of the projected position tolerance zone is determined by the application. It may be established by the thickness of the mating part through which the screws are to extend, or it may be determined by the thread hole depth when a thin mating part is involved, or any desired height which fulfills the part design requirements may be selected. When the projected tolerance zone is intended, it should be specified as shown with the symbol immediately following the stated tolerance and any material condition symbol in the feature control frame (see illustration). This system is also used on parts in which pins, studs, or other features are to be inserted and the critical assembled location is *above* the surface of the part.

The representative production gage shown on the preceding page also illustrates this principle. The gage is made up of the main gage body containing the datum B hole pin, the plate which contacts the datum surface A, a spring plate to provide orientation to datum C, four GO thread members per standard tolerances and, if desired, torquing flats to represent appropriate screw tightening and setting. The plate contains four holes which are .003 (the positional tolerance) larger than the four GO thread pins which are inserted into the threaded holes of the part through the gage plate. CMM simulation using computer soft-gaging techniques can be used in lieu of a hard-gage process.

The relationship of the GO thread member projections and the holes in the gage plate simulate the functional assembly requirement of the two mating parts and the four screws.

Gage-maker's tolerances would, of course, also apply but are not shown.

## MMC Related to MMC Datum Feature

The illustration on the opposite page shows a circular pattern of 8 holes located by position tolerancing with respect to the center hole.

The requirement states "the eight holes are to be located at position within Ø.002 at .403 MMC size of the holes with respect to datum A and datum hole B (at 1.025 MMC)."

As the actual size of the holes increases in production from MMC .403 to the high limit of .405, .002 is added to the positional tolerance. As the datum hole B increases from MMC of 1.025 to 1.027, .002 is added to the hole pattern's positional tolerance as a group. Depending on the size increase of the holes, the total tolerance may be .006 instead of .002.

A production gage for this part is also shown. The pins in the pattern are spaced at 45° intervals on a 3.000 circle, with the pin size at .401 or at MMC size of the part hole .403 minus the positional tolerance of .002. The center pin is 1.025 (MMC size of the datum hole).

Gage-maker's tolerances also apply.

**Example**

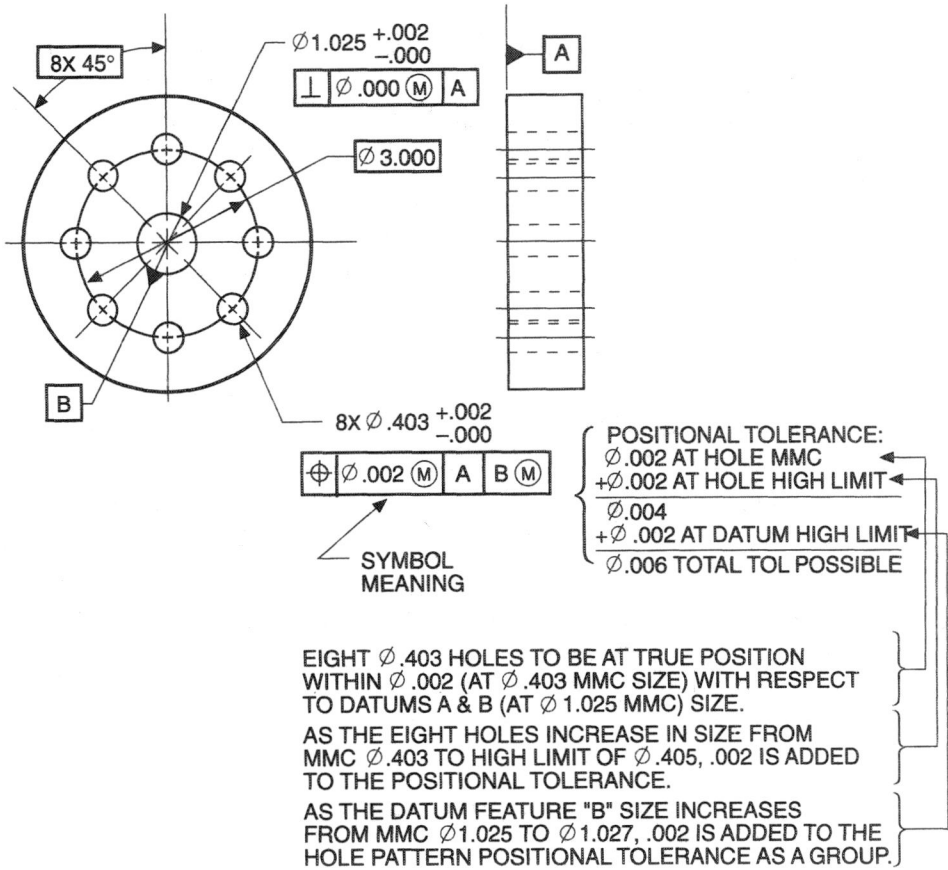

Ø 1.025 +.002 / −.000

⊥ | Ø .000 Ⓜ | A

Ø 3.000

8X Ø .403 +.002 / −.000

⌖ | Ø .002 Ⓜ | A | B Ⓜ

SYMBOL
MEANING

POSITIONAL TOLERANCE:
Ø .002 AT HOLE MMC
+Ø .002 AT HOLE HIGH LIMIT
Ø .004
+Ø .002 AT DATUM HIGH LIMIT
Ø .006 TOTAL TOL POSSIBLE

EIGHT Ø .403 HOLES TO BE AT TRUE POSITION
WITHIN Ø .002 (AT Ø .403 MMC SIZE) WITH RESPECT
TO DATUMS A & B (AT Ø 1.025 MMC) SIZE.

AS THE EIGHT HOLES INCREASE IN SIZE FROM
MMC Ø .403 TO HIGH LIMIT OF Ø .405, .002 IS ADDED
TO THE POSITIONAL TOLERANCE.

AS THE DATUM FEATURE "B" SIZE INCREASES
FROM MMC Ø 1.025 TO Ø 1.027, .002 IS ADDED TO THE
HOLE PATTERN POSITIONAL TOLERANCE AS A GROUP.

8X 45°

A

B

**Gage**

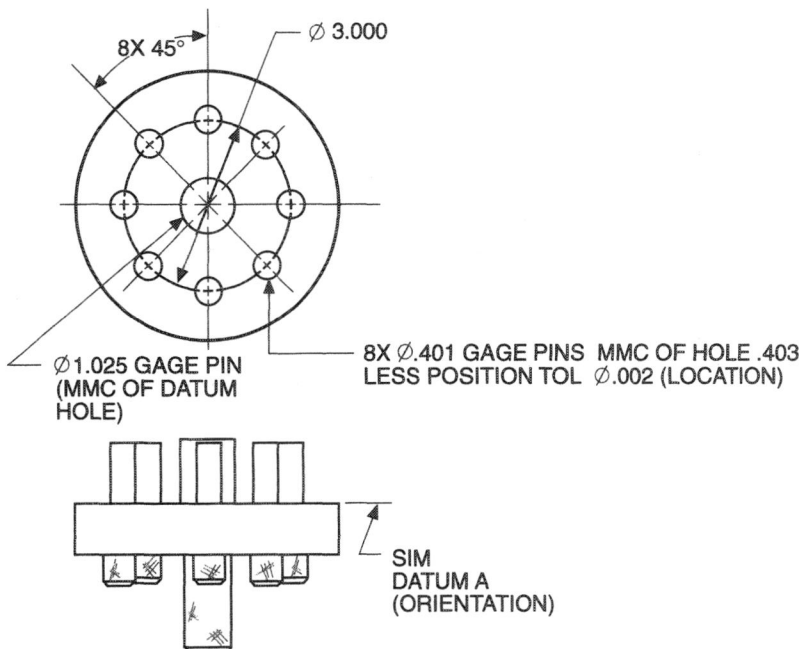

8X 45°     Ø 3.000

Ø 1.025 GAGE PIN
(MMC OF DATUM
HOLE)

8X Ø .401 GAGE PINS  MMC OF HOLE .403
LESS POSITION TOL  Ø .002 (LOCATION)

SIM
DATUM A
(ORIENTATION)

## MMC Related to MMC Datum Feature

This example shows the use of positional tolerancing on two locating pins on a part. The two pins are to take reference from the center hole which accommodates a shaft mechanism of the mating part (not shown). The center hole is selected as the datum reference. This example reminds us that positional tolerancing is not restricted to holes. It can also be applied to pins or to any feature on which a center (axis) is the basis for location.

The meaning of the position requirement is, "the two pins are to be located at true position within Ø.002 tolerance zone at .125 MMC size of the pins with respect to datums A, B (at MMC .500), and C."

Note that MMC of the pins is the high limit of size. As the two pins reduce in actual size from MMC of .125 to the

low limit (LMC) of .122, .003 is added to the positional tolerance. As the datum feature B (the center hole) increases from MMC of .500 to .502 high limit, .002 is added to the two pins' positional tolerance as a group.

Depending on the production sizes of the bosses and datum hole, the total positional tolerance on the boss location may vary from Ø.002 to .007.

A representative gage to check the position location and the relationship with the datum hole is also shown. Note that the holes to check the position location of the pins are to the high limit of MMC of the pins *plus* the position diameter tolerance. The datum hole pin is to the MMC size of the datum hole of .500.

Gage-maker's tolerances also apply.

**Example**

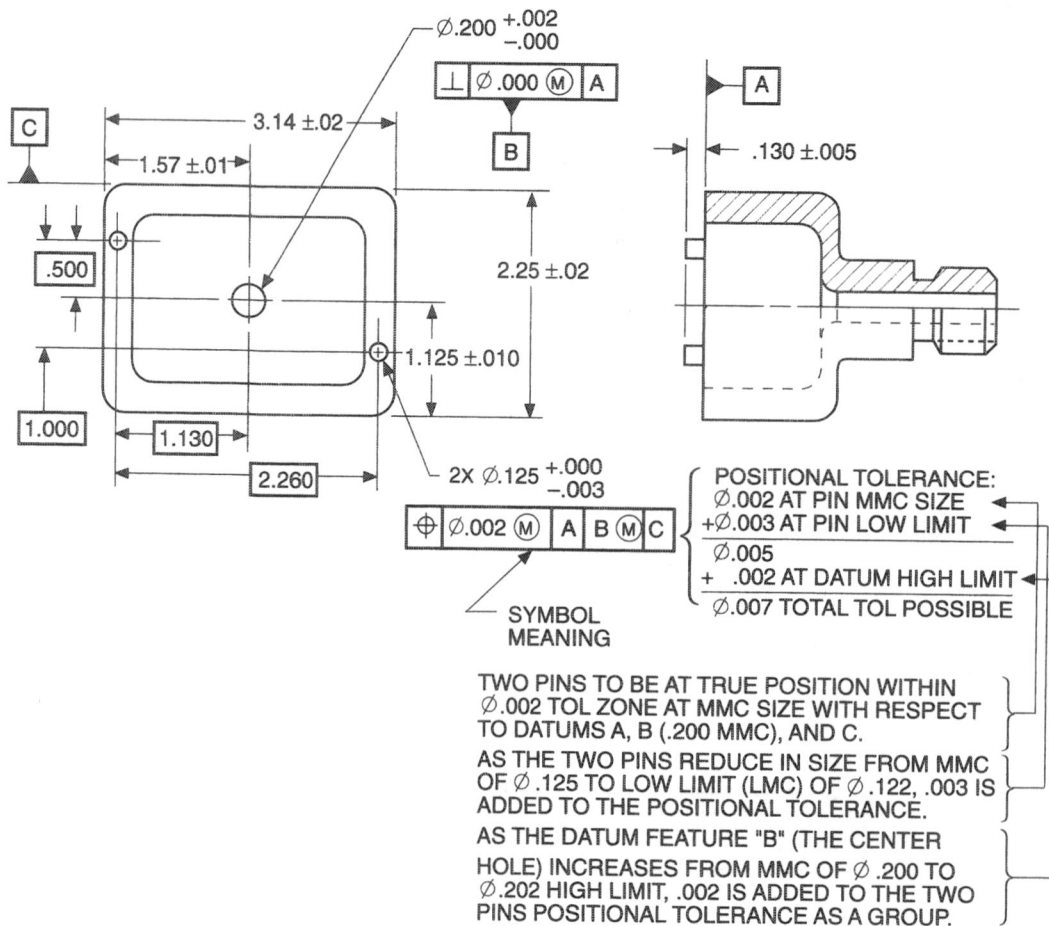

POSITIONAL TOLERANCE:
Ø.002 AT PIN MMC SIZE
+Ø.003 AT PIN LOW LIMIT
Ø.005
+  .002 AT DATUM HIGH LIMIT
Ø.007 TOTAL TOL POSSIBLE

TWO PINS TO BE AT TRUE POSITION WITHIN Ø.002 TOL ZONE AT MMC SIZE WITH RESPECT TO DATUMS A, B (.200 MMC), AND C.

AS THE TWO PINS REDUCE IN SIZE FROM MMC OF Ø.125 TO LOW LIMIT (LMC) OF Ø.122, .003 IS ADDED TO THE POSITIONAL TOLERANCE.

AS THE DATUM FEATURE "B" (THE CENTER HOLE) INCREASES FROM MMC OF Ø.200 TO Ø.202 HIGH LIMIT, .002 IS ADDED TO THE TWO PINS POSITIONAL TOLERANCE AS A GROUP.

**Gage**

SIM DATUM C (ROTATION)

.500
1.000

2X Ø.127 HOLE
(MMC OF
PIN .125
PLUS
POSITION
TOL Ø.002)

1.130
2.260

Ø.200 PIN (MMC OF
DATUM B HOLE)
(LOCATION)

SIM
DATUM A
(ORIENTATION)

## Position Tolerance—Least Material Condition (LMC)

Occasionally a method is required to control a situation which is essentially the reverse of the usual position relationship; that is, the stated position tolerance applies at the *least material condition*, LMC, of the feature or datum, instead of at MMC, and increases as the feature or datum *departs from* the least material condition.

**Definition.** Least Material condition (LMC) is the condition in which a feature of size contains the least amount of material within the stated limits of size: for example, maximum hole diameter, minimum shaft diameter. Least material condition is the condition opposite to MMC. For example, a shaft is at least material condition when it is at its *low* limit of size and a hole is at least material condition when it is at its *high* limit of size.

This method is applicable to special design requirements that will not permit MMC or that do not warrant the exacting requirements of RFS. It can be used to maintain critical wall thickness or critical center locations of features for which accuracy of location can be relaxed (position tolerance increased) when the feature leaves least material condition and approaches MMC. The amount of increase of positional tolerance permissible is equal to the feature actual size departure from least material condition.

The term "least material condition" and the abbreviation LMC have been used instead of "minimum material condition" (which is synonymous) to avoid confusion, since the abbreviation would be the same as that for maximum material condition. The symbol modifier Ⓛ is used to indicate the LMC requirement applicable to feature or datum.

Although the use of LMC does impose exacting requirements on both manufacturing and inspection, it permits additional tolerances.

Whenever least material condition (LMC) or Ⓛ is specified on a drawing, the position tolerance applies only when the feature is produced at its LMC size. See Fig. 6-110.

Additional positional tolerance is permissible but is dependent on, and equal to, the difference between the actually produced feature size (within its size tolerance) and LMC. See Fig. 6-111.

Position Tolerance—Least Material Condition
(LMC)

**Example**

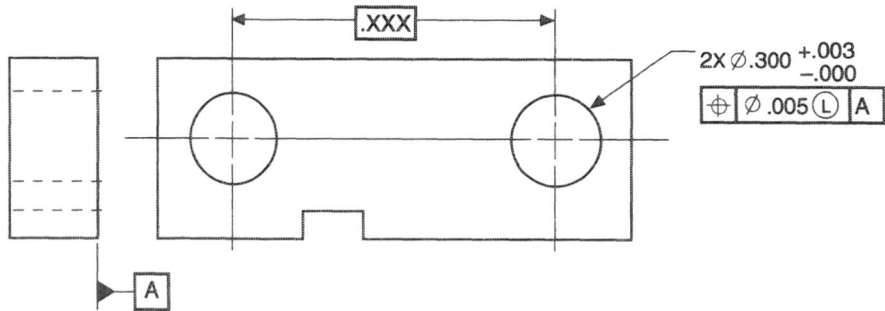

**Meaning**

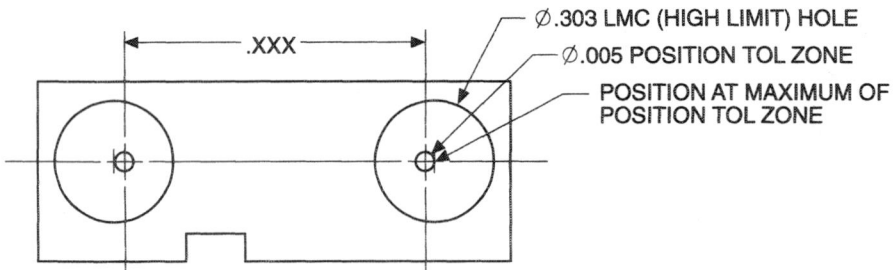

**Figure 6-110**

**Figure 6-111**

## LMC Related to RFS Datum Feature

When a pattern of holes is dimensioned relative to the location of another hole, as shown on the following illustration, this hole is identified as a datum and the hole pattern is located dimensionally with respect to it. In this case, as contrasted with some earlier examples of similar parts using MMC, the holes are controlled on an LMC basis with a more critical relationship (RFS) to the datum center hole. In a situation of this kind the effect of the MMC principle could be detrimental to the necessary precision of the design requirement, so the design specifies the reverse principle, LMC. Such applications as gear centers, or other critical component interfaces, could require such considerations. LMC provides a compensating effect which states in essence that the hole location criterion is at its LMC size ("sloppiest") and can permit an increase in position tolerance as it (the hole) reduces in actual mating size (toward MMC) and "improves" the component/fit relationship. This also provides a dynamic alternative to specifying RFS to the four hole location.

In the example, the four holes are related to the center datum hole. As the position of the center datum hole shifts, the position of the four hole pattern itself must follow as dictated by the function of the part. Imagine that this part has a mating part with a shaft and four pins which must assemble with the five holes in the illustrated part.

Note that the .500 datum center hole is also located from datums (surfaces) A, B, and C. It is given a position tolerance of Ø.010 at MMC and a refined orientation, or perpendicularity, tolerance of Ø.003 at MMC. This properly controls the center hole relationship to the edge surfaces with a rather lenient position tolerance while the orientation to the primary datum A is maintained to closer tolerance. The center hole is identified as datum D so that the four .380 holes can be located with respect to it. Under MMC principles, the position and perpendicularity tolerances increase an amount equal to the actual mating hole size departure from MMC as shown in the illustration.

Wherever the datum hole D position varies in the design considerations, or in actual production, the four .380 hole pattern must follow. Note that the positional pattern dimensions originate at datum D to carry out this intent.

The four .380 holes are located by a position tolerance of Ø.005 at LMC with respect to datum A (for orientation), datum D regardless of feature size (RFS) (for location), and datum B (for rotation). Reference to the lower portion of the illustration under Meaning will assist in the understanding of the effect of the three datums and the importance of datum precedence.

The calculations of the Ø.005 positional tolerance for the Ø.380 holes proceeds as follows. Imagine the mating part pins as four Ø.370 MMC size pins. The fixed fastener method, as described in numerous places earlier in this text where MMC principles are applied is used; e.g.,

$$T = \frac{H - P}{2} \quad T = \frac{\varnothing.380 - \varnothing.370}{2} \quad T = \frac{\varnothing.010}{2}$$

T = Ø.005 on each part.

## Added Step to Apply LMC Ⓛ Principle to Position Tolerance

Reduce mating part pin size an amount equal to the size tolerance of the hole (e.g. in this case Ø.003). Mating part pin size is now Ø.367 at MMC, (if ⌖ Ø.005 Ⓜ to VC of Ø.372 of pin, if ⌖ Ø.005 to OB (Outer Boundary) of Ø.372).

In this case, since the LMC principle reverses the usual size/position relationship, and departure from LMC toward MMC of the hole (e.g. Ø.383 to Ø.380), a Ø.003 reduction of the pin size must be made to Ø.367 (to achieve a VC or OB of Ø.372) to accommodate the added shift of the axis of the Ø.380 hole of Ø.003 as a departure from LMC to MMC occurs. Otherwise, there could be, at worse case, a Ø.003 interference of the Ø.380 MMC size hole with the pin size at MMC and when permitted to shift within an Ø.008 tolerance zone (i.e. Ø.005 + Ø.003 + Ø.008).

*Reasoning:*

The MMC and LMC principles normally strive for different opposed objectives; that is, the MMC principle ensures a prescribed "clearance" between mating part features; whereas the LMC principle deals with "cross sectional material strength or mass" in a sort of compensating situation or for alignment (not usually a "fit"). Thus, when the principle of LMC as applied here for wall thickness concern or alignment, it infringes upon the "clearance principle" of MMC. There must be a compensating adjustment of the pin size, and its resulting VC or OB of the imagined mating part to restore a hole/pin "go" relationship but yet retain the LMC advantage on the subject part holes.

Because of the compensating necessity required on the part illustrated (with 4X Ø.380 + .003/-.000 holes), it is not recommended that the LMC principle be also invoked on the mating part. If such to be done, additional complex and special further compensating would be required as before explained. The value of the LMC concept must be balanced against its added complexity brought about by the application to mating parts. Retain the MMC principle on the mating part if at all possible for maximum advantage and practicality. If deemed necessary, an RFS selection for the position tolerance on the mating part could be an option. This minimizes some of the complication. The design requirements must be balanced with the manufacturing advantages as the best optimum situation achievable. If determined desirable to invoke LMC on the mating part pin features as well, consideration as before stated, e.g. the compensation to the amount of that feature's size tolerance,

**Example**

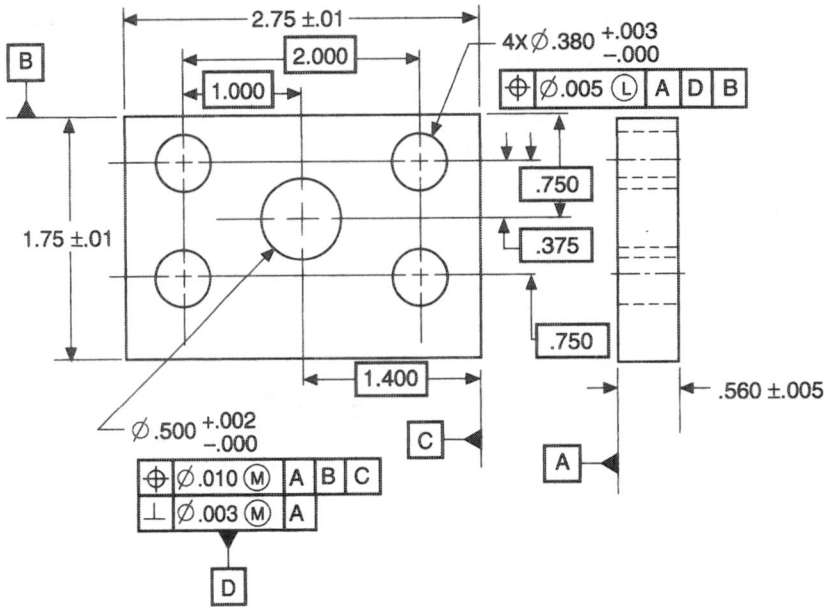

**Meaning**

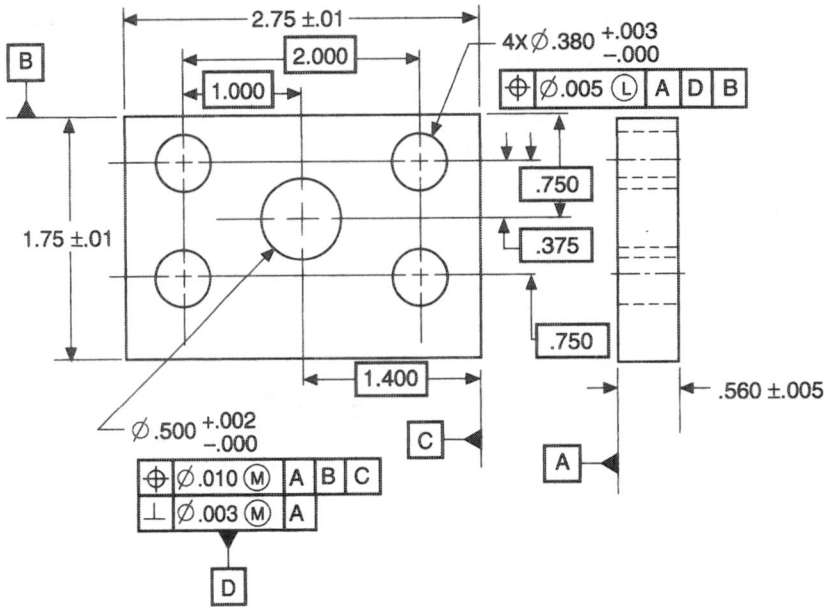

would need to be done in determining the final size values. In this example, increase the mating part hole size an amount equal to the size tolerance if the pin (imagine Ø.367 + .000/-.003 pin size) to hole size Ø.383 + .003/-.000. This adjustment would be necessary to ensure assembly of the mating part feature.

This will ensure that the part features will assemble as per routine positional tolerancing at MMC. By then reversing the principle on the illustrated part on page 490 to LMC from MMC as shown, the positional tolerance of Ø.005 applies at the LMC size limit of Ø.383 and increases an amount equal to the decrease in the actual mating envelope size toward MMC, Ø.380. Thus, as the produced hole reduces in its actual mating size from LMC size, Ø.383, down to Ø.382, the positional tolerance is increased by Ø.001; and so on down to Ø.380 (MMC), whereby its positional tolerance is then Ø.008 as seen in the tabulation at the lower part of the illustration. What has then happened is that the criterion has simply been reversed and the reduction of the actual mating size of the hole permits its added displacement an equal amount. This compensating effect always ensures the mating feature's assembly essentially the same as if MMC had been specified. However, it permits the hole positional displacement to increase *only* "inward" (toward true position). In contrast, if MMC had been stated, as in earlier examples of this same part, the permitted positional displacement of the hole is always "outward." Thus, the compensating effect of LMC is evident as a valuable optional approach to protect wall thickness, strength, alignment, etc., and yet permit an equal amount of positional tolerance where MMC may have been specified instead.

The datum feature D is referenced on an RFS basis. This neutralizes the effect of size departure from MMC of the datum feature to the related hole pattern movement. It must be located from the axis of datum D and cannot move with respect to that axis. This could be in keeping with the precision of the part interface, the LMC requirement, and also could be to restrict the pattern movement as a concern for wall thickness between the outer hole surface limits relative to the outside surfaces. Where deemed necessary to design intent, LMC could also be applied to datum D in its established relationship to the small hole pattern. In which case, the four hole pattern, as a group, could shift an amount equal to the amount of such departure from LMC of the actual mating size of the datum hole D. In such an application, added compensating considerations as before described would be required. This can become very complex in the details relative to "all features" now involved as a pattern. Retention of an RFS relationship to the datum feature could be the most logical choice to avoid complexity if this option is possible.

Functional gages and comparable techniques could be applied in the verification of the Ø.500 datum hole requirements as earlier illustrated on a similar part. However, functional gaging, of the physical variety, could *not* be performed on the four hole LMC requirement. It could be done with optical methods, graphic analysis, mathematically, etc., where the functional principles of LMC can be simulated as a gaging operation.

The advantages of LMC are (1) that it provides an often needed tool and (2) that it also provides an alternative other than RFS, where the MMC method is not desirable.

# C H A P T E R 7

# Working Drawings

## 7-1 INTRODUCTION

This chapter explains how to create assembly drawings, parts lists, and detail drawings. It includes guidelines for titles, revisions, tolerances, and release blocks. The chapter shows how to create a design layout, then use the layout to create assembly and detail drawings using the **Layer** command.

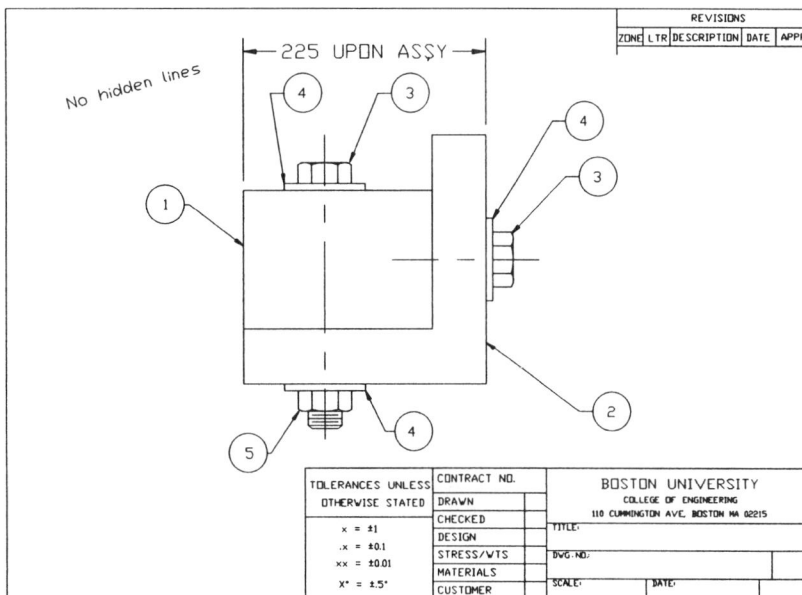

Figure 7-1

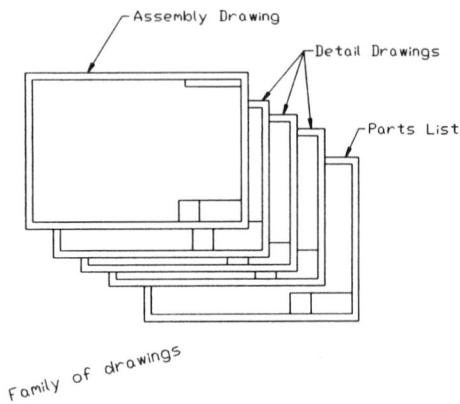

Figure 7-2

## 7-2 ASSEMBLY DRAWINGS

*Assembly drawings* show how objects fit together. See Figure 7-1. All information necessary to complete the assembly must be included on the drawing. This information may include specific assembly dimensions, torque require-

ments for bolts, finishing and shipping instructions, and any other appropriate company or customer specifications.

Assembly drawings are sometimes called *top drawings* because they are the first of a series of drawings used to define a group of parts that are to be assembled together. The group of drawings, referred to as a *family of drawings*, may include subassemblies, modification drawings, detail drawings, and a parts list. See Figure 7-2.

Assembly drawings do *not* contain hidden lines. A sectional view may be used to show internal areas critical to the assembly. Specific information about the internal surfaces of objects that make up the assembly can be found on the detail drawings of the individual objects. See Figure 7-3.

Each part of an assembly is identified by an assembly, or item, number. Assembly numbers are enclosed in a circle or ellipse, and a leader line is drawn between the assembly number and the part. Assembly numbers are unique to each assembly drawing; that is, a part that is used in several different assemblies may have a different assembly number on each assembly drawing.

If several of the same part are used in the same assembly, each part should be assigned the same assembly number. A leader line should be used to identify each part unless the

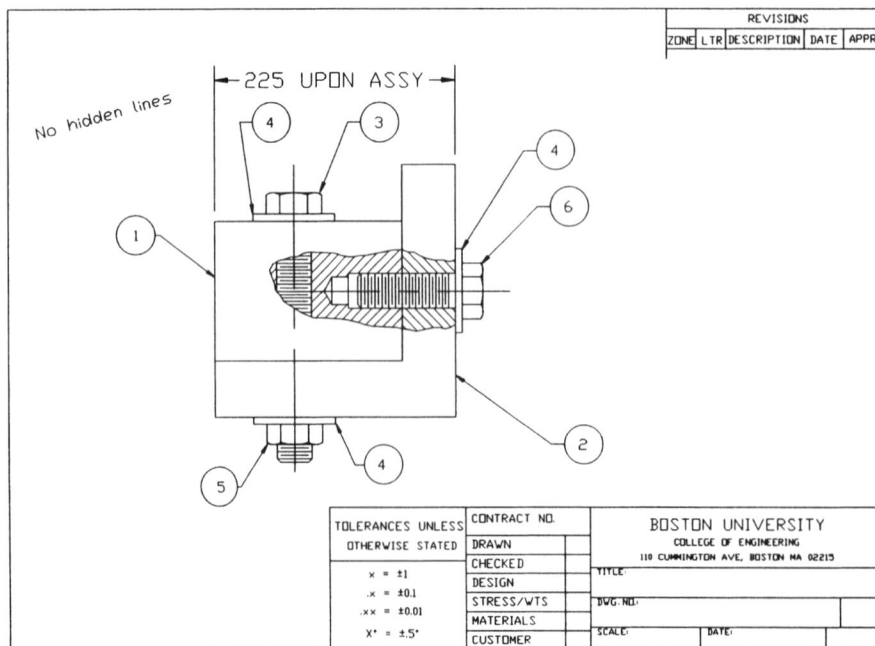

Figure 7-3

**Figure 7-4**

**Figure 7-6**

differences between the parts are obvious. In Figure 7-4 the difference between head sizes for the fasteners is obvious, so all parts need not be identified.

Assemblies should be shown in their natural or neutral positions. Figure 7-5 shows a clamp in a slightly open position. It is best to show the clamp jaws partially open, not fully closed or fully open. In general, a drawing should show an assembly in its most common position.

The range of motion for an assembly is shown using phantom lines. See Figure 7-6. Note how the range of motion for the hinged top piece is displayed using phantom lines.

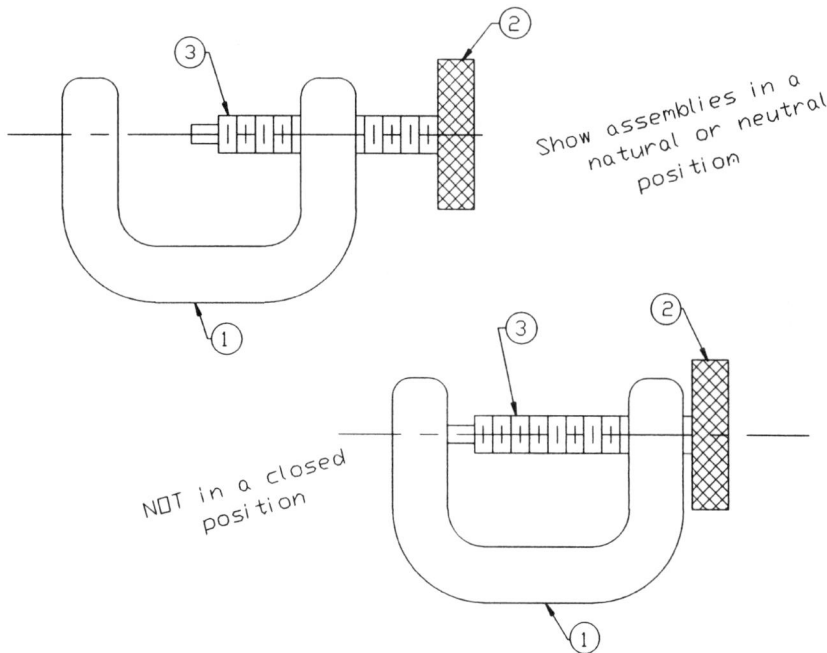

**Figure 7-5**

Figure 7-7

## 7-3 DRAWING FORMATS (TEMPLATES)

Figure 7-7 shows a general format for an assembly drawing. The format varies from company to company and with the size of the drawing paper.

AutoCAD includes many different drawing templates that conform to ANSI and ISO standards. Templates automatically include a drawing border, title block, revision block, and a partial release block.

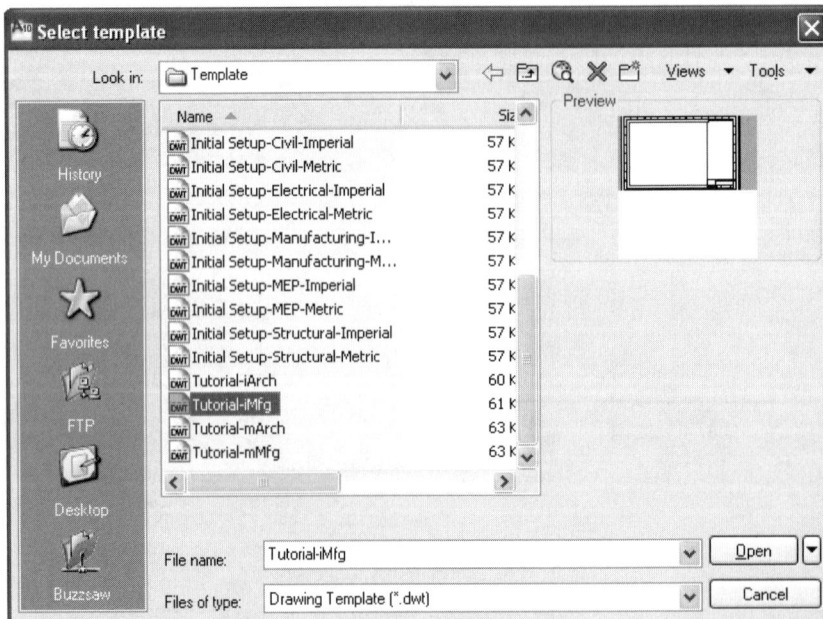

Figure 7-8

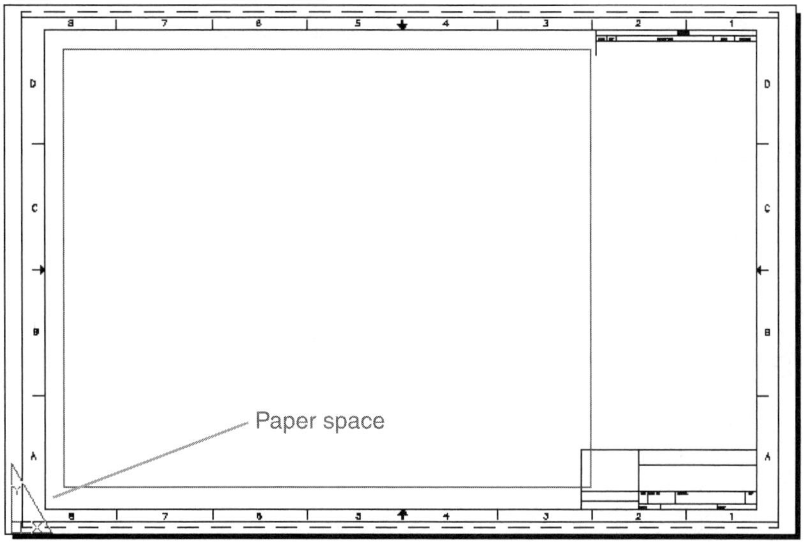

Figure 7-9

**To add a drawing template**

1. Click the **New** tool.

The **Select template** dialog box will appear. See Figure 7-8.

2. Select the **Tutorial-iMfg** template; click the **Open** box.

The template will appear on the screen. See Figure 7-9. Drawings may be created on the template just as they are with a blank, no template, drawing screen.

Figure 7-10 shows a completed title block for an iMfg template. It was completed using the **Text** tool. An A-size drawing sheet is 8.5 × 11 inches. If an A-size template is selected, the screen units will automatically be fitted to the sheet size and will be in inches.

The closest equivalent to an A-size inch drawing is an A4-size drawing, which is 210 × 297 millimeters. Figure 7-11 shows the **Select template** dialog box set to select a **Tutorial-mMfg** template. Figure 7-12 shows the resulting template. The screen units will be in millimeters.

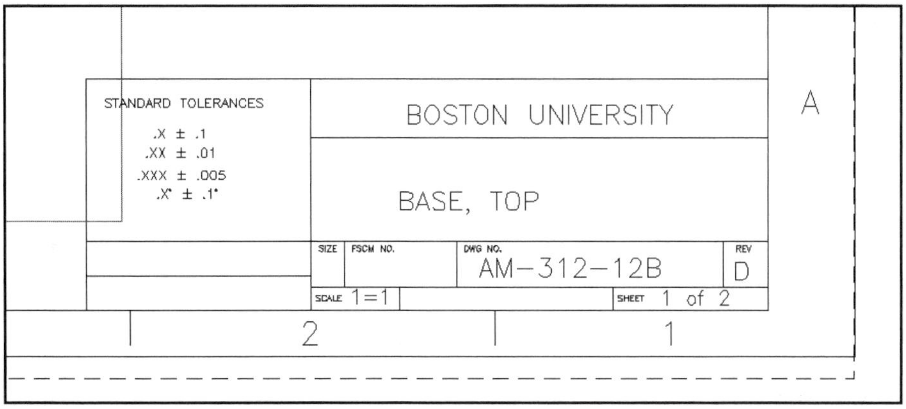

Figure 7-10

**Figure 7-11**

Template can also be accesed by clicking the **Insert** heading at the top of the screen and selecting **Layout,** then the **Layout from Template** option. See Figure 7-12.

## 7-4 TITLE BLOCK

Title blocks are located in the lower right corner of a drawing and include, at a minimum, the drawing's name and number, the company's name, the drawing scale, the release date of the drawing, and the sheet number of the drawing. Other information may be included. Figure 7-13 shows a sample title block created from an **iMfg Layout** template. The text was added using the **Multiline text** tool and positioned using the **Move** tool.

Figure 7-14 shows the title block from an mMfg template. The title block will appear with a series of XXX inputs. Use the **Explode** command and explode the title block.

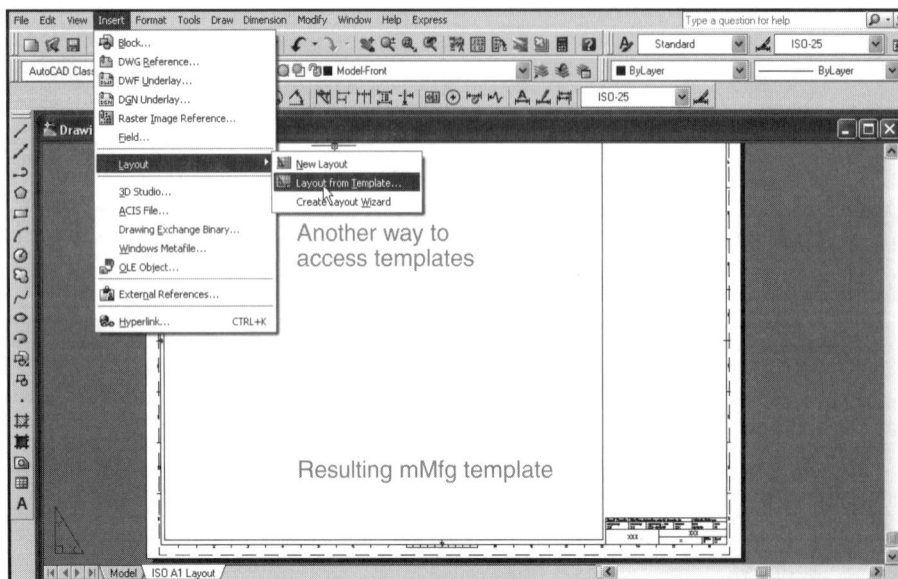

**Figure 7-12**

The XXX inputs will be replaced with words. Erase the unwanted words and replace them with the appropriate names and numbers.

## Drawing titles (names)

Drawing titles should be chosen so that they clearly define the function of the part. They should be presented in the following word sequence:

Noun, modifier, modifying phrase

For example,

SHAFT, HIGH SPEED, LEFT HAND
GASKET, LOWER

Noun names may be two words if normal usage includes the two words.

GEAR BOX, COMPOSITE
SHOCK ABSORBER, LEFT

| | | |
|---|---|---|
| | | REV |
| **BOSTON UNIVERSITY** | | |
| SPACER, RIGHT | | |
| SIZE A / FSCM NO. | DWG NO. AM311-1 | REV C |
| SCALE FULL | SHEET 1 of 2 | |

**Figure 7-13**

## Drawing numbers

Drawing numbers are assigned by companies according to their usage requirement. The numbering system varies

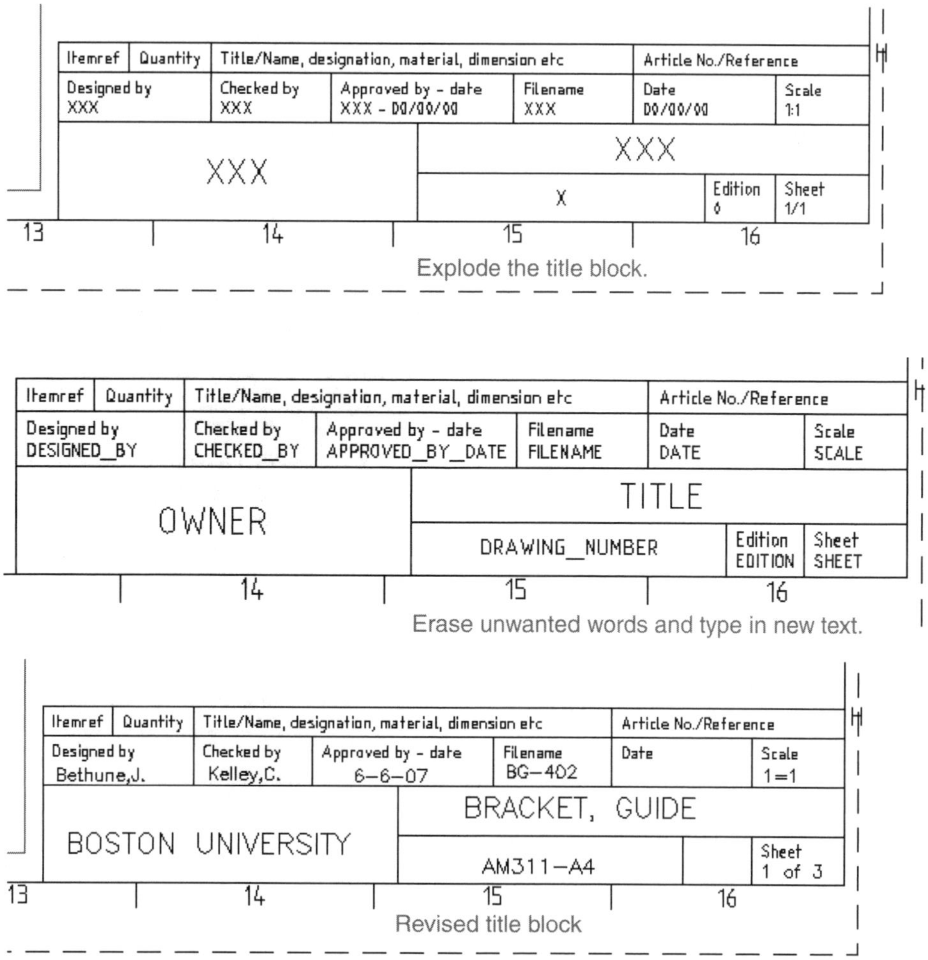

Explode the title block.

Erase unwanted words and type in new text.

Revised title block

**Figure 7-14**

greatly. Usually, drawing numbers are recorded in a log book to prevent duplication of numbers.

## Company name

The company's name and logo are preprinted on drawing paper or are included as a wblock so that they can be inserted on each drawing.

## Scale

Define the scale of the drawing.

SCALE: FULL, or 1 = 1

The drawing size is the same as the object size.

SCALE: 2 = 1

The drawing size is twice as large as the object size.

SCALE: 1 = 2

The drawing size is half as large as the object size.

## Release date

A drawing is released only after all persons required by the company's policy have reviewed it and added their signatures to the release block. Once released, a drawing becomes a legal document. Drawings that have not been officially released are often stamped with statements such as "NOT RELEASED" or "FOR REFERENCE ONLY."

## Sheet

The number of the sheet relative to the total number of sheets that make up the drawing should be stated clearly.

SHEET 2 OF 3

or

SH 2 OF 3

## 7-5  REVISION BLOCK

Drawings used in industry are constantly being changed. Products are improved or corrected, and drawings must be changed to reflect and document these changes. Figure 7-15 shows a sample revision block.

Drawing changes are listed in the revision block by letter. Revision blocks are located in the upper right corner of the drawing. See Figure 7-7. Figure 7-16 shows a revision block from an mMfg template. The text was added using **Mtext,** and the chart lines using **Line.**

Each drawing revision is listed by letter in the revision block. A brief description of the change is also included.

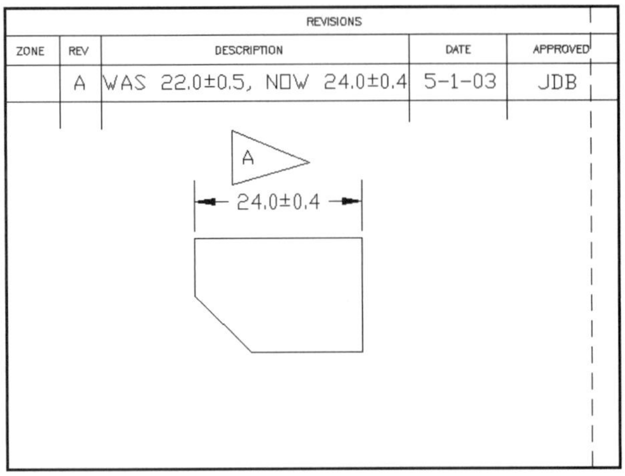

**Figure 7-15**

It is important that the description be as accurate and complete as possible. Revisions are often used to check drawing requirements on parts manufactured before the revisions were introduced.

The revision letter is also added to the field of the drawing in the area where the change was made. The letter is located within a "flag" to distinguish it from dimensions and drawing notes. The flag serves to notify anyone reading the drawing that revisions have been made.

Most companies have systems in place that allow engineers and designers to make quick changes to drawings. These change orders are called *engineering change orders* (ECOs), *change orders* (COs), or *engineering orders* (EOs), depending on the company's preference. Change orders are documented on special drawing sheets that are usually stapled to a print of the drawing. Figure 7-17 shows a change order attached to a drawing.

After a group of change orders accumulate, they are incorporated into the drawing. This process is called a *drawing revision,* which is different from a revision to the drawing. Drawing revisions are usually indicated by a letter located somewhere in the title block. The revision letters may be included as part of the drawing number or in a separate box in the title block. Whenever you are working on a drawing, make sure you have the latest revision and all

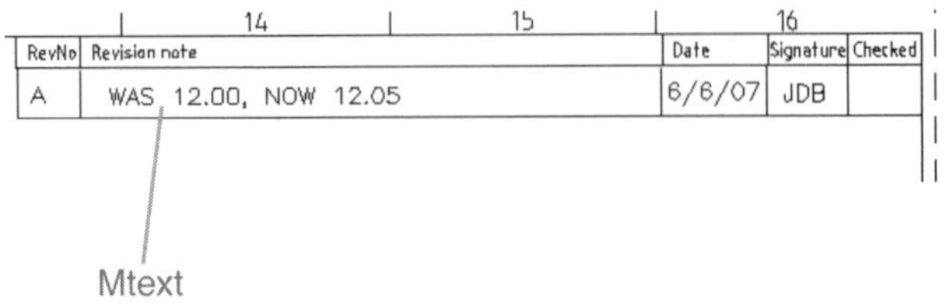

**Figure 7-16**

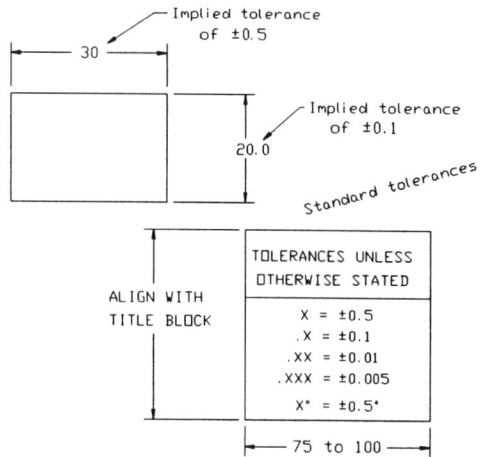

**Figure 7-17**

**Figure 7-19**

appropriate change orders. Companies have recording and referencing systems for listing all drawing revisions and drawing changes.

millimeter values. In Figure 7-18 the dimension 2.00 has an implied tolerance of ±.01. In Figure 7-19 the 20.0 dimension has an implied tolerance of ±0.1.

## 7-6 TOLERANCE BLOCK

Most drawings include a ***tolerance block*** next to the title block that lists the standard tolerances that apply to the dimensions on the drawing. A dimension that does not include a specific tolerance is assumed to have the appropriate standard tolerance.

Figure 7-18 shows a sample tolerance block for inch values, and Figure 7-19 shows a sample tolerance block for

## 7-7 RELEASE BLOCK

A ***release block*** contains a list of approval signatures or initials required before a drawing can be released for production. See Figure 7-20. The required signatures are generally as follows.

**Drawn**

Person that created the drawing.

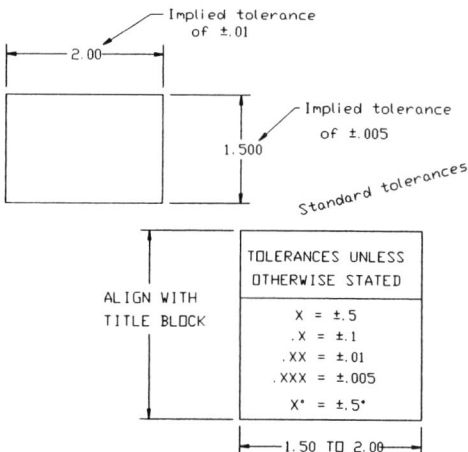

**Figure 7-18**

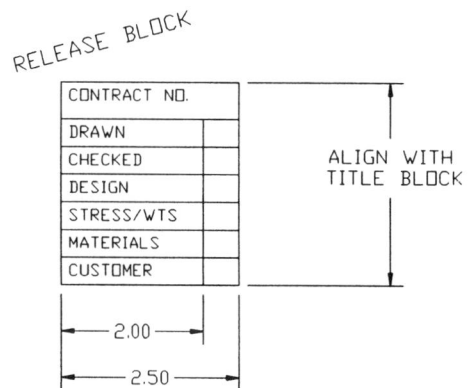

**Figure 7-20**

## Checked

Drawings are checked for errors and compliance with company procedures and conventions. Some large companies have checking departments, whereas smaller companies have drawings checked by a senior person or the drafting supervisor.

## Design

The engineer in charge of the design project. The designer and the drafter may be the same person.

## Stress/Wts

The department or person responsible for the stress analysis of the design.

## Materials

Usually, a person in the production department checks the design and makes sure that the necessary materials and the machine times for the design are available. This person may also schedule production time.

## Customer

The customer for the design may have a representative on site at the production facility to check that its design requirements are being met. For example, it is not unusual for the Air Force to assign an officer to a plant that manufactures its fighter aircraft.

## 7-8 PARTS LIST (BILL OF MATERIALS)

A *parts list* is a listing of all parts used on an assembly. See Figure 7-21. The parts list was created using the **Table** tool. Parts lists may be located on an assembly drawing above the title block or on a separate sheet of paper. Assembly drawings done using AutoCAD often include the parts list on a separate layer within the drawing. Use only capital letters on a parts list.

Figure 7-22 shows two sample parts list formats, including dimensions. Parts list formats vary greatly from company to company. The dimensions are given in inches. They may be converted to millimeters using the conversion factor 1.00 inch = 25.4 millimeters.

A parts list serves as a way to cross-reference detail drawing numbers to assembly item numbers. It also provides a list of the materials needed for production and is very helpful for scheduling and materials purchasing.

Parts purchased from a vendor and used exactly as they are supplied, without any modification, will not have detail drawings but are included on the parts list. The washers, bolts, and screws listed on the parts list shown in Figure 7-21 would not have detail drawings. This means that the information on the parts list must be sufficient for a purchaser to know exactly what size washers, bolts, and screws to buy. The information used to define an object on the drawing, the drawing callout, is also used on the parts list. See Chapter 6 for an explanation of drawing callouts for fasteners.

| PARTS LIST | | | | | |
|---|---|---|---|---|---|
| NO. | DESCRIPTION | PART NO. | MATL | NOTE | QTY |
| 1 | BLOCK, TOP | BU311S1 | SAE 1020 | | 1 |
| 2 | BLOCK, BOTTOM | BU311S2 | SAE 1020 | | 1 |
| 3 | M16×24 HEX HEAD BOLT | SPM16H | STEEL | 2 | 1 |
| 4 | 20×40×3 PLAIN WASHER | | STEEL | | 3 |
| 5 | M16 HEX NUT | 2P M16N | STEEL | | 1 |
| 6 | M16×20 HEX HEAD SCREW | | STEEL | | 1 |

Figure 7-21

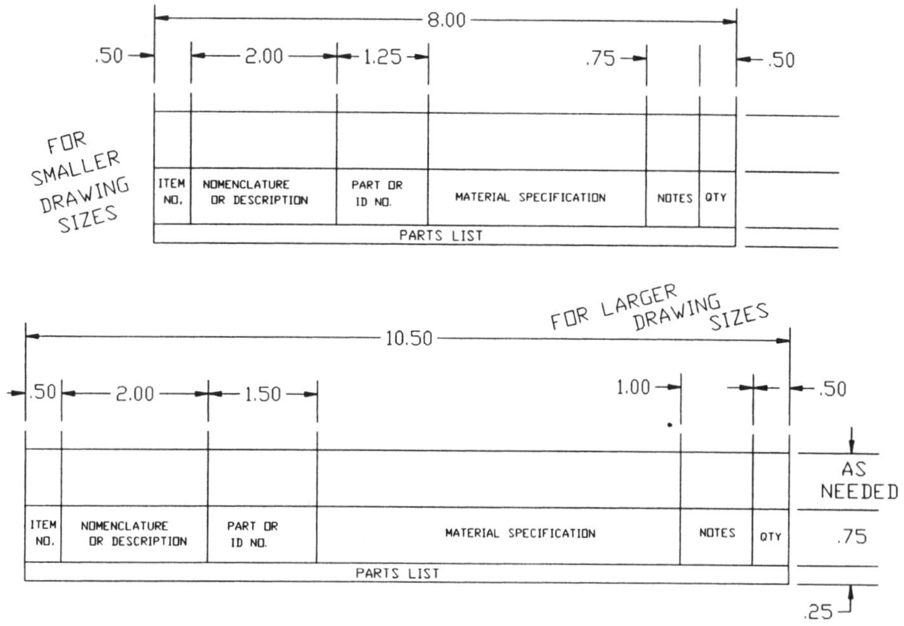

**Figure 7-22**

## 7-9 DETAIL DRAWINGS

A *detail drawing* is a drawing of a single part. The drawing should include all the information necessary to accurately manufacture the part, including orthographic views with all appropriate hidden lines, dimensions, tolerances, material requirements, and any special manufacturing requirements. Figure 7-23 shows a sample detail drawing.

Detail drawings include title, release, tolerance, and revision blocks located on the drawing in the same places as they are found on assembly drawings.

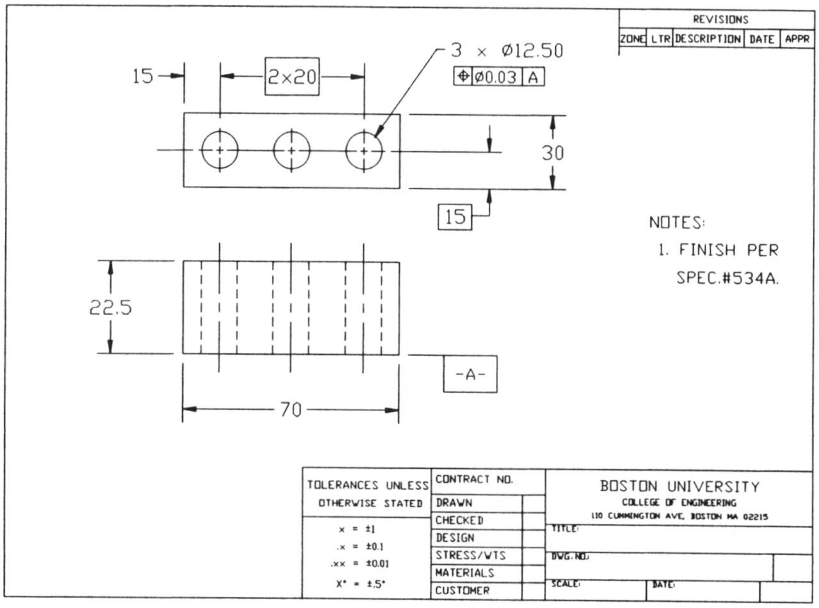

**Figure 7-23**

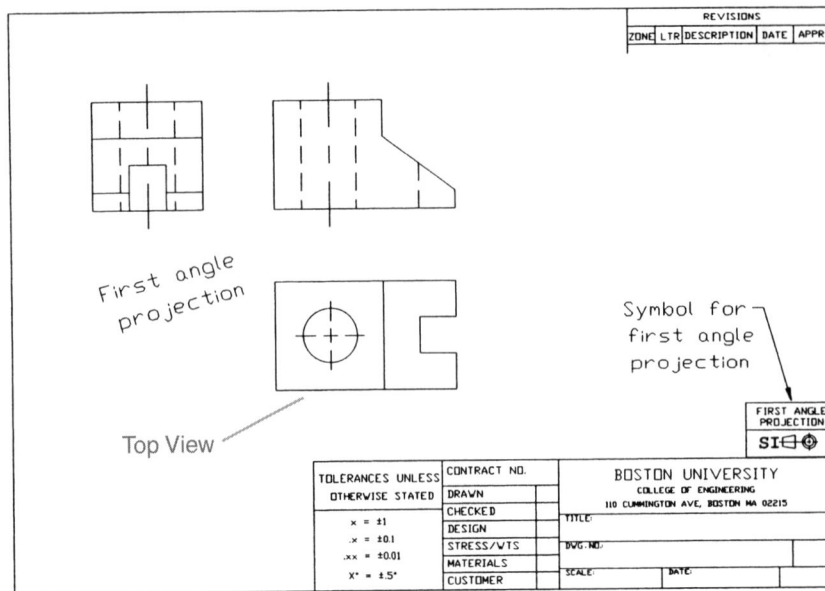

**Figure 7-24**

## 7-10 FIRST-ANGLE PROJECTION

The instructions for the creation of orthographic views as presented in this book are based on third-angle projection. See Chapter 1. Third-angle projection is used in the United States, Canada, and Great Britain, among other countries. Many other countries such as Japan use first-angle projection to create orthographic views.

Figure 7-24 shows an object and the orthographic views of the object created in first-angle projection. Note that the top view in the first-angle projection is the same as the top view in the third-angle projection, but it is located below the front view, not above the front view, as in the third-angle projection. Right-side views are also the same but are located to the left of the front view in first-angle projections and to the right in third-angle projections.

Many companies now do business internationally. They can have manufacturing plants in one country and assembly plants in another. Drawings can be prepared in several different countries. It is important to indicate on a drawing whether first-angle or third-angle projections are being used. Figure 7-25 shows the SI (International System of Units) symbols for first-angle projections. The SI symbol is included on a drawing in the lower right corner above the title block, as shown in Figure 7-24.

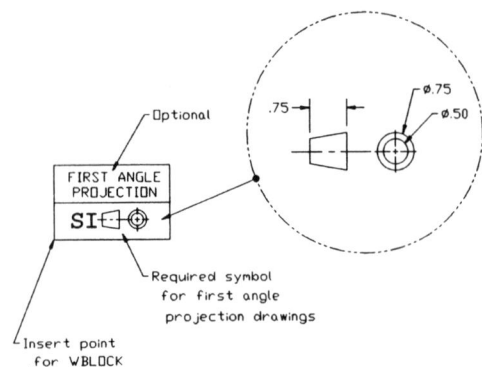

**Figure 7-25**

## 7-11  DRAWING NOTES

Drawing notes are used to provide manufacturing information that is not visual, for example, finishing instructions, torque requirements for bolts, and shipping instructions.

Drawing notes are usually located above the title block on the right side of the drawing. See Figure 7-26. Drawing notes are listed by number. If a note applies to a specific part of the drawing, the note number is enclosed in a triangle. The note numbers enclosed in triangles are also drawn next to the corresponding areas of the drawing.

## 7-12  DESIGN LAYOUTS

A design layout is not a drawing. It is like a visual calculation sheet used to size and locate parts as a design is developed. A design layout allows you to "build the assembly" on paper.

When drawings are created on a drawing board an initial layout is made locating and sizing the parts. Then the individual detail drawings and the assembly drawing are traced from the layout.

The same procedure can be followed using AutoCAD. First, create a design layout on one layer, then either transfer the individual parts and assembly drawing to other layers or create a new drawing from the layout drawing using the **Save As** command.

The sample problem in the next section demonstrates how to create a design layout and then use the layout to create the required detail and assembly drawings.

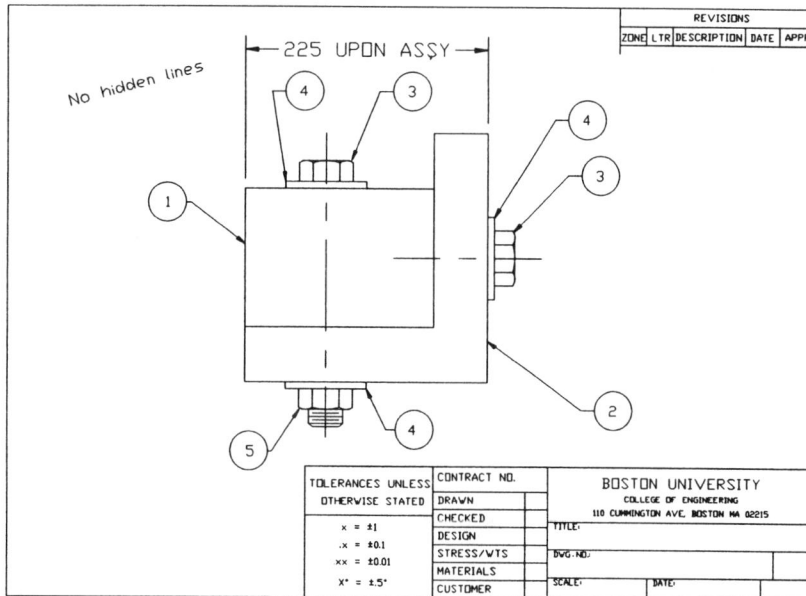

Figure 7-26

Figure 7-27

## 7-13 SAMPLE PROBLEM SP7-1

Figure 7-27 shows an engineer's sketch for a design problem. Prepare an assembly drawing and a detail drawing of the PN123 BASE and PN124 CENTER PLATE based on the following information.

1. Parts PM107S END BRACKET are existing parts and can be used as is. Figure 7-28 shows a detail drawing of the part.
2. Place the mounting holes for the two END BRACKETS 200 millimeters apart, as shown in the engineer's sketch.
3. Establish the size of the BASE so that it aligns with the END BRACKETS and is 40 millimeters thick.

4. Determine the sizes for the CENTER BRACKET so that it just fits between the end plates and has a 25-millimeter-diameter hole centered in its top surface. The CENTER BRACKET should extend 10 millimeters above the END BRACKETS and 5 millimeters above the BASE. The CENTER BRACKET should be 10 millimeters thick all around.
5. Mount the END BRACKETS to the BASE using M12 FLAT HEAD screws, and attach the END BRACKETS to the CENTER PLATE using M20 bolts with appropriate nuts.
6. Use only standard length fasteners.

**To create the design layout (See Figure 7-29.)**

1. Draw horizontal and vertical lines that define the top of the BASE and the distance between the mounting holes in the END BRACKETS.
2. Draw front and top views of the END BRACKETS using the vertical lines drawn in step 1. The vertical lines should align with the centerlines of the mounting holes.
3. Draw the BASE 40 millimeters thick.
4. Size the CENTER BRACKET.

Note that at this stage of the layout all lines are drawn using the same pattern. Lines can be changed to hidden lines or centerlines when the detail and assembly drawings are created.

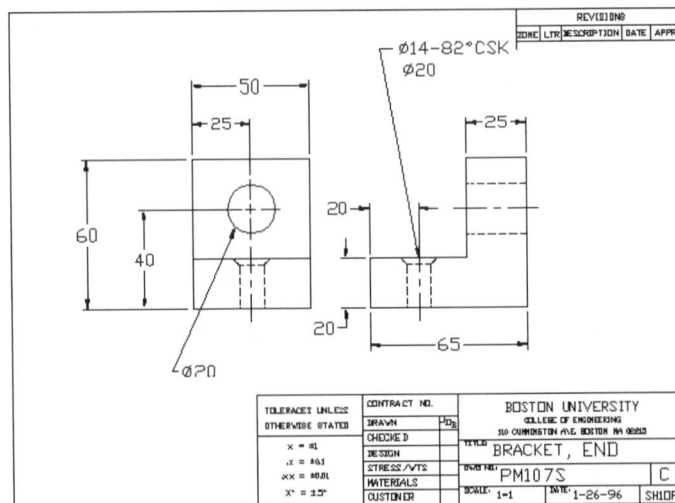

Figure 7-28

5. Add the M12 FLAT HEAD screws. A length of 20 millimeters was chosen from the tables in the appendix. The threaded holes in the BASE are M12 to match the screws. The tap drill size is included on the layout for reference.

6. Add the M20 × 60 bolts and nuts.

The bolt length is determined by adding 25 (END BRACKET) + 10 (CENTER BRACKET) + 15 (nut thickness = .75D) + 2 thread pitches = 52 millimeters. The next largest standard bolt length, as listed in the appendix, is 60, so an M20 × 60 bolt is specified.

7. The diameter of the clearance holes was chosen as 14 and 22 and is listed on the layout.

The design layout is complete. It may be used to create the required assembly drawing and detail drawings in one of two ways: using the **Layer** command and including all the drawings under one file name, or by using the **Save As** command to create separate drawings.

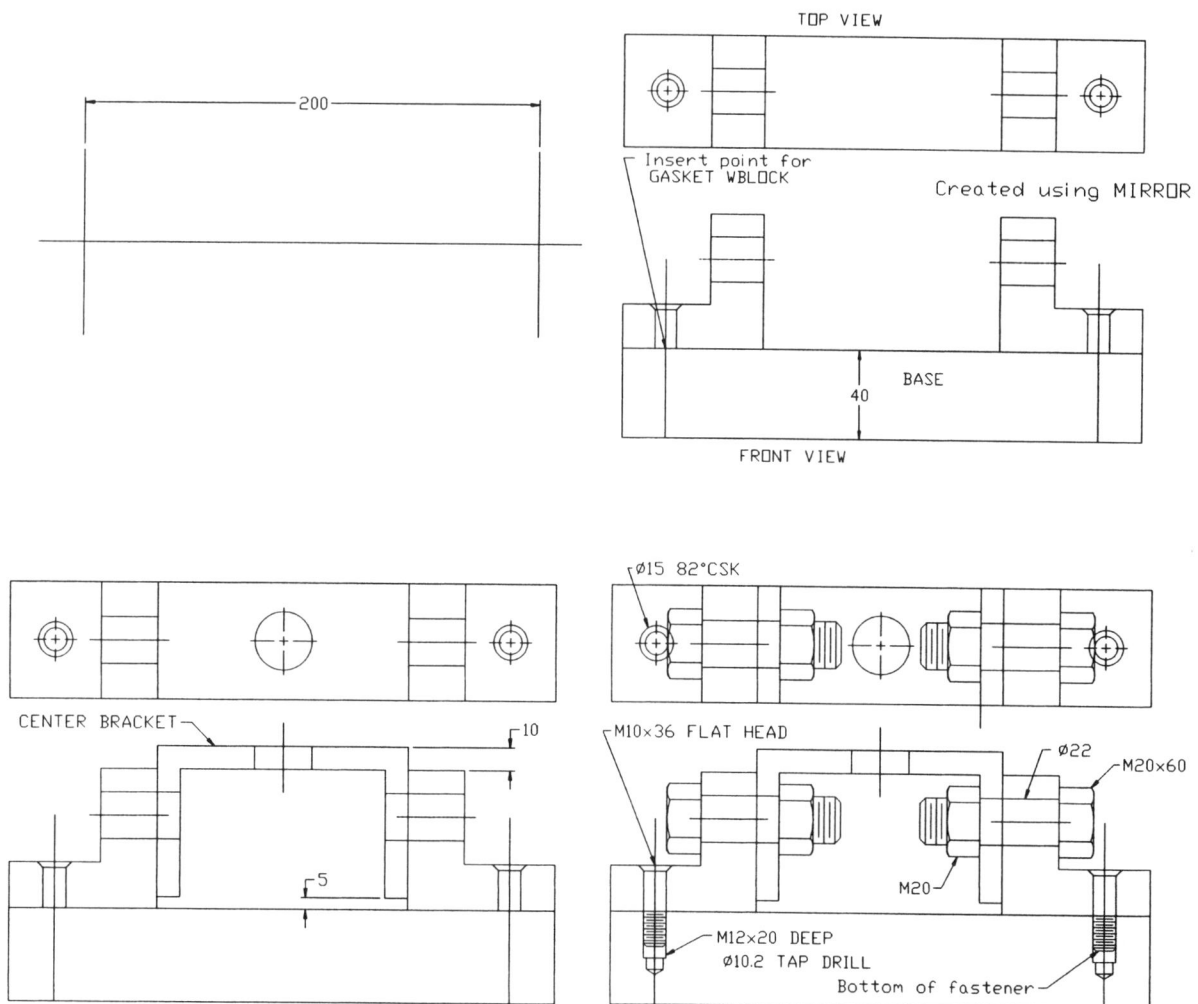

**Figure 7-29**

## To create a drawing using layers

When lines are moved from one layer to another the lines will disappear from the original layer. This will, in essence, erase the layout drawing because all the lines will have been moved to other layers. The layout may be preserved by first making a copy of the layout, or that part of the layout you wish to transfer, then using the copy for the transfers.

To make a copy of the layout, use the **Copy** command. Select the entire screen and select a base point. Select the exact same point as the second point of displacement, and AutoCAD will copy the layout exactly over itself. You can then move the appropriate lines to different layers.

*Note:* Explode all blocks before transferring them between layers. A block cannot be edited once it is moved from its original layer.

Create a detail drawing of the BASE by transferring the BASE from the design layout. Respond to the prompts as follows.

1. Copy the layout onto itself.

   *COMMAND:_Copy*
   *Select objects:*

2. Window the entire layout; press **Enter**.

   *<Base point or displacement>/Multiple:*

3. Select a point on the screen.

   *Second point of displacement:*

4. Select exactly the same point as that used in the previous step.

5. Create the following layers. See Section 3-24.

   ASSEMBLY
   END
   END-DIM
   BASE
   BASE-DIM
   CENTER
   CENTER-DIM
   PARTSLIST

The "DIM" layers will be used for the dimensions for the detail drawings. This will allow the dimensions to be shut off if they are not needed.

6. Transfer the BASE to a layer called **BASE.** Turn off all the other layers. The resulting transfer should look like Figure 7-30.

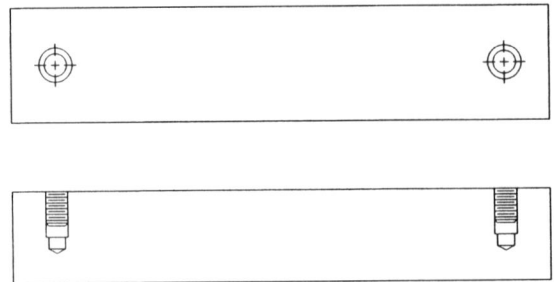

**Figure 7-30**

It will be difficult to make a perfectly clean transfer, that is, only the lines of the BASE. If other lines appear on the transferred BASE, erase and trim them as necessary.

7. Move the views closer together, if necessary.
8. Turn on the **BASE-DIM** layer and make it the current layer. Add the appropriate dimensions, tolerances, and callouts.
9. Create the drawing format and add the necessary information (drawing title, etc.).

The title block, and so forth, may be saved as a block or as part of a prototype drawing. Figure 7-31 shows the resulting drawing.

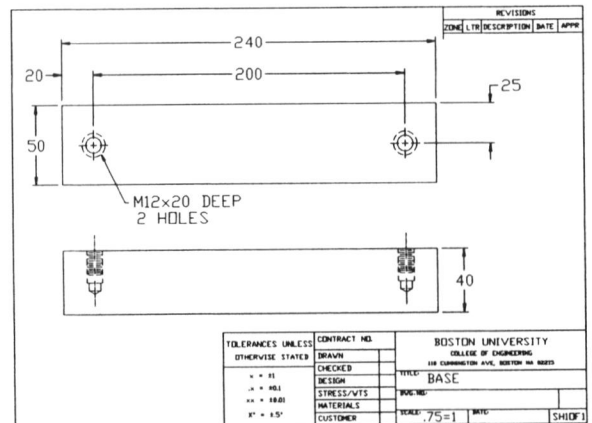

**Figure 7-31**

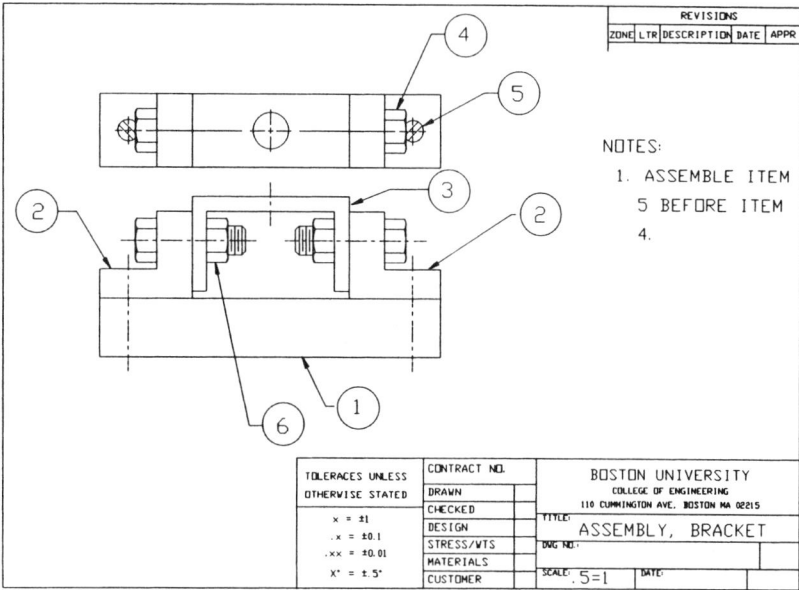

Figure 7-32

## To create a drawing from a layout

Figure 7-32 shows an assembly drawing that was created from the design layout. The procedure is as follows.

1. Save the design layout.
2. Create the assembly drawing from the layout drawing, but use the **Save As** command to save it under a different drawing name.
3. Add the appropriate drawing format and notes.
4. Save the assembly drawing using its new name.

Figure 7-33 shows a parts list for the assembly.

| PARTS LIST | | | |
|---|---|---|---|
| ITEM NO | DESCRIPTION | MATL | QTY |
| 1 | BASE | STEEL | 1 |
| 2 | END BRACKET - PM107S | STEEL | 2 |
| 3 | CENTER PLATE | STEEL | 1 |
| 4 | M20×60 HEX HEAD BOLT | STEEL | 2 |
| 5 | M12×36 FLAT HEAD SCREW | STEEL | 2 |
| 6 | M20 NUT | STEEL | 2 |

Figure 7-33

## 7-14  SAMPLE PROBLEM SP7-2

Figure 7-34 shows an assembly made from four different parts: a Base, a Bracket, and two Posts. The detail drawing of each part is shown in Figure 7-35.

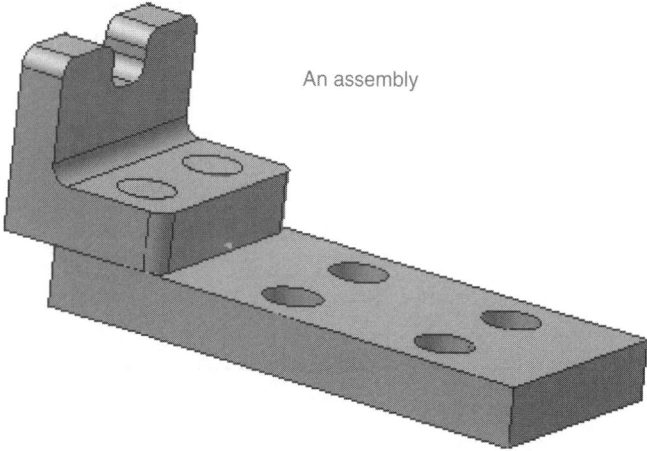

An assembly

Figure 7-34

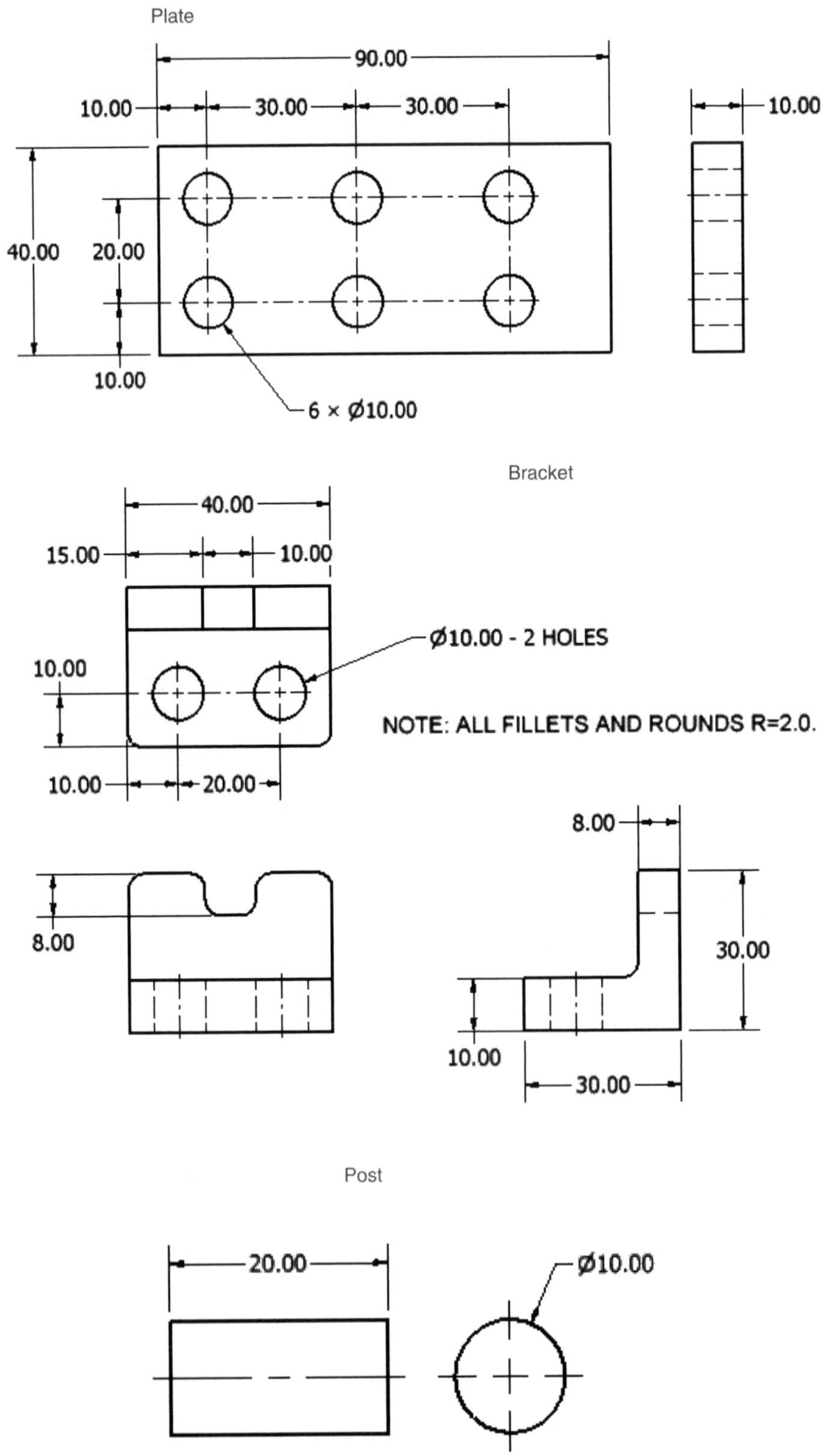

Plate

6 × Ø10.00

Bracket

Ø10.00 - 2 HOLES

NOTE: ALL FILLETS AND ROUNDS R=2.0.

Post

Figure 7-35

**To create an assembly drawing**

1. Draw orthographic views of the parts in their assembled position.

See Figure 7-36. In this example only a front and top view were drawn, as that is sufficient to define the assembly. The front view orientation was selected to display the profile of the assembled parts. Note that there are no hidden lines in assembly drawings.

2. Add assembly numbers to the drawing.

Remember that assembly numbers are different from part numbers. A part number is a number assigned to an individual part and remains with that part regardless of what assembly uses the part. An assembly number is unique to each assembly and does not transfer to other assemblies.

3. Create a parts list for the assembly.

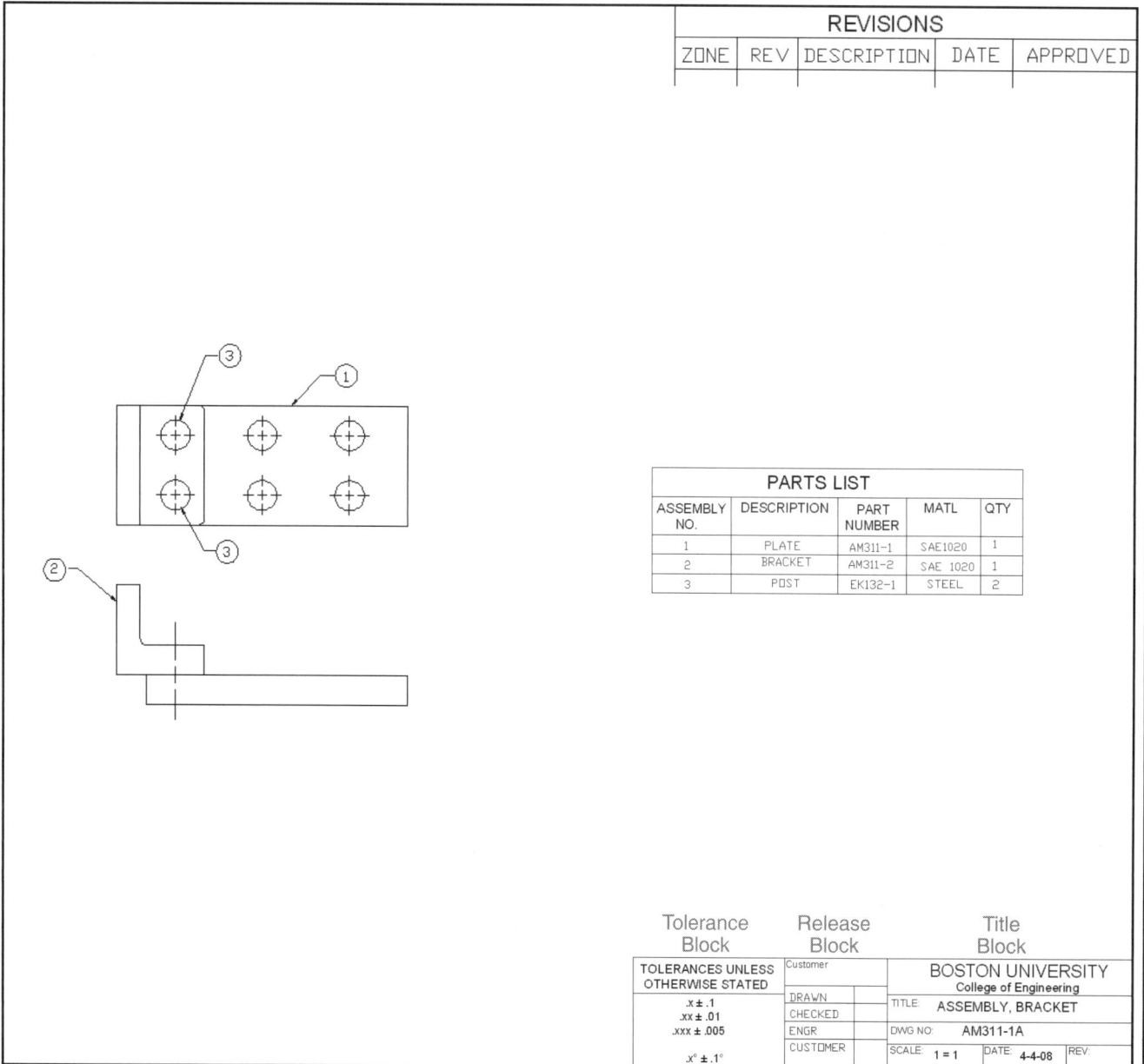

| | REVISIONS | | | |
|---|---|---|---|---|
| ZONE | REV | DESCRIPTION | DATE | APPROVED |
| | | | | |

| PARTS LIST | | | | |
|---|---|---|---|---|
| ASSEMBLY NO. | DESCRIPTION | PART NUMBER | MATL | QTY |
| 1 | PLATE | AM311-1 | SAE1020 | 1 |
| 2 | BRACKET | AM311-2 | SAE 1020 | 1 |
| 3 | POST | EK132-1 | STEEL | 2 |

Tolerance Block | Release Block | Title Block

TOLERANCES UNLESS OTHERWISE STATED
.X ± .1
.XX ± .01
.XXX ± .005
.X° ± .1°

Customer
DRAWN
CHECKED
ENGR
CUSTOMER

BOSTON UNIVERSITY
College of Engineering
TITLE ASSEMBLY, BRACKET
DWG NO: AM311-1A
SCALE 1 = 1 DATE: 4-4-08 REV:

**Figure 7-36**

As stated earlier, there is no standard format for parts lists. The example shown represents an average approach. Note the use of assembly numbers and part numbers.

4. Add a title block, a release block, and a tolerance block.

## 7-15 SAMPLE PROBLEM SP7-3

Figure 7-37 shows another example of an assembly drawing. In this example a front, a right-side, and a bottom view were used to define the assembly. The parts list is located on a second sheet. In general, drawings should not be overcrowded. Parts lists, sectional views, and other drawing information are often located on second or third sheets.

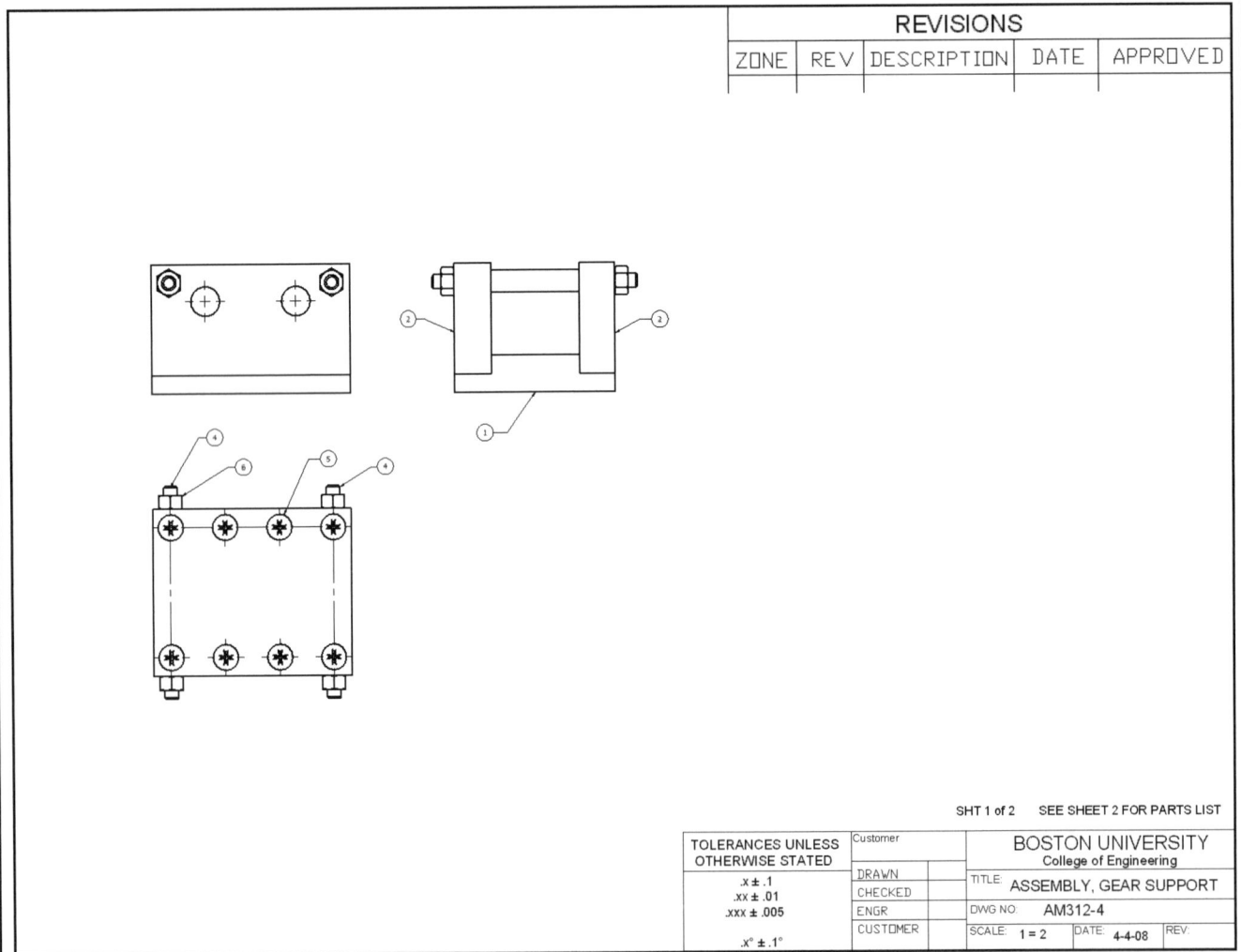

| REVISIONS | | | | |
|---|---|---|---|---|
| ZONE | REV | DESCRIPTION | DATE | APPROVED |
|  |  |  |  |  |

SHT 1 of 2     SEE SHEET 2 FOR PARTS LIST

| TOLERANCES UNLESS OTHERWISE STATED | Customer | BOSTON UNIVERSITY College of Engineering | | |
|---|---|---|---|---|
| .x ± .1 | DRAWN |  TITLE: ASSEMBLY, GEAR SUPPORT | | |
| .xx ± .01 | CHECKED | | | |
| .xxx ± .005 | ENGR | DWG NO: AM312-4 | | |
|  | CUSTOMER | SCALE: 1 = 2 | DATE: 4-4-08 | REV: |
| .x° ± .1° |  | | | |

Figure 7-37

| REVISIONS | | | | |
|---|---|---|---|---|
| ZONE | REV | DESCRIPTION | DATE | APPROVED |
| | | | | |

| Parts List | | | | |
|---|---|---|---|---|
| ITEM | DESCRIPTION | PART NUMBER | MATERIAL | QTY |
| 1 | PLATE, BASE | BU-311-A | Plexiglas | 1 |
| 2 | PLATE, SIDE | BU-311-B | Plexiglas | 2 |
| 3 | POST, GUIDE | BU-311-C | Steel | 2 |
| 4 | POST, THREADED | BU-311-D | Steel | 2 |
| 5 | Pozidriv ISO metric machine screws | AS 1427 - M8 x 20 | Steel, Mild | 8 |
| 6 | Metric Hex Nuts Styles 2 | ANSI B18.2.4.2M - M8x1.25 | Steel, Mild | 4 |

SHT 2 of 2

| TOLERANCES UNLESS OTHERWISE STATED | Customer | | BOSTON UNIVERSITY College of Engineering | |
|---|---|---|---|---|
| .x ± .1 .xx ± .01 .xxx ± .005 .x° ± .1° | DRAWN | | TITLE: ASSEMBLY, GEAR SUPPORT | |
| | CHECKED | | | |
| | ENGR | | DWG NO: AM312-4 | |
| | CUSTOMER | | SCALE: 1 = 2 | DATE: 4-4-08 | REV: |

Figure 7-37    Continued.

## 7-16  EXERCISE PROBLEMS

### EX7-1

A.  Draw an assembly drawing of the given objects.
B.  Draw detail drawings for all nonstandard parts.
C.  Prepare a parts list. Specify the length for the M10 hex head screw.

All parts made from SAE 1020 steel

M10 Hex head screw

12 x 18 x 3 washer 3 reqd

Link bar - 2 reqd 5 thick

Ø12 - 2 holes

R12 - both ends

40

30

20

M10 x 25 deep

15

40

45

Threaded block

### EX7-2

A.  Draw an assembly drawing of the given objects.
B.  Draw detail drawings for all nonstandard parts.
C.  Prepare a parts list.

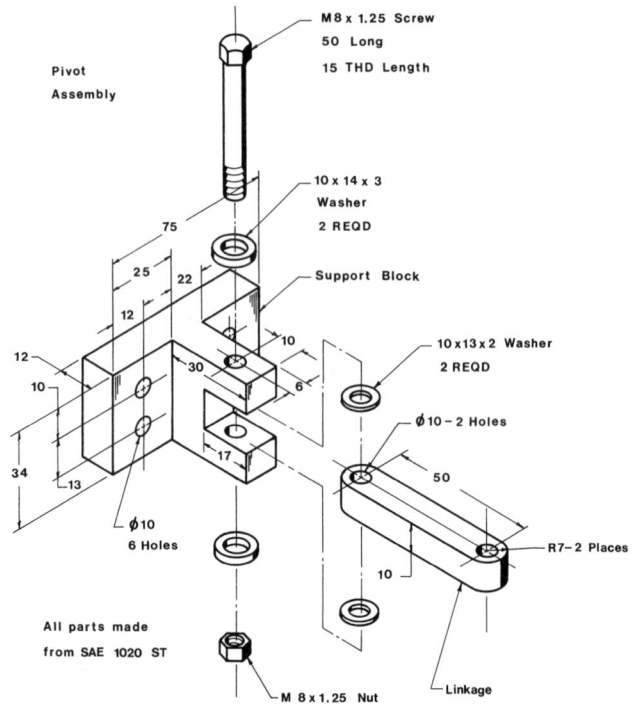

Pivot Assembly

M8 x 1.25 Screw
50 Long
15 THD Length

10 x 14 x 3 Washer
2 REQD

Support Block

10 x 13 x 2 Washer
2 REQD

Ø 10 - 2 Holes

50

75

25

22

12

10

6

30

12

10

17

34

13

Ø10 6 Holes

10

R7 - 2 Places

All parts made from SAE 1020 ST

M 8 x 1.25 Nut

Linkage

## EX7-3

A. Draw an assembly drawing of the given objects.
B. Draw detail drawings for all nonstandard parts.
C. Prepare a parts list.

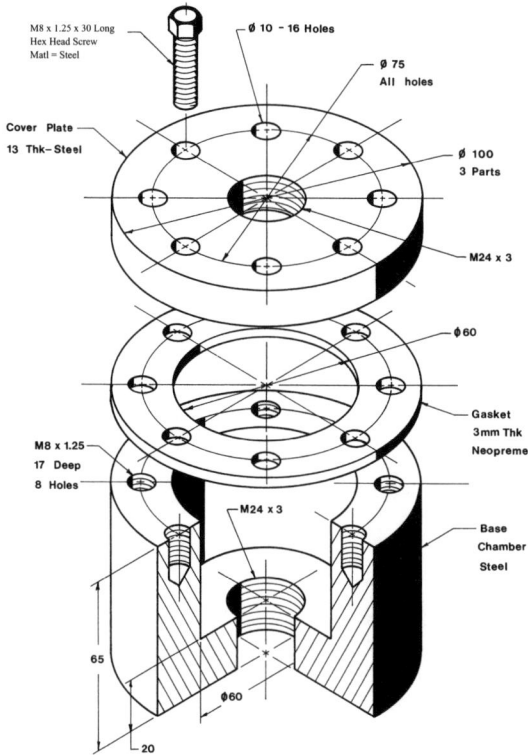

## EX7-4

A. Draw an assembly drawing of the given objects.
B. Draw detail drawings for all nonstandard parts.
C. Prepare a parts list.

## DESIGN

D. Replace part 3 with a bolt and appropriate nut. Add four washers.

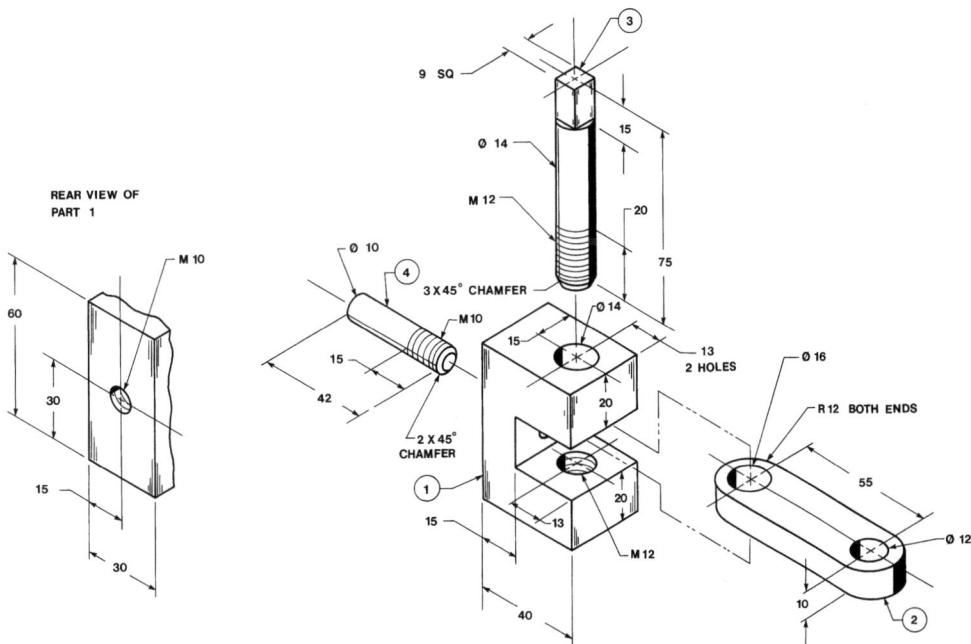

## EX7-5

A. Draw an assembly drawing of the given objects.
B. Draw detail drawings for all nonstandard parts.
C. Prepare a parts list.

## EX7-6

A. Draw an assembly drawing of the given objects.
B. Draw detail drawings for all nonstandard parts.
C. Prepare a parts list.

## DESIGN

D. Replace the rivets with the appropriate screws and nuts.

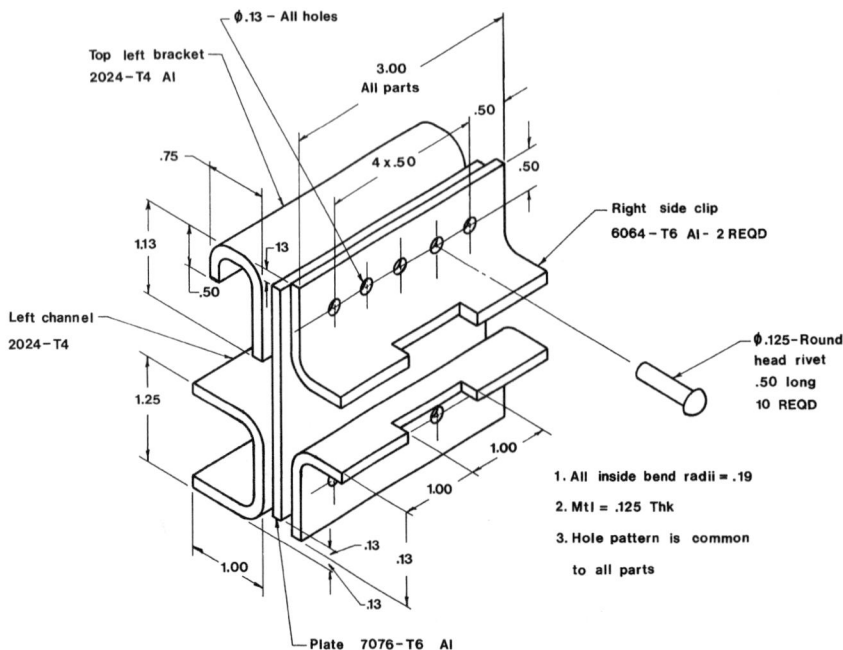

## EX7-7 MILLIMETERS

A. Draw an assembly drawing of the given objects.
B. Prepare a parts list.

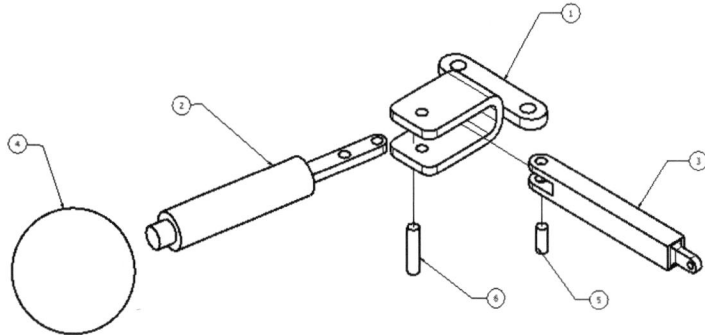

THROTTLE LINK
ASSEMBLY

| Parts List | | | | |
|---|---|---|---|---|
| ITEM | PART NUMBER | DESCRIPTION | MATERIAL | QTY |
| 1 | ENG-A43 | BOX,PIVOT | SAE1020 | 1 |
| 2 | ENG-A44 | POST,HANDLE | SAE1020 | 1 |
| 3 | ENG-A45 | LINK | SAE1020 | 1 |
| 4 | AM300-1 | HANDLE | STEEL | 1 |
| 5 | EK-132 | POST-Ø6x14 | STEEL | 1 |
| 6 | EK-131 | POST-Ø6x26 | STEEL | 1 |

PIVOT BOX

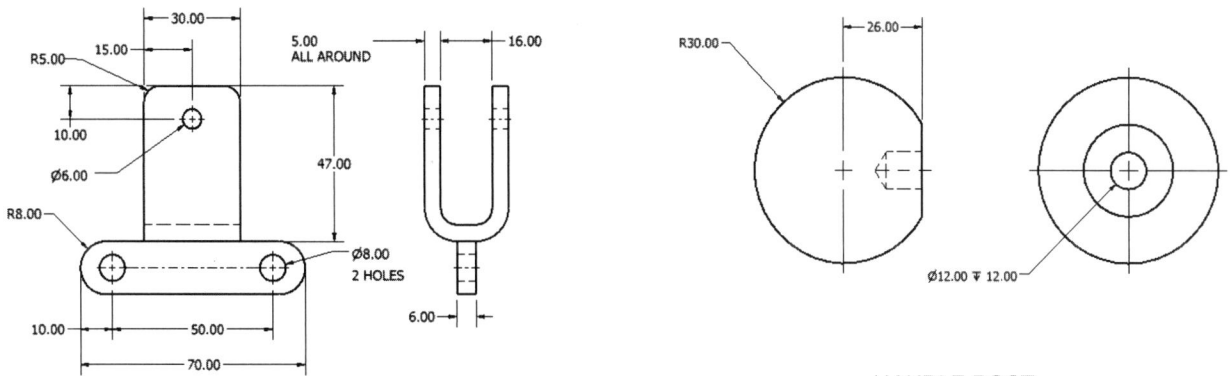

HANDLE POST

LINK

## EX7-8 INCHES

A. Draw an assembly drawing and parts list for the guide assembly.

ADJUSTABLE ASSEMBLY

| Parts List | | | | |
|---|---|---|---|---|
| ITEM | PART NUMBER | DESCRIPTION | MATERIAL | QTY |
| 1 | ENG-311 | BASE#4, CAST | Cast Iron | 1 |
| 2 | ENG-312 | SUPPORT, ROUNDED | SAE 1040 STEEL | 1 |
| 3 | ENG-404 | POST, ADJUSTABLE | Steel, Mild | 1 |
| 4 | BU-1964 | YOKE | Cast Iron | 1 |
| 5 | ANSI B18.8.2 1/4x1.3120 | Grooved pin, Type C - 1/4x1.312 ANSI B18.8.2 | Steel, Mild | 1 |
| 6 | ANSI B18.15 - 1/4 - 20. Shoulder Pattern Type 2 - Style A | Forged Eyebolt | Steel, Mild | 1 |
| 7 | ANSI B18.6.3 - 1/4 - 20 | Hex Machine Screw Nut | Steel, Mild | 1 |
| 8 | ANSI B18.2.2 - 3/8 - 16 | Hex Nut | Steel, Mild | 2 |

CAST BASE #4

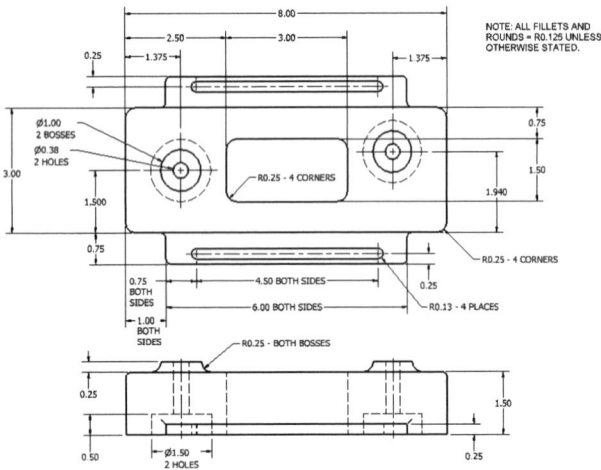

ROUNDED SUPPORT

ADJUSTABLE POST

YOKE

FORGED EYEBOLT

Thread = 1/4 = 20 UNC
0.03 X 45° Chamfer

## EX7-9

Redraw the given objects as an assembly drawing and add bolts with the appropriate nuts at the L and H holes. Add the appropriate drawing callouts. Specify standard bolt lengths.

A. Use inch values.
B. Use millimeter values.
C. Draw the front assembly view using a sectional view.
D. Draw the fasteners using schematic representations.
E. Draw the fasteners using simplified representations.
F. Prepare a parts list.
G. Prepare detail drawings for each part.

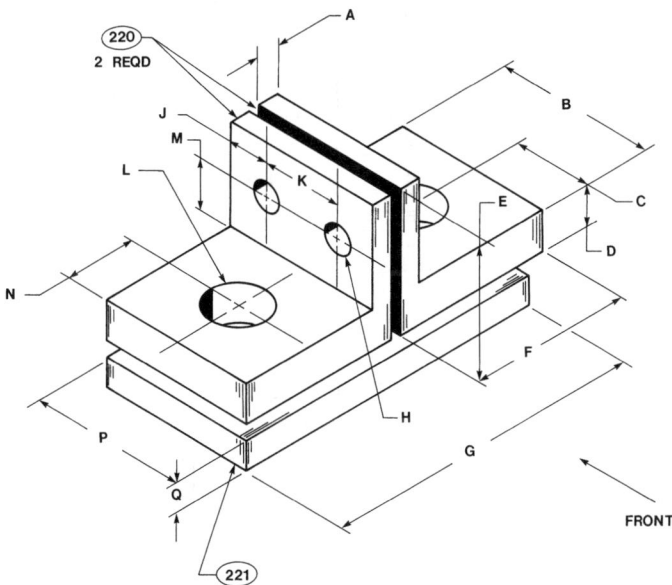

| DIMENSION | INCHES | mm |
|---|---|---|
| A | .25 | 6 |
| B | 2.00 | 50 |
| C | 1.00 | 25 |
| D | .50 | 13 |
| E | 1.75 | 45 |
| F | 2.00 | 50 |
| G | 4.00 | 100 |
| H | Ø.438 | Ø11 |
| J | .50 | 12.5 |
| K | 1.00 | 25 |
| L | Ø.781 | Ø19 |
| M | .63 | 16 |
| N | .88 | 22 |
| P | 2.00 | 50 |
| Q | .25 | 6 |

## EX7-10

Redraw the given views and add the appropriate hex head machine screws at M and N. Use standard length screws and allow at least two unused threads at the bottom of each threaded hole. Add a bolt with the appropriate nut at hole P.

A. Use inch values.
B. Use millimeter values.
C. Draw the front assembly view using a sectional view.
D. Draw the fasteners using schematic representations.
E. Draw the fasteners using simplified representations.
F. Prepare a parts list.
G. Prepare detail drawings for each part.

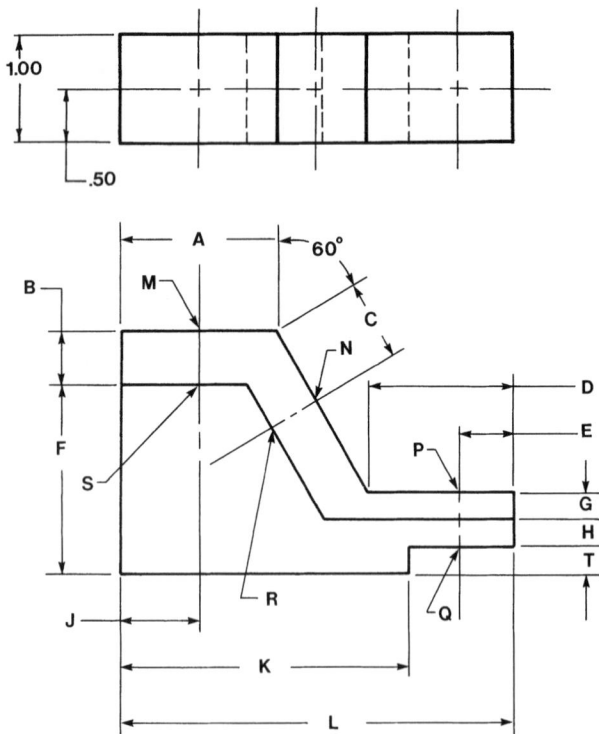

| DIMENSION | INCHES | mm |
|-----------|--------|-----|
| A | 1.50 | 38 |
| B | .50 | 13 |
| C | .75 | 19 |
| D | 1.38 | 35 |
| E | .50 | 13 |
| F | 1.75 | 44 |
| G | .25 | 6 |
| H | .25 | 6 |
| J | .75 | 19 |
| K | 2.75 | 70 |
| L | 3.75 | 96 |
| M | ø.31 | ø8 |
| N | ø.25 | ø6 |
| P | ø.41 | ø12 |
| Q | ø.41 | ø12 |
| R | .164-32 UNF X .50 DEEP | M4 X 14 DEEP |
| S | .250-20 UNC X 1.63 DEEP | M6 X 14 DEEP |
| T | .25 | 6 |

## EX7-11 MILLIMETERS

Draw an assembly drawing and parts list for the circular damper assembly shown.

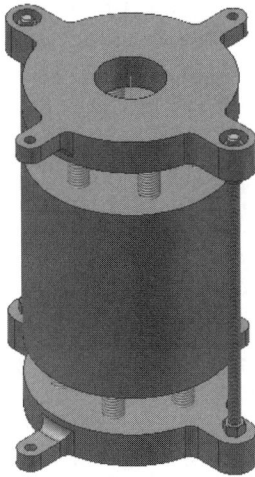

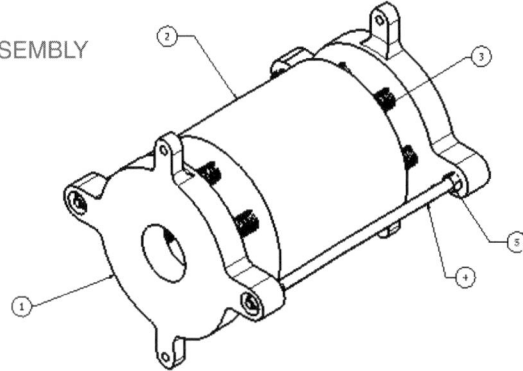

CIRCULAR DAMPER ASSEMBLY

| Parts List | | | | |
|---|---|---|---|---|
| ITEM | PART NUMBER | DESCRIPTION | MATERIAL | QTY |
| 1 | BU2008-1 | BASE, HOLDER | Steel | 2 |
| 3 | BU2008-2 | SPRING, COMPRESSION | Steel, Mild | 12 |
| 2 | BU2008-3 | COUNTERWEIGHT | Steel | 1 |
| 4 | AM312-12 | POST, THREADED | Steel | 2 |
| 5 | AS 1112 - M18  Type | HEX NUT | Steel, Mild | 8 |

Note: AM-312-12 Threaded Post is M18 x 560 long with 1 x 45° chamfers at each end

COUNTERWEIGHT

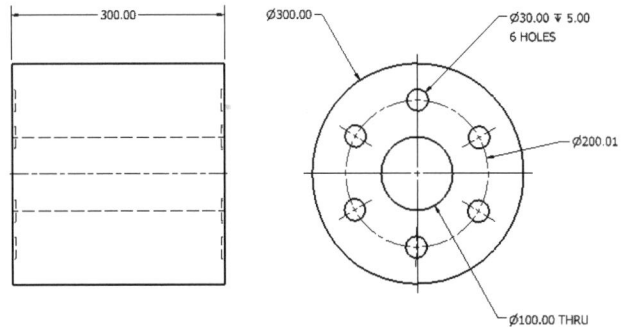

HOLDER BASE

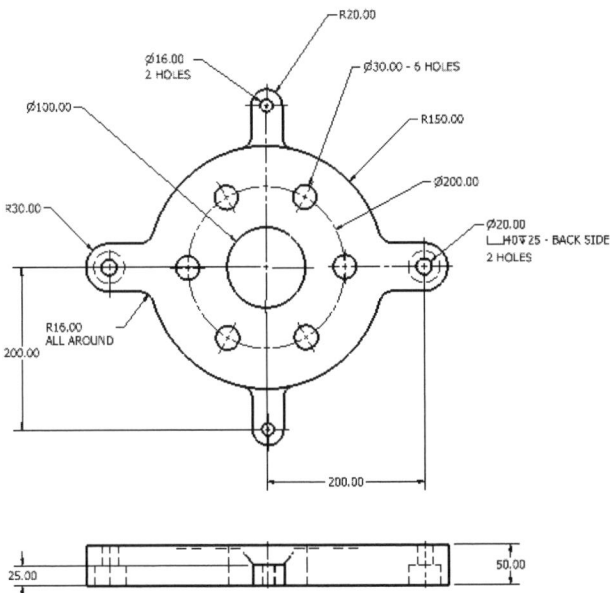

SPRING
Wire Ø = 3.0
Outside Ø = 28.0
Number of Coils = 16
Coil Direction = Right
Grind both ends

## EX7-12 INCHES

The given objects are to be assembled as shown. Select sizes for the parts that make the assembly possible. (Choose dimensions for the end blocks and then determine the screw and stud lengths.) The hex head screws (5) have a major diameter of either .375 inch or M10. The studs (3) are to have the same thread sizes as the screws and are to be screwed into the top part (2). The holes in the lower part (1) that accept the studs are to be clearance holes.

A. Draw an assembly drawing.
B. Draw detail drawings of each nonstandard part. Include positional tolerances for all holes.
C. Prepare a parts list.

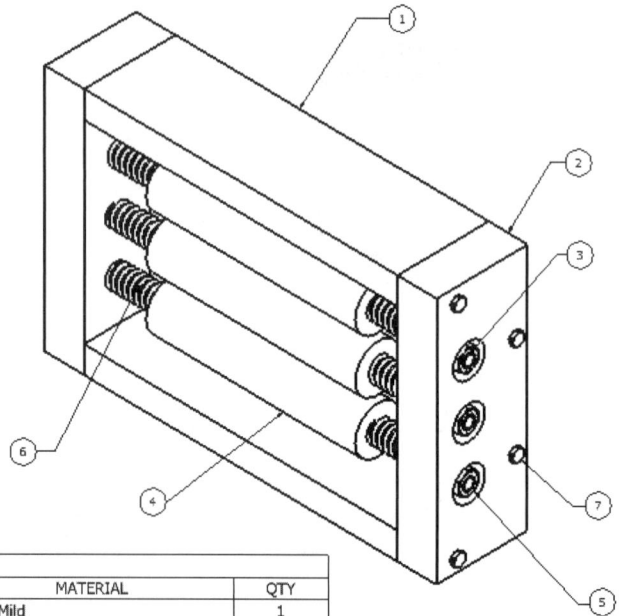

| Parts List | | | | |
|---|---|---|---|---|
| ITEM | PART NUMBER | DESCRIPTION | MATERIAL | QTY |
| 1 | AM311-1 | BASE | Steel, Mild | 1 |
| 2 | AM311-2 | PLATE,END | Steel, Mild | 2 |
| 3 | AM311-3 | POST, GUIDE | Steel, Mild | 3 |
| 4 | EK-152 | WEIGHT | Steel, Mild | 3 |
| 5 | AS 2465 - 1/2  UNC | HEX NUT | Steel, Mild | 6 |
| 6 | | COMPRESSION SPRING | Steel, Mild | 6 |
| 7 | AS 2465 - 1/4 x 2 1/2  UNC | HEX BOLT | Steel, Mild | 8 |

12.00

1.50

9.00

2.00

9.25

5.25

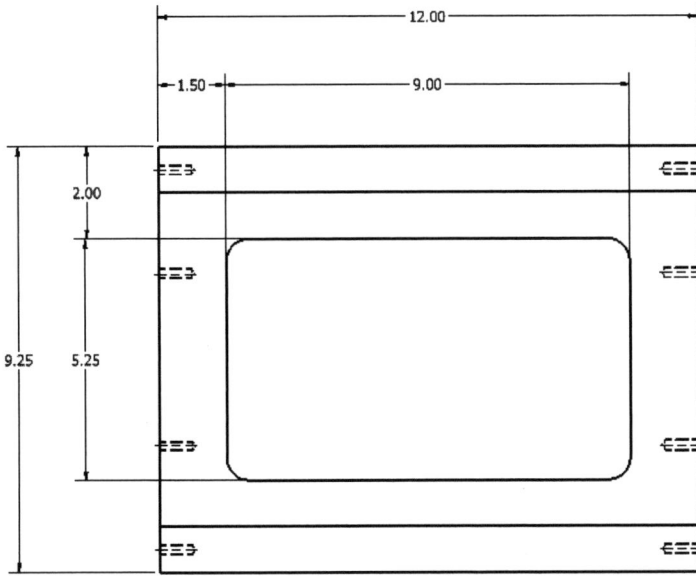

BASE
P/N AM311-1
MILD STEEL
1 REQD

0.75

1.00

0.50

2.75

1/4-20 UNC - 1B

2.75

1.00

0.50

0.75

2.50

WEIGHT
P/N EK-152
MILD STEEL
3 REQD

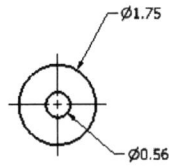

8.00

Ø1.75

Ø0.56

END

END PLATE
P/N AM311-2
MILD STEEL
2 REQD

Ø0.53 THRU - 3 HOLES

Ø1.25 ▼ 0.75
3 HOLES

Ø0.28 - 4 HOLES

GUIDE POST
P/N AM311-3
MILD STEEL
3 REQD

1/2-13 UNC - 1A
BOTH ENDS

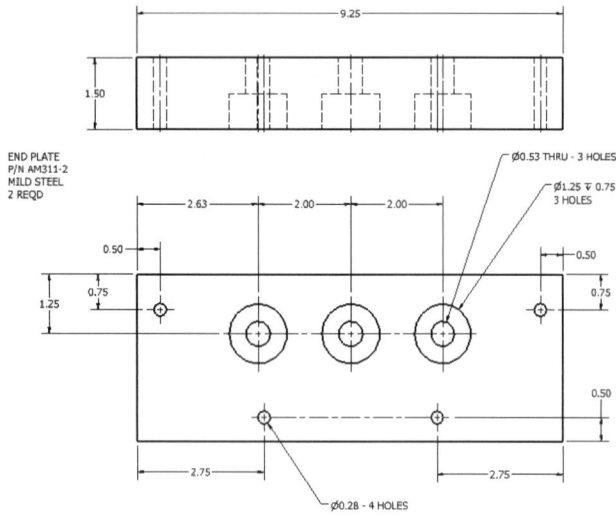

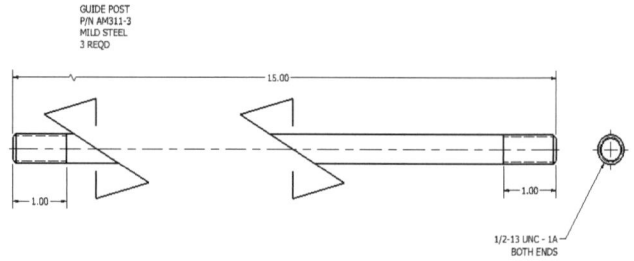

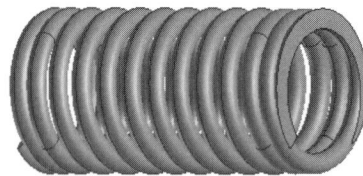

SPRING
Wire Ø = 0.125
Outside Ø = 1.000
Length = 2.00
Coil direction = light
Number of coils = 10
Grind both ends.

## EX7-13 INCHES

Draw an assembly drawing and parts list for the winding assembly.

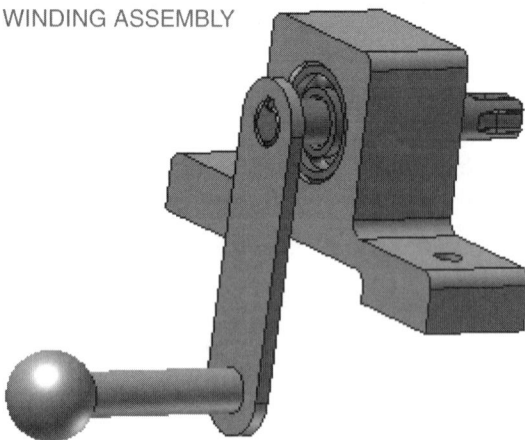

WINDING ASSEMBLY

| Parts List | | | | |
|---|---|---|---|---|
| ITEM | PART NUMBER | DESCRIPTION | MATERIAL | QTY |
| 1 | EK131-1 | SUPPORT | STEEL | 1 |
| 2 | EK131-2 | LINK | STEEL | 1 |
| 3 | EK131-3 | SHAFT,DRIVE | STEEL | 1 |
| 4 | EK131-4 | POST, THREADED | STEEL | 1 |
| 5 | EK131-5 | BALL | STEEL | 1 |
| 6 | BS 292 - BRM 3/4 | Deep Groove Ball Bearings | STEEL,MILD | 1 |
| 7 | 3/16x1/8x1/4 | RECTANGULAR KEY | STEEL | 1 |

SUPPORT

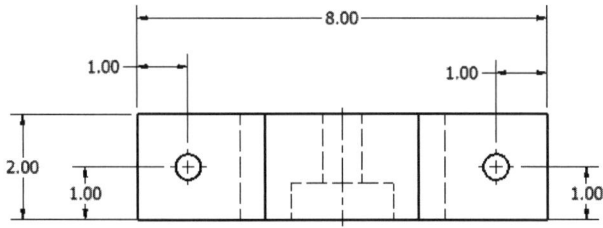

NOTE: ALL FILLETS AND ROUNDS R=0.250
UNLESS OTHERWISE STATED.

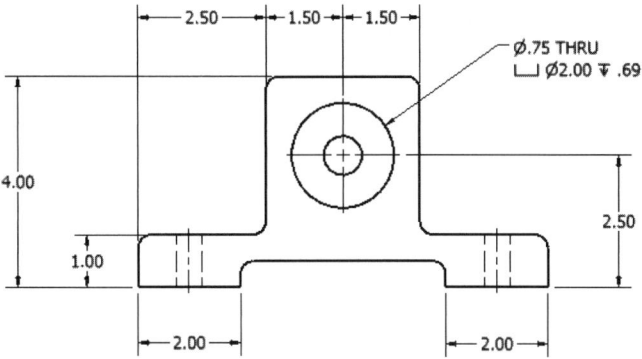

DRIVE SHAFT

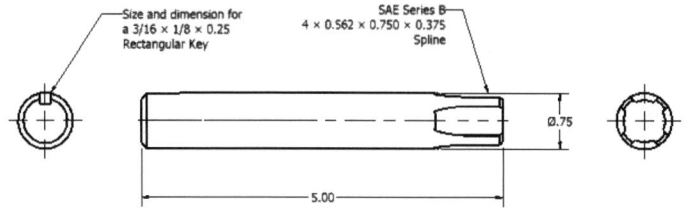

THREADED POST

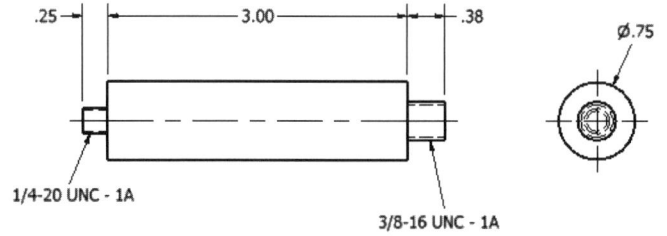

LINK

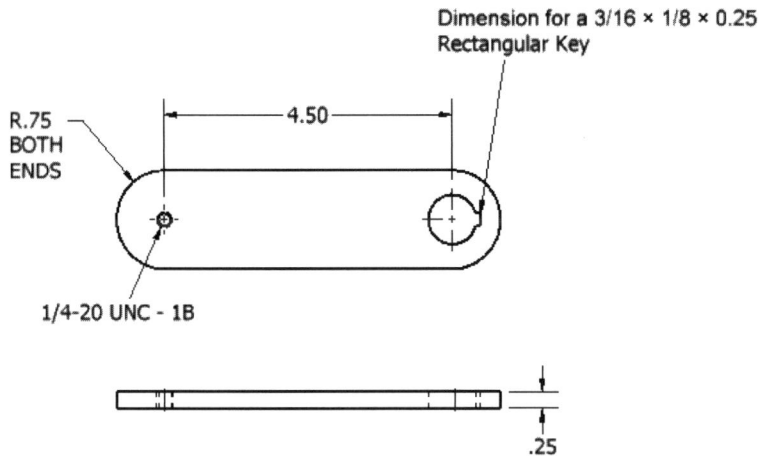

BALL

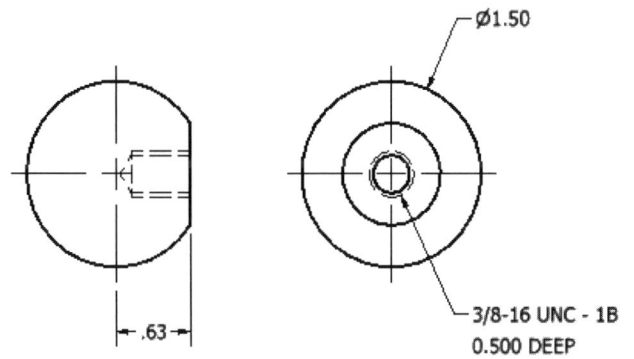

## EX7-14 INCHES

Design a hand-operated grinding wheel specifically for sharpening a chisel. The chisel is to be located on an adjustable rest while it is being sharpened. The mechanism should be able to be clamped to a table during operation using two thumbscrews.

A standard grinding wheel is Ø6.00 inch and 1/2 inch thick, and has an internal mounting hole with a 50.00 ±.03 bore.

### Prepare the following drawings.

A. Draw an assembly drawing.
B. Draw detail drawings of each nonstandard part. Include positional tolerances for all holes.
C. Prepare a parts list.

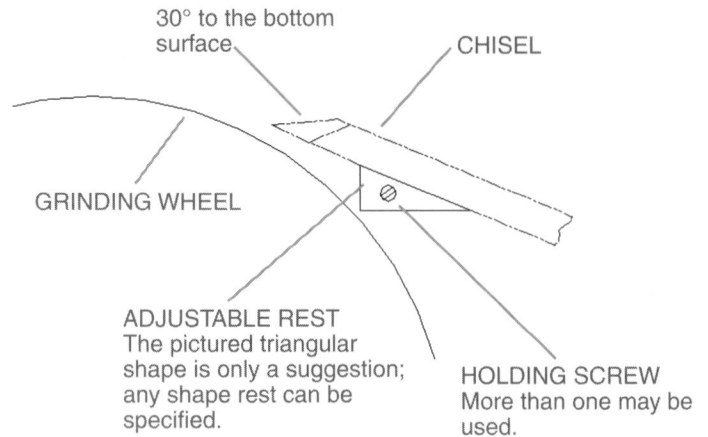

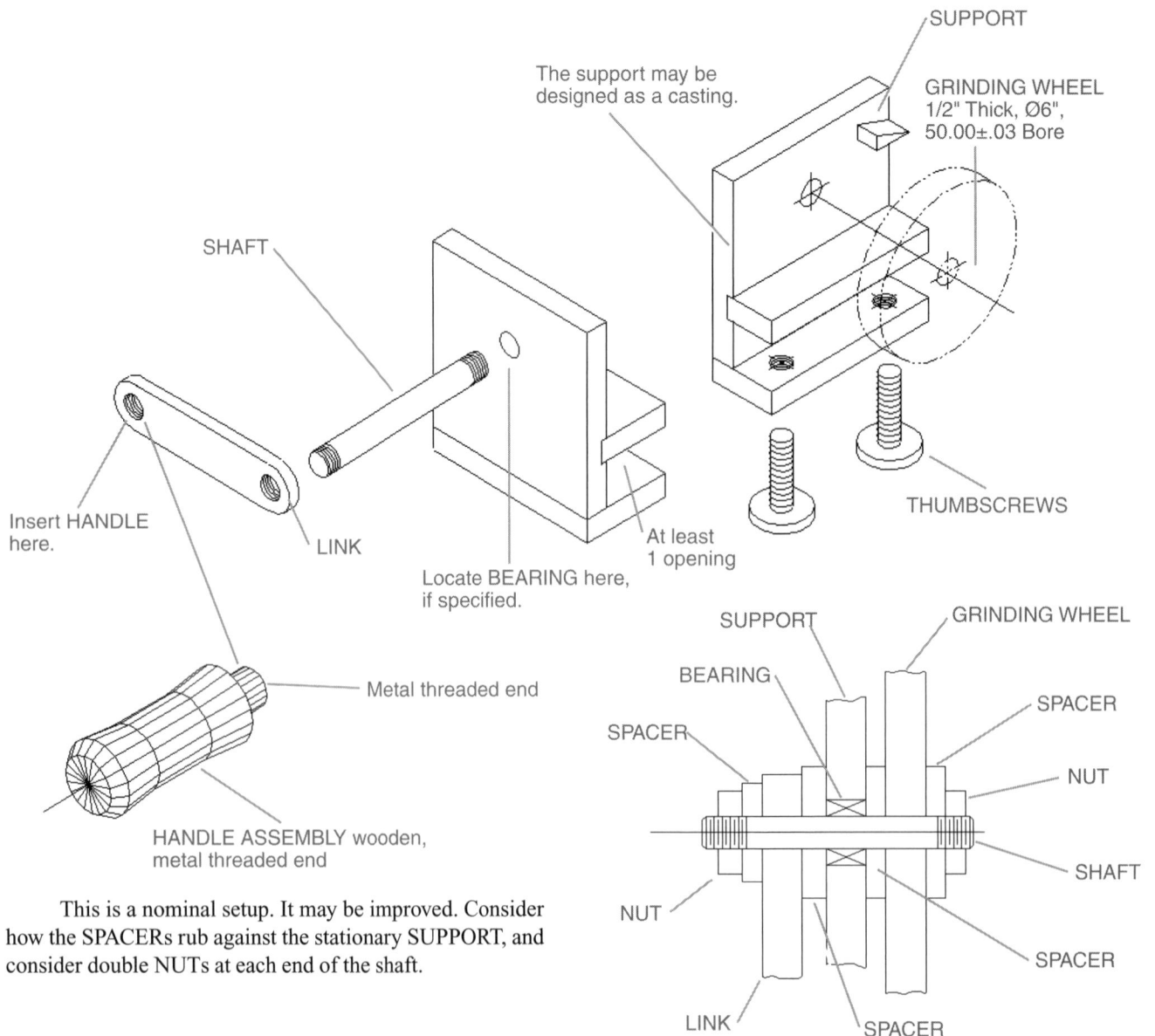

30° to the bottom surface

CHISEL

GRINDING WHEEL

ADJUSTABLE REST
The pictured triangular shape is only a suggestion; any shape rest can be specified.

HOLDING SCREW
More than one may be used.

SUPPORT

The support may be designed as a casting.

GRINDING WHEEL
1/2" Thick, Ø6", 50.00±.03 Bore

SHAFT

THUMBSCREWS

Insert HANDLE here.

LINK

Locate BEARING here, if specified.

At least 1 opening

Metal threaded end

HANDLE ASSEMBLY wooden, metal threaded end

SUPPORT    GRINDING WHEEL

BEARING    SPACER

SPACER    NUT

SHAFT

NUT    SPACER

LINK    SPACER

This is a nominal setup. It may be improved. Consider how the SPACERs rub against the stationary SUPPORT, and consider double NUTs at each end of the shaft.

## EX7-15 INCHES

A. Draw an assembly drawing of the Rocker Assembly.
B. Prepare a parts list.
C. Select appropriate fasteners to hold the assembly together.

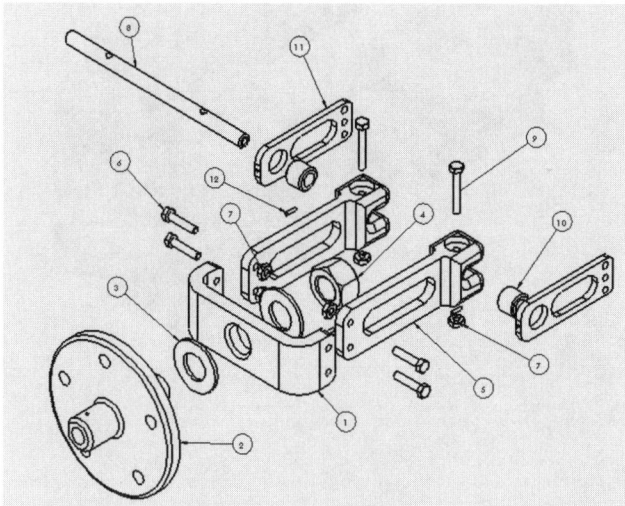

| ITEM NO. | PART NUMBER | DESCRIPTION | QTY. |
|---|---|---|---|
| 1 | ME 311-1 | WHEEL BRACKET | 1 |
| 2 | ME 311-2 | WHEEL SUPPORT | 1 |
| 3 | | 1.00 × 1.75 × .06 PLAIN WASHER | 2 |
| 4 | | 1 × 8 UNC HEX NUT | 1 |
| 5 | ME 311-3 | SUPPORT ARM | 2 |
| 6 | | 1/4 - 28 UNF × 1.25 HEX HEAD | 4 |
| 7 | | 1/4 - 28 UNF × 1.75 HEX HEAD | 2 |
| 8 | | 1/4 - 28 UNF HEX NUT | 6 |
| 9 | ME 311-4 | PIVOT SHAFT | 1 |
| 10 | | .500 × .875 .750 BEARING | 2 |
| 11 | ME 311-5 | STATIONARY ARM | 2 |
| 12 | | #6-32 × .560 UNC SET SCREW CONE POINT | 2 |

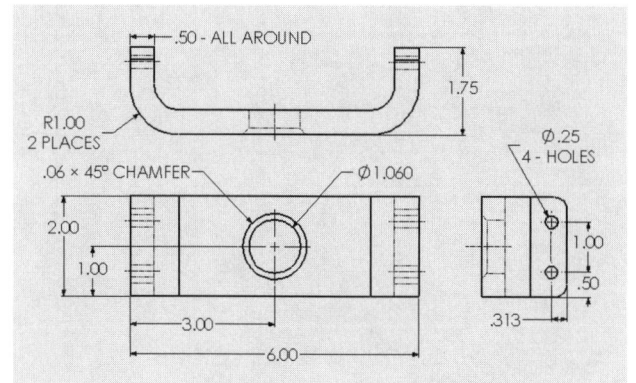

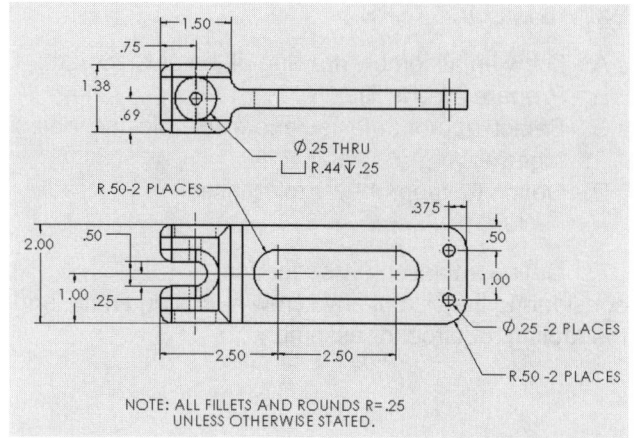

NOTE: ALL FILLETS AND ROUNDS R=.25 UNLESS OTHERWISE STATED.

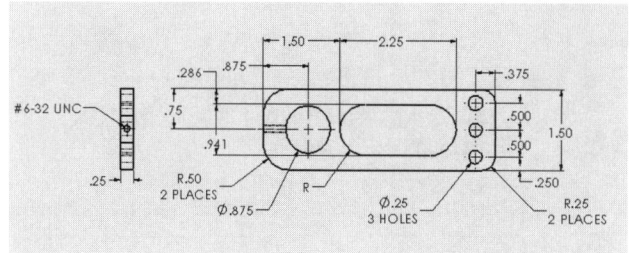

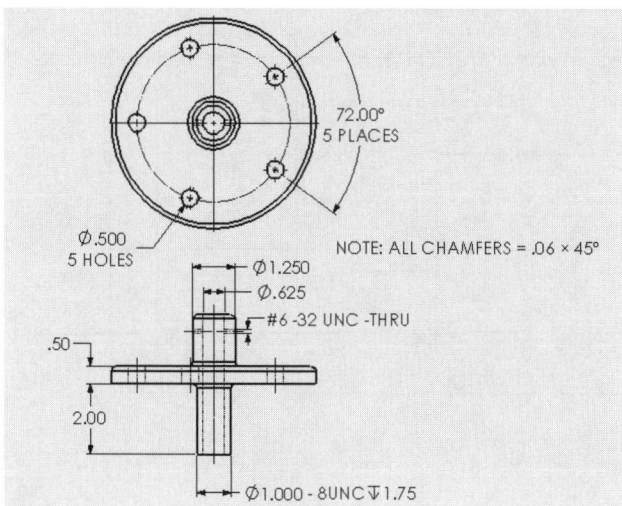

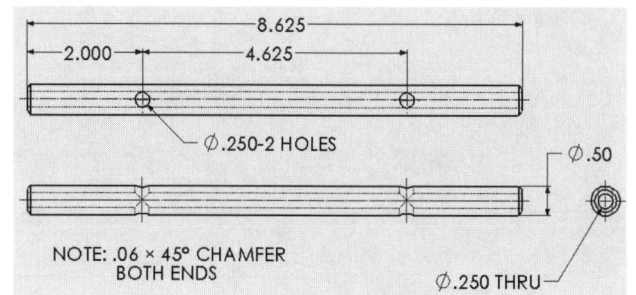

**EX7-16 MILLIMETERS**

A. Draw an assembly drawing of the given object.
B. Prepare a parts list.
C. Select appropriate fasteners to hold the object together.
D. Define the appropriate tolerances.

E. Define an assembly sequence considering the size of any screw heads and nuts, and the tooling required for assembly.

BRACKET ASSEMBLY

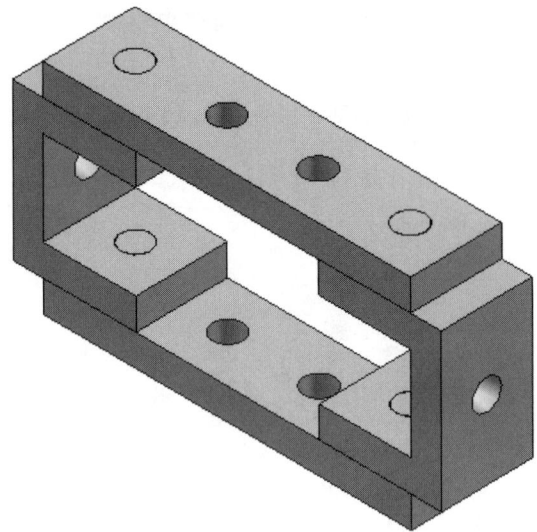

C-BRACKET, SAE1020 STEEL, 2 REQD
Part Number: AM311-1

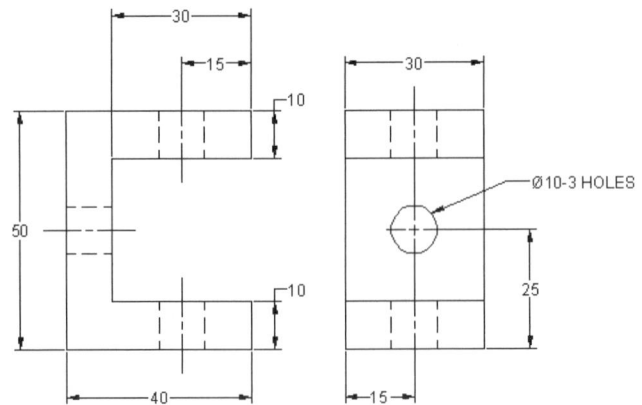

QUAD SPACER, SAE 1020 STEEL
2 REQD, Part Number: AM311-2

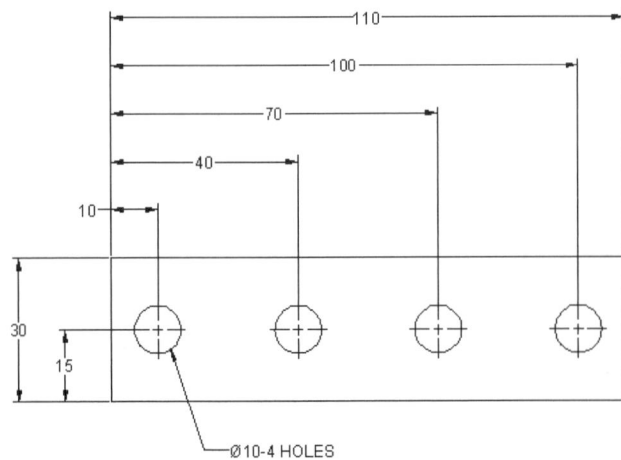

## EX7-17 MILLIMETERS

A. Draw an assembly drawing of the given object.
B. Prepare a parts list.
C. Select appropriate fasteners to hold the object together.
D. Define the appropriate tolerances.
E. Define an assembly sequence considering the size of any screw heads and nuts, and the tooling required for assembly.

CLIP ASSEMBLY

SUPPORT PLATE
SAE 1040 STEEL, 2 REQD
Part Number: AM312-3

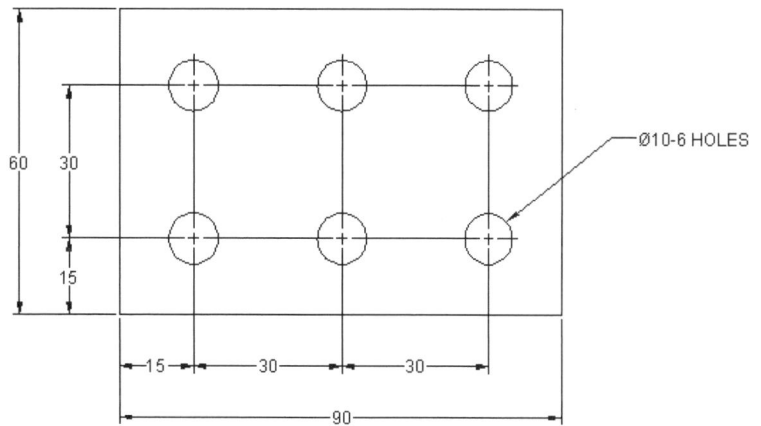

Ø10-6 HOLES

L-CLIP
SAE 1040 STEEL, 2 REQD
Part Number: AM312-4

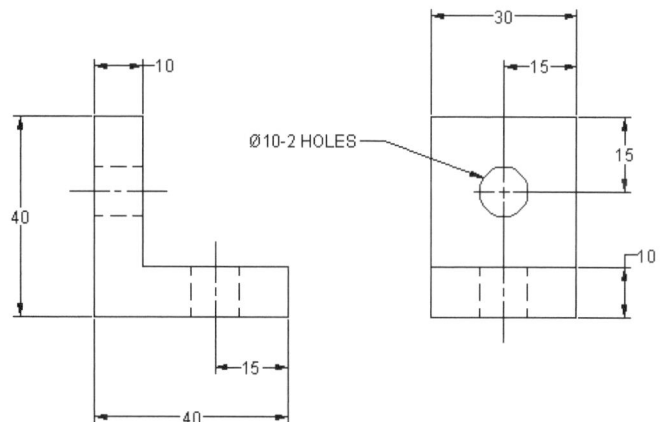

Ø10-2 HOLES

## EX7-18 MILLIMETERS

A. Draw an assembly drawing of the given object.
B. Prepare a parts list.
C. Select appropriate fasteners to hold the object together.
D. Define the appropriate tolerances.
E. Define an assembly sequence considering the size of any screw heads and nuts, and the tooling required for assembly.

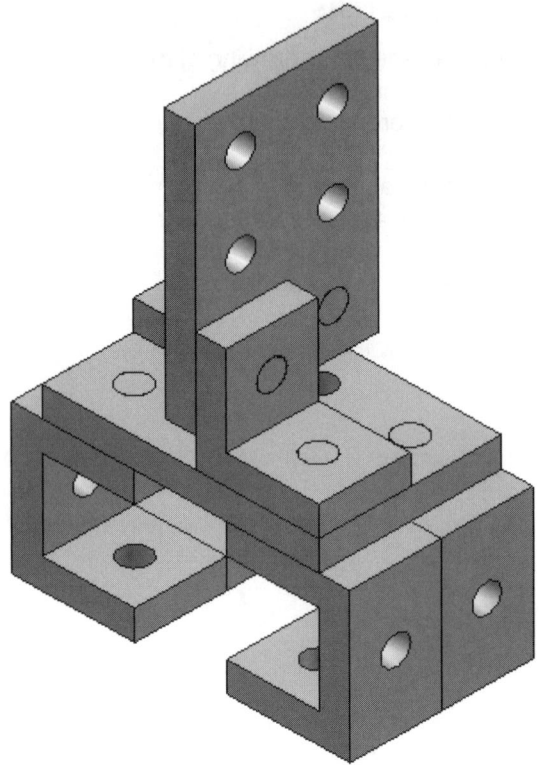

C-BRACKET
SAE 1020 STEEL, 4 REQD
Part Number: AM311-1

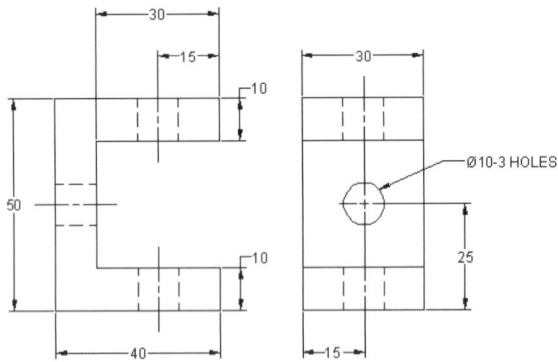

Ø10-3 HOLES

GUIDE ASSEMBLY

L-CLIP
SAE 1040 STEEL, 2 REQD
Part Number: AM312-4

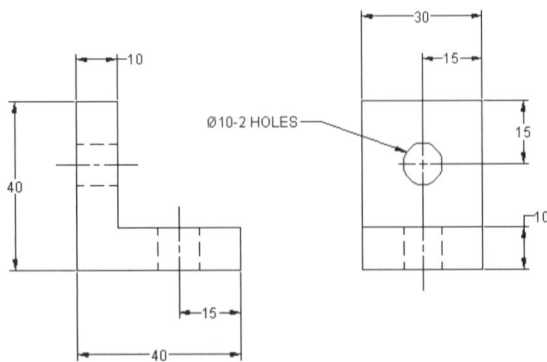

Ø10-2 HOLES

SUPPORT PLATE
SAE 1040 STEEL, 2 REQD
Part Number: AM312-3

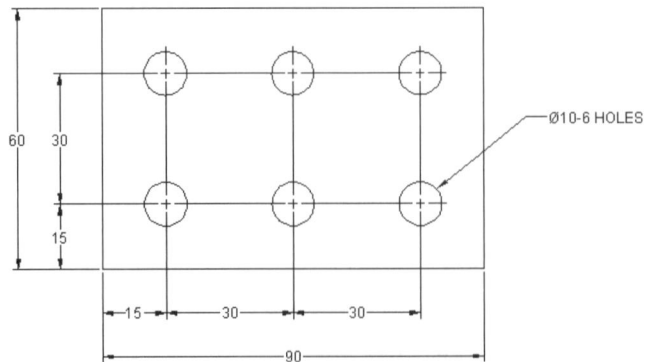

Ø10-6 HOLES

## EX7-19 MILLIMETERS

A. Draw an assembly drawing of the given object.
B. Prepare a parts list.
C. Select appropriate fasteners to hold the object together. Note that the fastener used to join the center link and the drive link passes over the web plate as the drive link rotates.
D. Define the appropriate tolerances.
E. Use phantom lines and define the motion of the center link and the rocker link if the drive link rotates 360°.

ROCKER ASSEMBLY

DRIVE LINK
Part Number:
AM311-22A

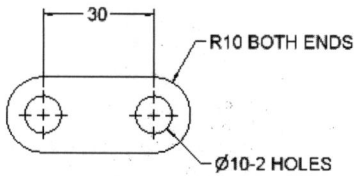

SAE 1040 STEEL
5mm THK

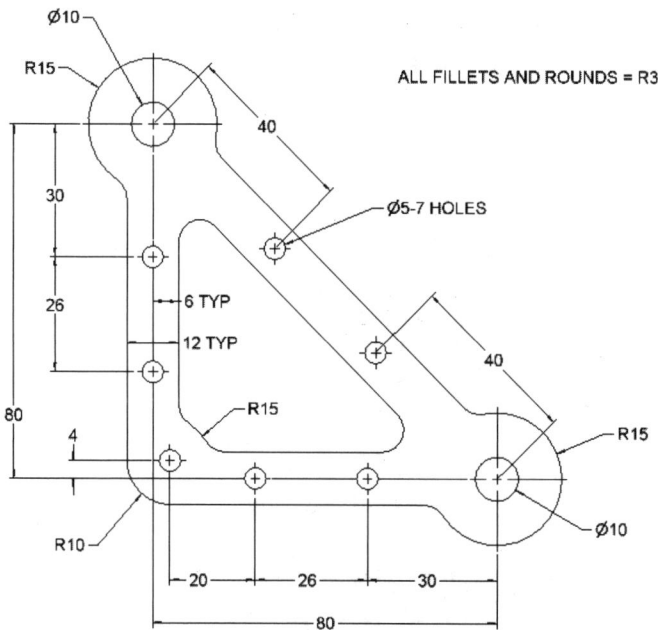

R10 BOTH ENDS
30
Ø10-2 HOLES

ALL FILLETS AND ROUNDS = R3

WEB PLATE
Part Number: AM311-22B
SAE 1040 STEEL
10mm THK

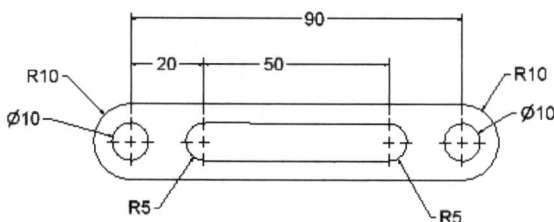

ROCKER LINK
Part Number: AM311-2C
SAE 1040 STEEL
5mm THK

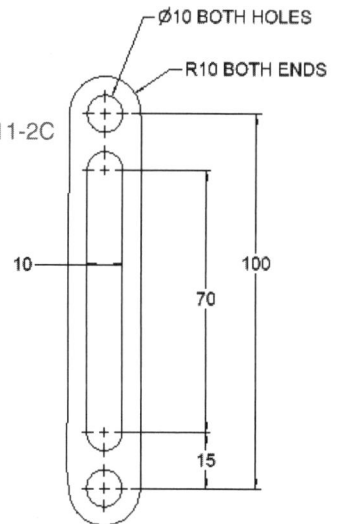

Ø10 BOTH HOLES
R10 BOTH ENDS

CENTER LINK
Part Number: AM311-22D
SAE 1040 STEEL
5mm THK

## EX7-20 MILLIMETERS

A. Draw an assembly drawing of the given object.
B. Prepare a parts list.
C. Select appropriate fasteners to hold the object together. Note that the fasteners used to join the side links to the holder arm pass over the holder base.
D. Define the appropriate tolerances. Use an H7/p6 tolerance between bushing-A and the holder base.
E. Use phantom lines to define the motion of the holder arm, side links, and cross link if the holder arm rotates between +30° and −30°.

LINK ASSEMBLY

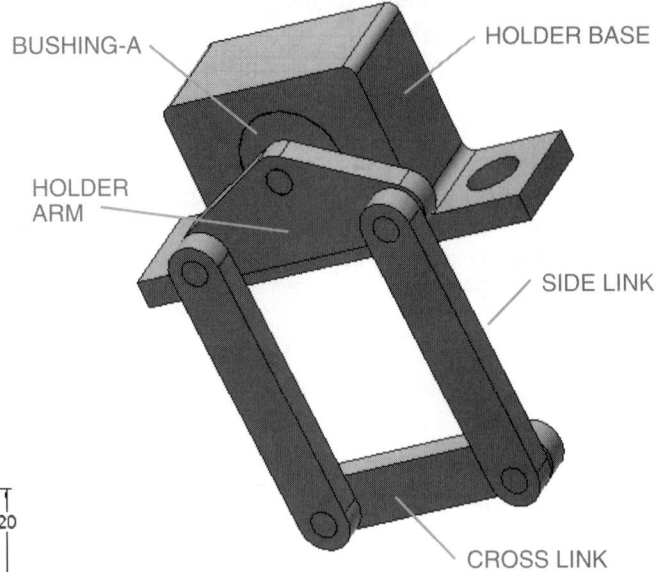

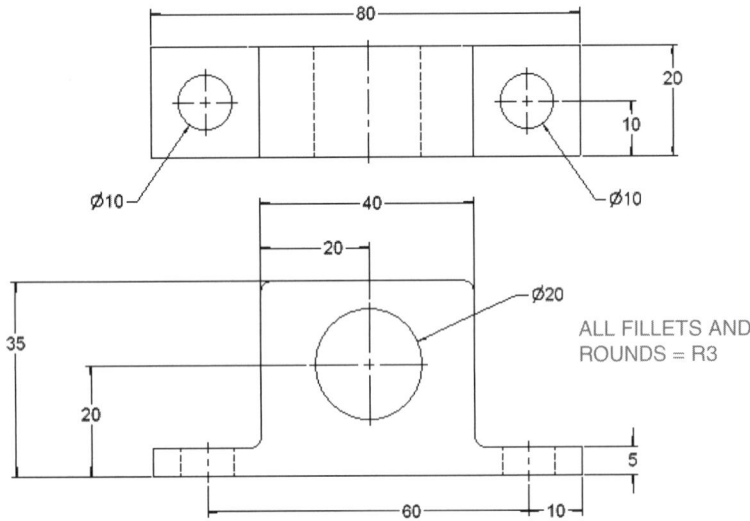

BUSHING-A

HOLDER BASE

HOLDER ARM

SIDE LINK

CROSS LINK

ALL FILLETS AND ROUNDS = R3

HOLDER ARM, BU100-2, 7075-T6, AL, 5mm THK

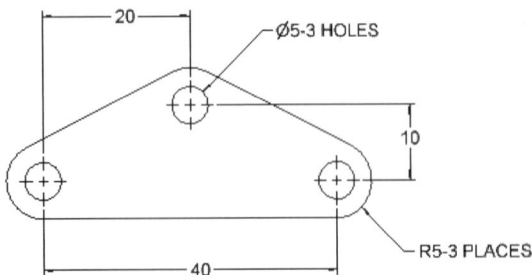

SIDE LINK
BU100-4
7075-T6 AL
5mm THK
2 REQD

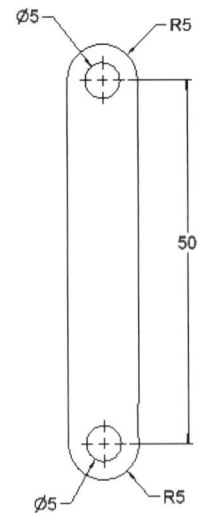

CROSS LINK, BU100-3, 7075-T6, AL, 5mm THK

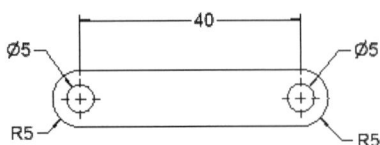

BUSHING-A
A/M CORP - B20-AD
TEFLON

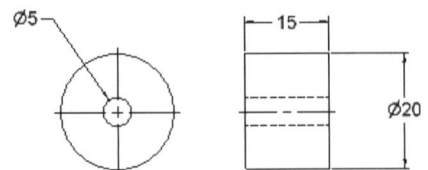

## EX7-21 MILLIMETERS

A. Draw an assembly drawing of the minivise.
B. Prepare a parts list.
C. Select the appropriate fasteners to hold the vise together.

D. Redesign the interface between the drive screw and the holder plate.

MINIVISE

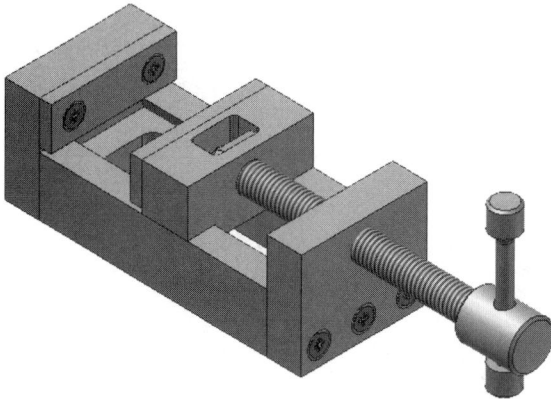

MINIVISE

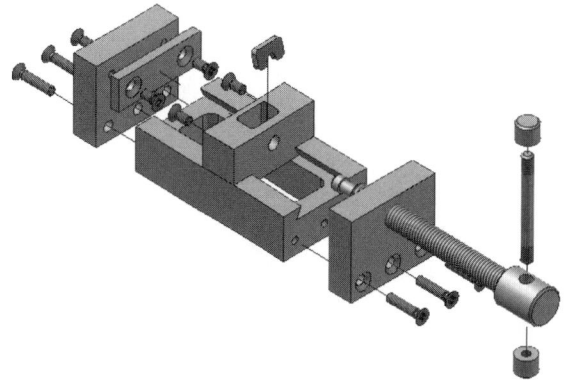

1. BASE
   SAE 1040 STEEL

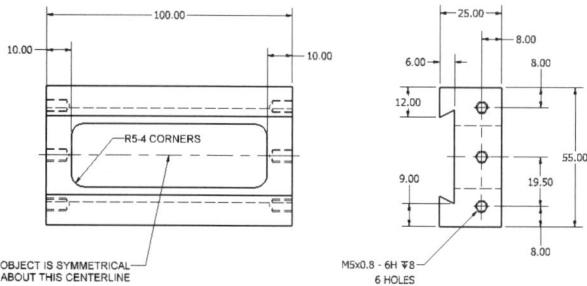

2. DRIVE SCREW
   SAE 1040 STEEL

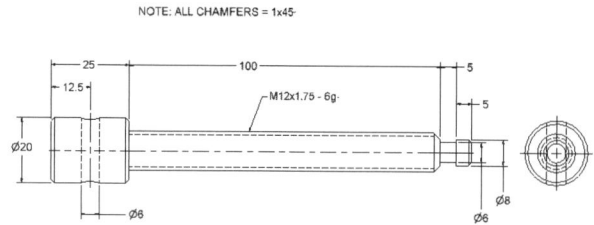

3. END PLATE
   SAE 1040 STEEL

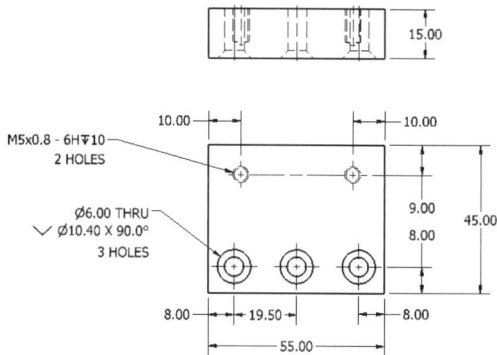

4. SLIDER
   SAE 1040 STEEL

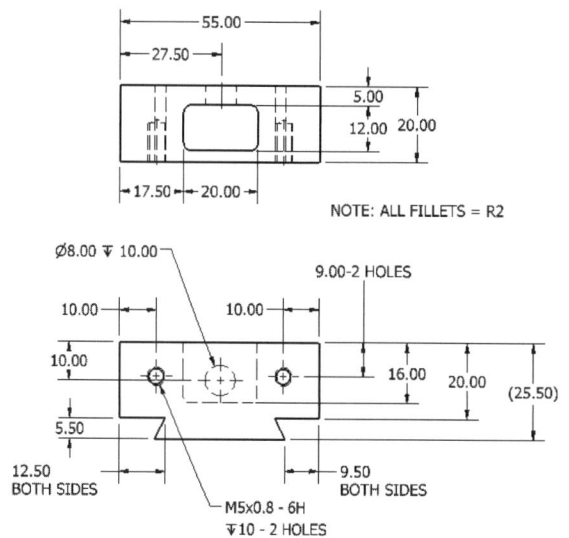

5. HOLDER PLATE
   SAE 1040 STEEL

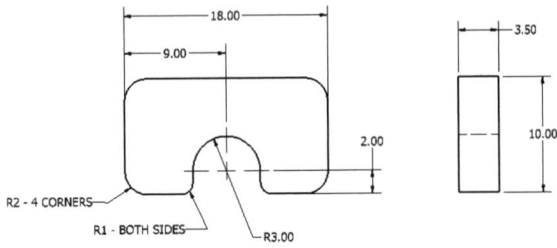

18.00
9.00
3.50
2.00
10.00
R2 - 4 CORNERS
R1 - BOTH SIDES
R3.00

6. FACE PLATE
   SAE 1040 STEEL

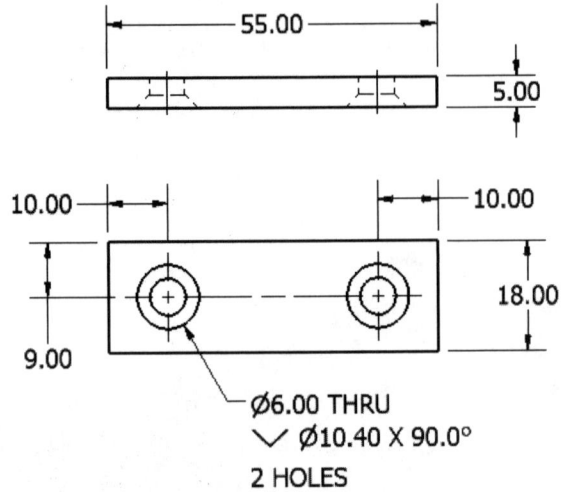

55.00
5.00
10.00
10.00
18.00
9.00
Ø6.00 THRU
⌵ Ø10.40 X 90.0°
2 HOLES

7. HANDLE
   SAE 1040 STEEL

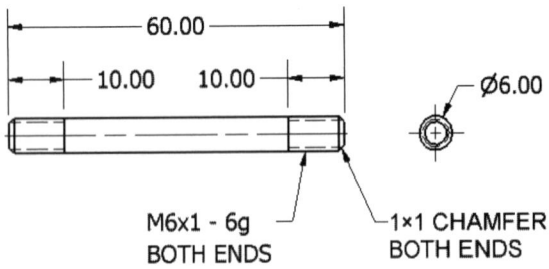

60.00
10.00
10.00
Ø6.00
M6x1 - 6g
BOTH ENDS
1×1 CHAMFER
BOTH ENDS

8. END CAP
   SAE 1040 STEEL

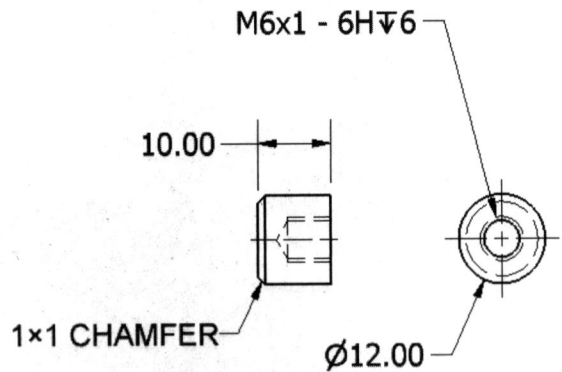

M6x1 - 6H⍌6
10.00
1×1 CHAMFER
Ø12.00

9. DRIVE PLATE
   SAE 1040 STEEL

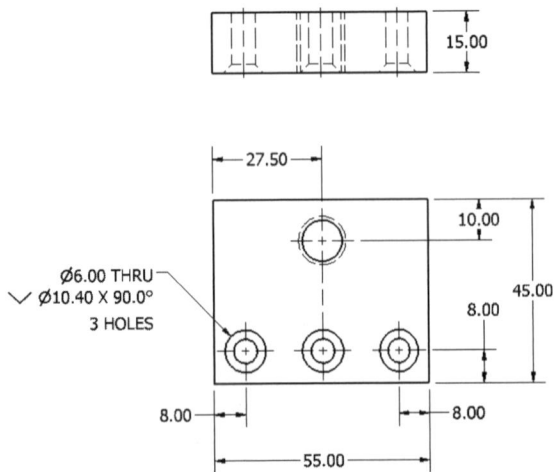

15.00
27.50
10.00
Ø6.00 THRU
⌵ Ø10.40 X 90.0°
3 HOLES
45.00
8.00
8.00
8.00
8.00
55.00

10. M5×10 RECESSED COUNTERSUNK
    HEAD-STEEL
11. M5×22 RECESSED COUNTERSUNK
    HEAD-STEEL

Upper level assembly drawing for a four wheel drive dynamometer (Inset shows 3D model). *Courtesy of Dynojet Research, Inc.*

# 7-17 WORKING DRAWINGS

Design professionals such as engineers and architects are creators or builders. They use graphics as a means to create, record, analyze, and communicate their design concepts or ideas so that they can be realized or made into real products or structures. The ability to communicate verbally, symbolically, and graphically is essential to building the teams necessary to create large scale projects.

Designs progress through five stages. Different types of drawings are required at each stage of the process. Early in the process, ideation sketches communicate and refine concepts for the project. Later, detailed layouts, analysis, and part drawings are created using 2D CAD or solid modeling techniques. Assembly drawings created in 3D CAD or using 2D methods show how multiple parts fit together. They describe the end result—how the individual pieces that must fit together to work.

Releasing and revising drawings is an important part of the design process. Revisions must be tracked, identified, logged, and saved for future reference. Understanding and using effective methods to manage paper and electronic documents is crucial to retain important information and prevent costly and even dangerous mistakes.

# 7-18 UNDERSTANDING DESIGN, DOCUMENTATION, AND WORKING DRAWINGS

## Top Down vs. Bottom Up Design

Methods of accomplishing the design process are sometimes described in terms of three general categories: top down, bottom up, and middle out.

*Top down* refers to starting the process of designing a product or system by considering the function of the entire system, then breaking that down into subassemblies or component

Check these Web sites for stock component CAD models to add to assemblies:

- http://www.carrlane.com
- http://parts.web2cad.de
- http://www.uspto.gov/web/offices/com/iip/pdf/brochure_05.pdf
- http://www.mycadmash.com/mash.asp lists lots of handy CAD sites

groups based on their major functions. Finally, each part that must be manufactured and assembled to create the design is defined. Layout drawings are often used to facilitate top down design by accurately showing the relationships between major functional items and how those may fit with existing equipment.

*Bottom up* refers to a design process starting at the part level. Individual components are sized and designed, then the final assembly is built around the design of the parts. This approach is helpful when the components are standardized parts.

*Middle out* refers to a combination of top down and bottom up design methods, where some parts are standardized and others are designed within the context of fitting into the design of the assembly.

You may prepare detail drawings first or assembly drawings first, depending on your process. Using solid modeling you may create a 2D or 3D layout first and then develop the models for the individual parts. Or you may create part models first and then assemble them together. Some companies use a fully digital documentation process and some create 2D detail drawings and assembly drawings that show how the parts fit together.

### Constraining 3D Assembly Models

With **constraint based modeling** software, you use assembly constraints to create relationships between modeled parts. The first part added to the assembly becomes the parent part. Other parts are mated to this **parent part** to build up the assembly. **Mating parts** have features that should fit together. Assembly constraints available in the 3D modeling software let you align mating parts. For example, in the 3D CAD assembly model shown in Figure 7-38, the parts are aligned with one another using assembly constraints, similar to the way you would build the actual device.

If you want two holes to line up, you can use an assembly constraint to align them. If a part changes, it will still be oriented in the assembly so that the holes align. Each software package will offer a similar set of constraint options, so you should become familiar with those available to you. Table 7-1 lists some common assembly constraints and their definitions. Assembly relationships can make your assembly model work for you. As you add parts, use constraints to orient the new part using relationships that will persist in the assembly.

A subordinate assembly, usually called a **subassembly**, is a group of components of a larger machine. Breaking products into subassemblies often makes it easier to coordinate when different designers are working on portions of the same device. Even if you are not using 3D modeling, structuring your drawings into subassemblies provides benefits such as making it easy to reuse subassemblies and track parts. The top down design process focuses on defining all of the subassembly requirements and how those interact in the assembly.

You can create a subassembly in much the same way you create an assembly: by making an assembly of the subassembly components. This subassembly can be added to the main assembly in the same way you add a part.

Organizing the model so that it comes together as it will on the assembly line can be useful in visualizing assembly difficulties. If a group of components are likely to be changed or replaced, linking all the subparts to a main component can make it easy to substitute an alternative design for that group of parts. Planning ahead is essential to creating assemblies efficiently and getting the most out of them.

### 3D Layouts and Skeleton Assemblies

Another method of creating an assembly using 3D CAD software is to start with an assembly framework in the form of an **assembly layout** or **skeleton** that can be used to define the locations of individual parts in the assembly. Using this method, parts are designed so they link to a skeleton framework in the assembly. A skeleton is a 3D drawing that defines major relationships in the assembly using lines, arcs, curves, and points, as shown in Figure 7-39. By creating the framework for each part up front, all parts do not have to be finished before they can be assembled. Parts can be assembled onto the skeleton at any stage of completion. By allowing the assembly to evolve as the parts are designed and refined, each designer can see the parts the others are creating—or at least the critical relationships between parts—by looking at the assembly.

### Working Drawings or Construction Drawings

The term **working drawings** describes a set of **assembly drawings** and detail drawings. A set of civil drawings with site, grading plans, and the many structural details for building a dam or bridge is an example of a set of working drawings.

Architectural drawings are another type of working (or construction) drawings (Figure 7-40). They are given to the contractor to show how to construct the building envisioned by the architect. Working drawings for machines include assembly drawings showing how parts fit together and detail drawings showing how to manufacture the parts. Weldments are a type of assembly drawing showing the welds that must be used to form an assembly from separate pieces of metal.

Drawings, models, and supporting documentation are the specifications for design manufacture. They are given to contractors to perform the work or manufacture individual parts, so they must represent the design accurately. The drawing is a legal document describing what work is to be performed or what parts are to be produced.

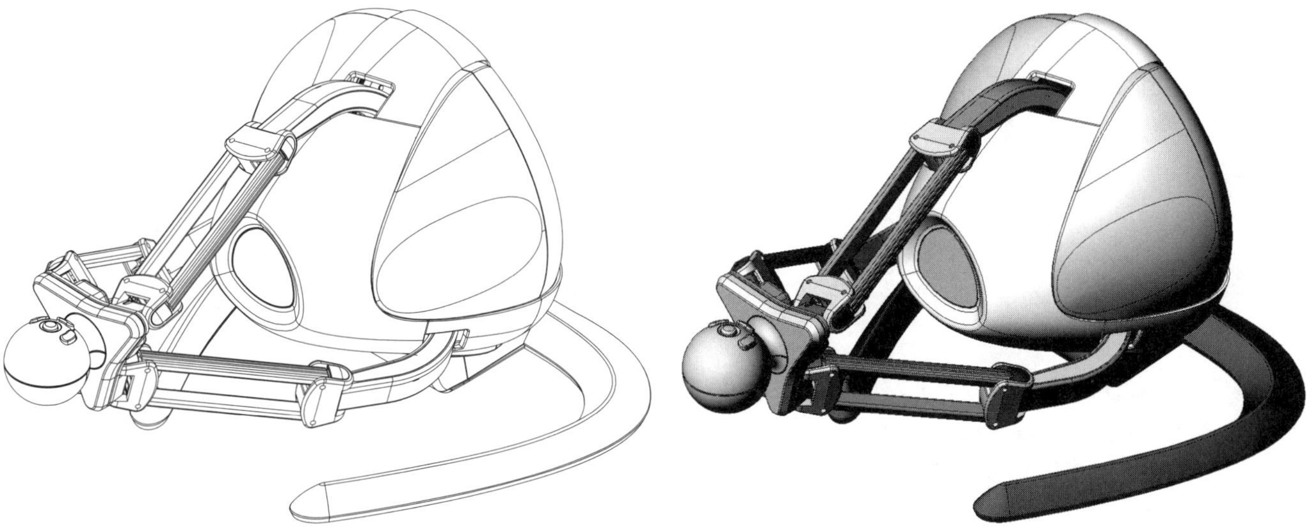

**Figure 7-38**   3D CAD Assembly Model for Lunar Design's Award Winning Design for a 3D Interactive Touch Device for the Home PC. The Novint Falcon lets users feel weight, shape, texture, dimension, and force effects when playing touch-enabled PC games. It accommodates a variety of controller grips, called end effectors, which allow users to more accurately engage with the experience of the game they are playing, such as feeling their stroke when they hit the golf ball on screen, or gaging how much edge they get on their snowboards as they fly down the slopes. To bring this sophisticated level of 3D touch technology device to market, Lunar Design worked closely with Novint and the commercial haptic developer Force Dimension. *Courtesy of Lunar.*

A careful process of checking and approving drawings and models helps prevent errors. Take preparing or approving drawings as a serious responsibility. Overlooking what may seem to be small or insignificant details may result in large amounts of wasted money or, worse yet, a person's injury or death.

## Assembly Drawings

An assembly drawing shows the assembled machine or structure, with all detail parts in their functional positions or as an **exploded view** where you can relate the parts to their functional positions.

There are different types of assembly drawings:

1. Design assemblies, or layouts.
2. General assemblies.
3. Detail assemblies.
4. Working drawing assemblies.
5. Outline or installation assemblies.
6. Inseparable assemblies (as in weldments, and others).

Assembly drawings are often generated from 3D CAD models. For example, the assembly drawing for the air brake in Figure 7-41 was generated from the 3D CAD model of the air brake shown in shaded view in Figure 7-42.

*Views*   Keep the purpose in mind when you select the views for an assembly drawing. The assembly drawing shows how the parts fit together and suggests the function of the entire unit. A complete set of orthographic views is not required. Often a single orthographic view will show all of the information needed when assembling the parts. The assembly drawing does not need to show how to make the parts, just how to put them together. The assembly worker receives the actual finished parts. The information for each individual part is shown on its detail drawing.

*Hidden Lines in Assembly Drawings*   Typically, hidden lines are not needed on assembly drawings. Keep in mind that the assembly drawing is often used by the worker who is putting the parts together. It needs to be easy to read and show the relationships between parts clearly. Hidden lines can make the drawing difficult to read, so use section or exploded views to show the interior parts in the assembly drawing.

**Table 7-1** Assembly Constraints for 3D Models.

| Name | Definition | Illustration |
|---|---|---|
| Mate | Mates two planar surfaces together. | |
| Mate offset | Mates two surfaces together so they have an offset between them. | |
| Align | Aligns two surfaces, datum points, vertices, or curve ends to be coplanar; also aligns revolved surfaces or axes to be coaxial. | |
| Align offset | Aligns two planar surfaces with an offset between them. | |
| Insert | Inserts a "male" revolved surface into a "female" revolved surface, aligning the axes. | |
| Orient | Orients two planar surfaces to be parallel and facing in the same direction. | |
| Coordinate system | Places a component into an assembly by aligning its coordinate system with an assembly coordinate system. | |

*Dimensions in Assembly Drawings*   Assembly drawings are not usually dimensioned except to show the relative positions of one feature to the next when that distance must be maintained at the time of the assembly, such as the maximum height of a jack, or the maximum opening between the jaws of a vise. When machining is required in the assembly operation, the necessary dimensions and notes may be given on the assembly drawing.

*Assembly Sections*   Since assemblies often have parts fitting into or overlapping other parts, 2D and 3D sections are useful views. For example, in Figure 7-43, try to imagine the right-side view drawn in elevation with interior parts represented by hidden lines.

Any kind of section may be used as needed. A broken-out section is shown in Figure 7-43. Half sections and removed sections are also frequently used. Pictorial sections are helpful in creating easy to read assembly drawings.

## Detail Drawings or Piece Part Drawings

Drawings of the individual parts are called **piece part drawings, part drawings,** or **detail drawings**. Detail drawings

contain all of the necessary information to manufacture any specific part being created for a product or design. Figures 7-44 and 7-45 show detail drawings. The information provided on detail drawings includes:

- All necessary drawing views or accurate 3D model information to fully define the shape.
- Dimensions that can be specified in a drawing or can be measured accurately from a 3D model.
- Tolerances either specified in a drawing or annotated in a 3D model so that how the tolerance applies can be clearly understood.
- The material for the manufactured part.
- Any general or specific notes including heat treatment, painting, coatings, hardness, pattern number, estimated weight, and surface finishes, such as maximum surface roughness.
- Approval or release and revision tracking, whether part of a 2D drawing title and revision block or part of a digital signature system.

## 7-19 SUBASSEMBLIES

A set of working drawings includes detail drawings of individual parts and the assembly drawing showing the assembled unit. Often an entire subassembly may be reused in a different design. It is easier to reuse the group of parts in a new design if they are grouped together logically and contained in

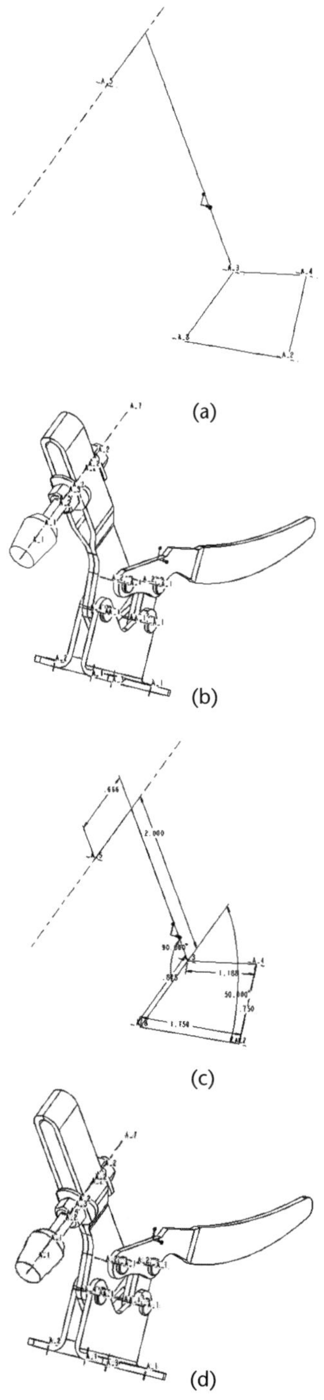

**Figure 7-39**  Skeleton Model for a Clamp

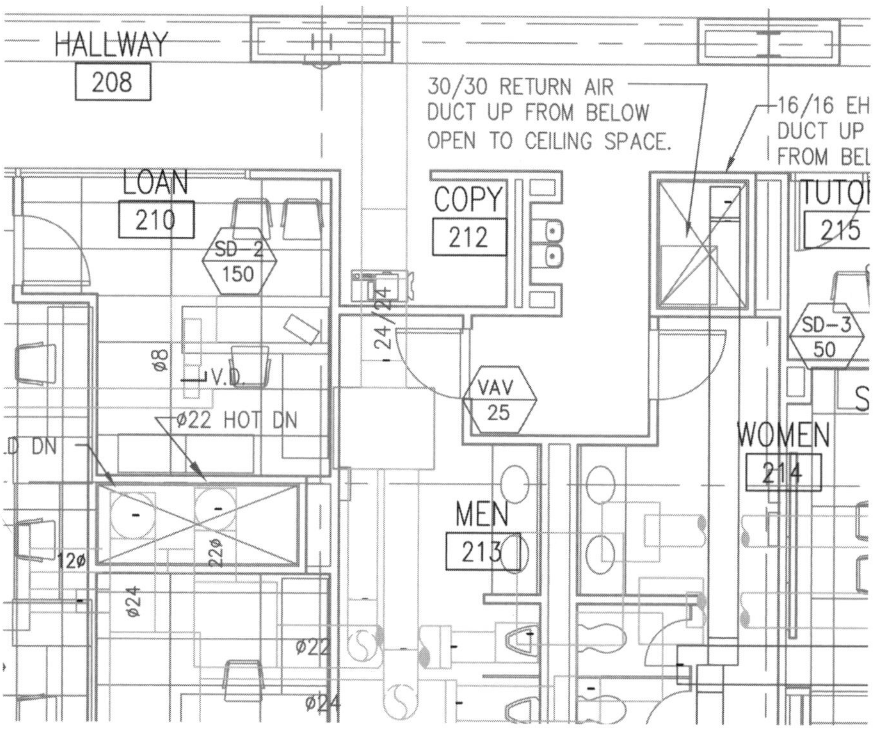

**Figure 7-40**  Portion of a Mechanical System for a Building. *Courtesy of Associated Construction Engineering, Inc.*

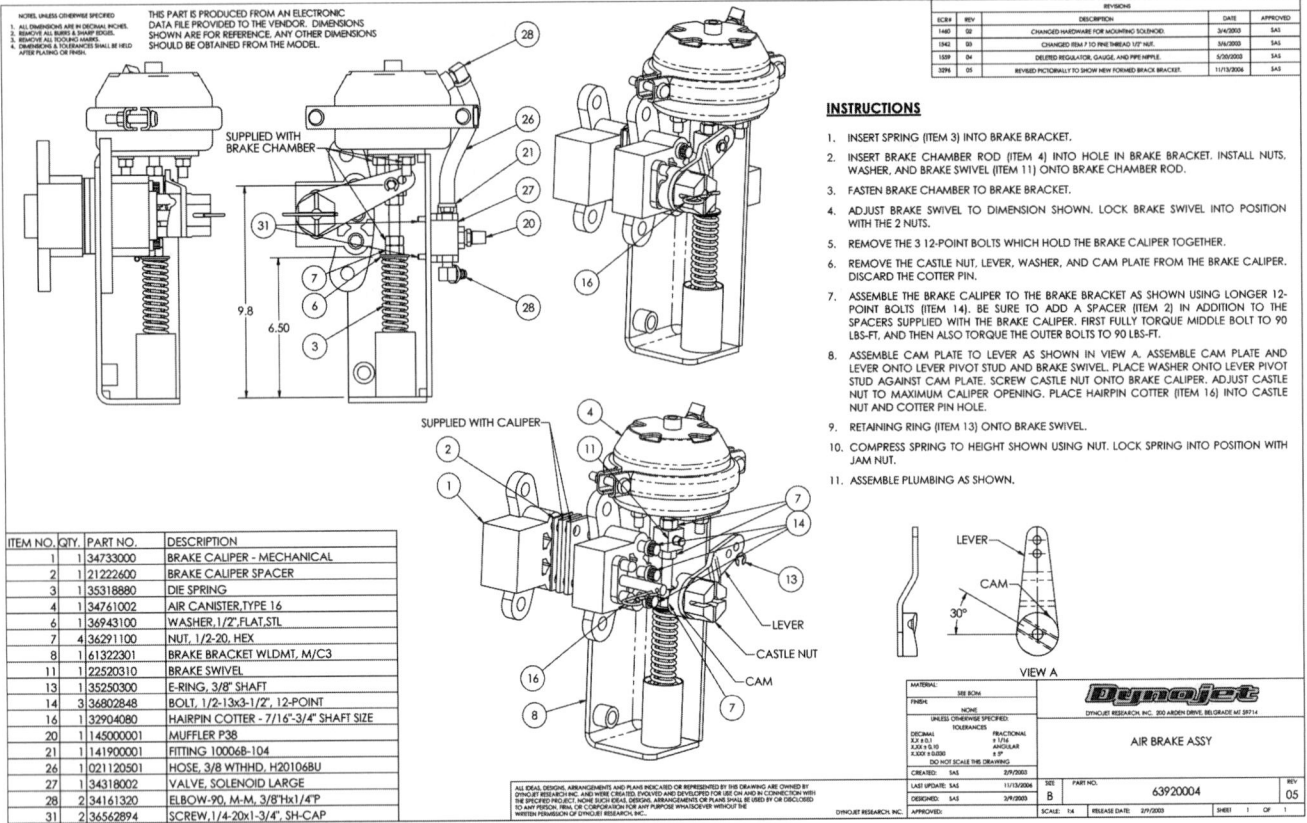

**Figure 7-41**   General Assembly Drawing for an Air Brake Created from a 3D CAD Model. *Courtesy of Dynojet Research, Inc.*

separate drawings. Your top level assembly drawing will appear cleaner if you keep subassemblies well organized, as the entire subassembly can be identified as a single item on a higher level assembly drawing. Fasteners for the subassembly that attach it to its mating parts in the next higher level assembly drawing are usually shown or listed on the bill of materials (sometimes referred to as BOM) at the higher level.

Structuring your product into assemblies and subassemblies requires thoughtful decision making in order to get the most advantage when retrieving part, subassembly, and assembly drawings later on. If your company uses a product data management system (PDM), planning is essential to seeing downstream results.

An example of a subassembly is shown in Figure 7-46. A subassembly is drawn the same way as an assembly drawing, just for a subgroup that assembles to other parts.

# 7-20 IDENTIFICATION

Use circled numbers called **balloon numbers** or **ball tags** to identify the parts in the assembly (Figure 7-47). Circles containing the part numbers are placed adjacent to the parts,

with leaders terminated by arrowheads touching the parts as shown in Figure 7-48.

The circles are placed in orderly horizontal or vertical rows and not scattered over the sheet. Leader lines should be drawn at an angle, not horizontally or vertically. Do not let leaders cross. Make adjacent leaders parallel or nearly so. For multiple small parts that are easily distinguished, a single leader may have multiple circle item numbers as shown in Figure 7-48.

The circled item number identifies each part. Show information for the part in the parts list that is usually included on the drawing sheet, but may also be a separate document.

Another method of identification is to letter the part names, numbers required, and part numbers, at the end of leaders as shown in Figure 7-46. More commonly, only the part numbers are given, together with standard straight-line leaders.

## Multidetail Drawings

When multiple detail drawings are shown on one sheet, identify each part are similar to those used in detail drawings where several details are shown on one sheet, as in Figure 7-49. Place circles containing the part numbers

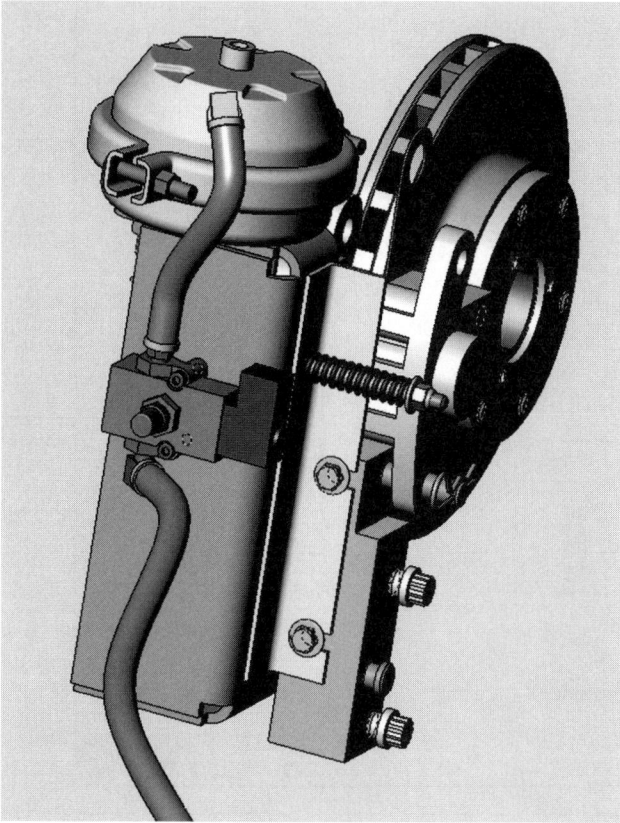

**Figure 7-42** 3D CAD Model for an Air Brake. *Courtesy of Dynojet Research, Inc.*

adjacent to the parts, with leaders terminated by arrowheads touching the parts as in Figure 7-48. A portion of a multidetail drawing is shown in Figure 7-49.

## 7-21 PARTS LISTS

**A parts list** or **bill of materials** itemizes the parts of a structure shown on an assembly drawing (ANSI Y14.34M–1996). The title strip alone is sufficient on detail drawings of only one part, but a parts list is necessary on assembly drawings or detail drawings of several parts. Parts can be listed in general order of size or importance or grouped by types.

Parts lists for machine drawings, Figure 7-50 contain:

- Part identification number (PIN).
- Description of each part.
- Quantity required in the assembly.
- The following abbreviations can be used to indicate quantities that are not exactly known: AR indicating *as required;* EST followed by a number for an *estimated quantity.*

For parts lists that contain application data, information for the next assembly level must be included.

Frequently other information is supplied in the parts list, such as material, CAGE code, pattern numbers, stock sizes of materials, and weights of parts.

### Automatic BOM Generation

CAD software often allows you to generate the parts list automatically or somewhat automatically. Figure 7-51 shows a dialog box used to automate generation of a parts list.

If you created a 3D assembly model by inserting CAD models for the parts, the software can query the assembly model for quantities and the file names that were inserted to generate the parts list. This is another good reason to use good file management practices and name your parts logically. Most software allows you to type in information, but overriding the information this way makes it harder to automatically update files, losing some of the advantage of using 3D CAD.

### Locating the Parts List

If the parts list rests on top of the title box or strip, the order of the items should be from the bottom upward so that new items can be added later, if necessary. If the parts list is placed in the upper-right corner, the items should read downward.

### Listing Standard Parts

Standard parts, whether purchased or company produced, are not drawn but are included in the parts list. Bolts, screws, bearings, pins, keys, and so on are identified by the part number from the assembly drawing and are specified by name and size or number.

## 7-22 ASSEMBLY SECTIONS

In assembly sections it is necessary not only to show the cut surfaces but also to distinguish between adjacent parts. Do this by drawing the section lines in opposing directions, as shown in Figure 7-52. The first large area is sectioned at 45° (Figure 7-52a). The next large area, (b), is sectioned at 45° in the opposite direction. Additional areas are then sectioned at other angles, such as 30° or 60° as shown at (c) or at other angles.

In small areas it is necessary to space the section lines closer together. In larger areas space section lining more widely or use outline section lining.

Use the general-purpose section lining for assemblies. You can also give a general indication of the materials used, through using symbolic section lining as shown in Figure 7-53.

In sectioning relatively thin parts in an assembly, such as gaskets and sheet metal parts, section lining is ineffective and should be left out or shown in solid black as in Figure 7-54.

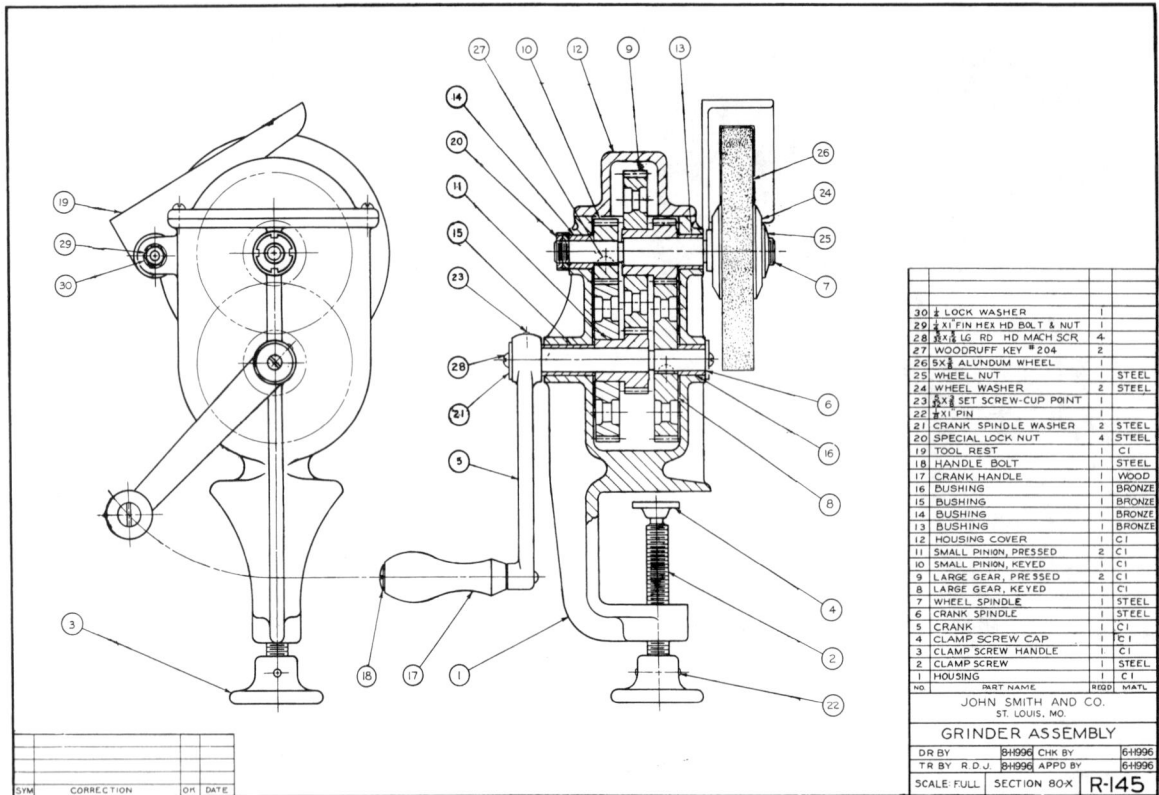

**Figure 7-43** Assembly Drawing of a Grinder

In architectural drawings, filling sectioned areas solidly is called **poche**, as shown in Figure 7-55. It is often used to show walls that have been cut through, as on floor plans.

Often solid objects, or parts that do not show required information, are sliced by the cutting plane. Leave these parts unsectioned, or "in the round." This includes bolts, nuts, shafts, keys, screws, pins, ball or roller bearings, gear teeth, spokes, and ribs, among others. See Figure 7-56.

## 7-23 WORKING DRAWING ASSEMBLY

A **working drawing assembly**, Figure 7-57, is a combined detail and assembly drawing. These drawings are often used in place of separate detail and assembly drawings when the assembly is simple enough for all of its parts to be shown clearly in the single drawing. In some cases, all but one or two parts can be drawn and dimensioned clearly in the assembly drawing, in which event these parts are detailed separately on the same sheet. This type of drawing is common in valve drawings, locomotive subassemblies, aircraft subassemblies, and drawings of jigs and fixtures.

## 7-24 INSTALLATION ASSEMBLIES

An assembly made specifically to show how to install or erect a machine or structure is an **installation assembly**. This type of drawing is also often called an outline assembly, because it shows only the outlines and the relationships of exterior surfaces. A typical installation assembly is shown in Figure 7-58. In aircraft drafting, an installation drawing (assembly) gives complete information for placing details or subassemblies in their final positions in the airplane.

## 7-25 CHECK ASSEMBLIES

After all detail drawings of a unit have been made, it may be necessary to make a **check assembly**, especially if a number of changes were made in the details. Such an assembly is shown accurately to scale in order to graphically check the correctness of the details and their relationship in assembly. After the check assembly has served its purpose, it may be converted into a general assembly drawing.

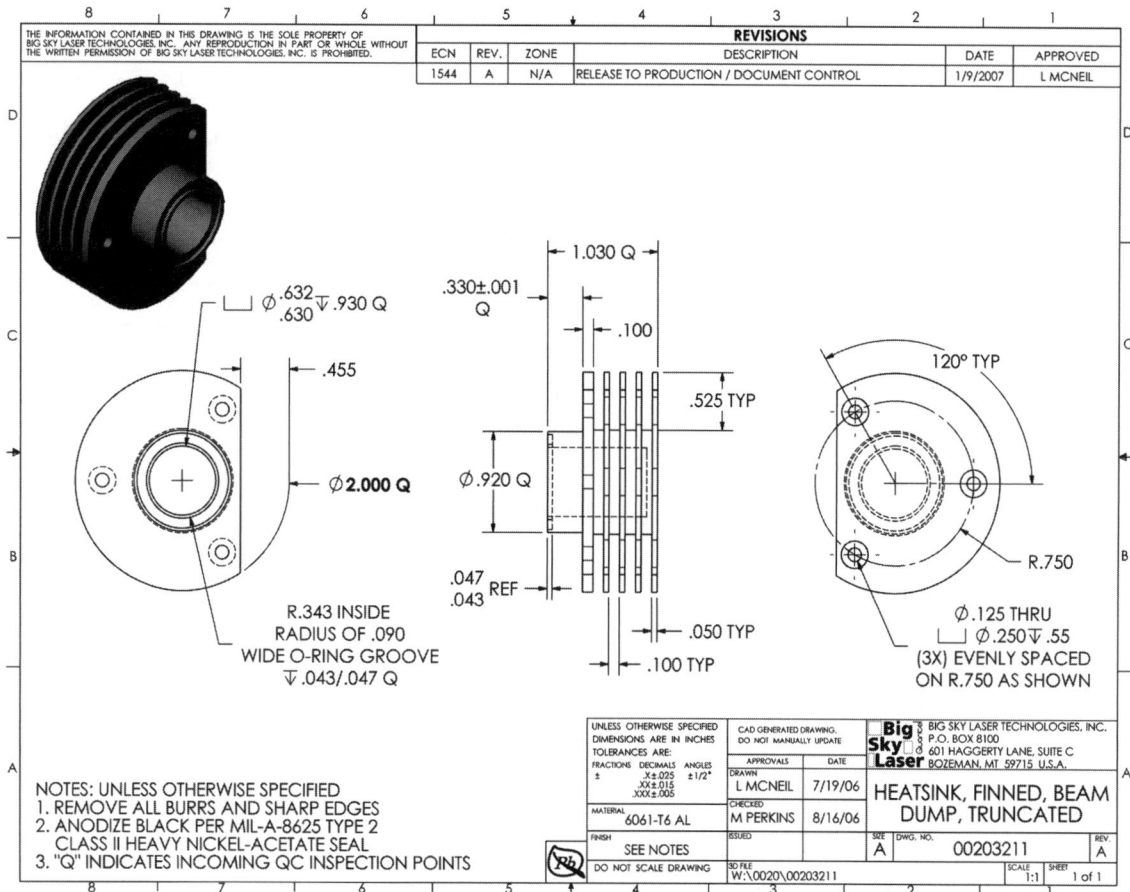

**Figure 7-44** Part Drawing for a Heat Sink. *Courtesy of Big Sky Laser.*

**Figure 7-45** A Detail Drawing

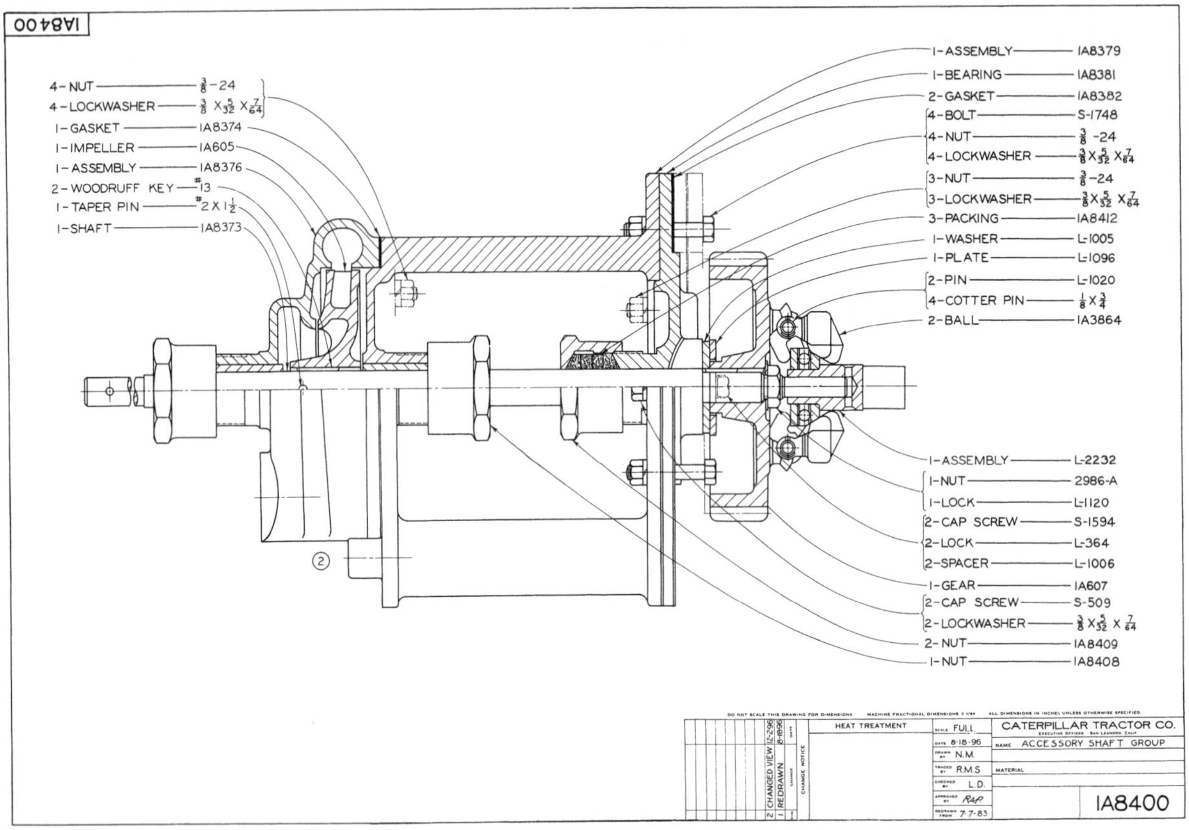

**Figure 7-46**   Subassembly of Accessory Shaft Group

# 7-26 WORKING DRAWING FORMATS

### Number of Details Per Sheet

There are two general methods for grouping detailed parts on sheets. Showing one detailed part per sheet is typically preferred because it is easier to repurpose drawings for other uses and to track revision data when the sheet does not contain extra parts.

Small machines or structures composed of few parts sometimes show all the details on one large sheet. Showing the assembly and all its details on one sheet can be convenient, but it is generally more difficult to revise and maintain. The same scale should be used for all details on a single sheet, if possible. When this is not possible, clearly note the scale under each dissimilar detail.

Most companies show one detail per sheet, however simple or small it may be. For many parts the basic 8.5 × 11" or 210 × 297 mm sheet works well. Since it is easy to lose a few drawings on smaller sheets from a set that is mostly on larger sheets, some companies use all 11 × 17" (or the equivalent metric size) for all parts.

### Digital Drawing Transmittal

Electronic file formats such as Portable Document Format (PDF), originally developed by Adobe Systems in 1993,

allow the originator to send a document that can be commented on without allowing the original document to be changed.

Several search engines allow you to search for text embedded in the PDF file. This means that PDF can provide advantages not just for storing, but retrieving the information later. Adobe Systems provides a useful document (in PDF format) on using PDF as an archiving standard. You can read it at http://www.adobe.com/products/acrobat/pdfs/pdfarchiving.pdf.

Using electronic files saves trees, makes it quicker to distribute and store documents, and allows others to review documents from various applications. Figure 7-59 shows a drawing stored in PDF format with comments and redlined markups.

### Title and Record Strips

Drawings constitute important and valuable information regarding products, so carefully designed, well-kept, systematic files are important.

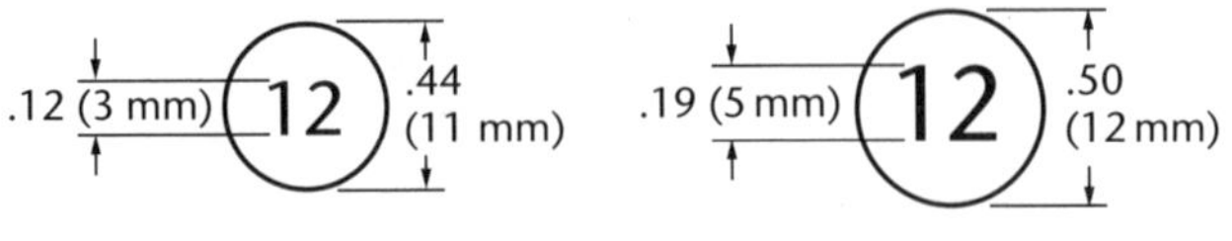

**Figure 7-47**   Identification Numbers

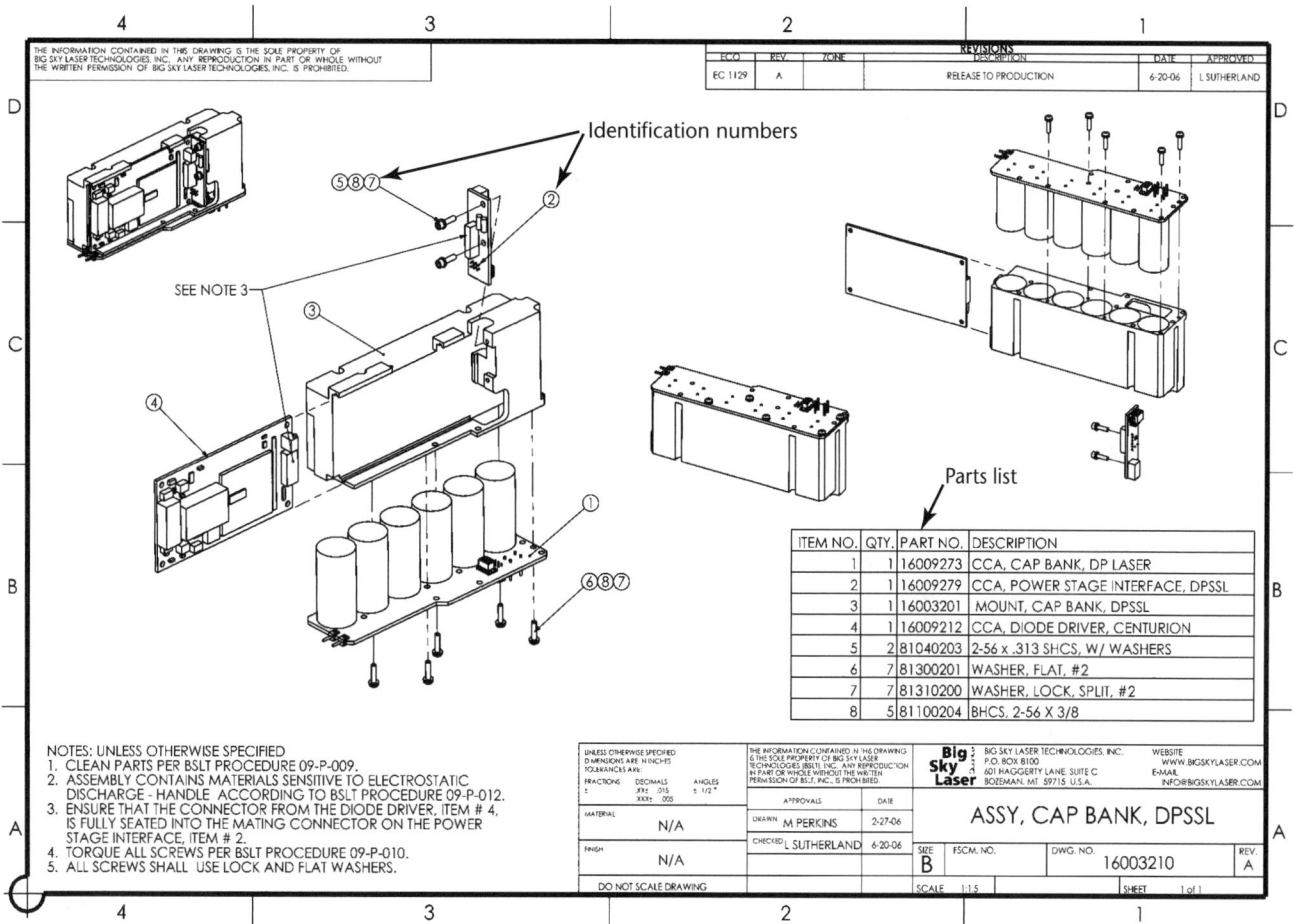

**Figure 7-48** Identification of Assembly Drawing Items with a Parts List. *Courtesy of Big Sky Laser.*

**Figure 7-49** Portion of a Drawing Showing Identification of Details with a Parts List

The function of the title and record strip is to show, in an organized way, all necessary information not given directly on the drawing with its dimensions and notes. The type of title used depends on the filing system in use, the manufacturing processes, and the requirements of the product. The following should generally be given in the title form:

1. Name of the object shown.
2. Name and address of manufacturer.
3. Name and address of the purchasing company, if any.
4. Signature of the person who made the drawing and date of completion.
5. Signature of the checker and date of completion.
6. Signature of the chief drafter, chief engineer, or other official, and the date of approval.
7. Scale of the drawing.
8. Number of the drawing.

| ITEM NO. | QTY. | PART NO. | DESCRIPTION |
|---|---|---|---|
| 1 | 1 | 16009273 | CCA, CAP BANK, DP LASER |
| 2 | 1 | 16009279 | CCA, POWER STAGE INTERFACE, DPSSL |
| 3 | 1 | 16003201 | MOUNT, CAP BANK, DPSSL |
| 4 | 1 | 16009212 | CCA, DIODE DRIVER, CENTURION |
| 5 | 2 | 81040203 | 2-56 x .313 SHCS, W/ WASHERS |
| 6 | 7 | 81300201 | WASHER, FLAT, #2 |
| 7 | 7 | 81310200 | WASHER, LOCK, SPLIT, #2 |
| 8 | 5 | 81100204 | BHCS, 2-56 X 3/8 |

**Figure 7-50**   Parts List. *Courtesy of Big Sky Laser.*

Other information may be included, such as material, quantity, heat treatment, finish, hardness, pattern number, estimated weight, superseding and superseded drawing numbers, symbol of machine, and so on, depending on the plant organization and unique aspects of the product. Some typical title blocks are shown in Figures 7-60, 7-61 and 7-62. See the inside back cover for traditional title forms and ANSI-approved sheet sizes.

The title form is usually placed along the bottom of the sheet or in the lower right-hand corner of the sheet, because drawings are often filed in flat, horizontal drawers, and the title must be easily found. However, as many filing systems are in use, the location of the title form depends on your company's organizational preference. Many companies adopt their own title forms or those preferred by ANSI.

To letter items in a title form:

- Use single-stroke vertical or inclined Gothic capitals.
- Letter items according to their relative importance. Use heavier, larger or more widely spaced lettering (or a combination of these) to indicate important items.
- Give the drawing number the most emphasis, followed by the name of the object and name of the company. (Date, scale, and originator's and checker's names are important, but do not need to be prominent.)

## 7-27 DRAWING NUMBERS

Every drawing should be numbered. Some companies use serial numbers, such as 6047, or a number with a prefix or suffix letter to indicate the sheet size, as A6047 or 6047-A. The size A sheet is a standard 8.5 × 11". In different numbering schemes, various parts of the drawing number indicate different things, such as model number of the machine and the nature or use of the part. In general, it is best to use a simple numbering system and not to load the number with too many indications.

The drawing number should be lettered 7 mm (.2500") high in the lower-right and upper-left corners of the sheet.

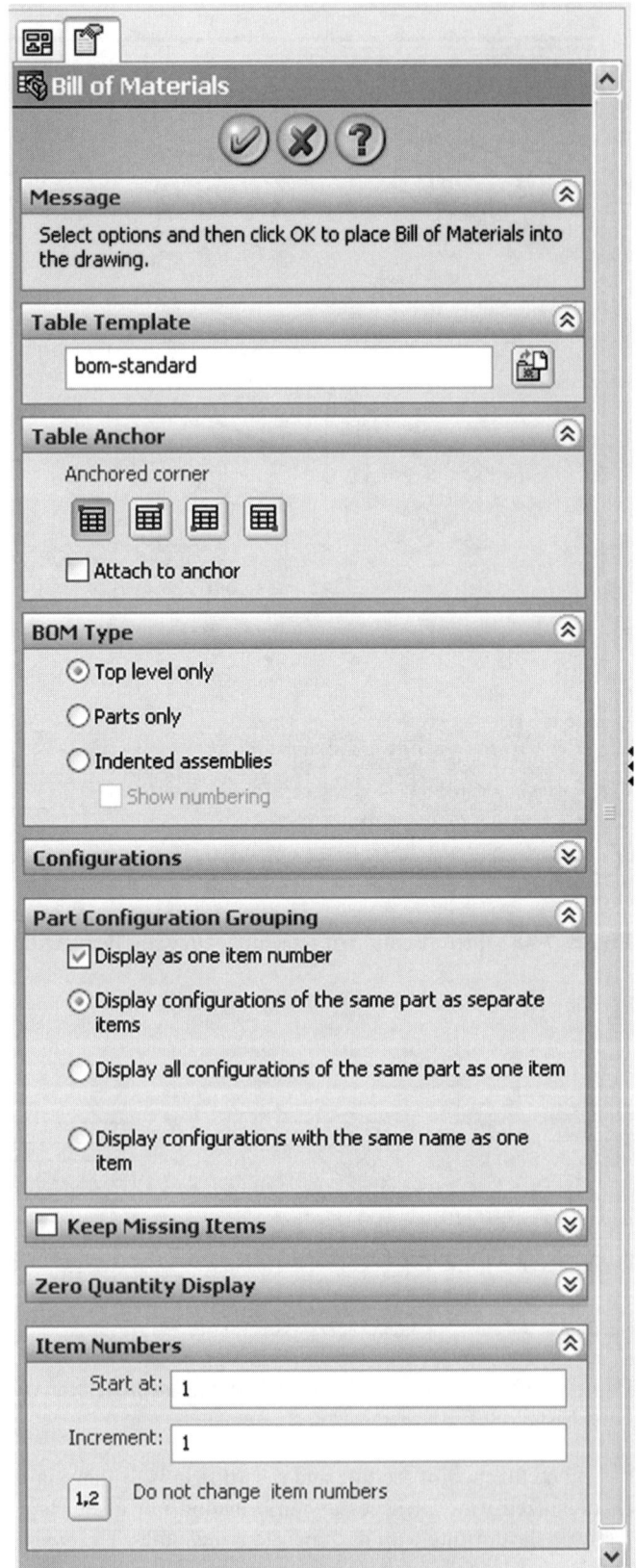

**7-51**   Solidworks Dialog Box Showing Options For Automatically Inserting a Bill of Materials Table. *Courtesy of Solidworks Corporation.*

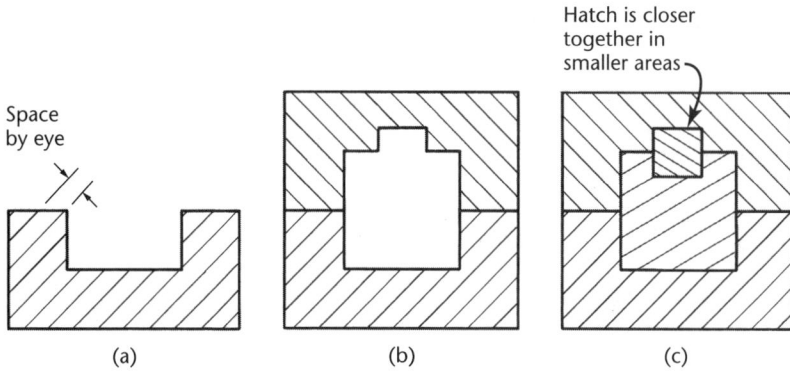

**Figure 7-52**    Section Lining in Assemblies (Full Size)

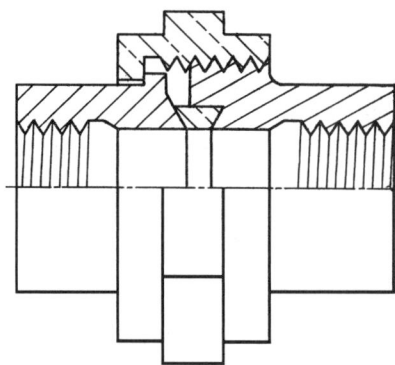

**Figurre 7-53**    Symbolic Section Lining

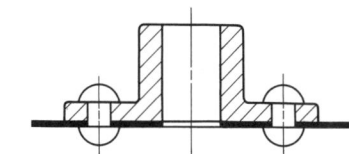

**Figure 7-54**    Solidly Hatching Small Parts

In order to benefit from a CAD system, you must be able to store and retrieve your drawings efficiently. Drawing tracking software allows users to search by part number or text items to retrieve drawing files and CAD models.

## 7-28 ZONING

To help people locate a particular item on a large or complex drawing, regular ruled intervals are labeled along the margins, often in the right and lower margins only. The intervals on the horizontal margin are labeled from right to left with numerals, and the intervals on the vertical margin are labeled from bottom to top with letters, similar to road maps. Note the zone letters and numbers around the border of Figure 7-63.

## 7-29 CHECKING DRAWINGS

The importance of accuracy in technical drawing cannot be overstated. Errors sometimes cause tremendous unnecessary expenditures. The signature on the drawing identifies who is responsible for its accuracy.

In small offices, checking is usually done by the designer or by one of the drafters. In large offices, experienced engineers may be employed to devote a major part of their time to checking drawings.

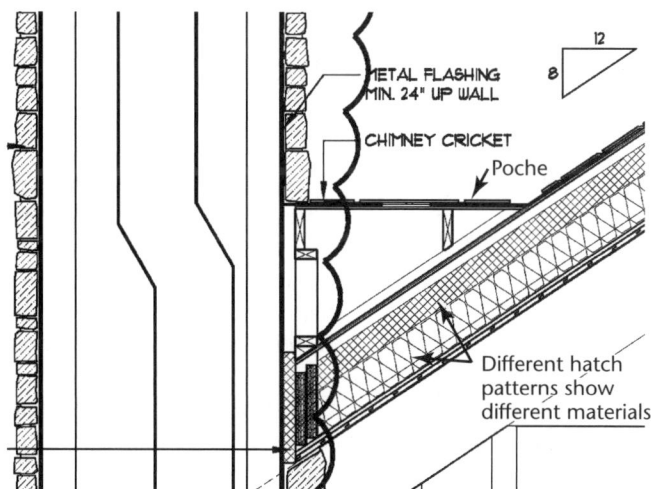

**Figure 7-55**    Use Hatch Patterns to Indicate Material and Poche Small Features *Detail of Drawing. Courtesy of Locati Architects.*

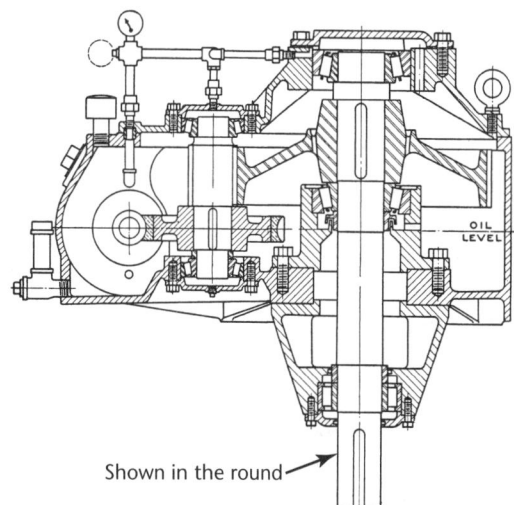

**Figure 7-56**    Assembly Section

*You can save time in creating assemblies by downloading stock parts. Many vendors have models for Web download.*

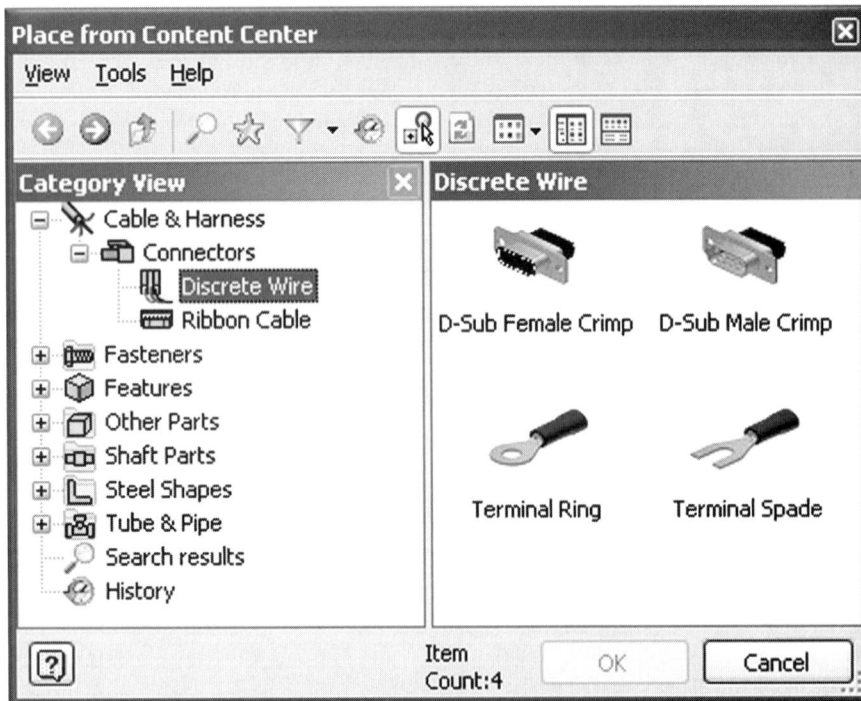

*Autodesk Inventor is an example of a software package that features stock parts that are useful for creating assemblies. With permission of Autodesk, Inc. © 2006-2007. All rights reserved.*

A drawing is carefully checked and signed by the person who made it. It is then checked by the lead designer for function, economy, practicability, fit, tolerances and so on. Corrections, if any, are then made by the original drafter.

The final checker should systematically review the drawing for any remaining errors. They should study the drawing with particular attention to:

1. Soundness of design, with reference to function, strength, materials, economy, manufacturability, serviceability, ease of assembly and repair, lubrication, and so on.
2. Choice of views, partial views, auxiliary views, sections, lettering, and so on.
3. Dimensions, with special reference to repetition, ambiguity, legibility, omissions, errors, and finish marks. Special attention should be given to tolerances.
4. Standard parts. In the interest of economy, as many parts as possible should be standard.
5. Notes, with special reference to clear wording and legibility.
6. Clearances. Moving parts should be checked in all possible positions to ensure freedom of movement.
7. Title form information.

## 7-30 DRAWING REVISIONS

Changes on drawings may be necessitated by changes in design, changes in tools, desires of customers, or errors in design or in production. An accurate record of all changes made to released drawings is tracked via a revision block. This is important so that the sources of all changes may be understood, verified, and approved.

The record of revisions should show the change, by whom, when, and why the change was made. An engineering change order (ECO) or

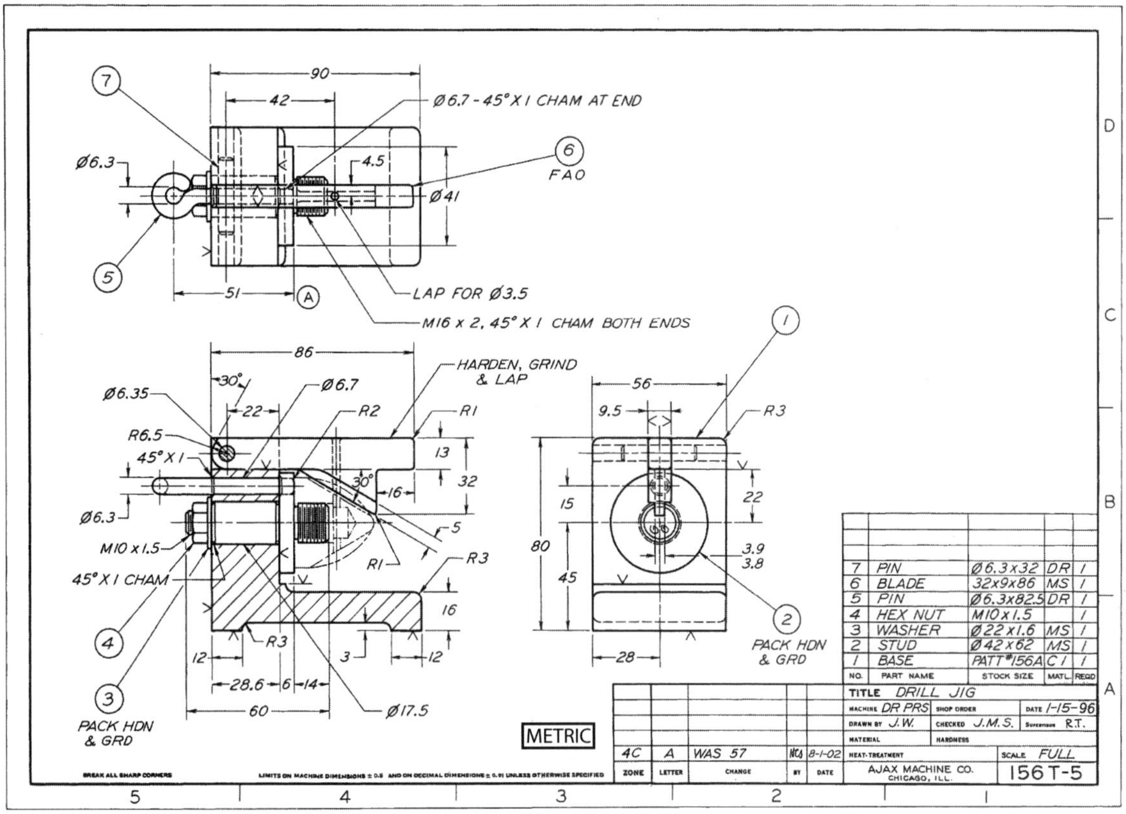

**Figure 7-57**   Working Drawing Assembly of Drill Jig.

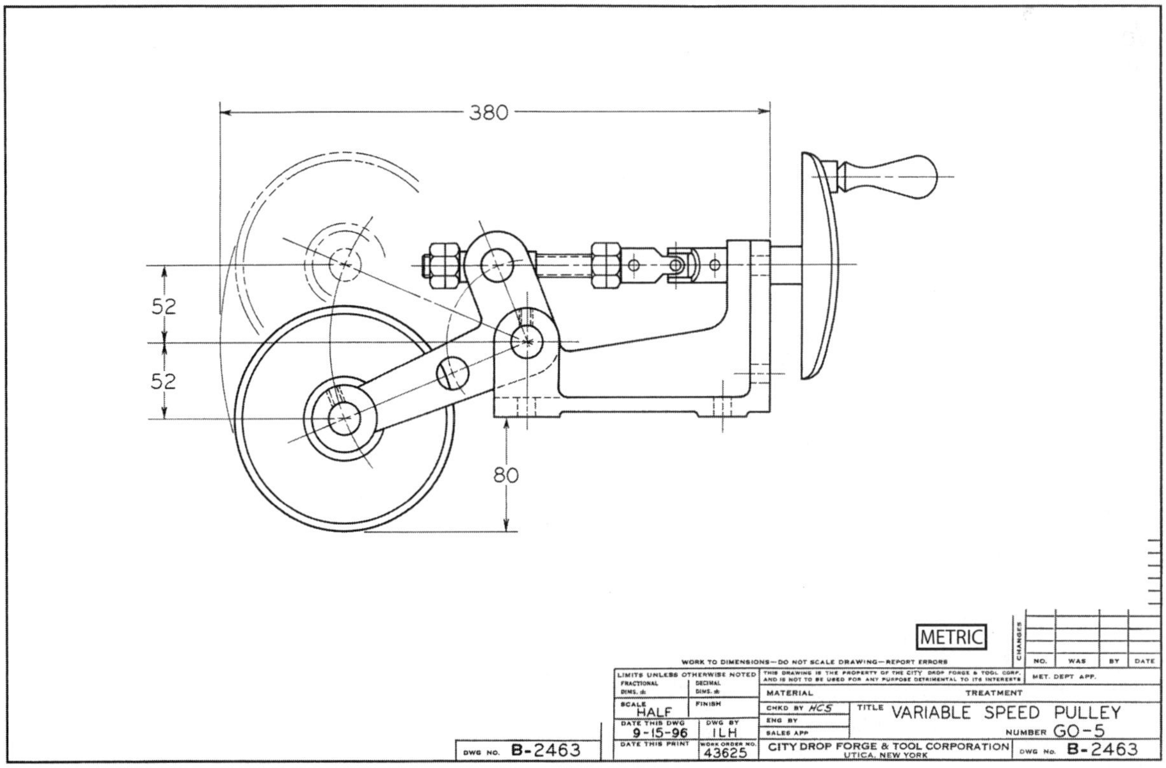

**Figure 7-58**   Installation Assembly

engineering change request (ECR) is processed to approve and track changes to drawings once they have been released for production. Some companies use a paper record for this and others manage it digitally.

Any changes or additions made to a drawing are tracked by a **revision number**. A symbol can be added to the drawing showing the item affected by the revision.

It is not recommended to remove information by crossing it out.

In rare cases when a dimension is not noticeably affected by a change, it may be underlined with a heavy line to indicate that it is not to scale.

It is important to keep prints or microfilms of each issue on file to show how the drawing appeared before the revision. Issue new prints to supersede old ones each time a change is made.

Digital systems absolutely must use careful backup procedures and, due to data loss concerns, are still not approved in some industries.

If considerable change on a drawing is necessary, it may be necessary to make a new drawing and stamp the old one OBSOLETE and store it in an "obsolete" file. In the title block of the old drawing, enter the words "SUPERSEDED BY" or "REPLACED BY" followed by the number of the new drawing. On the new drawing, under "SUPERSEDES" or "REPLACES," enter the number of the old drawing.

People use various methods to reference the area on a drawing where the change is made, with the entry in the revision block. The most common is to place numbers or letters in a small circle or triangle near each place where the changes were made and to use the same numbers or letters in the revision block, as shown in Figure 7-63. On zoned drawings the zone of the correction is shown in the revision block.

**7-59**    A Portion of a PDF File Showing Red-Lined Markups. *Courtesy of Dynojet Research, Inc.*

The change should also be described briefly, along with the date and initials of the person making the change.

## 7-31 SIMPLIFYING DRAWINGS

Drawing time is a considerable part of the total cost of a product. It makes sense to reduce drawing costs by using practices to simplify your drawings without losing clarity. For example, use partial views, half views, thread symbols,

---

## PAPER CONSERVATION

According to Worldwatch Institute, 40% of the trees harvested worldwide are used to make paper. The U.S. Environmental Protection Agency (EPA) estimates that paper makes up 38% of municipal solid waste.

The UNESCO Statistical Handbook estimated paper production in 1999 at 1,510 sheets of paper per inhabitant of the world. Even with digital data storage, paper consumption has only increased since then.

Web sites related to paper conservation include:
- http://www2.sims.berkeley.edu/research/projects/how-much-info-2003/print.html
- http://www.lesk.com/mlesk/ksg97/ksg.html

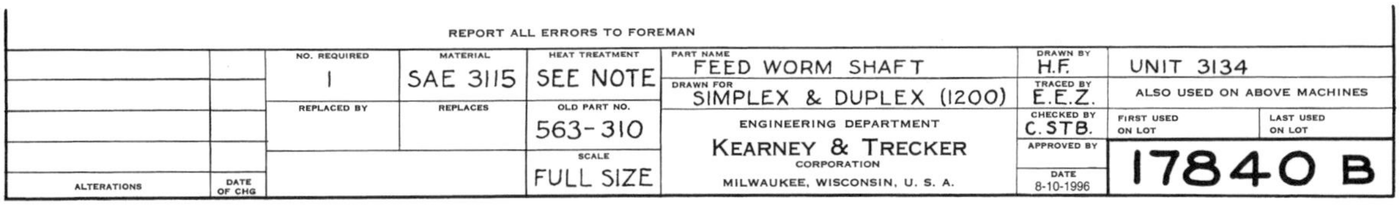

**Figure 7-60**   Title Strip

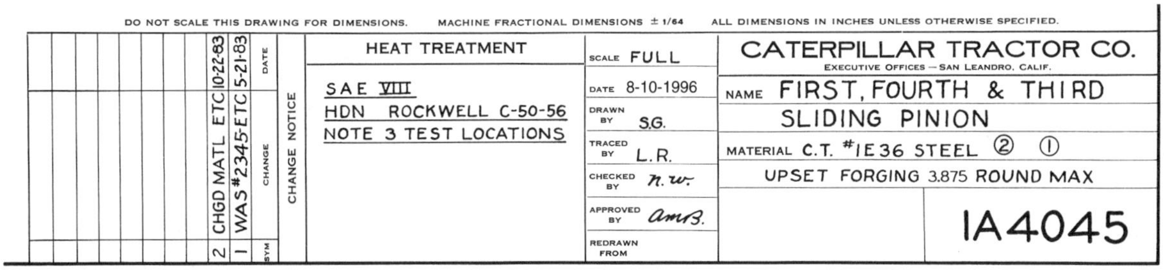

**Figure 7-61**   Title Strip. *Courtesy of Dynojet Research, Inc.*

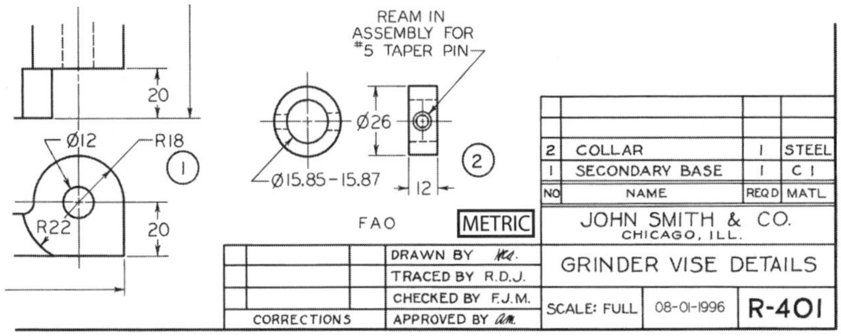

**Figure 7-62**   Identifying Details with a Parts List. *Courtesy of Big Sky Laser, Inc.*

piping symbols, and single-line spring drawings when appropriate. Omit lines or lettering on a drawing that are not needed for clarity. In addition to saving production time, this makes drawings easier to read. To simplify drawings:

1. Use word descriptions when practical.
2. Do not show unnecessary views.
3. Use standard symbols such as Ø and standard abbreviations.
4. Avoid elaborate, pictorial, or repetitive details. Use phantom lines to avoid drawing repeated features.
5. List rather than draw standard parts such as bolts, nuts, keys, and pins.
6. Omit unnecessary hidden lines.
7. Use outline section lining in large areas to save time and improve legibility.
8. Omit unnecessary duplication of notes and lettering.
9. Use symbolic representation for piping and thread.

10. Use CAD libraries and standard parts when feasible for design and drawings.

Some industries have simplified their drafting practices even more. Learn the practices appropriate to the industry for which you are creating drawings.

## 7-32 PATENT DRAWINGS

The patent application for a machine or device must include drawings to illustrate and explain the invention. All patent drawings must be mechanically correct and constitute complete illustrations of every feature of the invention claimed. The strict requirements of the U.S. Patent Office facilitate the examination of applications and the interpretation of patents issued. Examples of patent drawings are shown in Figure 7-64.

Drawings for patent applications are pictorial and explanatory in nature; therefore they are not as detailed as working drawings for production purposes. Centerlines,

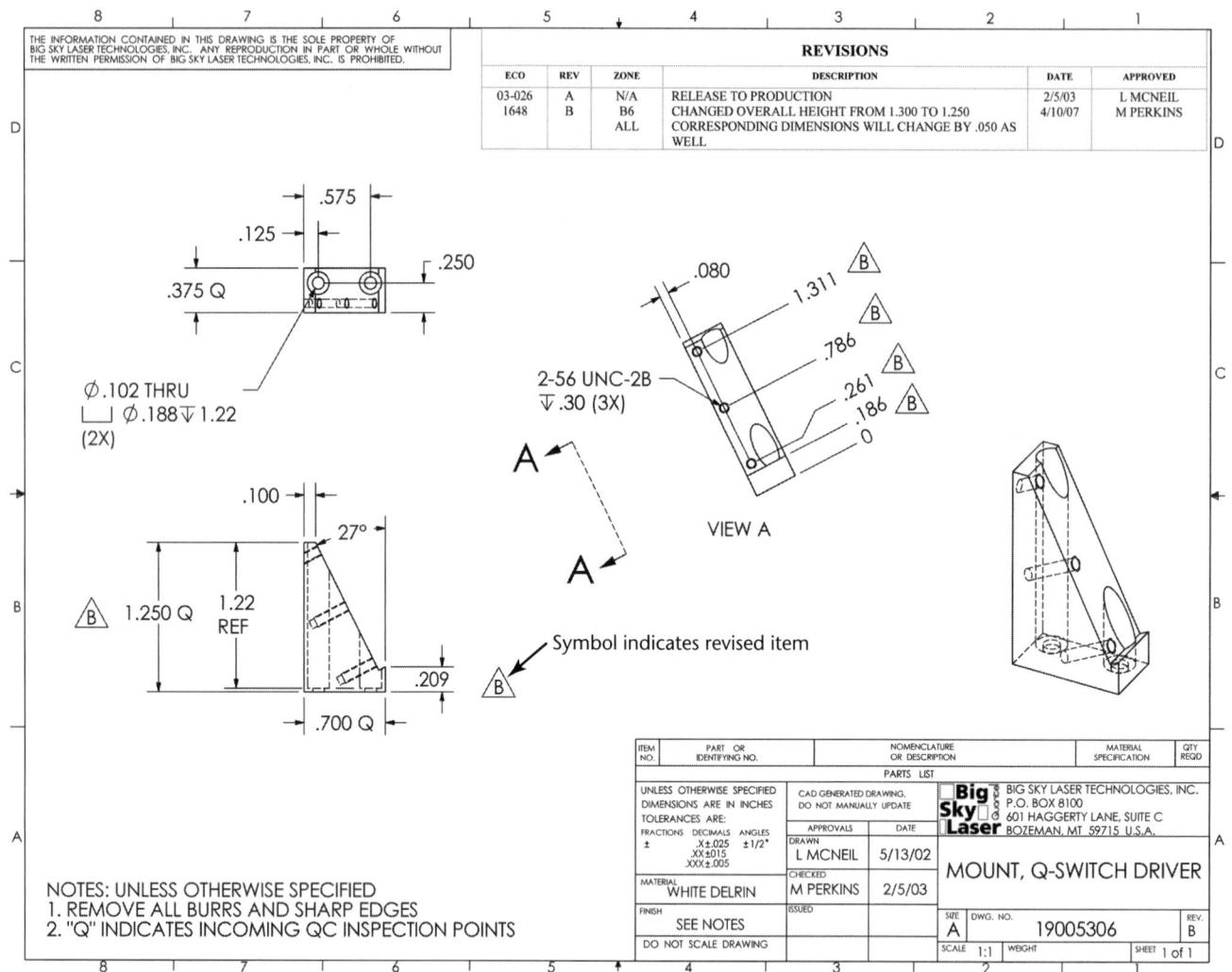

**Figure 7-63** Symbols Matching the Item in the Revision Block Indicate Revised Features on a Drawing. *Courtesy of Big Sky Laser, Inc.*

hidden lines, dimension notes, and so forth, are omitted, since specific dimensions, tolerances, and notes are often not required to patent the general design or innovation.

The drawings must contain as many figures as necessary to show the invention clearly. There is no limit on the number of sheets that may be submitted. The drawings can be produced by hand or from the same CAD database used to create the design documentation.

While most engineering drawings are produced with views in alignment on one sheet, patent drawings must show each separate view as one figure on a separate sheet. Figures should be numbered consecutively (i.e., Figure 1, Figure 2, Figure 3A, Figure 3B, etc.). Views, features, and parts are identified by numbers that refer to the descriptions and explanations given in the specification section of the patent application. The reference number for a part or feature should remain the same in every diagram.

Exploded isometric or perspective drawings with reference numbers identifying the parts (i.e., assembly drawings) are preferred. Centerlines are used to illustrate how parts are aligned in exploded views. While the drawing must show every feature that is listed in the patent claims, if standardized parts are used, they can be represented symbolically and do not have to be drawn in detail.

The figures may be plan, elevation, section, pictorial, and detail views of portions or elements, and they may be drawn to an enlarged scale if necessary.

The U.S. Patent Office has basic standards for drawings:

- All sheets within a single application must be the same size, and two sheet sizes are accepted:
  - U.S. size: 8.5 by 11" (216 mm × 279 mm).
  - International size: 210 mm × 297 mm.

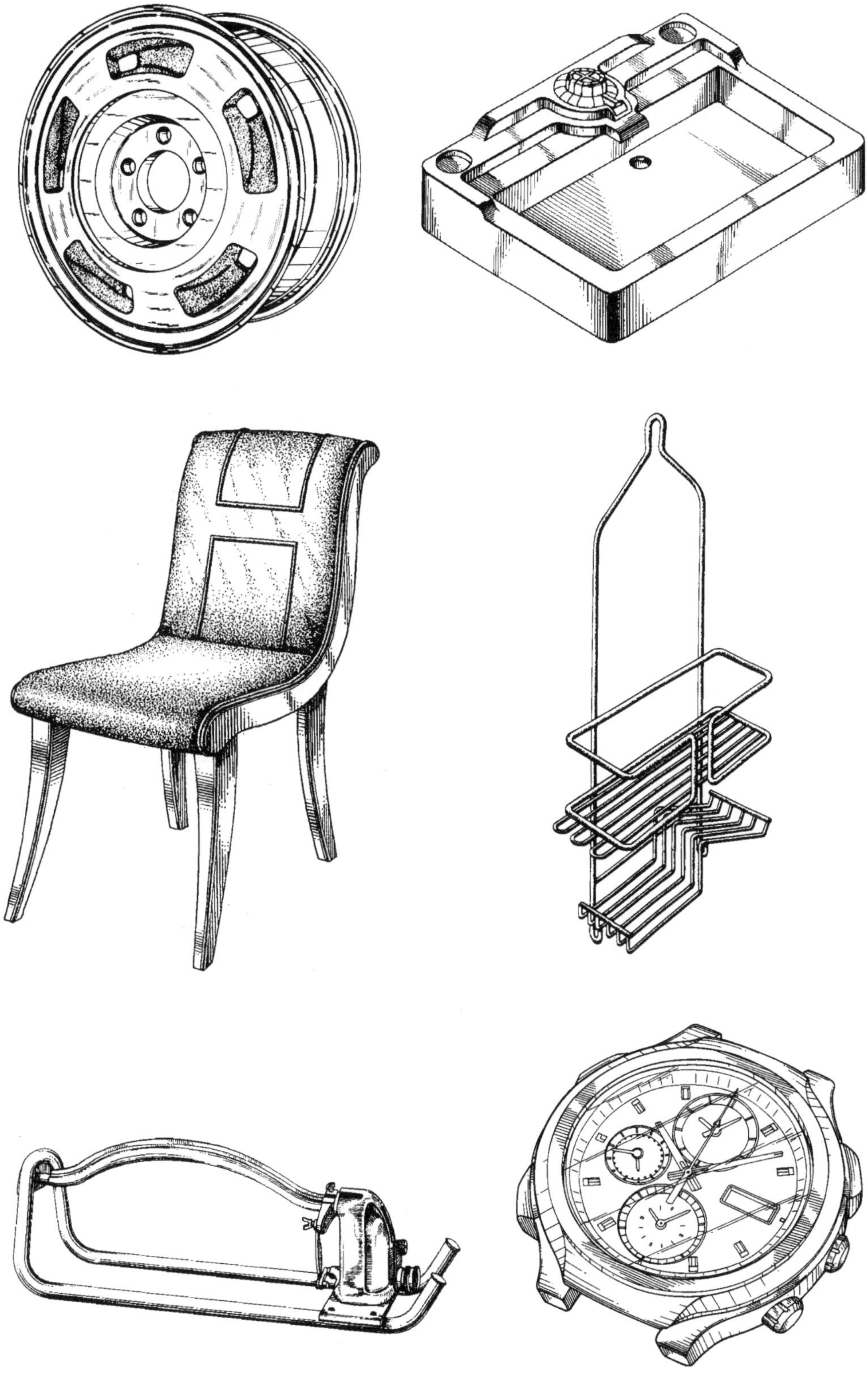

**Figure 7-64** Patent Drawing Examples *Although several examples are shown here, each drawing is shown on a separate sheet in the patent application. Courtesy of US. Patent and Trademark Office.*

- Paper must be single sided.
- Paper must be oriented vertically, so that the short side of the sheet is at the top (called portrait style in printing options).
- No border lines are permitted on the sheets.
- The following minimum margins must be maintained.
  - Top margin: 1" (25 mm).
  - Left margin: 1" (25 mm).
  - Right margin: .675" (15 mm).
  - Bottom margin: .375" (10 mm).
- No labels or drawing lines may extend into the margin except for the specific identification required at the top of each sheet and two scan target points.
- All drawings must be submitted in black and white—no color drawings or photos except in very limited cases.
- Lines must be solid black and suitable for reproduction at a smaller size.
- Shading (either cross hatch or stippling) is used whenever it improves readability. In rare cases when it is necessary to show a feature hidden behind a surface, a lighter solid line is used.
- Sketches are acceptable for the application process, but formal drawings will have to be created if accepted.

Photocopies are accepted since three copies of each drawing must be submitted. The drawings will not be returned so it is *not* a good idea to send an original with the initial patent application.

While the above gives you a basic idea of the standards for patent drawings, the strict requirements of the U.S. Patent Office are carefully documented on their Web site. Be sure to follow their requirements exactly if you are preparing drawings for a patent application.

For more information, log on to the U.S. Patent and Trademark Office's Web site at http://www.uspto.gov.

You can also consult the Guide for Patent Draftsmen, which can be obtained from the Superintendent of Documents, U.S. Government Printing Office, Washington, D.C. 20402.

# CAD AT WORK

## Assembly Drawings using Pro Engineer wildfire

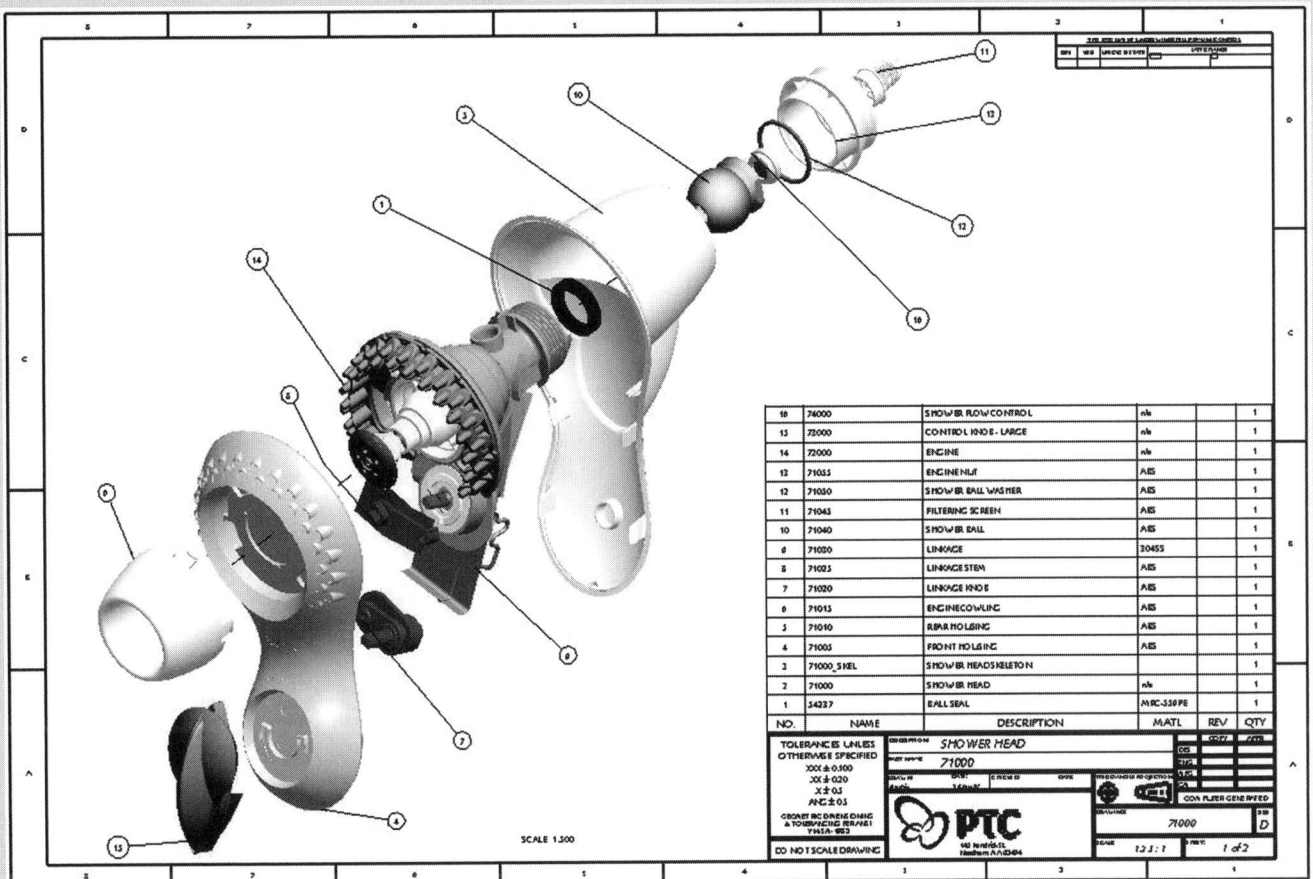

Example of a Color Shaded Exploded View Assembly from Pro Engineer Wildfire. *Courtesy of Parametric Technologies Corporation.*

Great looking assembly drawings are only one of the benefits of using 3D CAD for your designs. You can also check to see how parts fit together, perform tolerance studies, and even see how mechanisms you are designing will behave. Additionally, you can analyze the mass properties of your design, determine the volume and surface area of complicated shapes, and produce documentation drawings directly from the part models.

Software like Pro/Engineer Wildfire 3 allows you to use color shaded views of models in the exploded view assembly drawing, as shown in the figure above.

As you decide whether to use a color shaded drawing or a black and white line drawing for an assembly drawing, consider whether and how the drawing will be reproduced.

Shaded color views make it easy to identify and visualize the parts, but require color printing and copying to look their best on paper. This doesn't present a problem if you are distributing files electronically.

Even though color shaded drawings look great, there are times when black and white drawings are preferable, or required. For example, patent drawings must be black and white, showing visible lines and not hidden lines. Black and white drawings are also helpful in user manuals, which may be copied or printed in black and white.

With 3D CAD software, it is not difficult to switch between color shaded views, outlines, and views that show hidden lines, to suit the particular need that the drawing will meet.

## Portfolio

Assembly Drawing for Four Wheel Drive Pit Dyno. *Courtesy of Dynojet Research, Inc.*

This Cover for a Set of Architectural Plans Lists the Drawings in the Set and Abbreviations Used. *Courtesy of Locati Architects.*

Portfolio

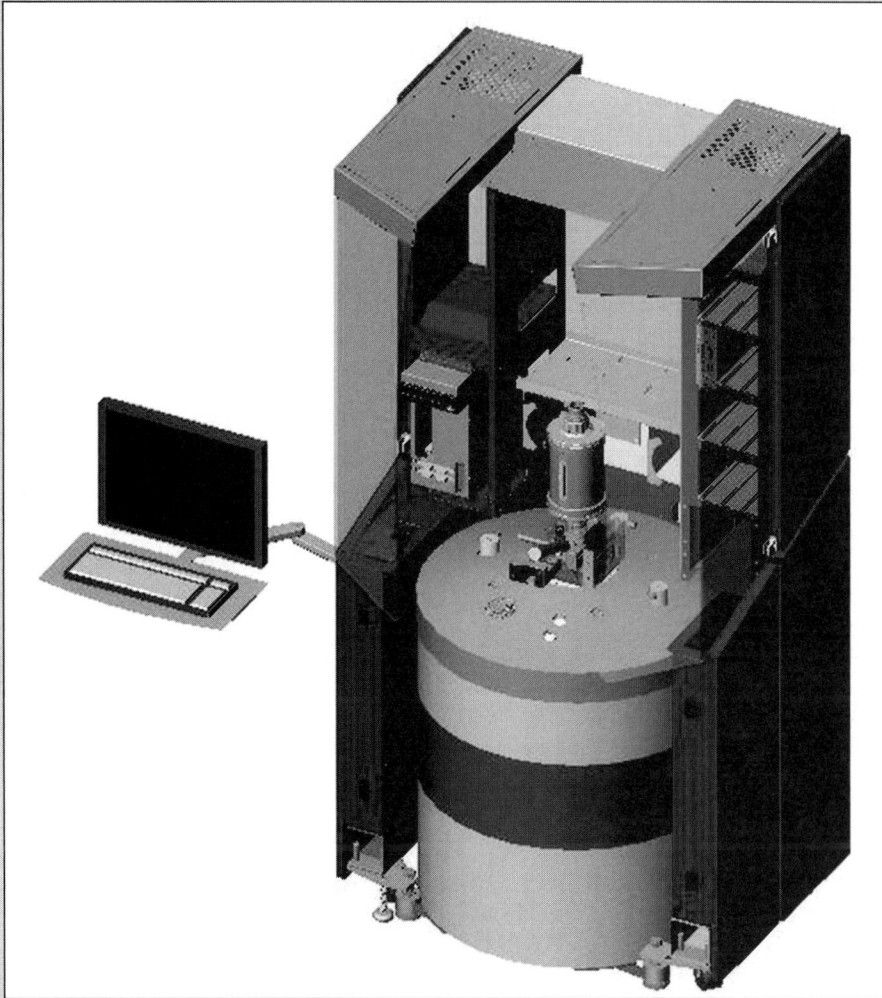

Fully Assembled 3D CAD Model Showing How Parts Fit in Assembly.
*Courtesy of Quantum Design.*

## 7-33 DESIGN AND WORKING DRAWING EXERCISES

### Design Exercises

The following suggestions for project assignments are of a general and very broad nature, and it is expected that they will help generate many ideas for specific design projects. Much design work is undertaken to improve an existing product or system by utilization of new materials, new techniques, or new systems or procedures. In addition to the design of the product itself, another large amount of design work is essential for the tooling, production, and handling of the product. You are encouraged to discuss with your instructor any ideas you may have for a project.

Each solution to a design problem, whether prepared by an individual student or formulated by a group, should be in the form of a report, which should be typed or carefully lettered, assembled, and bound. It is suggested that the report contain the following (or variations of the following, as specified by your instructor).

1. A title sheet. The title of the design project should be placed in approximately the center of the sheet, and your name or the names of those in the group in the lower right-hand corner. The symbol PL should follow the name of the project leader.
2. Table of contents with page numbers.
3. Statement of the purpose of the project with appropriate comments.
4. Preliminary design sketches, with comments on advantages and disadvantages of each, leading to the final selection of the best solution. All work should be signed and dated.
5. An accurately made pictorial and/or assembly drawing(s), using traditional drawing methods or CAD as assigned, if more than one part is involved in the design.
6. Detail working drawings, freehand, mechanical, or CAD-produced as assigned. The 8.5 ×11" sheet size is preferred for convenient insertion in the report. Larger sizes may be bound in the report with appropriate folding.

7. A bibliography or credit for important sources of information, if applicable.

### EX7-22

Design new or improved playground, recreational, or sporting equipment. For example, a new child's toy could be both recreational and educational. Create an assembly drawing.

### EX7-23

Design new or improved health equipment. For example, physically handicapped people need special equipment.

### EX7-24

Design a cup holder attachment to retrofit cars. It must accommodate a range of cup sizes from 8 oz to 64 oz size.

### EX7-25

Design a guitar stand to support either an acoustic or electric guitar. It should be convenient and stable, suitable for use on stage. Allow for quick change of guitars by the musician.

### EX7-26

Break up into design teams. See how many different ideas each team can come up with for a new layout of your classroom. Time limit is 20 minutes.

### EX7-27

Design a new or improved bike safety lock and chain. Integrate the locking devices into the bike's frame, if possible. Create an assembly drawing showing the features of your design.

## Working Drawing Exercises

The problems in Exercises 7.28–7.83 are presented to give you practice in making the type of regular working drawings used in industry. Many exercises, especially assemblies, offer an opportunity to exercise your ability to redesign or improve on the existing design. Due to the variations in sizes and in scales that may be used, you are required to select the sheet sizes and scales, when these are not specified, subject to the approval of the instructor. Standard sheet layouts are shown inside the front cover of this book. (Any of the title blocks shown inside the back cover of this book may be used, with modification if desired, or you may design the title block if assigned by the instructor.)

The statements for each problem are intentionally brief and your instructor may vary the requirements. Use the preferred metric system or the acceptable complete decimal inch system, as assigned.

In problems presented in pictorial form, the dimensions and finish marks are to provide you the information necessary to make the orthographic drawing or solid model. The dimensions given are in most cases those needed to make the parts, but due to the limitations of pictorial drawings they are not in all cases the dimensions that should be shown on the working drawing. In the pictorial problems, the rough and finished surfaces are shown, but finish marks are usually omitted. You should add all necessary finish marks and place all dimensions in the preferred places in the final drawings.

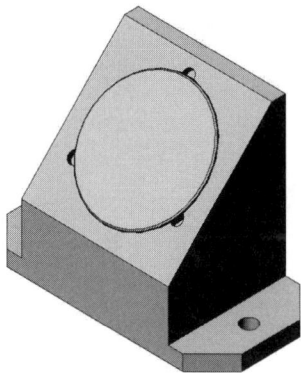

### EX7-28

Create part drawings and an assembly for the lens and mount. Maintain the critical distances and precise 45° angle for the lens.

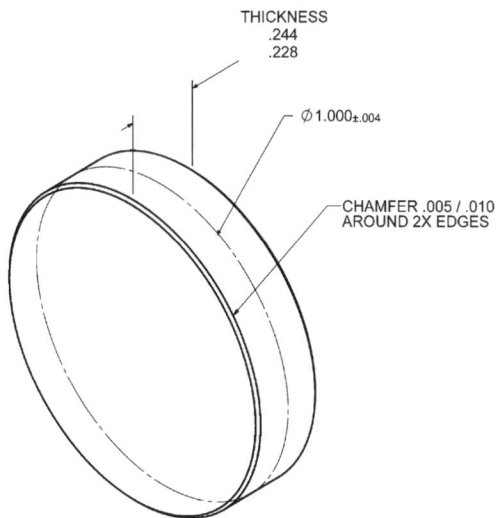

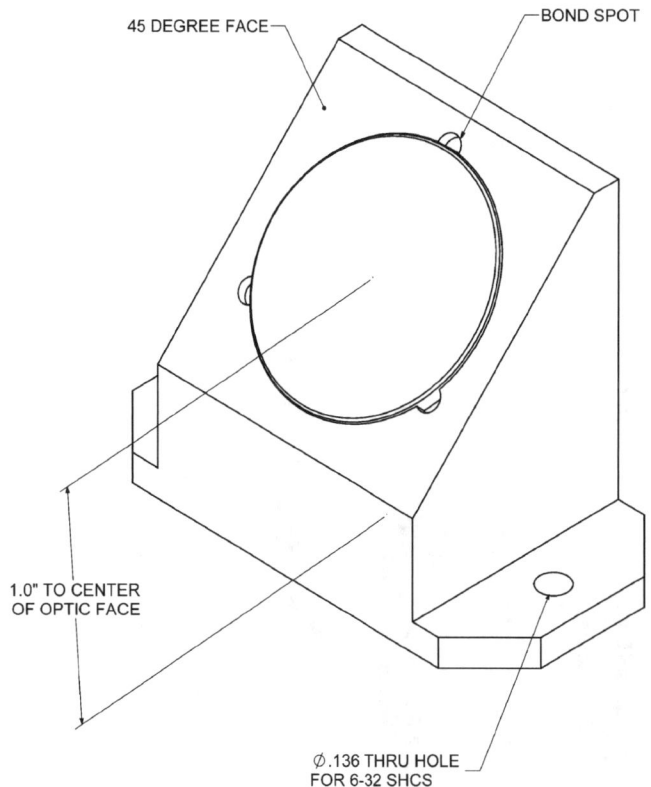

## EX7-29

Create an exploded assembly drawing for the gyroscope. Create detail drawings for the parts as assigned by your instructor. Dimensioned parts are shown on the facing page.

## EX7-30

Design the sheet metal housing for the power and D-sub connectors shown. Download stock models for standard parts. Create the flat patterns for the sheet metal if assigned by your instructor.

## EX7-31

Design a sheet metal drill bit case. Create detailed part and assembly drawings. Develop the flat patterns if assigned. Use "relations" in your model so that you can change the sizes for the holes and overall height, width and depth for the case and automatically update your design to different configurations.

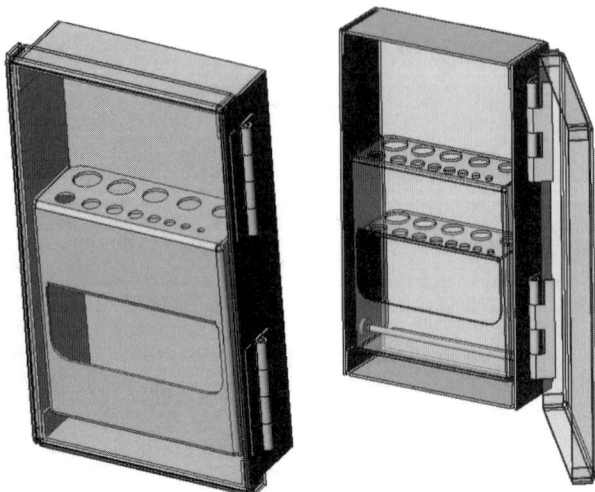

## EX7-32

Create an exploded assembly drawing for the clamp. Dimensioned parts are shown on page 535.

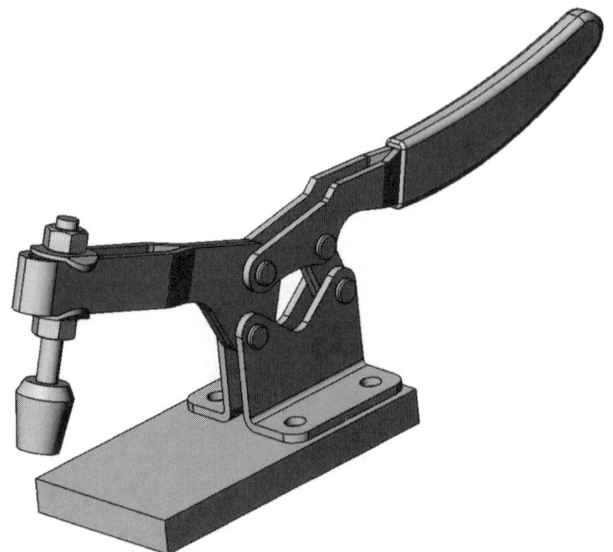

NOTES:
1. ALL DIMENSIONS ARE IN INCHES
2. SCALE FOR PARTS VARIES

OUTER RING

⌀ 2.625

⌀ .125 (CROSS-SECTION)

60°

R.75

3X R.05

DETAIL OF WEB CUT (7X EQUALLY SPACED AROUND ₵ )

⌀ .60

CENTER POST

.125

2X R.02

⌀ .063 THRU

.125

.250

2.500

INNER RING

⌀ 2.375

⌀ .125 (CROSS-SECTION)

⌀ .042 THRU (BOTH SIDES)

WHEEL

R.100

.310

.060

⌀ .125 THRU

6X R.02

R.125 TYP

.066

⌀ 1.925

⌀ .375

17°

R.03 AROUND PERIMETER OF WEB CUTS (BOTH SIDES)

BOTTOM PEG

.538 (TO ₵ OF R.0625 CUT)

.500

.125

R.063

⌀ .100

⌀ .188

2X R.05

BASE

1.520 (TO ₵ OF R.10 CIRCULAR CUT)

1.500 (OVERALL HEIGHT OF PART)

.332

⌀ .303

R.10

4.5° REF

.738

R.020

R.125

R.125

⌀ 1.125

TOP PEG

.538 (TO ₵ OF R.0625 CUT)

.500

.125

.031

90°

R.063

⌀ .100

⌀ .188

2X R.05

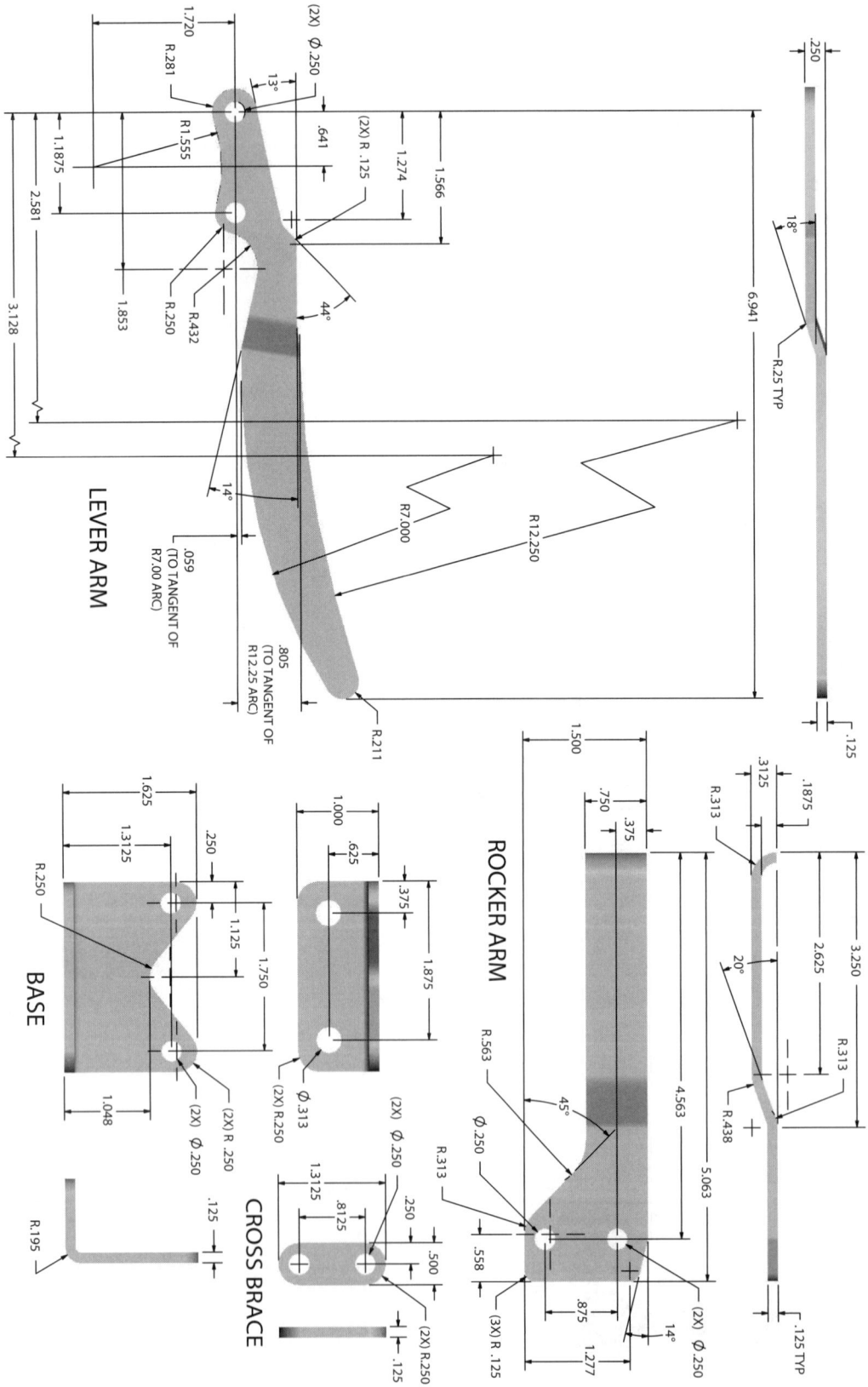

LEVER ARM

ROCKER ARM

BASE

CROSS BRACE

## EX7-33

Make detail drawing for the table bracket.

Ø45.12–45.23 THRU
1.5 SAW
M12 x 1.75
↧18
M12 x 1.75
TO SLOT
Ø72
R12
R32
R19
R20
Ø13.5
TO SLOT
CORE
22 WIDE
(Centered)
C1–1 REQD
FILLETS & ROUNDS R3
PARTIAL VIEW
IN DIRECTION OF
ARROW
METRIC

## EX7-34

Make detail drawing for the RH tool post. If assigned, convert dimensions to metric system.

2 x .625–11UNC–2B
45° CHAMFER x DEPTH OF THD
2 x Ø.625
⊔Ø.938 ↧.62
R.24
C1–1 REQD
ROUNDS R.06

## EX7-35

Make detail drawing for the drill press base. Use unidirectional metric or decimal-inch dimensions.

.375–16UNC–2B
Ø1.623–1.625
FILLETS & ROUNDS R.063
4 x .25–20UNC–2B
(Through)
R1.88
R.25
C1
1 REQD
PARTIAL BOTTOM VIEW
(REDUCED SCALE)
4 x Ø.375
R.438
R2

### EX7-36

Make detail drawing for the shifter fork. If assigned, convert dimensions to metric system.

### EX7-37

Make detail drawing for the idler arm.

### EX7-38

Make detail drawing for the drill press bracket. If assigned, convert dimensions to decimal inches or redesign the part with metric dimensions.

**EX7-39**

Make detail drawing for the dial holder. If assigned, convert dimensions to decimal inches or redesign the part with metric dimensions.

**EX7-40**

Make detail drawings half size for the rack slide. If assigned, convert dimensions to decimal inches or redesign the part with metric dimensions.

## EX7-41

Make detail drawing half size for the automatic stop box. If assigned, redesign the part with metric dimensions.

## EX7-42

Make detail drawings half size for the conveyer housing. If assigned, convert dimensions to decimal inches or redesign the parts with metric dimensions.

VIEW AT A          C I – I EACH REQD

**EX7-43**

For the spindle housing, draw as follows. Given: Front, left-side, and bottom views, and partial removed section. Required: Front view in full section, top view, and right-side view in half section on A-A. Draw half size. If assigned, dimension fully.

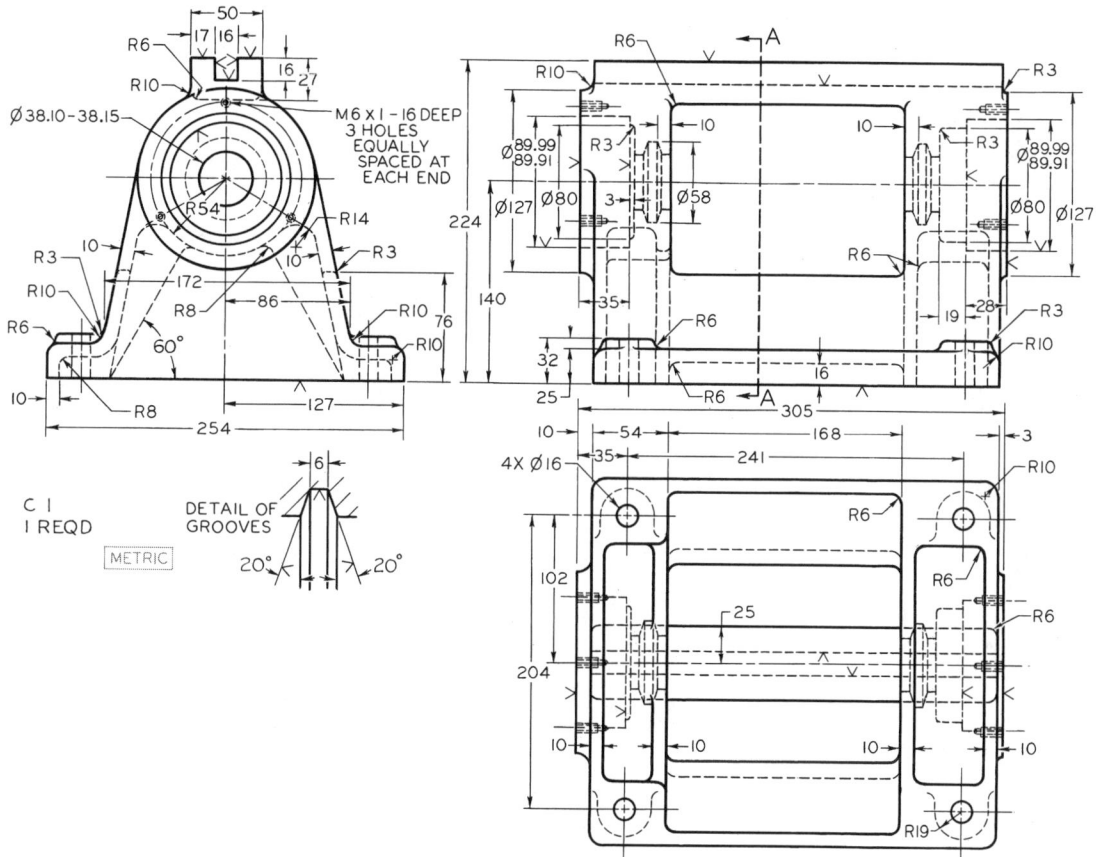

## EX7-44

For the arbor support bracket, draw the following. Given: Front and right-side views. Required: Front, left-side, and bottom views, and a detail section A-A. Use American National Standard tables for indicated fits and if required convert to metric values. If assigned, dimension in the metric or decimal inch system.

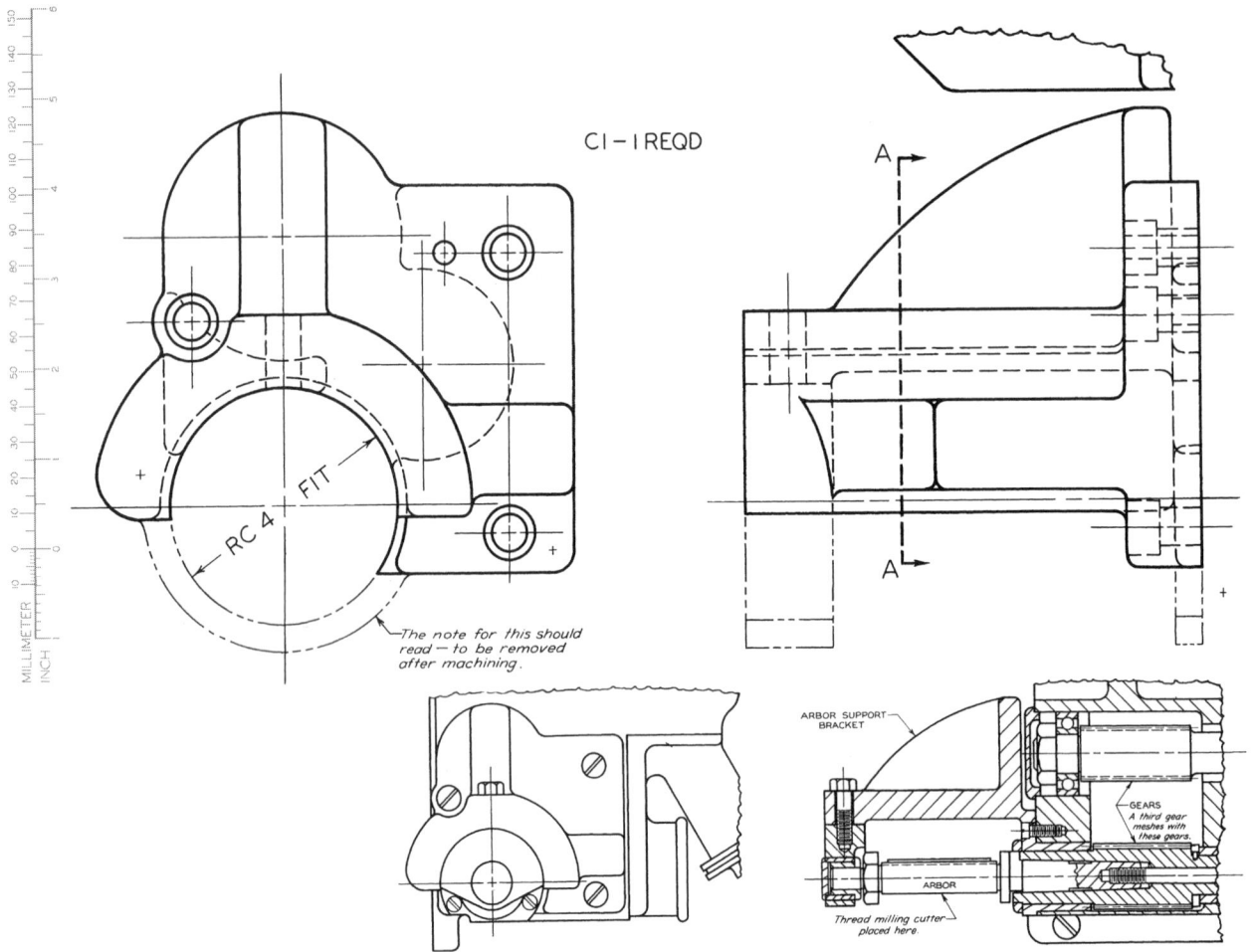

CI – I REQD

RC 4 FIT

The note for this should read – to be removed after machining.

ARBOR SUPPORT BRACKET

GEARS
A third gear meshes with these gears.

ARBOR

Thread milling cutter placed here.

## EX7-45

For the pump bracket for a thread milling machine, draw the following. Given: Front and left-side views. Required: Front and right-side views, and top view in section on A-A. Draw full size. If assigned, dimension fully.

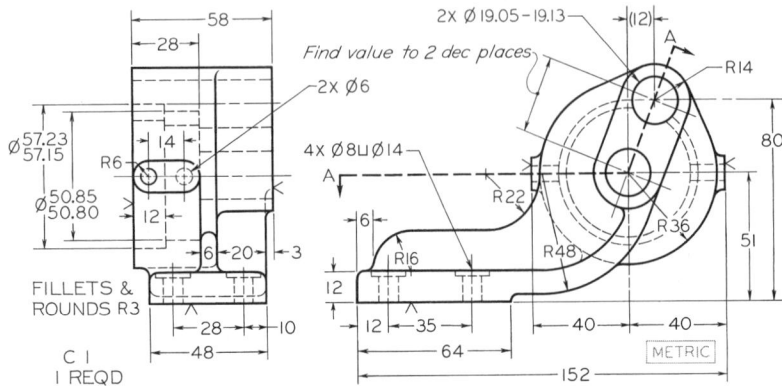

58
28
2x Ø 19.05-19.13
(12)
A
Find value to 2 dec places
2x Ø6
R14
Ø 57.23 / 57.15
R6
14
4x Ø8⌴Ø14
A
R22
80
Ø 50.85 / 50.80
12
R36
6
6
R16
R48
51
FILLETS & ROUNDS R3
6 20 3
12
R16
CI
I REQD
28 10
48
12 35 40 40
64
METRIC
152

## EX7-46

For the support base for planer, draw the following. Given: Front and top views. Required: Front and top views, left-side view in full section A-A, and removed section B-B. Draw full size. If assigned, dimension fully.

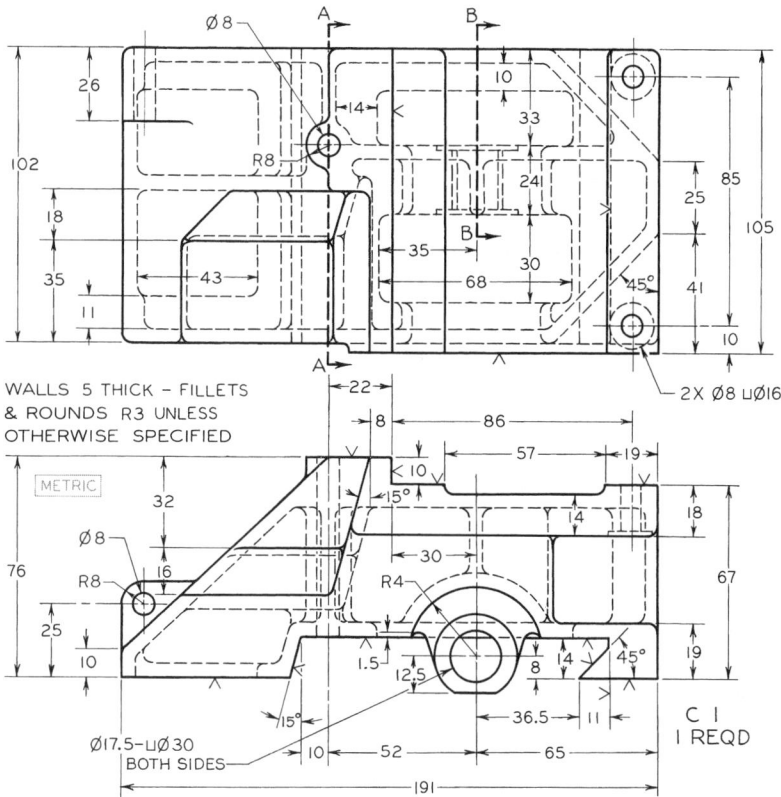

WALLS 5 THICK – FILLETS
& ROUNDS R3 UNLESS
OTHERWISE SPECIFIED

METRIC

C 1
1 REQD

Ø17.5–⊔Ø30
BOTH SIDES

## EX7-47

For the jaw base for chuck jaw, draw the following. Given: Top, right-side, and partial auxiliary views. Required: Top, left-side (beside top), front, and partial auxiliary views complete with dimensions, if assigned. Use metric or decimal inch dimensions. Use American National Standard tables for indicated fits or convert for metric values.

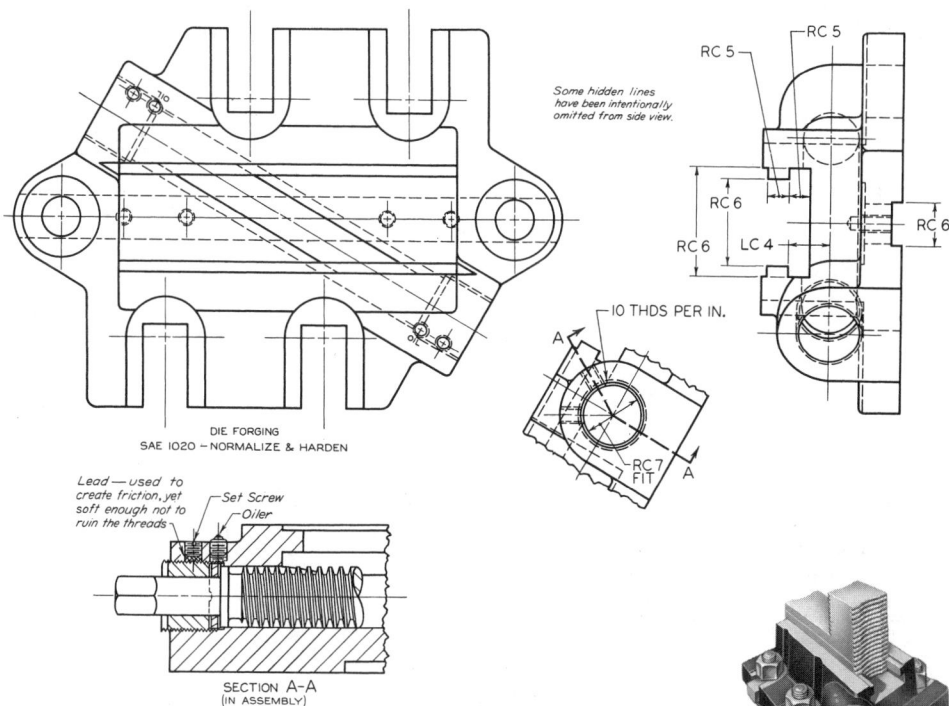

Some hidden lines
have been intentionally
omitted from side view.

RC 5

RC 5

RC 6

RC 6

RC 6

LC 4

RC 6

10 THDS PER IN.

A

A

RC 7
FIT

DIE FORGING
SAE 1020 – NORMALIZE & HARDEN

Lead — used to
create friction, yet
soft enough not to
ruin the threads

Set Screw
Oiler

SECTION A-A
(IN ASSEMBLY)

MILLIMETER
INCH

## EX7-48

For the fixture base for 60-ton vertical press, draw the following. Given: Front and right-side views. Required: Revolve front view 90° clockwise; then add top and left-side views. Draw half size. If assigned, complete with dimensions.

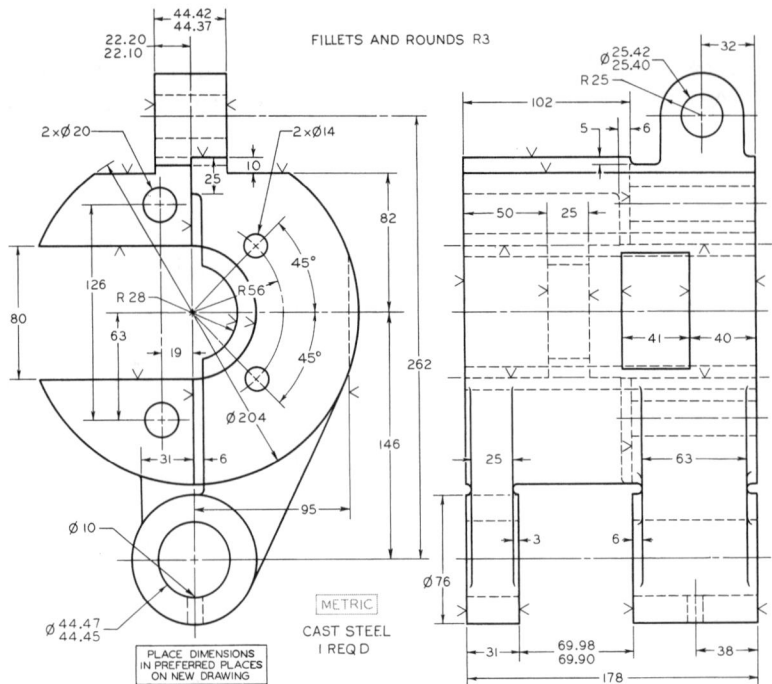

FILLETS AND ROUNDS R3

PLACE DIMENSIONS IN PREFERRED PLACES ON NEW DRAWING

METRIC

CAST STEEL
1 REQD

## EX7-49

For the bracket, draw the following. Given: Front, left-side, and bottom views, and partial removed section. Required: Make detail drawing. Draw front, top, and right-side views, and removed sections A-A and B-B. Draw half size. Draw section B-B full size. If assigned, complete with dimensions.

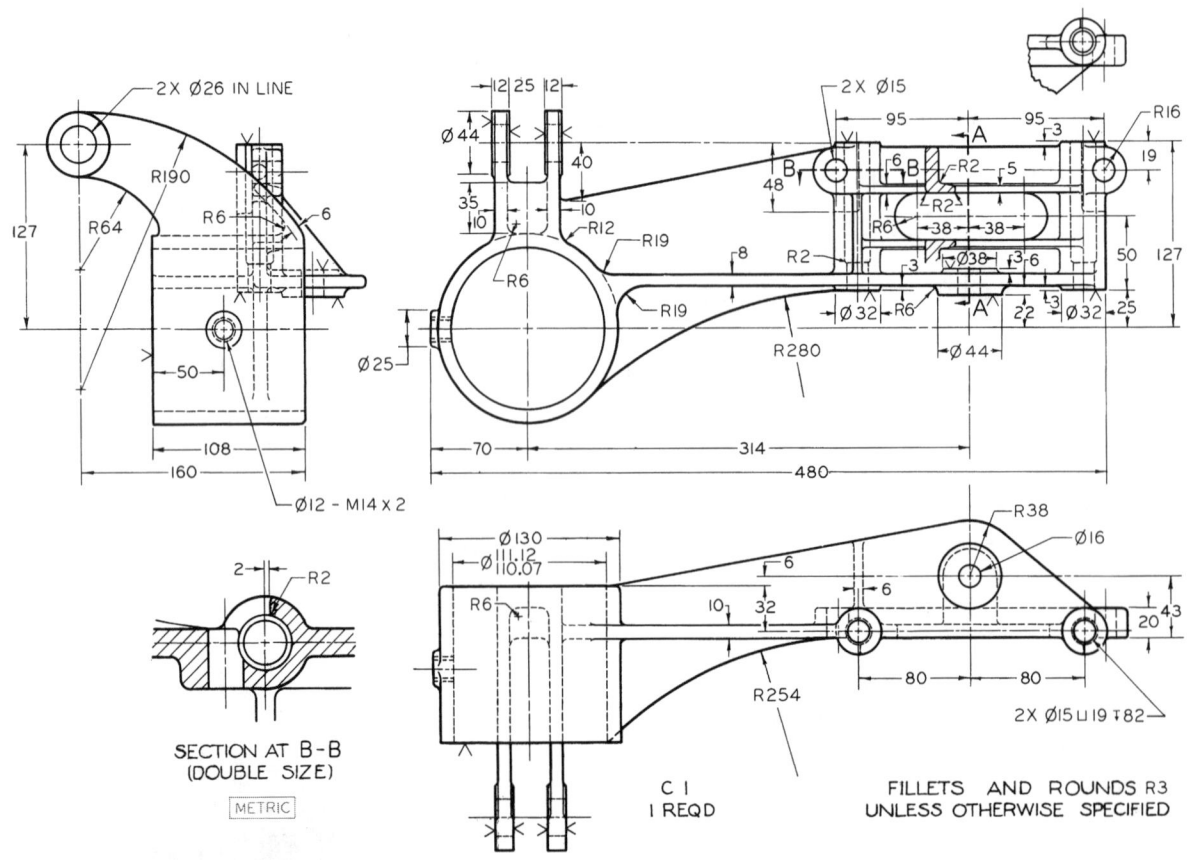

SECTION AT B-B
(DOUBLE SIZE)

METRIC

C 1
1 REQD

FILLETS AND ROUNDS R3
UNLESS OTHERWISE SPECIFIED

**EX7-50**

For the roller rest bracket for automatic screw machine, draw the following. Given: Front and left-side views. Required: Revolve front view 90° clockwise; then add top and left-side views. Draw half size. If assigned, complete with dimensions.

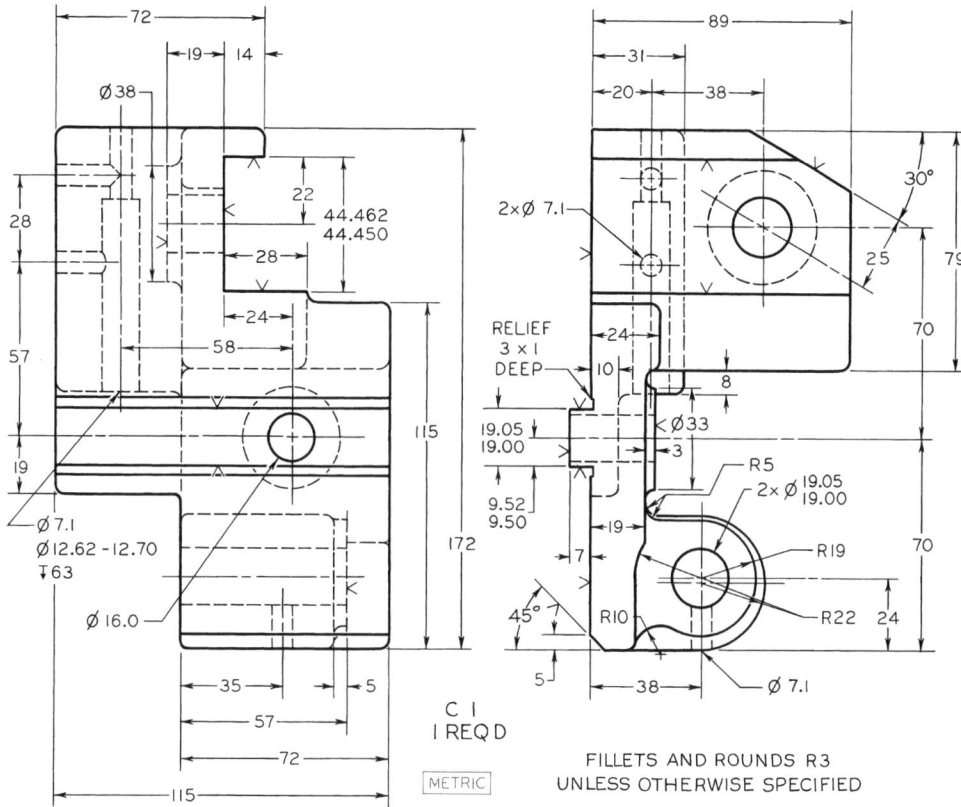

C 1
I REQ D

METRIC

FILLETS AND ROUNDS R3
UNLESS OTHERWISE SPECIFIED

**EX7-51**

For the guide bracket for gear shaper, draw the following. Given: Front and right-side views. Required: Front view, a partial right-side view, and two partial auxiliary views taken in direction of arrows. Draw half size. If assigned, complete with unidirectional dimensions.

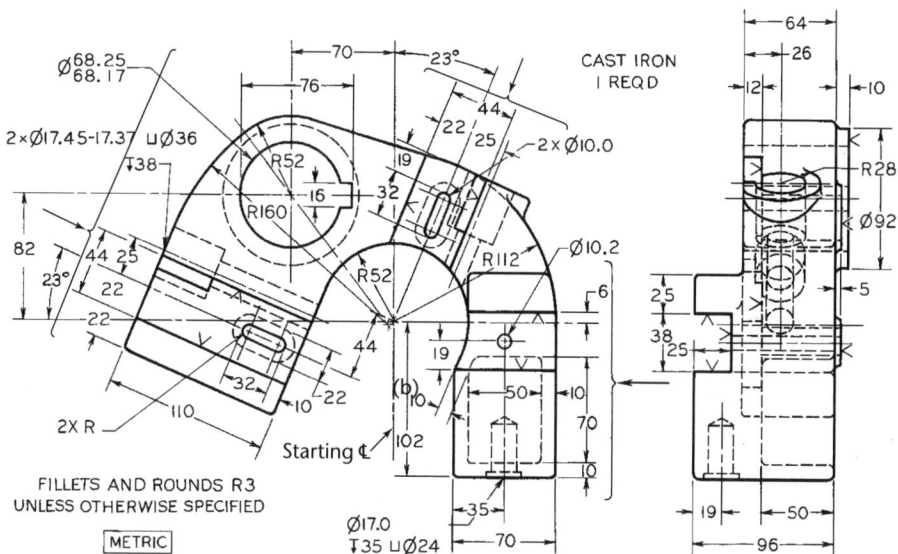

FILLETS AND ROUNDS R3
UNLESS OTHERWISE SPECIFIED

METRIC

CAST IRON
I REQ D

**EX7-52**

For the rear tool post, draw the following. Given: Front and left-side views. Required: Take left-side view as new top view; add front and left-side views, approx. 215 mm apart, a primary auxiliary view, then a secondary view taken so as to show true end view of 19 mm slot. Complete all views, except show only necessary hidden lines in auxiliary views. Draw full size. If assigned, complete with dimensions.

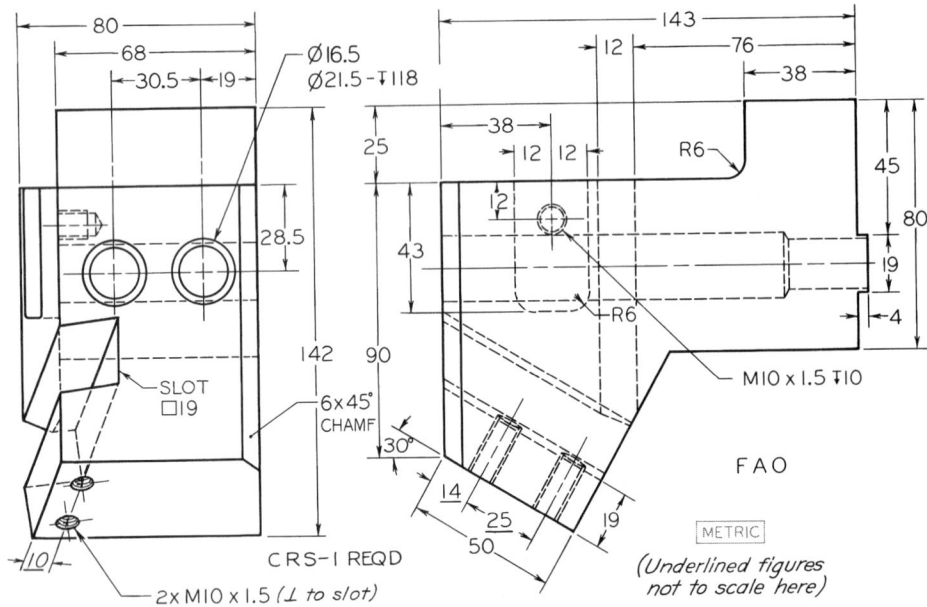

**EX7-53**

For the bearing for a worm gear, draw the following. Given: Front and right-side views. Required: Front, top, and left-side views. Draw full size. If assigned, complete with dimensions.

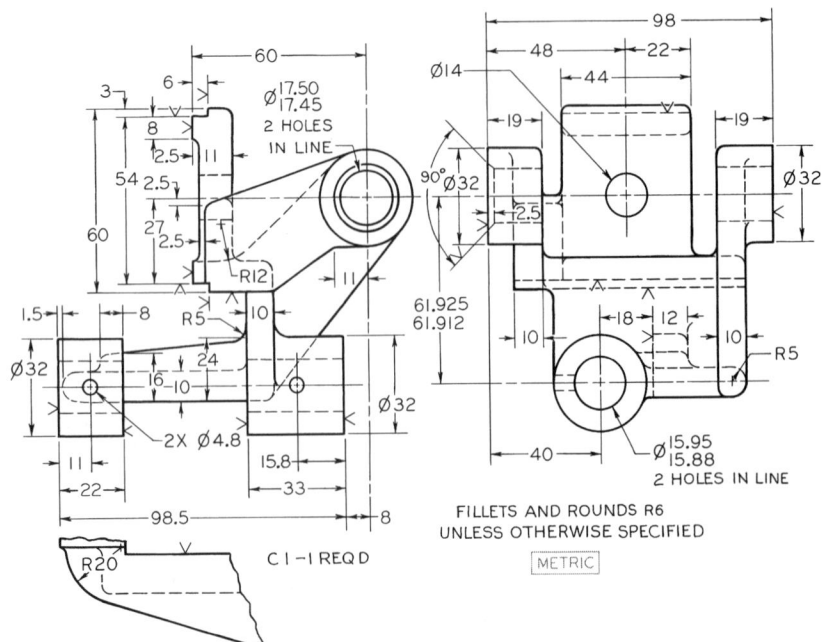

**EX7-54**

For the caterpillar tractor piston, draw the following. Make detail drawing full size. If assigned, use unidirectional decimal inch system, converting all fractions to two place decimal dimensions, or convert all dimensions to metric.

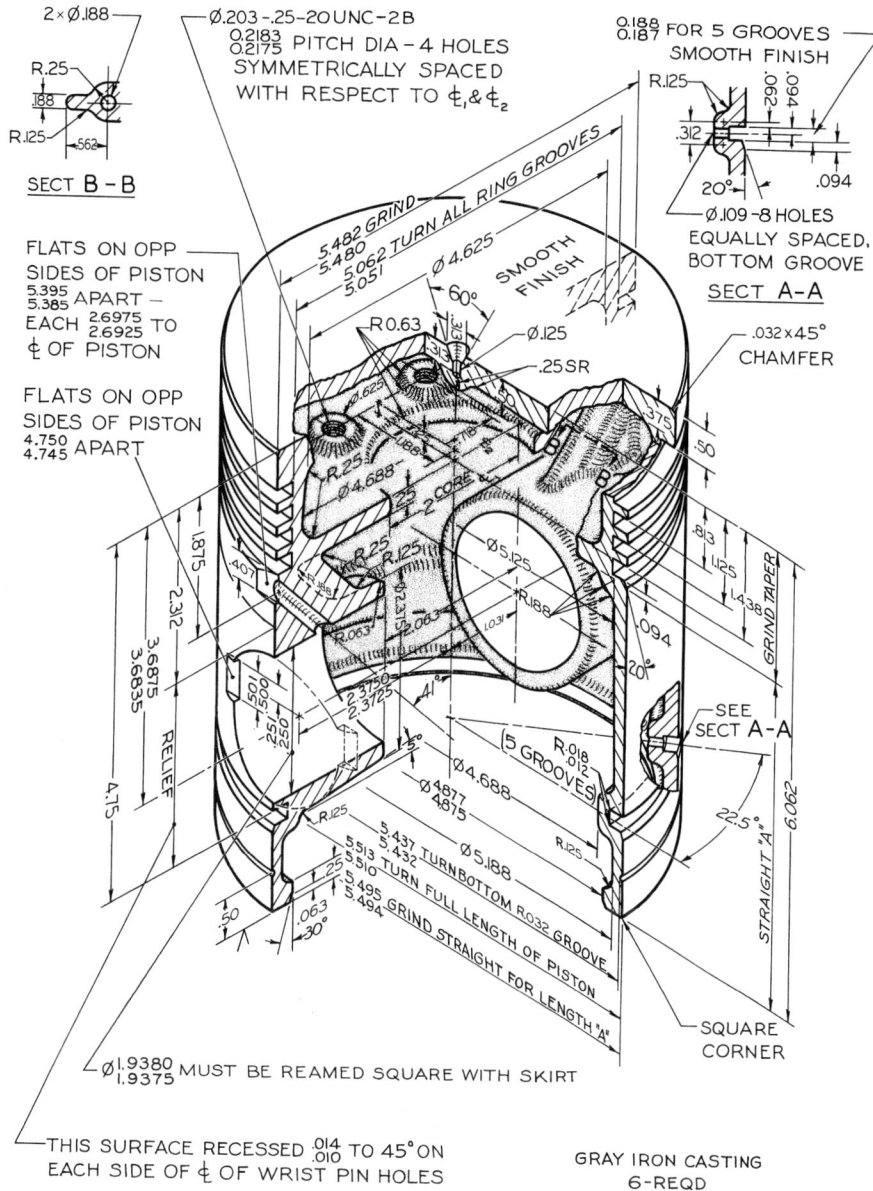

GRAY IRON CASTING
6-REQD

**EX7-55**

For the generator drive housing, draw the following. Given: Front and left-side views. Required: Front view, right-side view in full section, and top view in full section on A-A. Draw full size. If assigned, complete with dimensions.

**EX7-56**

For the machinist's clamp, draw the following. Draw details and assembly. If assigned, use unidirectional two place decimal inch dimensions or redesign for metric dimensions.

## EX7-57

For the hand rail column, draw the following. (1) Draw details. If assigned, complete with dimensions. (2) Draw assembly.

## EX7-58

For the drill jig, draw the following. (1) Draw details. If assigned, complete with dimensions. (2) Draw assembly.

## EX7-59

For the tool post, draw the following. (1) Draw details. (2) Draw assembly. If assigned, use unidirectional two place decimals for all fractional dimensions or redesign for all metric dimensions.

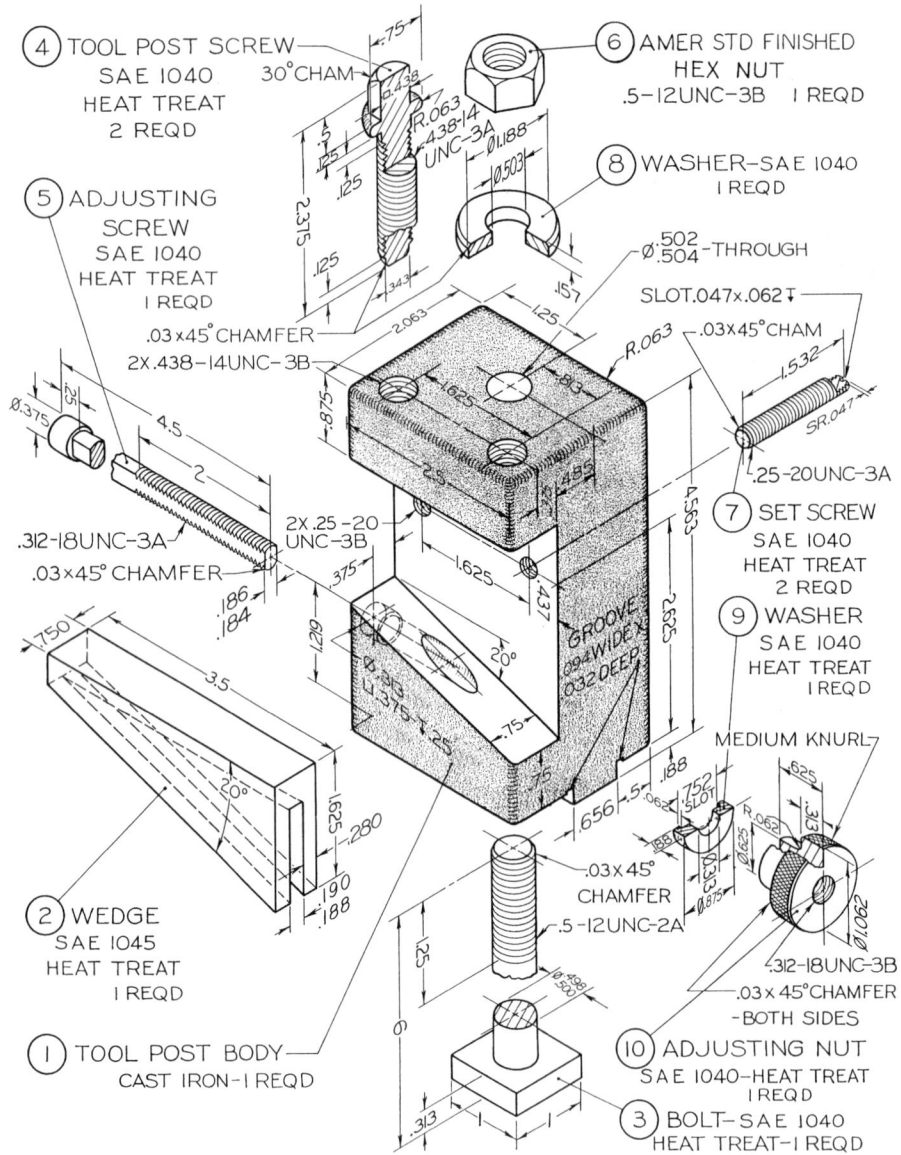

**EX7-60**

For the belt tightener, draw the following. (1) Draw details. (2) Draw assembly. It is assumed that the parts are to be made in quantity and they are to be dimensioned for interchangeability on the detail drawings. Design as follows. (a) Bushing fit in pulley: locational interference fit. (b) Shaft fit in bushing: free running fit. (c) Shaft fits in frame: sliding fit. (d) Pin fit in frame: free running fit. (e) Pulley hub length plus washers fit in frame: allowance 0.13 and tolerances 0.10. (f) Make bushing 0.25 mm shorter than pulley hub. (g) Bracket fit in frame: Same as (e).

## EX7-61

For the milling jack, draw the following. (1) Draw details. (2) Draw assembly. If assigned, convert dimensions to metric or decimal inch system.

## EX7-62

For the connecting bar, draw the following. (1) Draw details. (2) Draw assembly. If assigned, convert dimensions to metric or decimal inch system.

**EX7-63**

For the clamp stop, draw the following. (1) Draw details. (2) Draw assembly. If assigned, convert dimensions to decimal inch system or redesign for metric dimensions.

**EX7-64**

For the pillow block bearing, draw the following. (1) Draw details. (2) Draw assembly. If assigned, complete with dimensions.

## EX7-65

For the centering rest, draw the following. (1) Draw details. (2) Draw assembly. If assigned, complete with dimensions.

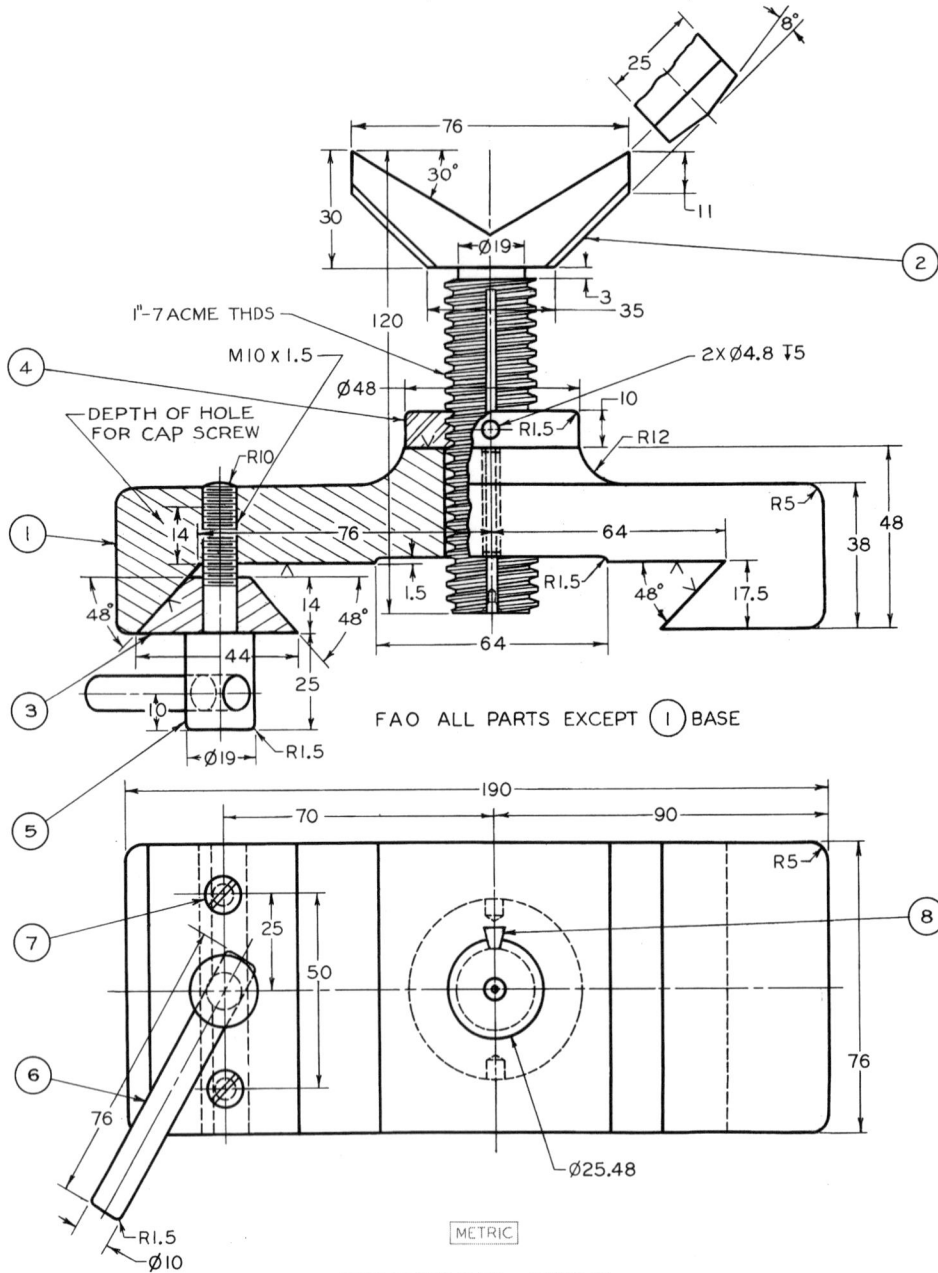

FAO ALL PARTS EXCEPT ① BASE

METRIC

CENTERING REST

| | PARTS LIST | | | | | | | |
|---|---|---|---|---|---|---|---|
| NO. | PART NAME | MATL | REQD | NO. | PART NAME | MATL | REQD |
| 1 | BASE | C I | 1 | 5 | CLAMP SCREW | SAE 1020 | 1 |
| 2 | REST | SAE 1020 | 1 | 6 | CLAMP HANDLE | SAE 1020 | 1 |
| 3 | CLAMP | SAE 1020 | 1 | 7 | M6X1-25LG FIL HD CAP SCREW | | 2 |
| 4 | ADJUSTING NUT | SAE 1020 | 1 | 8 | 5.5X5.5X3.2-25LG KEY | SAE 1030 | 1 |

## EX7-66

For the pipe vise, draw the following. (1) Draw details. (2) Draw assembly. To obtain dimensions, take distances directly from figure with dividers; then set dividers on printed scale and read measurements in millimeters or decimal inches as assigned. All threads are general purpose metric threads or unified coarse threads except the American National Standard pipe threads on handle and handle caps.

PIPE VISE

6 HANDLE — STD PIPE

7 HANDLE CAP — STD CAP

2 VISE SCREW — S A E 1045

9 HEX NUT — S A E 1020

10 GUIDE LINK — S A E 1020

3 GUIDE BAR — S A E 1020

1 VISE BASE — CAST STEEL

4 UPPER JAW — S A E 1045 HARDENED

5 LOWER JAW — S A E 1045 HARDENED

8 HEX CAP SCR — S A E 1020

MILLIMETERS
INCHES

**EX7-67**

For the tap wrench, draw the following. (1) Draw details. (2) Draw assembly. If assigned, use unidirectional two place decimals for all fractional dimensions or redesign for metric dimensions.

**EX7-68**

For the machinist's vise, draw the following. (1) Draw details. (2) Draw assembly. If assigned, use unidirectional two place decimals for all fractional dimensions or redesign for metric dimensions.

**EX7-69**

For the screw jack, draw the following. (1) Draw details. (2) Draw assembly. If assigned, convert dimensions to decimal inches or redesign for metric dimensions.

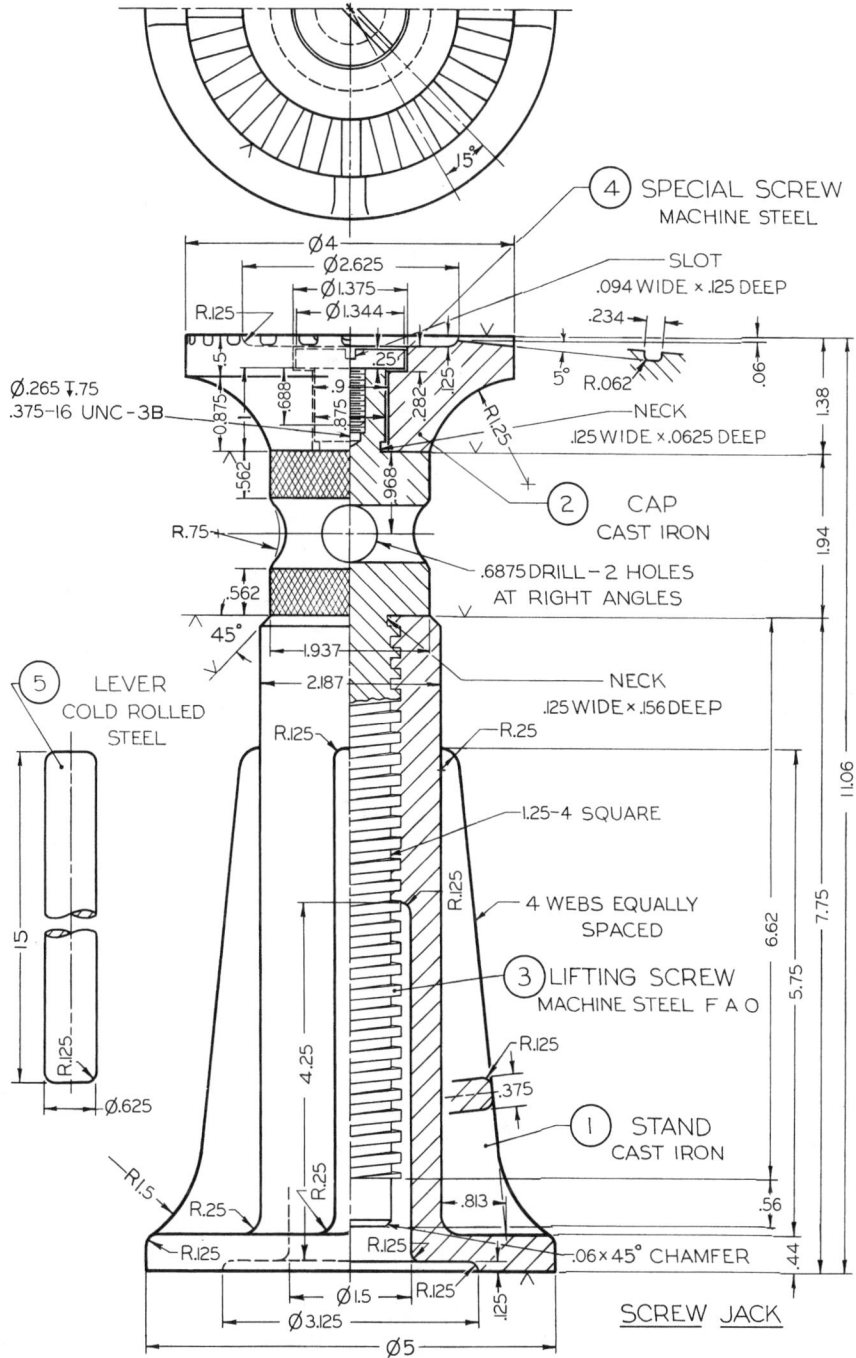

SCREW JACK

## EX7-70

For the stock bracket for cold saw machine, draw the following. (1) Draw details. (2) Draw assembly. If assigned, use unidirectional decimal dimensions or redesign for metric dimensions.

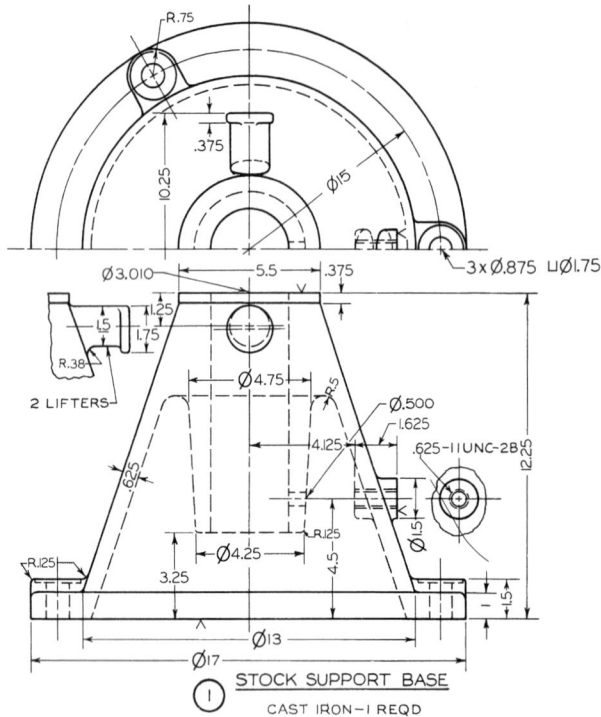

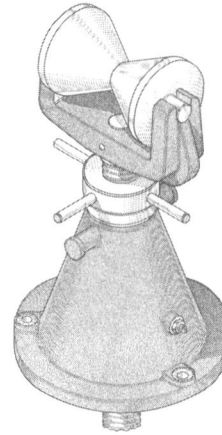

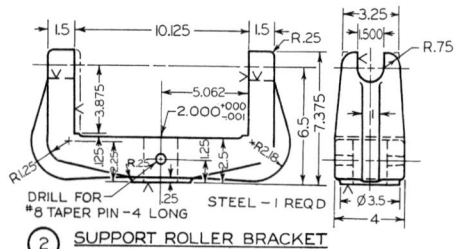

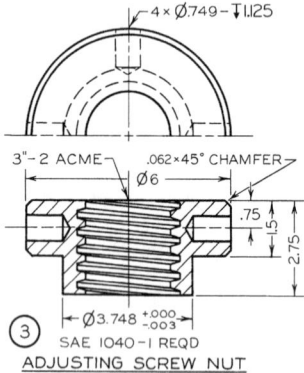

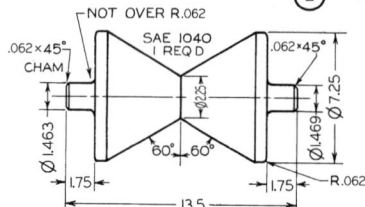

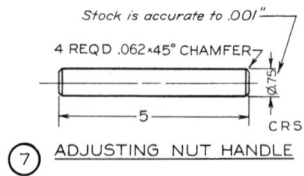

1 STOCK SUPPORT BASE
CAST IRON—1 REQD

2 SUPPORT ROLLER BRACKET
STEEL—1 REQD

3 ADJUSTING SCREW NUT
SAE 1040-1 REQD

4 STOCK SUPPORT ROLLER
SAE 1040 1 REQD

5 SUPPORT ADJUSTING SCREW
SAE 1040-1 REQD

6 ADJUSTING SCREW GUIDE
SAE 1020 1 REQD

7 ADJUSTING NUT HANDLE
4 REQD .062×45° CHAMFER
CRS

8 THRUST BEARING
.625 BALLS

STOCK ITEMS
9 1—#8 TAPER PIN 4" LENGTH
10 1—.625 FIN. HEX JAM NUT

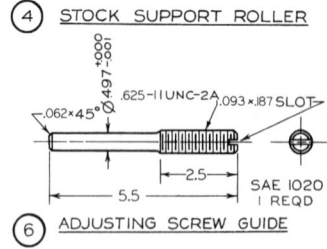

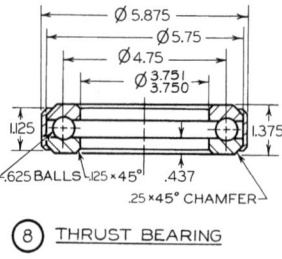

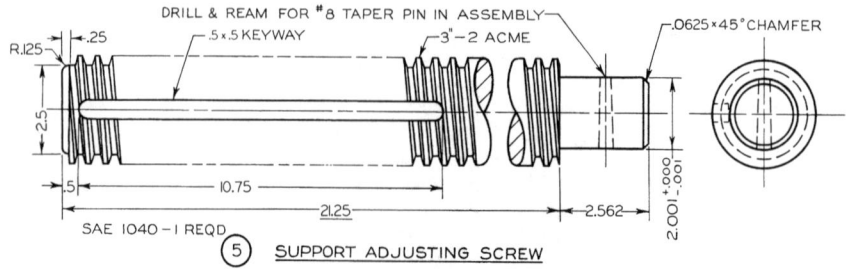

## EX7-71

For the front circular forming cutter holder, draw the following. (1) Draw details. (2) Draw assembly. To obtain dimensions, take distances directly from figure with dividers and set dividers on printed scale. Use metric or decimal inch dimensions as assigned.

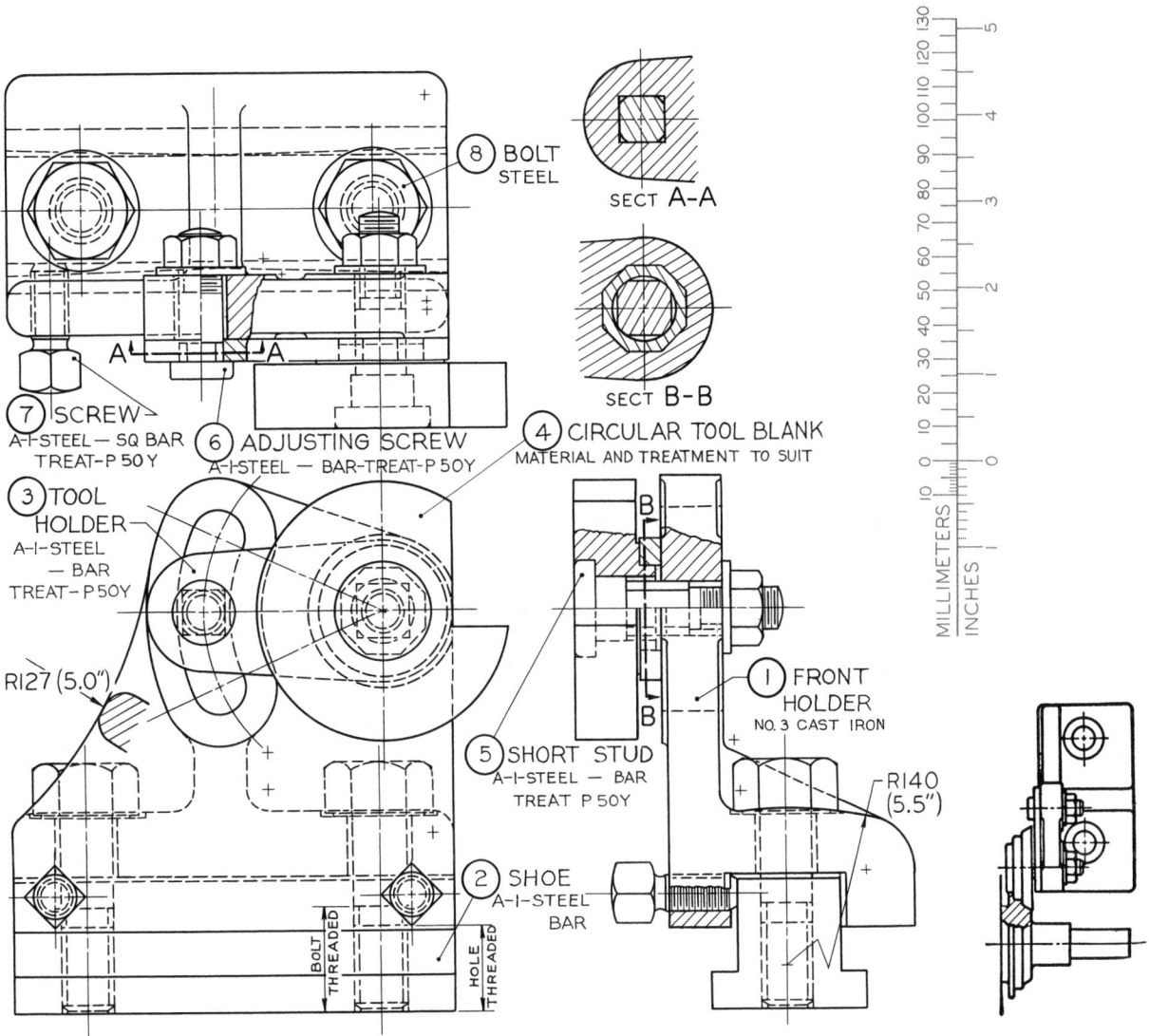

8 BOLT
STEEL

SECT A-A

SECT B-B

7 SCREW
A-I-STEEL — SQ BAR
TREAT-P 50 Y

6 ADJUSTING SCREW
A-I-STEEL — BAR-TREAT-P 50Y

4 CIRCULAR TOOL BLANK
MATERIAL AND TREATMENT TO SUIT

3 TOOL
HOLDER
A-I-STEEL
— BAR
TREAT-P 50Y

R127 (5.0")

5 SHORT STUD
A-I-STEEL — BAR
TREAT P 50Y

1 FRONT
HOLDER
NO. 3 CAST IRON

R140
(5.5")

2 SHOE
A-I-STEEL
BAR

BOLT THREADED

HOLE THREADED

MILLIMETERS
INCHES

130 120 110 100 90 80 70 60 50 40 30 20 10 0
5 4 3 2 1 0

## EX7-72

For the machine vise, draw the following. (1) Draw details. (2) Draw assembly. If assigned, convert dimensions to the decimal inch system or redesign with metric dimensions.

**EX7-73**

*Part (a)* For the grinder vise, draw the following. (1) Draw details. (2) Draw assembly. If assigned, convert dimensions to decimal inches or redesign with metric dimensions. See parts (b) and (c) on the following pages.

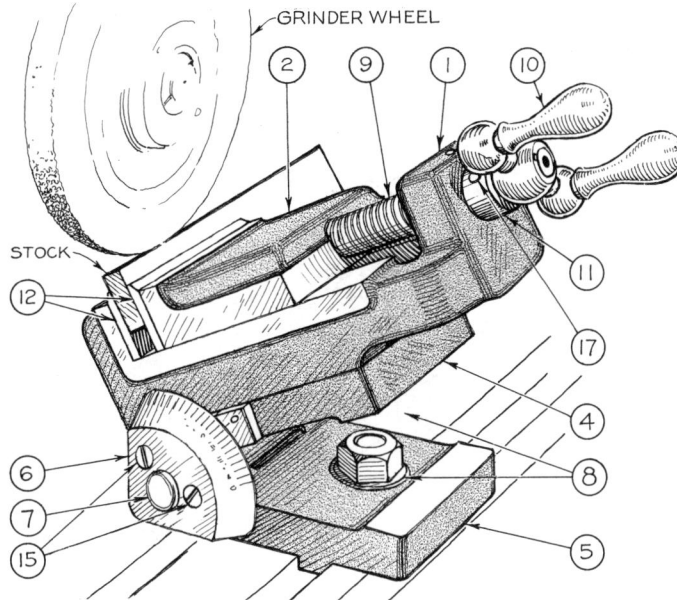

## EX7-73

*Part (b)* For the grinder vise, see EX7-73 *part (a)* for instructions.

VIEW IN DIRECTION
OF ARROW A

SPHERICAL CORNER

Ø.626
.625
⌴Ø1.062 FACE
ON BOTH SIDES
THICKNESS TO BE .750

2.5 MAX CUTTER

SHARP
CORNER
LOOKING
IN THIS
DIRECTION

FILLETS AND
ROUNDS R.125

4×.438−14UNC−2B
⊤.875

2×#10−32UNF−2B
⊤.437

R.375
R.25

CUTTER

① VISE BASE
CI I REQD

#10−32UNF−2A
.062×.062 SLOT
Ø.312

⑮ PROTRACTOR
SCREW SAE 1115
2 REQD

#10−32UNF−2A
.062×.031 SLOT
Ø.312
.062

⑯ JAW PLATE SCREW
SAE 1115 −4 REQD

Ø.20
⊤.5

PIN
(FN 1 fit)

.031×45°
CHAMFER
(PIN)

Ø.500
.499

SØ1

SØ.75

1.28
.875

CUT FOR ⑲
#204 WOODRUFF KEY

⑩ COMPOUND REST HANDLE
SAE 3140 − 1 REQD

.562−8 ACME
L H −CLEAR
THROUGH

PARTIAL
TOP VIEW

SPHERICAL
CORNER

SIDE
VIEW−
LOOKING
IN THE
DIRECTION
OF ARROW A

R.062

2×#10−32UNF−2B
⊤.438

2×.5−12UNC−2B ⊤.688
BOTTOM TAP

② SLIDING JAW
CI −1 REQD

90 80 70 60 50 40 30 20 10

40°

45°

20°

45°

R1.5

R.5

Ø.626
.625

2×Ø.219 ⌴Ø.344 ⊤.125

⑥ PROTRACTOR
SAE 3140−1 REQD

**EX7-73**

*Part (c)* For the grinder vise, see EX7-73 *part (a)* for instructions.

4 × Ø.469 ⌴Ø.656 ⊤.25

.687

R.125

1.5 — Ø1.312

3.25

1.5

1.125

.25

.125

R.2.25

LOCATE ZERO AND
MARK IN ASSEMBLY

R1.515

R.875

R.687

Ø.500
.501

4.687

.562

.75

1.749
1.748

.75

④ SECONDARY BASE
C I - I REQD

.031  30°  .093

Ø1.187

Ø.562

⑧  WASHER
S A E  1020 - 2 REQD

Ø.499
.498 — 12 UNC — 2A

USE ⑱
STANDARD
FIN HEX
NUT & WASHER
⑧

4.5

Ø.624
.623

R.062

45°
CHAMFER
TO BOTTOM
OF THD

⑦ HINGE BOLT
S A E  1040 - 1 REQD

REAM FOR ⑰
#0 TAPER PIN

.28

.437

Ø.626
.625

⑪  COLLAR
S A E  1020 - 1 REQD

45° CHAMFER TO BOTTOM
OF THD

.437 — 14 UNC — 2A

.062

Ø.625

.093

1.25

.25

⑬ SEC BASE SCREW
S A E  1020 - 4 REQD

2 × Ø.531 - ⌴Ø.781 - ⊤.25

.437

.625

1.25

.625  1.5

2.75

③ CLAMP PLATE
S A E  3140 - 1 REQD

45° CHAMFER TO BOTTOM
OF THD

.5 — 12 UNC — 2A

.062

Ø.75

.093

1.25

.25

⑭ CLAMP PLATE SCREW
S A E  3140  C H - 2 REQD

2 × Ø.218 - ⌴Ø.343 - ⊤.062

2.75  .5

3.5

.062

.375

45°

.187

1.125

⑫  JAW PLATE
S A E  1030 - 2 REQD

Ø.563 - CLEAR THROUGH
⌴Ø1.25 - ⊤.0625

(No tapped holes
on this side)

.687

1.5

3.25

R.125

R.125

.75

.25

.75

.375

1.375

.75

1.751
1.750

3.5

.75

R.125

2.25

Ø.500
.501

.625

.5

1.25

4.687

⑤ HINGE BASE
C I - I REQD

#10 - 32 UNF - 2B - .375 DEEP
2 HOLES ON 1" DIA BC

CUT FOR ⑲
#204 WOODRUFF KEY

REAM IN PLACE
WITH PART #11 FOR ⑰
#0 TAPER PIN

4.75

.75

.315

.498
.497

.624
.623

.937

2.375

.125

Ø.437

.562 — 8 ACME L H

.375

Ø.09 COMB.
DRILL AND
CSK — BOTH
ENDS

⑨ VISE SCREW
S A E  3140 - 1 REQD

## EX7-74

For the trolley, draw the following. (1) Draw details, omitting parts 7–14. (2) Draw assembly. If assigned, convert dimensions to decimal inches or redesign for metric dimensions.

.0625 X 45° CHAM

(4) AXLE
C R S – 2 R E Q D

0.625 × 45° CHAM

DETAIL SHOWING
BEARING ASSY

STOCK ITEMS:

(10) .875×8.25 SF HEX HD BOLT – 2 REQD
DRILL .187 COTTER PIN HOLE .187 FROM TIP

(11) .875 SF SLOTTED HEX NUT – 2 REQD

(12) .875 SAE PLAIN WASHER – 8 REQD

(13) .156 × 1.5 SAE COTTER PIN – 2 REQD

(14) #10(.190) X.625 ROUND HD MACH SCR – 4 REQD

8 BALLS Ø.312

(7) #99504 NEW
DEPARTURE BALL BEARING
2 REQ'D. (Stock item).

2 × Ø 0.218

(6) KEEPER PLATE
C R S – 2 REQD

2 × Ø.906

Ø.75 – ⊔Ø1.625 – ↧.4375

(3) CROSS HEAD
H R S – 1 REQ D

2 × Ø0.937

(5) ADJUSTING SHIM
H R S – 2 REQD

TUMBLE
TO RE-
MOVE
BURRS

Ø.5 CORE – 3 HOLES
EQUALLY SPACED

Ø.190 – 24 NC – 2 B
– 2 HOLES
(Through)

Ø1.117
1.116

(8) AXLE SNAP RING
SPRING STEEL – 2 REQD

(Stock item)

(1) TROLLEY WHEEL
C I – 2 REQD

(9) WHEEL SNAP RING
SPRING STEEL – 4 REQD

Ø2.000
Ø1.688

(Stock item)

.156

(2) SIDE PLATE
H R S – 2 REQD

2 × Ø.875

## EX7-75

For the arbor press, draw the following. (1) Draw details. (2) Draw assembly. If assigned, convert dimensions to decimal inches or redesign for metric dimensions.

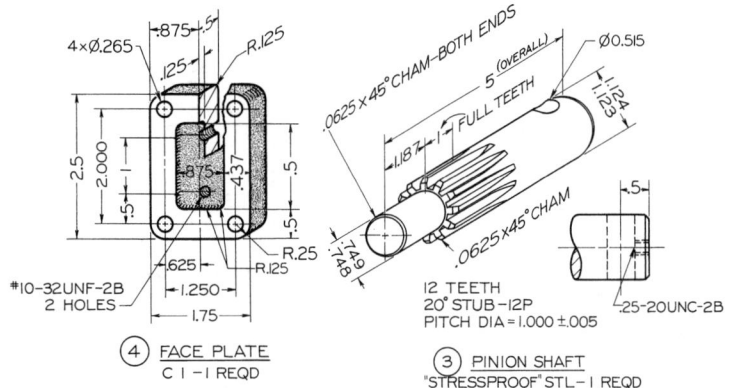

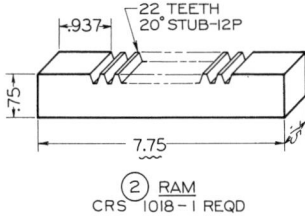

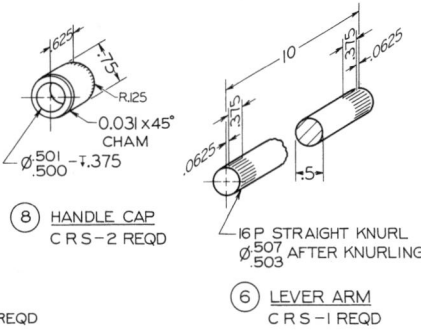

④ FACE PLATE
C I –I REQD

③ PINION SHAFT
"STRESSPROOF" STL–I REQD

② RAM
CRS 1018–I REQD

⑧ HANDLE CAP
C R S–2 REQD

⑥ LEVER ARM
C R S–I REQD

⑤ TABLE PLATE
C I –I REQD

⑫ .25–20 x.875 HEX HD CAP SCR–4 REQD
⑬ #10–32 x.625 HEX SOCK FL PT SET SCR–4 REQD
⑭ #10–32 x.187 SLOTTED FL PT SET SCR–I REQD
⑮ #10–32 S F HEX JAM NUT–4 REQD

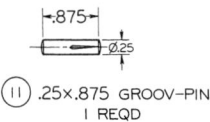

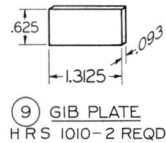

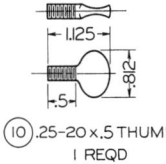

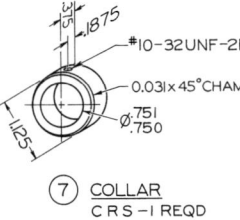

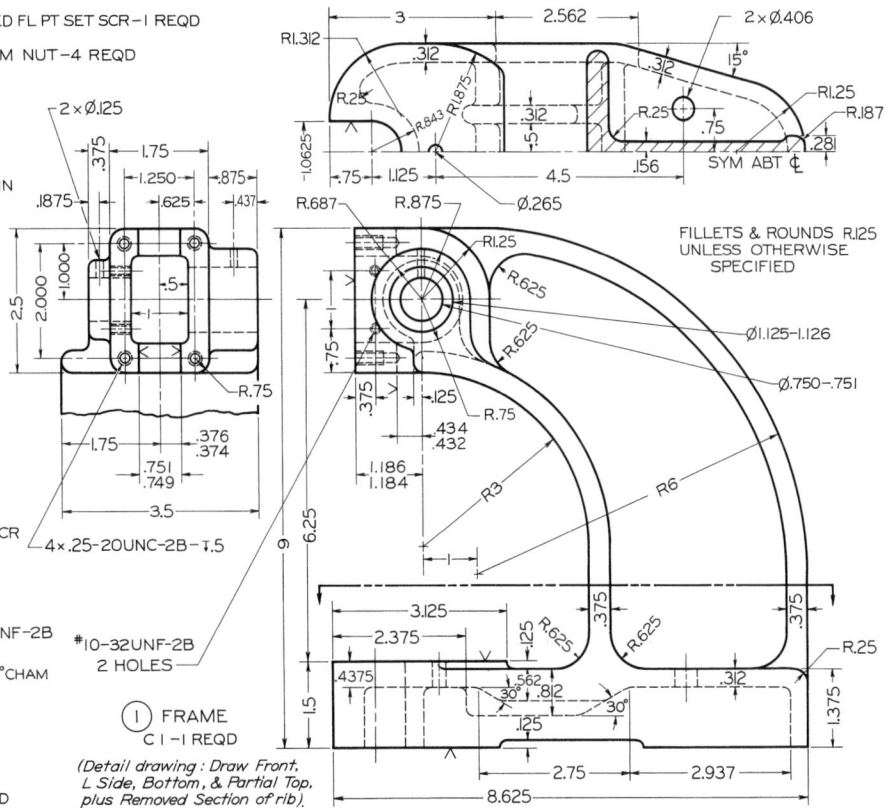

⑪ .25 x.875 GROOV–PIN
I REQD

⑨ GIB PLATE
H R S 1010– 2 REQD

⑩ .25–20 x .5 THUMB SCR
I REQD

① FRAME
C I –I REQD

*(Detail drawing : Draw Front,
L Side, Bottom, & Partial Top,
plus Removed Section of rib).*

⑦ COLLAR
C R S –I REQD

FILLETS & ROUNDS R.125
UNLESS OTHERWISE
SPECIFIED

**EX7-76**

For the forming cutter holder, draw the following. (1) Draw details using decimal or metric dimensions. (2) Draw assembly. Above layout is half size. To obtain dimensions, take distances directly from figure with dividers and double them. At left is shown the top view of the forming cutter holder in use on the lathe.

(8) SCREW
D-2-STEEL-.188 SQ BAR
TREAT-45Y-Z400

(6) BOLT
STEEL

SECT A-A

(5) STRAP
A-2-STEEL
1×1.750 BAR

(1) FRONT HOLDER
NO. 2-MALL IRON

(2) BLANK FORMING
TOOL
F-STEEL-1.125×3.125 BAR

(4) BOLT
A-1-STEEL-1.063 HEX BAR
TREAT-A55Z

(3) SHOE
A-3-STEEL
1.625×2 BAR

(7) SCREW
A-1-STEEL-.5 SQ BAR
TREAT-P50Y

## EX7-77

For the milling fixture for clutch arm, draw the following. (1) Draw details using the decimal inch system or redesign for metric dimensions, if assigned. (2) Draw assembly.

| Item | NAME | Amt | MATL | REMARKS | Item | NAME | Amt | MATL | REMARKS |
|---|---|---|---|---|---|---|---|---|---|
| 1 | BASE PLATE | 1 | CRS | 1×5×9.5 | 9 | SLEEVE | 1 | BRONZE | O D.718 – I D.640 |
| 2 | GAGE BLOCK | 1 | CRS | 1.5×2.875×4.875 | 10 | STUD | 1 | CRS | .625 DIA × 3 |
| 3 | LOCATING PLUG | 1 | CRS | 2.005 DIA × 2.25 | 11 | KEY | 2 | CRS | .5 × .812 × 1.5 |
| 4 | C-WASHER | 1 | CRS | 2.875 DIA ×.5 | 12 | SOC HD CAP SCR | 2 | STK | .312 ×.75 |
| 5 | REST BLOCK | 1 | CRS | 1.375 × 2 × 2.75 | 13 | PIN | 1 | DR | .375 ×2 |
| 6 | CLAMP | 1 | CRS | 1×1×3.625 | 14 | SOC HD CAP SCR | 3 | STK | .5 × 1.25 |
| 7 | .625 STD HEX NUT | 2 | STK | | 15 | DOWEL PIN | 2 | STK | .312 DIA × 1.5 |
| 8 | SPRING | 1 | MUSIC WIRE | WIRE .054 – O D.875 | 16 | SOC HD CAP SCR | 2 | STK | .5 × 1 |

## EX7-78

*Part (a)* For the drill speeder, draw the following. (1) Draw details. (2) Draw assembly. If assigned, convert dimensions to decimal inches or redesign with metric dimensions. See parts (b) and (c) on the following pages.

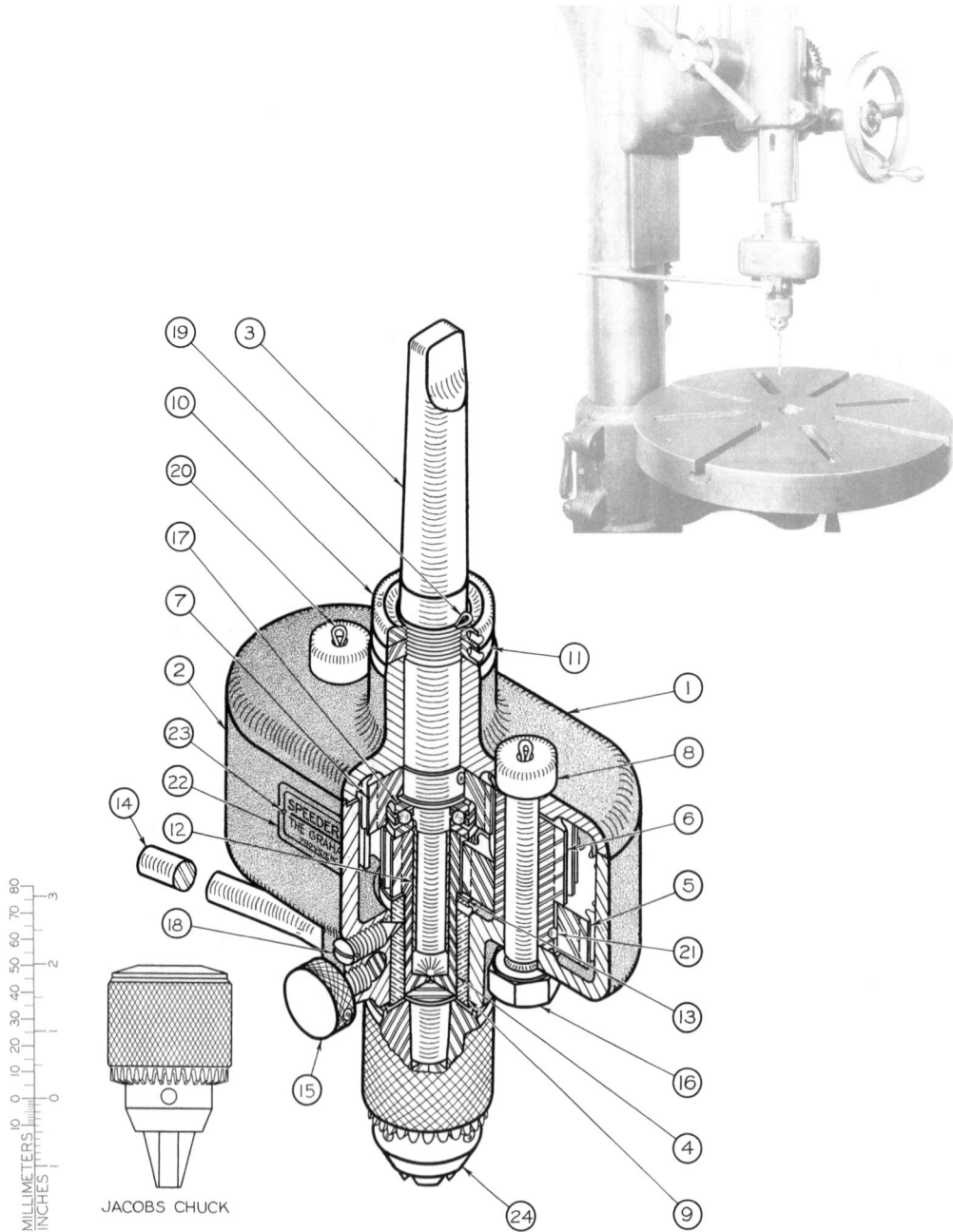

JACOBS CHUCK

**EX7-78**

*Part (b)* For the drill speeder, see Ex 7-78 part (a) for instructions.

∅1.063
1.062

R.062
∅3.125
.187
1.062
.312
R1.625
3.25
3.430
3.426

∅1.75
∅1.687
.187
2X ∅ .498 ⌴∅1.125
.500 FROM BOTTOM
1.75
.062
R.25
R.437
R.1875
∅1
∅1.75
ROUGH
.25
.312
.218

① BODY-UPPER HALF
C1 – 1 REQD

∅.125 SEAT FOR
㉑ BALL KEY
HARDEN TEETH
.312
∅1.416
1.415
15 T – 12 P
.500
.498
∅1.062
1.061
1.125
1.937

⑥ INTERMEDIATE PINION
AJAX STL–2 REQD

∅1.375
∅.937
.015

⑬ SPINDLE WASHER
CRS – 1 REQD

⑰ THRUST BEARING
1.344 OD × .625 ID × .562
STK – 1 REQD

2×∅ .500
.498
.187
.5
.5
R.062
30°
R.062
R1.625
∅3.125 PIN GAGE
2.625
3.25
3.430
3.426
∅1.438
1.437
#42 DRILL – 2 HOLES
FOR ㉓ #13 ESCUTCHEON PINS–2 REQD
.28
R.0312
.125
.687
1.937
2.312
3.5
2.062
1.625
.062×45° CHAMFER FOR OIL
R.125
∅1.125
TAPER INSIDE ONLY
WORK FROM
THIS SURFACE
R.125
R.375
1"SPOT
FACE
.375-16UNC-2B
R.375
.375-16 UNC –2B FOR ⑮ STOP ROD
SET SCREW
∅2

∅.125 FOR
㉑ BALL KEY
∅1.063
1.062
26 T – 12 P
∅2.334
2.333
.75

⑤ INTERMEDIATE GEAR
PAT. #337-C 1–2 REQD

② BODY-LOWER HALF
PAT. # 336 -C1-1 REQD

1
.5
.687
R.437
∅.453
1.562

## EX7-78

*Part (c)* For the drill speeder, see Ex  7-78 part (a) for instructions.

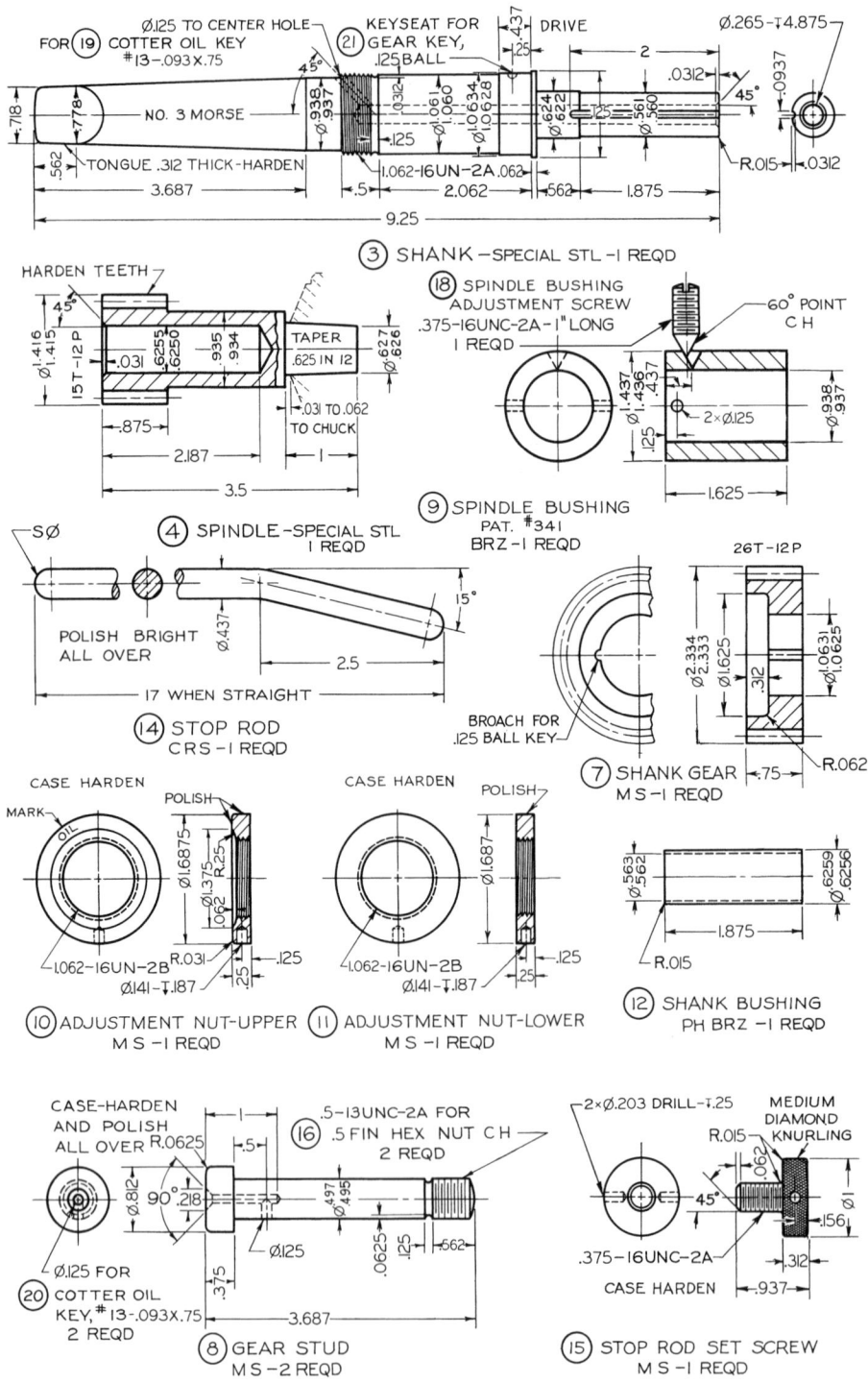

## EX7-79

*Part (a)* For the vertical slide tool, draw the following. (1) Draw details. If assigned, convert dimensions to decimal inches or re-design for metric system. (2) Draw assembly. Take given top view as front view in the new drawing; then add top and right-side views. If assigned, use unidirectional dimensions. See part (b) on the following page.

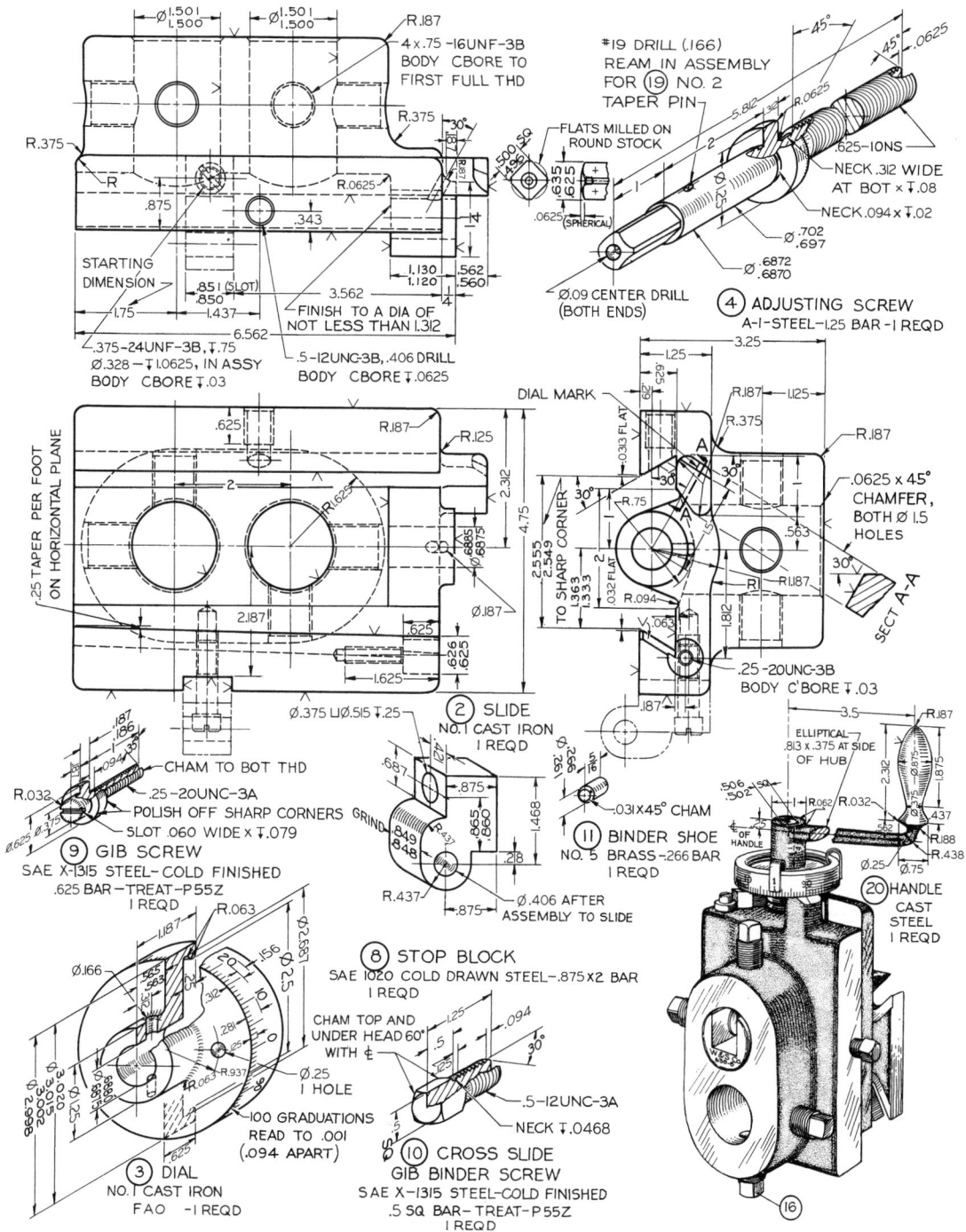

**EX7-79**

*Part (b)* For the vertical slide tool, see Ex 7-79 part (a) for instructions.

**EX7-80**

*Part (a)* For the slide tool, draw the following. Consult parts (b), (c), (d), and (e) on the following pages to: (1) Draw details using decimal inch dimensions or redesign with metric dimensions, if assigned. (2) Make an assembly drawing of this slide tool.

**EX7-80**

*Part (b)* Slide tool parts list.

| PARTS LIST | | NO. OF SHEETS 2 | SHEET NO. 1 | | | | MACHINE NO. M-219 | | | | |
|---|---|---|---|---|---|---|---|---|---|---|---|
| NAME NO. 4 SLIDE TOOL (SPECIFY SIZE OF SHANK REQ'D.) | | | | | | | LOT NUMBER | | | | |
| | | | | | | | NO. OF PIECES | | | | |

| TOTAL ON MACH. | NO. PCS. | NAME OF PART | PART NO. | CAST FROM PART NO. | TRACING NO. | MATERIAL | ROUGH WEIGHT PER PC. | DIA. | LENGTH | MILL | PART USED ON | NO. REQ. FINISH |
|---|---|---|---|---|---|---|---|---|---|---|---|---|
| | 1 | Body | 219-12 | | D-17417 | A-3-S D F | | | | | | |
| | 1 | Slide | 219-6 | | D-19255 | A-3-S D F | | | | | 219-12 | |
| | 1 | Nut | 219-9 | | E-19256 | #10 BZ | | | | | 219-6 | |
| | 1 | Gib | 219-1001 | | C-11129 | S A E 1020 | | | | | 219-6 | |
| | 1 | Slide Screw | 219-1002 | | C-11129 | A-3-S | | | | | 219-12 | |
| | 1 | Dial Bush. | 219-1003 | | C-11129 | A-1-S | | | | | 219-1002 | |
| | 1 | Dial Nut | 219-1004 | | C-11129 | A-1-S | | | | | 219-1002 | |
| | 1 | Handle | 219-1011 | | E-18270 | (Buy from Cincinnati Ball Crank Co.) | | | | | 219-1002 | |
| | 1 | Stop Screw (Short) | 219-1012 | | E-51950 | A-1-S | | | | | 219-6 | |
| | 1 | Stop Screw (Long) | 219-1013 | | E-51951 | A-1-S | | | | | 219-6 | |
| | 1 | Binder Shoe | 219-1015 | | E-51952 | #5 Brass | | | | | 219-6 | |
| | 1 | Handle Screw | 219-1016 | | E-62322 | X-1315 C F | | | | | 219-1011 | |
| | 1 | Binder Screw | 219-1017 | | E-63927 | A-1-S | | | | | 219-6 | |
| | 1 | Dial | 219-1018 | | E-39461 | A-1-S | | | | | 219-1002 | |
| | 2 | Gib Screw | 219-1019 | | E-52777 | A-1-S | | $\frac{1}{4}$-20 | 1 | | 219-6 | |
| | 1 | Binder Screw | 280-1010 | | E-24962 | A-1-S | | | | | 219-1018 | |
| | 2 | Tool Clamp Screws | 683-F-1002 | | E-19110 | D-2-S | | | | | 219-6 | |
| | 1 | Fill Hd Cap Scr | 1-A | | | A-1-S | | $\frac{3}{8}$ | $1\frac{3}{8}$ | | 219-6 219-9 | |
| | 1 | Key | No.404 Woodruff | | | | | | | | 219-1002 | |

**EX7-80**

*Part (c)* For the slide tool, see Ex 7-80 part (a) for instructions.

1—BODY —219—12
A-3 STL DROP FORGING

STARTING DIMENSION

GAGE — 300-B / 1619

DRILL JIG — 300-B / 1606

SECTION A-A
STARTING DIMENSION

Ø1.581 +.002 −.000

1.687—12 NS

FORGING DETAIL NO.219—12
A-3 STEEL

FILLETS & ROUNDS R.125 UNLESS
OTHERWISE SPECIFIED

1—SLIDE SCREW 219—1002
A-3-STEEL —1" BAR

.562—18UNF-2A

NECK FOR GRINDING -.063 W x .0156 DEEP

MILL FOR
#404 WOODRUFF
KEY

.5-10 ACME-LH

35° CHAMFER TO
BOTTOM OF THREAD

KEYWAY
.125 WIDE x .062 DEEP

NECK .0625 WIDE x .0156 DEEP

1— DIAL BUSH. 219—1003
SAE X—1315 STEEL — 1" BAR

**EX7-80**

*Part (d)* For the slide tool, see Ex 7-80 part (a) for instructions.

FILLETS & ROUNDS
R.125 UNLESS
OTHERWISE
SPECIFIED

GIB SCREW 219-1019
SAE X 1315. STEEL (COLD FINISHED) - .5 BAR
TREAT P 55Z

FORGING DETAIL NO. 219-6

A-3 STEEL

I-BINDER SHOE 219-1015
NO. 5 BRASS   .266 BAR

I-DIAL NUT 219-1004
SAE X-1315 STEEL-COLD FINISHED
.781 HEX BAR
TREAT - P 55Z

DRILL JIG 300-B/1608.

DOVE TAIL GAGE 300-B/3004

USE GAGE WITH
MASTER GIB 300-B/3009

I-SLIDE-219-6
A-3-STEEL DROP FORGING.

STAMP HERE
THE WARNER & SWASEY CO.
CLEVELAND, O. U.S.A.
M-219.

I-HANDLE SCREW-219-1016
SAE X-1315 STEEL-COLD FINISHED - .188 BAR
TREAT P55Z

COLLET 21-D/54
DRILL JIG 300-B/1610
TAPPING FIX 300-B/1611
SPECIAL TAP 300-B/1612

TO MATCH BODY

I-NUT-219-9
NO. 10 BRONZE

## EX7-80

*Part (e)* For the slide tool, see Exercise 7-80 part (a) for instructions.

COUNTERBORING JIG  300-B / 1673
COUNTERBORE  300-B / 1618

4.187

CUT OFF 1.5 LONGER THAN
DIMENSION GIVEN, TO ALLOW
FOR FITTING

.5    .25

.062    SLIDE

.125 ±.003

VIEW IN DIRECTION OF ARROW X

1 - GIB    219 - 1001
SAE 1020 STEEL - .375 × 1 BAR

NOTE – ON PLANE
A – A – THE TAPER
OF GIB IS .250
PER FT

.498 +.000 -.002

.5
.156
A    B    A
30°    B

NOTE – ON PLANE
B – B – THE TAPER
OF GIB IS .217
PER FT

NECK .046 DEEP

30°    45°    .436 +.000 -.002

.5    .437-14UNC-2A

.343
.125    .625    .015 × 45° CHAM
.375
1.625

2 - TOOL CLAMP SCREW  – 683-F-1002
D-2-STEEL
TREAT -45Y-Z 400

#21 (.159) DRILL, #10-32UNF-2B

#6
.375    .125    #00
.437    .093
.375
2.25    1.125

CINCINNATI BALL CRANK CO.
TO FURNISH WITH .375 ROUND HOLE
AND WE TO BROACH OUT
HOLE TO .375 SQUARE

.390

.75    2X S∅.75
.625
1.5

1 - COMPOUND REST HANDLE
219 - 1011

R.031    ∅.311 +.000 -.002
∅.687    .312-24UNF-2A
MEDIUM    .187    35° -0° +5°
KNURL         CHAMFER TO
.687         BOTTOM OF THREAD

1 - BINDER SCREW- 219-1017
SAE X-1315 STEEL-COLD FINISHED-.687 BAR
TREAT P55Z

SLOT .080 WIDE × .109 DEEP

.125    35° -0° +5°    .375-24UNF-2A    35° -0° +5°    .375 +.000 -.002
∅.375    45°
MED KNURL
R.031    .187    .125    4.437    CHAM TO
NECK             BOTTOM OF
.078 WIDE × .031 DEEP         THREAD

1 - STOP SCREW - 219 -1013
SAE X-1315 STEEL-COLD FINISHED - .625 BAR
TREAT -P 55Z

.031 × 45° CHAMFER    R.031    NO. 10-32 UNF-2A
∅.437    .190
MEDIUM    .28
KNURL    .687    35° -0° +5°
CHAMFER TO
BOTTOM OF THREAD

1 - BINDER SCREW - 280-1010
SAE X-1315 STEEL-COLD FINISHED -.437 BAR
TREAT P55Z

SLOT .080 WIDE × .109 DEEP

.125    35° -0° +5°    .375-24UNF-2A    35° -0° +5°    .375 +.000 -.002
∅.625    45°
MEDIUM    R.031
KNURL    .187    3.6875    CHAM TO
NECK             BOTTOM OF THREAD
.078 WIDE × .031 DEEP

1 - STOP SCREW  -219 -1012
SAE X-1315 STEEL-COLD FINISHED - .625 BAR
TREAT -P55Z

#21 (.159) DRILL, #10-32UNF-2B
90° FROM ZERO

.187
∅1.490 +.000 -.002    ∅.937 +.000 -.002    ∅.750 +.000 -.000    ∅1.187 +.001 -.000

100 DIVISIONS    R.062    .623 +.000 -.001
READS TO .001    .28
.047 APART    .406
.748 +.000 -.002
FAO

1-Dial 219-1018
A-1- STEEL  1.5 BAR

## EX7-81

*Part (a)* For the "any angle" tool vise, draw the following. (1) Draw details using decimal inch dimensions or redesign with metric dimensions, if assigned. (2) Draw assembly. See part (b) on the following page.

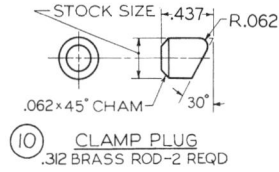

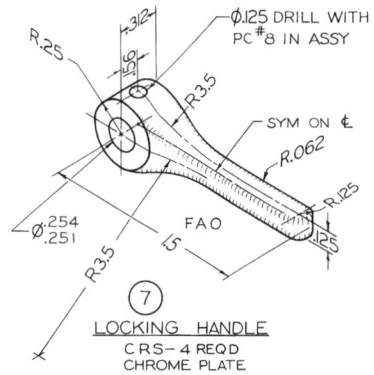

STOCK SIZE
.437
R.062
.062×45° CHAM
30°

⑩ CLAMP PLUG
.312 BRASS ROD—2 REQD

Ø.125 DRILL WITH
PC #8 IN ASSY

R.25
.312
.156
R3.5
SYM ON ₵
R.062
Ø.254
.251
FAO
R.125
.125
R3.5
L5

⑦ LOCKING HANDLE
CRS—4 REQD
CHROME PLATE

STANDARD PARTS

| NO.<br>REQD | |
|---|---|
| 4– | .375-16UNC-2A × 1.5 HEX SOCKET CUP PT SET SCR |
| 2– | .375-16UNC-2A × .562 HEX SOCKET FLAT PT SET SCR |
| 8– | .25-20UNC-2A × .625 FILLISTER HD CAP SCR |
| 12– | Ø.125 × .75 DRILL ROD |
| 2– | #4 (.112)–40UNC-2A × .25 ROUND HD MACH SCR |
| 4– | Ø.125 × .187 DRILL ROD |

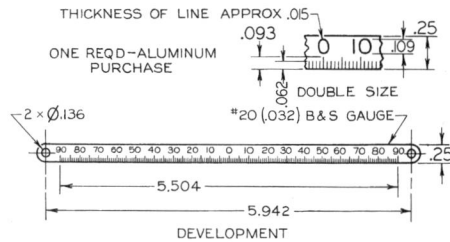

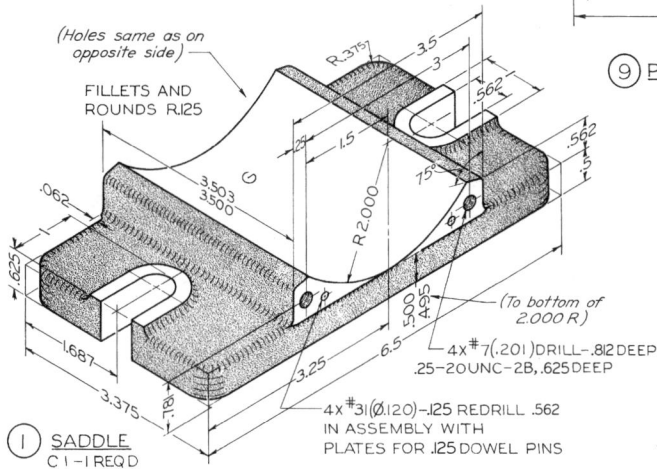

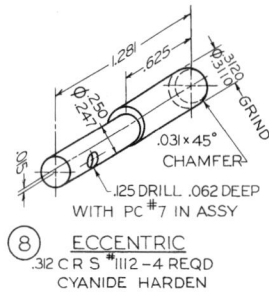

THICKNESS OF LINE APPROX .015
.093                     .25
ONE REQD–ALUMINUM          0  10  .109
PURCHASE                     .062
DOUBLE SIZE
2 × Ø.136          #20 (.032) B&S GAUGE
90 80 70 60 50 40 30 20 10  0  10 20 30 40 50 60 70 80 90   .25
5.504
5.942
DEVELOPMENT

⑨ PROTRACTOR SCALE

(Holes same as on
opposite side)

FILLETS AND
ROUNDS R.125

R.375
3.5
3
.562
.25
1.5
75°
.562
.562
.5
1.5
3.503
3.500
R 2.000
.062
.625
(To bottom of
2.000 R)
.500
.495
4x #7(.201)DRILL-.812 DEEP
.25-20UNC-2B,.625 DEEP
1.687
3.25
6.5
.181
3.375
4x #31(Ø.120)-.125 REDRILL .562
IN ASSEMBLY WITH
PLATES FOR .125 DOWEL PINS

① SADDLE
C1—1 REQD

1.281
.625
.3120
.3110 GRIND
Ø.250
.247
.031×45°
CHAMFER
.50
.125 DRILL .062 DEEP
WITH PC #7 IN ASSY

⑧ ECCENTRIC
.312 CRS #1112—4 REQD
CYANIDE HARDEN

**EX7-81**

*Part (b)* For the "any angle" tool vise, see Ex 7-81 part (a) for instructions.

SECTION A-A

*This surface flat; otherwise parts 5 and 6 are identical*

*(Dimensions as shown on PC. #5 at left)*

⑤ UPPER PLATE
C R S – 2 REQD

⑥ LOWER PLATE
C R S – 2 REQD

CYANIDE HARDEN–POLISH & BUFF ALL OVER

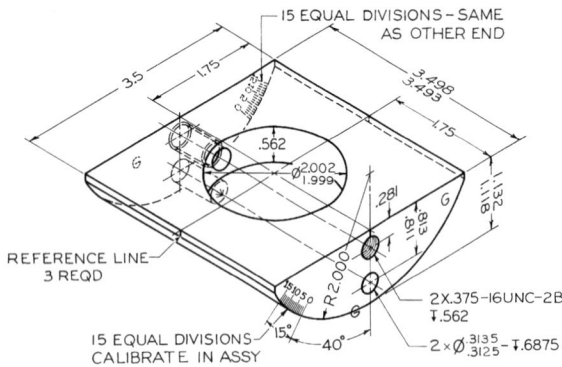

③ UPPER COMPOUND MEMBER
C R S – I REQD –CHROME PLATE

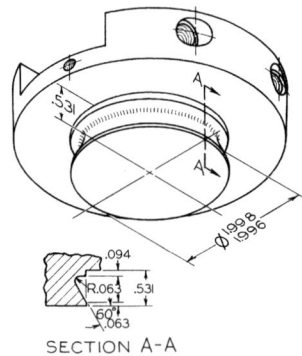

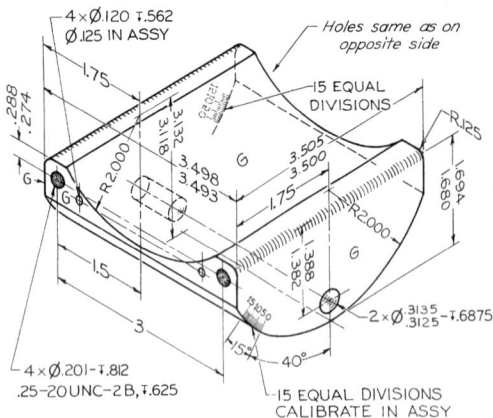

SECTION A-A

② COMPOUND CENTER MEMBER
C R S – I REQD –CHROME PLATE

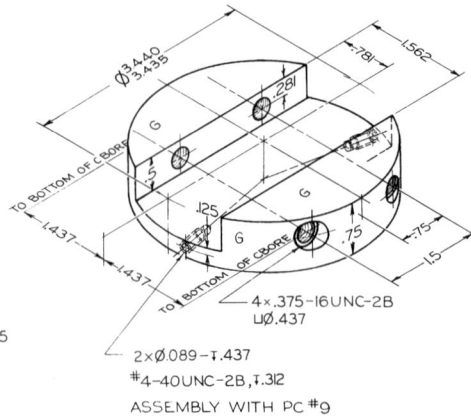

④ COMPOUND TOOL HOLDER
C R S – I REQD –CHROME PLATE

**EX7-82**

*Part (a)* For the fixture for centering connecting rod, draw the following. Consult Parts (b) and (c) on the following pages to: (1) Draw details using decimal inch dimensions or redesign with metric dimensions, if assigned. (2) Draw assembly. See parts (b) and (c) on the following pages.

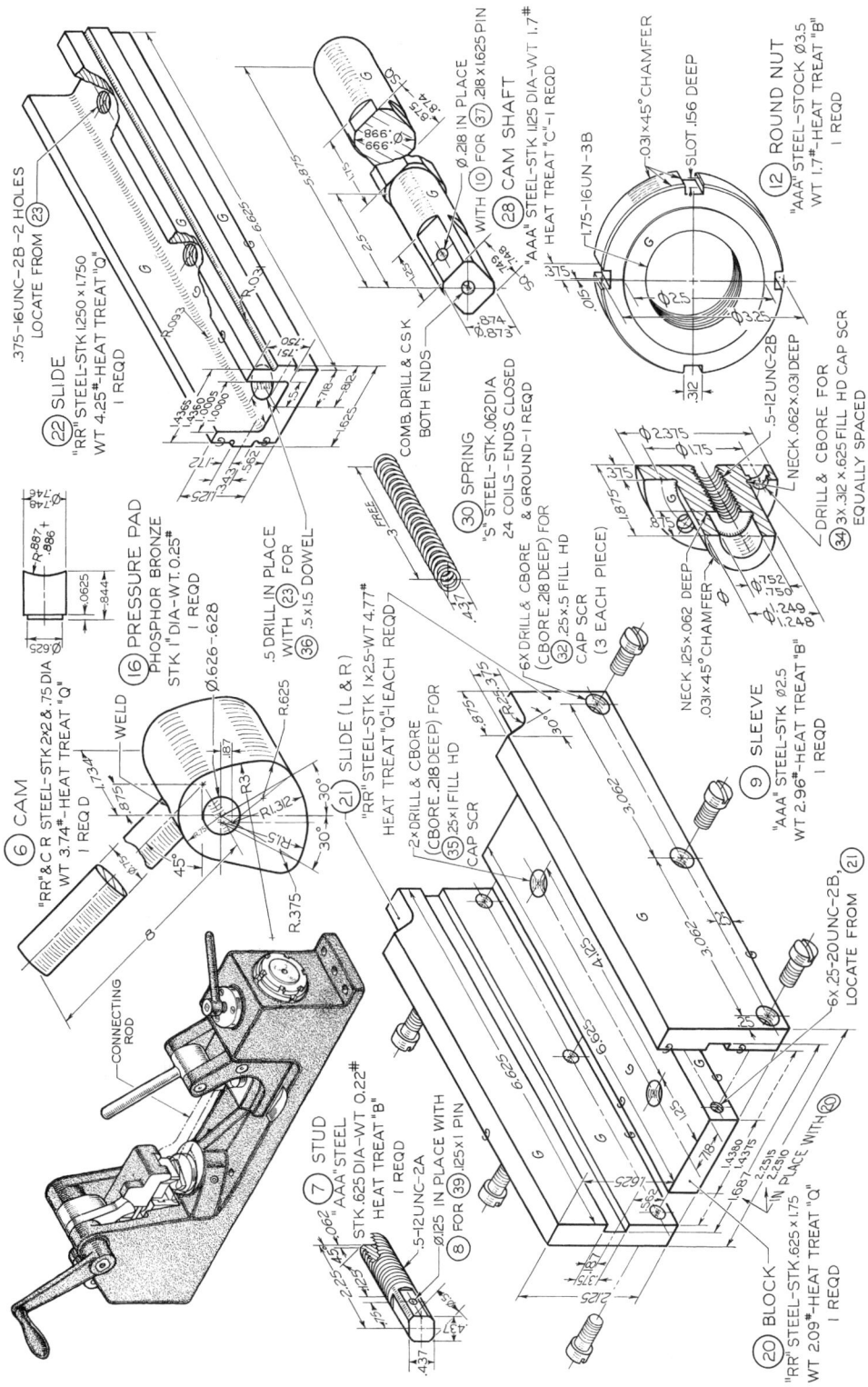

**EX7-82**

*Part (b)* For the fixture for centering connecting rod, see Ex 7-82 part (a) for instructions.

(23) COVER
"AAA" STEEL CASTING
WT 4#-HEAT TREAT "C"
1 REQD

2X DRILL & CBORE
(31) (CBORE .344 DEEP) FOR
.375x.125 FILL HD CAP SCR

.062x45°CHAMFER
.5 DRILL IN PLACE
WITH (22)

(22) BODY
C 1-WT 33#
1 REQD

FOR (11) DEEP GROOVE RADIAL BEARING
I D 1.7717-O D 3.3465-WIDTH .748
(8 BALLS .375 DIA-2 REQD

(-.5-12UNC-2B FOR (38) .5x.125
SOCKET HD SET SCR

FILLETS & ROUNDS R.25 UNLESS OTHERWISE SPECIFIED-DRAFT ANGLES 7°

3X .3l2-18UNC-2B,
.625DEEP LOCATE
FROM (9)

4×Ø.531
Ø1.125

3X.3l2-18UNC-2B, 2 HOLES (20)
LOCATE FROM

*Part I: Draw front, top, & right side views.*

NECK.062x.03I DEEP

COMB. DRILL &
CSK-BOTH ENDS

.28 DRILL IN PLACE WITH
(28) FOR (37) .28x.625 PIN .626-.628 REAM

.03Ix45° CHAMFER

(14) REST BUTTON
"RR" STEEL-STK .125DIA
HEAT TREAT "Q"-1 REQD
WT .75#

(10) HAND LEVER
(IN STOCK-PATT 19-Z-16-"A-B)
STEEL CASTING-WT 2.5#
1 REQD

SPHERICAL
END

6 SLOTS
EQUALLY
SPACED

.03Ix45°
CHAM

SLOT
.156 DEEP

.175-16UN-3B

.03Ix45°
CHAMFER-
BOTH SIDES

(13) ROUND NUT
PATT 19-Z-13)-"AAA" STEEL
HEAT TREAT "B"-1 REQD

(19) PLUNGER
MACHINE STEEL-STK
.625 DIA-WT 0.14#-HEAT
TREAT "Q"-1REQD

(4) LOCATER
"RR" STEEL-STK 1x2
WT 4.1#-HEAT TREAT "Q"
1 REQD

.125x45° CHAMF-BOTH SIDES

.625-BOTTOM

.25-20UNC-2B, 2 HOLES (20)
LOCATE FROM

**EX7-82**

*Part (c)* For the fixture for centering connecting rod, see Ex 7-82 part (a) for instructions.

(8) HAND LEVER
"AAA" STEEL
FORGING–WT 0.5#
I REQD

.125 DRILL IN PLACE WITH
(7) FOR (39) .125×1 PIN

(2) CRADLE
"AAA" STEEL CASTING
WT 22#–HEAT TREAT"B"
I REQD

(26) BUSHING
PHOSPHOR BRONZE
STK .15 DIA.×.75CORE–WT .63#
REQD
4×Ø.6240-.6245
FOR
(40) .625×3.5 PIN
2 REQD

(27) BUSHING
PHOSPHOR BRONZE
STK .15 DIA.×.75CORE–WT .8#
I REQD

Ø.624-.625  FOR
(41) .625×2.25 PIN

(24) CAM
"RR" STEEL-STK-.625×2.5
WT 2#–HEAT TREAT "Q"
I REQD

(18) SPRING
"S"STEEL-STK Ø.062
HEAT TREAT "SP"
8 COILS-ENDS CLOSED
& GROUND–I REQD

(17) SPRING
"S" STEEL-STK.093DIA
HEAT TREAT "SP"
I REQD

(25) CAM
"RR" STEEL-STK .15×2.75
WT 4#–HEAT TREAT "Q"
I REQD

(29) PLUNGER
"RR" STEEL-STK .75DIA
WT .28#–HEAT TREAT "Q"
I REQD

(3) CLAMP
"AAA" STEEL CASTING
WT 8.4#–HEAT TREAT "B"
I REQD

(15) ROUGH LOCATER
"RR" STEEL–STK Ø 2.5
WT 3.82#–HEAT
TREAT "Q"
I REQD

(5) CLAMP
"AAA" STEEL
STK .15×2
WT 3.5#–HEAT
TREAT "B"-I REQD

**EX7-83**

For the plastic open slot wiring duct, front and side views, draw the following. Redraw with metric dimensions reducing the size by 3.

## 7-34 DRAWING MANAGEMENT

Creating a drawing is part of the process of bringing the concept of a product or system to reality. Approval, management, retention, and storage of the drawing are other very important parts of the process that should not be overlooked.

Once a drawing has been created, you must be able to retrieve it to use it effectively.

Managing drawings and other design documentation is both a legal responsibility and an economic one. These records are construed as a contract between a company and the vendors or contractors that manufacture parts or build a structure or system.

The 400 Gigabyte hard drive in the upper right stores more than 100,000 drawings; about 1,000 times as much as the flat file shown at the left Western Digital. The flat file costs 10 times as much. *Courtesy of Mayline Group.*

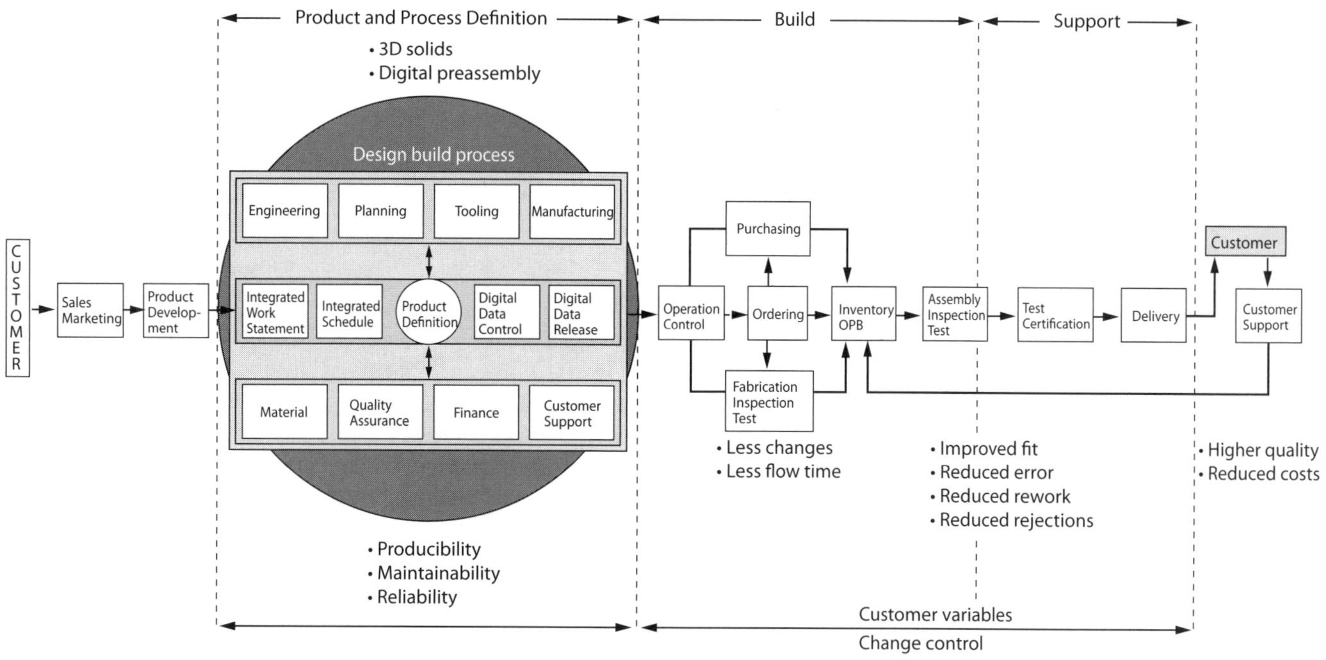

**Figure 7-65**    Boeing's preferred business process illustrates the concurrent nature of the design/build model at the heart of its design process. *Copyright © Boeing. Courtesy of Dynojet Research, Inc.*

For companies to compete in a fast paced marketplace they must be able to respond to changes and innovations that improve efficiency while keeping working documents and archives secure, organized, and conveniently accessible.

A well organized system for storing and retrieving past design efforts can be important for a company's continuing success.

# 7-35 MANAGING THE DESIGN PROCESS

Designing a product, structure, or system is an open ended process. A clear needs statement is the starting point. Recognizing the end point or achieving it can be challenging if you do not have a plan for the design process. Two critical items are time and cost.

In order to have a product to market by a particular date, you must allow time in the schedule for all of the stages in the design process, as well as coordinating for the product manufacture or construction. This may involve researching materials and suppliers, coordinating lead times for ordering materials, time for vendors to manufacture parts, incorporate changes, assemble, and market the product, produce user documentation, and a myriad of other details. Figure 7-65 shows a graphical representation of the Boeing Corporation's Preferred Business Process. Notice all of the stages before the product gets to the customer. If this is to happen on time, careful planning is needed.

Often a **Program Evaluation Review Technique** (PERT) chart or Gantt chart is used to track the time and activity for the project. By working backwards from the ending date, and allotting time to activities, you can develop a plan to meet the critical deadlines and predict the influence on one important activity if a deadline slips somewhere else.

The work of designing the product occurs near the beginning of the process, but it commits a large percentage of the total cost of the product. Figure 7-66 shows a graph of time versus the percentage of the project budget that is committed. Note that even though the conceptual design is not particularly costly, especially when compared with hand tooling for manufacturing, it is at this point that most of the cost of the project is committed. The difference between the product's success and failure may be determined early on.

Companies use design milestones and design reviews to help ensure that the design process accomplishes its goal in a timely fashion and to minimize the risk in proceeding with the design at that point. Typical milestones may include the following:

Product proposal The **product proposal** is usually driven by marketing and sales in response to consumer needs.

Design proposal The **design proposal** is a plan to meet that need and typically to show the profitability to the company in doing so.

Development plan The **development plan** includes all of the groups involved with the product, such as management, marketing, engineering, manufacturing, service, and sales.

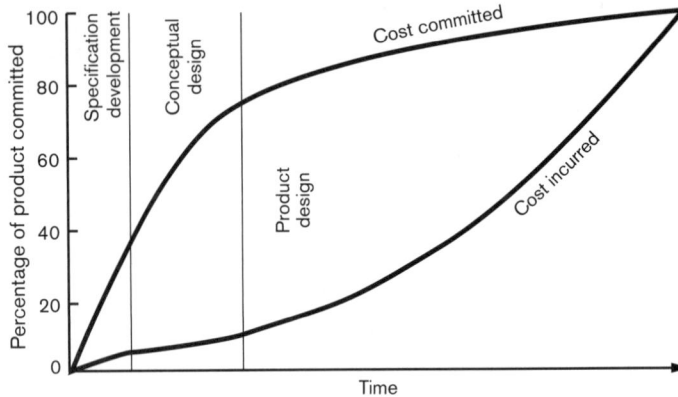

**Figure 7-66**    Early in the design process a large percentage of the life cycle cost of the product is committed. Late in the process, when costs are actually incurred, it may be impossible to reduce costs without major changes to the design. In order to produce products that are viable on the world market, you will want to consider as many alternative designs as possible early in the design process. *Reprinted with permission of the McGraw-Hill Companies from The Mechanical Design Process, Second Edition by David G. Ullman, 1997.*

Engineering release **Engineering release** ascertains that satisfactory design reviews have been completed for all of the components, modules, and systems, required in the device or system, and all questions have been answered to the point where it is ready for release. This usually includes several stages of design review both at a product and part level. After this initial engineering release of drawings or a digital database occurs, revisions are tracked and marked on the drawings.

Often companies use a letter to track design revisions before release. For example, at an initial prototype stage the drawing may be marked as revision A. Then as the design is refined a revision B may be indicated. At the initial engineering release, the drawings will then be updated to show release 1 (some companies use 0 as the initial release).

Product release    **Product release** requires that each part can be produced and function in the design and the product is now ready for the customer.

### Design Review

Entry level employees may be most involved in part reviews to determine if the part that they are working on will function correctly in the entire assembly and can be manufactured cost effectively. You may need to provide the documentation for design decisions; for example, the fastener selections, meeting company standards, analysis, tolerances, and other information.

# 7-36 UNDERSTANDING DRAWING MANAGEMENT

To control how drawings are released for manufacture, companies and industries have developed procedures to eliminate misunderstandings and costly mistakes. For example, if designs require changes, engineering change orders document these changes so they become a part of the permanent record of the design.

Regulatory bodies, standards organizations, and case law have all contributed to the rules for retaining and producing documentation for your engineering designs. Anyone who works with technical drawings should be familiar with the requirements for record retention in their industry.

Drawing management is obviously important, simply for the sake of convenience and to free designers to use their creative energy toward more lofty pursuits than shuffling through disorganized paperwork. More importantly, though, it plays a key role in providing legal documentation and contributing to a company's overall efficiency and profitability.

### Preserving Documentation

Drawings serve as an important part of an agreement between a designer and a producer. You can think of released drawings and 3D CAD models as a contract between you and the company building the structure or system, or the manufacturer producing the part or device. As such, it is

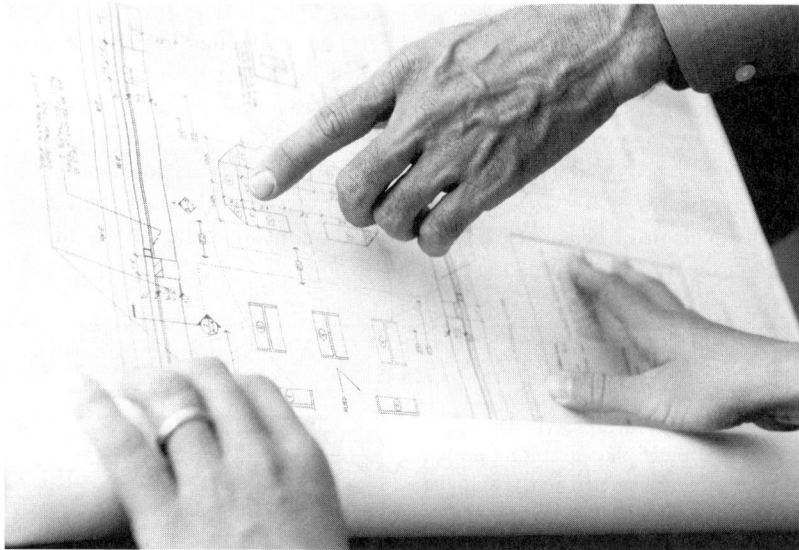

Documentation drawings can be construed as a contract with the manufacturers and clients, so management of both electronic and printed drawings is a key business practice in all industries that involve technical drawings. *Courtesy of Getty Images-Stock byte.*

crucial to be able to document what was actually provided in this drawing.

Companies may need to produce the **documentation** for various purposes, such as defending themselves against product safety liability and patent infringement lawsuits. In the event of legal issues, you must be able to produce the document as it was provided and in a method that is admissible in a court of law. Because electronically stored files can be altered, they may not be considered an acceptable method for documenting engineering designs.

Whether they are on paper, on mylar, or in electronic format, drawings are important company records and must be managed so that they can be retrieved, reproduced, revised, and retained effectively.

Most legal requests are for copies of all versions of the design, and all copies stored and used within the company. It can be costly and embarrassing if they are not well organized and correct, or if there are multiple versions in use that vary.

### Legal Standards for Drawings

Legal standards for how long engineering drawings need to be retained vary from state to state and industry to industry. For example, nuclear power plant drawings may be kept permanently or for hundreds of years, but the drawings for a medical device may be kept only a year or two after the product is obsolete (and no longer being manufactured or sold). In most cases, industry standards groups weigh the risk of record destruction in the context of product and public safety and make recommendations for its members.

The **American Records Management Association** (ARMA) standards for records retention are another source of information about engineering drawing retention. A

company—or an individual consultant—needs to understand these standards to determine how long to retain engineering drawing records. Even if the legal standard is less stringent, a court may rule that a firm should meet the industry's common practice to avoid a finding of negligence.

Industries that are regulated by the Food and Drug Administration (FDA) should also be aware of its Guidelines for Electronic Records and Signatures, a ruling that was finalized in early 1998. Regulatory agencies such as the FDA consider engineering drawings "specifications" or "documents" and have clarified their record keeping guidelines to include electronic forms such as graphics files. Companies that are undertaking records management guidelines should be aware of the most current rulings and efforts by industry standards and regulatory groups to encompass electronic media in their recommendations.

### Improving Efficiency

Efficient document management is also a key to the effective use of computer-aided design tools and concurrent engineering practices.

Effective storage and retrieval of engineering design documentation can make a difference in a company's ability to succeed in today's world marketplace. The effective use of a 3D design database can provide many benefits outside of just reduced drafting time or a shortened product development cycle.

Understanding the process that paper drawings go through for approval, release, and storage can also help you understand good practices for the approval, release, and storage of your electronic CAD data.

Concurrent engineering, a process that can improve efficiency and profitability by increasing the interplay between steps in the design process, depends on the team's ability to work together on interrelated tasks, often by using a design database. Access to current and accurate information is crucial to the team's ability to work simultaneously on different aspects of the same project.

On team efforts, the individuals involved can use a design database to streamline the design process by bringing the team together. The information kept in a 3D design database can be used either to produce paper drawings or to send as files for NC machining, mold design, or other manufacturing processes. It is also an important part of the engineering design record.

New software tools and processes to manage the flow of design information to and among members of the team are continually being improved.

## DRAWING MANAGEMENT IN ENGINEERING ETHICS

The Hyatt Regency Walkways Collapse provides a vivid example of the importance of accuracy and detail in engineering design and shop drawings (particularly regarding revisions), and the costly consequences of errors.

On July 17, 1981, the Hyatt Regency Hotel in Kansas City, Missouri, held a videotaped tea-dance party in their atrium lobby. With many party-goers standing and dancing on the suspended walkways, connections supporting the ceiling rods that held up the second and fourth-floor walkways across the atrium failed, and both walkways collapsed onto the crowded first-floor atrium below. The fourth-floor walkway collapsed onto the second-floor walkway, while the offset third-floor walkway remained intact. The collapse left 114 dead and in excess of 200 injured. Millions of dollars in costs resulted from the collapse, and thousands of lives were adversely affected.

The hotel had only been in operation for approximately one year at the time of the walkways' collapse, and the ensuing investigation of the accident revealed some unsettling facts.

During January and February, 1979, the design of the hanger rod connections was changed in a series of events and disputed communications between the fabricator (Havens Steel Company) and the engineering design team (G.C.E. International, Inc., a professional engineering firm). The fabricator changed the design from a one-rod to a two-rod system to simplify the assembly task, doubling the load on the connector, which ultimately resulted in the walkways' collapse.

The fabricator, in sworn testimony before the administrative judicial hearings after the accident, claimed that his company (Havens) telephoned the engineering firm (G.C.E.) for change approval. G.C.E. denied ever receiving such a call from Havens.

On October 14, 1979 (more than one year before the walkways collapsed), while the hotel was still under construction, more than 2700 square feet of the atrium roof collapsed because one of the roof connections at the north end of the atrium failed.

In testimony, G.C.E. stated that on three separate occasions they requested on-site project representation dur-

The fabricator of the failed walkway testified that his company had telephoned the engineering design team for change approval, but the engineering firm denied ever receiving such a call. *Courtesy of Texas A & M University.*

ing the construction phase; however, these requests were not acted on by the owner (Crown Center Redevelopment Corporation), due to additional costs of providing on-site inspection.

Even as originally designed, the walkways were barely capable of holding up the expected load, and would have failed to meet the requirements of the Kansas City Building Code.

Due to evidence supplied at the hearings, a number of principals involved lost their engineering licenses, a number of firms went bankrupt, and many expensive legal suits were settled out of court. The case serves as an excellent example of the importance of meeting professional responsibilities, and what the consequences are for professionals who fail to meet those responsibilities.

*Excerpted from "Negligence and The Professional Debate Over Responsibility For Design" A Case History of The Kansas City Hyatt Regency Walkways Collapse. Department of Philosophy and Department of Mechanical Engineering, Texas A&M University http://ethics.tamu.edu/ ethicscasestudies.htm*

## 7-37 DRAWING APPROVAL AND RELEASE

Once a drawing is determined to be complete, the title block on the drawing is used to document the change from a draft to a finished drawing. The drawing's creator signs and dates the "drawn by" block, perhaps a checker signs off and dates the "checked by" block, and the engineer approves the drawing for release by signing and dating the "approved by" block. A supervising engineer might also sign off to approve the drawing. On a set of paper drawings the signatures are written manually. When the drawings are approved digitally, approval is indicated by a "symbol of personal identification" according to the ASME Y1.14.42-2002 standard. The **approval indicator** can be any symbols, letters, numbers or a digital signature including bar codes, for example the initials in the title block shown in Figure 7-67.

Once approved, the drawing or set of drawings and contract are released to manufacturing or to the contractor to be produced. Copies of the approved drawings are circulated to various departments within the company as required, and one set of the printed drawings and contract is stored for the permanent record. Figure 7-68 shows a title block used to gather approval signatures on a paper drawing.

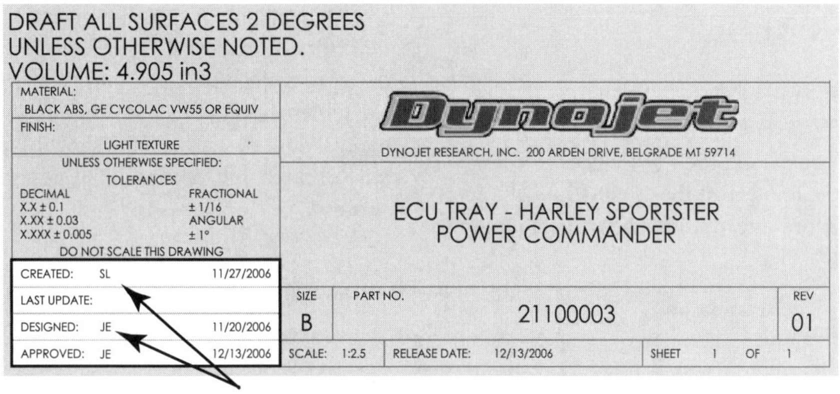

**Figure 7-67**  Electronic Signatures are Used on This Drawing. *Courtesy of Dynojet Research, Inc.*

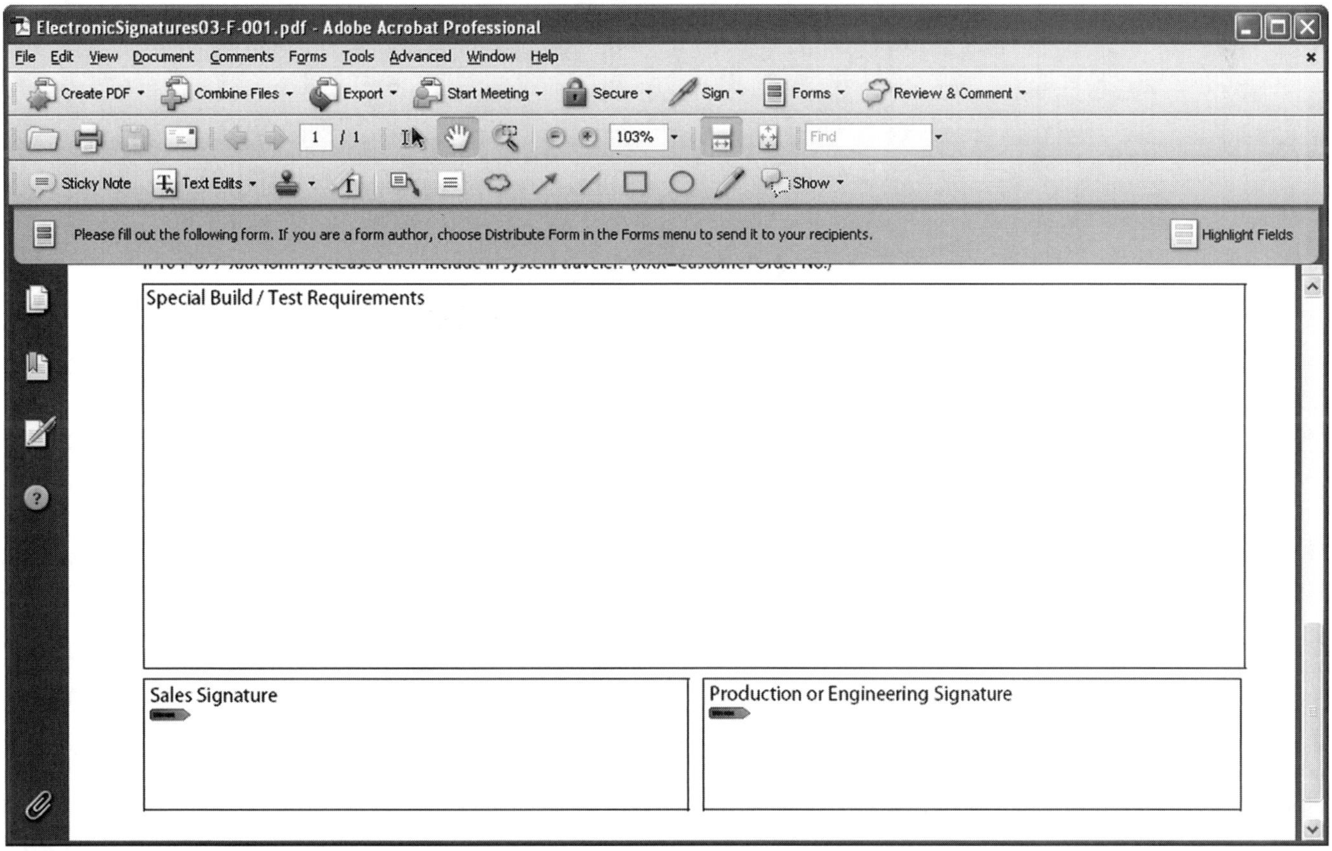

**Figure 7-68**  Electronic Signature Files are Used to Approve This Engineering Order Form. *Courtesy of Big Sky Laser.*

## 7-38 CHANGE ORDERS

In the imperfect real world, released drawings often require some type of correction during the process of constructing or manufacturing the product or system. After a drawing has been released, an **engineering change order** (ECO), shown in Figure 7-69, is used to document and approve drawing changes. An ECO, also called an engineering change notification (ECN) and sometimes an engineering change request (ECR), details the nature of the change in a separate document. After the ECO is approved, the drawing is revised and the revision noted on the drawing.

## 7-39 REVISION BLOCK

When revised, drawings are not replaced with a new drawing, instead a dated revision block, shown in Figure 7-70, is added to the drawing. A **revision block** describes the change and may also indicate the number of the engineering change order (which contains more information about the change). The revision block requires approval once again. A small number (contained in a triangle or circle) is added to call attention to the revision on the drawing. Some companies also included an easily visible revision number near the drawing number in the title block. Annotating the drawing number helps ensure that two people discussing the print from two different locations can verify that each is looking at the same revision of the drawing.

Once a drawing is revised and approved, a copy is stored for the permanent record. Companies that use printed drawings then must circulate new prints to all who received the previous print so they can update their files. Some companies require that old prints be collected to eliminate the possibility that some departments might continue to use them.

One advantage of a digital database is that when the system is well implemented and properly organized, it is much easier to track which revision is current. Obviously, a poorly implemented digital system is just as prone to confusion as a manual system. Figure 7-71 shows software written at Dynojet Research Inc. to track and approve engineering change requests.

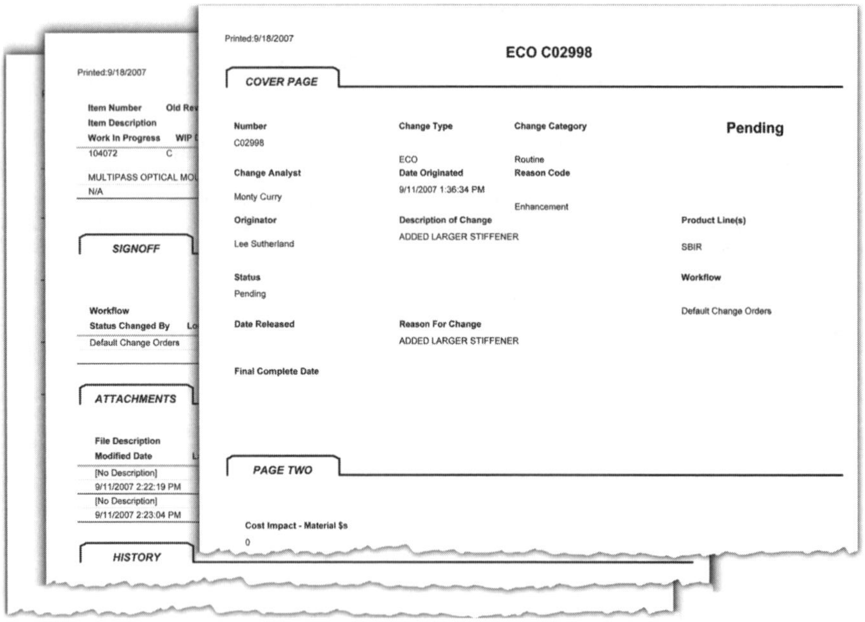
Printed:9/18/2007
ECO C02998
COVER PAGE
Number: C02998
Change Type: ECO
Change Category: Routine
Pending
Change Analyst: Monty Curry
Date Originated: 9/11/2007 1:36:34 PM
Reason Code: Enhancement
Originator: Lee Sutherland
Description of Change: ADDED LARGER STIFFENER
Product Line(s): SBIR
Status: Pending
Workflow: Default Change Orders
Date Released
Reason For Change: ADDED LARGER STIFFENER
Final Complete Date
PAGE TWO
Cost Impact - Material $s: 0

Figure 7-69 Engineering Change Order. *Courtesy of Zolo Technologies.*

| REVISIONS | | | | |
|---|---|---|---|---|
| ECR# | REV | DESCRIPTION | DATE | APPROVED |
| 1480 | 02 | INCREASED WIDTH OF KEY SLOTS TO EASE KEY INSTALLATION. | 4/1/2003 | SAS |
| | 02 | ADDED GDT TO CENTER KEYWAYS ON THE SHAFT. | | |
| 1813 | 03 | ADDED TOLERANCE TO Ø1.497. | 1/13/2004 | SAS |
| 3560 | 04 | ADDED GTD CONCENTRICITY TOLERANCE TO Ø1.497 | 2/1/2007 | JE |
| | 05 | ADDED REQUIREMENT FOR CENTER DRILLING ONE END OF SHAFT. | 5/2/2007 | SAS |

Figure 7-70 Revision Block. *Courtesy of Dynojet Research, Inc.*

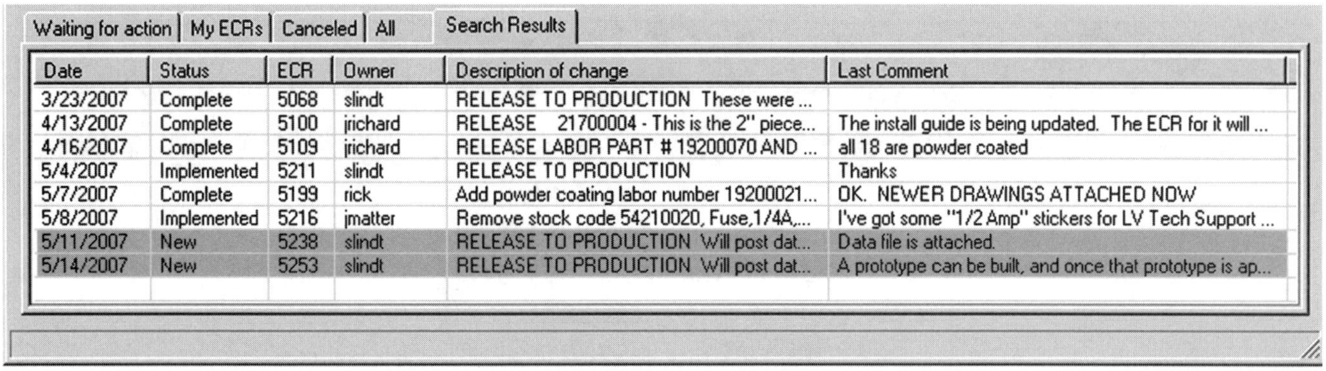

| Date | Status | ECR | Owner | Description of change | Last Comment |
|------|--------|-----|-------|----------------------|--------------|
| 3/23/2007 | Complete | 5068 | slindt | RELEASE TO PRODUCTION These were ... | |
| 4/13/2007 | Complete | 5100 | jrichard | RELEASE   21700004 - This is the 2'' piece... | The install guide is being updated.  The ECR for it will ... |
| 4/16/2007 | Complete | 5109 | jrichard | RELEASE LABOR PART # 19200070 AND ... | all 18 are powder coated |
| 5/4/2007 | Implemented | 5211 | slindt | RELEASE TO PRODUCTION | Thanks |
| 5/7/2007 | Complete | 5199 | rick | Add powder coating labor number 19200021... | OK.  NEWER DRAWINGS ATTACHED NOW |
| 5/8/2007 | Implemented | 5216 | jmatter | Remove stock code 54210020, Fuse,1/4A,... | I've got some "1/2 Amp" stickers for LV Tech Support ... |
| 5/11/2007 | New | 5238 | slindt | RELEASE TO PRODUCTION  Will post dat... | Data file is attached. |
| 5/14/2007 | New | 5253 | slindt | RELEASE TO PRODUCTION  Will post dat... | A prototype can be built, and once that prototype is ap... |

**Figure 7-71** Engineering Change Request Tracking Software. *Courtesy of Dynojet Research, Inc.*

## 7-40  A DRAWING AS A SNAPSHOT IN TIME

Each of the drawings archived and referenced in the process described above served to document a design at a particular point in time. Some companies continue to print and store a paper copy of the CAD file as their permanent record to provide this documentation. If the electronic CAD file is updated and no longer matches the drawing, the paper copy acts to preserve the design information. In this kind of system, the same approval and archival practices just described are applied to the paper drawings generated from the CAD database. This is a perfectly acceptable practice if you can retrieve these paper drawings later as needed.

Practices used today allow the same level of control for electronic files that store the design data. Quality standards in some industries allow for electronic documents to serve as the permanent record if they are properly controlled. The process of approval applies to CAD drawings that are "frozen" at each approval. Instead of storing a paper print, the company stores the electronic file and has backup procedures in place to ensure the future retrieval of the file.

The process just described covers only a portion of the design process and product information that a company needs to control. Sharing, controlling, and storing the electronic files used to document engineering design is important for getting the most value from a companies design effort.

## 7-41  GOOD PRACTICES FOR ELECTRONIC DRAWING STORAGE

Organized practices for storage, approval, retrieval, file naming, and tracking revision history for electronic CAD data are even more important than they are for paper-based systems. There are many benefits that motivate companies to invest in a 3D CAD database, such as improved communication between functional areas, computer integrated manufacturing, shorter design cycle time, cost savings, better access to information, and improved visualization capabilities. If the CAD files and documentation are not well organized, however, many of these benefits may not be realized. Chances are, you will work in a company that has developed and articulated its standards of data management, and you will be expected to adhere to them. The company may even have invested in data management software. In either case, much of the responsibility for managing data will be yours. Understanding the issues in personal file management will help you devise your own system, if you must, and help you appreciate the pros and cons of various approaches.

## 7-42  STORING ELECTRONIC FILES

The advent of personal computers on each designer's desktop contributes to the difficulty of managing electronic data. Each designer may organize the files he or she is working on differently, or keep multiple copies of a file in different directories. When others need to view or edit the file, it may be hard to be sure which is the current version. Without a thorough approval process for release of drawings, the engineer may neglect to track and store the revision history. Even when previous revisions of the drawing are stored electronically, they may not be useful because they don't satisfy the requirements for a static snapshot of the design at the time of release.

Many companies run into difficulties with their CAD data because they start out small and don't implement an organized system for managing the files. By the time they realize that they require better organization, they have thousands of poorly organized files and many CAD users with poor file storage habits. It is a very important part of your job to manage the product design records that you produce.

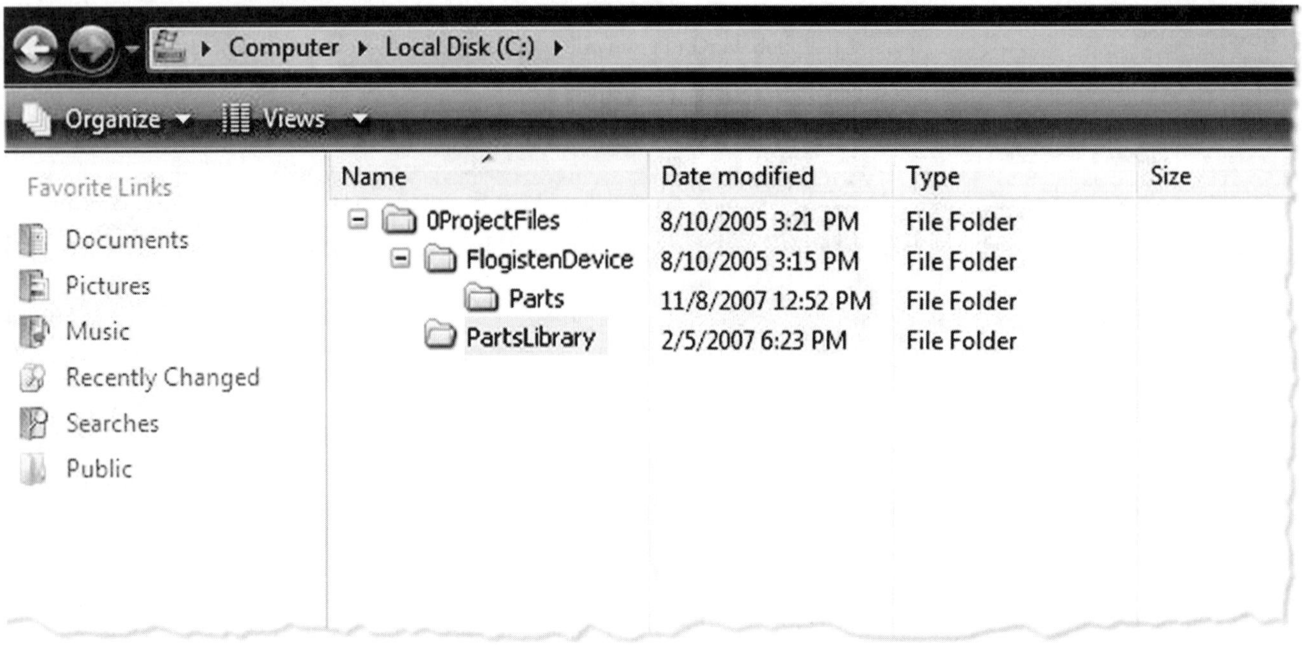

**Figure 7-72**    An Organized Directory Structure

## 7-43  ORGANIZED DIRECTORY STRUCTURES

Using an organized directory structure enables you to retrieve your CAD files and other electronically stored engineering data. Think of a directory on the disk as a kind of file folder. You would not put all of your paperwork loose in your file cabinet; neither should you scatter files over your hard disk.

Most people create a directory structure in a way that makes sense for the types of projects they do. A project-based directory structure, such as the one shown in Figure 7-72, allows you to store all files related to a particular project in the same directory.

Within each project, you may want to have subdirectories for different parts or different kinds of data associated with the project. A good rule of thumb is to think about situations when you (or others) will want to retrieve the data. Will you think of it in terms of the project? (Remember that a key to working with assemblies is the availability of the individual part files associated with it.) Or will you think of it in terms of another characteristic? Your CAD file directories may be project based, while other files may be organized differently. Making the directories work as you do will help you find what you need easily.

It is not a good idea to have a single "work" directory where you store all of your files, even if you intend to move them to another project-based directory later. This chore may not get done, or you may not remember which version of the file is current. Nor is it a good practice to store your work

files in the directory that contains the application software. (When you install software upgrades, you may accidentally overwrite or delete your work files.) Developing good habits that help you organize your files as you create them will eliminate confusion and save time later.

If you are using a networked computer system, you can store company documents on the network in a directory that others in the company can access. Because CAD files require large amounts of storage space, you need to manage the space on the hard disks and drives available to you. Keeping copies of all files on your personal system should be weighed against the frequency with which you need those files, the time involved in retrieving the files as needed, and the likelihood that the copy of the file on your system will still be current the next time you need it.

If you work for a company that has implemented a system to control their electronic data, you may be required to keep all of your CAD files in a workspace allotted to you on the network server. This makes it possible for the company to control (and regularly back up) this data, and it removes the burden of doing so from each individual employee.

## 7-44  FILE NAMING CONVENTIONS

File names assigned to your CAD files may be either random or semi random file names, or the name may provide information similar to that in a title block coded within its name. Whichever method you use, it is important to establish a procedure and name the files systematically. If a company-

**Figure 7-73** CAD files can use portions of the file name to code information telling you what the drawing is about. (a) The directory structure is organized around projects. Each filename includes information identifying the project. (b) The drawing title block includes the filename for reference. *Courtesy of LOCKHART, SHAWNA D.; JOHNSON, CINDY M., ENGINEERING DESIGN COMMUNICATION: CONVEYING DESIGN THROUGH GRAPHICS, 1st Edition, © 2000 pp. 60, 61, 99, 132,134, 388, 346. Reprinted by permission of Pearson Education, Inc., Upper Saddle River, NJ.*

**Figure 7-74** Part Number and Drawing Tracking Software Allows the User to Search the Description to Find a Drawing or Model. *Courtesy of Dynojet Research, Inc.*

wide policy for naming files does not exist, you should develop one for your own files to make them easy to find.

For names that use a code, even without opening the document or studying the model inside, you should be able to identify key information. Multipart file names may be used to indicate, for example, the part, the project it is part of, who created it, and a version number.

If you use numbers for your file names and assign them in a systematic way, users familiar with the system can search for kinds of documents, or for files by name. For example, you may have a sort of Dewey decimal system for types of drawings or documents your company typically produces, so that numbers starting with 100 are electrical schematics, 200s are assemblies, and so forth. The next set of numbers can further identify the type of file by another set of characteristics, such as project group numbers. A systematic alphanumeric code may work in a similar fashion. When the drawing name/number is not be the same as the CAD file name, it can become confusing. Another disadvantages is that for large numbers of drawings you can run out of values that fit your established code. It can be difficult to imagine a code that suits every type of project or part that may be undertaken.

When drawing names are assigned in a fairly random way, a company may maintain an organized database where users can look to find the particular drawing or file name. Sometimes drawing names are standardized only at the

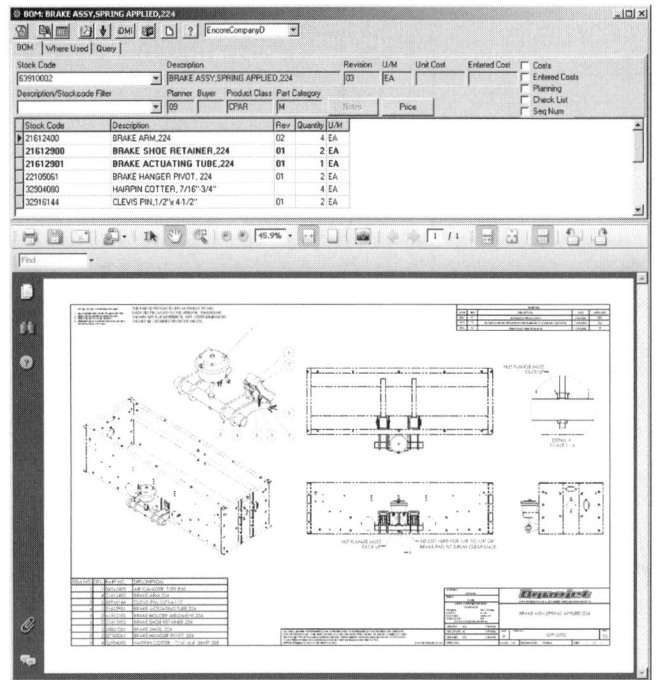

point that a part is approved and released to company archive. If so, the numerical system used for the CAD file name may also be used for the paper drawing name. Figure 7-73 shows a naming system you might use for CAD files. Figure 7-74 shows part number and drawing tracking software developed at Dynojet Research Inc. for its internal use in tracking drawings and CAD models.

Additional data may be added to the file name over time. The revision number can be added when revision drawings are approved. Once a company has file retention guidelines in place, file names can also incorporate a tag that indicates when it may be removed from long-term storage. This kind of tag is generally added by the group that manages the company's archives, but it demonstates the role of the file name in encapsulating information and aiding in file management.

Whatever system a company uses is definitely better than having each employee name drawings as he or she pleases.

## 7-45 DRAWING STANDARDS

Like file naming standards, **drawing standards** can help you work more productively and can contribute to the usefulness of the drawings as company records. Engineering drawings are the property of the company, not the individual

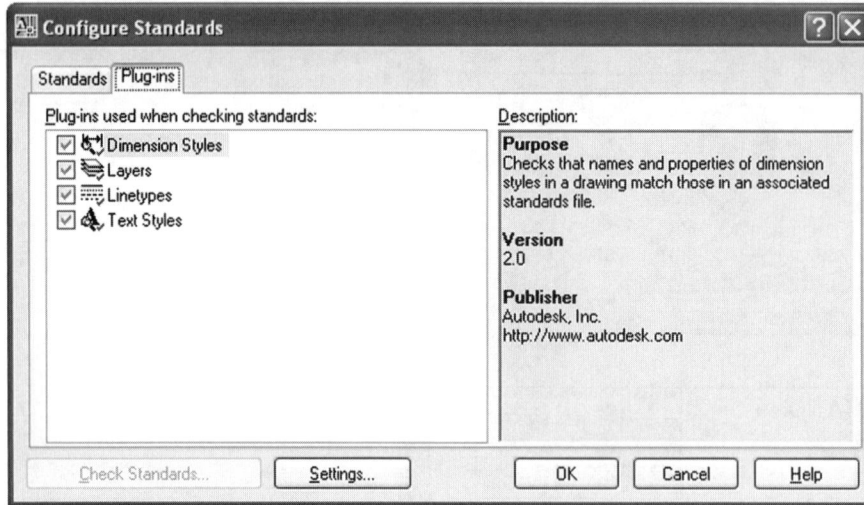

**Figure 7-75** AutoCAD 2008 Software's Configure Standards Dialog Box. *Certain images and materials contained in this publication were reproduced with the permission of Autodesk, Inc. © 2007. All rights reserved.*

Table 7-2  Standard Base Layers.* *Courtesy of National Oilwell Varco, L.P.*

| Line Type | Pen Number | Pen Thickness (AutoCAD) |
|---|---|---|
| Center | 1 | .003 |
| Section | 2 | .005 |
| Dim, Format | 3 | .005 |
| Object | 4 | .009 |
| Hatch | 5 | .003 |
| Bubbles | 6 | .004 |
| Material, Symbols, Text | 7 | .005 |
| Phantom | 8 | .003 |
| Hidden | 9 | .003 |

*Note: The layers used for certain drawing elements have been standardized company wide. Each element in the first column is placed on a separate layer in the drawing. Line type, color and line thickness for the various elements are set by layer in the prototype files used to start all drawings.

engineer. As such, their usefulness should not depend on a single individual being able to locate or interpret them.

Company standards may be in place to introduce consistency in the way drawings are constructed. Standardizing the layers in a drawing, for example, can make them more navigable. Architectural engineering firms use layers to organize the many systems, such as plumbing and electrical, that are part of building design. Each of these systems will be assigned a set of layers and layer names so that each drawing can be manipulated in the same way. Notes and drawing text are commonly stored on a separate layer across drawings so this information can be turned on and off as needed when working with the model. Because CAD software carries layer information along with parts, standardizing layer names can prevent unnecessary confusion when files are combined.

The CAD software itself may allow the user to select a CAD standard or check the drawing to see that it matches a CAD standard. Figure 7-75 shows the AutoCAD software's CAD standard dialog box.

The colors used for layers and different drawing elements can also be standardized, as shown in Table 7-2. Center lines may be red, for example. Because color is used to change pen widths on a pen plotter, this kind of standardization can facilitate printing the drawing with the appropriate line weights.

Many other aspects of creating a drawing can be standardized. The borders required for each different sheet size, the information contained in the title block (and how it is to appear), and the fonts and letter sizes to be used for different items are frequently spelled out in a company's drawing standards. The text of certain notes—such as manufacturing standards or safety control notices—are often standardized so that legally appropriate and consistent information is provided in all cases. In addition, libraries of symbols and stock parts may be provided to help you work more efficiently and produce engineering drawings that can easily be understood and reused by others.

The guidelines for drawing standards can be codified and stored with drawing archives as a navigational aid in the future. In some cases, the drafting group will help ensure that drawings meet company standards; in others, it is the responsibility of the designer to check the company's published standards. The standards may be enforced by the records management group responsible for archiving design documents, which will refuse to accept drawings that are not prepared according to standards. While it may be easy to rename a file after it is created to make it consistent with company standards, you should address drawing standards by starting new drawings from a prototype file or seed part that provides the company's common framework.

## 7-46 STORAGE OF PAPER OR MYLAR DRAWINGS

Paper or mylar drawings may be stored flat in large flat-drawer files or hung vertically in cabinets especially designed for the purpose (Figure 7-76). Exceptionally large drawings are often rolled and stored in tubes in racks or cabinets. Prints are often folded and stored in standard office file cases. Proper control procedures will enable the user of the drawing to find it in the file, to return it to its proper place, and to know where the drawing is when not in the file. Careful revision control processes must be used to ensure the most recent drawing version is being used and the old versions are stored to properly document the design history.

## 7-47 REPRODUCTION OF DRAWINGS

After the drawings of a machine or structure have been completed, it is usually necessary to supply copies to many different persons and firms. Exact, rapid, and economical reproduction methods must be used to transmit the information on the drawings to the group of drawing users, whether it is storing a PDF file version of the drawing on the Web or computer network or printing the drawing and distributing it.

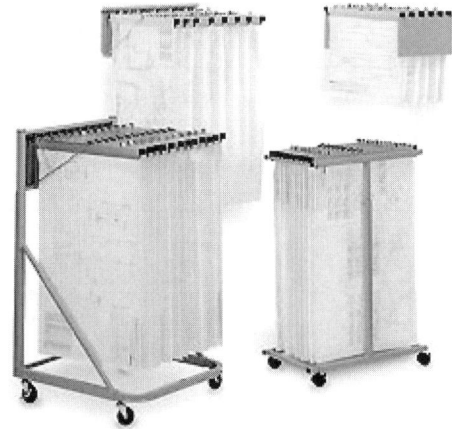

**Figure 7-76**   Paper or Mylar Drawings are Sometimes Stored Clamped into Hanging Racks. Courtesy of Mayline Group.

Even when a drawing is plotted from a CAD system, it is frequently faster to use a reproduction process to distribute the drawing. Plots can be created on mylar, or on specialty papers that allow prints to be made directly from the plot. Large format copy machines are also popular for creating distribution drawings of a print. Networks, groupware, Internet and Intranet, and modems are allowing many workplaces to go to a paperless office, where drawings are distributed electronically. However, when there is a need to take drawings into the field, printing is still frequently required.

## 7-48 PRINTING AND COPYING ENGINEERING DRAWINGS

Of the several processes in use for reproduction, the blueprint is the oldest process used for making prints of large drawings. It is essentially a photographic process in which the original drawing is the negative. The blueprint process, which was the common method used for the reproduction of drawings for many years, has now been replaced to a large extent by other more convenient and efficient processes.

### Engineering Printers

High-speed digital engineering printers, such as the one shown in Figure 7-77, are most widely used in the engineering and manufacturing industry. Such printers can produce output from either CAD files, hard copy, or a mixture of both. These printers can also produce multimedia products and digital sorting for automated set production.

### Diazo-Moist Prints

Diazo moist prints may be created on special black-print paper, cloth, or film producing a black-and-white print

**Figure 7-77** Xerox Engineering Printer/Copier Model 6024. *Courtesy of Xerox Corporation.*

to a blueprint. Moist diazo prints are fed through a special developer that dampens the coated side of the paper.

### The Diazo-Dry Process

The diazo-dry process is a contact method of reproduction which depends on light being transmitted through the original to produce a positive print. Pen or pencil lines, typewritten or any opaque printed matter or image can be reproduced this way. The diazo whiteprint method of reproduction consists of two steps—exposure and dry development by means of ammonia vapors.

### Xerography

Xerox prints are positive prints with black lines on a white background (Figure 7-78). A selenium-coated and electrostatically charged plate is used. A special camera is used to project the original onto the plate; hence, reduced or enlarged reproductions are possible. A negatively charged plastic powder is spread across the plate and adheres to the positively charged areas of the image. The powder is then transferred to paper by means of a positive electric charge and is baked onto the surface to produce the final print. Full-sized prints or reductions can be made inexpensively and quickly in the fully automated Xerox or other similar copy machines. Xerography allows for volume print making from original drawings or microfilms.

### Fax Technology

Fax machines (sometimes called telecopiers or facsimile machines) can receive or send documents (usually) over standard telephone lines in the office or in the field. A computer can send and receive documents directly, generally as a

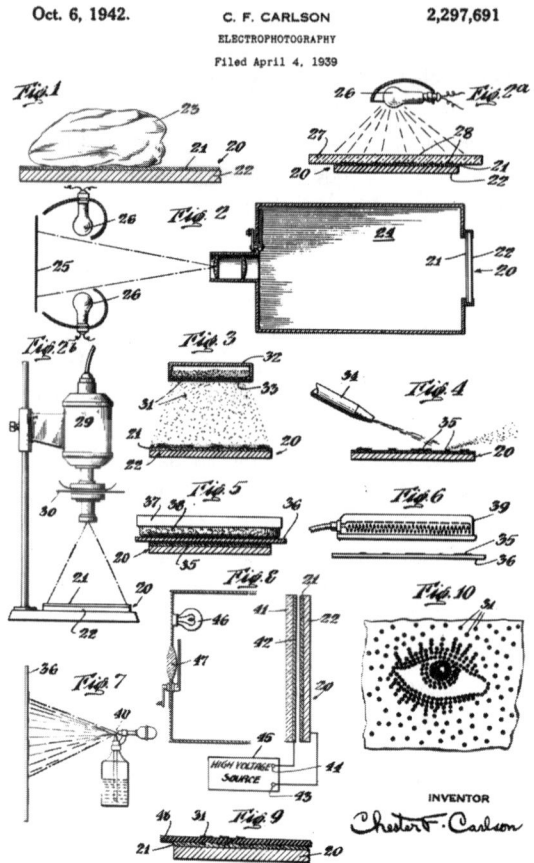

**Figure 7-78** Patent Drawing for Chester Carlson's Electrophotography Device. *Courtesy of Xerox Corporation.*

raster or bitmap type image, avoiding the need for a paper copy of the drawing. Many companies produce all in one copier, printer, and fax machine.

### Digital Image Processing

Modern digital techniques have made possible the direct production of drawings on a laser printer from a variety of input sources, including computers and electronic video equipment.

In addition to drawings produced with the assistance of a CAD program, hand-produced drawings may be stored digitally using digital scanning or by manually digitizing the drawing.

### Color Laser Printing/Copying

Color laser copiers can reproduce drawings in four colors, with black lettering (Figure 7-79), and some are available with optional built-in computer processing unit and monitor, video player, and film projector to permit convenient viewing and editing of drawings.

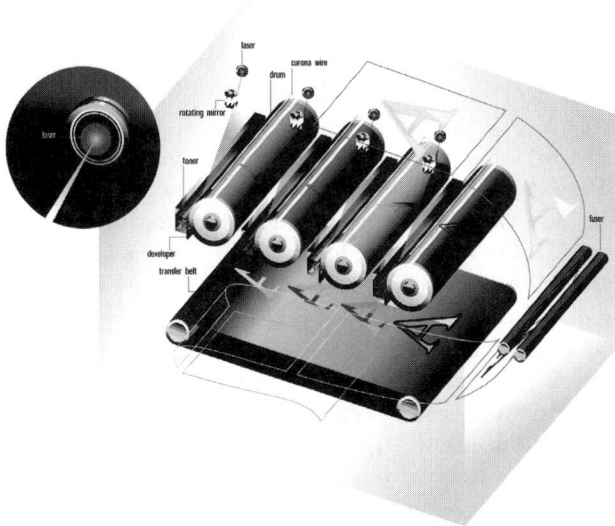

**Figure 7-79**  The single pass laser printer uses four separate photoconducting drums and four laser beams one for each color: cyan, magenta, yellow, and black. The four colors of powdery plastic toner are attracted to the positive charges that make up the image. *Courtesy of Xerox Corporation.*

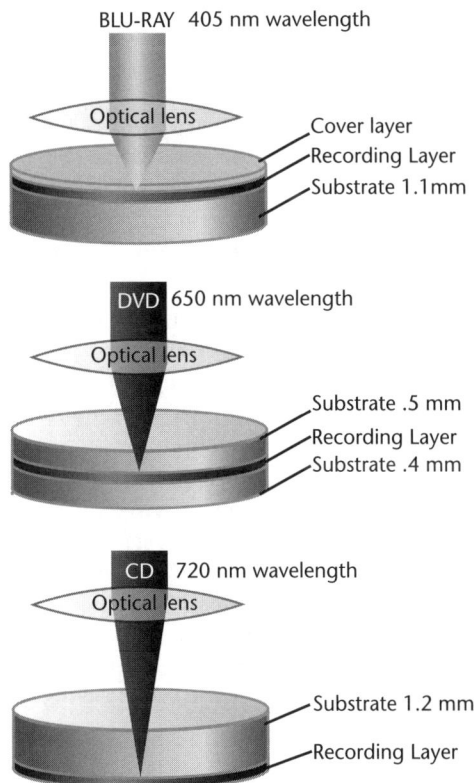

**Figure 7-80**  Blu-Ray disc system compared to DVD and CD Rom

# 7-49 DIGITAL STORAGE SYSTEMS

## CD-ROM

Recordable CD-ROM systems let you store digital information such as CAD drawings, digital audio and video, data, multimedia projects, and other digitally stored records. Write-once-read-many (WORM) CD-ROM storage devices provide excellent storage for CAD documentation. Once the CD has been written it cannot be erased or rewritten. The shelf life for most storage of the media if properly cared for claims to be at least 100 years, so it qualifies as an archival media for permanent storage. CDs are compact and easy to store, and CD-ROM players are standard equipment on many CAD systems. Another advantage of CD-ROM systems is that they are random access storage systems, so that you can go directly to the document you wish to retrieve, unlike tape systems, which must wind through all of the previous tape. Systems that can automatically retrieve from a selection of multiple CDs, called juke box systems, are available for quickly retrieving documents in a network storage situation.

## Optical Disk

Optical disk storage systems use optical magnetic media to store capacities up to 4.6 GB (gigabytes) on a single removable disk. They are rewritable media. This means that they are not suitable for archival storage of permanent records.

## DVD

Recordable DVD systems are similar to CD-ROM storage systems. Both use lasers to "read" microscopic bumps created in a spiral pattern on a layered polycarbonate substrate. A DVD holds more data than a CD because it can have more than one spiral in a separate layer. In addition, both sides of the disk can store data. This produces a range of 4.8 to 15.9 GB of storage on a disk.

## Blu-Ray

Blu-ray disks (BD) have a data layer that is nearer the surface and therefore closer to the "lens" of the laser. It uses a blue-violet laser to read bumps that are one-fifth the size of those read by the red lasers used by DVD systems. This allows for 50 GB of storage on a single disk (Figure 7-80).

## Network Drives

A terabyte (TB) is 1000 GB. Network drives that have fast seek times and allow multiple users to connect at one time are an effective method for storing data. Proper backup procedures are important to ensure that if the hard disk fails you do not lose your data.

## 7-50 DOCUMENT MANAGEMENT SOFTWARE

Specialized software is available to help manage document revision history, approval, storage, file naming, and other issues of managing digital documentation, such as that produced by CAD systems. Cyco's AutoManager TeamWork ™® is document management software geared to work groups of 5 to 15 people. Windchill™® is a full scale PDM/PLM solution from PTC corporation and there are many others in every range. In order to have an effective document management system, a lot of planning and setup needs to be done to ensure success. A software package alone will not provide instant success in managing the large number of files and meeting legal requirements for document storage. It requires setup time, training, and ongoing effort to make it effective. If you are unsure how long different documents must be retained, ARMA is a good place to go for information.

## 7-51 ELECTRONIC FILES AND THE INTERNET

Most engineering firms use some kind of CAD program, such as AutoCAD, Pro/Engineer Wildfire, SolidEdge, Catia, SolidWorks, or others, to produce the majority of their drawings. These drawings are created and then saved as an electronic file. Unlike hard copy diagrams, these electronic files can be manipulated, revised, and resaved on various storage devices and systems.

Saved electronic files are then categorized and downloaded into an electronic archive. Archived files can then be controlled and managed through a database. In this way, these files can be maintained, retrieved, reviewed, and revised whenever the need arises using this type of control management system.

One of the main advantages of using electronic files is that they can be shared easily with clients, designers, manufacturing staff, marketing management, purchasing agents, and suppliers through e-mail or via the Internet. E-mail (electronic mail) has become the most widely used electronic tool of the 21st century. It connects the user to the Internet and the World Wide Web. E-mail can be sent around the world or around the corner in a matter of seconds, thereby eliminating the need for phone conversations, mailings, or overnight courier deliveries. E-mails also provide the user with written documentation of all correspondence, which can be read, saved, or forwarded to other users. Electronic files may be attached to e-mails and sent to numerous people at once. This allows the user to communicate and share files with amazing speed. E-mail attachments are limited to a specific file size, based on the restrictions of the Internet provider (IP).

Web sites offer users instant access to the enormous amount of information available on the Web. Through an IP, such as America Online, MSN, Earthlink, ATT, and so on, just to name a few, the user gains access to all of the information placed on the Web. The user can also create his or her own Web site, placing any information or files he or she would want to share with the Internet community or colleagues. Many engineering firms create their own Web sites and post electronic files and images they wish to be viewed by clients, colleagues, and vendors. Such sites are usually password-protected (i.e., the user must provide a login name and password before gaining access to the site). Many Web sites are interactive (i.e., they respond to the user's commands).

Product designs can be communicated, shared, and interpreted quickly and easily through the Internet. This tremendous communication and design review tool helps shorten the design review process and helps eliminate productivity barriers such as incomplete data, slow fax machines, and overnight packages.

## 7-52 MICROFILM, MICROFICHE, AND COMPUTER-OUTPUT MICROFILM

Although electronic files have replaced the use of microfilm and microfiche, for the most part, some are still in use as data storage tools.

A microfilm is a photographic image of information, records, or drawings that is stored on film at a greatly reduced scale.

A microfiche is a cardlike film containing many rows of images or records or drawings. Card sizes used for storage are $3 \times 5$", $4 \times 6$", and $5 \times 8$". A typical $4 \times 6$" microfiche will contain the equivalent of 270 pages of information. The individual cards may be viewed on a reader and, if desired, a full-size copy may be made by using a reader-printer.

Computer-output microfilm (COM) refers to a process used to produce drawings and records on microfilm, with the aid of a computer. A COM unit will produce a microfilm from database information converted to an image on a high-resolution screen that is then photographed. The main advantages of COM are storage capability and speed.

# CAD AT WORK

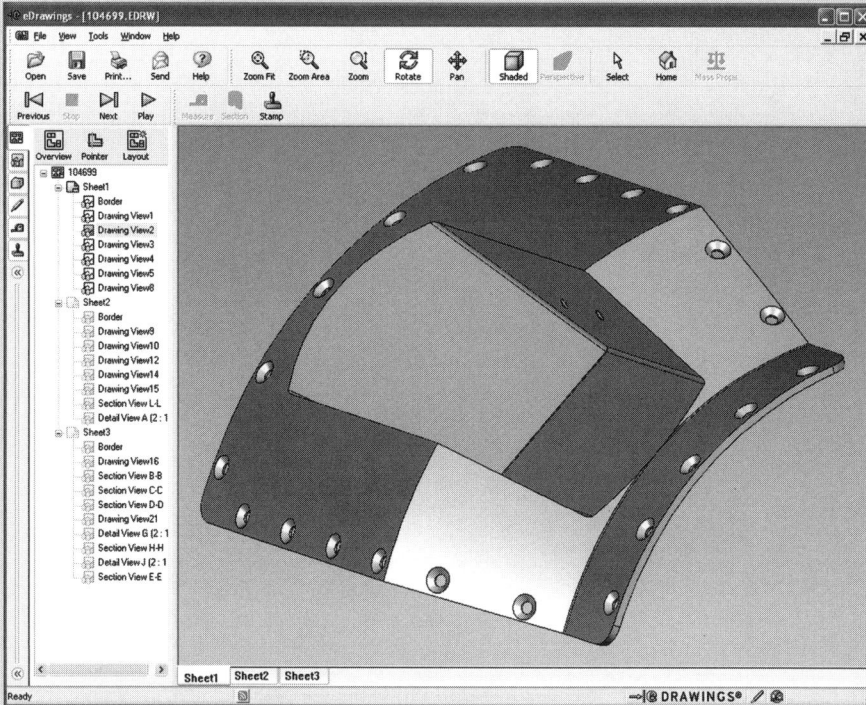

This 27 KB eDrawing can be easily e-mailed to a vendor, who can then view or even rotate the 3D part without any additional software. *Courtesy Zolo Technologies.*

## Communicating Your Drawings Electronically

Today, more and more offices are moving towards a totally electronic workplace. This means communicating with coworkers, clients, and vendors through the use of computers and electronic files. Previously, drawings would have been rendered by hand, copied, rolled up into shipping tubes, and then hand delivered or shipped to coworkers, clients, or vendors.

Modern offices no longer want to deal with hard copy diagrams, as they are bulky, costly to ship, take time to reach their destination, and can be damaged easily. Using electronic data to communicate designs saves time, money, and space. Most designers and clients are computer savvy and would prefer to look at designs on screen as opposed to looking at hard copy. Most computer users also want to be able to view designs as more than just static objects.

Communicating electronically to multiple users can pose challenges, though. Compatibility of files, hardware and software requirements, and varying Internet providers can cause communication problems.

SolidWorks offers eDrawings to answer many of these needs. eDrawings allows the user to share and interpret 2D and 3D product design data. With this new technology, the user can create review-enabled documents and send them to an unlimited number of recipients to mark up and measure via e-mail. The recipients do not need to purchase eDrawing Professional themselves. The user can embed eDrawings Viewer into the eDrawing files, allowing recipients the ability to view, mark up, and measure the drawings automatically. Recipients can create, edit, and save reviews by redlining 2D or 3D data and adding written comments;measure geometry in part, assembly, and drawing files when dimensions are omitted from the drawing; explode assemblies by dragging and dropping assembly components with the cursor; and move a cross-sectioning plane through a part or assembly to see design details hidden from view.

eDrawings also permits SolidWorks and AutoCAD integration by allowing the user to generate eDrawings instantly from within the SolidWorks or AutoCAD software programs.

Getting design information to vendors, suppliers, manufacturers, and coworkers is a large part of the challenge of collaboration. Communication problems often can delay a project. eDrawings offers better communication, smaller file size, and fewer interpretation mistakes by eliminating common communication barriers, by not requiring everyone in the review process to purchase additional software tools. eDrawings makes the files size as small as possible. The ability to rotate 3D models in an eDrawing aids in interpretation, reducing the chance of costly mistakes. EDrawings provide the file in a format that anyone can easily receive and immediately view. This new technology gives the user the capabilities needed to overcome many of the common barriers to effective design communications.

# C H A P T E R 8

# Gears, Bearings, and Cams

## 8-1 INTRODUCTION

This chapter explains how to draw and design with gears and bearings. The chapter does not discuss how to design specific gears and bearings but how to design using existing parts selected from manufacturers' catalogs. Various gear terms are defined, and design applications demonstrated.

The chapter also discusses how to design and draw a cam based on a displacement diagram. Different types of follower motion are explained as well as different types of followers.

## 8-2 TYPES OF GEARS

There are many types of gears, including spur, bevel, worm, helical, and rack. See Figure 8-1. Each type has its own terminology, drawing requirements, and design considerations.

## 8-3 GEAR TERMINOLOGY—SPUR

Following is a list of common spur gear terms and their meanings. Figure 8-2 illustrates the terms, and Figure 8-3 shows a listing of relative formulas.

**For spur gears using English units**

*Pitch diameter (PD)*—The diameter used to define the spacing of gears.

*Diametral pitch (DP)*—The number of teeth per inch or millimeter.

*Circular pitch (CP)*—The circular distance from a fixed point on one tooth to the same position on the next tooth as measured along the pitch diameter. The circumference of the pitch diameter divided by the number of teeth.

*Preferred pitches*—The standard sizes available from gear manufacturers. Whenever possible, use preferred gear sizes.

*Center distance (CD)*—The distance between the center points of two meshing gears.

*Backlash*—The difference between a tooth width and the engaging space on a meshing gear.

*Addendum (a)*—The height of a tooth above the pitch diameter.

*Dedendum (d)*—The depth of a tooth below the pitch diameter.

*Whole depth*—The total depth of a tooth. The addendum plus the dedendum.

*Working depth*—The depth of engagement of one gear into another. Equal to the sum of the two gear addendums.

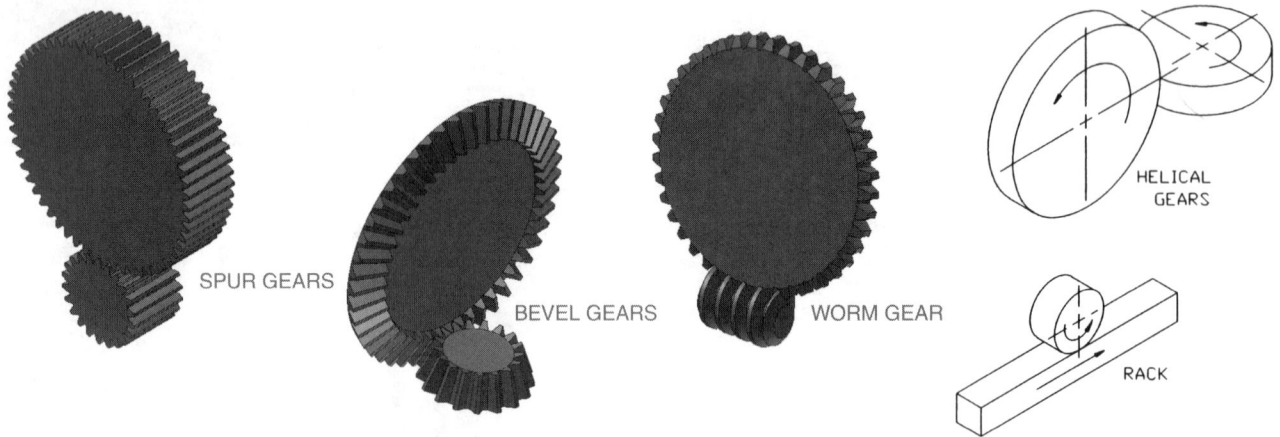

Figure 8-1

**Circular thickness**—The distance across a tooth as measured along the pitch diameter.

**Face width (FW)**—The distance from front to back along a tooth as measured perpendicular to the pitch diameter.

**Outside diameter (OD)**—The largest diameter of the gear. Equals the pitch diameter plus the addendum.

**Root diameter (RD)**—The diameter of the base of the teeth. The pitch diameter minus the dedendum.

**Clearance**—The distance between the addendum of a meshing gear and the dedendum of the mating gears.

**Pressure angle**—The angle between the line of action and a line tangent to the pitch diameter. Most gears have pressure angles of either 14.5° or 20°.

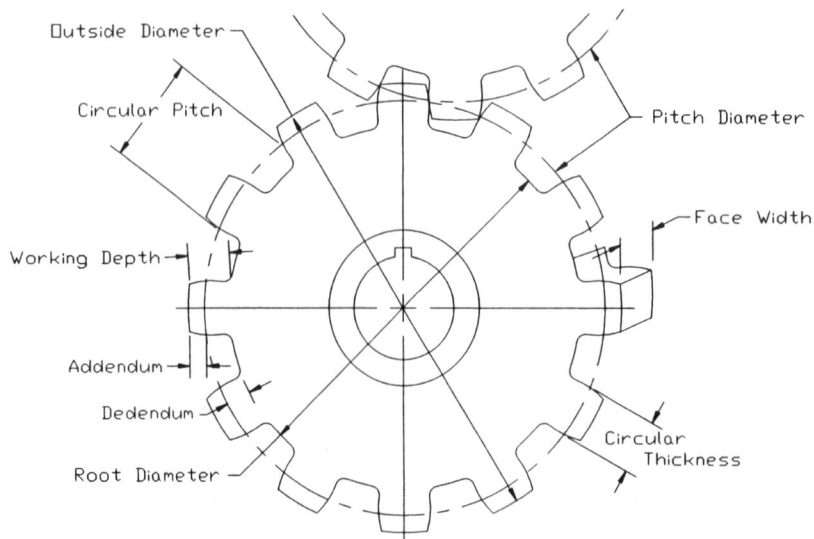

Figure 8-2

| Pitch Diameter (PD) | See catalogs |
|---|---|
| Circular Pitch (CP) | $CP = \dfrac{\pi}{DP}$ |
| Diametral Pitch (DP) | $DP = \dfrac{\pi}{CP}$ |
| Number of Teeth (N) | $N = (PD)(DP)$ |
| Outside Diameter (OD) | See catalogs |
| Addendum (a) | $a = \dfrac{1}{DP}$ |
| Dedendum (d) | $d = a + .125$ (for drawing purposes ONLY) |
| Root Diameter (RD) | $RD = PD - d$ |
| Circular Thickness (CT) | $CT = \dfrac{PD}{2N}$ |
| Face Width (F) | See catalogs |

**Figure 8-3**

For English units

$$\text{Pitch} = \frac{\text{Number of Teeth (N)}}{\text{Pitch Diameter (PD)}} = \frac{N}{PD}$$

For metric units

$$\text{Module} = \frac{\text{Pitch Diameter (PD)}}{\text{Number of Teeth (N)}} = \frac{PD}{N}$$

**Figure 8-4**

## For spur gears using metric units

The preceding definitions apply to both English unit and metric unit spur gears with the exception of pitch. For English unit gears, pitch is defined as the number of teeth per inch relative to the pitch diameter and is expressed by the formula shown in Figure 8-4. Gears made to metric specifications are defined in terms of the amount of pitch diameter per tooth, called the gear's *module.* Figure 8-4 shows the formula for calculating a gear's module. Metric gears also have a slightly different tooth shape that makes them incompatible with English unit gears.

## 8-4  SPUR GEAR DRAWINGS

Figure 8-5 shows a representation of spur gears. The individual teeth are not included in the front view but are represented by three phantom lines. The diameters of the three lines represent the outside diameter, the pitch diameter, and the root diameter.

The outside diameter and the pitch diameter are usually given in manufacturers' catalogs. The root diameter can be calculated from the pitch diameter using the formula presented in Figure 8-3.

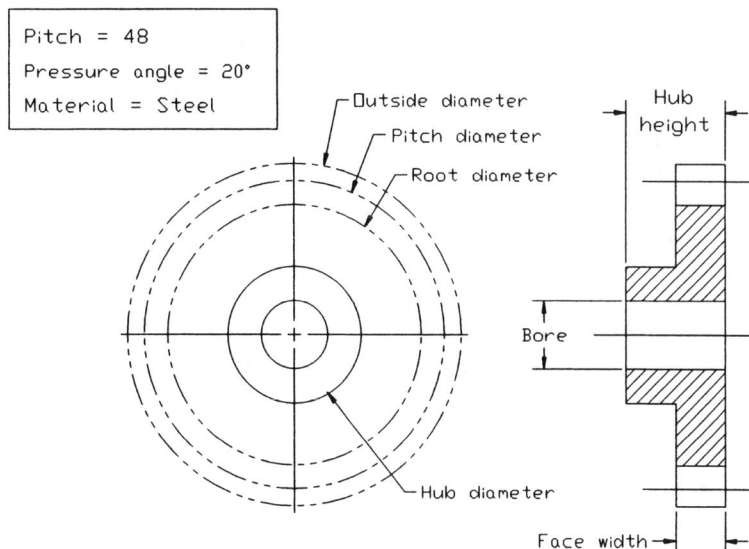

Pitch = 48
Pressure angle = 20°
Material = Steel

Outside diameter
Pitch diameter
Root diameter
Hub height
Bore
Hub diameter
Face width

**Figure 8-5**

**Figure 8-6**

**Figure 8-7**

The side view of the gear is drawn as a sectional view taken along the vertical centerline. The gear size is defined using the manufacturer's stock number, dimensions, and a list of appropriate design information.

Figure 8-6 shows the front representation view of three meshing gears. The gears are positioned so that the pitch circle diameters are tangent. Ideally, mating gears always mesh exactly tangent to their pitch circles.

Gear representations were developed because it was both difficult and time consuming to accurately draw individual teeth when creating drawings on a drawing board. AutoCAD can be used to create detailed gear drawings that include all teeth by using the **Array** and **Block** commands, among others; however, it is usually sufficient to show a few meshing teeth and use the representative centerlines for the remaining portions of both gears. See Figure 8-7.

Most gear tooth shapes are based on an involute curve. Involute-based teeth fit together well, transfer forces smoothly, and can use one cutter to generate all gear tooth variations within the same pitch. Standards for tooth proportions have been established by the American National Standards Institute (ANSI) and the American Gear Manufacturers Association (AGMA).

## 8-5 SAMPLE PROBLEM SP8-1

Draw a front view of a spur gear that has an outside diameter of 6.50 inches, a pitch diameter of 6.00 inches, and a root diameter of 5.25 inches. The gear has 12 teeth.

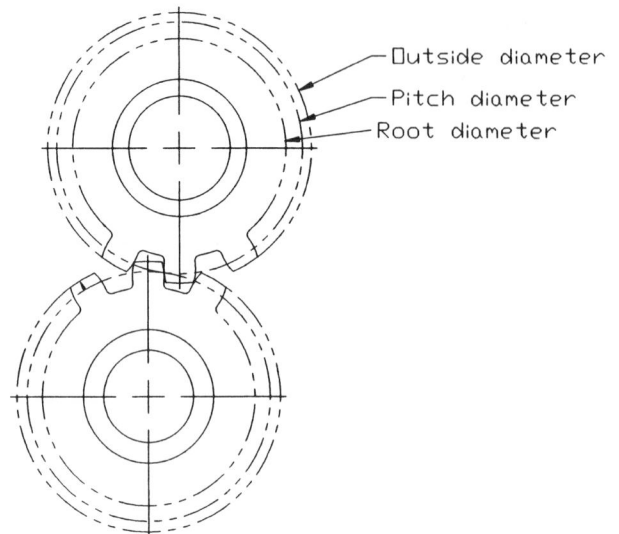

The method presented is a simplified method and is an acceptable representation for most drawing applications. See Figure 8-8.

1. Draw three concentric circles of diameter **6.50, 6.00,** and **5.25**. Use the **Center Mark** command modified to draw a centerline for the circles.
2. Use the **Array** command and create **48** ray lines as shown. Label three rays on each side of the top vertical centerline as shown. Zoom the labeled area.

The number of rays should equal four times the number of teeth to be drawn. Two adjoining sectors are used to define the width of the tooth, and the next two adjacent sectors are used to define the space between teeth.

3. Draw a circle whose center point is at the intersection of the pitch diameter and the ray labeled **2,** and whose radius is determined by the distance from the center point to the intersection of the pitch diameter and the ray labeled **−1.**
4. Repeat step 3 using the intersection of the pitch diameter and the point labeled **−2** as the center point.
5. Use the **Fillet** command to draw a radius at the base of the tooth. Select the side of the tooth as one of the lines for the fillet, and the root diameter as the other line.

Both the circle and ray line will be within the selected cursor, but AutoCAD will select the last entity drawn, so the diameter will be selected.

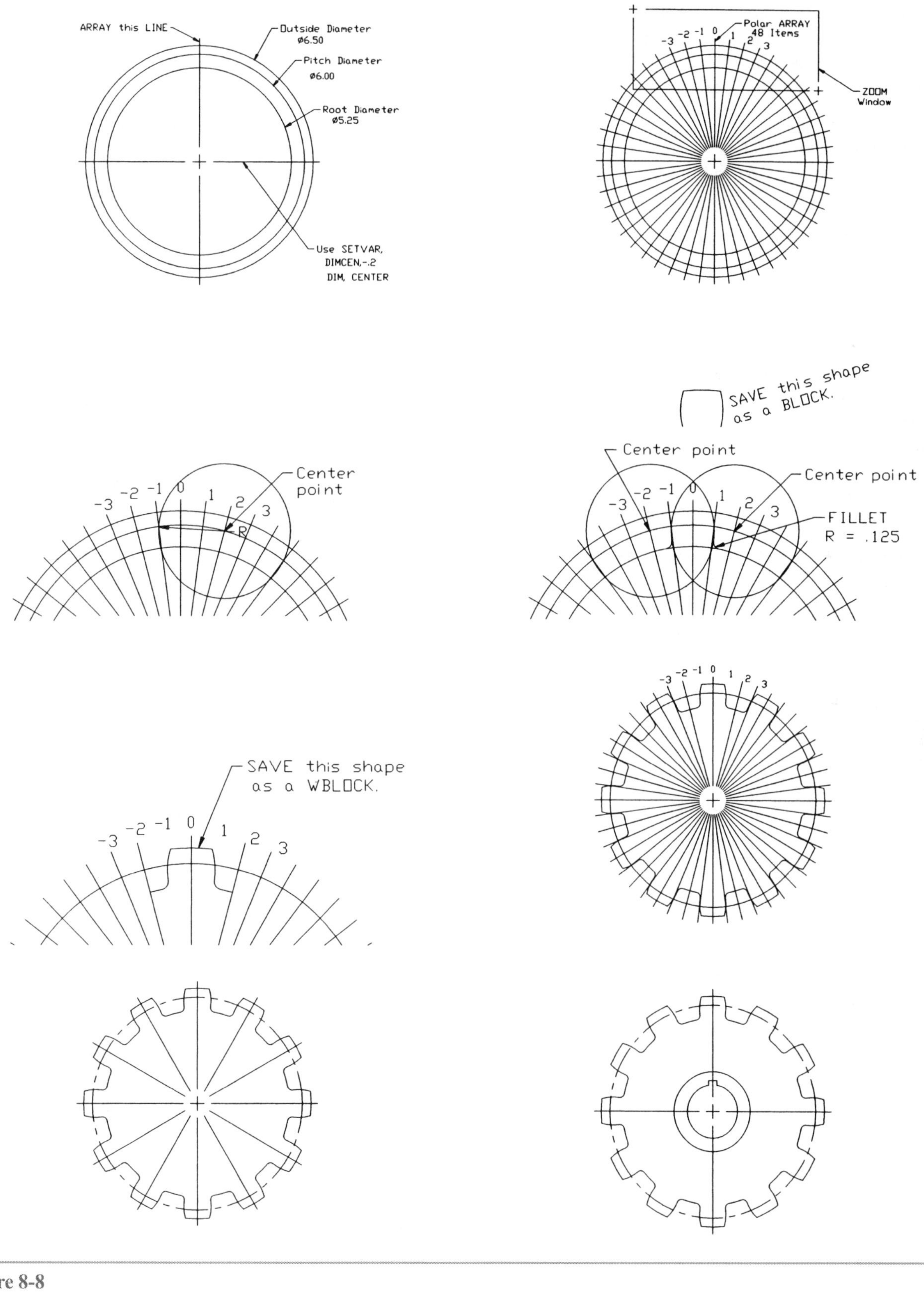

**Figure 8-8**

The tooth shape can be saved as a wblock named **TOOTH** and used when drawing other gears. Only the lines that represent the top of the gear and the two side sections need to be saved. A different size gear will have a different root diameter, and new fillets can be drawn that align with the new root diameter.

6. Use the **Trim** and **Erase** commands to remove excess lines and create the tooth shape between rays −**2** and **2** as shown.

This tooth shape may be saved as a wblock and used when drawing other gears.

7. Array the tooth shape about the gear's center point.
8. Erase the excess ray lines and change the pitch diameter to a centerline.
9. Draw the gear's bore hole or center hole and hub as required.

## 8-6 SAMPLE PROBLEM SP8-2

Figure 8-9A shows two meshing gears. In the example shown, the larger gear has a diameter of 6.00 inches with 24 teeth; the smaller gear has a diameter of 3.00 inches and 12 teeth. The drawing utilizes the wblock created in Sample Problem SP8-1 as follows.

The circular thickness of the wblock tooth as measured along the pitch diameter equals $\frac{1}{12}$ the circumference of the pitch diameter.

$(\frac{1}{12})(\pi\ PD)$

For the gear in SP8-1, where PD = 6

$(\frac{1}{12})(\pi\ 6) = \pi/2$
$= 1.57$ inches

The large gear requires 24 teeth on a pitch diameter of 6, or twice as many teeth as the gear used to create the wblock. The teeth on the 24-tooth gear must be half the size of the tooth created in the wblock. The teeth on the smaller gear must be the same size as the teeth on the larger gear.

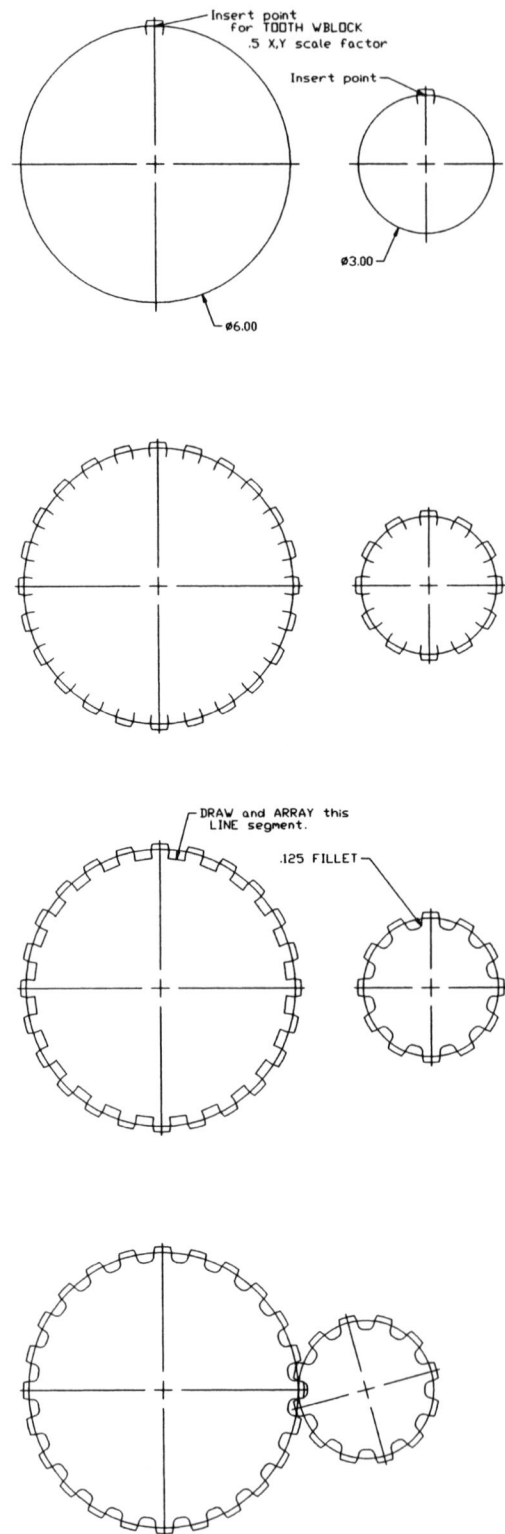

**Figure 8-9A**

**Figure 8-9B**

**To draw meshing spur gears (See Figure 8-9B.)**

1. Draw two circles of diameter **6.00** and **3.00** inches. Include the circles' centerlines. Use **Setvar, Dimcen,** set to **−.2,** or use a modified **Center Mark** tool.

The gears will be drawn separately, then meshed.

2. Insert the TOOTH wblock on both gears as shown on the pitch diameter. Use an X and Y scale factor of **.5**. Explode the wblock.
3. Use the **Array** command to create the required 24 and 12 teeth.
4. Draw a line between the roots of two of the teeth as shown. Use the **Array** command to create the root circle. Add the fillet to each tooth base.

The line in this example is a straight line acceptable for smaller gears. If more accuracy of shape is required, draw a complete root circle, then trim all of it away except the portion between two tooth roots. Array this sector between all the other teeth.

5. Use the **Rotate** command to orient the small gear with the large gear.

Each tooth on the large gear requires $360/24 = 15°$. Rotate the smaller gear 15°.

6. Use the **Move** command to position the small gear.

The pitch circles of the two gears should be tangent.

7. Rotate the smaller gear's centerline **−15°**.
8. Add the center hubs to both gears as shown.

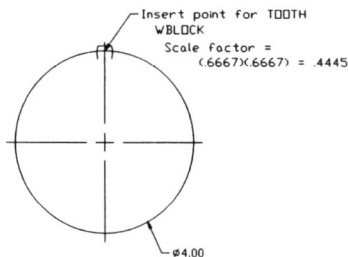

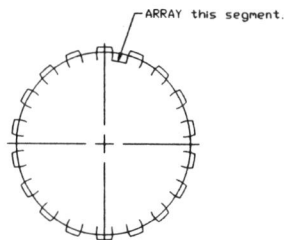

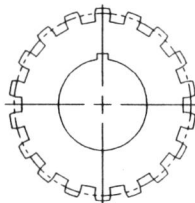

**Figure 8-10**

## 8-7 SAMPLE PROBLEM SP8-3

Figure 8-10 shows a gear that has a pitch diameter of 4.00 with 18 teeth. The TOOTH wblock can be used as follows.

The wblock is first reduced to accommodate the smaller diameter.

1. Draw a circle with a **4.00**-inch diameter. Include centerlines.
2. Insert the wblock using an X and Y scale factor of **.4445.**

The scale factor was derived by first considering the ratio between the number of teeth on the gear used to create the TOOTH wblock (12) and the number of teeth on the desired gear, 18, or 12/18 = .6667. The ratio between the diameters is also considered: 4.00/6.00 = .6667. The two ratios are multiplied together: (.6667)(.6667) = .4445.

3. Use the **Array** command to create the required 18 teeth.
4. Draw a line between the end lines of two of the teeth, then array the line **18** times around the gear.

The line may be drawn as a straight line for small gears or as an arc for larger gears. The arc may be created by drawing a circle, then using the **Trim** command to create the desired length.

5. Add the center hub as required.

The same wblock can be used for metric gears using the conversion factor 1.00 inch = 25.4 millimeters. It is probably easier to create a separate wblock for a metric tooth.

## 8-8 SELECTING SPUR GEARS

When two spur gears are engaged, the smaller gear is called the *pinion gear* and the larger gear is called simply the *gear*. The relationship between the relative speed of two mating gears is directly proportional to the gears' pitch diameters. Also, the number of teeth on a gear is proportional to the gear's pitch diameter. This means that the ratio of speed between two meshing gears is equal to the ratio of the number of teeth on the two gears. If one gear has 40 teeth and the other 20, the speed ratio between the two gears is 2:1.

For gears to mesh properly, they must have the same pitch and pressure angle. Gear manufacturers present their gears in charts that include a selection of gears with common pitches and pressure angles. Figure 8-11 shows a sample spur gear listing from the website of Stock Drive Products. The site allows you to select a diametral pitch and all other appropriate product details. Once a gear is selected, the gear information will be displayed. You may be able to generate an AutoCAD drawing of the gear if you have a compatible setup. Many manufacturers offer free product catalogs.

## 8-9 CENTER DISTANCE BETWEEN GEARS

The center distance between meshing spur gears is needed to align the gears properly. Ideally, gears mesh exactly on their pitch diameters, so the ideal center distance between two meshing gears is equal to the sum of the two pitch radii, or the sum of the two diameters divided by 2:

$$CD1 = (PD1 + PD2)/2$$

If two gears were chosen from the chart in Figure 8-11, and one had 30 teeth and a pitch diameter of .9375, and the other had 60 teeth and a pitch diameter of 1.8750, the center distance between the gears would be

$$CD1 = (.9375 + 1.8750)/2 = 2.8125$$

The center distance of gears is dependent on the tolerance of the gears' bores, the tolerance of the supporting shafts, and the feature and positional tolerance of the holes in the shaft's supporting structure. The following sample problem shows how these tolerances are considered and applied when matching two spur gears.

## 8-10 SAMPLE PROBLEM SP8-4

An electric motor generates power at 1750 rpm. Reduce this speed by a factor of 2 using steel metric gears with a module of 1.5 and a pressure angle of 20°. Figure 8-12 shows a list of gears. Determine the center distance between the gears, and specify the shaft sizes required for both gears.

Gear number A 1C22MYKW150 50A and number A 1C22MKYW150 100A were selected from the chart shown in Figure 8-12. The pinion gear has 50 teeth, and the large gear has 100, so if the pinion gear is mounted on the motor

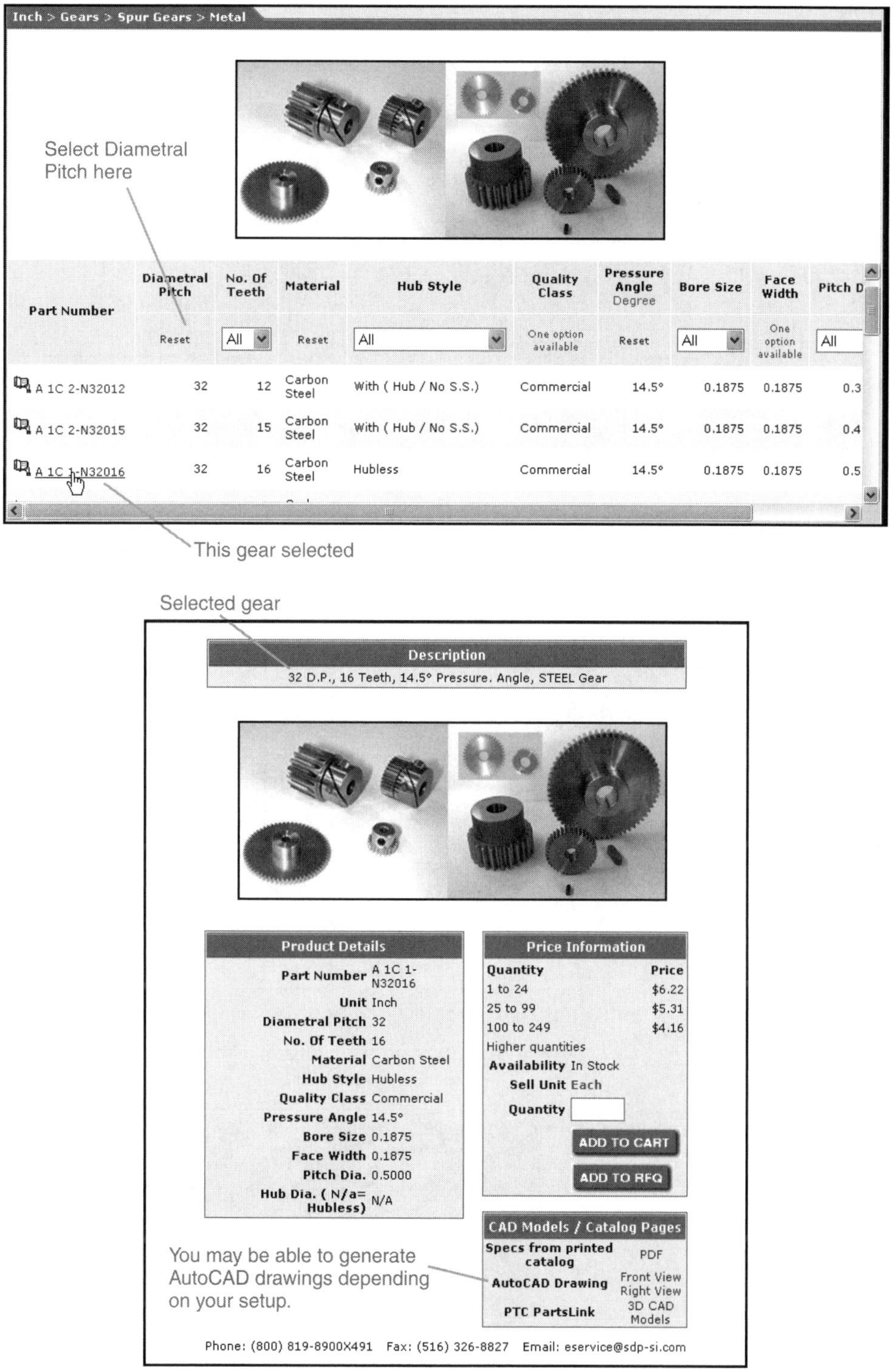

Inch > Gears > Spur Gears > Metal

Select Diametral Pitch here

| Part Number | Diametral Pitch | No. Of Teeth | Material | Hub Style | Quality Class | Pressure Angle Degree | Bore Size | Face Width | Pitch D |
|---|---|---|---|---|---|---|---|---|---|
| | Reset | All | Reset | All | One option available | Reset | All | One option available | All |
| A 1C 2-N32012 | 32 | 12 | Carbon Steel | With ( Hub / No S.S.) | Commercial | 14.5° | 0.1875 | 0.1875 | 0.3 |
| A 1C 2-N32015 | 32 | 15 | Carbon Steel | With ( Hub / No S.S.) | Commercial | 14.5° | 0.1875 | 0.1875 | 0.4 |
| A 1C 1-N32016 | 32 | 16 | Carbon Steel | Hubless | Commercial | 14.5° | 0.1875 | 0.1875 | 0.5 |

This gear selected

Selected gear

**Description**

32 D.P., 16 Teeth, 14.5° Pressure. Angle, STEEL Gear

**Product Details**

| | |
|---|---|
| Part Number | A 1C 1-N32016 |
| Unit | Inch |
| Diametral Pitch | 32 |
| No. Of Teeth | 16 |
| Material | Carbon Steel |
| Hub Style | Hubless |
| Quality Class | Commercial |
| Pressure Angle | 14.5° |
| Bore Size | 0.1875 |
| Face Width | 0.1875 |
| Pitch Dia. | 0.5000 |
| Hub Dia. ( N/a= Hubless) | N/A |

**Price Information**

| Quantity | Price |
|---|---|
| 1 to 24 | $6.22 |
| 25 to 99 | $5.31 |
| 100 to 249 | $4.16 |

Higher quantities

**Availability** In Stock

**Sell Unit** Each

**Quantity** [        ]

ADD TO CART

ADD TO RFQ

You may be able to generate AutoCAD drawings depending on your setup.

**CAD Models / Catalog Pages**

| | |
|---|---|
| Specs from printed catalog | PDF |
| AutoCAD Drawing | Front View Right View |
| PTC PartsLink | 3D CAD Models |

Phone: (800) 819-8900X491   Fax: (516) 326-8827   Email: eservice@sdp-si.com

**Figure 8-11**
Courtesy of Stock Drive Products.

FIG. 1                 FIG. 2

| Catalog Number | Fig No. | No. of Teeth | P.D. | O.D. | B* Bore (H7) | F Face Width | C Hub Dia | D Hub Proj | A Dia | W Dim | K Dim |
|---|---|---|---|---|---|---|---|---|---|---|---|
| A 1C22MYKW150 20A | | 20 | 30 | 33 | 14 | | 25 | | | | |
| A 1C22MYKW150 24A | | 24 | 36 | 39 | 16 | | 30 | | | 5 | 2.3 |
| A 1C22MYKW150 25A | | 25 | 37.5 | 40.5 | 18 | 18 | 32 | | | | |
| A 1C22MYKW150 28A | | 28 | 42 | 45 | | | 36 | | | | |
| A 1C22MYKW150 30A | | 30 | 45 | 48 | | | | | | | |
| A 1C22MYKW150 32A | 1 | 32 | 48 | 51 | | | | | | | |
| A 1C22MYKW150 36A | | 36 | 54 | 57 | | | | | | | |
| A 1C22MYKW150 40A | | 40 | 60 | 63 | | | | 14 | | 6 | 2.8 |
| A 1C22MYKW150 48A | | 48 | 72 | 75 | | | | | | | |
| A 1C22MYKW150 50A | | 50 | 75 | 78 | 20 | | 40 | | | | |
| A 1C22MYKW150 56A | | 56 | 84 | 87 | | | | | | | |
| A 1C22MYKW150 60A | | 60 | 90 | 93 | | 16 | | | 76 | | |
| A 1C22MYKW150 64A | | 64 | 96 | 99 | | | | | 82 | | |
| A 1C22MYKW150 70A | | 70 | 105 | 108 | | | | | 91 | | |
| A 1C22MYKW150 72A | 2 | 72 | 108 | 111 | | | | | 94 | | |
| A 1C22MYKW150 80A | | 80 | 120 | 123 | 25 | | 50 | | 106 | 8 | 3.3 |
| A 1C22MYKW150 100A | | 100 | 150 | 153 | | | | | 136 | | |

\* Gears with 14, 16, and 18mm bores have a tolerance of +.018, +.000.
20 and 25mm bores have a tolerance of +0.021/- 0

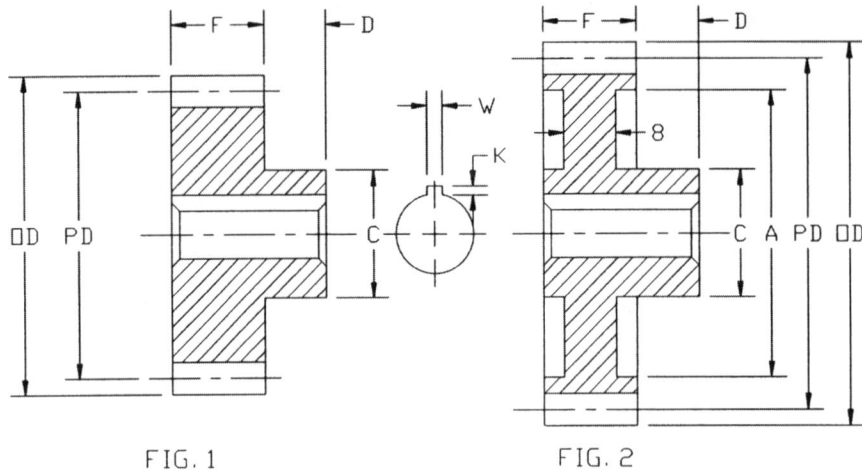

This information is available from the Stock Drive Products catalog or website.

| Part Number | Module (Pitch) | No. Of Teeth | Material | Hub Style | Quality Class | Pressure Angle Degree | Bore Size mm | Face Width mm | Pitch Dia. mm | Hub Dia. (0 = Hubless) mm |
|---|---|---|---|---|---|---|---|---|---|---|
| | One option available | One option available | One option available | One option available | One option available | One option available | One option available | One option available | One option available | One option available |
| A 1C22MYKW15050A | 1.5 | 50 | Carbon Steel | With Hub & Keyway | ISO 8 - ( AGMA 9 ) | 20° | 20.00 | 16.00 | 75.00 | 40.00 |

One part found
One part found

Metric > Gears > Spur Gears > Metal

From the website www.SDP-SI.com

**Figure 8-12**
Courtesy of Stock Drive Products.

shaft, the larger gear will turn at 875 rpm, or half the 1750 motor speed.

$$\frac{1}{2} = \frac{x}{1750}$$

$$x = \frac{1750}{2} = 875 \text{ rpm}$$

The specific design information for the selected gears is as follows:

PINION GEAR
  PD = 75
  OD = 78
  N = 50

  Bore = $20 \begin{smallmatrix} +.021 \\ -.000 \end{smallmatrix}$

  Tolerance = H7

LARGE GEAR
  PD = 150
  OD = 153
  N = 100

  Bore = $25 \begin{smallmatrix} +.021 \\ -.000 \end{smallmatrix}$

  Tolerance = H7

The center distance between the gears is equal to the sum of the two pitch diameters divided by 2:

$$\frac{PD1 + PD2}{2} =$$

$$\frac{75 + 150}{2} = 112.5$$

The manufacturer's catalog lists the bore tolerance as an H7. One gear has a nominal bore diameter of 20 and the other one of 25. Standard fit tolerances for metric values are discussed in Chapter 4, and appropriate tables are included in the appendix.

Tolerance values for a sliding fit (H7/g6) hole basis were selected for this design application. This means that the shaft tolerances are 19.993 and 19.980 for the pinion gear and 24.993 and 24.980 for the large gear.

## 8-11 COMBINING SPUR GEARS

Gears may be mounted on the same shaft. Gears on a common shaft have the same turning speed. Combining gears on the same shaft enables the designer to develop larger gear ratios within a smaller space.

Combining gears can also be used to help reduce the size of gear ratios between individual gears and the amount of space needed to create the reductions. Figure 8-13 shows a four-gear setup. Gears B and C are mounted on the same shaft. Gear A is the driver gear and is turning at a speed of 1750 rpm. The speed of gear D is determined as follows.

The ratio between gears A and B is

$$\frac{48}{72} = .6667$$

The speed of gear B is therefore

$$1750 \times .6667 = 1166.7 \text{ rpm}$$

Gears B and C are on the same shaft, so they have the same speed. Gear C is turning at 1166.7 rpm.

The ratio between gears C and D is

$$\frac{24}{48} = .5000$$

The speed of gear D is therefore

$$1166.7 \times .5000 = 583 \text{ rpm}$$

The ratio between gears A and D is 3:1.

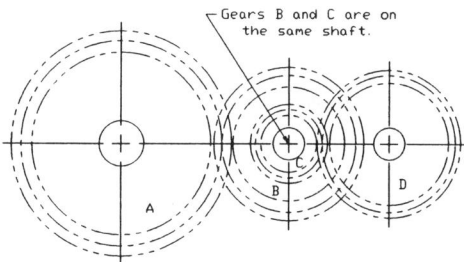

**Figure 8-13**

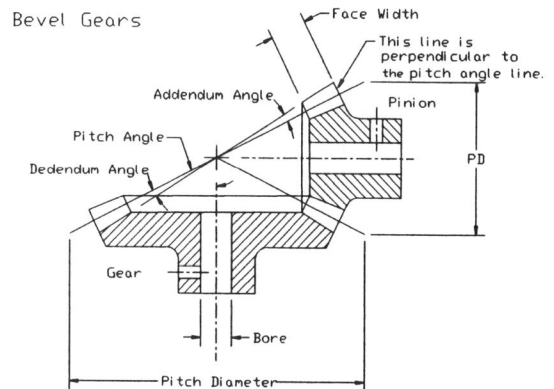

**Figure 8-14**
Courtesy of Stock Drive Products.

## Stock Drive Products

Metric Bevel Gears

20° Pressure Angle

Module 1.5, 2.0, 2.5, 3.0

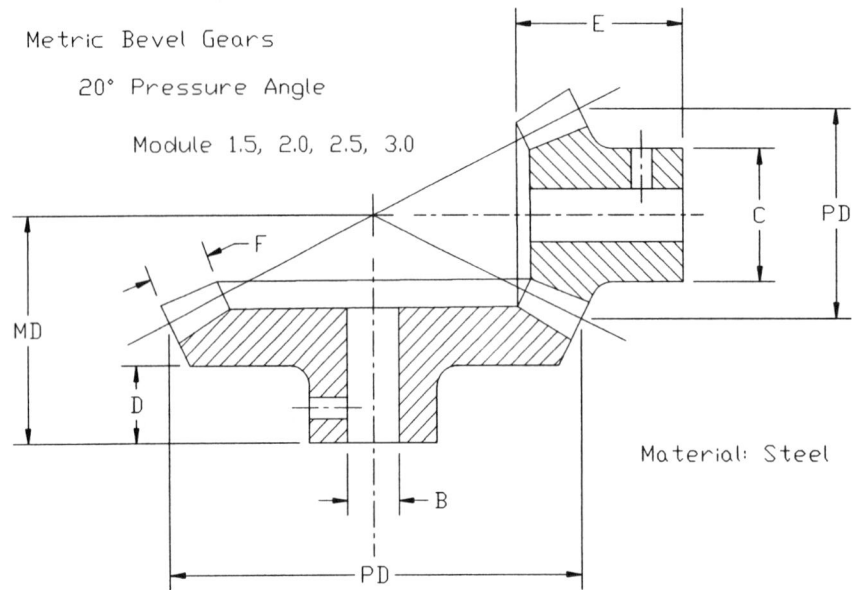

Material: Steel

| Catalog Number | Module | No. of Teeth | Ratio | PD | OD | B* Bore (H8) | F Face Width | E Length | C Hub Dia | D Hub Proj | MD Dim |
|---|---|---|---|---|---|---|---|---|---|---|---|
| A 1C 3MYK 15018 | 1:5 | 18 | | 27 | 29.7 | 8 | 9.8 | 23 | 22 | 12.5 | 40.74 |
| A 1C 3MYK 15036 | | 36 | | 54 | 55.4 | 10 | | 18.5 | 30 | 10 | 26.75 |
| A 1C 3MYK 20018 | 2:0 | 18 | | 36 | 39.6 | 9 | 12.6 | 29 | 28 | 15 | 53.12 |
| A 1C 3MYK 20036 | | 36 | 1:2 | 72 | 73.8 | 12 | | 24 | 36 | 13 | 35.21 |
| A 1C 3MYK 25018 | 2.5 | 18 | | 45 | 49.5 | 12 | 16.7 | 35 | 36 | 17 | 64.29 |
| A 1C 3MYK 25036H | | 36 | | 90 | 92.2 | 14 | | 29 | 50 | 15 | 42.55 |
| A 1C 3MYK 30018 | 3.0 | 18 | | 54 | 59.4 | 12 | 20 | 40 | 41 | 18 | 75.27 |
| A 1C 3MYK 30036H | | 36 | | 108 | 110.7 | 16 | | 36 | 60 | 19 | 52.32 |

* Gears with: 8, 9, 10mm bores have a tolerance of +0.022/−0
    12, 14, 16mm bores have a tolerance of +0.027/−0

This information is also available on the Stock Drive Products website www.SDP-SI.com.

**Figure 8-15**
Courtesy of Stock Drive Products.

# 8-12 GEAR TERMINOLOGY— BEVEL

*Bevel gears* align at an angle with each other. An angle of 90° is most common. Bevel gears use much of the same terminology as spur gears but with the addition of several terms related to the angles between the gears and the shape and position of the teeth. Figure 8-14 defines the related terminology.

Bevel gears must have the same pitch or module value and have the same pressure angle for them to mesh properly. Manufacturers' catalogs usually list bevel gears in matched sets designated by ratios that have been predetermined to fit together correctly. Figure 8-15 shows a sample list of matched set bevel gears using millimeter values, and Figure 8-16 shows a list for inch values.

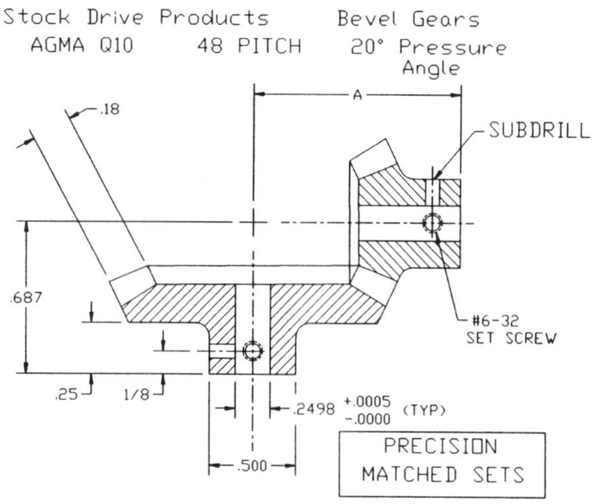

| Catalog Number | Ratio | No. of Teeth | PD | A | Material |
|---|---|---|---|---|---|
| S1346Z-48S30A030 | 1:1 | 30 30 | .625 | .687 | St Steel Aluminum |
| S1346Z-48S30S030 | 1:1 | 30 30 | .625 | .687 | St Steel |
| S1346Z-48A30A030 | 1:1 | 30 30 | .625 | .687 | Aluminum |
| S1346Z-48S30A045 | 1:1-1/2 | 30 45 | .625 .937 | .812 | St Steel Aluminum |
| S1346Z-48S30S045 | 1:1-1/2 | 30 45 | .625 .937 | .812 | St Steel |
| S1346Z-48A30A045 | 1:1-1/2 | 30 45 | .625 .937 | .812 | Aluminum |
| S1346Z-48S30A060 | 1:2 | 30 60 | .625 1.250 | .937 | St Steel Aluminum |
| S1346Z-48S30S060 | 1:2 | 30 60 | .625 1.250 | .937 | St Steel |
| S1346Z-48A30A060 | 1:2 | 30 60 | .625 1.250 | .937 | Aluminum |
| S1346Z-48S30A090 | 1:3 | 30 90 | .625 1.875 | 1.250 | St Steel Aluminum |
| S1346Z-48S30S090 | 1:3 | 30 90 | .625 1.875 | 1.250 | St Steel |
| S1346Z-48S30A090 | 1:3 | 30 90 | .625 1.875 | 1.250 | Aluminum |
| S1346Z-48S30A120 | 1:4 | 30 120 | .625 2.500 | 1.531 | St Steel Aluminum |
| S1346Z-48S30S120 | 1:4 | 30 120 | .625 2.500 | 1.531 | St Steel |
| S1346Z-48A30A120 | 1:4 | 30 120 | .625 2.500 | 1.531 | Aluminum |

**Figure 8-16**
Courtesy of Stock Drive Products.

# 8-13 HOW TO DRAW BEVEL GEARS

Bevel gears are usually drawn by working from dimensions listed in manufacturers' catalogs for specific matching sets. The gears are drawn using either a sectional view or a half-sectional view that shows the profiles of the two gears. Figure 8-17 shows a matching set of bevel gears that were drawn from the information given in Figure 8-15.

The procedure used to draw the gears, based on information given in manufacturers' catalogs, is as follows.

**To draw a matched set of beveled gears**

1. Draw a perpendicular centerline pattern and use the **Offset** command to define the pitch diameters of the pinion and gear.

2. Extend the pitch diameter lines and draw lines from the center point to the intersections as shown.

These lines are called the *face angle lines*. The ends of beveled gears are drawn perpendicular to the face angle lines. Gear manufacturers do not always include the outside

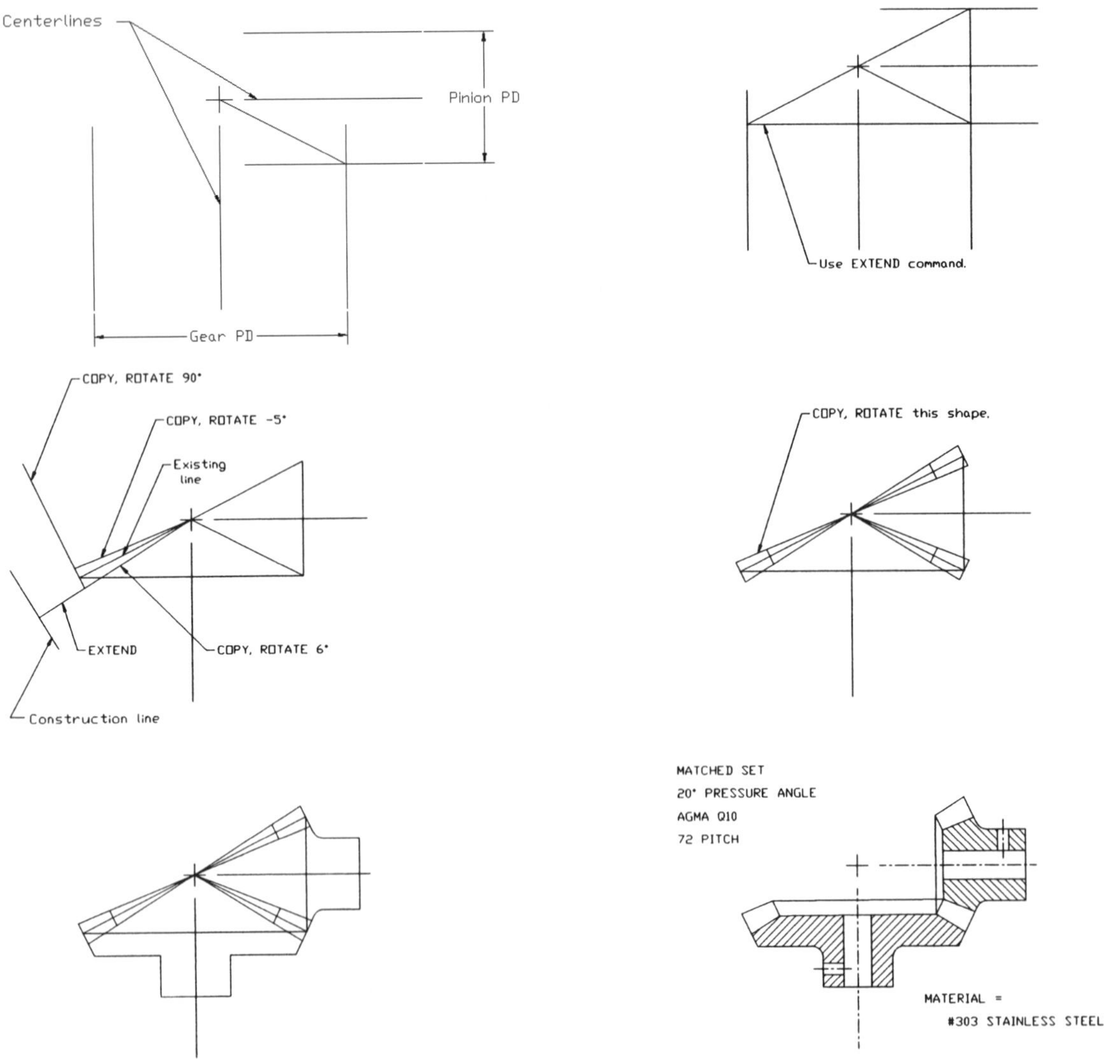

Figure 8-17

diameter values with matching sets of gears. The values are sometimes listed with data for the individual gears elsewhere in the catalog. If the outside diameter values are not given, they can be conservatively estimated by drawing the addendum angle approximately $-5.0°$ from the face angle and the dedendum $6.0°$ from the face angle.

The perpendicular end lines may be constructed using the **Copy** and **Rotate** commands. Copy the existing face angle line directly over the existing line, then rotate the line $90°$ from the face angle line.

Use the **Extend** command to extend the rotated lines as needed. Add a construction line to help define an intersection between the dedendum ray line and the face line perpendicular to the face angle line. Trim and erase any excess lines.

3. Use the **Copy** and **Rotate** commands to copy the face shape and rotate it into the two other positions shown.
4. Use the given dimensions to complete the profiles. Erase and trim lines as necessary.
5. Draw the bore holes and holes for the setscrews based on the manufacturer's specifications.
6. Use the **Hatch** command to draw the appropriate hatch lines.

The two gears should have hatch patterns at different angles. In this example the hatch pattern on the pinion is at $90°$ to the pattern on the gear.

## 8-14  WORM GEARS

A worm gear setup is created using a cylindrical gear called a *worm*, and a circular matching gear called a *worm gear.* See Figure 8-18. As with other types of gears, worm gears must have the same pitch and pressure angle to mesh correctly. Manufacturers list matching worms and worm gears together in their catalogs. Figure 8-20 shows a gear manufacturer's listing for a worm gear and the appropriate worms.

Worm gears are drawn using the representation shown in Figure 8-20 or using sectional views as shown in Figure 8-19. The worm teeth shown can be drawn using the procedure explained in Chapter 6 for acme threads.

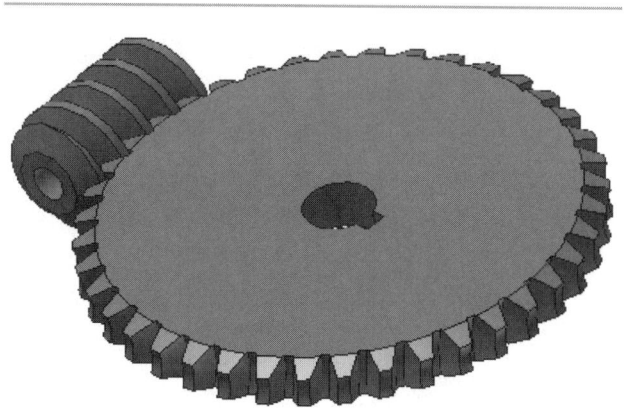

Figure 8-18

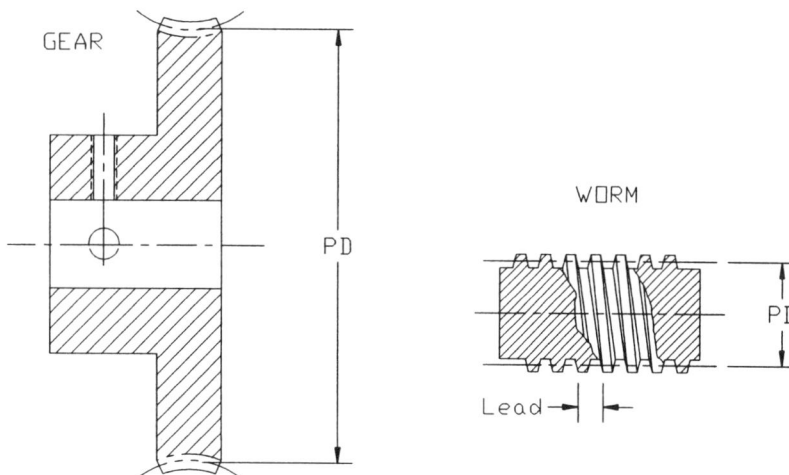

Figure 8-19

### BERG Precision Worm Gears

Bronze
QQ-B-637
Alloy 464

— Setscrew Supplied

— Spot Drill

32 Pitch
Right-Hand

— 3/16

7/64

1/2

B $^{+.0005}_{-.0000}$ PD

.22

— Worm

$$\text{Ratio} = \frac{\text{No. of Teeth}}{\text{Worm Threads}}$$

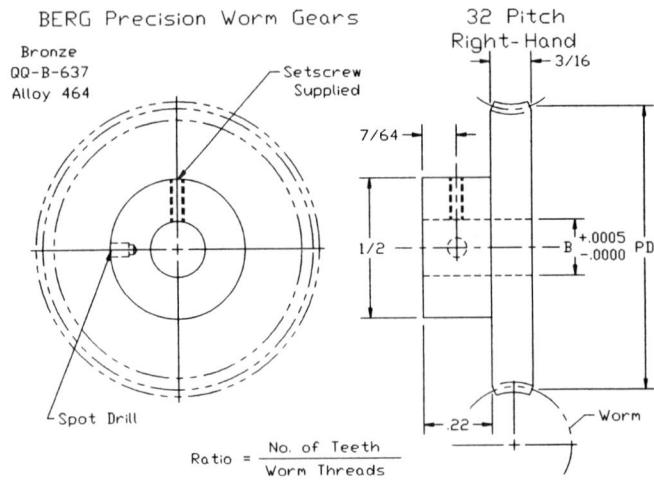

| FOR SINGLE THREAD WORM | FOR DOUBLE THREAD WORM | NO. OF TEETH | PITCH DIA |
|---|---|---|---|
| CIRCULAR PITCH .0982<br>HELIX ANGLE 4° - 5'<br>PRESSURE ANGLE 14-1/2° | CIRCULAR PITCH .1963<br>HELIX ANGLE 8° - 8'<br>PRESSURE ANGLE 20° | | |
| STOCK NUMBER | STOCK NUMBER | | |
| W32B29-S20 | W32B29-D20 | 20 | .625 |
| W32B29-S30 | W32B29-D30 | 30 | .938 |
| W32B29-S40 | W32B29-D40 | 40 | 1.250 |
| W32B29-S50 | W32B29-D50 | 50 | 1.562 |
| W32B29-S60 | W32B29-D60 | 60 | 1.875 |
| W32B29-S80 | W32B29-D80 | 80 | 2.500 |
| W32B29-S96 | W32B29-D96 | 96 | 3.000 |
| W32B29-S100 | W32B29-D100 | 100 | 3.125 |
| W32B29-S120 | W32B29-D120 | 120 | 3.750 |
| W32B29-S180 | W32B29-D180 | 180 | 5.625 |

### Precision Worms

303 Stainless
Steel

32 PITCH
3/16' Bore
Right-Hand

7/8

— Setscrew

.500 $^{+.000}_{-.003}$

.4375 PD (REF)

.1875 $^{+.0005}_{-.0000}$ 3/8

Counterbore relief on bore MFR's option

11/16

3/32

| STOCK NUMBER | THREAD | LEAD | LEAD ANGLE | PRESSURE ANGLE |
|---|---|---|---|---|
| W32S-3S | SINGLE | .0982 | 4°-5' | 14-1/2° |
| W32S-3D | DOUBLE | .1963 | 8°-8' | 20° |

**Figure 8-20**
Courtesy of W. M. Berg Inc.

The relationship between worm gears is determined by the *lead* of the worm thread. The lead of a worm thread is similar to the pitch of a thread discussed in Chapter 6. Worm threads may be single, double, or quadruple. If a worm has a double thread, it will advance the gear twice as fast as a worm with a single thread.

## 8-15 HELICAL GEARS

Helical gears are drawn as shown in Figure 8-21. The two gears are called the ***driver*** and the ***driven,*** as indicated. Figure 8-22 shows a manufacturer's listing of compatible helical gears.

Helical gears may be manufactured with either left- or right-hand threads. Left- and right-hand threads are used to determine the relative rotation direction of the gears.

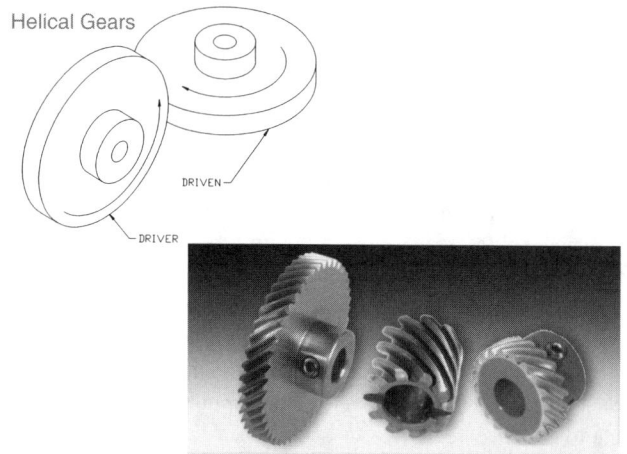

**Figure 8-21**
Courtesy of Stock Drive Products.

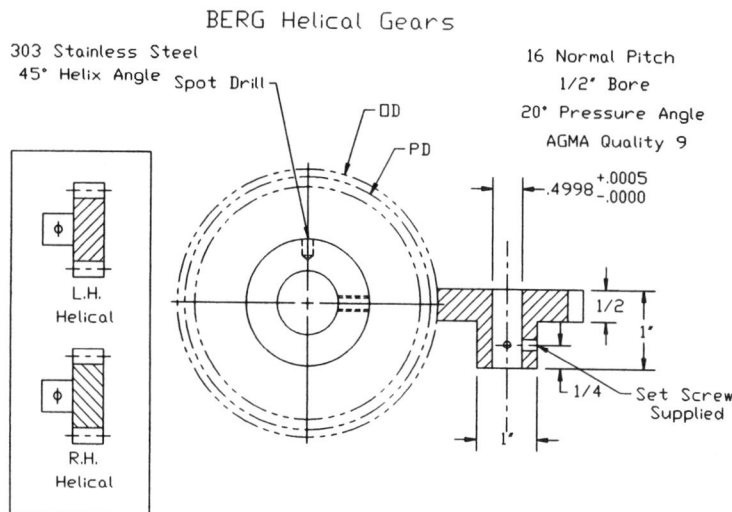

**Figure 8-22**
Courtesy of W. M. Berg Inc.

| STOCK NUMBER | | NO. OF TEETH | PITCH DIAMETER | OUTSIDE DIAMETER |
|---|---|---|---|---|
| RIGHT-HAND | LEFT-HAND | | | |
| H16S38-R 12 | H16S38-L 12 | 12 | 1.060 | 1.185 |
| H16S38-R 16 | H16S38-L 16 | 16 | 1.4142 | 1.539 |
| H16S38-R 20 | H16S38-L 20 | 20 | 1.7677 | 1.892 |
| H16S38-R 24 | H16S38-L 24 | 24 | 2.1213 | 2.246 |
| H16S38-R 32 | H16S38-L 32 | 32 | 2.8284 | 2.953 |
| H16S38-R 40 | H16S38-L 40 | 40 | 3.5355 | 3.660 |
| H16S38-R 48 | H16S38-L 48 | 48 | 4.2426 | 4.367 |

## 8-16 RACKS

***Racks*** are gears whose teeth are in a straight row. Racks are used to change rotary motion into linear motion. See Figure 8-23. Racks are usually driven by a spur gear called a *pinion*.

One of the most common applications of gear racks is the steering mechanism of an automobile. Rack-and-pinion steering helps create a more positive relationship between the rotation of the steering wheel and the linear input to the car's wheel than did the mechanical linkages used on older-model cars.

Figure 8-24 shows a manufacturer's list for racks. The rack and pinions must have the same pitch and pressure angle to mesh correctly.

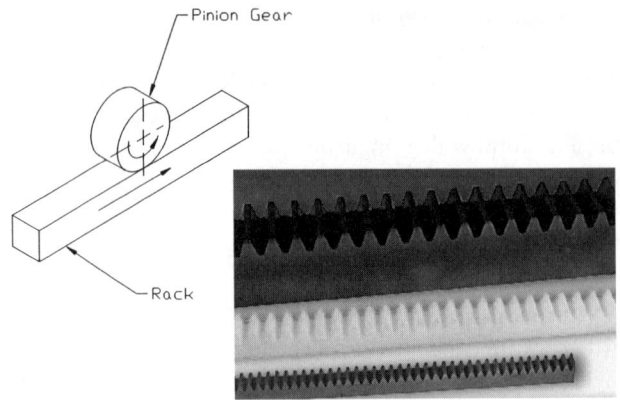

**Figure 8-23**
Courtesy of Stock Drive Products.

BERG Precision Racks

416 ST. Steel &
2024T4 Aluminum
Anodized

24 to 120 Pitch
20° Pressure Angle
AGMA Quality 10

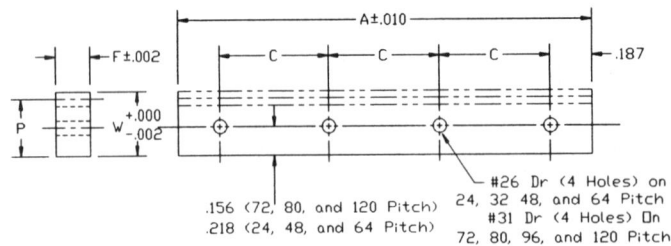

.156 (72, 80, and 120 Pitch)
.218 (24, 48, and 64 Pitch)

#26 Dr (4 Holes) on
24, 32 48, and 64 Pitch
#31 Dr (4 Holes) On
72, 80, 96, and 120 Pitch

| STOCK NUMBER | MATERIAL | PITCH | A | P | W | F | C |
|---|---|---|---|---|---|---|---|
| R4-5 R4-6 | STAINLESS ST ALUMINUM | 24 | 10' | .4383 | .480 | .230 | 3.208 |
| R4-9 R4-10 | STAINLESS ST ALUMINUM | 32 | 10' | .4487 | .480 | .230 | 3.208 |
| R4-11 R4-12 | STAINLESS ST ALUMINUM | 48 | 9' | .4592 | .480 | .230 | 2.879 |
| R4-15 R4-16 | STAINLESS ST ALUMINUM | 64 | 7' | .4644 | .480 | .230 | 2.208 |
| R4-17 R4-18 | STAINLESS ST ALUMINUM | 72 | 5' | .3411 | .355 | .167 | 1.541 |
| R4-19 R4-20 | STAINLESS ST ALUMINUM | 80 | 5' | .3425 | .355 | .167 | 1.541 |
| R4-21 R4-22 | STAINLESS ST ALUMINUM | 96 | 3' | .3446 | .355 | .167 | .875 |
| R4-23 R4-24 | STAINLESS ST ALUMINUM | 120 | 3' | .3467 | .355 | .167 | .875 |

**Figure 8-24**
Courtesy of W. M. Berg Inc.

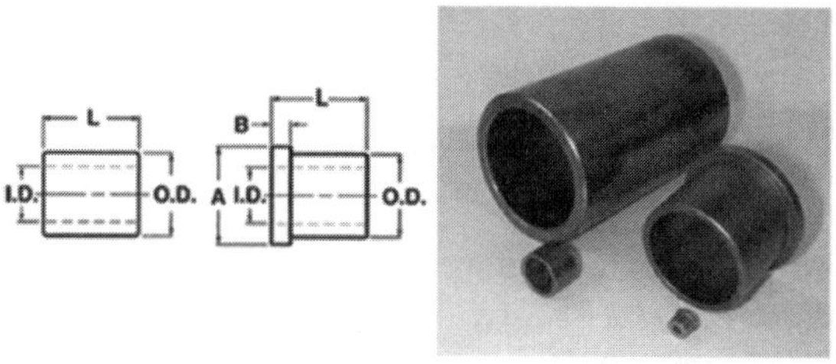

| Part Number | I.D. Inside Dia. Inch | Inside Dia. Tolerance Inch | Bearing Type | O.D. Outside Dia Inch | Outside Dia. Tolerance Inch | L Overall Length | A Flange Dia. | B Flange Thickness |
|---|---|---|---|---|---|---|---|---|
| | All | All | All | All | All | All | All | All |
| A 7B 4-F042 | 0.1245" | (+.001/-.000)" | Flanged | 0.2215" | (+.000/-.001)" | .122 - .115 | 0.34 | 0.050 |
| A 7B 4-F048 | 0.125" | (+.001/-.000)" | Flanged | 0.19" | (+.000/-.001)" | .228 - .216 | 0.25 | 0.035 |
| A 7B 4-F063 | 0.126" | (+.001/-.000)" | Flanged | 0.2215" | (+.000/-.001)" | .148 - .141 | 0.345 | 0.061 |
| A 7B 4-GF0404 | 0.126" | (+.001/-.000)" | Flanged | 0.252" | (+.000/-.001)" | 0.125 | 0.36 | 0.047 |
| A 7B 4-GF0408 | 0.126" | (+.001/-.000)" | Flanged | 0.252" | (+.000/-.001)" | 0.25 | 0.36 | 0.047 |

**Figure 8-25**
Courtesy of Stock Drive Products.

## 8-17 BALL BEARINGS

*Ball bearings* are used to help eliminate friction between moving and stationary parts. The moving and stationary parts are separated by a series of balls that ride in a *race*.

Figure 8-25 shows a listing for ball bearings that includes applicable dimensions and tolerances, which is from the Stock Drive Products website. There are many other types and sizes of ball bearings available.

Ball bearings may be drawn as shown in Figure 8-25 or by using one of the representations shown in Figure 8-26. It is recommended that the representations be drawn and saved as wblocks for use on future drawings.

The outside diameter of the bearings listed in Figure 8-25 has a tolerance of +.0000/−.0002. The same tolerance range applies to the center hole. These tight tolerances are manufactured because this type of ball bearing is usually assembled using a force fit. See Chapter 4 for an explanation of fits. Ball bearings are available that do not assemble using force fits.

The following sample problem shows how ball bearings can be used to support the gear's shafts.

## 8-18 SAMPLE PROBLEM SP8-5

Figure 8-27 shows two spur gears and a dimensioned drawing of a shaft used to support both gears. Design a support plate for the shafts. Use ball bearings to support the shafts. Specify dimensions and tolerances for the support plate and assume that the bearings are to be fitted into the support plate using an LN2 medium press fit. The tolerance for the center distance between the gears is to be +.001, −.000.

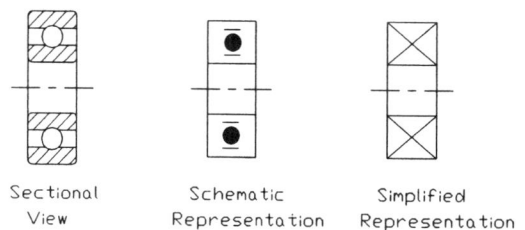

Sectional View   Schematic Representation   Simplified Representation

**Figure 8-26**

The maximum interference permitted for an LN2 medium press fit is .0011. See the fits tables in the appendix. For purposes of calculation, the bearing is considered to be the shaft and the holes in the support plate.

Maximum interference occurs when the shaft (bearing) is at its maximum diameter and the hole (support plate) is at its minimum. The maximum shaft diameter, the maximum diameter of the bearing, is .5000, as defined in the manufacturer's listing. This means that the minimum hole diameter should be .5000 − .0011 = .4989. An LN2 fit has a hole tolerance of +.0007, so the maximum hole size should be .4989 + .0007 = .4996.

The minimum interference, or the difference between the minimum shaft diameter and the maximum hole diameter, is .4998 − .4996 = .0002. There will always be at least .0002 interference between the ball bearing and the hole.

The bore of the selected bearing, listed in Figure 8-25, has a limit tolerance of .2500 − .2498. The shafts specified in Figure 8-27 have a limit tolerance of .2497 − .2495. This means that there will always be a slight clearance between the shaft and the bore hole.

The nominal center distance between the two gears is 3.000 inches. The given tolerance for the center distance is +.001, −.000. This tolerance can be ensured by assigning a positional tolerance of .0005 to each of the two holes applied at maximum material condition at the centerline. The base distance between the two holes is defined as 3.0000. The maximum center distance, including the positional tolerance, is 3.0000 + .0005 = 3.0005, and the minimum is 3.0000 − .0005 = 2.9995, or a total maximum tolerance of .001.

Figure 8-28 shows a detail drawing of the support plate.

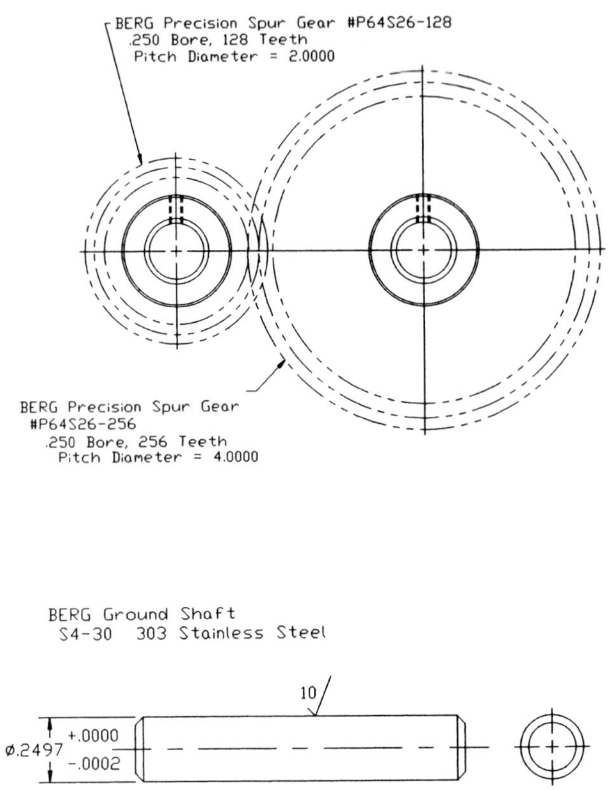

**Figure 8-27**
Courtesy of W. M. Berg Inc.

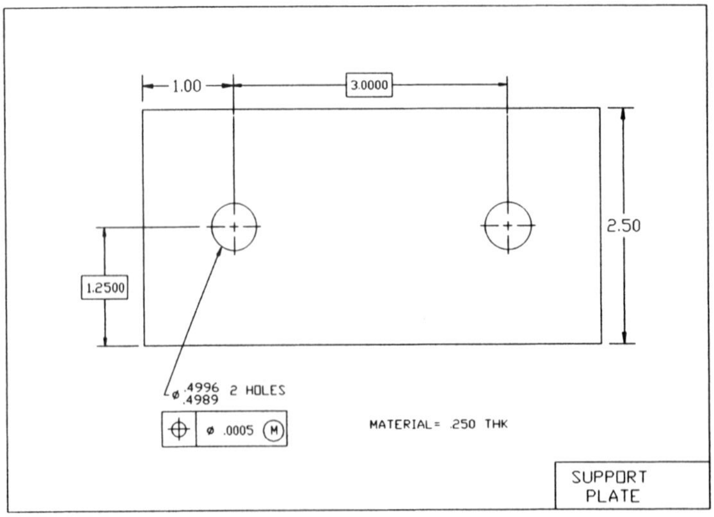

**Figure 8-28**

## 8-19 BUSHINGS

A *bushing* is a cylindrically shaped bearing that helps reduce friction between a moving part and a stationary part. Bushings, unlike ball bearings, have no moving parts. Bushings are usually made from oil-impregnated bronze or Teflon. Figure 8-29 shows a manufacturer's list of bronze bushings, and Figure 8-30 shows a list for Teflon bushings.

Bushings are cheaper than ball bearings but they wear over time, particularly if the application is high speed or one with heavy loading. Bushings are usually pressed into a supporting plate. Gear shafts must always have clearance from the inside diameters of bushings.

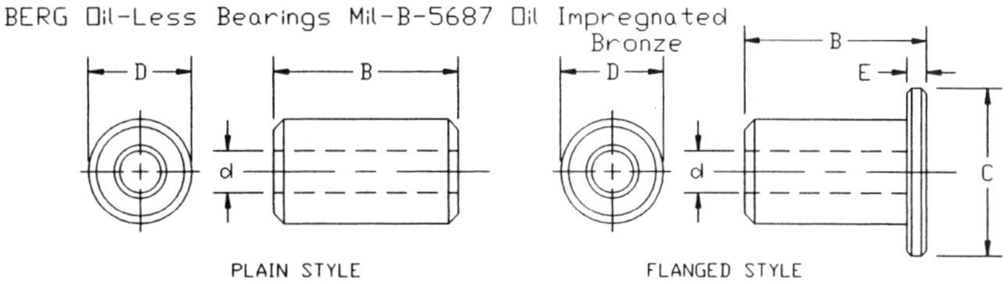

BERG Oil-Less Bearings Mil-B-5687 Oil Impregnated Bronze

PLAIN STYLE          FLANGED STYLE

| PLAIN STYLE | FLANGED STYLE | | | | | FLANGED ONLY | |
|---|---|---|---|---|---|---|---|
| STOCK NUMBER | STOCK NUMBER | SHAFT SIZE | d +.001 | D -.001 | B ±.005 | C ±.005 | E ±.003 |
| B6-2 | B7-31 | 1/8 | .126 | .252 | .250 | .375 | .047 |
| B6-7 | B7-35 | 3/16 | .188 | .314 | .500 | .375 | .047 |
| B6-11 | B7-11 | 1/4 | .251 | .377 | .500 | .562 | .047 |
| B6-15 | B7-40 | 5-16 | .313 | .439 | .625 | .562 | .062 |
| B6-18 | B7-44 | 3/8 | .376 | .502 | .625 | .625 | .062 |
| B6-23 | B7-49 | 1/2 | .501 | .627 | .750 | .875 | .062 |

**Figure 8-29**
Courtesy of W. M. Berg Inc.

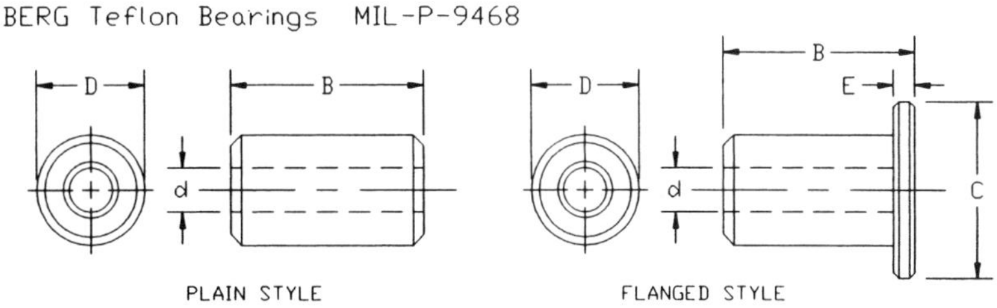

BERG Teflon Bearings  MIL-P-9468

PLAIN STYLE          FLANGED STYLE

| PLAIN STYLE | FLANGED STYLE | | | | | FLANGED ONLY | |
|---|---|---|---|---|---|---|---|
| STOCK NUMBER | STOCK NUMBER | SHAFT SIZE | d +.001 | D -.001 | B ±.005 | C ±.005 | E ±.003 |
|  | B9-3 | 1/8 | .126 | .252 | .250 | .312 | .047 |
| B8-5 | B9-6 | 3/16 | .188 | .315 | .250 | .375 | .047 |
| B8-11 | B9-11 | 1/4 | .251 | .377 | .500 | .500 | .047 |
| B8-14 | B9-15 | 5/16 | .313 | .439 | .500 | .562 | .093 |
| B8-19 | B9-19 | 3/8 | .376 | .502 | .625 | .687 | .093 |
| B8-23 | B8-29 | 1/2 | .501 | .628 | .750 | .875 | .125 |

**Figure 8-30**
Courtesy of W. M. Berg Inc.

## 8-20 SAMPLE PROBLEM SP8-6

Figure 8-31 shows two support plates used to support and align a matched set of bevel gears. The gears selected are numbered S1346Z–48S30S60 in the Berg listings presented in Figure 8-16. The calculations are similar to those presented earlier for Sample Problem SP8-5 but with the addition of tolerances for the holes and machine screws used to join the two perpendicular support plates together.

Figure 8-31 shows an assembly drawing of the two gears along with appropriate bushings, stock number B6-11 from Figure 8-29, and shafts and supporting parts 1 and 2. The figure also shows detail drawings of the two support plates with appropriate dimensions and tolerances.

The bushings are fitted into the support plates using an FN1 fit. The fit tables in the appendix define the maximum interference for an FN1 fit as .0075 and the minimum interference as .0001. The outside diameter of the bushing has a tolerance of .3770 − .3760, per Figure 8-29. The feature tolerances for the holes in the support parts are found as follows:

Shaft max − Hole min = Interference max
Shaft min − Hole max = Interference min

The feature tolerance for the hole is therefore .3750 − .3695.

The positional tolerance is determined as described in SP8-5 and is based on a tolerance of .001 between gear centers.

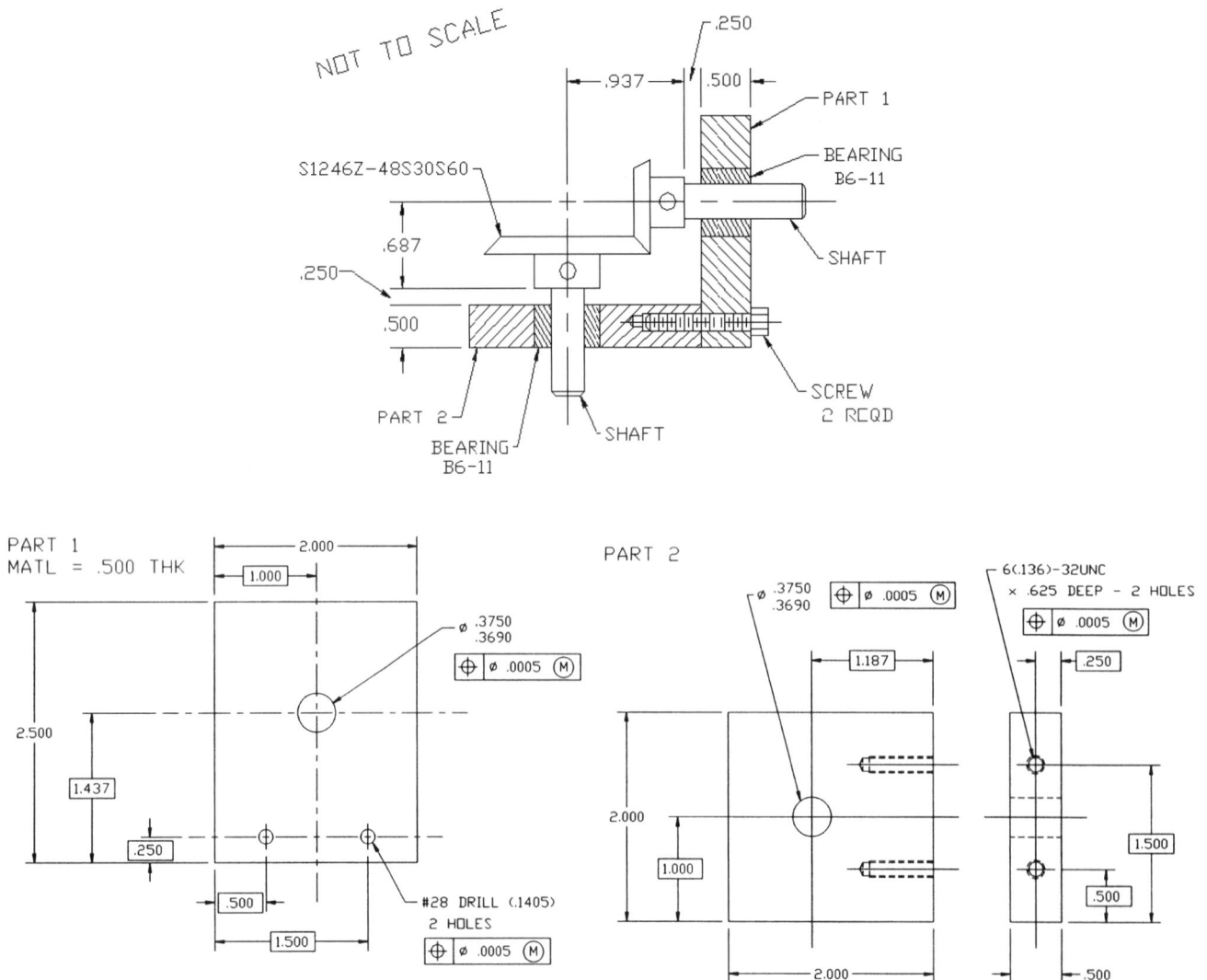

**Figure 8-31**

# 8-21  CAM DISPLACEMENT DIAGRAMS

A *displacement diagram* is used to define the motion of a cam follower. Displacement diagrams are set up as shown in Figure 8-32. The horizontal axis is marked off in 12 equal spaces that represent 30° on the cam. The vertical axis is used to define the linear displacement of the follower and is defined using either inches or millimeters.

The vertical axis of a displacement diagram must be drawn to scale because, once defined, the vertical distances are transferred to the cam's base circle to define the cam's shape. Figure 8-32 shows distances A, B, and C on both the displacement diagram and cam. The distances define the follower displacement at the 30°, 60°, and 90° marks, respectively.

The horizontal axis may use any equal spacing to indicate the angle because the vertical distances will be transferred to the cam along ray lines. The lower horizontal line represents the circumference of the base circle.

Figure 8-33 shows a second displacement diagram. Note how the distance between the 60° and 90° lines has been further subdivided. The additional lines are used to more accurately define the cam motion. Additional degree lines are often added when the follower is undergoing a rapid change of motion.

The term *dwell* means the cam follower does not move either up or down as the cam turns. Dwells are drawn as straight horizontal lines on a displacement diagram. Note the horizontal line between the 90° and 210° lines on the displacement diagram shown in Figure 8-33. Dwells are drawn as sectors of constant radius on the cam.

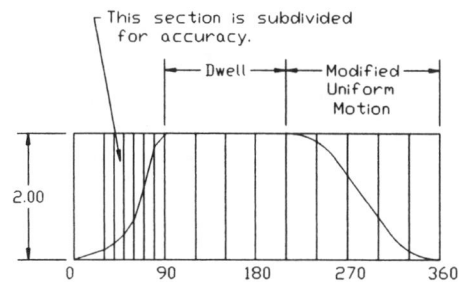

Figure 8-33

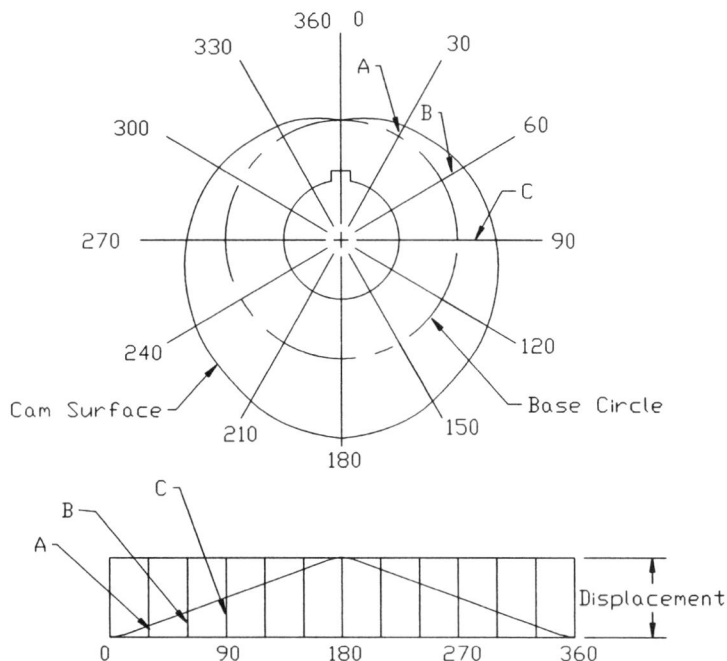

Figure 8-32

**To set up a displacement diagram**

See Figure 8-34. The given dimensions are in inches. The values in the brackets, [ ], are in millimeters.

1. Set **Grid = .5 [10]**
   **Snap = .25 [5]**
2. Draw a horizontal line **6 [120]** long.
3. Draw a vertical line **2 [40]** from the left end of the horizontal line as shown.

The length 2 [40] was chosen arbitrarily for this example. The vertical distance should be equal to the total displacement of the follower.

4. Use the **Array** command to create a rectangular array with **12** columns **.5 [10]** apart. Draw a horizontal line across the top of the diagram.
5. Label the horizontal axis in degrees, with each vertical line representing **30°**, and the vertical axis in inches [millimeters] of displacement.

## 8-22 CAM MOTIONS

The shape of a cam surface is designed to move a follower through a specific distance. The surface also determines the acceleration, deceleration, and smoothness of motion of the follower. It is important that a cam surface be shaped to maintain continuous contact with the follower. Several standard cam motions are defined next.

**Uniform motion**

Uniform motion is drawn as a straight line on a displacement diagram. See Figure 8-35. The follower rises the same distance for each degree of rotation by the cam.

**Modified uniform motion**

Modified uniform motion is similar to uniform motion but has a curved radius shape added to each end of the line to facilitate a smooth transition from the uniform motion to another type of motion or to a dwell section.

Figure 8-36 shows how to create a modified uniform motion on a displacement diagram. The procedure is as follows.

1. Set up a displacement diagram as presented in the preceding section.
2. Draw two arcs or circles of radius no greater than half the required displacement.

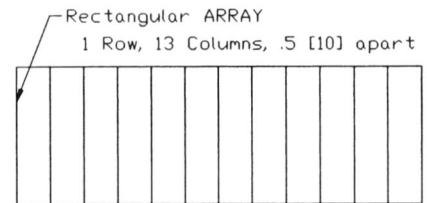

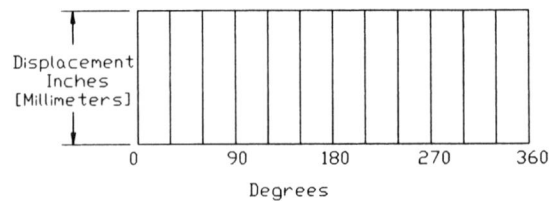

**Figure 8-34**

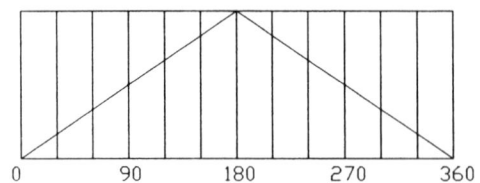

**Figure 8-35**

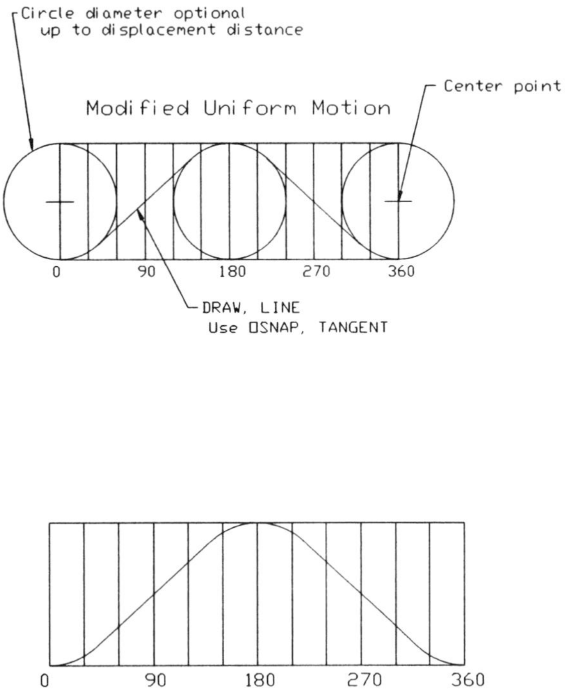

Figure 8-36

Any radius value can be used. In general, the larger the radius, the smoother the transition. In the example shown, a radius equal to half the total displacement was used.

3. Use **Osnap, Tangent,** and draw a line between the two arcs.
4. Erase and trim any excess lines.

### Harmonic motion

Figure 8-37 shows how to create a harmonic cam motion. The procedure is as follows.

1. Set up a displacement diagram as presented in Section 8-21.
2. Draw a semicircle aligned with the left side of the displacement diagram. The diameter of the semicircle equals the total height of the displacement.

3. Use the **Array** command or the **Line** command with relative coordinate inputs, and draw rays every **30°** on the circle as shown. Label the rays from 0° to 180° in 30° segments.
4. Use **Osnap, Intersection** with **Ortho** on, **<F8>**, and draw projection lines from the intersections of the rays with the circumference of the circle across the displacement diagram.
5. Draw a polyline starting at the lower left corner of the diagram and connecting the intersections of like angle lines.

For this example, the horizontal line from the circle's 30° increment intersects with the vertical line from the 30° mark on the displacement diagram.

6. Use **Edit Polyline, Fit** option, to change the straight polyline into a smooth curved line.
7. Project the same lines to the far side of the diagram to define the deceleration harmonic motion path.

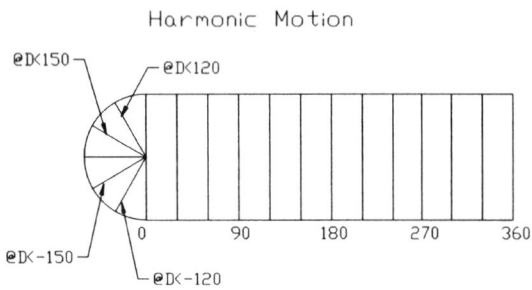

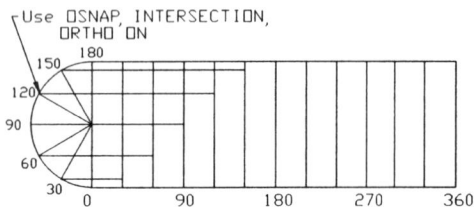

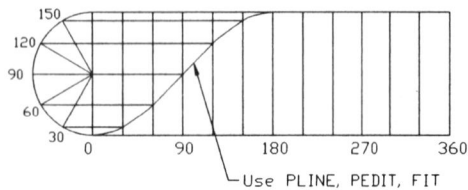

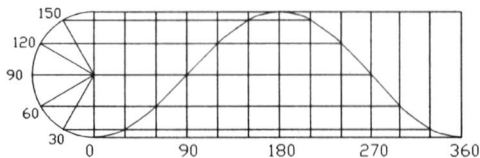

## Harmonic Motion

@DK150
@DK120
0   90   180   270   360
@DK-150
@DK-120

Use OSNAP, INTERSECTION, ORTHO ON
180
150
120
90
60
30
0   90   180   270   360

150
120
90
60
30
0   90   180   270   360
Use PLINE, PEDIT, FIT

150
120
90
60
30
0   90   180   270   360

**Figure 8-37**

## Uniform acceleration and deceleration

Uniform acceleration and deceleration are based on the knowledge that acceleration is related to distance by the square of the distance. Acceleration is measured in distance per second squared. Distances of units 1, 2, and 3 may be expressed as 1, 4, and 9, respectively.

Uniform acceleration and deceleration motions create smooth transitions between various displacement heights and are often used in high-speed applications.

Figure 8-38 shows how to create a uniform acceleration cam motion. The procedure is as follows.

1. Set up a displacement diagram as presented in Section 8-21.
2. Draw a horizontal construction line to the left and align it with the bottom horizontal line of the displacement diagram.
3. Use the **Array** command and create 1 column and **19** rows, **.1111** apart.

The 19 rows create 18 spaces. The uniform motion shape is created by combining six horizontal steps (30°, 60°, 90°, 120°, 150°, 180°) and the squares of six vertical steps (1, 4, 9, 4, 1, 0).

The vertical spacing is symmetrical about the centerline of the displacement diagram, so the original spacing is 1, 2, 3, 2, 1. The square of these values is used to create the uniform acceleration and deceleration.

The .1111 value was derived by dividing the displacement distance by the number of spaces, 2.00/18 = .1111.

4. Label the stack of vertical construction lines as shown.
5. Use the **Extend** command to extend the horizontal construction lines so that they intersect with the appropriate vertical degree line.
6. Use the **Edit Polyline** command's **Fit** option to create a smooth, continuous curve between the 0° and 180° lines.

The same line may be used to create a deceleration curve as shown.

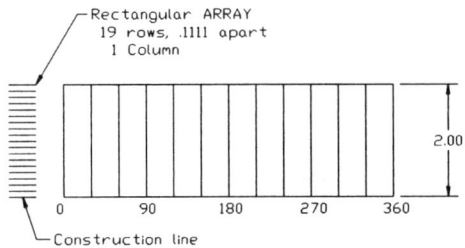

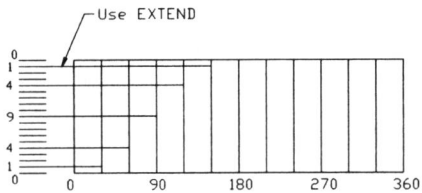

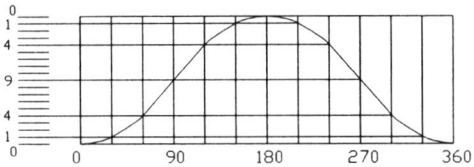

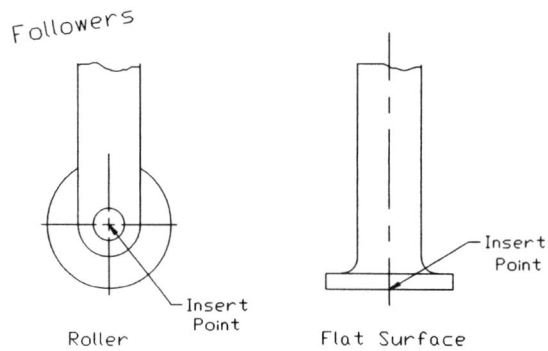

Rectangular ARRAY
19 rows, .1111 apart
1 Column

2.00

Construction line

Use EXTEND

PLINE, PEDIT, FIT

**Figure 8-38**

Followers

Insert Point

Roller

Insert Point

Flat Surface

**Figure 8-39**

## 8-23 CAM FOLLOWERS

There are two basic types of cam followers: those that roll as they follow the cam's surface, and those that have a fixed surface that slides in contact with the cam surface. Figure 8-39 shows an example of a roller follower and a fixed- or flat-surface follower. The flat-surface-type followers are limited to slow-moving cams with low force requirements.

Followers are usually spring-loaded to keep them in contact with the cam surface during operation. Springs were discussed in Section 6-24.

## 8-24 SAMPLE PROBLEM SP8-7

Design a cam that rises 2.00 inches over 180° using harmonic motion, dwells for 60°, descends 2.00 inches in 90° using modified uniform motion, and dwells the remaining 30°. The base circle for the cam is 3.00 inches in diameter, and the follower is a roller type with a 1.00-inch diameter. The cam will rotate in a counterclockwise direction. The center hole is .75 inch in diameter with a .125 × .875 keyway. See Figures 8-40 and 8-41.

1. Set up a displacement diagram as described in Section 8-21.
2. Define a path between the 0° and 180° vertical lines using the method described for harmonic motion. Draw the required circle on the left end of the diagram as shown.
3. Draw a horizontal line from the 180° to the 240° line.

This line defines the follower's dwell.

4. Draw two arcs of **.50** radius, one tangent to the top horizontal line of the displacement diagram and the second tangent to the bottom line.

Draw one arc on the 240° line and the other on the 330° line as shown. In this example the arcs have a radius equal to .25 of the total displacement. Any convenient radius value may be used.

5. Use the **Osnap, Tangent** command and draw a line tangent to the two arcs.
6. Erase and trim any excess lines and constructions.

This completes the displacement diagram. The follower distances are now transferred to the cam's base circle to define the surface shape. See Figure 8-41.

7. Draw two concentric circles of **3.00** and **8.00** diameter, respectively.

The 3.00 diameter is the base circle. The 8.00-diameter-circle value is derived from the radius of the base circle plus the maximum displacement plus the radius of the follower: 1.50 +2 +.5 =4.00 radius, or 8.00 diameter.

8. Use the **Setvar, Dimcen, −.2, Dim, Center** commands, or a modified **Center Mark** tool to draw the centerlines for the 8.00 circle.
9. Use the **Array** command, and polar array the top portion of the vertical centerline **12** times around the full 360°.
10. Label the ray lines as shown.

Note that the top vertical ray line is labeled both as 0 and 360.

11. Transfer the follower distances from the displacement diagram to the cam drawing.

Several different techniques can be used to transfer the distances: use the **Linear** dimension tool, the **Distance** command found under the **Tools** pull-down menu, or the **Inquiry** command. Use the **Osnap, Intersection** command to ensure accuracy. In this example the measured distance values are listed below the displacement diagram in Figure 8-40. The .50 addition to the bottom of the diagram is to account for the .50 follower radius. Note how the distance of 2.1000 was measured.

The distance values can be used to draw lines from the base circle along the appropriate ray line on the cam drawing using relative coordinate values. The values for this example are as follows:

| | |
|---|---|
| .5000-(0) | @.5000<90. |
| .6340-(30) | @.6340<60 |
| 1.0000-(60) | @1.0000<30 |
| 1.5000-(90) | @1.5000<0 |
| 2.0000-(120) | @2.0000<−30 |
| 2.3660-(150) | @2.3660<−60 |
| 2.5000-(180) | @2.5000<−90 |

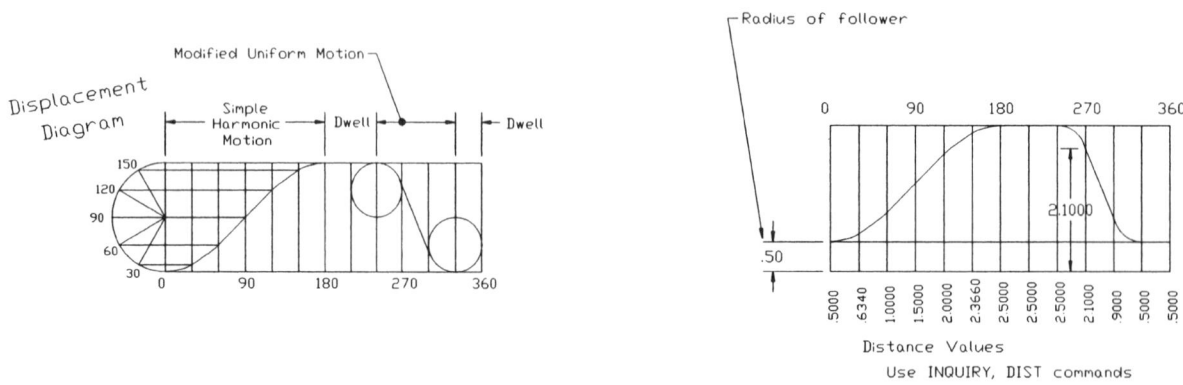

Figure 8-40

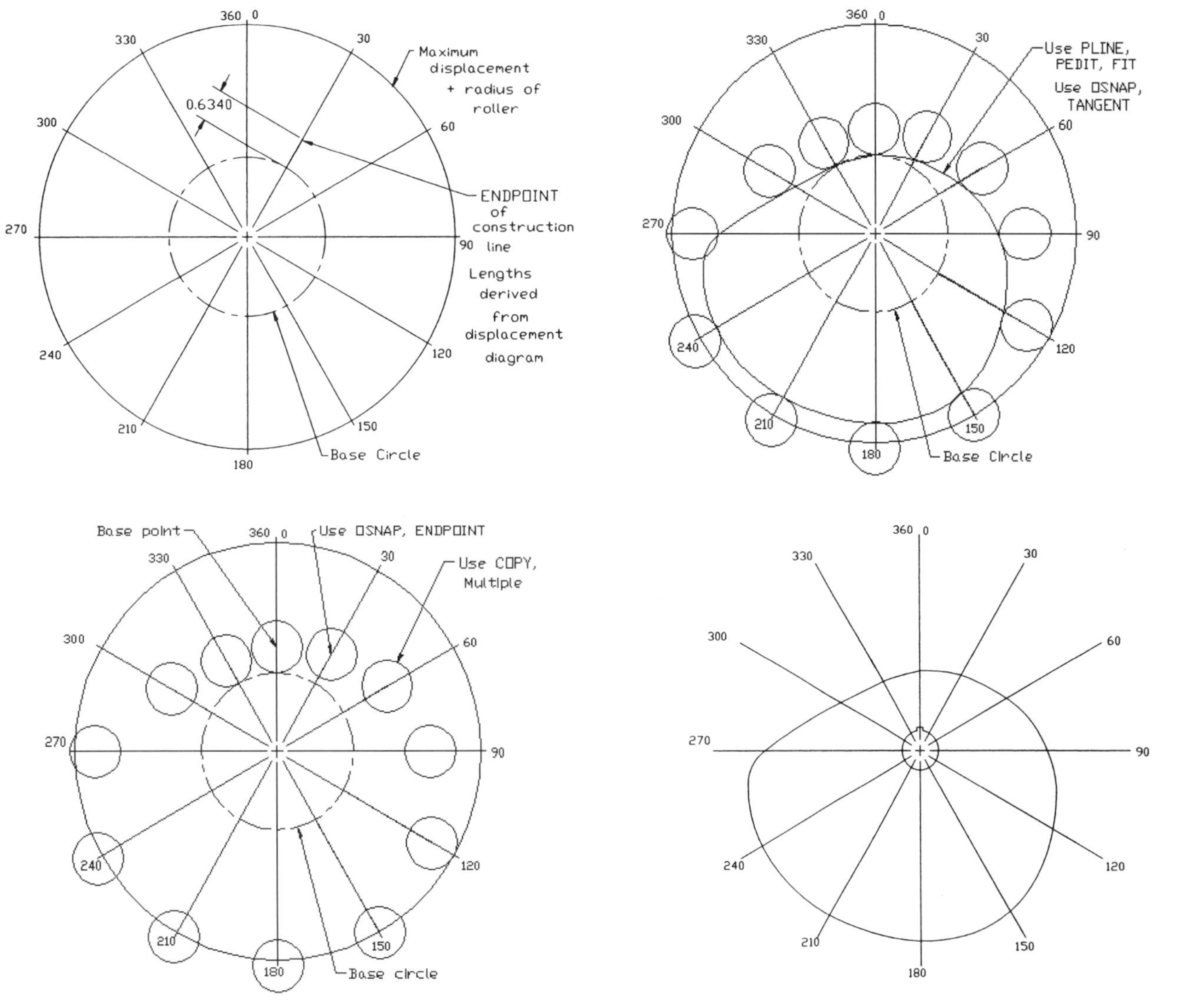

**Figure 8-41**

| | |
|---|---|
| 2.5000-(210) | @2.5000<-120 |
| 2.5000-(240) | @2.5000<-150 |
| 2.1000-(270) | @2.1000<180 |
| .9000-(300) | @.9000<150 |
| .5000-(330) | @.5000<120 |

Lines can be drawn over the existing vertical lines between the base line and the displacement path on the displacement diagram. The **Move, Rotate** or **Grips, Rotate** command can be used to transfer the lines to the base circle and appropriate ray line.

12. Draw circles of diameter **.50,** representing the roller follower, with their center points on the ends of the lines created in step 11.

Note that these lines and their endpoints are not visible on the screen because they are drawn directly over the existing ray lines. The **Osnap, Endpoint** command should be used to locate the circle's center point.

13. Use the **Draw, Polyline** command and draw a line tangent to the follower at each ray line.

The tangent point for each ray may not be exactly on the ray line.

14. Use the **Edit Polyline, Spline** command to create the cam surface.
15. Draw the center hole and keyway using the given dimensions.
16. Save the drawings, if desired.

# 8-25 EXERCISE PROBLEMS

## EX8-1 INCHES

Draw a spur gear with 24 teeth on a 4.00-pitch circle. The fillets at the base of each tooth have a radius of .0625. Locate a 1.00-diameter mounting hole in the center of the gear.

## EX8-2 MILLIMETERS

Draw a spur gear with 36 teeth on a 100-pitch circle. The fillets at the base of each tooth have a radius of 3. Locate a 20-diameter mounting hole in the center of the gear.

## EX8-3

Use the information presented in Figure 8-12 and draw front and side sectional views of gear number A 1C22MYKW150 32A.

## EX8-4

Use the information presented in Figure 8-12 and draw front and side sectional views of gear number A 1C22MYKW150 64A.

## EX8-5

Use the information presented in Figure 8-15 and draw a sectional view of bevel gears A 1C3MYK 30018 and A 1C3MYK 30036H.

## EX8-6

Use the information presented in Figure 8-15 and draw a sectional view of bevel gears A 1C3MYK 15018 and A 1C3MYK 15036.

## EX8-7

Use the information presented in Figure 8-16 and draw a sectional view of the matched set of bevel gears S1346Z−48S30A120.

## EX8-8

Use the information presented in Figure 8-20 and draw front and side views of a set of worm gears.

## EX8-9

Use the information presented in Figure 8-22 and draw a front view of a matched set of helical gears.

## EX8-10

Use the information presented in Figure 8-24 and draw a front view of a set of rack-and-pinion gears. Select a pinion gear from Figure 8-12.

## EX8-11 INCHES

Draw a displacement diagram and appropriate cam based on the following information.

Dwell for 60°, rise 1.00 inch using harmonic motion over 180°, dwell for 30°, then descend 1.00 inch using harmonic motion over 90°.

The cam's base circle is 4.00 inches in diameter. Include a 1.25-inch center mounting hole.

## EX8-12 MILLIMETERS

Draw a displacement diagram and appropriate cam based on the following information.

Dwell for 60°, rise 30 millimeters using harmonic motion over 180°, dwell for 30°, then descend 30 millimeters using harmonic motion over 90°.

The cam's base circle is 120 millimeters in diameter. Include a 20-millimeter center mounting hole.

## EX8-13 INCHES

Draw a displacement diagram and appropriate cam based on the following information.

Rise 1.25 inches over 180° using uniform acceleration motion, dwell for 60°, descend 1.25 inches over 90° using modified uniform motion, dwell for 30°.

The cam's base circle is 3.25 inches in diameter. Include a .75-inch-diameter center mounting hole.

## EX8-14 MILLIMETERS

Draw a displacement diagram and appropriate cam based on the following information.

Rise 20 millimeters over 180° using uniform acceleration motion, dwell for 60°, descend 20 millimeters over 90° using modified uniform motion, dwell for 30°.

The cam's base circle is 80 millimeters in diameter. Include a 16-millimeter-diameter center mounting hole.

## EX8-15 INCHES

Draw a displacement diagram and appropriate cam based on the following information.

Rise .60 inch in 90° using harmonic motion, dwell for 30°, rise .60 inch in 60° using modified uniform motion, dwell 60°, descend 1.20 inches using uniform deceleration in 120°.

The cam's base circle is 3.20 inches in diameter. Include a 1.75-inch-diameter center mounting hole.

## EX8-16 MILLIMETERS

Draw a displacement diagram and appropriate cam based on the following information.

Rise 16 millimeters in 90° using harmonic motion, dwell for 30°, rise 16 millimeters in 60° using modified uniform motion, dwell 60°, descend 32 millimeters using uniform deceleration in 120°.

The cam's base circle is 84 millimeters in diameter. Include a 20-millimeter-diameter center mounting hole.

## EX8-17 INCHES

Draw a displacement diagram and appropriate cam based on the following information. The cam's base circle is 2.00 inches. Include a ∅0.625 center mounting hole. Rise 0.375 inch in 90° using harmonic motion, dwell for 90°, rise 0.375 inch in 90° using harmonic motion, descend 0.750 inch using harmonic motion in 90°.

## EX8-18 MILLIMETERS

Draw a displacement diagram and appropriate cam based on the following information. The cam's base circle is 98 millimeters. Include a ∅18 center mounting hole. Rise 15 millimeters in 90° using harmonic motion, dwell for 90°, rise 15 millimeters in 90° using harmonic motion, descend 30 millimeters using harmonic motion in 90°.

## EX8-19 INCHES

Draw a displacement diagram and appropriate cam based on the following information. The cam's base circle is 2.00 inches. Include a ∅0.625 center mounting hole. Rise 0.375 inch in 90° using uniform acceleration and deceleration motion, dwell for 90°, rise 0.375 inch in 90° using uniform acceleration and deceleration motion, descend 0.750 inch using motion in 90°.

## EX8-20 MILLIMETERS

Draw a displacement diagram and appropriate cam based on the following information. The cam's base circle is 98 millimeters. Include a ∅18 millimeters center mounting hole. Rise 15 millimeters in 90° using uniform acceleration and deceleration motion, dwell for 90°, rise 15 millimeters in 90° using uniform acceleration and deceleration motion, descend 30 millimeters using motion in 90°.

**EX8-21 INCHES**

Draw an assembly drawing and parts list for the 2-gear assembly shown.

2-GEAR ASSEMBLY

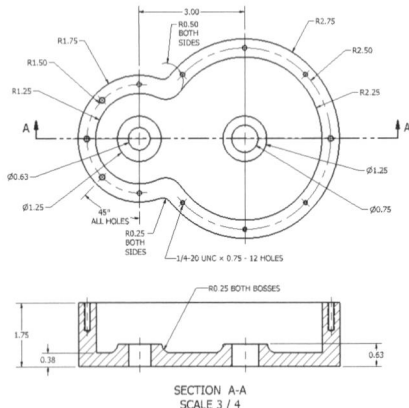

|  | Small<br>Gear | Large<br>Gear |
|---|---|---|
| Number of teeth | 48 | 96 |
| Face width | 0.50 | 0.50 |
| Diametral pitch | 24 | 24 |
| Pressure angle | 20 | 20 |
| Hub Ø | 0.750 | 1.00 |
| Base Ø | 0.500 | 0.625 |

| Parts List | | | | |
|---|---|---|---|---|
| ITEM | PART NUMBER | DESCRIPTION | MATERIAL | QTY |
| 1 | ENG-453-A | GEAR, HOUSING | CAST IRON | 1 |
| 2 | BU-1123 | BUSHING Ø0.75 | Delrin, Black | 1 |
| 3 | BU-1126 | BUSHING Ø0.625 | Delrin, Black | 1 |
| 4 | ASSEMBLY-6 | GEAR ASSEEMBLY | STEEL | 1 |
| 5 | AM-314 | SHAFT, GEAR Ø.625 | STEEL | 1 |
| 6 | AM-315 | SHAFT, GEAR Ø.0.500 | STEEL | 1 |
| 7 | ENG -566-B | COVER, GEAR | CAST IRON | 1 |
| 8 | ANSI B18.6.2 - 1/4-20 UNC<br>- 0.75 | Slotted Round Head Cap<br>Screw | Steel, Mild | 12 |

GEAR HOUSING

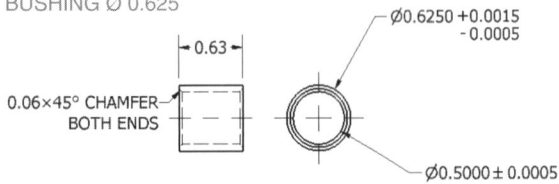

GEAR COVER

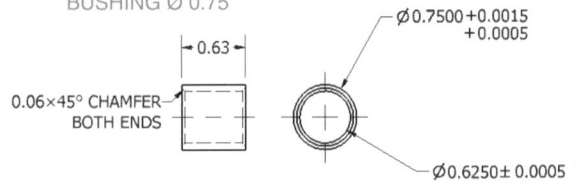

BUSHING Ø 0.625

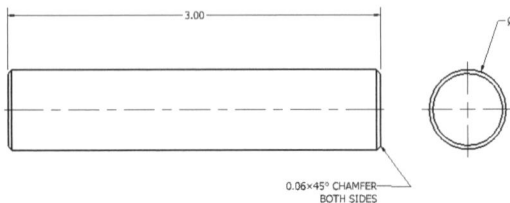

BUSHING Ø 0.75

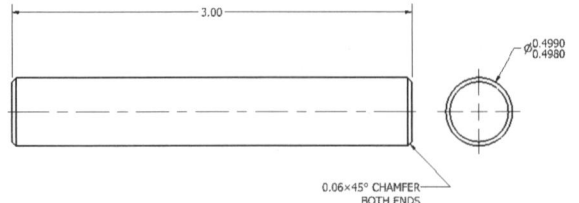

SHAFT, GEAR Ø 0.625

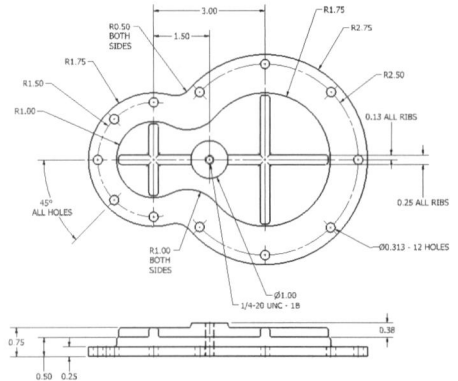

SHAFT, GEAR Ø 0.500

**EX8-22**

An electric motor operates at 3600 rpm. Design a gear that includes at least four spur gears (more may be used if needed) and reduces the motor speed to 200 rpm. Select gears and bearings from the tables in this book or from manufacturers' websites. Design each shaft as needed.

Enclose the gears in a box made from .50-inch [12-millimeter] plates, assembled using flat head screws. There should be at least three screws per edge on the box.

Extend the shafts for the input and output at least .50[12] outside the box. The other shafts should end at the edge of the box. Mount each shaft using two ball bearings, one mounted in each support plate.

Assume that the center distances have a tolerance of +.001, −.000 and the support shafts have tolerances of +.0000, −.0002, or their metric equivalent.

A. Draw an assembly drawing showing the gears in their assembled positions.
B. Support the gear shafts with ball bearings press fitted into the support plate. The support plate is to be .50 inch or 12 millimeters thick. The length and width dimensions are arbitrary, but there should be at least .25 [6] clearance between the gears and the support plate.
C. Draw detail drawings of the required support plates and shafts. Include dimensions and tolerances. Locate all support holes using positional tolerances.

Suggested websites:

www.wmberg.com
www.bostgear.com
www.newmantools.com

Do your own search for gears and bearings.

Top and end pieces omitted for clarity

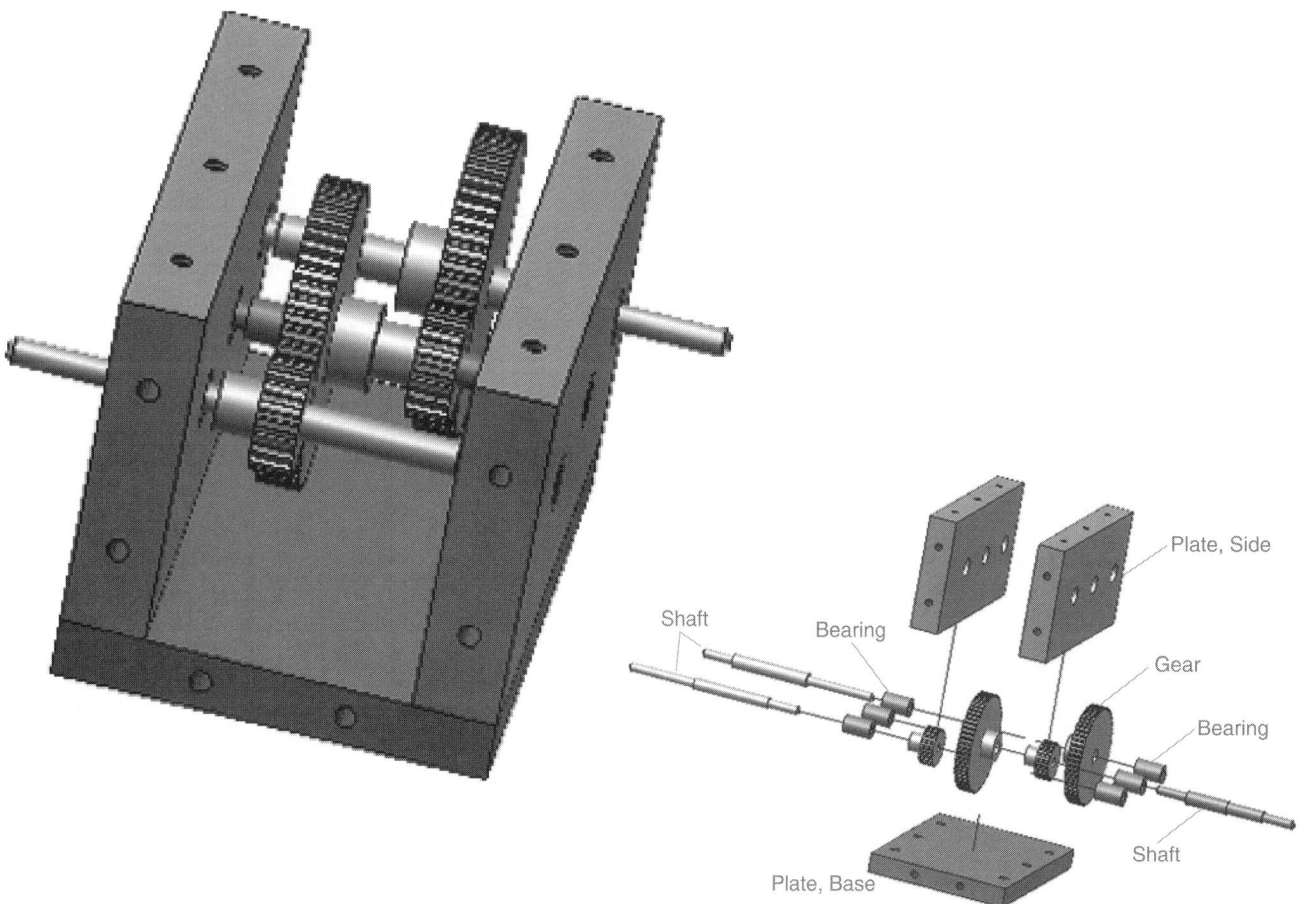

**EX8-23**

The following figure shows a general setup for matched bevel gears. Complete the design for a gear box with a ratio of 3:1 between the two gears. Select appropriate bearings and fasteners. Dimension and tolerance the shaft sizes and each of the six supporting plates. Extend the input and output shafts at least 1.00 inch [24 millimeters] beyond the surface of the box.

Assume that the center distances have a tolerance of +.001, −.000 and the support shafts have a tolerance of +.0000, −.0002, or their metric equivalent.

The figure shown represents half the gear box. The other half includes another three plates without the holes for the shafts. There should be at least three screws in each edge of the box.

A. Draw an assembly drawing showing the gears in their assembled positions.

B. Support the gear shafts with ball bearings press fitted into the support plate. The support plate is to be .50 inch or 12 millimeters thick. The length and width dimensions are arbitrary, but there should be at least .375 [10] clearance between the gears and the support plates.

C. Draw detail drawings of the required support plates. Include dimensions and tolerances. Locate all support holes using positional tolerances.

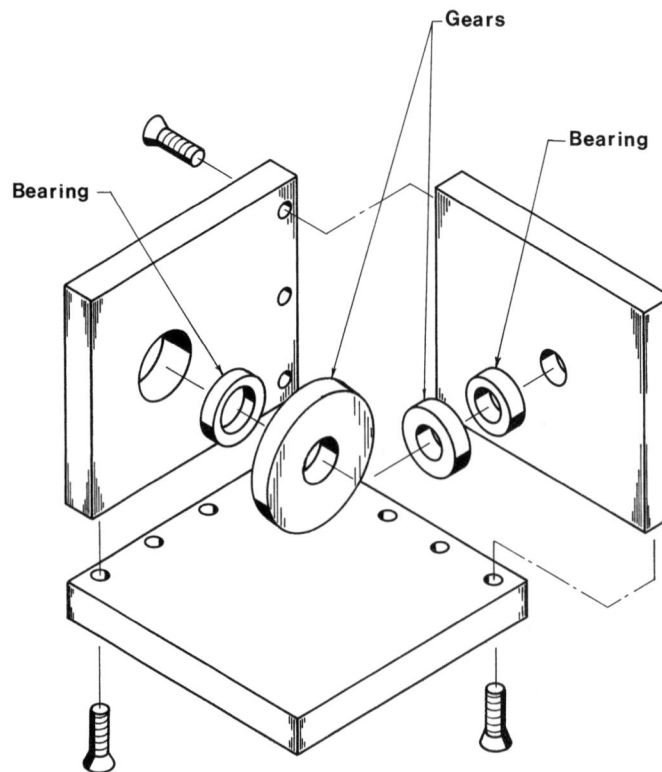

## EX8-24 MILLIMETERS

Draw an assembly drawing and parts list for the 4-gear assembly shown.

4-GEAR ASSEMBLY

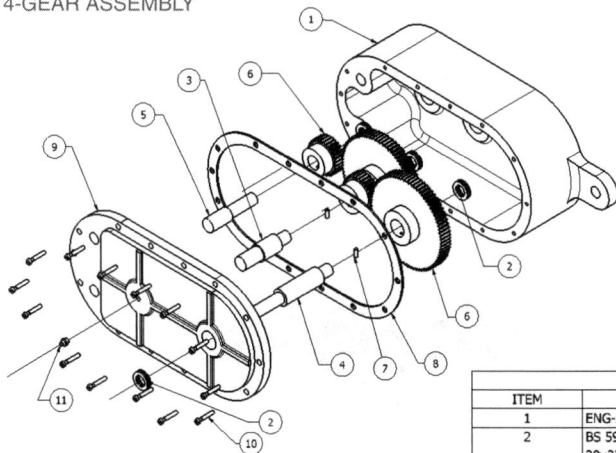

| Parts List | | | | |
|------|-------------|-------------|--------------|-----|
| ITEM | PART NUMBER | DESCRIPTION | MATERIAL | QTY |
| 1 | ENG-311-1 | 4-GEAR HOUSING | CAST IRON | 1 |
| 2 | BS 5989: Part 1 - 0 10 - 20x32x8 | Thrust  Thrust Ball Bearing | Steel, Mild | 4 |
| 3 | SH-4002 | SHAFT, NEUTRAL | STEEL | 1 |
| 4 | SH-4003 | SHAFT, OUTPUT | STEEL | 1 |
| 5 | SH-4004A | SHAFT,INPUT | STEEL | 1 |
| 6 | 4-GEAR-ASSEMBLY | | STEEL | 2 |
| 7 | CSN 02 1181 - M6 x 16 | Slotted Headless Set Screw - Flat Point | Steel, Mild | 2 |
| 8 | ENG-312-1 | GASKET | Brass, Soft Yellow | 1 |
| 9 | | COVER | | 1 |
| 10 | CNS 4355 - M 6 x 35 | Slotted Cheese Head Screw | Steel, Mild | 14 |
| 11 | CSN 02 7421 - M10 x 1coned short | Lubricating Nipple, coned Type A | Steel, Mild | 1 |

4-GEAR HOUSING

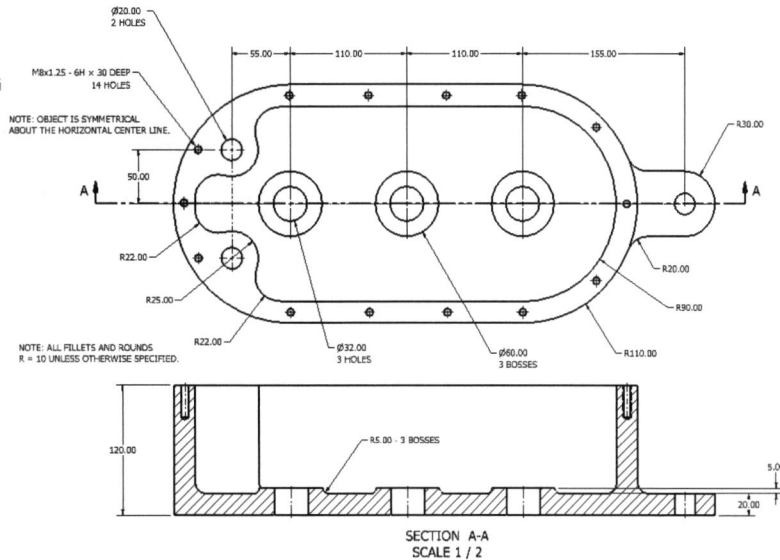

COVER

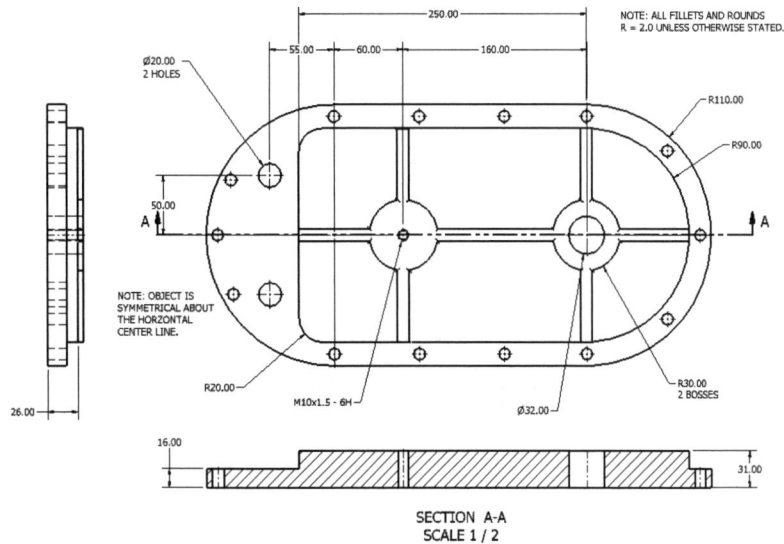

## GASKET

NOTE: HOLE PATTERN IS THE SAME FOR THE
GASKET, GEAR HOUSING, AND GEAR COVER.

NOTE: OBJECT IS SYMMETRICAL ABOUT
THE HORIZONTAL CENTER LINE.

220.00

74.00    74.00

R110.00
R100.00
R90.00

30.0°

45.0°

THICKNESS = 3

Ø10.00 - 14 HOLES

## GEAR ASSEMBLY

Large Gear
Module = 2
Pitch Ø = 160
Number of teeth = 80

Small Gear
Module = 2
Pitch Ø = 60
Number of teeth = 30

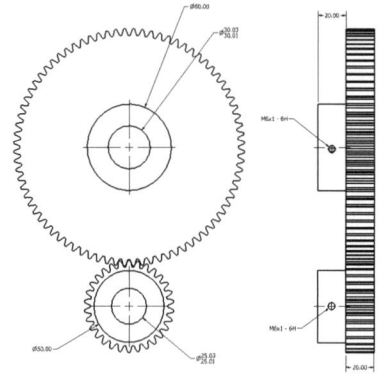

Ø90.00

M6x1 - 6H

M6x1 - 6H

Ø50.00

## OUTPUT SHAFT

## INPUT SHAFT

45.00    60.00

Ø$^{25.00}_{25.48}$

Ø$^{20.01}_{20.00}$

1 × 45° CHAMFER

25.00    85.00    40.00

Ø$^{30.00}_{29.98}$

Ø$^{20.01}_{20.00}$
BOTH
ENDS

1 × 45° CHAMFER

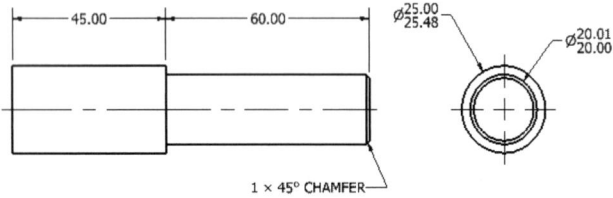

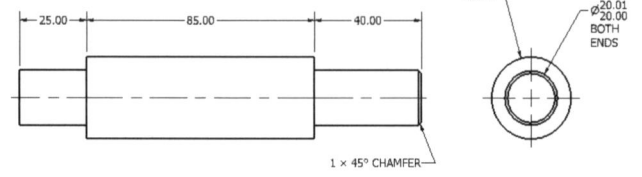

## NEUTRAL SHAFT

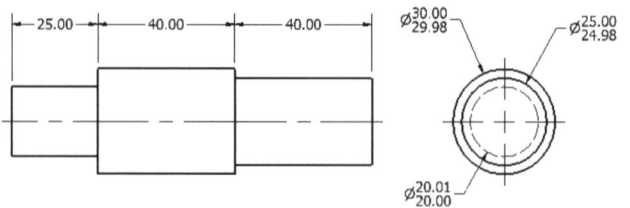

25.00    40.00    40.00

Ø$^{30.00}_{29.98}$

Ø$^{25.00}_{24.98}$

Ø$^{20.01}_{20.00}$

## LUBRICATING NIPPLE

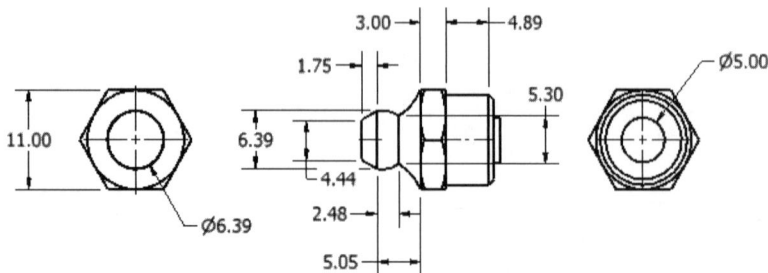

11.00

Ø6.39

3.00    4.89

1.75

6.39

4.44

2.48

5.05

5.30

Ø5.00

## EX8-25

Create the following for the slider assembly shown.

1. Select the appropriate fasteners.
2. Create an assembly drawing.
3. Create a parts list.
4. Specify the tolerance for the Guide Shaft/Bearings interface.

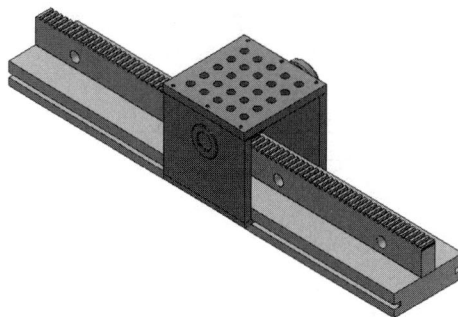

**RACK ASSEMBLY**

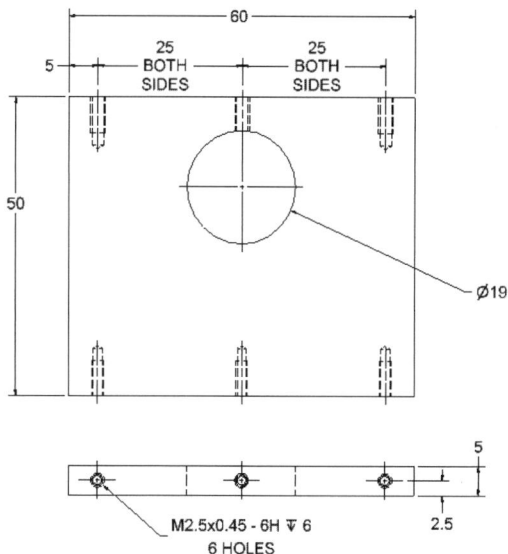

Hole Plate

Guide Shaft-1

Lock Plate-2

**RACK ASSEMBLY - 1**

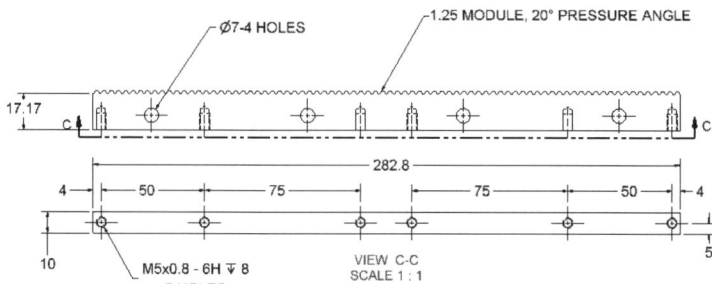

∅19

M2.5x0.45 - 6H ▼ 6
6 HOLES

**SIDE PLATE**

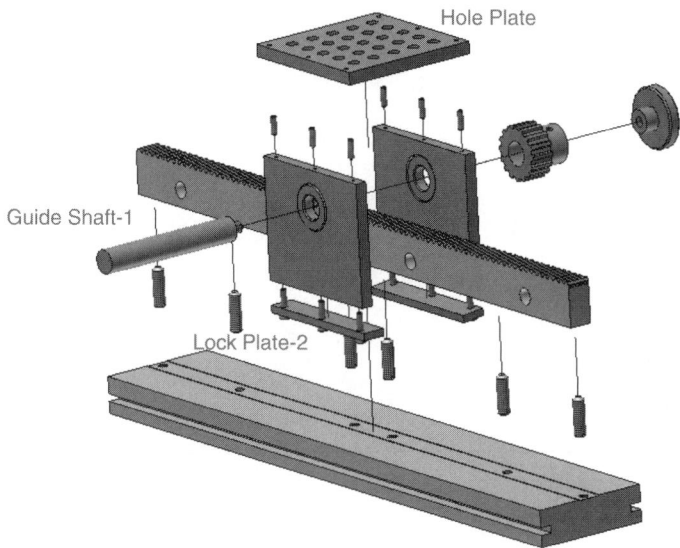

∅7-4 HOLES

1.25 MODULE, 20° PRESSURE ANGLE

17.17

282.8

4   50   75   75   50   4

10

M5x0.8 - 6H ▼ 8
6 HOLES

VIEW C-C
SCALE 1 : 1

BASED ON: ⊐PRECISION RACK R1M-2
BERG MANUFACTURING CO.

**RACK**

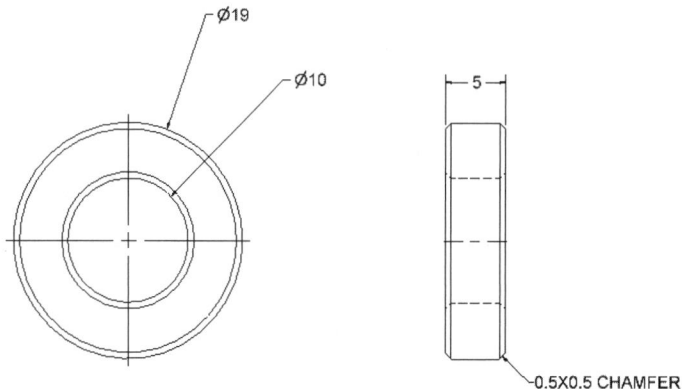

∅16
∅10 THRU
M3.5x0.6 - 6H ▼ 4

10   10

PITCH ∅ = 20.00

1.25 MODULE

PRECISION SPUR GEAR
BERG MANUFACTURING CO.
PART NUMBER: PBS86-16

**GEAR 16**

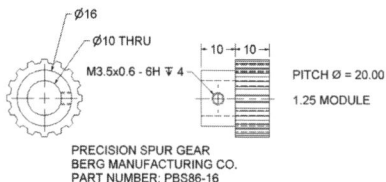

∅19
∅10
5

0.5X0.5 CHAMFER

BALL BEARING
BERG MANUFACTURING CO.
PART NUMBER: B11M-11

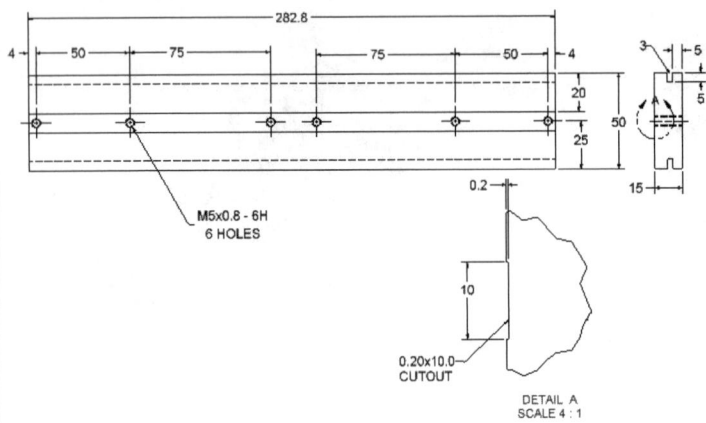

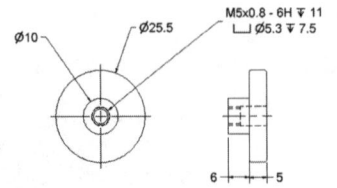

KNURLED THUMB NUT
BERG MANUFACTURING CO.
STOCK NUMBER: PD1M-15

**THUMB NUT**

**BASE SLIDER**

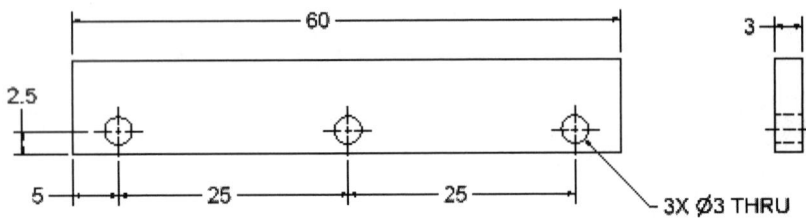

**LOCK PLATE - 2**

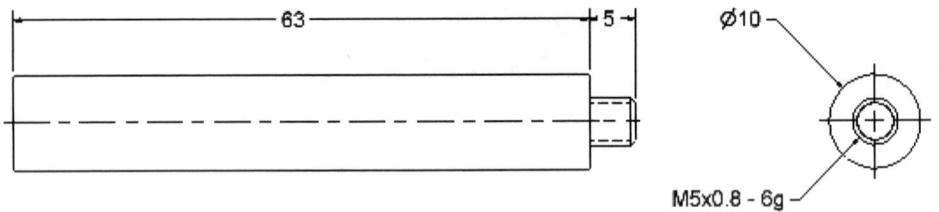

**GUIDE SHAFT - 1**

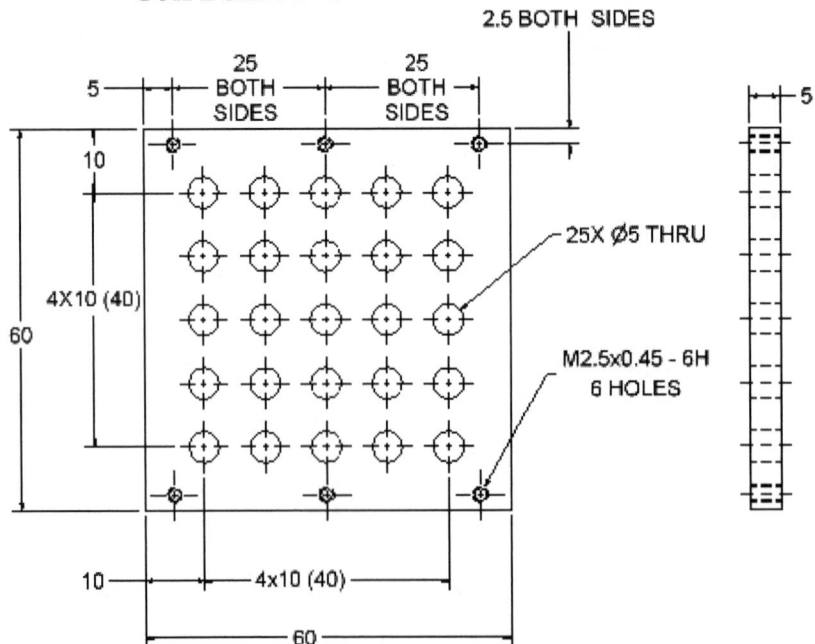

**HOLE PLATE**

## EX8-26 MILLIMETERS

Draw an assembly drawing and parts list for the cam asembly shown.

### Cam parameters

Base circle = $\varnothing$146
Face width = 16
Motion: rise 10 using harmonic motion in 90°, dwell for 180°, fall 10 in 90°
Bore = $\varnothing$16.0
Keyway = 2.3 × 5 × 16

Follower $\varnothing$ = 16
Follower width = 4
Square key = 5 × 5 × 16

### Bearing overall dimensions

DIN625-SKF 6203 (ID × OD × THK) 17 × 40 × 10
DIN625-SKF 634 4 × 13 × 4
GB 2273.2-87-7/70 8 × 18 × 5

### Spring parameters

Wire $\varnothing$ = 1.5
Inside $\varnothing$ = 9.0
Length = 20
Coil direction = Right
Active coils = 10
Grind both ends

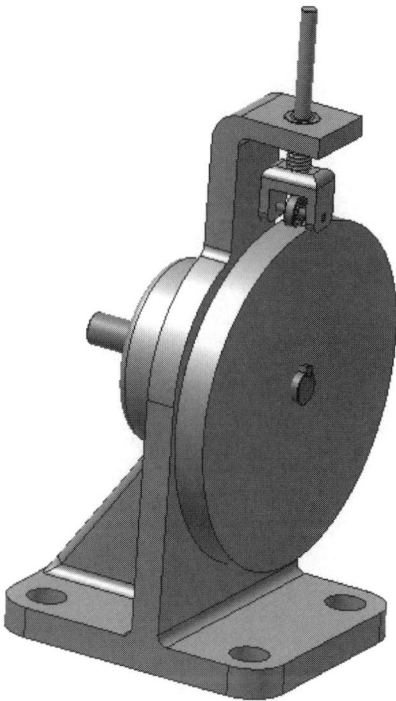

| Parts List | | | |
|---|---|---|---|
| ITEM | QTY | PART NUMBER | DESCRIPTION |
| 1 | 1 | ENG-2008-A | BASE, CAST |
| 2 | 1 | DIN625 - SKF 6203 | Single row ball bearings |
| 3 | 1 | SHF-4004-16 | SHAFT: Ø16×120, WITH 2.3×5×16 KEYWAY |
| 4 | 1 | | SUB-ASSEMBLY, FOLLOWER |
| 5 | 1 | SPR-C22 | SPRING, COMPRESSION |
| 6 | 1 | GB 273.2-87 - 7/70 - 8 x 18 x 5 | Rolling bearings - Thrust bearings - Plan of boundary dimensions |
| 7 | 1 | IS 2048 - 1983 - Specification for Parallel Keys and Keyways B 5 x 5 x 16 | Specification for Parallel Keys and Keyways |

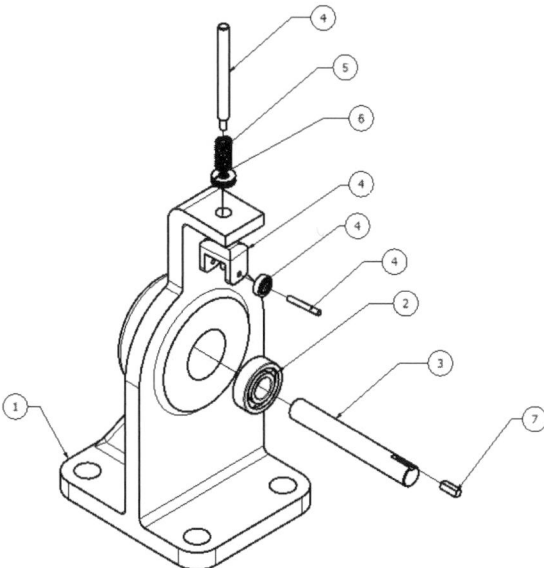

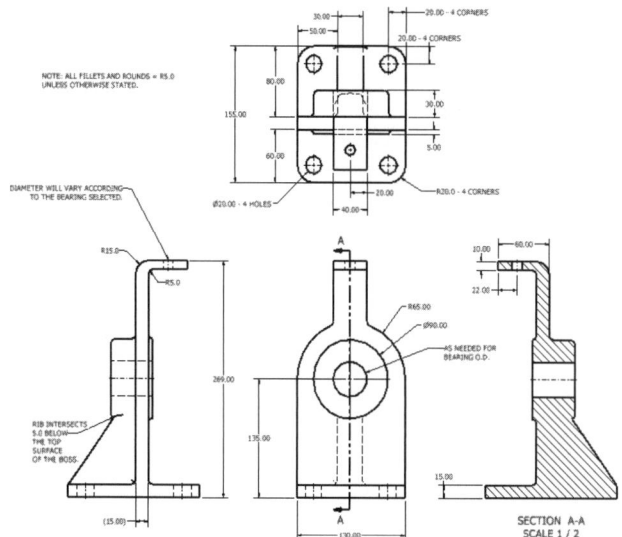

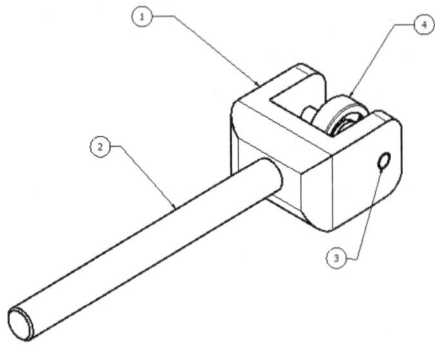

| Parts List | | | | |
|------|-------------|-------------|-------------|-----|
| ITEM | PART NUMBER | DESCRIPTION | MATERIAL | QTY |
| 1 | AM-232 | HOLDER | STEEL | 1 |
| 2 | AM-256 | POST, FOLLOWER | STEEL | 1 |
| 3 | BS 1804-2 - 4 x 30 | Parallel steel dowel pins - metric series | Steel, Mild | 1 |
| 4 | DIN625- SKF 634 | Single row ball bearings | Steel, Mild | 1 |

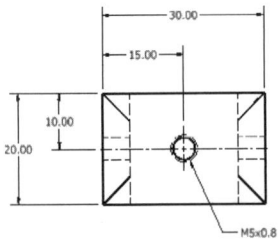

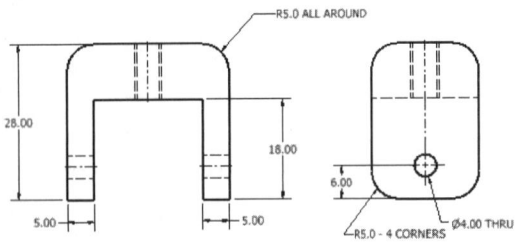

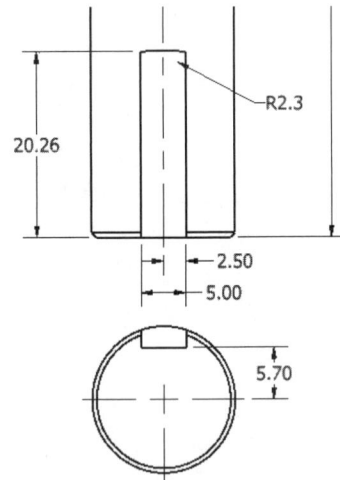

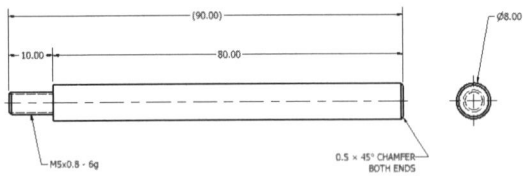

Four Spur Gears with Meshing Teeth. *Courtesy of Takeshi Takahara/Photo Researchers, Inc.*

## 8-26 GEARS AND CAMS

Gears, pulleys and belts, chains and sprockets, cams, linkages, and other devices are commonly used to transmit power and motion from one machine member to another. Understanding the function of these devices will help you create correct drawings and specifications for them.

Gears are one of the most common drive transfer mechanisms. Gears can change rotation direction, rotation speed, and axis orientation. The complex shape of a gear tooth is the result of mathematical computation. Drawing gears exactly on paper or with CAD requires considerable theoretical knowledge. However, often it is not necessary to represent details of gear teeth. Advanced CAD software programs often can generate gear models based on the parameters for the gear. Having an understanding of gear terminology before you start drawing a gear will help you provide the necessary information whether you draw by hand or use CAD.

Cams change rotational motion into reciprocating motion. Both gears and cams are commonly used in automobile engines and transmissions.

The American Gear Manufacturers Association (AGMA) was founded in 1916. AGMA produces standards for gear design and manufacture for American and global markets. They publish specific standards for gear types such as spur gears used in vehicles, plastic gears, bevel gears, and many others. AGMA 933-B03 is their standard for basic gear geometry. For more information when designing gears, check their Web site at www.AGMA.org for relevant standards.

## 8-27 UNDERSTANDING GEARS

**Gears** are used to transmit power and rotating or reciprocating motion from one machine part to another. They may be classified according to the position of the shafts that they connect. Parallel shafts, for example, may be connected by **spur gears, helical gears,** or **herringbone gears.** Intersecting shafts may be connected by **bevel gears** having either straight, skew, or spiral teeth. Nonparallel, nonintersecting

---

Search the following Web sites to learn more about designing gears (agma), stock drive products (sdp-si), and an interactive gear template generator (woodgears.ca):

- http://www.agma.org
- http://www.sdp-si.com/
- http://woodgears.ca/gear_cutting/template.html

Spur Gear. *Courtesy of Big Sky Laser.*

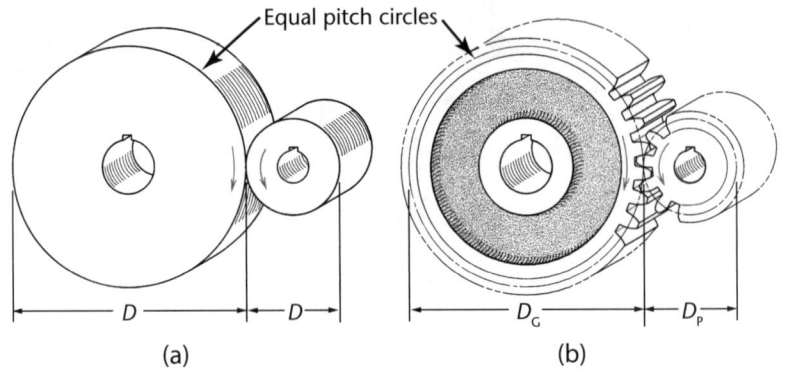

**Figure 8-42**    Friction Gears and Toothed Gears

shafts may be connected by **helical gears, hypoid gears,** or a worm and **worm gear.** A spur gear meshed with a rack will convert rotary motion to reciprocating motion.

ANSI/AGMA publishes detailed standards for gear design and drawing. Refer to these standards for current design specification and inspection practices for all of the gear types discussed in this chapter.

## Using Gears to Transmit Power

The **friction wheels** shown in Figure 8-42a transmit motion and power from one shaft to another parallel shaft. However, friction wheels are subject to slipping, and a great deal of pressure is required between them to create the necessary frictional force; therefore, they are usually used for low power applications, such as CD ROM drives. Spur gears (Figure 8-43b) have teeth on the cylindrical surfaces that fit together and transmit the same motion and power without slipping and with reduced bearing pressures.

If a friction wheel of diameter $D$ turns at $n$ **rpm** (revolutions per minute), the linear velocity, $v$, of a point on its periphery will be $\pi D n$, since $D n = c$ (circumference).

Example

Let                 $D = 3$ inches
                      $n = 100$ rpm
Then                $v = \pi D n = \pi(3)(100)$ in./min
                      $= 942$ in./min

or
$$\frac{942}{12} = 78.5 \text{ ft/min}$$

But the **pitch circles** of a pair of mating spur gears correspond exactly to the outside diameters of the friction wheels, and since the gears turn in contact without slipping, they must have the same linear velocity at the pitch line.

Therefore            $\pi D_G n_G = \pi D_P n_P$

or
$$\frac{D_G}{D_P} = \frac{n_P}{n_G} = m_G$$

where $D_G$ = **pitch diameter** of larger gear (called the wheel)

$D_P$ = pitch diameter of smaller gear (called the **pinion**)
$n_G$ = rpm of gear
$n_P$ = rpm of pinion
$m_G$ = **gear ratio**

The gear ratio is also expressed as $n_P/n_G$ or $D_G/D_P$.

Example

Let                 $D_G = 27''$    and    $D_P = 9''$
Then

Gear ratio $= m_G = 27:9$
                      $= 3:1$ (read as 3 to 1)

Example

For the same gear pair, let $n_P = 1725$ rpm.

Find $n_G$.

$$\frac{n_P}{n_G} = \frac{D_G}{D_P} \qquad \frac{n_G}{n_P} = \frac{D_P}{D_G}$$

$$n_G = n_P \frac{D_P}{D_G} = 1725 \cdot \frac{9}{27}$$

$$= \frac{1725}{3}$$

$$= 575 \text{ rpm}$$

The teeth on mating gears must be of equal width and spacing, so the number of teeth on each gear, $N$, is directly proportional to its pitch diameter, or

$$\frac{N_G}{N_P} = \frac{D_G}{D_P} = \frac{n_P}{n_G} = m_G$$

## Spur Gear Definitions and Formulas

Proportions and shapes of gear teeth are well standardized, and the terms illustrated and defined in Figure 8-43 are common to all spur gears. The dimensions relating to tooth height are for full-depth $14\frac{1}{2}°$ or $20°$ involute teeth.

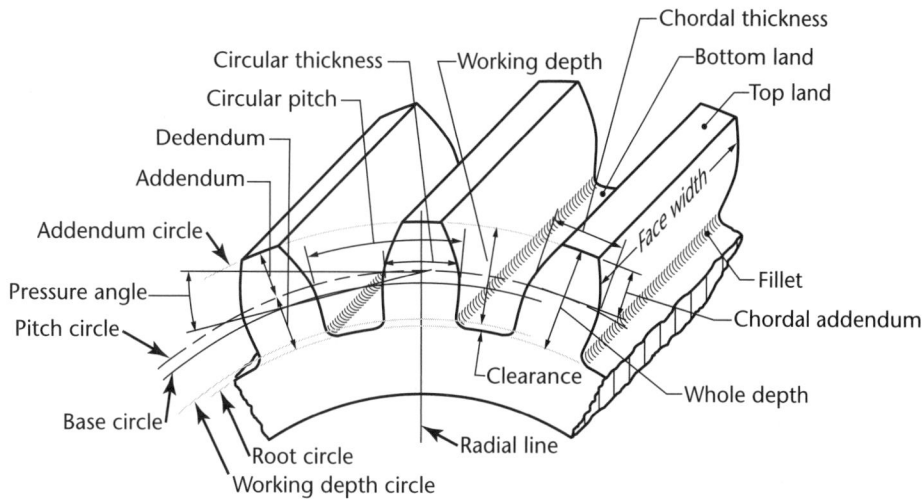

| Term | Symbol | Definition | Formula |
|------|--------|-----------|---------|
| Addendum | a | Radial distance from pitch circle to top of tooth. | $a = 1/P$ |
| Base circle | | Circle from which involute profile is generated. | |
| Chordal addendum | $\alpha_c$ | Radial distance from the top of a tooth to the chord of the pitch circle. | $a_c = a + \frac{1}{2}D\,[1\text{-}\cos(90°/N)]$ |
| Chordal thickness | $t_c$ | Thickness of a tooth measured along a chord of the pitch circle. | $t_c = D\sin(90°/N)$ |
| Circular pitch | p | Distance measured along pitch circle from a point on one tooth to corresponding point on the adjacent tooth; includes one tooth and one space. | $p = \pi D/N$ <br> $p = \pi/P$ |
| Circular thickness | t | Thickness of a tooth measured along the pitch circle: equal to one-half the circular pitch. | $t = p/2 = \pi/2P$ |
| Clearance | c | Distance between top of a tooth and bottom of mating space: equal to the dedendum minus the addendum. | $c = b - a = 0.157/P$ |
| Dedendum | b | Radial distance from pitch circle to bottom of tooth space. | $b = 1.157/P$ |
| Diametral pitch | P | A ratio equal to the number of teeth on the gear per inch of pitch diameter. | $P = N/D$ |
| Number of teeth | $N_G$ or $N_P$ | Number of teeth on the gear or pinion. | $N = P \times D$ |
| Outside diameter | $D_o$ | Diameter of addendum circle: equal to pitch diameter plus twice the addendum. | $D_o = D + 2a = (N + 2)/P$ |
| Pitch circle | | An imaginary circle that corresponds to the circumference of the friction gear from which the spur gear is derived. | |
| Pitch diameter | $D_G$ or $D_P$ | Diameter of pitch circle of gear or pinion. | $D = N/P$ |
| Pressure angle | $\phi$ | Angle that determines direction of pressure between contacting teeth and designates shape of involute teeth—e.g. $14\frac{1}{2}°$ involute; also determines the size of base circle. | |
| Root diameter | $D_R$ | Diameter of the root circle; equal to pitch diameter minus twice the dedendum. | $D_R = D - 2b = (N - 2.314)/P$ |
| Whole depth | $h_t$ | Total height of the tooth; equal to the addendum plus the dedendum. | $h_t = a + b = 2.157/P$ |
| Working depth | $h_k$ | Distance a tooth projects into mating space; equal to twice the addendum. | $h_k = 2a = 2/P$ |

**Figure 8-43**   Spur Gear Terminology

To make gears operate smoothly with a minimum of noise and vibration, the curved surface of the tooth profile uses a definite geometric form. The most common form in use today is the **involute** profile shown in Figure 8-43. (The word *involute* means "rolled inward.")

## 8-28 CONSTRUCTING A BASE CIRCLE

The involute tooth form depends on the pressure angle, which is ordinarily 14½° or 20°. This pressure angle determines the size of the **base circle;** from this the involute curve is generated.

To calculate the base circle for the spur gear as shown in Figure 8-44, follow these steps. At any point on the pitch circle, such as point *P* (the pitch point) draw a line tangent to the pitch circle; draw a second line through *P* at the required pressure angle (frequently approximated at 15° on the drawing). This line is called the line of contact. Next, draw a line perpendicular to the line of contact from the center, *C*. Then draw the base circle with radius *CJ* tangent to the line of contact at *J*.

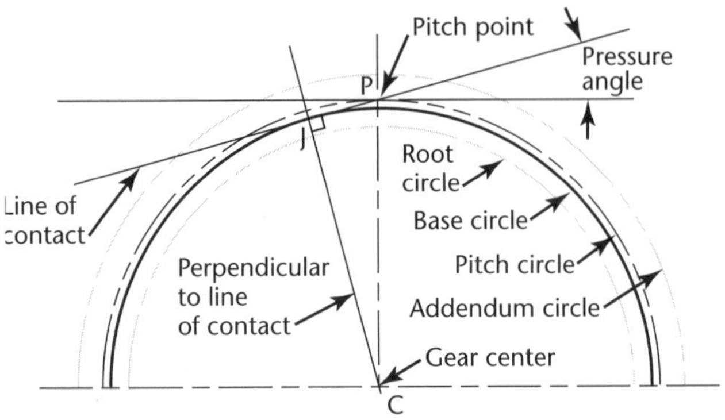

## 8-29 THE INVOLUTE TOOTH SHAPE

If the exact shape of the tooth is desired, the portion of the profile from the base circle to the **addendum** circle can be drawn as the involute of the base circle. In Figure 8-45, the tooth profile from *A* to *O* is an involute of the base circle. See Appendix 20 if you need to refer to the method of construction. The part of the profile below the base circle, line *OB*, is drawn as a radial line (a straight line drawn from the gear center) that terminates in the fillet at the **root circle.** The fillet should be equal in radius to one and one-half times the clearance from the tip of the tooth to the bottom of the mating space. See Figure 8-46.

## 8-30 APPROXIMATE INVOLUTE USING CIRCULAR ARCS

Involute curves can be closely approximated with two circular arcs, as shown in Figure 8-47. This method, originally devised by G. B. Grant, uses a table of arc radii, an involute odontograph, for gears with various numbers of teeth. To use this method, draw the base circle as described above, and set off the spacing of the teeth along the pitch circle. Then draw the face of the tooth from *P* to *A* with the face radius *R*, and draw the portion of the flank from *P* to *O* with the flank radius *r*. Draw both arcs from centers located on the base circle. The

**Figure 8-44** Construction of a Base Circle

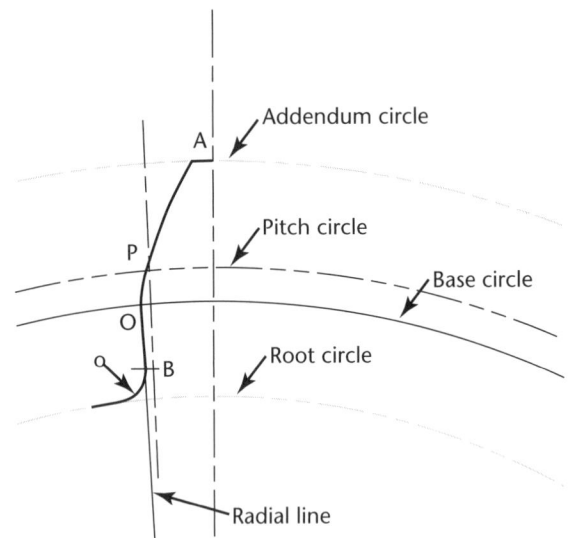

**Figure 8-45** The Involute Profile

**Figure 8-46** Shaded Model of Involute Shaped Spur Gear Teeth. *It is not typical to model gear teeth in detail as it creates an unnecessarily large and complex model.*

| No. of Teeth | 14½° | | 20° | |
|---|---|---|---|---|
| (N) | R (in.) | r (in.) | R (in.) | r (in.) |
| 12 | 2.87 | 0.79 | 3.21 | 1.31 |
| 13 | 3.02 | 0.88 | 3.40 | 1.45 |
| 14 | 3.17 | 0.97 | 3.58 | 1.60 |
| 15 | 3.31 | 1.06 | 3.76 | 1.75 |
| 16 | 3.46 | 1.16 | 3.94 | 1.90 |
| 17 | 3.60 | 1.26 | 4.12 | 2.05 |
| 18 | 3.74 | 1.36 | 4.30 | 2.20 |
| 19 | 3.88 | 1.46 | 4.48 | 2.35 |
| 20 | 4.02 | 1.56 | 4.66 | 2.51 |
| 21 | 4.16 | 1.66 | 4.84 | 2.66 |
| 22 | 4.29 | 1.77 | 5.02 | 2.82 |
| 23 | 4.43 | 1.87 | 5.20 | 2.98 |
| 24 | 4.57 | 1.98 | 5.37 | 3.14 |
| 25 | 4.70 | 2.08 | 5.55 | 3.29 |
| 26 | 4.84 | 2.19 | 5.73 | 3.45 |
| 27 | 4.97 | 2.30 | 5.90 | 3.61 |
| 28 | 5.11 | 2.41 | 6.08 | 3.77 |
| 29 | 5.24 | 2.52 | 6.25 | 3.93 |
| 30 | 5.37 | 2.63 | 6.43 | 4.10 |
| 31 | 5.51 | 2.74 | 6.60 | 4.26 |
| 32 | 5.64 | 2.85 | 6.78 | 4.42 |
| 33 | 5.77 | 2.96 | 6.95 | 4.58 |
| 34 | 5.90 | 3.07 | 7.13 | 4.74 |
| 35 | 6.03 | 3.18 | 7.30 | 4.91 |
| 36 | 6.17 | 3.29 | 7.47 | 5.07 |
| 37–39 | 6.36 | 3.46 | 7.82 | 5.32 |
| 40–44 | 6.82 | 3.86 | 8.52 | 5.90 |
| 45–50 | 7.50 | 4.46 | 9.48 | 6.76 |
| 51–60 | 8.40 | 5.28 | 10.84 | 7.92 |
| 61–72 | 9.76 | 6.54 | 12.76 | 9.68 |
| 73–90 | 11.42 | 8.14 | 15.32 | 11.96 |
| 91–120 | 0.118N | 0.156N | | |
| 121–180 | 0.122N | 0.165N | | |
| Over 180 | 0.125N | 0.171N | | |

**Figure 8-47**    Wellman's Involute Odontograph for Drawing Gear Teeth Using Circular Arcs

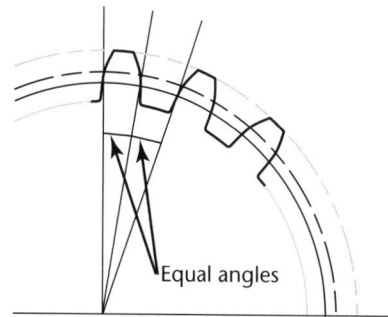

**Figure 8-48**    Spacing Gear Teeth

table in Figure 8-43 gives the correct face and flank radii for gears of one **diametral pitch.** For other pitches, divide the values in the table by the diametral pitch. For gears with more than 90 teeth, use a single radius (let $R = r$) computed from the appropriate formula given in Figure 8-47, then divide by diametral pitch. Below the base circle, complete the flank of the tooth with a radial line $OB$ and a fillet.

## 8-31  SPACING GEAR TEETH

Suppose that number of teeth ($N$) = 20 and diametral pitch ($P$) = 4 for a 14½° involute tooth. The values from Figure 8-47 are $R = 4.02$ and $r = 1.56$. These must be divided by $P$; yielding $r = 4.02/4 = 1.005''$ and $r = 1.56/4 = 0.39''$. Note that these values are for a full-size drawing.

Space the teeth around the periphery by laying out equal angles (Figure 8-48). The number of spaces should be $2N$, twice the number of teeth, to make the space between teeth equal to the tooth thickness at the pitch circle. In the example, since $N = 20$, $a = 360°/2N = 360°/40 = 9°$, the angle subtended by each tooth and each space.

When the pressure angle is 20° and the height of the tooth is reduced, the teeth are called **stub teeth.** Stub teeth are drawn in the same manner as other teeth except that $a = 0.8/P$, $b = 1/P$, and the pressure angle is 20°. The main advantage of stub teeth is that they are stronger than the 14½° standard full-depth teeth.

**TIP**

The Divide command in the AutoCAD software is very handy for dividing a circle or other geometry into any number of equal divisions. You can also use it to insert a block at the same time. A polar array is another useful command for creating gears.

Involute Gear Hob. Spur gears with involute tooth shapes are usually manufactured using a hobbing machine. This machine cuts gear teeth by rotating the gear blank and a cutter like the one shown at a fixed speed ratio. The cross sectional profile of the sides of the teeth on the cutter generate the involute tooth shape for the gear. Very small gears normally must be milled instead. *Courtesy of Hobsource.*

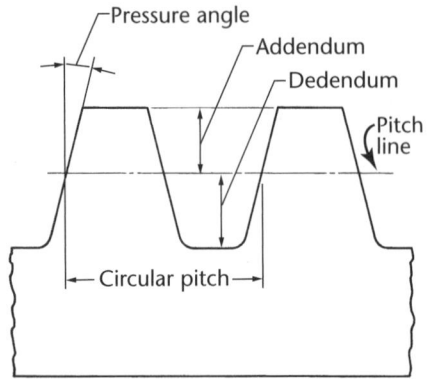

## 8-32  RACK TEETH

Gear teeth formed on a flat surface are called a **rack.** In the involute system, the sides of rack teeth are straight and are inclined at an angle equal to the pressure angle. To mesh with a gear, the linear pitch of the rack must be the same as the **circular pitch** of the gear, and the rack teeth must have the same height proportions as the gear teeth. See Figure 8-49.

## 8-33  WORKING DRAWINGS OF SPUR GEARS

A typical working drawing of a spur gear is shown in Figure 8-50. Since the teeth are cut to standard shape, it is not necessary to show individual teeth on the drawing. Instead,

**Figure 8-49**    Involute Rack Teeth

draw the addendum and root circles as phantom lines and the pitch circle as a centerline.

The drawing actually shows only a **gear blank**—a gear complete except for teeth. Since the machining of the blank and the cutting of the teeth are separate operations, the necessary dimensions are arranged in two groups: the blank dimensions are shown on the views, and the cutting data are given in a note or table.

Before laying out the working drawing, calculate the gear dimensions. For example, if the gear must have 48 teeth of 4 diametral pitch, with $14\frac{1}{2}°$ full-depth involute profile, as in Figure 8-50, calculate the following items in this order: pitch diameter, addendum, dedendum, outside diameter, root diameter, whole depth, chordal thickness, and chordal addendum.

The dimensions shown in Figure 8-50 are the minimum requirements for the spur gear. The chordal addendum

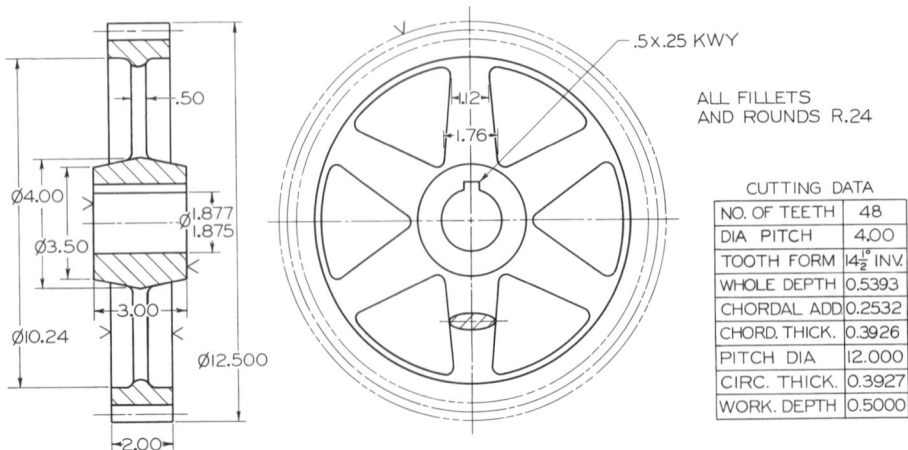

**Figure 8-50**    Working Drawing of a Spur Gear

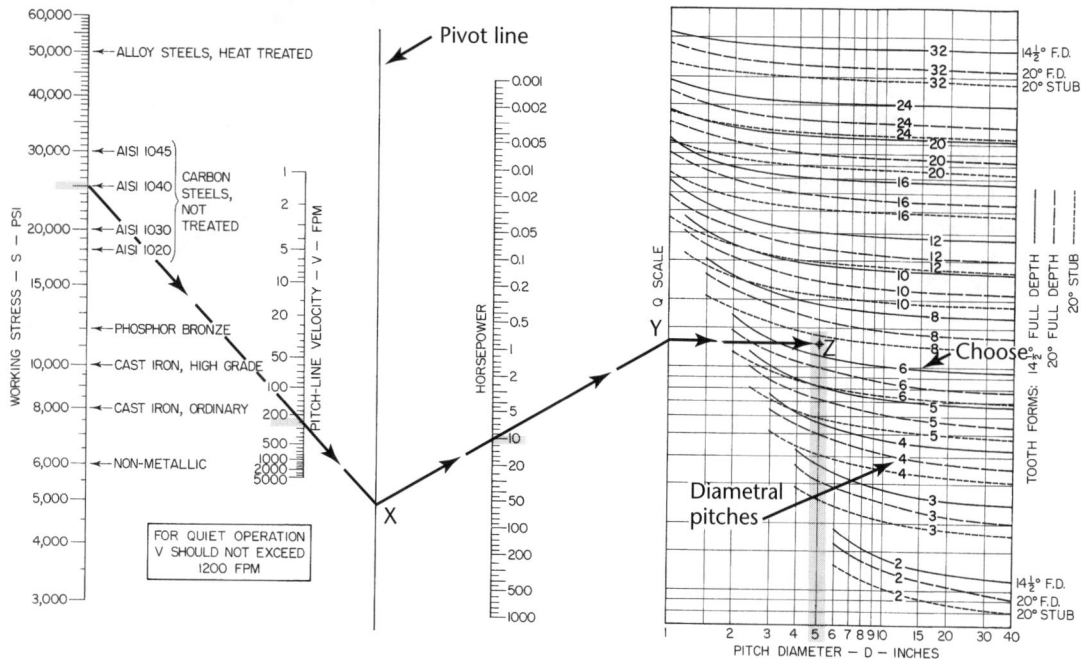

**Figure 8-51**    Chart for Design of Cut Spur Gears. *The chart is based on the Lewis equation, with the Barth velocity modification, and assumes gear face width equal to three times the circular pitch.*

and chordal thickness are given to aid in checking the finished gear. Other special data may be given in the table, according to the degree of precision required and the manufacturing method.

## 8-34  SPUR GEAR DESIGN

Spur gear design normally begins with selecting pitch diameters to suit the required speed ratio, center distance, and space limitations. The size of the teeth (the diametral pitch) depends on the gear speeds, gear materials, horsepower to be transmitted, and the selected tooth form. The complete analysis and design of precision gears is complex and beyond the scope of this textbook, but the chart in Figure 8-51 gives suitable diametral pitches for ordinary cut spur gears.

**Example**  To determine the diametral pitch and face width for a 5″ pitch diameter (G10400 carbon steel) pinion with 14½° full-depth teeth that must transmit 10 hp (horsepower) at 200 rpm.

Calculate the pitch in line velocity $V$ as follows:

$$V = \pi/12(D_p n_p)$$

$$= 0.262 \times 5 \times 200 = 262 \text{ fpm (ft/min)}$$

On the chart, draw a straight line from 25,000 psi on the working stress scale through 262 fpm on the velocity

scale to intersect the pivot line at $X$. From $X$, draw a second straight line through 10 on the horsepower scale to intersect the $Q$ scale at $Y$. From this point, enter the graph of pitches and go to the ordinate for 5″ pitch diameter. The junction point $Z$ falls slightly above the curve for 14½° full-depth teeth of 6 diametral pitch; hence, choose 6 diametral pitch for the gears.

For good proportions the face width of a spur gear should be about three times the circular pitch, and this proportion is incorporated in the chart. For 6 diametral pitch the circular pitch is $\pi/6$ or .5236″, and the face width is $3 \times .5236$ or 1.5708 inch. This width should be rounded to 1⅝ inch (1.625″).

When both the pinion and the gear are of the same material, the smaller gear is the weaker, and the design should be based on the pinion. When the materials are different, the chart should be used to determine the horsepower capacity of each gear.

## 8-35  WORM GEARS

Worm gears are used to transmit power between nonintersecting shafts that are at right angles to each other. A worm (Figure 8-52) is a screw with a thread shaped like a rack tooth. The worm wheel is similar to a helical gear that has been cut to conform to the shape of worm for more contact. Worm gearing offers a large speed ratio, since with one

**Figure 8-52** Worm

revolution, a single-thread worm advances the worm wheel only one tooth and a space.

Figure 8-53 shows a worm and a worm wheel engaged. The section taken through the center of the worm and perpendicular to the axis of the worm wheel shows that the worm section is identical to a rack and that the wheel section is identical to a spur gear. Consequently, in this plane the height proportions of thread and gear teeth are the same as for a spur gear of corresponding pitch.

**Pitch ( *p* )** The axial pitch of the worm is the distance from a point on one thread to the corresponding point on the next

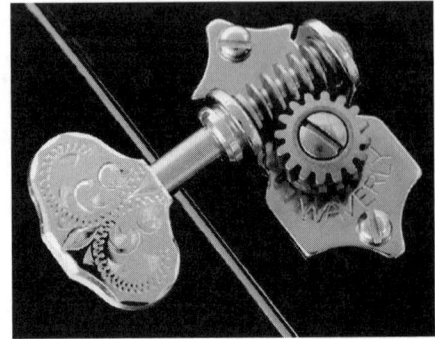

A Waverly Guitar Tuner. Worms are often designed so they can easily turn the gear, but the gear cannot turn the worm. Guitar machines, often called tuners, use worm gears so that the guitar can be tuned by turning the worm, but the strings are kept taut as the gear cannot turn the worm. *Courtesy of Stewart-MacDonald Company.*

thread measured parallel to the worm axis. The pitch of the worm must be exactly equal to the circular pitch of the gear.

**Lead ( *L* )** The lead is the distance that the thread advances axially in one turn. The lead is always a multiple of the pitch. Thus, for a single-thread worm, the lead equals the pitch; for a double-thread worm, the lead is twice the pitch, and so on.

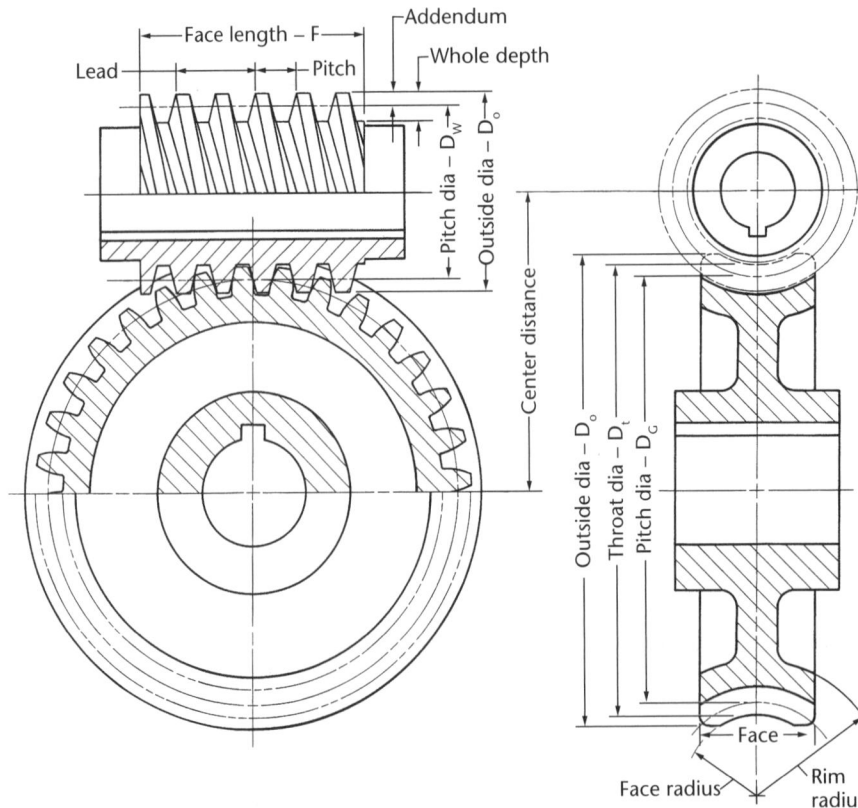

**Figure 8-53** Double-Thread Worm and Worm Gear

**Lead angle** ($\lambda$) The lead angle is the angle between a tangent to the helix at the pitch diameter and a plane perpendicular to the axis of the worm. The lead angle can be calculated from

$$\tan l = \frac{L}{pD_W}$$

where

$D_W$ = pitch diameter of the worm

The speed ratio of worm gears depends only on the number of threads on the worm and the number of teeth on the gear. Therefore,

$$m_G = \frac{N_G}{N_W}$$

where

$N_G$ = number of teeth on the gear
$N_W$ = number of threads on the worm

For $14\frac{1}{2}°$ standard involute teeth and single-thread or double-thread worms, the following proportions are the recommended practice of the AGMA. All formulas are expressed in terms of circular pitch $p$ instead of diametral pitch $P$. It is easier to machine the worm and the hob used to cut the gear if the circular pitch has an even rational value such as 5/8″.

**For the worm:**

| | |
|---|---|
| Pitch diameter | $D_w = 2.4p \times 1.1$ (recommended value, but it may vary) |
| Whole depth | $h_t = 0.686p$ |
| Outside diameter | $D_o = D_w + 0.636p$ |
| Face length | $F = p(4.5 + N_G/50)$ |

**For the gear:**

| | |
|---|---|
| Pitch diameter | $D_G = p(N_G/\pi)$ |
| Throat diameter | $D_t = D_G + 0.636p$ |
| Outside diameter | $D_o = D_t + 0.4775p$ |
| Face radius | $R_f = \frac{1}{2}D_w - 0.318p$ |
| Rim radius | $R_r = \frac{1}{2}D_w + p$ |
| Face width | $F = 2.38p \times 0.25$ |
| Center distance | $C = \frac{1}{2}(D_G + D_w)$ |

## 8-36 WORKING DRAWINGS OF WORM GEARS

In an assembly drawing, the engaged worm and gear can be shown as in Figure 8-53, but usually the gear teeth are omitted and the gear blank represented conventionally, as shown in the lower half of the circular view. On detail drawings, the worm and gear are usually drawn separately, as shown in Figures 8-54 and 8-55. Although their dimensioning depends on the production method, it is standard practice to dimension the blanks on the views and give the cutting data in a table, as shown. Note that dimensions that closely affect the engagement of the gear and worm have been given as three-place decimal or limit dimensions; other dimensions, such as rim radius, face lengths, and gear outside diameter, have been rounded to convenient two-place decimal values.

## 8-37 BEVEL GEARS

Bevel gears (Figure 8-56) are used to transmit power between shafts whose axes intersect. The analogous friction drive would consist of a pair of cones with a common apex at the point where their axes intersect, as in Figure 8-57. The axes may intersect at any angle, but right angles are most common. Bevel gear teeth have the same involute shape as teeth on spur gears, but they are tapered toward the cone apex; hence, the height and width of a bevel gear tooth vary with the distance from the cone apex. Spur gears are interchangeable (a spur gear of given pitch will run properly with any other spur gear of the same pitch and tooth form) but this is not true of bevel gears, which must be designed in pairs and will run only with each other.

The speed ratio of bevel gears can be calculated from the same formulas given for spur gears.

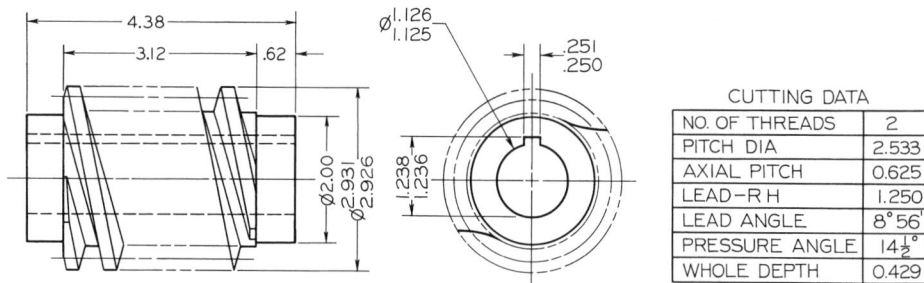

| CUTTING DATA | |
|---|---|
| NO. OF THREADS | 2 |
| PITCH DIA | 2.533 |
| AXIAL PITCH | 0.625 |
| LEAD-R H | 1.250 |
| LEAD ANGLE | 8° 56′ |
| PRESSURE ANGLE | $14\frac{1}{2}°$ |
| WHOLE DEPTH | 0.429 |

**Figure 8-54**  Working Drawing of a Worm

| CUTTING DATA | |
|---|---|
| NO. OF TEETH | 30 |
| PITCH DIA | 5.967 |
| ADDENDUM | 0.199 |
| WHOLE DEPTH | 0.429 |
| NO. OF THREADS | 2 |
| AXIAL PITCH | 0.625 |
| LEAD – R H | 1.250 |
| LEAD ANGLE | 8° 56' |
| PRESSURE ANGLE | $14\frac{1}{2}°$ |

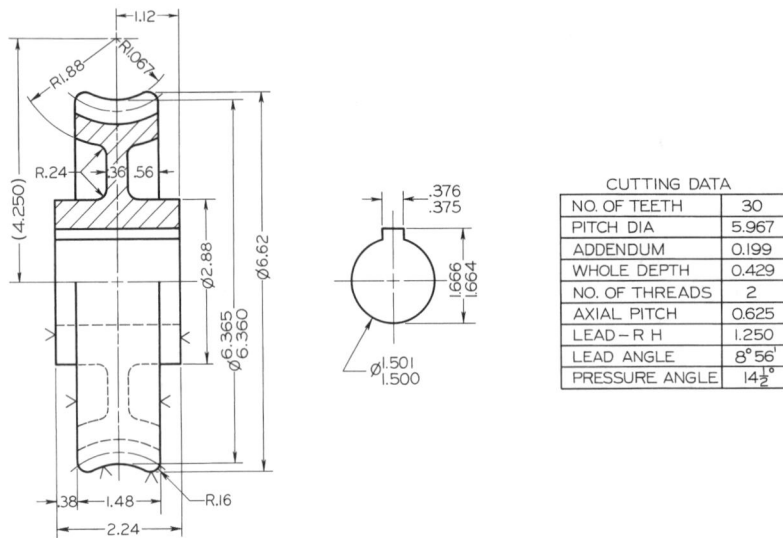

**Figure 8-55**   Working Drawing of a Worm Gear

**Figure 8-56**   Bevel Gears. *Courtesy of Stock Drive Products/ Sterling Instruments.*

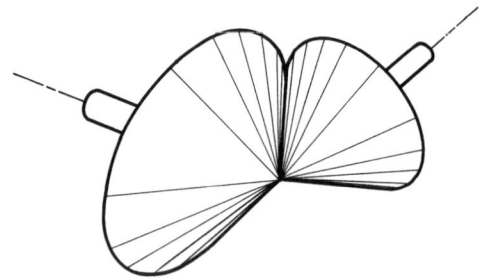

**Figure 8-57**   Friction Cones

## 8-38  BEVEL GEAR DEFINITIONS AND FORMULAS

Since the design of bevel gears is very similar to that of spur gears, many spur gear terms are applied with slight modification to bevel gears. Just as in spur gears, the pitch diameter *D* of a bevel gear is the diameter of the base of the pitch cone, the circular pitch *p* to the teeth is measured along this circle and the diametral pitch *P* is also based on this circle.

The important dimensions and angles of a bevel gear are illustrated in Figure 8-58. The pitch cone is shown as the triangle *AOB*, and the pitch angle is $\frac{1}{2}$ angle *AOB*. The root and face angle lines do not actually converge at point *O* but are often drawn as if they do for simplicity. Figure 8-58 shows that the pitch angle of each gear depends on the relative diameters of the gears. When the shafts are at right angles, the sum of the pitch angles for the two mating gears equals 90°.

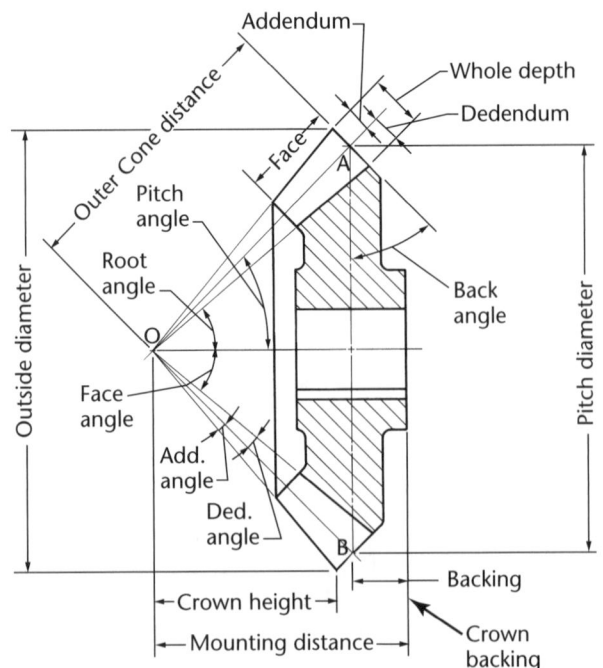

**8-58**   Bevel Gear Terminology

Therefore, the pitch angles, $\Gamma$ (gamma), are determined from the following equations:

$$\tan \Gamma_G = \frac{D_G}{D_P} = \frac{N_G}{N_P}$$

and

$$\tan \Gamma_G = \frac{D_P}{D_G} = \frac{N_P}{N_G}$$

In the simplified formulas given here, tooth proportions are assumed equal on both gears, but in modern practice bevel gears are often designed with unequal addenda and unequal tooth thicknesses to balance the strength of gear and pinion. Refer to AGMA standards for more details.

Terms for bevel gear definitions and formulas are given in Table 8-1.

Table 8-1    Bevel Gear Terms, Definitions, and Formulas.

| Term | Symbol | Definition | Formula |
|---|---|---|---|
| Addendum | $a$ | Distance from pitch cone to top of tooth; measured at large end of tooth. | $a = 1/P$ |
| Addendum angle | $\alpha$ | Angle subtended by addendum; same for gear and pinion. | $\tan \alpha = a/A$ |
| Back angle | | Usually equal to the pitch angle. | |
| Backing | $Y$ | Distance from base of pitch cone to rear of hub. | |
| Chordal addendum | $a_c$ | For bevel gears, the formulas given for spur gears can be used | |
| Chordal thickness | $t_c$ | if $D$ is replaced by $D/\cos \Gamma$ and $N$ is replaced by $N \cos \Gamma$. | |
| Crown backing | $Z$ | More practical than backing for shop use; dimension $Z$ given on drawings instead of $Y$. | $Z = Y + a \sin \Gamma$ |
| Crown height | $X$ | Distance, parallel to gear axis, from cone apex to crown of the gear. | $X = \frac{1}{2}D_o/\tan \Gamma_o$ |
| Dedendum | $b$ | Distance from pitch cone to bottom of tooth; measured at large end of tooth. | $b = 1.188/P$ |
| Dedendum angle | $\delta$ | Angle subtended by dedendum; same for gear and pinion. | $\tan \delta = b/A$ |
| Face angle | $\Gamma_o$ | Angle between top of teeth and the gear axis. | $\Gamma_o = \Gamma + \alpha$ |
| Face width | $F$ | Should not exceed $A$ or $10/P$, whichever is smaller. | |
| Mounting distance | $M$ | A dimension used primarily for inspection and assembly purposes. | $M = Y + \frac{1}{2}D/\tan \Gamma$ |
| Outer cone distance | $A$ | Slant height of pitch cone; same for gear and pinion. | $A = D/2 \sin \Gamma$ |
| Outside diameter | $D_o$ | Diameter of outside or crown circle of the gear. | $D_o = D + 2a \cos \Gamma$ |
| Pitch diameter | $D_G$ | Diameter of base of pitch cone of gear or pinion. | $D_G = N_G/P$ |
| | $D_P$ | | $D_P = N_P/P$ |
| Root angle | $\Gamma_R$ | Angle between the root of the teeth and the gear axis. | $\Gamma_R = \Gamma - \delta$ |

## 8-39 WORKING DRAWINGS OF BEVEL GEARS

Like those for spur gears, working drawings of bevel gears give only the dimensions of the gear blank. Data necessary for cutting the teeth are given in a note or table. A single sectional view will usually provide all necessary information (Figure 8-59). If a second view is required, only the gear blank is drawn, and the tooth profiles are omitted. Two gears are shown in their operating relationship. On detail drawings, each gear is usually drawn separately, as in Figure 8-58, and fully dimensioned. Placement of the gear-blank dimensions depends largely on the manufacturing methods used in producing the gear, but the scheme shown is commonly followed.

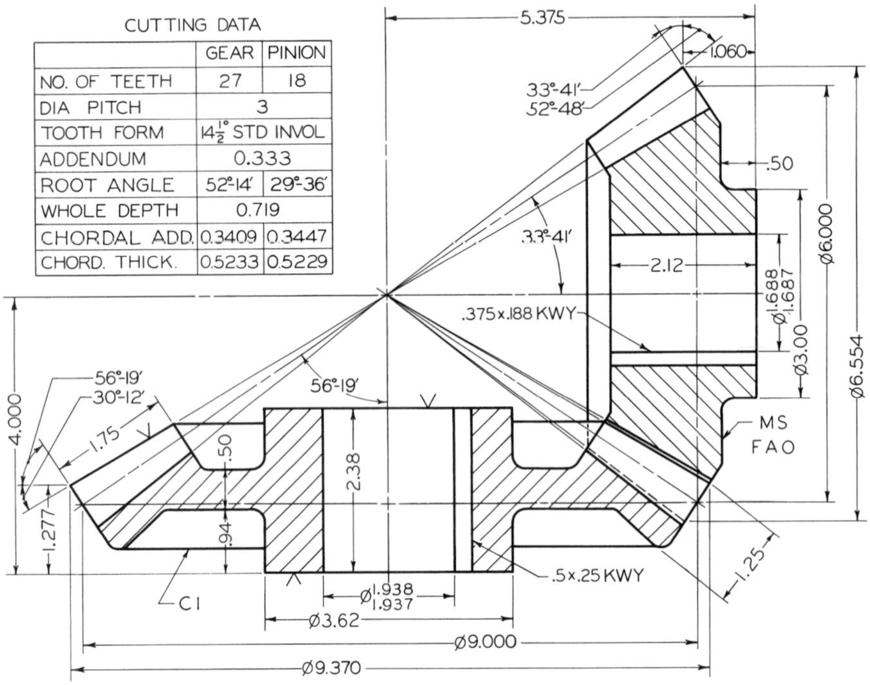
CUTTING DATA

| | GEAR | PINION |
|---|---|---|
| NO. OF TEETH | 27 | 18 |
| DIA PITCH | 3 | |
| TOOTH FORM | $14\frac{1}{2}$° STD INVOL | |
| ADDENDUM | 0.333 | |
| ROOT ANGLE | 52°-14′ | 29°-36′ |
| WHOLE DEPTH | 0.719 | |
| CHORDAL ADD. | 0.3409 | 0.3447 |
| CHORD. THICK. | 0.5233 | 0.5229 |



**Figure 8-59** Working Drawing of Bevel Gears

# CAD AT WORK

## Stock Gear Models And Drawings

A great Web site where you can find stock models for gears is the Stock Drive Products/Sterling Instruments site at https://sdp-si.com (Figure A). Once you have registered, you can search their site for useful standard parts that you might purchase for a design. After you have located a part, you can download 2D or 3D CAD data. Figure B shows the compatible file types available for download. Detailed data sheets are available for download in PDF format. You can even use a Web viewer to view and rotate the 3D model before you download it (Figure C).

When gears are ordered to use as standard parts in an assembly, a detailed gear drawing is not necessary. A table showing the gear information is often sufficient. Simplified representations of the shape—for example, the exterior envelope and/or base curves—may be used to represent the gear.

| Part Number | Diametral Pitch | No. Of Teeth | Material | Hub Style | Quality Class | Pressure Angle Degree | Bore Size | Face Width | Pitch Dia. |
|---|---|---|---|---|---|---|---|---|---|
| A 1B 1-N24008 | 24 | 8 | Brass | Hubless | Commercial | 14.5° | 0.1250 | 0.25 | 0.3330 |
| A 1B 1-N24014 | 24 | 14 | Brass | Hubless | Commercial | 14.5° | 0.1875 | 0.25 | 0.5830 |
| A 1B 1-N24018 | 24 | 18 | Brass | Hubless | Commercial | 14.5° | 0.1875 | 0.25 | 0.7500 |
| A 1B 1-N24020 | 24 | 20 | Brass | Hubless | Commercial | 14.5° | 0.1875 | 0.25 | 0.8330 |
| A 1B 1-N24024 | 24 | 24 | Brass | Hubless | Commercial | 14.5° | 0.1875 | 0.25 | 1.0000 |
| A 1B 1-Y24007 | 24 | 7 | Brass | Hubless | Commercial | 20° | 0.1250 | 0.25 | 0.2920 |
| A 1B 1-Y24008 | 24 | 8 | Brass | Hubless | Commercial | 20° | 0.1250 | 0.25 | 0.3330 |
| A 1B 1-Y24011 | 24 | 11 | Brass | Hubless | Commercial | 20° | 0.1880 | 0.25 | 0.4580 |
| A 1B 1-Y24012 | 24 | 12 | Brass | Hubless | Commercial | 20° | 0.1880 | 0.25 | 0.5000 |

(A) Stock Drive Product/Sterling Instruments provides downloadable CAD models for the parts they sell. *Courtesy of Stock Drive Products/Sterling Instruments.*

(B) Several CAD formats are available for download. *Courtesy of Stock Drive Products/Sterling Instruments.*

(C) An interactive viewer allows you to view and rotate the 3D model via the Web. *Courtesy of Stock Drive Products/Sterling Instruments.*

Camshafts. *Courtesy of Andrew Wakeford/Getty Images, Inc.-Photodisc.*

Billet Steel Cams for Ford 4.6/5.4L SOHC V-8. *Courtesy of Crane Cams, Inc.*

## 8-40 CAMS

**Cams** can produce unusual and irregular motions that would be difficult to produce otherwise. Figure 8-60a shows the basic principle of the cam. A shaft rotating at uniform speed carries an irregularly shaped disk called a cam; a reciprocating member, called the **cam follower,** presses a small roller against the curved surface of the cam. (The roller is held in contact with the cam by gravity or a spring.)

Rotating the cam causes the follower to reciprocate with a cyclic motion according to the shape of the **cam profile.**

Figure 8-60b shows an automobile valve cam which operates a flat-faced follower. Figure 8-60c shows a disk cam with the roller follower attached to a linkage to transmit motion to another part of the device.

The following sections discuss how to draw a cam profile that will cause the follower to produce the particular motion that is needed.

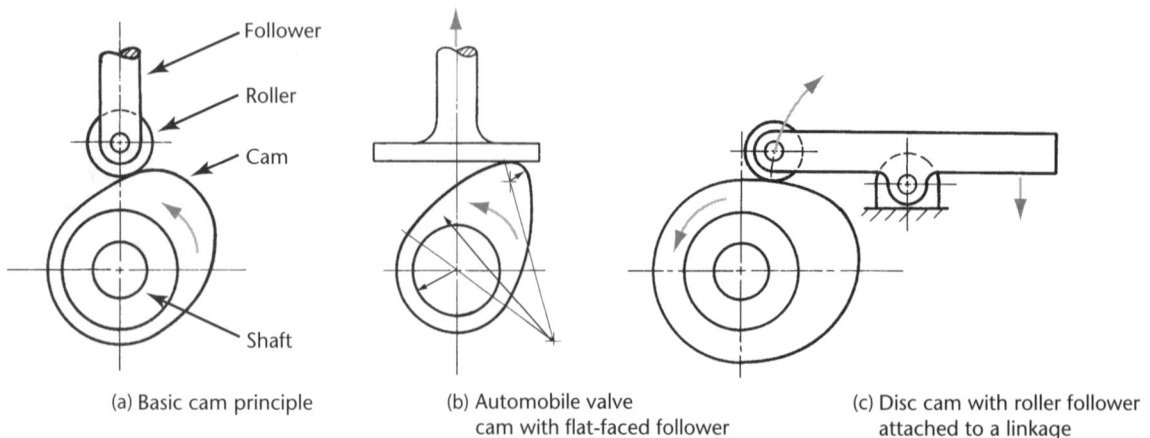

(a) Basic cam principle

(b) Automobile valve cam with flat-faced follower

(c) Disc cam with roller follower attached to a linkage

Figure **8-60** Disk Cams

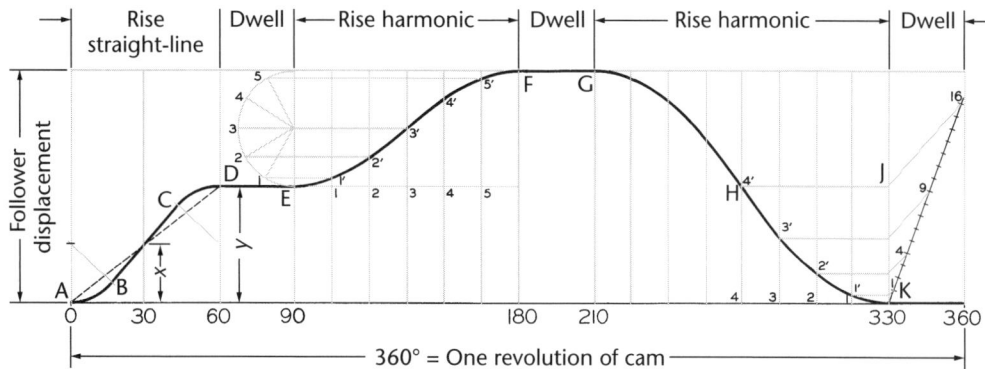

**Figure 8-61** Displacement Diagram with Typical Curves

## 8-41 DISPLACEMENT DIAGRAMS

Since the motion of the follower is your first concern, its rate of speed and its various positions should be carefully planned in a displacement diagram before the cam profile is constructed. A **displacement diagram** (Figure 8-61) is a curve showing the displacement of the follower as ordinates on a baseline that represents one revolution of the cam. Draw the follower displacement to scale, but you can use any convenient length to represent the 360° of cam rotation.

The motion of the follower as it rises or falls depends on the shape of the curves in the displacement diagram. In this diagram, four commonly employed types of curves are shown. If a straight line is used, such as the dashed line *AD* in Figure 8-61, the follower will move with a uniform velocity, but it will be forced to start and stop very abruptly, producing high acceleration and unnecessary force on the follower and other parts. This straight-line motion can be modified as shown in the curve *ABCD*, where arcs have been introduced at the beginning and at the end of the period.

The curve shown at *EF* gives harmonic motion to the follower. To construct this curve, draw a semicircle with a diameter equal to the desired rise. Divide the circumference of the semicircle into equal arcs. (The number of divisions should be the same as the number of horizontal divisions.) Then find points on the curve by projecting horizontally from the divisions on the semicircle to the corresponding ordinates.

The parabolic curve shown at *GHK* gives the follower constantly accelerated and decelerated motion. The half of the curve from *G* to *H* is exactly the reverse of the half from *H* to *K*. To construct the curve *HK*, divide the vertical height from *K* to *J* into distances proportional to $1^2$, $2^2$, $3^2$, or 1, 4, 9, and so on. The number of vertical divisions should be the same as the number of horizontal divisions. Find points on the curve by projecting horizontally from the divisions on the line *JK* to the corresponding ordinates. The parabolic curve and curves of higher polynomial equations produce smoother follower motion than those of the other curves discussed.

## 8-42 CAM PROFILES

The general method for constructing a cam profile is shown in Figure 8-62. The disk cam rotating counterclockwise on its shaft raises and lowers the roller follower, which is constrained to move along the straight line *AB*. The displacement diagram at the bottom of the figure shows the desired follower motion.

With the follower in its lowest or initial position, the center of the roller is at *A*, and *OA* is the radius of the base circle. The diameter of the base circle is determined from design parameters that are not discussed here.

Since the cam must remain stationary while it is being drawn, you can obtain an equivalent rotative effect by imagining that the cam stands still while the follower rotates about it in the opposite direction. Therefore, the base circle is divided into 12 equal divisions corresponding to the divisions used in the displacement diagram. These divisions begin at zero and are numbered in an opposite direction to the cam rotation.

The points 1, 2, and so on, on the follower axis *AB* indicate successive positions of the center of the roller and are located by transferring ordinates such as *x* and *y* from the displacement diagram. So, when the cam has rotated 60°, the follower roller must rise a distance *x* to position 2, and after 90° of rotation, a distance *y* to position 3, and so on.

Note that while the center of the roller moved from its initial position *A* to position 2, for example, the cam rotated 60° counterclockwise. Therefore, point *2* must be revolved clockwise about the cam center *O* to the corresponding 60° tangent line to establish point *2'*. In this position, the complete follower would appear as shown by the phantom outline. Points represent consecutive positions of the roller center, and a smooth curve drawn through these points is called the pitch curve. To obtain the actual cam profile, the roller must be drawn in a number of positions, and the cam profile drawn tangent to the roller circles, as shown. The best results are obtained by first drawing the pitch curve very carefully and then drawing several closely spaced roller circles with centers on the **pitch curve,** as shown between points *5'* and *6'*.

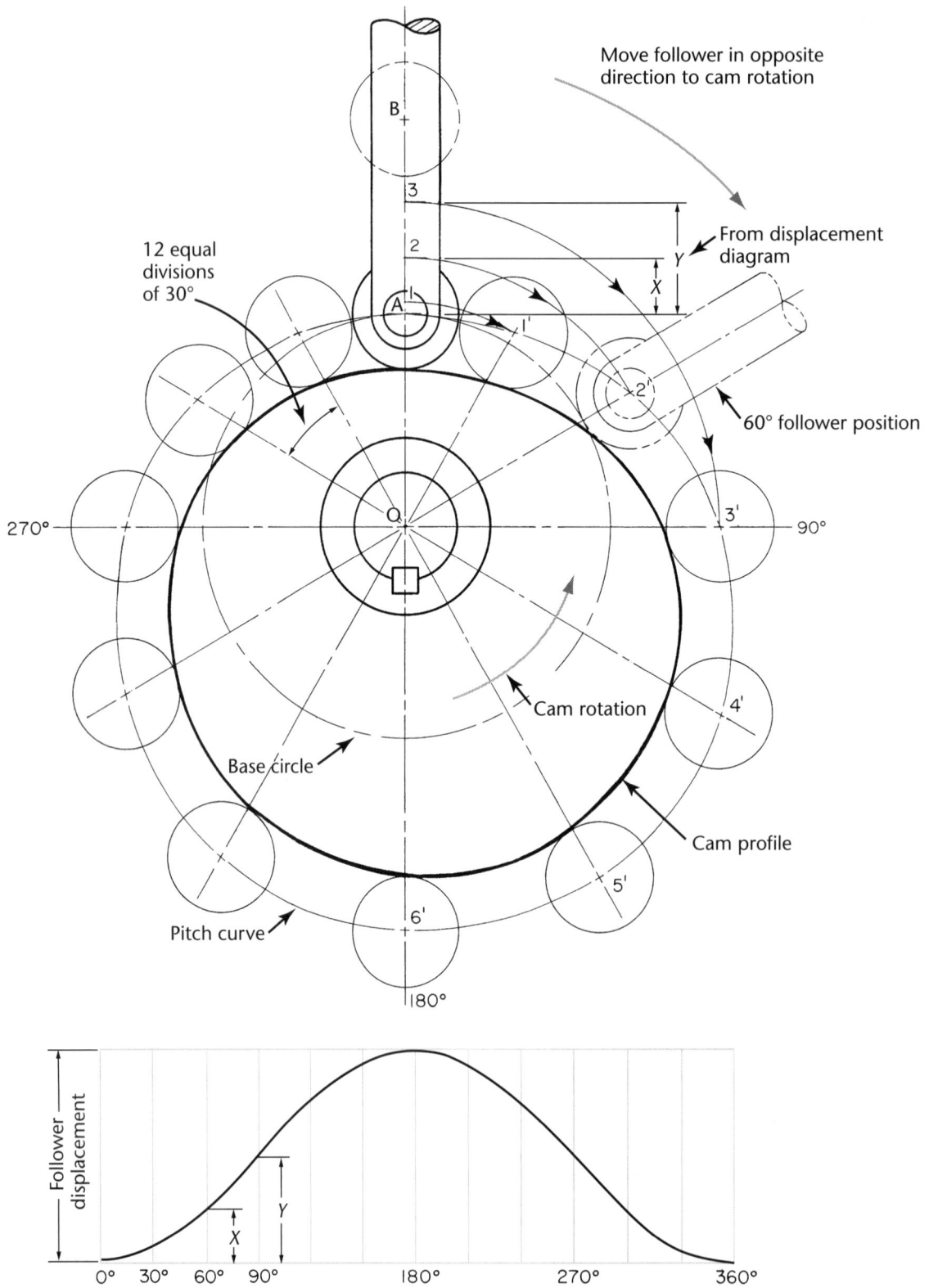

**Figure 8-62**    Disk Cam Profile Construction

## 8-43  OFFSET AND PIVOTED CAM FOLLOWERS

If the follower is offset as shown in Figure 8-63, an offset circle is drawn with center $O$ and radius equal to the amount of offset. As the cam turns, the extended centerline of the follower will always be tangent to this offset circle. The equiangular spaces are stepped off on the offset circle, and tangent lines are then drawn from each point on the offset circle, as shown.

If the roller is on a pivoted arm, as shown in Figure 8-64a, then the displacement of the roller center is along the circular arc $AB$. The height of the displacement diagram (not shown) should be made equal to the rectified length of arc $AB$. Ordinates from the diagram are then transferred to arc $AB$ to locate the roller positions $1, 2, 3, \ldots$. As the follower is revolved about the cam, pivot point $C$ moves in a circular path of radius $OC$ to the consecutive positions $C_1$, $C_2$, and so on. Length $AC$ is constant for all follower positions; hence, from each new position of point $C$, the follower arc of radius $R$ is drawn as shown at the 90° position. The roller centers $1$, $2$, $3$, are now revolved about the cam center $O$ to intersect the follower arcs at $1'$, $2'$, $3'$. After the pitch curve is completed, the cam profile is drawn tangent to the roller circles.

The construction for a flat-faced follower is shown in Figure 8-64b. The initial point of contact is at $A$, and points $1, 2, 3$, represent consecutive positions of the follower face. Then for the 90° position, point $3$ must be revolved 90°, as shown, to position $3'$, and the flat face of the follower is drawn through point $3'$ at right angles to the cam radius. When this procedure has been repeated for each position, the cam profile will be enveloped by a series of straight

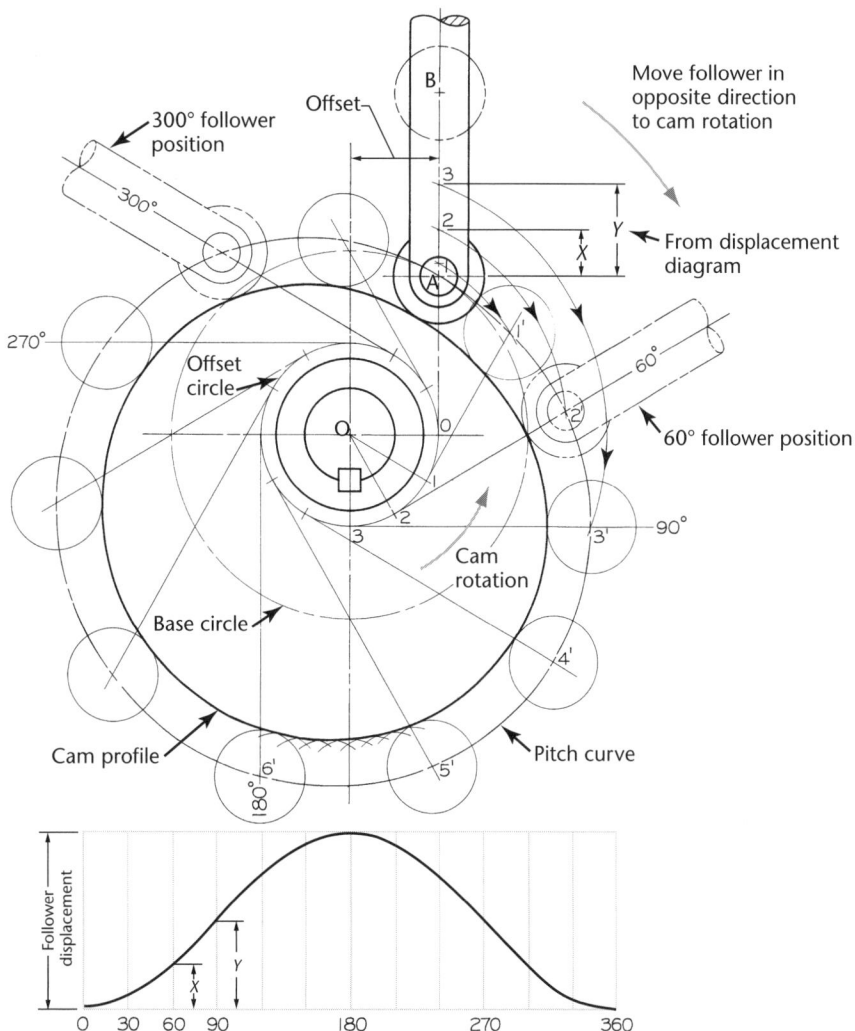

**Figure 8-63**   Disk Cam Profile with Offset Follower Construction

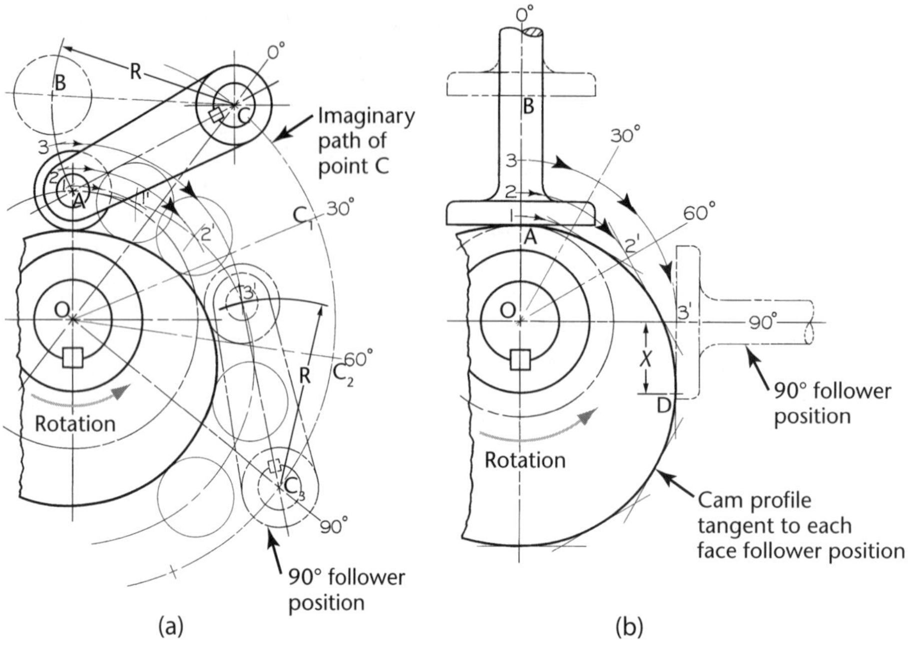

**Figure 8-64** Pivoted and Flat-Faced Followers

lines, and the cam profile is drawn inside and tangent to these lines. Note that the point of contact, initially at *A*, changes as the follower rises. At 90°, for example, contact is at *D*, a distance *X* to the right of the follower axis.

## 8-44 CYLINDRICAL CAMS

When the follower movement is in a plane parallel to the camshaft, some form of cylindrical cam must be employed. In Figure 8-65, for example, the follower rod moves vertically parallel to the cam axis as the attached roller follows the groove in the rotating cam cylinder.

A cylindrical cam of diameter *D* is required to lift the follower rod a distance *AB* with harmonic motion in 180° of cam motion and to return the rod in the remaining 180° with the same motion. The displacement diagram is drawn first and conveniently placed directly opposite the front view. The 360° length of the diagram must be made equal to $\pi D$ so that the resulting curves will be a true development of the outer surface of the cam cylinder. The pitch curve is drawn to represent the required motion, and a series of roller circles is then drawn to establish the sides of the groove tangent to these circles. This completes the development of the outer cylinder and actually provides all information needed for making the cam; hence, it is not uncommon to omit the curves in the front view.

To complete the front view, points on the curves are projected horizontally from the development. For example, at 60° in the development, the width of the groove measured parallel to the cam axis is *X*, a distance slightly greater than the actual roller diameter. This width *X* is projected to the front view directly below point *2*, the corresponding 60° position in the top view, to establish two points on the outer curves. The inner curves for the bottom of the groove can be established in the same manner, except that the groove width *X* is located below point *2'*, which lies on the inner diameter. The inner curves are only approximate because the width at the bottom of the groove is actually slightly greater than *X*, but the exact bottom width can only be determined by drawing a second development for the inner cylinder.

## 8-45 OTHER DRIVE DEVICES

When the distance between drive and driven shafts makes the use of gears impractical, devices such as belt and pulley or chain and sprocket drives are often employed. Layout and detailing standards vary somewhat from one company to another, but design procedures and specification methods may be obtained from almost any good textbook or handbook on mechanical design.

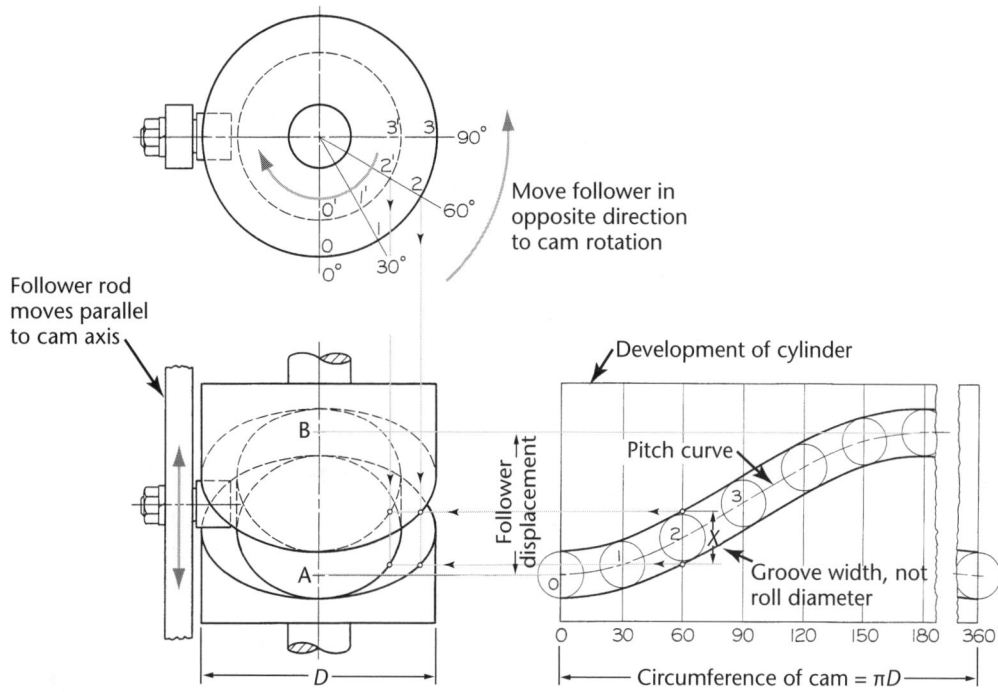

**Figure 8-65**    Cylindrical Cam

# CAD AT WORK

## Gears Using Autodesk Inventor Design Accelerator

When you know the design requirements for a gear pair, you can use a CAD package, such as Autodesk Inventor, to automatically generate the 3D models from the information. It can be very time-consuming to create drawings of gears and cams by hand, whereas using CAD frees you to spend more time on other aspects of your designs. Figure A shows the Autodesk Inventor software's Spur Gear Component Generator design tab which you can use to enter the defining details for the gear pair. Clicking the Preview option shows the values and a pictorial representation of the dimensional features for the gear as shown in Figure B.

Once you have entered the data, just select OK to generate the 3D models. They are automatically assembled so that the teeth mesh as shown in Figure C. Each gear is a separate part, for which you can list the data and generate the table of information necessary to manufacture the gear. You can also simulate motion to aid in inspecting how your design will function.

Gear drawings specify the information needed to manufacture the gear using a table. Figure D shows an example of the gear table that can be automatically generated using the software.

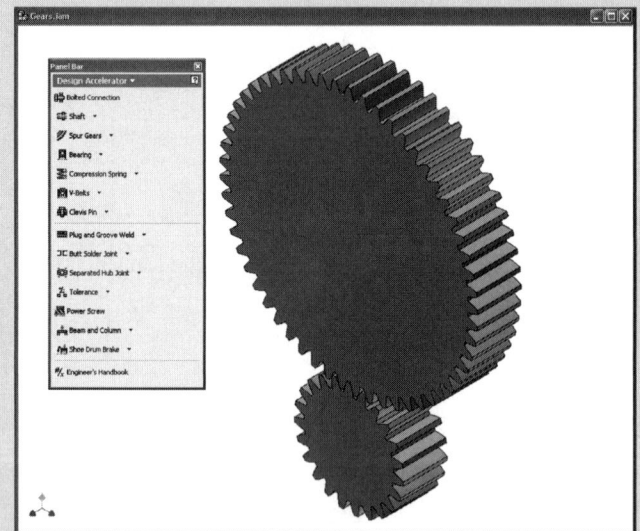

(A) Autodesk Inventor spur gear component generator design tab

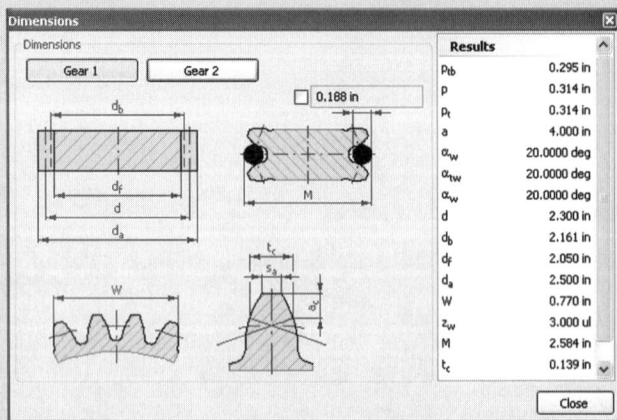

(C) 3D model

(B) Preview shows values and a representation of dimensional features

| | | Gear 1 | Gear 2 |
|---|---|---|---|
| Type of model | | Component | Component |
| Number of Teeth | N | 23 ul | 57 ul |
| Unit Correction | x | 0.0000 ul | 0.0000 ul |
| Pitch Diameter | d | 2.300 in | 5.700 in |
| Outside Diameter | $d_a$ | 2.500 in | 5.900 in |
| Root Diameter | $d_f$ | 2.050 in | 5.450 in |
| Base Circle Diameter | $d_b$ | 2.161 in | 5.356 in |
| Work Pitch Diameter | $d_w$ | 2.300 in | 5.700 in |
| Facewidth | b | 1.000 in | 1.000 in |
| Facewidth Ratio | $b_r$ | 0.4348 ul | 0.1754 ul |
| Addendum | $a^*$ | 1.0000 ul | 1.0000 ul |
| Clearance | $c^*$ | 0.2500 ul | 0.2500 ul |
| Root Fillet | $r_f^*$ | 0.3500 ul | 0.3500 ul |

(D) Table of information needed to manufacture the gear

## Portfolio

Post Screw
M3x5MM

Bronze Gear

Spring Washer

Stamped Base

String Post

Stainless Worm

Nylon Washer

Threaded Bushing

Brass Knob

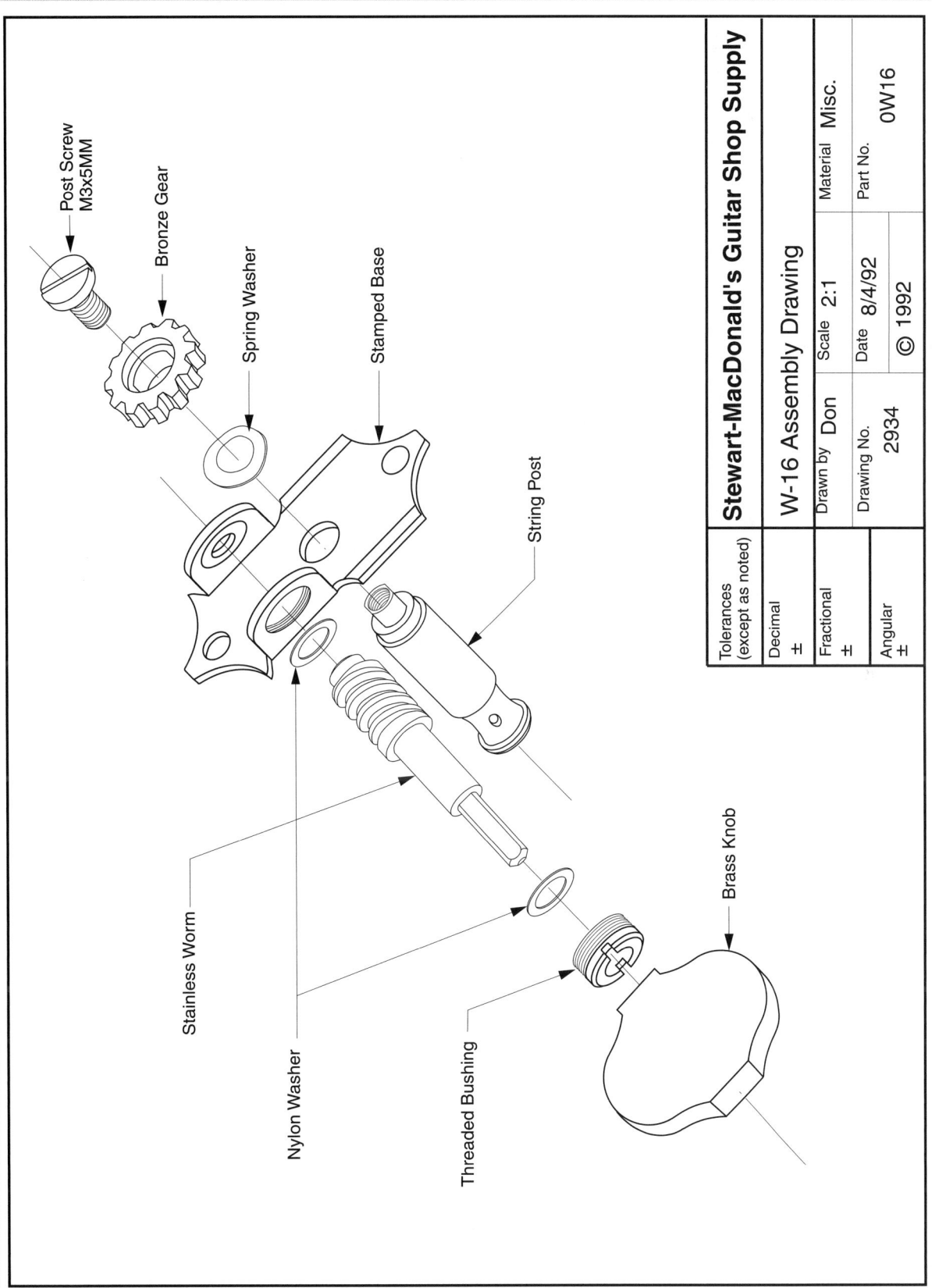

| Tolerances (except as noted) | **Stewart-MacDonald's Guitar Shop Supply** | | | |
|---|---|---|---|---|
| Decimal ± | W-16 Assembly Drawing | | Material Misc. | |
| Fractional ± | Drawn by Don | Scale 2:1 | Part No. | |
| Angular ± | Drawing No. 2934 | Date 8/4/92 | 0W16 | |
| | | © 1992 | | |

Assembly Drawing for Waverly Tuner. Courtesy of Stewart MacDonald Co.

# 8-46  EXERCISES

## Gearing and Cam Exercises

The following exercises provide practice in laying out and making working drawings of the common types of gears and cams. Where paper sizes are not given, select both scale and sheet layout. If assigned, convert the design layouts to metric dimensions. The instructor may choose to assign specific problems to be completed using CAD.

### *Gearing*
### EX 8-27

Draw the following by hand or using CAD, as assigned by your instructor.

a.  A 12-tooth IDP pinion engages a 15-tooth gear. Make a full-size drawing of a segment of each gear showing how the teeth mesh. Construct the 14½° involute teeth exactly, noting any points where the teeth appear to interfere. Label gear ratio $m_G$ on the drawing. Include dimensions, notes, and cutting data as assigned by your instructor.

b.  Follow the instructions for Part a but use 13 or 14 teeth, or any number of teeth as assigned by your instructor.

c.  Follow the instructions for Part a but use 20° stub teeth instead of involute teeth.

d.  Follow the instructions for Part a but use a rack in place of the 12-tooth pinion.

### EX 8-28

Make a display drawing of the pinion shown above. Show two views, drawing the teeth by the odontograph method shown in Figure 8-47. Draw double size. Omit dimensions.

### EX 8-29

A spur gear has 60 teeth of 5 diametral pitch. The face width is 1.50″. The shaft is 1.19″ in diameter. Make the hub 2.00″ long and 2.25″ in diameter. Calculate accurately all dimensions, and make a working drawing of the gear. Show six spokes, each 1.12″ wide at the hub, tapering to .75″ wide at the rim and .50″ thick. Use your own judgment for any dimensions not given. Draw half size.

### EX 8-30

Draw the following.

a.  Make a pictorial display of the intermediate pinion shown in Part 6 above. Show the gear as an oblique half section, similar to Ex 8-28. Draw four-times size, reducing the 30° receding lines by one-half, as in cabinet projection. Draw the teeth by the odontograph method.

b.  Make a working drawing of the intermediate pinion shown in Part 6 of the figure. Check the gear dimensions by calculation.

c.  Make a working drawing of the intermediate gear shown in Part 5 of the figure. Check the gear dimensions by calculation.

d.  Follow instructions for Part a, but use the pinion shown in Exercise EX8-28.

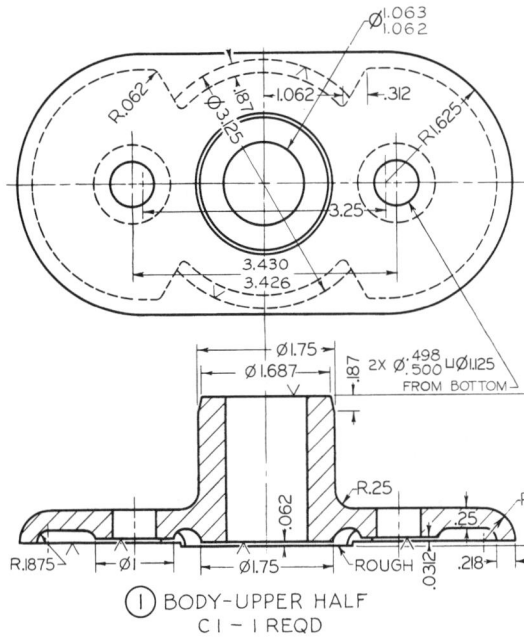

Ø1.063
1.062

R.062

.187

Ø3.125

—1.062—

—.312—

R1.625

3.25

3.430
3.426

Ø1.75
Ø1.687

.187

2X Ø .498
.500 ⊔Ø1.125
FROM BOTTOM

1.75

R.062

.062

R.25

.25

R.437

R.1875

Ø1

Ø1.75

ROUGH

.0312

.218

① BODY-UPPER HALF
C I - I REQD

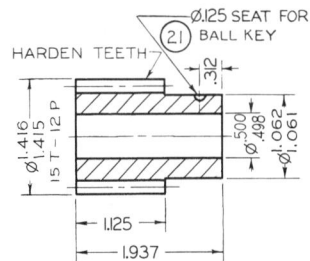

HARDEN TEETH

Ø.125 SEAT FOR
㉑ BALL KEY

.312

Ø1.416
1.415

15 T - 12 P

Ø.500
.498

Ø1.062
1.061

1.125

1.937

⑥ INTERMEDIATE PINION
AJAX STL - 2 REQD

Ø1.375
Ø.937

.015

⑬ SPINDLE WASHER
C R S - I REQD

⑰ THRUST BEARING
1.344 OD x .625 ID x .562
STK - I REQD

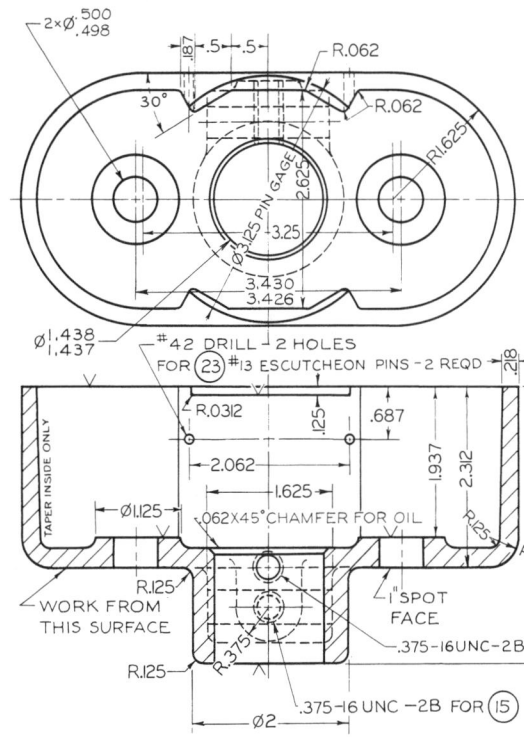

2 X Ø .500
.498

R.062

.187

.5

.5

30°

R.062

R1.625

Ø.125 PIN GAGE

2.625

3.25

3.430
3.426

Ø1.438
1.437

#42 DRILL - 2 HOLES
FOR ㉓ #13 ESCUTCHEON PINS - 2 REQD

TAPER INSIDE ONLY

R.0312

.125

.687

2.062

1.625

.062X45° CHAMFER FOR OIL

Ø1.125

R.125

R.375

WORK FROM
THIS SURFACE

1" SPOT
FACE

.375-16UNC-2B

R.125

1.937

2.312

3.5

R.375

.375-16 UNC -2B FOR ⑮ STOP ROD
SET SCREW

Ø2

② BODY-LOWER HALF
PAT. # 336 - C I - I REQD

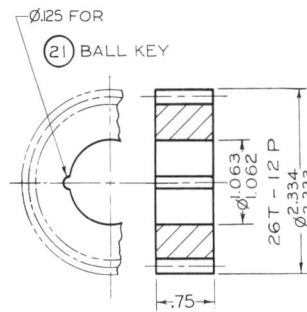

Ø.125 FOR
㉑ BALL KEY

Ø1.063
1.062

26 T - 12 P

Ø2.334
2.333

.75

⑤ INTERMEDIATE GEAR
PAT. #337- C I -2 REQD

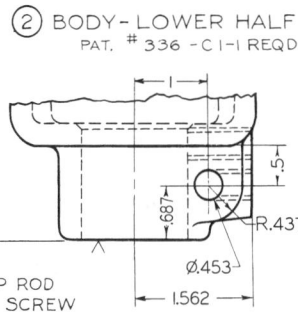

1

.5

.687

R.437

Ø.453

1.562

**EX 8-31**

Make a pictorial display of the spur gear shown above. Show the gear as an oblique half section, similar to Exercise 8-28. Draw four-times size, reducing the 30° receding lines by one-half, as in cabinet projection. Draw the teeth by the odontograph method.

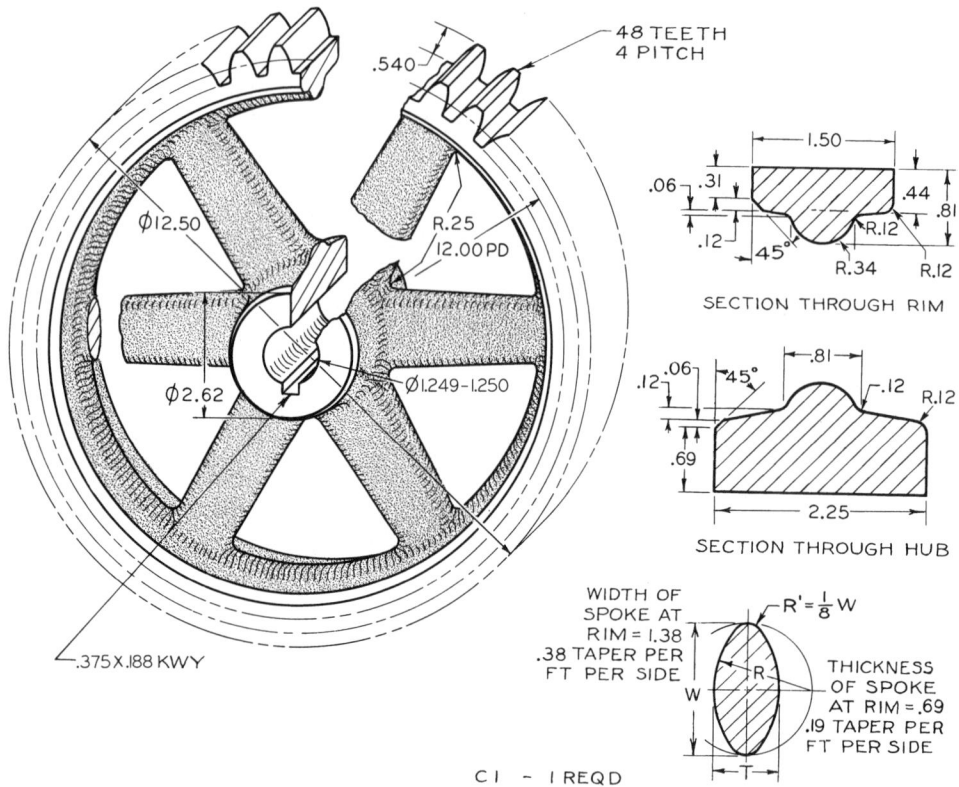

48 TEETH
4 PITCH

.540

Ø12.50

R.25
12.00 PD

Ø2.62

Ø1.249-1.250

.375 X .188 KWY

1.50
.06  .31
.44
.81
.12
45°
R.12
R.34  R.12
SECTION THROUGH RIM

.06  45°  .81
.12  .12 R.12
.69
2.25
SECTION THROUGH HUB

WIDTH OF
SPOKE AT
RIM = 1.38
.38 TAPER PER
FT PER SIDE
W
R' = ⅛ W
R
THICKNESS
OF SPOKE
AT RIM = .69
.19 TAPER PER
FT PER SIDE
T

CI - I REQD

**EX 8-32**

Draw the following by hand or using CAD, as assigned by your instructor.

a.  A pair of bevel gears has teeth of 4 diametral pitch. The pinion has 13 teeth, the gear 25 teeth. The face width is 1.12″. The pinion shaft is .94″ in diameter, and the gear shaft is 1.19″ in diameter. Calculate accurately all dimensions, and make a working drawing showing the gears engaged as in Figure 8-59. Make the hub diameters approximately twice the shaft diameters. The backing for the pinion must be .62″, and for the gear 1.25″. Use your own judgment for any dimensions not given.

b.  Make a working drawing of the pinion.
c.  Make a working drawing of the gear.
d.  Show two views of the pinion in Part b. Follow the general instructions in that problem.
e.  Follow instructions in Part a but use 5 diametral pitch and 30 and 15 teeth. The face is 1.00″. Shafts: pinion, 1.00″ diameter; gear, 1.50″ diameter. Backing: pinion .50″ gear, 1.00″.
f.  Follow instructions in Part a but use 4 diametral pitch, both gears 20 teeth. Select the correct face width. Shafts: 1.38″ diameter. Backing: 0.75″.

**EX 8-33**

Make a half-scale assembly drawing of the complete unit (a suggested layout with full size dimension for view placement is shown). The following full-size dimensions are sufficient to establish the position and general outline of the parts; the minor detail dimensions, web and rib shapes, and fillets and rounds should be designed with the help of the photograph in Part a of the figure

The gears are identical in size and are similar in shape to the one in Figure 8-48. There are 24 teeth of 3 diametral pitch in each gear. Face width, 1.50″. Shafts, 2.00″ diameter, extending 7.00″ beyond left bearing, 4.00″ beyond rear (break shafts as shown). Hub diameters, 2.75″. Backing, 1.00″. Hub lengths, 3.25″. The front gear is held by a square gib key and a .38″ set screw. The collar on front shaft next to right bearing is .75″ thick, 2.25″ outside diameter, with a .38″ set screw.

On the main casting, the split bearings are 2.75″ diameter, 3.00″ long, and 10.00″ apart. Each bearing cap is held by two 1/2″ bolts, 3.50″ apart, center to center. Oil holes have a layout for an assembly drawing of a similer unit. pipe tap for plug or grease cup. Shaft centerlines are 6.50″ above bottom surface, 8.00″ from the rear surface of casting. The main casting is 11.50″ high, and its base is 16.00″ long, 8.75″ wide, and .75″ thick. All webs, ribs, and walls are uniformly .50″ thick. In the base are eight holes (not shown in the photograph) for 1/2″ bolts, two outside each end, and two inside each end.

First block in the gears in each view, and then block in the principal main casting dimensions. Fill in details only after principal dimensions are clearly established.

(a) Counter shaft end used on a trough conveyor

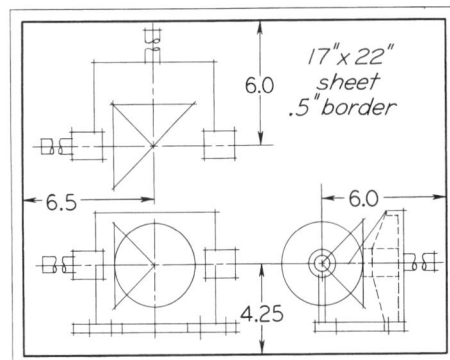

(b) Layout for an assembly drawing of a similar unit.

**EX 8-34**

a. The worm and worm gear shown in Figure 8-53 have a circular pitch of .62″, and the gear has 32 teeth of $14\frac{1}{2}°$ involute form. The worm is double threaded. Make an assembly drawing similar to Figure 8-53. Draw the teeth on the gear by the odontograph method of Figure 8-47. Calculate dimensions accurately, and use AGMA proportions. Shafts: worm, 1.12″ diameter; gear, 1.62″ diameter.

b. Make a working drawing of the worm in Part a.

c. Make a working drawing of the gear in Part a.

d. Make a working drawing of the worm in Part a, but make the worm single threaded.

**EX 8-35**

a. A single-thread worm has a lead of .75″. The worm gear has 28 teeth of standard form. Make a working drawing of the worm. The shaft is 1.25″ in diameter.

b. Make a working drawing of the gear in Part a. The shaft is 1.50″ in diameter.

**Cams**

**EX 8-36**

Draw the following for the 90° roller follower at right.

a. For the setup in the figure at right, draw the displacement diagram and determine the cam profile that will give the radial roller follower this motion: up 1.50″ in 120°, dwell 60°, down in 90°, dwell 90°. Motions are to be unmodified straight line and of uniform velocity. The roller is .75″ in diameter, and the base circle is 3.00″ in diameter. Note that the follower has zero offset. The cam rotates clockwise.

b. Instructions are the same as for Part a, except that the straight line motions are to be modified by arcs whose radii are equal to one-half the rise of the follower.

c. Instructions are the same as for Part a, except that the upward motion is to be harmonic and the downward motion parabolic.

d. Instructions are the same as for Part a, except that the follower is offset 1.00″ to the left of the cam centerline. Full size dimensions for suggested view placement are given in the example layout.

### EX 8-37

For the 90° flat-faced follower in the figure at right, draw the displacement diagram, and determine the cam profile that will give the flat-faced follower this motion: dwell 30°, up 1.50″ on a parabolic curve in 180°, dwell 30°, down with harmonic motion in 120°. The base circle is 3.00″ in diameter. After completing the cam profile, determine the necessary width of face of the follower by finding the position of the follower where the point of contact with the cam is farthest from the follower axis. The cam rotates counterclockwise.

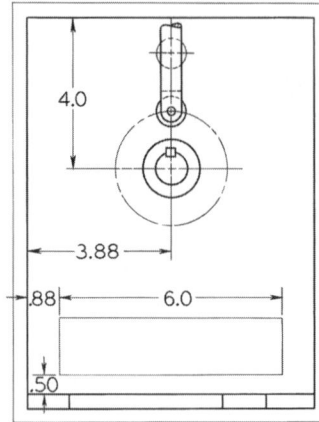

### EX 8-38

Draw the following for the 45° roller following at right.

a. For the setup in the figure at right, draw the displacement diagram, and determine the cam profile that will swing the pivoted follower through an angle of 30° with the same motion as prescribed in Exercise 8-52. The radius of the follower arm is 2.75″, and in its lowest position the center of the roller is directly over the center of the cam. The base circle is 2.50″ in diameter, and the roller is .75″ in diameter. The cam rotates counterclockwise.

b. Follow instructions for Part a, except that the motion is: up 1.50″ in 120°, dwell 60°, down in 90°, dwell 90°.

c. Follow instructions for Part a, except that the motion is to be: dwell 30°, up 1.50″ on a parabolic curve in 180°, dwell 30°, down with harmonic motion in 120°.

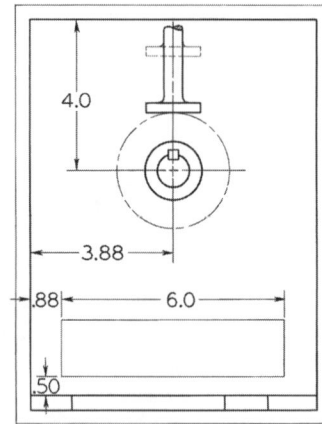

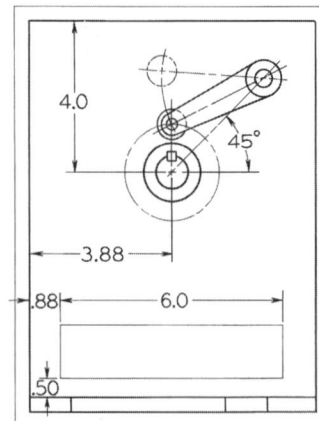

### EX 8-39

Draw the following by hand or using CAD, as given by your instructor.

a. Using an arrangement like Figure 8-65, construct a half development and complete the front view for the following cylindrical cam. Cam, 2.50″ diameter, 2.50″ high. Roller, .50″ diameter; cam groove, .38″ deep. Camshaft, .62″ diameter; follower rod, .62″ wide, .31″ thick. Motion: up 1.50″ with harmonic motion in 180°, down with the same motion in remaining 180°. Cam rotates counterclockwise. Assume lowest position of the follower at the front center of cam.

b. Follow instructions for Part a, but construct a full development for the following motion: up 1½″ with parabolic motion in 120°, dwell 90°, down with harmonic motion in 150°.

# Axonometric and Oblique Projections

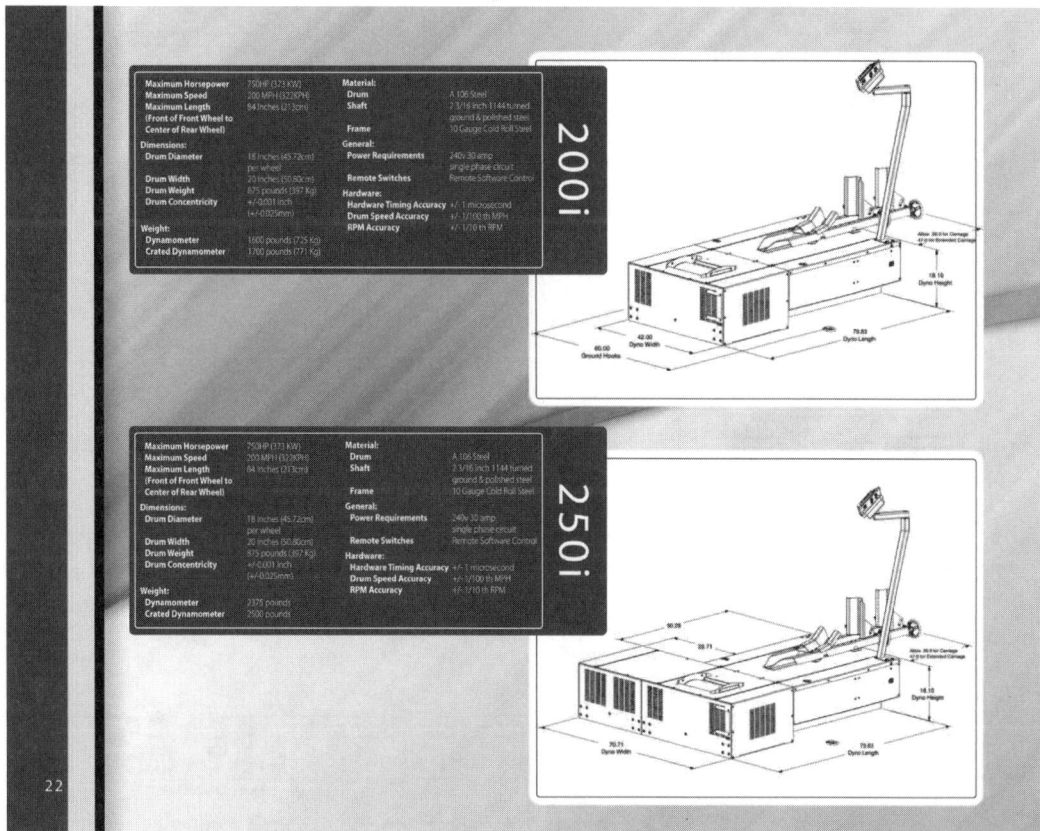

A Portion of a Sales Brochure Showing General Dimensions in Pictorial Drawings. *Courtesy of Dynojet Research, Inc.*

## 9-1 AXONOMETRIC PROJECTION

Multiview drawing makes it possible to accurately represent the complex forms of a design by showing a series of views and sections, but reading and interpreting this type of representation requires a thorough understanding of the principles of multiview projection. Although multiview drawings are commonly used to communicate information to a technical audience, they do not show length, width, and height in a single view and are hard for a layperson to visualize.

It is often necessary to communicate designs to people who do not have the technical training to interpret multiview projections. Axonometric projections show all three principal dimensions using a single drawing view, approximately as they appear to an observer. These projections are often called pictorial drawings because they look more like a picture than multiview drawings do. Since a pictorial drawing shows only the appearance of an object, it is not usually suitable for completely describing and dimensioning complex or detailed forms.

Pictorial drawings are also useful in developing design concepts. They can help you picture the relationships between design elements and quickly generate several solutions to a design problem.

## 9-2 UNDERSTANDING AXONOMETRIC DRAWINGS

Various types of pictorial drawings are used extensively in catalogs, sales literature, and technical work. They are often used in patent drawings; piping diagrams; machine, structural, architectural design, and furniture design; and for ideation sketching. The sketches for a wooden shelf in Figure 9-1 are examples of axonometric, orthographic, and perspective sketches.

The most common axonometric projection is **isometric**, which means "equal measure." When a cube is drawn in

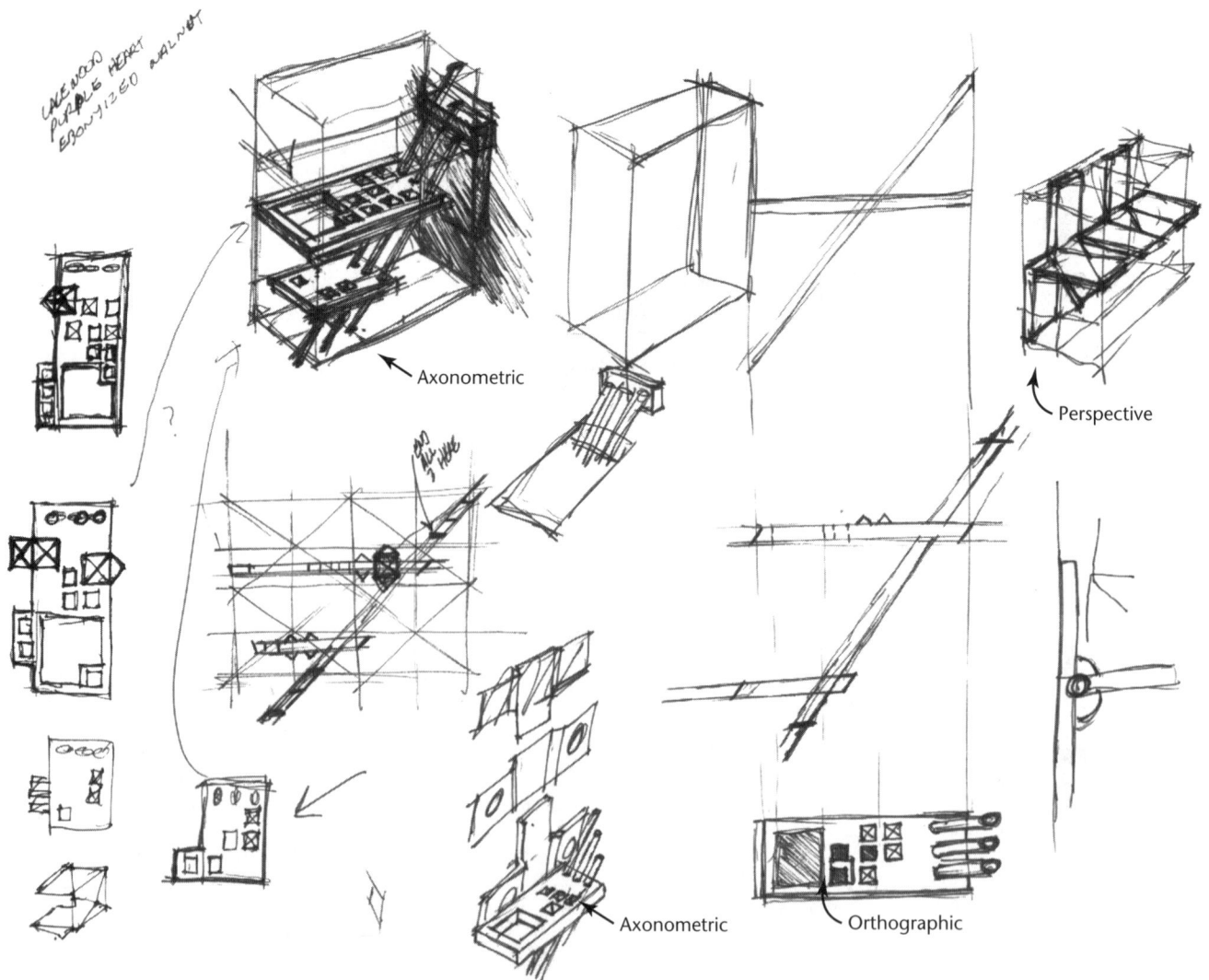

**Figure 9-1** Sketches for a Wooden Shelf using Axonometric, Orthographic, and Perspective Drawing Techniques—The Axonometric Projections in this Sketch are Drawn in Isometric. *Courtesy of Douglas Wintin.*

isometric, the axes are equally spaced (120° apart). Though not as realistic as perspective drawings, isometric drawings are much easier to draw. CAD software often displays the results of 3D models on the screen as isometric projections. Some CAD software allows you to choose between isometric, dimetric, trimetric, or perspective representation of your 3D models on the 2D computer screen. In sketching, dimetric and trimetric sometimes produce a better view than isometric but take longer to draw and are therefore used less frequently.

## Projection Methods Reviewed

The four principal types of projection are illustrated in Figure 9-2. All except the regular multiview projection (Figure 9-2a) are **pictorial** types since they show several sides of the object in a single view. In both **multiview projection** and **axonometric projection** the visual rays are parallel to each other and perpendicular to the plane of projection. Both are types of **orthographic projections** (Figure 9-2b).

In **oblique projection** (Figure 9-2c), the visual rays are parallel to each other but at an angle other than 90° to the plane of projection.

In perspective (Figure 9-2d), the visual rays extend from the observer's eye, or station point (SP), to all points of the object to form a "cone of rays" so that the portions of the object that are further away from the observer appear smaller than the closer portions of the object.

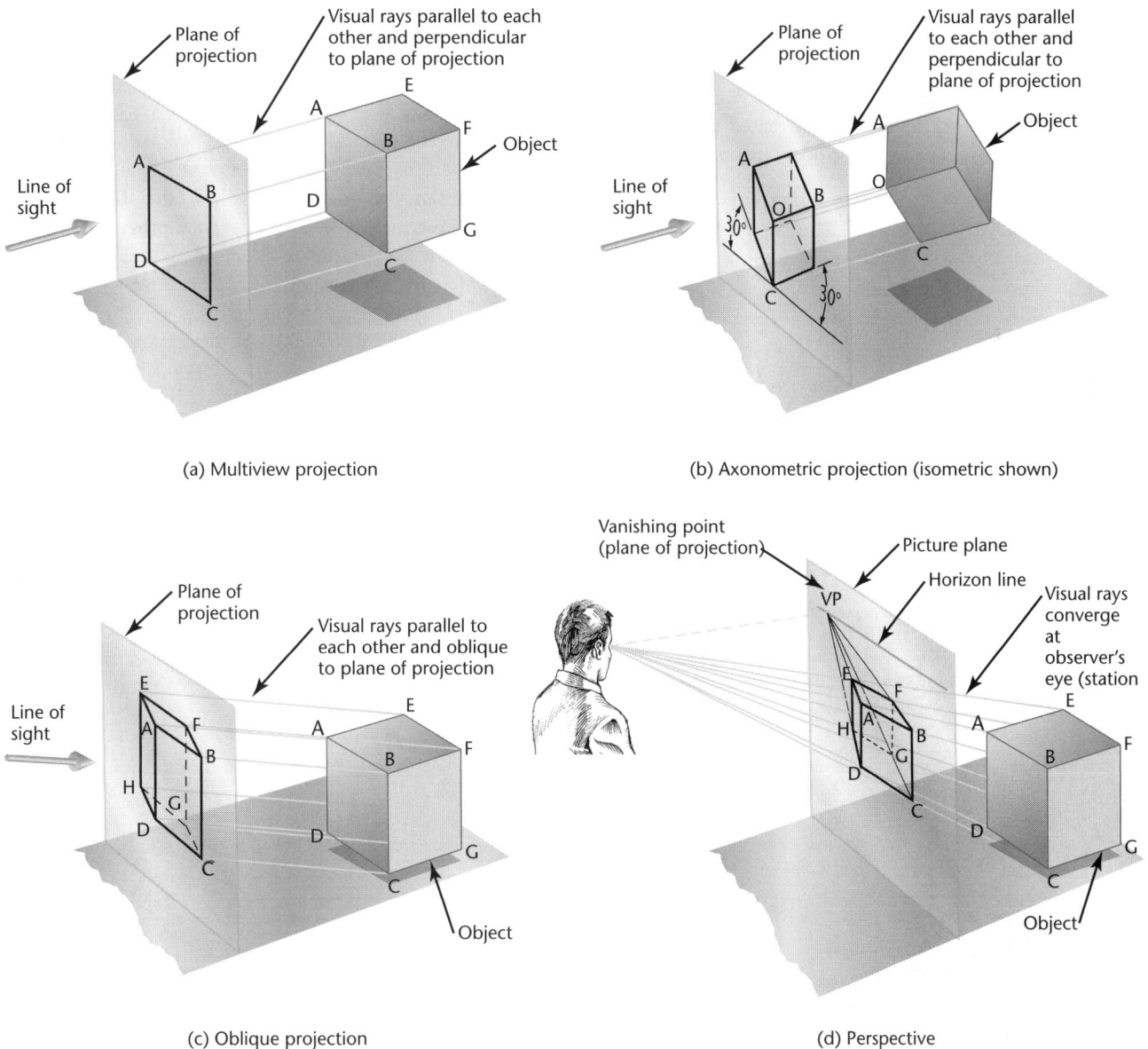

(a) Multiview projection

(b) Axonometric projection (isometric shown)

(c) Oblique projection

(d) Perspective

**Figure 9-2**    Four Types of Projection

## Types of Axonometric Projection

The feature that distinguishes axonometric projection from multiview projection is the inclined position of the object with respect to the planes of projection. When a surface or edge of the object is not parallel to the plane of projection, it appears foreshortened. When an angle is not parallel to the plane of projection, it appears either smaller or larger than the true angle.

To create an axonometric view, the object is tipped to the planes of projection so that all of the principal faces show in a single view. This produces a pictorial drawing that is easy to visualize. But, since the principal edges and surfaces of the object are inclined to the plane of projection, the lengths of the lines are foreshortened. The angles between surfaces and edges appear either larger or smaller than the true angle. There are an infinite variety of ways that the object may be oriented with respect to the plane of projection.

The degree of **foreshortening** of any line depends on its angle to the plane of projection. The greater the angle, the greater the foreshortening. If the degree of foreshortening is determined for each of the three edges of the cube that meet at one corner, scales can be easily constructed for measuring along these edges or any other edges parallel to them (Figure 9-3).

Use the three edges of the cube that meet at the corner nearest your view as the axonometric axes. In Figure 9-4, the axonometric axes, or simply the axes, are OA, OB, and OC. Figure 9-4 shows three axonometric projections.

**Isometric projection** (Figure 9-4a) has equal foreshortening along each of the three axis directions.

**Dimetric projection** (Figure 9-4b) has equal foreshortening along two axis directions and a different amount of foreshortening along the third axis. This is because it is not tipped an equal amount to all of the principal planes of projection.

**Trimetric projection** (Figure 9-4c) has different foreshortening along all three axis directions. This view is produced by an object that is not equally tipped to any of the planes of projection.

## Axonometric Projections and 3D Models

When you create a 3D CAD model, the object is stored so that vertices, surfaces, and solids are all defined relative to a 3D coordinate system. You can rotate your view of the object to produce a view from any direction. However, your computer screen is a flat surface, like a sheet of paper. The CAD software uses similar projection to produce the view transformations, creating the 2D view of the object on your computer screen. Most 3D CAD software provides a variety of preset isometric viewing directions to make it easy for you to manipulate the view. Some CAD software also allows for easy perspective viewing on screen.

After rotating the object you may want to return to a preset typical axonometric view like one of the examples shown in Figure 9-5. Figure 9-6 shows a 3D CAD model.

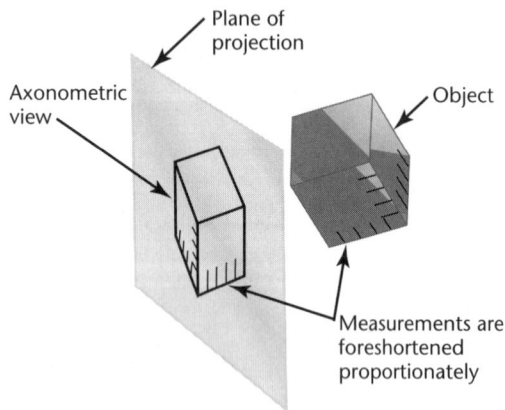

**Figure 9-3** Measurements are Foreshortened Proportionately based on Amount of Incline

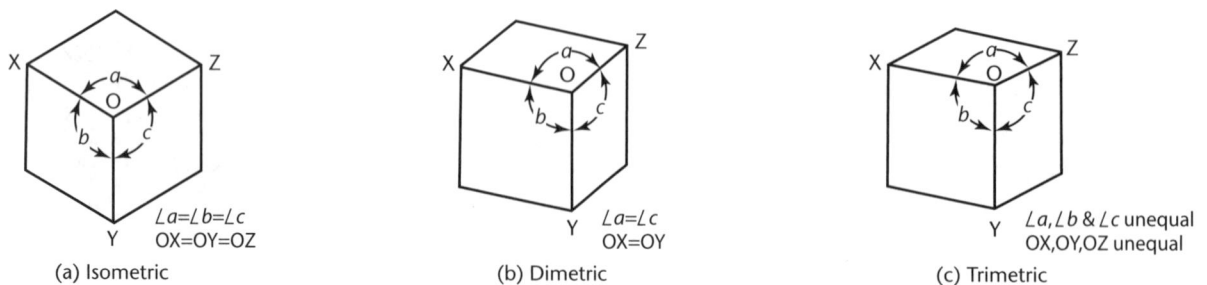

(a) Isometric    (b) Dimetric    (c) Trimetric

**Figure 9-4** Axonometric Projections

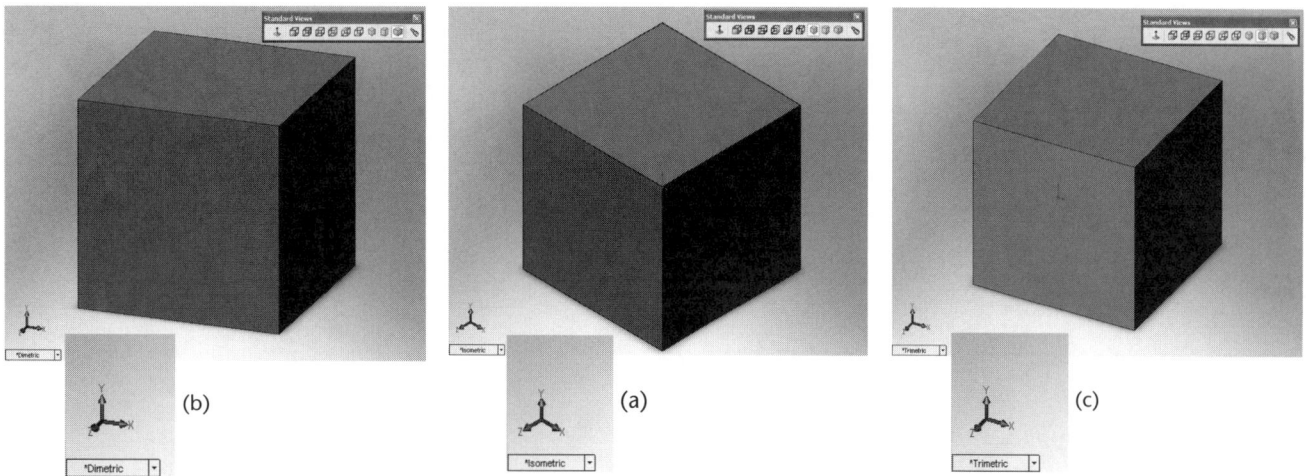

**Figure 9-5**    (a) Isometric View of a 1 inch Cube Shown in SolidWorks, (b) Dimetric View, (c) Trimetric View. *Courtesy of Solidworks Corporation.*

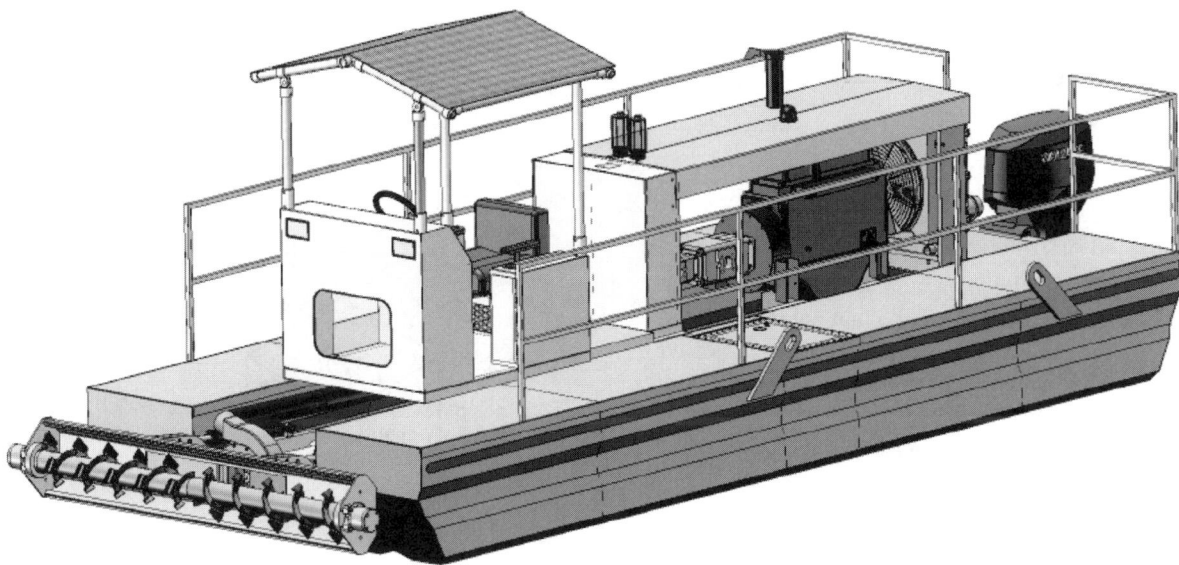

**Figure 9-6**    Complicated 3D CAD Models such as this Dredge from SRS Crisafulli Inc., are Often Viewed on Screen Using Isometric Display—Notice the Coordinate System Display in the Lower Left. *Courtesy of SRS Crisafulli, Inc.*

# 9-3  ISOMETRIC PROJECTION

In an isometric projection, all angles between the axonometric axes are equal. To produce an isometric projection (isometric means "equal measure"), you orient the object so that its principal edges (or axes) make equal angles with the plane of projection and are therefore foreshortened equally. Oriented this way, the edges of a cube would be projected so that they all measure the same and make equal angles with each other (of 120°) as shown in Figure 9-7.

## Creating Isometric Projections

Figure 9-8a shows a multiview drawing of a cube. Figure 9-8b shows the cube revolved 45° about an imaginary vertical axis. Now an auxiliary view in the direction of the arrow shows the diagonal of the cube as a point. This creates a true isometric projection. You can continue revolving the cube until the three edges *OX*, *OY*, and *OZ* make equal angles with the front plane of projection and show foreshortened equally. Again, a diagonal through the cube, in this case *OT*, appears as a point in the isometric view and the view

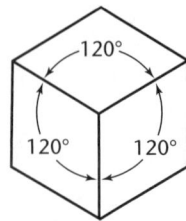

**Figure 9-7** Isometric Projection

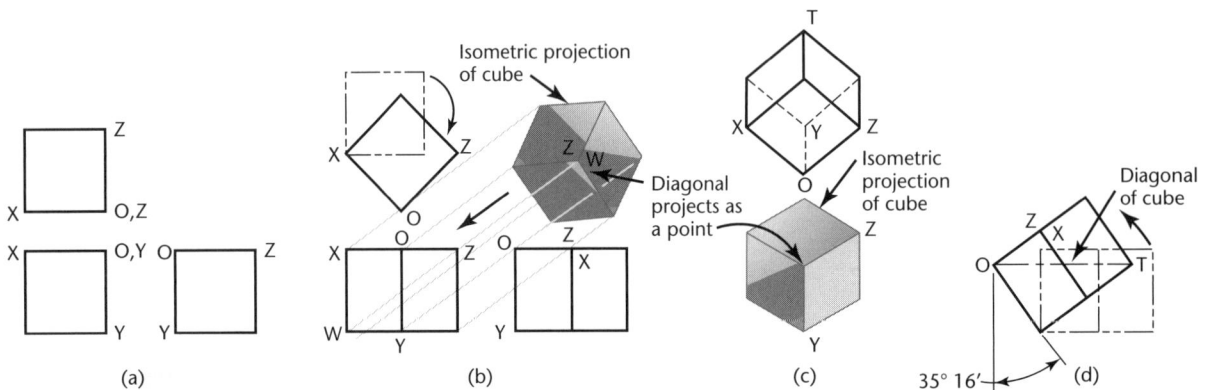

**Figure 9-8** Isometric Projection as a Second Auxiliary View

produced is a true isometric projection. In this projection the 12 edges of the cube make angles of about 35°16″ with the front plane of projection. The lengths of their projected edges are equal to the actual edge length multiplied by $\sqrt{\frac{2}{3}}$ or about 0.816. Thus the projected lengths are about 80 percent of the true lengths or about three-fourths of the true lengths.

## 9-4 ISOMETRIC AXES

The projections of the edges of a cube make angles of 120° with each other. You can use these as the **isometric axes** from which to make measurements. Any line parallel to one of these is called an isometric line. The angles in the isometric projection of the cube are either 120° or 60°, and all are projections of 90° angles. In an isometric projection of a cube, the faces of the cube, and any planes parallel to them, are called isometric planes. See Figure 9-9.

## 9-5 NONISOMETRIC LINES

Lines of an isometric drawing that are not parallel to the isometric axes are called **nonisometric lines** (Figure 9-10). Only lines of an object that are drawn parallel to the isometric axes are equally foreshortened. Nonisometric lines are drawn

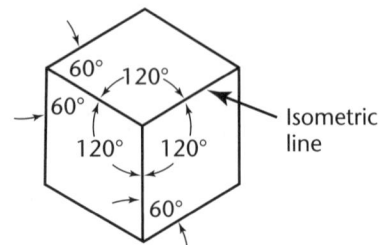

**Figure 9-9** Isometric Axes  a

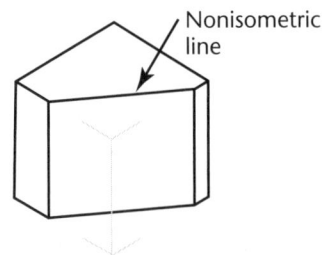

**Figure 9-10** Nonisometric Edges

at other angles and are not equally foreshortened. Therefore the lengths of features along nonisometric lines cannot be measured directly with the scale.

# 9-6 ISOMETRIC SCALES

An isometric scale can be used to draw correct isometric projections. All distances in this scale are $\sqrt{\frac{2}{3}} \times$ true size, or approximately 80 percent of true size. Figure 9-11a shows an isometric scale. More commonly, an isometric sketch or drawing is created using a standard scale, as in Figure 9-11b, disregarding the foreshortening that the tipped surfaces would produce in a true projection.

# 9-7 ISOMETRIC DRAWINGS

When you make a drawing using foreshortened measurements, or when the object is actually projected on a plane of projection, it is called an **isometric projection** (Figure 9-11a). When you make a drawing using the full length measurements of the actual object, it is an **isometric sketch** or **isometric drawing** (Figure 9-11b) to indicate that it lacks foreshortening.

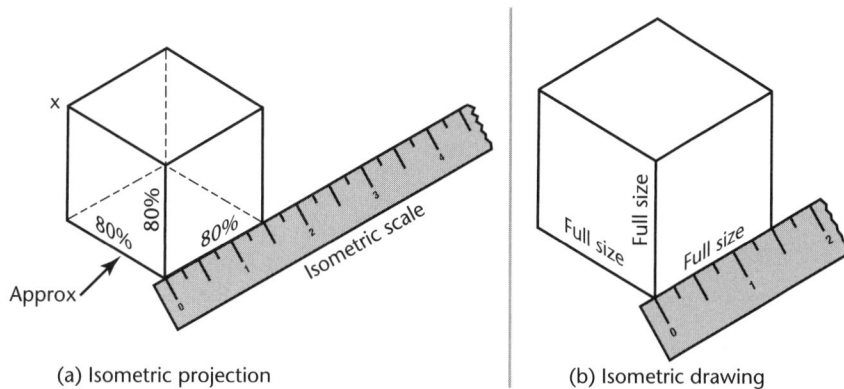

(a) Isometric projection    (b) Isometric drawing

**Figure 9-11**    Isometric and Ordinary Scales

## TIP

### Making an Isometric Scale

You can make an isometric scale from a strip of paper or cardboard as shown here by placing an ordinary scale at 45° to a horizontal line and the paper scale at 30° to the horizontal line. To mark the increments on the isometric scale, draw straight lines (perpendicular to the horizontal line) from the division lines on the ordinary scale. Alternatively, you can approximate an isometric scale. Scaled measurements of 9" = 1'-0, or three-quarter-size scale (or metric equivalent) can be used as an approximation.

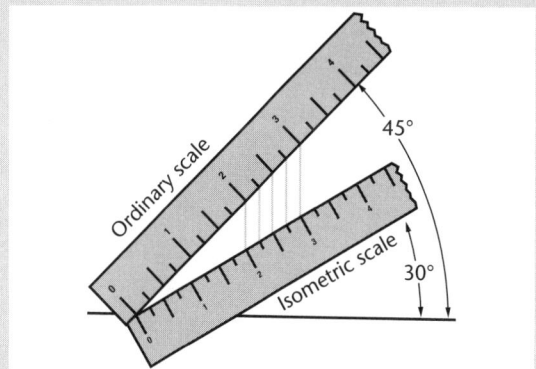

The isometric drawing is about 25 percent larger than the isometric projection, but the pictorial value is obviously the same in both. Since isometric sketches are quicker, as you can use the actual measurements, they are much more commonly drawn.

### Positions of the Isometric Axes

The first step in making an isometric drawing is to decide along which axis direction to show the height, width, and depth, respectively. Figure 9-12 shows four different orientations that you might start with to create an isometric

drawing of the block shown. Each is an isometric drawing of the same block, but with a different corner facing your view. These are only a few of many possible orientations.

You may orient the axes in any desired position, but the angle between them must remain 120°. In selecting an orientation for the axes, choose the position from which the object is usually viewed, or determine the position that best describes the shape of the object or better yet, both.

If the object is a long part, it will look best with the long axis oriented horizontally.

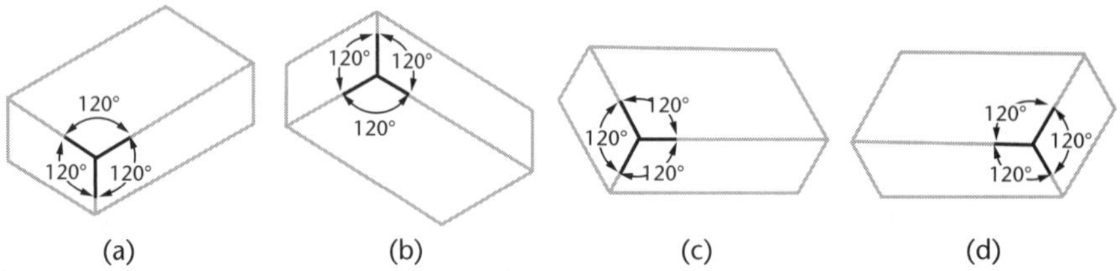

**Figure 9-12** Positions of Isometric Axes

## 9-8 MAKING AN ISOMETRIC DRAWING

Rectangular objects are easy to draw using box construction, which consists of imagining the object enclosed in a rectangular box whose sides coincide with the main faces of the object. For example, imagine the object shown in the two views in Figure 9-13 enclosed in a construction box, then locate the irregular features along the edges of the box as shown.

Figure 9-14 shows how to construct an isometric drawing of an object composed of all normal surfaces. Notice that all measurements are made parallel to the main edges of the enclosing box—that is, parallel to the isometric axes. No measurement along a nonisometric line can be measured directly with the scale as these lines are not foreshortened equally to the normal lines. Start at any one of the corners of the bounding box and draw along the isometric axis directions.

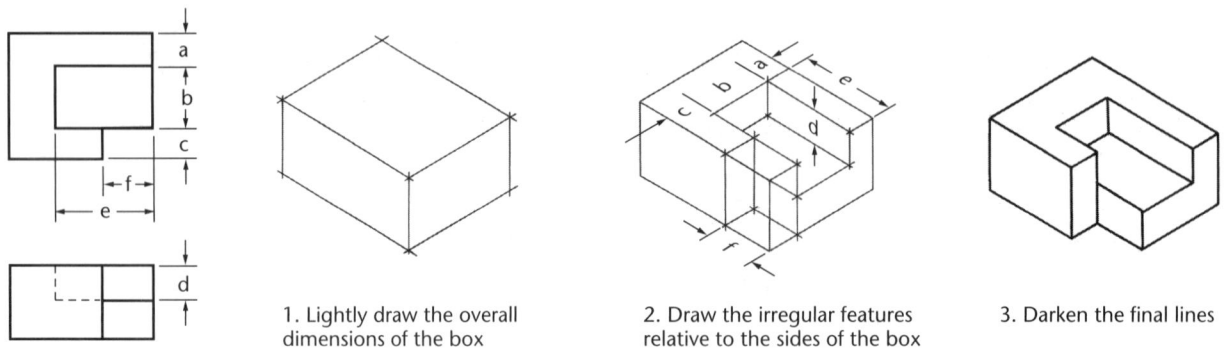

1. Lightly draw the overall dimensions of the box

2. Draw the irregular features relative to the sides of the box

3. Darken the final lines

**Figure 9-13** Box Construction

## 9-9 OFFSET LOCATION MEASUREMENTS

Use the method shown in Figure 9-15a and b to locate points with respect to each another. First draw the main enclosing block, then draw the offset lines (CA and BA) in the full size in the isometric drawing to located corner A of the small block or rectangular recess. These measurements are called **offset measurements**. Since they are parallel to edges of the main block in the multiview drawings, they will be parallel to the same edges in the isometric drawings (using the rule of parallelism).

Orthographic views are given

All measurements must be parallel to main edges of box

1. Select axes along which to block in height, weight and depth dimensions

2. Locate main areas to be removed from the overall block lightly sketch along isometric axes to define portion to be removed

3. Lightly block in any remaining major portions to be removed through the whole block

4. Lightly block in features to be removed from the remaining shape along isometric axes

5. Darken final lines

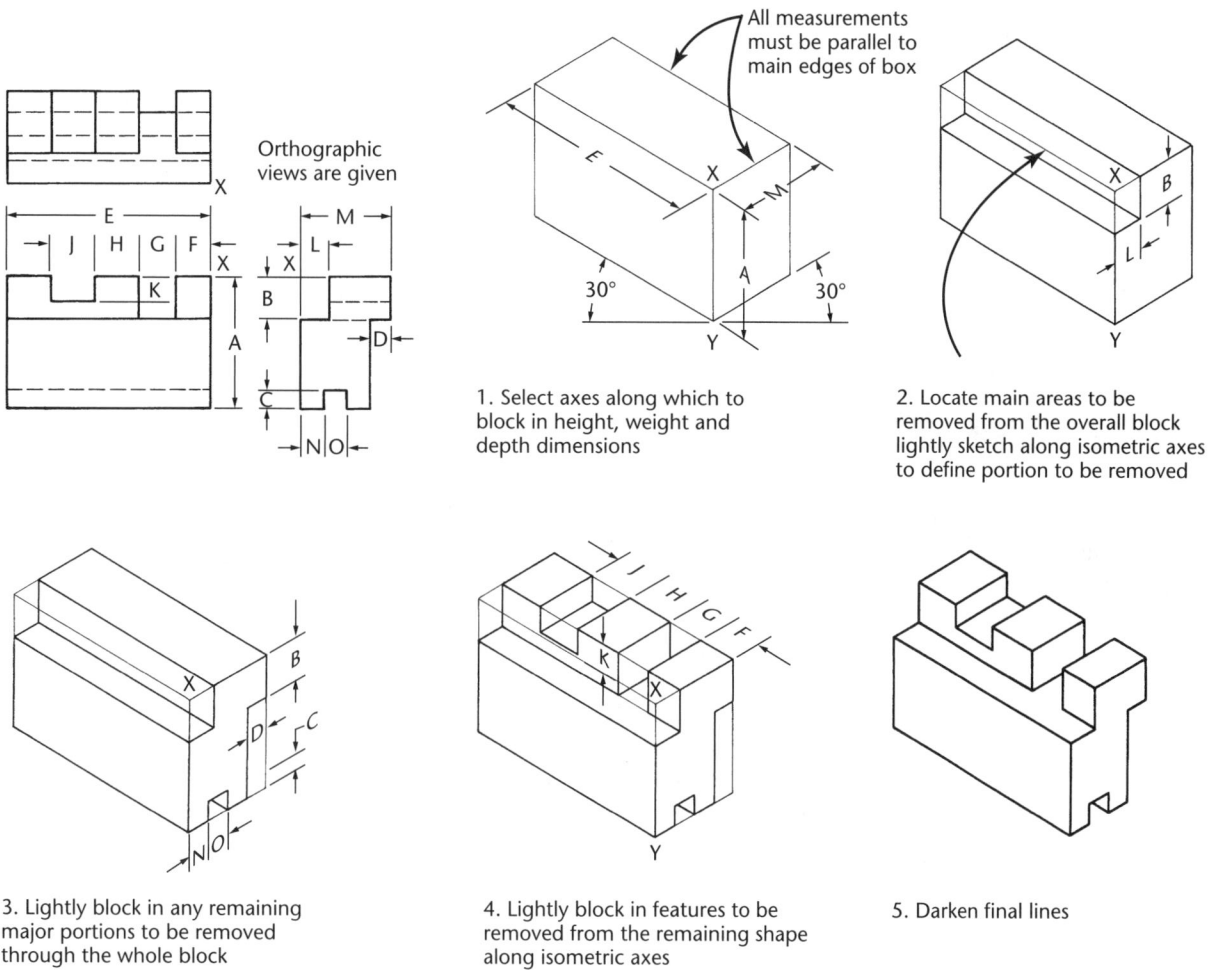

**Figure 9-14**    Steps in Making an Isometric Drawing of Normal Surfaces

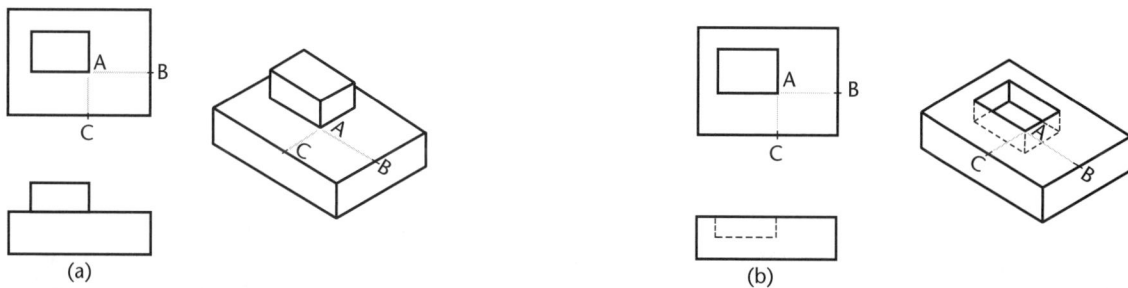

(a)

(b)

**Figure 9-15**    Offset Location Measurements

## 9-10  DRAWING NONISOMETRIC LINES

### STEP BY STEP

**How to Draw Nonisometric Lines**

The inclined lines BA and CA are shown true length in the top view (54 mm), but they are not true length in the isometric view. To draw these lines in the isometric drawing use a construction box and offset measurements.

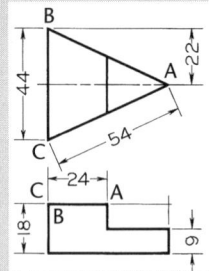

1. Directly measure dimensions that are along isometric lines (in this case, 44 mm, 18 mm, and 22 mm).

2. Since the 54 mm dimension is not along an isometric axis, it cannot be used to locate point A.

   Use trigonometry or draw a line parallel to the isometric axis to determine the distance to point A.

   Since this dimension is parallel to an isometric axis, it can be transferred to the isometric.

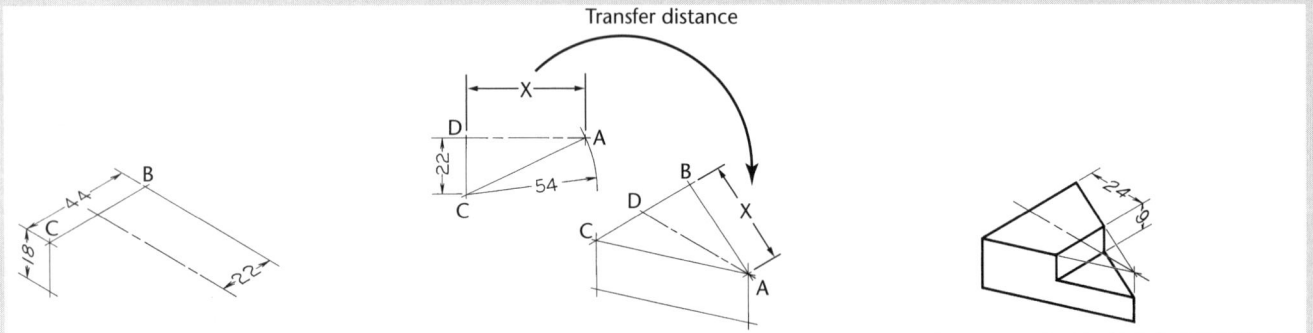

3. The dimensions 24 mm and 9 mm are parallel to isometric lines and can be measured directly.

Transfer distance

### TIP

To convince yourself that nonisometric lines will not be true length in the isometric drawing, use a scrap of paper and mark the distance BA (II) and then compare it with BA on the given top view in Figure 9-16a. Do the same for line CA. You will see that BA is shorter and CA is longer in the isometric than the corresponding lines in the given views.

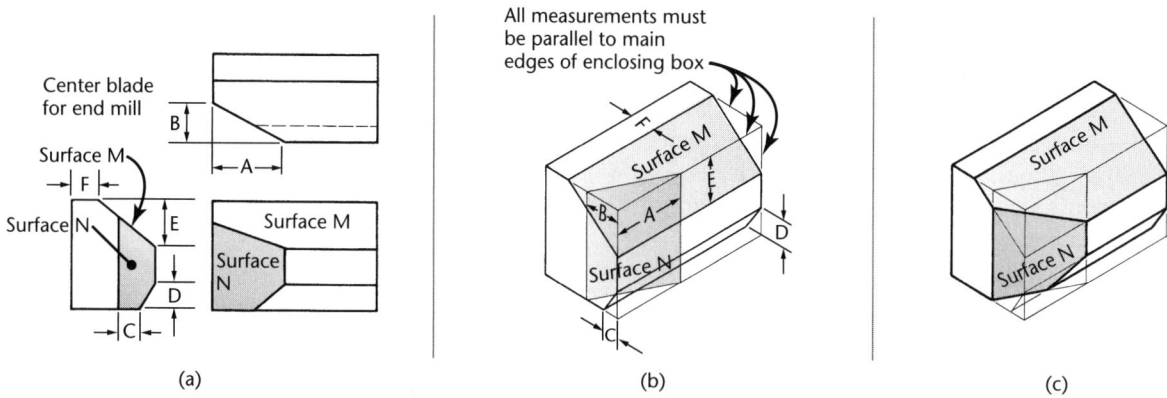

**Figure 9-16**   Inclined Surfaces in Isometric

## Isometric Drawings of Inclined Surfaces

Figure 9-16 shows how to construct an isometric drawing of an object that has some inclined surfaces and oblique edges. Notice that inclined surfaces are located by offset or coordinate measurements along the isometric lines. For example, dimensions *E* and *F* are measured to locate the inclined surface *M*, and dimensions *A* and *B* are used to locate surface *N*.

## 9-11   OBLIQUE SURFACES IN ISOMETRIC

# STEP BY STEP

### How to Draw Oblique Surfaces in Isometric

1. Find the intersections of the oblique surfaces with the isometric planes. Note that for this example, the oblique plane contains points A, B, and C.

2. To draw the plane, extend line AB to X and Y, in the same isometric plane as C. Use lines XC and YC to locate points E and F.

3. Finally, draw AD and ED using the rule that parallel lines appear parallel in every orthographic or isometric view.

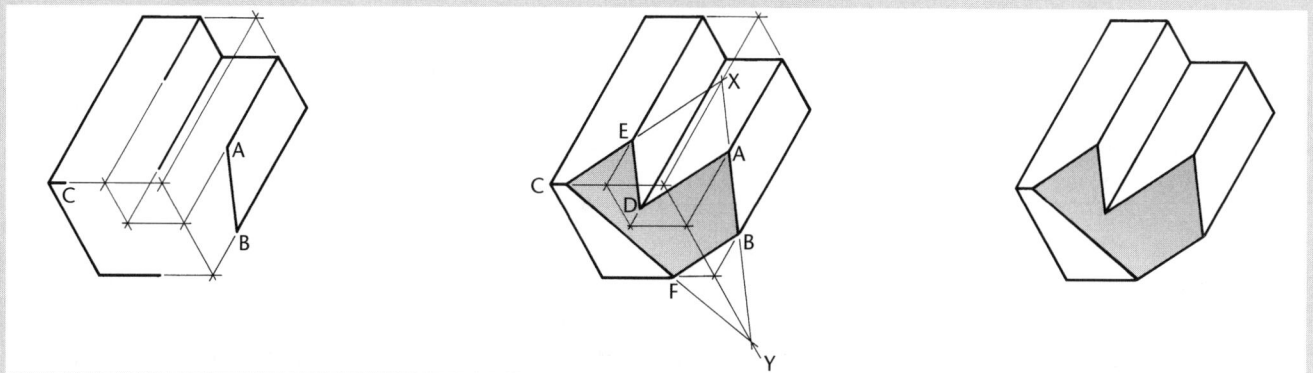

## 9-12 HIDDEN LINES AND CENTERLINES

Hidden lines are omitted unless they are needed to make the drawing clear. Figure 9-17 shows a case in which hidden lines are needed because a projecting part cannot be clearly shown without them. Sometimes it is better to include an isometric view from another direction than to try to show hidden features with hidden lines.

Draw centerlines if they are needed to indicate symmetry or if they are needed for dimensioning, but in general, use centerlines sparingly in isometric drawings. If in doubt, leave them out, as too many centerlines will look confusing.

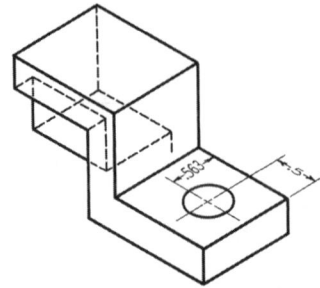

**Figure 9-17** Using Hidden Lines

## 9-13 ANGLES IN ISOMETRIC

Angles project true size only when the plane containing the angle is parallel to the plane of projection. An angle

## STEP BY STEP

**How to Draw Angles in Isometric**

The multiview drawing at left shows three 60° angles. None of the three angles will be 60° in the isometric drawing.

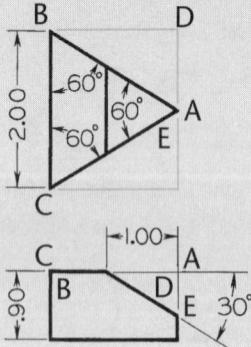

1. Lightly draw an enclosing box using the given dimensions, except for dimension X, which is not given.

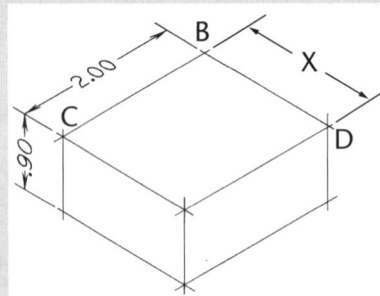

2. To find dimension X, draw triangle BDA from the top view full size, as shown.

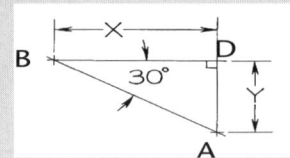

3. Transfer dimension X to the isometric to complete the enclosing box. Find dimension Y by a similar method and then transfer it to the isometric.

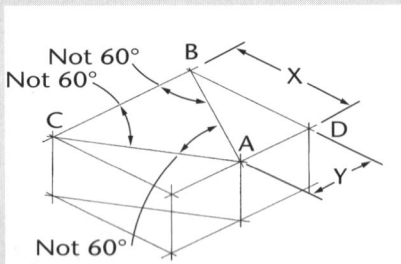

4. Complete the isometric by locating point E by using dimension K, as shown. A regular protractor cannot be used to measure angles in isometric drawings. Convert angular measurements to linear measurements along isometric axis lines.

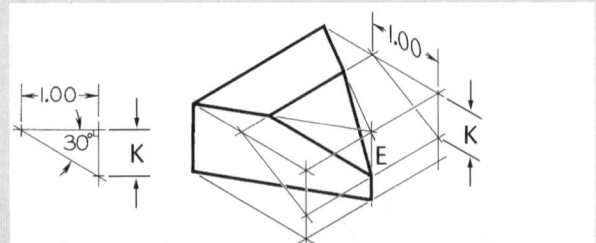

## TIP

**Checking Isometric Angles**

To convince yourself that none of the angles will be 60°, measure each angle in the isometric in Figure 9-17 with a protractor or scrap of paper and note the angle compared to the true 60°. None of the angles shown are the same in the isometric drawing. Two are smaller and one is larger than 60°.

**Estimating 30° angles**

If you are sketching on graph paper and estimating angles, an angle of 30° is roughly a rise of 1 to a run of 2.

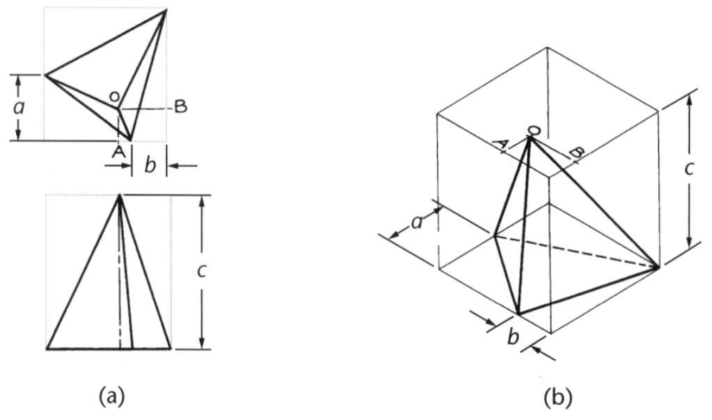

**Figure 9-18** Irregular Object in Isometric

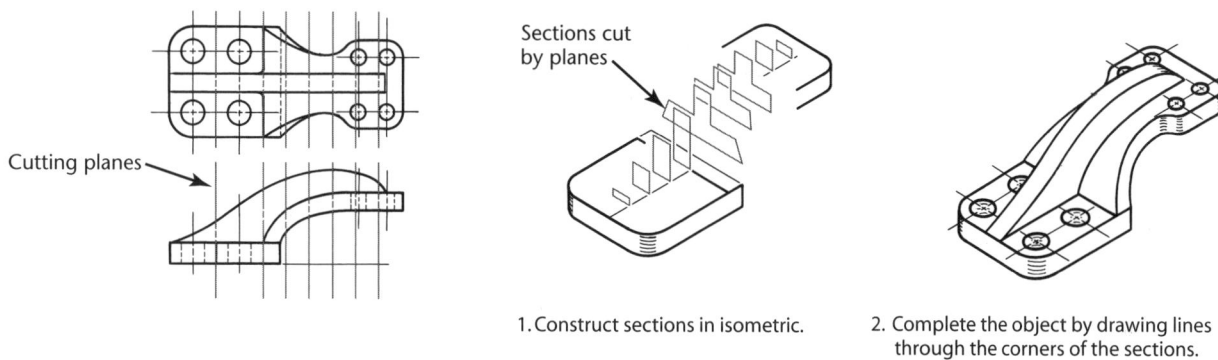

Sections cut by planes

1. Construct sections in isometric.

2. Complete the object by drawing lines through the corners of the sections.

**Figure 9-19** Using Sections in Isometric

may project to appear larger or smaller than the true angle depending on its position.

Since the various surfaces of the object are usually inclined to the front plane of projection, they generally will not be projected true size in an isometric drawing.

## 9-14 IRREGULAR OBJECTS

You can use the construction box method to draw objects that are not rectangular (Figure 9-18). Locate the points of the triangular base by offsetting **a** and **b** along the edges of the bottom of the construction box. Locate the vertex by offsetting lines **OA** and **OB** using the top of the construction box.

You can also draw irregular objects using a series of sections. The edge views of imaginary cutting planes are shown in the top and front views of the multiview drawing in Figure 9-19. In the example, all height dimensions are taken from the front view and all depth dimensions from the top view.

**TIP**

It is not always necessary to draw the complete construction box as shown in Figure 9-18b. If only the bottom of the box is drawn, the triangular base can be constructed as before. The orthographic projection of the vertex O' on the base can be drawn using offsets O'A and O'B, as shown, and then the vertical line O'O can be drawn, using measurement C.

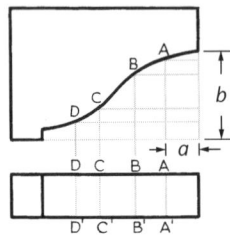

1. Use offset measurements *a* and *b* in the isometric to locate point A on the curve

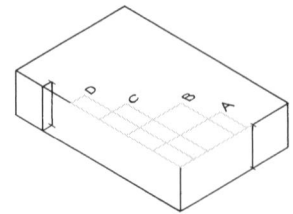

2. Locate points B, C, and D, and so on

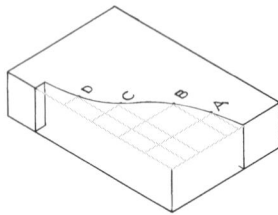

3. Sketch a smooth light freehand curve through the points

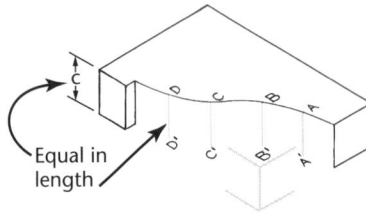

4. Draw a line vertically from point A to locate point A', and so on, making all equal to the height of block (c) then draw a light curve through the points

5. Darken the final lines

**Figure 9-20**   Curves in Isometric

## 9-15 CURVES IN ISOMETRIC

You can draw curves in isometric using a series of off-set measurements similar to those discussed in Section 9-9. Select any desired number of points at random along the curve in the given top view, such as points *A*, *B*, and *C* in Figure 9-20. Choose enough points to accurately locate the path of the curve (the more points, the greater the accuracy). Draw offset grid lines from each point parallel to the isometric axes and use them to locate each point in the isometric drawing as in the example shown in Figure 9-20.

## 9-16 TRUE ELLIPSES IN ISOMETRIC

If a circle lies in a plane that is not parallel to the plane of projection, the circle projects as an ellipse. The ellipse can be constructed using offset measurements.

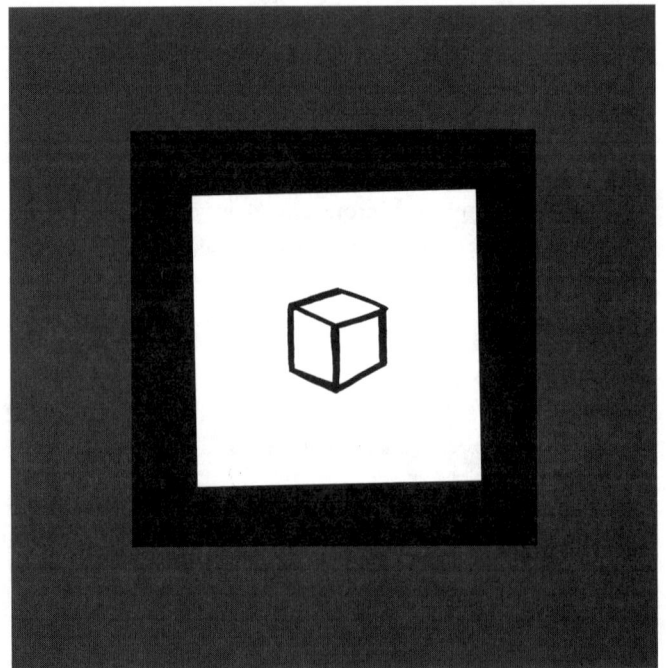

Tennis Ball (Factory Reject). *Cartoon by Roger Price. Courtesy of Droodles, "The Classic Collection."*

# STEP BY STEP

## Drawing an Isometric Ellipse By Offset Measurements

### Random Line Method

1. Draw parallel lines spaced at random across the circle.

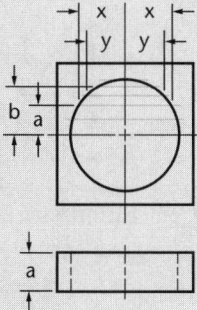

2. Transfer these lines to the isometric drawing. Where the hole exits the bottom of the block, locate points by measuring down a distance equal to the height $d$ of the block from each of the upper points. Draw the ellipse, part of which will be hidden, through these points. Darken the final drawing lines.

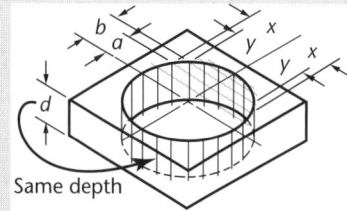

### Eight Point Method

1. Enclose the given circle in a square, and draw diagonals. Draw another square through the points of intersection of the diagonals and the circle as shown.

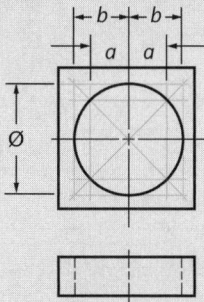

desired, add random parallel lines, as above.) The centerlines in the isometric are the conjugate diameters of the ellipse. The 45° diagonals coincide with the major and minor axes of the ellipse. The minor axis is equal in length to the sides of the inscribed square.

When more accuracy is required, divide the circle into 12 equal parts, as shown.

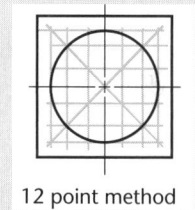

12 point method

2. Draw this same construction in the isometric, transferring distances a and b. (If more points are

### Nonisometric Lines

If a curve lies in a nonisometric plane, not all offset measurements can be applied directly. The elliptical face shown in the auxiliary view lies in an inclined nonisometric plane.

1. Draw lines in the orthographic view to locate points.

2. Enclose the cylinder in a construction box and draw the box in the isometric drawing. Draw the baseusing offset measurements and construct the

inclined ellipse by locating points and drawing the final curve through them.

Measure distances parallei to an isometric axis ($a$, $b$, $f$, etc.) in the isometric drawing on each side of the centerline X–X. Project those not parallel to any isometric axis ($e$, $f$, etc.) to the front view and down to the base, then measure along the lower edge of the construction box, as shown.

3. Darken final lines.

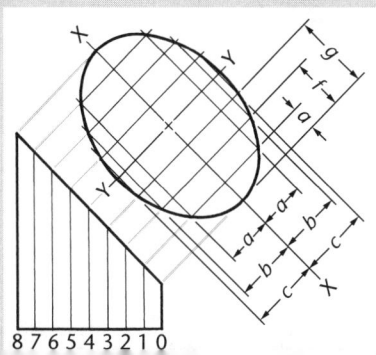

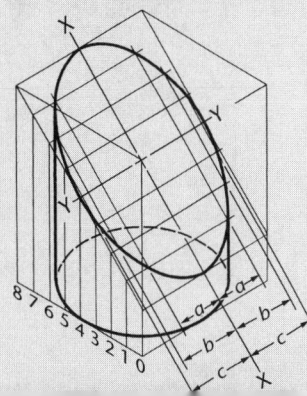

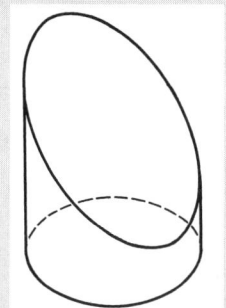

## 9-17 ORIENTING ELLIPSES IN ISOMETRIC DRAWINGS

Figure 9-21 shows a four center ellipses constructed on the three visible faces of a cube. Note that all of the diagonals are horizontal or at 60° with horizontal. Realizing this makes it easier to draw the shapes.

An approximate ellipse such as this, constructed from four arcs, is accurate enough for most isometric drawings. The four center method can be used only for ellipses in isometric planes. Earlier versions of CAD software, such as AutoCAD Release 10, used this method to create the approximate elliptical shapes available in the software. Current releases use an accurate ellipse.

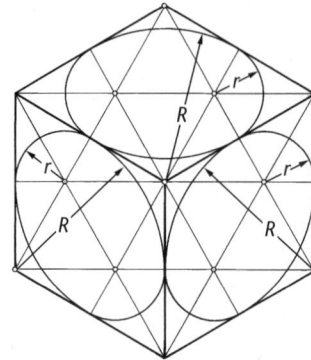

**Figure 9-21** Four Center Ellipses

## STEP BY STEP

**Drawing a Four Center Ellipse**

1. Draw or imagine a square enclosing the circle in the multiview drawing. Draw the isometric view of the square (an equilateral parallelogram with sides equal to the diameter of the circle).

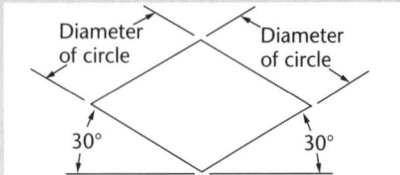

2. Create perpendicular bisectors to each side. They will intersect at four points, which will be centers for the four circular arcs.

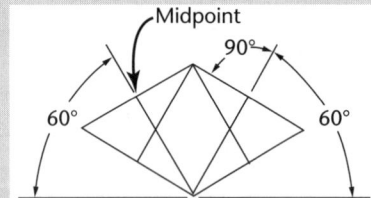

3. Draw the two large arcs, with radius R, from the intersections of the perpendiculars in the two closest corners of the parallelogram.

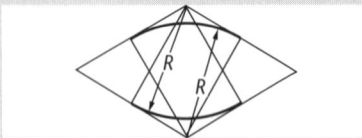

4. Draw the two small arcs, with radius r, from the intersections of the perpendiculars within the parallelogram, to complete the ellipse.

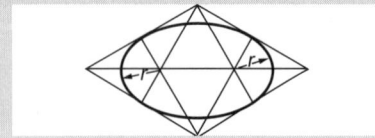

### TIP

Here is a useful rule. The major axis of the ellipse is always at right angles to the centerline of the cylinder, and the minor axis is at right angles to the major axis and coincides with the centerline.

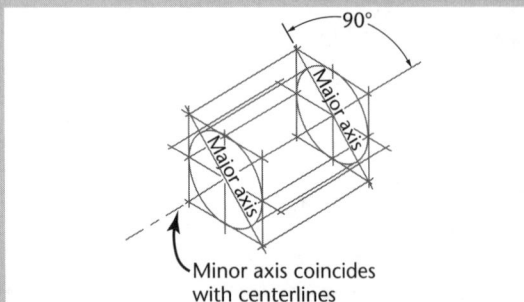

### TIP

As a check on the accurate location of these centers, you can draw a long diagonal of the parallelogram as shown in Step 4. The midpoints of the sides of the parallelogram are points of tangency for the four arcs.

## More Accurate Ellipses

The four center ellipse deviates considerably from a true ellipse. As shown in Figure 9-22a, a four center ellipse is somewhat shorter and "fatter" than a true ellipse. When the four center ellipse is not accurate enough, you can use a closer approximation called the Orth four center ellipse to produce a more accurate drawing.

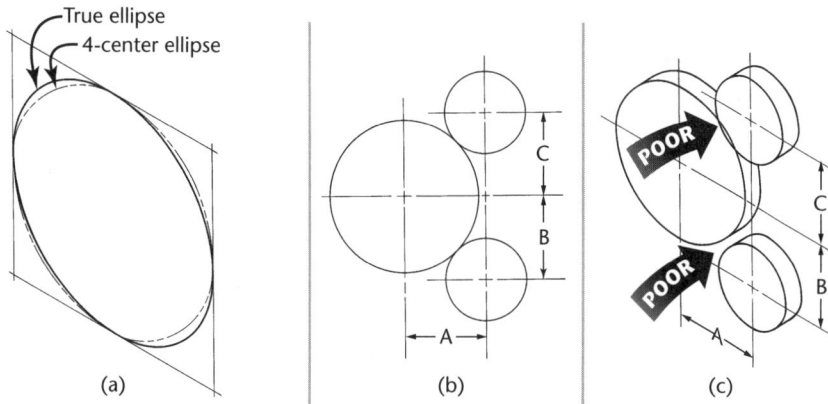

(a)    (b)    (c)

**Figure 9-22**    Inaccuracy of the Four Center Ellipse

## STEP BY STEP

### Drawing an Orth Four Center Ellipse

To create a more accurate approximate ellipse using the Orth method, follow the steps for these methods. The centerline method is convenient when starting from a hole or cylinder.

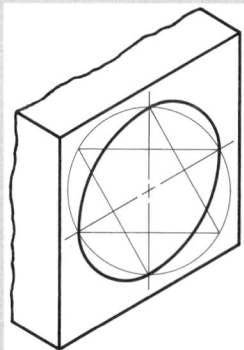

*Centerline Method*

1. Draw the isometric centerlines. From the center, draw a construction circle equal to the actual diameter of the hole or cylinder. The circle will intersect the centerlines at four points A, B, C, and D.
2. From the two intersection points on one centerline, draw perpendiculars to the other centerline. Then draw perpendiculars from the two intersection points on the other centerline to the first centerline.
3. With the intersections of the perpendiculars as centers, draw two small arcs and two large arcs.

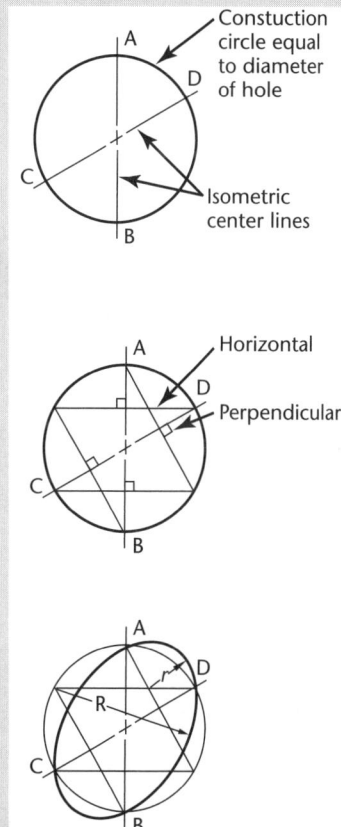

*Continued*

## Enclosing Rectangle Method

1.  Locate center and block in enclosing isometric rectangle.

2.  Use the midpoint of the isometric rectangle (the distance from *A* to *B*) to locate the foci on the major axis.

3.  Draw lines at 60° from horizontal through the foci (points *C* and *D*) to locate the center of the large arc *R*.

4.  Draw the two large arcs *R* tangent to the isometric rectangle. Draw two small arcs *r*, using foci points *C* and *D* as centers, to complete the approximate ellipse.

*Note that these steps are exactly the same as for the regular four center ellipse, except for the use of the isometric centerlines instead of the enclosing parallelogram. (When sketching, it works fine to just draw the enclosing rectangle and sketch the arcs tangent to its sides.)*

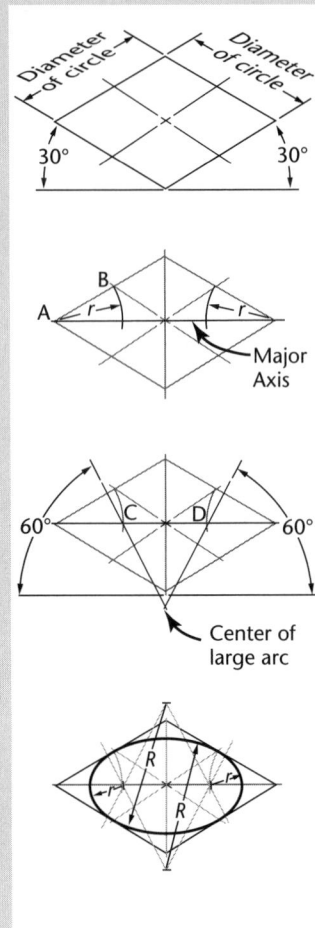

---

## TIP

### Isometric Templates

Special templates like this isometric template with angled lines and ellipses oriented in various isometric planes make it easy to draw isometric sketches.

The ellipses are provided with markings to coincide with the isometric centerlines of the holes—a convenient feature in isometric drawing.

You can also draw ellipses using an appropriate ellipse template selected to fit the major and minor axes.

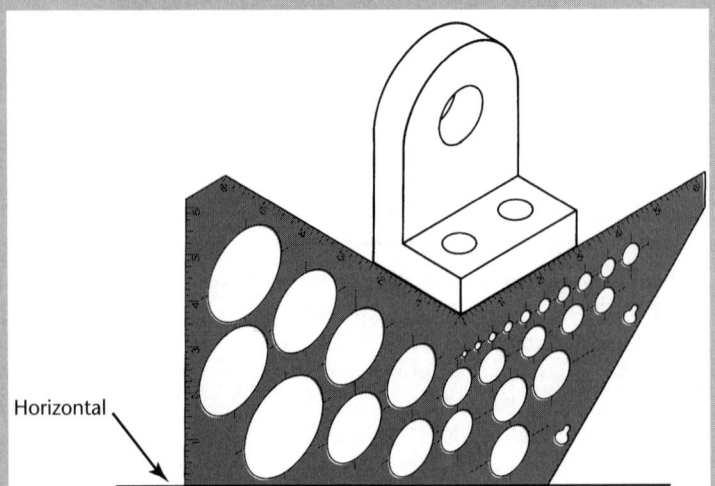

## 9-18 DRAWING ISOMETRIC CYLINDERS

A typical drawing with cylindrical shapes is shown in Figure 9-23. Note that the centers of the larger ellipse cannot be used for the smaller ellipse, though the ellipses represent concentric circles. Each ellipse has its own parallelogram and its own centers. Notice that the centers of the lower ellipse are drawn by projecting the centers of the upper large ellipse down a distance equal to the height of the cylinder.

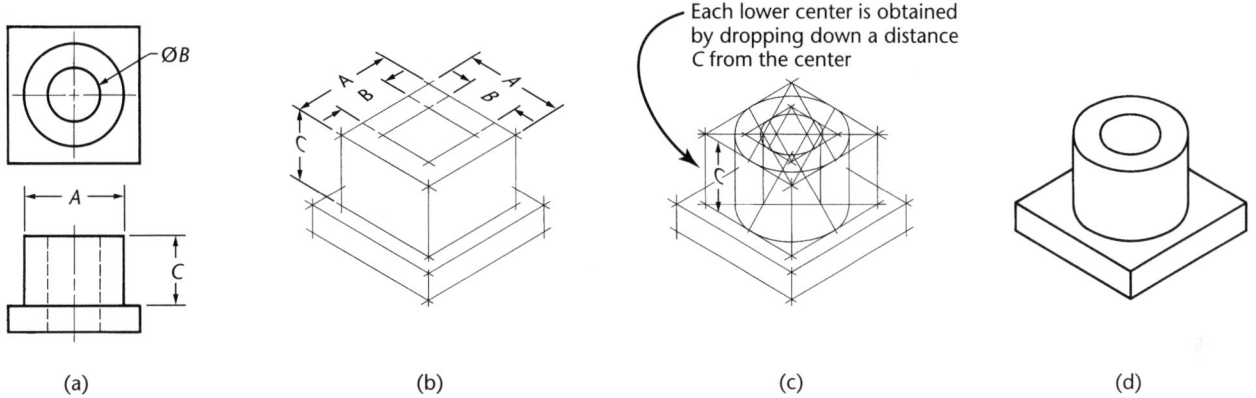

Each lower center is obtained by dropping down a distance C from the center

(a)    (b)    (c)    (d)

**Figure 9.23**    Isometric Drawing of a Bearing

## 9-19 SCREW THREADS IN ISOMETRIC

Parallel partial ellipses equally spaced at the symbolic thread pitch are used to represent only the crests of a screw thread in isometric (Figure 9-24). The ellipses may be sketched, drawn by the four center method, or created using an ellipse template.

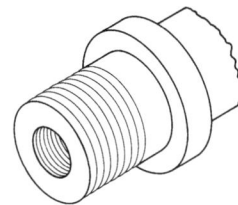

**Figure 9.24**    Screw Threads in Isometric

## 9-20 ARCS IN ISOMETRICS

The four center ellipse construction can be used to sketch or draw circular arcs in isometric. Figure 9-25a shows the complete construction. It is not necessary to draw the complete constructions for arcs, as shown in Figure 9-25b and c. Measure the radius R from the construction corner; then at each point, draw perpendiculars to the lines. Their intersection is the center of the arc. Note that the R distances are equal in Figure 9-25b and c, but that the actual radii used are quite different.

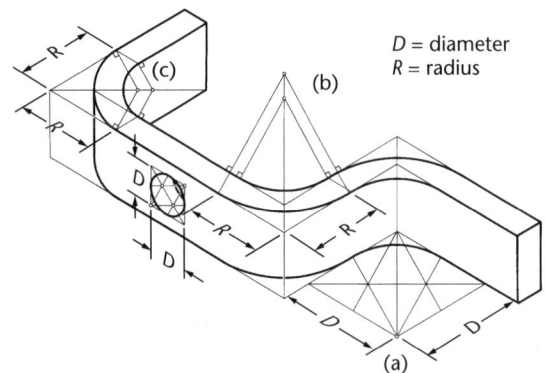

D = diameter
R = radius

**Figure 9-25**    Arcs in Isometric

## 9-21 INTERSECTIONS

To draw the elliptical intersection of a cylindrical hole in an oblique plane in isometric (Figure 9-26a), draw the ellipse in the isometric plane on top of the construction box (Figure 9-26b); then project points down to the oblique plane as shown. Each point forms a trapezoid, which is produced by a slicing plane parallel to a lateral surface of the block.

To draw the curve of intersection between two cylinders (Figure 9-27), use a series of imaginary cutting planes through the cylinders parallel to their axes. Each plane will cut elements on both cylinders that intersect at points on the curve of intersection (Figure 9-26b). As many points should be plotted as necessary to assure a smooth curve. For accuracy, draw the ends of the cylinders using the Orth four center construction, with ellipse guides, or by one of the true ellipse constructions.

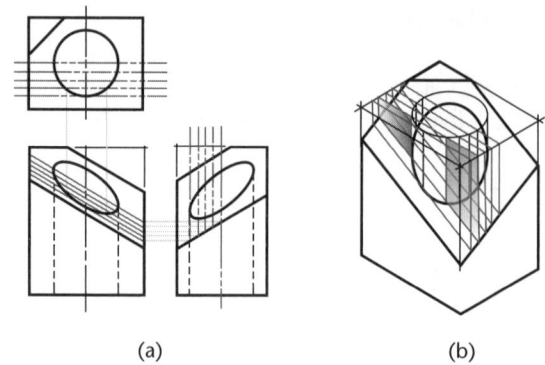

(a)         (b)

**Figure 9-26**   Oblique Plane and Cylinder

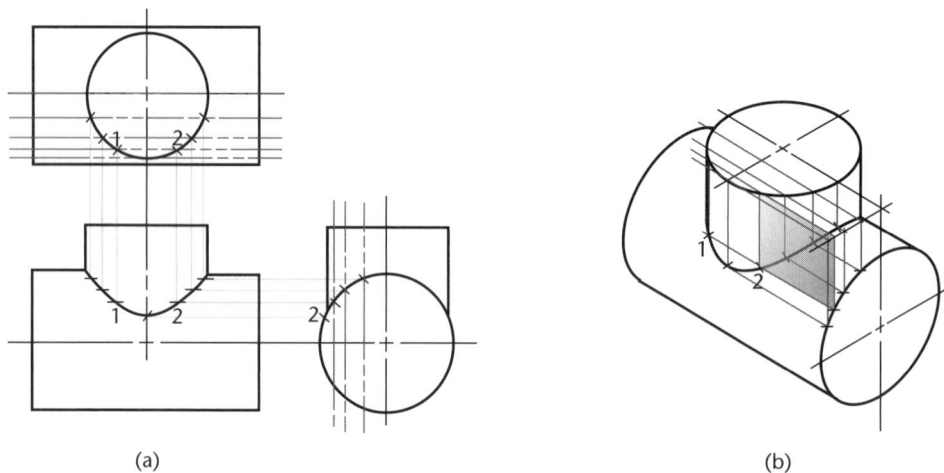

(a)         (b)

**Figure 9-27**   Intersection of Cylinders

## 9-22 SPHERES IN ISOMETRIC

The isometric drawing of any curved surface is the envelope of all lines that can be drawn on that surface. For spheres, select the great circles (circles cut by any plane through the center) as the lines on the surface. Since all great circles, except those that are perpendicular or parallel to the plane of projection, are shown as ellipses having equal major axes, their envelope is a circle whose diameter is the major axis of the ellipse.

Figure 9-28 shows two views of a sphere enclosed in a construction cube. Next, an isometric of a great circle is

drawn in a plane parallel to one face of the cube. There is no need to draw the ellipse, since only the points on the diagonal located by measurements *a* are needed to establish the ends of the major axis and thus to determine the radius of the sphere.

In the resulting isometric drawing the diameter of the circle is $\sqrt{\frac{2}{3}}$ times the actual diameter of the sphere. The isometric projection is simply a circle whose diameter is equal to the true diameter of the sphere.

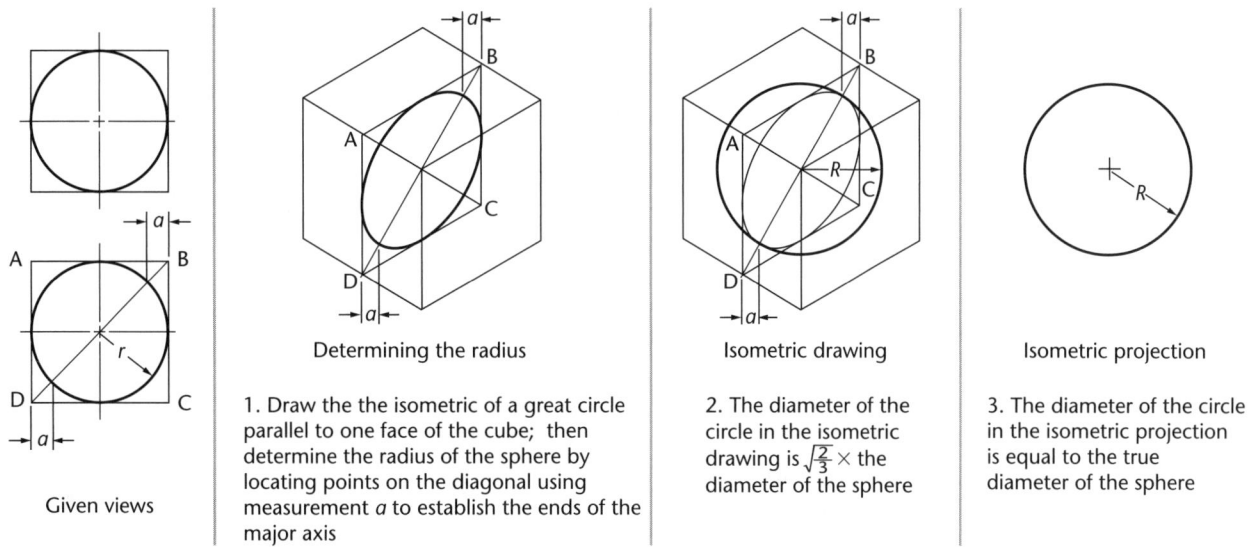

A

B

a

A

D

r

C

a

Given views

Determining the radius

A

B

a

D

a

C

1. Draw the the isometric of a great circle parallel to one face of the cube; then determine the radius of the sphere by locating points on the diagonal using measurement *a* to establish the ends of the major axis

Isometric drawing

A

B

a

R

C

D

a

2. The diameter of the circle in the isometric drawing is $\sqrt{\frac{2}{3}} \times$ the diameter of the sphere

Isometric projection

R

3. The diameter of the circle in the isometric projection is equal to the true diameter of the sphere

**Figure 9-28**    Isometric of a Sphere

## 9-23  ISOMETRIC SECTIONING

**Isometric sectioning** is useful in drawing open or irregularly shaped objects. Figure 9-29 shows an isometric full section. It is usually best to draw the cut surface first, then draw the portion of the object that lies behind the cutting plane.

To create an isometric half section, it is usually easiest to make an isometric drawing of the entire object, then add the cut surfaces as shown in Figure 9-30. Since only a quarter of the object is removed in a half section, the resulting pictorial drawing is more useful than a full section.

Isometric broken-out sections are also sometimes used. Section lining in isometric drawing is similar to that in multiview drawing. Section lining at an angle of 60° with horizontal as shown in Figures 9-29 and 9-30 is recommended, but change the direction if 60° would cause the lines to be parallel to a prominent visible line bounding the cut surface, or to other adjacent lines of the drawing.

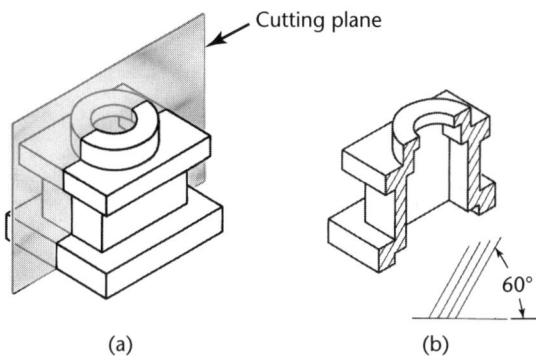

Cutting plane

(a)                 (b)

60°

**Figure 9-29**    Isometric Full Section

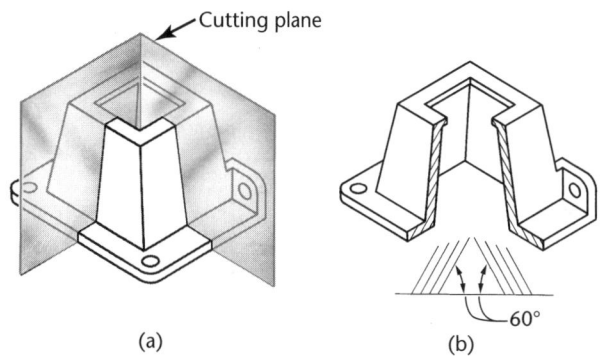

Cutting plane

(a)                 (b)

60°

**Figure 9-30**    Isometric Half Section

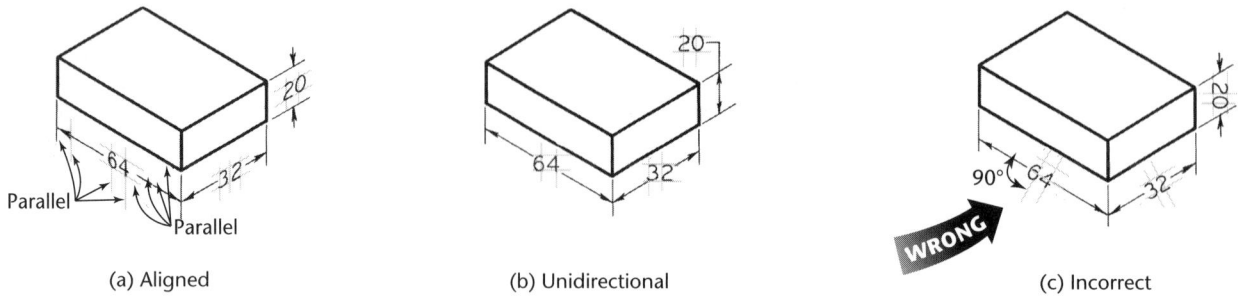

**Figure 9-31** Numerals and Arrowheads in Isometric (Metric Dimensions)

## 9-24 ISOMETRIC DIMENSIONING

**Isometric dimensions** are similar to ordinary dimensions used on multiview drawings but should match the pictorial style. Two methods of dimensioning are approved by ANSI—namely, the pictorial plane (aligned) system and the unidirectional system (Figure 9-31).

Note that vertical lettering is used for either system of dimensioning. Inclined lettering is not recommended for pictorial dimensioning. Figure 9-31a and b show how to draw numerals and arrowheads for the two systems.

In the aligned system, the extension lines, dimension lines, and lettering for the 64 mm dimension are all drawn in the isometric plane of one face of the object (Figure 9-31a). The "horizontal" guidelines for the lettering are drawn parallel to the dimension line, and the "vertical" guidelines are drawn parallel to the extension lines. The barbs of the arrowheads should line up parallel to the extension lines.

In the unidirectional system the extension lines and dimension lines for the 64 mm dimension are drawn in the isometric plane of one face of the object (Figure 9-31b). The lettering for the dimensions is vertical and reads from the bottom of the drawing. This simpler system of dimensioning is often used on pictorials for production purposes. Still, the barbs of the arrowheads should line up parallel to the extension lines, as in Figure 9-31a.

As shown in Figure 9-31c, the vertical guidelines for the letters should not be perpendicular to the dimension lines. The example in Figure 9-31c is incorrect because the 64 mm and 32 mm dimensions are not lettered in the plane of corresponding dimension and extension lines, nor are they in a vertical position to read from the bottom of the drawing. Note how the 20 mm dimension is awkward to read because of its position.

### Correct and Incorrect Isometric Dimensioning

Correct practice in isometric dimensioning using the aligned system of dimensioning is shown in Figure 9-32a.

Figure 9-32b shows several incorrect practices. The 3.125 dimension runs to a wrong extension line at the right, so the dimension does not lie in an isometric plane. Near the left side, a number of lines cross each other unnecessarily and terminate on the wrong lines. The upper .5 drill hole is located from the edge of the cylinder when it should be dimensioned from its centerline. Study these two drawings carefully to see additional mistakes in Figure 9-32b.

Isometric dimensioning methods apply equally to fractional, decimal, and metric dimensions.

Many examples of isometric dimensioning are given in the End of Chapter Exercises. Study these to find samples of almost any special case you may encounter.

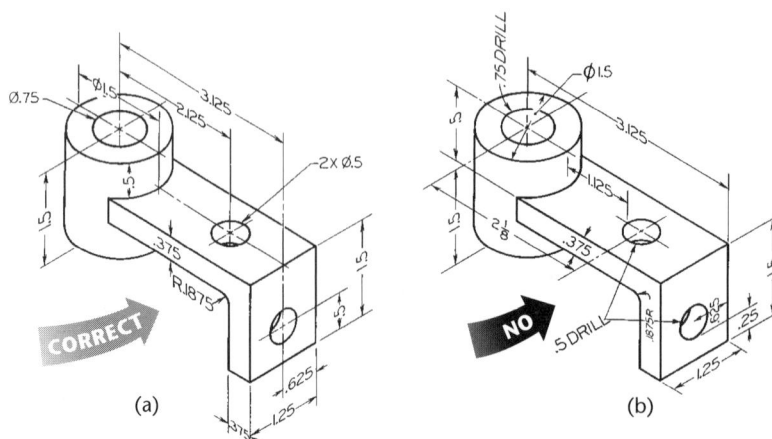

**Figure 9-32** Correct and Incorrect Isometric Dimensioning (Aligned System)

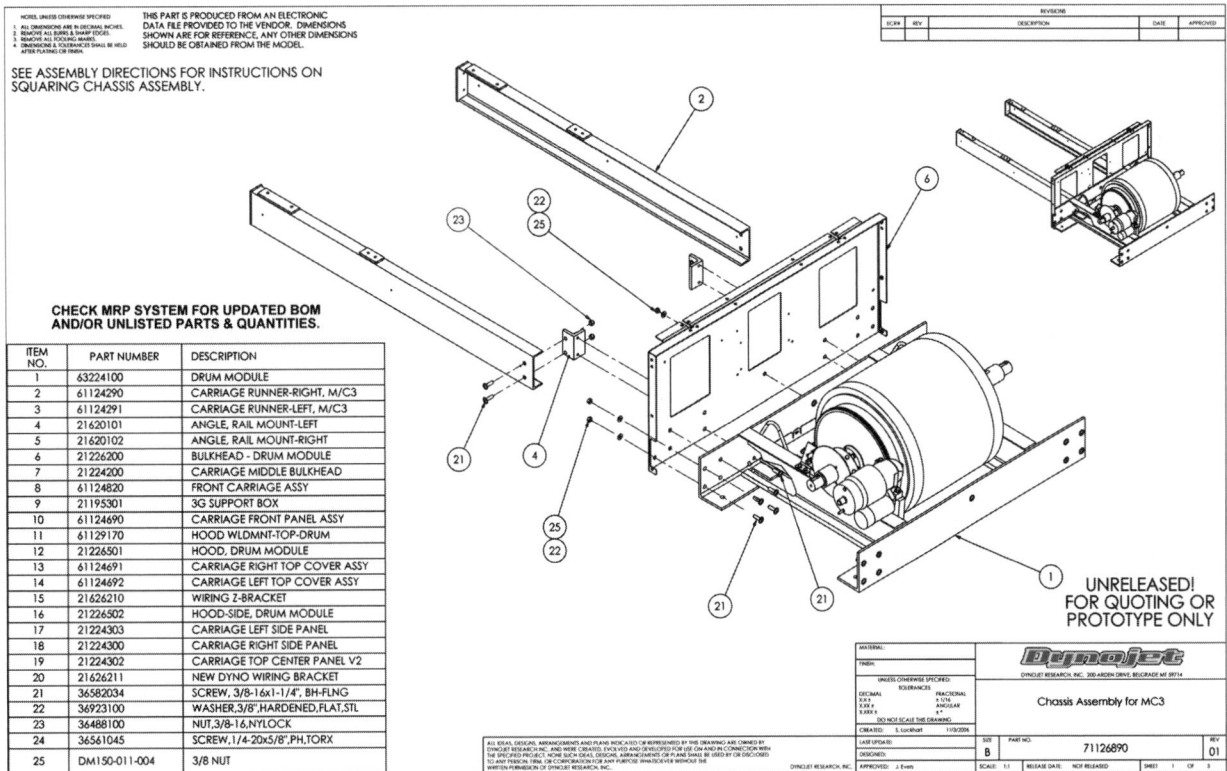

| ITEM NO. | PART NUMBER | DESCRIPTION |
|---|---|---|
| 1 | 63224100 | DRUM MODULE |
| 2 | 61124290 | CARRIAGE RUNNER-RIGHT, M/C3 |
| 3 | 61124291 | CARRIAGE RUNNER-LEFT, M/C3 |
| 4 | 21620101 | ANGLE, RAIL MOUNT-LEFT |
| 5 | 21620102 | ANGLE, RAIL MOUNT-RIGHT |
| 6 | 21226200 | BULKHEAD - DRUM MODULE |
| 7 | 21224200 | CARRIAGE MIDDLE BULKHEAD |
| 8 | 61124820 | FRONT CARRIAGE ASSY |
| 9 | 21195301 | 3G SUPPORT BOX |
| 10 | 61124690 | CARRIAGE FRONT PANEL ASSY |
| 11 | 61129170 | HOOD WLDMNT-TOP-DRUM |
| 12 | 21226501 | HOOD, DRUM MODULE |
| 13 | 61124691 | CARRIAGE RIGHT TOP COVER ASSY |
| 14 | 61124692 | CARRIAGE LEFT TOP COVER ASSY |
| 15 | 21626210 | WIRING Z-BRACKET |
| 16 | 21226502 | HOOD-SIDE, DRUM MODULE |
| 17 | 21224303 | CARRIAGE LEFT SIDE PANEL |
| 18 | 21224300 | CARRIAGE RIGHT SIDE PANEL |
| 19 | 21224302 | CARRIAGE TOP CENTER PANEL V2 |
| 20 | 21626211 | NEW DYNO WIRING BRACKET |
| 21 | 36582034 | SCREW, 3/8-16x1-1/4", BH-FLNG |
| 22 | 36923100 | WASHER,3/8",HARDENED,FLAT,STL |
| 23 | 36488100 | NUT,3/8-16,NYLOCK |
| 24 | 36561045 | SCREW,1/4-20x5/8",PH,TORX |
| 25 | DM150-011-004 | 3/8 NUT |

**Figure 9-33**   Exploded Isometric Assembly Drawing. *Courtesy of Dynojet Research, Inc.*

## 9-25  EXPLODED ASSEMBLIES

Exploded assemblies are often used in design presentations, catalogs, sales literature, and in the shop to show all the parts of an assembly and how they fit together. They may be drawn by any of the pictorial methods, including isometric (Figure 9-33).

## 9-26  PIPING DIAGRAMS

Isometric and oblique drawings are well suited for representation of piping layouts, as well as for all other structural work to be represented pictorially. An example is shown in Figure 9-34.

**Figure 9-34**   Portion of an Isometric Piping Diagram. *Courtesy of Associated Construction Engineering.*

## 9-27 DIMETRIC PROJECTION

A dimetric projection is an axonometric projection of an object where two of its axes make equal angles with the plane of projection and the third axis makes either a smaller or a greater angle (Figure 9-35). The two axes making equal angles with the plane of projection are foreshortened equally, while the third axis is foreshortened in a different proportion.

Usually the object is oriented so one axis is vertical. However, you can revolve the projection to any orientation if you want that particular view.

Do not confuse the angles between the axes in the drawing with the angles from the plane projection. These are two different, but related things. You can arrange the amount that the principal faces are tilted to the plane of projection any way that two angles between the axes are equal and over 90°.

The scales can be determined graphically, as shown in Figure 9-36a, in which *OP*, *OL*, and *OS* are the projections of the axes or converging edges of a cube. If the triangle *POS* is revolved about the axis line *PS* into the plane of projection, it will show its true size and shape as *PO'S*. If regular full-size scales are marked along the lines *O'P* and *O'S*, and the triangle is counterrevolved to its original position, the dimetric scales may be divided along the axes *OP* and *OS*, as shown.

You can use an architect's scale to make the measurements by assuming the scales and calculating the positions of the axes, as follows:

$$\cos a = -\frac{\sqrt{2h^2v^2 - v^4}}{2hv}$$

where *a* is one of the two equal angles between the projections of the axes, *h* is one of the two equal scales, and *v* is the third scale. Examples are shown in the upper row of Figure 9-35, where length measurements could be made using an architect's scale. One of these three positions of the axes will be found suitable for almost any practical drawing.

## 9-28 APPROXIMATE DIMETRIC DRAWING

Approximate dimetric drawings, which closely resemble true dimetrics, can be constructed by substituting for the true angles shown in the upper half of Figure 9-35 angles that can be obtained with the ordinary triangles and compass, as shown in the lower half of the figure. The resulting drawings will be accurate enough for all practical purposes.

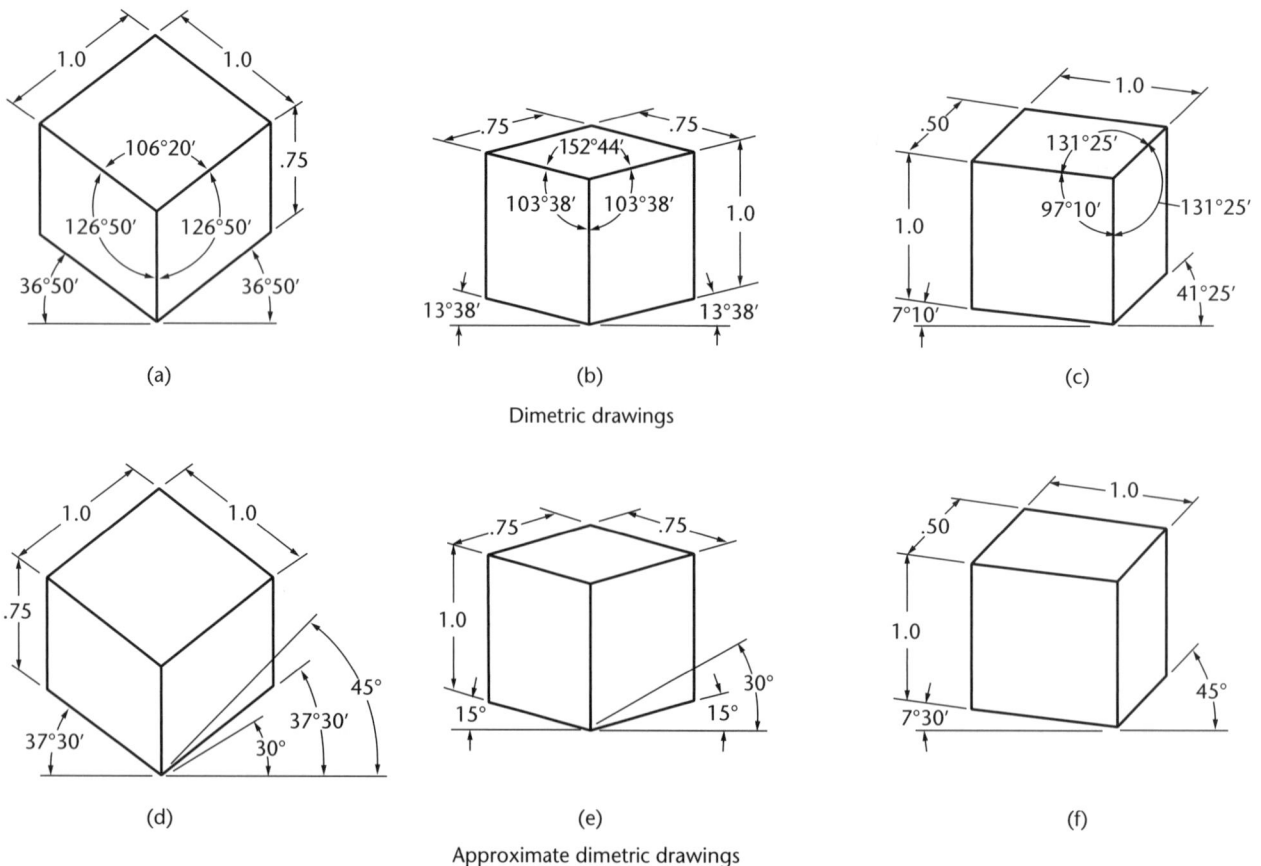

(a)  (b)  (c)

Dimetric drawings

(d)  (e)  (f)

Approximate dimetric drawings

**Figure 9-35**  Undertstanding Angles in Dimetric Projection

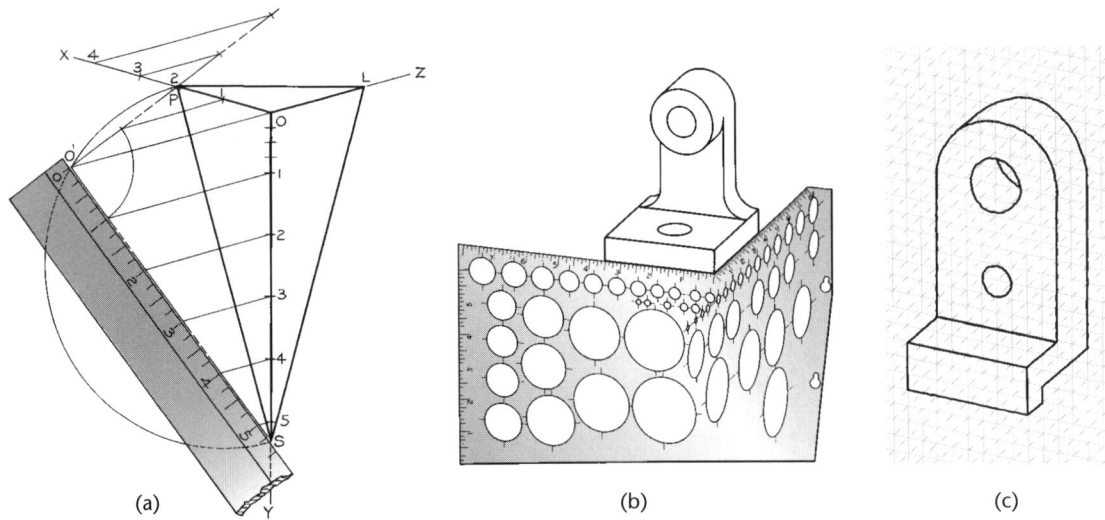

**Figure 9-36**  Dimetric Drawings

# STEP BY STEP

## How to Make Dimetric Drawings

1. To make a dimetric drawing for the views given, draw two intersecting axis lines at angles of 7.5° and 45° from horizontal. Draw the third axis direction vertically through them.

2. The dimensions for the principal face are measured full size. The dimension for the receding axis direction will be at half scale.

3. Block in the features relative to the surfaces of the enclosing box. The offset method of drawing a curve is shown in the figure.

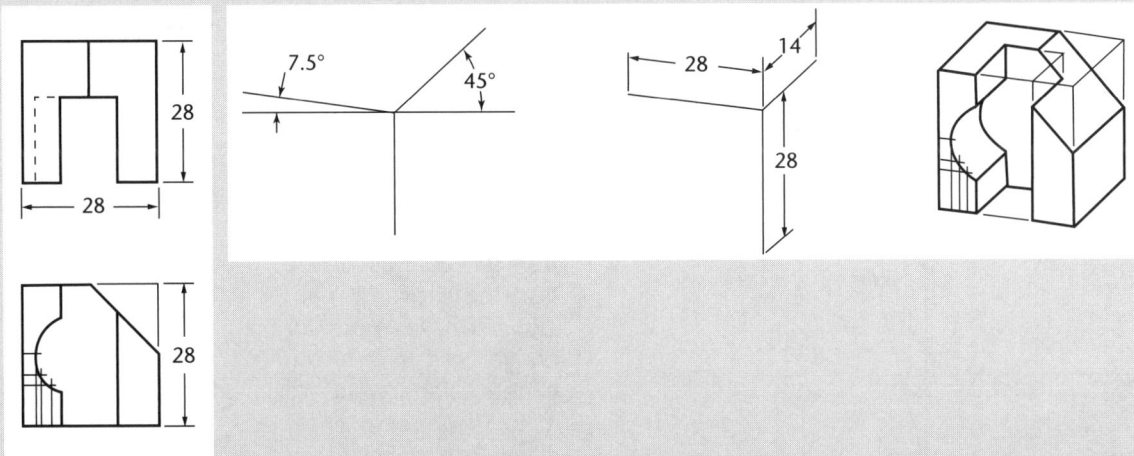

*Continued*

## An Approximate Dimetric Drawing

Follow these steps to make a dimetric sketch with the position similar to that in Figure 9-35e where the two angles are equal.

1. Using whichever angle produces a good drawing of your part, block in the dimetric axes. An angle of 20° from horizontal tends to show many parts well.

2. Block in the major features, foreshorten the dimensions along the two receding axes by approximately 75 percent.

3. Darken the final lines.

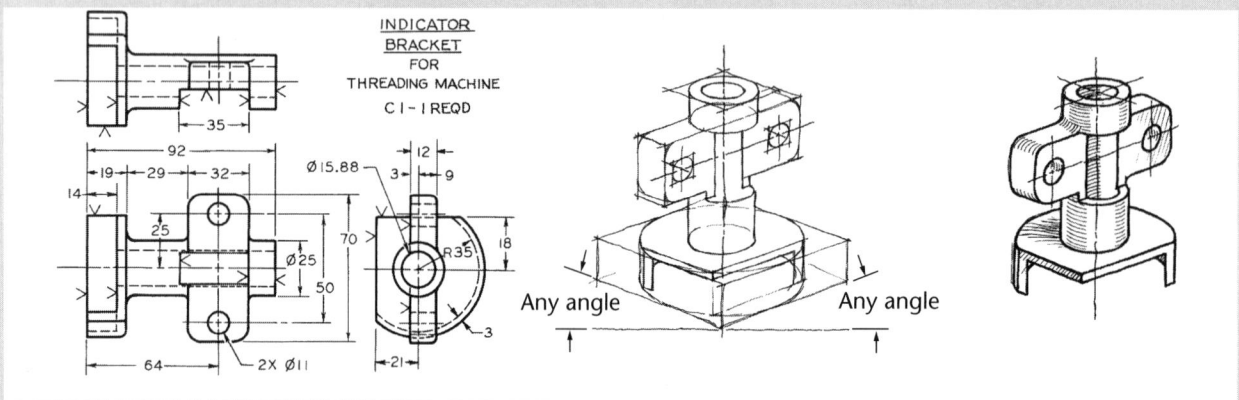

## 9-29  TRIMETRIC PROJECTION

A trimetric projection is an axonometric projection of an object oriented so that no two axes make equal angles with the plane of projection. In other words, each of the three axes, and the lines parallel to them, have different ratios of foreshortening. If the three axes are selected in any position on paper so that none of the angles is less than 90°,

and they are not an isometric nor a dimetric projection, the result will be a trimetric projection.

## 9-30  TRIMETRIC SCALES

Since the three axes are foreshortened differently, each axis will use measurement proportions different from the other two. You can select which scale to use as shown in Figure 9-37. Any two of the three triangular faces can be re-

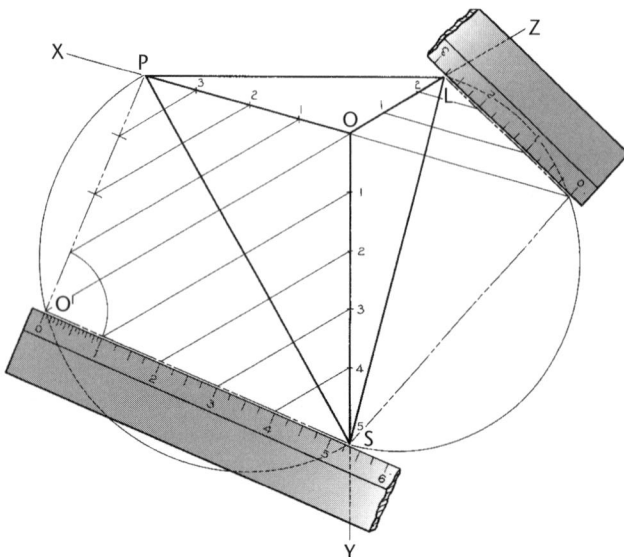

### TIP

You can make scales from thin card stock and transfer these dimensions to each card for easy reference. You might even want to make a trimetric angle from Bristol Board or plastic, as shown here, or six or seven of them, using angles for a variety of positions of the axes.

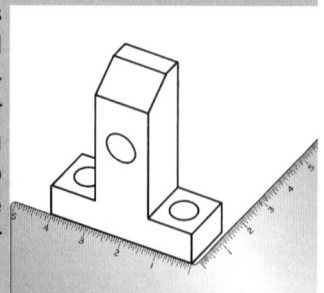

**Figure 9-37**   Trimetric Scales

volved into the plane of projection to show the true lengths of the three axes. In the revolved position, the regular scale is used to set off inches or fractions thereof. When the axes have been counterrevolved to their original positions, the scales will be correctly foreshortened, as shown.

## 9-31 TRIMETRIC ELLIPSES

The trimetric centerlines of a hole, or the end of a cylinder, become the conjugate diameters of an ellipse when drawn in trimetric. The ellipse may be drawn on the conjugate diameters.

One advantage of trimetric projection is the infinite number of positions of the object available. The angles and scales can be handled without too much difficulty, as shown in Sections 9-32 and 9-33. However, in drawing any axonometric ellipse, keep the following in mind:

1. On the drawing, the major axis is always perpendicular to the centerline, or axis, of the cylinder.
2. The minor axis is always perpendicular to the major axis; on the paper it coincides with the axis of the cylinder.
3. The length of the major axis is equal to the actual diameter of the cylinder.

The directions of both the major and minor axes, and the length of the major axis, will always be known, but not the length of the minor axis. Once it is determined, you can construct the ellipse using a template or any of a number of ellipse constructions. For sketching you can generally sketch an ellipse that looks correct by eye.

In Figure 9-38a, locate center $O$ as desired, and draw the horizontal and vertical construction lines that will contain the major and minor axes through $O$. Note that the major axis will be on the horizontal line perpendicular to the axis of the hole, and the minor axis will be perpendicular to it, or vertical.

Use the actual radius of the hole and draw the semicircle, as shown, to establish the ends $A$ and $B$ of the major axis. Draw $AF$ and $BF$ parallel to the axonometric edges $WX$ and $YX$, respectively, to locate $F$, which lies on the ellipse. Draw a vertical line through $F$ to intersect the semicircle at $F'$ and join $F'$ to $B'$, as shown. From $D'$, where the minor axis, extended, intersects the semicircle, draw $D'E$ and $ED$ parallel to $F'B$ and $BF$, respectively. Point $D$ is one end of the minor axis. From center $O$, strike arc $DC$ to locate $C$, the other end of the minor axis. On these axes, a true ellipse can be constructed, or drawn with an ellipse template.

In constructions where the enclosing parallelogram for an ellipse is available or easily constructed, the major and minor axes can be determined as shown in Figure 9-38b. The directions of both axes and the length of the major axis are known. Extend the axes to intersect the sides of the parallelogram at $L$ and $M$, and join the points with a straight line. From one end $N$ of the major axis, draw a line $NP$ parallel to $LM$. The point $P$ is one end of the minor axis. To find one end $T$ of the minor axis of the smaller ellipse, it is only necessary to draw $RT$ parallel to $LM$ or $NP$.

The method of constructing an ellipse on an oblique plane in trimetric is similar to that shown in the Step by Step in Section 9-17 for drawing an isometric ellipse by offset measurements.

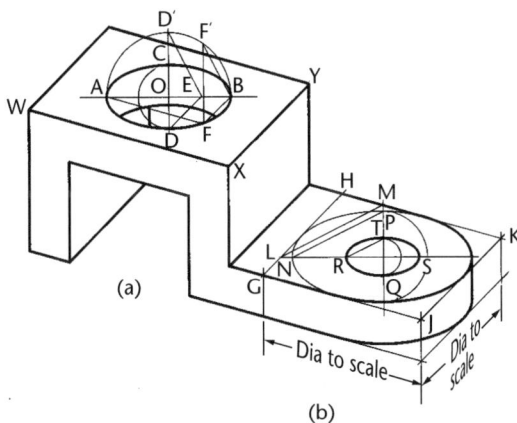

**Figure 9-38** Ellipses in Trimetric. *Method (b). Courtesy of Professor H. E. Grant.*

> **TIP**
>
> When you are creating a trimetric sketch of an ellipse, it works great to block in the trimetric rectangle that would enclose the ellipse and sketch the ellipse tangent to the midpoints of the rectangle.

## PRESENTATION DRAWING

The MARGE (Mars Autonomous Rover for Geoscience Exploration) aeroshell is part of a NASA Scout mission proposal developed by Malin Space Science Systems and the Raytheon Company in 2005 and 2006. The blunt, conical MARGE aeroshell is an integrated system providing safe delivery of its payload, two small, autonomous rovers, to the surface of Mars. The aeroshell is about 2.4 meters in diameter.

Shown here is the part of the system which provides aerobraking for the spacecraft's initial descent from orbit, the terminal rocket descent phase just before landing, and the final soft touchdown with the surface. With the protective backshell (where the parachute is located) and rovers removed, you can clearly see the components of the propulsion and control systems integrated into the rover egress deck, and color coded for clarity. In addition to aerobraking and rocket-powered descent, the MARGE aeroshell design incorporates crushable foam layers of increasing density to cushion the final touchdown with the planet surface. After the descent and landing phase is complete, clamps are disengaged and the rovers drive off the lip of the aeroshell under their own power.

### MARGE SUB-ASSY

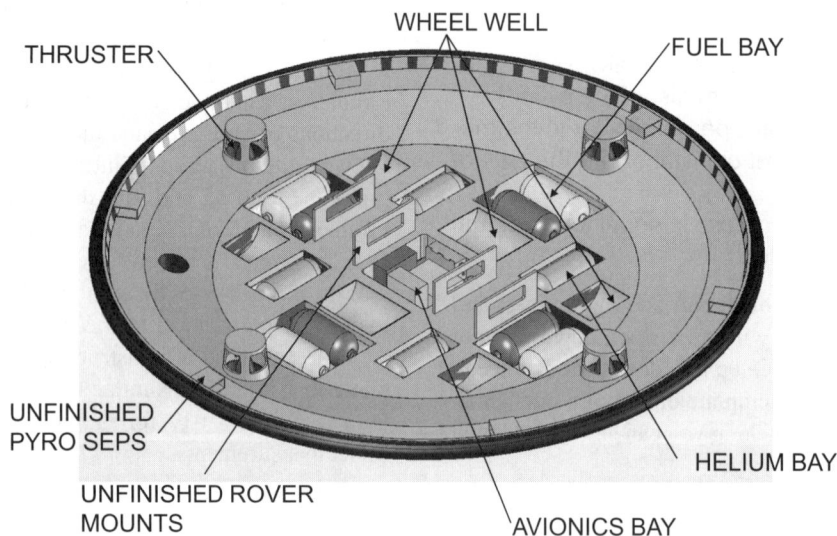

Shaded isometric views of 3D models are often used as presentation drawings. This isometric view of a proposed design for the MARGE Aeroshell was used as a presentation drawing to communicate the features of a concept developed by Malin Aerospace. *Courtesy of Malin Space Science Systems, Inc.*

## 9-32 AXONOMETRIC PROJECTION USING INTERSECTIONS

Before the advent of CAD engineering scholars devised methods to create an axonometric projection using projections from two orthographic views of the object. This method, called the method of intersections, was developed by Professors L. Eckhart and T. Schmid of the Vienna College of Engineering and was published in 1937.

To understand their method of axonometric projection, study Figure 9-39 as you read through the following steps. Assume that the axonometric projection of a rectangular object is given, and it is necessary to find the three orthographic projections: the top view, front view, and side view.

Place the object so that its principal edges coincide with the coordinate axes, and the plane of projection (the plane on which the axonometric projection is drawn) intersects the three coordinate planes in the triangle *ABC*.

From descriptive geometry, we know that lines *BC*, *CA*, and *AB* will be perpendicular, respectively, to axes *OX*, *OY*, and *OZ*. Any one of the three points *A*, *B*, or *C* may be assumed anywhere on one of the axes in order to draw triangle *ABC*.

To find the true size and shape of the top view, revolve the triangular portion of the horizontal plane *AOC*, which is in front of the plane of projection, about its base *CA*, into the plane of projection. In this case, the triangle is revolved inward to the plane of projection through the smallest angle made with it. The triangle would then be shown in its true size and shape, and you could draw the top view of the object

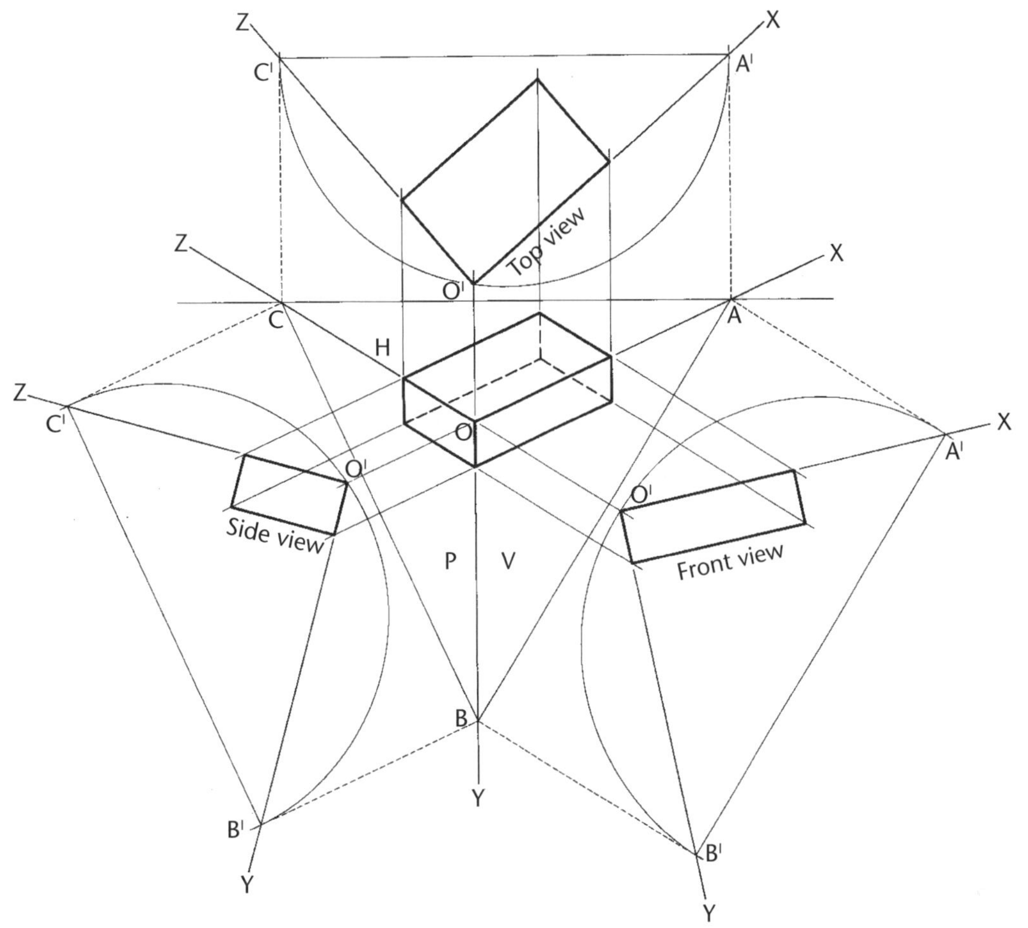

**Figure 9-39**  Views from an Axonometric Projection

in the triangle by projecting from the axonometric projection, as shown (since all width dimensions remain the same).

In the figure, the base *CA* of the triangle has been moved upward to *C'A'* so that the revolved position of the triangle will not overlap its projection.

The true sizes and shapes of the front view and side view can be found similarly, as shown in the figure.

Note that if the three orthographic projections, or in most cases any two of them, are given in their relative positions, as shown in Figure 9-39, the directions of the projections could be reversed so that the intersections of the projecting lines would determine the axonometric projection needed.

### Use of an Enclosing Box to Create an Isometric Sketch using Intersections

To draw an axonometric projection using intersections, it helps to make a sketch of the desired general appearance of the projection as shown in Figure 9-40. Even for complex objects the sketch need not be complete, just an enclosing box. Draw the projections of the coordinate axes *OX*, *OY*, and *OZ* parallel to the principal edges of the object, as shown in the sketch, and the three coordinate planes with the plane of projection.

Revolve the triangle *ABO* about its base *AB* as the axis into the plane of projection. Line *OA* will revolve to *O'A*, and this line, or one parallel to it, must be used as the baseline of the front view of the object. Draw the projecting lines from the front view to the axonometric parallel to the projection of the unrevolved Z-axis, as indicated in the figure.

Similarly, revolve the triangle *COB* about its base *CB* as the axis into the plane of projection. Line *CO* will revolve to *CO''*. Use this line, or one parallel to it, as the baseline of the side view. Make the direction of the projecting lines parallel to the projection of the unrevolved *X* axis, as shown.

Draw the front view baseline at a convenient location parallel to *O' X*. Use the parallel line you drew (*P3*) as the base and draw the front view of the object. Draw the side view baseline at a convenient location parallel to *O'' C*. Use it as the base (*P2*) for the side view of the object, as shown.

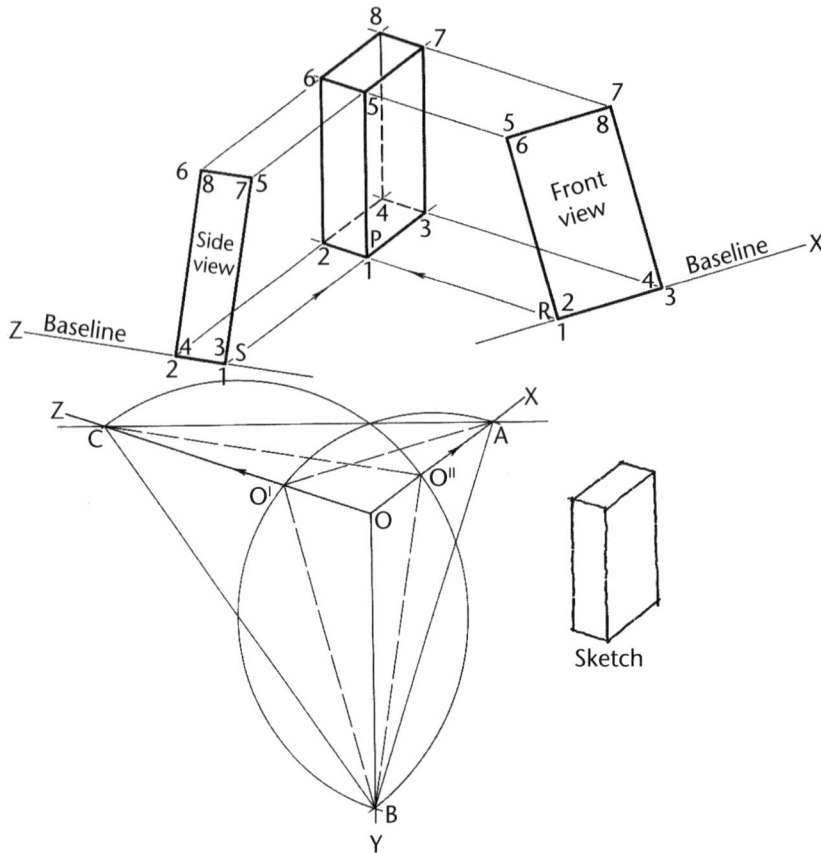

**Figure 9-40**    Axonometric Projection

From the corners of the front view, draw projecting lines parallel to $OZ$. From the corners of the side view, draw projecting lines parallel to $OX$. The intersections of these two sets of projecting lines determine the axonometric projection. It will be an isometric, a dimetric, or a trimetric projection, depending on the form of the sketch used as the basis for the projections.

If the angles formed by the three coordinate axes are equal, the projection is isometric; if two of them are equal, the projection is dimetric; and if none of the three angles are equal, the result is a trimetric projection.

To place the desired projection on a specific location on the drawing (Figure 9-40), select the desired projection $P$ of point 1, for example, and draw two projecting lines $PR$ ands $PS$ to intersect the two baselines and thereby to determine the locations of the two views on their baselines.

Another example of this method of axonometric projection is shown in Figure 9-41. In this case, it was only necessary to draw a sketch of the plan or base of the object in the desired position.

To understand how the axonometric projection in Figure 9-41 was created, examine the figure while reading through these steps.

Draw the axes with $OX$ and $OZ$ parallel to the sides of the sketch plan, and the remaining axis $OY$ in a vertical position.

Revolve triangles $COB$ and $AOB$, and draw the two baselines parallel to $O''C$ and $O'A$.

Choose point $P$, the lower front corner of the axonometric drawing, at a convenient place, and draw projecting lines toward the baselines parallel to axes $OX$ and $OZ$ to locate their positions. You can draw the views on the baselines or even cut them apart from another drawing and fasten them in place with drafting tape.

To draw the elliptical projection of the circle, use any points, such as $A$, on the circle in both front and side views. Note that point $A$ is the same altitude, $P$, above the baseline in both views. Draw the axonometric projection of point $A$ by projecting lines from the two views. You can project the major and minor axes this way, or by the methods shown in Figure 9-38.

True ellipses may be drawn by any of the methods shown in the Appendix or with an ellipse template. An approximate ellipse is fine for most drawings.

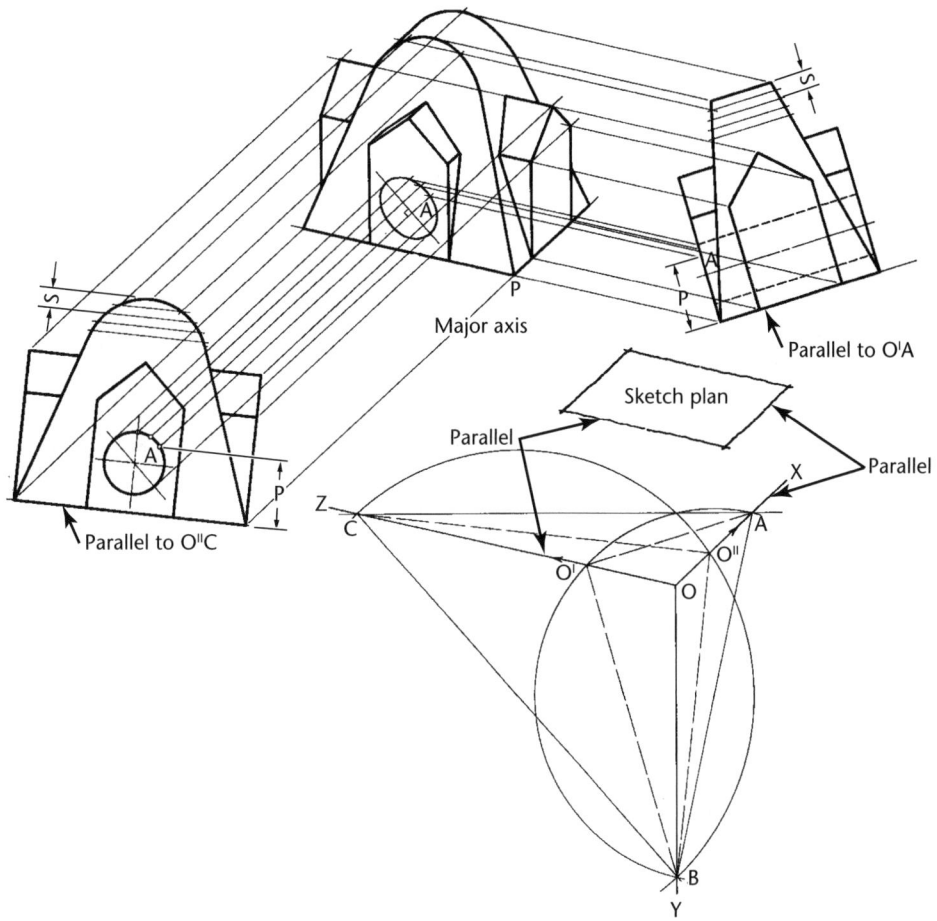

**Figure 9-41**    Axonometric Projection

## 9-33 COMPUTER GRAPHICS

Pictorial drawings of all sorts can be created using 3D CAD (Figures 9-42, 9-43). To create pictorials using 2D CAD, use projection techniques similar to those presented in this chapter. The advantage of 3D CAD is that once you make a 3D model of a part or assembly, you can change the viewing direction at any time for orthographic, isometric, or perspective views. You can also apply different materials to the drawing objects and shade them to produce a high degree of realism in the pictorial view.

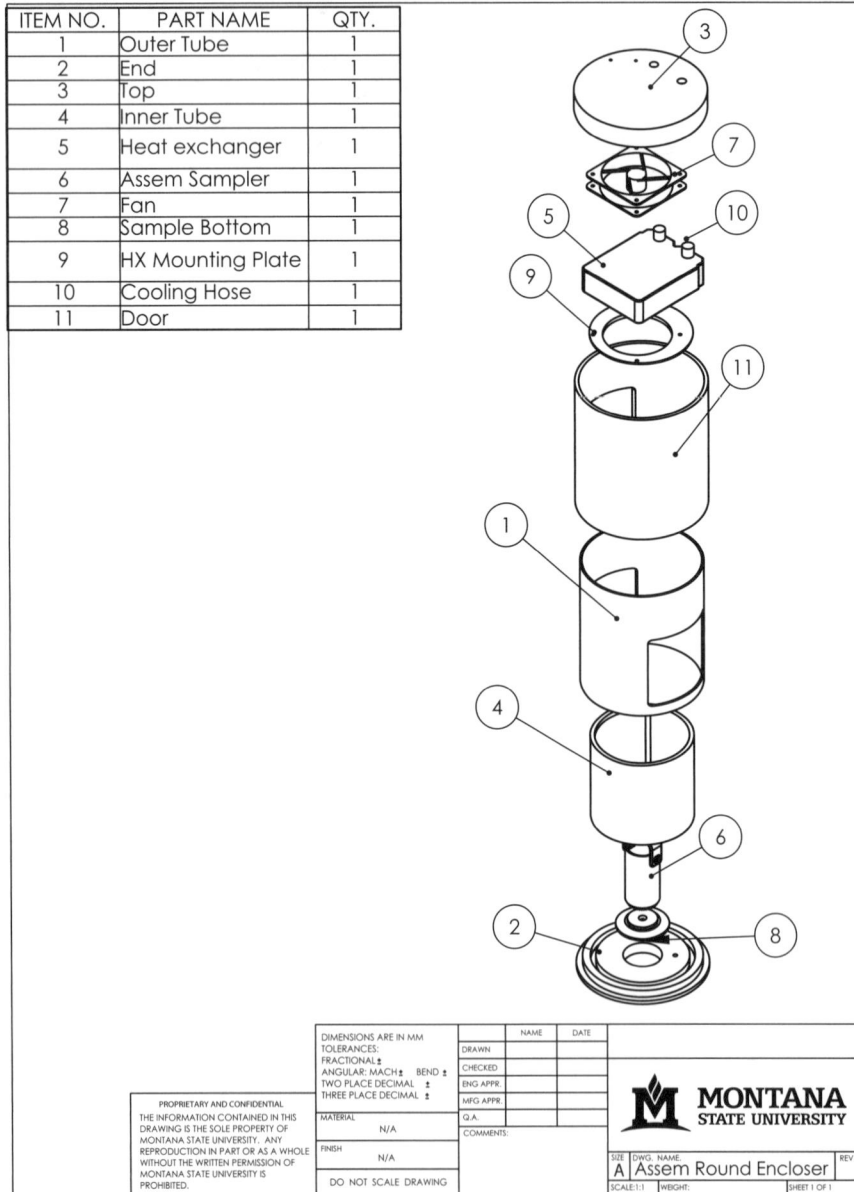

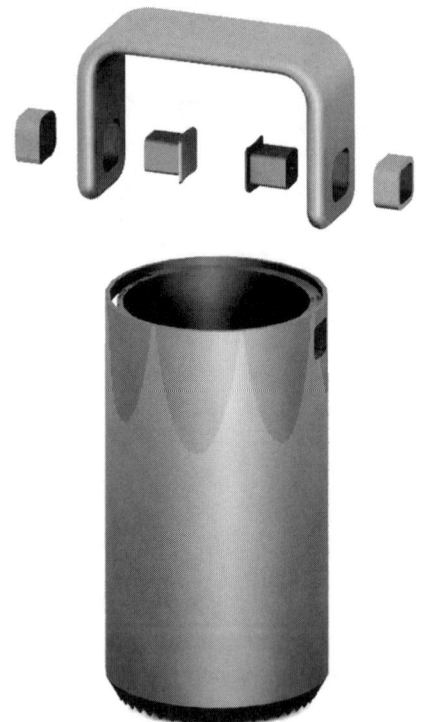

| ITEM NO. | PART NAME | QTY. |
|---|---|---|
| 1 | Outer Tube | 1 |
| 2 | End | 1 |
| 3 | Top | 1 |
| 4 | Inner Tube | 1 |
| 5 | Heat exchanger | 1 |
| 6 | Assem Sampler | 1 |
| 7 | Fan | 1 |
| 8 | Sample Bottom | 1 |
| 9 | HX Mounting Plate | 1 |
| 10 | Cooling Hose | 1 |
| 11 | Door | 1 |

**Figure 9-42** Shaded Dimetric Pictorial View from a 3D Model. *Courtesy of Robert Kincaid.*

| DIMENSIONS ARE IN MM | NAME | DATE | |
|---|---|---|---|
| TOLERANCES: | | | |
| FRACTIONAL ± | DRAWN | | |
| ANGULAR: MACH± BEND ± | CHECKED | | |
| TWO PLACE DECIMAL ± | ENG APPR. | | |
| THREE PLACE DECIMAL ± | MFG APPR. | | |
| | Q.A. | | |
| MATERIAL N/A | COMMENTS: | | |
| FINISH N/A | | | |

PROPRIETARY AND CONFIDENTIAL
THE INFORMATION CONTAINED IN THIS DRAWING IS THE SOLE PROPERTY OF MONTANA STATE UNIVERSITY. ANY REPRODUCTION IN PART OR AS A WHOLE WITHOUT THE WRITTEN PERMISSION OF MONTANA STATE UNIVERSITY IS PROHIBITED.

DO NOT SCALE DRAWING

**MONTANA STATE UNIVERSITY**

SIZE A | DWG. NAME. Assem Round Encloser | REV.
SCALE:1:1 | WEIGHT: | SHEET 1 OF 1

**Figure 9-43** Isometric Assembly Drawing. *Courtesy of PTC.*

# CAD AT WORK

## Isometric Sketches Using AutoCAD Software

Need a quick isometric sketch? AutoCAD software has special drafting settings for creating an isometric style grid.

Figure A shows the Drafting Settings dialog box in AutoCAD. When you check the button for Isometric Snap, the software calculates the spacing needed for an isometric grid. You can use it to make quick pictorial sketches like the example shown in Figure B. Piping diagrams are often done this way, although they can also be created using 3D tools.

Even though the drawing in Figure B looks 3D, it is really drawn in a flat 2D plane. You can observe this if you change the viewpoint so you are no longer looking straight onto the view.

The Ellipse command in AutoCAD has a special Isocircle option that makes drawing isometric ellipses easy. The isocircles are oriented in different directions depending on the angle of the snap cursor. Figure C shows isocircles and snap cursors for the three different orientations. In the software, you press CTRL and E simultaneously to toggle the cursor appearance.

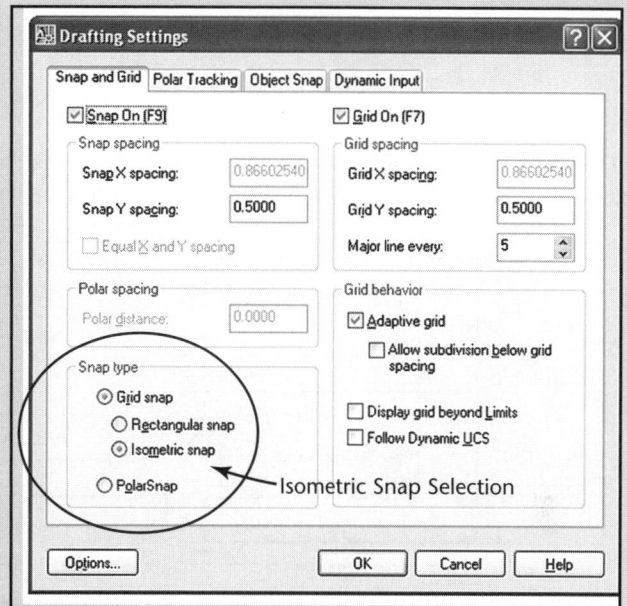

(A) Selecting isometric snap in the AutoCAD drafting settings dialog box.

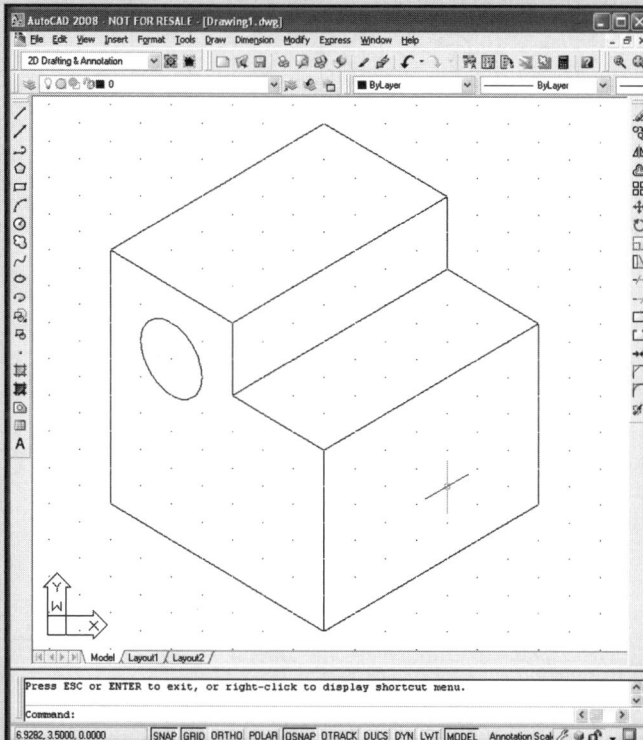

(B) A pictorial sketch created from a flat drawing using isometric snap.

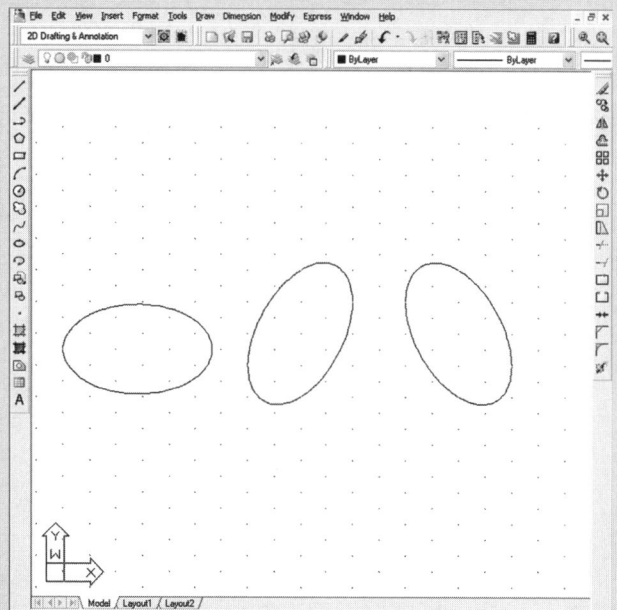

Center cursor    Right cursor    Left cursor

(C) Variously oriented isometric circles and the corresponding snap cursors used to create them.

## 9-34 EXERCISES

### Axonometric Problems

EX 9-1–9-9 are to be drawn axonometrically. The earlier isometric sketches may be drawn on isometric paper, and later sketches should be made on plain drawing paper.

**EX9-1**

(1) Make freehand isometric sketches. (2) Use CAD to make isometric drawings. (3) Make dimetric drawings. (4) Make trimetric drawings with axes chosen to show the objects to best advantage. Dimension your drawing only if assigned by your instructor.

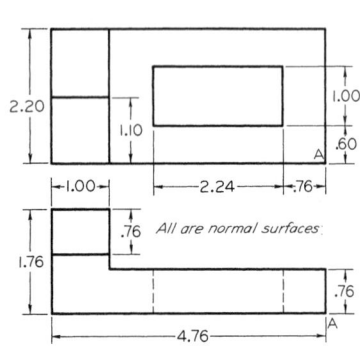

1 KEY PLATE

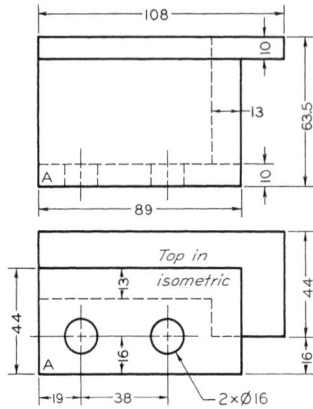

METRIC

2 BASE

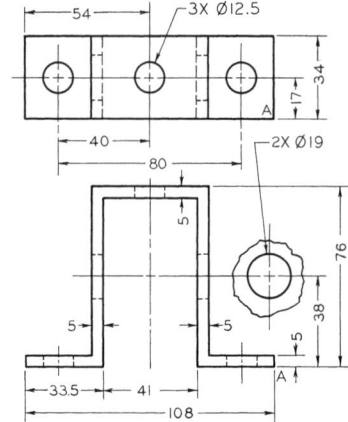

METRIC

3 STRAP

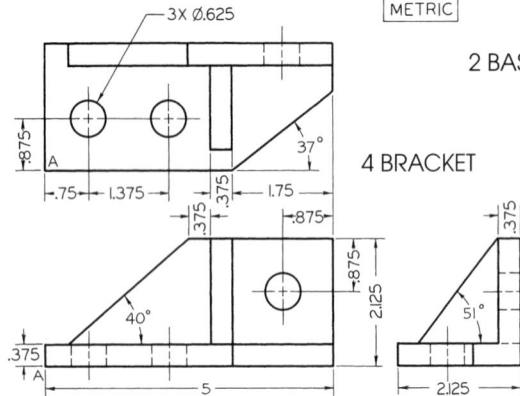

4 BRACKET

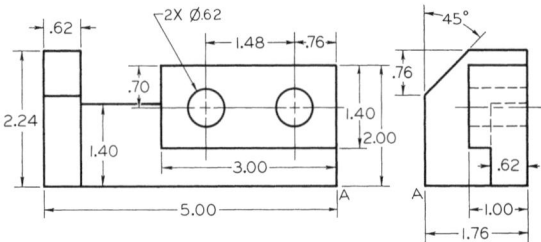

5 CUTTER BLOCK

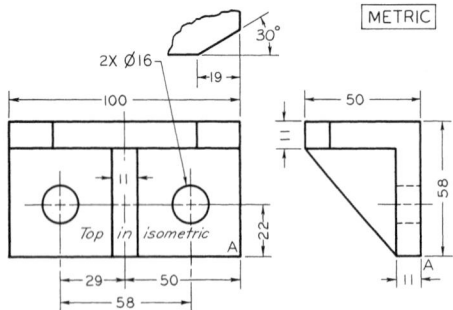

6 BRACKET

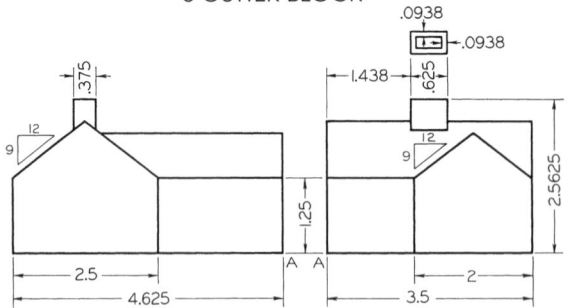

7 HOUSE MODEL

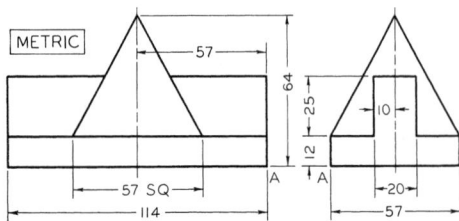

8 GUIDE BLOCK

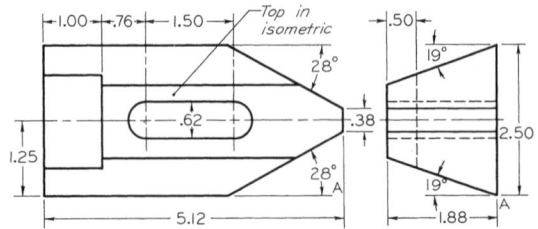

9 FINGER

**EX9-2**

(1) Make freehand isometric sketches. (2) Use CAD to make isometric drawings. (3) Make dimetric drawings. (4) Make trimetric drawings with axes chosen to show the objects to best advantage. Dimension your drawing only if assigned by your instructor.

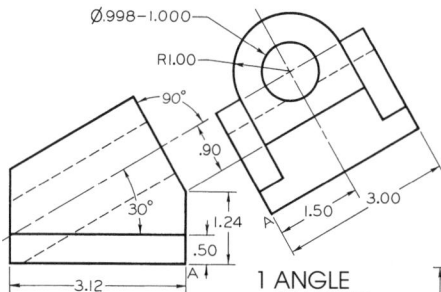

1 ANGLE BEARING

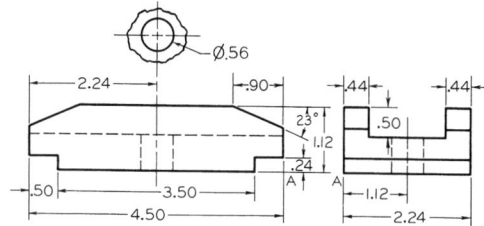

2 TAILSTOCK CLAMP

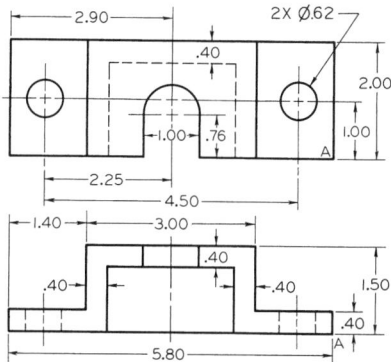

3 TABLE SUPPORT

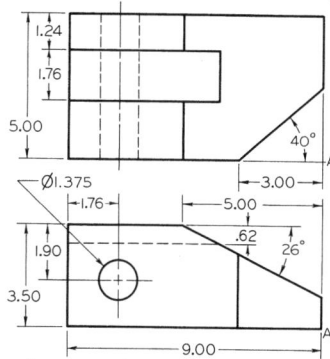

4 WEDGE

Draw 1:2 scale

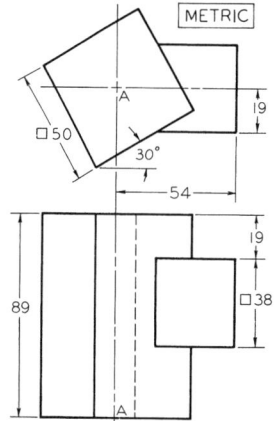

5 INTERSECTION

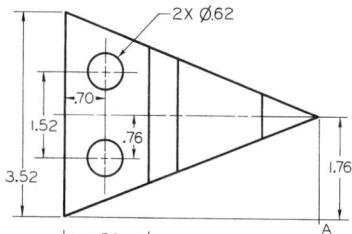

6 CONTROL BLOCK

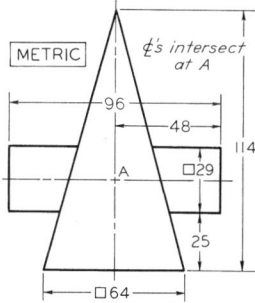

7 INTERSECTION

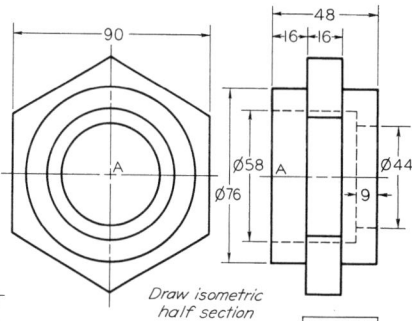

Draw isometric half section

8 HEX CAP

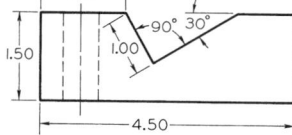

Draw 1:2 scale

9 BOOK END

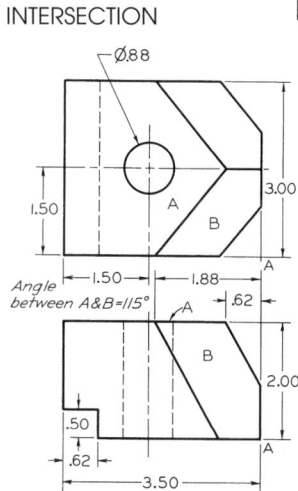

10 LOCATOR

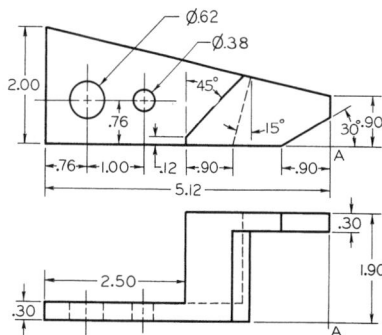

11 TRIP ARM

**EX9-3**

(1) Make freehand isometric sketches. (2) Use CAD to make isometric drawings. (3) Make dimetric drawings. (4) Make trimetric drawings with axes chosen to show the objects to best advantage. Dimension your drawing only if assigned by your instructor.

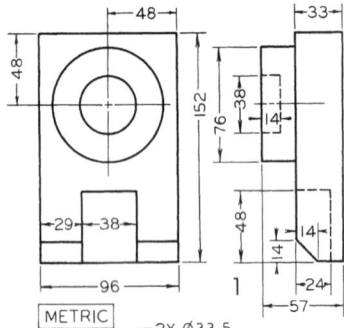

1

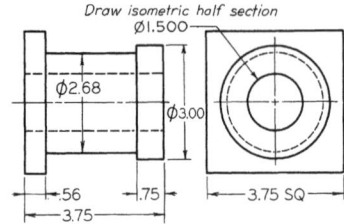

2

3

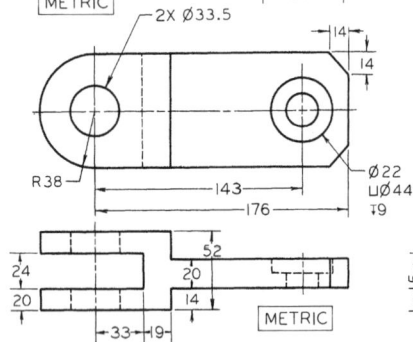

4

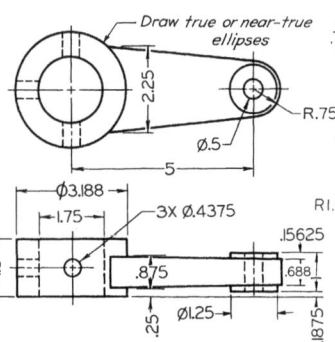

5

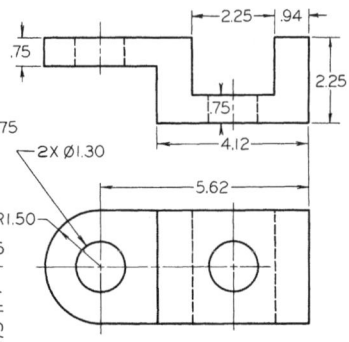

6

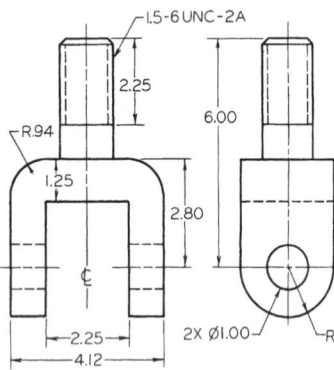

7

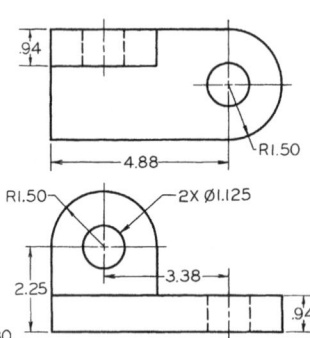

8

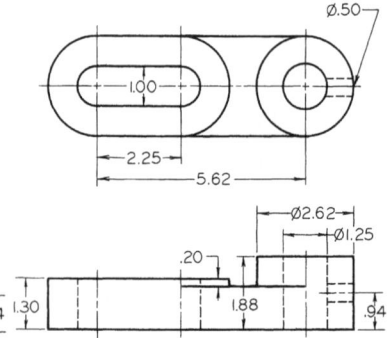

9

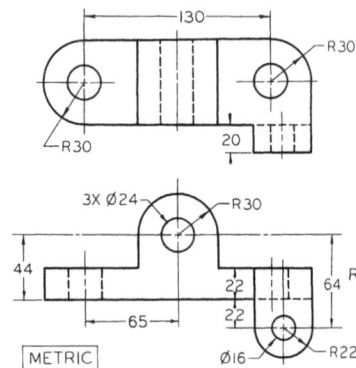

10

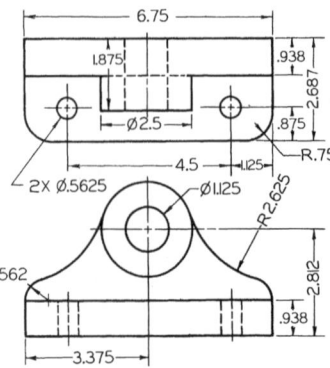

11

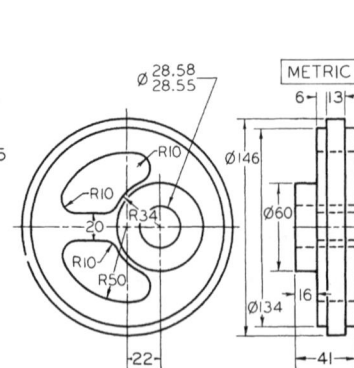

12

**EX9-4**

(1) Make freehand isometric sketches. (2) Use CAD to make isometric drawings. (3) Make dimetric drawings. (4) Make trimetric drawings with axes chosen to show the objects to best advantage. Dimension your drawing only if assigned by your instructor.

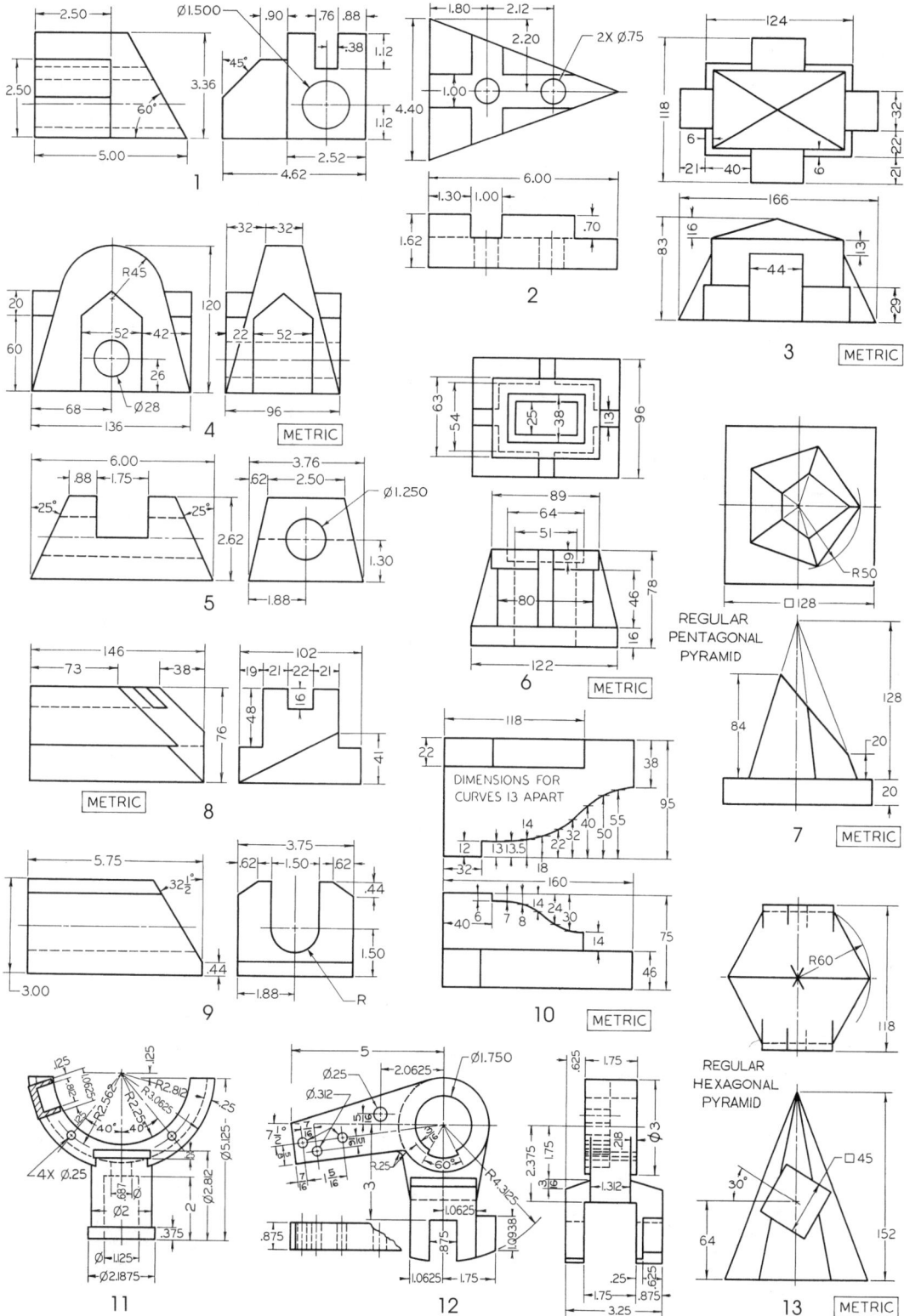

**EX9-5**

(1) Make freehand isometric sketches. (2) Use CAD to make isometric drawings. (3) Make dimetric drawings. (4) Make trimetric drawings with axes chosen to show the objects to best advantage. Dimension your drawing only if assigned by your instructor.

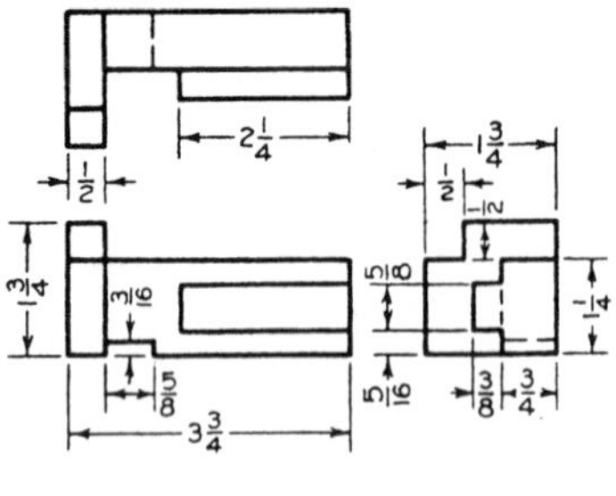

1

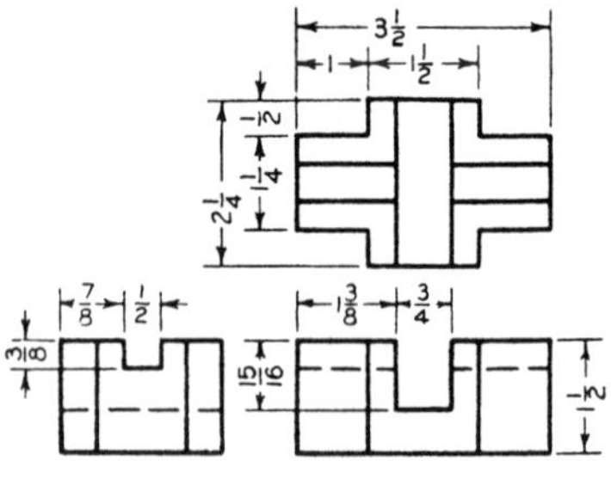

2

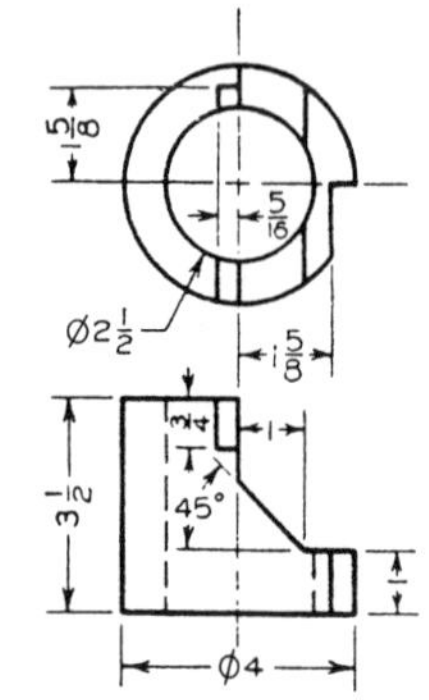

3

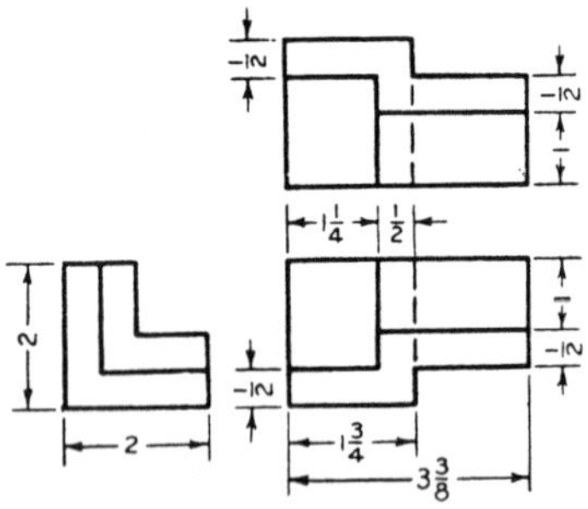

4

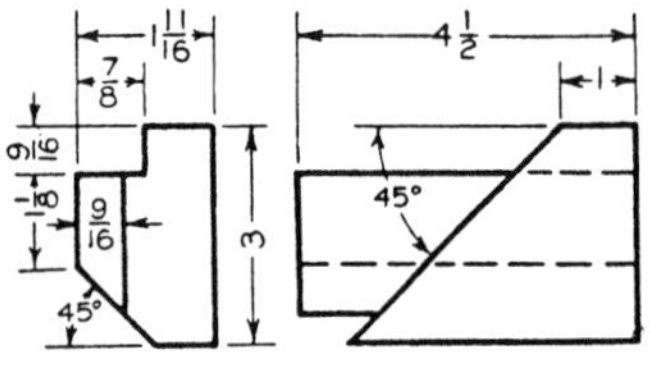

5

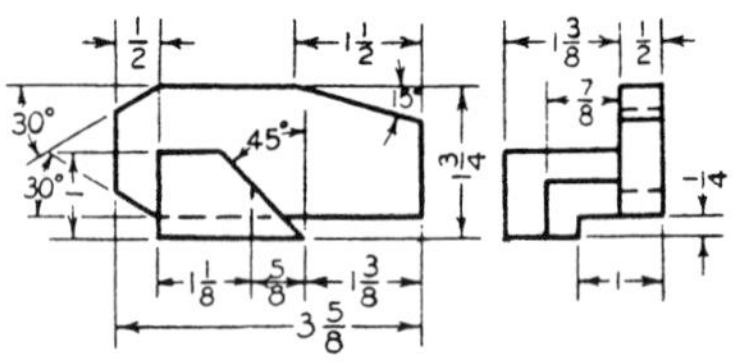

6

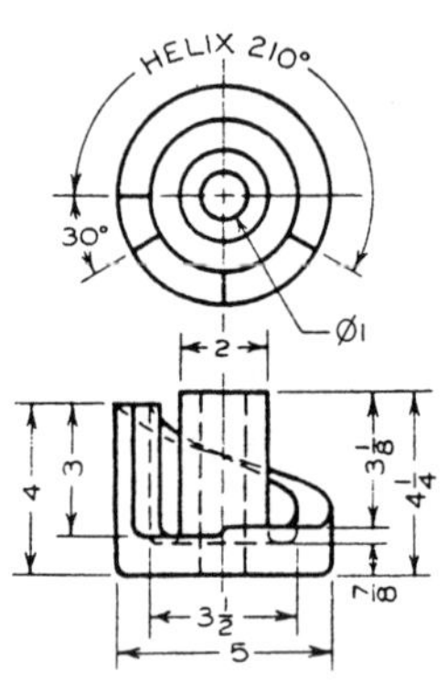

7

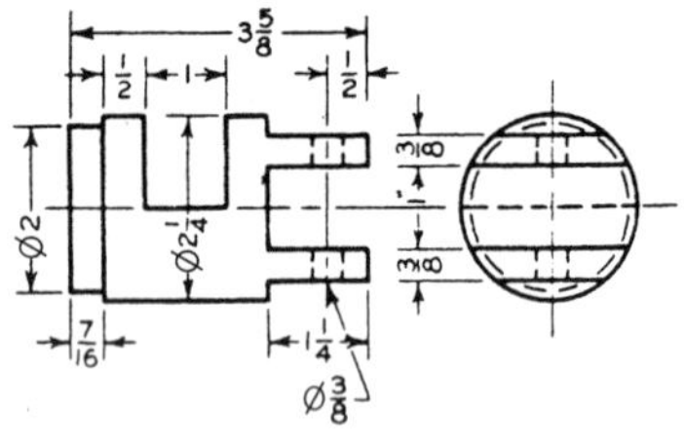

8

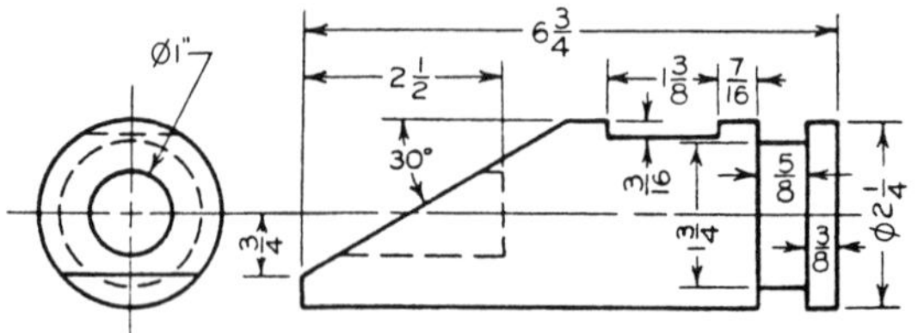

9

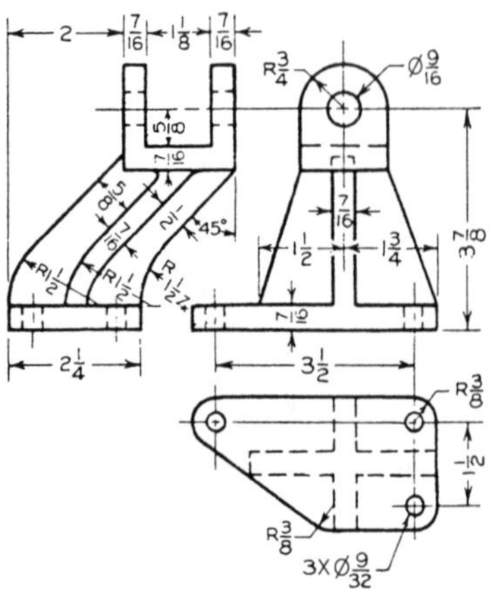

10

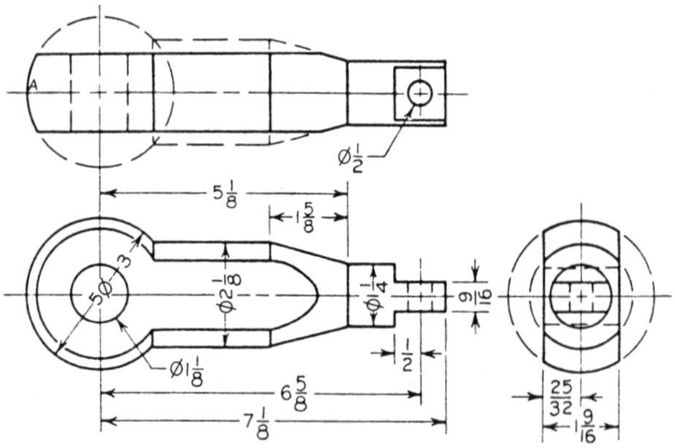

11

### EX9-6

Draw the nylon collar nut as follows. (1) Make an isometric freehand sketch. (2) Make an isometric drawing using CAD.

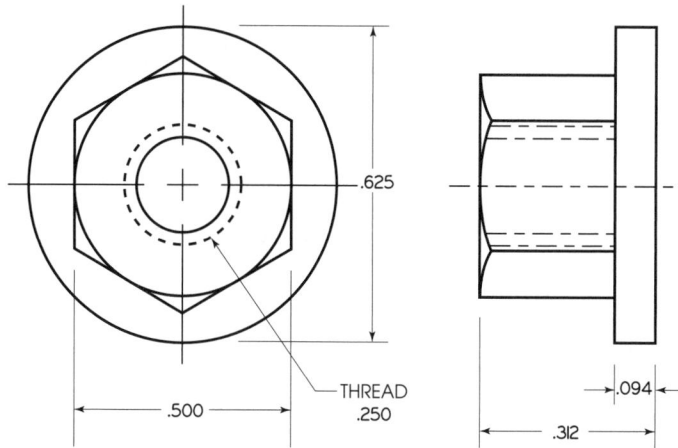

.625

THREAD
.250

.500

.094

.312

### EX9-7

Draw the plastic T-handle plated steel stud as follows. (1) Make a diametric drawing using CAD. (2) Make a trimetric drawing using CAD.

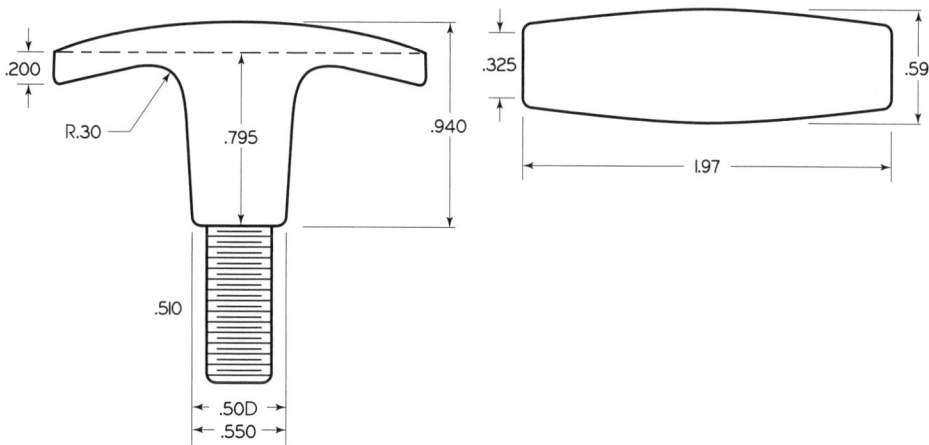

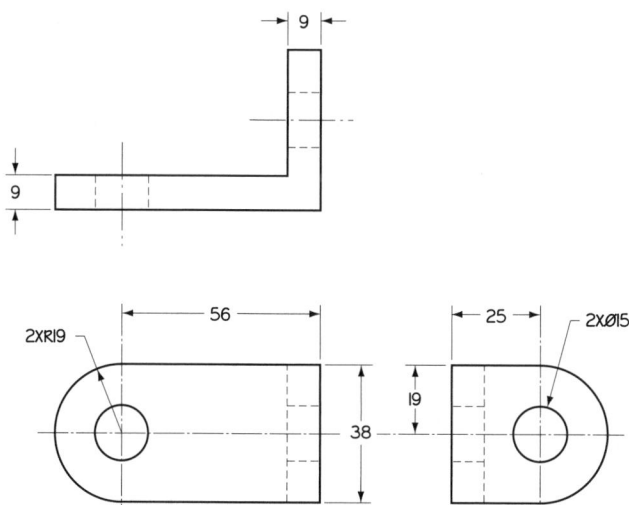

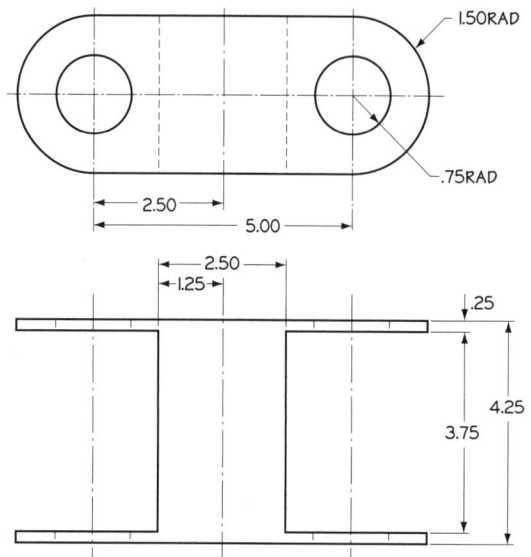

.200

R.30

.795

.940

.510

.50D

.550

.325

.59

1.97

9

9

1.50RAD

.75RAD

2.50

5.00

56

25

2XØ15

2XR19

19

38

2.50

1.25

.25

4.25

3.75

### EX9-8

Draw the mounting plate as follows. (1) Make an isometric freehand sketch. (2) Make isometric drawings using CAD.

### EX9-9

Draw the hanger as follows. (1) Make an isometric freehand sketch. (2) Make isometric drawings using CAD.

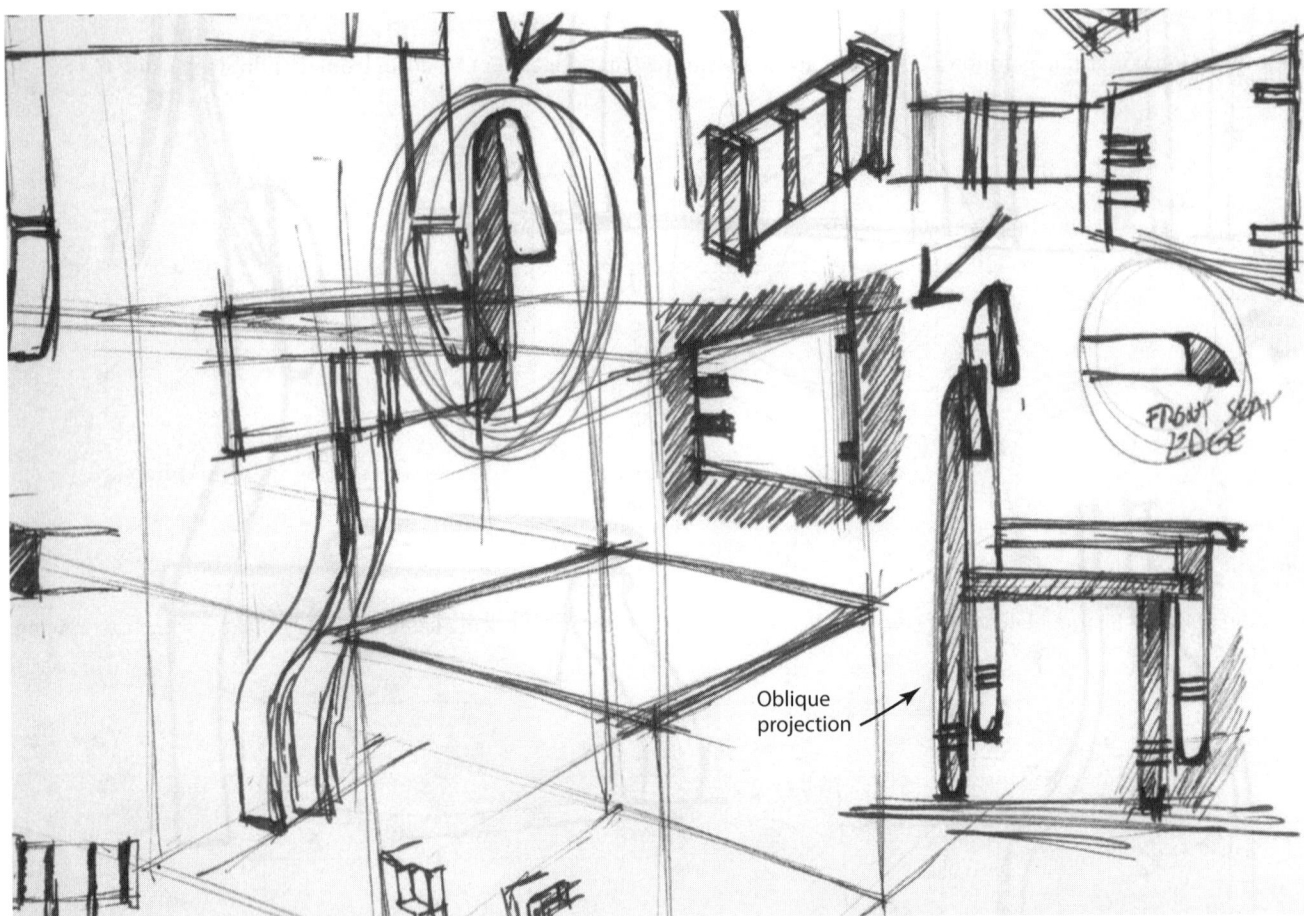

Oblique Projection in a Sketch. *Courtesy of Douglas Wintin.*

## 9-35 OBLIQUE PROJECTION

An oblique projection is a drawing where the projectors are parallel to each other but are at an angle other than 90° (oblique) to the plane of projection.

Oblique drawings are primarily used as a sketching technique. The front view in an oblique projection is the same as the front view in a multiview drawing, so circles and angles parallel to the front plane show as true size and shape and are therefore easy to draw. The downside is that while drawing circles in the front surface is easier, they are more difficult to draw in the top or side surfaces.

Oblique projection does not look as realistic as axonometric projection because the depth appears distorted. Using a proportion of the depth measurement makes the object's depth look more realistic. Oblique drawings are rarely created using CAD since it is more accurate to create a 3D model representing the actual shape and then show views of it using either isometric or perspective projection.

## 9-36 UNDERSTANDING OBLIQUE PROJECTIONS

In **oblique projections**, the projectors are parallel to each other but are not perpendicular to the **plane of projection.** To create an oblique projection, orient the object so that one of its principal faces is parallel to the plane of projection as illustrated in Figure 9-44. Bear in mind that the goal is to produce an informative drawing. Orient the surface showing the most information about the shape of the object so it is parallel to the plane of projection.

Figures 9-45, 9-46, and 9-47 show comparisons between oblique projection, orthographic projection, and isometric projection for a cube and a cylinder. In oblique projection, the front face is identical in the front orthographic view.

If an object is placed with one of its faces parallel to the plane of projection, the projected view will show the face true size and shape. This makes oblique drawings easier than isometric or other axonometric projection such as diametric or trimetric for many shapes. Surfaces that are not parallel to the plane of projection will not project in true size and shape.

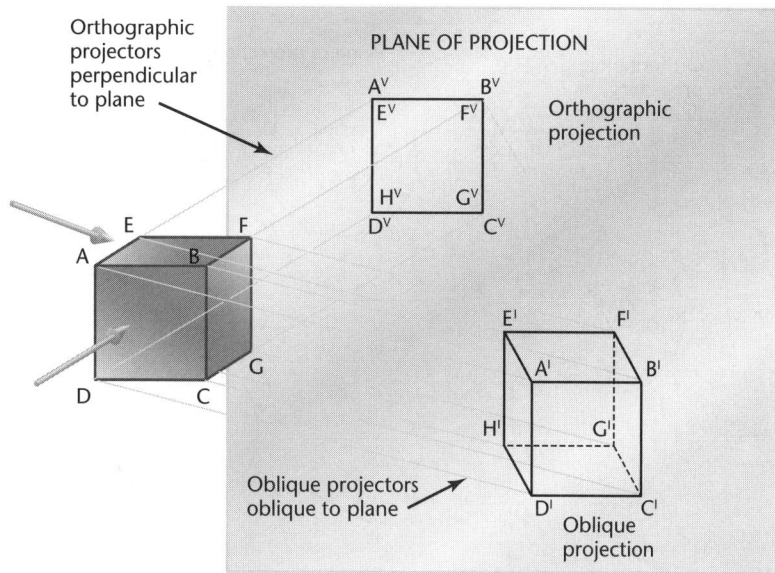

**Figure 9-44**    Comparison of Oblique and Orthographic Projections

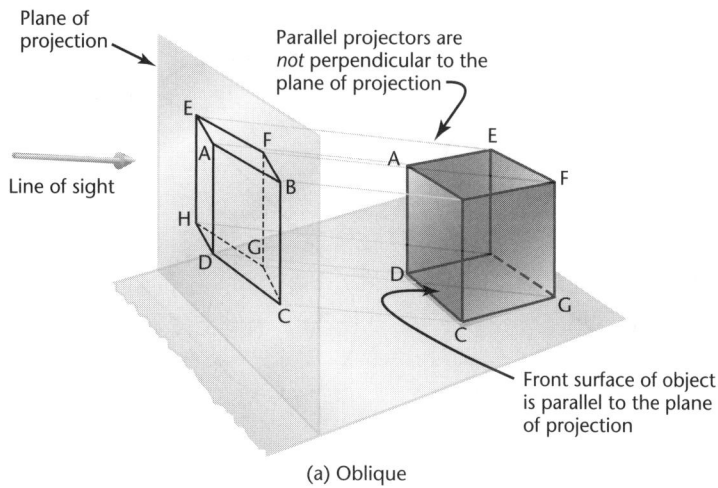

(a) Oblique

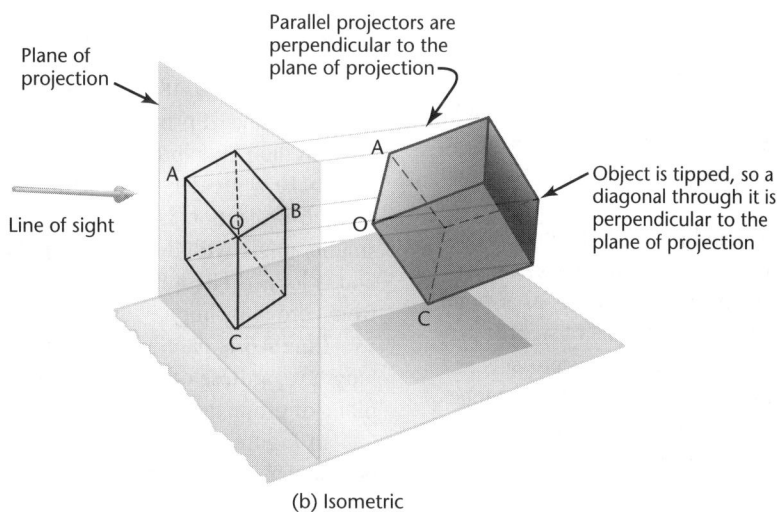

(b) Isometric

**Figure 9-45**    Comparison of Oblique and Isometric Projections.

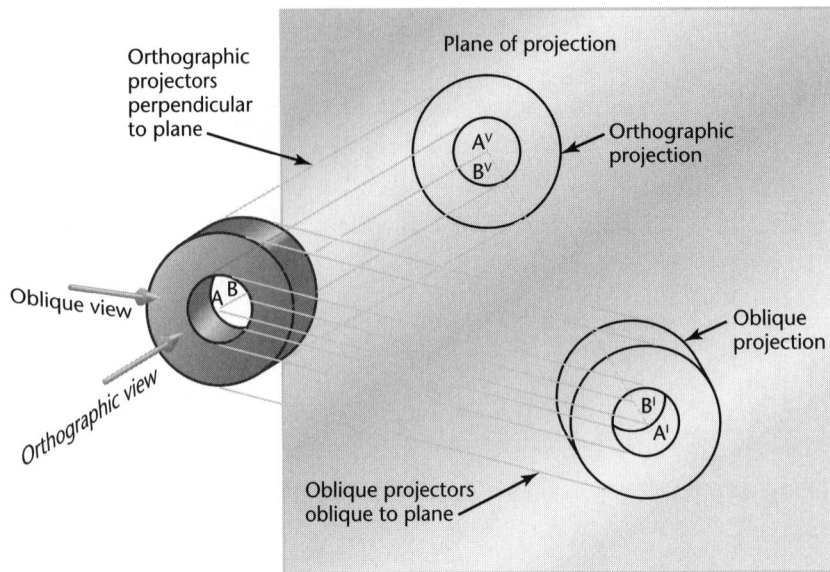

**Figure 9-46**    Circles Parallel to Plane of Projection

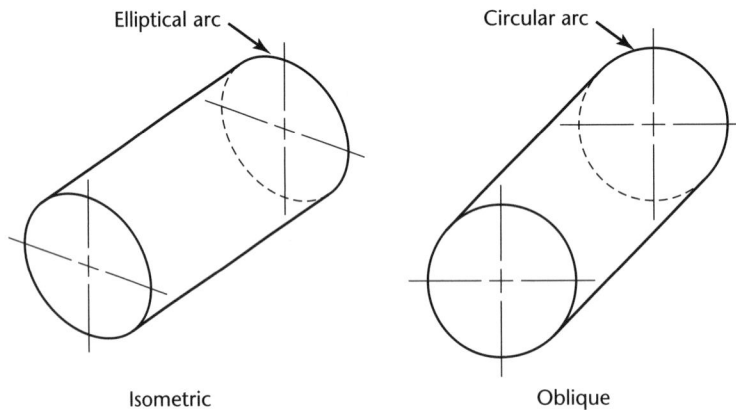

**Figure 9-47**    Comparison of Oblique and Isometric Projections for a Cylinder

In axonometric drawings, circular shapes nearly always project as ellipses because the principal faces are inclined to the viewing plane. If you position the object so that those surfaces are parallel to the viewing plane and draw an oblique projection, the circles will project as true shape and are easy to draw.

Oblique projections show the object from an angle where the projectors are not parallel to the viewing plane. An axis (like *AB* of the cylinder in Figure 9-46) projects as a point (*A^V B^V*) in the orthographic view where the line of sight is parallel to *AB*. But in the oblique projection, the axis projects as a line *A´B´*. The more nearly the direction of sight approaches being perpendicular to the plane of projection, the closer the oblique projection moves toward the orthographic projection, and the shorter *A´B´* becomes.

### Directions of Projectors

In Figure 9-48, the projectors make an angle of 45° with the plane of projection; so line *CD´*, which is perpendicular to the plane, projects true length at *C´D´*. If the projectors make a greater angle with the plane of projection, the oblique projection is shorter, and if the projectors make a smaller angle with the plane of projection, the oblique projection is longer. Theoretically, *CD´* could project in any length from zero to infinity.

Line *AB* is parallel to the plane and will project in true length regardless of the angle the projectors makes with the plane of projection.

In Figure 9-44, the lines *AE*, *BF*, *CG*, and *DH* are perpendicular to the plane of projection and project as parallel

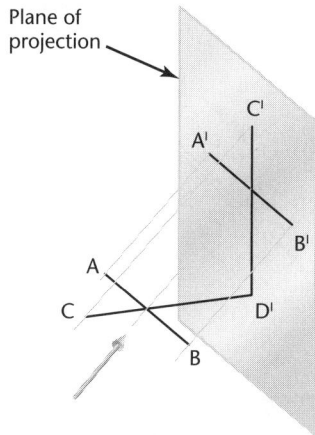

**Figure 9-48** Lengths of Projections

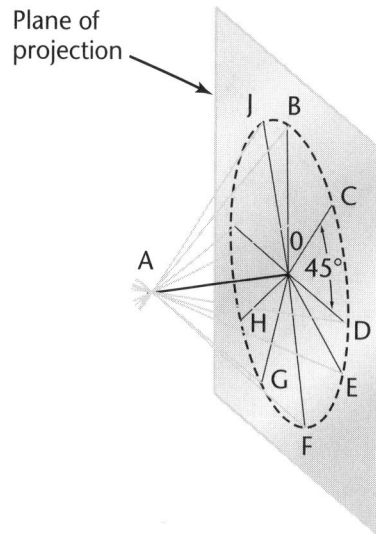

**Figure 9-49** Directions and Projections

inclined lines *A'E*, *B'F*, *C'G*, and *D'H'* in the oblique projection. These lines on the drawing are called the receding lines. They may be any length, depending on the direction of sight.

What angle will these lines be in the drawing measured from horizontal? In Figure 9-49, line *AO* is perpendicular to the plane of projection, and all the projectors make angles of 45° with it; therefore, all the oblique projections like *BO*, *CO*, and *DO* are equal in length to line *AO*. You can select the projectors at any angle and still produce any desired angle with the plane of projection. The directions of the projections *BO*, *CO*, *DO*, and so on, are independent of the angles the projectors make with the plane of projection. Traditionally, the angle used is 45° (*CO* in the figure), 30°, or 60° with horizontal.

## 9-37 LENGTH OF RECEDING LINES

Theoretically, **oblique projectors** can be at any angle to the plane of projection other than perpendicular or paral-

lel. The difference in the angle you choose causes **receding lines** of oblique drawings to vary in angle and in length from near zero to near infinity. However, many of those choices would not produce very useful drawings. Figure 9-50 shows a variety of oblique drawings with different lengths for the receding lines.

Since we see objects in perspective (where receding parallel lines appear to converge) oblique projections look unnatural to us. The longer the object in the receding direction, the more unnatural the object appears. For example, the object shown in Figure 9-50a is an isometric drawing of a cube where the receding lines are shown full length. They appear to be too long and they appear to widen toward the rear of the block. Figure 9-51b shows how unnatural the familiar pictorial image of railroad tracks leading off into the distance would look if drawn in an oblique projection. To give a more natural appearance, show long objects with the long axis parallel to the view, as shown in Figure 9-52.

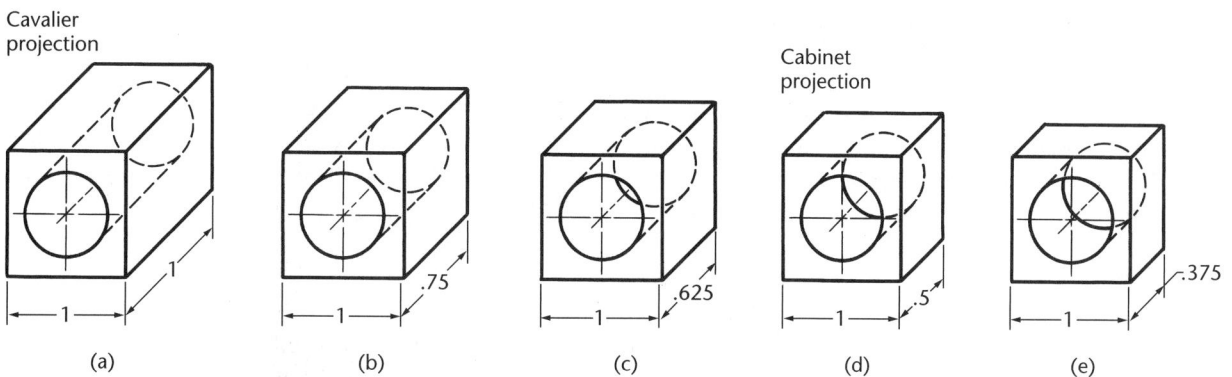

**Figure 9-50** Foreshortening of Receding Lines

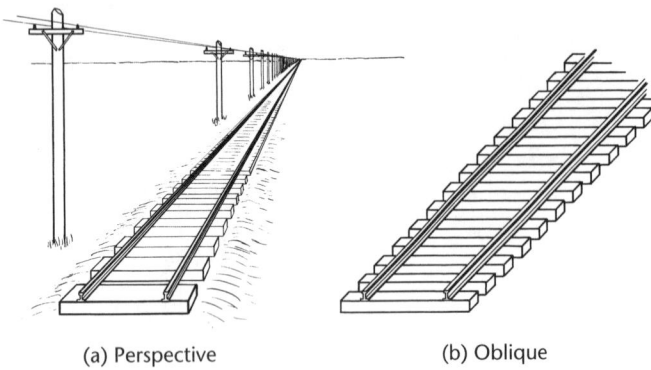

(a) Perspective                (b) Oblique

**Figure 9-51**    Unnatural Appearance of Oblique Drawing

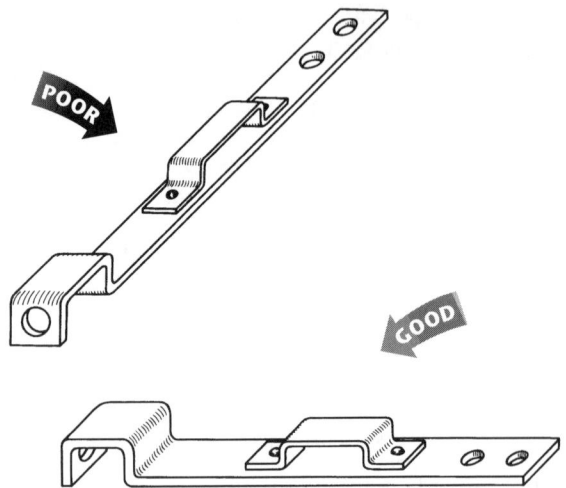

**Figure 9-52**    Long Axis Parallel to Plane of Projection

## Cavalier Projection

When the receding lines are true length—(the projectors make an angle of 45° with the plane of projection)—the oblique drawing is called a **cavalier projection** (Figure 9-50a). Cavalier projections originated in the drawing of medieval fortifications and were made on horizontal planes of projection. On these fortifications the central portion was higher than the rest, and it was called cavalier because of its dominating and commanding position.

## Cabinet Projection

When the receding lines are drawn to half size (Figure 9-50d), the drawing is known as a **cabinet projection**. This term is attributed to the early use of this type of oblique drawing in the furniture industries.

## 9-38  ANGLES OF RECEDING LINES

Typically 45° is a good angle to choose for receding lines, since this makes it is easy to sketch through the diagonal of squares on grid paper or using the snap feature in CAD. Other common angles for the receding lines are 30° or 60°, but any convenient angle may be used.

Choose an angle based on the shape of the object and the location of its significant features. For example, the larger angle used in Figure 9-53a shows a better view of the rectangular recess on the top of the object, while the smaller

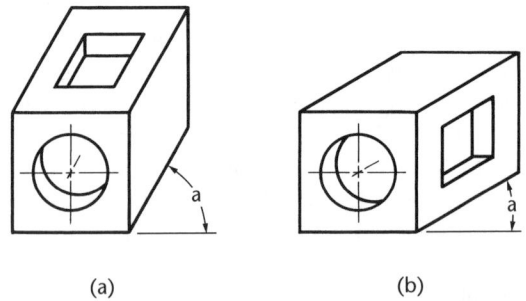

(a)                                (b)

**Figure 9-53**    Angle of Receding Axis

angle in Figure 9-53b shows a feature on the side of the object.

## 9-39  CHOICE OF POSITION

Orient the view so that important shapes are parallel to the viewing plane as shown in Figure 9-54. In Figure 9-54a and c, the circles and circular arcs are shown in their true shapes and are easy to draw. In Figure 9-54b and d they are not shown in true shape and must be plotted as free curves or ellipses.

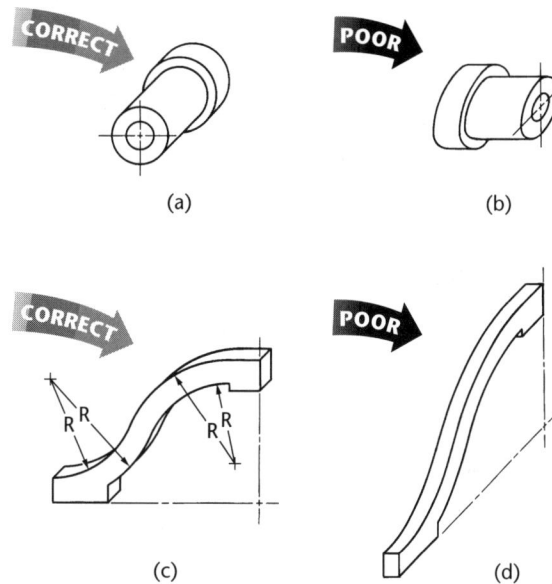

**Figure 9-54**    Essential Contours Parallel to Plane of Projection

# STEP BY STEP

### Using Box Construction To Create An Oblique Drawing

Follow these steps to draw a cavalier drawing of the rectangular object shown in the two orthographic views.

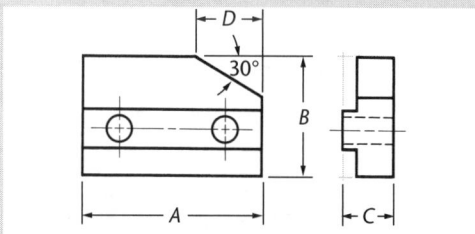

1. Lightly block in the overall width (*A*) and height (*B*) to form the enclosing rectangle for the front surface. Select an angle for the receding axis (*OZ*) and draw the depth (*C*) along it.

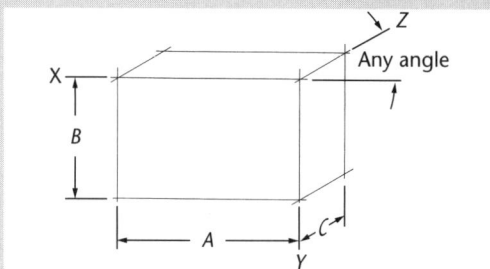

2. Lightly block in the details of the front surface shape including the two holes which will appear round. Add the details of the right side surface shape. Extend lines along the receding axis connecting the edges to form the remaining surface edges.

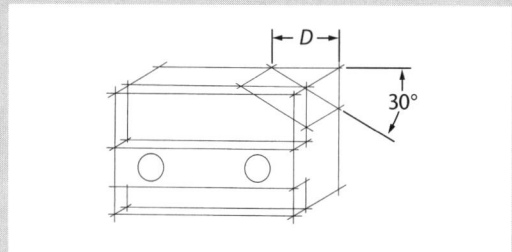

3. Darken the final lines.

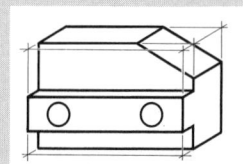

# STEP BY STEP

## Using Skeleton Construction In Oblique Drawing

Oblique drawings are especially useful for showing objects that have cylindrical shapes built on axes or centerlines. Construct an oblique drawing of the part shown using projected centerlines using these steps.

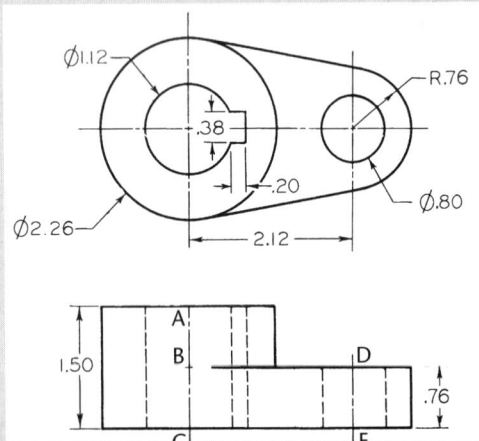

1. Position the object in the drawing so that the circles shown in the given top view are parallel to the plane of projection. This will show true shape in the oblique view. Draw the circular shape in the front plane of the oblique view and extend the center axis along the receding axis of the oblique drawing.

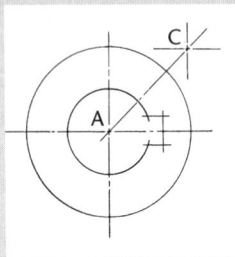

2. Add the centerline skeleton as shown.

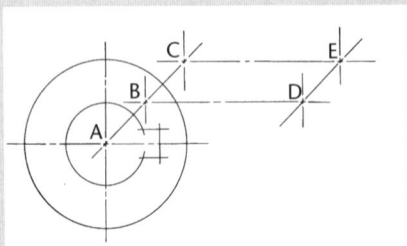

3. Build the drawing from the location of these centerlines.

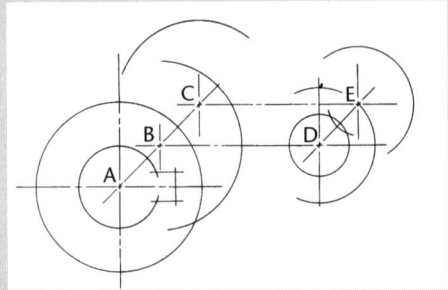

4. Construct all important points of tangency.

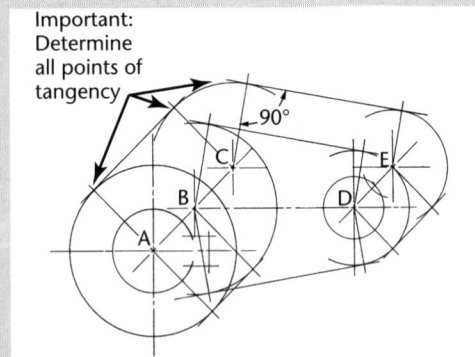

Important: Determine all points of tangency

5. Darken the final cavalier drawing.

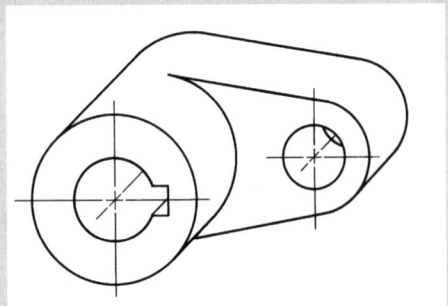

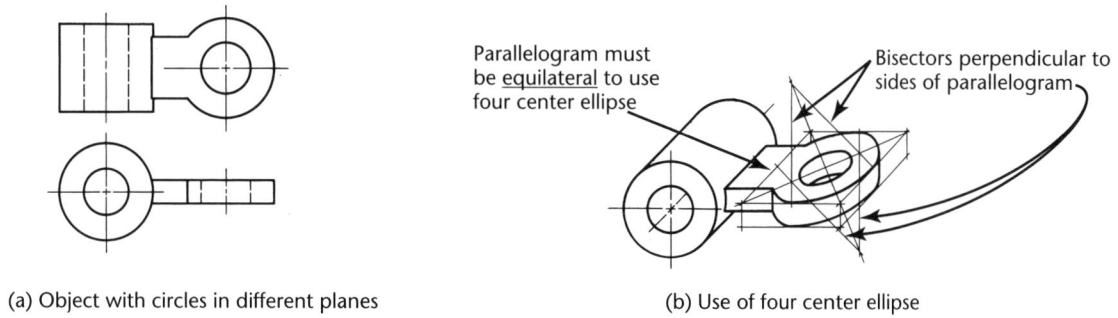

(a) Object with circles in different planes

Parallelogram must
be <u>equilateral</u> to use
four center ellipse

Bisectors perpendicular to
sides of parallelogram

(b) Use of four center ellipse

**Figure 9-55**    Circles and Arcs Not Parallel to Plane of Projection

## 9-40  ELLIPSES FOR OBLIQUE DRAWINGS

It is not always possible to orient the view of an object so that all of its rounded shapes are parallel to the plane of projection. For example, the object shown in Figure 9-55a has two sets of circular contours in different planes. Both cannot be simultaneously placed parallel to the plane of projection, so in the oblique projection, one of them must be viewed as an ellipse.

If you are sketching, you can just block in the enclosing rectangle and sketch the ellipse tangent to its sides. Using CAD, you can draw the ellipse by specifying its center and major and minor axes. In circumstances where a CAD system is not available, if you need an accurate ellipse, you can draw them by hand using one of the following methods.

### Alternate Four Center Ellipses

Normal four center ellipses can be made only in equilateral parallelograms so they cannot be used in an oblique

drawing where the **receding axis** is foreshortened. Instead, use this alternate four center ellipse to approximate ellipses in oblique drawings.

Draw the ellipse on two centerlines, as shown in Figure 9-56. This is the same method as is sometimes used in isometric drawings, but in oblique drawings it appears slightly different according to the different angles of the receding lines.

First, draw the two centerlines. Then, from the center, draw a construction circle equal to the diameter of the actual hole or cylinder. The circle will intersect each centerline at two points. From the two points on one centerline, draw perpendiculars to the other centerline. Then, from the two points on the other centerline, draw perpendiculars to the first centerline. From the intersections of the perpendiculars, draw four circular arcs, as shown.

### Four Center Ellipse for Cavalier Drawings

For cavalier drawings, you can use the normal four center ellipse method to draw ellipses (Figure 9-55b). This method can be used only in cavalier drawing because the

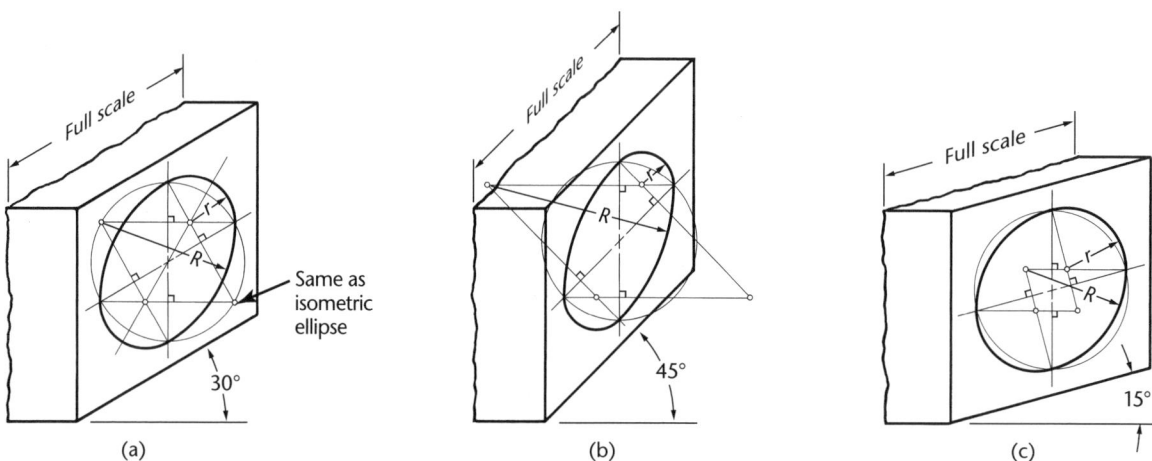

(a)    (b)    (c)

**Figure 9-56**    Alternate Four Center Ellipse

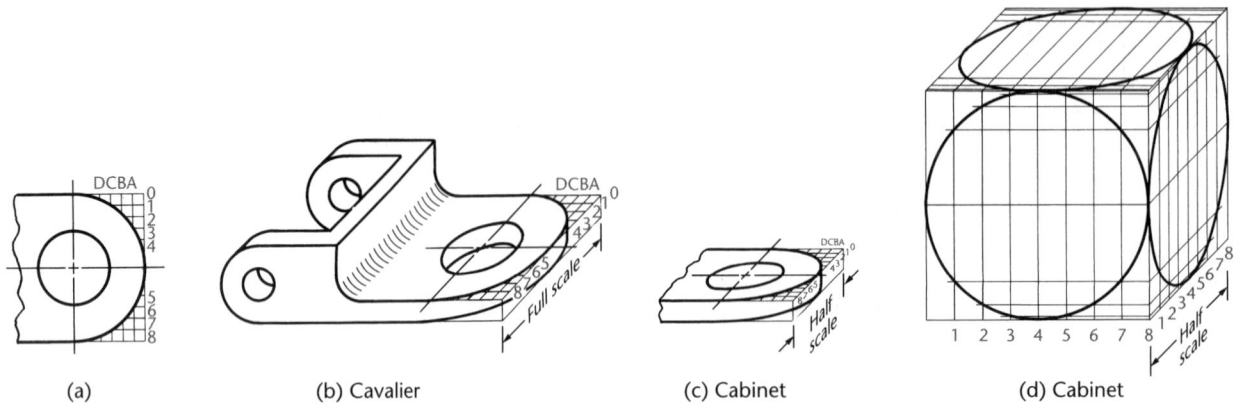

**Figure 9-57**    Use of Offset Measurements

receding axis is drawn to full scale forming an equilateral parallelogram.

To approximate the ellipse in the angled plane, draw the enclosing parallelogram. Then draw the perpendicular bisectors to the four sides of the parallelogram. The intersections of the perpendicular bisectors will be centers for the four circular arcs that form the approximate ellipse. If the angle of the receding lines is anything other than 30° from horizontal, as in this case, the centers of the two large arcs will not fall in the corners of the parallelogram.

When using a CAD system, you can quickly construct accurate ellipses and do not need these methods, but knowing them may be helpful for drawing in the field, or under other circumstances where CAD may not be readily available.

## 9-41  OFFSET MEASUREMENTS

Circles, circular arcs, and other curved or irregular lines can be drawn using **offset measurements,** as shown in Figure 9-57. Draw the offsets on the multiview drawing of the curve (Figure 9-57a), and transfer them to the oblique drawing (Figure 9-57b). In this case, the receding axis is full scale; therefore all offset measurements are drawn full scale. The four center ellipse could be used, but this method is more accurate.

In a cabinet drawing (Figure 9-57c) or any oblique drawing where the receding axis is at a reduced scale, the offset measurements along the receding axis must be drawn

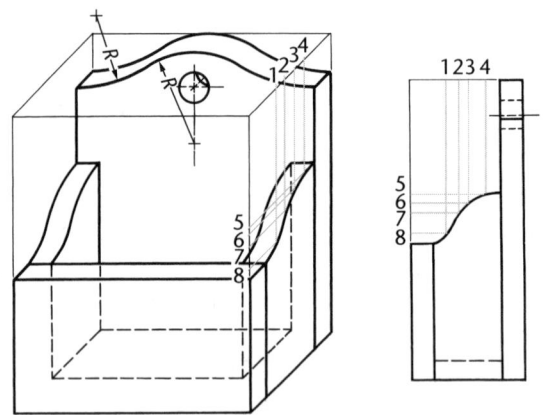

**Figure 9-58**    Use of Offset Measurements

to the same reduced scale. The four center ellipse cannot be used when the receding axis is not full scale. A method of drawing ellipses in a cabinet drawing of a cube is shown in Figure 9-57d.

Figure 9-58 shows a free curve drawn by means of offset measurements in an oblique drawing. This also illustrates hidden lines used to make the drawing clearer.

Offset measurements can be used to draw an ellipse in an inclined plane as shown in Figure 9-59. In Figure 9-59a, parallel lines represent imaginary cutting planes. Each plane cuts a rectangular surface between the front of the cylinder and the inclined surface. These rectangles are shown in the oblique drawing in Figure 9-59b. The curve is drawn through

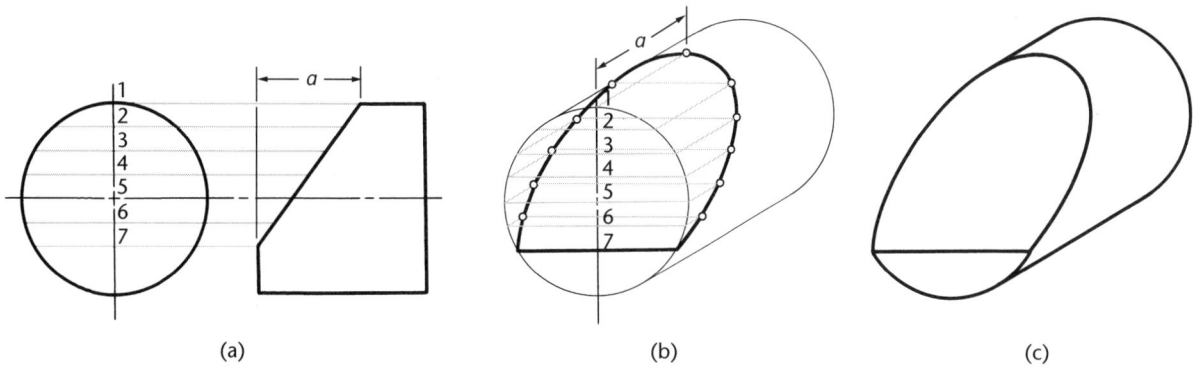

**Figure 9-59** Use of Offset Measurements

the corner points. The final cavalier drawing is shown in Figure 9-59c.

## 9-42 ANGLES IN OBLIQUE PROJECTION

When an angle that is specified in degrees lies in a receding plane, convert the angle into linear measurements to draw the angle in an oblique drawing. Figure 9-60a shows a drawing with an angle of 30° specified.

To draw the angle in the oblique drawing, you will need to know distance $X$. The distance from point $A$ to point $B$ is given as 32 mm. This can be measured directly in the cavalier drawing (Figure 9-60b). Find distance $X$ by drawing the right triangle $ABC$ (Figure 9-60c) using the dimensions given, which is quick and easy using CAD.

You can also use a mathematical solution to find the length of the side: The length of the opposite side equals the

tangent of the angle times the length of the adjacent side. In this case, the length is about 18.5 mm. Draw the angle in the cavalier drawing using the found distance.

Remember that all receding dimensions must be reduced to match the scale of the receding axis. Thus, in the cabinet drawing in Figure 9-60b, the distance $BC$ must be half the side $BC$ of the right triangle in Figure 9-60c.

## 9-43 OBLIQUE SECTIONS

Sections are often useful in oblique drawing, especially to show interior shapes. An oblique half section is shown in Figure 9-61. Oblique full sections are seldom used because they do not show enough of the exterior shapes. In general, all the types of sections for isometric drawing may be applied to oblique drawing.

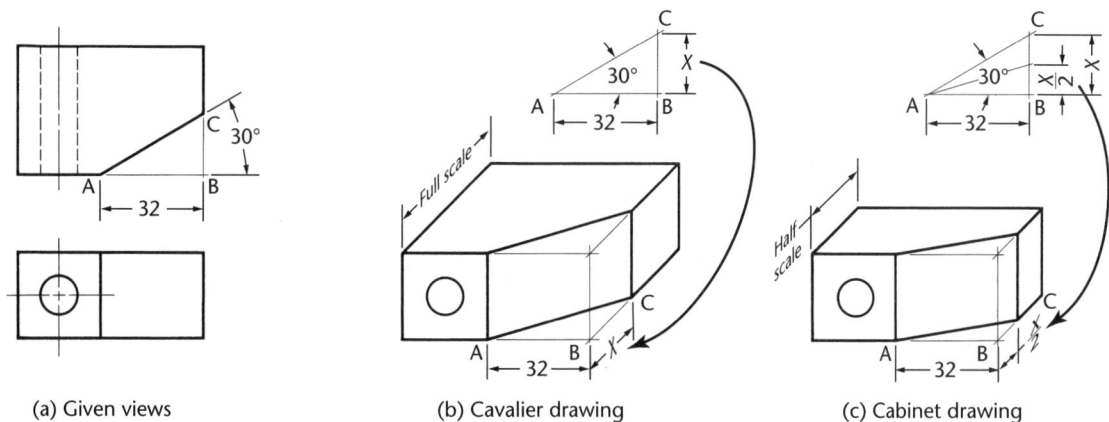

(a) Given views   (b) Cavalier drawing   (c) Cabinet drawing

**Figure 9-60** Angles in Oblique Projection

**Figure 9-61** Oblique Half Section

## 9-44 SCREW THREADS IN OBLIQUE

To show screw thread in a cavalier oblique projection, use parallel partial circles spaced equally to the symbolic thread pitch for only the crests (Figure 9-62). For a cabinet oblique projection, the space would be one half of the symbolic pitch. If you are not using CAD and the thread is positioned to require ellipses, you can draw them using an ellipse template or the four center method.

## 9-45 OBLIQUE DIMENSIONING

You can dimension oblique drawings in a way similar to that used for isometric drawings.

As shown in Figure 9-63, all dimension lines, extension lines, and arrowheads must lie in the planes of the object to which they apply. You should also place the dimension values in the corresponding planes when using the aligned dimensioning system (Figure 9-63a). For the preferred unidirectional system of dimensioning, all dimension figures are horizontal and read from the bottom of the drawing (Figure 9-63b). Use vertical lettering for all pictorial dimensioning.

Place dimensions outside the outlines of the drawing except when clarity is improved by placing the dimensions directly on the view.

## 9-46 COMPUTER GRAPHICS

Using CAD you can easily create oblique drawings by using a snap increment and drawing similarly to the method used on grid paper. If necessary, adjust for desired amount of foreshortening along the receding axis as well as the preferred direction of the axis. You can use CAD commands to draw curves, ellipses, elliptical arcs, and other similar features easily. However if you have a need for a complicated detailed pictorial, it is often easier and more accurate to create a 3D model rather than an oblique view.

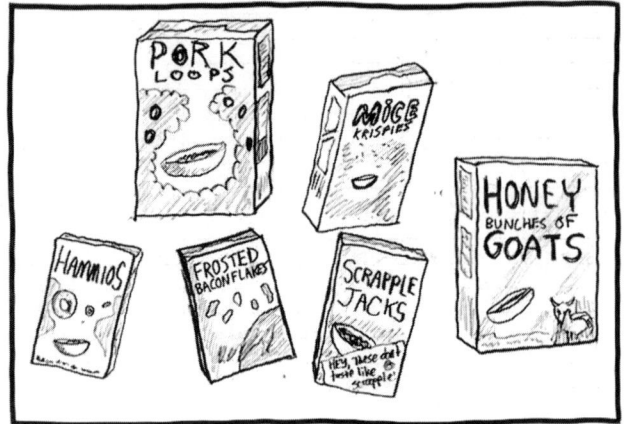

Meat Cereals. *Courtesy of Randall Munroe, xkcd.com.*

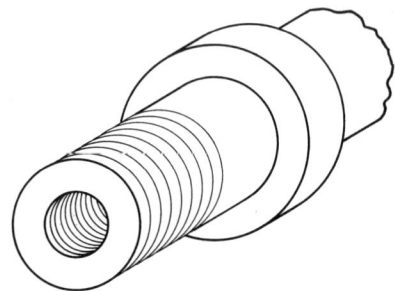

**Figure 9-62** Screw Threads in Oblique

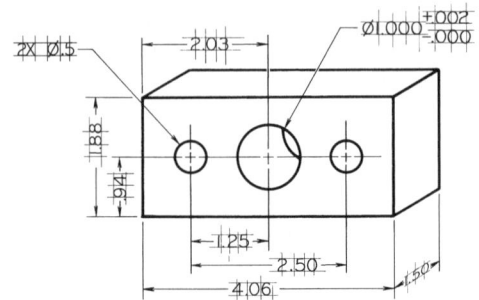

(a) Aligned

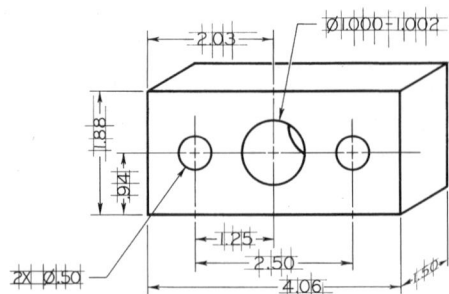

(b) Unidirectional

**Figure 9-63** Oblique Dimensioning

# CAD AT WORK

## Quick Oblique Drawing Using AutoCAD

You can use the snap tool available in the AutoCAD software to make a quick oblique drawing. An oblique drawing is a 2D drawing that gives the appearance of 3D by showing the object angled so that it shows all of the major surfaces in one pictorial view.

### *To Create a Quick Oblique Drawing*

Use methods similar to that for creating a sketch on paper:

1. Draw the front view of the object.
2. Use the snap increment or the polar tracking to draw one of the receding lines showing the depth of the object.
3. Copy the front surface to the back.
4. Add the receding edges.

3D solid models are rarely ever shown in oblique views. It is as quick and easy to create a 3D solid model and produce a view of that as it is to make an oblique sketch, but depending on your need, you may choose either option.

### *To Create a Solid Model*

1. Draw the front view.
2. Use the region command to create 2D areas from the drawing lines.
3. Subtract the areas that are holes from the exterior.
4. View the drawing from an angle so that you can see the results when you produce the 3D part.
5. Extrude the region to create a 3D solid.
6. Orbit the drawing to change your 3D viewpoint.

Which drawing appears most realistic?
Which do you think is most useful?

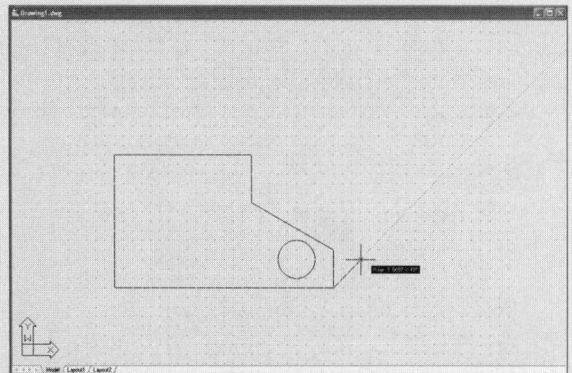

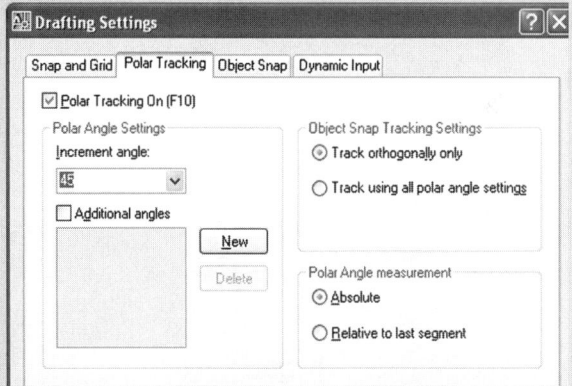

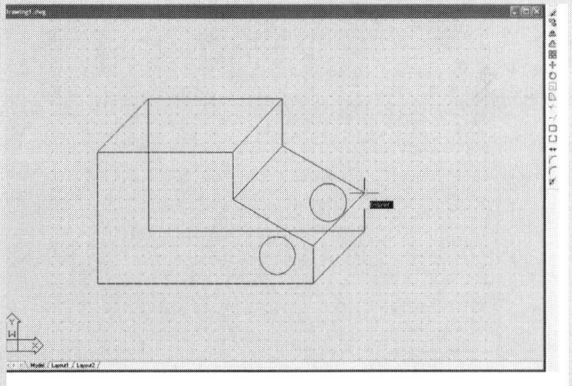

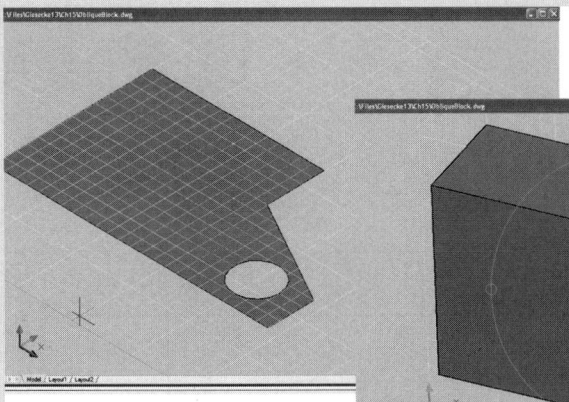

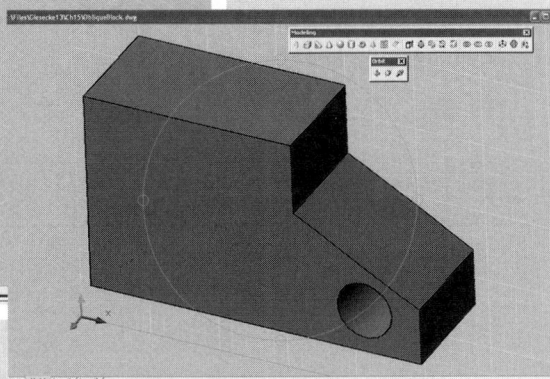

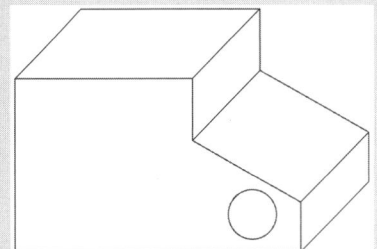

*Reproduced with the permission of Autodesk, Inc.*

## 9-47 EXERCISES

### Oblique Projection Problems

Exercises to be drawn in oblique—either cavalier or cabinet—are given in EX 9-10–9-14. They may be drawn freehand using graph paper or plain drawing paper as assigned by the instructor, or they may be drawn with instruments. In the latter case, all construction lines should be shown on the completed drawing.

Since many of the problems in this chapter are of a general nature, they can also be solved on most CAD systems. The instructor may ask you to use CAD on specific problems.

### Ex9-10

(1) Make freehand oblique sketches. (2) Make oblique drawings using CAD. Add dimensions to your drawing only if assigned by your instructor.

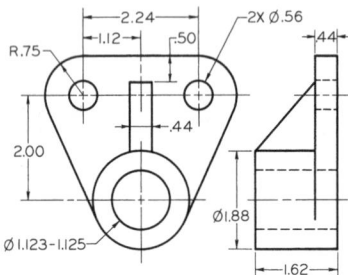
1 ROD GUIDE

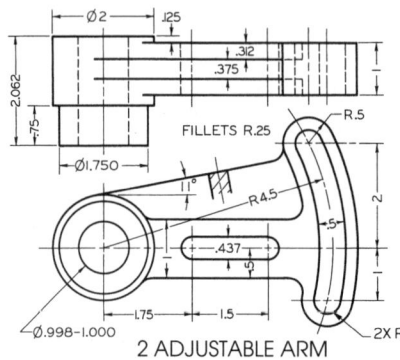
2 ADJUSTABLE ARM

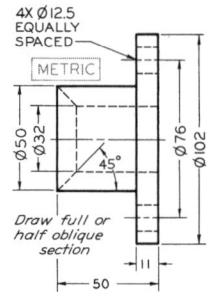
3 FOLLOWER

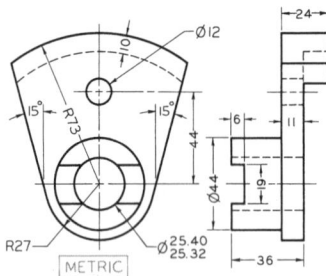
4 GUIDE ARM

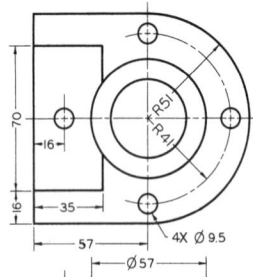
5 HOUSING CAP

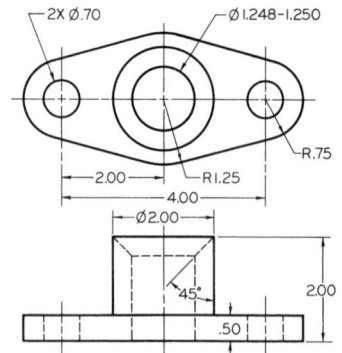
6 GLAND

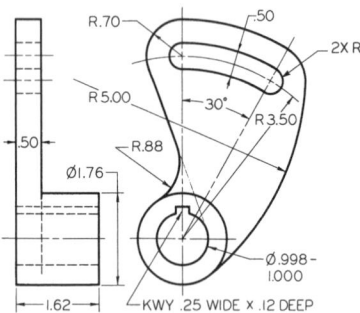
7 CONTROL ARM

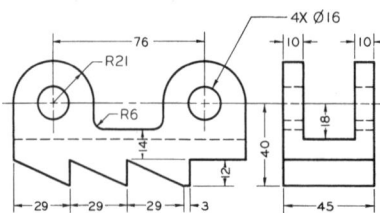
8 RACK

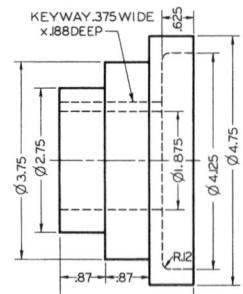
9 STEP CONE

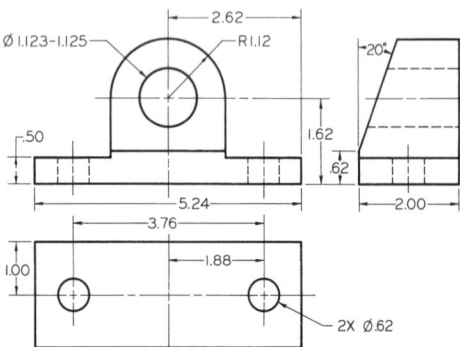
10 ANGLE BEARING

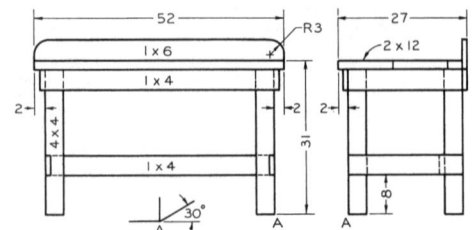
11 WORKBENCH

**Ex9-11**

(1) Make freehand oblique sketches. (2) Make oblique drawings using CAD. Add dimensions to your drawing only if assigned by your instructor.

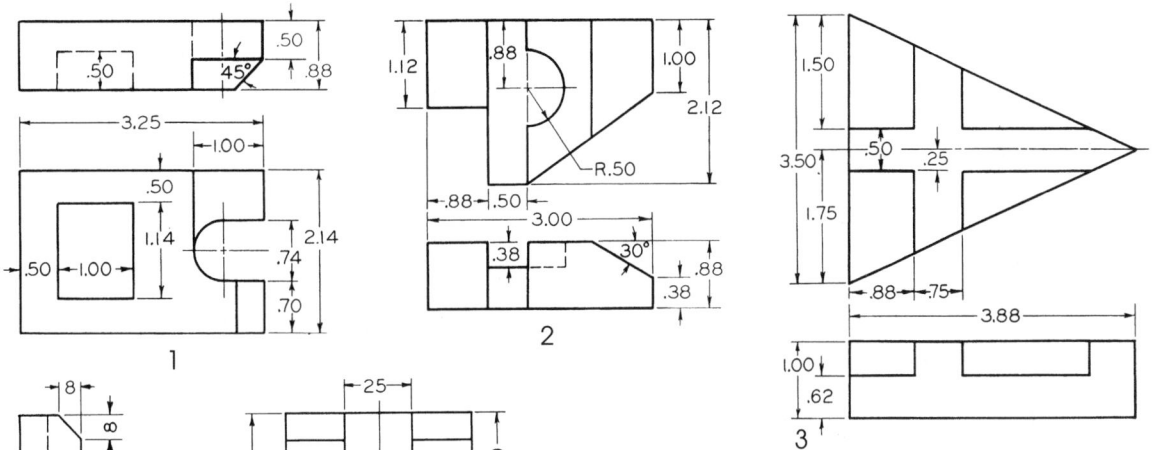

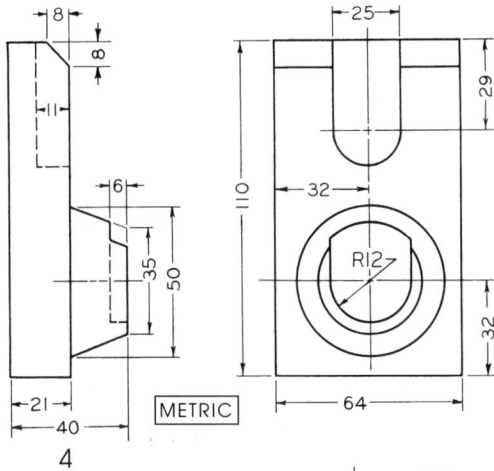

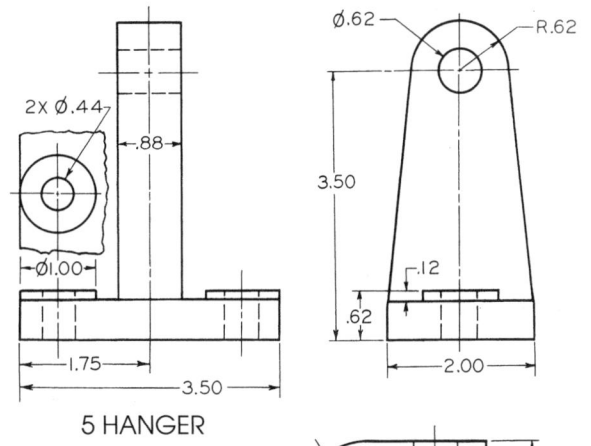

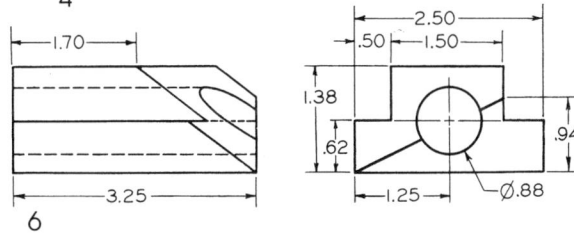

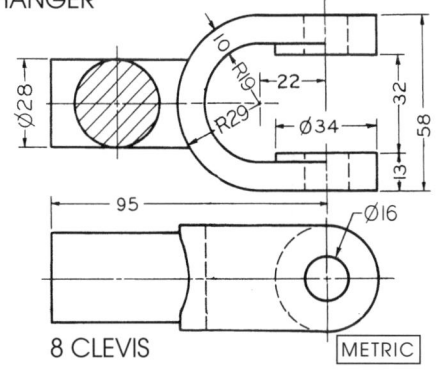

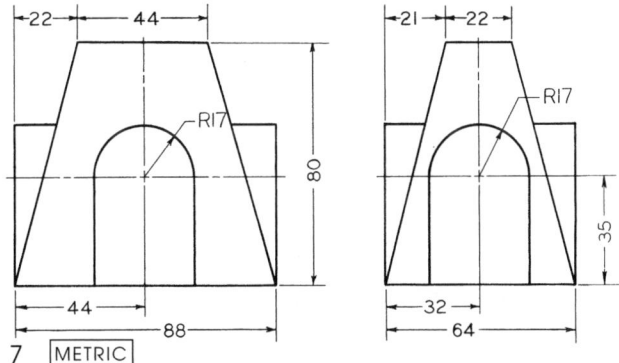

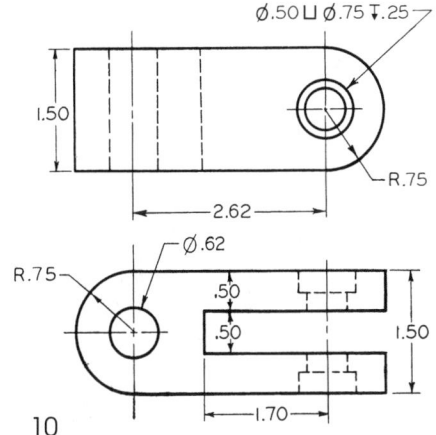

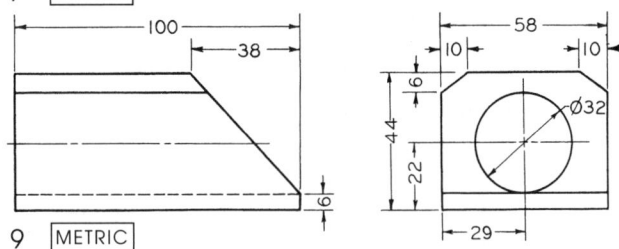

1

2

3

4    METRIC

5 HANGER

6

7    METRIC

8 CLEVIS    METRIC

9    METRIC

10

**Ex9-12**

(1) Make freehand oblique sketches. (2) Make oblique drawings using CAD. Add dimensions to your drawing only if assigned by your instructor.

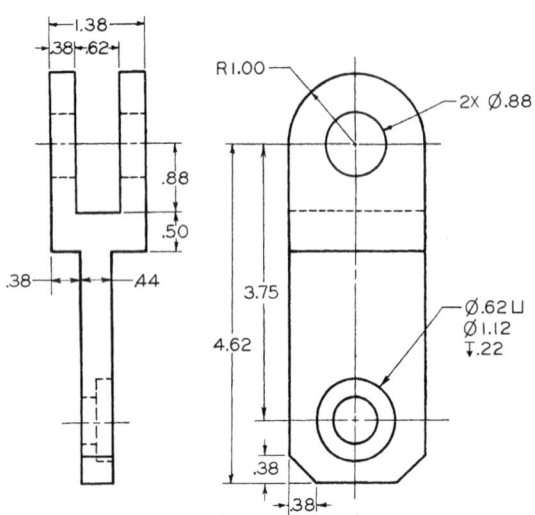

1 CLEVIS

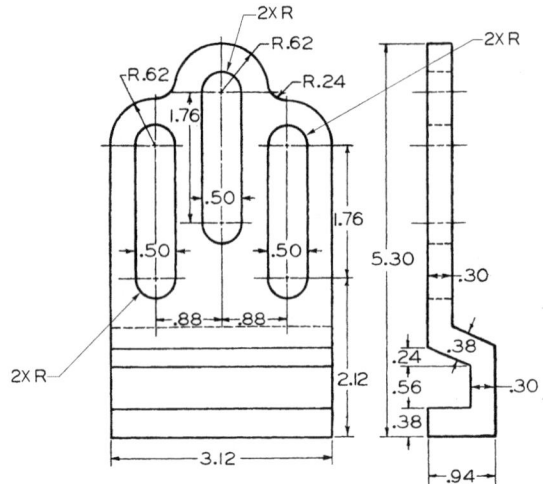

2 ADJUSTABLE ORDER

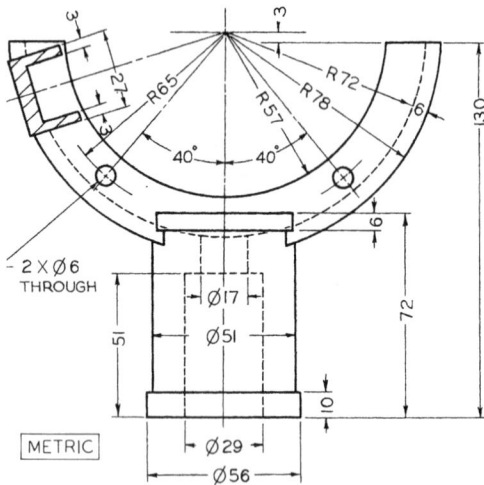

3 TURRET LATHE STOCK REST

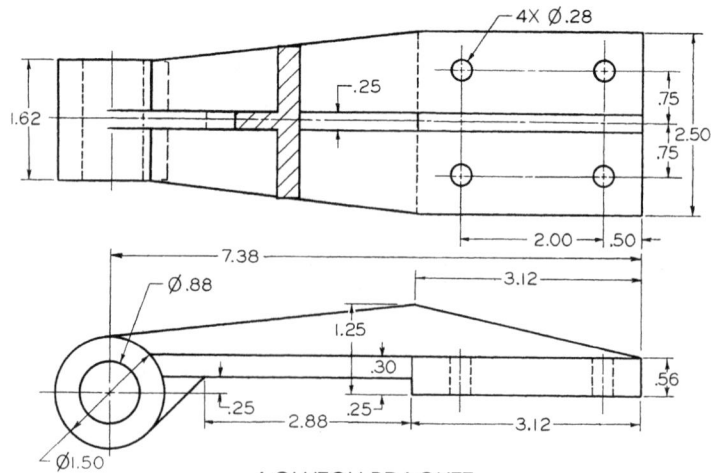

4 CLUTCH BRACKET

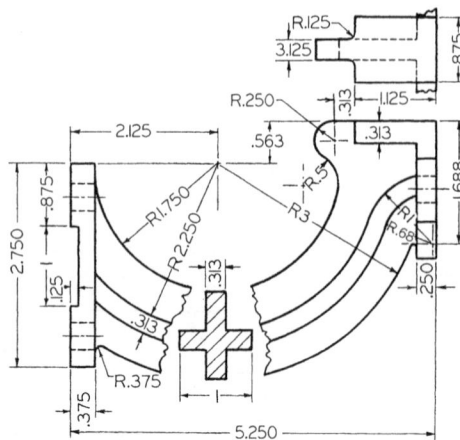

5 RAIL SUPPORT

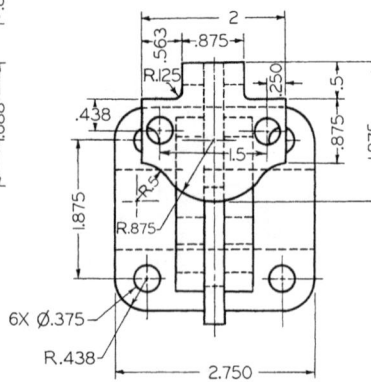

**Ex9-13**

(1) Make freehand oblique sketches. (2) Make oblique drawings using CAD. Add dimensions to your drawing only if assigned by your instructor.

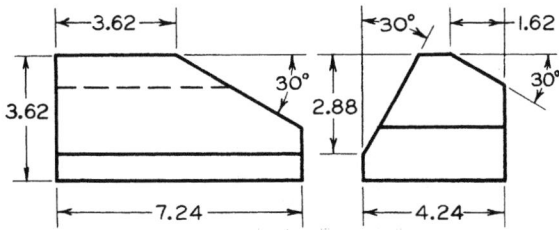

1 GUIDE

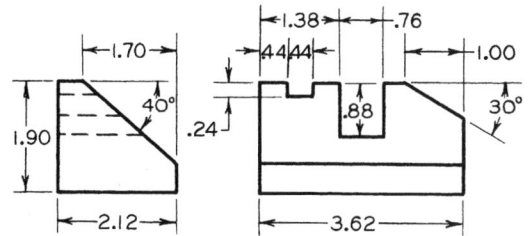

2 TERMINAL BLOCK

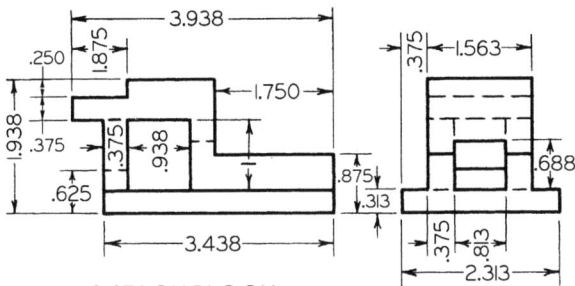

3 STACK BLOCK

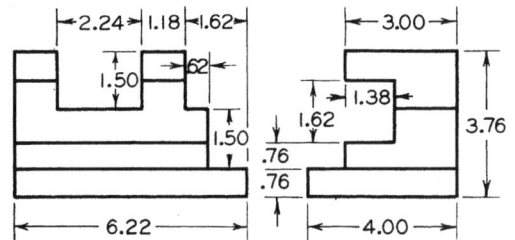

4 SLIDE

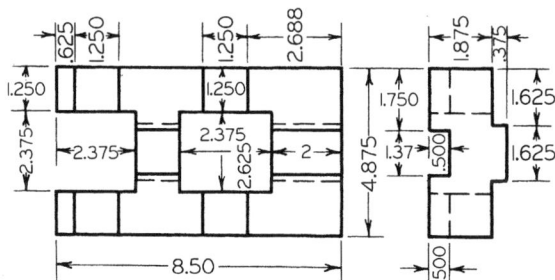

5 ADAPTER PLATE

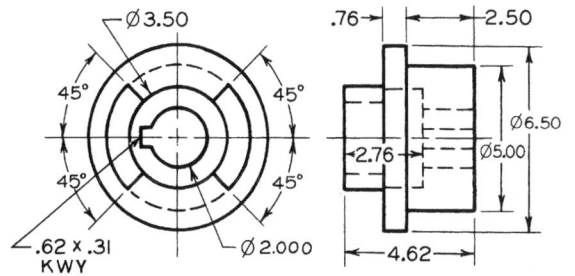

6 DRIVE SLEEVE

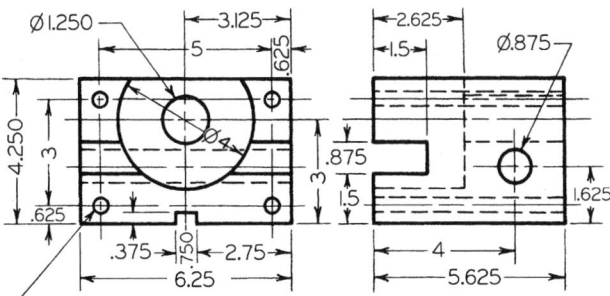

7 SAW GUIDE BLOCK

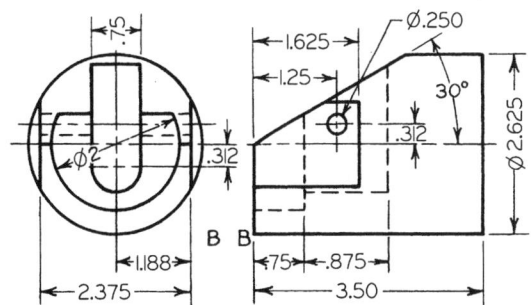

8 TRAVERSE STOP PISTON

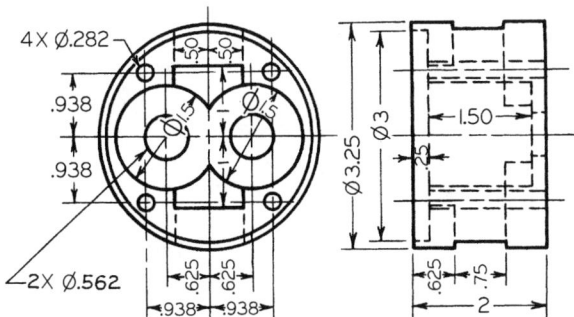

9 OIL PUMP BODY

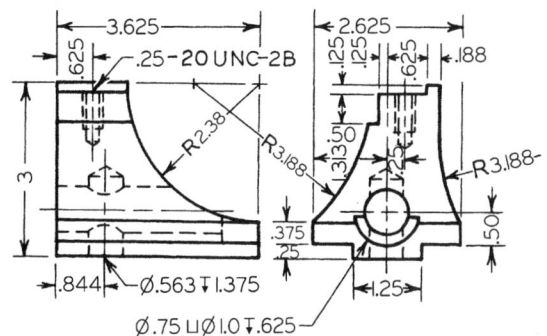

10 CUTTING OFF TOOL HOLDER

**Ex9-14**

For the linear actuator, make an oblique drawing. If requested by your instructor, add the overall dimensions.

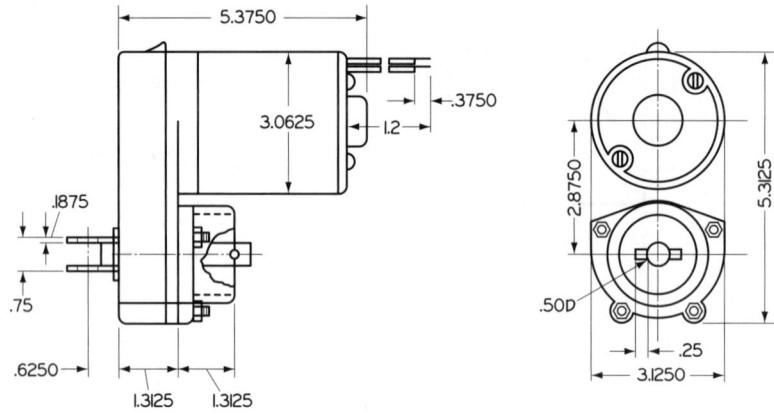

# CHAPTER 10

# Welding Representation

## 10-1 WELDING REPRESENTATION

For fastening parts together permanently rather than using bolts, screws, rivets, or other fasteners, welding is often the method of choice. Welding is widely used in fabricating machine parts or other structures that formerly would have been formed by casting or forging. Structural steel frames for buildings, ships, and other structures are often welded. Welding is one of few machine processes that adds material to a workpiece, rather than removing it.

Welding symbols on a mechanical drawing provide precise instructions for the welder. The weld type and the location of each weld must be clearly defined using standardized symbols. CAD libraries of welding symbols can simplify the drawing process. Welding templates can speed the process of drawing by hand.

Check the sites below for Web resources:
- Examples for students and links about ASME codes and standards: http://www.asme.org/Codes/About/Links/Links_Codes_ Standards.cfm
- Americal Welding Societies homepage: www.AWS.org

## 10-2 UNDERSTANDING WELDMENT DRAWINGS

Welding is used extensively and for a wide variety of attachment purposes. A series of "Standard Welding Symbols" were developed in 1947 to provide an accurate method of showing the exact types, sizes, and locations of welds on construction drawings of machines or structures. Before these standards were developed, notes on the drawing such as "To be welded throughout" or "To be completely welded," gave responsibility for welding control to the welding shop, which was dangerously vague and could be unnecessarily expensive, since shops would often "play it safe" by welding more than necessary.

### Welding Processes

The principal methods of welding are:

- Gas welding
- Arc welding
- Resistance welding

**Gas Welding** The oxyacetylene method is generally known as gas welding. Gas welding originated in 1895, when the French chemist Le Châtelier discovered that the combustion of acetylene gas with oxygen produced a flame hot enough to melt metals. This discovery was soon followed by the development of practical methods to produce and

An Automobile Frame is Welded on a Robotic Automobile Assembly Line. *Courtesy of Vladimir Pcholkin/Stone/ Getty Images.*

transport oxygen and acetylene and the construction of torches and welding rods.

**Arc Welding** the electric arc method is generally known as arc welding, In arc welding, the heat of an electric arc is used to fuse the metals that are to be welded or cut. Gas metal arc welding (**GMAW**) is often known by its subtypes metal inert gas welding (**MIG**) and metal active gas welding (**MAG**).

This type of welding uses a wire electrode that is fed along with a shielding gas through the welding gun. Gas tungsten arc welding (**GTAW**) often known as tungsten inert gas welding (**TIG**) and plasma arc welding are processes that conduct the arc through a heated gas called a plasma, to produce strong high-quality welds. Arc and gas welding are important construction processes in industry.

**Resistance Welding** Electric resistance welding is generally called resistance welding. In resistance welding, two pieces of metal are held together under some pressure, and a large amount of electric current is passed through the parts. The resistance of the metals to the passage of the current causes great heating at the junction of the two pieces, resulting in the welding of the metals.

## Standard Symbols

The text and illustrations of this chapter are based primarily on ANSI/AWS A2.4-2007 Standard Symbols for Welding, Brazing, and Nondestructive Examination. You may also want to refer to AWS A1.1-2001, Metric Practice Guide for the Welding Industry, and ANSI/AWS A3.0-1994, Standard Welding Terms and Definitions.

**Welding drawings** are a special type of assembly drawing as weldments are composed of a number of separate pieces fastened together as a unit. The welds themselves are not drawn but are clearly and completely indicated by the welding symbols. Figure 10-1 shows a drawing using standard welding symbols. Most CAD packages provide standard welding symbols that can quickly be inserted into your drawing.

The joints are all shown in the drawing as they would appear before welding. Dimensions are given to show the sizes of the individual pieces to be cut from stock. Each component piece is identified by encircled numbers and by specifications in the parts list.

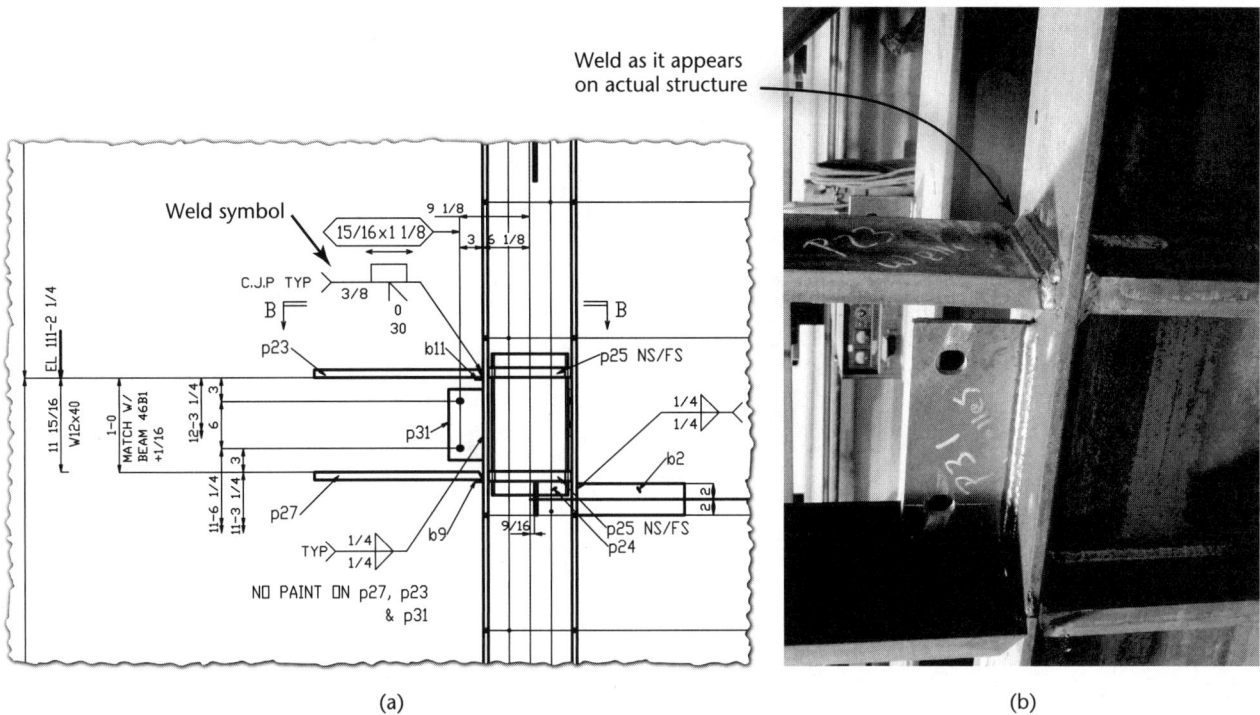

**Figure 10-1**    a. Portion of a weldment drawing. *Courtesy of Midwest Steel Industries*, b. Picture of welded structure (photo has been rotated to the same orientation as the drawing in part a.)

## 10-3  UNDERSTANDING A WELDING SYMBOL

A welding symbol added to the drawing has many different features that specify each detail of the weld. The items that can be specified are:

- Type of weld
- Process
- Depth of bevel, size or strength for some weld types
- Groove weld size
- Finishing designator
- Contour
- Groove angle

- Root opening
- Length of weld
- Number and **pitch** (center to center spacing) of welds
- Whether the weld is to be field welded (done on site)
- All around indicator
- Which side of the material is to be welded

That's a lot of information contained in one symbol. The next sections provide more detail about the information contained in the symbol you will add to your drawings. Figure 10-1 shows an example of a weld symbol used on a drawing and the actual welded part. Figure 10-2 points out features of a weld to which the symbol may refer.

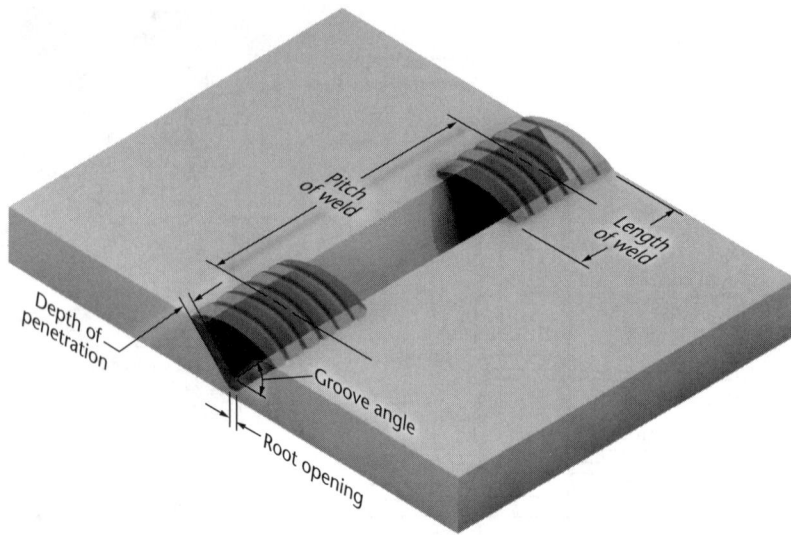

**Figure 10-2**   Features of a weld

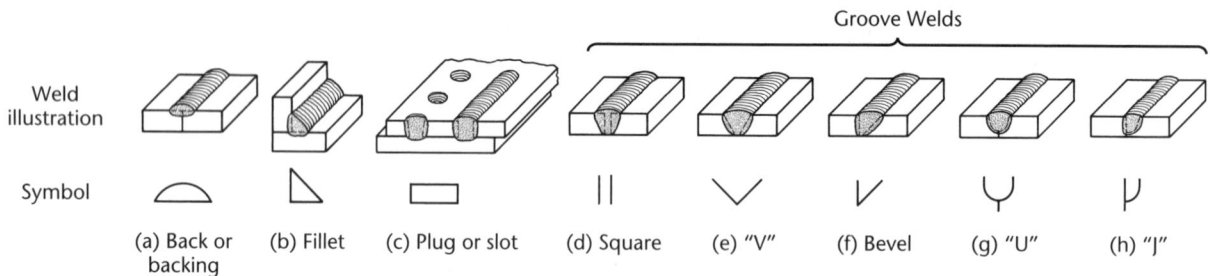

**Figure 10-3**   Basic Welds and Symbols

## 10-4  TYPES OF WELDED JOINTS

There are five basic types of welded joints: **butt joint, corner joint, T-joint, lap joint,** and **edge joint.** They are classified according to the positions of the parts being joined. Table 10-1 shows illustrations of the types of welded joints.

A number of different types of welds are applicable to each type of joint, depending on the thickness of metal, the strength of joint required, and other considerations.

## 10-5  TYPES OF WELDS

The four types of arc and gas welds are shown in Figure 10-3:

- Back or backing weld.
- Fillet weld.
- Plug or slot weld.
- Groove weld.

Groove welds are further classified as square, V, bevel, U, and J, as shown Figure 10-3d through h.

More than one type of weld may be applied to a single joint. For example, a V weld may be on one side and a back weld on the other side. Frequently, the same type of weld is used on opposite sides, forming such welds as a double-V, a double-U, or a double-J.

The four basic resistance welds are:

- Spot weld.
- Projection weld.
- Seam weld.
- Flash or upset weld.

Except for the flash or upset weld, the corresponding symbols for these welds and additional basic weld symbols for surfacing, groove, and flange joints are given in Figure

**Table 10-1** Basic Types of Welded Joints.

| Type of Joint | Example |
|---|---|
| Butt joint | |
| Corner joint | |
| T-joint | |
| Lap joint | |
| Edge joint | |

10-4. See Section 10-15 for the use of the square groove weld symbol for the flash or upset resistance weld. Supplementary symbols are shown in Figure 10-5. Depending on your field and the particular project, you may need only to use a simple symbol composed of the minimum elements (the arrow and the weld symbol) or you may need to use the additional components.

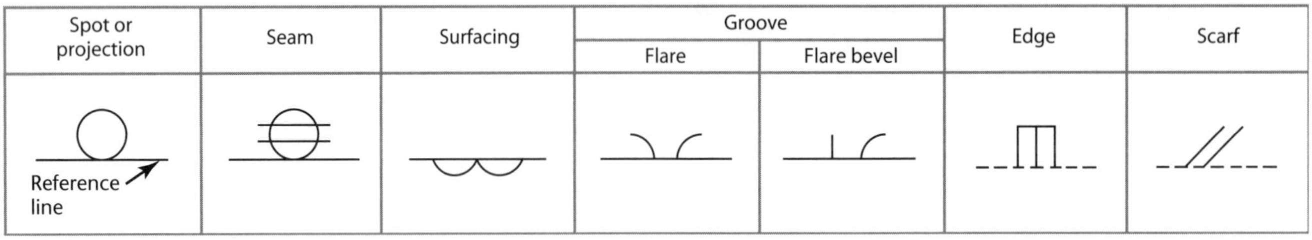

| Spot or projection | Seam | Surfacing | Groove | | Edge | Scarf |
|---|---|---|---|---|---|---|
| | | | Flare | Flare bevel | | |
| Reference line | | | | | | |

**Figure 10-4**   Additional Basic Weld Symbols

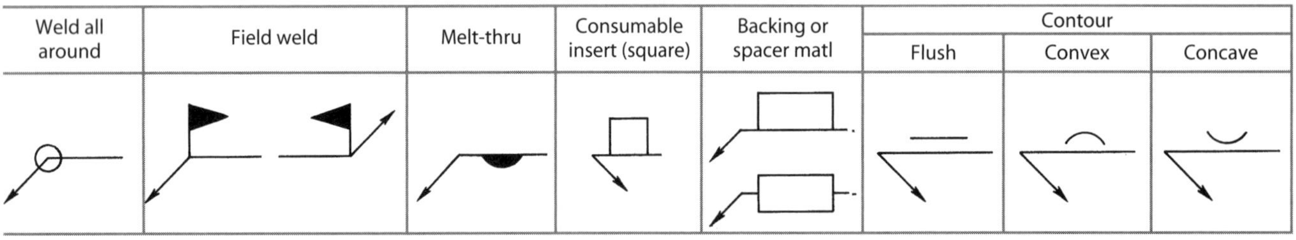

| Weld all around | Field weld | Melt-thru | Consumable insert (square) | Backing or spacer matl | Contour | | |
|---|---|---|---|---|---|---|---|
| | | | | | Flush | Convex | Concave |

**Figure 10-5**   Supplementary Symbols

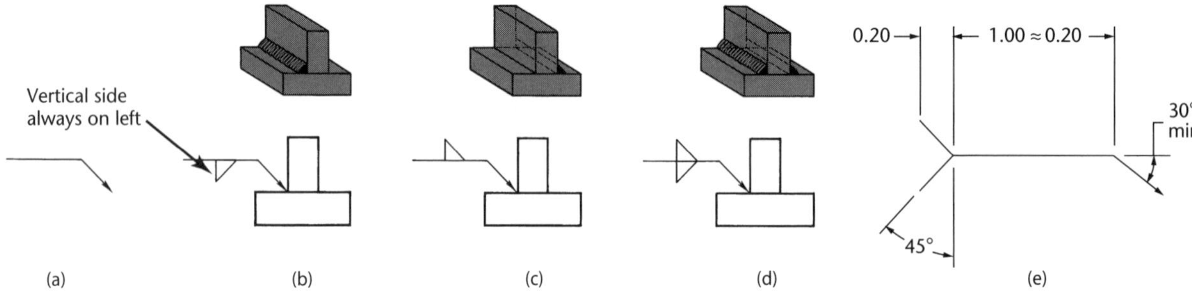

Vertical side always on left

0.20    1.00 ≈ 0.20    30° min    45°

(a)        (b)        (c)        (d)        (e)

**Figure 10-6**   Welding Symbols

## 10-6  WELDING SYMBOLS

The basic element of the symbol is the "bent" arrow, as shown in Figure 10-6a. The arrow points to the joint where the weld is to be made (Figure 10-6b). Attached to the reference line, or shank, of the arrow is the weld symbol for the desired weld. The symbol would be one of those illustrated in Figures 10-3 and 10-4. In this case, a fillet weld symbol has been used.

The weld symbol is placed below the reference line if the weld is to be on the **arrow side** of the joint, as in Figure 10-6b, or above the reference line if the weld is to be on the **other side** of the joint, as in Figure 10-6c. If the weld is to be on both the arrow side and the other side of the joint, weld symbols are placed on both sides of the reference line (Figure 10-6d). This rule for placement of the weld symbol is followed for all arc or gas weld symbols. Dimensions for creating a weld symbol are shown in Figure 10-6e.

When a joint is represented by a single line on a drawing, as in the top and side views of Figure 10-7b, the arrow side of the joint is regarded as the "near" side to the reader of the drawing, according to the usual conventions of technical drawing.

For the plug, slot, seam, and projection welding symbols, the arrow points to the outer surface of one of the members at the centerline of the weld. In such cases, the arrow side of the joint is the one to which the arrow points, or the "near" side to the reader (see Sections 10-11 and 10-15).

Note that for all fillet or groove symbols, the vertical side of the symbol is always drawn on the left as shown in Figure 10-6b.

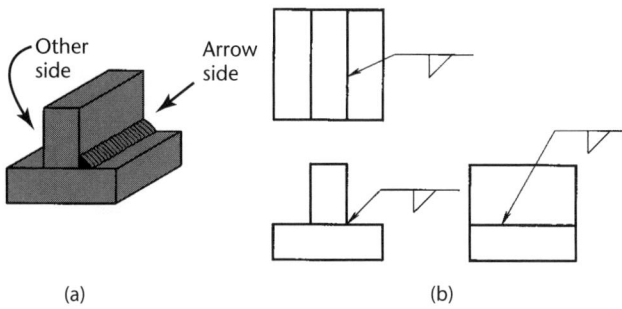

**Figure 10-7**  Arrow Side and Other Side

For best results, welding symbols should be drawn using CAD or a template, but in certain cases where necessary, they may be drawn freehand.

The complete welding symbol, enlarged, is shown in Figure 10-8.

Reference to a specification, process, or other supplementary information is indicated by any desired symbol in the tail of the arrow (Figure 10-9a). Otherwise, a general note may be placed on the drawing, such as

*UNLESS OTHERWISE INDICATED, MAKE ALL WELDS PER SPECIFICATION NO. XXX*

If no reference is indicated in the symbol, the tail may be omitted.

To avoid repeating the same information on many welding symbols on a drawing, general notes may be used, such as

*FILLET    WELDS    UNLESS    OTHERWISE INDICATED*

or

*ROOT OPENINGS FOR ALL GROOVE WELDS UNLESS OTHERWISE INDICATED*

Welds extending completely around a joint are indicated by an open circle around the elbow of the arrow (Figure 10-9b). When the weld all around symbol is not used, the welding symbol is understood to apply between abrupt changes in direction of the weld, unless otherwise shown. A short vertical staff with a solid triangular flag at the elbow of the arrow indicates a weld to be made "in the field" (on the site) rather than in the fabrication shop (Figure 10-9c).

Spot, seam, flash, or upset symbols usually do not have arrow-side or other-side significance and are simply centered on the reference line of the arrow (Figure 10-10a through c). Spot and seam symbols are shown on the drawing as indicated in Figure 10-10d and e. Note that the required process must be specified in the tail of the symbol (*RSW* = resistance spot weld, *EBW* = electron beam weld).

For bevel or J-groove welds, the arrow should point with a definite change of direction, or break, toward the member that is to be beveled or grooved (Figure 10-11a and b). In this case, the upper member is grooved. The break is omitted if the location of the bevel or groove is obvious.

Lettering for the symbols should be placed to read from the bottom or from the right side of the drawing in accordance with the aligned system (Figure 10-11c through e). Dimensions may be indicated on a drawing in the fractional, decimal inch, or metric system.

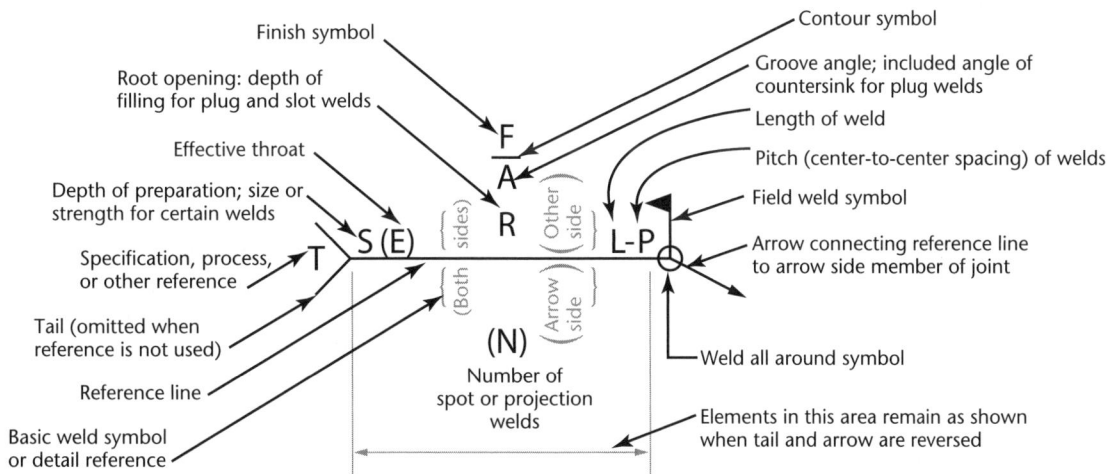

**Figure 10-8**  The Standard Locations of the Elements of the Welding Symbol

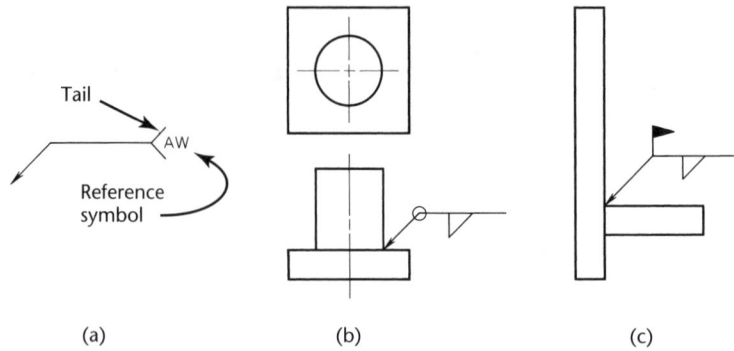

**Figure 10-9**    The Standard Locations of the Elements of the Welding Symbol

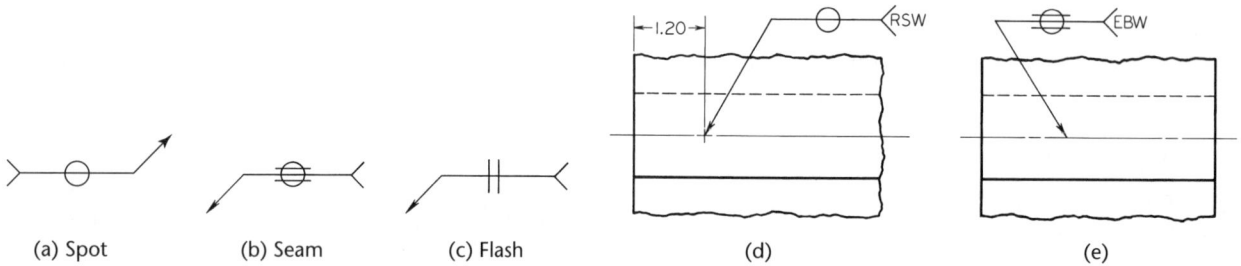

**Figure 10-10**    Spot, Seam, and Flash Welding Symbols

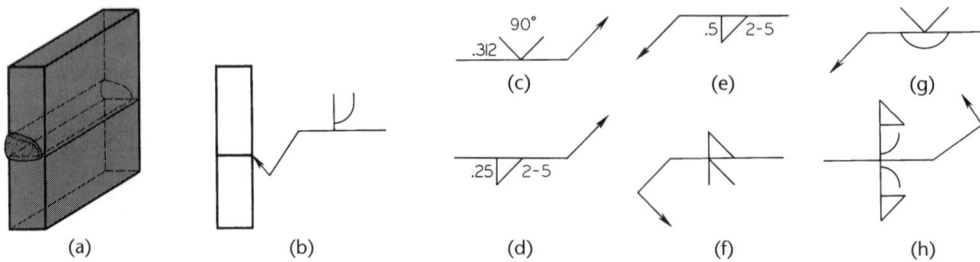

**Figure 10-11**    Welding Symbols

When a joint has more than one weld, the combined symbols are used (Figure 10-11f through h).

## 10-7  FILLET WELDS

The usual fillet weld has equal legs (Figure 10-12a). The size of the weld is the length of one leg, as indicated in Figure 10-12b by a dimension figure (fraction, decimal inch, or metric) at the left of the weld symbol. For fillet welds on both sides of a joint, the dimensions may be indicated on one side or both sides of the reference line (Figure 10-12c). The lengths of the welds and the pitch (center to center spacing of welds) are indicated as shown. When the welds on opposite sides are different in size, the sizes are given as shown in Figure 10-12d. If a fillet weld has unequal legs, the weld orientation is shown on the drawing, if necessary, and the lengths of the legs are given in parentheses to the left of the weld symbol, as in Figure 10-12e. If a general note is given on the drawing, such as

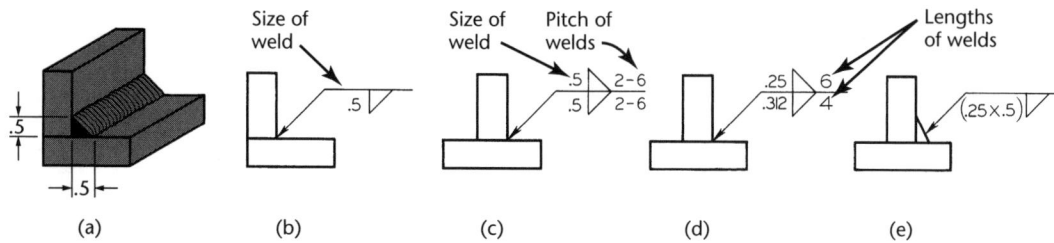

**Figure 10-12** Dimensioning of Fillet Welds

*ALL FILLET WELDS UNLESS OTHERWISE NOTED*

the size dimensions are omitted from the symbols.

No length dimension is needed for a weld that extends the full distance between abrupt changes of direction. For each abrupt change in direction, an additional arrow is added to the symbol, except when the weld all around symbol is used.

### Fillet Weld Length

Lengths of fillet welds may be indicated by symbols in conjunction with dimension lines (Figure 10-13a). The extent of fillet welding may be shown graphically with section lining if desired (Figure 10-13b).

### Intermittent Fillet Welding

Chain intermittent fillet welding is indicated as shown in Figure 10-14a. If the welds are staggered, the weld symbols are staggered (Figure 10-14b).

### Surface Contour and Fillet Welds

Unfinished flat-faced fillet welds are indicated by adding the **flush** symbol (see Figure 10-5), to the weld symbol (Figure 10-15a). If fillet welds are to be made flat-faced by mechanical means, add the flush-contour symbol and the user's standard finish symbol to the weld symbol (Figure 10-15b through d). These finish symbols indicate the method of finishing (*C* = chipping, *G* = grinding, *M* = machining, *R* = rolling, H = hammering) and not the degree of finish. If fillet welds are to be finished to a **convex** contour, the convex contour symbol is added, together with the finish symbol (Figure 10-15e).

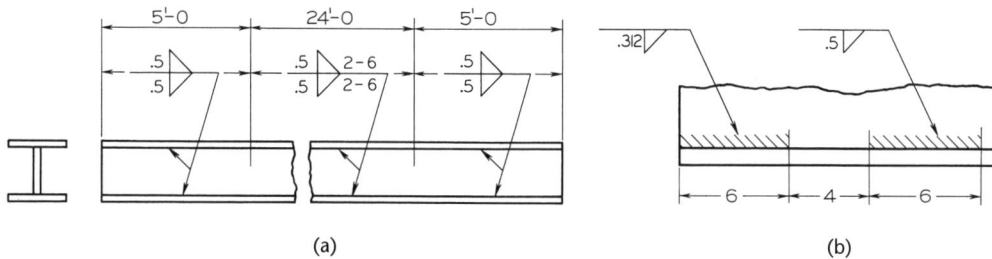

**Figure 10-13** Lengths of Fillet Welds

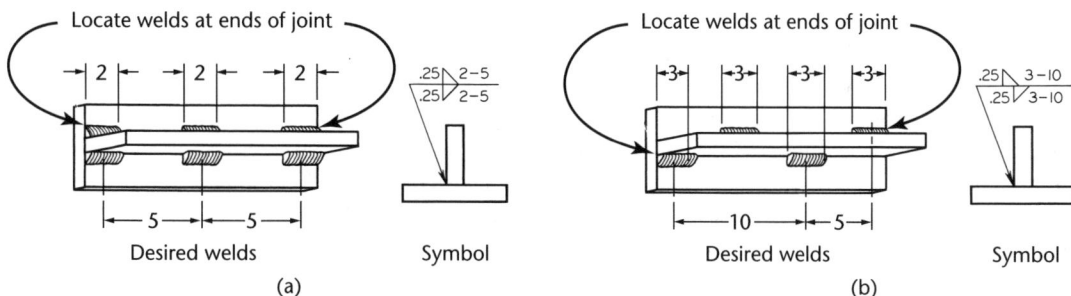

**Figure 10-14** Intermittent Welds

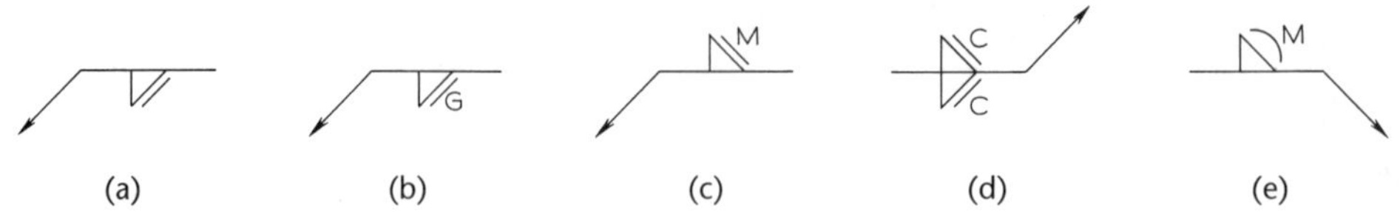

**Figure 10-15** Surface Contour of Fillet Welds

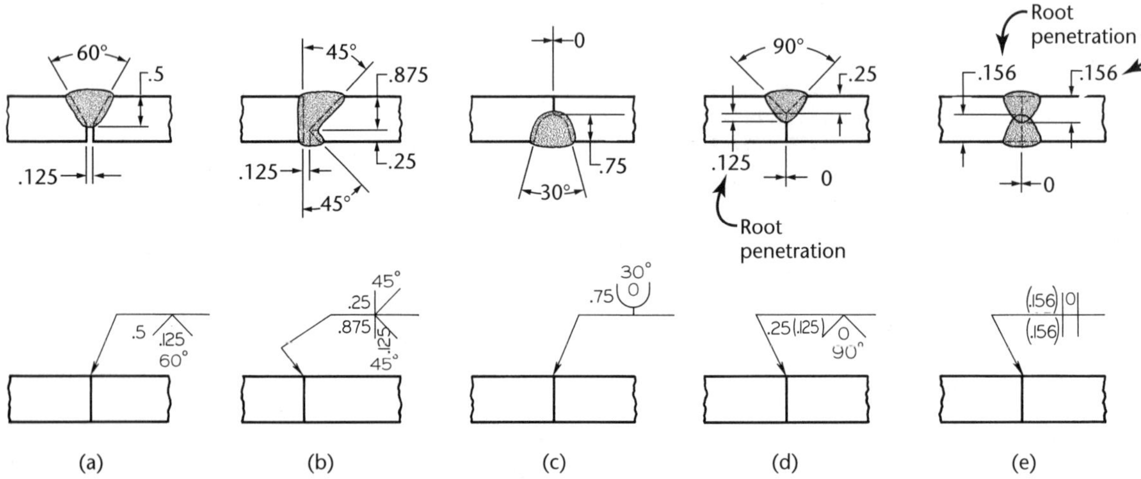

**Figure 10-16** Groove Welds

## 10-8 GROOVE WELDS

In Figure 10-16, various groove welds are shown above and the corresponding symbolic representations below. The sizes of the groove welds (depth of the V, bevel, U, or J) are indicated on the left of the weld symbol. For example, in Figure 10-16a, the size of the V-weld is .50″, in Figure 10-16b the sizes are .25″ and .875″, in Figure 10-16c the size is .75″, and in Figure 10-16d the size is .25″. For the symbol in Figure 10-16d the size is followed by .125″, which is the additional "root penetration" of the weld. In Figure 10-16e, the root penetration is .156" from zero, or from the outside of the members. Note the overlap of the root penetration in this case.

The root opening or space between members, when not covered by a company standard, is shown within the weld symbol. In Figure 10-16a and b, the root openings are .125″. In Figure 10-16c through e, the openings are zero.

The groove angles, when not covered by a company standard, appear just outside the openings of the weld symbols (Figure 10-16a and b).

A general note may be used on the drawing to avoid repeating the symbols, such as

*ALL V-GROOVE WELDS TO HAVE 60° GROOVE ANGLE UNLESS OTHERWISE SHOWN*

However, when the dimensions of one or both of two opposite welds differ from the general note, both welds should be completely dimensioned.

When single-groove or symmetrical double-groove welds extend completely through, the size need not be added to the welding symbol. For example, in Figure 10-16a, if the V-groove extended entirely through the joint, the depth or size would be simply the thickness of the stock and would not need to be indicated in the welding symbol.

### Surface Contour of Groove Welds

When groove welds are to be approximately flush without finishing, add the flush contour symbol (see Figure 10-5) to the weld symbols (Figs. 10-17a and b). If the welds are to be machined, add the flush contour symbol and the user's standard finish symbol to the weld symbol (Figure 10-17c and d). These finish symbols indicate the method of finishing (C = chipping, G = grinding, M = machining) and not the degree of finish. If a groove weld is to be finished with a convex contour, add the convex contour and finish symbols, as in Figure 10-17e.

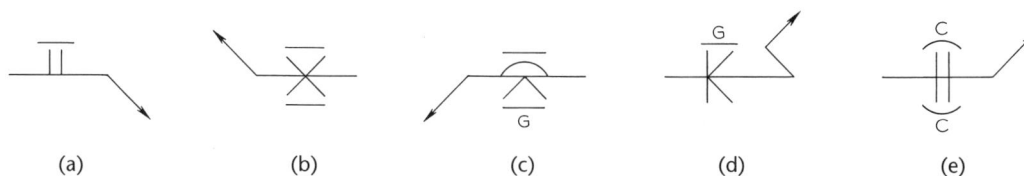

**Figure 10-17**   Surface Contour of Groove Welds

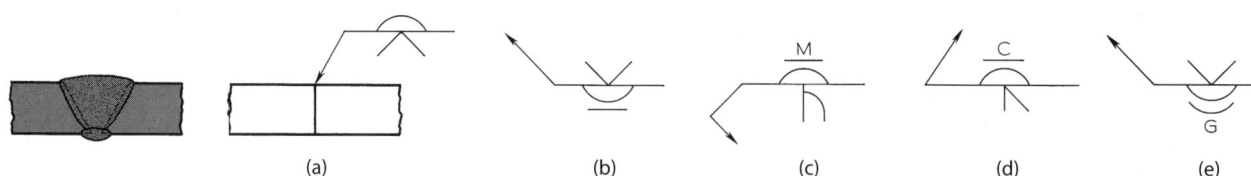

**Figure 10-18**   Back or Backing Weld Symbols

## 10-9  BACK OR BACKING WELDS

A back or backing symbol opposite the groove weld symbol indicates bead type welds used as back or backing welds on single groove welds (Figure 10-18a). Dimensions for back or backing welds are not shown on the symbol, but may be shown, if necessary, directly on the drawing.

A flush contour symbol included in the weld symbols indicates that the back or backing welds are to be approximately flush without machining (Figure 10-18b). If they are to be machined, the user's finish symbol is added (Figures 10-18c and d). If the welds are to be finished with a convex contour, the convex contour symbol and the finish symbol are included in the weld symbol, as shown in Figure 10-18e.

## 10-10  SURFACE WELDS

The surface weld symbol indicates a surface to be built up with single- or multiple-pass bead type welds (Figure 10-19). Since this symbol does not indicate a welded joint, there is no arrow-side or other-side significance, so the symbol is always drawn below the reference line. Indicate the minimum height of the weld deposit at the left of the weld symbol, except where no specific height is required. When a specific area of a surface is to be built up, give the dimensions of the area on the drawing.

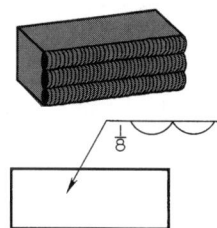

**Figure 10-19**   Surface Weld Symbol

## 10-11  PLUG AND SLOT WELDS

The same symbol is used for plug welds and slot welds. Figure 10-20a and d show the hole or slot that is made to receive the weld. If it is in the arrow-side member, place the weld symbol below the reference line as shown Figure 10-20b and c. If it is in the other-side member, place the weld symbol above the line as shown in Figure 10-20e and f.

Place the size of a plug weld (which is the smallest diameter of the hole, if countersunk) at the left of the weld symbol. If the included angle for the countersink of a plug welds is in accordance with the user's standard, omit it; otherwise, place it adjacent to the weld symbol as shown in Figure 10-20b and c.

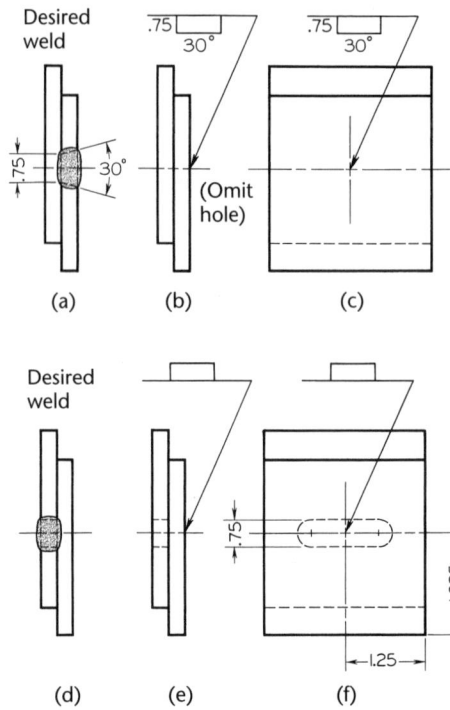

**Figure 10-20**   Plug and Slot Welds

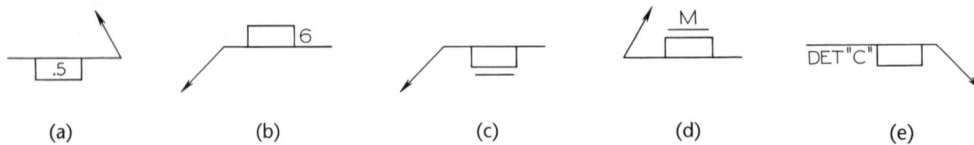

**Figure 10-21**   Plug Welds

A plug weld is understood to fill the depth of the hole unless its depth is indicated inside the weld symbol as shown in Figure 10-21a. The pitch of plug welds is shown at the right of the weld symbol (Figure 10-21b). If the weld is to be approximately flush without finishing, add the flush contour symbol as in Figure 10-21c. If the weld is to be made flush by mechanical means, a finish symbol is added (Figure 10-21d). Flush contour and finish symbols are used the same way for slot welds and for plug welds.

Indicate the depth of filling for slot welds the same way as for plug welds (Figure 10-21a). The size and location dimensions of slot welds cannot be shown on the welding symbol. Show them directly on the drawing (Figure 10-20f) or in a detail with a reference to it on the welding symbol as shown in Figure 10-21e.

## 10-12  SPOT WELDS

The spot weld symbol, with the required welding process indicated in the tail, may or may not have arrow-side or other-side significance. Show dimensions on the same side of the reference line as the symbol, or on either side when the symbol is centered on the reference line and no arrow-side or other-side significance is intended.

The size of a spot weld is its diameter. Show this value at the left of the weld symbol on either side of the reference line (Figure 10-22a). If you need to indicate the minimum acceptable shear strength in pounds per spot, instead of the size of the weld, place this value at the left of the weld symbol, and indicate pitch at the right of the weld symbol as shown in Figure 10-22b. In this case the spot welds are 3″ apart.

**Figure 10-22**   Spot Welds

If a joint requires a certain number of spot welds, give the number in parentheses above or below the symbol, as in Figure 10-22c. If the exposed surface of one member is to be flush, add the flush contour symbol above the symbol if it is the other-side member, and below it if it is the arrow-side member, as in Figure 10-22d. Figure 10-22e shows the welding symbol used in conjunction with ordinary dimensions.

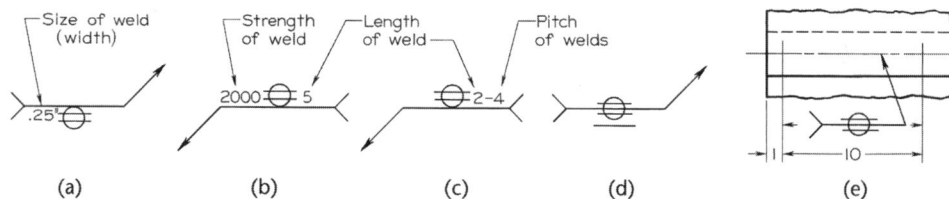

## 10-13  SEAM WELDS

The seam weld symbol, with the welding process indicated in the tail, may or may not have arrow-side or other-side significance. Dimensions are shown on the same side of the reference line as the symbol, or on either side when the symbol is centered on the reference line and no arrow-side or other-side significance is intended.

The size of the seam weld is its width. Show this value at the left of the weld symbol, on either side of the reference line (Figure 10-23a). If you need to indicate the minimum acceptable shear strength in pounds per linear inch, instead of the size of the weld, place this value at the left of the weld symbol, and show the length of a seam weld at the right of the weld symbol as in Figure 10-23b. In this case, the seam weld is 5″ long. If the weld extends the full distance between abrupt changes of direction, no length dimension in the symbol is given.

The pitch of intermittent seam welding is the distance between centers of lengths of welding. Show the pitch at the right of the length figure (Figure 10-23c). In this case, the welds are 2″ long and spaced 4″ center to center.

When the exposed surface of one member is to be flush, add the flush-contour symbol above the symbol if it is the other-side member and below it if it is the arrow-side member (Figure 10-23d). Figure 10-23e shows the welding symbol used in conjunction with ordinary dimensions.

**Figure 10-23**   Seam Welds

## 10-14  PROJECTION WELDS

In projection welding, one member is embossed in preparation for the weld (Figure 10-24a). When welded, the joint appears in section, as in Figure 10-24b. The weld symbols, in this case, are placed below the reference lines (Figure 10-24c) to indicate that the arrow-side member is the one that is embossed. The weld symbols would be placed above the lines if the other member were embossed.

Projection welds are dimensioned by either size or strength. The size is the diameter of the weld. This value is shown to the left of the weld symbol (Figure 10-24d). If you need to indicate the minimum acceptable shear strength in pounds per weld, place the value at the left of the weld symbol (Figure 10-24e). Indicate the pitch at the right of the weld symbol (Figure 10-24e). In this case, the welds are spaced 6″ (152 mm) apart. If the joint requires a definite number of welds, give the number in parentheses (Figure 10-24e). If the exposed

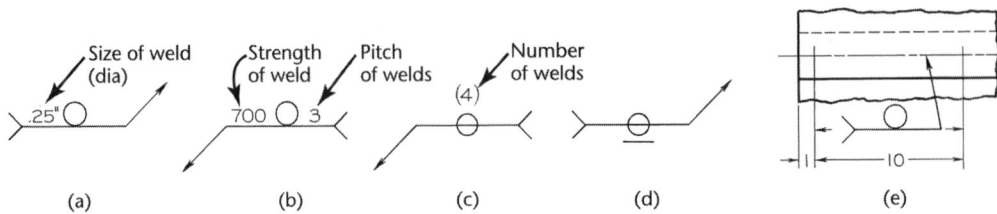

**Figure 10-24**    Projection Welds

surface of one member is to be flush, add the flush contour symbol (Figure 10-24f). Figure 10-24g shows the welding symbol used in conjunction with ordinary dimensions. The welding process reference is required in the tail of the symbol.

## 10-15  FLASH AND UPSET WELDS

Flash and upset weld symbols have no arrow-side or other-side significance, but the supplementary symbols do. A flash-welded joint is shown in Figure 10-25a, and an upset welded joint in Figure 10-25b. The joint after machining flush is shown in Figure 10-25c. The complete symbol (Figure 10-25d) includes the weld symbol together with the flush contour and machining symbols.

If the joint is ground to smooth contours (Figure 10-25e), the resulting welding drawing and symbol would be constructed as in Figure 10-25f, which includes convex-contour and grind symbols. In either Figure 10-25d or 10-25f, the joint may be finished on only one side, if desired, by indicating the contour and machining symbols on the appropriate side of the reference line. The dimensions of flash and upset welds are not shown on the welding symbol. Note that the process reference for flash welding (FW) or upset welding (UW) must be placed in the tail of the symbol.

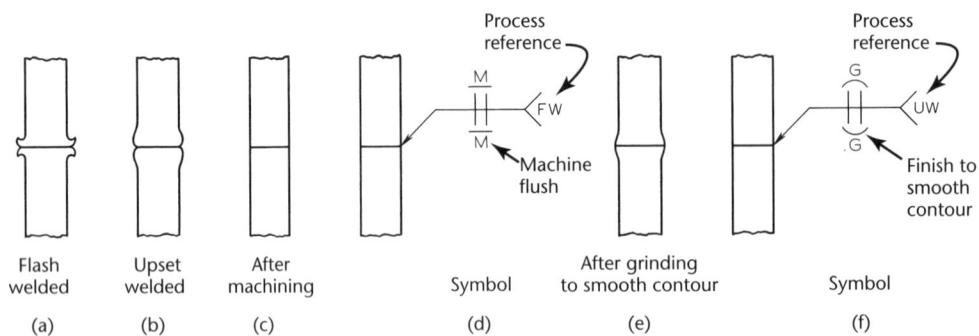

**Figure 10-25**    Flash and Upset Welds

## 10-16  WELDING APPLICATIONS

A typical example of welding fabrication for machine parts is shown in Figure 10-26. In many cases, especially when only one or a few identical parts are required, it is cheaper to produce by welding than to make patterns and sand castings and do the necessary machining. Thus, welding is particularly adaptable to custom-built constructions.

Welding is also suitable for large structures that are difficult or impossible to fabricate entirely in the shop, and it is coming into greater use for steel structures, such as building frames, bridges, and ships. A welded truss is shown in Figure 10-27. It is easier to place members in such a welded truss so that their center of gravity axes coincide with the working lines of the truss than is the case in a riveted truss.

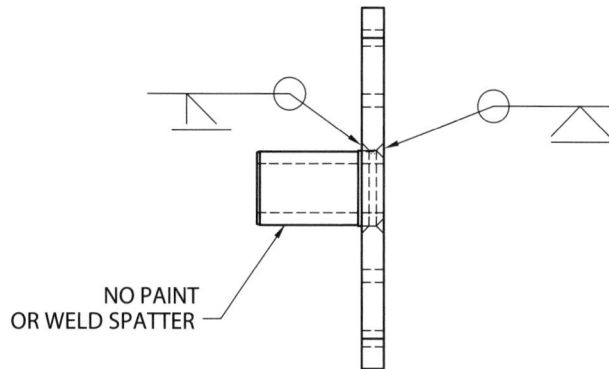

**Figure 10-26**    Application of welding fabrication for machine parts. *Courtesy of Dynojet Research, Inc.*

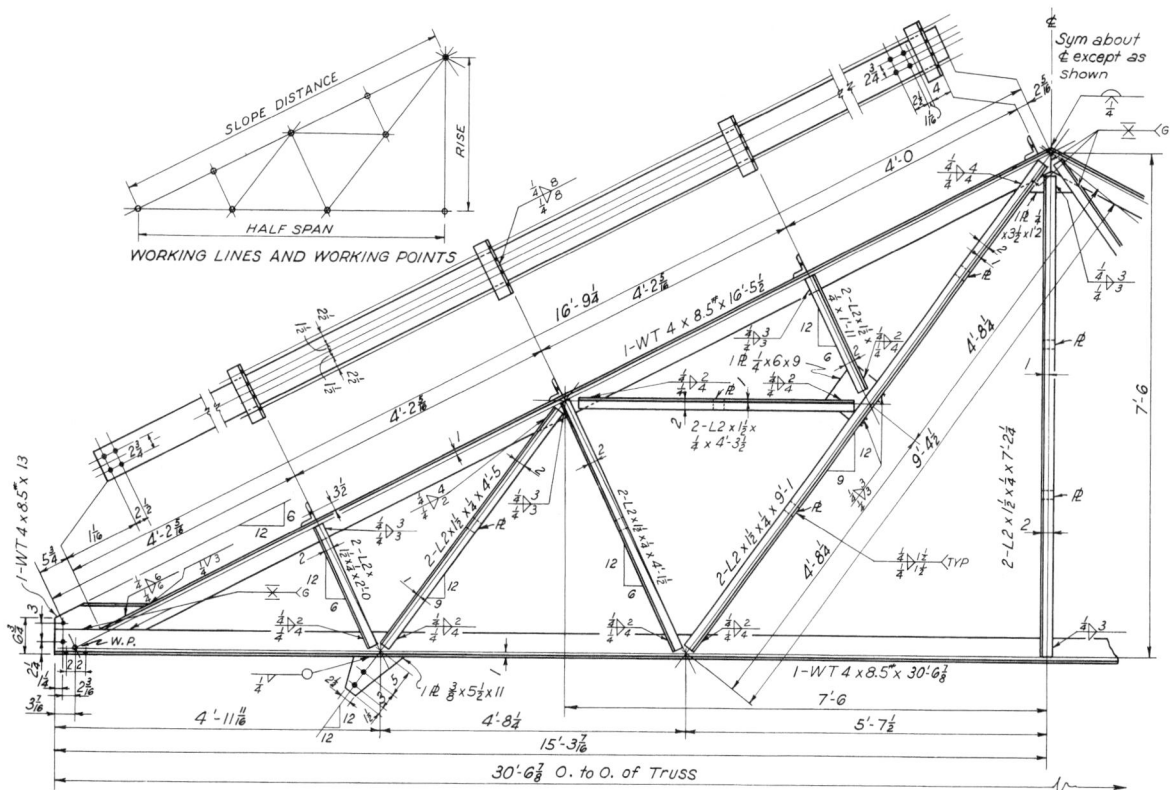

**Figure 10-27**    A Welded Truss

## 10-17  WELDING TEMPLATES

Welding templates can simplify drawing welding symbols by hand (which may be done in pencil or ink). They have all the forms needed for drawing the arrow, weld symbols, and supplementary symbols, as well as an illustration of the complete composite welding symbol for quick reference.

## 10-18  COMPUTER GRAPHICS

Welding symbols libraries available in CAD (Figure 10-28) allow for rapid application of accurate, uniform symbols that are in compliance with AWS standards (Figure 10-29). In addition to standard symbols, many CAD programs permit the operator to create custom symbols as required.

**Figure 10-28**    Computervision Production Drafting

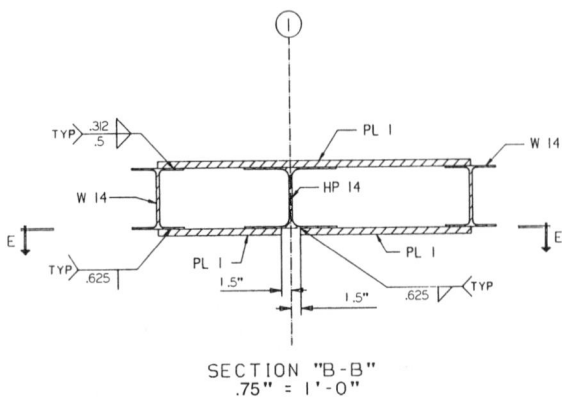

**Figure 10-29**    CAD-Generated Welded Structural Detail

# CAD AT WORK

## Weld Symbols from CAD

Most CAD systems provide a way to quickly generate weld symbols to place in your drawing. The Solidworks dialog box shown in Figure A allows you to select the symbol for type and size of weld, field placement, whether to add the all around symbol, special indications for contour and finish method. Once you have made you selections, the weld symbol as shown in Figure B is automatically generated based on your selections and you have only to click to place it in your drawing.

Specialized software like Design Data's SDS/2 allows you to design and model connections between members based on parameters that you define for each job. You can designate the material, edge distance, cope criteria and other connection specifications, and SDS/2 will automatically generate connections. You can model all of the details using the software's 3D modeler for example the fillet welds shown in figure C. 2D drawings can be taken directly from the model, providing an accurate fit in the field.

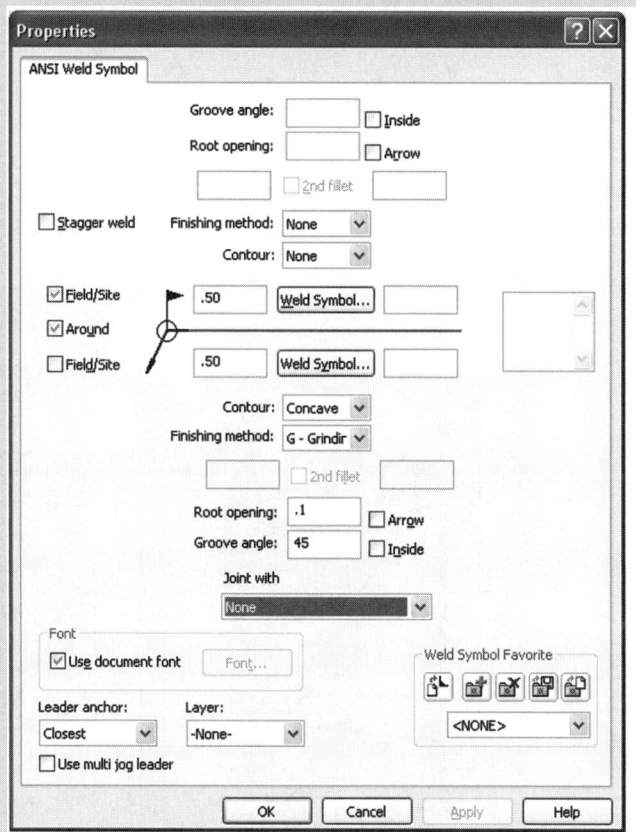

(A) SolidWorks Weld Symbol Dialog Box

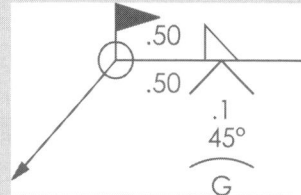

(B) Automatically Generated Weld Symbol

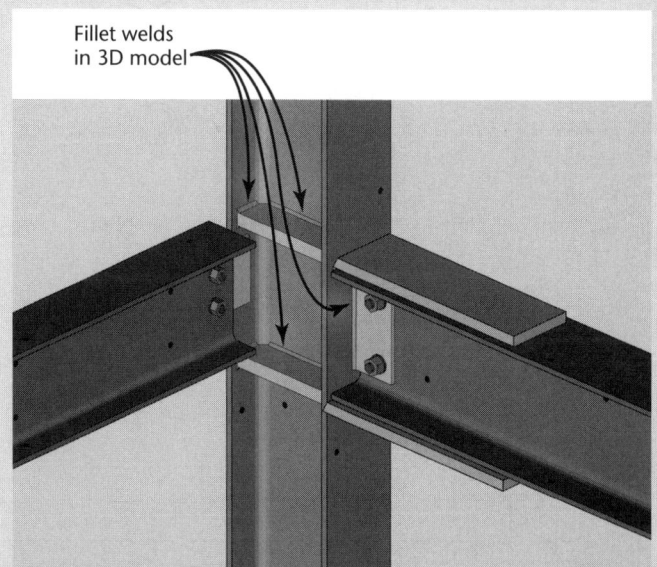

(C) Fillet welds in a 3D model. *Courtesy of Paul Hergett, Midwest Steel Industries.*

# Portfolio

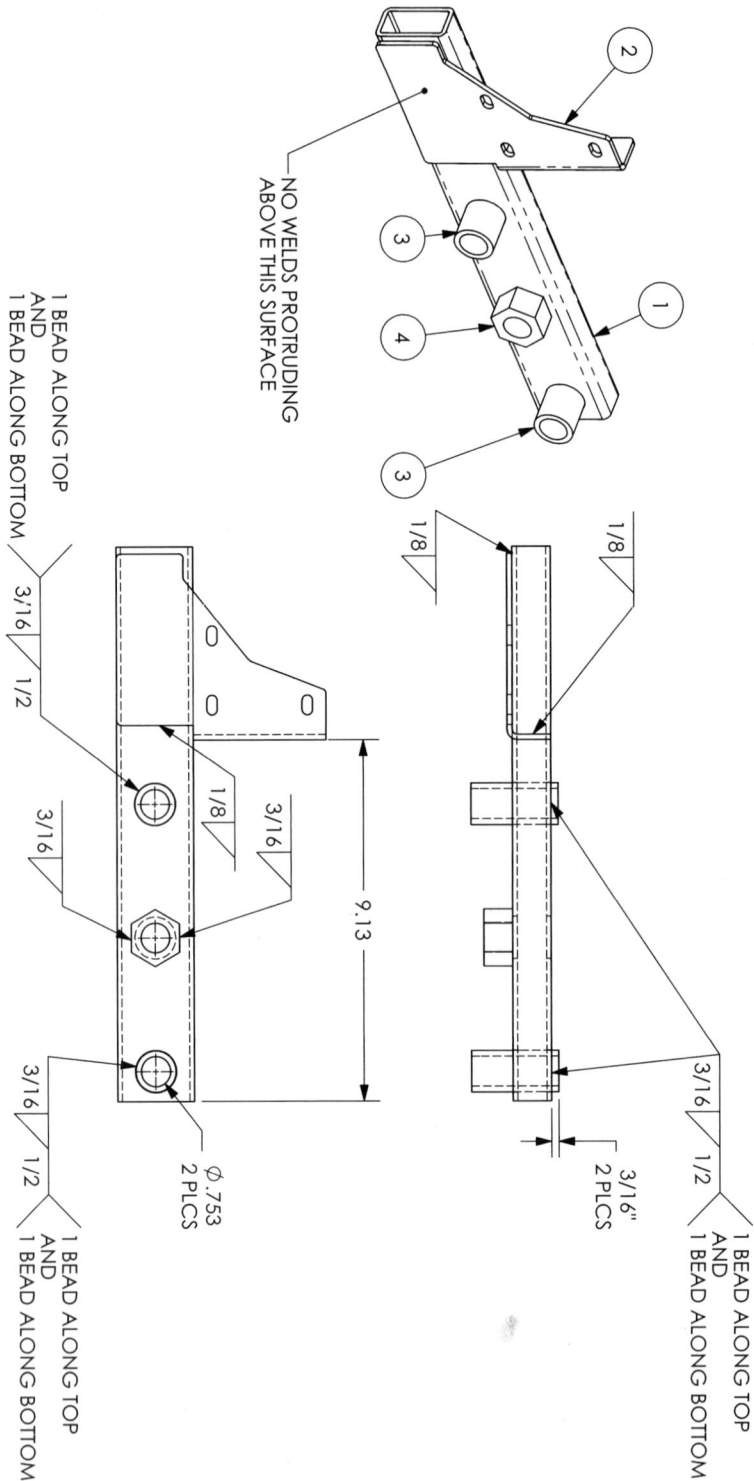

Welding symbols specify the welds for attaching the tubes, bushing, nut, and plate for the clamp arm. *Courtesy of Dynojet Research, Inc.*

# Portfolio

Welding symbols specify how plates attach to structural beam in this drawing created from a 3D model using SOSZ software. *Courtesy of Paul Hergett, Midwest Steel Industries.*

# 10-19  EXERCISES

## Welding Drawing Exercises

The following problems are given to familiarize the student with some applications of welding symbols to machine construction and to steel structures.

### EX10-1

Change to a welded part. Make working drawing, using appropriate welding symbols.

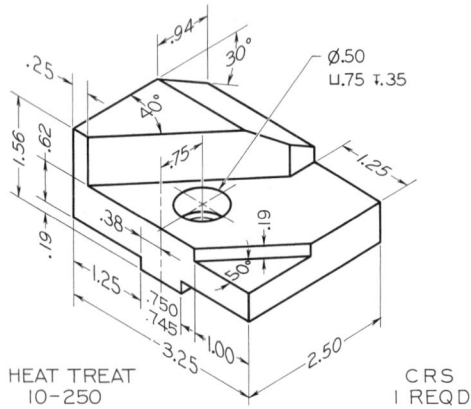

### EX10-2

Same instructions as for EX10-1.

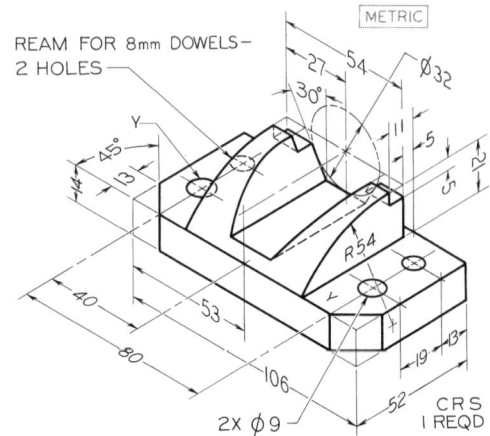

### EX10-3

Same instructions as for EX10-1.

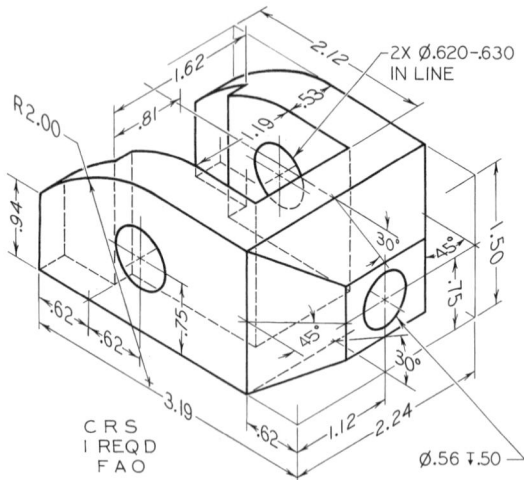

### EX10-4

Same instructions as for EX10-1.

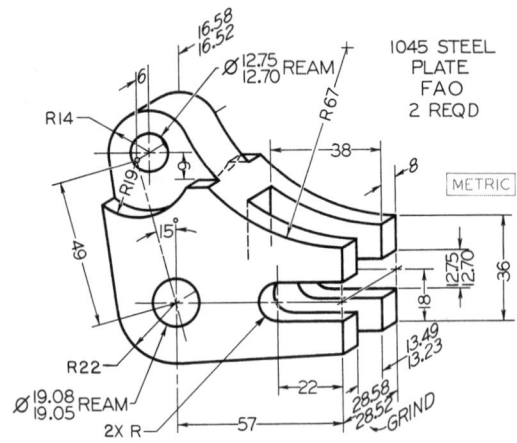

## EX10-5

Same instructions as for EX10-1.

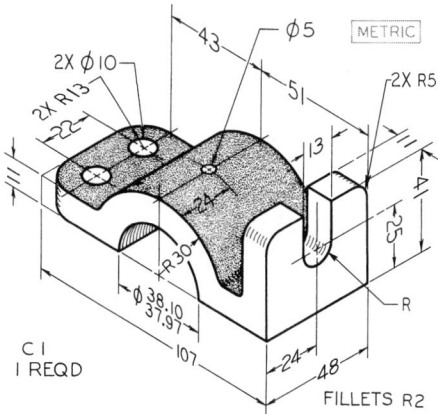

C I
I REQD

FILLETS R2

## EX10-6

Same instructions as for EX10-1.

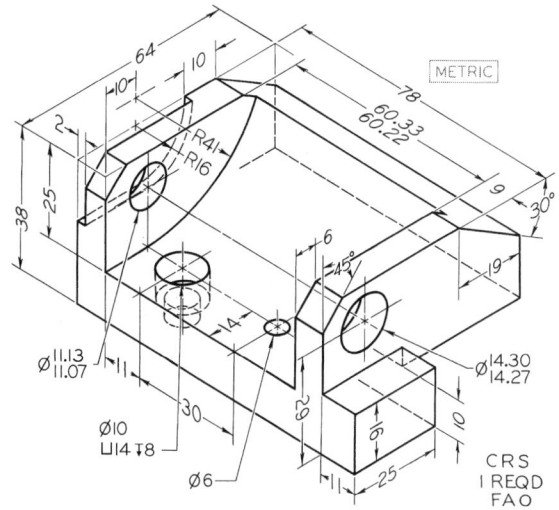

(Tongue
identical on
both sides)

(Slot clear
through)

## EX10-7

Same instructions as for EX10-1.

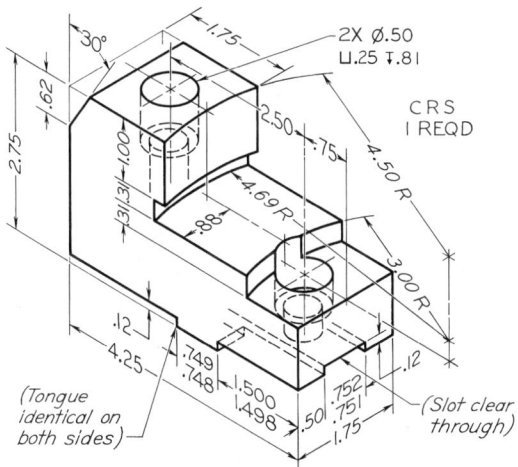

C I
I REQD

FILLETS & ROUNDS R3

## EX10-8

Same instructions as for EX10-1.

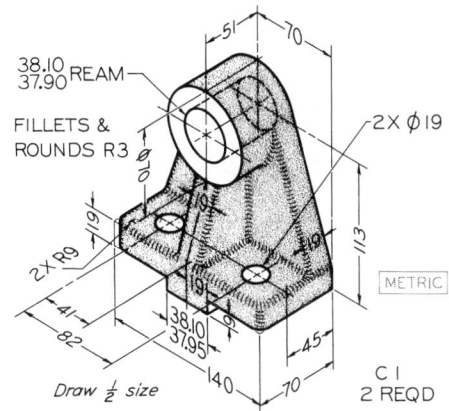

C R S
I REQD
FAO

## EX10-9

Same instructions as for EX10-1.

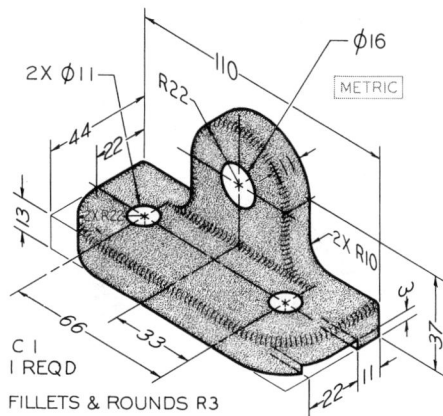

Draw ½ size

C I
2 REQD

## EX10-10

Same instructions as for EX10-1.

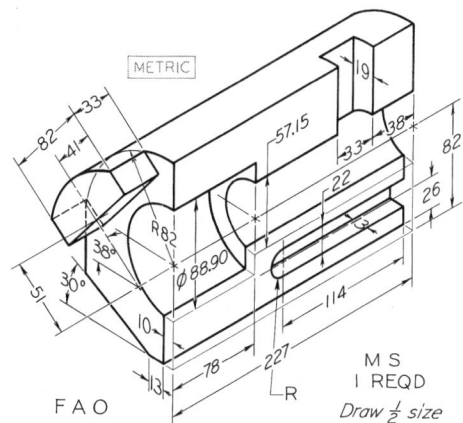

F A O

M S
I REQD
Draw ½ size

## EX10-11

Same instructions as for EX10-1.

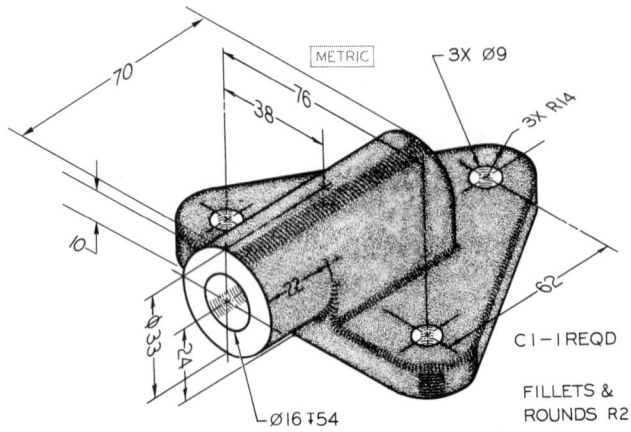

## EX10-12

Same instructions as for EX10-1.

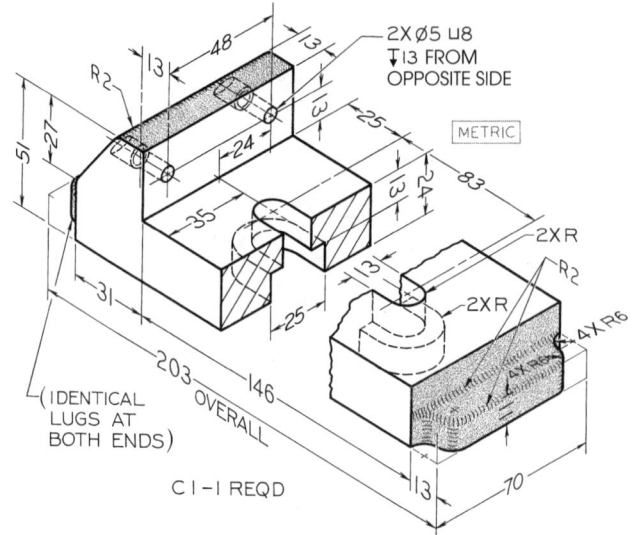

## EX10-13

Same instructions as for EX10-1.

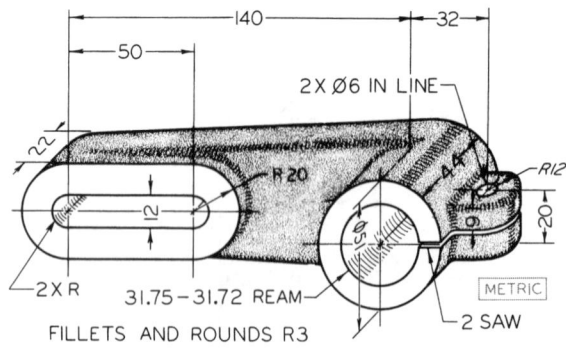

## EX10-14

Same instructions as for EX10-1.

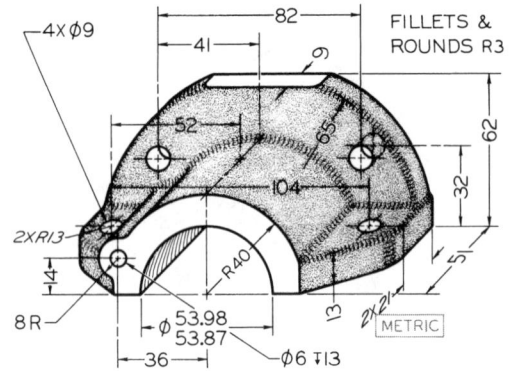

## EX10-15

Same instructions as for EX10-1.

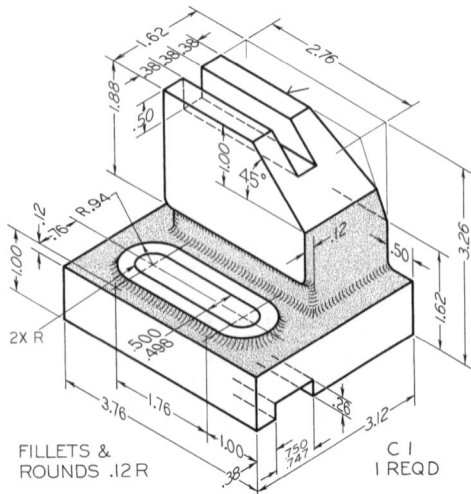

## EX10-16

Same instructions as for EX10-1.

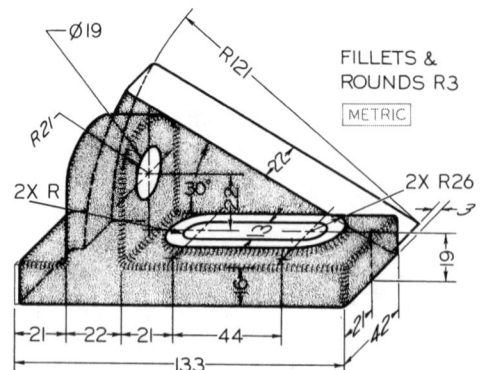

## EX10-17

Same instructions as for EX10-1.

## EX10-18

Same instructions as for EX10-1.

## EX10-19

Same instructions as for EX10-1.

## EX10-22

Make a half-size drawing of the joint at the center of the lower chord of the truss in Figure 10-27 where the chord is supported by two vertical angles. The chord is a structural tee, cut from an, $8 \times 5\frac{1}{4}$, 17 lb, wide-flange shape. Draw the

## EX10-20

Same instructions as for EX10-1.

## EX10-21

Same instructions as for EX10-1.

front and side views, and show the working lines, the two angles, the structural tee, and all welding symbols.

## EX10-23

Make a half-size front view, showing the welding symbols, of any joint of the truss in Figure 10-27 in which three or four members meet.

## EX10-24

Draw half-size front, top, and left-side views of the end joint of the truss in Figure 10-27 showing the welding symbols.

# Index

# H

Oval Head Setscrews, 381
Oval Point Setscrews, 382
Override Option, 204

# P

Pan Head Rivets, 385, 433, 434
Paper Conservation, 546
Paper Drawings, Storage of, 615
Paper Gaging, 465–66
Parallelism, 182, 193–96, 312–13,
　316
　Applicability of MMC, RFS, or
　　LMC to, 92
　Application of, 193
　Feature at MMC, Datum
　　Feature at RFS, 196
　Feature of Size at MMC, Datum
　　a Plane, 195
　Feature of Size at RFS, Datum
　　Feature at RFS, 196
　Feature of Size at RFS, Datum
　　Plane, 195
　Features of Size, 194
　Shape of Tolerance Zones for, 93
　Surface to Datum Plane, 194
　Symbol for, 78, 87, 92, 272
　Tolerance, 193
Parallelism Tolerance, 193
Parallel Shafts, 661
Parent Part, 532
Part Drawings, 535
Partial Surface Runout, 360
Partial Views
　Auxiliary, 161
　Sectional, 113
Parts List, 498–99, 537, 541, 542
Patch Bolt, 431
Patent Drawings, 547–50
Perfect Orientation, 90
Perpendicularity, 182–93, 312–13,
　314–15
　Applicability of MMC, RFS, or
　　LMC to, 92
　Application of, 183
　Cylindrical Feature at RFS,
　　Datum a Cylinder, 185
　Noncylindrical Feature at
　　MMC, Datum a Plane, 184
　Noncylindrical Feature at RFS,
　　Datum a Plane, 184
　Radial, 186–88
　Shape of Tolerance Zones for, 93
　Squareness, Normality, 182
　Symbol for, 78, 87, 92, 272
　Tolerance, 182–83
　Zero Tolerance at MMC, Max
　　Deviation, 189
Perspective, 689

Phantom Lines, 419
Phantom Styles, 101, 102
Pictorial Projections, 689
Piece Part Drawings, 535
Pinion, 662
Pipe Threads, 419–21
Pitch (p)—Worm Gears, 668
Pitch Circle of Gears, 662, 663
Pitch Curve of Cams, 675, 676
Pitch Diameter (PD) of Gears, 621,
　622, 662, 663, 671
Pitch Diameter of Threads, 404
Pitch Diameter Rule, 91
Pitch of Gears, 623
Pitch of Threads, 361, 404, 406–7
Pivoted Cam Followers, 677–78
Plain Washers, 383
Plane of Projection, 726
Plane Surfaces, Angularity of, 191
Plasma Arc Welding, 744
Plow Bolt, 431
Plug Welds, 746, 753, 754
Plus and Minus Tolerances
　Creating Using AutoCAD,
　　255–57
　Understanding, 255
Poche, 538
Points, Dimensioning to, 233
Polar Array, 665
Polar Dimensions, 230
Polyline Command
　Drawing Cam Followers, 649
　Drawing Cylinders, 24
　Drawing Hexagonal-Shaped
　　Heads, 374
　Drawing Intersections, 36, 38
　Drawing Irregular Surfaces, 27
　Drawing Square-Shaped Head,
　　376
Position, 447
　Applicability of MMC, RFS, or
　　LMC to, 92
　Choice of in Oblique
　　Projections, 730–31
　Composite Position
　　Tolerancing, 465–70
　Functional Gaging Principles,
　　472–73
　Mating Parts, Fixed Fasteners,
　　454–58
　Mating Parts, Floating
　　Fasteners, 451–53
　MMC Related to MMC Datum
　　Feature, 471–74, 478–82
　MMC Related to RFS Datum
　　Feature Functional Gaging
　　Principles, 475–76
　Relation to Datum Surface,
　　459–60

Relation to Specific Datum
　Surfaces, 458–64
RFS Related to RFS Datum
　Feature, 477–78
Shape of Tolerance Zone for, 93
Symbol for, 79, 87, 92
Positional Tolerances, 319–20,
　447–53
　Adding to Hole's Feature
　　Tolerance, 312
　Control of Coaxial Features,
　　349, 350
　Creation with AutoCAD, 308
　LMC, 483–87
　LMC Related to RFS Datum
　　Feature, 485–87
　Relation to Datum Surface,
　　469–70
Position System, 448
Position Theory, 447, 449–51
Power Springs, 436
Pratt & Whitney Keys, 384, 431,
　432
Precedence of Lines, 9
Precision Box
　Changing Number of Decimal
　　Places, 208, 311
　Creating Plus and Minus
　　Tolerances, 257
　Setting Precision of Angular
　　Dimensions, 260
Preferred Pitches of Gears, 621
Preferred Sizes, 270
Presentation Drawing, 714
Pressure Angle of Gears, 622, 663
Primary Units Option
　Changing Number of Decimal
　　Places, 208, 260, 311
　Creating Plus and Minus
　　Tolerances, 256
　Measurement Scale Box, 204–5
　Precision of Angular
　　Dimensions, 260
　Zero Suppression, 207
Principles
　Least Material Condition, 81
　Maximum Material Condition,
　　79–80
　Regardless of Feature Size
　　(RFS), 80–81
Printing Engineering Drawings,
　615–17
Product Proposals, 605
Product Release, 606
Profile, 291–99. *see also* Profile of a
　Line; Profile of a Surface
　Application of Tolerance, 291
　Geometric Tolerances, 316–17
　Method of Specifying, 291

# Appendix A
# Workbook

## GEOMETRIC TOLERANCING

Advances in technology have brought about the need for more preciseness in the design and reproduction of machined parts. To accommodate these needs, industry has adopted a system of geometric tolerance control symbols that have replaced lengthy drawing notes. As a blueprint reader, you must become acquainted with the various symbols and their interpretations. Study the table of symbols below, and learn to relate the characteristic with the symbol that it represents. Examples of usage will be shown on following pages.

| | TYPE OF TOLERANCE | CHARACTERISTIC | SYMBOL |
|---|---|---|---|
| FOR INDIVIDUAL FEATURES | FORM | STRAIGHTNESS | — |
| | | FLATNESS | ▱ |
| | | CIRCULARITY (ROUNDNESS) | ○ |
| | | CYLINDRICITY | ⌭ |
| FOR INDIVIDUAL OR RELATED FEATURES | PROFILE | PROFILE OF A LINE | ⌒ |
| | | PROFILE OF A SURFACE | ⌓ |
| FOR RELATED FEATURES | ORIENTATION | ANGULARITY | ∠ |
| | | PERPENDICULARITY | ⊥ |
| | | PARALLELISM | // |
| | LOCATION | POSITION | ⌖ |
| | | CONCENTRICITY | ◎ |
| | | SYMMETRY | ⩵ |
| | RUNOUT | CIRCULAR RUNOUT | ↗• |
| | | TOTAL RUNOUT | ↗↗• |
| • ARROWHEADS MAY BE FILLED OR NOT FILLED | | | |

Geometric Characteristic Symbols

*(Figures reprinted from ASME Y14.5M, by permission of the American Society of Mechanical Engineers. All rights reserved.)*

The symbols shown above are from the current ASME Y14.5M–1994 Dimensioning and Tolerancing Standard, adopted March 14, 1994.

These symbols also conform to the CAN/CSA–B78.2–M91 Dimensioning and Tolerancing of Technical Drawings, Basic Engineering Canadian Standard. Earlier versions of ASME Y14.5M–1994 were: ANSI Y14.5M–1982, ANSI Y14.5–1973, USASI Y14.5–1966, and ASA Y14.5–1957. Other previous standards consolidated were the MIL–STD–8C and the SAE Automotive Aerospace Drawing Standards. Since many drawings in existence today were created when earlier standards were in effect, those drawings might still contain some of the old symbols. Therefore, it will be necessary for you to also become familiar with the former symbols shown near the end of the workbook.

## FEATURE CONTROL FRAME

The feature control frame is divided into separate compartments. The geometric characteristic symbol will always appear in the first compartment of the frame. The second compartment will contain the tolerance. Where applicable, the tolerance is preceded by the diameter symbol and/or followed by a material condition symbol. See the examples shown below. (*Note:* the examples in this unit are expressed in millimeters.)

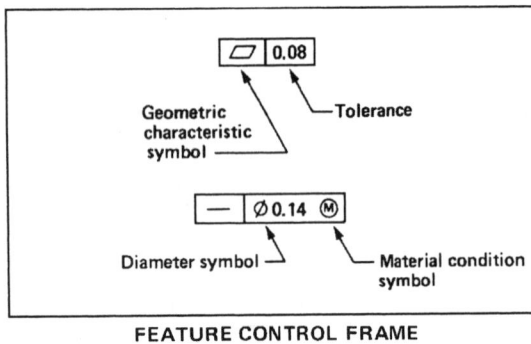

**FEATURE CONTROL FRAME**

   Where a geometric tolerance is related to a datum, the datum reference will occupy the compartment following the tolerance, inside the feature control frame. Earlier versions had the datum reference appearing between the symbol and the tolerance. (See the example below.)

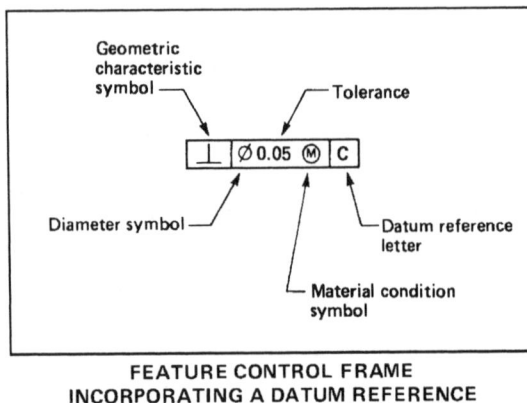

**FEATURE CONTROL FRAME**
**INCORPORATING A DATUM REFERENCE**

Where a datum is established by two datum features, such as an axis established by two datum diameters, both datum reference letters appear, separated by a dash. (See the example shown below.)

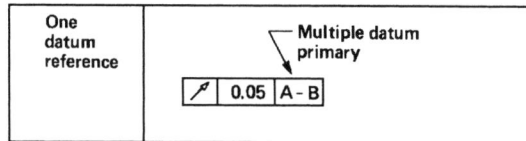

Where more than one datum is referenced, the letters (followed by a material condition symbol, where applicable) appear inside separate compartments in order of precedence, from left to right. (See the example below.)

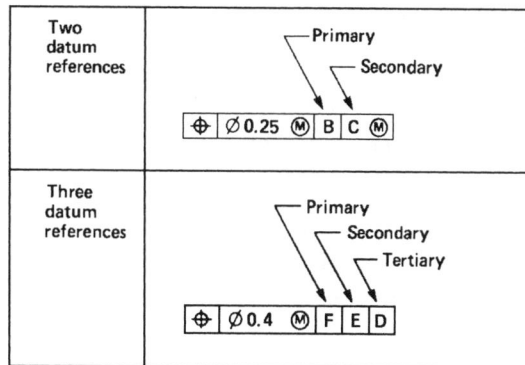

ORDER OF PRECEDENCE
OF DATUM REFERENCES

## MATERIAL CONDITION SYMBOLS

The symbols used to indicate "at maximum material condition," and "at least material condition" are shown below. They may be used in the feature control frame to indicate how the tolerance or datum reference is applied. Examples on this and the preceding page include material condition symbols.

| TERM | SYMBOL |
|---|---|
| AT MAXIMUM MATERIAL CONDITION | Ⓜ |
| AT LEAST MATERIAL CONDITION | Ⓛ |
| PROJECTED TOLERANCE ZONE | Ⓟ |

# PROJECTED TOLERANCE ZONE

Where a positional or an orientation tolerance is specified as a projected tolerance zone, the projected tolerance zone symbol is placed in the feature control frame, along with the dimension indicating the minimum height of the tolerance zone. This is to follow the stated tolerance and any modifier. (See the example below.)

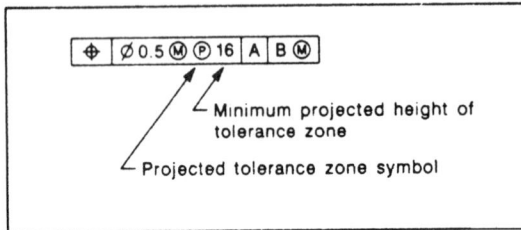

FEATURE CONTROL FRAME WITH A
PROJECTED TOLERANCE ZONE SYMBOL

# BASIC DIMENSION SYMBOL

A basic dimension is a theoretically exact value, with no tolerance. It is used as the basis from which permissible variations are established by tolerances in the feature control frames. The symbol is to enclose the dimension figure in a rectangular frame, as in the example shown below.

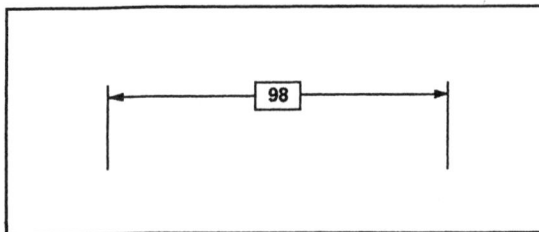

BASIC DIMENSION SYMBOL

## DATUM FEATURE SYMBOL

A datum feature is a physical feature of a part used to establish a datum. The symbol consists of a square frame containing the datum identifying letter and a leader line extending from the frame to the feature, terminating with a triangle. (See the example below.)

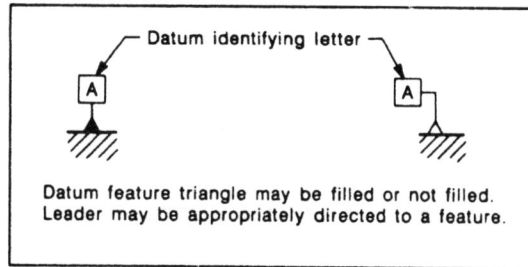

DATUM FEATURE SYMBOL

## COMBINED SYMBOLS

When a feature that is controlled by a geometric tolerance also serves as a datum feature, the feature control frame and the datum feature symbol may be combined. In the example shown below, the feature is controlled for position in relation to both datum A and datum B, and is identified as datum feature C. Whenever datum C is referenced elsewhere on the drawing, the reference applies only to datum C, not to datum A or B.

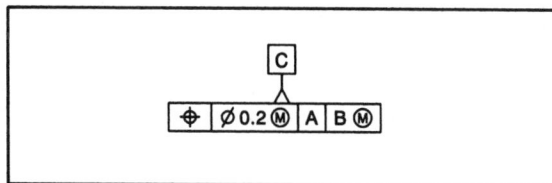

COMBINED FEATURE CONTROL FRAME AND
DATUM FEATURE SYMBOL

## COMPOSITE FRAMES

Where more than one tolerance is specified for the same geometric characteristic of a feature, the composite frame may be used. It will contain a single geometric characteristic symbol followed by each tolerance and datum requirement, one above the other. (See the example below.)

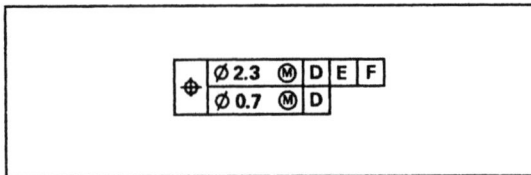

| ⊕ | ⌀2.3 Ⓜ | D | E | F |
|---|---|---|---|---|
|   | ⌀0.7 Ⓜ | D |   |   |

**COMPOSITE FEATURE CONTROL FRAME**

## DATUM TARGET SYMBOL

A datum target is a specified point, line, or area used to establish datum points, lines, planes, or areas for manufacturing or inspecting purposes. The datum target symbol is a circle divided into halves. The bottom half contains the datum identification letter followed by the target number. If the datum target is an area, the size may appear in the top half of the symbol. (See the example below.)

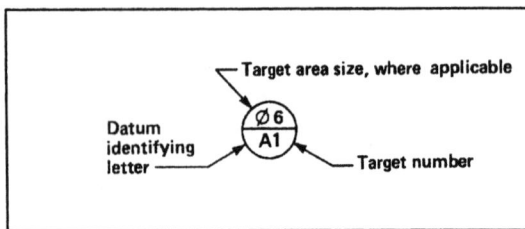

Target area size, where applicable

Datum identifying letter

⌀6
A1

Target number

**DATUM TARGET SYMBOL**

## ALL AROUND SYMBOL

A circle at the junction of the leader elbow is the symbol that indicates a profile tolerance applies to surfaces all around the part. (See the example below.)

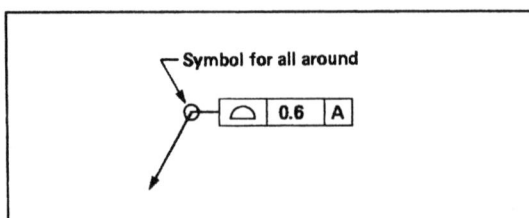

Symbol for all around

| ⌒ | 0.6 | A |
|---|---|---|

**SYMBOL FOR ALL AROUND**

## GD&T Symbol Worksheet

INSTRUCTIONS: Shown below are the various symbols introduced to you on the preceding pages. Without referring to those pages, try to match the symbols with the list of words, then write the correct words alongside each symbol.

1._____

2._____

3._____

4._____

5._____

6._____

7._____

8._____

9._____

10._____

11._____

12._____

13._____

14._____

15._____

16._____

17._____

18._____

19._____

20._____

21._____

22._____

23._____

24._____

All around

Angularity

At least material condition

At maximum material condition

Basic dimension

Circular runout

Circularity

Concentricity

Cylindricity

Datum feature

Datum target

Diameter

Dimension origin

Flatness

Parallelism

Perpendicularity

Position

Profile of a line

Profile of a surface

Projected tolerance zone

Reference dimension

Straightness

Symmetry

Total runout

INSTRUCTIONS: Identify each component of the feature control frame drawn below. Write your answers on the lines provided.

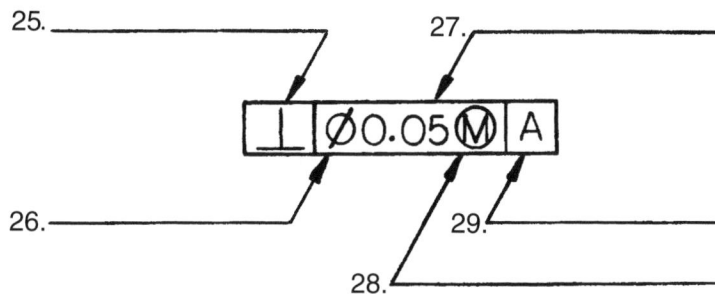

25._____

26._____

27._____

28._____

29._____

# FORM TOLERANCES

The starting point for all geometric dimensioning and tolerancing (GD&T) is with Rule #1. This rule states that the form of an individual feature is controlled by its limits of size to the extent that the surfaces of a feature shall not extend beyond a boundary (envelope) of perfect form at MMC. This boundary is the true geometric form represented by the drawing. With few exceptions, no variation in form is permitted if the feature is produced at its MMC limit of size. Where the actual size of a feature has departed from MMC toward LMC, a variation in form is allowed equal to the amount of such departure. Therefore, if a feature has a large size range, it is possible for this same feature to have a large form error if only a size is given. To control the amount of form error that a feature might have, form controls are used. A form tolerance specifies a zone within which the considered feature, its line elements, its axis, or its centerplane must be contained. Where the tolerance value represents the diameter of a cylindrical zone, it is preceded by the diameter symbol. Form tolerances are applicable to individual features only. Therefore, feature control frames associated with form controls will not contain datum reference letters. All form controls applied to a surface apply RFS (regardless of feature size). Only straightness, when it is applied to a feature of size, can use the Ⓜ modifier. The four form tolerances are straightness, flatness, circularity, and cylindricity.

## Straightness Tolerance

Straightness is a condition where an element of a surface or an axis is a straight line. A straightness tolerance specifies a tolerance zone. The line element or axis must lie within this zone. A straightness tolerance applies in the view where the elements to be controlled are shown as a straight line. Straightness can be applied to a surface. This results in a two-dimensional tolerance zone (two parallel straight lines apart by the tolerance value). It can also be applied to a feature of size, resulting in a three-dimensional tolerance zone that consists of either a cylinder (having the diameter of the specified tolerance zone), or two parallel planes apart by the tolerance value. All points of the derived median line must lie within the specified tolerance zone. Observe from the examples how the tolerance zone increases as the feature size deviates from MMC when Ⓜ is included in the feature control frame.

## THIS ON THE DRAWING

$$\varnothing \begin{matrix} 16.00 \\ 15.89 \end{matrix} (16h11)$$

## MEANS THIS

(a)

Ø16.00
MMC

0.02 wide tolerance zone

(b)

Ø16.00
MMC

0.02 wide tolerance zone

(c)

Ø16.00
MMC

Each longitudinal element of the surface must lie between two parallel lines (0.02 apart) where the two lines and the nominal axis of the part share a common plane. The feature must be within the specified limits of size and the boundary of perfect form at MMC (16.00).

Note: Waisting (b) or barreling (c) of the surface, though within the straightness tolerance, must not exceed the limits of size of the feature.

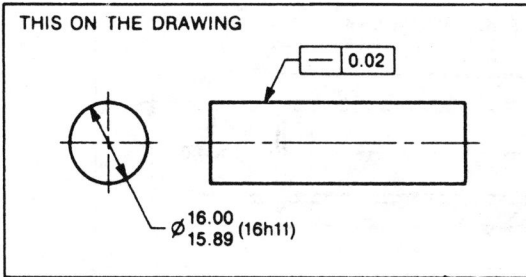

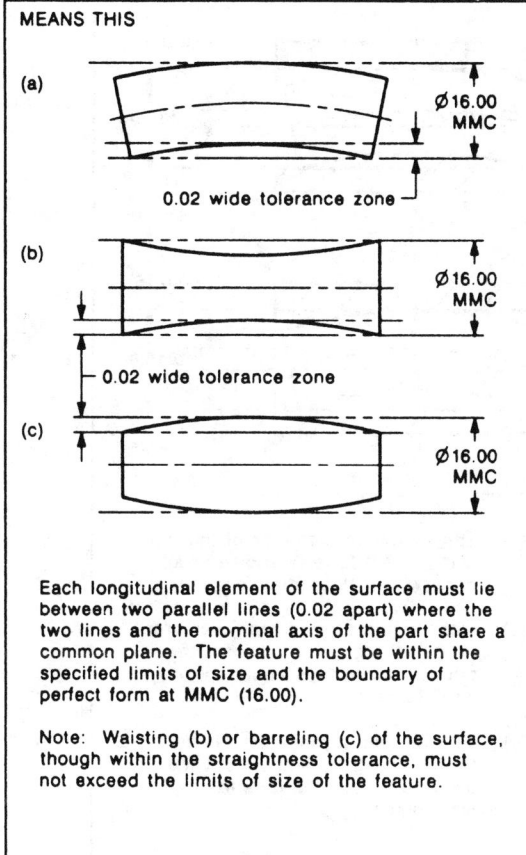

**SPECIFYING STRAIGHTNESS OF SURFACE ELEMENTS**

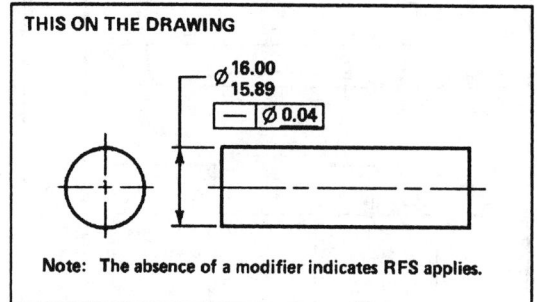

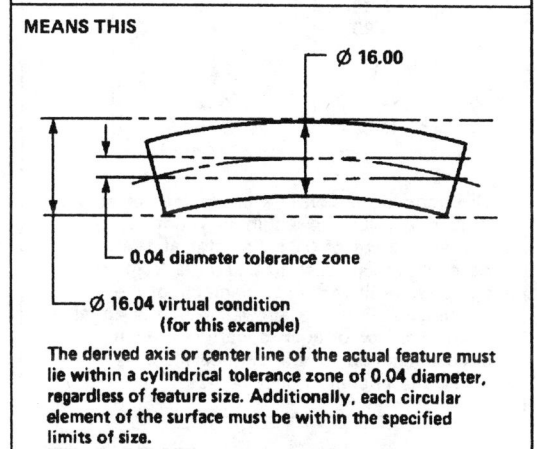

## THIS ON THE DRAWING

$$\varnothing \begin{matrix} 16.00 \\ 15.89 \end{matrix}$$

Note: The absence of a modifier indicates RFS applies.

## MEANS THIS

Ø 16.00

0.04 diameter tolerance zone

Ø 16.04 virtual condition
(for this example)

The derived axis or center line of the actual feature must lie within a cylindrical tolerance zone of 0.04 diameter, regardless of feature size. Additionally, each circular element of the surface must be within the specified limits of size.

**SPECIFYING STRAIGHTNESS RFS**

*(Reprinted from ASME Y14.5M, by permission of the American Society of Mechanical Engineers. All rights reserved.)*

$\varnothing^{16.00}_{15.89}$ (16h11)

| — | $\varnothing$ 0.04 Ⓜ |

$\varnothing$16.04
$\varnothing$16.00

(a)

**MEANS THIS**

$\varnothing$16.04 virtual condition

$\varnothing$16.00
$\varnothing$0.04
$\varnothing$16.04

(b)

$\varnothing$15.89
$\varnothing$0.15
$\varnothing$16.04

(c)

| Feature size | Diameter tolerance zone allowed |
|---|---|
| 16.00 | 0.04 |
| 15.99 | 0.05 |
| 15.98 | 0.06 |
| ↓ | ↓ |
| 15.90 | 0.14 |
| 15.89 | 0.15 |

The derived median line of the feature actual local sizes must lie within a cylindrical tolerance zone of 0.04 diameter at MMC. As each actual local size departs from MMC, an increase in the local diameter of the tolerance cylinder is allowed which is equal to the amount of such departure. Each circular element of the surface must be within the specified limit of size.

Meanings:

(a) The maximum diameter of the pin with perfect form is shown in a gage with a 16.04 diameter hole;

(b) with the pin at maximum diameter (16.00), the gage will accept the pin with up to 0.04 variation in straightness;

(c) with the pin at minimum diameter (15.89), the gage will accept the pin with up to 0.15 variation in straightness.

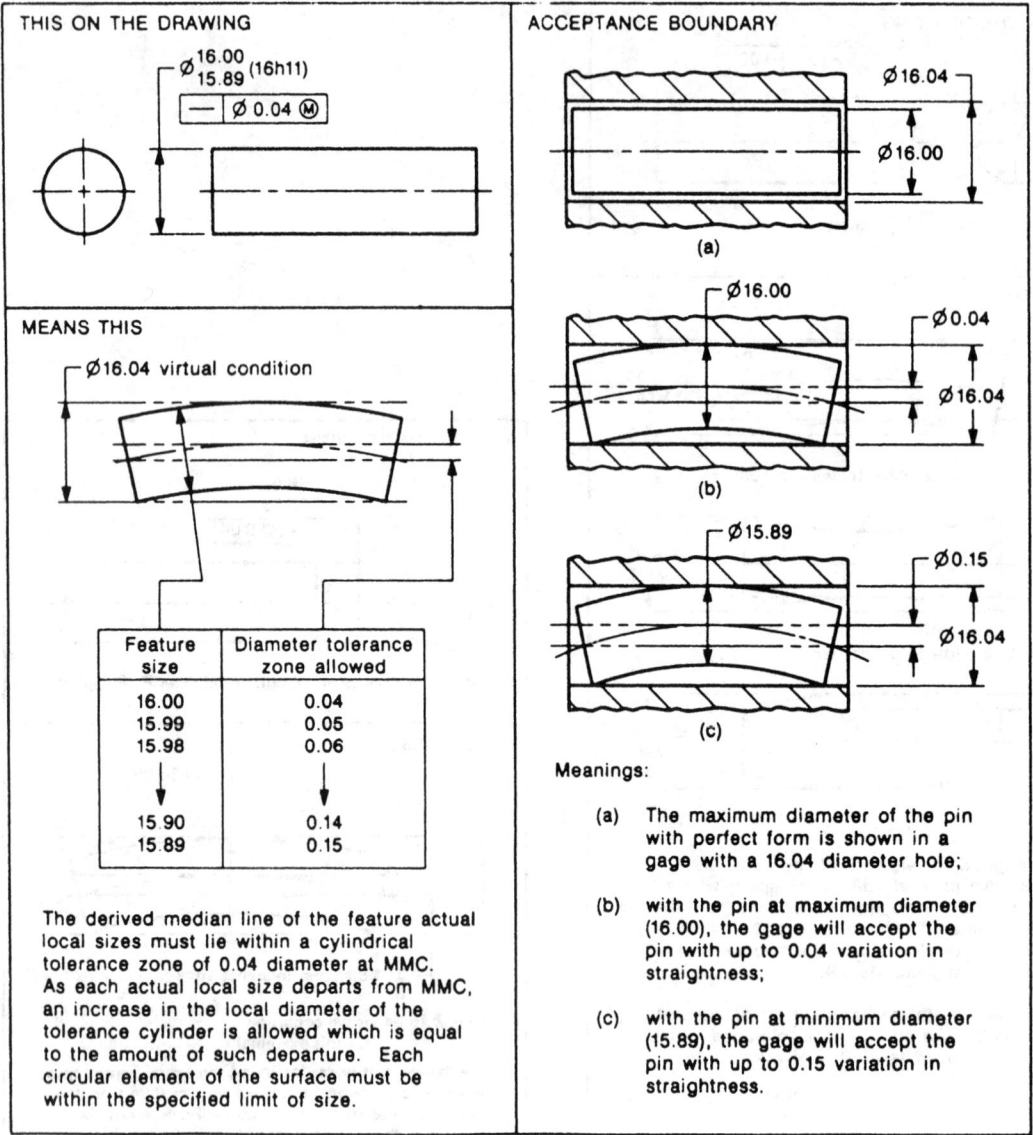

**SPECIFYING STRAIGHTNESS AT MMC**

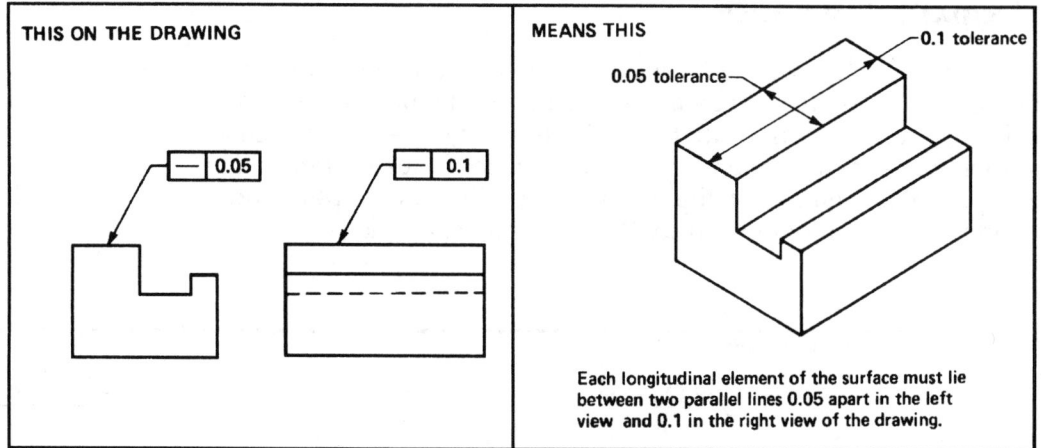

**SPECIFYING STRAIGHTNESS OF FLAT SURFACES**

## Flatness Tolerance

Flatness is the condition of a surface having all elements in one plane. A flatness tolerance (which is a three-dimensional tolerance zone) specifies a zone defined by two parallel planes within which the surface must lie. A flatness tolerance and its symbol will appear in the view where the surface elements to be controlled are represented by a line. Compare the example of flatness (three-dimensional) shown below with the application of straightness of flat surfaces (two-dimensional) shown in the example above.

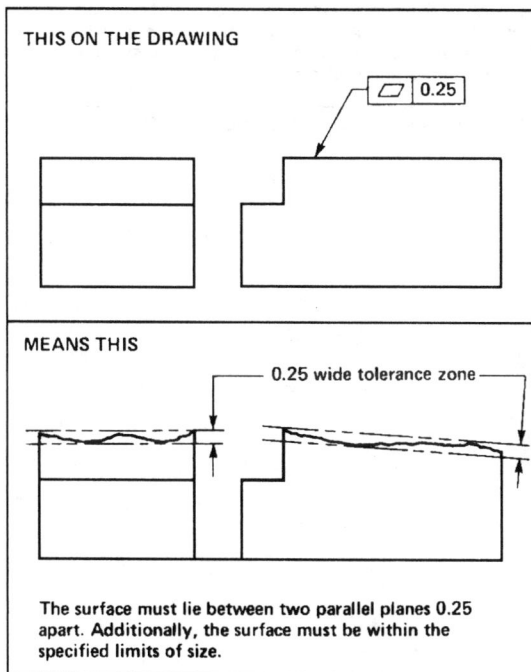

**SPECIFYING FLATNESS**

## Circularity Tolerance

Circularity is a condition of a surface of revolution where all points of the surface intersected by any plane perpendicular to a common axis are equidistant from that axis. A circularity tolerance specifies a tolerance zone bounded by two concentric circles within which each circular element of the surface must lie, and applies independently at each plane. (Study the following example.)

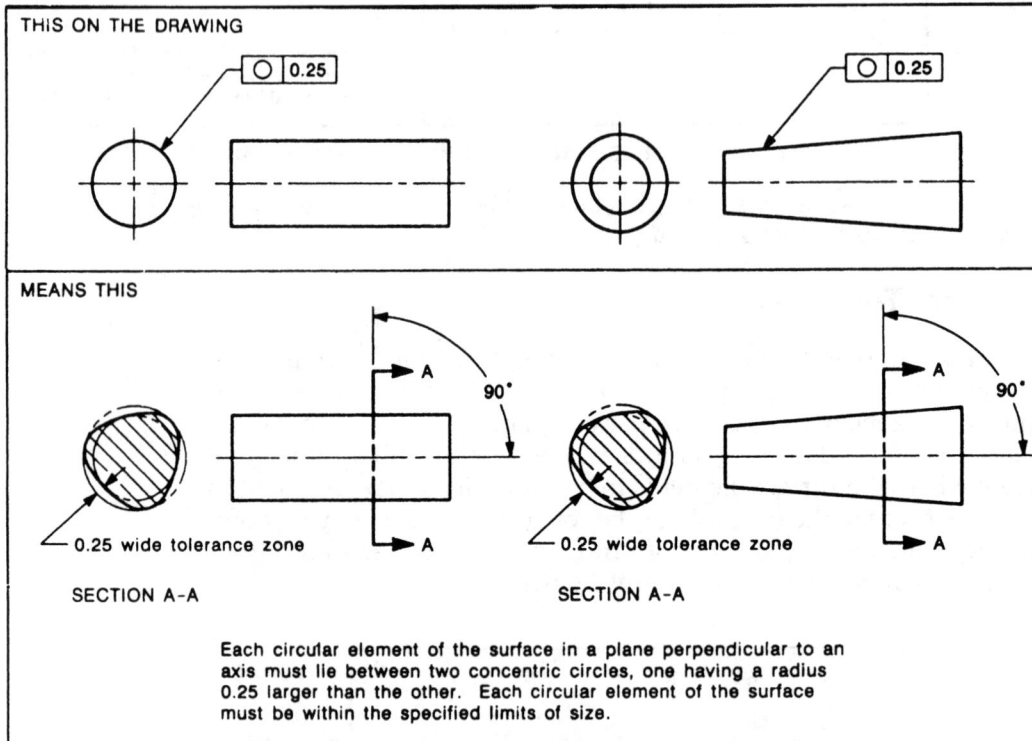

**THIS ON THE DRAWING**

⊙ 0.25    ⊙ 0.25

**MEANS THIS**

A    90°    A    90°

0.25 wide tolerance zone    0.25 wide tolerance zone

SECTION A-A    SECTION A-A

Each circular element of the surface in a plane perpendicular to an axis must lie between two concentric circles, one having a radius 0.25 larger than the other. Each circular element of the surface must be within the specified limits of size.

SPECIFYING CIRCULARITY FOR A CYLINDER OR CONE

## Cylindricity Tolerance

Cylindricity is a condition of a surface of revolution in which all points of the surface are equidistant from a common axis. A cylindricity tolerance specifies a tolerance zone bounded by two concentric cylinders within which the surface must lie. In the case of cylindricity, unlike that of circularity, the tolerance applies simultaneously to both circular and longitudinal elements of the surface.

*Note:* The cylindricity tolerance is a composite control of form that includes circularity, straightness, and taper of a cylindrical feature. (Study the following example.)

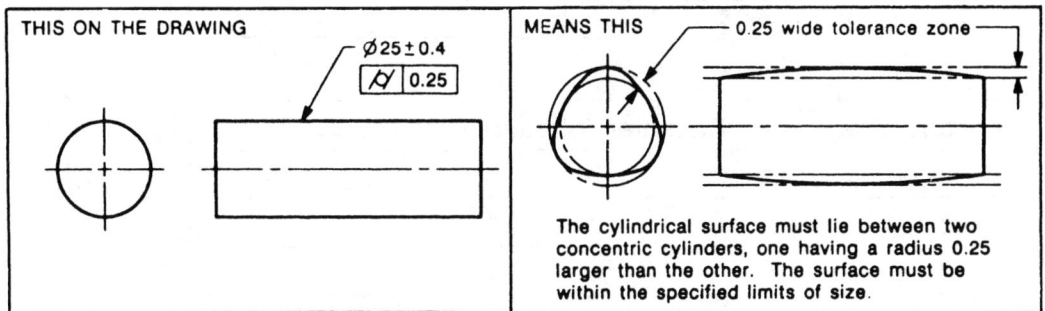

| THIS ON THE DRAWING | MEANS THIS |
|---|---|
| Ø25±0.4 [⌀\| 0.25] | 0.25 wide tolerance zone — The cylindrical surface must lie between two concentric cylinders, one having a radius 0.25 larger than the other. The surface must be within the specified limits of size. |

SPECIFYING CYLINDRICITY

## Tolerance Calculations

INSTRUCTIONS: Calculate the maximum diameter tolerance zone for each of the feature sizes listed below. Refer to the example on page 776.

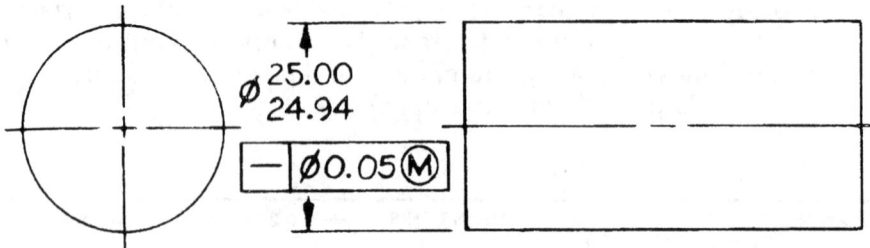

| Feature Size | Diameter Tolerance Zone Allowed |
|---|---|
| 25.00 | |
| 24.99 | |
| 24.98 | |
| 24.97 | |
| 24.96 | |
| 24.95 | |
| 24.94 | |

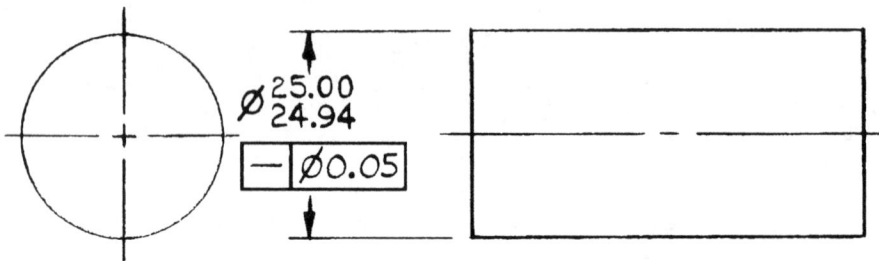

| Feature Size | Diameter Tolerance Zone Allowed |
|---|---|
| 25.00 | |
| 24.99 | |
| 24.98 | |
| 24.97 | |
| 24.96 | |
| 24.95 | |
| 24.94 | |

NOTE: DWG IS INCOMPLETE
BY INTENT

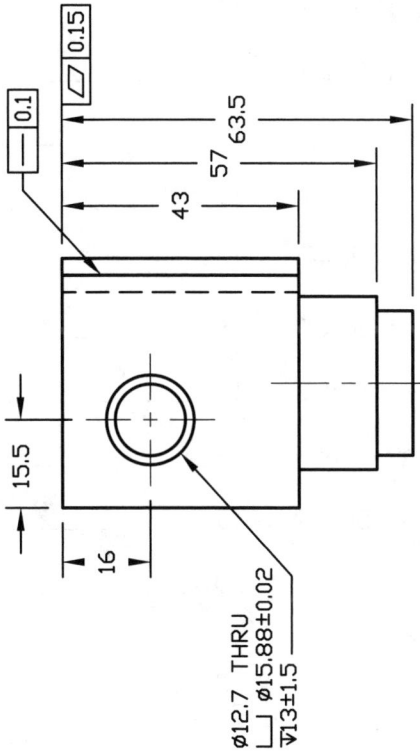

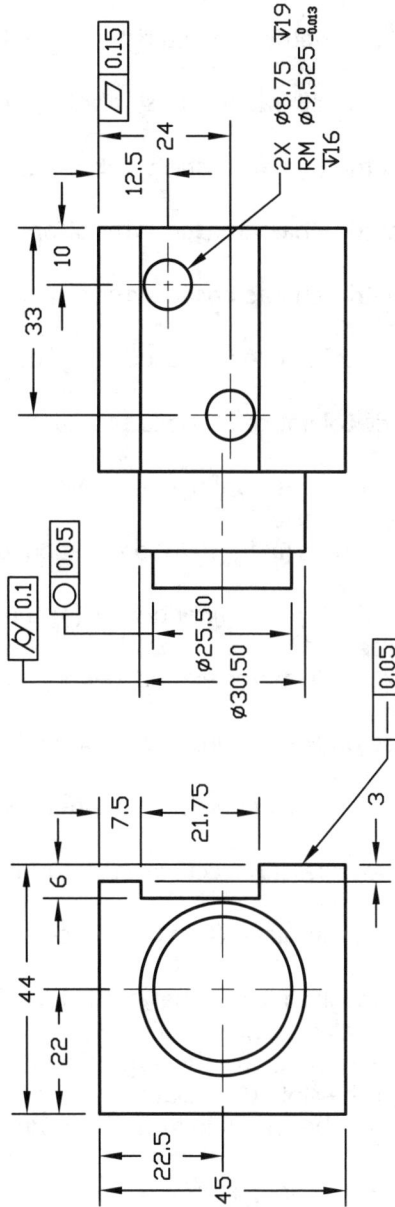

NOTE: FAO

METRIC

THIRD-ANGLE PROJECTION

DWG NO.

31A069

| UNSPECIFIED TOLERANCES: | MATERIAL | | | |
| ±0.25 | 50 X 50 X 65 HRS | | | |
| | DWN BY M.E. HOLLAND | REDRAWN BY M. ŽOKVIĆ | SCALE 1:1 | CHKD LS |
| ANGLES ±0° 30' | TITLE | | | |
| | SPL TEST BLK | | | |

INSTRUCTIONS: Refer to drawing 31A069 on page 781 to answer the following questions.

1. How many geometric tolerances are specified? (Total)

1. _____

2. How many *different* geometric characteristic symbols are shown?

2. _____

3. Interpret the geometric characteristic symbol appearing on the front view.

3. _____

4. Interpret the geometric characteristic symbol appearing on the large diameter.

4. _____

5. Interpret the geometric characteristic symbol appearing on the small diameter.

5. _____

6. Interpret the geometric characteristic symbol appearing on the top surface of the block.

6. _____

7. How much geometric tolerance is applied to cylindricity?

7. _____

8. How much geometric tolerance is applied to flatness?

8. _____

9. Do the geometric tolerances apply at (a) MMC, (b) LMC, or (c) RFS (regardless of feature size)?

9. _____

10. Are the geometric tolerances (a) form, (b) orientation, or (c) location?

10. _____

11. Are the dimensions in the top view arranged by (a) chain, (b) broken-chain, or (c) datum method?

11. _____

12. How far does the 25.5 diameter extend beyond the end of the 30.5 diameter?

12. _____

13. What is the tolerance on the counterbore diameter?

13. _____

14. What is the tolerance on the diameter of the reamed holes?

14. _____

15. Will either of the blind holes intersect with the counterbored hole? (Calculate.)

15. _____

16. How much tolerance accumulates (C to C) between the blind holes?

16. _____

17. With the tolerances assigned, can either of the blind holes overlap the side of the slot? (Calculate.)

17. _____

18. With the tolerances assigned, can the large diameter ever extend above the bottom of the slot? (Calculate.)

18. _____

19. How much material is to be removed from the length of the blank? (Raw material size)

19. _____

20. What does the note FAO abbreviate?

20. _____

# DATUMS AND THE DATUM REFERENCE SYSTEM

A part in space has six degrees of freedom. In order to restrict the six degrees of freedom, the part is placed in a simulated datum reference frame comprised of three mutually perpendicular planes. The first of these three mutually perpendicular planes will restrict three degrees of freedom: rotation around the X axis, rotation around the Y axis, and movement along the Z axis. The second of the three mutually perpendicular planes will restrict two degrees of freedom: rotation around the Z axis and movement along the Y axis. The third of the three mutually perpendicular planes will restrict the last degree of freedom: movement along the X axis.

With all six degrees of freedom restrained, the part can be inspected accurately and the repeatability of the inspection process is assured.

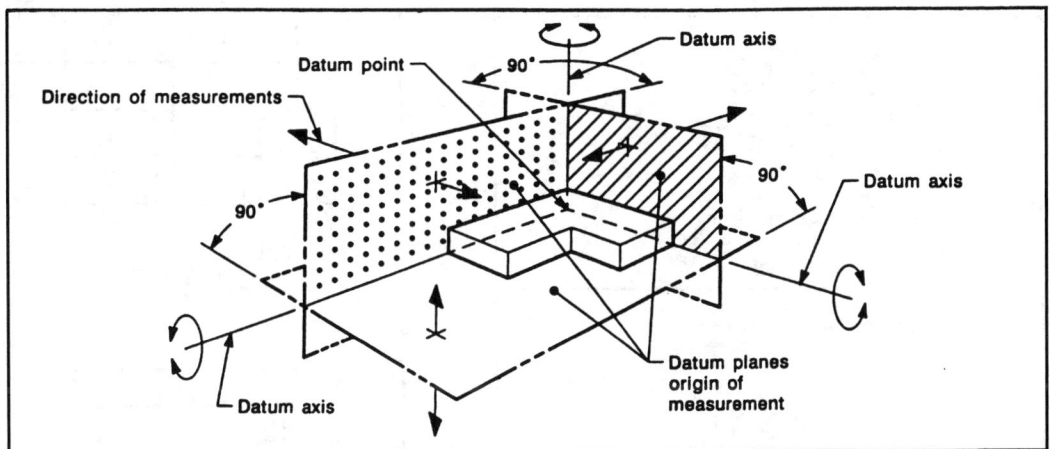

DATUM REFERENCE FRAME

A datum is a theoretically perfect point, plane, or axis from which measurement is made. It is simulated by the inspection equipment. A datum feature is a part feature that is designated as a datum. The datum feature is imperfect, but the datum simulator (such as a surface plate) is of well enough quality to be considered perfect for inspection requirements. When the part feature (datum feature) comes into contact with the surface plate (datum feature simulator), we have a simulated datum from which our measurements are made.

How the part is placed in the measuring equipment, and the order in which the measurements are taken, is determined by the datum order as given in the feature control frame (see p. 768). Selection of datum features is based on the functional requirements of the part. They are often features that orient or locate the part in the assembly. They are shown on the drawing by the use of the Datum Feature Symbol shown on page 771 and are referenced in the Feature Control Frame shown on page 769.

An engineering drawing will contain a large number of dimensions. However, only dimensions related to a datum reference frame with the use of geometric tolerances are measured by using a datum reference frame simulation. Features such as size are not measured in a simulated datum reference frame. For example, the size of a hole is not dependent upon its location on the part in order to be acceptable. Likewise, the location of the hole on the part is not dependent upon its size tolerance. However, for the part to be acceptable, the hole must meet both size and location requirements. Therefore, the size of the hole is determined first, then its location is verified. In order for the part to be acceptable, both the size and the location of the hole must be within the given tolerance.

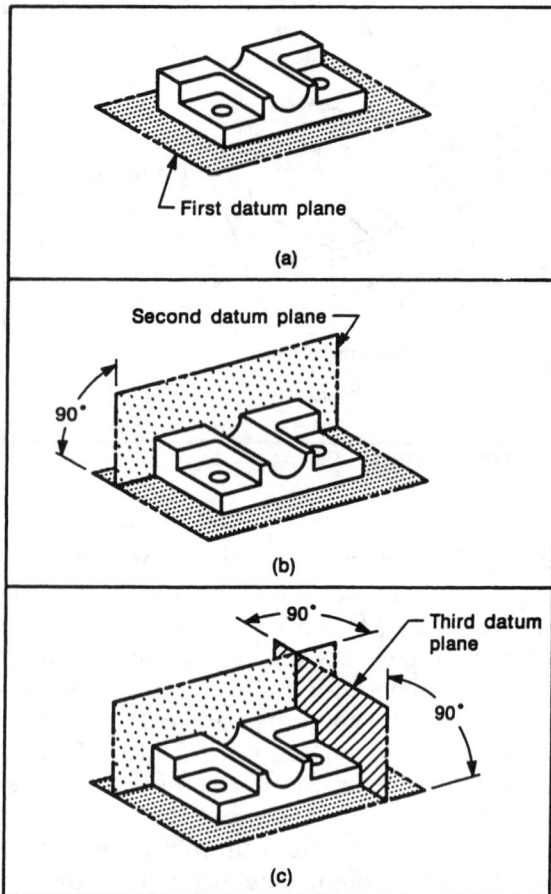

**SEQUENCE OF DATUM FEATURES RELATES THE PART TO THE DATUM REFERENCE FRAME**

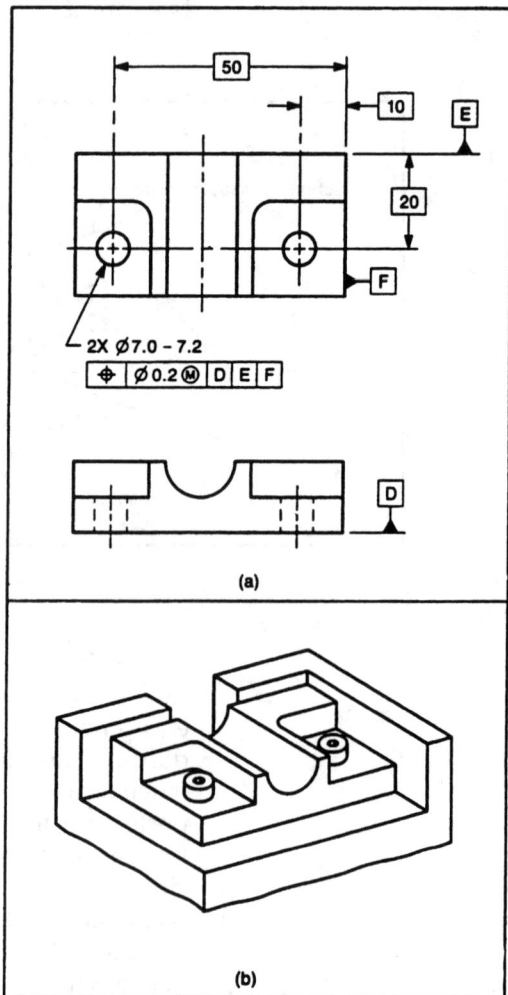

**PART WHERE DATUM FEATURES ARE PLANE SURFACES**

# ORIENTATION TOLERANCES

Angularity, parallelism, perpendicularity, and, in some instances, profile are orientation tolerances. All are applicable to related features, and will always include datum reference letters in their feature control frames. These tolerances control the orientation of features to one another. The considered feature may be related to more than one datum feature if required to stabilize the tolerance zone in more than one direction. When no modifying symbol appears in the feature control frame, RFS applies.

## Angularity Tolerance

Angularity is the condition of a surface, center plane, or axis at a specified angle (other than 90°) from a datum plane or axis. An angularity tolerance specifies a zone defined by two parallel planes. It also specifies a cylindrical tolerance zone at the basic angle from one or more datum planes, or a datum axis, within which the surface, center plane, or axis of the considered feature must lie. (See the following examples.)

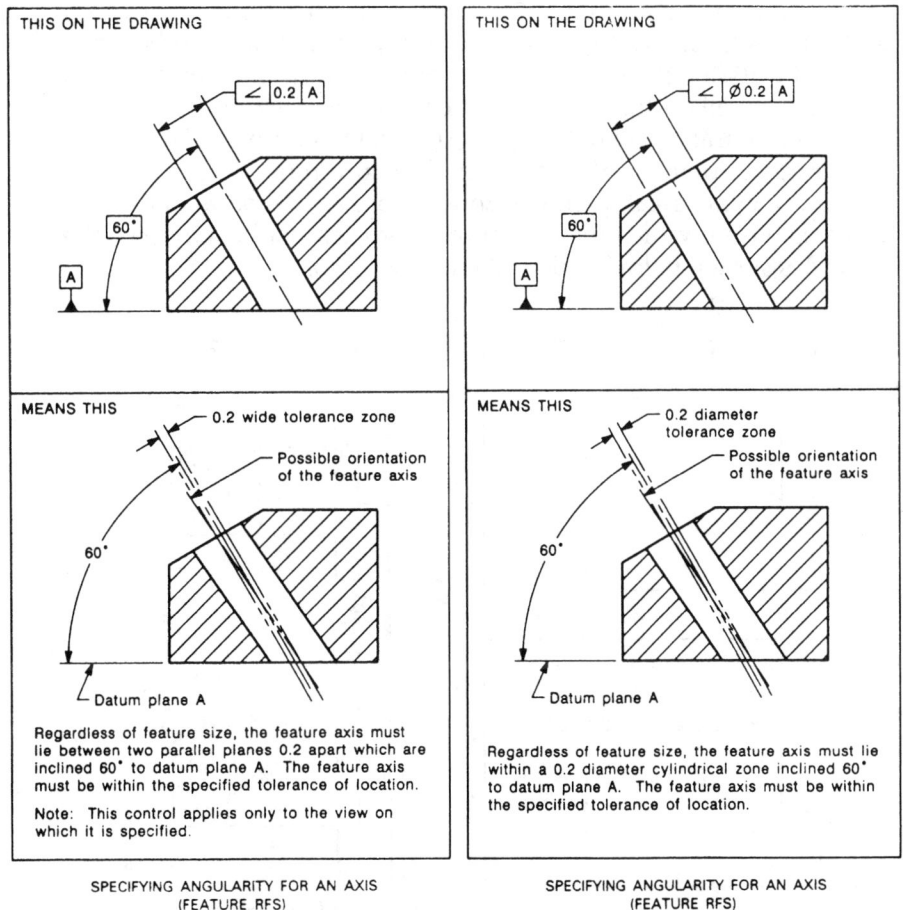

THIS ON THE DRAWING

∠ 0.2 A

60°

A

MEANS THIS

0.2 wide tolerance zone

Possible orientation of the feature axis

60°

Datum plane A

Regardless of feature size, the feature axis must lie between two parallel planes 0.2 apart which are inclined 60° to datum plane A. The feature axis must be within the specified tolerance of location.

Note: This control applies only to the view on which it is specified.

SPECIFYING ANGULARITY FOR AN AXIS
(FEATURE RFS)

THIS ON THE DRAWING

∠ Ø0.2 A

60°

A

MEANS THIS

0.2 diameter tolerance zone

Possible orientation of the feature axis

60°

Datum plane A

Regardless of feature size, the feature axis must lie within a 0.2 diameter cylindrical zone inclined 60° to datum plane A. The feature axis must be within the specified tolerance of location.

SPECIFYING ANGULARITY FOR AN AXIS
(FEATURE RFS)

*(Reprinted from ASME Y14.5M, by permission of the American Society of Mechanical Engineers. All rights reserved.)*

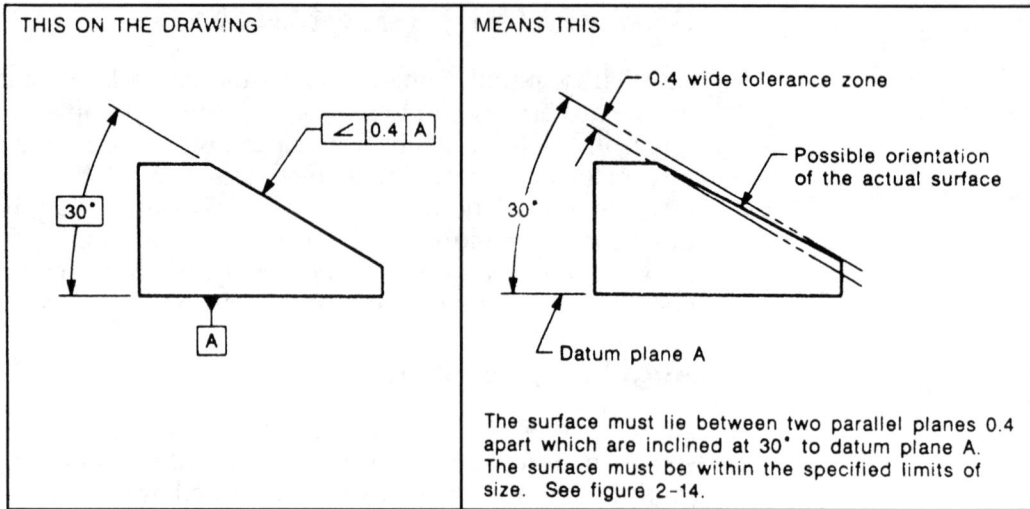

THIS ON THE DRAWING

MEANS THIS

0.4 wide tolerance zone

Possible orientation
of the actual surface

Datum plane A

The surface must lie between two parallel planes 0.4
apart which are inclined at 30° to datum plane A.
The surface must be within the specified limits of
size. See figure 2-14.

SPECIFYING ANGULARITY FOR A PLANE
SURFACE

## Parallelism Tolerance

Parallelism is the condition of a surface or center plane equidistant at all points from a datum plane. It also describes an axis, equidistant along its length from one or more datum planes or a datum axis. A parallelism tolerance specifies a tolerance zone defined by two planes parallel to a datum plane (or axis), within which the surface or center plane of the considered feature must lie. It may also establish a cylindrical tolerance zone whose axis is parallel to a datum axis within which the axis of the feature must lie. Observe the examples shown below and on the following page.

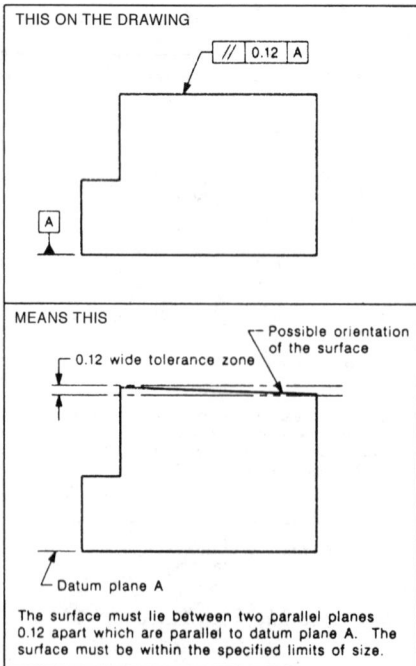

THIS ON THE DRAWING

MEANS THIS

0.12 wide tolerance zone

Possible orientation
of the surface

Datum plane A

The surface must lie between two parallel planes
0.12 apart which are parallel to datum plane A. The
surface must be within the specified limits of size.

SPECIFYING PARALLELISM FOR A PLANE
SURFACE

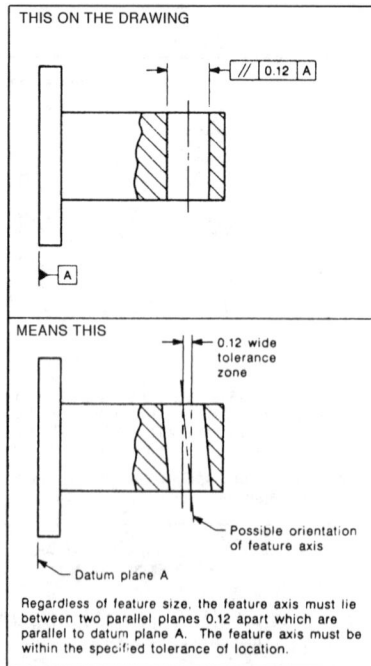

THIS ON THE DRAWING

MEANS THIS

0.12 wide
tolerance
zone

Possible orientation
of feature axis

Datum plane A

Regardless of feature size, the feature axis must lie
between two parallel planes 0.12 apart which are
parallel to datum plane A. The feature axis must be
within the specified tolerance of location.

SPECIFYING PARALLELISM FOR AN AXIS
(FEATURE RFS)

*(Figures on this page reprinted from ASME Y14.5M, by permission of the American Society of Mechanical Engineers. All rights reserved.)*

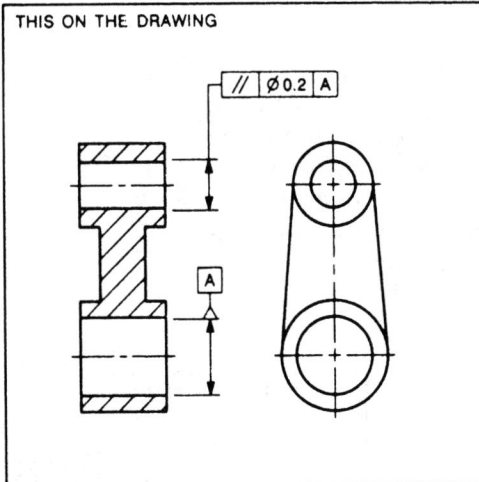

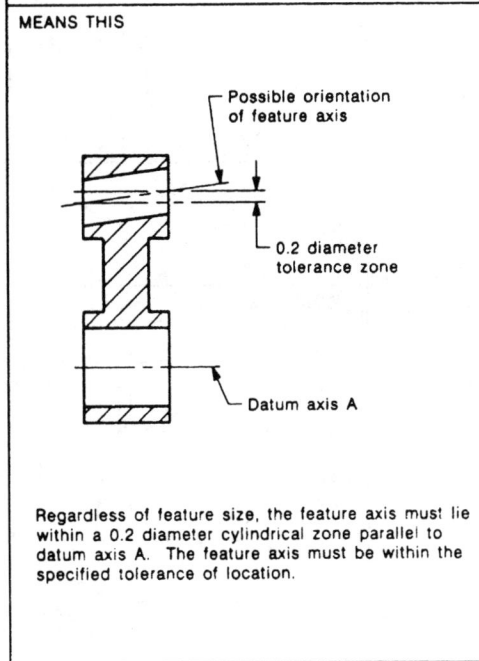

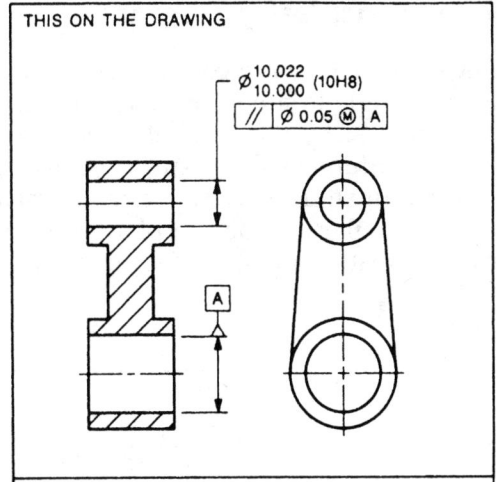

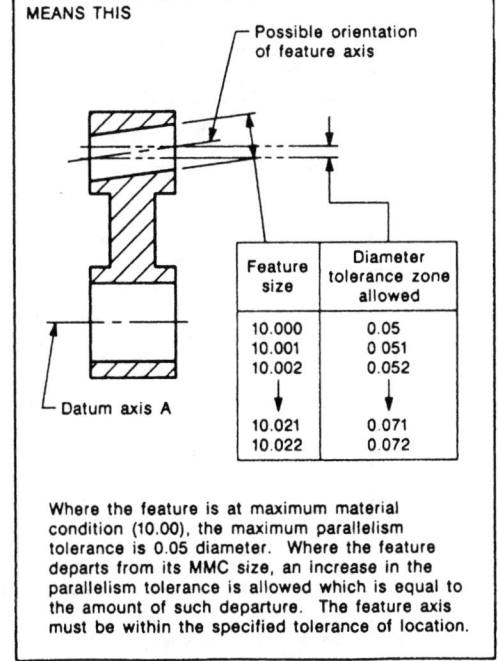

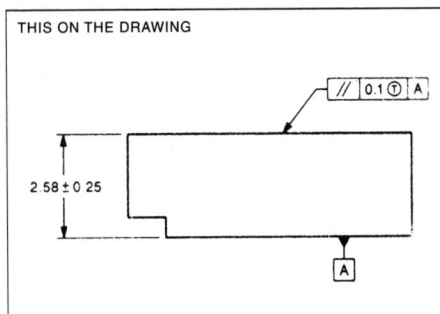

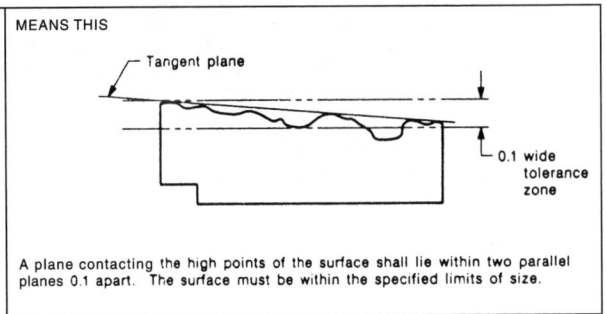

## Perpendicularity Tolerance

Perpendicularity is a condition of a surface, center plane, or axis at a right angle to a datum plane or axis. It may specify a tolerance zone defined by two parallel planes perpendicular to a datum plane or axis within which the surface, center plane, or axis of the considered feature must lie; or it may establish a cylindrical tolerance zone perpendicular to a datum plane within which the axis of the considered feature must lie. Observe the examples below and on the following pages.

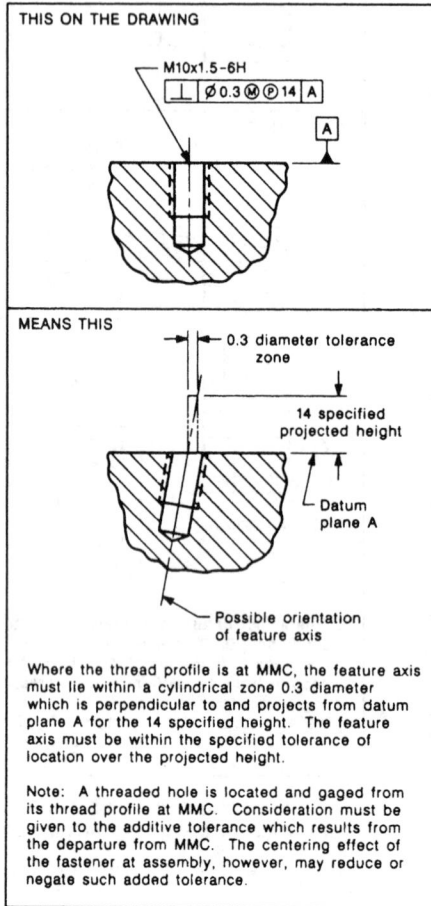

THIS ON THE DRAWING

M10x1.5-6H

⊥ | Ø0.3 Ⓜ Ⓟ 14 | A

A

MEANS THIS

0.3 diameter tolerance zone

14 specified projected height

Datum plane A

Possible orientation of feature axis

Where the thread profile is at MMC, the feature axis must lie within a cylindrical zone 0.3 diameter which is perpendicular to and projects from datum plane A for the 14 specified height. The feature axis must be within the specified tolerance of location over the projected height.

Note: A threaded hole is located and gaged from its thread profile at MMC. Consideration must be given to the additive tolerance which results from the departure from MMC. The centering effect of the fastener at assembly, however, may reduce or negate such added tolerance.

SPECIFYING PERPENDICULARITY FOR AN AXIS AT A PROJECTED HEIGHT (THREADED HOLE OR INSERT AT MMC)

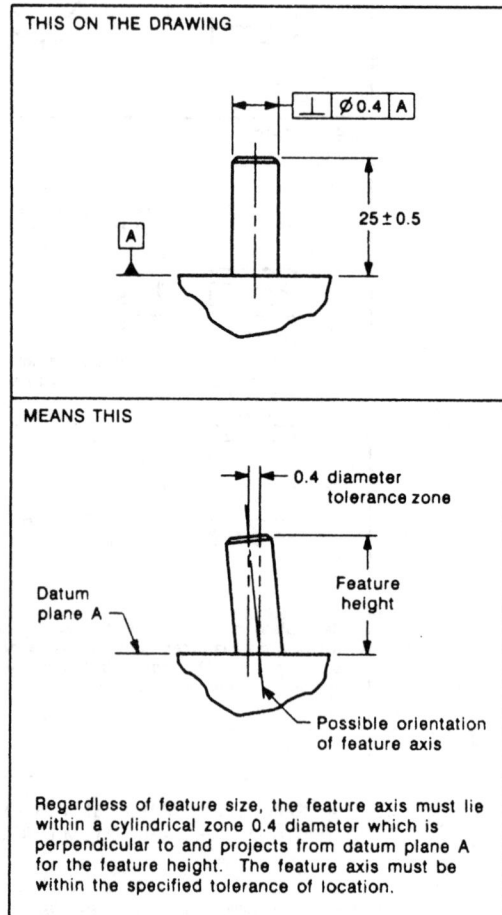

THIS ON THE DRAWING

⊥ | Ø0.4 | A

25 ± 0.5

A

MEANS THIS

0.4 diameter tolerance zone

Feature height

Datum plane A

Possible orientation of feature axis

Regardless of feature size, the feature axis must lie within a cylindrical zone 0.4 diameter which is perpendicular to and projects from datum plane A for the feature height. The feature axis must be within the specified tolerance of location.

SPECIFYING PERPENDICULARITY FOR AN AXIS (PIN OR BOSS RFS)

*(Reprinted from ASME Y14.5M, by permission of the American Society of Mechanical Engineers. All rights reserved.)*

**THIS ON THE DRAWING**

⊥ 0.12 A

A

**MEANS THIS**

Possible orientation of the surface

0.12 wide tolerance zone

Datum plane A

The surface must lie between two parallel planes 0.12 apart which are perpendicular to datum plane A. The surface must be within the specified limits of size.

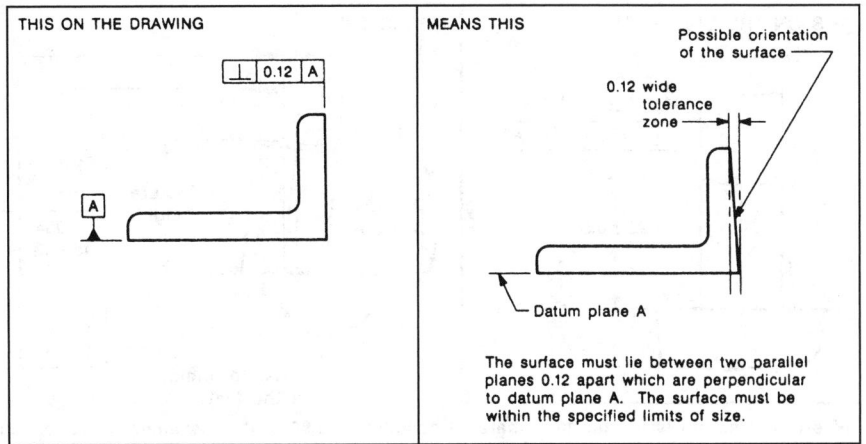

SPECIFYING PERPENDICULARITY FOR A
PLANE SURFACE

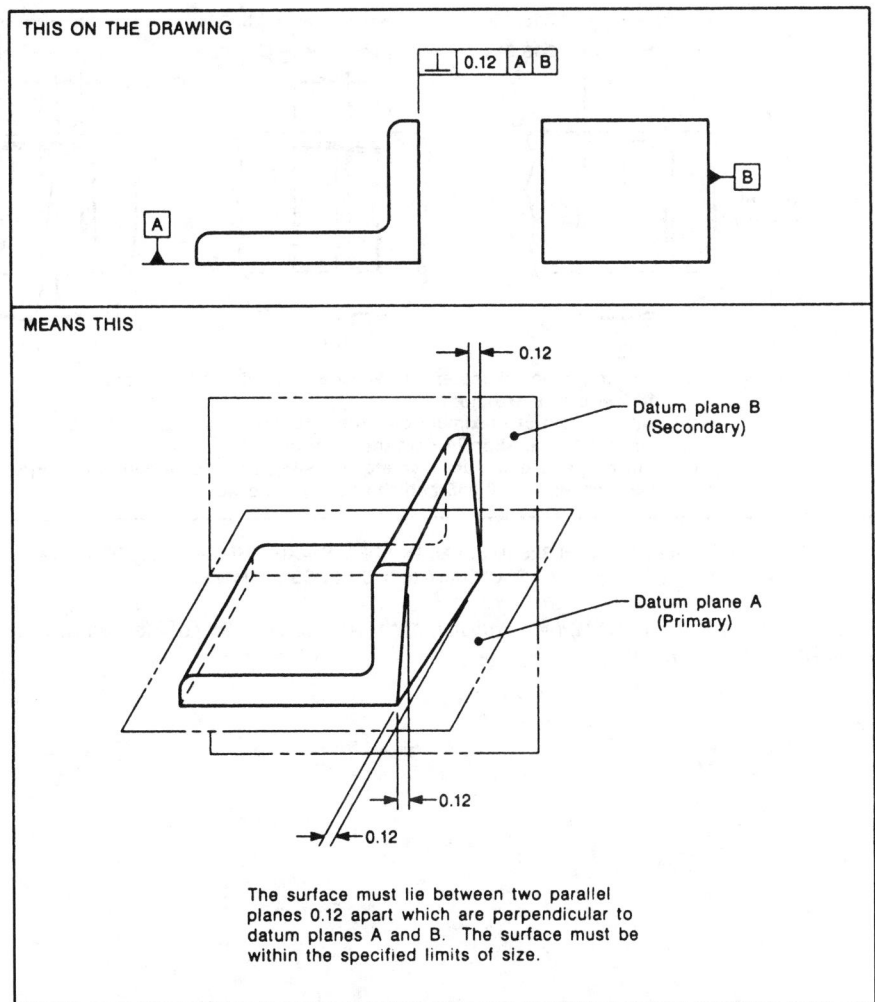

**THIS ON THE DRAWING**

⊥ 0.12 A B

A

B

**MEANS THIS**

0.12

Datum plane B (Secondary)

Datum plane A (Primary)

0.12

0.12

The surface must lie between two parallel planes 0.12 apart which are perpendicular to datum planes A and B. The surface must be within the specified limits of size.

SPECIFYING PERPENDICULARITY FOR A PLANE SURFACE RELATIVE TO TWO DATUMS

| THIS ON THE DRAWING | MEANS THIS | | |
|---|---|---|---|

$\varnothing\,^{15.984}_{15.966}$ (16f7)

⊥ | $\varnothing$ 0.05 Ⓜ | A

25 ± 0.5

A

Datum plane A

Feature height

Possible orientation of the feature axis

| Feature size | Diameter tolerance zone allowed |
|---|---|
| 15.984 | 0.05 |
| 15.983 | 0.051 |
| 15.982 | 0.052 |
| ↓ | ↓ |
| 15.967 | 0.067 |
| 15.966 | 0.068 |

Where the feature is at maximum material condition (15.984), the maximum perpendicularity tolerance is 0.05 diameter. Where the feature departs from its MMC size, an increase in the perpendicularity tolerance is allowed which is equal to the amount of such departure. The feature axis must be within the specified tolerance of location.

ACCEPTANCE BOUNDARY

$\varnothing$16.034
$\varnothing$15.984

$\varnothing$0.05

$\varnothing$16.034
$\varnothing$15.984

$\varnothing$0.068

$\varnothing$16.034
$\varnothing$15.966

Datum plane A

(a)　　　　(b)　　　　(c)

Meaning:　(a)　The maximum diameter pin with perfect orientation is shown in a gage with a 16.034 diameter hole;
　　　　(b)　with the pin at maximum diameter (15.984), the gage will accept the part with up to 0.05 variation in perpendicularity;
　　　　(c)　the pin is at minimum diameter (15.966), and the variation in perpendicularity may increase to 0.068 and the part will be acceptable.

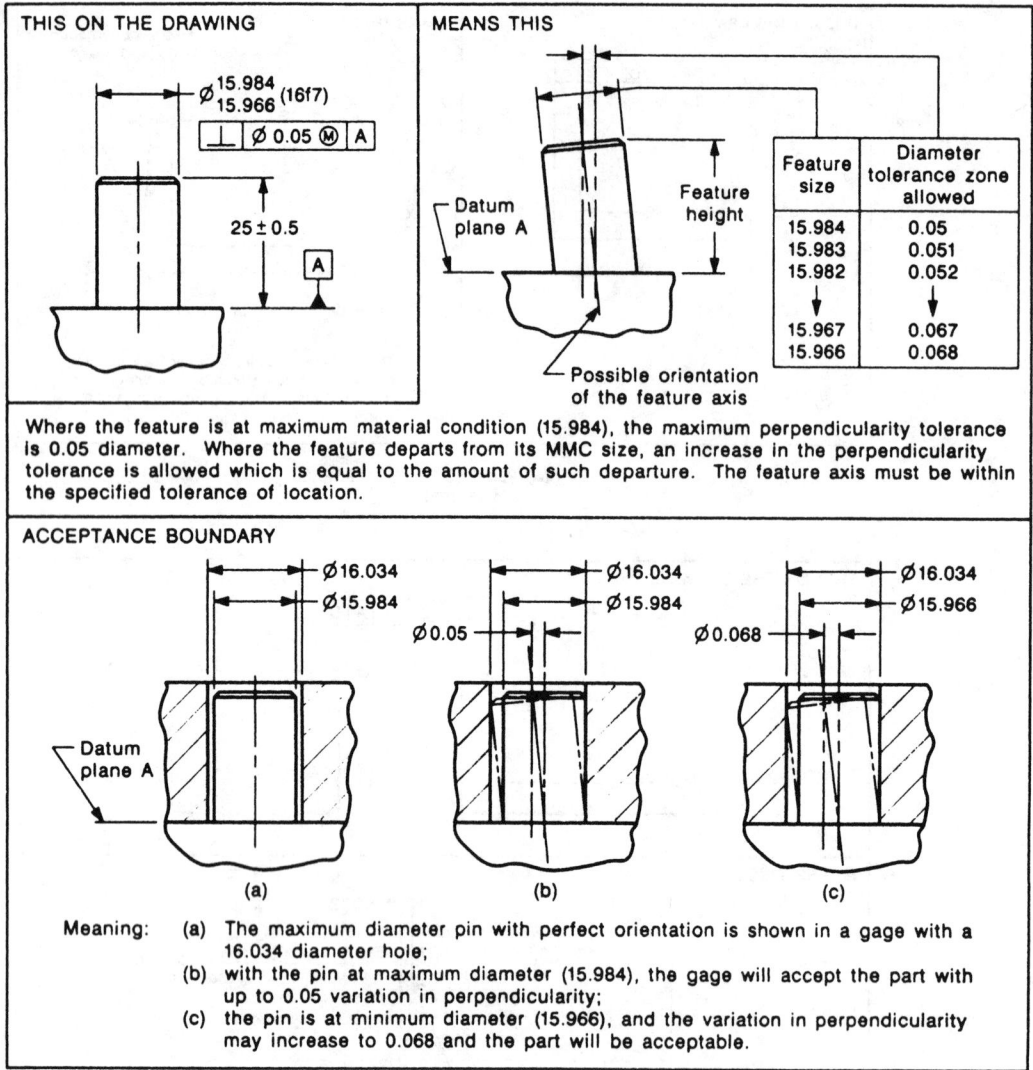

SPECIFYING PERPENDICULARITY FOR AN AXIS SHOWING ACCEPTANCE
BOUNDARY (PIN OR BOSS AT MMC)

**THIS ON THE DRAWING**

$\emptyset \, ^{50.16}_{50.00}$ (50H11)

| ⊥ | $\emptyset$ 0 Ⓜ | A |

**MEANS THIS**

Datum plane A

Possible orientation of the feature axis

| Feature size | Diameter tolerance zone allowed |
|---|---|
| 50.00 | 0 |
| 50.01 | 0.01 |
| 50.02 | 0.02 |
| ↓ | ↓ |
| 50.15 | 0.15 |
| 50 16 | 0.16 |

Where the feature is at maximum material condition (50.00), its axis must be perpendicular to datum plane A. Where the feature departs from MMC, a perpendicularity tolerance is allowed which is equal to the amount of such departure. The feature axis must be within the specified tolerance of location.

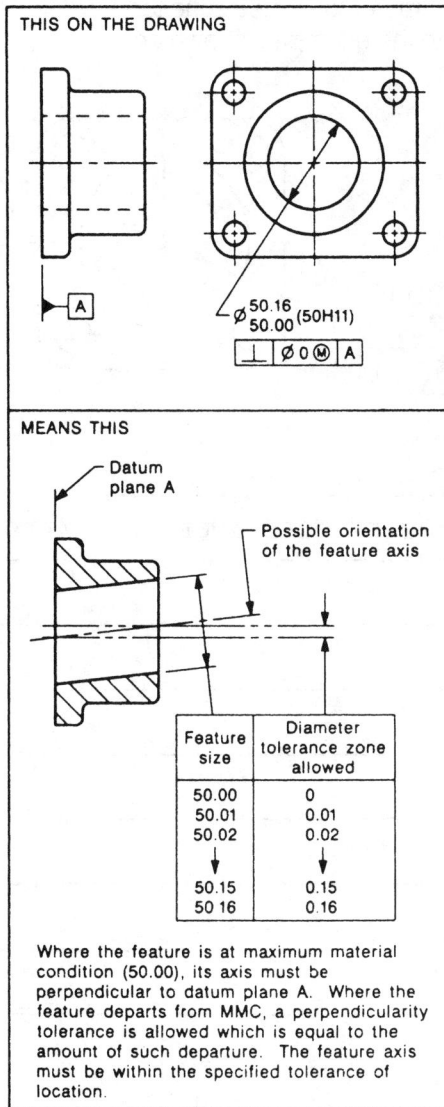

SPECIFYING PERPENDICULARITY FOR AN AXIS (ZERO TOLERANCE AT MMC)

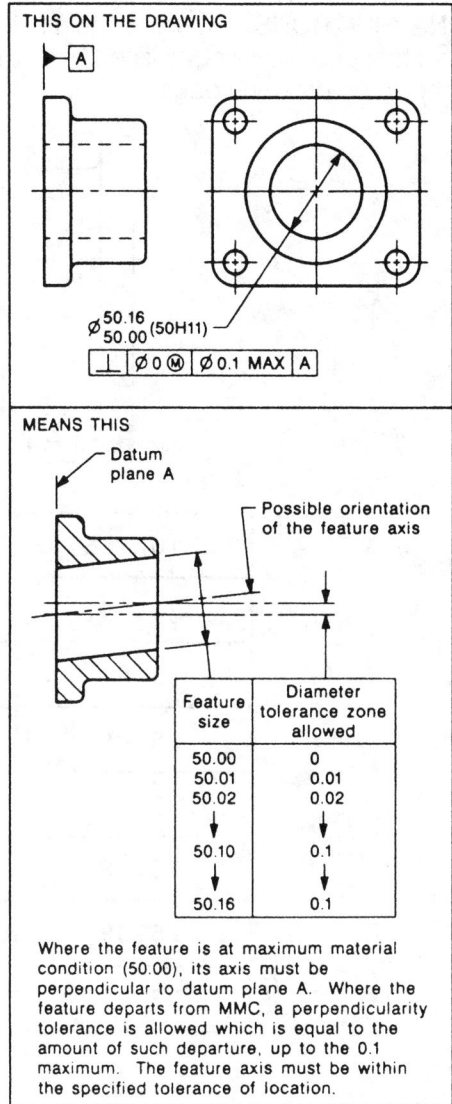

**THIS ON THE DRAWING**

A

$\emptyset \, ^{50.16}_{50.00}$ (50H11)

| ⊥ | $\emptyset$ 0 Ⓜ | $\emptyset$ 0.1 MAX | A |

**MEANS THIS**

Datum plane A

Possible orientation of the feature axis

| Feature size | Diameter tolerance zone allowed |
|---|---|
| 50.00 | 0 |
| 50.01 | 0.01 |
| 50.02 | 0.02 |
| ↓ | ↓ |
| 50.10 | 0.1 |
| ↓ | ↓ |
| 50.16 | 0.1 |

Where the feature is at maximum material condition (50.00), its axis must be perpendicular to datum plane A. Where the feature departs from MMC, a perpendicularity tolerance is allowed which is equal to the amount of such departure, up to the 0.1 maximum. The feature axis must be within the specified tolerance of location.

SPECIFYING PERPENDICULARITY FOR AN AXIS (ZERO TOLERANCE AT MMC WITH A MAXIMUM SPECIFIED)

## Tolerance Calculations

INSTRUCTIONS: Calculate the maximum diameter tolerance zone for each of the feature sizes listed below. Refer to the examples shown on shown on previous page.

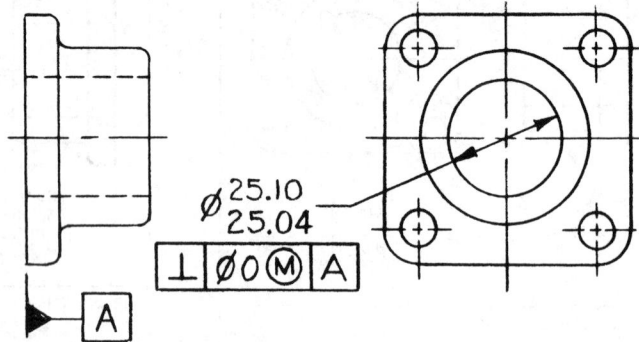

| FEATURE SIZE | DIAMETER TOLERANCE ZONE ALLOWED |
|:---:|:---|
| 25.04 | |
| 25.05 | |
| 25.06 | |
| 25.07 | |
| 25.08 | |
| 25.09 | |
| 25.10 | |

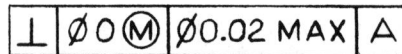

| FEATURE SIZE | DIAMETER TOLERANCE ZONE ALLOWED |
|:---:|:---|
| 25.04 | |
| 25.05 | |
| 25.06 | |
| 25.07 | |
| 25.08 | |
| 25.09 | |
| 25.10 | |

NOTE: DWG IS INCOMPLETE
BY INTENT

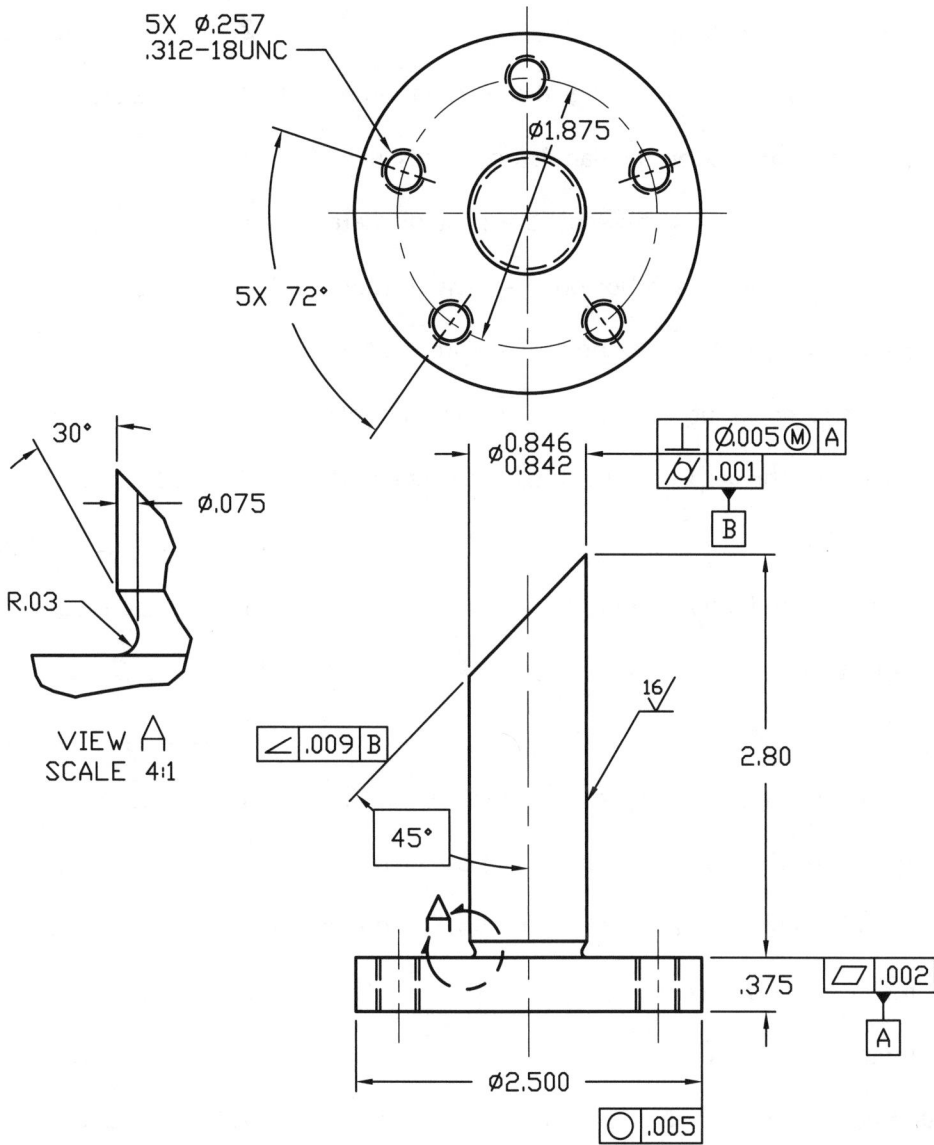

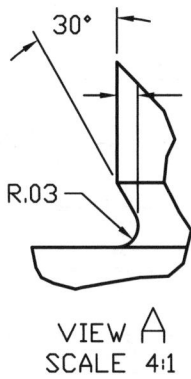

5X Ø.257
.312–18UNC

Ø1.875

5X 72°

30°

Ø.075

R.03

VIEW A
SCALE 4:1

Ø0.846
0.842

⊥ | Ø.005Ⓜ | A
⌗ | .001

B

16

2.80

∠ | .009 | B

45°

A

▱ | .002

A

.375

Ø2.500

◯ | .005

NOTE: FAO

| UNSPECIFIED TOLERANCES: | MATERIAL | | | THIRD–ANGLE PROJECTION |
| 2-PL DEC IN.(.xx)±.01 | AISI 1040 FORGING STL | | | |
| 3-PL DEC IN.(.xxx)±.005 | DWN BY R. BRAINERD | REDRAWN BY M. ŽOKVIC | SCALE 1:1 | CHKD LS |
| DRILLED HOLES +.010 -.002 | TITLE FLANGE POST | | DWG NO. 31A070 | |
| ANGLE ±0° 30′ | | | | |

INSTRUCTIONS: Refer to drawing 31A070 on page 811 to answer the following questions.

1. How many geometric tolerances are specified?      1. _____

2. How many surfaces serve as datum features?      2. _____

3. What does the geometric symbol on the flange diameter represent?      3. _____

4. What does the geometric symbol on the flange face represent?      4. _____

5. What does the upper geometric symbol on the post diameter represent?      5. _____

6. What does the lower geometric symbol on the post diameter represent?      6. _____

7. What does the geometric symbol on the post face represent?      7. _____

8. What does the rectangle enclosing the 45° dimension represent?      8. _____

9. What does the modifier on the perpendicularity tolerance represent?      9. _____

10. What is the flatness tolerance?      10. _____

11. What is the circularity (roundness) tolerance?      11. _____

12. What is the cylindricity tolerance?      12. _____

13. What is the angularity tolerance?      13. _____

14. What is the datum reference for the perpendicularity tolerance?      14. _____

15. What is the perpendicularity tolerance when the post diameter is .846?      15. _____

16. What is the perpendicularity tolerance when the post diameter is .842?      16. _____

17. What is the diameter of the bolt circle?      17. _____

18. What is the diameter of the neck?      18. _____

19. How many full threads does each hole contain?      19. _____

20. What is the accumulated tolerance on the overall height?      20. _____

**Optional Exercise Using Tolerance of Position:**

1a. After completing the workbook, students may return to this drawing and create a feature control frame for the 5 × .312-18UNC threaded holes that states the following conditions:

The axis of each threaded hole is to be located within a .010 diameter cylindrical tolerance zone at RFS that is oriented perpendicular to datum A, and located relative to datum B at MMC. Designate the proper dimensions as basic by enclosing them inside a rectangular frame.

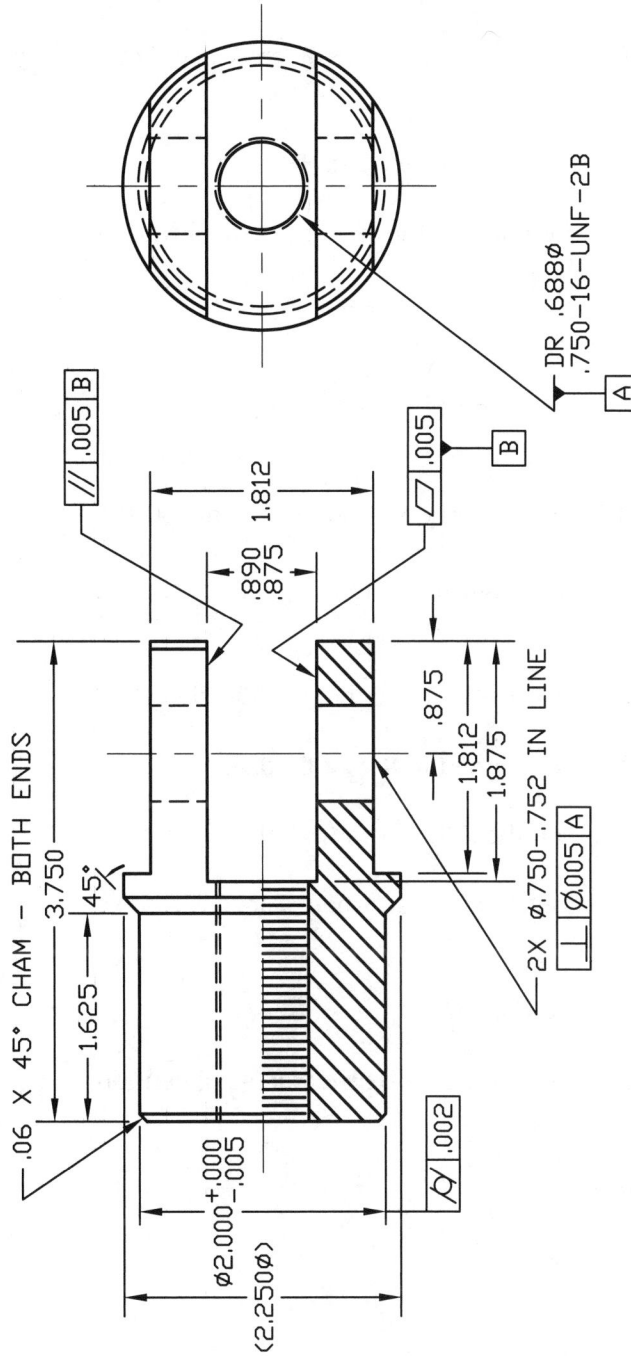

NOTE: DWG INCOMPLETE BY INTENT

DR .688∅
.750-16-UNF-2B

// .005 B

1.812

.890
.875

∠ .005 B

.875

1.812
1.875

2X ∅.750-.752 IN LINE

⊥ ∅.005 A

.06 X 45° CHAM - BOTH ENDS

3.750

45°

1.625

⌀ .002

∅2.000 +.000 -.005

(2.250∅)

| UNSPECIFIED TOLERANCES: | MATERIAL | ∅2.250 X 3.88 AISI 1117 CDS | | | |
|---|---|---|---|---|---|
| 2-PL DEC IN.(.xx)±.03 | DWN BY | REDRAWN BY | SCALE | CHKD | |
| 3-PL DEC IN.(.xxx)±.015 | A. L. WHITE | M. ŽOKVIC | NTS | LS | |
| DRILLED HOLES +.010 -.002 | TITLE | | | | |
| ANGLE ±3° | | CLEVIS | | | |

THIRD-ANGLE PROJECTION

DWG NO.
31A071

INSTRUCTIONS: Refer to drawing 31A071 on page 813 to answer the following questions.

1. How many geometric tolerances are specified?

1. _____

2. How many of the geometric tolerances are considered an orientation type of tolerance?

2. _____

3. Do the geometric tolerances apply at MMC or RFS?

3. _____

4. What do the letters at the ends of the feature control frames represent?

4. _____

5. How many surfaces serve as datum features?

5. _____

6. Interpret the geometric characteristic symbol appearing on the 2.000 diameter.

6. _____

7. Interpret the geometric characteristic symbol appearing on the .750 clearance holes.

7. _____

8. Interpret the geometric characteristic symbols appearing on the flat surfaces.

8. _____

9. How much geometric tolerance is assigned to the 2.000 diameter?

9. _____

10. How much size tolerance is assigned to the 2.000 diameter?

10. _____

11. What is the size tolerance on the .750 clearance holes?

11. _____

12. What is the maximum metal thickness where each .750 hole passes through?

12. _____

13. What type of section view is drawn?

13. _____

14. What do the parentheses around the large-diameter dimension represent?

14. _____

15. Is the clevis symmetrical?

15. _____

NOTE: DWG IS INCOMPLETE
BY INTENT

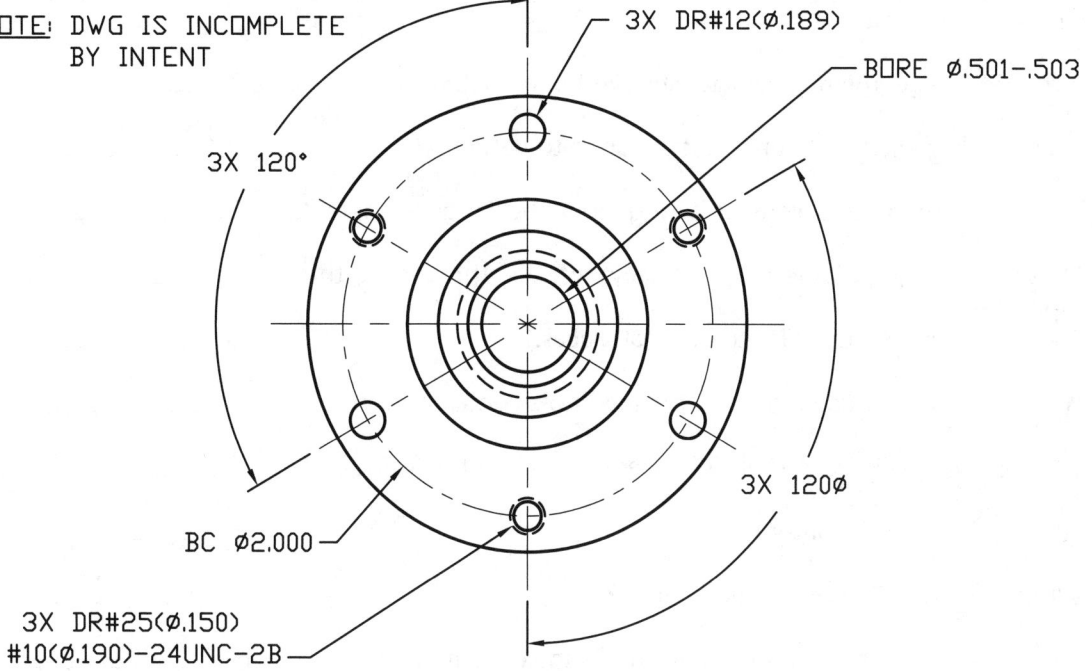

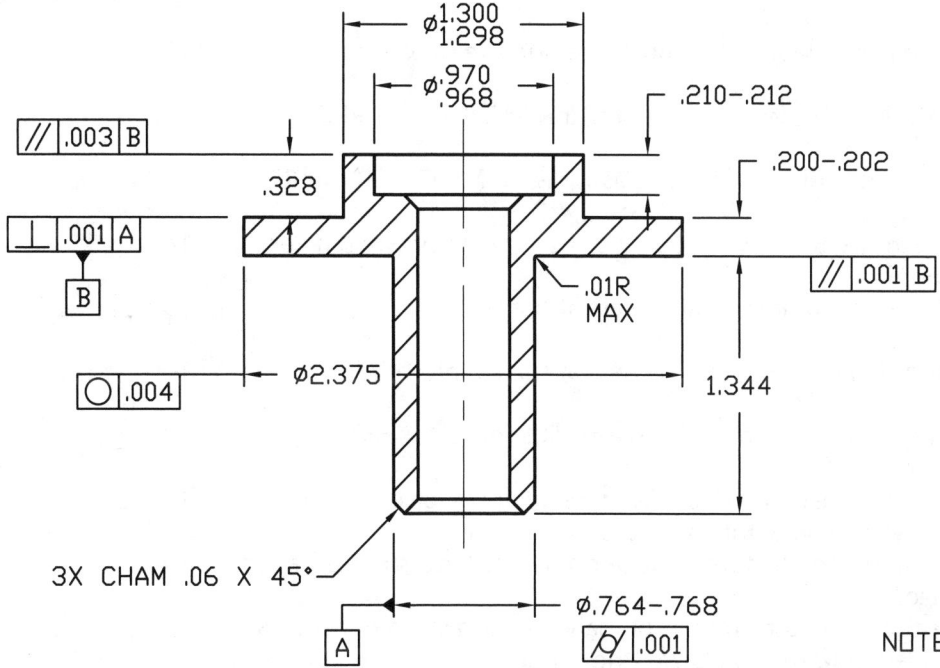

3X DR#12(⌀.189)

BORE ⌀.501-.503

3X 120°

3X 120⌀

BC ⌀2.000

3X DR#25(⌀.150)
#10(⌀.190)-24UNC-2B

⌀1.300
1.298

⌀.970
.968

.210-.212

.200-.202

// .003 B

.328

⊥ .001 A

B

// .001 B

.01R
MAX

○ .004

⌀2.375

1.344

⌀.764-.768

3X CHAM .06 X 45°

A

⌀ .001

NOTE: FAO

| UNSPECIFIED TOLERANCES: | MATERIAL C1020 HRS | | | | THIRD-ANGLE PROJECTION |
|---|---|---|---|---|---|
| 2-PL DEC IN.(.xx)±.01 | DWN BY R.B. GONZALEZ | REDRAWN BY M. ZOKVIC | SCALE 1:1 | CHKD LS | |
| 3-PL DEC IN.(.xxx)±.005 | TITLE | | | | DWG NO. |
| DRILLED HOLES +.010 -.002 | FR FLG FITTING | | | | 31A072 |
| ANGLE ±0° 30' | | | | | |

INSTRUCTIONS: Refer to drawing 31A072 on page 815 to answer the following questions.

1. Interpret the geometric characteristic symbol on datum A.

1. _____

2. How much geometric tolerance is assigned to datum A?

2. _____

3. How much geometric tolerance is assigned to the flange diameter?

3. _____

4. Show the major diameter (decimally) of the threads in the tapped holes.

4. _____

5. Show the thread pitch. (Three decimal places.)

5. _____

6. How many *full* threads will each threaded hole contain?

6. _____

7. What is the maximum permissible overall length of the fitting?

7. _____

8. What is the MMC of datum A?

8. _____

9. What is the MMC of the bore diameter?

9. _____

10. Show the upper and lower limits of the flange diameter.

10. _____

11. Show the percentage of carbon content in the steel specified.

11. _____

12. Is the parallelism tolerance assigned at MMC, LMC, or RFS?

12. _____

13. Is the perpendicularity tolerance assigned at MMC, LMC, or RFS?

13. _____

14. How much size tolerance is permitted on the unthreaded holes in the flange?

14. _____

15. What is the minimum wall thickness at the bore?

15. _____

16. What is the minimum wall thickness at the counterbore?

16. _____

17. What is the total size tolerance permissible on datum A?

17. _____

18. How many of the geometric tolerances do not involve a datum reference? What type of tolerance are they?

18. _____

19. How many of the feature control frames include an orientation tolerance?

19. _____

20. What is the minimum amount of material permissible between an unthreaded hole and the OD of the flange?

20. _____

NOTE: Students may return to this drawing (after completing the workbook) to add the proper positional tolerance controls for the three (3) clearance holes and the three (3) threaded holes.

# LOCATION TOLERANCES

Location tolerances include tolerance of symmetry, concentricity, and position. They are used to control the center distances, location, coaxiality, and concentricity or symmetry between features such as holes, slots, bosses, and tabs.

## Symmetry Tolerance

Symmetry is the condition where the median points of all opposed elements of two or more feature surfaces are located within a tolerance zone of two parallel planes equally disposed on either side of the axis or center plane of the datum reference. The symmetry tolerance and the datum reference can only be applied at RFS. (See the example below.)

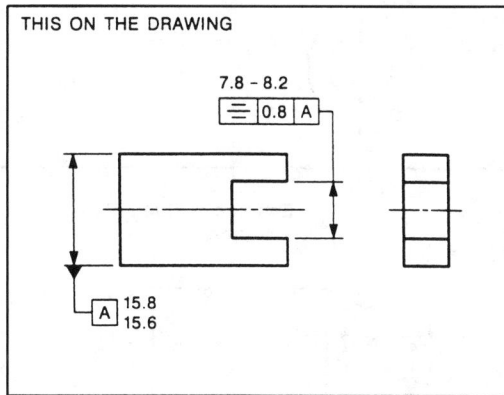

(a)

*(Reprinted from ASME Y14.5M, by permission of the American Society of Mechanical Engineers. All rights reserved.)*

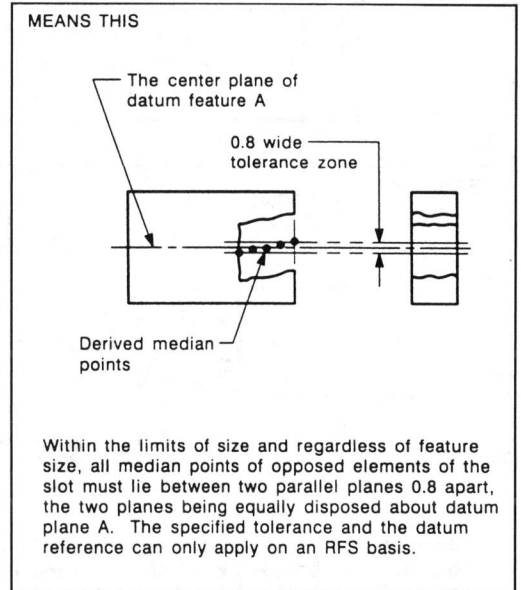

Within the limits of size and regardless of feature size, all median points of opposed elements of the slot must lie between two parallel planes 0.8 apart, the two planes being equally disposed about datum plane A. The specified tolerance and the datum reference can only apply on an RFS basis.

SYMMETRY TOLERANCING

(b)

## Concentricity Tolerance

Concentricity is the condition where the median points of all diametrically opposed elements of a cylindrical feature are located within a cylindrical tolerance zone equally disposed around the axis of the datum reference. The concentricity tolerance and the datum reference can only be applied at RFS.

Both concentricity and symmetry are very expensive controls to use. Therefore, it is preferred to use other controls such as position, runout, or profile in place of concentricity, and to use position in place of symmetry.

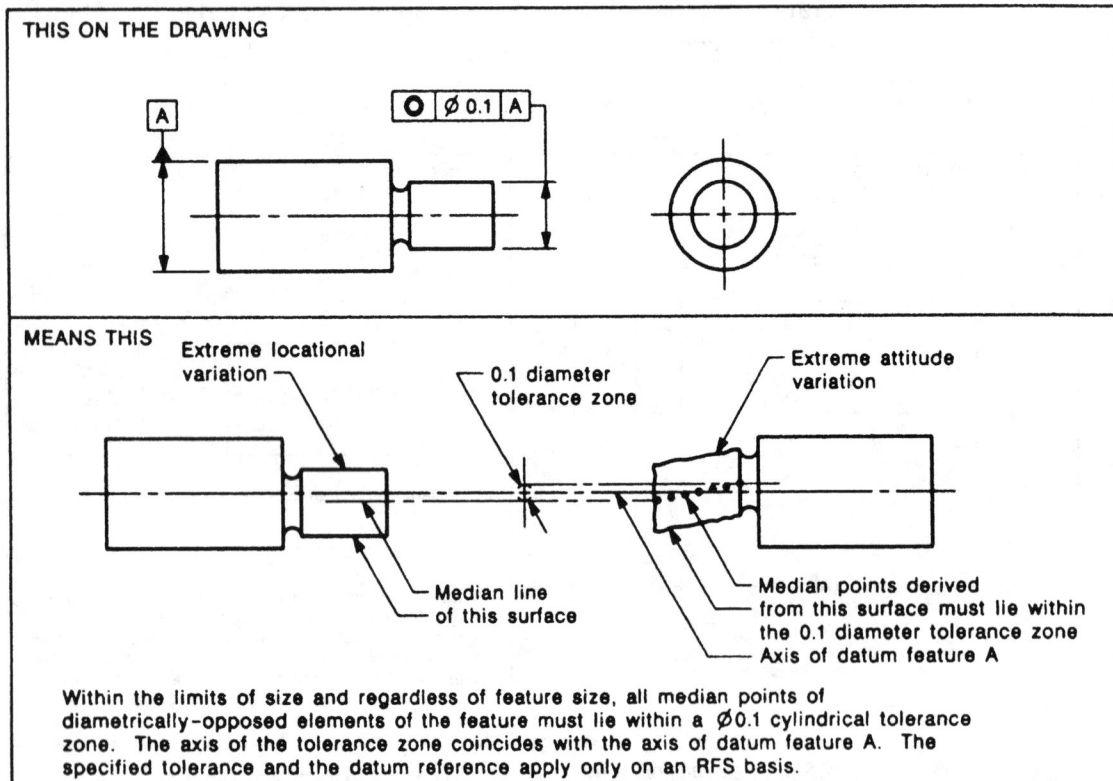

THIS ON THE DRAWING

MEANS THIS

Extreme locational variation

0.1 diameter tolerance zone

Extreme attitude variation

Median line of this surface

Median points derived from this surface must lie within the 0.1 diameter tolerance zone

Axis of datum feature A

Within the limits of size and regardless of feature size, all median points of diametrically-opposed elements of the feature must lie within a Ø0.1 cylindrical tolerance zone. The axis of the tolerance zone coincides with the axis of datum feature A. The specified tolerance and the datum reference apply only on an RFS basis.

CONCENTRICITY TOLERANCING

*(Reprinted from ASME Y14.5M, by permission of the American Society of Mechanical Engineers. All rights reserved.)*

## Positional Tolerance

A positional tolerance defines a zone within which the center, axis, or center plane of a feature of size is permitted to vary from true (theoretically exact) position. Basic dimensions establish the true position from specified datum features and between interrelated features.

The advantages of positional tolerancing can be clearly seen in the examples that follow. Observe the .010 square tolerance zone that results from coordinate plus and minus tolerancing. This tolerancing method permits a location of .007 from the true center if it occurs in the corners of the .010 square zone (1.4 × .005). An increase of 57% more tolerance zone can be obtained by creating a round tolerance zone. This is accomplished by enclosing the location dimensions in rectangles, thus designating them as basic dimensions with theoretically exact numerical values. The round tolerance zone is then centered at the intersection of these basic dimensions.

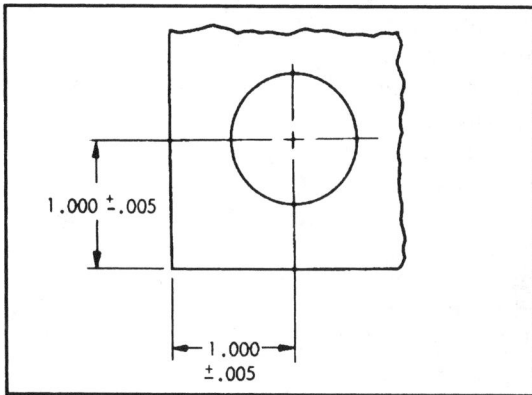

Coordinate tolerancing.

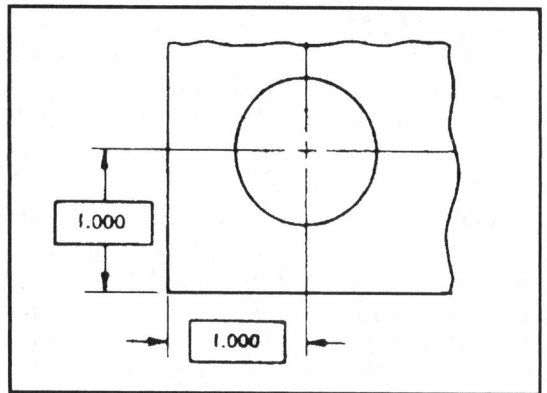

Positional tolerancing.

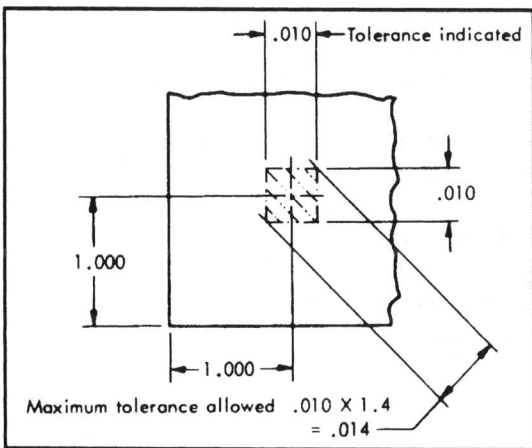

Coordinate tolerance zone (square).

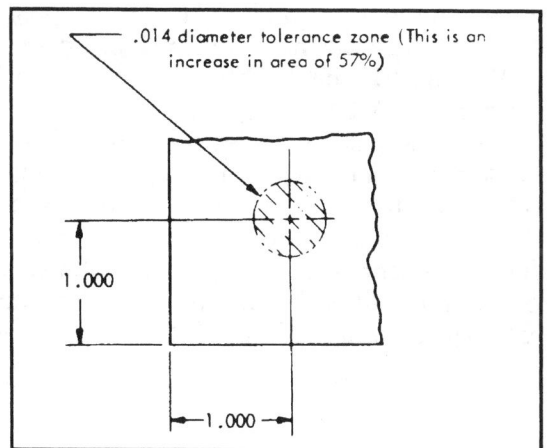

Positional tolerance zone (round).

## Positional Tolerance Modifiers

With the publication of the ASME Y14.5M–1994 standard, a new method of applying positional tolerancing was introduced. The previous ANSI Y14.5M standard stated that all position controls had to have the modifier MMC, LMC, or RFS in either the tolerance portion or the datum reference portion of the feature control frame. Now, in the ASME Y14.5M–1994 standard, when no material condition modifier appears (either Ⓜ or Ⓛ) then the control automatically applies at RFS. RFS is referred to as the default condition for all tolerance controls or datum references (Rule #2).

Maximum material condition means that internal features such as holes and slots would be at their minimum allowable size, whereas external features such as shafts would be at their maximum allowable size. For example, MMC of a .500–.505 diameter hole is at .500, while MMC of a .500–.505 diameter shaft is at .505 diameter. When positional tolerance is specified at MMC, perfect form is required (Rule #1) and the tolerance zone size is dependent on the size of the considered feature. Where the actual size of the feature has departed from MMC, an increase in the tolerance is allowed equal to the amount of the departure. For example, if a .014 dia. tolerance zone is applied to a .500–.505 hole at MMC, the tolerance zone would be .014 diameter, if the hole measured exactly .500 diameter. If the hole measured .501, the tolerance would increase to .015 dia., .016 dia. for .502, .017 dia. for .503, and on up to .019 dia. tolerance zone for a hole measuring .505 diameter.

Specifying positional tolerance at LMC (least material condition) would mean that the stated positional tolerance applies when the feature contains the least amount of material permitted by its toleranced size dimension. Specification of positional tolerance at LMC requires that the feature have perfect form at LMC, but perfect form at MMC is not required. Where the feature departs from its LMC size, an increase in positional tolerance is allowed equal to the amount of the departure. Using the .500–.505 dia. hole as an example, when the hole is at LMC (.505), the positional tolerance zone is .014 dia. If the hole measured .504 dia., the tolerance would increase to .015 dia., .016 dia. for .503, and on up to .019 dia. for a hole measuring .500 (MMC).

When a feature is referenced at RFS (no Ⓜ or Ⓛ with the datum letters), then the tolerance zone remains the same as long as the feature size falls within the acceptable size range.

When using positional tolerancing, the tolerance zone may be of any convenient size, as long as it contains an Ⓜ or an Ⓛ material condition modifier that will allow the tolerance zone to grow as the feature size changes. Observe the examples on the following pages.

## Zero Positional Tolerance at MMC

In applications where it is necessary to provide greater than normal tolerance within functional limits, the principle of positional tolerancing at MMC may be extended. This is accomplished by adjusting the minimum size of a hole to the absolute minimum required for insertion of a fastener located at true position, and specifying a zero positional tolerance at MMC.

Figures 1 and 2 below illustrate the same part, one with zero positional tolerance at MMC, the other with conventional positional tolerance at MMC. Note that the maximum size limit of the clearance hole remains the same, but the minimum was adjusted to correspond with a 14 mm diameter fastener. This results in an increase in the size tolerance for the clearance holes, the increase being equal to the conventional positional tolerance specified in Fig. 2. Although the positional tolerance specified in Fig. 1 is zero at MMC, the positional tolerance allowed is in direct proportion to the actual clearance hole size, as shown in the table that follows.

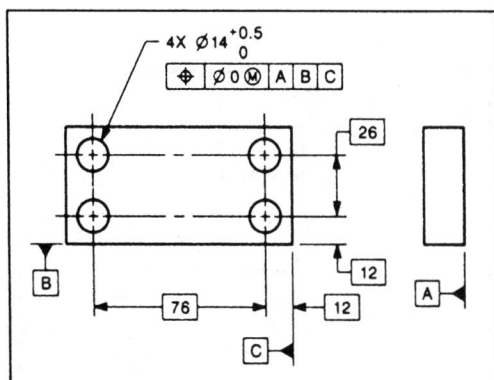

FIG. 1 ZERO POSITIONAL TOLERANCING AT MMC

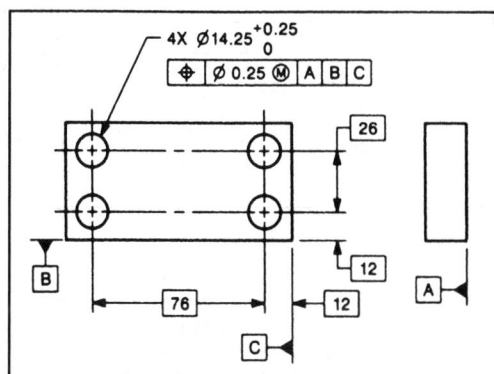

FIG. 2 CONVENTIONAL POSITIONAL TOLERANCING AT MMC

*(Reprinted from ASME Y14.5M, by permission of the American Society of Mechanical Engineers. All rights reserved.)*

| Clearance Hole Diameter (Feature Actual Mating Size) | Positional Tolerance Diameter Allowed (Zero Positional Tolerancing at MMC) | Positional Tolerance Diameter Allowed (Conventional Positional Tolerancing at MMC) |
|---|---|---|
| 14 | 0 | Part would function, but |
| 14.1 | 0.1 | would have to be rejected |
| 14.2 | 0.2 | because of size violation. |
| 14.25 | 0.25 | 0.25 |
| 14.3 | 0.3 | 0.3 |
| 14.4 | 0.4 | 0.4 |
| 14.5 | 0.5 | 0.5 |

Zero positional tolerancing allows for all functional parts to be accepted.

## Positional Tolerancing for Symmetrical Relationships

Positional tolerancing for symmetrical relationships is that condition where the center plane of the actual mating envelope of one or more features is congruent with the axis or center plane of a datum feature within specified limits. MMC, LMC, or RFS may be specified to apply to both the tolerance and the datum feature.

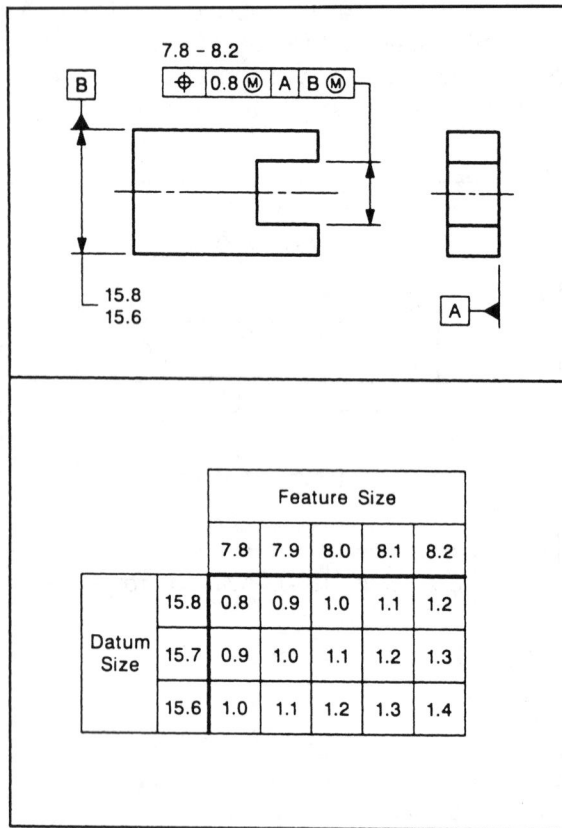

7.8 – 8.2

B    ⊕ | 0.8 Ⓜ | A | B Ⓜ

15.8
15.6

A

|  |  | Feature Size | | | | |
|---|---|---|---|---|---|---|
|  |  | 7.8 | 7.9 | 8.0 | 8.1 | 8.2 |
| Datum Size | 15.8 | 0.8 | 0.9 | 1.0 | 1.1 | 1.2 |
|  | 15.7 | 0.9 | 1.0 | 1.1 | 1.2 | 1.3 |
|  | 15.6 | 1.0 | 1.1 | 1.2 | 1.3 | 1.4 |

POSITIONAL TOLERANCING AT MMC FOR
SYMMETRICAL FEATURES

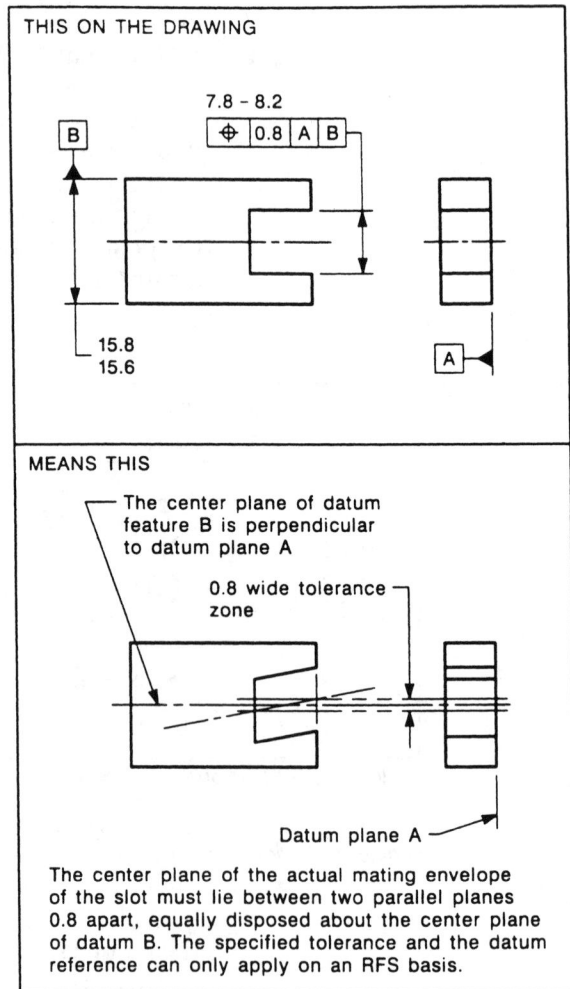

THIS ON THE DRAWING

7.8 – 8.2

B    ⊕ | 0.8 | A | B

15.8
15.6

A

MEANS THIS

The center plane of datum
feature B is perpendicular
to datum plane A

0.8 wide tolerance
zone

Datum plane A

The center plane of the actual mating envelope
of the slot must lie between two parallel planes
0.8 apart, equally disposed about the center plane
of datum B. The specified tolerance and the datum
reference can only apply on an RFS basis.

POSITIONAL TOLERANCING RFS–RFS FOR
SYMMETRICAL FEATURES

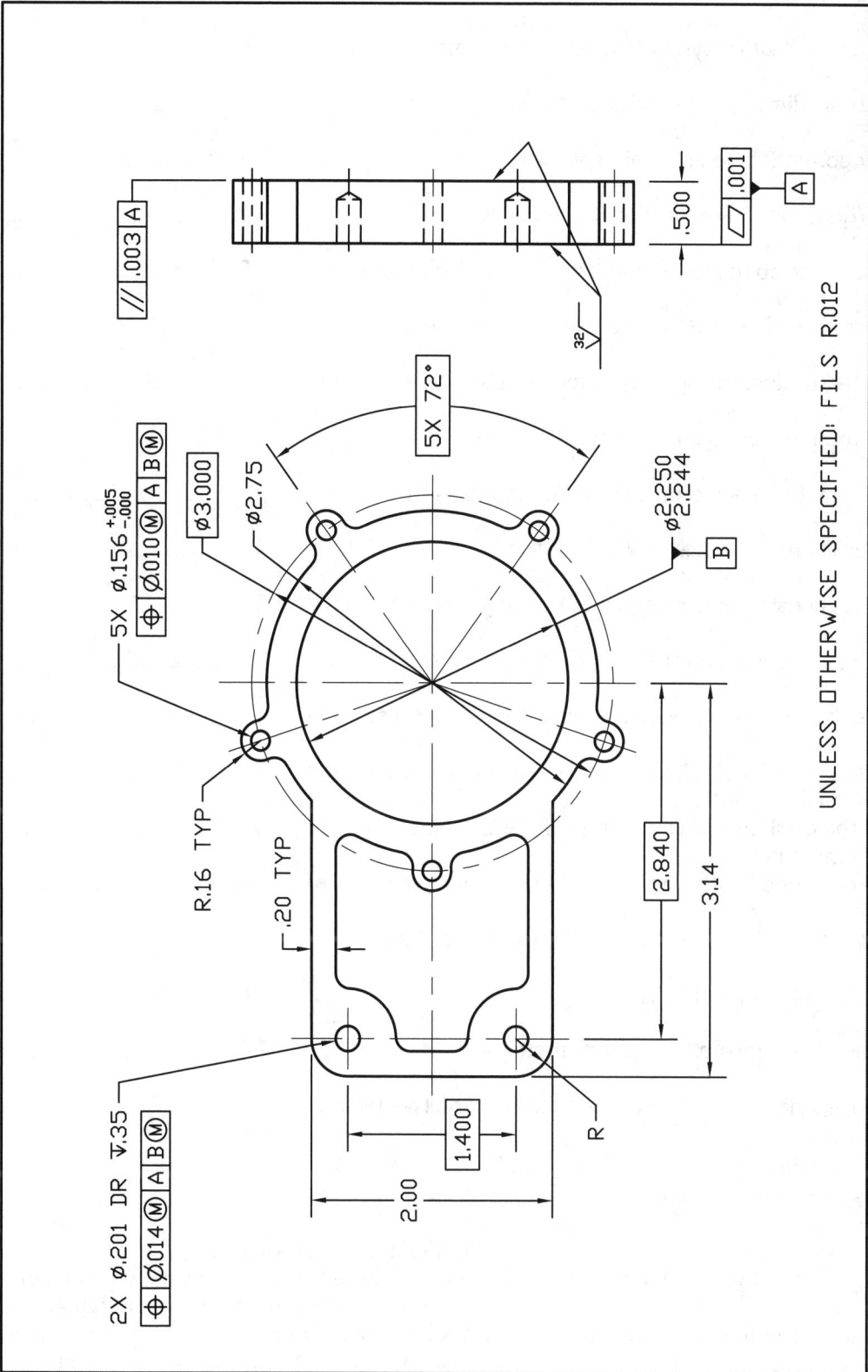

UNLESS OTHERWISE SPECIFIED: FILS R.012

| | MATERIAL | SAE 903 ALLOY | | | | | |
|---|---|---|---|---|---|---|---|
| UNSPECIFIED TOLERANCES: | DWN BY K.M. KAIDER | REDRAWN BY M. ŽOKVIC | | | THIRD-ANGLE PROJECTION | | |
| 2-PL DEC IN.(.xx)±.01 | | | | | | | |
| 3-PL DEC IN.(.xxx)±.005 | TITLE | | | | | | |
| DRILLED HOLES +.010 -.002 | RESERVOIR ADAPTER | | | | | | |
| ANGLE ±0° 30' | | SCALE 1:1 | CHKD LS | | DWG NO. 31A073 | | |

**INSTRUCTIONS:** Refer to drawing 31A073 on page 823 to answer the following questions.

1. How many datum feature symbols appear on the drawing?          1. _____

2. How many basic dimensions are designated?          2. _____

3. How many geometric tolerances are specified?          3. _____

4. How many (question 3) are locational tolerances?          4. _____

5. What characteristic do the locational tolerance symbols represent?          5. _____

6. What characteristic does the form tolerance symbol represent?          6. _____

7. What characteristic does the orientation tolerance symbol represent?          7. _____

8. What is the total tolerance on the surface of datum A?          8. _____

9. What is the total tolerance on the diameter of datum B?          9. _____

10. What is the positional tolerance on the .156 diameter holes at MMC?          10. _____

11. What is the positional tolerance on the .201 diameter holes at MMC?          11. _____

12. What is the size tolerance on the .201 diameter holes?          12. _____

13. What size is the positional tolerance zone if the five holes measure .156?          13. _____

14. What size is the positional tolerance zone if the five holes measure .157? If they measure LMC?          14. _____

15. What size is the positional tolerance zone if the two holes measure .201? If they measure LMC?          15. _____

16. Is the tolerance zone for the hole locations (a) square-shaped or (b) round-shaped?          16. _____

17. What roughness height is specified for the two flat surfaces?          17. _____

18. What size bolt circle is specified?          18. _____

19. How far apart angularly are the .156 diameter holes?          19. _____

20. What is the overall length measured along the horizontal center line?          20. _____

EXTRA: Add a Perpendicularity Control to datum B. Make the 2.250 2.244Ø hole perpendicular within Ø .005 at MMC to datum A.

Two 8.75 dia. reamed holes:
Indicate the bottom of the 21.75 slot shown in the front view of datum A.
Indicate the top surface in the right-side view of datum B.
Indicate the right side of the right-side view of datum C.
Locate the axis of each hole in a 0.05 cylindrical zone at MMC that is perpendicular to datum A, and located from datum B and datum C. Make the dimensions used for location basic.

12.7 dia. counterbored hole:
Indicate the left surface in the top view of datum D.
The upper surface in the top view is datum B.
Locate the axis of the 12.7 dia. hole in a 0.1 cylindrical tolerance zone at MMC that is perpendicular to datum B and located from datum D and datum C. Make the dimensions used for location basic.

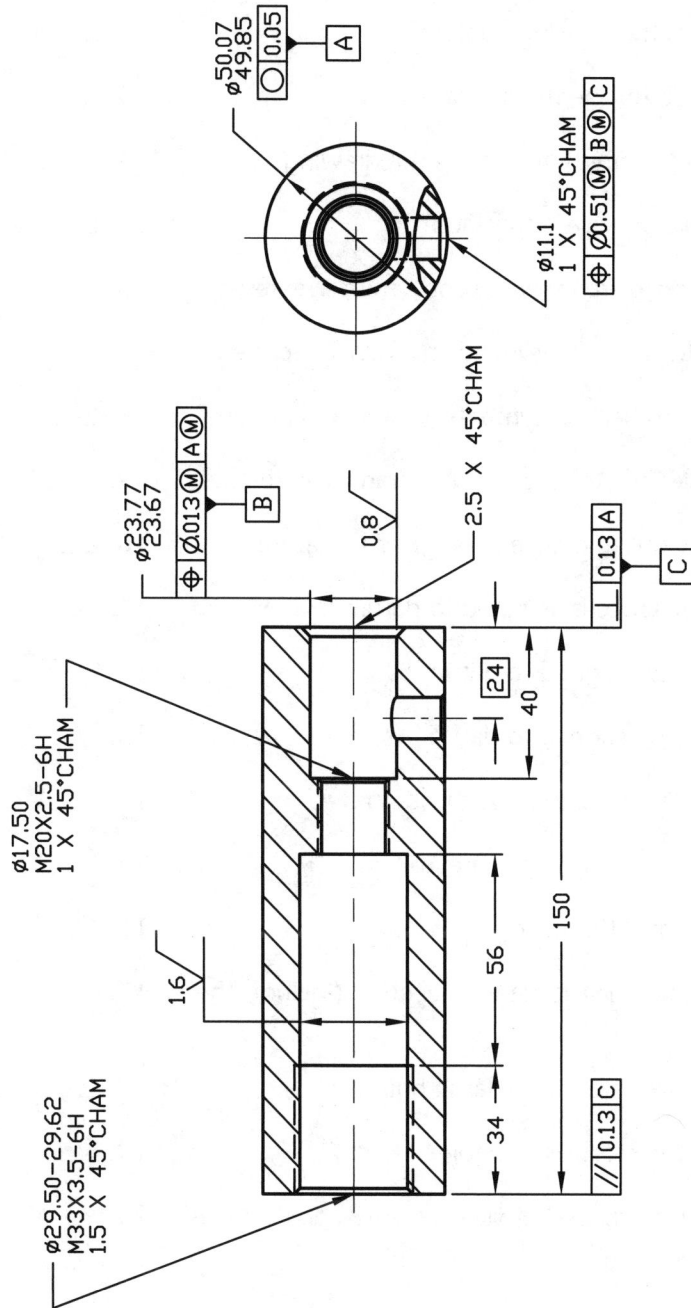

NOTE: DWG IS INCOMPLETE BY INTENT

Ø50.07
49.85
○ 0.05   A

Ø11.1
1 X 45°CHAM
⊕ Ø0.51 Ⓜ B Ⓜ C

Ø23.77
23.67
⊕ Ø013 Ⓜ A Ⓜ
B

Ⓜ A Ⓜ

2.5 X 45°CHAM

0.8

24

40

⊥ 0.13 A   C

Ø17.50
M20X2.5-6H
1 X 45°CHAM

1.6

150

56

34

Ø29.50-29.62
M33X3.5-6H
1.5 X 45°CHAM

// 0.13 C

METRIC

THIRD-ANGLE PROJECTION

DWG NO.
31A074

UNSPECIFIED TOLERANCES:
±0.5

ANGLES ±0° 30'

MATERIAL  C F STL RDS AISI 1018

DWN BY E.A. OTOADESE   REDRAWN BY M. ZOKVIC   SCALE 1:2   CHKD LS

TITLE

VALVE HSG

INSTRUCTIONS: Refer to drawing 31A074 on page 825 to answer the following questions.

1. What measurement units were used to dimension the drawing?  1. _____

2. How many chamfers does the housing contain?  2. _____

3. What tolerance is specified for the thread class of fit?  3. _____

4. How many datum feature symbols appear on the drawing?  4. _____

5. How many geometric tolerances are specified?  5. _____

6. What characteristic does the tolerance symbol on datum A represent?  6. _____

7. What characteristic does the tolerance symbol on datum B represent?  7. _____

8. What characteristic does the tolerance symbol on datum C represent?  8. _____

9. What does the modifier Ⓜ in the feature control frames represent?  9. _____

10. How much roughness height tolerance is assigned to datum B?  10. _____

11. How much geometric tolerance is assigned to datum B?  11. _____

12. How much size tolerance is assigned to datum B?  12. _____

13. How much size tolerance is assigned to datum A?  13. _____

14. How much geometric tolerance is assigned to datum A?  14. _____

15. What is the major diameter of the large threads?  15. _____

16. What is the pitch of the small threads?  16. _____

17. What is the length of the small threaded section? (Include the chamfer.)  17. _____

18. What is the unthreaded diameter of the large hole?  18. _____

19. What is the diameter of the cross-drilled hole?  19. _____

20. What is the maximum wall thickness between datum B and the OD?  20. _____

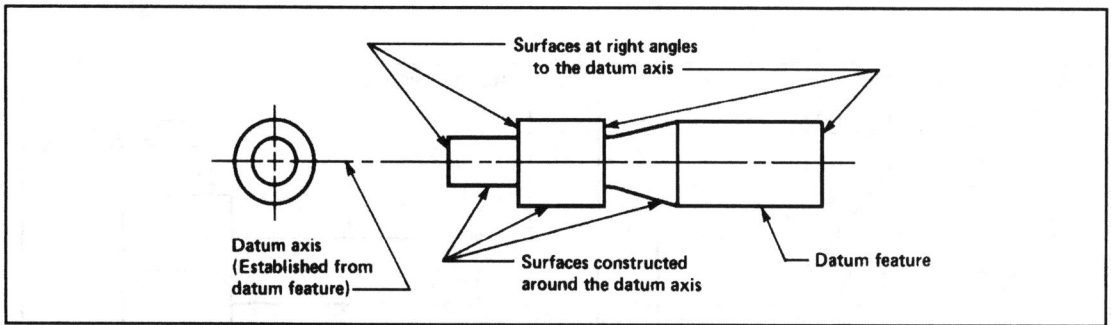

Surfaces at right angles
to the datum axis

Datum axis
(Established from
datum feature)

Surfaces constructed
around the datum axis

Datum feature

**FEATURES APPLICABLE TO RUNOUT TOLERANCING**

## RUNOUT TOLERANCES

Runout is a composite tolerance used to control the functional relationship of one or more features of a part to a datum axis. The types of features controlled by runout tolerances include those surfaces constructed around a datum axis and those constructed at right angles to a datum axis. (See the illustration above.)

The datum axis for a runout control is established in one of three ways. First, a datum axis can be established by a diameter of sufficient length to be repeatable. Second, two diameters having sufficient axial separation can be used. Third, a diameter and a face at right angles to it can be used and the datums are specified separately to indicate datum preference. In other words, it could be a face and a diameter, or a diameter and a face.

Each considered feature must be within its runout tolerance when rotated about the datum axis. The tolerance specified is the total tolerance, or full indicator movement (FIM). There are two types of runout control, circular runout and total runout.

### Circular Runout

Circular runout provides control of circular elements of a surface. The tolerance is applied independently at any circular measuring position as the part is rotated 360°. Where applied to surfaces constructed around a datum axis, circular runout may be used to control the cumulative variations of circularity and coaxiality. Where applied to surfaces constructed at right angles to the datum axis, circular runout controls circular elements of a plane surface (wobble).

### Total Runout

Total runout provides composite control of all surface elements. The tolerance is applied simultaneously to all circular and profile measuring positions as the part is rotated 360°. Where applied to surfaces constructed around a datum axis, total runout is used to control cumulative variations of circularity, straightness, coaxiality, angularity, taper, and profile of a surface. Where applied to surfaces constructed at right angles to a datum axis, total runout controls cumulative variations of perpendicularity (to detect wobble) and flatness (to detect concavity or convexity).

THIS ON THE DRAWING

MEANS THIS

At any measuring position, each circular element of these surfaces must be within the specified runout tolerance (0.02 full indicator movement) when the part is rotated 360° about the datum axis with the indicator fixed in a position normal to the true geometric shape. The feature must be within the specified limits of size.

(This controls only the circular elements of the surfaces, not the total surfaces.)

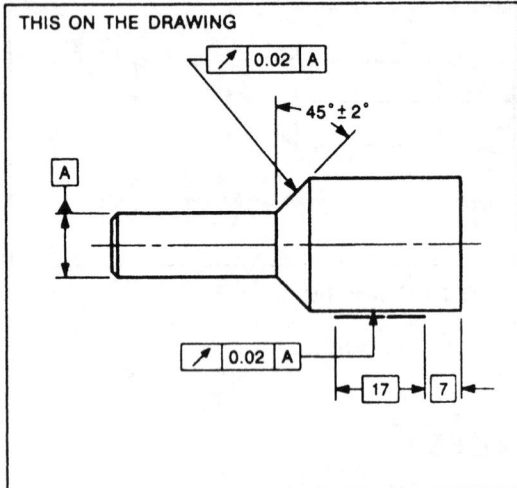

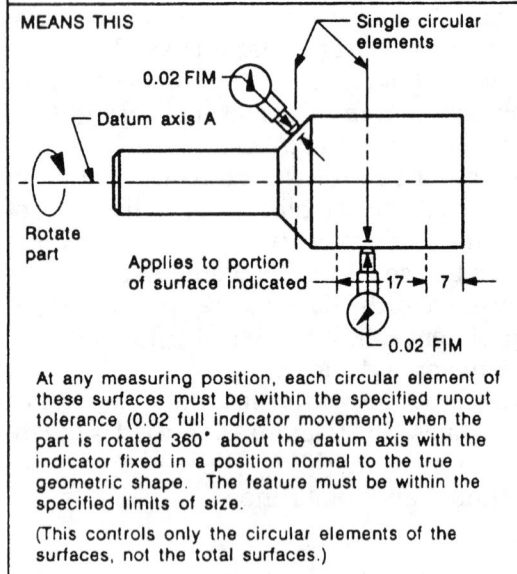

SPECIFYING CIRCULAR RUNOUT RELATIVE TO
A DATUM DIAMETER

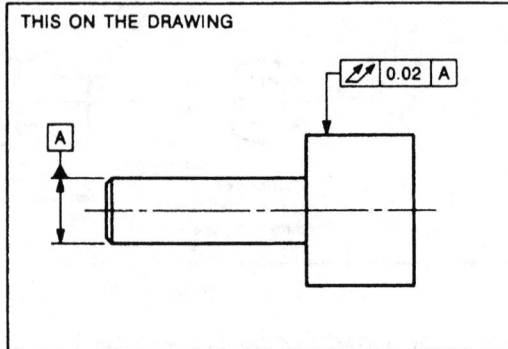

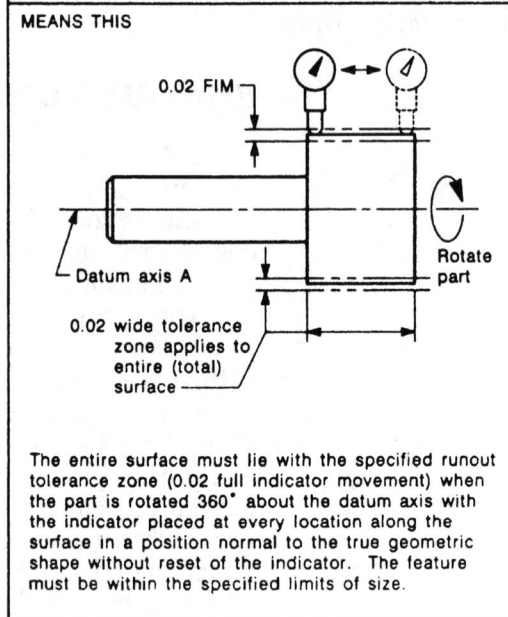

THIS ON THE DRAWING

MEANS THIS

The entire surface must lie with the specified runout tolerance zone (0.02 full indicator movement) when the part is rotated 360° about the datum axis with the indicator placed at every location along the surface in a position normal to the true geometric shape without reset of the indicator. The feature must be within the specified limits of size.

SPECIFYING TOTAL RUNOUT RELATIVE TO A
DATUM DIAMETER

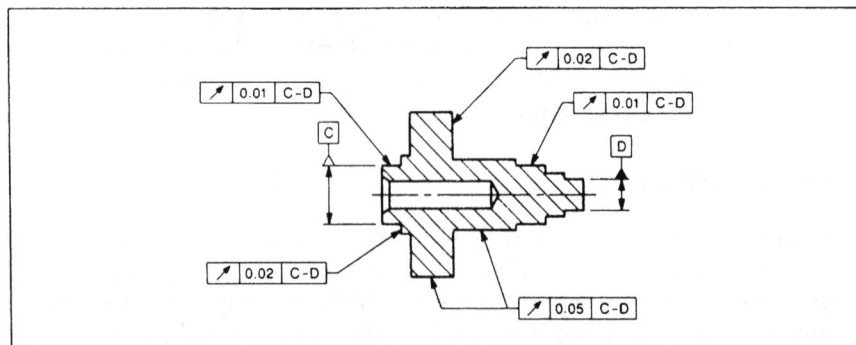

SPECIFYING RUNOUT RELATIVE TO TWO DATUM DIAMETERS

*(Reprinted from ASME Y14.5M, by permission of the American Society of Mechanical Engineers. All rights reserved.)*

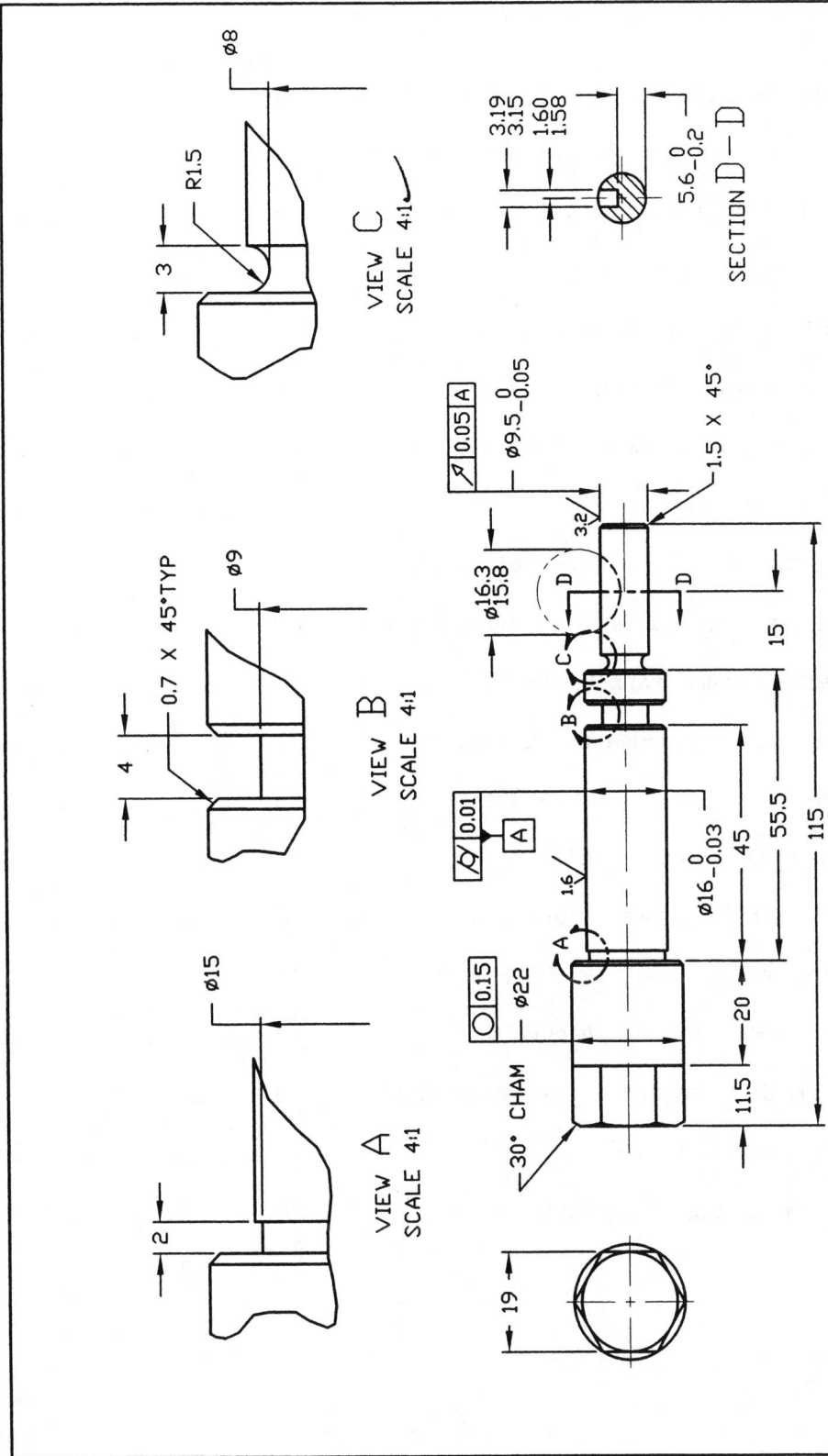

VIEW A
SCALE 4:1

ø15

2

VIEW B
SCALE 4:1

0.7 X 45°TYP

ø9

4

VIEW C
SCALE 4:1

ø8

R1.5

3

SECTION D-D

3.19
3.15

1.60
1.58

5.6 0 -0.2

⌖ 0.05 A

ø9.5 0 -0.05

1.5 X 45°

3.2

ø16.3
15.8

⌀ 0.01   A

1.6

ø16 0 -0.03

45

55.5

115

15

30° CHAM

ø22

○ 0.15

19

11.5

20

HT NOTES:
CARB RC 56 MIN
0.5 MIN DP

| UNSPECIFIED TOLERANCES: | MATERIAL | ø25 X 120 CF RD AISI 1018 | | |
|---|---|---|---|---|
| ±0.25 | DWN BY W.W. CHURCHILL | REDRAWN BY M. ZOKVIC | SCALE 1:1 | CHK'D LS |
| ANGLES ±0° 30' | TITLE | SPL TEST BLK | | |

THIRD ANGLE PROJECTION

METRIC

DWG NO.
31A075

INSTRUCTIONS: Refer to drawing 31A075 on page 829 to answer the following questions.

1. What metric units do the dimensional tolerances represent?

1. _____

2. What metric units do the surface roughness tolerances represent?

2. _____

3. Which diameter (Ø16 or Ø9.5) requires the smoothest surface?

3. _____

4. What is the minimum width of the keyseat?

4. _____

5. What is the maximum diameter of the keyseat cutter?

5. _____

6. What is the minimum Rockwell hardness permissible?

6. _____

7. What is the dimension across the flat sides of the hexagon?

7. _____

8. How many chamfers does the shaft contain?

8. _____

9. What characteristic does the tolerance symbol on the Ø22 represent?

9. _____

10. What characteristic does the tolerance symbol on the Ø16 represent?

10. _____

11. What characteristic does the tolerance symbol on the Ø9.5 represent?

11. _____

12. Which one of the above diameters is identified as datum A?

12. _____

13. Which one of the diameters uses datum A as a reference?

13. _____

14. How deep is the annular groove in datum A?

14. _____

15. What is the width of the annular groove in datum A?

15. _____

16. What is the width of the neck at the left side of datum A?

16. _____

17. What is the width of the neck at the left side of Ø9.5?

17. _____

18. What is the length of the Ø9.5 after necking and chamfering?

18. _____

19. How much size tolerance applies to Ø22?

19. _____

20. How much geometric tolerance applies to Ø22?

20. _____

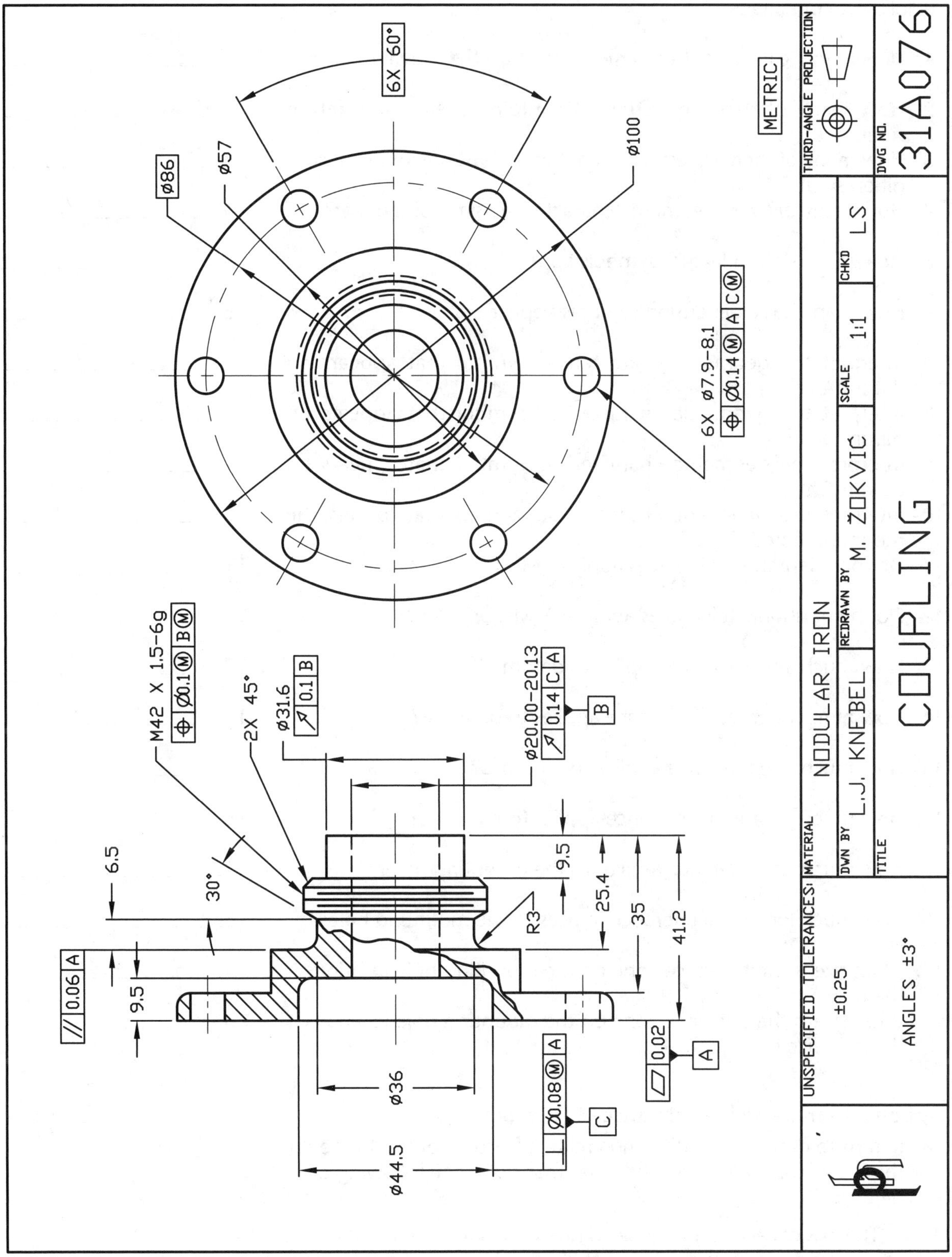

6X 60°

Ø57

Ø86

Ø100

6X Ø7.9-8.1
⊕ Ø0.14 Ⓜ A C Ⓜ

M42 X 1.5-6g
⊕ Ø0.1 Ⓜ B Ⓜ

2X 45°

Ø31.6
↗ 0.1 B

Ø20.00-20.13
↗ 0.14 C A

B

6.5

30°

9.5

// 0.06 A

9.5

25.4

35

41.2

R3

Ø36

Ø44.5

⊥ Ø0.08 Ⓜ C

▱ 0.02

A

METRIC

THIRD-ANGLE PROJECTION

DWG NO.
31A076

| UNSPECIFIED TOLERANCES: | MATERIAL | NODULAR IRON | | |
|---|---|---|---|---|
| ±0.25 | DWN BY L.J. KNEBEL | REDRAWN BY M. ZOKVIC | SCALE 1:1 | CHKD LS |
| ANGLES ±3° | TITLE | COUPLING | | |

1. How many feature control frames appear on the drawing?　　　　1. _____

2. How many of them (question 1) contain at least one datum reference?　　　　2. _____

3. How many of them (question 1) contain at least one material condition symbol?　　　　3. _____

4. How many *different* geometric characteristic symbols appear?　　　　4. _____

5. How many datum feature symbols appear?　　　　5. _____

6. How many basic dimension symbols appear?　　　　6. _____

7. Interpret the geometric characteristic symbol that appears on datum A.　　　　7. _____

8. Interpret the geometric characteristic symbol that appears on datum B.　　　　8. _____

9. Interpret the geometric characteristic symbol that appears on datum C.　　　　9. _____

10. Interpret the geometric characteristic symbol that locates the mounting holes.　　　　10. _____

11. Do the orientation tolerances apply at MMC or RFS?　　　　11. _____

12. Do the positional tolerances apply at MMC or RFS?　　　　12. _____

13. How much size tolerance applies to datum B?　　　　13. _____

14. How much geometric tolerance applies to datum B?　　　　14. _____

15. How much size tolerance applies to datum C?　　　　15. _____

16. How much geometric tolerance applies to datum C?　　　　16. _____

17. How much size tolerance applies to the mounting holes?　　　　17. _____

18. How much locational tolerance applies to the mounting holes?　　　　18. _____

19. What size is the tolerance zone on the mounting hole locations if the hole size is exactly 7.9?　　　　19. _____

20. What size is the tolerance zone on the mounting hole locations if the hole size is exactly 8.1?　　　　20. _____

**Optional Exercise Using Tolerance of Position:**

1a. Return to drawing 31A072 and create a feature control frame for the 3 3 #10-24UNC threaded holes that states the following conditions:

　　The axis of each of the threaded holes is located within a .005 diameter cylindrical tolerance zone at RFS that is oriented perpendicular to datum B, and located relative to datum C at MMC. Designate datum C as the 1.300/1.298 diameter. Designate the proper dimensions as basic by enclosing them inside a rectangular frame.

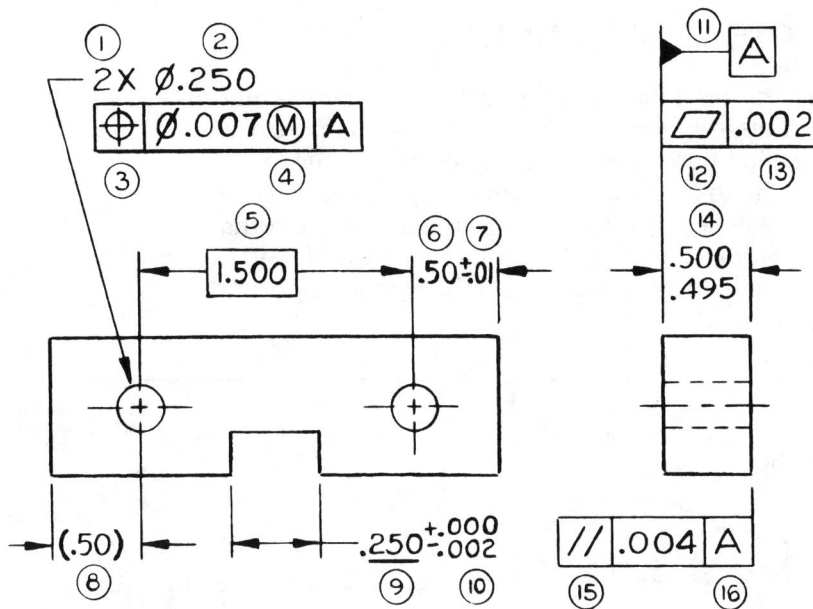

## Symbol Quiz

INSTRUCTIONS: Match the descriptions listed below with the proper symbols and tolerances shown in the illustration above. Select the correct answers from the encircled numbers adjacent to the features.

 1. Bilateral tolerance

 2. Unilateral tolerance

 3. Geometric tolerance

 4. Datum reference

 5. Datum feature symbol

 6. Basic dimension symbol

 7. Material condition symbol

 8. Form tolerance symbol

 9. Orientation tolerance symbol

10. Location tolerance symbol

11. Reference dimension

12. Dimension not-to-scale

13. Limit dimension

14. Number of times symbol

15. Nominal size

16. Unspecified tolerance

 1. _____

 2. _____

 3. _____

 4. _____

 5. _____

 6. _____

 7. _____

 8. _____

 9. _____

10. _____

11. _____

12. _____

13. _____

14. _____

15. _____

16. _____

# PROJECTED TOLERANCE ZONE

This concept can be applied where the variation in perpendicularity of threaded or press-fit holes could cause fasteners such as screws, studs, or pins to interfere with mating parts. See the examples shown below. Note that it is the variation in perpendicularity of the portion of the fastener passing through the mating part that is significant. The location and perpendicularity of the threaded hole is of importance only insofar as it affects the extended portion of the engaging fastener.

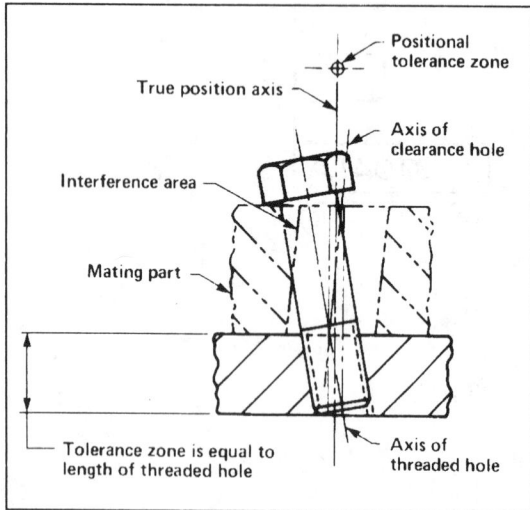

INTERFERENCE DIAGRAM, FASTENER
AND HOLE

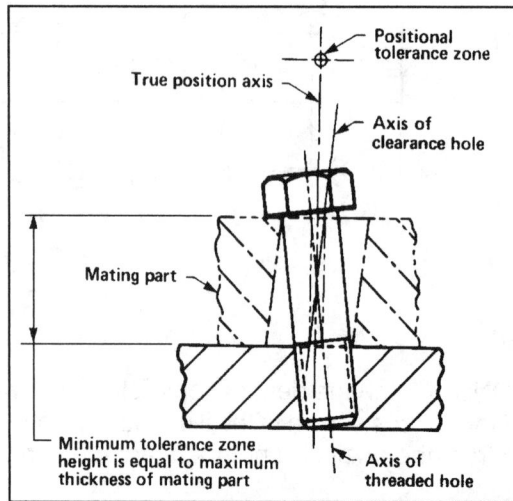

BASIS FOR PROJECTED TOLERANCE ZONE

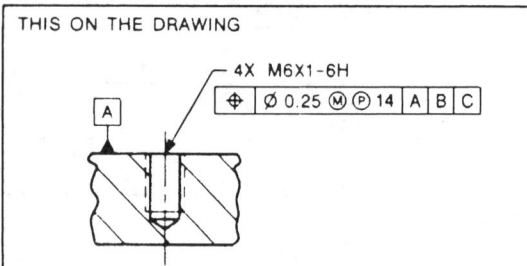

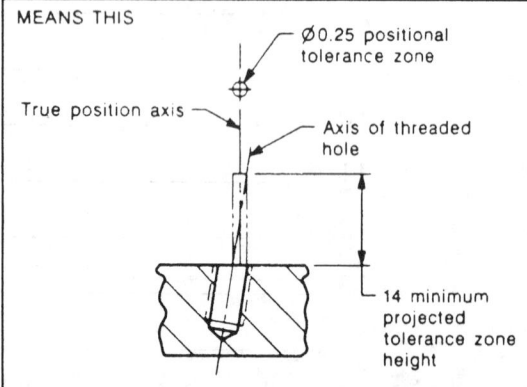

PROJECTED TOLERANCE ZONE SPECIFIED

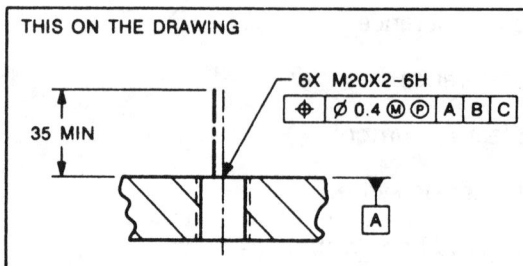

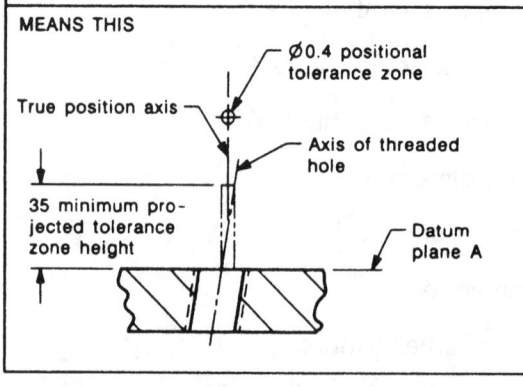

PROJECTED TOLERANCE ZONE INDICATED
WITH CHAIN LINE

Ø4.47

Ø3.500

5X Ø.531 +.003/-.000

⊕ Ø.014 Ⓜ .56 B C Ⓜ
Ⓟ

5X 72°

R.005 MAX

VIEW A
SCALE 3:1

Ø1.858±.005
⌀ .005 A

R.05 MAX

VIEW C
SCALE 3:1

Ø1.874±.001
⌀ .003 A B

.06
.44

30°

Ø2.44

VIEW B
SCALE 3:1

30°
R.03

.50

2.50

.12

1.322±.002
.070±.002

B

1.750

30°

30°

.06

125

Ø1.562±.016

Ø1.7505±.0005
◯ .0005  A

Ø2.5006±.0006
⌀⌀ .003 A B  C

NOTES:
.12 FINISH ALLOW.
.19 RADII
2° DRAFT
FAO

THIRD-ANGLE PROJECTION

DWG NO. 31A077

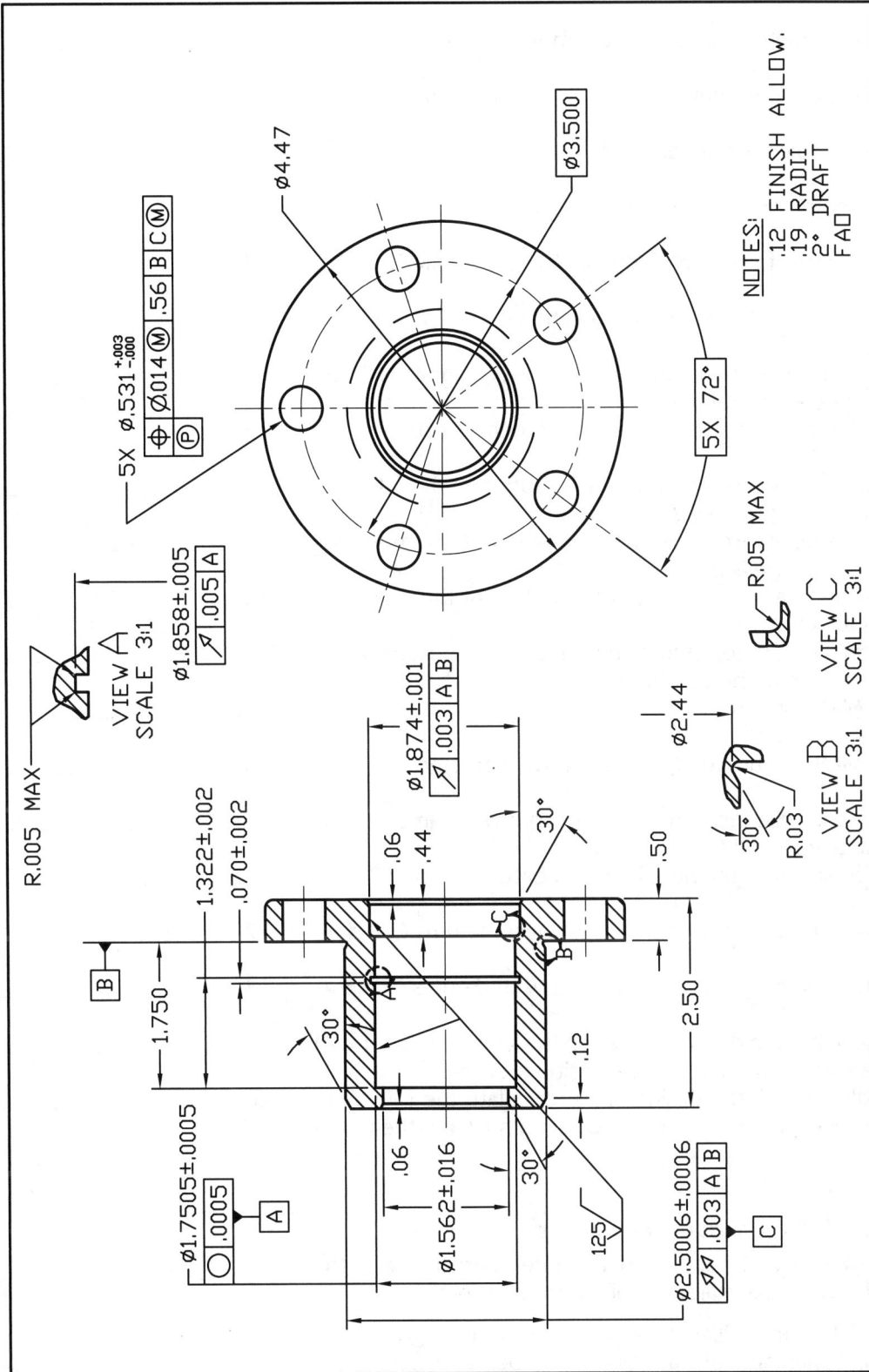

| UNSPECIFIED TOLERANCES: | MATERIAL | | |
|---|---|---|---|
| 2-PL DEC IN.(.xx)±.03 | GRAY IRON – ASTM 48-76 CLASS 25B | | |
| 3-PL DEC IN.(.xxx)±.005 | REDRAWN BY M. ŽOKVIĆ | SCALE 1:2 | CHKD LS |
| DRILLED HOLES +.010/-.002 | DWN BY J.W. DIETZ | | |
| ANGLE ±0° 30' | TITLE BRG CARRIER | | |

INSTRUCTIONS: Refer to drawing 31A077 on page 835 to answer the following questions.

1. Interpret the geometric characteristic symbol on datum A.

1. _____

2. Interpret the geometric characteristic symbol on datum C.

2. _____

3. What is the size tolerance on datum C?

3. _____

4. What is the geometric tolerance on datum A?

4. _____

5. Does the geometric tolerance on datum A apply at MMC or RFS?

5. _____

6. Does the geometric tolerance on the location of the five holes apply at MMC or RFS?

6. _____

7. What would the material condition symbol Ⓛ signify if it appeared in a feature control frame?

7. _____

8. What is the shape of the tolerance zone for the location of the mounting holes?

8. _____

9. What two basic dimensions establish the location of the tolerance zones for the mounting holes?

9. _____

10. What size is the tolerance zone for the location of the mounting holes if their exact size is .531Ø?

10. _____

11. What size is the tolerance zone for the location of the mounting holes if their exact size is .533Ø?

11. _____

12. What is the minimum projected height of the positional tolerance zone for the mounting hole locations?

12. _____

13. What is the MMC of datum C?

13. _____

14. What is the MMC of the (.070) inside groove diameter?

14. _____

15. Is the geometric tolerance on the inside groove diameter total runout or circular runout?

15. _____

16. What is the diameter of the neck between datums C and B?

16. _____

17. Calculate the minimum wall thickness between datums C and A.

17. _____

18. Calculate the maximum wall thickness between datums C and A.

18. _____

19. Calculate the minimum distance between the left side of the 1.874 counterbore and the right side of the inside groove.

19. _____

20. Taking *all* tolerances into consideration, calculate the minimum possible web between the side of a mounting hole and the OD of the flange.

20. _____

## Optional Exercise Using Tolerance of Position:

1a. Return to drawing 31A072 and create a feature control frame for the 3 × #12 drilled holes that states the following conditions:

The axis of each of the drilled holes is located within a .010 diameter cylindrical tolerance zone at MMC that is perpendicular to datum D and located relative to datum A at MMC. Designate datum D as the opposite side from datum B. Designate datum A as the .766–.764 diameter. Designate the proper dimensions as basic by enclosing them inside a rectangular frame.

# PROFILE TOLERANCES

The profile tolerance specifies a uniform boundary along the true profile within which all the elements of the surface must lie. It can be used to control the form, location, orientation, and size of a part feature, either singularly or in a combination. Profile of a line or surface can be used to tolerance simple or complex shapes.

The true profile must be defined with basic dimensions, basic angular dimensions, and/or basic radii. The true profile may be located with either basic or toleranced dimensions to the datums called out in the feature control frame. The default condition for a profile tolerance is equal bilateral. The tolerance can be assigned as either bilateral (equal or unequal) to both sides of the true profile, or unilaterally to either one side or the other. For unilateral tolerances, a short phantom line is drawn parallel to the true profile to indicate if the tolerance zone boundary is inside or outside of the true profile. Observe the examples on the following pages.

Profile tolerances are applicable to individual features, or to related features that require datum references. Profile tolerance must always be applied RFS to the surface being toleranced. However, the datum references that are features of size can be referenced at MMC, LMC, or RFS depending on the requirements.

## Profile of a Line

The tolerance zone established by the profile of a line tolerance is *two dimensional*, extending along the length of the considered feature. This applies to the profiles of parts having a varying cross section (such as the tapered wing of an aircraft), or to random cross sections of parts (as in the illustration on the following page, where it is not desired to control the entire surface of the feature as a single entity).

THIS ON THE DRAWING

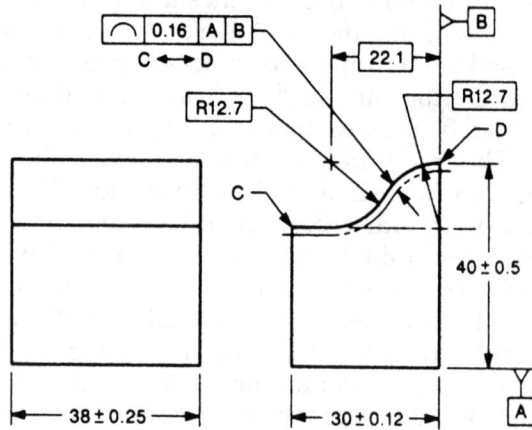

⌒ | 0.16 | A | B
C ↔ D

R12.7

B

22.1

R12.7

D

C

40 ± 0.5

38 ± 0.25

30 ± 0.12

A

MEANS THIS

22.1

Datum plane B

R12.7

R12.7

0.16 wide tolerance zone

Profile tolerance zone

22.1

R12.7

R12.7

Datum plane B

R12.7

R12.7

40.5

39.5

Datum plane A

40 ± 0.5 Size tolerance zone

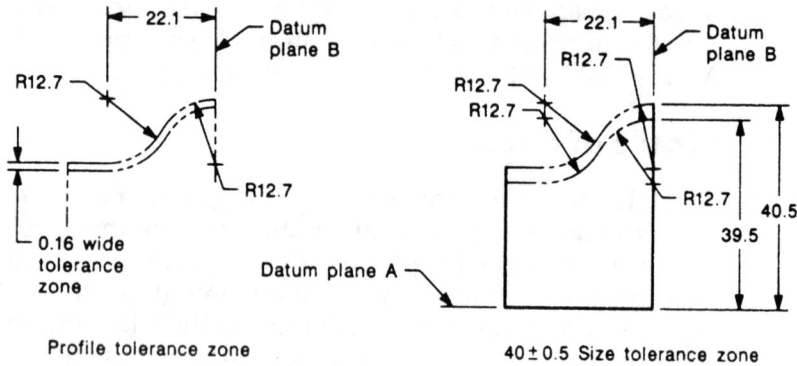

Each line element of the surface between points C and D, at any cross section, must lie between two profile boundaries 0.16 apart in relation to datum planes A and B. The surface must be within the specified limits of size.

PROFILE OF A LINE AND SIZE CONTROL

(Reprinted from ASME Y14.5M, by permission of the American Society of Mechanical Engineers. All rights reserved.)

THIS ON THE DRAWING

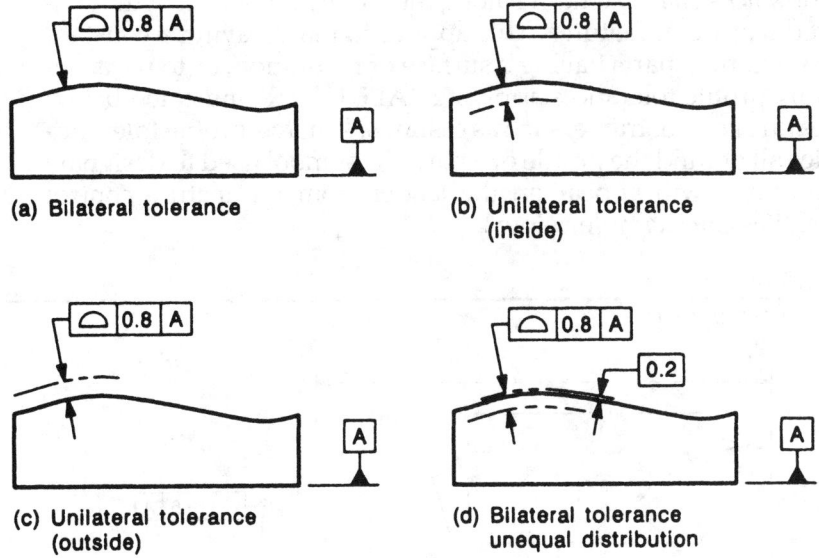

(a) Bilateral tolerance

(b) Unilateral tolerance (inside)

(c) Unilateral tolerance (outside)

(d) Bilateral tolerance unequal distribution

MEANS THIS

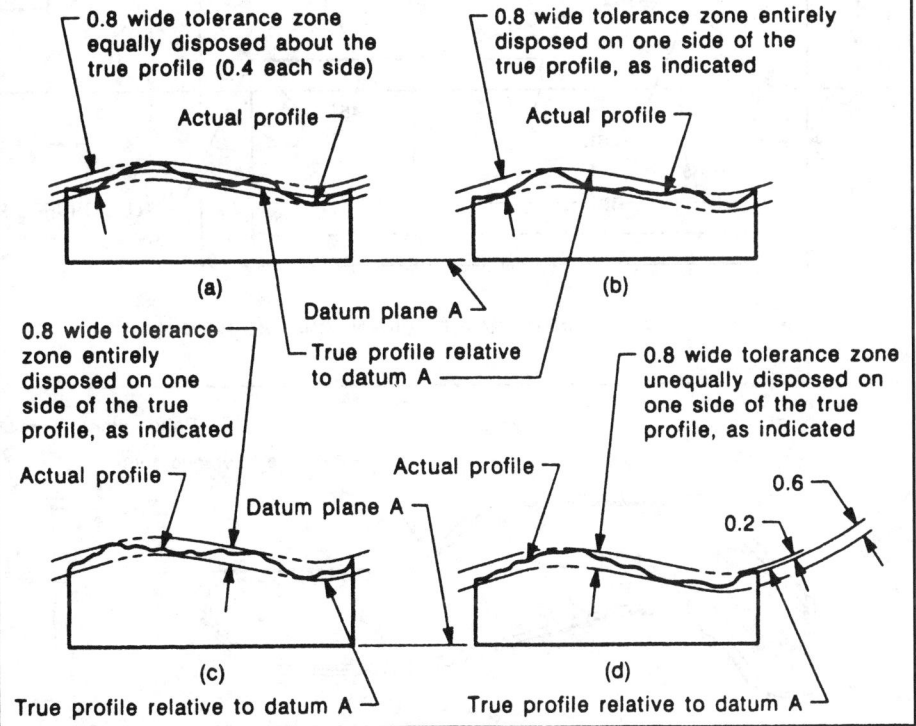

0.8 wide tolerance zone equally disposed about the true profile (0.4 each side)

Actual profile

(a)

Datum plane A

True profile relative to datum A

0.8 wide tolerance zone entirely disposed on one side of the true profile, as indicated

Actual profile

(b)

0.8 wide tolerance zone entirely disposed on one side of the true profile, as indicated

Actual profile

Datum plane A

(c)

True profile relative to datum A

0.8 wide tolerance zone unequally disposed on one side of the true profile, as indicated

Actual profile

0.6

0.2

(d)

True profile relative to datum A

*(Reprinted from ASME Y14.5M, by permission of the American Society of Mechanical Engineers. All rights reserved.)*

## Profile of a Surface

The tolerance zone established by the profile of a surface tolerance is *three dimensional*, extending along the length and width of the considered feature or features. This applies to parts having a constant cross section, to parts having a surface of revolution, or to parts defined by profile tolerances applying "ALL OVER" indicated below the feature control frame, such as castings. Where a profile tolerance applies all around the profile of a part, the symbol used to designate "all around" will appear on the leader from the feature control frame. (See the example below.)

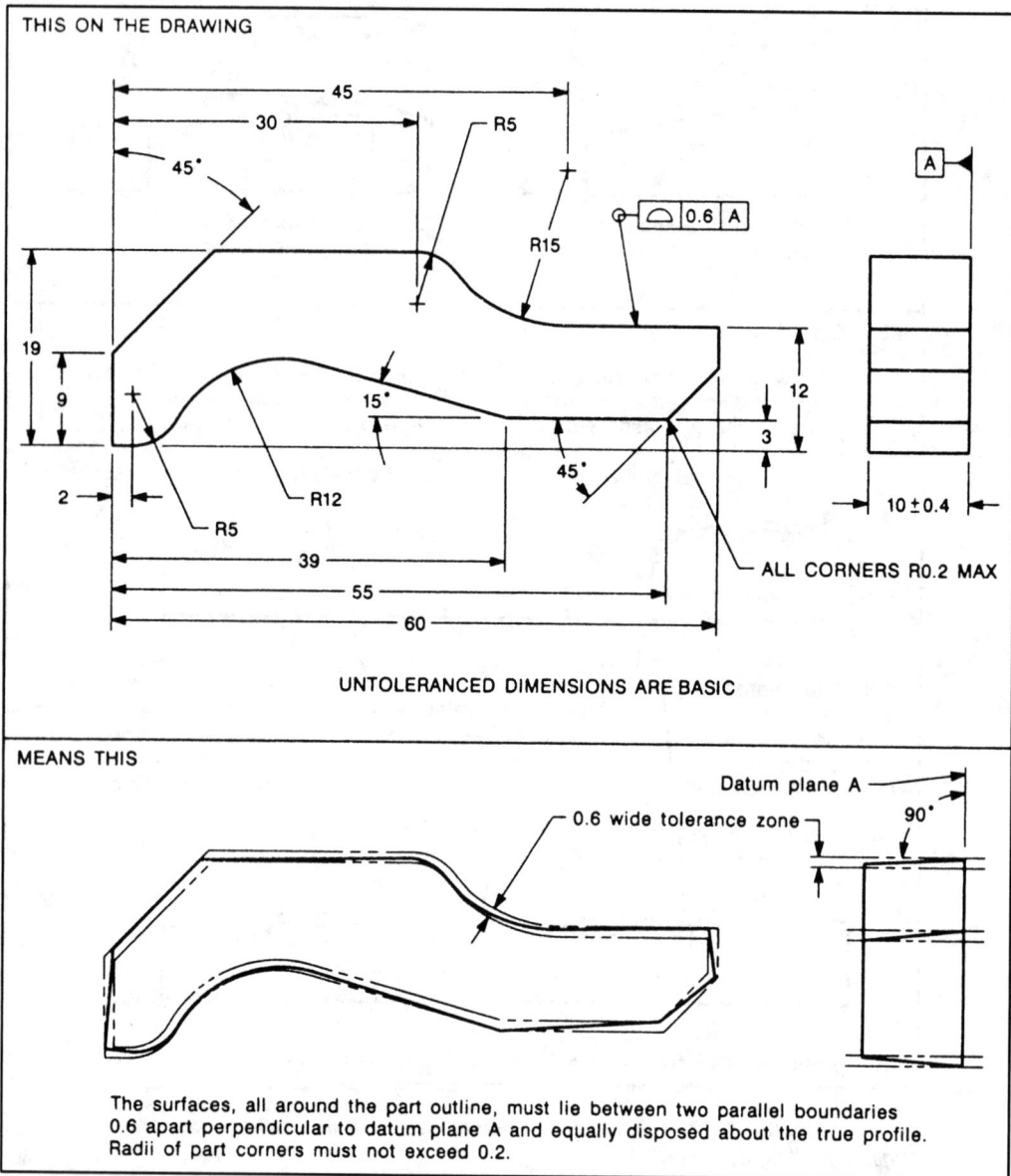

THIS ON THE DRAWING

UNTOLERANCED DIMENSIONS ARE BASIC

MEANS THIS

The surfaces, all around the part outline, must lie between two parallel boundaries 0.6 apart perpendicular to datum plane A and equally disposed about the true profile. Radii of part corners must not exceed 0.2.

SPECIFYING PROFILE OF A SURFACE ALL AROUND

*(Reprinted from ASME Y14.5M, by permission of the American Society of Mechanical Engineers. All rights reserved.)*

Using the "between" symbol underneath the feature control frame with the profile tolerancing control enables the designer to place different amounts of profile control to different segments of the same part. (See the example below.)

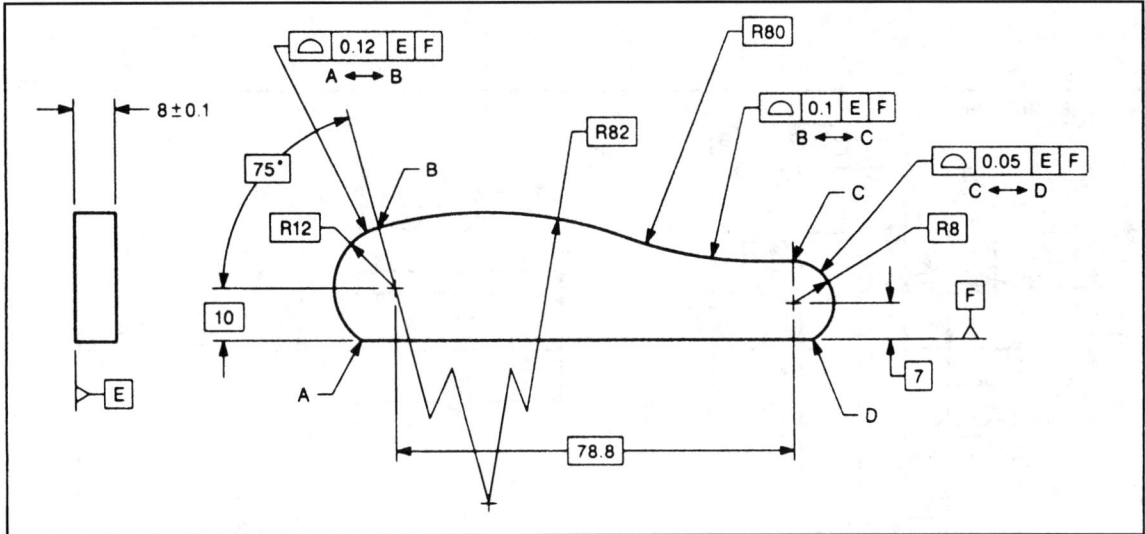

SPECIFYING DIFFERENT PROFILE TOLERANCES ON SEGMENTS OF A PROFILE

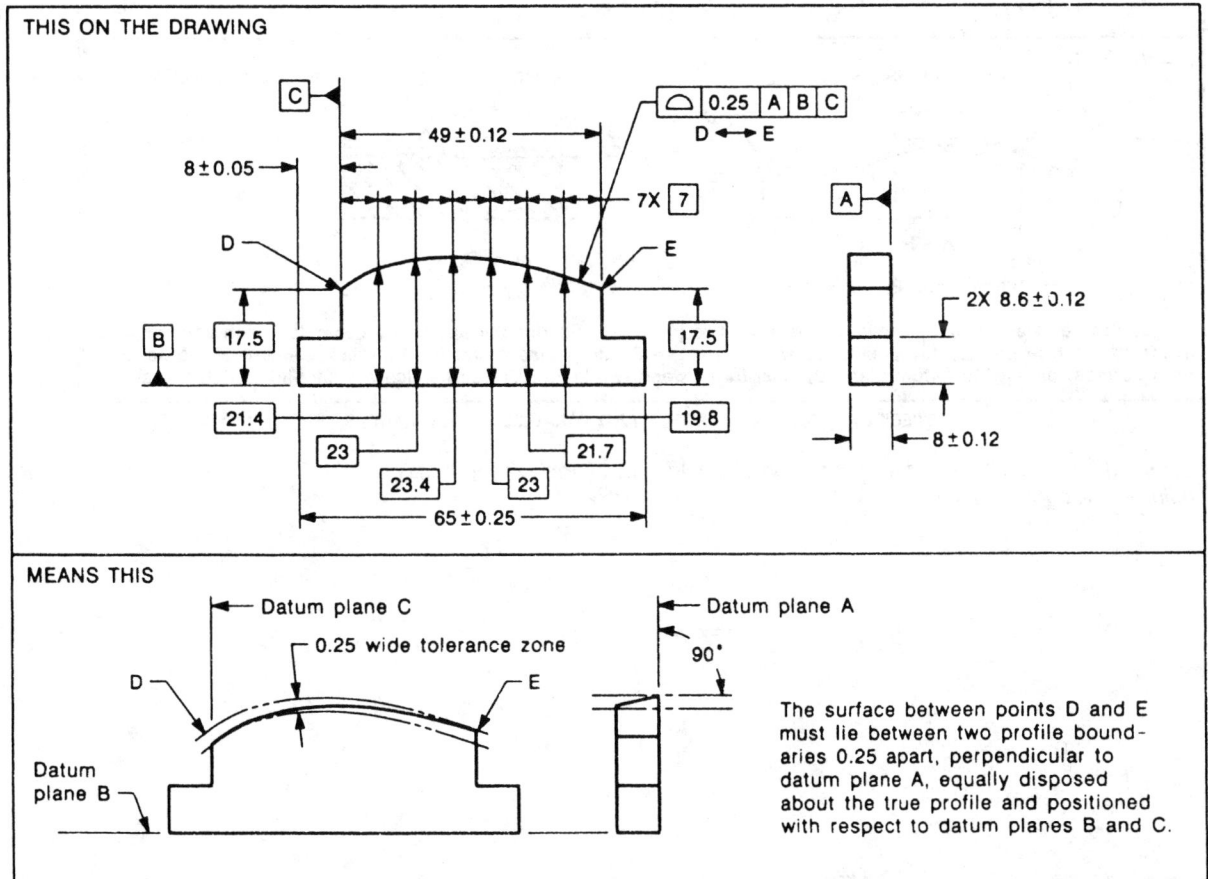

SPECIFYING PROFILE OF A SURFACE BETWEEN POINTS

*(Reprinted from ASME Y14.5M, by permission of the American Society of Mechanical Engineers. All rights reserved.)*

Profile of a surface control may be combined with other controls in order to provide specific types of control for a given feature. (See the example below.)

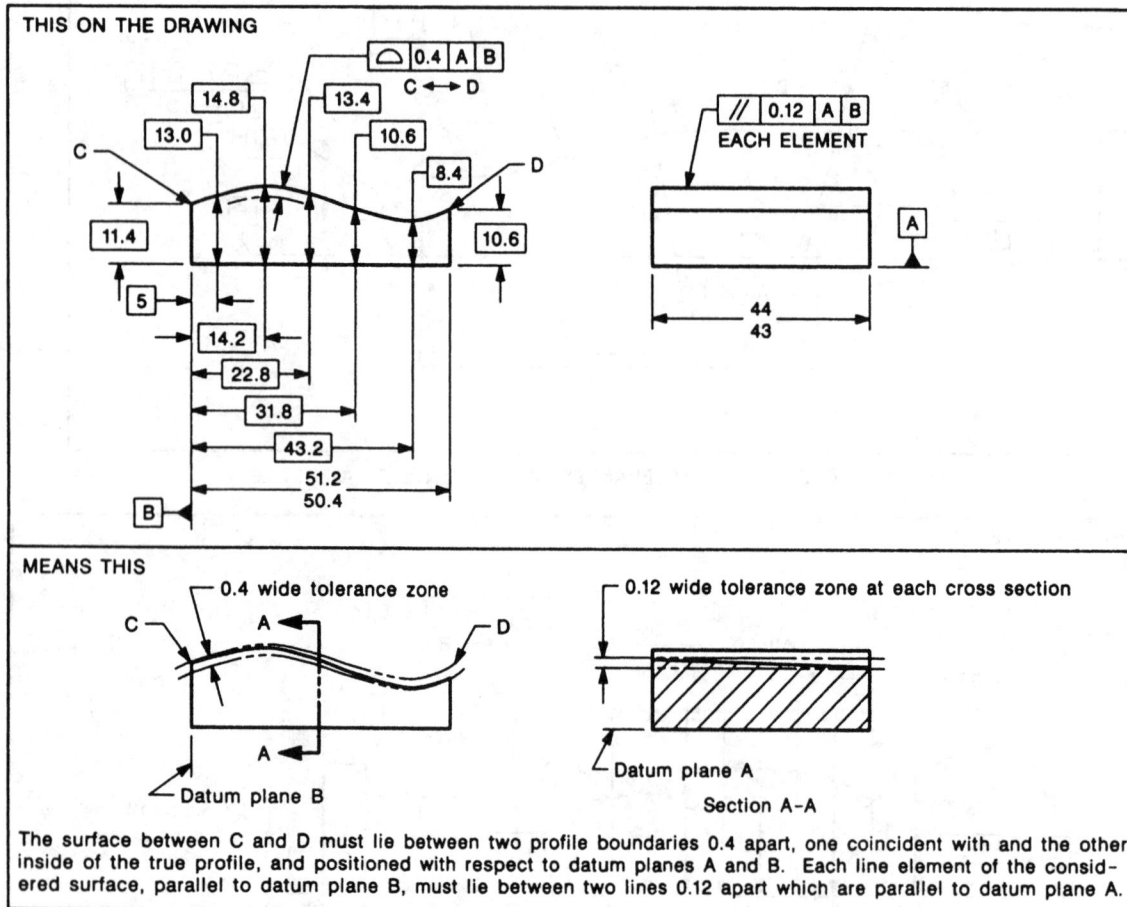

THIS ON THE DRAWING

MEANS THIS

0.4 wide tolerance zone

Datum plane B

0.12 wide tolerance zone at each cross section

Datum plane A

Section A-A

The surface between C and D must lie between two profile boundaries 0.4 apart, one coincident with and the other inside of the true profile, and positioned with respect to datum planes A and B. Each line element of the considered surface, parallel to datum plane B, must lie between two lines 0.12 apart which are parallel to datum plane A.

SPECIFYING COMBINED PROFILE AND PARALLELISM TOLERANCES

## Profile Tolerance for Coplanar Surfaces

A profile of a surface tolerance may be used where it is desired to treat two or more surfaces as a single interrupted or noncontinuous surface. In this case, a control is provided similar to that achieved by a flatness tolerance applied to a single plane surface. The profile of a surface tolerance establishes a tolerance zone defined by two parallel planes within which the considered surfaces must lie. No datum reference is stated in the figure below, as in the case of flatness, since the orientation of the tolerance zone is established from contact of the part against a reference standard; the plane is established by the considered surfaces themselves. Where two or more surfaces are involved, it may be desirable to identity which specific surface(s) are to be used as the datum feature(s). Datum feature symbols are applied to these surfaces with the appropriate tolerance for their relationship to each other. The datum reference letters are added to the feature control frame for the features being controlled. The tolerance zone thus established applies to all coplanar surfaces including datum surfaces. See the figure on page 844.

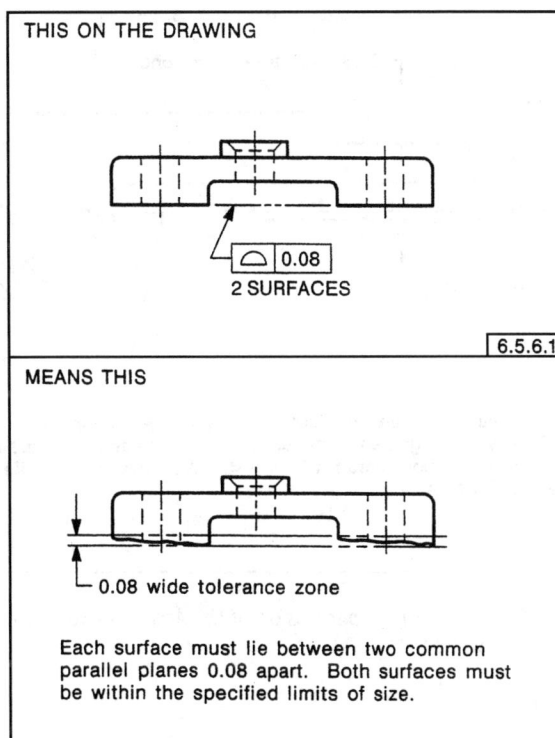

THIS ON THE DRAWING

0.08

2 SURFACES

6.5.6.1

MEANS THIS

0.08 wide tolerance zone

Each surface must lie between two common parallel planes 0.08 apart. Both surfaces must be within the specified limits of size.

*(Reprinted from ASME Y14.5M, by permission of the American Society of Mechanical Engineers. All rights reserved.)*

THIS ON THE DRAWING

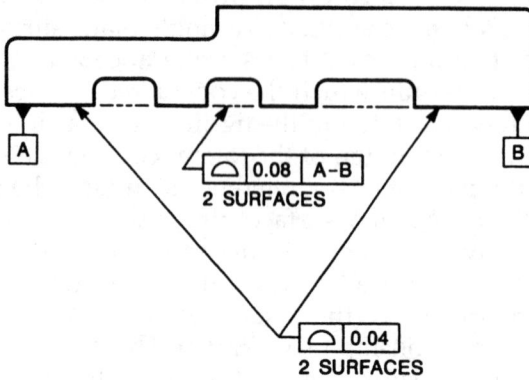

△ | 0.08 | A-B
2 SURFACES

△ | 0.04
2 SURFACES

6.5.6.1

MEANS THIS

Simulated datum A-B

0.08 wide tolerance zone

0.04

Datum plane A-B

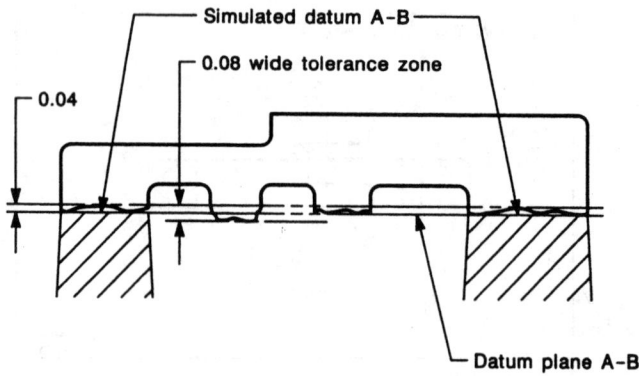

The datum features A and B must lie between two common planes 0.04 apart. The two designated surfaces must lie between two parallel planes equally disposed about datum plane A-B. All surfaces must lie within the specified limits of size.

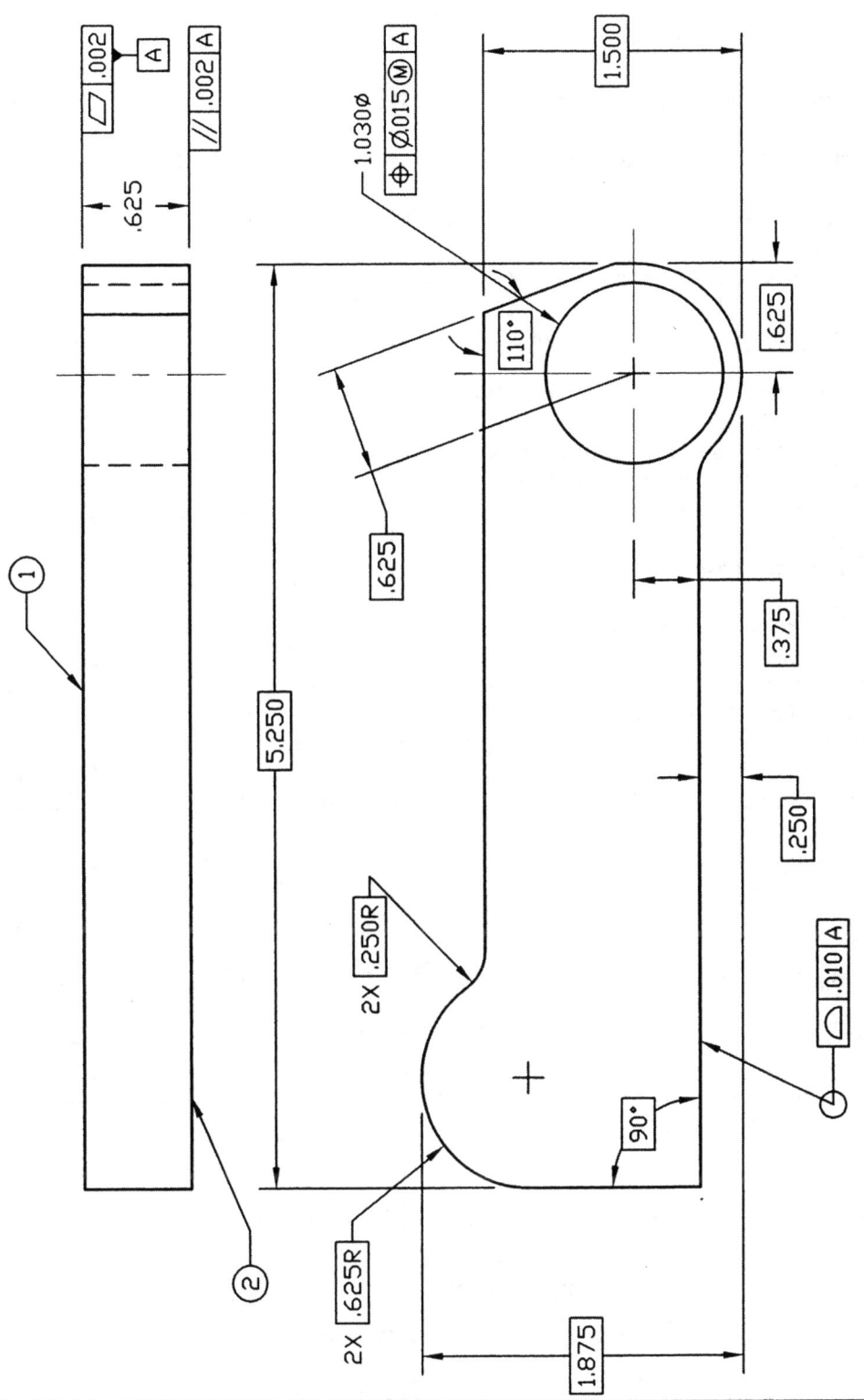

DRAWING IS INCOMPLETE BY INTENT

.002 A

// .002 A

.625

①

②

2X .625R

1.875

5.250

2X .250R

90°

1.030⌀

⊕ ⌀015 Ⓜ A

110°

.625

1.500

.625

.375

.250

◁ .010 A

| UNSPECIFIED TOLERANCES | | MATERIAL | STEEL, HOT ROLLED, .625 THK. STK., M1010-1020 | |
|---|---|---|---|---|
| TOLERANCES ON | | DWN BY M.M. OLUJIĆ | REDRAWN BY M. ŽOKVIĆ | SCALE 1:1 | CHKD LS |
| 3 PLACES | ±.005 | TITLE | | |
| ANGLES | ±1° | HATCH LEVER | | |
| CHAMFERS | ±5° | | | |

THIRD-ANGLE PROJECTION

DWG NO. 31B001

INSTRUCTIONS: Refer to drawing 31B001 on page 845 to answer the following questions.

1. How many datum feature identification symbols appear on the drawing?

1. _____

2. How many bais dimensions are designated?

2. _____

3. How many geometric tolerances are specified?

3. _____

4. How many Profile tolerances are specified?

4. _____

5. How many Form tolerances are specified?

5. _____

6. Is the Profile tolerance two dimensional or three dimensional?

6. _____

7. Is the Profile tolerance unilateral or equal bilateral?

7. _____

8. Does the Profile tolerance apply to only a portion of the outside of the part, or all around the outside of the part?

8. _____

9. If the Flatness control on Surface 1 was removed, what would be the maximum flatness error allowed on Surface 1?

9. _____

10. What is the flatness of Surface 2 limited to?

10. _____

11. What is the maximum length of the part?

11. _____

12. What is the minimum width of the part?

12. _____

13. If the Parallelism control was removed, what would be the maximum allowable taper between Surfaces 1 and 2?

13. _____

14. What is the size of the large radius at the lower right end of the part?

14. _____

15. What is the maximum allowable flatness error on the right end of the part?

15. _____

16. What is the size range of the 1.030 diameter hole?

16. _____

17. What size is the position tolerance zone if the 1.030 diameter hole is at LMC?

17. _____

18. What size is the pin on the gage used to verify the hole location?

18. _____

19. What is the maximum amount of datum possible for the position tolerance callout?

19. _____

20. What is the minimum wall thickness allowable between the 1.030 dia. hole and the right end of the part?

20. _____

**Optional Exercise (31B001):**

Make the bottom surface flat within .002 and designate it datum B.
Make the right end flat within .002 and designate it datum C.
Position the 1.030 dia. hole relative to datums A, B, and C.

**Optional Exercise Using Profile Tolerance and Tolerance of Position**

1a. Return to the clevis pin print (drawing 31A071) on page 795 and, using the Profile tolerance, make the two surfaces on the right end of the part coplanar within .002. Designate the right end as datum C and then locate the 2× 0.750–.752 diameter in line holes within 0.005 relative to datum A at RFS and datum C. Make the .875 dimension basic for the Position callout.

# GEOMETRIC CHARACTERISTIC SYMBOLS

The current dimensioning and tolerancing standards, ASME Y14.5M–1994 and CAN/CSA-B78.2-M91, were followed exclusively throughout this textbook. However, many older drawings still in use today contain geometric characteristic symbols from earlier standards. Therefore, it is imperative that you learn the old as well as the new symbols. Shown below are the symbols from the ANSI Y14.5–1973 standard. Study them, along with their footnotes.

| | | CHARACTERISTIC | SYMBOL | NOTES |
|---|---|---|---|---|
| INDIVIDUAL FEATURES | FORM TOLERANCES | STRAIGHTNESS | — | 1 |
| | | FLATNESS | ▱ | 1 |
| | | ROUNDNESS (CIRCULARITY) | ○ | |
| | | CYLINDRICITY | ⌭ | |
| INDIVIDUAL OR RELATED FEATURES | | PROFILE OF A LINE | ⌒ | 2 |
| | | PROFILE OF A SURFACE | ⌓ | 2 |
| RELATED FEATURES | | ANGULARITY | ∠ | |
| | | PERPENDICULARITY (SQUARENESS) | ⊥ | |
| | | PARALLELISM | // | 3 |
| | LOCATION TOLERANCES | POSITION | ⊕ | |
| | | CONCENTRICITY | ◎ | 3,7 |
| | | SYMMETRY | ≡ | 5 |
| | RUNOUT TOLERANCES | CIRCULAR | ↗ | 4 |
| | | TOTAL | ↗ | 4,6 |

Note:
1) The symbol ∿ formerly denoted flatness.
The symbol ⌒ or — formerly denoted flatness and straightness.
2) Considered "related" features where datums are specified.
3) The symbol ‖ and ◉ formerly denoted parallelism and concentricity, respectively.
4) The symbol ↗ without the qualifier "CIRCULAR" formerly denoted total runout.
5) Where symmetry applies, it is preferred that the position symbol be used.
6) "TOTAL" must be specified under the feature control symbol.
7) Consider the use of position or runout.

*(Reprinted from ASME Y14.5, by permission of the American Society of Mechanical Engineers. All rights reserved.)*

# FORMER PRACTICES

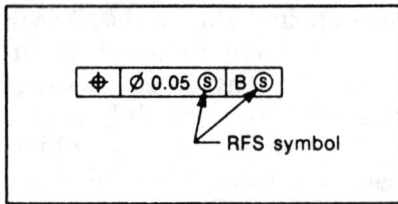

FORMER RFS SYMBOL APPLIED TO A
FEATURE AND DATUM

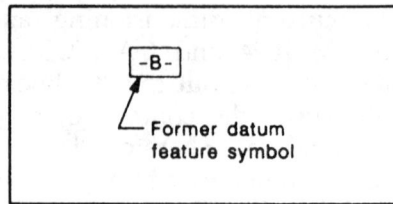

FORMER DATUM FEATURE SYMBOL

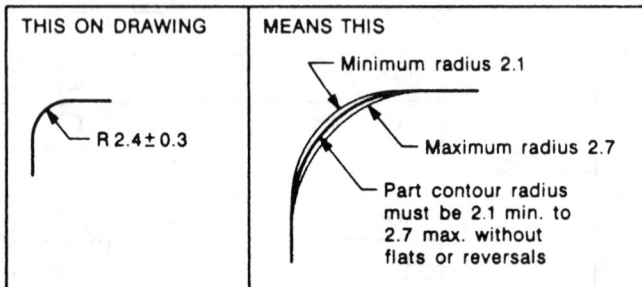

FORMER INTERPRETATION OF THE
TOLERANCE ZONE CREATED BY THE SYMBOL R

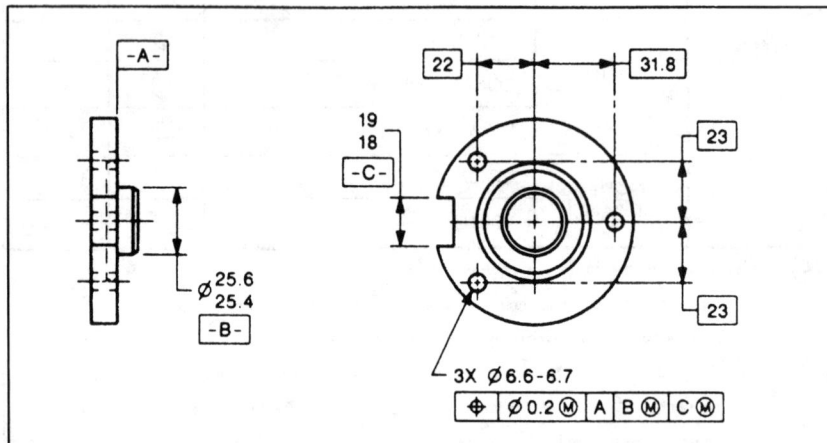

EXAMPLE OF FORMER DATUM FEATURE SYMBOL APPLICATIONS

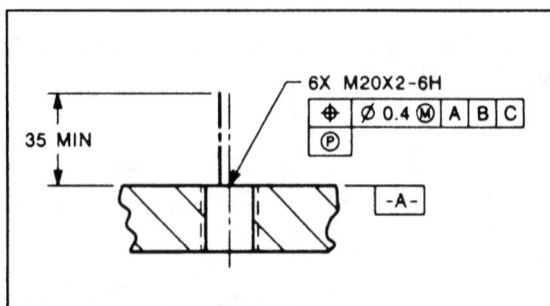

FORMER METHOD OF INDICATING A
PROJECTED TOLERANCE ZONE